University of Colorado at Boulder

MEGAN DONAHUE
Space Telescope Science Institute

NICHOLAS SCHNEIDER
University of Colorado at Boulder

MARK VOIT
Space Telescope Science Institute

The Cosmic Perspective

An Imprint of Addison-Wesley Longman, Inc.
Menlo Park, California • Reading, Massachusetts • New York • Harlow, England •
Don Mills, Ontario • Sydney • Mexico City • Madrid • Amsterdam

Acquisitions Editor: *Sami Iwata*
Publisher: *Robin J. Heyden*
Market Developer: *Andy Fisher*
Marketing Manager: *Gay Meixel*
Publishing Associates: *Bridget Biscotti Bradley, Paige Farmer*
Developmental Editor: *Laura Maria Bonazzoli*
Production Coordination: *Joan Marsh*
Production: *Mary Douglas, Rogue Valley Publications*
Photo Research: *Myrna Engler*
Technical Artist: *John Goshorn/Techarts*
Copyeditor: *Mary Roybal*
Text and Cover Designer: *Andrew Ogus*
Partial Cover Photo: Teenagers on Bicyles: *Juan Silva/The Image Bank*
Composition: *Thompson Type*
Film: *H&S Graphics/Tom Anderson, Michelle Kessel, and Rich Stanislawski*
Cover Printer: *Phoenix Color Corporation*
Printer and Binder: *Von Hoffmann Press*

Library of Congress Cataloging-in-Publication Data
The cosmic perspective / Jeffrey O. Bennett . . . [et al.].
p. cm.
Includes index.
ISBN 0-201-87878-X (alk. paper)
1. Astronomy. I. Bennett, Jeffrey O.
QB43.2.C684 1998
520—dc21 98-48585
CIP

ISBN: 0-201-87878-X

3 4 5 6 7 8 9 10—VHP—02 01 00

The Cosmic Perspective

About the Cover

The front cover is based on a remarkable image, called the Hubble Deep Field, taken by the Hubble Space Telescope in December, 1995. Although the cover shows the image filling the sky, in reality the entire image represents a point in the sky no larger in appearance than a grain of sand held at arm's length. The Hubble Space Telescope made this image by collecting 10 days' worth of light from this tiny piece of the sky. Nearly all of the many colorful objects in the image are entire galaxies, each consisting of billions of stars. You can read more about the Hubble Deep Field in Chapter 19 and see the complete Hubble Deep Field in Figure 19.1.

We shall not cease from exploration
And the end of all our exploring
Will be to arrive where we started
And know the place for the first time.
T. S. ELIOT

Dedication

TO ALL WHO HAVE EVER WONDERED about the mysteries of the universe. We hope this book will answer some of your questions—and that it will also raise new questions in your mind that will keep you curious and interested in the ongoing human adventure of astronomy.

And, especially, to the members of the "baby boom" that has occurred among the authors and editors during the writing of this book: Michaela, Emily, Rachel, Sebastian, Elizabeth, and Grant. The study of the universe begins at birth, and we hope that you will grow up in a world with far less poverty, hatred, and war so that all people will have the opportunity to contemplate the mysteries of the universe into which they are born.

BRIEF CONTENTS

DETAILED CONTENTS

PART II

KEY CONCEPTS FOR ASTRONOMY 95

4 The Science of Astronomy 96

5 A Universe of Matter and Energy 118

6 Universal Motion: from Copernicus to Newton 134

7 Light: the Cosmic Messenger 160

Appendices

PREFACE

We humans have gazed into the sky for countless generations, wondering how our lives are connected to the Sun, Moon, planets, and stars that adorn the heavens. Today, through the science of astronomy, we know that these connections go far deeper than our ancestors ever imagined. This book tells the story of modern astronomy and the new perspective—*The Cosmic Perspective*—with which it allows us to view ourselves and our planet. It is written for anyone who is curious about the universe, but it is designed primarily as a textbook for college students who are not planning to major in mathematics or science.

Approach

This book grew out of our experience teaching astronomy both to college students and to the general public over the past 20 years. During this time, a flood of new discoveries fueled a revolution in our understanding of the cosmos, but the basic organization and approach of most astronomy textbooks remained unchanged. We felt that the time had come to rethink how to organize and teach the major concepts in astronomy. This book is the result. Several major innovations in organization set this book apart.

A big picture approach. To help students achieve a lasting appreciation of astronomy, we have tried to ensure that astronomical facts are always expressed in the context of the big picture of the universe. The first three chapters set the stage by providing a broad overview of modern astronomy, ensuring that students understand the structural hierarchy and general history of the universe (Chapter 1), the scale of space and time (Chapter 2), and the motions of Earth in space (Chapter 3). The rest of the book builds upon this framework, and each chapter ends with a section called "The Big Picture" that reinforces how the concepts covered fit into the broader context of astronomy.

True comparative planetology. Departing from the traditional planet-by-planet approach to studying our solar system, we focus on the physical processes that determine the similarities and differences among planets. This approach helps students see the relevance of planetary science to their own lives by enabling them to gain a far deeper appreciation of our unique world.

We begin our section on comparative planetology (Part III of the text) by discussing solar system formation (Chapter 8), which provides students with the context needed to make planetary comparisons. We then turn to the study of the terrestrial worlds with chapters on planetary geology (Chapter 9) and planetary atmospheres (Chapter 10). Our chapter on the jovian worlds (Chapter 11) similarly focuses on processes, with emphasis on the atmospheric processes of the jovian planets themselves and on the "ice geology" of the major jovian moons. In Chapter 12, we focus on asteroids, comets, and Pluto (because it is now known to share much more in common with the comets of the Kuiper belt than with the other planets of our solar system). We also discuss the important role of impacts in Earth's past and the threat of possible impacts in the future. Finally, in Chapter 13, we bring all the previous lessons from other planets to bear on the study of our home world—and then use what we learn about Earth to investigate the prospects for finding life elsewhere in the universe.

An evolving vision of galaxies and dark matter. Traditionally, galaxies have been covered from a classification standpoint, rather like the way that stars were studied in the early part of the 20th century. In just the past decade or two, our understanding of galaxy evolution has grown enormously. As a result, it is now possible to discuss the entire subject of galaxies and cosmology (Part VI of the text) within a unifying evolutionary context. Chapter 18 begins this discussion by focusing on the Milky Way as a dynamic system that cycles gas through stars, chemically enriching the galaxy as a whole. We also discuss some of the key clues that our galaxy offers about galactic evolution, such as the differences between stars in the disk and halo and the evidence that dark matter dominates the total mass of the galaxy. Chapter 19 emphasizes how we can study galaxies in the context of the universe by introducing the cosmic distance scale, Hubble's law, and the concepts of lookback time and the cosmological horizon. We are then ready in Chapter 20 to explore the emerging story of galactic evolution as it is known today. We place starburst galaxies and quasars in their proper context as phases in the evolution of otherwise normal galaxies, rather than as separate classes of abnormal galaxies. We devote Chapter 21 to the important topic of dark matter, which is integral to the process of galaxy formation and evolution as well as to questions of the fate of the universe. With the evolutionary context well established, students are better able to understand the importance of studying the Big Bang itself in Chapter 22.

A focus on the universality of physics. We believe that one of the primary goals of any astronomy course should be to help students understand how we are able to learn so much about the universe. We therefore emphasize the "universality of physics"—the idea that knowing a few key principles that govern matter, energy, light, and motion can help them better understand both the phenomena of their daily lives and the mysteries of the cosmos. Each of our chapters covering fundamental ideas of science (Part II of the text) begins with a section on science in everyday life in which we use everyday experiences to help students build a scientific view of the world. We then develop formal concepts of physics by building upon these familiar experiences. Also, we devote a chapter (Chapter 4) to the development of science and how it functions in the modern world and another entire chapter (Chapter 5) to the concepts of matter and energy, with emphasis on conservation of energy. In addition, because student difficulties often stem from prior misconceptions, we address many misconceptions head on with our boxed features called "Common Misconceptions," which are found throughout the text.

Supplementary chapters on relativity and quantum mechanics. It is impossible to teach a class in astronomy without receiving questions such as: Why can't we go faster than the speed of light? What is the universe expanding into? What is a black hole? The common thread to these questions is that the answers, or possible answers, can be understood only through relativity. We believe that the ideas of relativity can be made accessible to anyone, and we have therefore included two chapters (Chapters S3 and S4) covering basic ideas of special and general relativity. We have also included a chapter on quantum mechanics and its astronomical implications (Chapter S5). These three chapters will enable students to understand stars, galaxies, and cosmology in much greater depth than they could otherwise. And from our teaching experience, students will find the topics of relativity and quantum mechanics to be among their favorite topics in the course. However, recognizing that many instructors will not have the time or inclination to cover these chapters, we have made them supplemental chapters—covering them will enhance student understanding of the chapters that follow, but they are not prerequisite to the remaining chapters.

Designed for Flexibility

We have designed this book so that it can be used as a text for introductory astronomy courses with many different areas of emphasis (such as the solar system, stars and galaxies, or both) and with varying course lengths from one quarter to one year. In particular, although we have tried to weave a storyline through the book from beginning to end, we have also made each chapter as self-contained as possible. This structure allows instructors to pick and choose among topics or rearrange the order of coverage. The supplementary chapters and the Mathematical Insights provide a variety of options for tailoring a course.

Art That Teaches

Textbook art and photos are a balance of beauty and teaching. The art and photo program must not only engage the eye, most importantly it must illustrate and reinforce the material within the text discussion. With this goal in mind, we have worked side by side with our artists, weaving together the connection between the book's words and illustrations. All photos have been chosen not only for their eye-catching quality but for how well they complement the text. Our aim was to achieve a new level of integration between content and art that would help students better understand astronomy.

Additional Features

Supplementary ("S") Chapters. The book includes six supplementary chapters, designated S1 through S6. These chapters cover material that is of interest to many students and instructors but that often must be skipped because of length constraints in courses—celestial timekeeping and navigation (Chapter S1), telescopes and spacecraft (Chapter S2), relativity and quantum ideas (Chapters S3–S5), and interstellar travel (Chapter S6). The chapters are placed where they would make the most sense if one were following the book from beginning to end.

Mathematical Insights. We have strived to explain astronomical concepts without resorting to mathematics except in a very few instances where the equations are absolutely necessary to the explanations. However, in many cases mathematics can provide deeper insight into astronomical ideas. We therefore feature "Mathematical Insights" throughout the book—boxes in which we go into a little more mathematical detail than we do in the main flow of the text. Instructors can tailor the level of mathematical rigor in their courses by choosing which Mathematical Insights to require, and students can decide whether to take advantage of other Mathematical Insights depending on their own interests.

Time Out to Think. Appearing throughout every chapter, the "Time Out to Think" features pose a short conceptual question designed to help students reflect on important new ideas. They also serve as excellent starting points for classroom discussions.

Common Misconceptions. These boxes address and correct popularly held but incorrect ideas related to the chapter material. Examples of topics covered in these boxes include misconceptions about the seasons (in Chapter 3), about why astronauts are weightless in orbit (in Chapter 6), and about what would happen to a spaceship passing near a black hole (in Chapter 17).

Thinking About. These boxes contain supplementary discussion of topics related to the chapter material but not prerequisite to the continuing discussion. Examples of topics covered in these boxes include deductive versus inductive arguments (in Chapter 4), weather and chaos (in Chapter 10), and the cosmological constant (in Chapter 22).

End-of-Chapter Questions and Problems. The following three sections appear at the end of each chapter:

- *Review Questions*: Every important concept in a chapter is covered by at least one review question, and the questions can be answered by rereading relevant sections of the chapter. For students, these questions are more useful than a standard chapter summary because they require interaction, as opposed to passive reading. Instructors can help students focus their studying by telling them which questions they are responsible to know for exams.
- *Discussion Questions*: These questions are meant to be particularly thought-provoking and generally do not have objective answers. As such, they are ideal for in-class discussion.
- *Problems*: Each chapter includes several problems of varying levels of difficulty that are designed to be assigned as written homework. Problem titles help explain the topic of a problem, and explicitly state when problems require outside research or activity.

Cross-References. Cross-references in brackets help students find the relevant discussion and create connections between chapters.

Stay Current With Our Web Site: *www.astrospot.com*

New discoveries in astronomy are occurring at a remarkable rate. To stay current, visit our companion web site at www.astrospot.com. Astrospot includes many features for both students and instructors including:

- chapter-by-chapter study aids including chapter summaries, immediate feedback quizzes, and a hyperlinked glossary.
- current articles, weekly updates on new discoveries, and interviews with scientists
- animations and interactive simulations to clarify key concepts
- weekly sky calendar
- links to other on-line astronomical resources, sorted and organized to help users make sense of the vast amount of available information
- additional resources for instructors, including an on-line version of the Instructor's Resource Guide.
- A complimentary subscription to Astrospot is packaged with each new copy of the text.

Supplements for Instructors

In addition to the Instructor's Resource Area at www.astrospot.com, the following additional resources are available to qualified adopters; contact your Addison Wesley Longman sales representative for more information.

- *Instructor's Resource Guide (0-201-38414-0).* Includes an overview of the philosophy of the text, suggested course syllabi, interactive teaching methods, a chapter-by-chapter commentary, and solutions to end-of-chapter problems.
- *Addison Wesley Longman Course Management System.* Allows the instructor to offer on-line quizzing, create individualized syllabi, conduct threaded discussion groups, and provide a format for administering on-line content.
- *Instructor's Presentation CD-ROM (0-201-38419-1).* Contains all of the full-color line art figures and hundreds of photos from the text for use in lecture presentations. Images may be exported into other programs, such as PowerPoint.
- *Transparency Acetates (0-201-38416-7).* A selection of over 200 figures from the text printed on full-color transparency acetates.
- *Test Bank (0-201-38415-9).* Includes multiple-choice, true/false, and free-response questions for each chapter.
- *TestGen-EQ with QuizMaster-EQ (Windows: 0-201-38418-3; Mac: 0-201-38417-5).* Allows the instructor to view and edit questions, add questions, and create and print different versions of tests. A built-in question editor creates graphs, imports graphics, and inserts text. QuizMaster-EQ automatically grades exams, stores results on

disk, and allows the instructor to view or print a variety of reports.

- *Supplementary Texts for Courses or Labs.* For students who wish to learn more about the sky, *Norton's Star Atlas* provides an unmatched resource with extensive star charts, information about what to see in the sky, and hints on observing techniques, use of telescopes, and astrophotography. It is also an excellent resource for astronomy labs. Contact your bookstore or your Addison Wesley Longman sales representative for ordering information. *Norton's Star Atlas and Reference Handbook,* 19th edition, edited by Ian Ridpath. ISBN: 0-582-31283-3 (hardcover), ISBN: 0-582-35655-5 (softcover).

Acknowledgments

It has taken the intense efforts of many people to bring this book to publication. We could not possibly list everyone who has helped, but we would like to call attention to a few people who have played particularly important roles. First, we thank the people who helped turn our writing into an actual book: our editors and friends at Addison Wesley Longman who have stuck with us through thick and thin, including Bill Poole, Robin Heyden, Linda Davis, Joan Marsh, Andy Fisher, Gay Meixel, Paige Farmer, Bridget Biscotti Bradley, and especially Sami Iwata; our production team, including Mary Douglas, Myrna Engler, Mary Roybal, and Karen Stough; our designer, Andrew Ogus; our web team, including Kim Askew and Kim Dow; and the artists who have created the outstanding illustrations, Joe Bergeron, John Goshorn, and Stan Maddock. Special thanks to Laura Bonazzoli, our developmental editor, who helped us clarify the writing in countless places throughout the book.

We've also been fortunate to have an outstanding group of reviewers, whose extensive comments and suggestions helped us shape the book. We thank the following people who reviewed drafts of the book in various stages:

John Beaver, University of Wisconsin at Fox Valley
Dipak Chowdhury, Indiana University–Purdue University at Fort Wayne
Robert Egler, North Carolina State University at Raleigh
Robert A. Fesen, Dartmouth College
Sidney Freudenstein, Metropolitan State College of Denver
Richard Gray, Appalachian State University
David Griffiths, Oregon State University
David Grinspoon, University of Colorado
Bruce Jakosky, University of Colorado
Kristine Larsen, Central Connecticut State University
Larry Lebofsky, University of Arizona
Michael LoPresto, Henry Ford Community College
William R. Luebke, Modesto Junior College
Marie Machacek, Massachusetts Institute of Technology
Steven Majewski, University of Virginia
John Safko, University of South Carolina
James A. Scarborough, Delta State University
James Schombert, University of Oregon
Joslyn Schoemer, Challenger Center for Space Science Education
Dale Smith, Bowling Green State University
John Spencer, Lowell Observatory
John Stolar, West Chester University
Jack Sulentic, University of Alabama
C. Sean Sutton, Mount Holyoke College
J. Wayne Wooten, Pensacola Junior College
Dennis Zaritsky, University of California, Santa Cruz

In addition, we thank the following colleagues who helped us clarify technical points or checked the accuracy of technical discussions in the book:

Thomas Ayres, University of Colorado
Cecilia Barnbaum, Valdosta State University
Rick Binzel, Massachusetts Institute of Technology
Humberto Campins, University of Florida
Robin Canup, Southwest Research Institute
Mark Dickinson, Space Telescope Science Institute
Harry Ferguson, Space Telescope Science Institute
Andrew Hamilton, University of Colorado
Todd Henry, Harvard–Smithsonian Center for Astrophysics
Dave Jewitt, University of Hawaii
Hal Levison, Southwest Research Institute
Mario Livio, Space Telescope Science Institute
Mark Marley, New Mexico State University
Bob Pappalardo, Brown University
Michael Shara, Space Telescope Science Institute
Glen Stewart, University of Colorado
John Stolar, West Chester University
Dave Tholen, University of Hawaii
Nick Thomas, MPI/Lindau (Germany)
Don Yeomans, Jet Propulsion Laboratory

Finally, we thank the many people who have greatly influenced our outlook on education and our perspective on the universe over the years, including Tom Ayres, Fran Bagenal, Forrest Boley, Robert A. Brown, George Dulk, Erica Ellingson, Timothy Ferris, Katy Garmany, Jeff Goldstein, David Grinspoon, Don Hunten, Catherine McCord, Dick McCray, Dee Mook, Cheri Morrow, Charlie Pellerin, Carl Sagan, Mike Shull, John Spencer, and John Stocke.

Jeff Bennett
Megan Donahue
Nick Schneider
Mark Voit

HOW TO SUCCEED IN YOUR ASTRONOMY COURSE

Most readers of this book are enrolled in a college course in introductory astronomy. If you are one of these readers, we offer you the following hints to help you succeed in your astronomy course.

Using This Book

Before we address general strategies for studying, here are a few guidelines that will help you use *this* book most effectively.

- Read assigned material twice.
  - Make your first pass *before* the material is covered in class. Use this pass to get a "feel" for all the material and to identify concepts that you may want to ask about in class.
  - Read the material for the second time shortly after it is covered in class. This will help solidify your understanding and allow you to make notes that will help you study for exams later.
- Take advantage of the features that will help you study.
  - Always read the *Time Out to Think* features, and use them to help you absorb the material and to identify areas where you may have questions.
  - Use the *cross-references* indicated in brackets in the text and the *glossary* at the back of the book to find more information about terms or concepts that you don't recall.
  - Go to the text web site at **www.astrospot.com** to find additional study aids, including chapter summaries and quizzes with answers.
- It's your book, so don't be afraid to make notes in it that will help you study later.
  - Don't highlight—underline! Using a pen or pencil to underline material requires greater care than highlighting and therefore helps keep you alert as you study. And be selective in your underlining—for purposes of studying later, it won't help if you underlined everything.
  - There's plenty of "white space" in the margins and elsewhere, so use it to make notes as you read. Your own notes will later be very valuable when you are doing homework or studying for exams.
- After you complete the reading, and again when you study for exams, make sure you can answer the *Review Questions* at the end of each chapter. If you are having difficulty with a review question, reread the relevant portions of the chapter until the answer becomes clear.

Budgeting Your Time

One of the easiest ways to ensure success in any college course is to make sure you budget enough time for studying. A general rule of thumb for college classes is that you should expect to study about 2 to 3 hours per week *outside* class for each unit of credit. For example, based on this rule of thumb, a student taking 15 credit hours should expect to spend 30 to 45 hours each week studying outside of class. Combined with time in class, this works out to a total of 45 to 60 hours spent on academic work—not much more than the time a typical job requires, and you get to choose your own hours. Of course, if you are working while you attend school, you will need to budget your time carefully.

As a rough guideline, your studying time in astronomy might be divided as follows:

If Your Course is:	Time for Reading the Assigned Text (per week)	Time for Homework Assignments (per week)	Time for Review and Test Preparation (average per week)	Total Study Time (per week)
3 credits	2 to 4 hours	2 to 3 hours	2 hours	6 to 9 hours
4 credits	3 to 5 hours	2 to 4 hours	3 hours	8 to 12 hours
5 credits	3 to 5 hours	3 to 6 hours	4 hours	10 to 15 hours

If you find that you are spending fewer hours than these guidelines suggest, you can probably improve your grade by studying more. If you are spending more hours than these guidelines suggest, you may be studying inefficiently; in that case, you should talk to your instructor about how to study more effectively.

General Strategies for Studying

- Don't miss class. Listening to lectures and participating in discussions is much more effective than reading someone else's notes. Active participation will help you retain what you are learning.
- Budget your time effectively. An hour or two each day is more effective, and far less painful, than studying all night before homework is due or before exams.
- If a concept gives you trouble, do additional reading or studying beyond what has been assigned. And if you still have trouble, ask for help: You surely can find friends, colleagues, or teachers who will be glad to help you learn.
- Working together with friends can be valuable in helping you understand difficult concepts. However, be sure that you learn *with* your friends and do not become dependent on them.
- Be sure that any work you turn in is of *collegiate quality*: neat and easy to read, well organized, and demonstrating mastery of the subject matter. Although it takes extra effort to make your work look this good, the effort will help you solidify your learning and is also good practice for the expectations that future professors and employers will have.

Preparing for Exams

- Study the review questions, and rework problems and other assignments; try additional questions to be sure you understand the concepts. Study your performance on assignments, quizzes, or exams from earlier in the term.
- Try the quizzes available on the text web site at **www.astrospot.com.**
- Study your notes from lectures and discussions. Pay attention to what your instructor expects you to know for an exam.
- Reread the relevant sections in the textbook, paying special attention to notes you have made on the pages.
- Study individually *before* joining a study group with friends. Study groups are effective only if every individual comes prepared to contribute.
- Don't stay up too late before an exam. Don't eat a big meal within an hour of the exam (thinking is more difficult when blood is being diverted to the digestive system).
- Try to relax before and during the exam. If you have studied effectively, you are capable of doing well. Staying relaxed will help you think clearly.

ABOUT THE AUTHORS

Jeffrey Bennett received a B.A. in biophysics from the University of California at San Diego in 1981 and a Ph.D. in astrophysics from the University of Colorado in 1987. His thesis research focused on Sun-like stars, but he now specializes in mathematics and science education. He has extensive teaching experience at the elementary and secondary levels and has taught more than fifty college courses in astronomy, physics, mathematics, and education. During a 2-year term as a Visiting Senior Scientist at NASA headquarters, he guided the creation of NASA's Initiative to Develop Education through Astronomy (IDEA). He is also the author of a mathematics textbook (Bennett and Briggs, *Using and Understanding Mathematics,* Addison Wesley Longman, 1999) and is currently working with the Smithsonian Institution and the Challenger Center to build a scale-model solar system near the National Mall. When not working, he enjoys participating in Masters Swimming and hiking the trails of Boulder, Colorado, with his wife, Lisa, son, Grant, and dog, Max.

Megan Donahue is an astronomer at the Space Telescope Science Institute in Baltimore, Maryland. After growing up in rural Nebraska, she obtained an S.B. in physics from the Massachusetts Institute of Technology in 1985 and a Ph.D. in astrophysics from the University of Colorado in 1990. Her thesis on intergalactic gas and clusters of galaxies won the Robert J. Trumpler Award (1993). She continued her research as a Carnegie Fellow at the Observatories of the Carnegie Institution in Pasadena, California, and later was an Institute Fellow at the Space Telescope Science Institute. She is an active observer, using ground-based telescopes, the Hubble Space Telescope, and orbiting X-ray telescopes. Her research focuses on questions of galaxy evolution, the nature of intergalactic space, large-scale structure formation, and dark matter and the fate of the universe. She married Mark Voit while in graduate school and they are the parents of two children, Michaela and Sebastian.

Nicholas Schneider is an associate professor in the Department of Astrophysical and Planetary Sciences at the University of Colorado and a researcher in the Laboratory for Atmospheric and Space Physics. He received his B.A. in physics and astronomy from Dartmouth College in 1979 and his Ph.D. in planetary science from the University of Arizona in 1988. In 1991, he received the National Science Foundation's Presidential Young Investigator Award. His research interests include planetary atmospheres and planetary astronomy, with a focus on the odd case of Jupiter's moon Io. He enjoys teaching at all levels and is active in efforts to improve undergraduate astronomy education. Off the job, he enjoys exploring the outdoors with his family and figuring out how things work.

Mark Voit is an astronomer in the Office of Public Outreach at the Space Telescope Science Institute. He earned his A.B. in physics at Princeton University in 1983 and his Ph.D. in astrophysics at the University of Colorado in 1990. He continued his studies at the California Institute of Technology, where he was Research Fellow in theoretical astrophysics. NASA then awarded him a Hubble Fellowship, under which he conducted research at the Johns Hopkins University. His research interests range from interstellar processes in our own galaxy to the clustering of galaxies in the early universe. Occasionally he escapes to the outdoors, where he and his wife, Megan Donahue, enjoy running, hiking, orienteering, and playing with their children.

ABOUT THE ARTIST

Joe Bergeron is a space artist who collaborated with the authors on many of the illustrations in this text. His decades as an amateur astronomer, combined with his imagination and ability to analyze the way things look, form his unique mixture of artistic skills. His science fiction work has been included in books by Isaac Asimov, Piers Anthony, and Gregory Benford. He was a contributor to the Time-Life Books series *Voyage Through the Universe*.

Bergeron's artwork is influenced by his background as an amateur astronomer. Many of his most memorable sightings were revealed by small telescopes, or even the naked eye. He has been privileged to see the sky painted in bold colors by the aurora borealis, the glory of the solar corona revealed during an eclipse, a golden star flickering behind the rings of Saturn, and pepper-black spots deposited on the clouds of Jupiter by a rain of comets—and, of course, light streaming like water from comets Hyakutake and Hale–Bopp.

Bergeron still paints, and he also creates digital images using a Macintosh computer. All of his illustrations that appear in this text were created on the computer. His work is not restricted to space themes but includes wildlife, landscapes, fantasy, and figures. For more about him, visit his web site at: http://members.aol.com/jabergeron

The Cosmic Perspective

PART I
DEVELOPING A PERSPECTIVE

I

The Universe Discovers Itself

FAR FROM CITY LIGHTS ON A CLEAR NIGHT, YOU CAN GAZE UPWARD at a sky filled with stars. If you lie back and watch for a few hours, you will observe the stars marching steadily across the sky. Confronted by the seemingly infinite heavens, you might wonder how the Earth and the universe came to be. With these thoughts, you will be sharing an experience common to humans around the world and in thousands of generations past. Remarkably, modern science offers answers to many fundamental questions about the universe and our place within it. We now know the basic content and scale of the universe. We know the age of the Earth and the approximate age of the universe. And, although much remains to be discovered, we are rapidly learning how the simple constituents of the early universe developed into the incredible diversity of life on Earth.

Above all, modern science shows that we are intimately connected to the stars that fill the night sky, in ways that our ancestors never imagined. We are part of the universe, and hence, through our science, the universe is discovering itself.

1.1 The Earth Is *Not* the Center of the Universe

If you observe the sky carefully, you can see why most of our ancestors believed that the heavens revolved about the Earth. The Sun, Moon, planets, and stars *appear* to circle around our sky each day, and we cannot feel the constant motion of the Earth as it rotates on its axis and orbits the Sun. The ancient belief in an Earth-centered, or **geocentric,** universe was shattered only about 400 years ago, when the work of Copernicus, Kepler, and Galileo finally established that the Earth is just one of many planets orbiting the Sun.

Recognition that the Earth is *not* the center of the universe heralded the birth of modern astronomy. We have learned far more about the cosmos[1] since that time. The primary purpose of this text is to tell the story of our vast and incredible universe to the extent that we know it today.

The scientific study of our universe is not without its social implications. New discoveries sometimes challenge conventional thinking. Vatican authorities put Galileo under house arrest in 1633 for his claims that the Earth orbits the Sun. Although the Church soon recognized that Galileo was right, he was not formally vindicated until Pope John Paul II issued an official statement in 1992.

These days, astronomical discoveries rarely create direct social conflict. Instead, their most direct social implications stem from the links between science and technology. The same discoveries that help us understand our universe spur the creation of techniques and instruments that can be used in medicine, engineering, and weaponry. One of the central issues of our time is whether these new technologies ultimately will improve our lives or facilitate our destruction. Could the study of the universe influence this issue?

While attempts to answer such a question generally fall beyond the bounds of science, and hence beyond the scope of a science textbook, you might ponder at least one person's answer. In the late 1600s, the Dutch scientist Christiaan Huygens made numerous astronomical discoveries, including the first reasonable estimates of the distances to the stars. He thereby recognized that the planets and the stars constitute real *worlds,* in many cases worlds far larger than the Earth. Huygens's hope that this knowledge might change human behavior for the better is reflected in the following quotation (c. 1690):

> *How vast those Orbs must be, and how inconsiderable this Earth, the Theatre upon which all our mighty Designs, all our Navigations, and all our Wars are transacted, is when compared to them. A very fit consideration, and matter of Reflection, for those Kings and Princes who sacrifice the Lives of so many People, only to flatter their Ambition in being Masters of some pitiful corner of this small Spot.*

[1]In astronomy, the term *cosmos* is synonymous with *universe.*

1.2 Our Cosmic Address

If we are not the center of the universe, then what *is* our place? In very general terms, the answer is shown in Figure 1.1. The Earth is the third (in order of distance) of nine planets that orbit the Sun. The Earth is orbited by the Moon, and most of the other planets also are orbited by one or more moons. Our **solar system** consists of the Sun, the planets and their moons, and the myriad of smaller objects, such as asteroids and comets, that orbit the Sun.

Our Sun is a star, just like the countless stars in our night sky. The Sun and all the stars we can see with the naked eye make up only a small part of a huge, disk-shaped collection of stars called the **Milky Way Galaxy.** A galaxy is a great island of stars in

Basic Astronomical Definitions

Star A large, glowing ball of gas that generates energy through nuclear fusion in its core. The term *star* is sometimes applied to objects that are in the process of becoming true stars (e.g., protostars) and to the remains of stars that have died (e.g., neutron stars).

Planet An object that orbits a star and that, while much smaller than a star, is relatively large in size. There is no "official" minimum size for a planet, but the nine planets in our solar system are all at least 2,000 kilometers in diameter. Planets may be rocky, icy, or gaseous in composition, and they shine primarily by reflecting light from their star.

Moon **(or *satellite*)** An object that orbits a planet. The term *satellite* is also used more generally to refer to *any* object orbiting another object.

Asteroid A relatively small, rocky object that orbits a star. Asteroids are sometimes called *minor planets* because they are similar to planets but smaller.

Comet A relatively small, icy object that orbits a star.

the Local Supercluster

the Local Group

GURE 1.1 Our place in the universe.

Earth

the Milky Way galaxy
the solar system

space, containing from a few hundred million to a trillion or more stars. The Milky Way Galaxy is relatively large, containing more than 100 billion stars. Our solar system is located about two-thirds of the distance from the galactic center to the edge of the galactic disk.

Many galaxies congregate in small **groups** or larger **clusters** that may contain hundreds or thousands of galaxies. The Milky Way Galaxy belongs to a group of 30 or so galaxies called the **Local Group.**

On the largest scale, groups, clusters, and isolated individual galaxies appear to be loosely associated into giant chains and sheets that span great distances across the universe. Between these vast structures lie huge **voids** containing few if any galaxies. Some of the galaxy structures appear more tightly grouped than others and are called **superclusters.** The supercluster to which our Local Group belongs is called, not surprisingly, the **Local Supercluster.**

Finally, the **universe** is the sum total of all matter and energy. That is, it encompasses the superclusters and voids, and everything within them. As you progress through this text, you will learn much more about the different levels of structure in the universe. To keep the hierarchy straight, you might imagine how a faraway friend would address a postcard to someone here on Earth (Figure 1.2).

1.3 Our Cosmic Origins

Having briefly discussed our place in the universe, we turn next to another fundamental question: How did we get here? Remarkably, modern science is beginning to piece together parts of the answer. Much of the rest of this text explains in considerable detail the scientific evidence concerning our cosmic origins. This evidence will help you understand how various astronomical processes are related to our lives on Earth. For now, however, let's look at a quick overview of this scientific story of creation, as summarized in Figure 1.3 (p. 8–9).

Telescopic observations of distant galaxies show that the entire universe is **expanding.** That is, average distances between galaxies are increasing with time. If the universe is expanding, we can logically conclude that everything was closer together in the past. From the observed rate of expansion, we estimate that the expansion must have started somewhere between 10 and 16 billion years ago. This beginning is commonly referred to as the **Big Bang.**

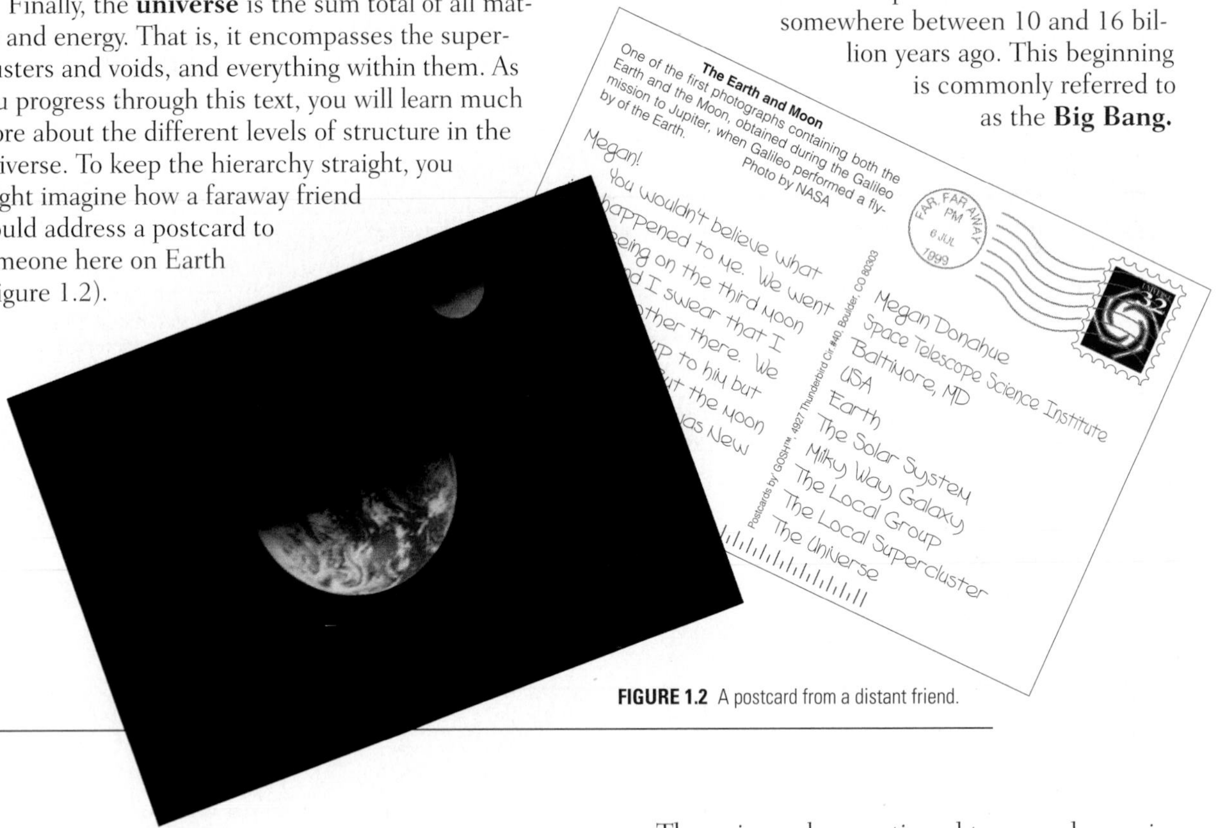

FIGURE 1.2 A postcard from a distant friend.

TIME OUT TO THINK *Some people think that our tiny physical size in the vast universe makes us insignificant. Others think that our ability to learn about the wonders of the universe gives us significance despite our small size. What do you think?*

The universe has continued to expand ever since the Big Bang, but not without some very important changes taking place. In particular, the force of *gravity* that attracts all objects to all other objects has presumably slowed the overall expansion, and in some relatively small regions of the universe gravity has halted the expansion altogether. Within a billion years of the Big Bang, gravity had created the local concentrations of matter that became the galaxies we see today, including our own Milky Way.

Gravity also drives the collapse of smaller clumps of gas and dust within galaxies, thereby forming stars. A typical star-forming cloud may give birth to hundreds of thousands of stars over a period of a few million years. As a fragment of the gas from a star-forming cloud collapses to make an individual star, the gas tends to swirl into a disk around the newly forming star. Chunks of metal, rock, or ice can solidify and combine in this swirling disk, slowly building into planets. We believe that our entire solar system, including the Sun and the Earth, formed in this way about 4.6 billion years ago—or roughly $\frac{1}{2}$ to $\frac{2}{3}$ of the way through the 10- to 16-billion-year history of the universe.

Humans have appeared only very recently in the Earth's history, but we can trace the origins of the materials from which we are made all the way back to the Big Bang. The Big Bang produced only the two simplest elements, *hydrogen* and *helium,* along with a trace amount of the element lithium. All other elements were manufactured by stars through **nuclear fusion,** in which light elements fuse (i.e., their nuclei join together) to form heavier ones. The energy released by nuclear fusion is what makes stars shine. Stars shine for most of their lives by fusing hydrogen into helium, but near the ends of their lives advanced fusion reactions in the more massive stars produce elements such as carbon, oxygen, nitrogen, and iron. Many of these massive stars die in titanic explosions that release the heavy elements into space, where they mix with other gases and dust in the galaxy. In a sense, galaxies function as giant recycling plants, recycling material expelled from dying stars into subsequent generations of stars.

The processes of heavy element production and cosmic recycling had already been taking place for several billion years by the time our solar system formed about 4.6 billion years ago. The cloud that gave birth to our solar system still was about 98% hydrogen and helium, but the other 2% contained all the other chemical elements. The small rocky planets, including the Earth, were made from a small part of this 2%. We do not know exactly how the elements on the Earth's surface developed into the first forms of life, but life was already flourishing on Earth more than 3 billion years ago. Biological evolution took over once life arose, leading to the great diversity of life on Earth today—and to us.

In summary, all the material from which we and the Earth are made (except hydrogen and most helium) was created inside stars that died before the birth of our Sun and recycled into our solar system through the dynamics of the Milky Way Galaxy. We are intimately connected to the stars because we are the products of stars. In the words of astronomer Carl Sagan (1934–96), we are "star stuff."

1.4 Images of Time

We study the universe by studying light from distant stars and galaxies. Light travels extremely fast by earthly standards: The speed of light is 300,000 kilometers per second, a speed at which it would be possible to circle the Earth nearly eight times in just 1 second! Nevertheless, even light takes substantial amounts of time to travel the vast distances in space. For example, light takes about 1 second to travel from the Moon to the Earth, and about 8 minutes to travel from the Sun to the Earth.

Light from the stars takes many years to reach us, and distances to the stars are measured in units called **light-years.** One light-year is the distance that light can travel in 1 year, which is about 10 trillion kilometers, or 6 trillion miles. Note that a light-year is a unit of *distance,* not of time.

Common Misconceptions: Light-Years

A recent advertisement illustrated a common misconception by claiming, "It will be light-years before anyone builds a better product." This advertisement makes no sense, because light-years are a unit of *distance,* not a unit of time. If you are unsure whether the term *light-years* is being used correctly, try testing the statement by remembering that 1 light-year is approximately 10 trillion kilometers, or 6 trillion miles. The advertisement then reads, "It will be 6 trillion miles before anyone builds a better product," which clearly does not make sense.

The brightest star in the night sky, Sirius, is about 8 light-years from our solar system, which means that it takes light from Sirius about 8 years to reach us. Thus, when we look at Sirius, we see light that left the star about 8 years ago. The Orion Nebula, a star-forming region visible to the naked eye as a small, cloudy patch in the sword of the constellation Orion, lies about 1,500 light-years from Earth. Light therefore takes about 1,500 years to travel from the Orion Nebula to the Earth, so we see the Orion Nebula as it looked about 1,500 years ago—at about the time of the fall of the Roman Empire. We cannot see how the Orion Nebula looks today because the light it is emitting right now will not reach us for another

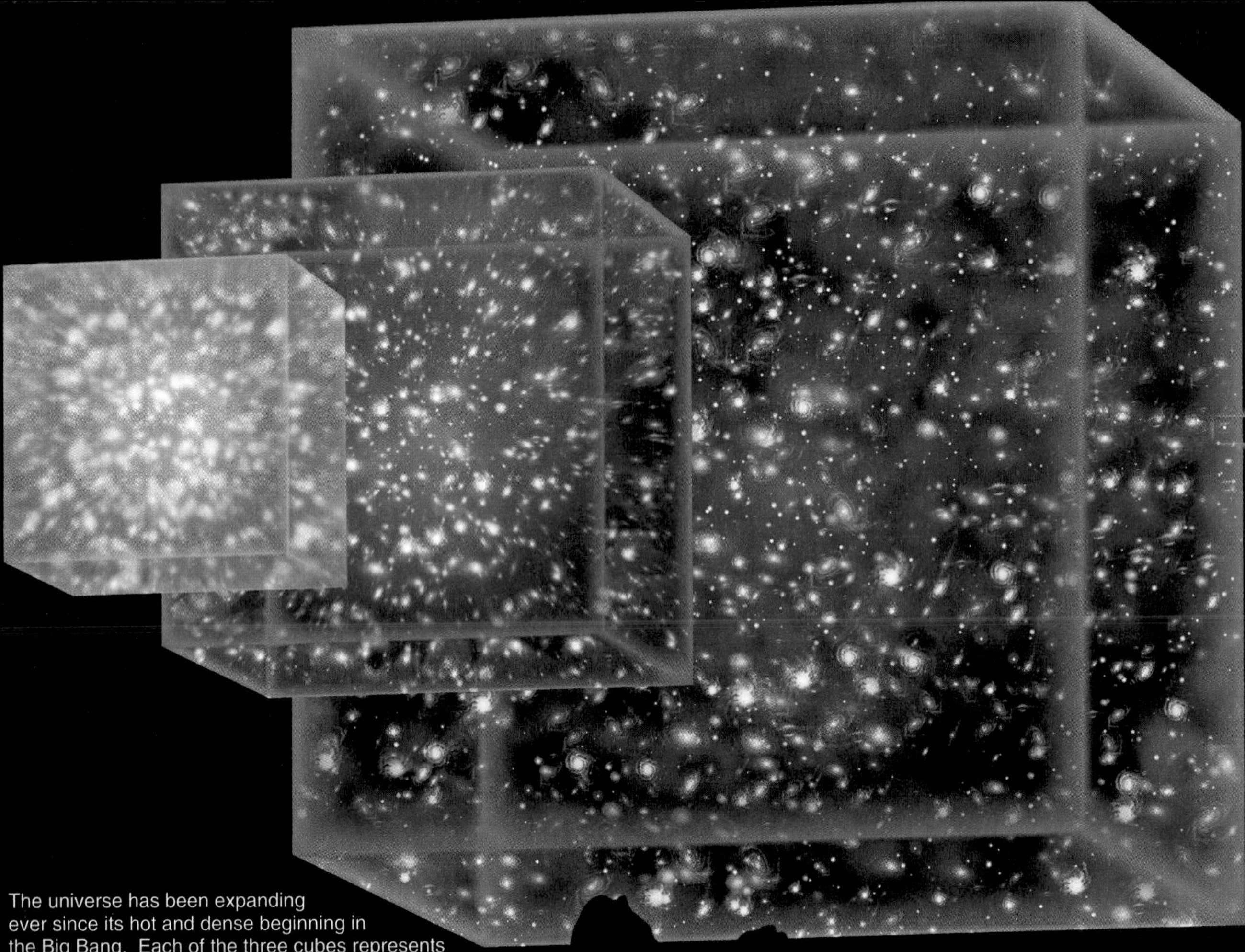

The universe has been expanding ever since its hot and dense beginning in the Big Bang. Each of the three cubes represents the same region of the universe, showing how the region expands with time.

FIGURE 1.3 Our cosmic origins: All the matter and energy in the universe was created in the Big Bang. This sequence of paintings shows the progression of that matter and energy from the Big Bang to human life. Note that the elements from which we are made were produced in stars that shined long ago, and these elements formed the Earth, thanks to the recycling role played by our galaxy.

The Earth was built with elements produced in stars that lived and died in the Milky Way before our solar system formed.

Within a few billion years after the Big Bang, gravity caused local concentrations of matter to collapse into galaxies even while the universe as a whole continued to expand.
Galaxies like the Milky Way act as cosmic recycling plants: stars are made from the material in clouds of gas and dust within the galaxy, and stars return material to interstellar space when they die.
A star forms at the center of a collapsing cloud of gas and dust, and planets may form in the spinning disk that surrounds the young star.
Stars shine with the energy produced by nuclear fusion in their cores; the fusion also creates heavier elements from lighter ones.
Massive stars explode when they die, scattering the elements they've produced into space.

1,500 years. In fact, if any major events have occurred in the Orion Nebula in the past 1,500 years, we cannot yet know about them because the light from these events has not yet reached us.

Because it takes time for light to travel through space, *the farther away we look in distance, the further back we look in time.* If we observe the center of our galaxy, which is about 28,000 light-years away, we are observing events that took place 28,000 years ago. If we look to a galaxy that lies 10 million light-years away, we are seeing it as it was 10 million years ago. If we observe a distant cluster of galaxies that lies 1 billion light-years away, we are seeing the cluster as it was 1 billion years ago.

This fact—that when we look out in space we look back in time—provides us with an incredible ability: We can observe the evolution of the universe. Although the actual observations are difficult, the idea is simple and depends only on two assumptions: Galaxies everywhere are more or less alike, and they all formed at about the same time in the early universe. (As we will see later, both assumptions are supported by observations.) For example, when we observe galaxies that are 1 billion light-years away, we see them as they were 1 billion years ago—and hence when they were 1 billion years younger than they are today. If we want to know what galaxies looked like 5 billion years ago, we need only look at galaxies 5 billion light-years away. By observing many galaxies at many distances—and hence at many times in the past—we observe how galaxies as a group have changed as the universe has aged.

Ultimately, the speed of light limits the portion of the universe that we can see. For example, if the universe is 12 billion years old, then light from galaxies more than 12 billion light-years away would not have had time to reach us.[2] We would therefore say that the **observable universe** extends 12 billion light-years in all directions from Earth. Objects that lie beyond 12 billion light-years would not be part of our *observable* universe since we could not *observe* them. Furthermore, if we looked to objects that are *nearly* 12 billion light-years away, we would see them as they were nearly 12 billion years ago—shortly after the Big Bang. (If the universe is less or more than 12 billion years old, then the size of the observable universe is correspondingly smaller or larger.)

It is also amazing to realize that any "snapshot" of a distant galaxy or cluster of galaxies is a picture of *both* space *and* time. For example, the Great Galaxy in Andromeda, also known as M31, lies about 2.5 million light-years from Earth. Figure 1.4 therefore is a picture of how M31 looked about 2.5 million years ago, when early humans were first walking the Earth. Moreover, the *diameter* of M31 is about 100,000 light-years, so light from the far side of the galaxy required 100,000 years longer to reach us than light from the near side. Thus, the picture of M31 shows 100,000 years of time: This single photograph captured light that left the near side of the galaxy some 100,000 years later than did the light it captured from the far side. When we study the universe, it is impossible to separate space and time.

[2]This explanation is somewhat oversimplified, but it captures the basic point that we can see only to a certain distance that is dependent on the age of the universe (see Chapter 19).

Mathematical Insight 1.1 How Far Is a Light-Year?

It's easy to calculate the distance represented by a light-year if you recall that

$$\text{distance} = \text{speed} \times \text{time}$$

For example, if you travel at a speed of 50 kilometers per hour for 2 hours, you will travel 100 kilometers. A light-year is the distance covered by light, traveling at a speed of 300,000 kilometers per second, in a time of 1 year. In the process of multiplying the speed and time, you must convert the year to seconds in order to arrive at a final answer in units of kilometers.

$$\begin{aligned}1 \text{ light-year} &= (\text{speed of light}) \times (1 \text{ yr}) \\ &= \left(300{,}000 \frac{\text{km}}{\text{s}}\right) \times \left(1 \text{ yr} \times \frac{365 \text{ days}}{1 \text{ yr}} \times \frac{24 \text{ hr}}{1 \text{ day}} \times \frac{60 \text{ min}}{1 \text{ hr}} \times \frac{60 \text{ s}}{1 \text{ min}}\right) \\ &= 9{,}460{,}000{,}000{,}000 \text{ km}\end{aligned}$$

That is, "1 light-year" is just an easy way of saying "9.46 trillion kilometers" or "almost 10 trillion kilometers."

FIGURE 1.4 M31, the Great Galaxy in Andromeda. This photograph, taken with a camera attached to a small telescope, shows a galaxy some 100,000 light-years in diameter containing more than 100 billion stars.

1.5 Discovering the Universe for Yourself

In these first few pages, we have discussed facts about the universe that would have seemed incredible to our ancestors. But we are just beginning to learn how we fit into the cosmos. Most of our knowledge has been acquired relatively recently, and the present time is particularly exciting—advances in technology are spurring an unprecedented rate of astronomical discovery. A new generation of telescopes, both on the ground and in space, is probing the depths of the universe. Increasingly sophisticated space probes are collecting new data about the planets and other objects in our solar system. And rapid advances in computing technology are helping scientists analyze the vast amounts of new data and model the processes that occur in planets, stars, galaxies, and the universe.

One of the goals of this book is to help *you* share in the ongoing human adventure of astronomical discovery. One of the best ways to become a part of this adventure is to discover the universe for yourself by doing what other humans have done for thousands of generations: Go outside, observe the sky around you, and contemplate the awe-inspiring universe of which you are a part. You will then be more able to appreciate the many discoveries we will discuss throughout this book. To help you get started, let's take an imaginary stargazing trip and see what we can see.

The Dome of the Sky

Imagine standing in an open field, looking at the sky above you. If no obstacles (e.g., buildings, trees, or hills) affect your view, you see a clear boundary between Earth and sky called your **horizon.** This clear division between Earth and sky makes it *appear* that you are standing on a flat surface and that the sky is a great dome encompassing the world. Many ancient peoples mistook this appearance for reality, imagining the Earth to be flat and lying under a great celestial dome.

In general, you can pinpoint the position of any object in your sky by stating its **direction** along your horizon[3] and its **altitude** above the horizon. Figure 1.5 shows a person pointing to a star located in a southeasterly direction at an altitude of 60°. Note that an object on the horizon is defined to have an altitude of 0° and an object at your **zenith**—the point directly above your head—has an altitude of 90°. The dividing line between east and west in your sky is your **meridian,** a half-circle extending from your horizon (altitude 0°) due south, through your zenith, to your horizon due north.

[3]A more technical way to specify direction along the horizon is with *azimuth* (or *bearing*), which is measured clockwise around the horizon from due north. For example, the azimuth of due north is 0°, due east is 90°, due south is 180°, and due west is 270°.

FIGURE 1.5 Definitions in your local sky.

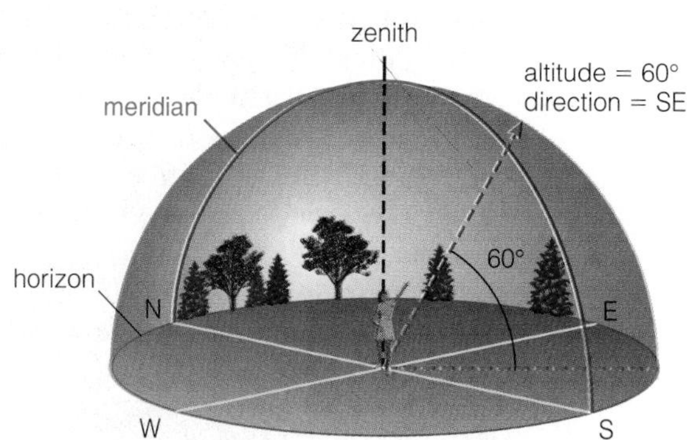

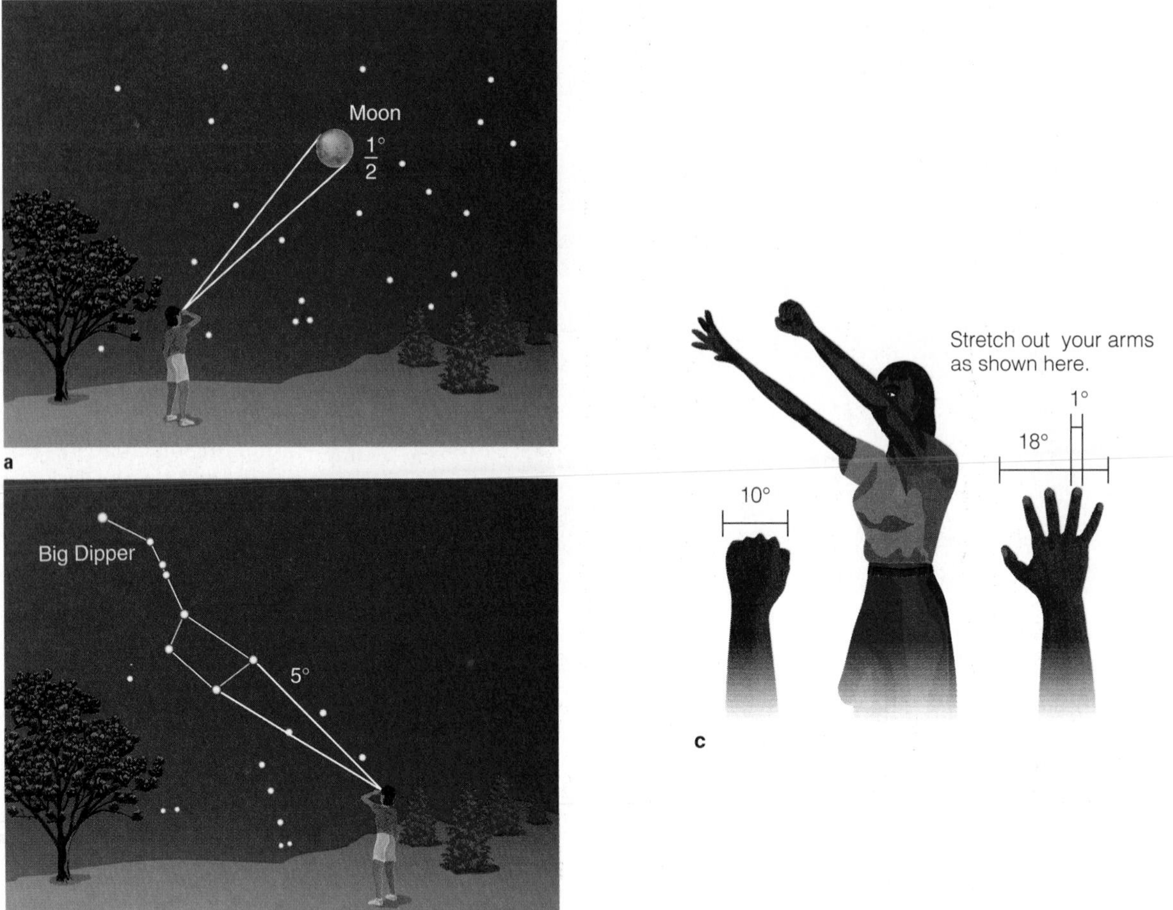

FIGURE 1.6 (**a**) The angular size of the Moon is about $\frac{1}{2}$°. (**b**) The angular distance between the pointer stars of the Big Dipper is about 5°. (**c**) You can estimate angular sizes or distances with your outstretched hand.

It's important to realize that, when we look at objects in the sky, we have no immediate way of knowing their actual size or how distant one object is from another. The Moon and Sun appear to be about the same size in our sky, but this is only because the Moon is much closer than the Sun. The diameter of the Sun is actually about 400 times larger than that of the Moon. The stars Sirius and Betelgeuse appear similar in brightness and relatively near each other in the sky, giving us no hint of the fact that they are actually separated by some 500 light-years.

Because we cannot immediately determine actual sizes or distances, we can measure only *angles* when we look at objects in the sky. We measure these angles by extending imaginary lines outward from our eyes. For example, Figure 1.6a shows that the **angular size** of the Moon is about $\frac{1}{2}$°, which is also the angular size of the Sun. Figure 1.6b shows that the **angular distance** between the two pointer stars of the Big Dipper is about 5°. You can make rough estimates of angular sizes or distances by using your outstretched hand as a simple measuring device (Figure 1.6c). For more precise astronomical measurements, we subdivide each degree into 60 **arcminutes** and further subdivide each arcminute into 60 **arcseconds:**

$$1° = 60 \text{ arcminutes} = 3{,}600 \text{ arcseconds}$$

When writing angles, we abbreviate arcminutes by using the symbol ′ and arcseconds by using the symbol ″. For example, 35°27′15″ is read "35 degrees, 27 arcminutes, and 15 arcseconds."

The Celestial Sphere

Knowing that the Earth is round, you might imagine the dome of your sky to be part of a great **celestial sphere** that surrounds the Earth (Figure 1.7). The sky looks like a dome because only *half* the celestial sphere is visible above your horizon at any particular

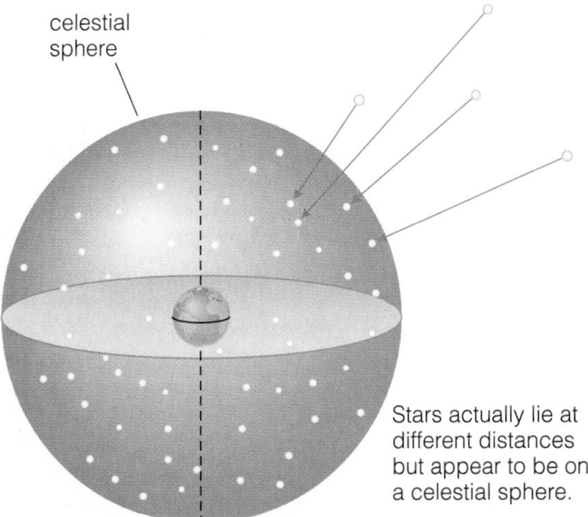

FIGURE 1.7 The Earth is *imagined* to be at the center of a celestial sphere.

moment. Different stars lie at different distances from Earth, but all stars are so far away that we lack any sense of depth perception when we look at them. Thus, the stars all *appear* to reside on the celestial sphere. To the naked eye, even the objects in our own solar system—the Sun, Moon, and planets—appear to travel among the stars on the same celestial sphere. Thus, even though the celestial sphere has no *physical* reality, the idea of a celestial sphere can be a useful tool for learning about the sky.

Patterns in the Sky

Shortly after sunset, as daylight fades to darkness, the sky appears to fill slowly with stars. On clear, moonless nights far from city lights, as many as 2,000–3,000 stars may be visible to your naked eye. As you look at the stars, your mind might group them into many different patterns. If you observe the sky night after night or year after year, you will recognize the same patterns of stars.

People of nearly every culture have given names to patterns in the sky.[4] The pattern that the Greeks named Orion, the hunter (Figure 1.8), was seen as a supreme warrior called *Shen* by the ancient Chinese. Hindus in ancient India also saw a warrior, called *Skanda,* who rode a peacock as the general of a great celestial army. The three stars of Orion's belt were seen as three fishermen in a canoe by Aborigines of northern Australia, an idea remarkably similar to that of the Wasco Indians, who lived along the Columbia River in Oregon. As seen from southern California, the stars of Orion's belt climb almost straight up into the sky as they rise in the east, which may explain why the Chemehuevi Indians of the California desert saw these stars as a line of three sure-footed mountain sheep. These are but a few of the many names, each accompanied by a rich folklore, given to the pattern of stars that we call Orion.[5]

The patterns of stars seen in the sky usually are called *constellations.* In astronomy, however, the term **constellation** refers to a *region* of the sky. Any place you point to in the sky belongs to some constellation;

FIGURE 1.8 The pattern of stars in the constellation Orion. In this photograph, brighter stars appear larger because they are overexposed.

familiar patterns of stars merely help locate particular constellations. For example, the constellation Orion includes all the stars in the familiar pattern of the hunter, *along with* the region of the sky in which these stars are found.

The official borders of the constellations were set in 1928 by members of the International Astronomical Union (IAU), an association of astronomers from around the world. The IAU divided the celestial sphere into 88 constellations whose borders correspond roughly to the star patterns recognized by

[4]The rare exceptions primarily were peoples who inhabited dense tropical forests, such as on the island of Papua New Guinea in the Pacific or in the Amazon forest in South America, and therefore rarely saw a clear night sky.

[5]These and other stories about the constellation Orion are described in greater detail in E. C. Krupp, *Beyond the Blue Horizon* (Oxford University Press, 1991).

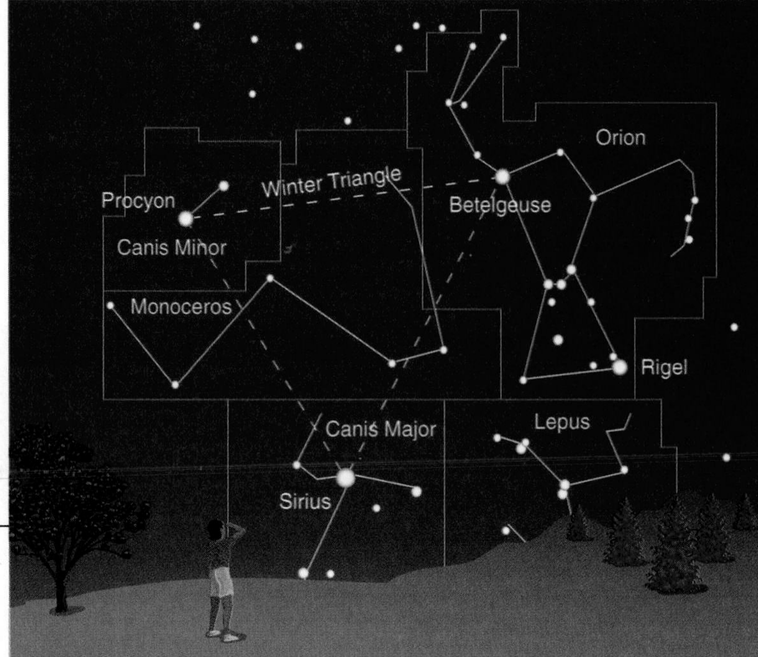

FIGURE 1.9 Red lines mark official borders of several constellations near Orion. Yellow lines connect recognizable patterns of stars within constellations. Sirius, Procyon, and Betelgeuse form a pattern spanning several constellations and called the Winter Triangle; it is easy to find on clear winter evenings.

Europeans (Figure 1.9). Thus, despite the wide variety of names given to patterns of stars by different cultures, the official names of constellations visible from the Northern Hemisphere can be traced back to the ancient Greeks and to other cultures of Southern Europe, the Middle East, and Northern Africa. No one knows exactly when these constellations were first named, although it probably was at least 5,000 years ago. The official names of the constellations visible from the Southern Hemisphere are primarily those given by seventeenth-century European explorers.

Learning your way around the constellations is no more difficult than learning the way around your neighborhood. The patterns of the stars do not change over the course of a human lifetime, and recognizing the patterns of just 20–40 constellations is enough to make the entire sky seem familiar. The best way to learn the constellations is just to go out and practice; it may help to start with a few visits to a planetarium and to purchase an inexpensive star chart to help you identify the patterns you see.

The Milky Way

As your eyes adapt to darkness at a dark site, you'll begin to see the whitish band of light called the *Milky Way*. It is from this band of light that our Milky Way Galaxy gets its name. You can see only part of the Milky Way at any particular time, but it stretches all the way around the celestial sphere, passing through more than a dozen constellations. If you look carefully, you will notice that the Milky Way is narrow in some places and wider in others and has dark fissures running through it. Like the patterns of stars in the sky, the patterns of light and dark in the Milky Way remain fixed among the constellations. The widest and brightest parts of the Milky Way are most easily seen from the Southern Hemisphere (Figure 1.10), which probably explains why the Aborigines of Australia gave names to the dark patterns they saw within the Milky Way in the same way that other cultures gave names to patterns of stars.

The band of light called the Milky Way bears an important relationship to the Milky Way Galaxy: *It traces the galactic plane as it appears from our location in the outskirts of the galaxy.* The Milky Way Galaxy is shaped like a thin pancake with a bulge in the middle (Figure 1.11). We view the universe from our location within this "pancake," about two-thirds of the distance outward from its center. When we look in any direction *within* the plane of the galaxy, we see countless stars, along with interstellar gas and dust. It is these stars and glowing clouds of gas that form the band of light we call the Milky Way. The densest interstellar clouds of gas and dust are dark and obscure our view of the stars behind them. These dense clouds generally are confined to a particularly thin, central layer of the galactic plane, forming the main narrow fissure that we see running through the Milky Way. The central bulge of the galaxy makes the Milky

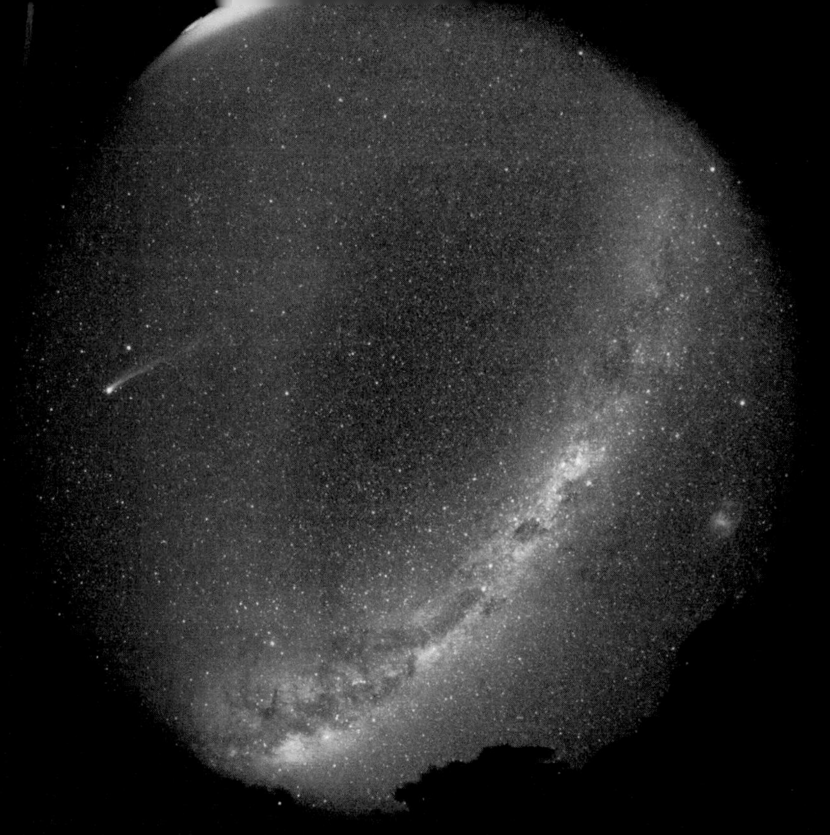

FIGURE 1.10 A "fish eye" view of the Milky Way in the Australian sky. Comet Hale–Bopp is also visible in this 1997 photo.

FIGURE 1.11 Artist's conception of the Milky Way galaxy from afar, showing how the galaxy's structure affects our view from Earth. When we look *into* the galactic plane in any direction, our view is blocked by stars, gas, and dust. Thus, we see the galactic plane as the band of light we call the Milky Way, stretching a full 360° around our sky (i.e., around the celestial sphere). We have a clear view to the distant universe only when we look away from the galactic plane.

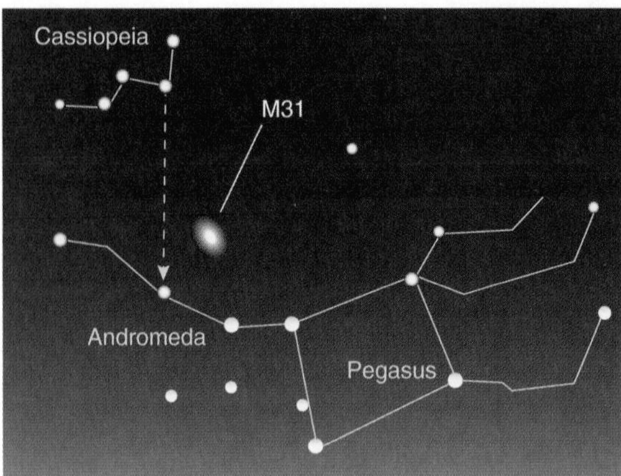

FIGURE 1.12 Location of M31 in the night sky.

Way slightly wider in the direction of the galactic center, which is the direction of the constellation Sagittarius in our sky.

When we look in directions pointing *outward* from our location within the galactic plane, we see fewer stars, and there is relatively little gas and dust to obscure our view. Thus, we have a clear view to the far reaches of the universe, limited only by what our eyes or instruments allow. On a clear, dark night away from city lights, you may see a fuzzy patch in the constellation Andromeda (Figure 1.12). You might think this patch is just a small cloud. Actually, you are seeing far beyond the boundaries of the Milky Way Galaxy to the Great Galaxy in Andromeda (M31), which lies 2.5 million light-years from Earth.

TIME OUT TO THINK *Contemplate the fact that light from M31 journeyed through space for some 2.5 million years to reach you and is the combined light of hundreds of billions of stars. Do you think it possible that among some of those stars there are students like yourself gazing outward in amazement at the Milky Way Galaxy?*

Common Misconceptions: Stars in the Daytime

Because we don't see stars in the daytime, some people believe that the stars disappear in the daytime and "come out" at night. In fact, the stars are always present. The only reason you cannot see stars in the daytime is that their dim light is overwhelmed by the bright daytime sky. You *can* see bright stars in the daytime with the aid of a telescope, and you may see stars in the daytime if you are fortunate enough to observe a total eclipse of the Sun. Astronauts also see stars in the daytime: Above the Earth's atmosphere, where no air is present to scatter sunlight through the sky, the Sun appears as a bright disk against a dark sky filled with stars.

The Sun, Moon, and Planets

Unlike the stars and the Milky Way, the Sun, Moon, and planets do not always appear in the same places among the constellations. As we discuss in Chapter 3, everything in the universe is in constant motion. The stars are actually moving at surprisingly high speeds, but this motion is undetectable to the naked eye because the stars are so far away. It's much easier to notice the motion of the Sun, Moon, and planets, but even these objects appear to remain *nearly* stationary among the stars on any particular day or night. Of course, it's difficult to see the Sun among the constellations because whenever the Sun is above the horizon (that is, in the daytime), its bright light drowns out the faint light of other stars. But on many nights you will see the Moon and some of the five planets that are visible to the naked eye: Mercury, Venus, Mars, Jupiter, and Saturn.[6]

On any particular night, the locations of the planets among the constellations depend on the relative positions of the Earth and the planets in their orbits around the Sun. Mercury can be seen only infrequently, and then only just after sunset or before sunrise because it is so close to the Sun. Newspapers and astronomy magazines may alert you to times when you have a good chance of seeing Mercury. Venus often shines brightly in the early evening in the west or before dawn in the east; if you see a very bright "star" in the early evening or early morning, it probably is Venus. Jupiter, when it is visible at night, is the brightest object in the sky besides the Moon and Venus. Mars is recognizable by its red color, but be careful not to confuse it with a bright red star. Saturn also is easy to see with the naked eye, but because many stars are just as bright as Saturn, it helps to know where to look. (Also, planets tend not to twinkle as much as stars.) Sometimes several planets and the Moon may appear close together in the sky, offering a particularly beautiful sight (Figure 1.13).

The Circling Sky

If you remain outside for a few hours, you will notice that the entire celestial sphere appears to be rotating, carrying the Sun, Moon, planets, and stars across the sky. In reality, it is the *Earth* that is rotating on its axis once each day. The Earth rotates from *west to east,* so the celestial sphere appears to rotate around

[6]Uranus also is sometimes visible to the naked eye; it is so faint and moves so slowly among the stars that it was not recognized as a planet by ancient astronomers.

FIGURE 1.13 This photograph shows Venus, Mars, and Jupiter (the three white dots, from left to right) appearing close together and near the Moon in November, 1995.

us in the opposite direction, from *east to west* (Figure 1.14). By remembering that the Sun rises in the east and sets in the west, you can always figure out that the Earth rotates from west to east (which is *counter-clockwise* as viewed from above the North Pole).

Two special points remain stationary in the sky as the celestial sphere appears to turn: the **north celestial pole** (**NCP**) and the **south celestial pole** (**SCP**). The celestial poles are the points on the celestial sphere that represent extensions of the Earth's rotational axis into space. All other points on the celestial sphere appear to circle around this axis once each day. The largest circle is the **celestial equator** (**CE**), which represents an extension of the Earth's equator onto the celestial sphere.

In the Northern Hemisphere, you can find the north celestial pole in your sky because it lies within 1° of the *North Star,* Polaris. You can locate Polaris with the aid of the *pointer stars* at the end of the bowl of the Big Dipper (Figure 1.15a). If you face north and watch for a while, you will see that all the stars move in *counterclockwise* circles around the north celestial pole. A full circle requires a full day to complete; for example, in 6 hours ($\frac{1}{4}$ day) the sky rotates 90° around the north celestial pole. You can capture an image of the circling sky on film with a time-exposure photograph (Figure 1.15b). Stars that make relatively small circles around the north celestial pole remain continuously above your horizon and are said to be **circumpolar** in your sky. All other stars (along with the Sun, Moon, and planets) have only a portion of their daily circle above your horizon and hence rise in the east and set in the west. In the Southern Hemisphere, you can watch the stars circle slowly *clockwise* around the south celestial pole. Although no bright star marks the south celestial pole, you can find it with the aid of the Southern Cross (Figure 1.16).

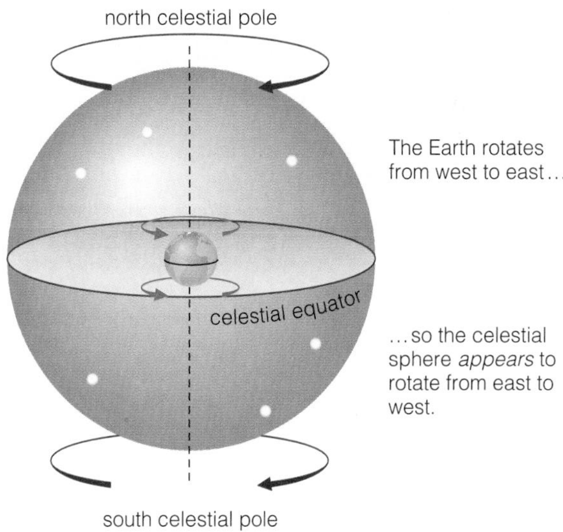

FIGURE 1.14 Apparent rotation of the celestial sphere.

TIME OUT TO THINK *If it is nighttime and clear, go outside and find the Big Dipper and the north celestial pole (or the Southern Cross and the south celestial pole if you are in the Southern Hemisphere). Where do you expect the Big Dipper (or Southern Cross) to appear in a few hours? Set an alarm to remind yourself to go back out and check your prediction in 3–6 hours.*

The fact that you can see the north celestial pole only from the Northern Hemisphere and the south celestial pole only from the Southern Hemisphere points out the fact that your location on Earth affects what you see in the sky. The ancient Greek scientist Anaximander (c. 610–546 B.C.) used reports from

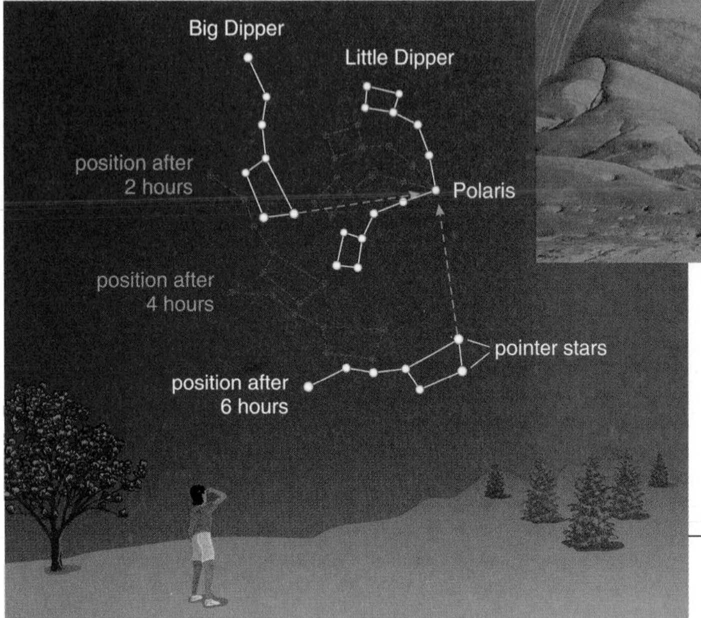

FIGURE 1.15 (**a**) Positions of the Big Dipper and Little Dipper are shown over a period of 6 hours. Note that the sky appears to turn counterclockwise around the north celestial pole. (**b**) This time-exposure photograph, taken at Arches National Park in Utah, shows how the Earth's rotation causes stars to make daily circles around our sky.

travelers about how the sky changes with travel north or south to conclude that the Earth is *not* flat; he thought it might be a cylinder curved in the north-south direction. By about 500 B.C., Pythagoras (c. 560–480 B.C.) had concluded that the variation in the sky comes about because the Earth is a sphere. Thus, the Greeks knew that the Earth is round some 2,000 years before the time of Columbus.

The location of any point on Earth can be described by its north or south **latitude** and its east or west **longitude** (Figure 1.17). Latitude is defined to be 0° at the equator, so the North Pole and South Pole have latitude 90°N and 90°S, respectively. Note that "lines of latitude" actually are circles running parallel to the equator. "Lines of longitude" are semicircles extending from the North Pole to the South Pole. The line of longitude passing through Greenwich, England, is defined to have a longitude of 0°; the choice of Greenwich was made by international treaty in

FIGURE 1.16 Positions of the Southern Cross are shown over a period of 6 hours. Note that the sky appears to turn clockwise around the south celestial pole.

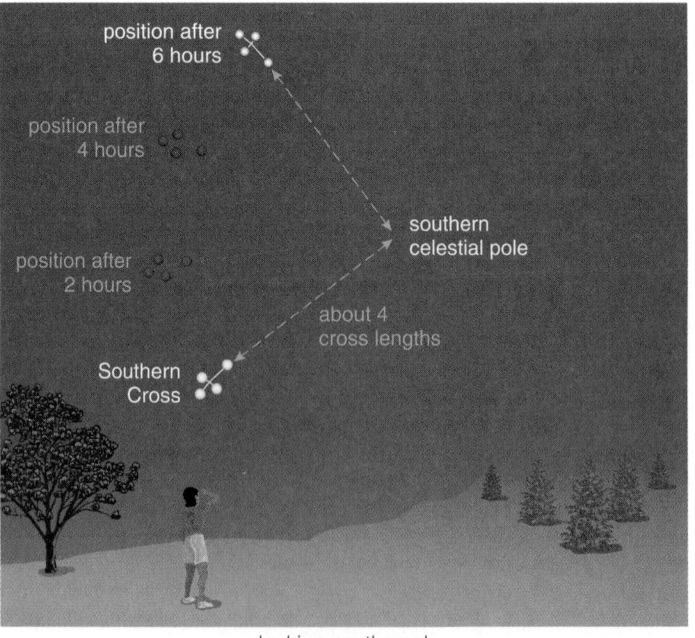

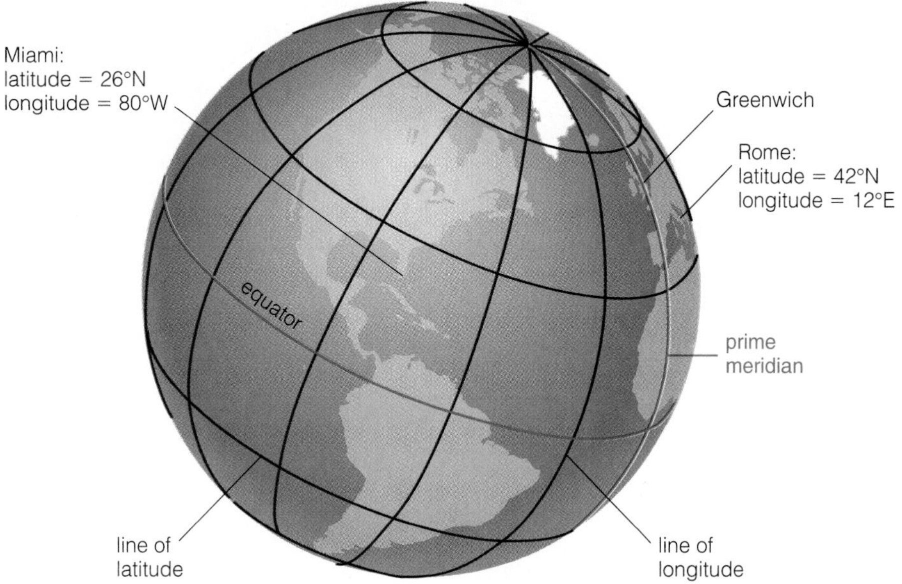

FIGURE 1.17 Depictions of latitude and longitude.

1884. Stating a latitude and a longitude pinpoints a location. For example, Figure 1.17 shows that Rome lies at about 42°N latitude and 12°E longitude and that Miami lies at about 26°N latitude and 80°W longitude.

TIME OUT TO THINK *Lines of longitude are sometimes called* meridians, *and the line of 0° longitude is called the* prime meridian. *What is the relationship between a meridian of longitude and* your *meridian that goes from due north on your horizon, through your zenith, to due south on your horizon?* (Hint: *Suppose the line of longitude passing through your location were projected into the sky. Where would it appear?*)

A detailed description of how the sky varies with latitude is presented in Chapter S1, but you can get the basic idea by examining Figure 1.18. Your latitude determines your orientation on Earth relative to the celestial sphere—"down" is toward the center of the Earth, and "up" (toward your zenith) is directly away from the center of the Earth. If you're in the Northern Hemisphere, Earth's rotation makes stars near the north celestial pole circumpolar, while stars near the south celestial pole never rise above your horizon (Figure 1.18a). The opposite is true in the Southern Hemisphere (Figure 1.18b). Noncircumpolar stars rise in the east and set in the west no matter where you are located. If you study the geometry

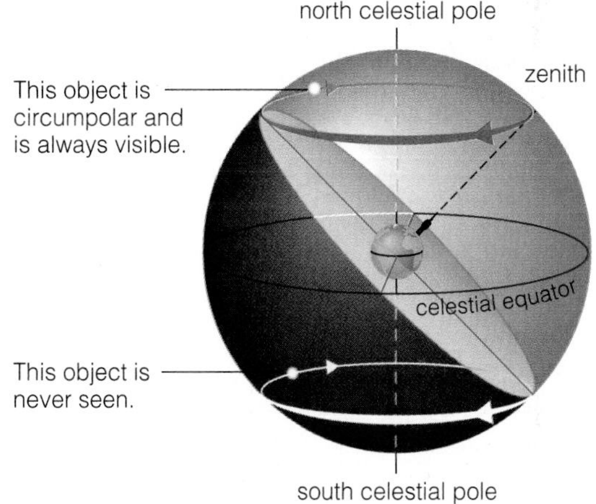

a Northern Hemisphere

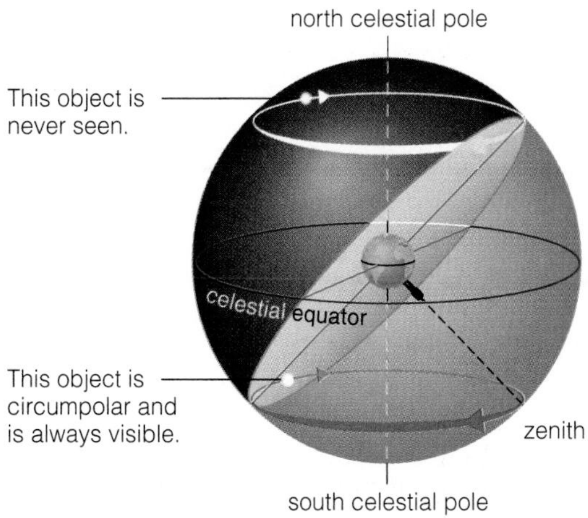

b Southern Hemisphere

FIGURE 1.18 Your latitude determines the portion of the celestial sphere visible in your sky. (**a**) A Northern Hemisphere sky. (**b**) A Southern Hemisphere sky.

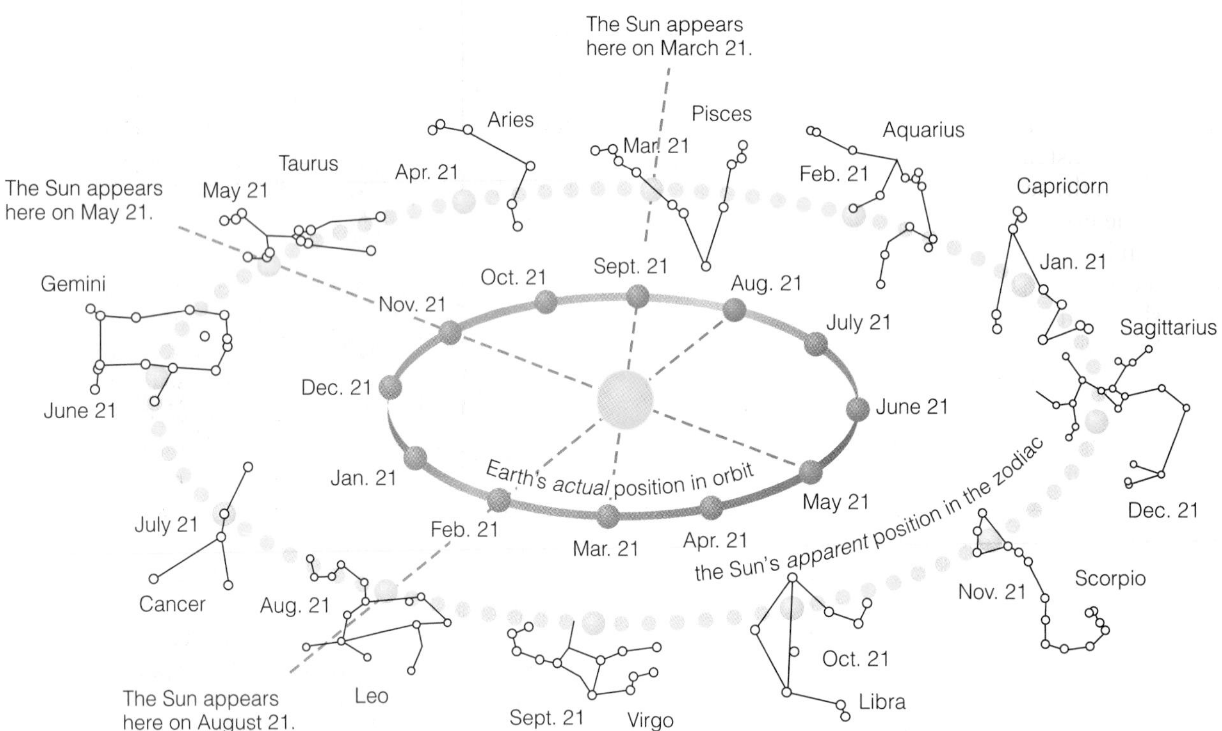

FIGURE 1.19 This diagram shows why the Sun appears to move steadily eastward around the ecliptic each year, passing through the 12 constellations that make up the zodiac. If you look at the Earth's position in its orbit on May 21, for example, you'll see that the Sun appears to be in the constellation Taurus. Similarly, from the Earth's position on August 21, the Sun appears to be in Leo. You can use this diagram to determine the constellation in which the Sun appears on various dates of the year.

carefully, you may notice another useful fact: *The altitude of the celestial pole in your sky is equal to your latitude.* For example, if the north celestial pole appears in your sky 40° above your horizon, your latitude is 40°N.

Seasonal Changes in the Night Sky

The basic patterns of motion in the sky remain the same from one day to the next: The Sun, Moon, planets, and stars trace daily circles around the sky, and those bodies that are not circumpolar from your

Full Moon

Follows new moon by about 14 days

Rises at 6 P.M.
Sets at 6 A.M.

Waning Gibbous

Follows new moon by about 17 days

Rises at 9 P.M.
Sets at 9 A.M.

Third Quarter

Follows new moon by about 21 days

Rises at 12 A.M.
Sets at 12 P.M.

Waning Crescent

Follows new moon by about 24 days

Rises at 3 A.M.
Sets at 3 P.M.

FIGURE 1.20 Phases of the Moon, along with approximate rise and set times for each phase. The photo for new moon shows only a blue sky because it is above the horizon in daylight, but it cannot be seen as it is very close to the Sun in the sky.

latitude rise in the east and set in the west. However, if you observe the sky night after night, you will notice changes that cannot be seen in a single night. One pattern that becomes obvious after a few months is that the constellations visible at a particular time change with the seasons. For example, Orion is prominent in the evening sky in February, but in September it can be seen only shortly before dawn.

You can understand the seasonal changes in the night sky by remembering that the Earth orbits the Sun. The Earth completes one orbit of the Sun each year, always moving in a counterclockwise direction as viewed from above the North Pole. From our viewpoint on the Earth, the Sun therefore appears to move steadily *eastward* around the celestial sphere over the course of the year (Figure 1.19). The Sun's apparent annual path among the constellations is called the **ecliptic.** The ecliptic passes through the 12 constellations that we collectively call the **zodiac.**

The constellations visible at a particular time of night and year depend on the Sun's apparent location among the zodiac constellations. In late August, for example, the Sun appears to be in the constellation Leo, so you won't see the stars of Leo at all—they're above your horizon only during the daytime. As the Earth rotates, Leo sets with the Sun, followed soon thereafter by Virgo. Thus, as the sky darkens, you'll see the stars of Libra come into view near your western horizon. If you want to see the stars of Gemini or Cancer in late August, you'll need to get up shortly before dawn—Gemini and Cancer will rise in the east only shortly before the Sun rises with Leo.

TIME OUT TO THINK *Based on Figure 1.19 and today's date, where does the Sun currently appear among the zodiac constellations? What zodiac constellation will you see in the west shortly after sunset? What zodiac constellation will you see in the east shortly before sunrise?*

The Lunar Cycle

Like all objects on the celestial sphere, each day the Moon rises in the east and sets in the west. But the Moon's appearance and the time at which it rises and sets all change with the cycle of **lunar phases.** Each complete cycle of phases takes about $29\frac{1}{2}$ days—hence the origin of the word *month*[7] (think of "moonth"). Figure 1.20 shows the appearance of the Moon at different phases, along with the approximate times of its rising and setting. You've probably noticed that the Moon is sometimes visible in the daytime, as the rise and set times make clear. For example, first-quarter moon is visible in the afternoon before sunset, and third-quarter moon is visible in the morning after sunrise. We'll discuss the cause of the lunar cycle in Chapter 3.

The Moon orbits the Earth in a little less than a month, going in the same direction in which the Earth orbits the Sun and nearly in the same plane. The Moon therefore appears to complete a circuit around the celestial sphere in a little less than a month, always moving steadily eastward through the

[7]More precisely, the approximately $29\frac{1}{2}$-day cycle of lunar phases defines a *lunar month.* Our calendar months vary between 28 days and 31 days.

constellations of the zodiac. You can notice this motion in the course of just a few hours if the Moon happens to be near a bright star in the sky.

TIME OUT TO THINK *Many daily newspapers provide information about the phases of the Moon, often on the weather page. Find a newspaper that does so, and look up the current phase of the Moon. Based on the approximate rise and set times shown in Figure 1.20, should the Moon be visible in your sky right now? If so, go outside and try to find it.*

Common Misconceptions: What Makes the North Star Special?

Most people are aware that the North Star, Polaris, is a special star. Contrary to a relatively common belief, however, it is *not* the brightest star in the sky; more than 50 other stars are either considerably brighter or comparable in brightness. Polaris is special because it so closely marks the direction of due north and because its altitude in your sky is nearly equal to your latitude on Earth, making it very useful in navigation.

The Ancient Mystery of the Planets

The word *planet* comes from the Greek for "wandering star" and reflects the mystery that surrounded planets in ancient times. Like the Sun and the Moon, the planets appear to move slowly through the constellations of the zodiac. Mars, for example, takes a little less than 2 years to complete a circuit through the zodiac constellations, while the more distant Jupiter takes about 12 years. However, while the Sun and the Moon always appear to move eastward relative to the stars, the planets occasionally reverse course and appear to move *westward* through the zodiac. A period during which a planet appears to move westward relative to the stars is called a period of **apparent retrograde motion** (*retrograde* means "backward"). Figure 1.21 shows the apparent position of Jupiter among the stars over the course of a year. Note that it appears to drift eastward among the stars for most of the year but spends about 4 months in apparent retrograde motion. All the other planets also have periods of apparent retrograde motion, although the duration of the period and the time from one period to the next are different for each planet.

Ancient astronomers who believed in an Earth-centered universe could easily "explain" the daily paths of the stars through the sky by imagining that the celestial sphere was real and that it rotated around the Earth each day. The apparent motions of the Sun and the Moon were only slightly more mysterious: Some ancient Greeks explained these motions by imagining that the Sun and the Moon each resided on its own sphere that turned at a slightly different rate from that of the sphere of the stars. But the apparent retrograde motion of the planets posed a far greater mystery. Even the addition of five more rotating spheres for the five known planets could not adequately explain why the planets sometimes appear to move backward.

We'll see in Chapter 3 that apparent retrograde motion is not difficult to understand once you know that the Earth is one of the planets orbiting the Sun. Indeed, it was the quest to understand apparent planetary motion that eventually led to the abandonment of belief in an Earth-centered universe and to the discovery of the laws governing *real* planetary motion and gravity. In many ways, the modern technological society that we take for granted can be traced back directly to the scientific revolution that began because of the quest to explain the slow wandering of the planets among the stars in our sky.

Meteors and More

At this point in the text, your principal goal in observing the sky should be simply to enjoy what you see and let your mind wander to thoughts about the universe and your place within it. As you watch the sky, you'll soon become familiar with the motions of the Sun, Moon, and planets. Occasionally, you may also notice other celestial phenomena.

On almost any clear night, you are likely to see a few **meteors,** or "shooting stars," flashing through the sky. Meteors are created by particles from interplanetary space entering and burning up in the Earth's atmosphere. Surprisingly, most meteors are produced by particles no larger than a pea. These small particles crash into the atmosphere at high speed, heating the surrounding air so much that it glows. The flash of a meteor typically lasts only a second or less. Occasionally, a larger particle makes an especially bright meteor called a **fireball;** the fireball may end with a brilliant burst as the heated particle explodes. On very rare occasions, a particle is large enough to survive entry into the atmosphere and strike the ground. The surviving rock is called a

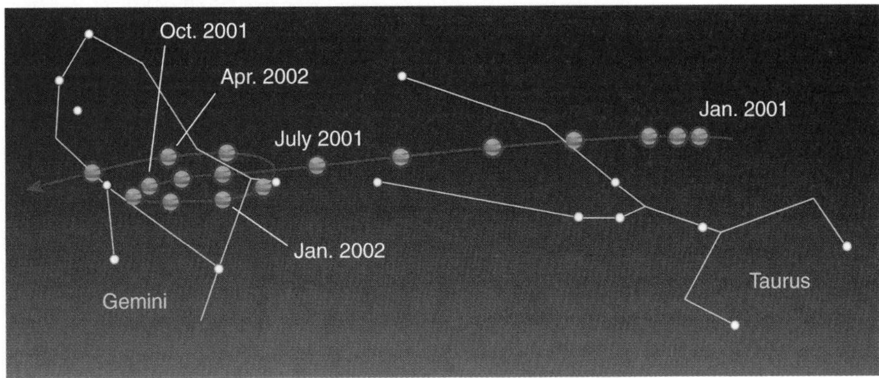

Dots represent Jupiter's position at 1-month intervals.

FIGURE 1.21 Jupiter's position among the stars in our sky in the year 2001–2002.

meteorite—it is a piece of rock from space, and you can touch meteorites in many science museums.

On some nights, unusually large numbers of meteors streak through the sky. These **meteor showers** usually occur when the Earth passes through a stream of interplanetary particles (left behind by a comet) in its orbit, and they recur on about the same date each year. Table 1.1 lists the major annual meteor showers; the Perseids and the Leonids tend to be particularly spectacular.

It is also easy to spot some of the hundreds of *artificial satellites* that orbit the Earth, most launched by the United States and the former Soviet Union. A particular satellite usually is visible only for a few minutes, moving straight and steadily across your sky. It's generally easier to notice some of the many satellites with *polar orbits* (i.e., orbits that pass over the North and South poles); these satellites move nearly due north or due south through your sky. Because satellites shine by reflected sunlight, they are best seen shortly after dusk or before dawn.

At high latitudes (north or south), the sky may dance with the lights of an **aurora** (Figure 1.22), called the *aurora borealis* (or "northern lights") in the Northern Hemisphere and the *aurora australis* in the Southern Hemisphere. Auroras are caused by the interaction of charged particles ejected from the Sun with the Earth's atmosphere and magnetic field. Occasionally, the Sun erupts with a spectacular flare that produces unusually large numbers of particles. Then the aurora may be seen even at mid-latitudes. Scientists often study these spectacular solar events, and when they occur the local news may alert you to watch for an aurora.

Table 1.1 Major Annual Meteor Showers

Shower Name	Approximate Date
Quadrantids	January 3
Lyrids	April 22
Eta Aquarids	May 5
Delta Aquarids	July 28
Perseids	August 12
Orionids	October 22
Taurids	November 3
Leonids	November 17
Geminids	December 14
Ursids	December 23

Other celestial events are much rarer. *Eclipses* (lunar or solar) occur a few times each year, but each eclipse is visible only from selected places on Earth. If you stay near home, you may see a total *lunar* eclipse every couple of years but a total *solar* eclipse only once in a lifetime. *Comets* are balls of ice and dust (sometimes described as "dirty snowballs") that come from the outer solar system. The ice evaporates to form a bright "head" and a long "tail" as the comet swings around the Sun. A comet appearance as spectacular as that of comet Hale–Bopp in 1997 (Figure 1.23) typically occurs only once every few decades. On even rarer occasions, humans have witnessed fantastic events beyond our solar system, such as the explosion of a distant star. If you keep your eyes on the sky, there is no end to the wonders you may see.

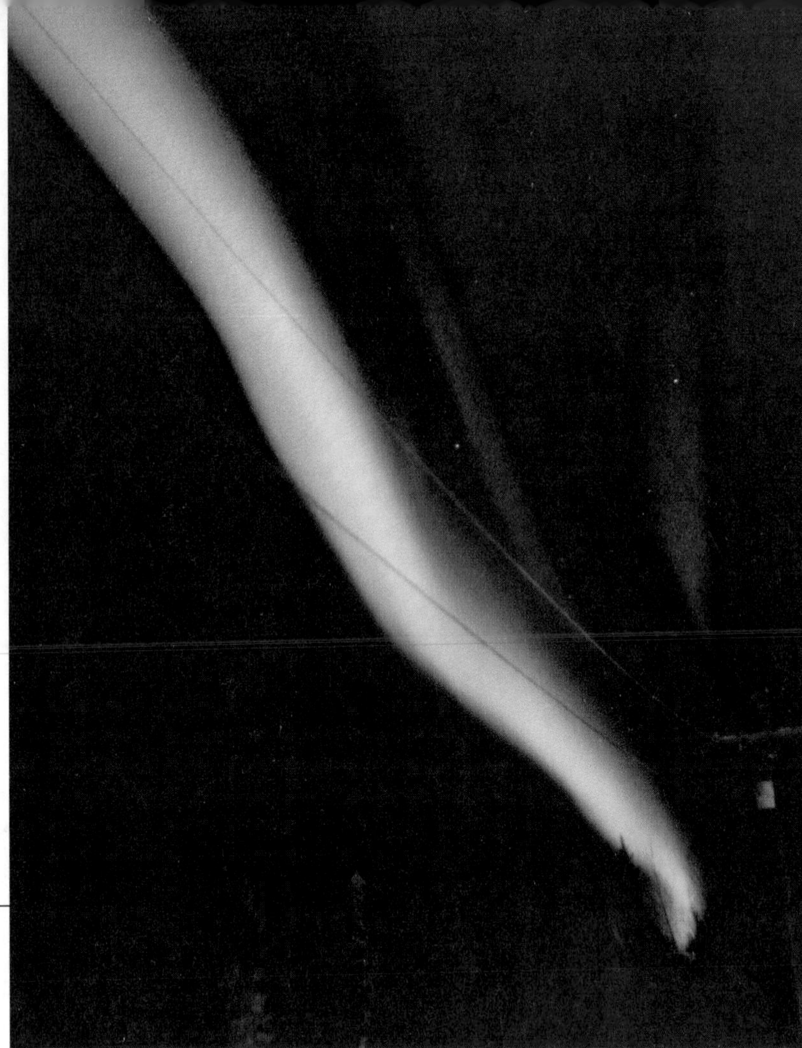

FIGURE 1.22 This photograph shows the aurora borealis, which is created when charged particles from the Sun crash into the Earth's atmosphere. Because the Earth's magnetic field directs these particles toward the poles, the aurora borealis is generally visible only at northerly latitudes. A similar phenomenon, the aurora australis, occurs at high southerly latitudes.

For those who have seen the Earth from space, the experience most certainly changes your perspective. The things that we share in the world are far more valuable than those which divide us.

DONALD WILLIAMS, ASTRONAUT (USA)

THE BIG PICTURE

In this first chapter, we have developed a broad overview of the universe and how to appreciate the universe by observing the sky. It is not necessary for you to remember all the details—everything presented in this chapter will be covered in greater detail later in the book. However, you should understand enough so that the following "big picture" ideas are clear:

- The Earth is not the center of the universe, but the apparent motions of the sky can make it appear to be. Learning the true place of the Earth in the universe required the combined efforts of numerous people through many generations.
- The universe contains physical structures at many levels, ranging from the tiniest subatomic particles to superclusters of galaxies.
- The universe has evolved since its beginning in the Big Bang. We can observe how parts of the universe looked at different ages simply by looking to different distances: The farther away we look in distance, the further back we look in time.
- We are intimately connected to the cosmos because the atoms from which we are made were forged in stars, released into space when these stars died, and incorporated into our solar system when it formed some 4.6 billion years ago.
- You can enhance your enjoyment of learning astronomy by spending time outside observing the sky. The more you learn about the appearance and apparent motions of objects seen in the sky, the more you will appreciate what you can see in the universe.

FIGURE 1.23 Comet Hale–Bopp, photographed over Boulder, Colorado, during its appearance in 1997.

Review Questions

1. What do we mean by a *geocentric* universe? When was the geocentric viewpoint shown to be incorrect? Name the three scientists who played the primary roles in overturning the geocentric viewpoint.
2. Describe the major levels of structure in the universe, beginning with planets like the Earth and working up to the universe itself.
3. What do we mean when we say that the universe is *expanding?* Why does an expanding universe imply a beginning called the *Big Bang?*
4. Explain what Carl Sagan meant when he said that we are "star stuff."
5. Explain the statement *"The farther away we look in distance, the further back we look in time."*
6. What do we mean by the *observable universe?* Why does its size depend on the age of the universe?
7. Indicate whether each of the following statements is true or false, and explain why.
   a. Our solar system is located in the center of the Milky Way Galaxy.
   b. The Local Group contains many billions of stars.
   c. A typical supercluster contains no more than about 10,000 stars.
   d. No galaxies existed before the Big Bang.
   e. The observable universe is the same size today as it was a few billion years ago.
8. Why does your sky look like a dome? Define your *horizon, zenith,* and *meridian.*
9. Explain how you can describe the position of any object in your sky by stating an *altitude* and a *direction.* If you look right now to an altitude of 30° in a northwesterly direction, what are you looking at?
10. Explain why we can describe *angular* sizes and distances when we look at objects in the sky, but not *physical* sizes and distances. Describe how you can estimate angular sizes and distances with your outstretched hand. What are *arcminutes* and *arcseconds?*
11. What is the *celestial sphere?* Does it really exist? Explain.
12. What is a *constellation?* How is a constellation related to a pattern of stars in the sky?
13. Describe how the Milky Way would look in the sky of someone observing from a planet around a star in the Great Galaxy in Andromeda.

14. Explain why we don't see any other galaxies when we look into the band of light in our sky that we call the Milky Way. How is this band of light related to the Milky Way Galaxy?
15. Which planets are sometimes visible to the naked eye? How can you identify them?
16. Define the *north celestial pole, south celestial pole,* and *celestial equator.* How is each of these related to the poles and equator of the Earth?
17. Explain how you can find the north celestial pole with the aid of the Big Dipper, and the south celestial pole with the aid of the Southern Cross.
18. Describe the general daily pattern of motion of stars in the sky. What are *circumpolar* stars?
19. Suppose that Orion is rising in your eastern sky. Where will it be in 6 hours? How do you know?
20. Describe how we measure latitude and longitude. What is your current latitude and longitude?
21. Suppose that the north celestial pole is located due north at an altitude of 33° in your sky. What is your latitude? How do you know?
22. What is the name of the North Star? What makes this star special?
23. True or false: Columbus was the first person to discover that the Earth is round. Explain.
24. What is the *ecliptic?* How is it related to the constellations of the *zodiac?*
25. What do we mean by the *lunar cycle?* How is it related to a *month?* Name the eight basic phases that the Moon goes through in the lunar cycle.
26. What is *apparent retrograde motion* of the planets? Explain why apparent retrograde motion *does not* mean that planets sometimes rise in the west and set in the east. Why was apparent retrograde motion more mysterious to ancient astronomers than the rising and setting of the Sun, Moon, and stars?
27. What is a *meteor?* What is a *meteor shower?* Explain how meteors are related to *fireballs* and to *meteorites.*
28. What is an *aurora?* Where are you likely to see one?
29. Did you see comet Hale–Bopp in 1997? How often should we expect to see other spectacular comets?

Discussion Questions

1. *Geocentric Language.* Many common phrases reflect the ancient Earth-centered view of our universe. For example, the phrase "the Sun rises each day" implies that the Sun is really moving over the Earth. In fact, the Sun only *appears* to rise as the rotation of the Earth carries us to a place where we can see the Sun in our sky. Identify at least three more common phrases that imply an Earth-centered viewpoint, and briefly describe the reality behind each one.
2. *"How Vast Those Orbs Must Be."* The quotation from Christiaan Huygens on page 3 suggests that humans might be less inclined to wage war if everyone appreciated the Earth's place in the universe.
   a. Do you think human behavior would generally improve if everyone knew more about the Earth's physical place in the universe? Explain and defend your opinion.
   b. A 1987 survey found that approximately 20% of Americans did not know that the Earth orbits the Sun; the proportion is likely even higher in countries that have lower standards of education. Do you think this fact has any bearing at all on human behavior? Do you think it is important for people to know that the Earth is not the center of the universe? If so, what steps could we take in society to improve education and public discussion of the Earth's place in space and time?
3. *From Pythagoras to Columbus.* The ancient Greeks were aware that the Earth was round; they even had a good idea of its size. Why, then, do we often hear that Columbus discovered that the Earth is round?

Problems

1. *Sensible Statements?* Evaluate each of the following statements and decide whether it is sensible. Explain.

   Example: I walked east from our base camp at the North Pole.

   Solution: The statement does not make sense because *east* has no meaning at the North Pole—all directions are south from the North Pole.

   a. The universe is between 10 and 16 billion years old.
   b. The universe is between 10 and 16 billion light-years old.
   c. It will take me light-years to complete this homework assignment!
   d. Someday, we may build spaceships capable of traveling at a speed of 1 light-minute per hour. (Think carefully about this one!)
   e. If you had a very fast spaceship, you could travel to the celestial sphere in about 100 years.
   f. When I looked into the dark fissure of the Milky Way with my binoculars, I saw what must have been a cluster of distant galaxies.
   g. I enjoyed seeing the Southern Cross during my last trip to Argentina.
   h. Last night I saw Jupiter in the constellation Ursa Major. (*Hint:* Is Ursa Major part of the zodiac?)

i. Last night I saw Mars move westward through the sky in its apparent retrograde motion.

j. Although all the known stars appear to rise in the east and set in the west, we might someday discover a star that will appear to rise in the west and set in the east.

2. *Distances by Light.* Just as a light-year is the distance that light can travel in 1 year, we can define a light-second as the distance that light can travel in 1 second, a light-minute as the distance that light can travel in 1 minute, and so on. Following the method presented in Mathematical Insight 1.1, calculate the distances represented by each of the following: 1 light-second; 1 light-minute; 1 light-hour; and 1 light-day.

3. *Proxima Centauri.* The nearest star to the Sun, Proxima Centauri, is about 4.3 light-years from Earth. How far is this in kilometers? In miles?

4. *Your View of the Sky.*
   a. Find your latitude and longitude, and state the source of your information.
   b. Describe the altitude and direction in your sky at which the north or south celestial pole appears.
   c. Is Polaris a circumpolar star in your sky? Explain.
   d. Describe the location of the meridian in your sky. (*Hint:* Remember that the meridian traces a semicircle through your sky.)
   e. Describe where the celestial equator is located in your sky. Note that the celestial equator is a circle, not a point. (*Hint:* Study Figure 1.18.)

5. *Sky Information.* Many newspapers publish regular sky information such as the phase of the Moon, times of sunrise and sunset, and interesting things to watch for in the sky.
   a. Look through your local newspaper in detail, including both daily and Sunday editions, and make a list of all the types of sky information that are regularly published.
   b. Repeat part (a) for at least one national newspaper, such as the *New York Times* or *USA Today.*

6. *Observing Project: The Circling Sky.* Pick a particular time of night (e.g., 10 P.M.) at which you can go outside periodically throughout the next couple of months. At least once a week, go outside, find the Big Dipper, and draw a diagram showing its orientation relative to Polaris and your northern horizon. (In the Southern Hemisphere, do this observation for the Southern Cross relative to the south celestial pole and the southern horizon.) Observe how the position of the Big Dipper (or Southern Cross), for a given time of night, changes with the date. (*Note:* If clocks are changed between observations [e.g., from daylight saving time to standard time], you will need to change your observing time. For a switch from daylight saving time to standard time, move your observing time 1 hour earlier [e.g., from 10 P.M. to 9 P.M.]; for a switch from standard time to daylight saving time, move your observing time 1 hour later.)

7. *Observing Project: Seasonal Changes in the Night Sky.* Identify a few major constellations that currently are visible in your sky. At least once a week during the next couple of months, go outside at the same time of night and note the altitude and direction of each of these constellations. Record your findings. How do the locations of the constellations in your sky change from night to night? From week to week, do the same constellations rise earlier or later? What else can you conclude about how the constellations visible in your sky change with the seasons?

8. *Observing Project: Follow the Sun.* This activity presents some simple ways to notice how the Sun's path through your sky varies with the seasons.
   a. Use your fist to estimate the altitude of the Sun in your sky when it is on or close to your meridian (around noon on standard time or 1 P.M. on daylight saving time). Record the date and altitude. Repeat this observation at least once a week, noticing changes over a period of a few months.
   b. Find a convenient spot from which you can observe sunrise and/or sunset at least once a week for the next several months. For each observation, note the place where the Sun rises or sets along your horizon in relation to landmarks. Draw a simple picture showing the point of sunrise or sunset, and record the date and time. Note how the location and time of sunrise and/or sunset change over a few months.
   c. Find a post (or other vertical object that is narrow and just a few feet tall) and, at some particular time of day, measure the length and direction (use a compass) of its shadow. Record your results, and repeat at the same time of day at least once a week. Notice and explain the changes over a few months.

9. *Observing Project: Motion and Phases of the Moon.*
   a. Based on the current phase of the Moon (check a newspaper), figure out the time at which it will be convenient to observe it at night. Find the Moon in the sky, and identify its phase and the zodiac constellation in which it appears. Then compare the position of the Moon relative to the stars at times a few hours apart. Can you detect the motion of the Moon relative to the stars? Explain.
   b. Observe the Moon again 29 days after your first observation. Has it returned to its original phase? Is it in the same zodiac constellation? Discuss your observations.

10. *Observing Project: Apparent Planetary Motion.* Find out what planets are currently visible in your evening sky. At least once a week, observe the planets and draw a diagram showing the position of each visible planet relative to stars in a zodiac constellation. From week to week, note how the planets are moving relative to the stars. Can you see any of the apparently "erratic" features of planetary motion? Explain.

2

The Scale of the Universe

How does Earth compare in size to the other planets? How far apart are the planets? How far away are the stars? How big is the universe? How does a human lifetime compare with the age of the universe?

Each of these questions can be answered numerically, and you can find the answers in almost any encyclopedia. However, because they are literally astronomical, the numbers are meaningless unless they are somehow put into perspective.

One of the best ways to put the universe into perspective is by imagining or building various scale models. This chapter is devoted to giving meaning to incredible cosmic distances and times through scale models and other scaling techniques.

2.1 A Walking Tour of the Solar System

How would you describe an airplane to someone who has never seen one? It might be easiest to begin by showing a scale model of the airplane. In the same way, a scale model provides one of the best ways to develop perspective on our solar system. A convenient scale for a model of the solar system is 1 to 10 billion; that is, lengths (e.g., diameters or distances) in the model are *one ten-billionth,* or 10^{-10}, of actual length, making every 100,000 kilometers in the real solar system become 1 centimeter in the model. (If you are unfamiliar with powers of 10 notation or need a review, see Appendix A.)

Table 2.1 lists both real diameters and distances in our solar system and the diameters and distances for a model of the solar system on a scale of 1 to 10 billion. Note that all the model planets lie within 0.6 kilometer (less than $\frac{1}{2}$ mile) of the Sun, so we could walk from the Sun to Pluto in just a few minutes. The model Sun is about the size of a grapefruit, and the model planets range in size from dust speck–size Pluto to marble-size Jupiter (Figure 2.1).

Many colleges and science museums feature scale model solar system exhibits. Let's imagine a walk through such a model solar system, stopping

FIGURE 2.1 The sizes of the Sun and the planets on the 1-to-10-billion scale.

Table 2.1 Solar System Sizes and Distances, 1-to-10-Billion Scale

Object	Real Diameter	Real Distance from Sun (average)	Model Diameter	Model Distance from Sun
Sun	1,392,500 km	—	139 mm = 13.9 cm	—
Mercury	4,880 km	57.9 million km	0.5 mm	6 m
Venus	12,100 km	108.2 million km	1.2 mm	11 m
Earth	12,760 km	149.6 million km	1.3 mm	15 m
Mars	6,790 km	227.9 million km	0.7 mm	23 m
Jupiter	143,000 km	778.3 million km	14.3 mm	78 m
Saturn	120,000 km	1,427 million km	12.0 mm	143 m
Uranus	52,000 km	2,870 million km	5.2 mm	287 m
Neptune	48,400 km	4,497 million km	4.8 mm	450 m
Pluto	2,260 km	5,900 million km	0.2 mm	590 m

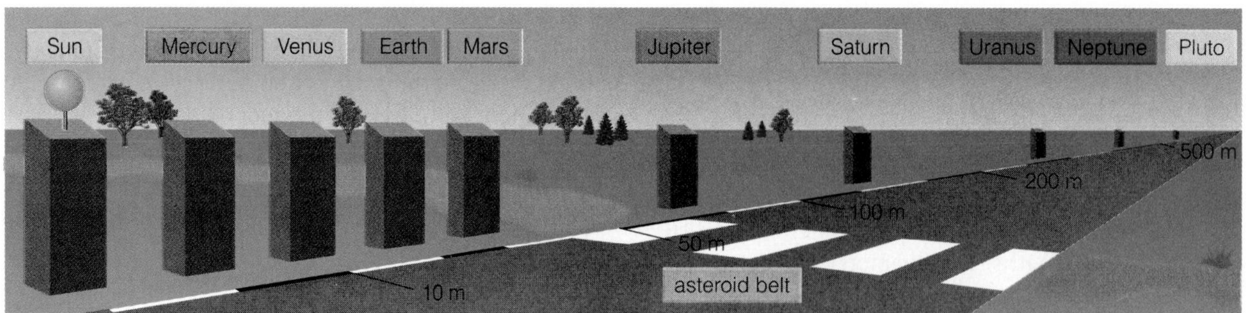

FIGURE 2.2 Pedestals for planet locations in a 1-to-10-billion scale model solar system.

along the way to become acquainted with the Sun and the nine planets (Figure 2.2). Keep in mind that, while our model lines up all the planets in a straight line from the Sun, each planet actually orbits the Sun independently. In reality, the planets line up in a reasonably straight line only once every few hundred years.

TIME OUT TO THINK *To help yourself picture the scale, make a quick model for the Earth and the Sun. Find a grapefruit or similar-size object to represent the Sun, and a pinhead to represent the Earth. In a long hallway or outdoors, place your Sun on a chair or table and take your tiny Earth about 15 meters from your Sun. Explain why the angular size of your model Sun as seen from your model Earth is about the same as that of the real Sun from the real Earth (about 0.5°).*

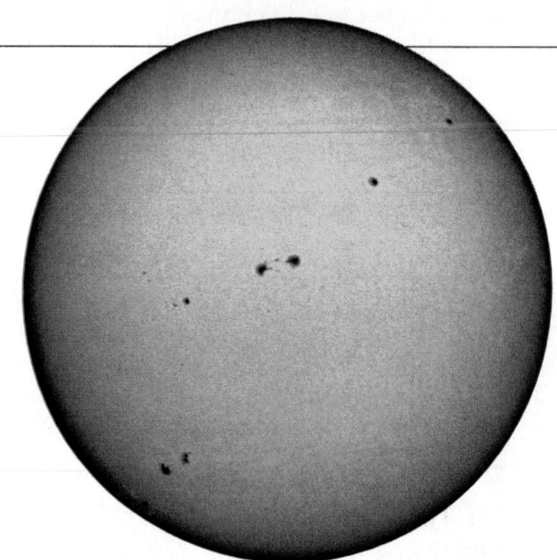

FIGURE 2.3 A photograph of the Sun. The dark splotches are sunspots.

The Sun

The Sun is a star. It is the central object of our solar system and contains more than 99.9% of the solar system's total mass; that is, the Sun outweighs everything else in the solar system combined by a thousand times. The Sun essentially is a great ball of hydrogen and helium, with small amounts of other elements mixed in.

The Sun generates energy by converting hydrogen into helium deep in its core through a process called **nuclear fusion;** each fusion reaction converts the nuclei of four hydrogen atoms into the nucleus of one helium atom. Overall, 600 million tons of hydrogen are converted into 596 million tons of helium every second. The "missing" 4 million tons of mass is converted to energy in accord with Einstein's famous equation, E 5 mc^2. In this equation, E stands for the amount of energy released when an amount of mass m is converted into energy, and c is the speed of light [Section 5.2].

Although the Sun constantly generates energy in its core, the energy released at any particular moment typically takes a million years to work its way outward to the seething 6,000°C (about 10,000°F) surface of the Sun (Figure 2.3). Once there, it is radiated into space in the form of light. About 8 minutes later, a tiny fraction of this energy is intercepted by the Earth, where it provides the light and heat that sustain life.

Mercury

Only about six steps (assuming 1 meter per step) from the grapefruit-size Sun in our model solar system lies the innermost planet, Mercury (Figure 2.4). With a diameter of 0.5 millimeter, the model Mercury is about the size of the period at the end of this sentence, making it the second smallest planet in the solar system (after Pluto). Mercury is a desolate, cratered, and virtually airless world, looking much like our Moon. Because it is so close to the Sun, temperatures on Mercury's daylight side reach about 400°C (about 750°F)—hot enough to melt lead. But with no atmosphere to retain heat during its long

FIGURE 2.4 Mercury. This picture is a mosaic of photographs from the *Mariner 10* spacecraft (1974). The areas labeled "no data" were not photographed.

nights (which last about 3 months), nighttime temperatures plummet to below −150°C (about −240°F), giving Mercury the most extreme temperature range of any planet.

Venus

A few steps beyond Mercury lies the second planet from the Sun, Venus (Figure 2.5). Because Venus and Earth are nearly identical in size—both looking like pinheads in the model solar system—Venus is

Mathematical Insight 2.1 A 1-to-10-Billion Scale

Calculating any length (e.g., diameter or distance) on a 1-to-10-billion scale simply requires dividing the real length by 10 billion (10^{10}):

$$\text{model length} = \frac{\text{actual length}}{10^{10}}$$

Example 1: The Earth's real diameter is 12,760 km (1.276×10^4). What is its diameter on the 1-to-10-billion scale?

Solution: We divide the real diameter by 10 billion to find the scaled diameter. Remember that when we divide powers of 10, we *subtract* the exponents:

$$\text{model Earth diameter} = \frac{1.276 \times 10^4 \text{ km}}{10^{10}}$$
$$= 1.276 \times 10^{4-10} \text{ km} = 1.276 \times 10^{-6} \text{ km}$$

We can round our answer to 1.3×10^{-6} km. However, because it's difficult to form a mental picture of 10^{-6} km, we should convert the answer to something more meaningful, such as millimeters. We do this by multiplying the answer in kilometers by 1,000 meters per kilometer (10^3 m/km) and then by 1,000 millimeters per meter (10^3 mm/m):

$$1.3 \times 10^{-6} \text{ km} \times \underbrace{\frac{10^3 \text{ m}}{1 \text{ km}}}_{\substack{\text{conversion} \\ \text{to meters}}} \times \underbrace{\frac{10^3 \text{ mm}}{1 \text{ m}}}_{\substack{\text{conversion} \\ \text{to millimeters}}} = 1.3 \text{ mm}$$

Thus, the model Earth diameter is about 1.3 mm.

Example 2: The Earth's real distance from the Sun is about 150 million km, or 1.5×10^8 km. What is its scaled distance from the Sun?

Solution: We divide the real distance by 10 billion to find the scaled distance:

$$\text{Earth–Sun distance in model} = \frac{1.5 \times 10^8 \text{ km}}{10^{10}} = 1.5 \times 10^{-2} \text{ km}$$

Again, we can make the answer easier to visualize if we convert units; in this case, we convert from kilometers to meters:

$$1.5 \times 10^{-2} \text{ km} \times \frac{10^3 \text{ m}}{1 \text{ km}} = 1.5 \times 10^1 \text{ m} = 15 \text{ m}$$

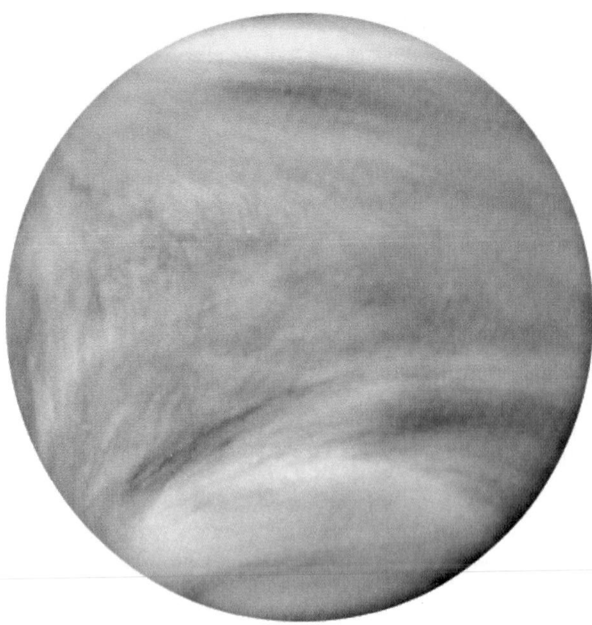

FIGURE 2.5 Venus. This picture was taken by the *Pioneer Venus Orbiter* with cameras sensitive to ultraviolet light; with visible light, cloud features cannot be distinguished from the general haze.

a As a spacecraft orbits the Moon, Earth appears to rise over the lunar landscape.

FIGURE 2.6 Views of the Earth.

sometimes called our "sister planet." Clouds enshroud the real Venus, and past generations of scientists speculated that it might have an atmosphere similar to that of the Earth. Although sunlight is more intense at Venus than at Earth, its clouds reflect most of this sunlight back to space, so scientists in the past suggested that the planet should be similar in temperature to the Earth. Science fiction writers imagined Venus as a lush, tropical world.

When twentieth-century astronomical technology finally allowed study of the Venusian surface, visions of a tropical paradise were shattered. Rather than being Earth-like, Venus's atmosphere is thick with carbon dioxide, which causes an extreme *greenhouse effect* that heats the surface to an incredible 450°C (about 850°F). Day and night, Venus is hotter than a pizza oven. The weight of the atmosphere makes the surface pressure 90 times that of Earth. Besides the crushing pressure and searing temperature, a visitor to Venus would also feel the corrosive effects of sulfuric acid and other toxic chemicals in its atmosphere. Far from being a beautiful sister planet to Earth, Venus resembles a traditional view of hell.

Earth

Beyond Venus, and about 15 steps from the Sun, Earth is the next planet on our walking tour. From space, Earth is striking in its beauty (Figure 2.6a). Blue oceans cover nearly three-fourths of the surface and are broken by the continental land masses and scattered islands. The polar caps are white with snow and ice, and white clouds are scattered above Earth's surface. At night, the glow of artificial lights clearly reveals the presence of an intelligent civilization (Figure 2.6b).

Earth is the first planet on our tour with a moon. On our scale of 1 to 10 billion, our Moon orbits the Earth at a distance of about 4 centimeters, so you could easily cover both the Earth and the Moon with your thumb; the Moon's diameter is about one-fourth Earth's diameter (Figure 2.6c). The Moon is the only world besides Earth on which humans have ever stepped: 12 people walked on the Moon from 1969 through 1972 as part of NASA's Apollo program.

TIME OUT TO THINK *Hold your pinhead-size model Earth in your hand, and look toward your model Sun 15 meters away. Aside from the tiny Earth that you hold, there is nowhere else in the solar system—and nowhere that we yet know of in the universe—where humans can survive outside the artificial environment of a spacecraft or spacesuit. Does holding the model Earth affect your perspective on human existence in any way? Explain.*

Mars

A few steps beyond Earth lies the model Mars, with about half the diameter of Earth. Mars has two tiny moons, Phobos and Deimos, but they are so small as to be microscopic on the 1-to-10-billion scale.

b A composite photograph showing the nighttime glow of lights from human activity across the Earth. The lights, which come from cities and from agricultural, oil, and gas fires around the world, clearly outline the continents. The "curtain" of light in the upper left is the aurora borealis.

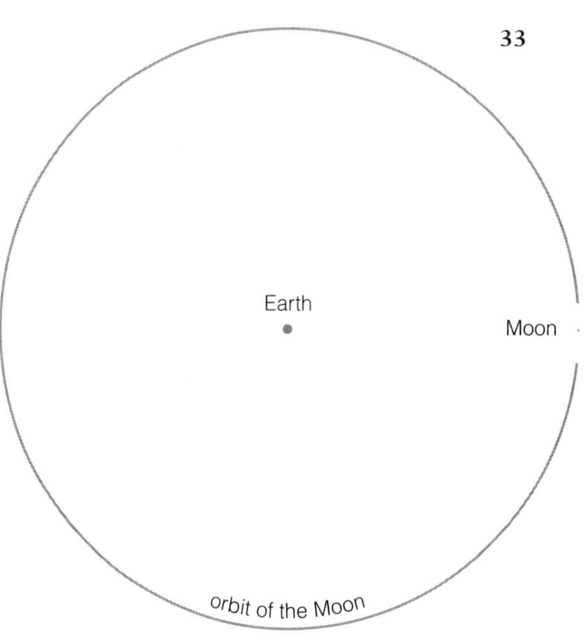

c This sketch shows the Earth-Moon system on the 1-to-10 billion scale.

Mars is an intriguing world (Figure 2.7), with extinct volcanoes that dwarf the largest mountains on Earth, a great canyon that runs nearly one-fifth of the way around the planet, and polar caps made of water ice[1] and frozen carbon dioxide ("dry ice"). Although all of the water on Mars is frozen today, the presence of dried-up riverbeds indicates that water once flowed on the Martian surface. Hence, although life on Mars is highly unlikely today, Martian conditions may have been suitable for life in the past.

A long-standing dream of space exploration is to send humans to study the incredible landscape of Mars and perhaps to prospect for fossil evidence of ancient life. The trip from Earth to Mars, which is just a few steps in our model solar system, would take about 8 months in each direction with present-day spacecraft. Nevertheless, it is quite possible that humans will walk on the surface of Mars within your lifetime.

[1]In science, the term *ice* can refer not only to frozen water but also to things such as frozen carbon dioxide, frozen methane, and frozen ammonia.

FIGURE 2.7 (**a**) Mars. (**b**) The surface of Mars as seen from the *Pathfinder* lander in 1997.

a The asteroid Gaspra (16 km across), photographed by the Galileo spacecraft that continued on to Jupiter.

FIGURE 2.8

b Meteor Crater in Arizona was created about 50,000 years ago by an asteroid impact. The crater is more than a kilometer across and almost 200 meters deep, but the asteroid that made it was only about 50 meters across.

The Asteroid Belt

As we cross the gap between Mars and Jupiter, we pass through the **asteroid belt,** where thousands of asteroids orbit the Sun. Asteroids are smaller than planets, making them extremely difficult to see on a 1-to-10-billion scale. Although the largest asteroids are spherical, many small asteroids have irregular shapes (Figure 2.8a).

Some asteroids are found outside the asteroid belt. Among the most important asteroids, at least to humans, are those whose orbits cross the orbit of the Earth. Dozens of craters on the Earth attest to past impacts of asteroids, although such craters can be difficult to detect because of erosion by wind and rain (Figure 2.8b). The impact of an asteroid (or possibly of a comet) is thought to have precipitated the extinction of the dinosaurs some 65 million years ago [Section 12.6]. If a similar impact occurred today, its effects could be catastrophic for human civilization. Fortunately, the probability of a major asteroid impact during our lifetime is small; some astronomers monitor asteroids in hopes of providing enough warning of an impending impact to allow us to somehow avert it.

As we walk through the asteroid belt, we will notice one of the most important characteristics of our solar system. In contrast to the few steps separating the four inner planets in the model solar system, the walk from Mars to Jupiter covers more than 50 meters, or more than half the length of a football field. As we look ahead, we will notice that increasingly large distances separate the remaining planets. Thus, the solar system has two major regions: the *inner solar system,* where Mercury, Venus, Earth, and Mars orbit the Sun relatively closely; and the *outer solar system,* where the remaining planets are widely separated.

Jupiter

Upon reaching the model Jupiter, we will find a marble-size planet, making it a giant in comparison to Earth and the other three planets of the inner solar system. Indeed, Jupiter (Figure 2.9a) is more massive than all the rest of the planets *combined,* and its gravity influences the orbits of several other planets and countless asteroids and comets. At least 16 moons orbit Jupiter. The four largest—Io, Europa, Ganymede, and Callisto (Figure 2.9b)—are worlds in their own right, easily visible on the scale of our model solar system. A thin set of rings also orbits Jupiter, but these rings are extremely faint and do not show up in most photographs. Jupiter's rings, like all planetary rings, are made of countless individual particles ranging in size from dust specks to large boulders.

FIGURE 2.9 (a) Jupiter. (b) A montage of photographs showing Jupiter's four largest moons, Io, Europa, Ganymede, and Callisto. The montage is arranged to show size comparison to Jupiter's Great Red Spot.

TIME OUT TO THINK *To get a better sense of the Jupiter system, get out a ruler and make a quick model on paper. Draw a circle 14 millimeters in diameter to represent the model Jupiter. Here are the data you'll need for its four largest moons (the distances should be measured from the* center *of the model Jupiter):*

	Model Diameter	*Model Distance*
Io	*0.36 mm*	*4.2 cm*
Europa	*0.31 mm*	*6.7 cm*
Ganymede	*0.53 mm*	*10.7 cm*
Callisto	*0.48 mm*	*18.8 cm*

How does the Jupiter system resemble the entire solar system in miniature?

Like the Sun, Jupiter is made primarily of hydrogen and helium and has no solid surface. If you descended into Jupiter's atmosphere, you would be crushed by the growing gas pressure long before you ever reached the planet's core.

Jupiter's rapid rotation—a day is less than 10 hours long—helps spawn incredible winds and storms. The Great Red Spot, a huge storm with winds of 500 km/hr (300 mi/hr), is some three times the size of the Earth. Like the king of the gods in Roman mythology whose name it bears, Jupiter reigns as king of the planets.

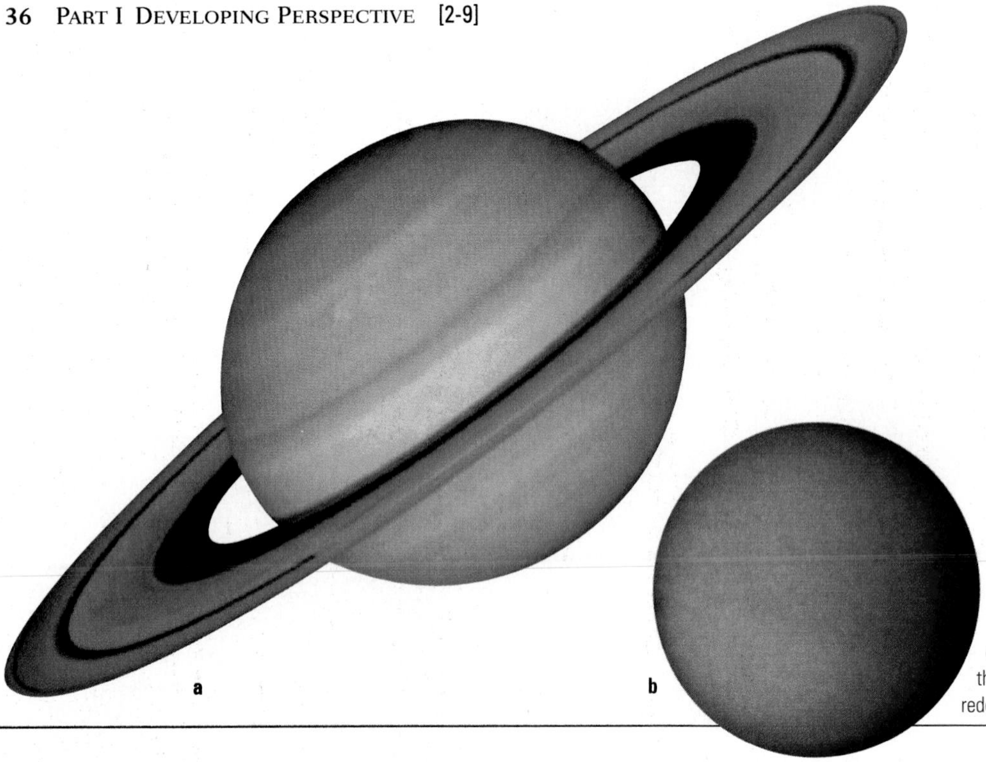

a

b

FIGURE 2.10 (**a**) Saturn. (**b**) Saturn's moon Titan is enshrouded by a thick atmosphere that makes it look like a hazy, reddish sphere in photographs.

Saturn

The walk from Jupiter to Saturn is about 65 meters, or nearly three-fourths the length of a football field, again showing the very different nature of the outer solar system compared with the inner solar system. Like the larger Jupiter, Saturn is a giant world made primarily of hydrogen and helium and orbited by rings and many moons (at least 20 are known).

The beautiful rings are Saturn's most distinguishing feature (Figure 2.10a). Saturn's rings are more spectacular than the rings of Jupiter, Uranus, or Neptune because they contain more material and reflect far more sunlight.

The largest of Saturn's moons, Titan, has a thick atmosphere that hides the surface from view (Figure 2.10b). Although it is far colder than Earth's atmosphere and contains no oxygen, Titan's atmosphere bears two striking similarities to that of Earth: It is made mostly of nitrogen, and the atmospheric pressure is similar. Many scientists believe that Titan is a rich natural laboratory for organic chemistry—a few even speculate that it might harbor life.

Uranus

We have a long walk of about 144 meters from the model Saturn to the model Uranus, equivalent to the length of about $1\frac{1}{2}$ football fields. Although Uranus is much smaller than Jupiter or Saturn, it still is far larger than Earth. Uranus is made largely of hydrogen and helium and of hydrogen compounds such as methane (CH_4). Its pale blue-green color is due to the methane in its atmosphere (Figure 2.11a). Uranus has rings and at least 15 moons.

Perhaps the most unusual feature of Uranus is that, compared to all the other planets except Pluto, it appears to be "tipped" onto its side. That is, Uranus rotates with its axis nearly in the same plane as its orbit (Figure 2.11b). This unusual orientation may be the result of a cataclysmic collision suffered by Uranus as it was forming, some 4.6 billion years ago.

Although Uranus is faintly visible to the naked eye, its 84-year orbit of the Sun makes its motion among the constellations difficult to detect. As a result, Uranus was not discovered until 1781.[2] Its discoverer, William Herschel, first suggested naming the planet *Georgium Sidus,* Latin for "George's star," in honor of his patron, King George III of England. Fortunately, the idea of "Planet George" never caught on. Instead, many eighteenth- and nineteenth-century astronomers referred to the new planet as *Herschel,* after its discoverer. The modern name Uranus—after the mythological father of Saturn—was first suggested by one of Herschel's contemporaries, the astronomer Johann Bode, and was generally accepted by the mid-nineteenth century.

Neptune

The eighth model planet, Neptune, lies more than 160 meters beyond the model of Uranus. At least 10 moons orbit Neptune, along with a set of dark rings that are extremely difficult to photograph. Neptune looks nearly like a twin of Uranus, with very similar

[2] Uranus appears on earlier star charts, but William Herschel was the first to recognize it as a planet.

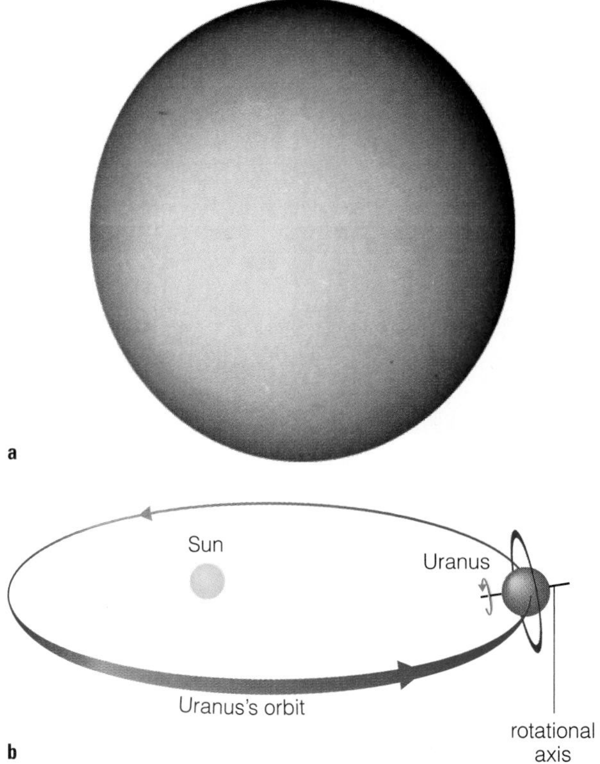

FIGURE 2.11 (**a**) Uranus. (**b**) The unusual orientation of Uranus's rotation axis relative to its orbit.

size and composition, although it is more strikingly blue (Figure 2.12).

The story of Neptune's discovery is particularly interesting. By the mid-nineteenth century, scientists were convinced of the validity of the theory of gravity developed by Isaac Newton in the late 1600s. However, careful observations of Uranus had shown its orbit to be slightly inconsistent with that predicted by Newton's theory. In the early 1840s, a 24-year-old Englishman named John Adams suggested that the inconsistency resulted from the gravitational influence of a previously unknown planet. Making mathematical calculations based on Newton's theory of gravity, he predicted a location for the unknown planet. Unfortunately, Adams was a student at the time and was unable to convince British astronomers to look for the predicted planet.

In the summer of 1846, French astronomer Urbain Leverrier independently made the same calculation. Leverrier had connections, however, and he sent a letter suggesting a search for the presumed planet to Johann Galle at the Berlin Observatory. On the night of September 23, 1846, Galle pointed his telescope to the position suggested by Leverrier. There, within 1° of its predicted position, he saw the planet Neptune. Hence, Neptune's discovery truly was made with mathematics and only confirmed with a telescope.

Before we walk on to Pluto, let us consider an interesting side note to this story. Leverrier's faith in Newton's law of gravity also led him to suggest that known discrepancies in Mercury's orbit were caused by an unknown planet. He suggested that this planet, which he called Vulcan, had not yet been seen because it was so close to the Sun. Leverrier, who died in 1877, surely would have been disappointed to learn that Vulcan does not exist and that the discrepancies in Mercury's orbit are, in fact, due to inherent limitations of Newton's theory of gravity. Today, Mercury's orbit is successfully explained by a more accurate theory of gravity, Einstein's general theory of relativity (see Chapter S4).

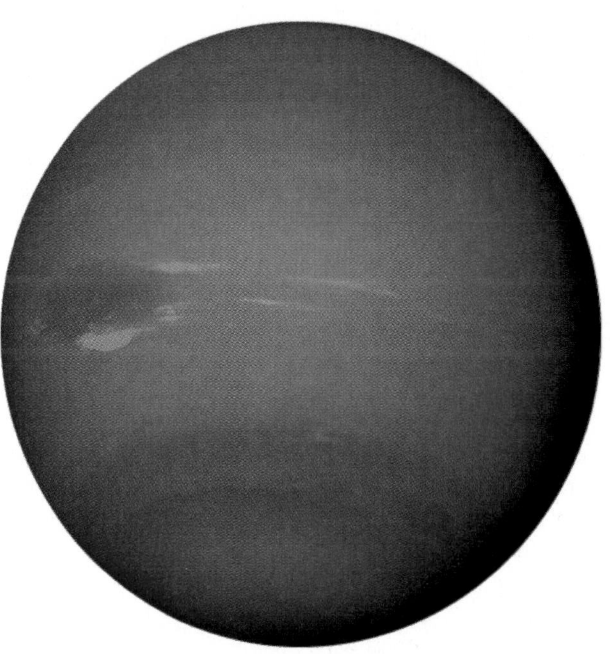

FIGURE 2.12 Neptune.

Pluto

At its average distance from the Sun, the model Pluto is a long walk from Neptune. The grapefruit-size Sun, more than $\frac{1}{2}$ kilometer away, appears tiny from here. It is easy to imagine that the world of Pluto must be cold and dark. One look at the model Pluto, along with its moon Charon, shows it to be out of character with the rest of the planets. Whereas the other planets of the outer solar system are gaseous giants, Pluto is made mostly of ices and is the smallest of all the planets (Figure 2.13a).

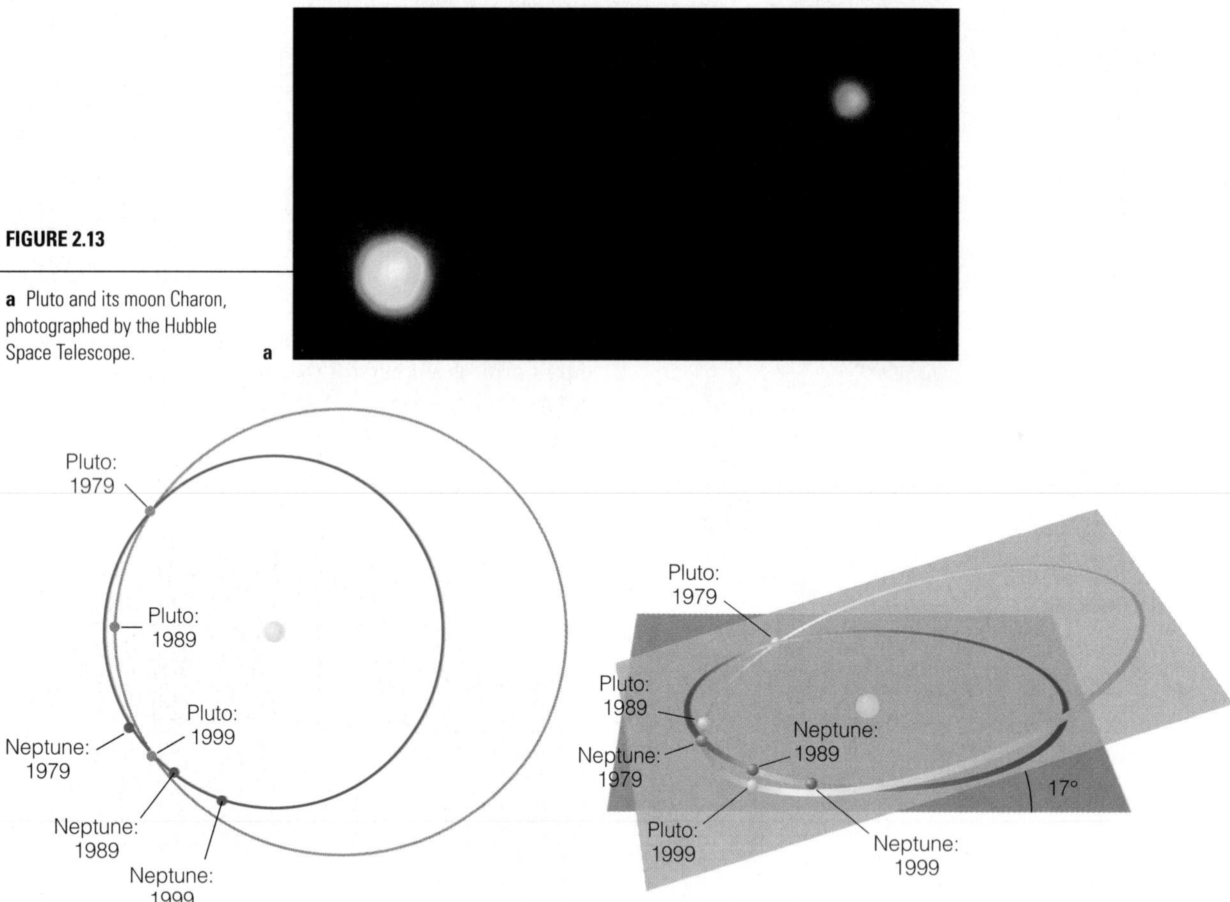

FIGURE 2.13

a Pluto and its moon Charon, photographed by the Hubble Space Telescope.

b Pluto's orbit sometimes brings it closer to the Sun than Neptune (left diagram). However, because Pluto's orbit is inclined 17° to the ecliptic plane (right diagram), the two orbits do not actually cross. (The orbits appear to cross in this diagram only because of the perspective.)

Pluto's orbit also is unusual. While all the other planets orbit the Sun along nearly circular paths, Pluto's orbit is highly elongated, and it is substantially inclined relative to the orbits of the other planets (Figure 2.13b). In each 248-year orbit of the Sun, Pluto is the most distant planet in all but 20 years; Pluto was closer to the Sun than Neptune between 1979 and 1999.

Pluto was discovered in 1930 by Clyde Tombaugh. By that time, astronomers had many more observations of the orbit of Uranus and thought that it still showed a few discrepancies, even after accounting for Neptune's influence. They initially assumed that Pluto was the cause of these discrepancies, but we now know that Pluto is too small to have a noticeable effect on Uranus. The supposed discrepancies turned out to be measurement errors. Recent discoveries have suggested that Pluto may be just one of many large, icy bodies in the outer solar system. In fact, if Pluto had not been discovered until today, it might be considered the largest known of these objects rather than the smallest planet. Some astronomers have even proposed demoting Pluto from its status as a planet. The controversy over Pluto's "planethood" only makes it a more interesting object and illustrates the often fuzzy boundaries between categories of astronomical objects.

Comets

Once we pass Pluto, we leave the realm of the planets. However, the most numerous objects in the solar system still lie ahead: balls of ice and dust called *comets* (Figure 2.14a). You are probably familiar with the occasional appearance of a comet in the inner solar system, where the heat of the Sun evaporates some of its ice and it grows a beautiful tail (see Figure 1.23). Astronomers have calculated the orbits of many comets that have visited the inner solar system and have determined that most come from far beyond the orbit of Pluto (Figure 2.14b). Our present technology does not allow us to detect comets when they are very far from the Sun. However, in order to

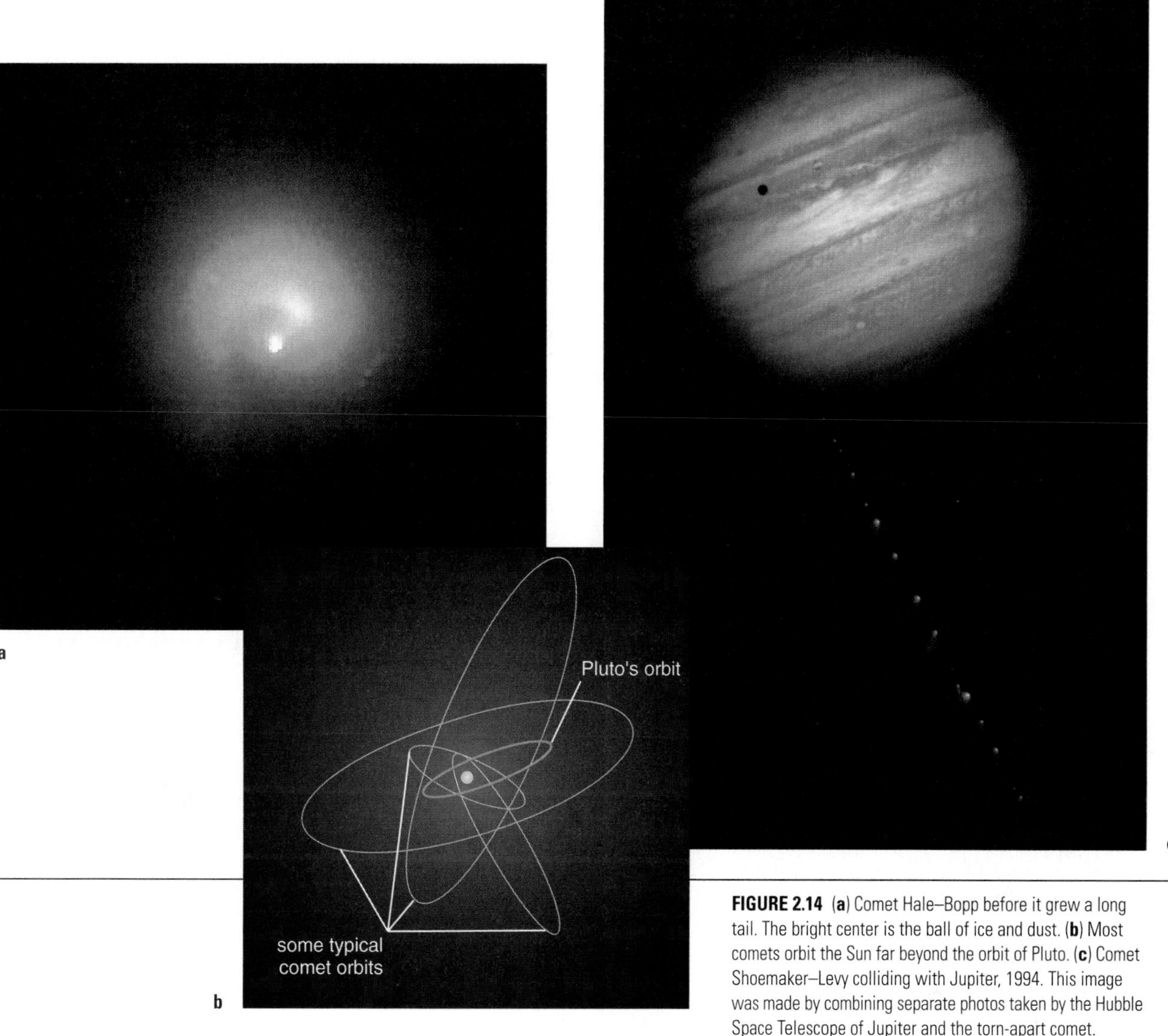

FIGURE 2.14 (**a**) Comet Hale–Bopp before it grew a long tail. The bright center is the ball of ice and dust. (**b**) Most comets orbit the Sun far beyond the orbit of Pluto. (**c**) Comet Shoemaker–Levy colliding with Jupiter, 1994. This image was made by combining separate photos taken by the Hubble Space Telescope of Jupiter and the torn-apart comet.

account for the frequency with which we see comets in the inner solar system, there must be in the neighborhood of a *trillion* (10^{12}) comets inhabiting the outskirts of our solar system.

Like asteroids, comets occasionally smash into planets. We were fortunate to witness the collision of a comet with Jupiter in 1994 (Figure 2.14c). Similar collisions of comets with Earth may have played a profound role in our existence: Some theoretical models of the formation of our solar system suggest that the Earth had very little water when it formed. If that is true, then most of the water in our oceans may have come from the melted ice of comets that struck the Earth long ago.

Looking Back

Having completed our walking tour of the solar system, let's take a brief look back at the trip. Perhaps the most striking feature of the solar system is its *emptiness*. Although we imagined walking to model planets laid out in a straight line from the Sun, a better model would show their orbits extending all the way around the Sun. Such a model on a scale of 1 to 10 billion would require an area of some 1.5 square kilometers (0.6 square mile), equivalent to some 300 football fields. Aside from the grapefruit-size Sun at the center, all we would find in the rest of this area would be the nine planets and their moons, plus scattered asteroids and comets—no object larger than marble-size Jupiter, and most objects far smaller than pinhead-size Earth.

TIME OUT TO THINK *Although robotic spacecraft have visited all the planets except Pluto, human beings have traveled no farther than the Moon. Compare the size of the Earth–Moon system to that of the entire solar system. Based on the much greater distances to the planets, do you think that people will ever visit any of them? If so, when? If not, why not?*

2.2 Onward to the Stars

The star system nearest to our own is the three-star system *Alpha Centauri.* The largest of its three stars, Alpha Centauri A, is about the same size as our Sun—another grapefruit on our 1-to-10-billion scale. Having completed the less than 1 kilometer walk from the Sun to Pluto, which takes about 10 minutes, how much farther would we have to walk to reach Alpha Centauri A?

TIME OUT TO THINK ***Before reading on, make a guess. Look at the grapefruit-size model Sun that you placed nearby. Where do you think you'll find the next grapefruit-size star? Within a few blocks? Within your city? Farther away?***

Recall that a light-year is about 10 trillion kilometers. On the scale of 1 to 10 billion, a light-year is one ten-billionth of this real distance, which we can calculate by dividing 10 trillion (10^{13}) kilometers by 10 billion (10^{10}):

$$\underbrace{10^{13}\text{ km}}_{\substack{\text{real}\\\text{light-year}}} \div \underbrace{10^{10}}_{\text{scale}} = \underbrace{10^{3}\text{ km}}_{\substack{\text{scaled}\\\text{light-year}}}$$

That is, a light-year on the scale of 1 to 10 billion is about 10^3, or 1,000, kilometers. Alpha Centauri A actually is about 4.4 light-years distant, so it is about 4,400 kilometers (2,700 miles) away on our model scale—roughly equivalent to the distance from New York to Los Angeles!

Simply trying to *see* Alpha Centauri A from the Earth is like trying to look from New York at a grapefruit in Los Angeles (neglecting the problems introduced by the curvature of the Earth). It may seem incredible that we can see it at all, but the bright light of this star *is* visible in the night sky of Earth; in fact, for those at southerly latitudes who can see it, Alpha Centauri A is one of the brightest stars in the sky. Nevertheless, it appears as a mere point of light with no apparent size, even to our most powerful telescopes.

Seeing the other two stars of Alpha Centauri is far more difficult because they are much dimmer. On our model scale, the second largest of the three stars, Alpha Centauri B, is about the size of a baseball and is located about the same distance from Alpha Centauri A as Pluto is from the Sun. Because this separation between the A and B stars of Alpha Centauri is so small compared to their distance from Earth, they appear as a single point of light to the naked eye. The third star, *Proxima Centauri,* orbits the other two stars of Alpha Centauri at a real distance of about 0.1 light-year, or about 100 kilometers on our scale. Because it takes thousands of years to complete a single orbit and because it happens to be nearer to us at present than the other two stars of Alpha Centauri, Proxima Centauri is the nearest star to Earth besides the Sun. However, Proxima Centauri is too dim to be seen with the naked eye.

If seeing a *star* is difficult, consider the difficulty of seeing *planets* beyond our solar system. Imagine a grapefruit-size model of Alpha Centauri A as a bright light bulb, orbited by a pinhead-size model planet, like Earth, at a distance of 15 meters. Now imagine looking from New York and trying to see this pinhead in Los Angeles. You probably won't be surprised to learn that no Earth-like planets have yet been detected around any other star. Indeed, the bigger surprise may be that we *have* discovered Jupiter-size planets around other stars,[3] with the first definitive evidence for the existence of planets around Sun-like stars reported in late 1995. In all likelihood, the answer to the question of whether Earth-like planets orbit nearby stars will be known within a couple of decades.

TIME OUT TO THINK ***Contemplate the fact that, throughout all of human history, no one knew for certain whether planets exist around other Sun-like stars until 1995. If an Earth-like planet is discovered in the near future, do you think it will have any impact on how we view our place in the universe? If so, how? If not, why not?***

Although our telescopic technology allows us to learn much about the stars, *travel* to the stars is another story. Consider the *Voyager 2* spacecraft. Launched in 1977, *Voyager 2* flew past Jupiter in 1979, Saturn in 1981, Uranus in 1986, and Neptune in 1989. (*Voyager*'s trajectory could not take it past Pluto.) *Voyager 2* is now bound for the stars at a speed near 50,000 kilometers per hour (30,000 miles per hour)—about as fast as anything ever built by humans. At this speed, *Voyager 2* would take about 100,000 years to reach Alpha Centauri—if it were headed in that direction, which it isn't. Clearly, convenient interstellar travel remains well beyond our present technology.

2.3 The Milky Way Galaxy

The vast separation between our solar system and Alpha Centauri is typical of the separations among

[3]These planets have been detected by their gravitational effects on their stars; current technology does not allow us to obtain actual images of these planets [Section 8.8].

star systems in the outskirts of the Milky Way Galaxy. Thus, the 1-to-10-billion scale is useless for modeling even just a few dozen of the nearest stars, because they could not all be spaced properly on the entire surface of the Earth. Visualizing the entire galaxy is even more difficult and clearly requires a new scale.

Let's reduce our scale further by a factor of 1 billion (making it a scale of 1 to 10^{19}). On this new scale, each light-year becomes 1 millimeter; for example, the 4.4-light-year separation between the Sun and Alpha Centauri A becomes just 4.4 millimeters on this scale—smaller than the width of your little finger. The stars themselves are microscopic on this scale; they are about the size of individual atoms.

Aside from the fact that stars become microscopic, this new scale makes it easy for us to visualize the Milky Way Galaxy. Its diameter of 100,000 light-years becomes 100 meters on this scale, or about the length of a football field.

TIME OUT TO THINK ***Go to a football field or visualize standing on one with our scale model of the galaxy centered over midfield. Standing on the 20-yard line puts you at about the correct location for our solar system. Hold your thumb and forefinger about 4 millimeters apart to represent the separation between the Sun and Alpha Centauri, and remember that the stars themselves are microscopic on this scale. Can you capture the enormity of the Milky Way Galaxy in words? If so, how?***

Another way to put the galaxy in perspective is to consider its number of stars. Based on star counts and gravitational effects, the number of stars in the Milky Way Galaxy is known to be somewhat more than 100 billion. Let's be conservative and assume that there are 100 billion stars. Imagine that, tonight, you are having difficulty falling asleep (perhaps because you are contemplating the enormity of the universe). Instead of counting sheep, you decide to count stars. If you are able to count about one star each second, on average, how long would it take you to count the 100 billion stars in the Milky Way?

Clearly, the answer is 100 billion (10^{11}) seconds. But how long is that? Let's convert it to years:

$$10^{11}\text{ s} \times \left(\frac{1\text{ min}}{60\text{ s}}\right) \times \left(\frac{1\text{ hr}}{60\text{ min}}\right) \times \left(\frac{1\text{ day}}{24\text{ hr}}\right) \times \left(\frac{1\text{ yr}}{365\text{ days}}\right) = 3{,}171\text{ yr}$$

You would need more than three thousand years just to *count* to 100 billion. And that assumes that you never take a break—no sleeping, no eating, and absolutely no dying!

Common Misconceptions: Confusing Very Different Things

Most people are familiar with the terms *solar system* and *galaxy,* but many people sometimes mix them up. Notice how incredibly different our solar system is from our galaxy. The solar system is a *single* star system consisting of our Sun and the various objects that orbit it, including Earth and eight other planets. The galaxy is a collection of some 100 billion star systems—so many that it would take thousands of years just to *count* them. Thus, confusing the terms *solar system* and *galaxy* represents a mistake by a factor of 100 billion—a fairly big mistake!

2.4 Beyond the Milky Way

As incredible as the scale of our galaxy may seem, the Milky Way is only one of some 50–100 billion galaxies in the observable universe. Just as it would take thousands of years to count all the stars in the Milky Way, it would also take thousands of years to count all the galaxies.

We could try to change our scale again in hopes of visualizing the size of the universe, but instead let's focus on the number of stars. Assuming that 100 billion stars is typical for a galaxy and that there are about 100 billion galaxies in the observable universe, we can calculate the approximate number of stars in the observable universe:

$$100\text{ billion} \times 100\text{ billion} = 10^{11} \times 10^{11} = 10^{22}$$

This number, 10^{22}, is so large that it does not even have a common name. How big is it? Visit a beach. Run your hands through the fine-grained sand (Figure 2.15). Try to imagine counting every one of the tiny grains of sand as they slip through your fingers. Then imagine counting every grain of sand on the beach where you are sitting. Next think about counting *all* the grains of sand on *all* the beaches everywhere on Earth. The number you would count would be *less than* 10^{22}. That is, the number of stars in the universe is larger than the total number of grains of sand on all the beaches on Earth.

2.5 The Scale of Time

Now that we have developed some perspective on the scale of space, we can do the same for the scale of time. Imagine the entire history of the universe, from the Big Bang to the present, compressed into a

FIGURE 2.15 The number of stars in the universe is larger than the total number of grains of sand on all the beaches on Earth.

single year. We can represent this history with a *cosmic calendar,*[4] on which the Big Bang takes place at the first instant of January 1 and the present day is just before the stroke of midnight on December 31 (Figure 2.16).

[4]This scheme for illustrating the scale of time with a cosmic calendar is beautifully presented (with a slightly different scale) by Carl Sagan in several of his books and in the *Cosmos* video series.

The uncertainty about the age of the universe poses a slight difficulty in constructing the cosmic calendar. The age is somewhere between about 10 billion and 16 billion years. We will arbitrarily choose 12 billion years as the approximate age of the universe so that each month on the cosmic calendar represents 1 billion years.

On this scale, the Milky Way Galaxy probably formed sometime in February. Many generations of stars lived and died in the subsequent cosmic months, enriching the galaxy with the heavier elements from which we and the Earth are made. Not until about August 13, which represents a time 4.6 billion years ago, did our solar system form from a cloud of gas and dust in the Milky Way. All the planets in our solar system formed at roughly the same time, so August 13 also is the Earth's birth date on the cosmic calendar.

No one knows exactly when the earliest life arose on Earth, but it certainly was within the first billion years, or by mid-September on the cosmic calendar. However, for most of Earth's history, living organisms remained relatively primitive and microscopic in size. Larger invertebrate life arose only about 600 million years ago, or about December 13 on the cosmic calendar.

FIGURE 2.16 The cosmic calendar compresses the history of the universe into 1 year; this version assumes that the universe is 12 billion years old.

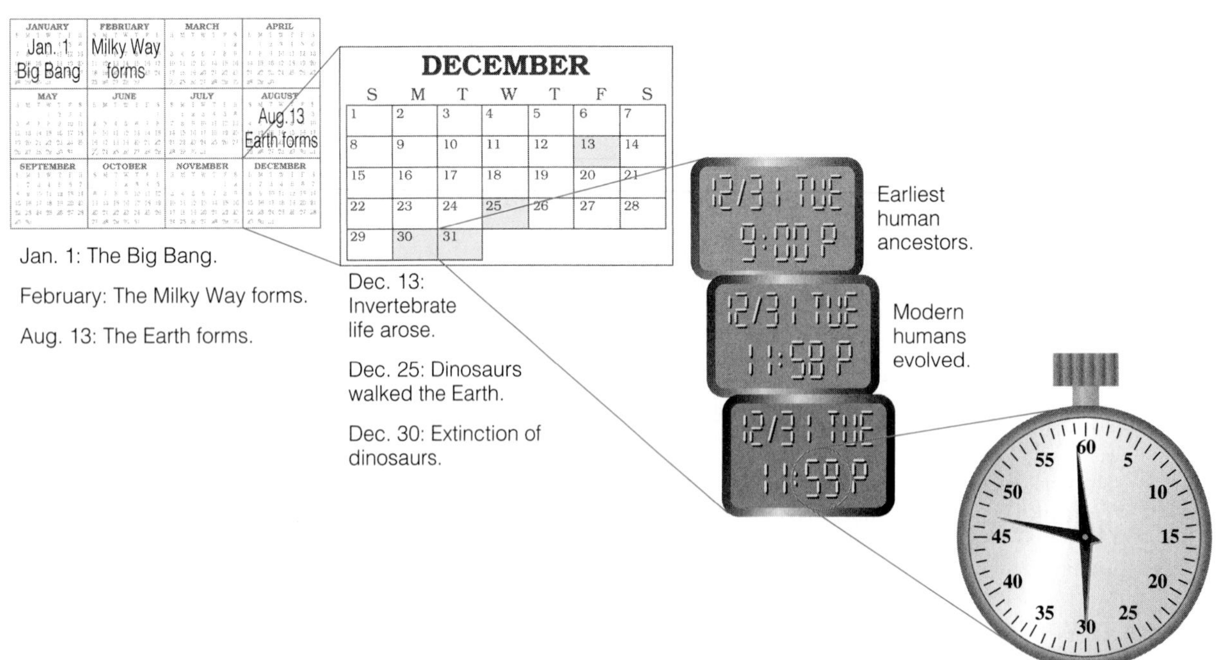

The earliest dinosaurs roamed the Earth about 230 million years ago, or shortly after midnight on Christmas (December 25) on the cosmic calendar. The last dinosaurs (and most other species living at the time) suddenly became extinct about 65 million years ago—around 2 A.M. on December 30 of the cosmic calendar.

With the dinosaurs gone, small furry mammals inherited the Earth. Some 60 million years later, or around 9 P.M. on December 31 of the cosmic calendar, the earliest hominids (human ancestors) walked upright. Humans slowly evolved, developing larger brains and the ability to use tools and communicate. By about 11:58 P.M.—2 minutes before midnight on the cosmic calendar, or about 40,000 years ago in real time—our ancestors essentially were modern humans.

As shown on the stopwatch in Figure 2.16, most of the major events of human history took place within the final seconds of the final minute of the final day on the cosmic calendar. Agriculture arose about 30 seconds ago. The Egyptians built the pyramids about 13 seconds ago. It was only about 1 second ago on the cosmic calendar that Kepler and Galileo first proved that the Earth is a planet. The average college student was born less than 0.1 second ago, around 11:59:59.9 P.M. on December 31. On the scale of cosmic time, the human species is the youngest of infants, and a human lifetime is a mere blink of an eye.

TIME OUT TO THINK ***In the last few tenths of a second before midnight on December 31 of the cosmic calendar, we have developed an incredible civilization and learned a great deal about the universe, but we also have developed technology through which we could destroy ourselves. The midnight bell is striking, and the choice for the future is ours. How far into the next cosmic year do you think our civilization will survive? Defend your opinion.***

Have you noticed that the astronomers and mathematicians are much the most cheerful people of the lot? I suppose that perpetually contemplating things on so vast a scale makes them feel either that it doesn't matter a hoot anyway, or that anything so large and elaborate must have some sense in it somewhere.

DOROTHY L. SAYERS (1893–1957)

Mathematical Insight 2.2 Using the Cosmic Calendar

The cosmic calendar is set up so that 1 year represents 12 billion years of real time. In other words, it represents time on a scale of 1 to 12 billion. Thus, using the cosmic calendar is similar to using a scale model.

Example: The solar system formed about 4.6 billion (4.6×10^9) years ago. When was this on the cosmic calendar?

Solution: We convert from real time to cosmic calendar time by dividing by 12 billion (1.2×10^{10}):

$$\text{cosmic calendar time} = \frac{\text{real time}}{1.2 \times 10^{10}} = \frac{4.6 \times 10^9 \text{ yr}}{1.2 \times 10^{10}} = 0.383 \text{ yr}$$

It is easier to work in days than in fractions of a year, so we convert 0.383 year to days by multiplying by 365 days per year:

$$0.383 \text{ yr} \times \frac{365 \text{ days}}{1 \text{ yr}} = 140 \text{ days}$$

Thus, the Earth formed about 140 days ago on the cosmic calendar. The months of August through December have a total of 153 days (31 days each for August, October, and December and 30 days each for September and November), so 140 days before December 31 is 13 days after the beginning of August, or August 13.

THE BIG PICTURE

This chapter was all about "the big picture"—how to develop perspective on the scale of space and time in the universe. Throughout the rest of the book, we will return many times to the particular perspectives developed here. Keep them in mind.

- Despite their relatively small size on the scale of the solar system, each of the planets is a unique world. Although you needn't memorize detailed facts about each planet at this point, you should be able to associate at least one or two unique characteristics with each of the planets.
- In comparison with distances within our solar system, the stars are incredibly far away. On the same scale on which we can walk from the Sun to Pluto in less than 1 kilometer, it is a journey of thousands of kilometers to even the nearest stars besides the Sun.
- As difficult as it is to imagine the distances to the stars, these distances pale in comparison to the scale of the Milky Way Galaxy. There are so many stars in the Milky Way that it would take thousands of years just to count them. Moreover, there may be nearly as many galaxies in the universe as stars in the Milky Way.
- The scale of time is no less incredible than the scale of space. If we imagine that the history of the universe spans a 1-year cosmic calendar, a human lifetime is a mere blink of an eye.

Review Questions

1. Imagine that you are describing the 1-to-10-billion scale model solar system to friends or family members. In your own words, give them a feel for the layout of the solar system and the sizes of the Sun and planets. For each planet, give a few facts that you find particularly interesting.
2. What is the Sun made of? How does its mass compare with the mass of the planets?
3. What is nuclear fusion? Where does it occur in the Sun? Why is it important?
4. Which planet has the highest average surface temperature? Why?
5. Why do many people believe that conditions on Mars may have been suitable for life in the past?
6. Why do we group planets into those of the *inner* solar system and those of the *outer* solar system? Which planets belong to each group?
7. List several ways in which Pluto is a "misfit" compared to other planets of the outer solar system.
8. In general terms, describe the differences between asteroids and comets.
9. On the 1-to-10-billion scale, how far is it to the nearest stars? Use this fact to describe the difficulty of seeing planets around other stars and the difficulty of traveling to the stars.
10. Why can't we use the 1-to-10-billion scale to represent the entire Milky Way Galaxy?
11. Describe the Milky Way Galaxy on a scale on which 1 millimeter represents 1 light-year. Where is our solar system located?
12. How many stars are in the Milky Way Galaxy? How long would it take to count them at a rate of one per second?
13. How many galaxies are in the universe? How many stars are in the universe? Describe the number by comparing it to grains of sand.
14. Imagine that you are describing the cosmic calendar to friends or family members. In your own words, give them a feel for how the human race fits into the scale of time.

Discussion Questions

1. *The Greenhouse Effect.* Why is the greenhouse effect a topic of environmental debate on Earth? Why might the study of Venus help us understand the greenhouse effect on Earth?
2. *Crossing the Asteroid Belt.* Do you think that a spacecraft sent through the asteroid belt is likely to suffer a collision with an asteroid? Why or why not?
3. *Voyager 2.* Do you think that *Voyager 2* will ever send back data about its visits to other stars? Why or why not?

Problems

1. *Tour Report.* Imagine that you have just completed a walking tour of the 1-to-10-billion scale model solar system described in this chapter. Upon returning home, imagine that friends or family members ask you the following questions. Answer each question in one or a few sentences.
   a. Is the Sun really much bigger than the Earth?
   b. I've heard that the Sun uses nuclear energy. Is this true?
   c. Would it be much harder to send humans to Mars than to the Moon?
   d. I heard that Neptune is farther from the Sun than Pluto. Can you explain the real story?
   e. Why is Jupiter called a giant planet?
   f. Why weren't there any stars besides the Sun in the scale model?
   g. What was the most interesting thing you learned during your tour?
2. *Driving to the Planets (and Stars).* Another method for gaining perspective on the size of the solar system is to imagine that you could drive your car in space. Assume that you can drive at a constant speed of 100 kilometers per hour (about 62 miles per hour), and refer to the data in Table 2.1. (In reality, the law of gravity would make driving through space at a constant speed all but impossible.)
   a. If you could drive around the Earth's equator, how long would it take to circle the Earth? (*Hint:* circumference of a circle $= 2 \times \pi \times$ radius $= \pi \times$ diameter)
   b. Suppose you started driving from the Sun. Make a table showing how long it would take, in years, to reach each of the nine planets at a speed of 100 kilometers per hour.
   c. How long would it take to drive the 4.3 light-years to Proxima Centauri at 100 kilometers per hour? (*Hint:* Start by converting Proxima Centauri's distance to kilometers.)
   d. The *Voyager 2* spacecraft is on its way out of our solar system at a speed of about 50,000 kilometers per hour. If it were headed in the correct direction (which it is not), how long would it take *Voyager 2* to reach Proxima Centauri?
3. *Communication to Mars.* Robotic spacecraft on Mars receive instructions from Earth and transmit their data back to Earth by radio. Radio waves are a form of light and therefore travel through space at the speed of light ($c = 300{,}000$ km/s).
   a. Using the average distances from the Sun given in Table 2.1, calculate the approximate distance from Earth to Mars when both lie in a straight line from the Sun. (*Hint:* This calculation requires only *subtracting* two numbers from the table.) About how long would it take for radio signals to be transmitted from Mars to Earth or vice versa?
   b. Calculate the approximate distance from Earth to Mars when the two planets are on opposite sides of the Sun. How long would it now take for radio signals to be transmitted from Mars to Earth or vice versa?
   c. Suppose you are designing a robotic lander with wheels that will drive around on Mars. Based on your answers to parts (a) and (b), explain why you must program your robot to make decisions when it encounters obstacles such as rocks or cliffs, rather than controlling it remotely from Earth.
4. *Talking to Galileo and Cassini.* The Galileo spacecraft began orbiting Jupiter in late 1995, and the Cassini spacecraft is scheduled to reach Saturn in 2004.
   a. Following the basic procedure outlined in problem 3, calculate the approximate amount of time that it takes for radio signals to travel from Earth to Jupiter when the two planets are closest together and when they are farthest apart.
   b. Repeat part (a) for Saturn.
   c. Suppose that, someday, humans are sent to explore the moons of Jupiter or Saturn. Would it be possible for a TV station on Earth to broadcast a *live* interview with the astronauts? Why or why not?
5. *Scaling the Local Group of Galaxies.* Both the Milky Way Galaxy and the Great Galaxy in Andromeda (M31) have diameters of about 100,000 light-years. The distance between the two galaxies is about 2.5 million light-years. Using a scale on which 1 centimeter represents 100,000 light-years, draw a sketch showing both galaxies and the distance between them to scale.
6. *Number of Civilizations.* Another means of gaining perspective on the galaxy is to consider whether other stars might harbor civilizations like ours. In reality, no one knows whether *any* such civilizations exist, let alone how common they might be if they do exist. However, with so many stars, many people believe it likely that there are many other civilizations. In this problem, assume that there are 100 billion stars in the Milky Way Galaxy.
   a. Suppose that if you pick a star at random the odds that it harbors a civilization are about the same as your odds of winning big in a state lottery: about 1 in 1 million. How many civilizations would exist in the Milky Way Galaxy? How does the number change if the odds are 1 in 1 billion? 1 in 10,000?
   b. Write a short essay (two or three paragraphs) detailing and defending your personal belief as to whether other civilizations exist in the Milky Way Galaxy.

7. *Project: Build Your Own Model Solar System.* Using the model diameters and distances shown in Table 2.1 and a long, straight path (e.g., on a campus walkway or along a park trail or a road), lay out a 1-to-10-billion scale model of the solar system. A few hints:
   - For the planets, use appropriate-size small spheres (e.g., ball bearings or Styrofoam) or make the small spheres from bits of clay. For the Sun, use an appropriate-size piece of fruit or Styrofoam sphere.
   - You can measure the distances roughly by practicing a 1-meter or 2-meter stride and counting stride lengths to find the locations for each planet. For greater accuracy, use a "measuring wheel" like those used in marking a football field.
   - Place your planets on a post or a dish so you won't lose them when you are looking for them later.

   After laying out your model, take a few friends on a tour through it. If the necessary resources are available, you might want to construct your model with materials strong enough to be left in place throughout the semester (or even longer) so you can revisit it as you learn more about the solar system.

8. *Project: Make Your Own Time Line.* As an alternative to the *cosmic calendar* for showing the scale of time, make a 1-meter-long time line of the history of the universe. Show the locations of important events in cosmic history and in the Earth's history.

3

Spaceship Earth

WHEREVER YOU ARE AS YOU READ THIS BOOK, YOU PROBABLY have the feeling that you're "just sitting here." Nothing could be further from the truth.

In fact, you are being spun in circles as the Earth rotates, you are racing around the Sun in the Earth's orbit, and you are careening with the Milky Way Galaxy through the cosmos. In the words of the noted inventor and philosopher R. Buckminster Fuller (1895–1983), you are a traveler on *spaceship Earth.*

In this chapter, we'll discuss the real motions of the Earth and the Moon in our solar system and use these motions to explain the *apparent* phenomena of the sky that so mystified our ancestors. We'll then discuss the remarkable journey of our solar system within the Milky Way Galaxy and conclude by examining the motion of the Milky Way Galaxy relative to other galaxies in the universe.

3.1 Rotation and Revolution

The Earth **rotates** on its axis once each *day.* As a result, you are whirling around the Earth's rotation axis at a speed that depends on your latitude (Figure 3.1). Note that, in most populated regions, the speed of rotation is 1,000 kilometers per hour (600 miles per hour) or more—faster than most airplanes travel.

The Earth also **revolves** around the Sun once each *year,* following an *orbit* that is not quite circular (Figure 3.2). The deviation from a perfect circle is so slight that your eye may have a difficult time seeing it in a diagram. The *average* distance of the Earth from the Sun is called an **astronomical unit (AU)**:[1]

$$1 \text{ AU} \approx 150{,}000{,}000 \text{ km } (93{,}000{,}000 \text{ mi})$$

The Earth travels slightly faster in its orbit when it is nearer to the Sun and slightly slower when it is farther from the Sun. But, at all times, the Earth is carrying you around the Sun at a speed in excess of 100,000 kilometers per hour (60,000 miles per hour)!

The plane of the Earth's orbit around the Sun is called the **ecliptic plane.** The Earth's rotation axis happens to be tilted[2] by $23\frac{1}{2}°$ from a line *perpendicular* to the ecliptic plane, pointing nearly in the direction of Polaris, the North Star (Figure 3.3). If you could view the Earth's motion from Polaris, you would see that the Earth both rotates and revolves counterclockwise.

TIME OUT TO THINK *Suppose you viewed the Earth's motion from far above the* South Pole *(e.g., from one of the stars in the Southern Cross) rather than from above the North Pole. Which way would the Earth appear to rotate and revolve? Explain.*

[1]Technically, an astronomical unit is the *semimajor axis* of the Earth's orbit, which is the mean of its minimum (perihelion) and maximum (aphelion) distances from the Sun.

[2]A technical term for this tilt is *obliquity.* Thus, the Earth's obliquity is $23\frac{1}{2}°$.

The Length of a Day

How do the characteristics of the Earth's rotation and revolution affect what we see in the sky? Let's start by examining rotation. The Earth makes one complete rotation about every 23 hours 56 minutes (actually 23 hours 56 minutes 4.09 seconds). From our viewpoint on the Earth, the stars therefore seem to spin around us with this same period, which is called a **sidereal day** (pronounced sy-dear-ee-al)—*sidereal* means "related to the stars." You can confirm this fact by starting a stopwatch at the moment when any particular star is on your *meridian* [Section 1.5] and stopping the watch the next day when the star again is on your meridian; the watch will read 23 hours 56 minutes.

FIGURE 3.1 Rotation carries you around the Earth's axis at a speed that depends on your latitude.

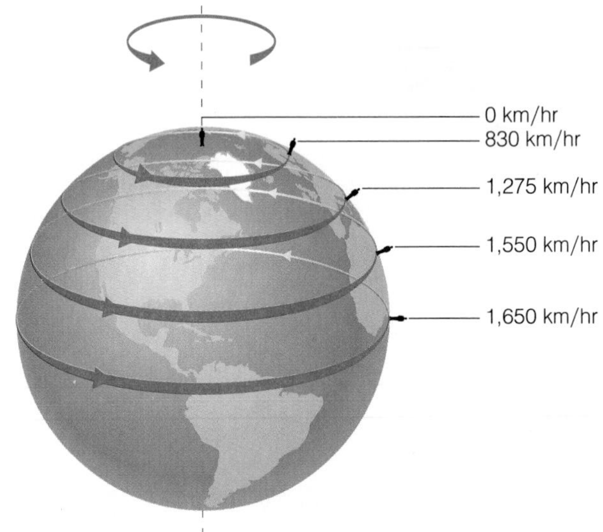

FIGURE 3.2 Basic properties of the Earth's orbit.

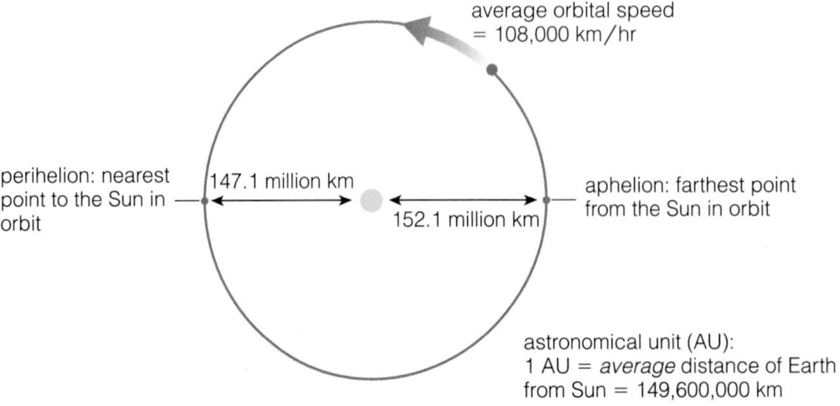

You may be surprised that the Earth's rotation period is about 4 minutes short of 24 hours. In fact, 24 hours is a **solar day:** the average time between successive meridian crossings of the Sun.[3]

The Sun takes about 4 minutes longer than the stars to circle your sky because the Earth is orbiting the Sun at the same time that it is rotating. This will become clear with a simple demonstration. Set an object on a chair or table to represent the Sun while you stand a few steps away to represent the Earth. Now imagine that a distant star appears in the same direction as the Sun (Figure 3.4a). If you rotate exactly once (counterclockwise) while standing in place, you will again be pointing in the direction of both the Sun and the star. Next, repeat your single rotation while at the same time taking a couple of steps counterclockwise around the Sun to represent the Earth's orbit (Figure 3.4b). At the end of this rotation, you'll again be pointing in the direction of the distant star, but you will *not* be pointing toward the Sun. You'll need to rotate a little bit more before you again are pointing toward the Sun; while one rotation represents a sidereal day, you need slightly more than one rotation to make a solar day.

[3]The actual time varies slightly over the course of the year because of the Earth's varying orbital speed and the tilt of its axis.

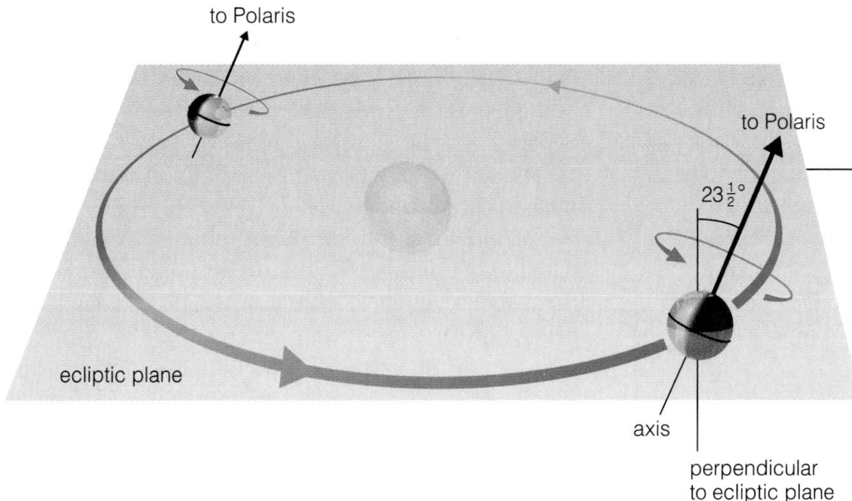

FIGURE 3.3 The Earth's axis is tilted by $23\frac{1}{2}°$ from a line *perpendicular* to the ecliptic plane. Note that, as viewed from above the North Pole, the Earth rotates *and* revolves counter clockwise.

FIGURE 3.4 A simple demonstration shows why a solar day is slightly longer than a sidereal day. (**a**) One full rotation represents a sidereal day. (**b**) The simultaneous motion of the Earth around the Sun means that a solar day requires slightly more than one full rotation. (**c**) In reality, the Earth travels about 1° per day around its orbit. Thus, a sidereal day is one full (360°) rotation, while a solar day requires about 361° of rotation. (*Note:* Drawing is not to scale.)

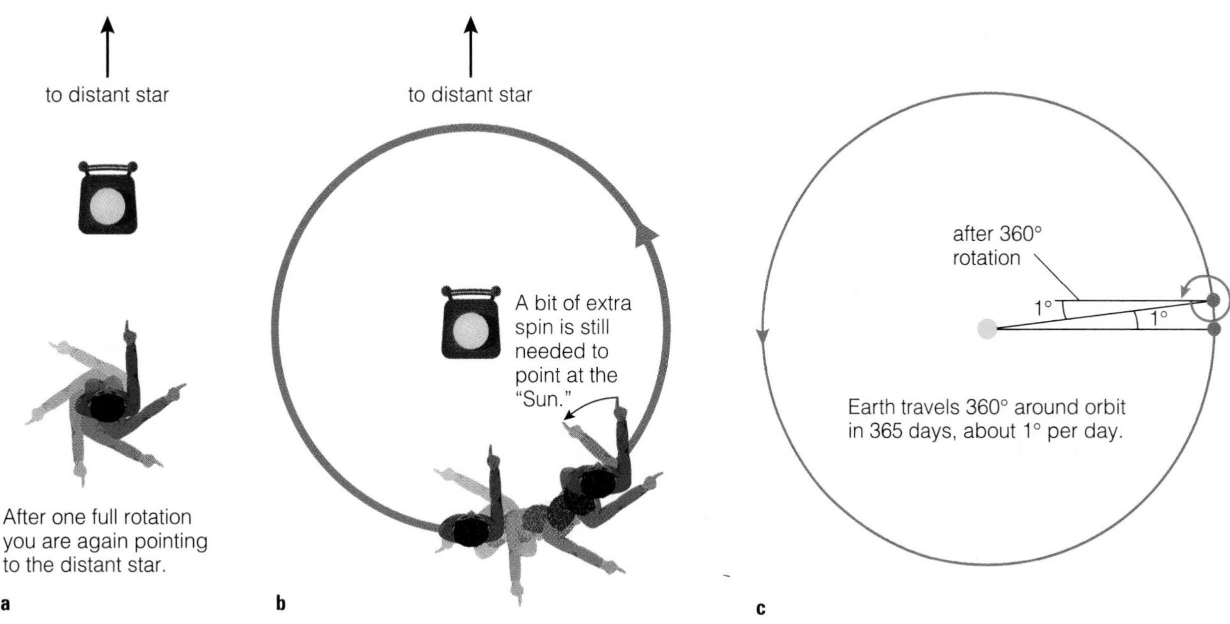

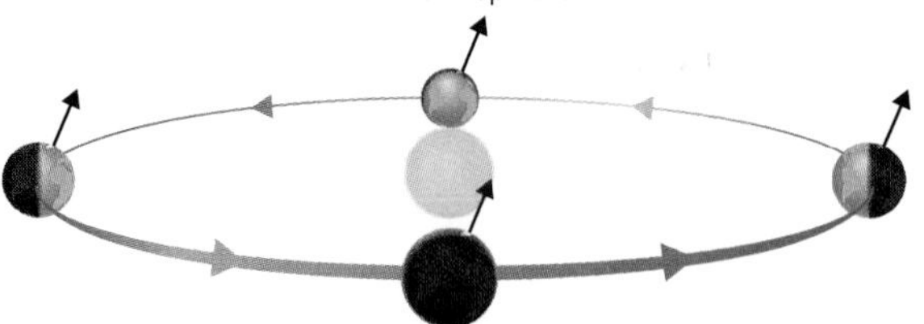

FIGURE 3.5 Seasons occur because, even though the Earth's axis remains pointed toward Polaris throughout the year, the orientation of the axis *relative to the Sun* changes as the Earth orbits the Sun. Around the time of the summer solstice, the Northern Hemisphere has summer because it is tipped toward the Sun, and the Southern Hemisphere has winter because it is tipped away from the Sun. The situation is reversed around the time of the winter solstice when the Northern Hemisphere has winter and the Southern Hemisphere has summer. At the equinoxes, both hemispheres receive equal amounts of light.

The Earth takes about 365 days (1 year) to complete one revolution, so it moves about 1/365 of the way around its orbit each day—just under 1° (Figure 3.4c). Thus, although any particular star will reappear on your meridian with each 360° rotation (one sidereal day), about 361° of rotation are required between successive passes of the Sun across your meridian. The extra 1° of rotation takes the extra 4 minutes by which the solar day is longer than the sidereal day.

Common Misconceptions: The Cause of Seasons

When asked the cause of the seasons, many people mistakenly answer that they are caused by variations in Earth's distance from the Sun. Indeed, in a famous survey, even many Harvard graduates gave this incorrect answer. By knowing that the Northern and Southern hemispheres experience opposite seasons, you can immediately realize that Earth's varying distance from the Sun *cannot* be the cause of the seasons; if it were, both hemispheres would have summer at the same time. Although Earth's distance from the Sun *does* vary slightly over the course of a year (Earth is closest to the Sun in January and farthest away in July), this effect is greatly overwhelmed by the way the tilt of the rotation axis causes the Northern and Southern hemispheres to alternately receive more or less direct sunlight. That is, Earth's varying distance from the Sun has no noticeable effect on our seasons. (Interestingly, this is not necessarily the case for other planets; seasons on Mars, for example, *are* affected by its varying distance from the Sun.)

The Seasons

The Earth's rotation axis remains pointed in the same direction (toward Polaris) throughout the year. The orientation of the axis *relative to the Sun* therefore changes over the course of the year, which is the cause of the *seasons* (Figure 3.5). The more directly a part of the Earth is receiving sunlight, the longer and higher is the Sun's path through the sky, making daylight hours longer and generally warmer. Note that the seasons are opposite in the Northern and Southern hemispheres.

Four points in the Earth's orbit have special names related to the seasons. The Earth is at point 1 in Figure 3.5 on about March 21 each year, which represents the **spring equinox** (or *vernal equinox*). Both hemispheres receive equal amounts of sunlight at this time; it is the beginning of spring for the Northern Hemisphere but the beginning of fall for the Southern Hemisphere.[4] The Earth reaches point 2, which represents the **summer solstice,** around June 21—the day on which the Northern Hemisphere receives its most direct sunlight and the Southern Hemisphere receives its least direct sunlight. Point 3 represents the **fall equinox** (or *autumnal equinox*), which occurs around September 21; both hemispheres again receive the same amount of sunlight, but it is the beginning of fall in the Northern Hemisphere and the beginning of spring in the Southern Hemisphere. Finally, point 4 represents the **winter solstice,** which occurs around December 21, the day on which the Northern Hemisphere receives its least direct sunlight and the Southern Hemisphere receives its most direct sunlight.

[4]The solstices and equinoxes were named by people living in the Northern Hemisphere, so the names reflect the Northern Hemisphere seasons.

TIME OUT TO THINK *Use the fact that the seasons are opposite in the Northern and Southern hemispheres to explain why round-trip airfare from the United States to Australia generally is much higher in the months of December through February than in the months of June through August. During which months would you expect airfares from the United States to Europe to be highest? Why?*

Apparent Planetary Motion

Ancient astronomers who believed the Earth to be at the center of the universe had no trouble understanding the apparent motion of the Sun or the stars through the sky. Instead of realizing that the Earth is rotating, they simply imagined the Earth to be standing still while the stars rotated around us. The fact that the Sun takes 4 minutes longer than the stars to complete its daily path through the sky simply meant that the Sun went around the Earth slightly more slowly than the stars. But the nightly changes in the positions of the planets among the stars—especially the periods of *apparent retrograde motion* during which planets appear to reverse course relative to the stars for a few months [Section 1.5]—were far more mysterious to ancient scientists.

The apparent retrograde motion is very simply explained in a Sun-centered solar system. You can demonstrate it for yourself with the help of a friend (Figure 3.6). Go outside to a large, open field (such as a football or soccer field) and pick a spot in the center to represent the Sun. You can represent the Earth, walking counterclockwise around the Sun, while your friend represents a more distant planet (e.g., Mars, Jupiter, or Saturn) by walking counterclockwise around the Sun at a greater distance. Your friend should walk more slowly than you because more distant planets orbit the Sun more slowly. As you walk, watch how your friend appears to move relative to buildings or trees in the distance. Although both of you always walk counterclockwise around the Sun, your friend will *appear* to you to move backward against the background during the part of your "orbit" at which you catch up and pass him or her. To understand the apparent retrograde motions of Mercury and Venus, which are closer to the Sun than is the Earth, simply switch places with your friend and repeat the demonstration.

TIME OUT TO THINK *Don't just take our word for it. Find a friend, go outside, and try this demonstration. It takes only a couple of minutes, and it should forever remove the mystery from the apparent retrograde motion of the planets. Describe your results.*

The apparent retrograde motion demonstration applies directly to the planets. For example, because Mars takes about 2 years to orbit the Sun (actually 1.88 years), it covers about half its orbit during the 1 year in which Earth makes a complete orbit. If you trace lines of sight from Earth to Mars from different points in their orbits, you will see that the line of sight usually moves eastward relative to the stars but moves westward during the time when Earth is passing Mars in its orbit (Figure 3.7). Like your friend in the demonstration, Mars never really changes direction; it only *appears* to change direction from our perspective on Earth.

Stellar Parallax

If the apparent retrograde motion of the planets is so readily explained by recognizing that the Earth is a planet, why wasn't this idea accepted in ancient

FIGURE 3.6 The retrograde motion demonstration. Watch how your friend usually appears to you to move forward against the background of buildings in the distance but appears to move backward as you catch up with and pass him or her in your "orbit."

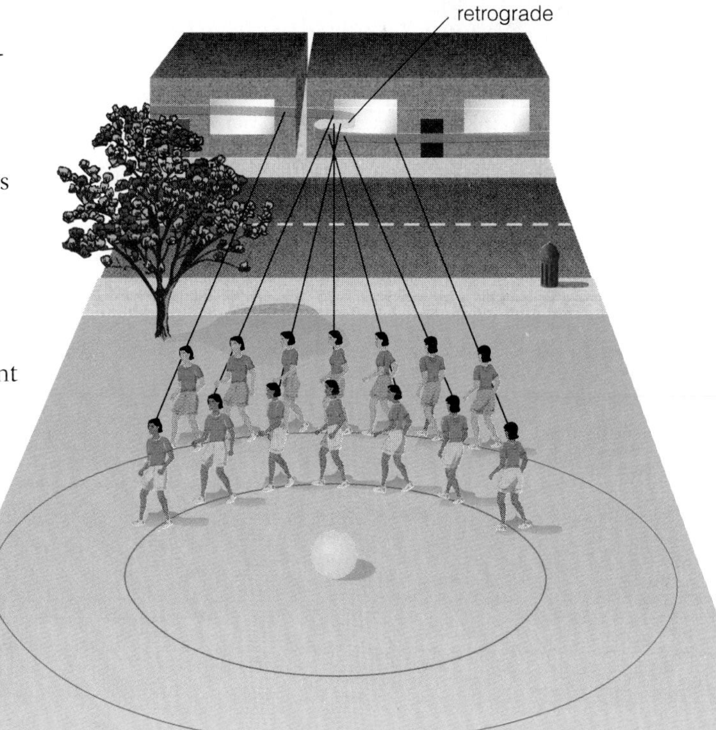

Aristarchus

Until the early 1600s, nearly everyone believed that the Earth was the center of the universe. Yet Aristarchus (c. 310–230 B.C.) argued otherwise almost 2,000 years earlier. How did he reach his conclusion?

In about 260 B.C., Aristarchus pointed out that the apparent retrograde motion of the planets could be explained much more simply if Earth and other planets orbited the Sun. To account for the lack of detectable stellar parallax, Aristarchus suggested that the stars were extremely far away.

He further strengthened his argument by estimating the sizes of the Moon and the Sun. By observing the shadow of the Earth on the Moon during a lunar eclipse, Aristarchus estimated the Moon's diameter to be about one-third of the Earth's diameter—only slightly higher than the actual value. He then used a geometric argument, based on measuring the angle between the Moon and the Sun at first- and third-quarter phases, to conclude that the Sun must be larger than the Earth. (His measurements were imprecise, so he estimated the Sun's diameter to be about 7 times the Earth's, rather than the correct value of about 100 times.) Since the Sun is larger than the Earth, he argued that it was more natural for the Earth to go around the Sun than vice versa.

Like most scientific work, Aristarchus's work built upon the work of others. In particular, Heracleides (c. 388–315 B.C.) had suggested that the Earth rotates, an idea that Aristarchus drew from to explain the apparent daily rotation of the stars in our sky. Heracleides also was the first to suggest that *not all* heavenly bodies circle the Earth; based on the fact that Mercury and Venus always are close to the Sun in the sky, he argued that these two planets must orbit the Sun. Thus, in suggesting that all the planets orbit the Sun, Aristarchus was extending the ideas of Heracleides and others before him. Alas, Aristarchus's arguments were not widely accepted in ancient times and were revived only with the work of Copernicus some 1,800 years later.

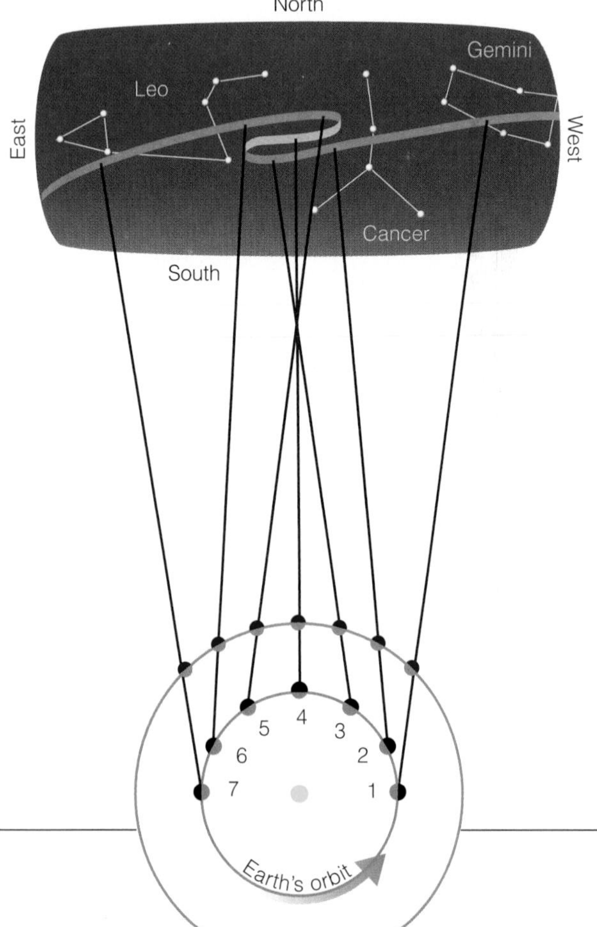

FIGURE 3.7 The explanation for apparent retrograde motion. Follow the lines of sight from Earth to Mars in numerical order. The period where the lines of sight shift *westward* relative to the distant stars is the period during which we observe apparent retrograde motion for Mars.

times? In fact, the idea that the Earth goes around the Sun was suggested as early as 260 B.C. by the Greek astronomer Aristarchus and likely was debated many times thereafter. The German scholar Nicholas of Cusa wrote that the Earth goes around the Sun in a book published in 1440,[5] and the desire for a simple explanation for apparent retrograde motion was the primary motivation of Copernicus when he again put forth Aristarchus's idea in the early 1500s[6] [Section 6.3].

Nevertheless, the idea that the Earth goes around the Sun did not gain wide acceptance among scientists until the work of Kepler and Galileo in the early 1600s. Although there were many reasons for the historic reluctance to abandon the idea of an Earth-centered universe, perhaps the most prominent involved the inability of ancient peoples to detect something called *stellar parallax.*

[5]Interestingly, while Galileo was punished by the Church for holding the same belief two centuries later, Nicholas was named a Cardinal of the Church in 1448.

[6]Copernicus probably was not aware of the book by Nicholas of Cusa, but he *was* aware of the claim by Aristarchus.

Extend your arm and hold up one finger. If you keep your finger still but alternately close your left eye and right eye, your finger will *appear* to jump back and forth against the background. This apparent shifting in the position of your finger demonstrates **parallax** and occurs simply because your two eyes view your finger from opposite sides of your nose. Note that if you move your finger closer to your face, the parallax increases. In contrast, if you look at a distant tree or flagpole instead of your finger, you probably cannot detect any parallax by alternately closing your left eye and right eye. Thus, parallax depends on distance, with nearer objects exhibiting greater parallax than more distant objects.

Now imagine that your two eyes represent the Earth at opposite sides of its orbit around the Sun and that your finger represents a distant star—this is the idea of **stellar parallax.** That is, because we view the stars from different places in our orbit at different times of year, nearby stars should *appear* to shift back and forth against distant stars in the background during the course of the year (Figure 3.8).

FIGURE 3.8 Stellar parallax is an apparent shift in the position of a nearby star as we look at it from different places in the Earth's orbit. This figure is greatly exaggerated; in reality, the amount of shift is far too small to detect with the naked eye.

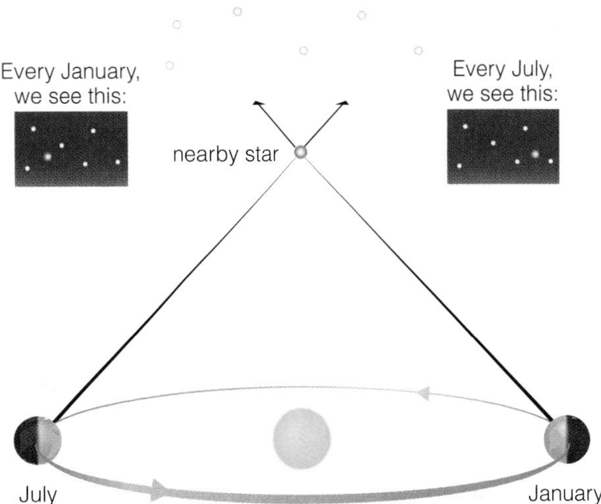

THINKING ABOUT . . .

And Yet It Moves

On June 22, 1633, Galileo was brought before a Church inquisition in Rome and ordered to recant his heretical view that the Earth goes around the Sun rather than vice versa. Fearing for his life, Galileo recanted as ordered. However, legend has it that, as he rose from his knees following his renunciation, he whispered under his breath, *Eppur si muove*—Italian for "And yet it moves." The evidence supporting the idea that the Earth moves was quite strong by the mid-1600s, but it was still indirect. Today, we have much more direct proof.

French physicist Jean Foucault provided the first direct proof of *rotation* in 1851. Foucault built a large pendulum that he carefully started swinging. Because any pendulum tends to swing always in the same plane, the Earth's rotation made Foucault's pendulum appear to twist slowly in a circle. Today, *Foucault pendulums* are a popular attraction at many science centers and museums. A second direct proof that the Earth rotates is provided by the *Coriolis effect,* first described by French physicist Gustave Coriolis (1792–1843). The Coriolis effect, which would not happen if the Earth were not rotating, is responsible for things such as the swirling of hurricanes and the fact that missiles shot great distances on the Earth deviate from straight-line paths [Section 10.3].

Stellar parallax provides direct proof that the Earth orbits the Sun, and it was first measured by German astronomer Friedrich Bessel in 1838. However, another direct proof that the Earth orbits the Sun actually was found about a century earlier by English astronomer James Bradley (1693–1762). To understand Bradley's proof, imagine that starlight is like rain, falling straight down. If you are standing still you should hold your umbrella straight over your head, but if you are walking through the rain you should tilt your umbrella forward, because your motion makes the rain appear to be coming down at an angle. Bradley discovered that observing light from stars requires that telescopes be slightly tilted—just like the umbrella; this effect is called the *aberration of starlight.*

However, no matter how hard they searched, ancient astronomers could find no sign of stellar parallax. They therefore concluded that one of the following must be true:

1. The stars are so far away that stellar parallax is undetectable to the naked eye.
2. There is no stellar parallax because the Earth is the center of the universe.

Unfortunately, with notable exceptions such as Aristarchus, Nicholas of Cusa, and Copernicus, until the 1600s most astronomers rejected the correct answer (1) because they could not imagine that the stars could be *that* far away.

TIME OUT TO THINK ***How far apart are opposite sides of the Earth's orbit? How far away are the nearest stars? Describe the challenge of detecting stellar parallax. It may help to visualize the Earth's orbit and the distance to the stars on the 1-to-10-billion scale used in Chapter 2.***

Although stellar parallax is undetectable with the naked eye, today it can be accurately measured for stars several hundred light-years away with the aid of powerful telescopes and photographs. Careful measurements of stellar parallax are the most reliable means of measuring distances to nearby stars [Section 15.2].

3.2 Precession

Besides rotating on its axis and orbiting the Sun, the Earth moves within the solar system in several subtler ways. The most important of these motions is one you're familiar with if you've ever played with a top: As it spins rapidly on its axis, a top also wobbles more slowly around an axis perpendicular to the floor (Figure 3.9a). This wobbling motion is called **precession.**

Like a spinning top, the rotating Earth also precesses[7]—but much more slowly. Each cycle of precession takes about 26,000 years (Figure 3.9b). Moreover, whereas friction slows a top's rotation until it falls over, the Earth's tilt remains about $23\frac{1}{2}°$ throughout its precession cycle. Precession slowly changes the direction the Earth's axis points in space. Today the axis points toward Polaris, but about 13,000 years from now it will point nearly in the direction of the star Vega. During most of the precession cycle, the axis does not point very near to any bright star.

Precession changes not only the location of the north celestial pole among the stars, but also the locations in the Earth's orbit at which the equinoxes and solstices occur. For example, the spring equinox today occurs when the Sun appears in the direction of the constellation Pisces, but 2,000 years ago it occurred when the Sun appeared in Aries. (That is why the spring equinox is sometimes called "the first point of Aries.") During a complete 26,000-year precession cycle, the Sun's apparent location among the stars on the spring equinox slowly drifts all the way around the zodiac. As a result, the time from one spring equinox to the next is not exactly the same as the Earth's true orbital period around the Sun. The time required for the Earth to complete exactly one orbit is called a **sidereal year.** The time from one spring equinox to the next, called a **tropical year,** is about 20 minutes shorter than the sidereal year. Our calendar is based on the tropical year, ensuring that spring always

FIGURE 3.9 (**a**) A spinning top slowly wobbles, or *precesses.* (**b**) The Earth's axis also precesses.

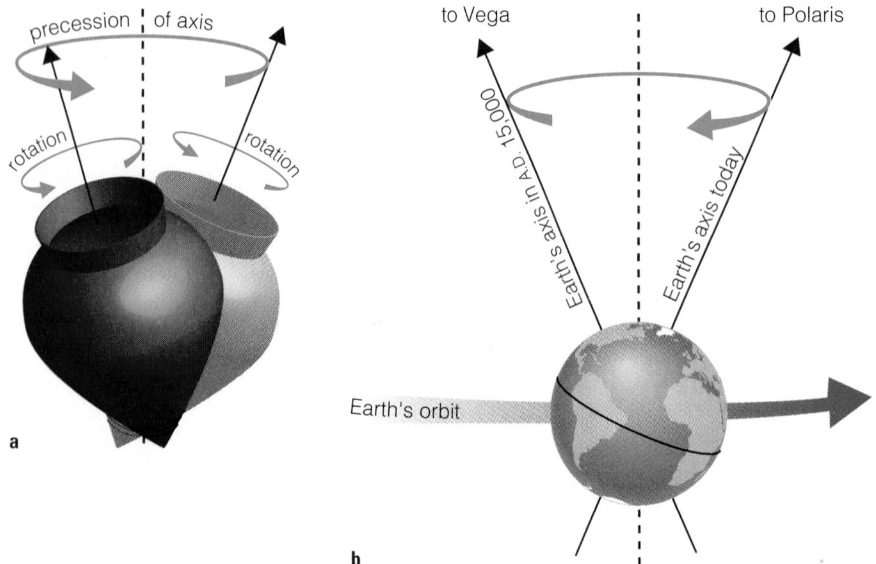

[7]The causes of the Earth's precession are forces (torques [Section 6.1]) exerted by the gravitational attractions of the Moon and the Sun on the Earth's equatorial bulge.

Common Misconceptions: Sun Signs

You probably know your astrological "sign." When astrology began a few thousand years ago, your sign was supposed to represent the constellation in which the Sun appeared on your birth date. However, for most people, this no longer is the case. For example, if your birthday is the spring equinox, March 21, a newspaper horoscope will show that your sign is Aries, but the Sun appears in Pisces on that date. In fact, your astrological sign corresponds to the constellation in which the Sun *would have appeared* on your birth date if you had lived about 2,000 years ago. Why? Because the astrological signs are based on the positions of the Sun among the stars as described by the Greek scientist Ptolemy in his book *Tetrabiblios,* which was written in about A.D. 150 [Section 4.6].

begins in March in the Northern Hemisphere, rather than having the first day of spring slowly change its date due to precession.

3.3 The Moon, Our Constant Companion

In discussing the motions of the Earth as a planet, we've explained all the apparent motions of the sky visible to the naked eye except lunar phases and eclipses. Understanding these phenomena requires understanding the motion of the Moon, the constant companion of spaceship Earth.

The Moon orbits the Earth at an average distance of about 380,000 kilometers (235,000 miles). It travels around the Earth at an average speed of about 3,680 kilometers per hour (2,270 miles per hour), completing each orbit of the Earth in one **sidereal month,** which is about $27\frac{1}{4}$ days[8] (Figure 3.10).

Lunar Phases

The easiest way to understand the lunar phases is with a simple demonstration, using a small ball to represent the Moon while your head represents the Earth. If it's daytime and the Sun is shining, take your ball outside and watch how you see phases as you move the ball around your head; if it's dark or cloudy, you can place a flashlight a few meters away to represent the Sun. As you hold your ball at various places in its "orbit" around your head, you'll observe that phases result from just two basic facts:

1. At any particular time, half of the ball faces the Sun (or flashlight) and therefore is bright, while the other half faces away from the Sun and therefore is dark.
2. As you look at the ball, you see some combination of its bright and dark faces. This combination is the phase of the ball.

We see lunar phases for the same reason: Half of the Moon is always illuminated by the Sun, but the amount of this illuminated half that we see from Earth depends on the Moon's position in its orbit (Figure 3.11). Note that the time of day during which the Moon is visible also depends on its phase. For example, full moon occurs when the Moon is opposite the Sun in the sky, so the full moon rises around sunset, reaches the meridian at midnight, and sets around sunrise.

TIME OUT TO THINK *Suppose you go outside in the* morning *and notice that the visible face of the Moon is half light and half dark. Is this a first-quarter or third-quarter moon? How do you know?* (Hint: *Study Figure 3.11.*)

Each complete cycle of lunar phases (e.g., from new moon to new moon) takes about $29\frac{1}{2}$ days, which is called a **lunar month** (or a *synodic month*). The lunar month is longer than the sidereal month for essentially the same reason that a solar day is

FIGURE 3.10 A schematic diagram of the Moon's orbit.

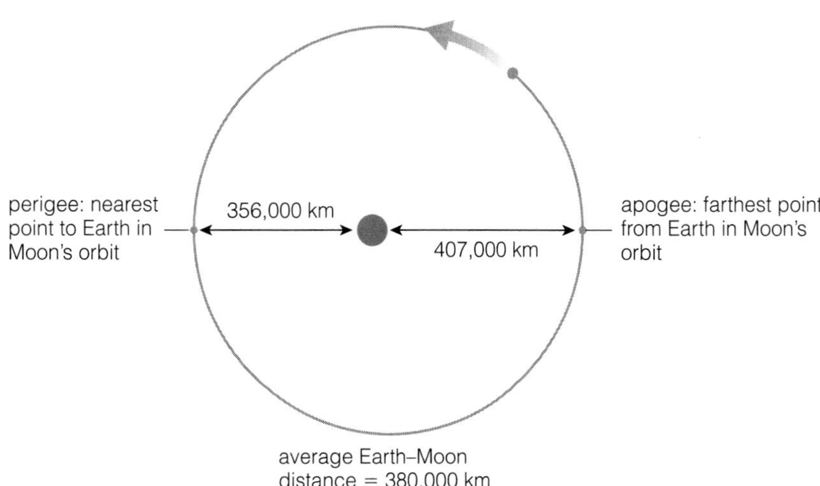

[8]The length of the sidereal month (and also of the lunar month) varies because of variations in the Moon's orbital speed.

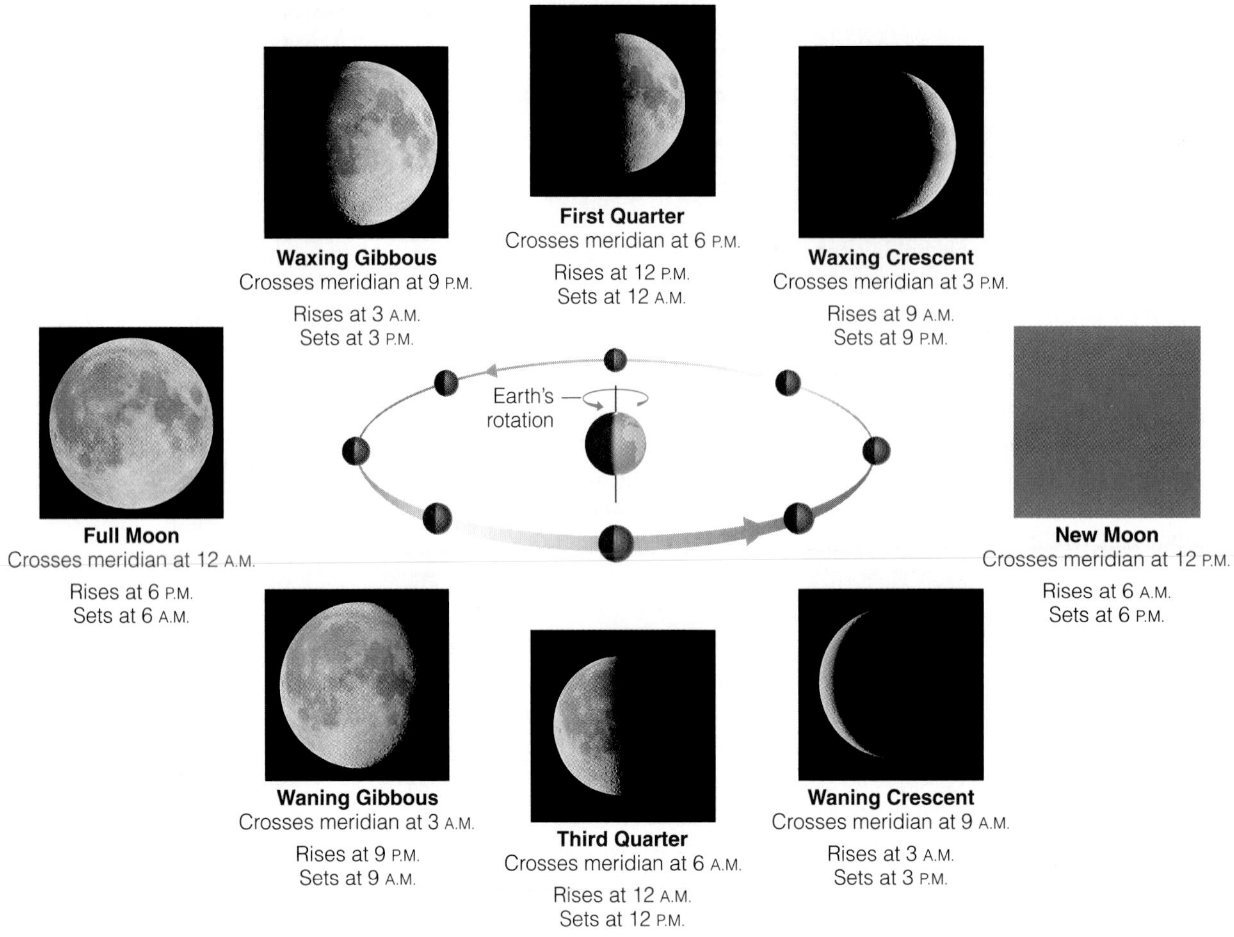

Waxing Gibbous
Crosses meridian at 9 P.M.
Rises at 3 A.M.
Sets at 3 P.M.

First Quarter
Crosses meridian at 6 P.M.
Rises at 12 P.M.
Sets at 12 A.M.

Waxing Crescent
Crosses meridian at 3 P.M.
Rises at 9 A.M.
Sets at 9 P.M.

Earth's rotation

Full Moon
Crosses meridian at 12 A.M.
Rises at 6 P.M.
Sets at 6 A.M.

New Moon
Crosses meridian at 12 P.M.
Rises at 6 A.M.
Sets at 6 P.M.

Waning Gibbous
Crosses meridian at 3 A.M.
Rises at 9 P.M.
Sets at 9 A.M.

Third Quarter
Crosses meridian at 6 A.M.
Rises at 12 A.M.
Sets at 12 P.M.

Waning Crescent
Crosses meridian at 9 A.M.
Rises at 3 A.M.
Sets at 3 P.M.

FIGURE 3.11 The phases of the Moon depend on its location relative to the Sun in its orbit. To understand rise and set times, imagine yourself standing on the rotating Earth and looking at the Moon in a particular phase. (The photo for the new moon shows only a blue sky because a new moon is above the horizon in daylight, but cannot be seen because it is very close to the Sun in the sky.)

longer than a sidereal day: The Earth's motion around the Sun means the Moon must complete more than one full orbit of the Earth from one new moon to the next (Figure 3.12).

Although we see many *phases* of the Moon, we do *not* see many *faces*. In fact, from the Earth we always see the same face of the Moon.[9] This tells us that the Moon must rotate once on its axis in the same time that it makes a single orbit of the Earth, as you can see with another simple demonstration. Place a ball on a table to represent the Earth, while *you* represent the Moon. Start by facing the ball. If you do not rotate at all as you orbit the ball, you'll be looking away from it by the time you are halfway around your orbit (Figure 3.13a). The only way you can face the ball at all times is by completing exactly one rotation while you complete one orbit (Figure 3.13b).

Earth travels about 30° per month around the Sun, so the Moon must orbit around Earth about 360° + 30° = 390° from new moon to new moon.

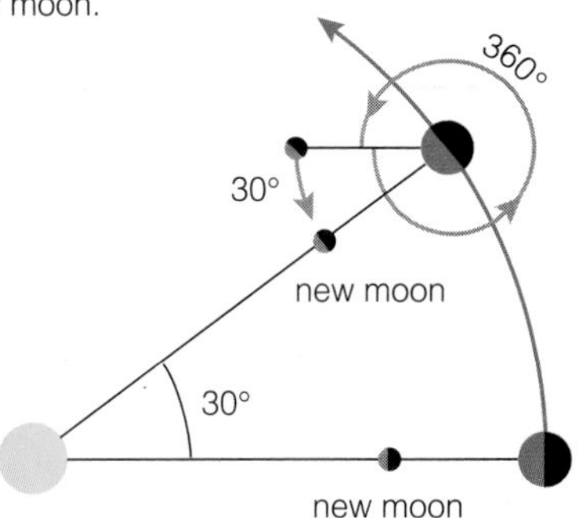

FIGURE 3.12 The Moon completes one 360° orbit in about $27\frac{1}{4}$ days (a sidereal month), but the time from new moon to new moon is about $29\frac{1}{2}$ days (a lunar month).

[9]Because the Moon's orbital speed varies while its rotation rate is steady, the Moon's visible face appears to wobble slightly back and forth as it orbits the Earth. This effect, called *libration*, allows us to see a total of about 59% of the Moon's surface by viewing it at different times of the month. (Of course, we see only 50% of the Moon at any single time.)

FIGURE 3.13 You can demonstrate why the Moon's rotational period must be the same as its orbital period by walking around a model of Earth. (**a**) If you do not rotate while walking around the model, you will not always face it. (**b**) You can face the model at all times only if you rotate exactly once during each orbit.

The View from the Moon

A good way to solidify your understanding of the lunar phases is to imagine that you live on the side of the Moon that faces the Earth. Look again at Figure 3.11. Note that at new moon you would be facing the day side of the Earth. Thus, you would see *full earth* when people on Earth see new moon. Similarly, at full moon you would be facing the night side of the Earth. Thus, you would see *new earth* when people on Earth see full moon. In general, you'd always see the Earth in a phase opposite the phase of the Moon that people on Earth see. Moreover, your solar day would be the same length as a lunar month, so you'd have about 2 weeks of daylight and 2 weeks of darkness as you watched the Earth go through its cycle of phases.

TIME OUT TO THINK *On Earth, we see the Moon rise and set in our sky each day. If you lived on the Moon, would you see the Earth rise and set? Why or why not?* (Hint: *Remember that the Moon always keeps the same face toward Earth.*)

Common Misconceptions: Moon on the Horizon

You've probably noticed that the full moon appears far larger when it is near the horizon than when it is high in your sky. However, this appearance is an illusion. If you measure the angular size of the full moon on a particular night, you'll find it is the same whether it is near the horizon or high in the sky. In fact, the Moon's angular size in the sky depends only on its distance from the Earth. Although this distance varies over the course of the Moon's monthly orbit, it does not change enough to cause a noticeable effect on a single night. (You can eliminate the illusion by viewing the Moon upside down between your legs when it is on the horizon.)

Common Misconceptions: The "Dark Side" of the Moon

The term *dark side of the Moon* really should be used to mean the side facing away from the Sun. Unfortunately, *dark side* traditionally meant what would better be called the *far side*—the hemisphere that never can be seen from the Earth. Many people still refer to the far side as the "dark side," even though this side is not necessarily dark. For example, during new moon the far side faces the Sun and hence is completely sunlit; the only time the far side actually is completely dark is at full moon, when it faces away from both the Sun and the Earth.

Thinking about the view from the Moon clarifies another interesting feature of the lunar phases: The dark portion of the lunar face is not *totally* dark. Imagine that you are standing on the Moon when it is in a crescent phase. Because it's nearly new moon as seen from Earth, you will see nearly full earth in your sky. Just as we can see at night by the light of the Moon, the light of the Earth illuminates your night moonscape. (In fact, Earth is much larger and brighter in the lunar sky than the full moon is in

FIGURE 3.14 This painting represents the ecliptic plane as the surface of a pond. The Moon's orbit is slightly tilted to the ecliptic plane, thus, in this illustration, the Moon spends half of each orbit above the pond surface and half below the surface. Eclipses occur only when the Moon is both crossing the ecliptic plane (splashing through the pond surface) *and* has a phase of either new moon or full moon—as is the case with the lower left and top right orbits shown. At all other times new moons and full moons occur above or below the ecliptic plane, so no eclipse is possible.

Earth's sky.) This faint light illuminating the "dark" portion of the Moon's face is often called the *ashen light* or *earthshine;* it is the light that enables us to see the outline of the full face of the Moon even when the Moon is not full.

Eclipses

Look once more at Figure 3.11. If this figure told the whole story of the lunar phases, a new moon would always block our view of the Sun, and the Earth would always prevent sunlight from reaching a full moon. More precisely, this figure makes it look as if the Moon's shadow should fall on Earth during new moon and that Earth's shadow should fall on the Moon during full moon. Any time one astronomical object casts a shadow on another, we say that an **eclipse** is occurring. Thus, Figure 3.11 makes it look as if we should have an eclipse with every new moon and every full moon—but we don't.

The missing piece of the story in Figure 3.11 is that the Moon's orbit is inclined to the ecliptic plane by about 5°. An easy way to visualize this inclination is to imagine the ecliptic plane as the surface of a pond, as shown in Figure 3.14. Because of the inclination of its orbit, the Moon spends most of its time either above or below this surface. It crosses *through* this surface only twice during each orbit: once coming out and once going back in. The two points in each orbit at which the Moon crosses the surface (which represents the ecliptic plane) are called the **nodes** of the Moon's orbit.

Figure 3.14 shows the position of the Moon's orbit at several different times of year. Note that the nodes are aligned the same way in each case (diagonally on the page). As a result, the nodes lie in a straight line with the Earth and the Sun only about twice each year. (For reasons we'll discuss shortly, it is not *exactly* twice each year.) Because an eclipse

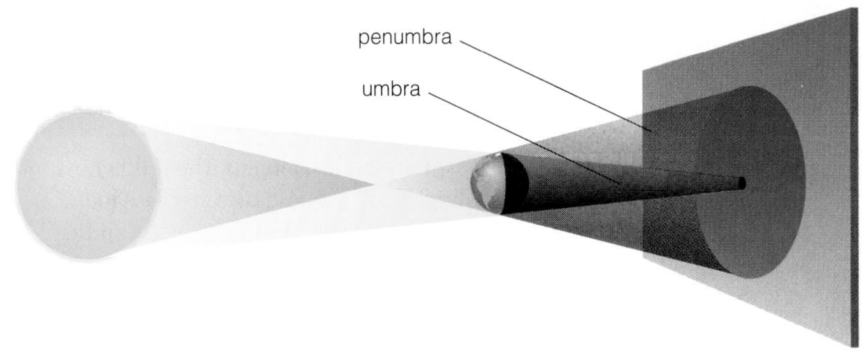

FIGURE 3.15 The shadow cast by an object in sunlight: Sunlight is fully blocked in the umbra and partially blocked in the penumbra.

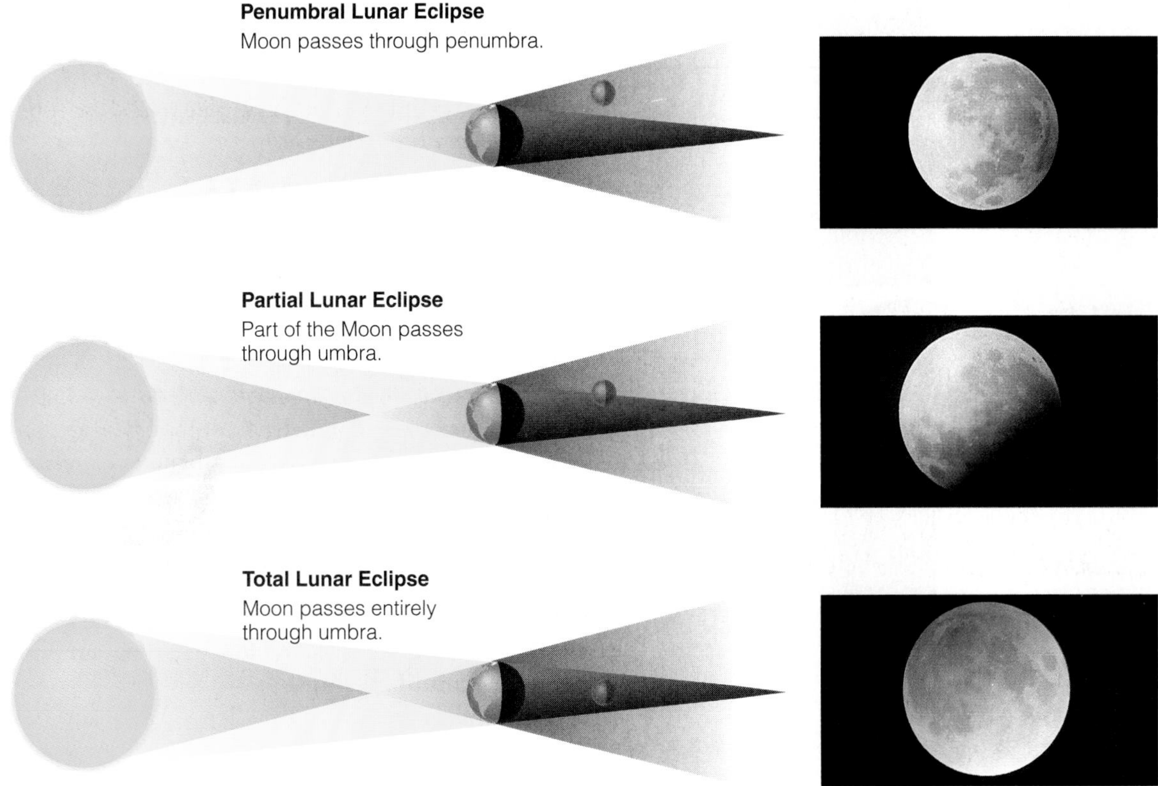

FIGURE 3.16 The three types of lunar eclipse.

can occur only when the Earth and the Moon lie in a straight line with the Sun, two conditions must be met simultaneously for an eclipse to occur:

1. The nodes of the Moon's orbit must be nearly aligned with the Earth and the Sun.[10]
2. The phase of the Moon must be either new or full.

Types of Eclipse There are two basic types of eclipse. A **lunar eclipse** occurs when the Moon passes through the Earth's shadow and therefore can occur only at *full moon*. A **solar eclipse** occurs when the Moon's shadow falls on the Earth and therefore can occur only at *new moon*. But the full story of eclipse types is more complex, because the shadow of the Moon or the Earth consists of two distinct regions: a central **umbra** where sunlight is completely blocked and a surrounding **penumbra** where sunlight is only partially blocked (Figure 3.15).

A lunar eclipse begins at the moment that the Moon's orbit first carries it into the Earth's penumbra. After that, we will see one of three types of lunar eclipse (Figure 3.16). If the Sun, the Earth, and the Moon are nearly perfectly aligned, the Moon will pass through the Earth's umbra, and we will see a **total lunar eclipse.** If the alignment is somewhat less perfect, only part of the full moon will pass

[10]The alignment does not have to be perfect, just close enough so that the Earth and the Moon lie at least *partially* on a straight line with the Sun.

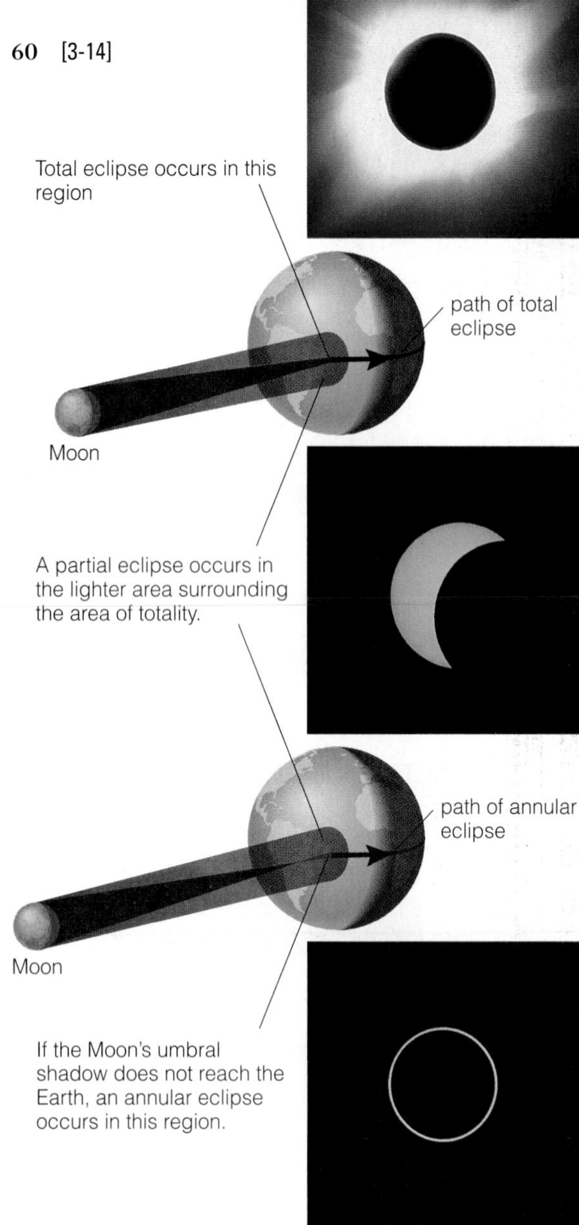

FIGURE 3.17 The three types of solar eclipse.

through the umbra (with the rest in the penumbra), and we will see a **partial lunar eclipse.** If the Moon passes *only* through the Earth's penumbra, we will see a **penumbral lunar eclipse.**

Penumbral eclipses are the most common type of lunar eclipse, but they are difficult to notice because the full moon darkens only slightly. Partial lunar eclipses are easier to see because the Earth's umbral shadow clearly darkens part of the Moon's face. Note that the Earth's umbra casts a curved shadow on the Moon, a fact that helped convince the ancient Greeks that the Earth is round. A total lunar eclipse is particularly spectacular, because the Moon becomes dark and eerily red during *totality* (the time during which the Moon is entirely engulfed in the umbra)—dark because it is in shadow, and red because the Earth's atmosphere bends some of the red light from the Sun around the Earth and toward the Moon. A total lunar eclipse can last several hours from the time the Moon first touches the Earth's penumbra to the time it fully emerges from the penumbra, but the maximum possible duration of totality is about 1 hour 40 minutes.

TIME OUT TO THINK *When a total lunar eclipse occurs, it can be seen by people living on only a little more than half of the Earth. Which half? Why?* (Hint: *You must be able to see the full moon to see a lunar eclipse.*)

We can also see three types of solar eclipse (Figure 3.17). Remember that the Moon's distance from the Earth varies in its orbit. If a solar eclipse happens to occur when the Moon is relatively close to the Earth, the Moon's umbra touches a small area of the Earth (no more than about 270 kilometers in diameter). Anyone within this area will see a **total solar eclipse.** Surrounding this region of totality is a much larger area (typically about 7,000 kilometers in diameter) that falls within the Moon's penumbral shadow; anyone within this region will see a **partial solar eclipse.** If the eclipse happens to occur when the Moon is relatively far from the Earth, the umbra may not reach the Earth at all. In that case, anyone in the small region of the Earth directly behind the umbra will see an **annular eclipse,** in which a ring of sunlight surrounds the disk of the Moon. Again, anyone in the surrounding penumbral shadow will see a partial solar eclipse. During any solar eclipse, the combination of the Earth's rotation and the orbital motion of the Moon causes the circular umbral and penumbral shadows to race across the face of the Earth at a typical speed of about 1,700 kilometers per hour (relative to the ground). As a result, the umbral (or annular) shadow traces a narrow path across the Earth, and totality (or annularity) never lasts more than a few minutes in any particular place.

A total solar eclipse is a spectacular sight. It begins when the disk of the Moon first appears to touch the Sun. Over the next couple of hours, the Moon appears to take a larger and larger "bite" out of the Sun. As totality approaches, the sky darkens and temperatures fall. Birds head back to their nests, and crickets begin their nighttime chirping. During the few minutes of totality, the Moon completely blocks the normally visible disk of the Sun, allowing the faint *corona* to be seen (Figure 3.18). The surrounding sky takes on a twilight glow, and planets and bright stars become visible in the daytime. As totality ends, the Sun slowly emerges from behind the Moon

FIGURE 3.18 This multiple-exposure photograph shows the progression of a total solar eclipse. Totality (central image) lasts only a few minutes.

over the next couple of hours. However, because your eyes have adapted to the darkness, totality *appears* to end far more abruptly than it began.

Predicting Eclipses Few phenomena have so inspired and humbled humans throughout the ages as eclipses. For many cultures, eclipses were mystical events associated with fate or the gods, and countless stories and legends surround eclipses. For example, legend has it that the Greek philosopher Thales (c. 624–546 B.C.) successfully predicted the year (but presumably not the precise time) when a total eclipse of the Sun would be visible in the area where he lived, which is now part of Turkey. Coincidentally, the eclipse occurred as two opposing armies (the Medes and the Lydians) were massing for battle. The eclipse so frightened the armies that they put down their weapons, signed a treaty, and returned home. Because modern research shows that the only eclipse visible in that part of the world at about that time occurred on May 28, 585 B.C., we know the precise date on which the treaty was signed—the first historical event that can be dated precisely.

Much of the mystery of eclipses probably stems from the fact that they are relatively difficult to predict. Look again at Figure 3.14. The two periods each year when the nodes of the Moon's orbit are nearly aligned with the Sun are called **eclipse seasons.** Each eclipse season lasts a few weeks, so some type of lunar eclipse occurs during each eclipse season's full moon, and some type of solar eclipse occurs

THINKING ABOUT . . .

The Moon and Human Behavior

From myths of werewolves to stories of romance under the full moon, human culture is filled with claims that our behavior is influenced by the phase of the Moon. Can we say anything scientific about such claims?

The Moon clearly has important influences on Earth. The most noticeable is the tides, for which the Moon is primarily responsible [Section 6.5]. However, *tidal forces* are felt only by relatively large objects, such as Earth, and we can rule out the possibility that tidal forces from the Moon affect small objects such as people.

If a physical force from the Moon cannot affect human behavior, could we be influenced in other ways? Certainly, anyone who lives near the oceans is influenced by the rising and falling of the tides. For example, fishermen and boaters must follow the tides. Thus, although the Moon is not *directly* influencing their behavior, it is doing so indirectly through its effect on the oceans.

Many physiological patterns in countless species appear to follow the lunar phases, and the human menstrual cycle is so close in length to a lunar month that it is difficult to believe the similarity is mere coincidence. Nevertheless, aside from the obvious physical cycles and the influence of tides on people who live near the oceans, claims that the lunar phase affects human behavior are much more difficult to verify scientifically. For example, although it is possible that the full moon brings out certain behaviors, it may also simply be that some behaviors are easier to exhibit when the sky is bright. While a beautiful full moon may bring out your desire to walk on the beach under the moonlight, there is no scientific evidence that the full moon would affect you the same way if you lived in a cave and couldn't see it.

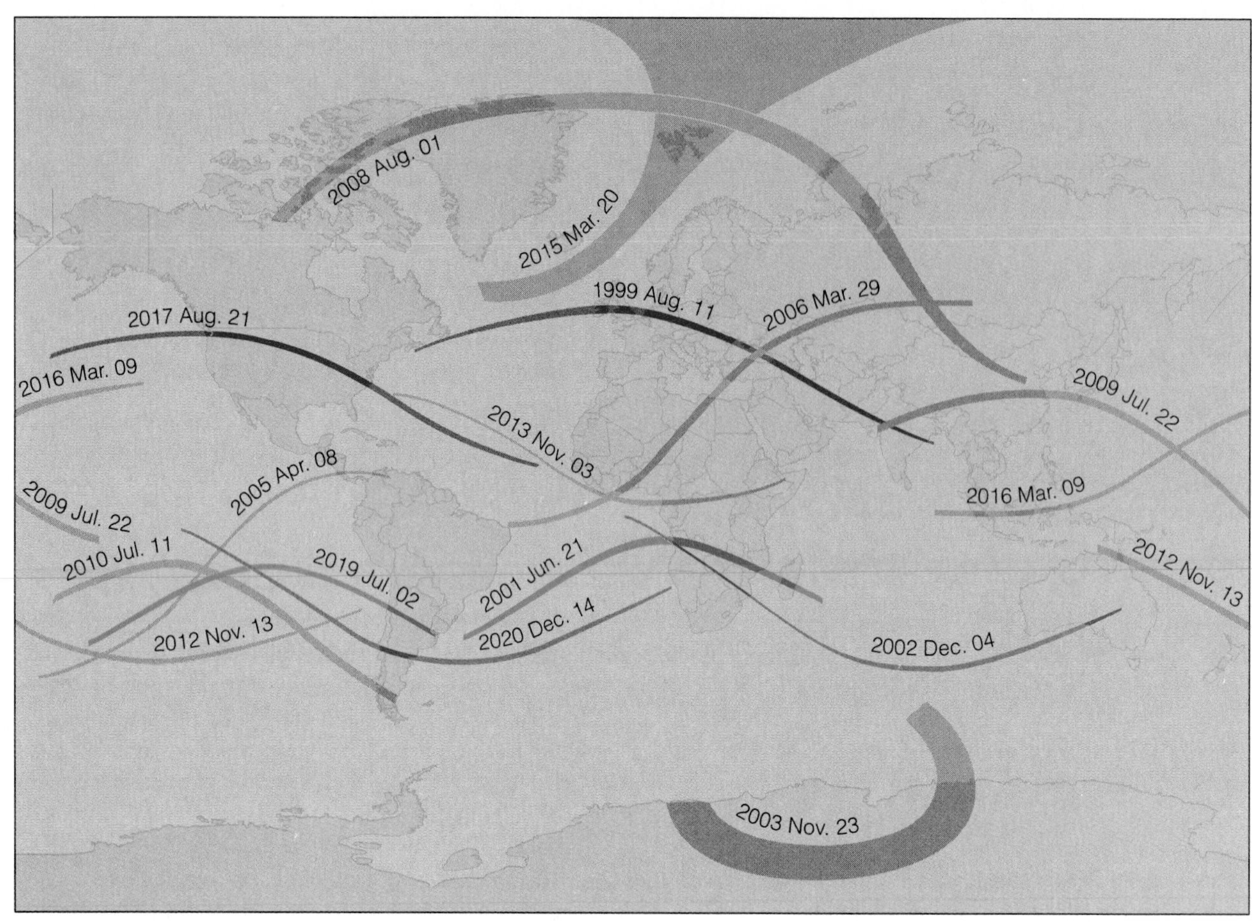

FIGURE 3.19 Solar eclipses, 1999–2020. You can recognize the saros cycle by noting, for example, that the 2017 eclipse occurs 18 years 11 days after the 1999 eclipse.

Table 3.1 Total Lunar Eclipses 1999–2004

January 21, 2000
July 16, 2000
January 9, 2001
May 16, 2003
November 9, 2003
May 4, 2004
October 28, 2004

during its new moon. If Figure 3.14 told the whole story, eclipse seasons would occur every 6 months and predicting eclipses would be easy. For example, if eclipse seasons always occurred in January and July, eclipses would occur only on the dates of new and full moons in those months. But Figure 3.14 does not show one important fact about the Moon's orbit: The nodes slowly precess around the orbit. As a result, eclipse seasons actually occur slightly less than 6 months apart (about 173 days apart) and therefore do not recur in the same months year after year.

The combination of the changing dates of eclipse seasons and the $29\frac{1}{2}$-day cycle of lunar phases turns out to make eclipses recur in a cycle of about 18 years $11\frac{1}{3}$ days. For example, if a solar eclipse occurs today, another will occur 18 years $11\frac{1}{3}$ days from now. This roughly 18-year cycle is called the **saros cycle.**

Astronomers in many ancient cultures identified the saros cycle and therefore could predict *when* eclipses would occur. However, the saros cycle does not account for all the complications involved in predicting eclipses. If a solar eclipse occurs today, the one that occurs 18 years $11\frac{1}{3}$ days from now will not be visible from the same places on the Earth and may not be of the same type (e.g., one may be total and the other only partial). As a result, no ancient culture achieved the ability to predict eclipses in every detail. Today, eclipses *can* be predicted because we know the precise details of the orbits of the Earth and the Moon. In fact, *you* can predict the dates of eclipses far into the future with the aid of one of many inexpensive astronomical software packages. Table 3.1 lists upcoming total lunar eclipses, and Figure 3.19 shows paths of totality for upcoming total solar eclipses. It is well worth your while to plan a trip to see a total solar eclipse.

3.4 Traveling in the Milky Way Galaxy

Within the solar system, our travels on spaceship Earth all follow relatively simple, repetitive patterns: We rotate daily around the Earth's axis, we orbit the Sun on the same path each year, and we are accompanied by a Moon that orbits the Earth monthly. But this is only the beginning of our journeys, as our entire solar system is moving through the Milky Way Galaxy and our galaxy is moving relative to other galaxies in the universe. We can begin to understand these journeys by considering how our solar system is moving relative to other nearby stars.

The Motion of Stars in the Local Solar Neighborhood

If you study the patterns of stars in the constellations night after night, you will not notice any changes. Throughout your life, they still will not appear to change. Few things in life appear as fixed and unchanging as the patterns of the stars . . . and few appearances are so deceiving!

You are probably familiar with traffic police measuring automobile speeds with radar. Imagine that you could use a police officer's radar device to measure a star's speed. If you measured the speeds of nearby stars in the local solar neighborhood, including all those visible to the naked eye, you would find them to be moving at an average speed of about 70,000 kilometers per hour (about 40,000 miles per hour). At that speed, a star moves a distance equivalent to that between the Earth and the Moon about every 6 hours.

All stars are in constant motion, and it is as valid to say that our solar system is moving with respect to a particular star as it is to say that the star is moving with respect to us. That is, all the stars in the local solar neighborhood are moving with respect to one another at typical speeds of 70,000 kilometers per hour.

If stars move at such high speeds, why don't we see them racing around the sky? The answer lies in their vast distances. Even at speeds of 70,000 kilometers per hour, the positions of nearby stars in the sky shift by only a few degrees in a few thousand years. That explains why the patterns of the constellations have hardly changed since they were named a few thousand years ago. Nevertheless, if you could keep watching, the patterns eventually *would* change. In 10,000 years, the constellations will be noticeably different from those we see today. In 500,000 years, they will be unrecognizable. Likewise, the patterns seen by our ancestors 500,000 years ago were very different from those we see today.

Imagine watching a time-lapse movie, made over millions of years, of the stars in the local solar neighborhood. You would see the stars racing around relative to one another. Some stars would leave your field of view, while others would enter it (Figure 3.20).

TIME OUT TO THINK *Despite the chaos of motion in the local solar neighborhood over millions of years, collisions between star systems are extremely unlikely. Explain why.* (Hint: *Consider the sizes of star systems, such as the solar system, relative to the distances between them.*)

FIGURE 3.20 As shown in this painting, the local solar neighborhood is only a tiny portion of the Milky Way Galaxy. The stars in the local solar neighborhood actually move quite fast relative to our solar system, but the enormous distances between stars make this motion barely detectable on human time scales. If you could watch a time-lapse movie made over a time period of millions of years, you would see the stars in the local solar neighborhood racing around in seemingly random directions.

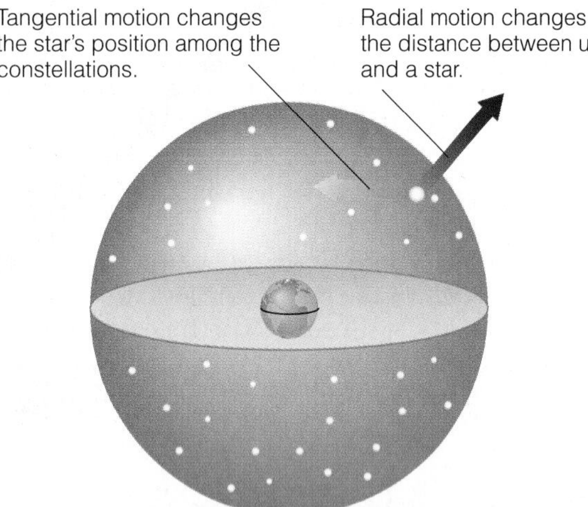

FIGURE 3.21 The true motion of a star can be broken down into radial and tangential components.

Detecting Stellar Motion

If the motions of stars are undetectable to the naked eye, how do we know stars are moving? We begin by breaking down a star's motion into two components: a **radial motion** directed toward or away from us and a **tangential motion** directed across our line of sight (Figure 3.21).

Only the tangential motion slowly changes the star's position among the constellations. Although this motion is not noticeable to the naked eye, we *can* measure it for many nearby stars by comparing telescopic photographs taken years or decades apart.[11] In general, the more distant a star, the more difficult it is to detect its tangential motion, and our current technology generally can measure tangential motion only for stars within a few thousand light-years of Earth.

TIME OUT TO THINK *Explain why, if two stars have the same speed of tangential motion, it is more difficult to detect the motion of the more distant star.* (Hint: *Suppose two cars are moving across your field of view, both traveling at 100 kilometers per hour. If one car is twice as far from you as the other, which car will move out of your field of view sooner?*)

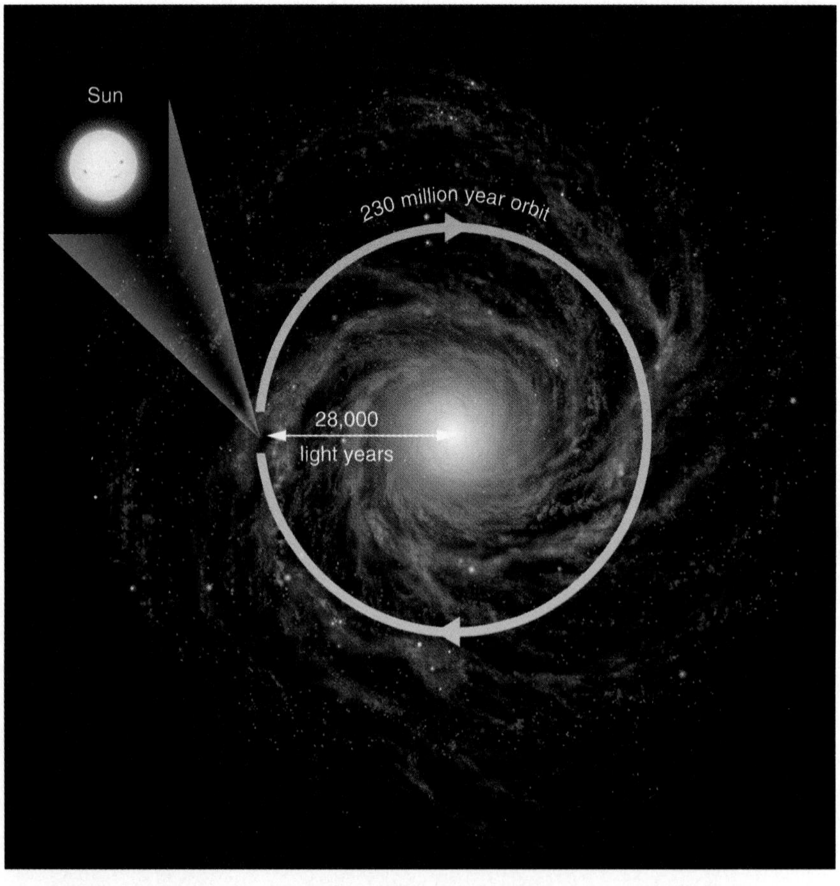

FIGURE 3.22 The rotation of the Milky Way Galaxy. Our solar system completes one nearly circular orbit around the galactic center every 230 million years.

Because radial motion does not affect a star's position in our sky, you might expect measuring radial motion to be extremely difficult. However, through a gift of nature called the *Doppler effect,* measuring radial velocities turns out to be relatively easy with modern technology. The Doppler effect causes a change in the color of a star (or any other object) that depends on how fast it is moving toward or away from us. We'll discuss exactly how the Doppler effect is measured in Chapter 7, but

[11]Such photographic position comparisons tell us the *angular* rate at which a star is moving, which astronomers call its *proper motion* (e.g., if the star moved 0.1° across the sky in 100 years, its proper motion would be 0.1° per 100 years, or 1° per 1,000 years). Calculating the star's *tangential speed* requires knowing both its proper motion and its distance.

Most of the galaxy's light comes from stars and gas in the galactic disk...

...but most of the galaxy's mass lies above and below the disk in the *halo.*

FIGURE 3.23 Careful study of galactic rotation indicates that most of the mass in the galaxy is in the form of mysterious *dark matter.* (The dark matter may actually extend quite far out; see Figure 21.1.)

the key point is this: *By measuring the Doppler effect, we can determine the speed of radial motion for any object, no matter how far away it is.*

We can combine measurements of radial and tangential motions to determine the overall motion of nearby stars. For more distant objects, we can generally determine only radial motion.

Rotation of the Galaxy

If you look closely at leaves floating in a stream, their motions relative to one another might appear to be random and chaotic, just like the motions of stars in the local solar neighborhood. As you widen your view, you will see that all the leaves are being carried in the same general direction by the downstream current. In the same way, if you widened your view beyond the local solar neighborhood, you would eventually see that all the stars generally are moving together around the center of the Milky Way Galaxy.

Figure 3.22 depicts the rotation of the Milky Way Galaxy, which was discovered by analysis of the tangential and radial motions of stars and clouds of interstellar gas. Our solar system, along with other star systems located at the same 28,000-light-year distance from the galactic center, completes one revolution around the galactic center in about 230 million years. Even if you could look at the Milky Way Galaxy from outside it, this motion would be unnoticeable to your naked eye. However, if you calculate the speed of our solar system as we orbit the center of the galaxy, you will find that it is close to 1 million kilometers per hour (600,000 miles per hour)!

Before we discuss larger-scale motion, it is worth noting that the galactic rotation reveals one of the greatest mysteries in science—one that we will study in depth later in the book. The speed at which stars orbit the galactic center depends on the strength of gravity, and the strength of gravity depends on how mass is distributed throughout the galaxy. Thus, careful study of the galactic rotation allows us to determine the distribution of mass in the galaxy. Remarkably, we find that the stars we see in the disk of the galaxy represent only the "tip of the iceberg" compared to the mass of the entire galaxy (Figure 3.23). That is, most of the mass of the galaxy is located outside the visible disk, in the galaxy's *halo.* We don't know the nature of this mass because we have not detected any light coming from it, and we therefore call it **dark matter.** Studies of other galaxies show that they are also made mostly of dark matter. In fact, most of the mass in the universe appears to be made of this mysterious dark matter, even though we do not know what it is.

3.5 Galaxies in Motion

Just as our solar system is moving within the Milky Way Galaxy, our galaxy is moving among the other galaxies of the universe. The precise dynamics of these motions are difficult to determine because, for objects beyond our own galaxy, we usually can measure only *radial* velocities. Nevertheless, we have discovered some astonishing facts about the motion of galaxies in the universe.

Motion in the Local Group

The entire Milky Way Galaxy is moving in relation to the 30 or so other galaxies in the Local Group (Figure 3.24). Again, although the speeds are enormous by earthly standards, the motion is unnoticeable to the naked eye. For example, the Milky Way is moving toward the Great Galaxy in Andromeda (M31) at a speed of about 300,000 kilometers per hour. At that rate, it will take us more than 90 million years to move just 1% of the 2.5-million-light-year distance to M31.

If you could watch a time-lapse movie taken over billions of years, you would see the galaxies in the Local Group swirl about one another. Although we don't know the precise dynamics of this motion, a few aspects of it seem clear. For example, two small galaxies—called the *Large Magellanic Cloud* and the *Small Magellanic Cloud*—orbit the Milky Way. M31 is accompanied by at least four small galaxies that probably orbit it. In addition, based on gravitational considerations, it seems likely that the Milky Way and M31 orbit each other; if so, each single orbit must take billions of years. Our Milky Way and the other galaxies of the Local Group are like members of an expedition, bound together by gravity as we travel through the universe.

The Expansion of the Universe

If we look at stars in the local solar neighborhood, we find that some stars are moving toward us and others are moving away. Similarly, if we look at galaxies in

FIGURE 3.24 The Local Group consists of our galaxy (the Milky Way) and about 30 or so other galaxies that are all bound together by gravity. This artist's painting is roughly to scale except that the sizes of the galaxies are exaggerated so they are easier to see.

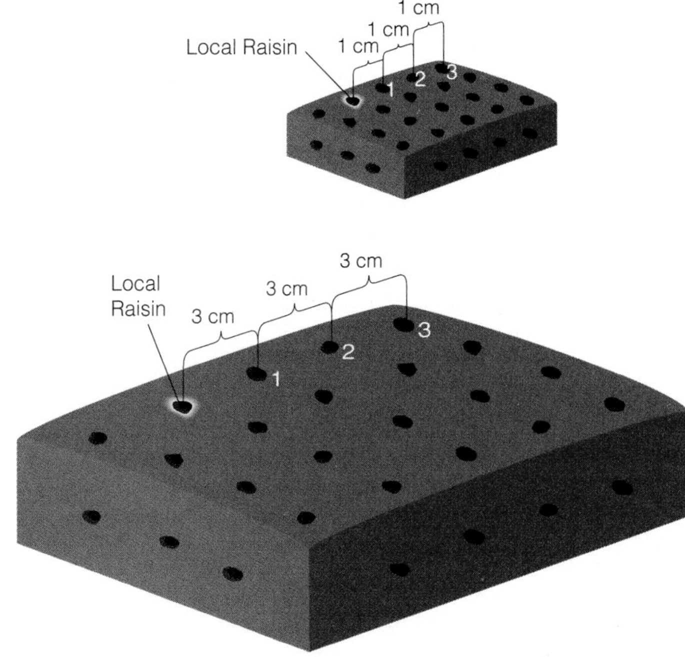

FIGURE 3.25 An expanding raisin cake illustrates basic principles of the expansion of the universe.

the Local Group, we find that some are moving toward us and others are moving away. When we look outside the Local Group, however, we find two astonishing facts that were first discovered by Edwin Hubble, for whom the Hubble Space Telescope was named:

1. Virtually every galaxy outside the Local Group is moving *away* from us.
2. The more distant the galaxy, the faster it appears to be racing away from us.

Upon first hearing of these two facts, you might be tempted to conclude that our Local Group suffers a cosmic case of chicken pox that leads other galaxies to race away from us. However, there is a much more natural explanation: *The entire universe is expanding*. Although discussion of the details of the expansion of the universe will have to wait until later in the book, we can explain the basic principle by using the analogy of baking a raisin cake.

Imagine that you make a raisin cake and that the distance between adjacent raisins happens to be about 1 centimeter. You place the cake in the oven, where it expands as it bakes. After 1 hour, you remove the cake, which has expanded so that the distance between adjacent raisins is 3 centimeters (Figure 3.25).

Table 3.2 Distances and Speeds As Seen from the Local Raisin in Figure 3.25

Raisin Number	Distance Before Baking	Distance After Baking (1 hour later)	Speed
1	1 cm	3 cm	2 cm/hr
2	2 cm	6 cm	4 cm/hr
3	3 cm	9 cm	6 cm/hr
4	4 cm	12 cm	8 cm/hr
⋮	⋮	⋮	⋮
10	10 cm	30 cm	20 cm/hr
⋮	⋮	⋮	⋮

Pick any raisin (it doesn't matter which one), call it the Local Raisin, and identify it in the pictures of the cake both before and after baking; Figure 3.25 shows one possible choice for the Local Raisin, with several other nearby raisins labeled. Note that, *before* baking, Raisin 1 is 1 centimeter away from the Local Raisin, Raisin 2 is 2 centimeters away, and Raisin 3 is 3 centimeters away. *After* baking, Raisin 1 is 3 centimeters away from the Local Raisin, Raisin 2 is 6 centimeters away, and Raisin 3 is 9 centimeters away. Nothing should seem surprising so far, as the new distances simply reflect the fact that the cake expanded uniformly everywhere.

Now suppose you live *inside* the Local Raisin. From your vantage point, the other raisins appear to move away from you as the cake expands. Raisin 1 is 1 centimeter away before baking and 3 centimeters away after baking; it therefore appears to move 2 centimeters during the hour, making its speed 2 centimeters per hour as seen from the Local Raisin. Raisin 2 is 2 centimeters away before baking and 6 centimeters away afterward, so it appears to move 4 centimeters during the hour of baking, giving it a speed of 4 centimeters per hour as seen from the Local Raisin. Table 3.2 lists the distances of other raisins before and after baking, along with their speed as seen from the Local Raisin. Note that, although the entire cake expanded uniformly, from the vantage point of the Local Raisin each subsequent raisin moved away at increasingly faster speeds. Furthermore, because selection of the Local Raisin was arbitrary, you will find the same results no matter which raisin you choose to represent your Local Raisin.

TIME OUT TO THINK *To confirm that selection of the Local Raisin is arbitrary, choose a different raisin to represent the Local Raisin in Figure 3.25, and then label its subsequent raisins Raisin 1, Raisin 2, and so on. Compare the distances of these raisins from the new Local Raisin before and after baking to confirm that you still find the same results shown in Table 3.2.*

If you now imagine the Local Raisin to represent our Local Group of galaxies, and the other raisins to represent more distant galaxies, you have a basic picture of the expansion of the universe.[12] That is, *space* itself is growing, and more distant galaxies move away from us faster because they are carried along with this expansion like raisins in an expanding cake. The same effects of expansion would be seen from any place in the universe.

Because of expansion, galaxies relatively near the Local Group are moving away from us at speeds of a few million kilometers per hour. The speeds grow increasingly greater with distance, and the most distant galaxies discovered to date are receding from us at more than 95% of the speed of light.

One more important point about our motion with respect to distant galaxies needs to be mentioned. Because gravity causes galaxies and clusters to attract one another, some objects are not moving away from us as fast as we would expect from the expansion of the universe alone. For example, although the expansion of the universe is carrying our Local Group away from the Virgo Cluster of galaxies, which lies near the center of our Local Supercluster, we are not moving away as fast as the expansion of the universe alone would suggest. Thus, we conclude that gravity is attracting us toward the direction of the Virgo Cluster, despite the fact that it is receding from us. By closely studying differences between expected and actual velocities away from distant galaxies, we have discovered large-scale motions in the universe that are superimposed on the general expansion.

[12]The raisin cake model is not perfect; in particular, the cake has a center and edges, but the universe does not. We will discuss a somewhat better model in Chapter 19.

TIME OUT TO THINK *At first, you might think it strange that we can be moving* away *from the Virgo Cluster even though we are gravitationally attracted* toward *it. To understand how this is possible, imagine throwing a ball up into the air. As the ball is on its way upward, it is moving away from the center of the Earth. But in which direction is gravity acting on the ball? How does gravity affect the speed of the ball as it travels upward? Explain how the effect of gravity on the upward-moving ball is similar to the effect of gravity on our motion relative to the Virgo Cluster of galaxies.*

Will the universe keep expanding forever, or will it someday stop expanding and begin to collapse? The answer to this question is not yet known; we will explore the current state of knowledge in Chapter 21. In brief, the fate of the universe hinges on the total mass of the universe—including the mass of the mysterious dark matter. If the total mass of the universe is large enough, the mutual gravitation of all this mass will eventually stop the expansion. If the total mass is not large enough, the expansion will continue forever.

> *The most important fact about Spaceship Earth: An instruction book didn't come with it.*
>
> R. BUCKMINSTER FULLER (1895–1983)

THE BIG PICTURE

This chapter covered the major motions of spaceship Earth and its companion, the Moon, as we journey through the cosmos, as well as how these motions explain the apparent phenomena of the sky. Before continuing, be sure you understand the following "big picture" ideas from this chapter:

- All of us are being carried on a remarkable journey through the cosmos on spaceship Earth. Many of the speeds at which spaceship Earth carries us relative to other objects in the universe are far higher than the speeds at which we ever travel on the Earth.
- We originally learned of our motions by studying changes in the position and appearance of objects in our sky. Today, we can use our understanding of the real motions of the Earth to better understand what we see in the sky.

- Not only is the Earth in motion, but so is everything in the cosmos. Indeed, just as there is a hierarchy of structure in the universe (from planet to solar system to galaxy and so on), there is also a hierarchy of motion (from rotation to revolution to galactic rotation and so on).

Review Questions

1. What do we mean by the *rotation* of the Earth? Based on Figure 3.1, approximately how fast are *you* moving with rotation?
2. Describe the Earth's *revolution* around the Sun. What is an *astronomical unit?* What is the *ecliptic plane?*
3. What is the difference between a *sidereal day* and a *solar day?* Why are they different?
4. Describe the cause of the Earth's seasons and explain why the seasons are opposite in the Northern and Southern hemispheres. Define the *spring equinox, summer solstice, fall equinox,* and *winter solstice.*
5. Why is the *apparent retrograde motion* of the planets difficult to explain in an Earth-centered system? Describe its natural explanation in a Sun-centered solar system.
6. What is *stellar parallax?* Why were ancient peoples unable to detect it? Explain why the existence of stellar parallax constitutes proof that the Earth really does orbit the Sun.
7. Describe the Earth's 26,000-year cycle of *precession.* How does precession explain the fact that the spring equinox is often called the "first point of Aries," even though it is located in the constellation Pisces?
8. What is the difference between a *sidereal year* and a *tropical year?* Why are they different? Which one is our calendar based on?
9. Describe the basic characteristics of the orbit of the Moon. What is the difference between a *lunar month* and a *sidereal month?* Why are they different?
10. Describe the Moon's cycle of phases and explain *why* we see phases of the Moon. Also explain why someone living on the Moon would see the Earth go through phases.
11. Explain why the fact that we always see the same *face* of the Moon means that the Moon's orbital and rotational periods are the same.
12. What is the *ashen light* (or *earthshine*)? How does it affect our view of the Moon?
13. Why don't we have an eclipse at every new and full moon? Explain why an eclipse can occur only when the *nodes* of the Moon's orbit are aligned with the Sun. At what phase is a *lunar eclipse* possible? A *solar eclipse?*
14. When discussing a shadow cast by the Earth or the Moon, what do we mean by the *umbra* and the *penumbra?*
15. Describe and differentiate between a *penumbral, partial,* and *total lunar eclipse.*
16. Describe and differentiate between a *partial, total,* and *annular solar eclipse.*
17. What are *eclipse seasons?* Why don't they occur at precisely 6-month intervals?
18. What is the *saros cycle?* Why isn't knowing the saros cycle enough to make complete predictions of eclipses?
19. Describe the general motion of stars in the local solar neighborhood. Given the relatively high speeds of stars in the local solar neighborhood, why don't we notice them racing around the sky?
20. Describe how the true motion of a star can be broken into *radial* and *tangential* components and how each can be measured.
21. Describe the overall pattern of rotation of the Milky Way Galaxy. About how long does it take our solar system to complete one orbit?
22. Briefly explain how analysis of the Milky Way's rotation leads us to conclude that most of its mass is in the form of mysterious *dark matter.*
23. Describe the raisin cake model of the universe and explain how it shows that a *uniform* expansion would lead us to see more distant galaxies moving away from us at higher speeds.

Discussion Questions

1. *Varying Planetary Brightness.* The apparent retrograde motion of the planets was not the only planetary mystery in ancient times. Another mystery concerned the fact that planets vary in brightness as they move through the zodiac. Explain why planets farther from the Sun than Earth appear brightest in the *middle* of their periods of apparent retrograde motion. (*Hint:* Study Figure 3.7.)
2. *Mercury and Venus.* Explain why the planets Mercury and Venus can be seen only either shortly before sunrise or shortly after sunset. Which of these planets is easier to see, and why? At what points in their orbits do Mercury and Venus appear brightest in the sky? Based on your answers, discuss why Venus is sometimes called *the morning star* and sometimes called *the evening star.* (*Hint:* Draw a diagram similar to Figure 3.7, but for Venus or Mercury.)

Problems

1. *Motion Summary.* Summarize the motions of spaceship Earth by making a three-column table. In column 1, list each of the motions of the Earth discussed in this chapter. In column 2, state the approximate speed (or range of speeds) of each motion listed in column 1. In column 3, list the apparent motions associated with each real motion.

2. *Annual Changes in the Night Sky.*
   a. Imagine that, today, both the Sun and a distant star appear on your meridian at exactly 12:00 noon (of course, it will be difficult to see the star in the daylight!). At what time will the star cross your meridian tomorrow? Two days from now? One week from now? (*Hint:* Remember that a solar day is 4 minutes *longer* than a sidereal day.)
   b. No matter where you live, the constellation Orion crosses your meridian at about midnight in mid-December. At what time does Orion cross your meridian in mid-January? Mid-February? Mid-June? Based on your answers, briefly explain how the difference between a solar and a sidereal day explains why the stars visible at night change with the seasons.

3. *No Axis Tilt.* Suppose that, rather than having a $23\frac{1}{2}°$ axis tilt, the Earth's axis had no tilt (i.e., suppose the axis were perpendicular to the ecliptic plane). Would we still have seasons? Why or why not?

4. *View from the Moon.* Suppose you lived on the Moon and your home was located near the center of the face that we see from the Earth.
   a. During the phase of full moon, what phase would you see for the Earth (e.g., new earth, full earth, first-quarter earth, etc.)? Would it be day or night at your home? Explain.
   b. Repeat part (a) for the phase of new moon.
   c. At what phase of the Moon would you see sunset? What phase of the Earth would you see at this time? Explain.
   d. At what phase of the Moon would you see sunrise? What phase of the Earth would you see at this time? Explain.
   e. What would you see if you were on the Moon during a total lunar eclipse?
   f. What would you see if you were on the Moon during a total solar eclipse?

5. *Earthshine.* You can ignore eclipses in answering the questions in this problem.
   a. If you lived on the side of the Moon facing the Earth, would there ever be a time when you would have neither sunlight nor earthshine to illuminate your home? If so, when? If not, why not?
   b. Repeat part (a) for the case in which you lived on the far side of the Moon.

6. *Sensible Statements?* Evaluate each of the following statements and decide whether it is sensible. Explain.
   a. I saw the Sun rising in the west.
   b. In South Africa, it's usually quite warm around the time of the winter solstice and quite cool around the time of the summer solstice.
   c. January is the best month to ski in New Zealand.
   d. It was midnight, and the crescent moon was on my meridian.
   e. Tomorrow the Moon will be in first-quarter phase, and today I saw a total lunar eclipse.
   f. When I was a child living in southern Florida, I once saw total solar eclipses from my home 2 years in a row.
   g. In the morning light, I saw the third-quarter moon near my western horizon.
   h. I was up early and saw the full moon rising just before sunrise.

7. *A Farther Moon.* Suppose the distance to the Moon were twice its actual value. Would it still be possible to have a total solar eclipse? An annular eclipse? A total lunar eclipse? Explain your answers clearly in one or two paragraphs.

8. *A Smaller Earth.* Suppose the Earth were smaller in size. Would solar eclipses be any different? If so, how? What about lunar eclipses? Explain your answers clearly in one or two paragraphs.

9. *361° of Rotation.*
   a. Knowing that the Earth rotates through 360° in a sidereal day, how long does it take for the Earth to rotate 1°? Give your answer in minutes and seconds. (*Hint:* Begin by converting the length of a sidereal day to seconds.)
   b. Your answer to part (a) should be close to the roughly 4-minute difference between a solar day and a sidereal day. Briefly explain why.
   c. Explain why the number of sidereal days in a year is *one* more than the number of solar days (there are about 366.25 sidereal days in a year and 365.25 solar days in a year).

10. *Racing Around the Axis.* Imagine that you live on the equator. Calculate how fast you are traveling around the center of the Earth. Give your answer in both kilometers per hour and miles per hour. Facts you will need: (i) The equatorial radius of Earth is about 6,380 kilometers. (ii) The circumference of a circle is $2 \times \pi \times$ radius. (*Hint:* Begin by calculating the distance you travel around the axis with each daily rotation, then recall that speed = distance ÷ time.)

11. *Racing Around the Sun.* By considering the Earth's orbit to be a circle with a radius of 1 AU, calculate

the speed of the Earth in its revolution around the Sun. Give your answer in both kilometers per hour and miles per hour.

12. *Racing Around the Galaxy.* Calculate how fast we are traveling around the center of the galaxy. Assume a circular orbit with radius 28,000 light-years and that each orbit takes 230 million years. Give your answer in both kilometers per hour and miles per hour.

13. *Raisin Cake Universe.* Suppose that, before baking, all the raisins in a cake are 1 centimeter apart and after baking they are all 4 centimeters apart.
    a. Draw diagrams to represent your cake before and after baking.
    b. Identify one raisin as the Local Raisin on your diagrams. Construct a table similar to Table 3.2 showing the distance and speed of other raisins as seen from the Local Raisin.
    c. In your own words, explain how your expanding cake is similar to the expanding universe.

14. *Sun Sign.* According to a newspaper astrology column, what is your sign? Does your birthday fall near the beginning, middle, or end of that sign? Now find out the constellation in which the Sun actually appears on your birthday. (You'll need a star chart or a model of the celestial sphere that lists dates along the ecliptic.) How far off is your astrological sign from your actual Sun sign? Explain the discrepancy.

15. *Project: A Connecticut Yankee.* Find the book *A Connecticut Yankee in King Arthur's Court* by Mark Twain. Read the portion that deals with the Connecticut Yankee predicting an eclipse (or read the entire book). In a one- to two-page essay, summarize the episode and describe how it helps the Connecticut Yankee gain power, and comment on how astrologers and others who learned to predict eclipses in ancient times might have used their knowledge to acquire power.

S1

Celestial Timekeeping and Navigation

WHAT DO A.M. AND P.M. MEAN? WHY DO WE HAVE LEAP YEARS? How can sailors find their way at sea? You may be surprised to learn that all these questions are tied to the apparent motions of the Sun, Moon, and stars in the sky.

The principles of celestial timekeeping and navigation are easy to understand. However, we must investigate the appearance of the celestial sphere in a bit more detail than we have up to this point, including how the sky varies with latitude. This chapter begins with this detailed investigation and then moves on to its primary purpose: explaining the origins of our clocks and calendars and the principles of navigating by the stars.

S1.1 A Map of the Celestial Sphere

Let's begin with a brief review of what we already know about the appearance of the sky from Earth. Every star is located at a unique distance from Earth, but all stars are so far away that they appear to be part of a great *celestial sphere* that surrounds us (see Figure 1.7 [Section 1.5]). Stars are moving through space at fairly high speeds relative to our solar system, but their great distances make them appear fixed on the celestial sphere over the time scale of a human lifetime. The patterns of the fixed stars mark the locations of the 88 distinct constellations recognized by the International Astronomical Union. Thus, we can make a map of the celestial sphere by marking a globe with the patterns of bright stars and the official borders of the constellations (Figure S1.1).

Because we look outward from Earth, we appear to be in the center of the celestial sphere. Thus, when we look at the Sun, Moon, or planets, we see them positioned against the backdrop of the constellations. As we orbit the Sun each year, to us the Sun appears to move steadily around the celestial sphere, following the path called the *ecliptic* through the constellations of the zodiac (see Figure 1.19). The ecliptic is marked with yellow dots in Figure S1.1. If each dot represented the Sun's apparent position among the constellations on a particular day of the year, we would need 365 closely spaced dots to mark the Sun's circular path around the ecliptic. Note that the ecliptic is the projection of the Earth's orbital plane (the *ecliptic plane*) into space. Just as the orbital plane is inclined by $23\frac{1}{2}°$ to the Earth's equator because of the tilt of the Earth's axis, the ecliptic is inclined by $23\frac{1}{2}°$ to the *celestial equator.*

We know that the Earth is rotating on its axis, but because we do not feel this motion the celestial sphere appears to rotate around us each day. The Earth rotates counterclockwise as viewed from above the North Pole, so the celestial sphere appears to rotate clockwise around the Earth (see Figure 1.14). The axis of this apparent rotation defines the *north and south celestial poles* and the *celestial equator.*

The best way to study the celestial sphere is to use a model like the one shown in Figure S1.1. However, when working with a flat page in a book, it is easier to show the celestial sphere without the stars as in Figure S1.2, which shows Earth in the center and the locations of the celestial equator, celestial poles, and ecliptic.

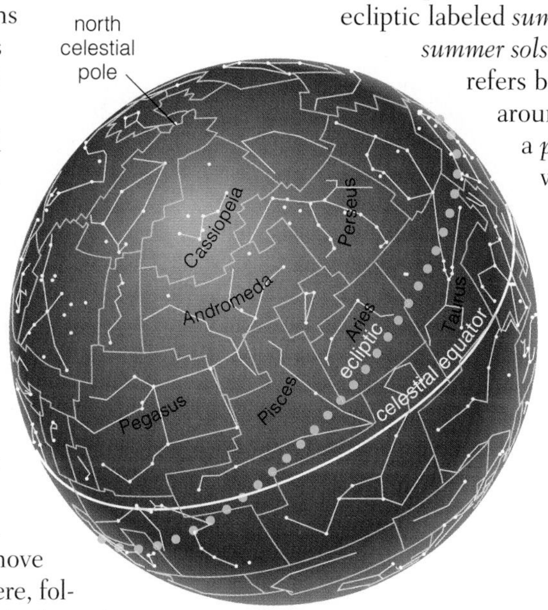

FIGURE S1.1 A map of the celestial sphere.

Four Special Points: Solstices and Equinoxes

The *summer solstice* is the moment each year when the Northern Hemisphere receives the most direct sunlight and the Southern Hemisphere receives the least [Section 3.1]. From Figure S1.2, it should be clear that the Sun shines most directly on the Northern Hemisphere when it is at the point along the ecliptic labeled *summer solstice.* Thus, the term *summer solstice* has a dual meaning: It refers both to a *moment* that occurs around June 21[1] each year and to a *point* on the celestial sphere at which the Sun appears to be located at that moment. Similarly, the point called the *winter solstice* represents the Sun's apparent location at the moment of the winter solstice, and the points called the *spring equinox*

[1] The exact times and dates of the solstices and equinoxes vary slightly from year to year.

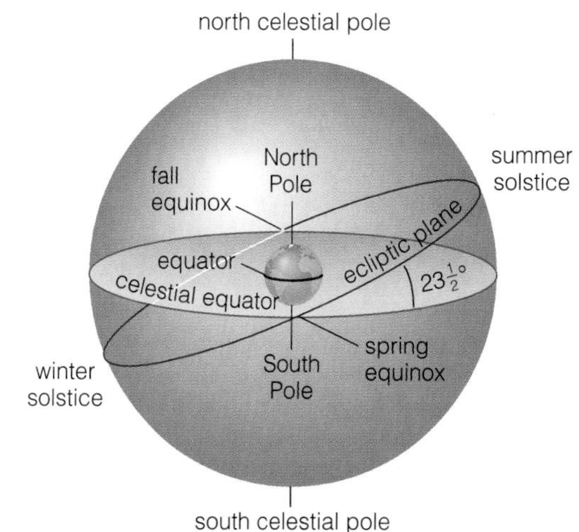

FIGURE S1.2 The inclination of the ecliptic to the celestial equator defines the locations of the equinoxes and solstices.

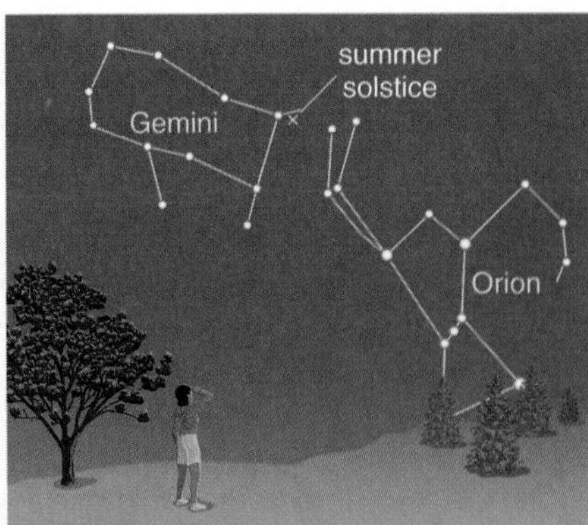

a looking southward

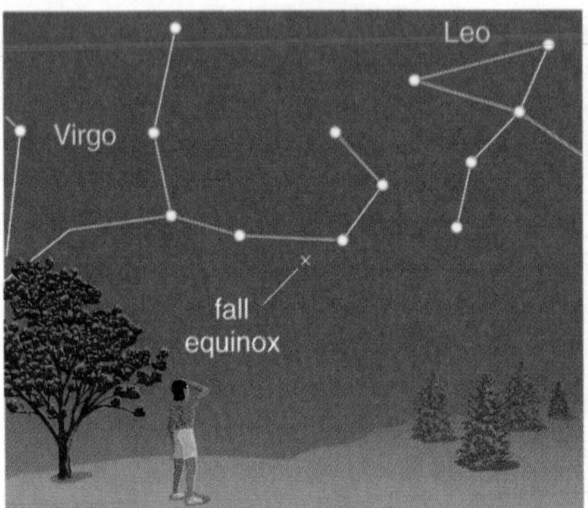

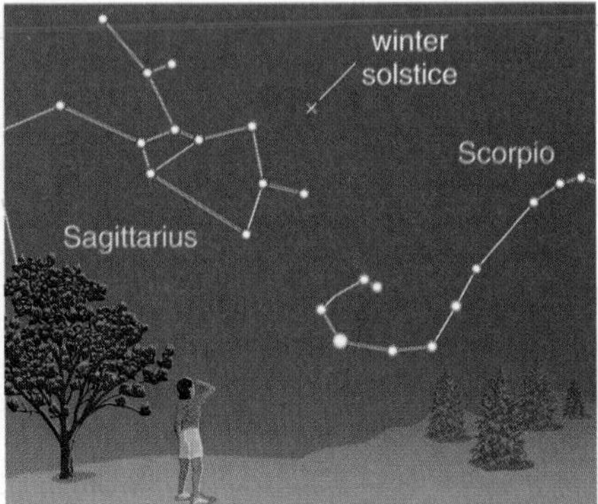

b looking southward

FIGURE S1.3 Finding the equinoxes and solstices on the celestial sphere.

and *fall equinox* represent the Sun's apparent locations at the moments of the spring equinox and fall equinox, respectively.

No bright stars mark the locations of the solstices and equinoxes among the constellations, but you can find these points in the sky with the aid of nearby bright stars (Figure S1.3). For example, the spring equinox is located in the constellation Pisces and can be found with the aid of the four bright stars in the Great Square of Pegasus.

TIME OUT TO THINK ***It is much easier to visualize the celestial sphere if you make a model with a ball; the inexpensive plastic balls available at most convenience stores work well. Use a felt-tip pen to mark the north and south celestial poles on your ball; then mark the celestial equator and the ecliptic. Label the equinoxes and solstices. Can you see how your model corresponds to the diagram in Figure S1.2?***

Celestial Coordinates

Up to this point in the book, we've worked with two distinct coordinate systems. First, we saw how you can pinpoint the location of an object in your local sky by specifying its coordinates of *altitude* and *direction* (Figure S1.4a). Note that these coordinates are fixed in your local sky; for example, you always know what it means to point to an altitude of 60° in a direction of southeast. Second, we've used the coordinates of *latitude* and *longitude* to pinpoint the locations of places on Earth (Figure S1.4b). These coordinates are fixed on the Earth; for example, if someone asks you to go to latitude 42°N and longitude 12°E, you'll know they want you to go to Rome.

In a similar way, it's useful to have a system of **celestial coordinates** that can pinpoint locations on the celestial sphere. These coordinates should be fixed on the celestial sphere so that a particular star's coordinates tell you exactly where it is located among the constellations. The coordinates we use on the celestial sphere are called *declination* and *right ascension* (Figure S1.4c). Despite their odd-sounding names, these coordinates are easy to understand.

Declination (**dec**) on the celestial sphere is very similar to *latitude* on Earth:

- Just as circles of latitude are parallel to the Earth's equator, circles of declination are parallel to the celestial equator.
- Just as the latitude of the Earth's equator is 0°, the declination of the celestial equator is 0°.
- Latitudes are labeled north or south, but declinations are labeled positive or negative. For example, the Earth's North Pole has latitude 90°N, and the South Pole has latitude 90°S; the north celestial pole has declination +90°, and the south celestial pole has declination −90°.

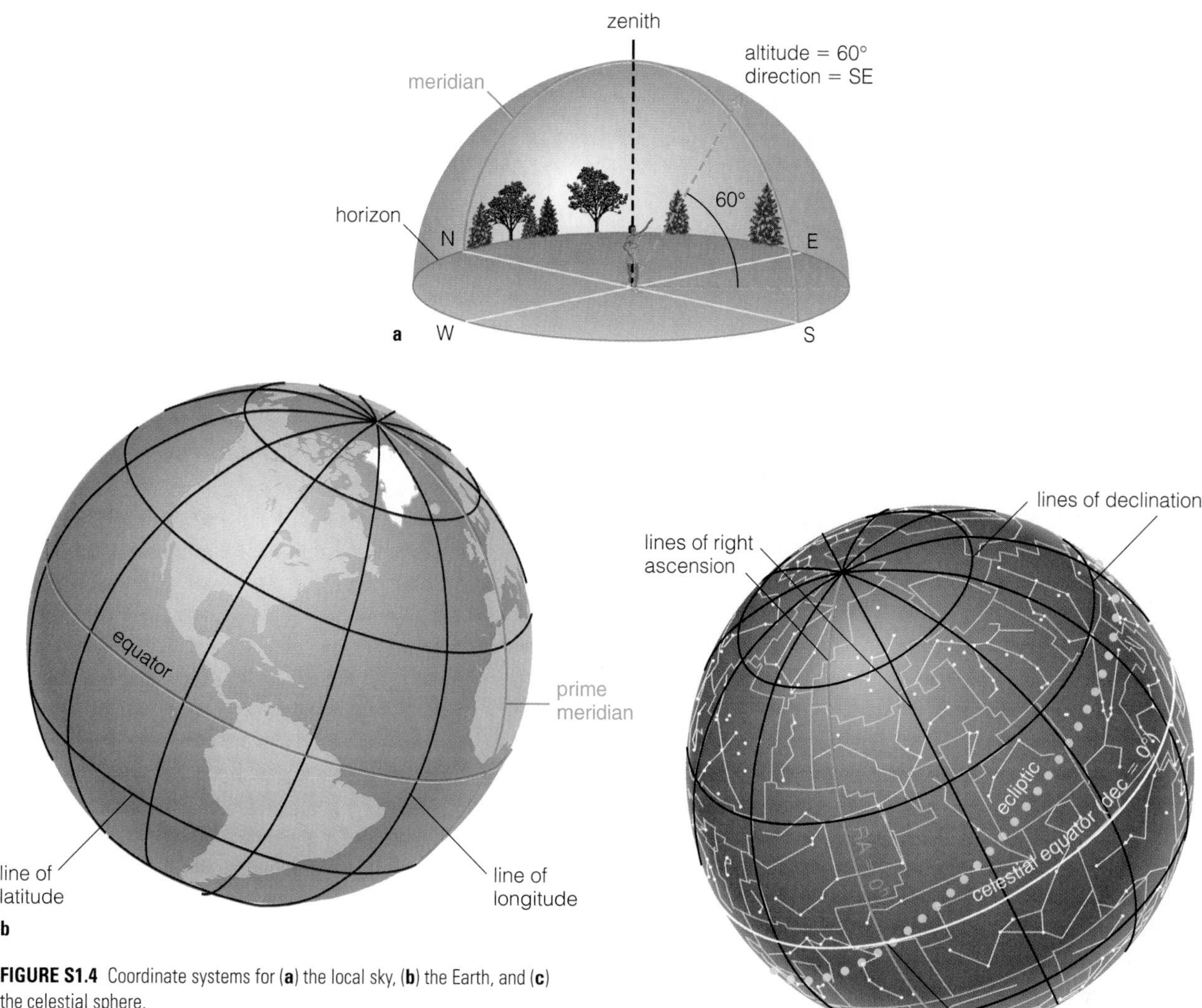

FIGURE S1.4 Coordinate systems for (**a**) the local sky, (**b**) the Earth, and (**c**) the celestial sphere.

Right ascension (**RA**) on the celestial sphere is very similar to *longitude* on Earth:

- Just as lines of longitude are semicircles extending from the Earth's North Pole to its South Pole, lines of right ascension are semicircles extending from the north celestial pole to the south celestial pole.
- The starting point for measuring longitude on Earth is the line of longitude running through Greenwich, England. On the celestial sphere, the starting point for measuring right ascension is the line of right ascension that runs through the spring equinox.
- Longitude is usually measured in degrees east or west of Greenwich, but right ascension is usually measured in hours (and minutes and seconds) east of the spring equinox. Because the celestial sphere appears to rotate around us once each day, 24 hours of right ascension[2] is equivalent to 360°. Thus, each hour of right ascension represents an angle of 360° ÷ 24 = 15°.

We can identify every star on the celestial sphere by its celestial coordinates.[3] For example, the star Vega (in the constellation Lyra) has dec = +38°44′ and RA = 18^h35^m (Figure S1.5). The declination tells us that Vega is 38°44′ north of the celestial equator, and the right ascension tells us that it is about 279°

[2]The astute reader will recognize that the celestial sphere appears to rotate in a *sidereal* day, which is 23 hours 56 minutes. Thus, the celestial sphere appears to rotate through an *angle* of 24 hours of right ascension in a *time* of 23 hours 56 minutes.

[3]Celestial coordinates actually change slowly because they are measured relative to the celestial equator, which changes its position among the stars with the Earth's 26,000-year precession cycle.

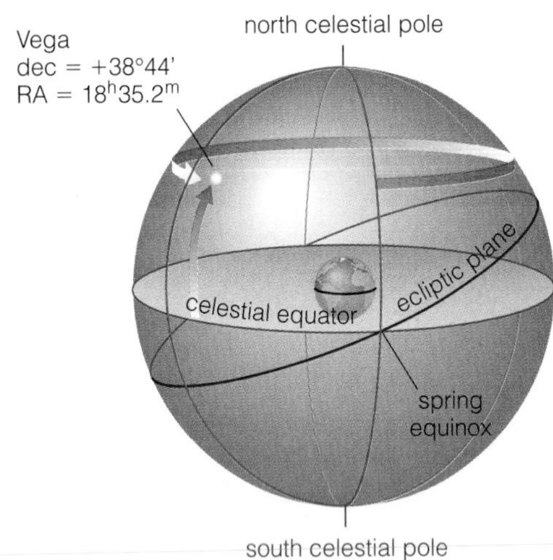

FIGURE S1.5 The celestial coordinates of Vega.

east of the spring equinox. (The 18 hours represent $18 \times 15° = 270°$, and the 35 minutes represent $\frac{35}{60} \times 15° \approx 9°$.)

TIME OUT TO THINK *Add the following to your ball model of the celestial sphere: circles of declination for dec = +30°, dec = +60°, dec = −30°, and dec = −60°; lines of right ascension for RA = 0^h, RA = 6^h, RA = 12^h, and RA = 18^h. Then place a dot at the appropriate coordinates to represent Vega.*

Celestial Coordinates of the Sun

Because the Sun appears to move along the ecliptic, its celestial coordinates change over the course of the year. Let's begin with the *moment* when the Sun is located along the ecliptic at the *point* of the spring equinox, which occurs each year around March 21. Figure S1.4c shows clearly that the celestial coordinates of the spring equinox are RA = 0^h and dec = 0°. Over the next 12 months, the Sun moves around the ecliptic through all 24 hours of right ascension. Therefore, the Sun's right ascension increases by about 2 hours per month. That is, the Sun has RA = 2^h around April 21, RA = 4^h around May 21, and so on.

Meanwhile, the Sun's declination gradually increases from 0° on the spring equinox to $+23\frac{1}{2}°$ by the time of the summer solstice, decreases through 0° on the fall equinox to $-23\frac{1}{2}°$ on the winter solstice, and returns to 0° by the next spring equinox. Table S1.1 shows the Sun's annual celestial coordinates at 1-month intervals.

TIME OUT TO THINK *On your ball model of the celestial sphere, add dots along the ecliptic to show the Sun's apparent position on each of the dates listed in Table S1.1. You should now be able to estimate the Sun's celestial coordinates on any day of the year. What are the Sun's celestial coordinates on your birthday?*

Celestial Coordinates of the Moon and Planets

Like the Sun, the Moon and planets *appear* to move among the constellations of the zodiac. However, unlike the Sun, the Moon and planets do not follow the same precise path over and over. Consider the Moon: As the Moon orbits the Earth each month, it appears to make a complete circuit through the zodiac. Because the Moon's orbit is inclined about 5° to the ecliptic plane [Section 3.3], the Moon always appears to be within 5° of the ecliptic on the celestial sphere. However, because the *nodes* of the Moon's orbit (the two points where the Moon's orbit crosses the ecliptic plane) slowly precess, the Moon's precise path through the zodiac appears to shift from one month to the next.

The apparent paths of the planets are even more complex: From one night to the next, planets usually appear to drift slowly eastward through the zodiac. However, during their periods of apparent retrograde motion [Section 3.1], the planets reverse course and appear to drift westward through the zodiac from night to night.

Despite the complexity of their apparent motions, the real motions of the Moon and planets are quite predictable. Many astronomical software packages can calculate the precise celestial coordinates of the

Table S1.1 The Sun's Approximate Celestial Coordinates at 1-Month Intervals

Approximate Date	RA	Dec	Approximate Date	RA	Dec
Mar. 21 (Spring equinox)	0 hr	0°	Sept. 21 (Fall equinox)	12 hr	0°
Apr. 21	2 hr	+12°	Oct. 21	14 hr	−12°
May 21	4 hr	+20°	Nov. 21	16 hr	−20°
June 21 (Summer solstice)	6 hr	$+23\frac{1}{2}°$	Dec. 21 (Winter solstice)	18 hr	$-23\frac{1}{2}°$
July 21	8 hr	+20°	Jan. 21	20 hr	−20°
Aug. 21	10 hr	+12°	Feb. 21	22 hr	−12°

Moon or planets for any date—whether today or thousands of years in the past or future. You can also purchase astronomical charts and tables that show the celestial coordinates of the Moon and planets in particular years.

S1.2 The Circling Sky

If we view the celestial sphere from the outside, the daily paths of the stars are extremely simple. Every star appears to make a daily circle around the Earth, and each star's circle is determined only by its declination. Figure S1.6 shows the daily paths for a few representative stars.

However, because we live *on* the Earth, the planet beneath us always blocks our view of half the celestial sphere. As a result, we see only half the celestial sphere at any one moment, and the simple daily paths of the stars look somewhat more complex than they would appear if viewed from outside the celestial sphere. In Chapter 1, we saw that your view of the celestial sphere depends on your latitude [Section 1.5]. Now we are ready to study this dependence in greater detail. It's easiest to begin by considering the sky as it appears from the Earth's North Pole.

The North Pole

Imagine standing at the North Pole (Figure S1.7a). Your "up" points toward the north celestial pole, which therefore appears at your zenith. The ground blocks your view of the half of the celestial sphere south of the celestial equator. Thus, the celestial equator runs along your horizon (Figure S1.7b). The daily circles traced by the stars keep them at constant altitudes above or below your horizon. That is, any object with a positive declination is circumpolar, remaining above your horizon at all times, while objects with negative declinations are never visible to you. Note that, as you look up, the stars trace *counterclockwise* circles around the north celestial pole. Note also that all directions are *south* from the North Pole. Thus, there is no meridian at the North Pole.

TIME OUT TO THINK *Imagine that you are at the North Pole. Visualize the daily paths through your sky of stars with various declinations. What is the path of Polaris, which has a declination of about +89°? What is the path of Vega, with a declination of about +39°? If you are having difficulty, rotating your ball model of the celestial sphere may help.*

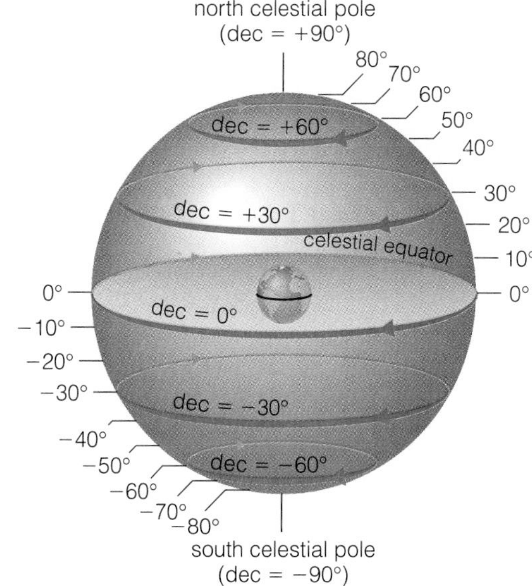

FIGURE S1.6 The daily paths of stars around the celestial sphere depend only on their declinations.

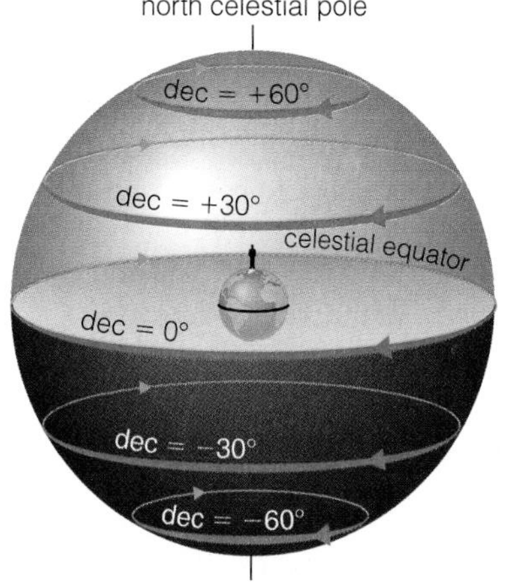

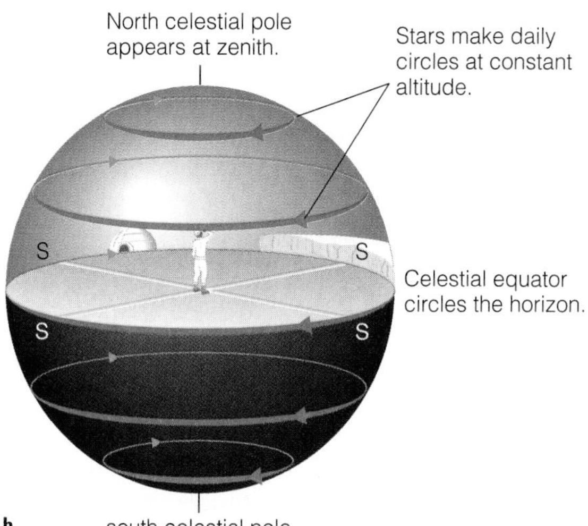

FIGURE S1.7 (**a**) Your orientation relative to the celestial sphere from the North Pole. (**b**) The local sky at the North Pole.

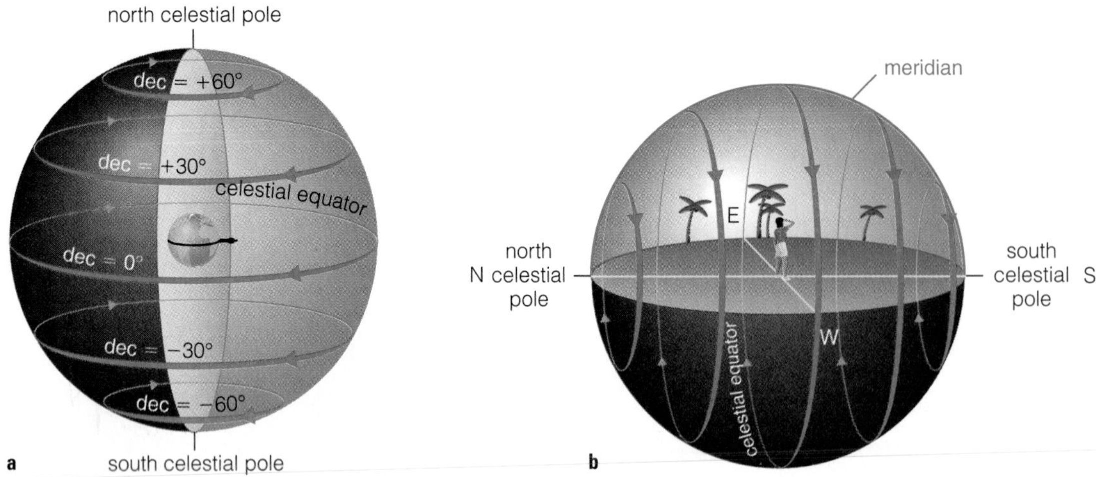

FIGURE S1.8 (**a**) Your orientation relative to the celestial sphere from the Earth's equator. (**b**) The local sky at the equator.

The Equator

Next imagine that you are standing somewhere on the Earth's equator (latitude = 0°), such as in Ecuador, in Kenya, or on the island of Borneo. Figure S1.8a shows that, from the equator, your "up" points directly away from (i.e., perpendicular to) the Earth's rotation axis.

Figure S1.8b shows how you view half the celestial sphere at any one moment from the equator. Note that the figure shows the entire celestial sphere turned so that your "up" is also up on the page. The north celestial pole always appears due north on your horizon, and the south celestial pole always appears due south on your horizon. The celestial equator always extends from the horizon due east, through the zenith, to the horizon due west. As is true from any place on Earth (except the poles), your *meridian* extends from your horizon due south, through your zenith, to your horizon due north (see Figure 1.5b). Aside from these constants, you'll see the stars rise and set as the celestial sphere appears to turn. Note the following key features of the equatorial sky shown in Figure S1.8:

- Stars with dec = 0° lie *on* the celestial equator and therefore rise due east, cross the meridian at the zenith, and set due west.
- Stars with *positive* declinations rise north of due east and set north of due west, and stars with *negative* declinations rise south of due east and set south of due west. The exact rise and set points of stars, and their altitudes when they cross the meridian, depend on their declinations. For example, a star with dec = +30° crosses the meridian 30° to the north of the zenith, which means at an *altitude* of 60°. (Be sure you don't confuse a star's *altitude* in your local sky with its *declination* on the celestial sphere.)
- Exactly half of any star's daily circle lies above the horizon, so *every* star is above the horizon for half of each sidereal day, or about 12 hours.

TIME OUT TO THINK *Imagine that you are at the equator, and picture the daily paths of stars with various declinations. Are any stars circumpolar at the equator? Is there any portion of the celestial sphere that you never see at the equator? Rotating your ball model of the celestial sphere while holding it in the orientation shown in Figure S1.8b may help you.*

40°N Latitude

For our third example, imagine that you are at latitude 40°N, such as in Denver, Indianapolis, Philadelphia, or Beijing. Figure S1.9a shows that, because you are located 40° north of the Earth's equator, your "up" makes a 40° angle with both the Earth's equator and the celestial equator.

Figure S1.9b shows how you view half the celestial sphere at any one moment. For convenience, altitudes and directions along the meridian are also shown. Note that no direction is indicated for the zenith (altitude 90°) because, if you look straight up, you are not facing in any particular direction along the horizon. If you study Figure S1.9b carefully, you

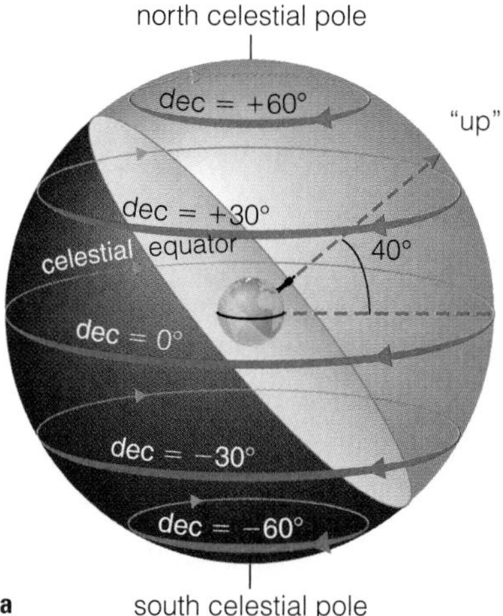

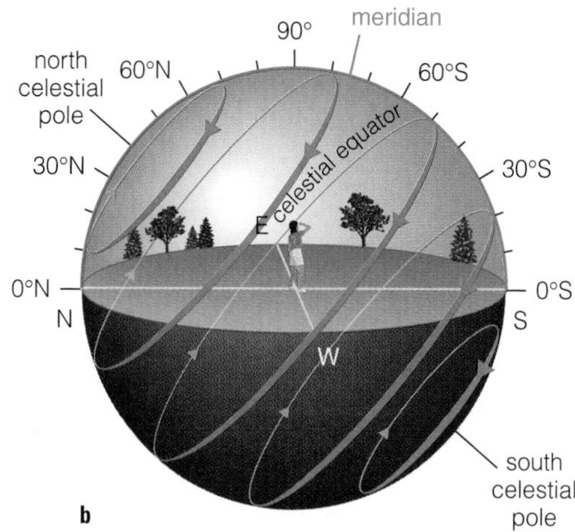

FIGURE S1.9 (**a**) Your orientation relative to the celestial sphere from 40°N latitude. (**b**) The local sky at 40°N latitude. Altitudes and directions are shown along the meridian.

should notice the following key features of the sky at 40°N latitude:

- The north celestial pole appears 40° above the due north horizon. That is, the *altitude* of the north celestial pole is the same as your *latitude*.
- The celestial equator extends from the horizon due east, up to an altitude of 50° on the meridian due south, and down to the horizon due west. Note that the 50° altitude on the meridian is 90° *minus* your latitude: 90° − 40° = 50°.
- Stars with dec = 0° lie *on* the celestial equator and therefore follow the path of the celestial equator through your sky. That is, they rise due east, cross the meridian at 50° in the south, and set due west. Exactly half the celestial equator is above the horizon, so stars with dec = 0° are above your horizon for about 12 hours each day.
- Stars that lie on the celestial sphere *within* 40° of the north celestial pole are circumpolar, tracing daily *counterclockwise* circles around the north celestial pole. Stars that lie on the celestial sphere *within* 40° of the south celestial pole never rise above the horizon.
- All other stars with *positive* declinations rise north of due east and set north of due west. Note that *more* than half the daily circle for any of these stars is above the horizon, so they are above the horizon for *more* than 12 hours each day. The greater the star's declination, the closer it is to being circumpolar.
- All other stars with *negative* declinations rise south of due east and set south of due west. Note that *less* than half the daily circle for any of these stars is above the horizon, so they are above the horizon for *less* than 12 hours each day. The more negative the star's declination, the shorter its daily path through your sky.

TIME OUT TO THINK *Imagine that you are at 40°N latitude, and picture the daily paths of stars with various declinations. Again, it may help to rotate your ball model of the celestial sphere while holding it in the orientation shown in Figure S1.9b. Where does a star with dec = +30° cross your meridian? What about a star with dec = −30°?*

All Other Latitudes

Generalizing the procedure we used for 40°N latitude, you can take the following two steps to get a picture of daily star paths at any other latitude:

Step 1. Begin with a model or picture of the celestial sphere that shows the daily paths of stars as they appear from outside the celestial sphere. Then add a person at the latitude you wish to study, as shown in Figure S1.10a for latitude 30°S.

Step 2. Draw the local sky for that latitude by rotating the picture so that the person's "up" is also up on the page, removing the Earth from the center, cutting off the portion of the celestial sphere that falls below the horizon, labeling directions along the horizon, and labeling the meridian with altitudes and directions. The result for latitude 30°S is shown in Figure S1.10b.

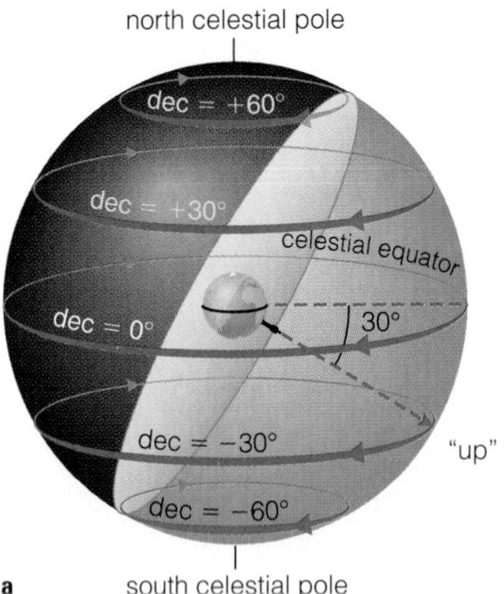

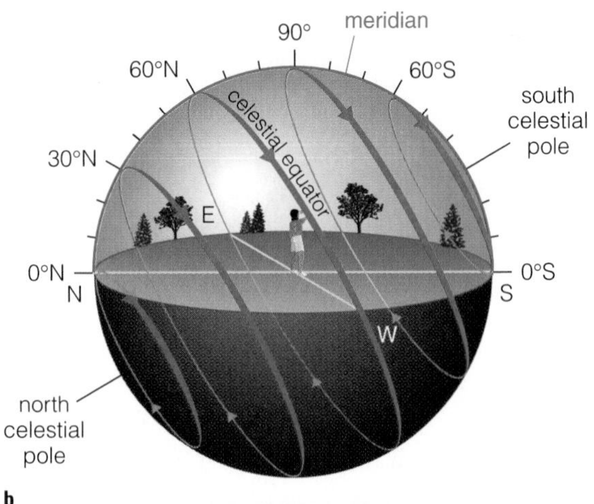

FIGURE S1.10 To picture the sky for any particular latitude: (**a**) Place a person at the correct latitude on a picture like this one, which shows latitude 30°S. (**b**) Then rotate and relabel the picture so it looks like a local sky. It helps to label altitudes and directions along the meridian.

Note that the following rules always hold for all locations in the Northern Hemisphere:

- The north celestial pole appears in the sky in a direction of due north and at an altitude equal to the latitude.
- The celestial equator stretches from the horizon due east, up to an altitude of 90° minus the latitude on the meridian in the south, and back down to the horizon due west.

Similarly, the following rules always hold for all locations in the Southern Hemisphere:

- The south celestial pole appears in the sky in a direction of due south and at an altitude equal to the latitude.
- The celestial equator stretches from the horizon due east, up to an altitude of 90° minus the latitude on the meridian in the north, and back down to the horizon due west.

Mathematical Insight S1.1 Recipe for Star Tracks

You can visualize the path of any particular star through the sky at any latitude by drawing diagrams similar to those in Figure S1.9b and Figure S1.10b. However, some people prefer to use a more mathematical "recipe," such as the one that follows. Before you begin, you must know the *declination* (*dec*) of the star whose path you wish to find and the *latitude* (*lat*) for which you wish to find it; note that lat is always positive (or 0° for the equator) in the equations that follow.

1. Check to see whether the star is either circumpolar or never above your horizon.
   a. For the Northern Hemisphere:
      i. The star is circumpolar if it has dec $\geq$ (90° − lat). In that case, it will trace a daily, *counterclockwise* circle around the north celestial pole with a radius of (90° − dec).
      ii. The star is never above your horizon if it has dec $\leq$ (lat − 90°).
   b. For the Southern Hemisphere:
      i. The star is circumpolar if it has dec $\leq$ (lat − 90°). In that case, it will trace a daily, *clockwise* circle around the south celestial pole with a radius of (90° + dec).
      ii. The star is never above your horizon if it has dec $\geq$ (90° − lat).

TIME OUT TO THINK *Look again at Figures S1.9b and S1.10b, and confirm that the rules for finding the celestial poles and celestial equator hold for the latitudes shown. Then follow the two-step procedure to draw a representation of* your *local sky.*

S1.3 The Path of the Sun

We've seen that the path of any star through your sky depends only on your latitude and the star's declination. The same is true for the path of the Sun: On any particular day, the path of the Sun depends only on your latitude and the Sun's declination. However, because the Sun's declination changes over the course of the year, its daily path also changes over the course of the year. By using the declinations of the Sun shown in Table S1.1, you can easily figure out the Sun's path on various dates. For example, the Sun has declination 0° on the equinoxes and hence follows the same path as a star with declination 0° on March 21 and September 21. The Sun has declination $+23\frac{1}{2}°$ on the summer solstice and therefore follows the same path as a star with dec = $23\frac{1}{2}°$. On April 21 and August 21, when the Sun has declination +12°, it follows the path of a star with dec = 12°. And so on.

As an example, consider again the sky at 40°N latitude shown in Figure S1.9. The celestial equator goes from due east on the horizon, through the meridian at an altitude of 50° in the south, to due west on the horizon. We can find the path of the Sun on various dates as follows:

- On the equinoxes (March 21 and September 21): The Sun has dec = 0° and therefore rises due east, crosses the meridian at an altitude of 50° in the south, and sets due west.
- On the summer solstice (June 21): The Sun has dec = $23\frac{1}{2}°$ and therefore rises well north of due east and sets well north of due west. It crosses the meridian $23\frac{1}{2}°$ higher than the celestial equator, or at an altitude of $50° + 23\frac{1}{2}° = 73\frac{1}{2}°$ in the south.
- On the winter solstice (December 21): The Sun has dec = $-23\frac{1}{2}°$ and therefore rises well south of due east and sets well south of due west. It crosses the meridian $23\frac{1}{2}°$ lower than the celestial equator, or at an altitude of $50° - 23\frac{1}{2}° = 26\frac{1}{2}°$ in the south.
- On April 21 and August 21: The Sun has dec = +12° and therefore rises north of due east and sets north of due west. It crosses the meridian 12° higher than the celestial equator, or at an altitude of 50° + 12° = 62° in the south.

Figure S1.11 shows the Sun's path for latitude 40°N at monthly intervals. You can follow a similar

---

2. If the star is sometimes above your horizon but not circumpolar, it rises and sets along a path described as follows:
   a. For the Northern Hemisphere:
      i. If the star has dec = 0°, it is on the celestial equator and therefore rises due east, crosses the meridian at an altitude of (90° − lat) in the south, and sets due west.
      ii. If the star has dec < 0°, it rises *south of* due east, crosses the meridian at an altitude of (90° − lat + dec) in the *south,* and sets *south of* due west.*
      iii. If the star has dec > 0°, it rises *north of* due east and sets *north of* due west.* To find its meridian-crossing altitude, go *northward* along the meridian by an amount equal to the declination, starting from the celestial equator at (90° − lat) in the south.

(continued)

*Calculating exactly where the star rises and sets along the horizon requires mathematics beyond the scope of this book. Many astronomical software packages can perform these calculations automatically.

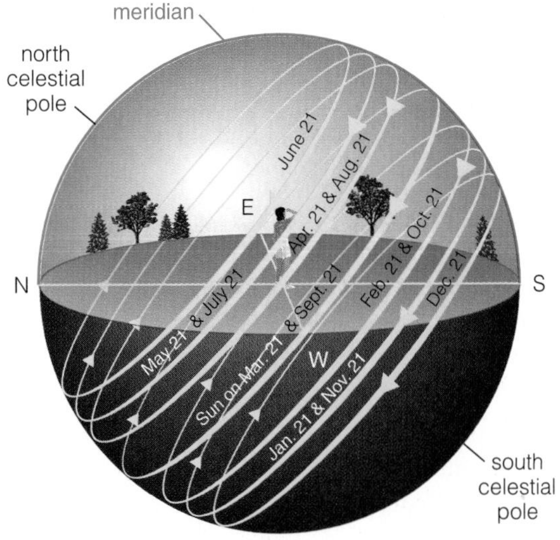

FIGURE S1.11 The Sun's daily path for 40°N latitude at monthly intervals.

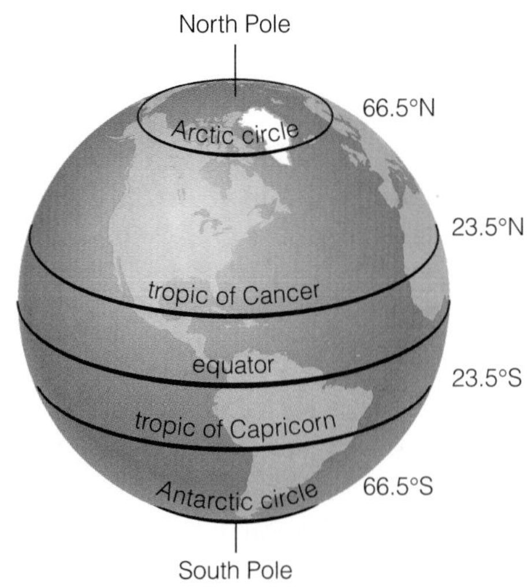

FIGURE S1.12 Special latitudes defined by the Sun's path through the sky.

process to determine the Sun's path for any other latitude. However, the $23\frac{1}{2}°$ tilt of the Earth's axis makes these patterns particularly interesting at the special latitudes shown in Figure S1.12.

The North and South Poles

Recall that the celestial equator (dec = 0°) circles the horizon at the North Pole. Because the Sun appears *on* the celestial equator on the day of the spring equinox, the Sun circles the north polar sky *on the horizon* on March 21 (Figure S1.13); note that it completes a full circle of the horizon in 24 hours (1 solar day). The Sun's declination slowly increases during the next 3 months, so its daily circles slowly increase in altitude. It reaches its highest point on the summer solstice (June 21), when its declination of $+23\frac{1}{2}°$ means that the Sun circles the north polar sky at an altitude of $23\frac{1}{2}°$. Over the next 3 months, the Sun's daily circles gradually fall lower, again reaching the horizon on the fall equinox (September 21). The Sun's declination is negative for the next 6 months (until the following spring equinox), so it

Mathematical Insight S1.1 ***(continued)***

b. For the Southern Hemisphere:
   i. If the star has dec = 0°, it is on the celestial equator and therefore rises due east, crosses the meridian at an altitude of (90° − lat) in the north, and sets due west.
   ii. If the star has dec > 0°, it rises *north of* due east, crosses the meridian at an altitude of (90° − lat − dec) in the *north,* and sets *north of* due west.*
   iii. If the star has dec < 0°, it rises *south of* due east and sets *south of* due west.* To find its meridian-crossing altitude, go *southward* along the meridian by an amount −dec, starting from the celestial equator at (90° − lat) in the north.

Example 1: Find the path of a star with dec = +30° for latitudes (a) 55°N and (b) 25°N.

Solution: (a) For latitude 55°N: By step 1a, the star is *not* circumpolar because its declination is *not* greater than 90° − lat = 90° − 55° = 35°, but it is sometimes visible. By step 2a.iii, we find its meridian-crossing altitude by going dec = +30° northward along the meridian from the celestial equator, which crosses the meridian at 90° − lat = 35° in the *south.* Thus, the star rises north of due

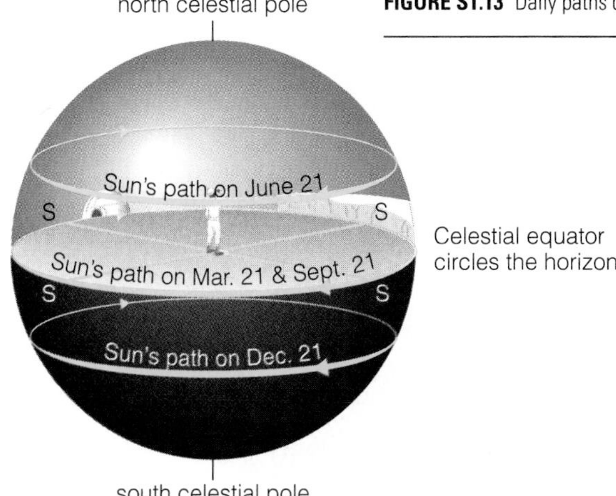

FIGURE S1.13 Daily paths of the Sun at the North Pole.

remains below the north polar horizon.[4] Thus, the North Pole essentially has 6 months of daylight and 6 months of darkness, with an extended twilight that lasts a few weeks beyond the fall equinox and an extended dawn that begins a few weeks before the spring equinox.

The situation is opposite at the South Pole. There the Sun's daily circles slowly rise from the horizon on the fall equinox to a maximum altitude of $23\frac{1}{2}°$ on the *winter* solstice and slowly fall back to the horizon on the spring equinox. Thus, the South Pole has the Sun above the horizon from the fall equinox on September 21 to the spring equinox on March 21, and below the horizon for the following 6 months.

[4]If the Earth had no atmosphere, the Sun would be above and below the north polar horizon for exactly half a year each. But the atmosphere bends light enough so that the Sun *appears* above the horizon for slightly more than half a year. (Near the horizon, the Sun appears about 1° higher than it would without an atmosphere.)

The Equator

At the Earth's equator, the *celestial equator* extends from the horizon due east, through the zenith, to the horizon due west. The Sun therefore follows this path on the dates of the spring and fall equinoxes, reaching the zenith at noon on these two dates (Figure S1.14). In the months between the spring and fall equinoxes, the Sun follows a track through the northern half of the sky; during the other half of the year, it follows a track through the southern half of the sky.

Note that, no matter what the date, half the Sun's daily circle is above the horizon, and the other half is below the horizon. Thus, the Sun is always above the horizon for 12 hours and below it for 12 hours at the equator. Moreover, the Sun appears equally high in the equatorial sky on the summer and winter solstices. Thus, equatorial regions do not have four seasons as do regions at temperate northern or southern latitudes. Instead, equatorial regions generally have a

east, crosses the meridian at 35° + 30° = 65° in the south, and sets north of due west.

(b) For latitude 25°N: Here, the celestial equator crosses the meridian at 90° − lat = 90° − 25° = 65° in the south. Thus, by step 2a.iii, we find the star's meridian-crossing altitude by going dec = +30° northward along the meridian from 65° in the south. Going 25° northward takes us to the zenith (altitude 90°), so going 30° northward takes us to an altitude of 85° in the *north*. The star rises north of due east, crosses the meridian at 85° in the north, and sets north of due west.

Example 2: Find the path of a star with dec = −35° for latitudes (a) 60°N and (b) 60°S.

Solution: (a) For latitude 60°N: By step 1a.ii, the star is never above the horizon because its declination of −35° is *less* than lat − 90° = 60° − 90° = −30°.

(b) For latitude 60°S: By step 1b.i, the star is circumpolar because its declination of −35° is *less* than lat − 90° = 60° − 90° = −30°. It traces a daily, *clockwise* circle around the south celestial pole with a *radius* of 90° + dec = 90° + −35° = 55°.

*Calculating exactly where the star rises and sets along the horizon requires mathematics beyond the scope of this book. Many astronomical software packages can perform these calculations automatically.

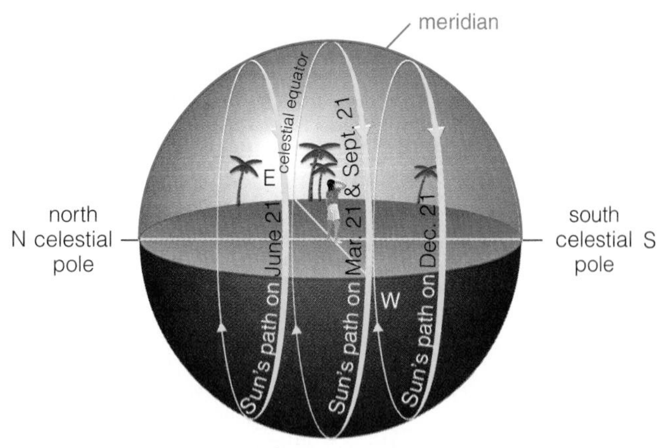

FIGURE S1.14 Daily paths of the Sun at the equator.

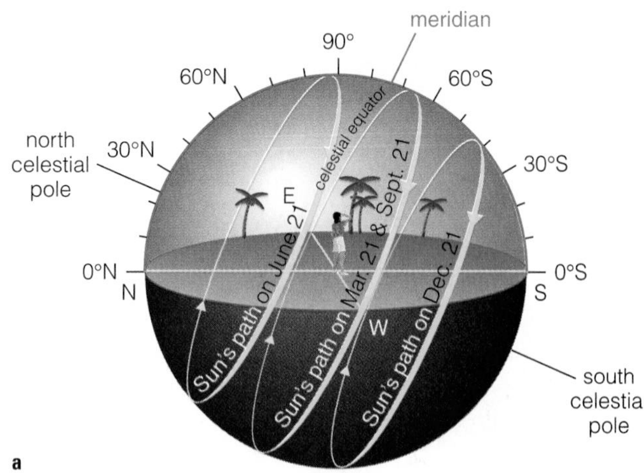

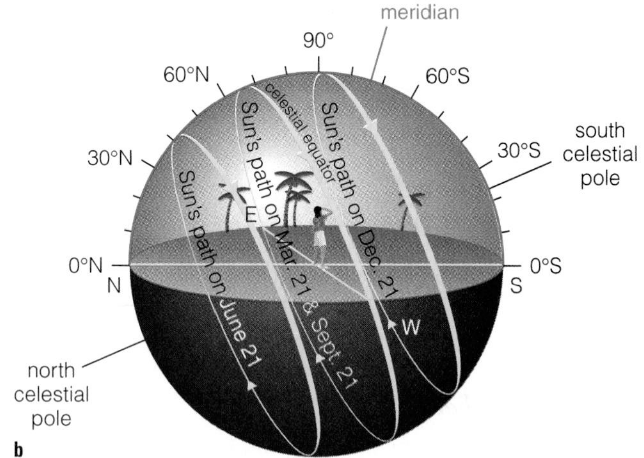

FIGURE S1.15 (**a**) Daily paths of the Sun at the tropic of Cancer. (**b**) Daily paths of the Sun at the tropic of Capricorn.

rainy season and a dry season, with the weather determined by wind patterns on the Earth.

The Tropics of Cancer and Capricorn

The circle on the Earth with latitude 23.5°N is called the **tropic of Cancer,** and the circle with latitude 23.5°S is called the **tropic of Capricorn.** The region between these two circles is called the *Tropics.*

Figure S1.15 shows the sky and the path of the Sun for the tropics of Cancer and Capricorn. Note that, at the tropic of Cancer, the Sun reaches the zenith at noon on the summer solstice and at no other time during the year. At the tropic of Capricorn, the Sun reaches the zenith only at noon on the winter solstice. Thus, the tropics of Cancer and Capricorn are the extreme latitudes where the Sun can appear at the zenith. Between the tropics of Cancer and Capricorn, the Sun passes through the zenith twice a year at noon (the dates vary with latitude). Outside the Tropics, the Sun never appears directly overhead.

Common Misconceptions: High Noon

When is the Sun directly overhead in your sky? If you ask this question of friends and acquaintances, you'll probably find that many people answer "at noon." It's true that the Sun reaches its *highest* point each day when it crosses the meridian (hence the term "high noon"), but unless you live in the Tropics the Sun is *never* directly overhead. In fact, any time you can see the Sun as you walk around, you can be sure it is *not* at your zenith; unless you are lying down, seeing objects at the zenith requires cranking your neck back into a very uncomfortable position.

The Arctic and Antarctic Circles

The $23\frac{1}{2}°$ tilt of the Earth's axis also defines special latitudes that are $23\frac{1}{2}°$ away from each pole, or at latitudes $90° - 23.5° = 66.5°$ north and south. The circle of latitude at 66.5°N is called the **Arctic Circle,** and the circle at 66.5°S is called the **Antarctic Circle.**

Figure S1.16 shows the sky and the path of the Sun for the Arctic and Antarctic circles. Note that, at the Arctic Circle, the Sun never sets on the summer solstice: It is circumpolar, reaching a maximum altitude of 47° in the south at noon and skimming the northern horizon at midnight. Similarly, at the Antarctic Circle, the Sun is circumpolar on the winter solstice. Places near the Arctic Circle or Antarctic Circle are sometimes called *land of the midnight Sun* because the Sun is visible at midnight during part of each year (Figure S1.17). Of course, the name "land of the noon darkness" is more appropriate in their respective winters, when the Sun peeks above the horizon only momentarily at noon.

TIME OUT TO THINK *The Arctic Circle has one day on which the Sun never sets—the summer solstice. The North Pole has 6 months during which the Sun never sets, with the summer solstice occurring in the middle of these 6 months. Based on these facts, describe the patterns of daylight and darkness that you would expect for latitudes* between *the Arctic Circle and the North Pole.*

S1.4 Clocks and Calendars

Our clocks and calendars are based on the Sun's apparent movement through the sky. You are already familiar with the basic principles. For example, you know that the Sun is highest in the sky around noon, and you know that March 21 is the day on which the Sun appears at the position of the spring equinox on the celestial sphere. However, several subtleties make the details of timekeeping somewhat more complex.

Daily Timekeeping

The most basic way of measuring time is by the Sun's position in *your* local sky, which you can measure with a sundial (Figure S1.18). We call this "sundial time" the **apparent solar time.** We define *noon* in apparent solar time to be the moment when the Sun is *on* the meridian, and hence when the sundial casts its shortest shadow. We say that the apparent solar time is *ante meridian,* or A.M., when the Sun is on its way toward the meridian (*ante* means "before"). For example, if the Sun will reach your meridian 2 hours from now, the apparent solar time is 10 A.M. Similarly, we say that the apparent solar time is *post meridian,* or P.M., after the Sun has crossed the meridian (*post* means "after"). If the Sun crossed the meridian 3 hours ago, the apparent solar time is 3 P.M. Note that noon and midnight are *neither* A.M. nor P.M.; thus, we must specify whether we mean 12:00 noon or 12:00 midnight. (Alternatively, we can use a 24-hour clock on which midnight is 0:00, noon is 12:00, 1 P.M. is 13:00, and so on.)

TIME OUT TO THINK *Is it daytime or nighttime at 12:01 A.M.? At 12:01 P.M.? Explain.*

Surprisingly, it is more difficult to determine the apparent solar time with a watch than with a sundial. A sundial *always* reads noon when the Sun crosses the meridian. In contrast, suppose you set your watch to read precisely 12:00 when a sundial reads noon today. Tomorrow, you'll find that your watch reads a few seconds before or after 12:00 when the sundial reads noon. Throughout the year, you'll find that your watch can read anywhere within about a half-hour range around 12:00 when the sundial reads noon!

Why doesn't your watch stay synchronized with the sundial? In Chapter 3, we learned that 24 hours is the *average* length of a solar day. However, the tilt

FIGURE S1.16 (**a**) Daily paths of the Sun at the Arctic Circle. (**b**) Daily paths of the Sun at the Antarctic Circle.

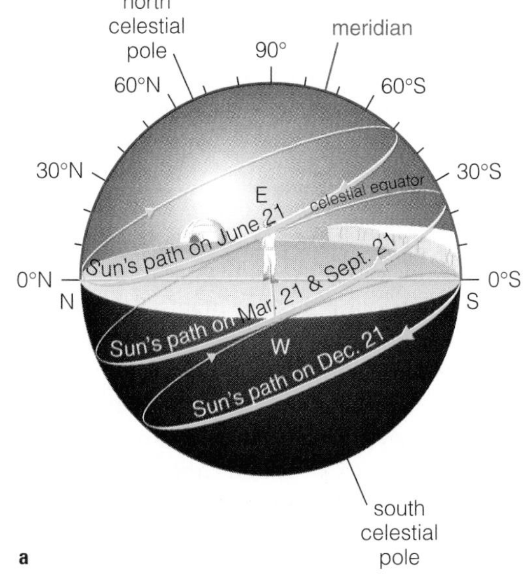

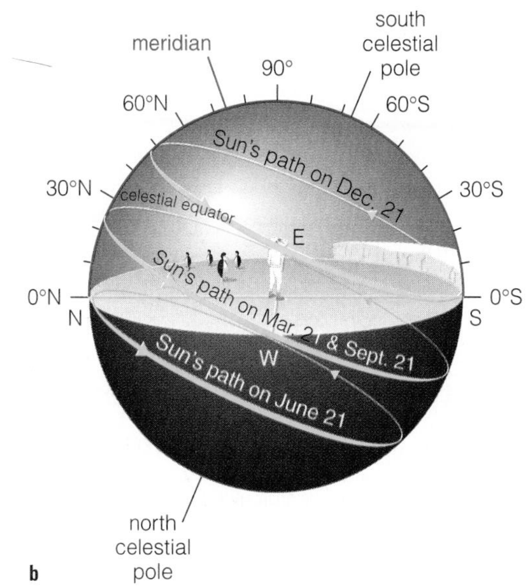

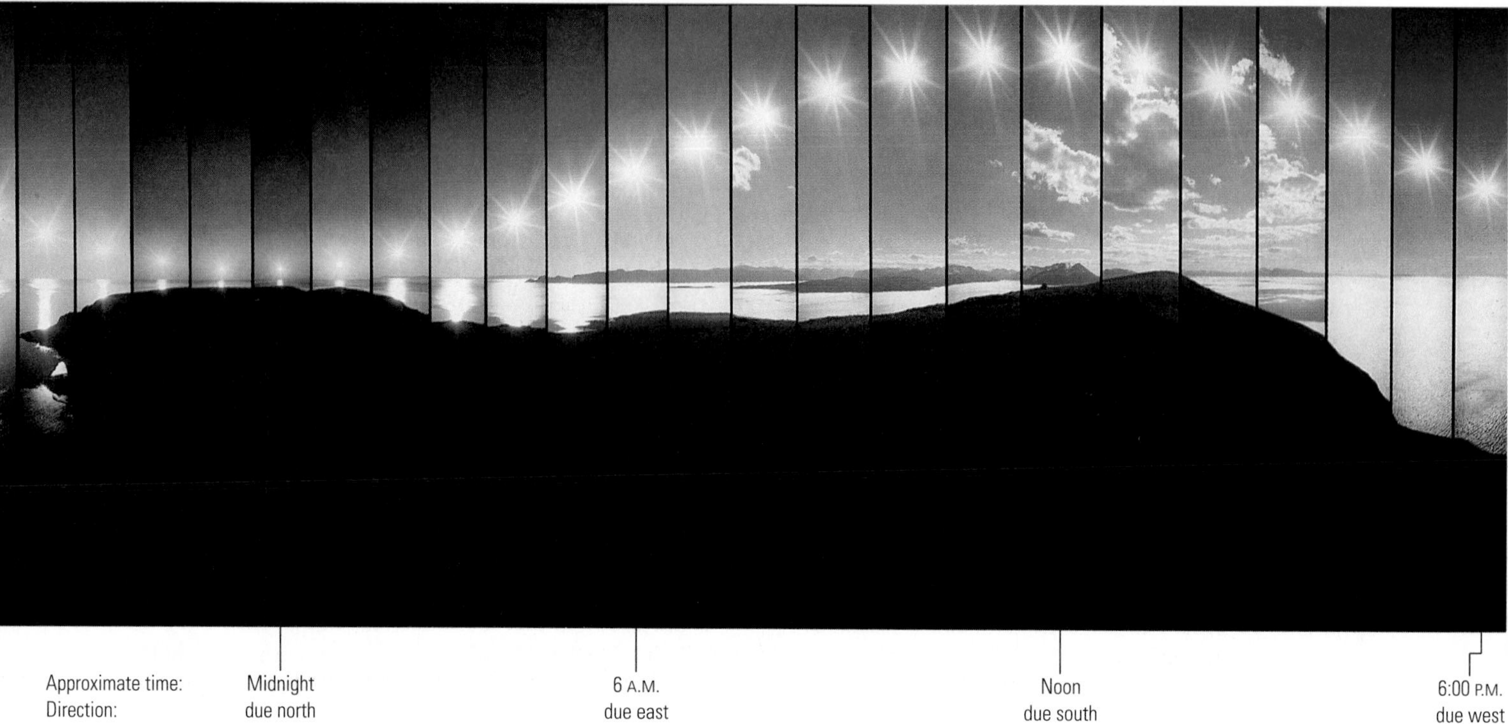

FIGURE S1.17 This sequence of photos shows the progression of the Sun around the horizon on the summer solstice at the Arctic Circle.

FIGURE S1.18 A sundial shows the apparent solar time.

of the Earth's axis and the Earth's varying speed in its orbit around the Sun cause the actual time between successive meridian crossings of the Sun to vary over the course of the year: On a particular day, the time may be as much as 25 seconds shorter or longer than 24 hours. These daily discrepancies between the actual and average lengths of the solar day accumulate at some times of year, thereby explaining the discrepancies between watch time and sundial time.

If you set your watch so that it reads precisely 12:00 when the Sun is on the meridian on an "average" day, then your watch is set to read **mean solar time** (*mean* is another word for *average*). If you photograph the Sun each day at the same mean solar time, you'll see its position trace a figure 8 over the course of a year (Figure S1.19), reflecting the difference between apparent and mean solar time.

Mean solar time is more convenient than apparent solar time because, once set, a reliable mechanical or electronic clock can tell you the mean solar time even when it is cloudy or night. However, mean solar time is still a *local* measure of time and therefore varies with longitude because of the Earth's west-to-east rotation. You're probably familiar with this idea; for example, you probably know that clocks in New York are set 3 hours ahead of clocks in Los Angeles. However, clocks reading mean solar time vary even over relatively short east-west distances.

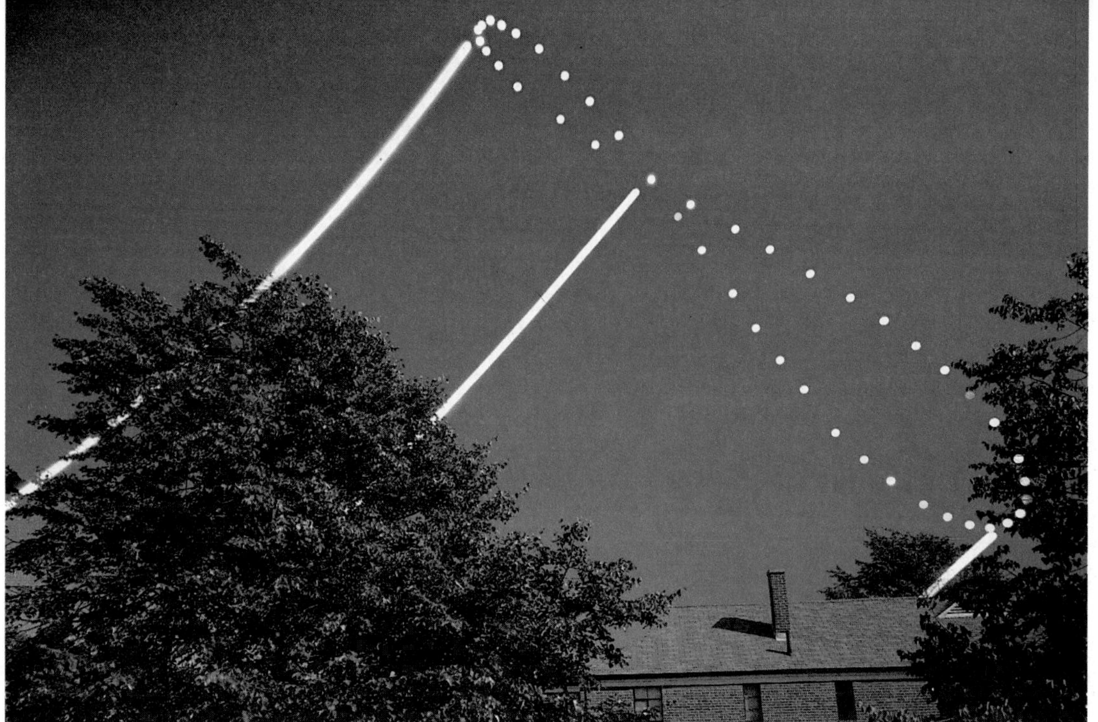

FIGURE S1.19 This composite photograph shows images of the Sun snapped at 10-day intervals over an entire year, always from the same location and at the same mean solar time. The three bright streaks show the path of the Sun's rise on three particular dates.

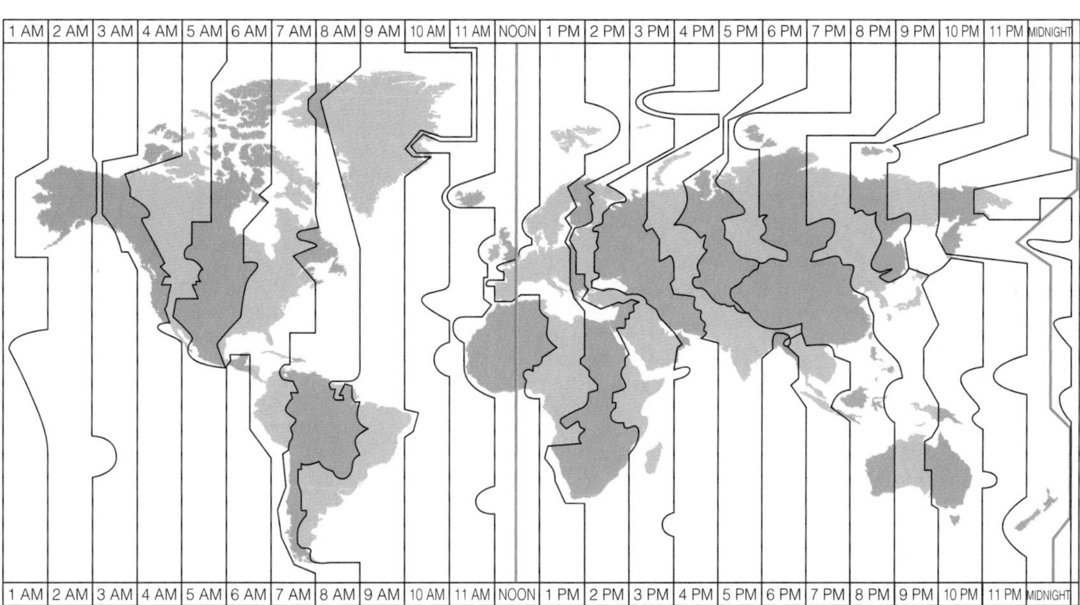

FIGURE S1.20 Standard time zones around the world.

For example, mean solar clocks in west Los Angeles are about 2 minutes behind mean solar clocks in Pasadena (about 30 miles to the east).

Clocks reading mean solar time were common during the early history of the United States. However, by the late 1800s, widespread railroad travel made the use of mean solar time increasingly problematic. Some states had dozens of different "official" times, usually corresponding to mean solar time in dozens of different cities, and each railroad company made schedules according to its own "railroad time." The many time systems made it difficult for passengers to follow the scheduling of trains. On November 18, 1883, the railroad companies agreed to a new system that divided the United States into four time zones, setting all clocks within each zone to the same time. That was the birth of **standard time,** in which the world today is divided into 24 time zones (Figure S1.20). Depending on where you live within a

THINKING ABOUT . . .

Eratosthenes Measures the Earth

In a remarkable ancient feat, the Greek astronomer and geographer Eratosthenes (c. 276–196 B.C.) estimated the size of the Earth in about 240 B.C. He did it by comparing the altitude of the Sun on the summer solstice in the Egyptian cities of Syene (modern-day Aswan) and Alexandria.

Syene lies on the tropic of Cancer, and Eratosthenes knew that the Sun passes directly overhead in Syene on the summer solstice. He also knew that, in the city of Alexandria to the north, the Sun came within only 7° of the zenith on the summer solstice. He therefore concluded that Alexandria must be 7° of latitude to the north of Syene (Figure S1.21). Because 7° is $\frac{7}{360}$ of a circle, he concluded that the north-south distance between Alexandria and Syene must be $\frac{7}{360}$ of the circumference of the Earth.

Eratosthenes estimated the north-south distance between Syene and Alexandria to be 5,000 stadia (the *stadium* was a Greek unit of distance). Thus, he concluded that

$$\frac{7}{360} \times \text{circumference of Earth} = 5{,}000 \text{ stadia}$$

from which he found the Earth's circumference to be about 250,000 stadia.

Today, we don't know exactly what distance a stadium meant to Eratosthenes; however, based on the actual sizes of Greek stadiums, it must have been about $\frac{1}{6}$ kilometer. Thus, Eratosthenes estimated the circumference of the Earth to be about $\frac{250{,}000}{6}$ kilometers, or about 42,000 kilometers—remarkably close to the modern value of just over 40,000 kilometers.

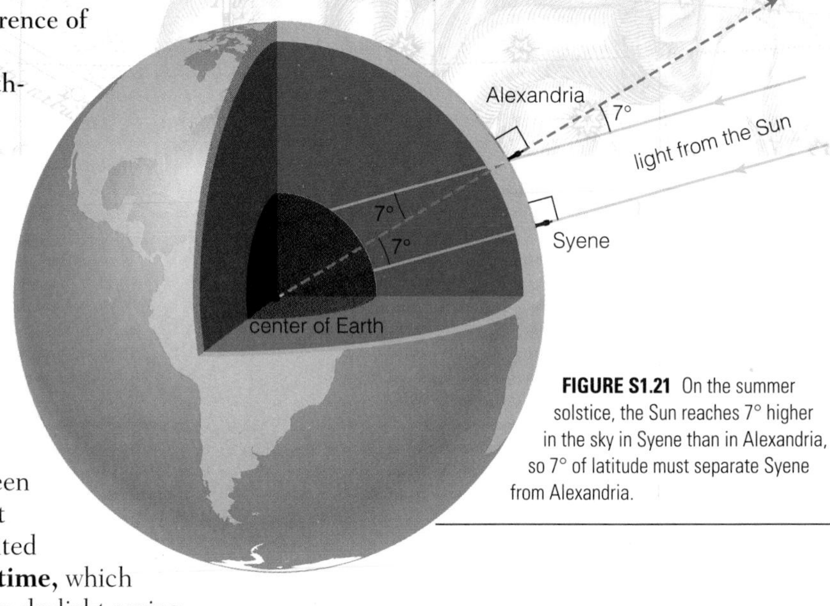

FIGURE S1.21 On the summer solstice, the Sun reaches 7° higher in the sky in Syene than in Alexandria, so 7° of latitude must separate Syene from Alexandria.

time zone, your standard time can vary by up to about a half-hour from your mean solar time.[5]

In most parts of the United States, clocks are set to standard time for only part of the year. Between the first Sunday in April and the last Sunday in October, most of the United States changes to **daylight saving time,** which is 1 hour ahead of standard time. On daylight saving time, clocks read around 1 P.M., rather than around noon, when the Sun is on the meridian.

As we will see, it is useful to have a single time for the entire Earth for purposes of navigation and astronomy. For historical reasons, this "world" time was chosen to be the mean solar time in Greenwich, England—the place that also defines longitude 0°. Today, this *Greenwich mean time* (*GMT*) is often called **universal time** (**UT**).

The Calendar

Our modern calendar is based on the length of the tropical year, which extends from the moment of one spring equinox to the next [Section 3.2]. The origins of our calendar go back to ancient Egypt. By 4200 B.C., the Egyptians were using a calendar that counted 365 days in a year, which they divided into 12 months of 30 days each, with 5 extra days at the end.

Because the tropical year actually is closer to $365\frac{1}{4}$ days, the Egyptian calendar slowly drifted out of synchronization with the seasons by about 1 day every 4 years. For example, if the spring equinox occurred on March 21 one year, 4 years later it would occur on March 22, 4 years after that on March 23,

[5]In principle, the standard time in a particular time zone is the mean solar time for the *center* of the time zone. In that case, local mean solar time always would be within $\frac{1}{2}$ hour of standard time. However, time zones often have unusual shapes in order to conform to social, economic, and political realities, so larger variations between standard and mean solar time sometimes occur.

and so on. Over many centuries, the spring equinox would occur in many different months. To keep the seasons and the calendar synchronized, in 46 B.C. Julius Caesar decreed the adoption of a new calendar. This **Julian calendar** introduced the concept of a **leap year:** Every fourth year has 366 days, rather than 365, so that the average length of the calendar year is $365\frac{1}{4}$ days. The Julian calendar also eliminated the Egyptians' 5 extra days at the end of each year by lengthening some of the months from 30 to 31 days.

The Julian calendar originally set the date of the spring equinox at March 21. If it had been perfectly synchronized with the tropical year, this calendar would have ensured that the spring equinox came on the same date every 4 years (i.e., every leap year cycle). Unfortunately, the Julian calendar didn't work perfectly because the actual length of a tropical year is about 11 minutes short of $365\frac{1}{4}$ days. As a result, the moment of the spring equinox slowly advanced by an average of 11 minutes per year. By the late 1500s, the Julian calendar had the spring equinox occurring on March 11, rather than March 21.

In 1582, Pope Gregory XIII decided to introduce a new calendar that was designed to keep the spring equinox on March 21 after every 4-year cycle. This **Gregorian calendar** is the one we still use today. It is based on some adjustments to the Julian calendar. First, because the spring equinox was occurring on March 11, Pope Gregory ordained that the day in 1582 following October 4 would be October 15.[6] By eliminating the dates between October 5 and October 14, 1582, he restored the date of the spring equinox to March 21 in 1583. Second, the Gregorian calendar uses modified rules for leap year: While leap year usually occurs every 4 years, the Gregorian calendar does not have leap year when the century changes *unless* that year is divisible by 400. For example, 1900 was *not* a leap year because it was the start of a new century and 1900 is not divisible by 400; but note that 2000 *is* a leap year because it is divisible by 400. The net result of these adjustments is to make the average length of the Gregorian calendar year almost exactly the same as the actual length of a tropical year, which ensures that the spring equinox will occur every fourth year on March 21 for thousands of years to come.

[6]The Pope's decree was not immediately accepted everywhere. For example, the Gregorian calendar was not adopted in England or the American colonies until 1752, and it was not adopted in Russia until 1919. In parts of the world that still used the Julian calendar, the dates October 5 through October 14 *did* occur in 1582, and the 10 days were removed from the calendar at some later time.

S1.5 Principles of Celestial Navigation

Imagine that you're on a ship at sea, far from any landmarks. How can you figure out where you are? It's easy, at least in principle, if you understand the apparent motions of the sky that we have discussed in this chapter.

Latitude

Determining your latitude is particularly easy if you can find the north or south celestial pole: Your latitude is equal to the altitude of the celestial pole in your sky. In the Northern Hemisphere at night, you can determine your approximate latitude by measuring the altitude of Polaris. Because Polaris has a declination within 1° of the north celestial pole, its altitude is within 1° of your latitude. For example, if Polaris has altitude 17°, your latitude is between 16°N and 18°N.

If you want to be more precise (or if you are in the Southern Hemisphere) you can determine your latitude from the altitude of *any* star as it crosses your meridian. For example, suppose Vega happens to be on your meridian at the moment, and it appears in your southern sky at an altitude of 78°44′. Because Vega has dec = +38°44′ (see Figure S1.5), it crosses your meridian 38°44′ northward of the celestial equator. As shown in Figure S1.22a, you can therefore conclude that the celestial equator is crossing your meridian at an altitude of 40°00′ in the south. Your latitude must be 50°00′N, because the

Common Misconceptions: Compass Directions

Most people determine direction with the aid of a compass rather than the stars. However, to many people's surprise, a compass needle doesn't actually point to true geographic north. Instead, the compass needle responds to the Earth's magnetic field and points to *magnetic* north, which can be substantially different from true north. As a result, if you want to navigate precisely with a compass, you need a special map that shows local variations in the Earth's magnetic field. Such maps are available at most camping stores; however, they are not perfectly reliable because the magnetic field also varies with time. In general, celestial navigation is much more reliable than a compass for determining direction.

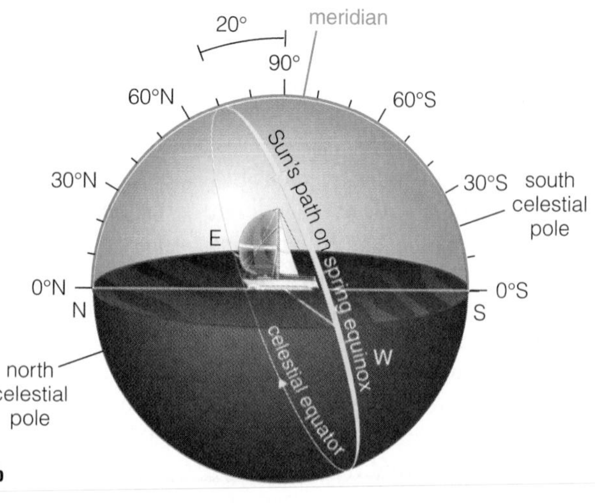

FIGURE S1.22 (**a**) Finding your latitude from the meridian-crossing altitude of Vega. (**b**) Finding your latitude from the meridian-crossing altitude of the Sun on the spring equinox.

celestial equator always crosses the meridian at an altitude of 90° minus the latitude, and it crosses in the south for latitudes in the Northern Hemisphere.

In the daytime, you can find your latitude from the Sun's altitude on your meridian, as long as you know the date. For example, suppose the date is March 21 and the Sun crosses your meridian at an altitude of 70° in the north (Figure S1.22b). Because the Sun has dec = 0° on March 21, you can conclude that the celestial equator also crosses your meridian in the *north* at an altitude of 70°. Thus, you are in the Southern Hemisphere at latitude 20°S (because 90° − 20° = 70°).

Longitude

Determining your longitude requires comparing the current positions of objects in your sky with their positions as seen from some known longitude. As a simple example, suppose you are at sea and you use a sundial to determine that the apparent solar time is 1 P.M. You immediately radio a friend in England and learn that it is 3 P.M. in Greenwich. Thus, your time is 2 hours behind the time in Greenwich, which means you are 2 hours west of Greenwich. (An earlier time means you are *west* of Greenwich because the Earth rotates from west to east.) Each hour corresponds to 15° of longitude (360° ÷ 24 = 15°), so "2 hours west of Greenwich" means longitude 30°W.

The process for finding your longitude at night is similar. For example, suppose Vega is on your meridian and a radio call to your friend reveals that it crossed the meridian 6 hours ago in Greenwich. This means you now are 6 hours east of Greenwich, or at longitude 90°E (6 × 15° = 90°).

Celestial Navigation in Practice

Although celestial navigation is easy in principle, there are several important details to consider in practice. First, measuring either latitude or longitude requires some method for precisely measuring angles in the sky. The ancient Greeks invented a device called an *astrolabe* for this purpose (Figure S1.23a); it was significantly improved by Islamic scholars during the Middle Ages. Medieval sailors often measured angles with a simple pair of calibrated perpendicular sticks called a *cross-staff* or *Jacob's staff* (Figure S1.23b). A more modern device called a *sextant* allows much more precise angle determinations by incorporating a small telescope for sightings (Figure S1.23c). Sextants are still used for celestial navigation on many ships. If you want to practice celestial navigation yourself, you can buy an inexpensive plastic sextant at many science-oriented stores.

Next, knowing the paths of different objects through the sky requires knowing their celestial coordinates. At night, you'll therefore need a table listing the celestial coordinates of bright stars. In addition, you must either know the constellations and bright stars extremely well or carry star charts to help identify them. For navigating by the Sun in the daytime, you'll need a table listing the Sun's celestial coordinates on each day of the year.

Finally, to determine your longitude, you need to know the current position of the Sun or a particular star in a known location, such as Greenwich, England. Although you could do this by calling a friend who lives there, it's more practical to carry a clock set to universal time (i.e., Greenwich mean time). In the

a

b

c

FIGURE S1.23 (**a**) An astrolabe in use (left) and close-up (right). (**b**) A cross-staff. (**c**) A sextant.

daytime, the clock makes it very easy to determine your longitude. If apparent solar time is 1 P.M. in your location and your clock tells you that the time in Greenwich is 3 P.M., then your longitude is 30°W. The task is more difficult at night, because you must compare the position of a *star* in your sky to its current position in Greenwich. Astronomical tables are available that allow you to determine the current position of any star in the Greenwich sky from the date and the universal time.

Note that, in the absence of a clock that tells you the time in Greenwich (or any particular place), there is no simple way to determine your longitude. Most of the European voyages of discovery beginning in the 1400s relied on little more than guesswork about longitude, although some sailors learned complex mathematical techniques for estimating longitude through observation of the lunar phases. More accurate longitude determination, upon which the development of extensive ocean commerce and travel depended, required the invention of a clock that would remain accurate on a ship rocking in the ocean swells. By the early 1700s, solving this problem was considered so important that the British government offered a substantial monetary prize for its solution. The prize was claimed in 1761 by John Harrison, who designed a clock that lost only 5 seconds during a 9-week voyage to Jamaica.[7]

[7]The story of the difficulties surrounding the measurement of longitude at sea and how the problem finally was solved by Harrison is chronicled in Dava Sobel, *Longitude* (Walker and Company, 1995).

THINKING ABOUT . . .

The Global Positioning System

Although celestial navigation can be very precise, it is possible only when the sky is clear. Today, a new type of celestial navigation is in wide use: a system of navigation by satellites that function like artificial stars. This **global positioning system (GPS)** involves about two dozen satellites orbiting the Earth at an altitude of 20,000 kilometers. Each satellite transmits a radio signal that can be received by small radio receivers (some small enough to hold in your hand)—rain or shine, and day or night. GPS receivers have a built-in computer that calculates your precise position on Earth by comparing the signals received from several GPS satellites.

The United States originally built GPS in the late 1970s to give its military troops an advantage over adversaries, and the system was designed so that secret information was needed to use the system to its fullest. Thus, while civilians (and adversaries) could use GPS to determine their location on Earth to within about 100 meters, U.S. military personnel could pinpoint their location to within 1 meter. However, civilian scientists eventually found ways to measure position with far better precision than the designers of GPS had imagined possible—even without the secret information. For example, GPS has been used by geologists to measure *millimeter-scale* changes in the Earth's crust. Today, the many applications of GPS include automobile navigation, helping airplanes land safely, guiding the blind around town, and helping lost hikers find their way home. More applications are being developed, and GPS will likely play an increasingly prominent role in our lives.

SOCRATES: *Shall we make astronomy the next study? What do you say?*

GLAUCON: *Certainly. A working knowledge of the seasons, months, and years is beneficial to everyone, to commanders as well as to farmers and sailors.*

SOCRATES: *You make me smile, Glaucon. You are so afraid that the public will accuse you of recommending unprofitable studies.*

PLATO, *REPUBLIC*

THE BIG PICTURE

In this chapter, we built upon concepts from Chapters 1–3 to form a clear picture of how the sky appears from different latitudes and how the path of the Sun varies with the seasons. We were thereby able to discuss the principles of our clocks and calendars and of celestial navigation. As you look back at what we've covered, keep in mind the following "big picture" ideas:

- Our modern systems of timekeeping are rooted in the apparent motions of the Sun through the sky. Although it's easy to forget these roots when you look at a clock or a calendar, remember that the sky was the only guide to time for most of human history.
- The term *celestial navigation* sounds a bit mysterious, but it involves simple principles that allow you to determine your location on the Earth. Even if you're never lost at sea, you may find the basic techniques of celestial navigation useful to orient yourself at night (e.g., on your next camping trip).
- If you understand the apparent motions of the sky discussed in this chapter and also learn the constellations and bright stars, you'll feel very much "at home" under the stars at night.

Review Questions

1. Why is the Earth placed in the center of models of the celestial sphere? Explain the relationships between the Earth's equator and the celestial equator, the Earth's poles and the north and south celestial poles, and the Earth's orbit around the Sun and the ecliptic.

2. Explain what we mean when we say that the terms *spring equinox, summer solstice, fall equinox,* and *winter solstice* refer both to *moments* in time and to *points* on the celestial sphere.
3. Describe the system of locating points on the celestial sphere by *declination* and *right ascension.* How are these *celestial coordinates* similar to latitude and longitude on Earth?
4. How do the Sun's celestial coordinates change over the course of each year? Why?
5. Why don't we show the paths of the Moon or planets on the celestial sphere?
6. Suppose you are standing at the North Pole. Where is the celestial equator in your sky? Where is the north celestial pole? Describe the daily motion of the sky.
7. Repeat question 6 for the South Pole.
8. Repeat question 6 for the Earth's equator. Explain why, from the Earth's equator, we see exactly half the daily circle of any star.
9. Repeat question 6 for latitude 40°N. Explain why, from latitude 40°N, objects with positive declinations are above the horizon for more than half the day and objects with negative declinations are above the horizon for less than half the day.
10. Describe the two-step process for drawing a picture of the daily paths of stars at any latitude. What is the general rule for finding the north or south celestial pole at any latitude? What is the general rule for finding the celestial equator?
11. Explain how, once you've determined the daily paths of stars at a particular latitude, you can determine the daily path of the Sun at different times of year.
12. Starting on the spring equinox, describe the motion of the Sun in the north polar sky over the next year.
13. Repeat question 12 for the South Pole.
14. Repeat question 12 for the Earth's equator.
15. What is special about the *tropics of Cancer* and *Capricorn?* Describe the motion of the Sun at each on the summer and winter solstices.
16. What are the *Tropics?* Why don't the Tropics experience four seasons as do other areas of the Earth?
17. What is special about the *Arctic Circle* and *Antarctic Circle?* Describe the motion of the Sun at each on the summer and winter solstices.
18. What is *apparent solar time?* Why is it different from *mean solar time?*
19. Why did it become necessary to create *time zones* in the late 1800s? Explain how *local* mean solar time is related to *standard time* and *daylight saving time.*
20. What is *universal time* (UT)? Why is it useful?
21. Describe the origins of the Julian and Gregorian calendars. Is 2000 a leap year? Is 2100? Explain.
22. Briefly describe how you can find your latitude by finding the north or south celestial pole.
23. Briefly describe how you can find your latitude from the meridian altitude of any star or of the Sun.
24. Briefly describe how you can find your longitude from the position of the Sun and a UT clock.

Discussion Questions

1. *Northern Chauvinism.* Why is the solstice in June called the *summer solstice,* when it marks winter for places like Australia, New Zealand, and South Africa? Why is the writing on maps and globes usually oriented so that the Northern Hemisphere is at the top, even though there is no up or down in space? Discuss.
2. *When Do Seasons Begin?* It is usually said that the moments of the equinoxes and solstices mark the *beginnings* of the seasons. For example, it is said that the summer solstice marks the beginning of summer. Based on the annual progression of the Sun's celestial coordinates, do you think it is accurate to say that the equinoxes and solstices mark beginnings? Why or why not?
3. *Seasonal Weather.* The summer solstice occurs in June, but the months of July and August generally have warmer temperatures in the Northern Hemisphere. Similarly, the winter solstice occurs in December, but the months of January and February generally have cooler temperatures in the Northern Hemisphere. Why do you think this is the case?

Problems

1. *Sensible Coordinates?* Consider each of the following statements. Which ones make sense, and which are nonsense? Why?
   a. Last night around 8 P.M., I saw Jupiter at an altitude of 45° in the south.
   b. The latitude of Orion's belt is about 5°.
   c. Today the Sun is at an altitude of 10° on the celestial sphere.
   d. Today the Sun is at a declination of 10° on the celestial sphere.
   e. The latitude of the Earth's South Pole is −90°.

f. Los Angeles is west of New York by about 3 hours of right ascension.

g. The summer solstice is east of the vernal equinox by 6 hours of right ascension.

2. *Fundamentals of Your Local Sky.* Answer each of the following questions for *your* latitude.
   a. Where is the north (or south) celestial pole in your sky?
   b. Describe the path of the meridian in your sky; be sure to specify its shape and at least three distinct points along it (such as the points where it meets your horizon and its highest point).
   c. Describe the path of the celestial equator in your sky; be sure to specify its shape and at least three distinct points along it (such as the points where it meets your horizon and crosses your meridian).
   d. Does the Sun ever appear at your zenith? If so, when? If not, why not?
   e. What is the range of declinations that make a star circumpolar in your sky? Explain.
   f. What is the range of declinations that you can never see in your sky? Explain.

3. *Path of the Sun in Your Sky.* Describe the path of the Sun through your sky for each of the following days.
   a. The day of the spring equinox.
   b. The day of the summer solstice.
   c. The day of the fall equinox.
   d. The day of the winter solstice.
   e. Today. (*Hint:* Estimate the right ascension and declination of the Sun for today's date by using the data in Table S1.1.)

4. *The Sun at the North Pole.* Repeat problem 3 for the sky at the North Pole.

5. *The Sun at the South Pole.* Repeat problem 3 for the sky at the South Pole.

6. *The Sun at the Equator.* Repeat problem 3 for the sky at the equator.

7. *The Sun in Australia.* Repeat problem 3 for the sky in Sydney, Australia (latitude 34°S).

8. *Lost at Sea I.* During an upcoming vacation, you decide to take a solo boat trip. While contemplating the universe, you lose track of your location. Fortunately, you have some astronomical tables and instruments, as well as a UT clock. You put together the following description of your situation:
   - It is the day of the spring equinox.
   - The Sun is on your meridian at an altitude of 75° in the south.
   - The UT clock reads 22:00.

   a. What is your latitude? How do you know?
   b. What is your longitude? How do you know?
   c. Consult a map. Based on your position, where is the nearest land? Which way should you sail to reach it?

9. *Lost at Sea II.* Repeat problem 8, but this time based on the following description of your situation:
   - It is the day of the summer solstice.
   - The Sun is on your meridian at an altitude of 67.5° in the north.
   - The UT clock reads 06:00.

10. *Lost at Sea III.* Repeat problem 8, but this time based on the following description of your situation:
    - Your local time is midnight.
    - Polaris appears at an altitude of 67° in the north.
    - The UT clock reads 01:00.

11. *Lost at Sea IV.* Repeat problem 8, but this time based on the following description of your situation:
    - Your local time is 6 A.M.
    - From the position of the Southern Cross, you estimate that the south celestial pole is at an altitude of 33° in the south.
    - The UT clock reads 11:00.

12. *Daylight Hours.* You've probably noticed that daylight hours remain long for about 2 months around the summer solstice and short for about 2 months around the winter solstice. In contrast, the length of daylight changes dramatically in the couple of months around the times of each equinox. Explain why. (*Hint:* Study Table S1.1 and Figure S1.11 to see how the Sun's declination and path through the sky change from month to month.)

13. *Changing Stars with Changing Seasons.*
    a. Suppose it is the spring equinox. What is the right ascension of stars that are on your meridian at midnight? How do you know? (*Hint:* What is the right ascension of the Sun on the spring equinox, and where is the Sun at midnight?)
    b. Repeat part (a) for the summer solstice, fall equinox, and winter solstice.
    c. Based on your answers to part (a) and part (b), explain why we see different stars in different seasons.

14. *Evening and Morning Skies.*
    a. Suppose it is the spring equinox. What zodiac constellation rises just ahead of the Sun in the morning? What zodiac constellation sets just after sunset in the evening?
    b. Repeat part (a) for the summer solstice, fall equinox, and winter solstice.
    c. Based on your answers to part (a) and part (b), explain why, in general, stars of the summer evening are also stars of the predawn winter, and vice versa.

15. *Research: International Date Line.* What is the international date line? Explain why, in principle, it corresponds to longitude 180°. Then research why, in reality, it makes a number of zigs and zags around the line of longitude 180°. Summarize your findings in one page or less.

PART II
KEY CONCEPTS FOR ASTRONOMY

4
The Science of Astronomy

TODAY WE KNOW THAT THE EARTH IS A PLANET ORBITING A RATHER ordinary star, in a galaxy of a hundred billion or more stars, in an incredibly vast universe. We know that the Earth, along with the entire cosmos, is in constant motion. We know that, on the scale of cosmic time, human civilization has existed only for the briefest moment. Yet we have acquired all this knowledge only recently in human history. How have we managed to learn these things?

It wasn't easy. Astronomy is the oldest of the sciences, with roots extending as far back as recorded history allows us to see. But while our current understanding of the universe rests on foundations laid long ago, the most impressive advances in knowledge have come in just the last few centuries.

In this chapter, we will study the development and nature of the science of astronomy. We'll explore what we mean by *science* and the *scientific method* and trace how modern astronomy arose from its roots in ancient observations.

4.1 Everyday Science

A common stereotype holds that scientists are men and women in white lab coats who somehow think differently from other people. In reality, scientific thinking is a fundamental part of human nature.

Think about how a baby behaves. By about a year of age, she notices that objects fall to the ground when she drops them. She lets go of a ball; it falls. She pushes a plate of food from her high chair; it falls too. She continues to drop all kinds of objects, and they all plummet to Earth. Through her powers of observation and trial-and-error tests, the baby has learned something about the physical world: Things fall when they are unsupported. Eventually, she becomes so certain of this fact that, to her parents' delight, she no longer needs to test it continually.

Then one day somebody gives the baby a helium balloon. She releases it; to her surprise, it rises to the ceiling! Her rudimentary understanding of physics must be revised. She now knows that the principle "all things fall" does not represent the whole truth, although it still serves her quite well in most situations. It probably will be years before she learns enough about the atmosphere, the force of gravity, and the concept of density to understand *why* the balloon rises when most other objects fall. For the moment, the baby probably will be delighted that she has learned something new and unexpected.

The baby's experience with falling objects and balloons exemplifies scientific thinking, which, in essence, is a way of learning about nature through careful observation and trial-and-error experiments. Rather than thinking differently from other people, modern scientists simply are trained to organize this everyday thinking in a way that makes it easier for them to share their discoveries with others and thereby employ the collective wisdom of many people in the search for new knowledge. This organizational scheme is what we call the *scientific method;* it is the method by which humanity has acquired its present knowledge of the universe.

TIME OUT TO THINK ***When was the last time you used trial and error to learn something? Describe a few cases where you have learned by trial and error in cooking, participating in sports, fixing something, or any other situation.***

Just as it takes years for a child to learn to communicate through language, art, or music, it took humanity a long time to develop the principles of the scientific method. In its modern form, the scientific method requires painstaking attention to detail, relentless testing of each piece of information to ensure its reliability, and a willingness to give up old beliefs that are not consistent with observed facts about the physical world. For professional scientists, the demands of the scientific method are the "hard work" part of the job. At heart, professional scientists are like the baby with the balloon, delighted by the unexpected and motivated by those rare moments when they—and all of us—learn something new about the universe.

4.2 Ancient Observations

We will discuss the modern scientific method shortly, but first we will explore how it arose from the observations of ancient peoples. Our exploration begins in central Africa, where some old traditions still live. There, people of many indigenous societies can predict the weather with reasonable accuracy by making careful observations of the Moon. The Moon begins its monthly cycle as a crescent in the western sky just after sunset. Through long traditions of sky watching, many African cultures learned that the orientation of the crescent "horns" relative to the horizon is closely tied to rainfall patterns. For example, the horns of the waxing crescent moon as seen from central Nigeria tilt to the right at the start of the annual rainy season and to the left as the rainy season ends (Figure 4.1).

Central African cultures passed knowledge from one generation to the next primarily through traditions and spoken stories, so no one knows when they first developed the ability to predict weather using lunar cycles—it might date back thousands of years. In fact, the earliest known written astronomical record comes from central Africa, near the border of modern-day Congo and Uganda. It consists of an animal bone etched with patterns that appear to represent part of a lunar calendar, probably carved around 6500 B.C.[1] If these ancient Africans already possessed lunar calendars, they may have learned to predict the weather long before the rise of written records elsewhere in the world.

Why did ancient Africans, and other people around the world, bother to make such careful and detailed observations of the sky? Biologically speaking, the human species has remained essentially unchanged for more than 10,000 years, so ancient

[1]This etched bone is known as the *Ishango bone.* There is some controversy regarding the age and interpretation of its markings, but most archaeologists believe it was carved around 6500 B.C.

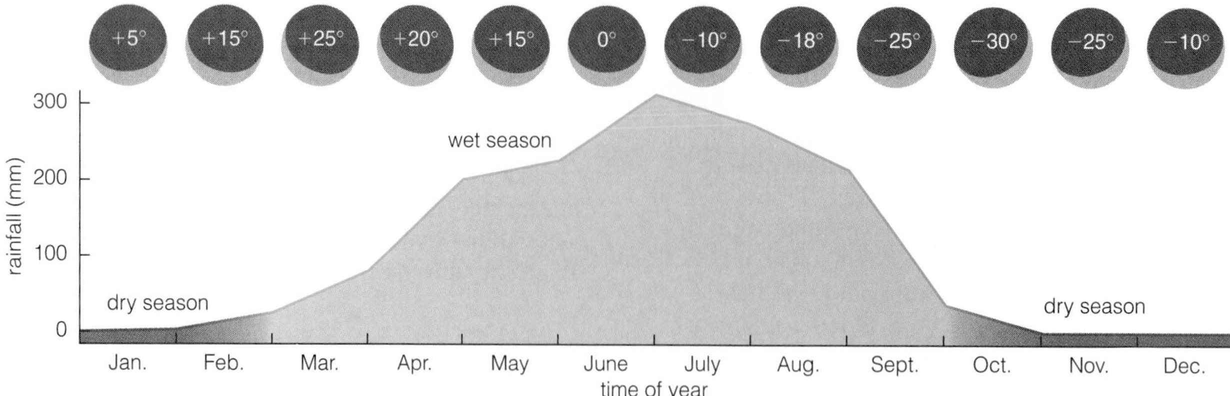

FIGURE 4.1 The graph depicts the annual rainfall pattern in central Nigeria, which is characterized by a wet season and a dry season, like most tropical latitudes. The rainfall is correlated with annual changes in the orientation of the crescent "horns" relative to the western horizon as seen from central Nigeria when the Moon is in waxing crescent phase. (The angles in the figure are measured relative to 0°, when the two horns line up parallel to the horizon.) Thus, observations of the waxing crescent moon can be used to determine the time of year and hence the expected rainfall.

peoples were just as curious and intelligent as we are today. In the daytime, they surely recognized the importance of the Sun to their lives. At night, without our modern-day luxury of retreating to lighted rooms, they were much more aware of the starry sky than we are today. Under these circumstances, it's not surprising that they paid attention to patterns of motion in the sky and developed ideas and stories to explain what they saw.

Astronomy played a practical role in ancient societies by enabling them to keep track of time and seasons. The ability of central Africans to predict the weather from the lunar cycle was a means of tracking the dry and rainy seasons, a crucial skill for people who depended on agriculture for survival. This ability to track the time and seasons may seem quaint today, when digital watches tell us both the precise time and the date, but it required considerable knowledge and skill in ancient times, when the only clocks and calendars were in the sky.

Our modern measures of time still reflect their ancient astronomical roots. Our 24-hour day is the time it takes for the Sun to circle our sky. The length of a month comes from the lunar cycle, and our calendar year is based on the cycle of the seasons. The days of the week are named after the seven naked-eye objects that appear to move among the constellations: the Sun, the Moon, and the five planets recognized in ancient times (Table 4.1).

Table 4.1 The Seven Days of the Week and the Astronomical Objects They Honor The correspondence between objects and days is easy to see in French and Spanish. In English, the correspondence becomes clear when we look at the names of the objects used by the Teutonic tribes who lived in the region of modern-day Germany.

Object	Teutonic Name	English	French	Spanish
Sun	Sun	Sunday	dimanche	domingo
Moon	Moon	Monday	lundi	lunes
Mars	Tiw	Tuesday	mardi	martes
Mercury	Woden	Wednesday	mercredi	miércoles
Jupiter	Thor	Thursday	jeudi	jueves
Venus	Fria	Friday	vendredi	viernes
Saturn	Saturn	Saturday	samedi	sábado

Determining the Time of Day

In the daytime, ancient peoples could tell time by observing the Sun's path through their sky—as long as it wasn't cloudy. They knew it was morning when the Sun was in the east, and afternoon when the Sun was in the west. For greater precision, many ancient cultures probably used sundials, which at their simplest consist of nothing more than a stick whose shadow length changes with the Sun's position in the sky (Figure 4.2) [Section S1.4].

Estimating the time on a clear night requires slightly more familiarity with the sky. The Moon's position and phase give an indication of the time [Section 3.3]. For example, a first-quarter moon sets around midnight, so it is not yet midnight if the first-quarter moon is still above the western horizon. The positions of the stars also indicate the time if you know the approximate date. For example, in December the constellation Orion rises around sunset, reaches the meridian around midnight, and sets around sunrise. Hence, if you know it is winter and you see Orion setting, you know dawn is approaching. Most ancient peoples probably were adept at estimating the time of night, although written evidence is sparse.

Our modern system of dividing the day into 24 *hours* arose in ancient Egypt some 5,000 years ago. The Egyptians divided the daylight into 12 equal parts,[2] and we still break the 24-hour day into 12 hours each of a.m. and p.m. However, the Egyptian "hours" were not a fixed amount of time because the amount of daylight varies during the year. For example, "summer hours" were longer than "winter hours," because $\frac{1}{12}$ of the daylight lasts longer in summer than in winter. Only much later in history did the hour become a fixed amount of time, subdivided into 60 equal minutes each consisting of 60 equal seconds.

FIGURE 4.2 An ancient Roman sundial.

The Egyptians also divided the night into 12 equal parts, and early Egyptians used the stars to determine the time at night. Egyptian *star clocks,* often found painted on the coffin lids of Egyptian pharaohs, essentially cataloged where particular stars appear in the sky at particular times of night and particular times of year. By knowing the date from their calendar and observing the positions of particular stars in the sky, the Egyptians could use the star clocks to estimate the time of night.

By about 1500 B.C., the Egyptians had abandoned the use of the star clocks in favor of clocks that measured time by the flow of water through an opening of a particular size, just as hourglasses measure time by the flow of sand through a narrow neck.[3] These *water clocks* had the advantage of being useful even when the sky was cloudy, and they eventually became the primary timekeeping instruments for many cultures, including the Greeks, Romans, and Chinese. The water clocks, in turn, were replaced by mechanical clocks in the late 1600s and by electronic clocks in the twentieth century. Despite the availability of other types of clocks, sundials remained in use throughout ancient times and are still popular today both for their decorative value and as reminders that the Sun and stars once were our only guides to time.

Determining the Time of Year

Knowing the approximate time of year was very important to ancient peoples who needed to know when to plant and harvest their crops, and many cultures built structures to help them mark the seasons. One of the oldest standing human-made structures served such a purpose: Stonehenge in southern England, which was constructed in stages spanning a period of some 1,200 years beginning in about 2750 B.C. (Figure 4.3). Observers standing in its center see the Sun rise directly over the "Heel Stone" only on the summer solstice. Stonehenge also served as a social gathering place and probably as a religious site; no one knows whether its original purpose was social or astronomical. Perhaps it was built for both: In ancient times, social rituals and practical astronomy probably were deeply intertwined.

One of the most spectacular structures used to mark the seasons was a building known as the Templo Mayor in the ancient Aztec city of Tenochtitlán, located on the site of modern-day Mexico City (Figure 4.4). Twin temples surmounted a flat-topped, 150-foot-high pyramid. From the location of a royal observer watching from the opposite side of the plaza, the Sun rose directly through the notch between the twin temples on the equinoxes. Like Stonehenge, the Templo Mayor served important

[2]No one knows why the Egyptians chose to divide the daylight into 12 parts. Some historians speculate that the choice was related to their division of the year, in which their calendar had 360 days divided into twelve 30-day months [Section S1.4].

[3]Hourglasses using sand were not invented until about the eighth century A.D., long after the advent of water clocks. Natural sand grains vary in size, so making accurate hourglasses required technology for manufacturing uniform grains of sand.

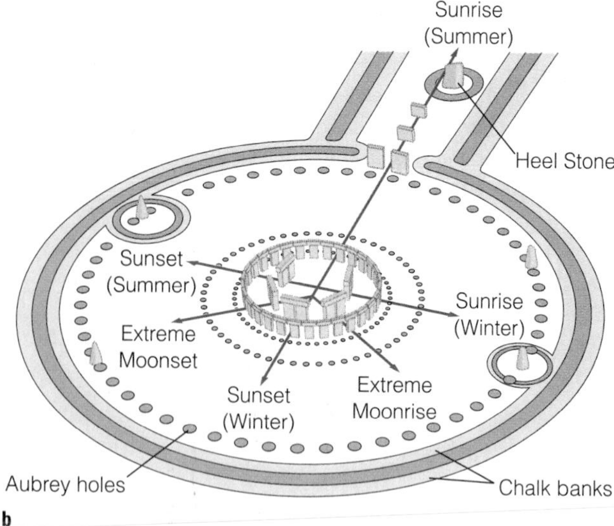

a

b

FIGURE 4.3 (**a**) Stonehenge today. (**b**) A sketch showing how archaeologists believe Stonehenge looked when construction was completed around 1550 B.C. Note, for example, that observers standing in the center would see the Sun rise directly over the Heel Stone on the summer solstice.

FIGURE 4.4 This scale model shows the Templo Mayor and the surrounding plaza as they looked before Aztec civilization was destroyed.

social and religious functions in addition to its astronomical role. Before it was destroyed by the Conquistadors, other Spanish visitors reported stories of elaborate rituals, sometimes including human sacrifice, that took place at the Templo Mayor at times determined by astronomical observations.

Many ancient cultures aligned their buildings with the cardinal directions (north, south, east, and west), enabling them to mark the rising and setting of the Sun relative to the building orientation. Some even created monuments with a single special astronomical purpose. Perhaps in a practical form of ancient art, someone among the ancient Anasazi people carved a spiral known as the Sun Dagger on a vertical cliff face near the top of a butte in Chaco Canyon, New Mexico (Figure 4.5). The Sun's rays are shaped into a dagger of sunlight by the tall stones in front of the carved spiral. The dagger of sunlight pierces the center of the spiral only once each year—at noon on the summer solstice.

FIGURE 4.5 (**a**) The Sun Dagger is an arrangement of rocks that shape the Sun's light into a dagger, with a spiral carved on a rock face behind them. The dagger pierces the center of the carved spiral each year at noon on the summer solstice. (**b**) The Sun Dagger is located on a vertical cliff face high on this butte in New Mexico.

Lunar Cycles

Many ancient civilizations paid particular attention to the lunar cycle, often using it as the basis for lunar calendars. The months on lunar calendars generally have either 29 or 30 days, chosen to make the average agree with the approximately $29\frac{1}{2}$-day lunar cycle. A 12-month lunar calendar has only 354 or 355 days, or about 11 days less than a calendar based on the tropical year. Such a calendar is still used in the Muslim religion, which is why the month-long fast of Ramadan (the ninth month) begins about 11 days earlier with each subsequent year.

Other lunar calendars take advantage of the fact that 19 years is almost precisely 235 lunar months. Thus, every 19 years, we get the same lunar phases on about the same dates. Because this fact was discovered in 432 B.C. by the Babylonian astronomer Meton, the 19-year period is called the **Metonic cycle.** A lunar calendar can be synchronized to the Metonic cycle by adding a thirteenth month to 7 of every 19 years (making exactly 235 months in each 19-year period), thereby ensuring that "new year" comes on approximately the same date every nineteenth year. The Jewish calendar still follows the Metonic cycle, adding a thirteenth month in the third, sixth, eighth, eleventh, fourteenth, seventeenth, and nineteenth years of each cycle. This also explains why the date of Easter changes each year: The New Testament ties the date of Easter to the Jewish festival of Passover, which has its date set by the Jewish lunar calendar. In a slight modification of the original scheme, most Christians today celebrate Easter on *the first Sunday after the first full moon after March 21;* if the full moon falls on Sunday, Easter is the following Sunday.

Some ancient cultures learned to predict eclipses by recognizing the 18-year saros cycle [Section 3.3]. In the Middle East, the ancient Babylonians achieved remarkable success in predicting eclipses more than 2,500 years ago and passed their knowledge along to the ancient Greeks. Perhaps the most successful predictions of eclipses prior to modern times were made by the Mayans in Central America. Unfortunately, we know little about the extent of Mayan knowledge because the Spanish Conquistadors burned most Mayan writings, but one surviving manuscript contains detailed information about eclipses. In addition, the Mayan calendar featured a sacred cycle that almost certainly was related to eclipses.[4]

[4]This Mayan cycle, called the *sacred round,* lasted 260 days––almost exactly $1\frac{1}{2}$ times the 173.32 days between successive eclipse seasons [Section 3.3].

FIGURE 4.6 The Moon rising between two stones of the 4,000-year-old sacred stone circle at Callanish, Scotland (on the Isle of Lewis in the Scottish Hebrides). The full moon rises in this position only once every 18.6 years.

FIGURE 4.7 The Mayan observatory at Chichén Itzá.

The complexity of the Moon's orbit leads to other long-term patterns in the Moon's appearance. For example, the full moon rises at its most southerly point along the eastern horizon only once every 18.6 years, a phenomenon that can be vividly observed from the 4,000-year-old sacred stone circle at Callinish, Scotland (Figure 4.6).

Observations of Planets and Stars

Many cultures also made careful observations of the planets and stars. Mayan observatories in Central America, such as the one still standing at Chichén Itzá (Figure 4.7), had windows strategically placed for observations of Venus, which held a sacred place in Mayan religion.

Because the visible stars change with the seasons, some cultures used observations of stars to determine the time of year. More than 2,000 years ago, the early Aztec people marked the beginning of their year when the group of stars called the Pleiades first rose in the east after having been hidden from view for 40 days by the glare of the Sun. On the other side of the world, southern Greeks also used the Pleiades to time their planting and harvests, as described in this excerpt from an epic poem (*Works and Days*) composed by the Greek farmer Hesiod in about 800 B.C.:

> *When you notice the daughters of Atlas, the Pleiades, rising, start on your reaping, and on your sowing when they are setting. They are hidden from your view for a period of forty full days, both night and day, but then once again, as the year*

a

b

FIGURE 4.8 (**a**) More than 800 lines, some miles long, are etched in the Nazca desert sand of Peru. (**b**) The desert also features many large figures of animals, such as the hummingbird shown here.

moves round, they reappear at the time for you to be sharpening your sickle.

Observational aids could also mark the rising and setting of the stars. More than 800 lines, some stretching for miles, are etched in the dry desert sand of Peru between the Ingenio and Nazca rivers (Figure 4.8). Many of these lines may simply have been well-traveled pathways, but others are aligned in directions that point to places where bright stars rose,[5] or where the Sun rises at particular times of year. In addition to the many straight lines, the desert also features many large figures of animals, which may be representations of constellations made by the Incas who lived in the region.

TIME OUT TO THINK ***The striking animal figures in the Peruvian desert do not show up clearly unless seen from above. As a result, some UFO enthusiasts argue that the patterns could not have been made by the people who lived there but must have been created by aliens. What do you think of this argument? Defend your opinion.***

Although the astronomical roots of the Nazca lines are debatable, the great Incan cities of South America clearly were built with astronomy in mind. Entire cities appear to have been designed so that particular arrangements of roads, buildings, or other human-made structures would point to places where bright stars rise or set, or where the Sun rises and sets at particular times of year.[6]

Structures for astronomical observation also were popular in North America. Lodges built by the Pawnee people in Kansas featured strategically placed holes for observing the passage of constellations that figured prominently in their folklore. In the northern plains of the United States, Native American "Medicine Wheels" probably were designed for astronomical observations. Like many other structures in the western United States, the "spokes" of the Medicine Wheel at Big Horn, Wyoming, were aligned with the rising and setting of bright stars when it was built, as well as with the rising and setting of the Sun on the equinoxes and solstices (Figure 4.9). The 28 spokes of the Medicine Wheel probably relate to the month of the Native Americans, which they measured as 28 days (rather than 29 or 30) because they did not count the day of the new moon.

Perhaps the people most dependent on knowledge of the stars were the Polynesians, who lived and traveled among the many islands of the mid- and South Pacific. Because the next island in a journey usually was too distant to be seen, poor navigation meant becoming lost at sea. As a result, the most esteemed position in Polynesian culture was that of the Navigator, a person who had acquired the detailed

[5]Because of precession, stars rise and set in different places today than they did hundreds or thousands of years ago.

[6]A detailed discussion of the layout of Incan cities can be found in the excellent book *Ancient Astronomers,* by Anthony F. Aveni (Smithsonian Books, St. Remy Press and Smithsonian Institution, 1993). Aveni's book also describes in greater detail many of the other observational techniques of ancient peoples that are mentioned in this chapter.

FIGURE 4.9 Aerial and ground views of the Big Horn Medicine Wheel in Wyoming. Note the 28 "spokes" radiating out from the center. These spokes probably relate to the month of the Native Americans.

FIGURE 4.10 A traditional Polynesian navigational instrument.

knowledge necessary to successfully navigate great distances among the islands. Although modern knowledge of the techniques of the Navigators is limited, we do know that they employed a combination of detailed knowledge of astronomy with equally impressive knowledge of the patterns of waves and swells around different islands (Figure 4.10). The stars provided their broad navigational sense, pointing them in the correct direction of their intended destination. As they neared a destination, the wave and swell patterns would guide them to their precise landing point. The Navigator memorized all his skills and passed them from generation to generation through a well-developed program for training future Navigators. Unfortunately, with the advent of modern navigational technology, many of the skills of the Navigators have been lost.

Mathematical Insight 4.1 The Metonic Cycle

It is easy to verify that the 19-year Metonic cycle is almost precisely 235 lunar months. A year is about 365.25 days, so 19 years is equivalent to:

$$19 \text{ yr} \times \frac{365.25 \text{ days}}{1 \text{ yr}} = 6{,}939.75 \text{ days}$$

The average length of a lunar month is 29.53 days, so 235 lunar months is:

$$235 \text{ months} \times \frac{29.53 \text{ days}}{1 \text{ month}} = 6{,}939.55 \text{ days}$$

Thus, the difference between 19 years and 235 lunar months is only 0.2 day, or about 5 hours.

The dates of lunar phases repeat with each Metonic cycle. For example, there was a new moon on October 1, 1998, and we will have a new moon on October 1, 2017. (The dates may be off by 1 day on the 19-year cycle because of the 5-hour "error" in the Metonic cycle and also because 19 years on our calendar may be either 6,939 days or 6,940 days, depending on the leap-year cycle.)

FIGURE 4.11 Instruments at the ancient astronomical observatory in Beijing.

From Observation to Science

Before a structure such as Stonehenge could be built, careful observations had to be made and repeated over and over to ensure their validity. Careful, repeatable observations also underlie the modern scientific method. To this extent, elements of modern science were present in many early human cultures.

The degree to which scientific ideas developed in different societies depended on practical needs, social and political customs, and interactions with other cultures. Because all of these factors can change, it should not be surprising that different cultures were more scientifically or technologically advanced than others at different times in history. The ancient Chinese kept remarkably detailed records of astronomical observations beginning at least 5,000 years ago, and these records enabled them to discover many patterns of sky motion hundreds of years before their counterparts in other parts of the world. The Chinese also developed advanced instrumentation to aid their observations, which may have spurred their general technological development.

Yet, by the end of the 1600s, Chinese science and technology had fallen behind that of Europe. Some historians argue that a primary reason for this decline was that the Chinese tended to regard their science and technology as state secrets, a fact suggested by the fortresslike appearance of the ancient astronomical observatory in Beijing (Figure 4.11). The ruling class insisted on collecting and keeping secret the voluminous written records that helped the Chinese achieve their many historical firsts in science and technology. But this same secrecy may eventually have slowed Chinese

---

Twelve lunar months is less than 365 days:

$$12 \text{ months} \times \frac{29.53 \text{ days}}{1 \text{ month}} = 354.36 \text{ days}$$

Thus, the new year on a 12-month lunar calendar advances by about 11 days every year relative to a 365-day solar calendar. Thus, if a 12-month lunar calendar has its new year on September 30 one year, the next new year will be around September 19, the following one around September 8, and so on.

To keep the date of new year from moving too much relative to a solar calendar, a lunar calendar must have 235 months in any 19-year period. But 19 years of 12 months each is only $19 \times 12 = 228$ months. Thus, a lunar calendar needs an extra 7 months over the 19-year period to make the total 235 months. One way to accomplish this is to have 7 years with a thirteenth lunar month in every 19-year cycle.

scientific development by preventing the broad-based collaborative science that fueled the European advance beginning in the Renaissance.

In Central America, the ancient Mayans also were ahead of their time in many ways. For example, their system of numbers and mathematics looks distinctly modern; their invention of the concept of zero preceded by some 500 years its introduction in the Eurasian world (by Hindu mathematicians, around A.D. 600). The Aztecs, Incas, Anasazi, and other ancient peoples of the Americas may also have been quite advanced in many other areas, but few written records survive to tell the tale.

It appears that virtually all cultures employed scientific thinking to varying degrees. Had the circumstances of history been different, any one of these many cultures might have been the first to develop what we consider to be modern science. But, in the end, history takes only one of countless possible paths, and the path that led to modern science emerged from the ancient civilizations of the Mediterranean and the Middle East.

4.3 The Modern Lineage

By 3000 B.C., civilization was well established in two major regions of the Middle East: Egypt and Mesopotamia[7] (Figure 4.12). Their geographical location placed these civilizations at a crossroads for travelers, merchants, and armies of Europe, Asia, and Africa. This mixing of cultures fostered creativity, and the broad interactions among peoples ensured that new ideas spread throughout the region. Over the next 2,500 years, numerous great cultures arose. For example, the ancient Egyptians built the Great Pyramids between 2700 and 2100 B.C., using their astronomical knowledge to orient the Pyramids with the cardinal directions. They also invented papyrus scrolls and ink-based writing. The Babylonians invented methods of writing on clay tablets and developed arithmetic to serve in commerce and later in astronomical calculations. Many more of our modern principles of commerce, law, religion, and science originated with the cultures of Egypt and Mesopotamia.

The development of principles of modern science accelerated with the rise of Greece as a power in the Middle East, beginning around 500 B.C. In 330 B.C., Alexander the Great led the expansion of the Greek empire throughout the Middle East, absorbing all the former empires of Egypt and Mesopotamia. Alexander had a keen interest in science and education, perhaps fueled by his association with Aristotle [Section 6.3], who was his personal tutor. He encouraged the pursuit of knowledge and respect for foreign cultures. On the delta of the Nile in Egypt, he founded the city of Alexandria, which he hoped would become a center of world culture.

[7]Mesopotamia was the ancient Greek name for the region between the Tigris and Euphrates rivers, but modern historians use the name for the entire region today occupied by Iraq. Three major cultures thrived at various times in ancient Mesopotamia: Sumer, Babylonia, and Assyria.

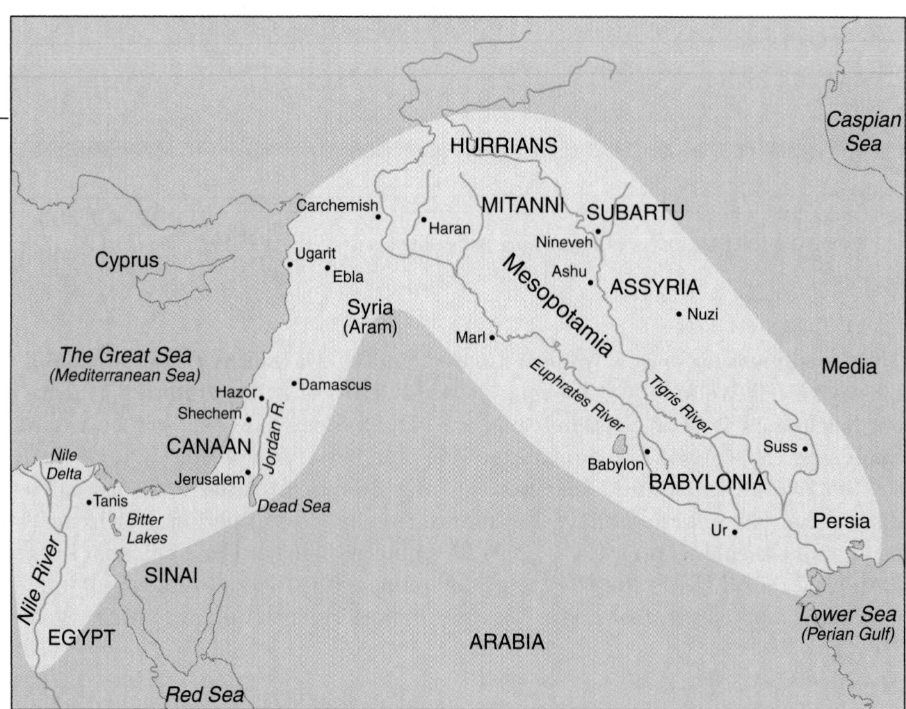

FIGURE 4.12 The Middle East in ancient times.

a

b

FIGURE 4.13 These renderings show an artist's reconstruction, based on scholarly research, of how (**a**) the Great Hall and (**b**) a scroll room of the Library of Alexandria might have looked.

Although Alexander died at the age of 35, his dream came true. His successors continued to build Alexandria, including the construction of a great library and research center around 300 B.C. (Figure 4.13). The Library of Alexandria was the world's preeminent center of research for the next 700 years. Its end is associated with the death of Hypatia, perhaps the most prominent female scholar of the ancient world. Hypatia was a resident scholar of the Library of Alexandria, director of the observatory in Alexandria, and one of the leading mathematicians and astronomers of her time. Unfortunately, she lived during a time of rising sentiment against free inquiry and was murdered by anti-intellectual mobs in A.D. 415. The final destruction of the Library of Alexandria followed not long after her death.

At its peak, the Library of Alexandria held more than a half million books, hand-written on papyrus scrolls. As the invention of the printing press was far in the future, most of the scrolls probably were original manuscripts or the single copies of original manuscripts. When the library was destroyed, most of its storehouse of ancient wisdom was lost forever.

TIME OUT TO THINK *How old is the school that you attend? Compare its age to the 700 years that the Library of Alexandria survived. Estimate the number of books you've read in your life. Compare this number to the half million books, most lost forever, that once were housed in the Library of Alexandria.*

The Ptolemaic Model of the Universe

Perhaps the most important scientific idea developed by the ancient Greeks was that of creating **models** of nature. Just as a model airplane is a representation of a real airplane, a model of nature seeks to represent some aspect of nature. The model can then be used to explain and predict real phenomena, without any need to invoke myth, magic, or the supernatural. Scientific models usually are conceptual models rather than miniature representations. The ancient Greek idea of a celestial sphere is an example of a scientific model: The celestial sphere is the model, and it can be used to explain and predict the apparent motions of stars in our sky.

By combining the concept of modeling with advances in logic and mathematics, the ancient Greeks practiced a pursuit of knowledge very much like modern science. They used their models to make predictions, such as predicting where the planets should appear among the constellations of the night sky. If the predictions proved inaccurate, they changed and refined their models. In this way, the Greeks soon learned that the simple idea of a celestial sphere could not explain all the phenomena they saw in the sky (particularly the apparent retrograde motion of the planets). They therefore added additional spheres for the Sun, the Moon, and each of the planets and then continued to change their ideas of how the spheres moved in an ongoing attempt to make the model agree with reality (Figure 4.14).

The culmination of this ancient modeling was the work of Claudius Ptolemy (c. A.D. 100–170), pronounced *tol-e-mee*. At the Library of Alexandria, Ptolemy synthesized and further developed many earlier Greek models of the universe. His ultimate model preserved the Greek idea of an Earth-centered universe, yet it was sufficiently accurate in its predictions to remain in use for the next 1,500 years.

Ptolemy's model is a tribute to human ingenuity in that it made reasonably accurate predictions despite being built on the flawed premise of an Earth-centered universe. Following a Greek tradition dating back to Plato (428–348 B.C.), Ptolemy imagined that

FIGURE 4.14 The heavenly spheres, an ancient Greek model of the universe (c. 200 B.C.).

all heavenly motions must be in perfect circles. To explain the apparent retrograde motion of planets, Ptolemy used an idea first suggested by Apollonius (c. 240–190 B.C.) and further developed by Hipparchus (c. 190–120 B.C.). This idea held that planets moved along small circles that, in turn, moved around a larger circle (Figure 4.15). As seen from Earth, this "circle upon circle" motion meant that a planet usually moved eastward relative to the stars but sometimes moved westward in apparent retrograde motion.

Ptolemy selected the sizes of the circles and the rates of motion of the planets along the circles so that his model reproduced the apparent retrograde motion of all the planets. He also incorporated more complex ideas into his model, such as allowing the Earth to be slightly off-center from the circles of a particular planet, so that it could forecast future planetary positions to within a few degrees—which was considered quite accurate at that time.

The Islamic Role

Most of the great scholarly work of the ancient Greeks was lost with the fall of the Roman Empire and the destruction of the Library of Alexandria in the fifth century A.D. Much more would have been lost had it not been for the rise of a new center of intellectual achievement in Baghdad (Iraq). While European civilization fell into the period of intellectual decline known as the Dark Ages, scholars of the new religion of Islam sought knowledge of mathematics and astronomy in hopes of better understanding the wisdom of Allah. During the eighth and ninth centuries, the Muslim empire centered in Baghdad translated and thereby saved many of the ancient Greek works.

Around A.D. 800, an Islamic leader named Al-Mamun (A.D. 786–833) established a "House of Wisdom" in Baghdad comparable in scope to the destroyed Library of Alexandria. Founded in a spirit of great openness and tolerance, the House of Wisdom employed Jews, Christians, and Muslims, all working together in scholarly pursuits. Using the translated Greek scientific manuscripts as building blocks, these scholars of the Islamic world developed the mathematics of algebra and many new instruments and techniques for astronomical observation and calculation. Most of the official names of constellations and stars come from Arabic because of the work of the scholars at Baghdad. If you look at a star chart, you will see that the names of many bright stars begin with *al* (e.g., Aldebaran, Algol), which simply means "the" in Arabic. Ptolemy's 13-volume

FIGURE 4.15 Ptolemy's model of the universe.

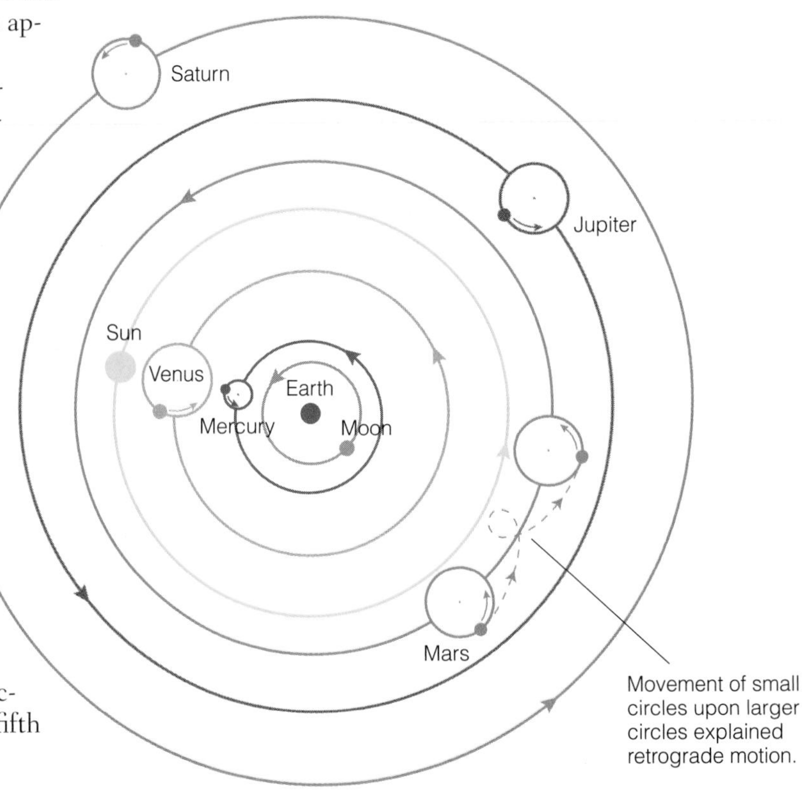

manuscript describing his astronomical system is also known by its Arabic name, *Almagest,* from Arabic words meaning "the greatest compilation."

The Islamic world of the Middle Ages had frequent contact with Hindu scholars from India, who in turn brought knowledge of ideas and discoveries from China. Hence, the intellectual center in Baghdad achieved a synthesis of the surviving work of the ancient Greeks with that of the Indians and Chinese. The accumulated knowledge of the Arabs spread throughout the Byzantine Empire (the eastern part of the former Roman Empire) during the Middle Ages. When the Byzantine capital of Constantinople (modern-day Istanbul) fell to the Turks in 1453, many Eastern scholars headed west to Europe, bringing with them the knowledge that helped ignite the European Renaissance.

4.4 Modern Science and the Scientific Method

As we have seen, many different cultures independently developed scientific modes of thinking, and the ancient Greeks developed principles of modeling very similar to those underlying the modern scientific method. But there are important differences between these ancient ideas and modern science. For example, Ptolemy's writings suggest that he designed his complex, Earth-centered model of the universe to aid in calculations, with little concern over whether it reflected reality. While modern scientists also employ purely computational models when necessary, understanding the underlying reality of nature is an important goal of modern science.

The principles that govern modern science coalesced during the European Renaissance, which was marked by a spirit of open inquiry. Within 100 years after the fall of Constantinople, Copernicus began the process of overturning the Earth-centered, Ptolemaic model, setting into motion a burst of scientific discovery that continues to this day. Before we can fully appreciate the Copernican revolution, which we'll discuss in Chapter 6, we must investigate a more basic question: What is modern science?

The word **science** comes from the Latin *scientia,* meaning "knowledge" or "to know"; it is also the xroot of the words *conscience* (related to the idea of self-knowledge or self-awareness) and *omniscience* (the quality of being all-knowing). Today, science connotes a special kind of knowledge: the kind that makes predictions that can be confirmed by rigorous observations and experiments. In brief, the modern **scientific method** is an organized approach to explaining observed facts with a model of nature, subject to the constraint that any proposed model must be testable and the provision that the model must be modified or discarded if it fails these tests.

In its most idealized form, the scientific method begins with a set of observed facts. A fact is supposed to be a statement that is objectively true. For example, we consider it a fact that the Sun rises each morning, that the planet Mars appeared in a particular place in our sky last night, and that the Earth rotates. Facts are not always obvious, as illustrated by the case of the Earth's rotation—for most of human history, the Earth was assumed to be stationary at the center of the universe. In addition, our interpretations of facts often are based on beliefs about the world that others might not share. For example, when we say that the Sun rises each morning, we assume that it is the *same* Sun day after day—an idea that might not have been accepted by ancient Egyptians, whose mythology held that the Sun died with every sunset and was reborn with every sunrise. Nevertheless, facts are the raw material that scientific models seek to explain, so it is important that scientists agree on the facts. In the context of science, a fact must therefore be something that anyone can verify for himself or herself, at least in principle.

Once the facts have been collected, a model can be proposed to explain them. A useful model must also make predictions that can be tested through experiments or further observations. Ptolemy's model of the universe was useful because it predicted future locations of the Sun, Moon, and planets in the sky. However, although the Ptolemaic model remained in use for nearly 1,500 years, it eventually became clear that its predictions could not be made to match actual observations—a key reason why the Earth-centered model of the universe finally was discarded.

In summary, the idealized scientific method proceeds as follows:

- *Observation:* The scientific method begins with the collection of a set of observed facts.
- *Hypothesis:* A model is proposed to explain the observed facts and to make new predictions. A proposed model is often called a **hypothesis,** which essentially means an *educated guess.*
- *Experiment:* The model's predictions are tested through experiments or further observations. When a prediction is verified, we gain confidence that the model truly represents nature. When a prediction fails, we recognize that the model is flawed, and we therefore must refine or discard the model.

THINKING ABOUT . . .

Logic and the Scientific Method

In science, we attempt to learn new knowledge through logical *reasoning*. The process of reasoning is carried out by constructing *arguments* in which we begin with a set of *premises* and try to draw appropriate *conclusions*. Note that this concept of a logical argument differs somewhat from the definition of argument in everyday life; in particular, a logical argument need not imply any animosity or dissension. There are two basic types of argument: *deductive* and *inductive*. Both are important in the scientific method.

The following is a simple example of a *deductive argument:*

PREMISE: All planets orbit the Sun in ellipses with the Sun at one focus.
PREMISE: Earth is a planet.
CONCLUSION: Earth orbits the Sun in an ellipse with the Sun at one focus.

Note that, as long as the two premises are true, the conclusion *must* also be true; this is the essence of deduction. In addition, note that the first premise is a general statement that applies to all planets, whereas the conclusion is a specific statement that applies only to Earth. As this example suggests, deduction is often used to *deduce* a specific prediction from a more general theory. If the specific prediction proves to be false, then we know there must be something wrong with the theoretical premises from which it was deduced. If it proves true, then we've acquired a piece of evidence in support of the theory.

Contrast the deductive argument above with the following example of an *inductive argument:*

PREMISE: Birds fly up into the air but eventually come back down.
PREMISE: People or animals that jump into the air fall back down.
PREMISE: Rocks thrown into the air come back down.
PREMISE: Balls thrown into the air come back down.
CONCLUSION: What goes up must come down.

Imagine that you lived in ancient Greece, when the Earth was thought to be a realm distinct from the sky. Because each premise supports the conclusion, you might agree that this is a strong inductive argument and that its conclusion probably is true. However, no matter how many more examples you consider of objects that go up and come down, you could never *prove* that the conclusion is true—only that it seems *likely* to be true. A single counterexample could prove the conclusion to be false—for example, the fact that some rockets are launched into space and never return to Earth.

Inductive arguments essentially are designed to *generalize* from specific facts to a broader model or theory. That is why theories can never be proved true beyond all doubt—they can only be shown to be consistent with ever larger bodies of evidence. Theories can be proved false, however, if they fail to account for observed or experimental facts.

- *Theory:* A model must be continually challenged with new observations and experiments by many different scientists. A model achieves the status of a **scientific theory** only after a broad range of its predictions have been repeatedly verified. Note that, while we can have great confidence that a scientific theory truly represents nature, we can never prove a theory to be true *beyond all doubt*. Therefore, even well-established theories must be subject to continuing challenges through further observations and experiments.

In reality, scientific discoveries rarely are made by a process as mechanical as the idealized scientific method described above. For example, Johannes Kepler, who discovered the laws of planetary motion in the early 1600s, tested his model against observations that had previously been made, rather than verifying new predictions of his model [Section 6.3]. Moreover, like most scientific work, Kepler's work involved intuition, collaboration with others, moments of insight, and luck. Nevertheless, with hindsight we can look back at Kepler's theory and see that other scientists eventually made plenty of observations to verify the planetary positions predicted by his model. In that sense, the scientific method represents an ideal prescription for judging objectively whether a proposed model of nature is close to the truth.

TIME OUT TO THINK *Look up* theory *in a dictionary. How does its definition in ordinary English differ from that given here for a* scientific theory? *If you hear someone talking about a "theory," how can you decide whether it has been well tested?*

Science, Nonscience, and Pseudoscience

People often seek knowledge in many ways that do not follow the basic tenets of the scientific method and therefore are not science. When we leave the realm of science, claims of knowledge become far more subjective and difficult to verify. Nevertheless, it can be useful to categorize such claims of knowledge as either *pseudoscience* or *nonscience*.

By **pseudoscience** we mean attempts to search for knowledge in ways that may at first seem scientific but do not really adhere to the testing and verification requirements of the scientific method; the prefix *pseudo* means "false." For example, at the beginning of each year you can find tabloid newspapers offering predictions made by people who claim to be able to "see" the future. Because they make specific predictions, we can test their claims by checking whether their predictions come true, and numerous studies have shown that their predictions come true no more often than would be expected by pure chance. But these seers seem unconcerned with such studies. The fact that they make testable claims but then ignore the results of the tests marks their claimed ability to see the future as pseudoscience.

By **nonscience** we mean knowledge sought through pure intuition, societal traditions, ancient scriptures, and other processes that make no claim to adhere to the scientific method. In nonscience, experimental testing is irrelevant because nonscientific beliefs are based on things such as faith, political conviction, and tradition. Because nonscience makes no pretense of following the scientific method, science can say nothing about the validity of nonscience.

TIME OUT TO THINK *Can the scientific method be applied in any way to questions of faith, prayer, emotion, or values? Defend your view.*

Is Science Objective?

The boundaries between science, nonscience, and pseudoscience are sometimes blurry. In particular, because science is practiced by human beings, individual scientists carry their personal biases and beliefs with them in their scientific work. These biases can influence how a scientist proposes or tests a model, and in some cases scientists have been known to cheat—either deliberately or subconsciously—to obtain the results they desire. For example, in the late nineteenth and early twentieth centuries, some astronomers claimed to see networks of "canals" in their blurry telescopic images of Mars, and they hypothesized that Mars was home to a dying civilization that used the canals to transport water to thirsty cities [Section 9.5]. Because no such canals actually exist, these astronomers apparently were allowing their beliefs to influence how they interpreted blurry images. It was, in essence, a form of cheating—even though it certainly was not intentional.

Sometimes, bias can show up even in the thinking of the scientific community as a whole. Thus, some valid ideas may not be considered by any scientist because the ideas fall too far outside the general patterns of thought, or the **paradigm,** of the time. One such example is the development of Einstein's theory of relativity. Many other scientists had gleaned hints of this theory in the decades before Einstein but did not investigate them, at least in part because they seemed too outlandish.

The beauty of the scientific method is that it encourages continued testing by many people. Thus, even if personal biases affect some results, tests by others eventually will discover the mistakes. Similarly, even when a new idea falls outside the accepted paradigm, sufficient testing and verification of the idea will eventually force a change in the paradigm. Thus, although individual scientists rarely follow the scientific method in its idealized form, the collective action of many scientists over many years generally *does* follow the basic tenets of the scientific method. That is why, despite the biases of individual scientists, science *as a whole* is objective.

4.5 Astronomy Today

The modern scientific method has fueled a dramatic rise in human knowledge of the universe. It took humanity some 6,000 years to progress from the early lunar astronomy recorded in Africa to the ancient Greek idea of creating models of the universe, and another 2,000 years to finally recognize that the Earth orbits the Sun. In contrast, the subsequent 400 years have seen discovery upon discovery, and we have come to know that the universe is filled with wonders far beyond the wildest imagination of our ancestors.

Today, the science of astronomy continues to develop at an ever-growing rate. The advent of new telescopic technologies, along with the positioning of many telescopes in space, is fueling a new renaissance in astronomy. In the 1990s alone, we have made the first discoveries of planets around other stars, witnessed the birth of galaxies near the edge of the observable universe, and accumulated strong evidence for the existence of black holes. If we measure information as a computer does, in *bytes*—where 1 byte represents the information content of one typed character and 1 megabyte represents the information in a 500-page book—then the information recorded at a single major observatory in a single night exceeds the total amount of astronomical information recorded by all ancient peoples combined.

The abundance of new data is disseminated worldwide via computer networks, and new data are collected at international observatories in many different countries and in space. The result is that, after a few centuries during which advances in astronomy were made primarily in Europe and the United States, astronomy today is a worldwide endeavor—just as it was for most of human history. Modern astronomy is studied in nearly identical form in schools everywhere in the world. The International Astronomical Union has thousands of members, from nearly every country in the world, engaged in active astronomical research. Important astronomical discoveries are being made by scientists of every race and culture. Just as the ancient civilizations of Egypt and Mesopotamia benefited from the cross-fertilization of ideas brought by people of many different cultures, astronomy today is benefiting from the great variety of perspectives brought by scientists from all over the world.

We might imagine the scientific method to be the trunk of a great tree of knowledge. If we follow the roots back in time, we see tendrils reaching out to ancient cultures throughout Africa, Asia, and Europe. If we look upward, we see the tree branching into the international endeavor of science today, with new shoots sprouting in nearly every country on Earth, promising further growth in human knowledge of the universe.

4.6 Astrology

Although the terms *astrology* and *astronomy* sound very similar, today they describe very different practices. In ancient times, however, astrology and astronomy often went hand in hand, and astrology played an important role in the historical development of astronomy.

In brief, the basic tenet of astrology is that human events are influenced by the apparent positions of the Sun, Moon, and planets among the stars in our sky. The origins of this idea are easy to understand. After all, there is no doubt that the position of the Sun in the sky influences our lives—it determines the seasons and hence the times of planting and harvesting, of warmth and cold, and of daylight and darkness. Similarly, the Moon determines the tides, and the cycle of lunar phases coincides with many biological cycles. Because the planets also appear to move among the stars, it seemed reasonable to imagine that planets also influence our lives, even if these influences were much more difficult to discover.

Ancient astrologers hoped that they might learn *how* the positions of the Sun, Moon, and planets influence our lives. They charted the skies, seeking correlations with events on Earth. For example, if an earthquake occurred when Saturn was entering the constellation Leo, might Saturn's position have been the cause of the earthquake? If the king became ill when Mars appeared in the constellation Gemini and the first-quarter moon appeared in Scorpio, might it mean another tragedy for the king when this particular alignment of the Moon and Mars next recurred? Surely, the ancient astrologers thought, the patterns of influence eventually would become clear. Thus, the astrologers hoped that they might someday learn to forecast human events with the same reliability with which astronomical observations of the Sun could forecast the coming of spring.

Astrology's Role in Astronomical History

Because forecasts of the seasons and forecasts of human events were imagined to be closely related, astrologers and astronomers usually were one and the same in the ancient world. For example, Ptolemy published a treatise on astrology called *Tetrabiblios* in addition to his books on astronomy. This work remains the foundation for much of astrology today. Interestingly, Ptolemy himself recognized that astrology stood on a far shakier foundation than astronomy. In the introduction to *Tetrabiblios,* Ptolemy compared astronomical and astrological predictions:

> *[Astronomy], which is first both in order and effectiveness, is that whereby we apprehend the aspects of the movements of sun, moon, and stars in relation to each other and to the earth. . . . I shall now give an account of the second and less sufficient method [of prediction (astrology)] in a proper philosophical way, so that one whose aim is the truth might never compare its perceptions with the sureness of the first, unvarying science. . . .*

Other ancient scientists probably likewise recognized that their astrological predictions were far less reliable than their astronomical ones. Nevertheless, if there was even a slight possibility that astrologers could forecast the future, no king or political leader would dare to be without one. Astrologers therefore held esteemed positions as political advisers in the ancient world and were provided with the resources they needed to continue charting the heavens and history. Much of the development of ancient astronomy was made possible through the support of astrology by wealthy political leaders.

Throughout the Middle Ages and into the Renaissance, many astronomers continued to practice astrology. For example, Kepler cast numerous **horoscopes**—the predictive charts of astrology—even as he was discovering the laws of planetary motion. However, given Kepler's later description of astrology as "the foolish stepdaughter of astronomy" and "a dreadful superstition," he may have cast the horoscopes solely as a source of much-needed income. Galileo is famed among modern-day astrologers for having cast a horoscope for the Grand Duke of Tuscany around the same time that he began his observations with the telescope. However, his horoscope predicted a long and fruitful life for the duke, who died just a few weeks later.

The scientific triumph of Kepler and Galileo in showing the Earth to be a planet orbiting the Sun heralded the end of the long linkage between astronomy and astrology. Nevertheless, astrology remains popular today around the world. Many more people earn income by casting horoscopes than by doing scientific astronomical research, and books and articles on astrology often outsell all but the most popular books on astronomy.

Scientific Tests of Astrology

Today, different astrologers follow different practices, which makes it difficult even to define astrology. Some astrologers no longer claim any ability to make testable predictions and therefore are practicing a form of *nonscience*—which modern science can say nothing about. However, for most astrologers, the business of astrology is casting horoscopes. Horoscopes are often cast for individuals and either predict future events in the person's life or describe characteristics of the person's personality and life. If the horoscope predicts future events, it can be evaluated by whether the predictions come true. If it describes the person's personality and life, we can check whether the description is accurate.

A *scientific* test of astrology requires evaluating many horoscopes and comparing their accuracy to what would be expected by pure chance. For example, suppose a horoscope states that a person's mother and father are separated. Because that is true of roughly half the population in the United States, an astrologer who casts 100 such horoscopes would be expected by pure chance to get it right around 50 times. Thus, we would be impressed with the predictive ability of the astrologer only if he or she were right much more often than 50 times out of 100. In hundreds of scientific tests, astrological predictions have never proved accurate by a substantially greater margin than expected from pure chance.[8] Similarly, in tests in which astrologers are asked to cast horoscopes for people they have never met, the horoscopes fail to correctly match personalities any more often than expected by chance. The verdict is clear: The methods of astrology are useless for predicting the past, the present, or the future.

What about newspaper horoscopes, which often appear to ring true? If you read them carefully, you will find that these horoscopes generally make predictions that are so vague as to be unverifiable. For example, a prediction that "it is a good day to spend time with your friends" doesn't offer much for testing. Indeed, if you read the horoscopes for all 12 astrological signs, you'll probably find that several of them apply equally well to you.

TIME OUT TO THINK ***Look in a local newspaper for today's weather forecast and for your horoscope. Contrast the nature of their predictions. By the end of the day, you will know if the weather forecast was accurate. Will you have any way of knowing whether your horoscope was accurate? Explain.***

Does It Make Sense?

In science, observations and experiments are the ultimate judge of any idea. No matter how outlandish an idea might appear, it cannot be dismissed if it successfully meets observational or experimental tests. The idea that the Earth rotates and orbits the Sun at one time seemed outlandish, yet today it is so strongly supported by the evidence that we consider it a fact. The idea that the positions of the Sun, Moon, and planets among the stars influence our lives might sound outlandish today, but if astrology were to make predictions that came true, adherence to the scientific method would force us to take it seriously. However, given that scientific tests of astrology have never found any evidence that its predictive methods work, it is worth looking back at its premises to see whether they make sense. Might there be a few kernels of wisdom buried within the lore of astrology?

Let's begin with one of the key premises of astrology: There is special meaning in the patterns of the

---

[8]An excellent summary of scientific tests of astrology can be found in Roger B. Culver and Philip A. Ianna, *Astrology: True or False?* (Buffalo, New York: Prometheus Books, 1988).

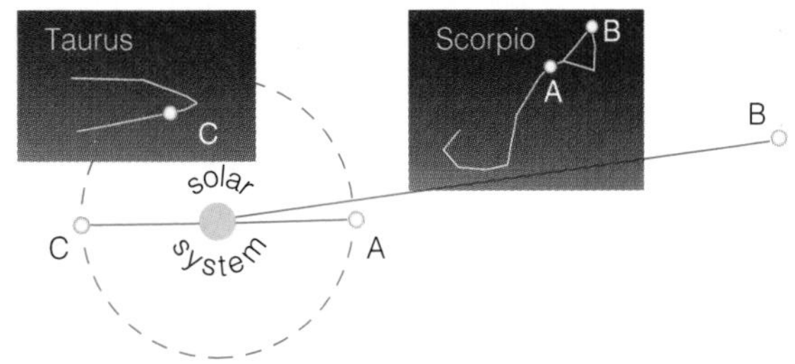

FIGURE 4.16 Two stars in different constellations (A and C) may be physically closer than two stars in the same constellation (A and B).

stars in the constellations. This idea may have seemed quite reasonable in ancient times, when the stars were assumed to be fixed on an unchanging celestial sphere. But today we know that the patterns of the stars in the constellations are accidents of the moment; long ago, the constellations did not look the same, and they will look still different far in the future [Section 3.4]. Moreover, the stars in a constellation don't necessarily have any *physical* association. Two stars that appear on opposite sides of our sky might well be closer together in reality than two stars in the same constellation (Figure 4.16). Constellations are only *apparent* associations of stars, with no more physical reality than the water in a desert mirage.

Astrology also places great importance on the positions of the planets in the sky. Again, this idea might have seemed quite reasonable in ancient times, when it was thought that the planets truly wandered among the stars. But today we know that the planets only *appear* to wander among the stars. In reality, the planets are in our own solar system, while the stars are vastly farther away. It is difficult to see how mere appearances could have profound effects on our lives.

Many other ideas at the heart of astrology are equally suspect. For example, most astrologers claim that a proper horoscope must account for the positions of *all* the planets. Does that mean that all horoscopes cast before the discovery of Pluto in 1930 were invalid? If so, why didn't astrologers notice that something was wrong with their horoscopes and thereby predict Pluto's existence? Given that several moons in the solar system are larger than Pluto, with two larger than Mercury, should astrologers also be tracking the positions of these moons? What about asteroids, which orbit the Sun like planets? What about planets orbiting other stars? With seemingly unanswerable questions like these, there seems little hope that astrology will ever meet its ancient goal of being able to forecast human events.

We especially need imagination in science.

MARIA MITCHELL (1818–1889), Astronomer and first woman elected to American Academy of Arts and Sciences

THE BIG PICTURE

In this chapter, we turned from the big picture of the universe presented in Chapters 1–3 to the topic of how we have learned so much about the universe. Key "big picture" concepts from this chapter include the following:

- The basic ingredients of scientific thinking—careful observation and trial-and-error testing—are a part of everyone's experience. The modern scientific method simply provides a way of organizing this everyday thinking to facilitate the learning and sharing of new knowledge.
- Although knowledge about the universe is growing rapidly today, each new piece rests upon foundations of older discoveries. The foundations of astronomy reach back farther in history than any other science and are intertwined with the general development of human culture and civilization.
- The concept of making models of nature lies at the heart of modern science. Models are created by generalizing from many specific facts. They then yield further predictions that allow them to be tested. Models that withstand repeated testing rise to the status of scientific theories, while models that fail must be modified or discarded.

- Although *astronomy* and *astrology* once developed hand in hand, today they represent very different things. The predictions of astrology either fail rigorous testing or are so vague as to be untestable. Astronomy is a science and is the primary means by which the hur..an species learns about the physical universe.

Review Questions

1. In what ways is scientific thinking a fundamental part of human nature? What characteristics does the modern scientific method possess that go beyond this everyday type of thinking?
2. Explain how the people of central Africa predicted the weather by observing the Moon.
3. How are the names of the seven days of the week related to astronomical objects? Based on astronomical considerations, why are there seven days, rather than some other number?
4. How did the meaning of an *hour* differ in ancient times from its meaning today? Why?
5. Briefly explain how the ancient Egyptians used star clocks to estimate the time of night.
6. Briefly describe each of the following, and give a brief description of its astronomical uses: Stonehenge, the Templo Mayor, the Sun Dagger, the Mayan observatory at Chichén Itzá, lines in the Nazca desert, Pawnee lodges, the Big Horn Medicine Wheel.
7. Explain why the Muslim fast of Ramadan occurs earlier with each subsequent year.
8. What is the *Metonic cycle?* Explain how a lunar calendar can take advantage of the Metonic cycle to stay *nearly* synchronized with a solar calendar. How does the Jewish calendar follow the Metonic cycle? How does this influence the date of Easter?
9. What ancient cultures had the greatest known success in predicting eclipses? Why do we know so little about the extent of Mayan knowledge and abilities?
10. What role did the Navigator play in Polynesian society? Why did the Navigator need great familiarity with the stars?
11. What was the Library of Alexandria? When was it built? When was it destroyed? What role did it play in the ancient empires of Greece and Rome?
12. What do we mean by a *model* of nature? Explain how the celestial sphere represents such a model.
13. Who was Ptolemy? When did he live? Why is he so important in the history of astronomy?
14. Briefly describe how the Ptolemaic model of the universe explains apparent retrograde motion while preserving the idea of an Earth-centered universe.
15. Briefly describe the role played by Islamic scholars in preserving ancient Greek knowledge and laying the groundwork for the European Renaissance.
16. Describe the basic process of the scientific method: observation, hypothesis, experiment, and theory. What is the role of a *model* in this method? What do we mean by *facts* when discussing science?
17. Briefly describe how science differs from *pseudoscience* and *nonscience.*
18. Briefly explain, with examples, how it is possible for science as a whole to be objective despite the fact that all individual scientists have personal biases and beliefs.
19. Briefly describe the international character of modern astronomy.
20. What is the basic tenet of *astrology?* Why might this idea have seemed quite reasonable in ancient times?
21. Briefly describe how and why astronomy and astrology went hand in hand in the ancient world. How did the development of astronomy benefit from this tie with astrology?
22. How do we make scientific tests of astrology? What have such tests found about the validity of astrological predictions?
23. Briefly discuss some of the ideas that lie at the heart of astrology and how these ideas made sense according to ancient beliefs. Do they still make sense today? Explain.

Discussion Questions

1. *The Impact of Science.* The modern world is filled with ideas, knowledge, and technology that developed through science and application of the scientific method. Discuss some of these things and how they affect our lives. Which of these impacts do you think are positive? Which are negative? Overall, do you think that science has benefited the human race? Defend your opinion.

2. *The Importance of Ancient Astronomy.* Why was astronomy important to people in ancient times? Discuss both the practical importance of astronomy and the importance it may have had for religious or other traditions. Which do you think was more important in the development of ancient astronomy, its practical or its philosophical role? Defend your opinion.

3. *Secrecy and Science.* The text mentioned that some historians believe that Chinese science and technology fell behind that of Europe because of the Chinese culture of secrecy. Do you agree that secrecy can hold back the advance of science? Why or why not? For the past 200 years, the United States has allowed a greater degree of free speech than most other countries in the world. How much of a role do you think this has played in making the United States the world leader in science and technology? Defend your opinion.

4. *Lunar Cycles.* We have now discussed *four* distinct lunar cycles: the $29\frac{1}{2}$-day cycle of phases, the 18-year 11-day saros cycle, the 19-year Metonic cycle, and the 18.6-year cycle over which a full moon rises at its most southerly place along the horizon. Discuss how each of these cycles can be observed and the role that each cycle has played in human history.

Problems

1. *Data Bits and Bytes.* Information can be measured in *bits,* where one bit (short for "binary digit") represents the answer to a single yes or no question. A single bit has only two (binary) possibilities. Because one bit of information has two possibilities, two bits have $2^2 = 4$ possibilities, three bits have $2^3 = 8$ possibilities, four bits have $2^4 = 16$ possibilities, and so on. One *byte* is made up of eight bits and therefore has $2^8 = 256$ possibilities.
   a. How many different *characters* are there in the English language? (Be sure to count both lower- and uppercase letters, numbers, and punctuation characters.) Is 1 byte sufficient to represent a single character? Explain.
   b. A typical page of text in a typical book contains about 2,000 characters. How many pages of text can be stored with 1 *megabyte?*
   c. A color picture can be built from many closely spaced dots. Suppose a picture contains 1,000 dots on a side, for a total of $1{,}000 \times 1{,}000 = 1$ million dots. If each dot requires 1 byte of information (to represent, say, 1 of 256 possible colors), how many bytes are required to represent the entire picture? By the measure of bits and bytes, is it true that a picture represents a thousand words? Explain.
   d. The text stated that a single night of observation at a modern observatory can collect more information than was recorded (i.e., written) by all ancient peoples combined. Do you think this is a fair comparison? Why or why not?

2. *Cultural Accomplishments in Astronomy.* Choose a particular culture of interest to you, and research the astronomical knowledge and accomplishments of that culture. Write a two- to three-page summary of your findings.

3. *Astronomical Structures.* Choose an ancient astronomical structure of interest to you (e.g., Stonehenge, Nazca lines, Pawnee lodges, etc.), and research its history. Write a two- to three-page summary of your findings. If possible, build a scale model of the structure, or create detailed diagrams to illustrate how the structure was used.

4. *Venus and the Mayans.* The planet Venus apparently played a particularly important role in Mayan society. Research the evidence for the substantial role of Venus, and write a one- to two-page summary of current knowledge about the role of Venus in Mayan society.

5. *Easter.* Find out when Easter is celebrated by different Christian churches. Then research how and why different groups set different dates for Easter. Summarize your findings in a one- to two-page written report.

6. *Time Line for the Ancient Middle East.* Create a time line marking the rise and fall of the many civilizations of the Middle East from about 3000 B.C. through the fall of the Roman Empire. Also include significant historical events on your time line, especially events relating in some way to science or astronomy.

7. *Greek Astronomers.* Many ancient Greek scientists had ideas that, in retrospect, seem well ahead of their time. Choose one or more of the following ancient Greek scientists, and learn enough about his or her work in science and astronomy to write a one- to two-page "scientific biography." Be sure to include discussion of whether and how their work influenced modern astronomy and any areas in which their ideas seemed particularly prescient.

Thales	Seleucus	Eudoxus
Democritus	Pythagoras	Eratosthenes
Callipus	Plato	Hypatia
Hipparchus	Archimedes	Empedocles
Anaximander	Ptolemy	Aristottle
Meton	Anaxagoras	Apollonius
Aristarchus		

8. *The Ptolemaic Model.* This chapter gave only a very brief description of Ptolemy's model of the universe. Investigate the model in greater depth. Using diagrams and text as needed, give a two- to three-page detailed description of the model.

9. *Scientific Tests of Astrology.* Find out about at least two scientific tests that have been conducted to test the validity of astrology. Write a short summary of each test, including how the test was conducted, how its results were evaluated, and the conclusions reached.

10. *Project: Devise Your Own Test.* Devise your own scientific test of astrology. Clearly define the methods you will use in your test and how you will evaluate the results. Then carry out the test.

5

A Universe of Matter and Energy

IN THIS AND THE NEXT THREE CHAPTERS, WE TURN OUR ATTENtion to the scientific concepts that lie at the heart of modern astronomy. We begin by investigating the nature of matter and energy, the fundamental stuff from which the universe is made. The history of the universe essentially is a story about the interplay between matter and energy since the beginning of time. Interactions between matter and energy govern everything from the creation of hydrogen and helium in the Big Bang to the processes that make Earth a habitable planet. Understanding the universe therefore depends on familiarity with how matter responds to the ebb and flow of energy.

The concepts of matter and energy presented in this chapter will enable you to understand most of the topics in this book. Some of the concepts and terminology may already be familiar to you. If not, don't worry. We will go over them again as they arise in different contexts, and you can refer back to this chapter as needed during the remainder of your studies.

5.1 Matter and Energy in Everyday Life

The meaning of **matter** is obvious to most people, at least on a practical level. Matter is simply material, such as rocks, water, or air. You can hold matter in your hand or put it in a box. The meaning of **energy** is not quite as obvious, although we certainly talk a lot about it. We pay energy bills to the power companies, we use energy from gasoline to run our cars, and we argue about whether nuclear energy is a sensible alternative to fossil fuels. On a personal level, we often talk about how energetic we feel on a particular day. But what *is* energy?

Broadly speaking, energy is what makes matter move. For Americans, the most familiar way of measuring energy is in units of Calories,[1] which we use to describe how much energy our bodies can draw from food. A typical adult uses about 2,500 Calories of energy each day. Among other things, this energy keeps our hearts beating and our lungs breathing, generates the heat that maintains our 37°C (98.6°F) body temperature, and allows us to walk and run.

Just as there are many different units for measuring height, such as inches, feet, and meters, there are many alternatives to Calories for measuring energy. If you look closely at an electric bill, you'll probably find that the power company charges you for electrical energy in units called *kilowatt-hours*. If you purchase a gas appliance, its energy requirements may be labeled in *British thermal units,* or BTUs. In science and internationally, the favored unit of energy is the **joule,** which is equivalent to about 1/4,000 of a Calorie. Thus, the 2,500 Calories used daily by a typical adult is equivalent to about 10 million joules. For purposes of comparison, some energies given in joules are listed in Table 5.1.

Although energy can always be measured in joules, it comes in many different forms. Already we have talked about food energy, electrical energy, the energy of a beating heart, heat energy represented by our body temperature, and more. Fortunately, the many forms of energy can be grouped into three basic categories.

First, whenever matter is moving, it has energy of motion, or **kinetic energy** (*kinetic* comes from a Greek word meaning "motion"). Falling rocks, the moving blades on an electric mixer, a car driving down the highway, and the molecules moving in the air around us are all examples of objects with kinetic energy.

The second basic category of energy is **potential energy,** or energy being stored for later conversion into kinetic energy. A rock perched on a ledge has *gravitational* potential energy because it will fall if it slips off the edge. Gasoline contains *chemical* potential energy, which a car engine converts to the kinetic energy of the moving car. Power companies supply *electrical* potential energy, which we use to run dishwashers and other appliances.

The third basic category is energy carried by light, or **radiative energy** (the word *radiation* often is used as a synonym for *light*). Plants directly convert the radiative energy of sunlight into chemical potential energy through the process called *photosynthesis.* Radiative energy is fundamental to astronomy, because telescopes collect the radiative energy of light from distant stars.

TIME OUT TO THINK ***We all buy energy in many different forms for many different purposes. For example, we buy chemical potential energy in the form of food to fuel our bodies. Describe several other forms of energy that your family or you commonly buy, and some of the uses to which you put each form of energy.***

Understanding how energy changes from one form to another helps us understand many common phenomena. For example, a diver standing on a 10-meter platform has gravitational potential energy

Table 5.1 Energy Comparisons

Item	Energy (joules)
Average daytime solar energy striking Earth, per m^2 per second	1×10^3
Energy released by metabolism of one average candy bar	1×10^6
Energy needed for 1 hour of walking (adult)	1×10^6
Kinetic energy of average car traveling at 60 mi/hr	1×10^6
Daily energy needs of average adult	1×10^7
Energy released by burning 1 liter of oil	1.2×10^7
Energy released by fission of 1 kg of uranium-235	5.6×10^{13}
Energy released by 1-megaton H-bomb	5×10^{15}
Energy released by major earthquake (magnitude 8.0)	2.5×10^{16}
U.S. annual energy consumption	10^{20}
Annual energy generation of Sun	10^{34}
Energy released by supernova (explosion of a star)	10^{44}–10^{46}

[1]One calorie (lowercase *c*) is the amount of energy needed to raise the temperature of 1 gram of water by 1°C. A food Calorie (uppercase C) is 1,000 calories, which is equivalent to 4,184 joules.

FIGURE 5.1 Understanding energy can help us understand the graceful movements of a diver.

owing to her height above the water and chemical potential energy stored in her body tissues. She uses the chemical potential energy to flex her muscles in such a way as to initiate her dive and then to help her execute graceful twists and spins (Figure 5.1). Meanwhile, her gravitational potential energy becomes kinetic energy of motion as she falls toward the water.

Following how energy changes from one form to another also is important to understanding the universe. For example, the particles in a collapsing cloud of interstellar gas convert gravitational potential energy into kinetic energy as they fall inward, and their motion generates heat that can eventually ignite a star. But before we study such astronomical phenomena, we first need ways to describe energy quantitatively.

5.2 A Scientific View of Energy

Over the past several centuries, scientists have discovered ways to quantify energy in all its different forms. In this section, we discuss a few ways of quantifying different forms of kinetic and potential energy; we'll save discussion of radiative energy for Chapter 7, in which we study properties of light.

Kinetic Energy

Not surprisingly, a fast-moving car has more kinetic energy than a slow-moving car, and a large truck has more kinetic energy than a car traveling at the same speed. You can calculate the kinetic energy of any moving object with a very simple formula:

$$\text{kinetic energy} = \frac{1}{2}mv^2$$

where m is the mass of the object and v is its speed (v for *velocity*). If you measure the mass in kilograms and the speed in meters per second, the resulting answer will be in joules.[2] The kinetic energy formula is easy to interpret. The m in the formula tells us that kinetic energy is proportional to mass: A 5-ton truck has 5 times the kinetic energy of a 1-ton car moving at the same speed. The v^2 tells us that kinetic energy increases with the *square* of the velocity: If you double your speed (e.g., from 30 km/hr to 60 km/hr), your kinetic energy becomes $2^2 = 4$ times greater.

Thermal Energy Suppose we want to know about the kinetic energy of the countless tiny particles (atoms and molecules) inside a rock, or of the countless particles in the air or in a distant star. Each of these tiny particles has its own motion relative to surrounding particles, and these motions are constantly changing as the particles jostle one another. The result is that the particles inside a substance appear to be moving randomly: Any individual particle may be moving in any direction with any of a wide range of speeds. Despite the seemingly random motion of particles within a substance, it's easy to measure the *average* kinetic energy of the particles—which we call **temperature.** A higher temperature simply means that, on average, the particles have more kinetic energy. Because more kinetic energy means faster average speed,[3] raising the temperature of something means making its component particles move faster (Figure 5.3, p. 122). The speeds of particles within a substance can be surprisingly fast. For example, the air molecules around you move at typical speeds of about 500 meters per second (about 1,000 miles per hour).

The energy contained *within* a substance as measured by its temperature often is called **thermal energy.** Thus, thermal energy represents the collective kinetic energy of the many individual particles moving within a substance.

---

[2]That is, energy has units of a mass times a velocity squared, and a joule is defined such that 1 joule = 1 (kg × m^2/s^2).

[3]This assumes particles of the same mass, in which case particles with greater kinetic energy must be moving faster.

Mathematical Insight 5.1 Temperature Scales

Three temperature scales are commonly used today (Figure 5.2). People in the United States usually measure temperature on the **Fahrenheit** scale, defined so that water freezes at 32°F and boils at 212°F. Internationally, temperature is usually measured on the **Celsius** scale, which places the freezing point of water at 0°C and the boiling point at 100°C.

Scientists measure temperature on the **Kelvin** scale, which is the same as the Celsius scale except for its zero point. A temperature of 0 K is the coldest possible temperature, known as **absolute zero**; it is equivalent to −273.15°C. (The degree symbol ° is not necessary when writing temperatures on the Kelvin scale.) Thus, any particular temperature has a Kelvin value that is numerically 273.15 larger than its Celsius value:

$$\text{temperature (Kelvin)} = \text{temperature (°C)} + 273.15$$

$$\text{temperature (°C)} = \text{temperature (Kelvin)} - 273.15$$

To find the conversion between Fahrenheit and Celsius, note that the Fahrenheit scale has 180° (212°F − 32°F = 180°F) between the freezing and boiling points of water, whereas the Celsius scale has only 100° between these points. Each Celsius degree therefore represents a temperature change equivalent to 1.8 Fahrenheit degrees. Furthermore, the freezing point of water is numerically 32 larger on the Fahrenheit scale than on the Celsius scale. Combining these two facts we find the conversion between Fahrenheit and Celsius:

$$\text{temperature (°C)} = \frac{[\text{temperature (°F)}] - 32\text{°F}}{1.8\frac{\text{°F}}{\text{°C}}}$$

$$\text{temperature (°F)} = 32\text{°F} + \left(1.8\frac{\text{°F}}{\text{°C}}\right) \times [\text{temperature (°C)}]$$

Example: Convert human body temperature of 98.6°F into Celsius and Kelvin.

Solution: First, we convert 98.6°F to Celsius:

$$\text{body temperature (°C)} = \frac{98.6\text{°F} - 32\text{°F}}{1.8\frac{\text{°F}}{\text{°C}}} = \frac{66.6\text{°F}}{1.8\frac{\text{°F}}{\text{°C}}} = 37.0\text{°C}$$

Next, we convert from Celsius to Kelvin:

$$\text{body temp. (K)} = \text{body temp. (°C)} + 273.15 = 37.0 + 273.15 = 310.15\text{ K}$$

Thus, human body temperature of 98.6°F is 37.0°C or 310.15 K.

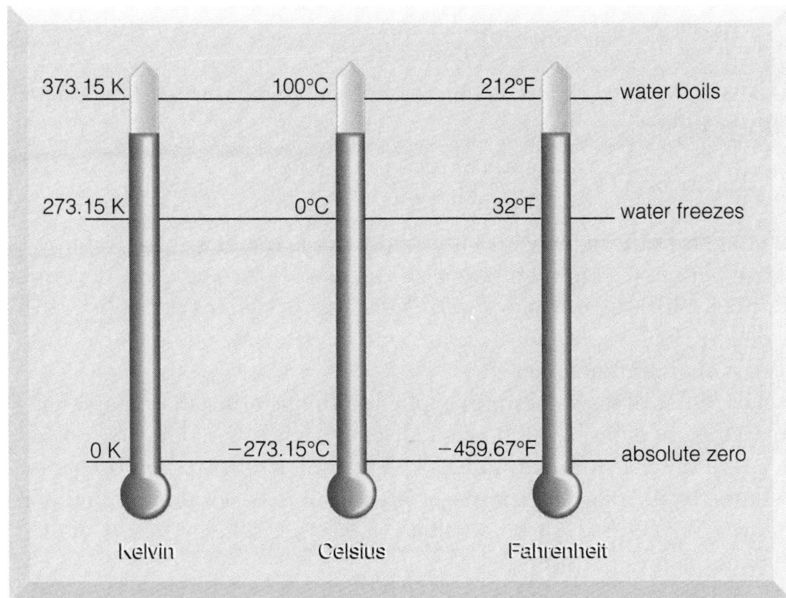

FIGURE 5.2 Three common temperature scales: Kelvin, Celsius, and Fahrenheit.

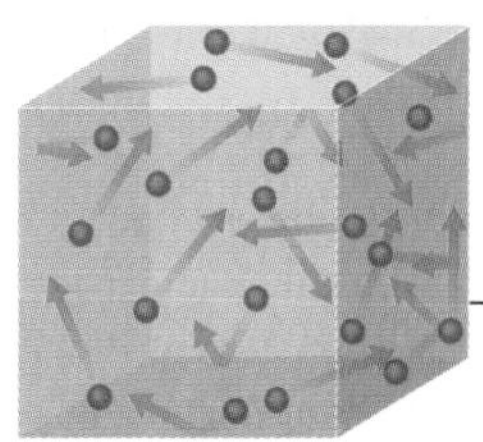

FIGURE 5.3 The particles in the box on the right have a higher temperature because their average speeds are higher (assuming that both boxes contain particles of the same mass).

Temperature and Heat The concepts of *temperature* and *heat* are not quite the same. To understand the difference, imagine the following experiment (but don't try it!). Suppose you heat your oven to 500°F. Then you open the oven door, quickly thrust your arm inside (without touching anything), and immediately remove it. What will happen to your arm? Not much! Now suppose you boil a pot of water. Although the temperature of boiling water is only 212°F, you would be badly burned if you put your arm in the pot, even for an instant. Thus, the pot of water transfers more heat to your arm than does the air in the oven, even though it has a lower temperature. That is, your arm heats more quickly in the water. This is because heat content depends on both the temperature and the total number of particles, which is much larger in a pot of water (Figure 5.4).

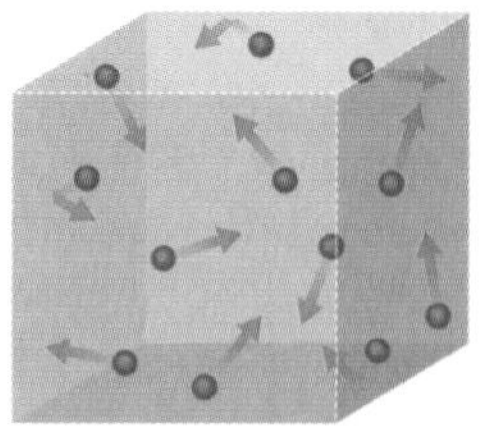

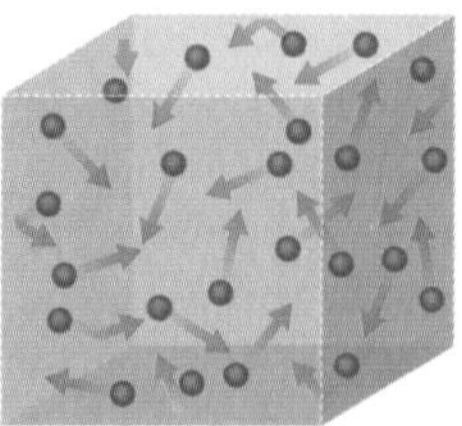

FIGURE 5.4 Both boxes have the same temperature, but the box on the right contains more *heat* because it contains more particles.

Let's look at what is happening on the molecular level. If air or water is hotter than your body, molecules striking your skin transfer some of their thermal energy to your arm. The high temperature in a 500°F oven means that the air molecules strike your skin harder, on average, than the molecules in a 212°F pot of boiling water. However, because the *density* is so much higher in the pot of water, many

Mathematical Insight 5.2 Density

The term *density* usually refers to *mass density*, which quantifies how much *mass* is packed into each unit of *volume*. The more tightly matter is packed, the higher its density. In science, the most common unit of density is grams per cubic centimeter (a cubic centimeter is about the size of a sugar cube):

$$\text{density} = \frac{\text{mass (in g)}}{\text{volume (in cm}^3\text{)}}$$

For example, if a 40-gram rock has a volume of 10 cubic centimeters, we calculate its density as follows:

$$\text{density of rock} = \frac{\text{mass of rock}}{\text{volume of rock}} = \frac{40 \text{ g}}{10 \text{ cm}^3} = 4\frac{\text{g}}{\text{cm}^3}$$

A useful guide for putting densities in perspective is the density of water, which is 1 gram per cubic centimeter; this isn't a coincidence—it's how the gram was defined. Rocks, like the one above with a density of 4 grams per cubic centimeter, are denser than water and therefore sink in water. Wood floats because its density is less than that of water.

The concept of density is sometimes applied to things other than mass. For example, an average of about 25,000 people live on each square kilometer of Manhattan, so we say that Manhattan has a *population density* of 25,000 people per square kilometer. As another example, 1 liter of oil releases about 12 million joules of energy when burned, so we say that the *chemical energy density* of oil is about 12 million joules per liter.

more molecules strike your skin each second. Each molecular collision transfers a little less thermal energy, but the total amount of heat flowing into your arm is so large that your skin burns rapidly.

TIME OUT TO THINK *If the air or water is colder than your body temperature, then the molecules in your skin have more average kinetic energy than the molecules of air or water. As a result,* your *thermal energy is transferred to the surrounding cold air or water, rather than vice versa. Using this fact, explain why falling into a 32°F (0°C) lake is much more dangerous than standing naked outside on a 32°F day.*

The environment in space provides another example of the difference between temperature and heat. Surprisingly, the temperature in low Earth orbit is several thousand degrees. However, astronauts working in Earth orbit (e.g., outside the Space Shuttle) tend to get very cold and therefore use heated space suits and gloves. Why do the astronauts feel cold despite the high temperature? Because the extremely low density of space means that relatively few particles are available to transfer heat to an astronaut.[4]

Potential Energy

Potential energy can be stored in many different forms and is not always easy to quantify. For example, the only practical way to measure the chemical potential energy in oil is to burn it and measure how much heat (thermal energy) it releases. Fortunately, it is easy to describe two types of potential energy that we use frequently in astronomy.

Gravitational Potential Energy Gravitational potential energy is extremely important in astronomy. The conversion of gravitational potential energy into kinetic (or thermal) energy helps explain everything from the speed at which objects fall to the ground to the formation processes of stars and planets. The mathematical formula for gravitational potential energy can take a variety of forms, but in words the idea is simple: *The amount of gravitational potential energy released as an object falls depends on its mass, the strength of gravity, and the distance it falls.*

This statement explains the obvious fact that falling from a 10-story building hurts more than falling out of a chair. Your gravitational potential energy is much greater on top of the 10-story building than in your chair because you can fall much farther. Because your gravitational potential energy will be converted to kinetic energy when you fall, you'll have a lot more kinetic energy by the time you hit the ground after falling from the building than from the chair, which means you'll hit the ground much harder.

Gravitational potential energy also helps us understand how the Sun became hot enough to sustain nuclear fusion. Before the Sun formed, it was a large, cold, diffuse cloud of gas. Most of the individual gas particles were far from the center of this large cloud and therefore had considerable amounts of gravitational potential energy. As the cloud collapsed under its own gravity, the gravitational potential energy of these particles was converted to thermal energy, eventually making the center of the cloud so hot that it could ignite nuclear fusion.

Mass-Energy Although matter and energy seem very different in daily life, they are intimately connected. Einstein's special theory of relativity shows that mass itself is a form of potential energy, often called **mass-energy.** Einstein demonstrated that the mass-energy of any piece of matter is given by the following formula:

$$E = mc^2$$

where E is the amount of potential energy, m is the mass of the object, and c is the speed of light. If the mass is measured in kilograms and the speed of light in meters per second, the resulting mass-energy has units of joules. Note that the speed of light is a large number ($c = 3 \times 10^8$ m/s) and the speed of light squared is much larger still ($c^2 = 9 \times 10^{16}$ m^2/s^2). Thus, Einstein's formula implies that a relatively small amount of mass represents a huge amount of mass-energy.

Mass-energy can be converted to other forms of energy, but noticeable amounts of mass become other forms of energy only under special but important circumstances. The process of nuclear fusion in the core of the Sun converts some of the Sun's mass into energy, ultimately generating the sunlight that makes life on Earth possible. On Earth, nuclear reactors and nuclear bombs also work in accord with Einstein's formula. In nuclear reactors, the splitting (fission) of elements such as uranium or plutonium converts some of the mass-energy of these materials into heat, which is then used to generate electrical power. In an H-bomb, nuclear fusion similar to that in the Sun converts a small amount of the mass-energy in hydrogen into the bomb's destructive en-

---

[4]You may wonder how the astronauts become cold since the low density also means that the astronauts cannot transfer much of their own thermal energy to the particles in space. It turns out that they lose their body heat by emitting *thermal radiation,* which we discuss in Chapter 7.

ergy. Incredibly, a 1-megaton H-bomb that could destroy a major city requires the conversion of only about 0.1 kilogram (about 3 ounces) of mass into energy (Figure 5.5).

Just as $E = mc^2$ tells us that mass can be converted into other forms of energy, it also tells us that energy can be transformed into mass. In *particle accelerators,* scientists accelerate subatomic particles to extremely high speeds so that they have a great deal of kinetic energy. When these particles collide with one another or with a barrier, some of the energy released in the collision spontaneously turns into mass, appearing as a shower of subatomic particles. These showers of particles allow scientists to test theories about how matter behaves at extremely high temperatures such as those that prevailed in the universe during the first fraction of a second after the Big Bang. Among the most powerful particle accelerators in the world are Fermilab in Illinois, the Stanford Linear Accelerator in California, and CERN in Switzerland. If you ever have a chance to visit one of these facilities, it is well worth a trip.

TIME OUT TO THINK *Einstein's formula $E = mc^2$ is probably the most famous physics equation of all time. Considering its role as described in the preceding paragraphs, do you think its fame is well deserved? If someone asked you for a one-sentence description of why this formula is important, what would you say?*

Conservation of Energy

We have seen that energy takes a variety of forms and that it can change from one form to another under certain circumstances. A fundamental principle in science is that, regardless of how we change the form of energy, the total quantity of energy never changes. That is, energy can be neither created nor destroyed; it can only change from one form to another. This principle is called the **law of conservation of energy.** It has been carefully tested in many experiments, and it is a pillar upon which modern theories of the universe are built. Because of this law, the story of the universe is a story of the interplay of energy and matter: All actions in the universe involve exchanges of energy or the conversion of energy from one form to another.

Mathematical Insight 5.3 Mass-Energy

It's easy to calculate mass-energies with Einstein's formula $E = mc^2$. Once we calculate an energy, we can compare it to other known energies.

Example: Suppose a 1-kilogram rock were completely converted into energy. How much energy would it release? Compare this to the energy released by burning 1 liter of oil.

Solution: The total mass-energy of the rock is $E = mc^2$, where m is the 1-kg mass and $c = 3 \times 10^8$ m/s:

$$\begin{aligned} E = mc^2 &= 1 \text{ kg} \times \left(3 \times 10^8 \frac{\text{m}}{\text{s}}\right)^2 \\ &= 1 \text{ kg} \times \left(9 \times 10^{16} \frac{\text{m}^2}{\text{s}^2}\right) \\ &= 9 \times 10^{16} \frac{\text{kg} \times \text{m}^2}{\text{s}^2} \\ &= 9 \times 10^{16} \text{ joules} \end{aligned}$$

To compare this energy to the 12 million joules of chemical potential energy released by burning 1 liter of oil (see Table 5.1), we divide:

$$\frac{\text{mass-energy of 1-kg rock}}{\text{chemical potential energy of 1 liter oil}} = \frac{9 \times 10^{16} \text{ joules}}{1.2 \times 10^7 \text{ joules}} = 7.5 \times 10^9$$

That is, if the rock could be converted completely to energy, its mass would supply as much energy as 7.5 billion liters of oil—which is roughly the amount of oil used by *all* cars in the United States in a week. However, no technology available now or in the foreseeable future can release all the mass-energy from a rock.

For example, imagine that you've thrown a baseball so that it is moving and hence has kinetic energy. Where did this kinetic energy come from? The baseball got its kinetic energy from the motion of your arm as you threw it; that is, some of the kinetic energy of your moving arm was transferred to the baseball. Your arm, in turn, got its kinetic energy from the release of chemical potential energy stored in your muscle tissues. Your muscles got this energy from the chemical potential energy stored in the foods you ate. The energy stored in the food came from sunlight, which plants convert into chemical potential energy through photosynthesis. The radiative energy of the Sun was generated through the process of nuclear fusion, which releases some of the mass-energy stored in the Sun's supply of hydrogen. Thus, the ultimate source of the energy of the moving baseball is the mass-energy stored in hydrogen—which was created in the Big Bang.

TIME OUT TO THINK *If you read the preceding paragraph carefully, you'll see that our description of how the baseball got its energy assumes that you are a vegetarian. How is the energy's pathway different if you got your food energy from meat?*

We have described where the baseball got its kinetic energy. Where will this energy go? As the baseball moves through the air, some of its energy is transferred to molecules in the air, generating heat or sound. If someone catches the baseball, its energy will cause his or her hand to recoil and perhaps will also generate some heat and sound. Ultimately, the energy of the moving baseball will be converted to a barely noticeable amount of heat (thermal energy) in the air, the ground, or a person's hand, making it extremely difficult to track. Nevertheless, the energy will never disappear. The total energy content of the universe was determined in the Big Bang. It remains the same today and will stay the same forever into the future.

5.3 The Material World

Now that we have seen how energy animates the matter in the universe, it's time to consider matter itself in greater detail. You are familiar with two basic properties of matter on Earth from everyday experience. First, matter can exist in different **phases:** as a **solid,** such as ice or a rock; as a **liquid,** such as flowing water or oil; or as a **gas,** such as air. Second, even in a particular phase, matter comes in a great variety of different substances.

FIGURE 5.5 The energy from this H-bomb comes from converting only about 0.1 kg of mass into energy.

But what *is* matter, and why does it have so many different forms? We can begin to answer the question by following the lead of the ancient Greek philosopher Democritus (c. 470–380 B.C.), who wondered what would happen if we broke a piece of matter, such as a rock, into ever smaller pieces. Democritus believed that the rock would eventually break into particles so small that nothing smaller could be possible. He called these particles *atoms,* a Greek term meaning "indivisible."[5] Building upon the beliefs of earlier Greek scientists, Democritus thought that all materials were composed of just four basic *elements:* fire, water, earth, and air. He proposed that the different properties of the elements could be explained by the physical characteristics of their atoms. Democritus suggested that atoms of water were smooth and round, so water flowed and had no fixed shape, and that burns were painful because atoms of fire were thorny. He imagined atoms of earth to be rough and jagged, like pieces of a three-dimensional jigsaw

[5]By modern definitions, atoms are *not* indivisible because they are composed of smaller particles.

puzzle, so that they could stick together to form a solid substance. He even explained the creation of the world with an idea that sounds uncannily modern, suggesting that the universe began as a chaotic mix of atoms that slowly clumped together to form the Earth.

Although Democritus was wrong about there being only four types of atoms and about their specific properties, his general idea was right. Today, we know that all ordinary matter is composed of **atoms** and that each different type of atom corresponds to a different chemical **element.** Ninety-one different elements are found naturally on Earth,[6] and more than 20 others have been created in laboratories. Among the most familiar chemical elements are hydrogen, helium, carbon, oxygen, silicon, iron, gold, silver, lead, and uranium.

The number of different material substances is far greater than the number of chemical elements because atoms can combine to form **molecules.** The chemical properties of a molecule are different from those of its individual atoms. For example, water behaves very differently from pure hydrogen or pure oxygen, even though water molecules are composed of two hydrogen atoms and one oxygen atom, as indicated by the familiar symbol H_2O.

Atomic Structure

Atoms are incredibly small: Millions could fit end-to-end across the period at the end of this sentence, and the number in a single drop of water (10^{22} to 10^{23} atoms) exceeds the number of stars in the observable universe. Yet atoms are composed of even smaller particles. Nearly all of an atom's mass is contained in its **nucleus,** which is surrounded by one or more **electrons.** The nucleus, in turn, is composed of two types of particles: **protons** and **neutrons.**

The properties of an atom depend mainly on the amount of **electrical charge** in its nucleus. Electrical charge is a fundamental physical property that is always conserved, just as energy is always conserved. Electrons carry a negative electrical charge of -1, protons carry a positive electrical charge of $+1$, and neutrons are electrically neutral. You may recall that oppositely charged particles attract one another and similarly charged particles repel one another. The attraction between the positively charged protons in the nucleus and the negatively charged electrons that surround it are what hold an atom together.[7] Ordinary atoms have identical numbers of electrons and protons, making them electrically neutral overall.

Although electrons can be thought of as tiny particles, they are not quite like tiny grains of sand, and they don't really orbit the nucleus as planets orbit the Sun. Instead, the electrons in an atom are "smeared out," forming a kind of cloud that surrounds the nucleus and gives the atom its apparent size. The electrons aren't really cloudy; it's just that it is impossible to pinpoint their positions (for reasons discussed in Chapter S5). An atomic nucleus is very tiny, even compared to the atom itself: If we imagine an atom on a scale on which its nucleus is the size of your fist, its electron cloud will be many miles wide. The structure of a typical atom is represented in Figure 5.6.

Common Misconceptions: The Illusion of Solidity

Bang your hand on a table. Although the table feels solid, it is made almost entirely of empty space! Nearly all the mass of the table is contained in the nuclei of its atoms. But the volume of an atom is more than a trillion times the volume of its nucleus, so relatively speaking the nuclei of adjacent atoms are nowhere near to touching one another. The solidity of the table comes about from a combination of electrical interactions between the charged particles in its atoms and the strange quantum laws governing the behavior of electrons. If we could somehow pack all the table's nuclei together, the table's mass would fit into a microscopic speck. Although *we* cannot pack matter together in this way, nature can and does—in *neutron stars,* which we will study in Chapter 17.

Each different chemical element contains a different number of protons in its nucleus called its **atomic number**. For example, a hydrogen nucleus contains just one proton, so its atomic number is 1; a helium nucleus contains two protons, so its atomic number is 2.

The *combined* number of protons and neutrons in an atom is called its **atomic weight** (or **atomic mass).** The atomic weight of ordinary hydrogen is 1 because its nucleus is just a single proton. Helium usually has two neutrons in addition to its two protons, giving it an atomic weight of 4. Carbon usually has six protons and six neutrons, giving it an atomic weight of 12.

---

[6]These are 91 of the first 92 elements on the periodic table; the exception is technetium (atomic number 43). See Appendix B.

[7]You may wonder why electrical repulsion doesn't cause the positively charged protons in a nucleus to fly apart from one another. It tries, but it is overcome by an even stronger force that holds nuclei together, called the *strong nuclear force* (see Chapter S5).

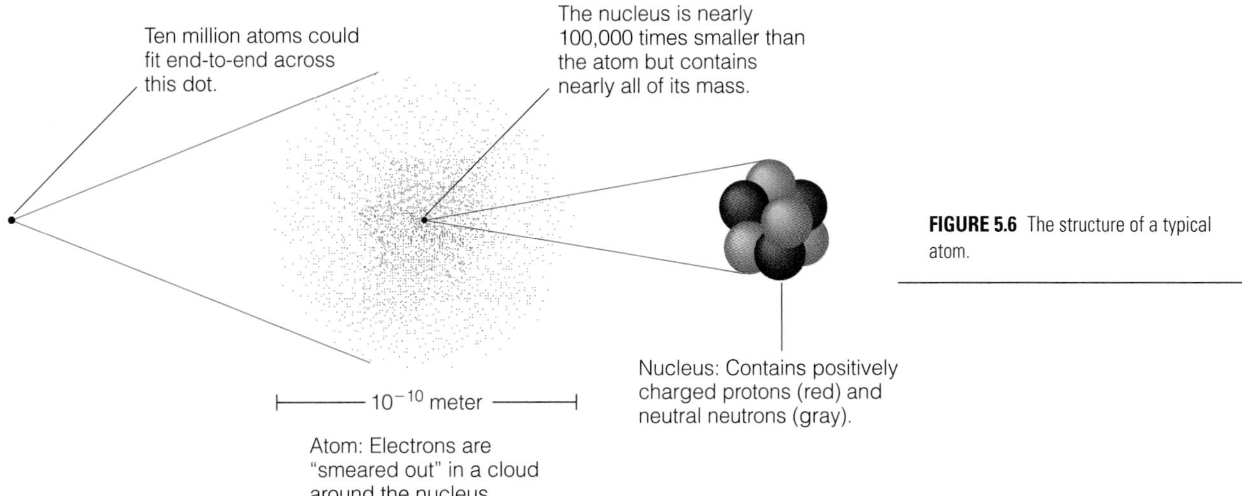

FIGURE 5.6 The structure of a typical atom.

Sometimes, the same element can have varying numbers of neutrons. For example, in rare cases a hydrogen nucleus contains a neutron in addition to its proton, making its atomic weight 2; hydrogen with atomic weight 2 is often called *deuterium*. The different forms of hydrogen are called **isotopes** of one another.[8] Each isotope of an element has the *same* number of protons but a *different* number of neutrons.

We usually denote different isotopes by showing the atomic weight to the upper left of the chemical symbol. For example, most carbon atoms contain six neutrons in addition to their six protons, making their atomic weight $6 + 6 = 12$; we therefore represent this isotope of carbon by writing ^{12}C. The symbol ^{14}C represents a rarer isotope of carbon that has atomic weight 14 and hence eight neutrons rather than six. Figure 5.7 shows examples illustrating the terminology of atomic number, atomic weight, and isotopes.

[8]A third isotope of hydrogen, tritium (^3H), contains two neutrons in addition to its one proton. Tritium is used in hydrogen bombs.

TIME OUT TO THINK ***The symbol ^4He represents helium with an atomic weight of 4; this is the most common form, containing two protons and two neutrons. What does the symbol ^3He represent?***

Phases of Matter

Everyday experience tells us that, depending on the temperature, the same substance can exist in different phases. The main difference between different phases of matter is how tightly neighboring particles are bound together. As a substance is heated, the average kinetic energy of its particles increases, enabling the particles to break the bonds holding them to their neighbors. Each change in phase corresponds to the breaking of a different kind of bond. Although

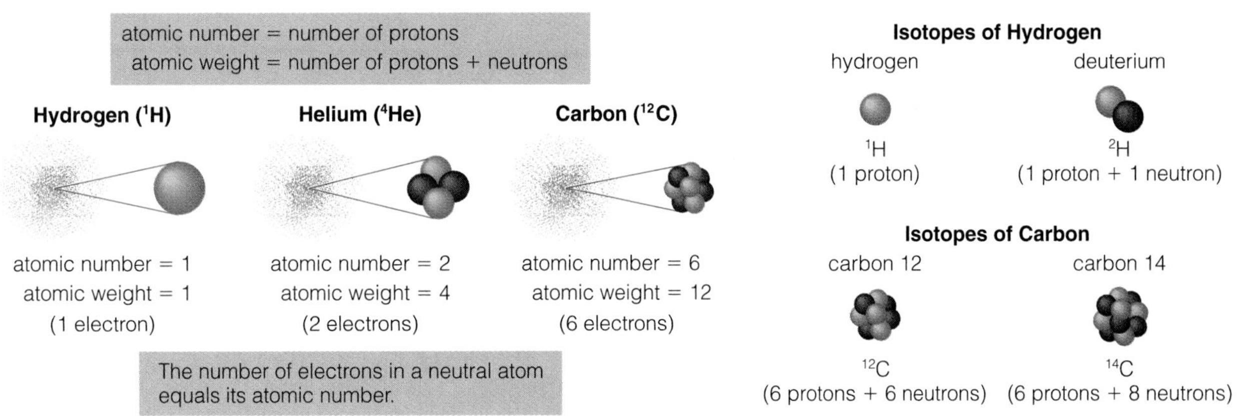

FIGURE 5.7 Terminology of atoms.

phase changes occur in all substances, let's consider what happens when we heat water, starting from ice:

- Below 0°C (32°F), water molecules have a relatively low average kinetic energy, and each molecule is bound tightly to its neighbors, making the *solid* structure of ice.
- As the temperature increases (but remains below freezing), the rigid arrangement of the molecules in ice vibrates more and more. At 0°C, the molecules have enough energy to break the solid bonds of ice. The molecules can then move relatively freely among one another, allowing the water to flow as a *liquid*. Even in liquid water, a type of bond between adjacent molecules still keeps them close together.
- When a water molecule breaks free of all bonds with its neighbors, we call it a molecule of water vapor, which is a *gas*. Molecules in the gas phase move independently of other molecules. Even at low temperatures at which water is a solid or liquid, a few molecules will have enough energy to enter the gas phase. We call the process **evaporation** when molecules escape from a liquid and **sublimation** when they escape from a solid. As temperatures rise, the rates of sublimation and evaporation increase, eventually changing all the solid and liquid water into water vapor.

What happens if we continue to raise the temperature of water vapor? As the temperature rises, the molecules move faster, making collisions among them more and more violent. These collisions eventually split the water molecules into their component atoms of hydrogen and oxygen. The process in which the bonds that hold the atoms of a molecule together are broken is called **molecular dissociation.** At still higher temperatures, collisions can break the bonds holding electrons around the nuclei of individual atoms, allowing the electrons to go free. The loss of one or more negatively charged electrons leaves a remaining atom with a net positive charge. Such positively charged atoms are usually called **ions,**[9] and the process of stripping electrons from atoms is called **ionization.** Thus, at high temperatures, what once was water becomes a hot gas consisting of freely moving electrons and positively charged ions of hydrogen and oxygen. This type of hot gas, in which atoms have become ionized, is called a **plasma,** sometimes referred to as "the fourth phase of matter."

Note that neutral hydrogen contains only one electron, which balances the single positive charge of the one proton in its nucleus. Thus, hydrogen can be ionized only once, and the remaining hydrogen ion, designated H^+, is simply a proton. Oxygen, with atomic number 8, has eight electrons when it is neutral, so it can be ionized multiple times. *Singly ionized* oxygen is missing one electron, so it has a charge of +1 and is designated O^+. *Doubly ionized* oxygen, or O^{+2}, is missing two electrons; *triply ionized* oxygen, or O^{+3}, is missing three electrons; and so on. At extremely high temperatures, oxygen can be *fully ionized,* in which case all eight electrons are stripped away and the remaining ion has a charge of +8.

Other chemical substances go through similar phase changes, but the temperatures at which phase changes occur depend on the type of atom or molecule involved. A chunk of iron, for example, is solid at room temperature. If you heat the iron, it will melt into a liquid at 1,535°C and then boil into a gas at about 3,000°C. Increasing the temperature further will ionize the iron atoms, forming a plasma. As the temperature continues to rise, the iron will become doubly ionized, then triply ionized, and so on. At a temperature of about 100 million Kelvin, the iron will be fully ionized: All 26 of its electrons will be stripped away, leaving ions of iron with a positive charge of +26 (designated Fe^{+26}). Figure 5.8 illustrates the general progression of phase changes.

Common Misconceptions: One Phase at a Time?

In daily life, we usually think of H_2O as being in the phase of either solid ice, liquid water, or water vapor, with the phase depending on the temperature. In reality, two or even all three phases can exist at the same time. In particular, some sublimation *always* occurs over solid ice, and some evaporation *always* occurs over liquid water. Thus, the phases of solid ice and liquid water never occur alone, but instead occur in conjunction with water vapor. The *amount* of water vapor increases with the temperature, which is why you see more and more steam rising from a kettle even before it boils.

TIME OUT TO THINK *Although plasma is rare at the temperatures that exist naturally on Earth, it is actually the most common phase of*

[9]The term *ion* is also used in reference to negatively charged atoms, that is, atoms possessing more electrons than protons. However, in astronomy, you can assume that *ion* refers to a positively charged atom unless explicitly stated otherwise.

ordinary matter in the universe. Can you explain why? (Hint: *What is the most massive object in the solar system, and how hot is it?*)

5.4 Energy in Atoms

We've said that the story of the universe is a story of interactions between energy and matter. Having discussed the concepts of energy and matter separately, our next step is to put the concepts together so we can understand their interactions. Because atoms are the components of matter, we need to understand how atoms store, release, and exchange energy.

So far, we've seen two different ways in which atoms have energy. First, by virtue of their mass, they possess mass-energy in the amount mc^2, where m is the mass and c is the speed of light. However, this mass-energy can be released only under special circumstances, such as in nuclear reactions, so it has no appreciable effect on most atomic interactions. Second, we know that a moving atom has kinetic energy in the amount $\frac{1}{2}mv^2$, where v is its speed. The temperature of a substance is a measure of the average kinetic energy of its atoms (or molecules), and the individual atoms can exchange kinetic energy through collisions.

Atoms can also contain energy in a third way: as *electric potential energy* in the distribution of their electrons around their nuclei. The simplest case is that of hydrogen, which has only one electron. Remember that an electron tends to be "smeared out" into a cloud around the nucleus. When the electron is "smeared out" to the minimum extent that nature allows, the atom contains its smallest possible amount of electric potential energy; we say that the atom is in its **ground state** (Figure 5.9). If the electron somehow gains energy, then it becomes "smeared out" over a greater volume, and we say that the atom is in an **excited state.** If the electron gains enough energy, it can escape the atom completely, in which case the atom has been *ionized.*

Perhaps the most surprising aspect of atoms was discovered in the 1910s, when scientists realized that electrons in atoms can have only *particular* energies (which correspond to particular sizes and shapes of the electron cloud). A simple analogy can illuminate this idea. Suppose you're washing windows on a building. If you use an adjustable platform to reach high windows, you can stop the platform at any height above the ground (Figure 5.10a). But if you use a ladder, you can stand only at *particular* heights—the heights of the rungs of the ladder—and not at any height in between (Figure 5.10b). The possible

FIGURE 5.8 The general progression of phase changes.

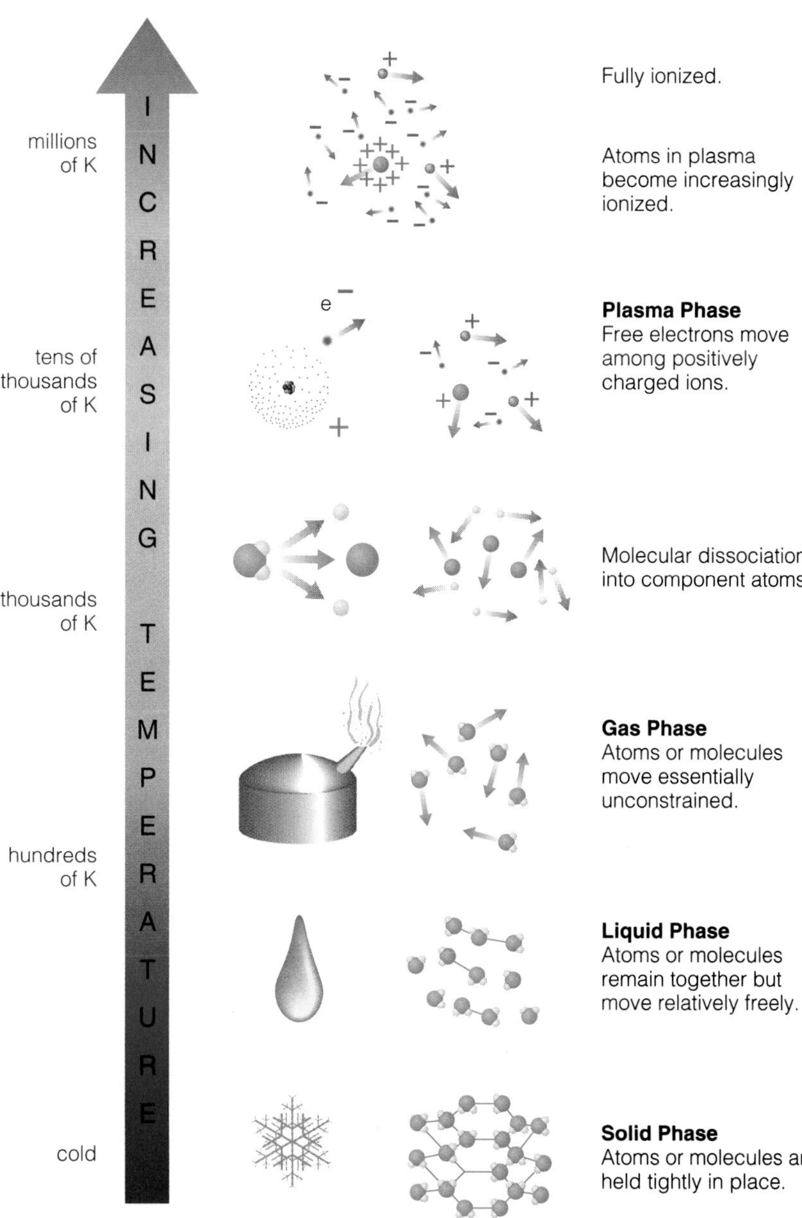

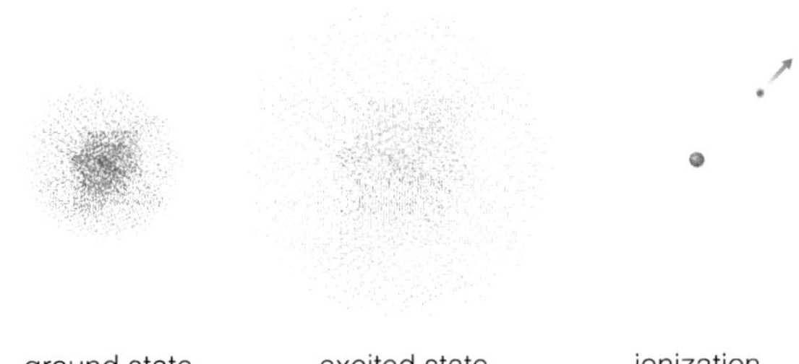

FIGURE 5.9 In its ground state, an electron is "smeared out" to the minimum extent allowed by nature. Adding energy can raise the electron to an excited state that occupies a larger volume. Adding enough energy can ionize the atom.

energies of electrons in atoms are like the possible heights on a ladder. Only a few particular energies are possible, and energies between these special few are not allowed.

The possible energy levels of the electron in hydrogen are represented like the steps of a ladder in Figure 5.11. The ground state, labeled level 1, is the bottom rung of the ladder; its energy is labeled zero because the atom has no excess electrical potential energy to lose. Each subsequent rung of the ladder represents a possible excited state for the electron and is labeled with the electron's energy above the ground state. Because the energies involved in a single atom are so small, it is easier to represent them with a unit of energy called the **electron-volt,** or **eV,** than with joules:

$$1 \text{ eV} = 1.60 \times 10^{-19} \text{ joule}$$

For example, the energy of an electron in energy level 2 is 10.2 eV greater than that of an electron in the ground state; that is, an electron must gain 10.2 eV of energy to "jump" from level 1 to level 2. Similarly, jumping from level 1 to level 3 requires gaining 12.1 eV of energy.

Note that, unlike with a ladder built for climbing, the rungs on the electron's energy ladder are closer together near the top. The top itself represents the energy of ionization—if the electron gains this much energy, or 13.6 eV above the ground state in the case of hydrogen, the electron breaks free from the atom. (Any excess energy beyond that needed for ionization becomes kinetic energy of the free-moving electron.)

Because energy is always conserved, an electron cannot jump to a higher energy level unless its atom gains the energy from somewhere else. Generally, the atom gains this energy either from the kinetic energy of another particle colliding with it or from the absorption of energy carried by light. Similarly, when an electron falls to a *lower* energy level, it either transfers its energy to another particle through a collision or emits light that carries the energy away. The key point is this: *Electron jumps can occur only with the particular amounts of energy representing differences between possible energy levels.*

The result is that electrons in atoms can absorb or emit only particular amounts of energy and not other amounts in between. For example, if you attempt to provide a hydrogen atom in the ground state with 11.1 eV of energy, the atom won't accept it because it is too high to boost the electron to level 2 but not high enough to boost it to level 3.

TIME OUT TO THINK *We will see in Chapter 7 that light comes in "pieces" called* photons *that carry specific amounts of energy. Can a hydrogen atom absorb a photon with 11.1 eV of energy? Why or why not? Can it absorb a photon with 10.2 eV of energy? Explain.*

a

b

FIGURE 5.10 (**a**) A window washer on an elevator can be at any height. (**b**) A window washer on a ladder can be at only particular heights. Similarly, electrons in an atom can have only particular energy levels.

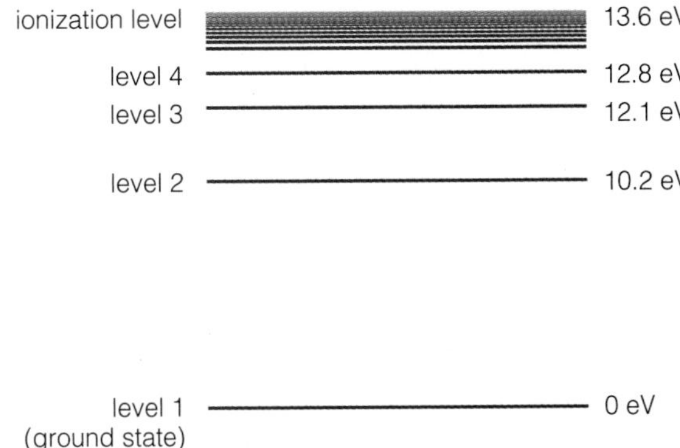

FIGURE 5.11 Energy levels for the electron in a hydrogen atom. There are many more closely spaced energy levels between level 4 and the ionization level.

If you think about it, the idea that electrons in atoms can jump only between particular energy levels is quite bizarre. It is as if you had a car that could go only particular speeds and not other speeds in between. How strange it would seem if your car suddenly jumped from 5 miles per hour to 20 miles per hour without gradually passing through a speed of 10 miles per hour! In scientific terminology, the electron's energy levels are said to be *quantized,* and the study of the energy levels of electrons (and other particles) is called *quantum mechanics.*

Electrons have quantized energy levels in all atoms, not just in hydrogen. Moreover, the allowed energy levels differ from element to element and even from one ion of an element to another ion of the same element. (In general, atoms with more electrons have larger and more complex sets of energy levels.) For example, the energy levels allowed for the electrons in carbon are different from those allowed for electrons in helium, which in turn are different from those allowed for electrons in hydrogen. This fact holds the key to the study of distant objects in the universe. As we will see in Chapter 7 when we study light, the different energy levels of different elements allow light to carry "fingerprints" that can tell us the chemical composition of distant objects.

Common Misconceptions: Orbiting Electrons?

Most people have been taught to think of electrons in atoms as "orbiting" the nucleus like tiny planets orbiting a tiny sun. But this representation is simply not true. Electrons and other subatomic particles do not behave at all like baseballs, rocks, or planets. In fact, the behavior of subatomic particles is so strange that human minds may simply be incapable of visualizing it. Nevertheless, the *effects* of electrons are easy to observe, and one of the most important effects is that electrons give atoms their size. Thus, we say that electrons in atoms are "smeared out" into an "electron cloud" around the nucleus. This description is rather vague, but it is far more accurate than the misleading picture of electrons circling like tiny planets.

The eternal mystery of the world is its comprehensibility. The fact that it is comprehensible is a miracle.

ALBERT EINSTEIN

THE BIG PICTURE

In this chapter, we discussed the concepts of energy and matter in some detail. Key "big picture" ideas to draw from this chapter include the following:

- Energy and matter are the two basic ingredients of the universe. Therefore, understanding energy and matter, and the interplay between them, is crucial to understanding the universe.
- Energy can be neither created nor destroyed. That is, energy always is conserved, and we can understand many processes in the universe by following how energy changes from one form to another in its interactions with matter. Don't forget that mass itself is a form of potential energy, called mass-energy.
- The strange laws of quantum mechanics govern the interactions of matter and energy on an atomic level. Electrons in atoms can have only particular energies and not energies in between, and each element has a different set of allowed energy levels.

Review Questions

1. List several different units used to measure energy. How is a *joule* related to a *Calorie?*
2. Briefly describe and differentiate between *kinetic energy, potential energy,* and *radiative energy.*
3. What is the formula for the kinetic energy of an object? Based on this formula, explain why (a) a 4-ton truck moving at 100 km/hr has four times as much kinetic energy as a 1-ton car moving at 100 km/hr; (b) a 1-ton car moving at 100 km/hr has the *same* kinetic energy as a 4-ton truck moving at 50 km/hr.
4. What does *temperature* measure? How is it related to kinetic energy? What is *thermal energy?*
5. Briefly describe the Fahrenheit, Celsius, and Kelvin temperature scales. What is absolute zero?
6. In your own words, briefly describe how a 500°F oven and a pot of boiling water illustrate the difference between *temperature* and *heat.* How is this difference important to astronauts in Earth orbit?
7. What is *gravitational potential energy?* Explain why (a) a bowling ball perched on a cliff ledge has more gravitational potential energy than a baseball perched on the same ledge; (b) a diver on a 10-meter platform has more gravitational potential energy than a diver on a 3-meter diving board; (c) a 100-kg satellite orbiting Jupiter has more gravitational potential energy than a 100-kg satellite orbiting the Earth, assuming both satellites orbit at the same distance from the planet centers.
8. What do we mean by mass-energy? Explain the meaning of the formula $E = mc^2$ and what it has to do with the Sun, nuclear bombs, and particle accelerators.
9. What is the law of conservation of energy? Your body is using energy right now to keep you alive. Where does this energy come from? Where does it go?
10. Briefly define *atom, element,* and *molecule.* How was Democritus's idea of *atoms* similar to the modern concept of atoms? How was it different?
11. What is *electrical charge?* What type of electrical charge is carried by *protons, neutrons,* and *electrons?* Under what circumstances do electrical charges attract? Under what circumstances do they repel?
12. Briefly describe the structure of an atom. How big is an atom? How big is the *nucleus* in comparison to the entire atom?
13. Test your understanding of atomic terminology. (a) The most common form of iron has 26 protons and 30 neutrons in its nucleus. State its *atomic number,* its *atomic weight,* and its number of electrons if it is electrically neutral. (b) Consider the following three atoms: Atom 1 has seven protons and eight neutrons; atom 2 has eight protons and seven neutrons; atom 3 has eight protons and eight neutrons. Which two are *isotopes* of each other? (c) Oxygen has atomic number 8. How many times must an oxygen atom be ionized to create an O^{+5} ion? How many electrons are in an O^{+5} ion?
14. What do we mean by a *phase* of matter? Briefly describe how the idea of *bonds* between atoms (or molecules) explains the difference between the *solid, liquid,* and *gas* phases.
15. Briefly explain why a few atoms (or molecules) are always in gas phase around any solid or liquid. Then explain how *sublimation* and *evaporation* are similar, and how they are different.
16. Explain why, at sufficiently high temperatures, molecules undergo *molecular dissociation* and atoms undergo *ionization.*
17. What is a *plasma?* Explain why, at very high temperatures, all matter is in this "fourth phase of matter."
18. Describe the three basic ways in which atoms can contain energy. In what way does an atom in an *excited state* contain more energy than an atom in the *ground state?*
19. How are the possible energy levels of electrons in atoms similar to the possible gravitational potential energies of a person on a ladder? How are they different?
20. What is an *electron-volt* (eV)? Why is it more convenient to describe energies in atoms in units of eV than in units of joules?
21. How can an electron gain the energy needed to jump to a higher energy level? How can it rid itself of energy to fall to a lower energy level? Why can electrons in atoms absorb or emit only particular amounts of energy?

Discussion Questions

1. *Knowledge of Mass-Energy.* Einstein's discovery that energy and mass are equivalent has led to technological developments both beneficial and dangerous. Discuss some of these developments. Overall, do you think the human race would be better or worse off if we had never discovered that mass is a form of energy? Defend your opinion.
2. *Perpetual Motion Machines.* Every so often, someone claims to have built a machine that can generate energy perpetually from nothing. Why isn't this possible according to the known laws of nature? Why do you think claims of perpetual motion machines sometimes get substantial media attention?
3. *Indoor Pollution.* Given that sublimation and evaporation are very similar processes, why is sublimation generally much more difficult to notice? Discuss how sublimation, particularly from plastics and other human-made materials, can cause "indoor pollution."

Problems

1. *Calculating Densities.* Find the average density of the following objects in grams per cubic centimeter.
   a. A rock with volume 15 cm^3 and mass of 0.25 kg.
   b. Earth, with its mass of 6×10^{24} kg and radius of about 6,400 km. (*Hint:* The formula for the volume of a sphere is $\frac{4}{3} \times \pi \times \text{radius}^3$.)
   c. The Sun, with its mass of 2×10^{30} kg and radius of about 700,000 km.
2. *Energy Comparisons.* Use the data in Table 5.1 to answer each of the following questions.
   a. Compare the energy of a 1-megaton hydrogen bomb to the energy released by a major earthquake.
   b. If the United States obtained all its energy from oil, how much oil would be needed each year?
   c. Compare the Sun's annual energy output to the energy released by a supernova.
3. *Moving Candy Bar.* Metabolizing a candy bar releases about 10^6 joules. How fast must the candy bar travel to have the same 10^6 joules in the form of kinetic energy? (Assume the candy bar mass is 0.2 kg.) Is your answer faster or slower than you expected?
4. *Einstein's Famous Formula.*
   a. What is the meaning of the formula $E = mc^2$? Be sure to define each variable.
   b. How does this formula explain the generation of energy by the Sun?
   c. How does this formula explain the destructive power of nuclear bombs?
5. *Spontaneous Human Combustion.* Suppose that, through a horrific act of an angry god (or a very powerful alien, if you prefer), all the mass in your body was suddenly converted into energy according to the formula $E = mc^2$. How much energy would be produced? Compare this to the energy of a nuclear bomb (see Table 5.1). What effect would your disappearance have on the surrounding region?
6. *The Fourth Phase of Matter.*
   a. Explain why nearly all the matter in the Sun is in the plasma phase.
   b. Based on your answer to part (a), explain why plasma is the most common phase of matter in the universe.
   c. Given that plasma is the most common phase of matter in the universe, why is it so rare on Earth?
7. *Atomic Terminology Practice.*
   a. Consider fluorine atoms with 9 protons and 10 neutrons. What are the atomic number and atomic weight of this fluorine? Suppose we could add a proton to this fluorine nucleus. Would the result still be fluorine? Explain. What if we added a neutron to the fluorine nucleus?
   b. The most common isotope of oxygen has atomic number 8 and atomic weight 16. Another isotope of oxygen has two extra neutrons. What are the atomic number and atomic weight of this isotope?
   c. The most common isotope of gold has atomic number 79 and atomic weight 197. How many protons and neutrons does the gold nucleus contain? Assuming the gold is electrically neutral, how many electrons does it have? If the gold is triply ionized, how many electrons does it have?
   d. The most common isotope of uranium is ^{238}U, but the form used in nuclear bombs and nuclear power plants is ^{235}U. Given that uranium has atomic number 92, how many neutrons are in each of these two isotopes of uranium?
8. *Energy Level Transitions.* The labeled transitions below represent an electron moving between energy levels in hydrogen. Answer each of the following questions and explain your answers.

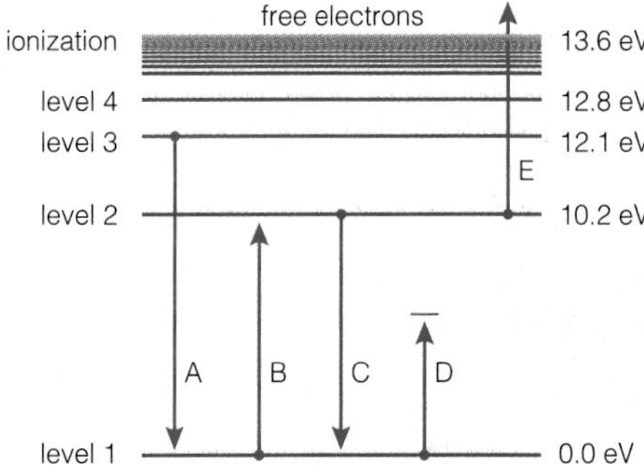

   a. Which transition could represent an electron that *gains* 10.2 eV of energy?
   b. Which transition represents an electron that *loses* 10.2 eV of energy?
   c. Which transition represents an electron that is breaking free of the atom?
   d. Which transition, as shown, is *not* possible?
   e. Describe the process taking place in transition A.
9. *Research: Democritus and the Path of History.* Besides his belief in atoms, Democritus held several other strikingly modern notions. For example, he maintained that the Moon was a world with mountains and valleys and that the Milky Way was composed of countless individual stars—ideas that weren't generally accepted until the time of Galileo, more than 2,000 years later. Unfortunately, we know of Democritus's work only secondhand because none of the 72 books he is said to have written survived the destruction of the Library of Alexandria. Do research to learn what we know about Democritus and how we know it. Write a two- to three-page essay discussing your findings and your opinion of how history might have been different if the work of Democritus had not been lost.

6

Universal Motion: from Copernicus to Newton

EVERYTHING IN THE UNIVERSE IS IN CONSTANT Motion, from the random meandering of molecules in the air to the large-scale drifting of galaxies in super clusters. Remarkably, just a few physical laws govern all this motion. The elucidation of these laws over the past several centuries is surely one of the greatest scientific triumphs of all time.

We have two primary goals in this chapter: understanding the laws that govern the motion of celestial objects and understanding how those laws were discovered. Achieving these goals requires that we first explore motion and the laws that govern it more generally. Once we develop this background, you'll be able to understand the extraordinary story of how the ancient belief in an Earth-centered universe was finally overthrown. As before, much of the subject matter of this chapter may be familiar to you already. Again, don't worry if this is not the case. The material is not difficult, and studying it carefully will greatly enhance your understanding of the astronomy in the rest of the book as well as many everyday phenomena.

6.1 Describing Motion: Examples from Daily Life

Think about what happens when you throw a ball to a dog: The dog runs and catches it. Now think about the complexity of this trick. The ball leaves your hand traveling in some particular direction with some particular amount of kinetic energy. As the ball rises, gravity converts some of its kinetic energy into potential energy, slowing the ball's rise until it reaches the top of its trajectory. Then gravity transforms the potential energy of the ball back into kinetic energy, bringing it back toward the ground. Meanwhile, the ball may lose some of its kinetic energy to air resistance or may be pushed by gusts of wind. Despite this complexity, the dog still catches the ball.

We humans can perform an even better trick: We have learned how to figure out where the ball will land even before throwing it, and we can perform this trick with extraordinary precision. Understanding how we perform this trick and applying it to problems of motion throughout the universe require understanding the laws that govern motion. We'll study these laws shortly, but first we need to discuss the scientific concepts used to describe motion.

We all have a great deal of experience with motion and natural intuition as to how motion works, so we begin our discussion of motion with some familiar examples. Indeed, you probably are familiar with all the terms defined in this section, although their scientific definitions may differ subtly from those you use in casual conversation.

Speed, Velocity, and Acceleration

The concepts we use in determining the trajectory of a ball, a rocket, or a planet are familiar to you from driving a car. The speedometer indicates your **speed,** usually in units of both miles per hour (mi/hr) and kilometers per hour (km/hr); for example, 100 km/hr is a speed. Your **velocity** is your speed in a certain direction; "100 km/hr going due north" describes a velocity. It is possible to change your velocity without changing your speed by, for example, holding to a steady 60 km/hr as you drive around a curve. Because your direction is changing as you round the curve, your *velocity* is also changing—even though your *speed* is constant.

Whenever your velocity is changing, you are experiencing **acceleration.** You are undoubtedly familiar with the term *acceleration* as it applies to increasing speed, such as when you accelerate away from a stop sign while driving your car. In science, we also say that you are accelerating when you slow down or turn. Slowing occurs when your acceleration is in a direction opposite to your motion; we say that your acceleration is *negative* and causes your velocity to decrease. Turning changes your *direction,* which means a change in velocity (and thus involves acceleration) even if your speed remains constant.

Note that you don't feel anything when you are traveling at *constant velocity,* which is why you don't feel any sensation of motion when you're traveling in an airplane on a smooth flight. In contrast, you can often feel acceleration: As you speed up in a car you feel yourself being pushed back into your seat, as you slow down you feel yourself being pulled forward from the seat, and as you drive around a curve you lean outward because of your acceleration.

One of the most important types of acceleration is that caused by gravity, which makes objects accelerate as they fall. In a famous (though probably apocryphal) experiment that involved dropping weights from the Leaning Tower of Pisa, Galileo demonstrated that gravity accelerates all objects by the same amount, regardless of their mass. This fact may be surprising because it seems to contradict everyday experience: A feather floats gently to the ground, while a rock plummets. However, this difference is caused by air resistance. If you dropped a feather and a rock on the Moon, where there is no air, both would fall at exactly the same rate.

TIME OUT TO THINK ***Find a piece of paper and a small rock. Hold both at the same height, one in each hand, and let them go at the same instant. The rock, of course, hits the ground first. Next crumple the paper into a small ball and repeat the experiment. What happens? Explain how this experiment suggests that gravity accelerates all objects by the same amount.***

The acceleration of a falling object is called the **acceleration of gravity,** abbreviated g. On Earth, the acceleration of gravity causes falling objects to fall faster by 9.8 meters per second (m/s), or about 10 m/s, with each passing second. For example, suppose you drop a rock from a tall building. At the moment you let it go, its speed is 0 m/s. After 1 second, the rock will be falling downward at about 10 m/s. After 2 seconds, it will be falling at about 20 m/s. In the absence of air resistance, its speed will continue to increase by about 10 m/s each second until it hits the ground (Figure 6.1).

Mathematically, acceleration is the *rate of change* in velocity; that is, it is how much velocity changes

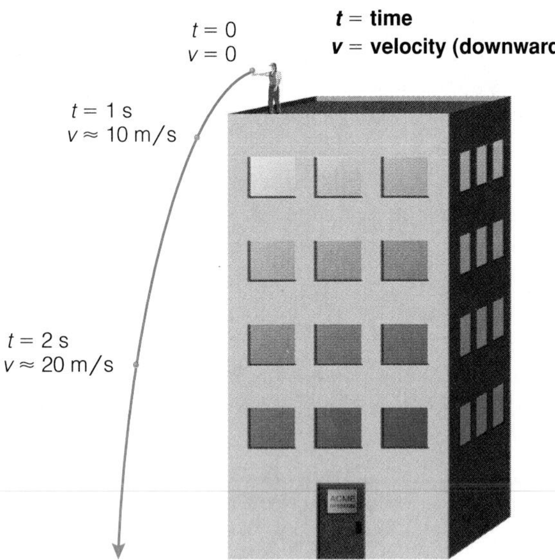

FIGURE 6.1 On Earth, gravity causes falling objects to accelerate downward at about 10 m/s^2. That is, a falling object's velocity increases by about 10 m/s with each passing second.

in a particular amount of time. Because the acceleration of gravity makes a falling object's velocity change by about 10 m/s with each second that elapses, the numerical value of g is:

$$g = \frac{\text{change in velocity}}{\text{elapsed time}} \approx \frac{10\,\frac{\text{m}}{\text{s}}}{1\text{ s}} = \frac{10\text{ m}}{\text{s}} \times \frac{1}{1\text{ s}} = 10\,\frac{\text{m}}{\text{s}^2}$$

Note that the units of acceleration are m/s^2, which we read "meters per second *per* second" or "meters per second squared." More precisely, the acceleration of gravity is $g = 9.8$ m/s^2 (downward).

Momentum and Force

Imagine that you're stopped innocently in your car at a red light when suddenly a bug flying at a velocity of 30 km/hr due south slams into your windshield. What will happen to your car? Not much, except perhaps a bit of a mess on your windshield. Next imagine that a 2-ton truck runs the red light and hits you head-on with the same velocity as the bug. Clearly, the truck will cause far more damage.

Scientifically, we say that the truck imparts a much larger jolt than the bug because it transfers more **momentum** to you. Momentum describes a combination of mass and velocity. We can describe the momentum of the truck before the collision as "2 tons moving due south at 30 km/hr," while the momentum of the bug is perhaps "1 gram moving due south at 30 km/hr." Mathematically, momentum is defined as the product of mass and velocity:

$$\text{momentum} = \text{mass} \times \text{velocity}$$

In transferring some of its momentum to your car, the truck (or bug) has exerted a **force** on your car. More generally, a force is anything that can cause a change in momentum. You are familiar with many types of force besides collisional force. For example, if you shift into neutral while driving along a flat stretch of road, the forces of air resistance and road friction will continually sap your car's momentum (transferring it to molecules in the air and pavement), slowing your velocity until you come to a stop. You probably are also familiar with forces arising from gravity, electricity, and magnetism.

Note that the mere presence of a force does not always cause a change in momentum. For example, if the engine works hard enough, a car can maintain constant velocity—and hence constant momentum—despite air resistance and road friction. In this case, the force generated by the engine precisely offsets the forces of air resistance and road friction, and we say that no **net force** is acting on the car.

As long as an object is not shedding (or gaining) mass, a change in momentum necessarily means a change in velocity (because momentum = mass × velocity), which is an acceleration. Hence, any net force will cause acceleration, and all accelerations must be caused by a force. That is why you feel forces—pushing you back into your seat as you speed up[1] and so on—when you accelerate in your car.

Mass and Weight

Up until now, we've been glossing over one key term: *mass*. Your **mass** refers to the amount of matter in your body, which is different from your *weight*. Imagine standing on a scale in an elevator (Figure 6.2). When the elevator is stationary or moving at constant velocity, the scale reads your "normal" weight. When the elevator is accelerating upward, the floor exerts an additional force that makes you feel heavier, and the scale verifies your greater weight. Note that your weight is greater than its "normal" value only while the elevator is *accelerating*, not while it is moving at constant velocity. When the elevator accelerates downward, the floor and scale are dropping away, so your weight is reduced. Thus, your weight varies with

[1]A technicality: Force acts in the same direction as the acceleration it causes. For example, when the car's velocity is increasing, the direction of the force is forward; however, you *feel* yourself pushed *backward* because the seat (and the whole car) is pushing forward against you.

the elevator's motion, while your mass remains the same. (You can verify these facts by taking a small bathroom scale with you on an elevator.)

If the cable breaks so that the elevator is in **free-fall,** the floor drops away at the same rate that you fall. You lose contact with the scale, becoming **weightless.** In fact, you are in free-fall whenever nothing is *preventing* you from falling. For example, you are in free-fall when you jump off a chair or spring from a diving board or trampoline. Surprising as it may seem, you have therefore experienced weightlessness many times in your life and can experience it right now simply by jumping off your chair. Of course, your weightlessness lasts for only the very short time until you hit the ground.

More precisely, your **weight** describes the *force* that acts on your mass, and it depends on the strength of gravity and other forces acting upon you (such as the force caused by the elevator's acceleration). Although your mass would be the same whether you were on the Earth, on the Moon, or in the Space Shuttle, you would weigh less on the Moon because of its weaker gravity and would be weightless in the Space Shuttle because it is in a constant state of free-fall as it orbits the Earth.

If you're surprised that the Space Shuttle is in free-fall, think of it like this: An object is in free-fall any time it is falling solely because of gravity. The faster the object is moving over the Earth's surface, the farther it travels before hitting the ground. We can see this by imagining a cannon that shoots a ball horizontally from a tall mountain; the faster it shoots the ball, the farther the ball goes (Figure 6.3). If the

FIGURE 6.3 The faster the cannonball is shot, the farther it goes before hitting the ground. If it goes fast enough, it will continually "fall around," or orbit, the Earth. With an even faster speed, it may escape the Earth's gravity altogether.

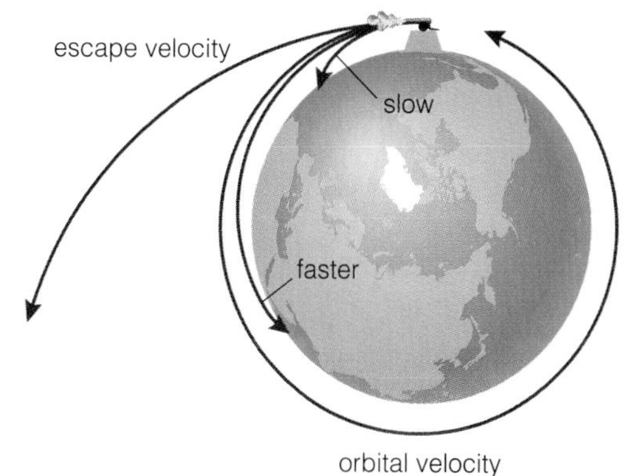

FIGURE 6.2 Mass is not the same as weight. The man's mass never changes, but his weight does.

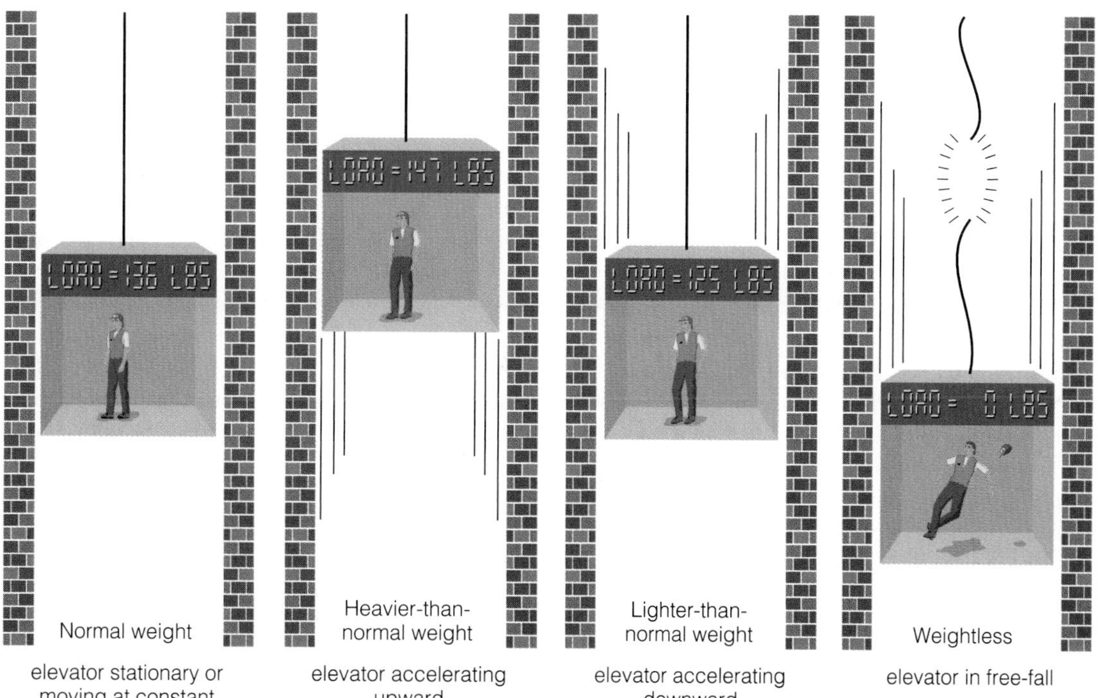

cannonball goes fast enough, its motion keeps it constantly "falling around" the Earth so that it never hits the ground (as long as we neglect air resistance). That is, an object orbiting the Earth is constantly in free-fall. Figure 6.3 also shows that an object launched at the **escape velocity** (or faster) will completely escape the Earth's gravity. The escape velocity from the Earth's surface is about 40,000 km/hr (25,000 mi/hr).

TIME OUT TO THINK *In the* Hitchhiker's Guide to the Galaxy *books, author Douglas Adams says that the trick to flying is to throw yourself at the ground and miss. Although this phrase does not really explain flying, which involves lift from air, it does describe an orbit fairly well. Explain.*

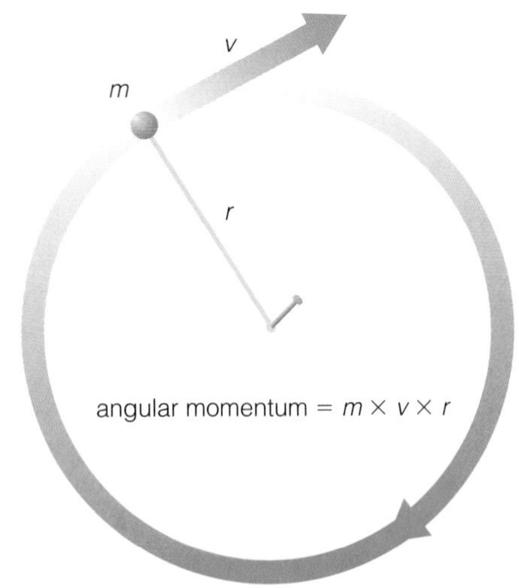

FIGURE 6.4 The angular momentum of an object moving in a circle is $m \times v \times r$.

Moving in Circles

Think about an ice skater spinning in place. She isn't going anywhere, so she has no overall velocity. Nevertheless, she has some kind of kinetic energy because every part of her body is moving in circles. How can we describe her motion?

The speed and direction in which she is spinning make up her *angular velocity.* For example, if she spins all the way around twice each second, her angular velocity is 2 revolutions per second (for simplicity, we won't worry about the direction of angular quantities). Suppose you walk out onto the ice to try to stop her spin. Clearly, you're going to have to apply some kind of force; for example, you might push against her outstretched arms. Because a force is involved, she must have some kind of momentum that can be changed, even though it is not forward momentum because she isn't going anywhere. The type of momentum carried by the spinning skater is called **angular momentum** because it depends on her angular velocity.

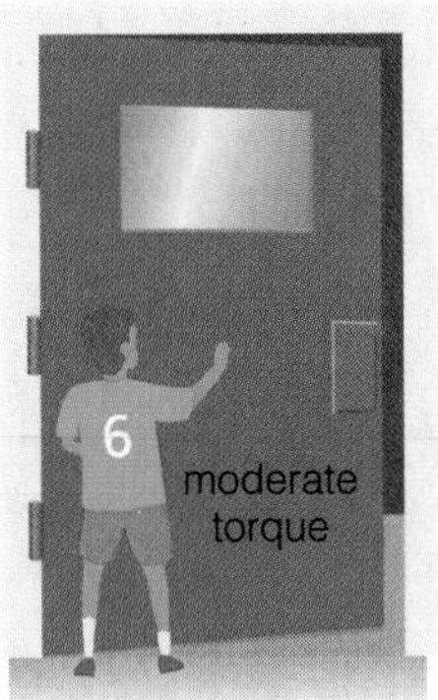

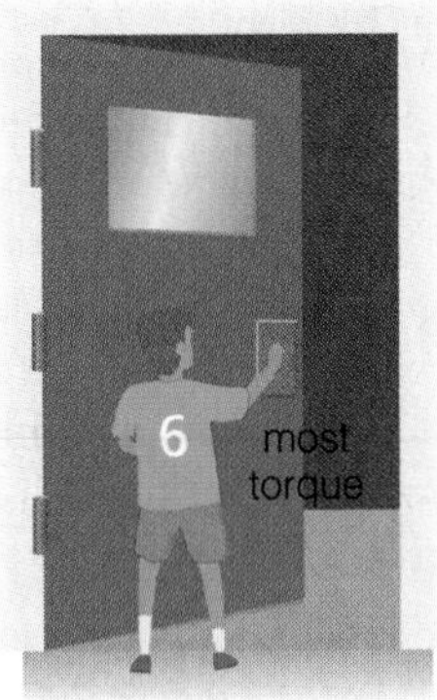

FIGURE 6.5 Opening a door requires applying a torque. Given the same amount of force, the torque on the door is greater if you push farther from the hinges (the door's rotation axis).

Any object that is spinning or orbiting has angular momentum. For example, the Earth has two kinds of angular momentum: It has *rotational* angular momentum because it spins on its axis once each day, and it has *orbital* angular momentum because it is going around the Sun. Quantitatively, the angular momentum of an object going in a circle is easy to calculate; the formula is:

$$\text{angular momentum} = m \times v \times r$$

where m is the object's mass, v is its speed around the circle, and r is the radius of the circle (Figure 6.4).

The type of force that can change an object's angular momentum is often called a *twisting force,* or a **torque.** For example, opening a door requires rotating the door on its hinges. Making it rotate means giving it some angular momentum, which you can do by applying a torque (Figure 6.5). Note that the torque depends not only on how much force you use to push the door, but also on *where* you push. The farther out from the hinges you push, the more torque you can apply and the easier it is to open the door.

TIME OUT TO THINK *Suppose an ice skater wants to* start *spinning. Explain why she won't start spinning if she simply stomps her foot straight down on the ice. How should she push off against the ice to start spinning? Why? What should she do when she wants to stop spinning?*

6.2 Understanding Motion: Newton's Laws

The human "trick" of being able to figure out where a ball will land before it is thrown or to predict how other motions will unfold requires understanding precisely how forces affect objects in motion. At first, the complexity of motion in daily life might lead you to guess that the laws governing how forces affect motion are also quite complex. For example, if you watch a falling piece of paper waft lazily to the ground, you'll see it rock irregularly back and forth in a seemingly unpredictable pattern. However, the complexity of this motion arises because the paper is affected by a variety of forces, including gravity and the changing forces caused by air currents. If you could analyze the forces individually, you'd find that each force affects the paper's motion in a simple, predictable way. Three simple laws, known as **Newton's laws of motion,** describe how forces affect motion.[2] Let's investigate them, again using examples from everyday life.

Newton's First Law of Motion

Newton's first law of motion deals with situations in which no net force is acting on an object. In particular, it states:

In the absence of a net force, an object moves with constant velocity.

Thus, objects at rest (velocity = 0) tend to remain at rest, and objects in motion tend to remain in motion with no change in either their speed or their direction.

The idea that an object at rest should remain at rest is rather obvious; after all, if a car is parked on a flat street, it won't suddenly start moving for no reason. But what if the car is traveling along a flat, straight road? Newton's first law says that the car should keep going forever *unless* a force acts upon it, but you know that the car eventually will come to a stop if you take your foot off the gas pedal. We must conclude that some forces are stopping the car—in this case forces arising from friction and air resistance.[3] If the car were in space, and therefore unaffected by friction or air, it would keep moving forever (though gravity would eventually alter its speed and direction). That is why interplanetary spacecraft, once launched into space, need no fuel to keep going.

Although friction cannot be eliminated on Earth, it sometimes can be minimized enough so that Newton's first law becomes more evident. For example, friction is low on ice, so a single push-off allows an ice skater to glide for a long time. Popular arcade games like *air hockey* also minimize friction so that a puck can travel for a long time before friction finally stops it.

Newton's first law also explains why you don't feel any sensation of motion when you're traveling in an airplane on a smooth flight. As long as the plane is traveling at constant velocity, no net force is acting on it or on you. Therefore, you feel no different from how you would feel at rest: You can walk around the cabin, play catch with a person a few rows forward, or relax and go to sleep just as though you were "at rest" on the ground.

Newton's Second Law of Motion

Newton's second law of motion tells us what happens to an object when a net force *is* present. We have already said that a net force changes an object's momentum, accelerating it in the direction of the force. Newton's second law quantifies this relationship, which can be stated in two equivalent ways:

$$\text{force} = \text{rate of change in momentum}$$

$$\text{force} = \text{mass} \times \text{acceleration} \quad (F = ma)$$

This law explains why you can throw a baseball farther than you can throw a shot-put. For both the baseball and the shot-put, the force delivered by your arm equals the product of mass and acceleration. Because the mass of the shot-put is greater than that of the baseball, the same force from your arm gives the shot-put a smaller acceleration. Due to its smaller acceleration, the shot-put leaves your hand with less speed than the baseball and thus travels a shorter distance before hitting the ground.

Newton's Third Law of Motion

Think for a moment about standing still on the ground. The force of gravity acts downward on you, so if this force were acting alone Newton's second law would demand that you be accelerating downward. The fact that you are not falling means that the ground must be pushing back up on you with exactly the right amount of force to offset gravity. How does

---

[2]Credit for discovering these laws really should go jointly to Galileo and Newton. They are called Newton's laws because he enumerated them in his book *Principia.*

[3]Some readers might wonder why gravity does not help stop the car. A force can affect motion only if it is acting along the direction of motion. Because gravity acts downward, it cannot affect the motion of a car traveling on a flat road.

the ground know how much force to apply? The answer is given by Newton's third law of motion:

For any force, there always is an equal and opposite reaction force.

According to this law, your body exerts a gravitational force on the Earth identical to the one the Earth exerts on you, except that it acts in the opposite direction. In this mutual pull, the ground just happens to be caught in the middle. Because other forces (between atoms and molecules in the Earth) keep the ground stationary with respect to the center of the Earth, the opposite, upward force is transmitted to you by the ground, holding you in place. Newton's third law also explains how rockets work: Engines generate an explosive force driving hot gas out the back, which creates an equal and opposite force propelling the rocket forward.

Conservation of Momentum

If you look more closely at Newton's laws, you will see that they all reflect aspects of a deeper principle: the *conservation of momentum.* Like the amount of energy, the total amount of momentum in the universe is conserved—that is, it does not change. Newton's first law says that an individual object's momentum will not change at all if the object is left alone. Newton's second law says that a force can change the object's momentum, but Newton's third law says that another equal and opposite force simultaneously changes some other object's momentum by a precisely opposite amount. Total momentum always remains unchanged.

On a pool table, momentum conservation is rather obvious. When one ball hits another ball "dead on," the first ball stops and the second one takes off with the speed of the first (Figure 6.6).

Common Misconceptions: What Makes a Rocket Launch?

If you've ever watched a rocket launch, it's easy to see why many people believe that the rocket "pushes off" the ground. In fact, the ground has nothing to do with the rocket launch. The rocket takes off because of momentum conservation. Rocket engines are designed to expel hot gas with an enormous amount of momentum. To balance the explosive force driving gas out the back of the rocket, an equal and opposite force must propel the rocket forward, keeping the total momentum—gas plus rocket—unchanged. Thus, rockets can be launched horizontally as well as vertically, and a rocket can be "launched" in space (e.g., from a space station) with no need for any nearby solid ground.

Mathematical Insight 6.1 Units of Force, Mass, and Weight

Newton's second law, $F = ma$, shows that the units of force are equal to a unit of mass multiplied by a unit of acceleration. For example, if a mass of 1 kg accelerates at 10 m/s^2, the magnitude of the responsible force is:

$$\text{force} = \text{mass} \times \text{acceleration} = 1\ \text{kg} \times 10\,\frac{\text{m}}{\text{s}^2} = 10\,\frac{\text{kg} \times \text{m}}{\text{s}^2} = 10\ \text{newtons}$$

Thus, the standard unit of force is the *kilogram-meter per second squared,* which is called the **newton** for short.

We now can further clarify the difference between mass and weight. When you stand on a scale, it records the downward force that you exert on it, which is equal and opposite to the upward force it exerts on you. If the scale were suddenly pulled out from under your feet, you would begin accelerating downward with the acceleration of gravity. Thus, when you are standing still, the scale must be supporting you with a force equal to your mass times the acceleration of gravity. Your weight must also equal this force (but in an opposite direction):

$$\text{weight when not accelerating} = \text{mass} \times \text{acceleration of gravity}$$

Like any force, weight has units of mass times acceleration. Thus, although we commonly speak of weights in *kilograms,* this usage is not technically correct: Kilograms are a unit of mass, not of force. You may safely ignore this technicality as long as you are dealing with objects on the Earth that are not accelerating. In elevators, or on other planets, the distinction between mass and weight is important and cannot be ignored.

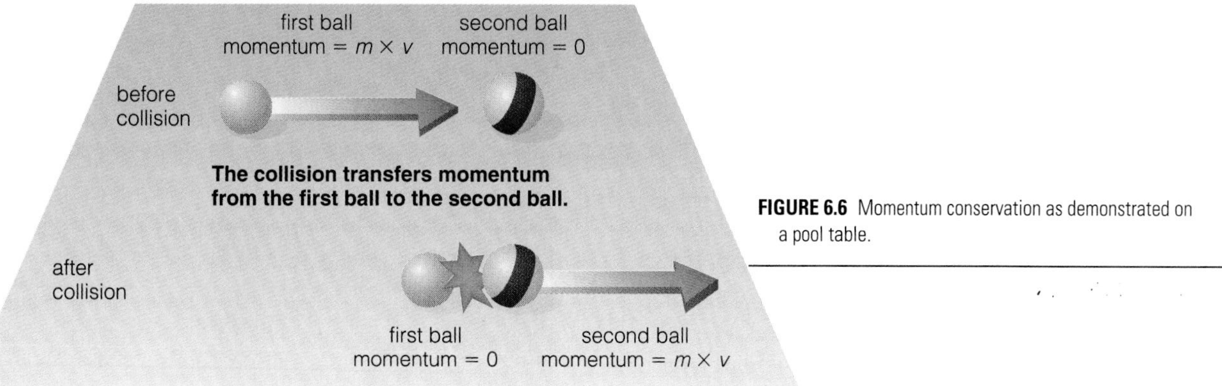

FIGURE 6.6 Momentum conservation as demonstrated on a pool table.

Sometimes, momentum conservation is less apparent. When you jump into the air, how do you get your upward momentum? As your legs propel you skyward, they are actually pushing the Earth in the other direction, giving the Earth's momentum an equal and opposite kick. However, the Earth's huge mass renders its acceleration undetectable. During your brief flight, the gravitational forces between you and the Earth return your upward momentum to the Earth. Again, the total momentum remains the same at all times.

In astronomy, *angular* momentum conservation is particularly important. Twisting forces, or torques, can transfer angular momentum from one object to another, but the total always remains the same. When no net torque is present, a law very similar to Newton's first law applies—the **law of conservation of angular momentum:**

In the absence of net torque (twisting force), the total angular momentum of a system remains constant.

Because gravity usually is not a twisting force, angular momentum almost always is conserved when we are dealing with astronomical motion. We will see countless applications of this law throughout the book, but in everyday life it is most familiar to us in the example of an ice skater. Since there is so little friction on ice, a spinning ice skater essentially keeps a constant angular momentum. Recall that angular momentum can be expressed as the product $m \times v \times r$. When the skater pulls in her extended arms, she effectively decreases her radius. Therefore, for the product $m \times v \times r$ to remain unchanged, her velocity of rotation must increase (Figure 6.7).

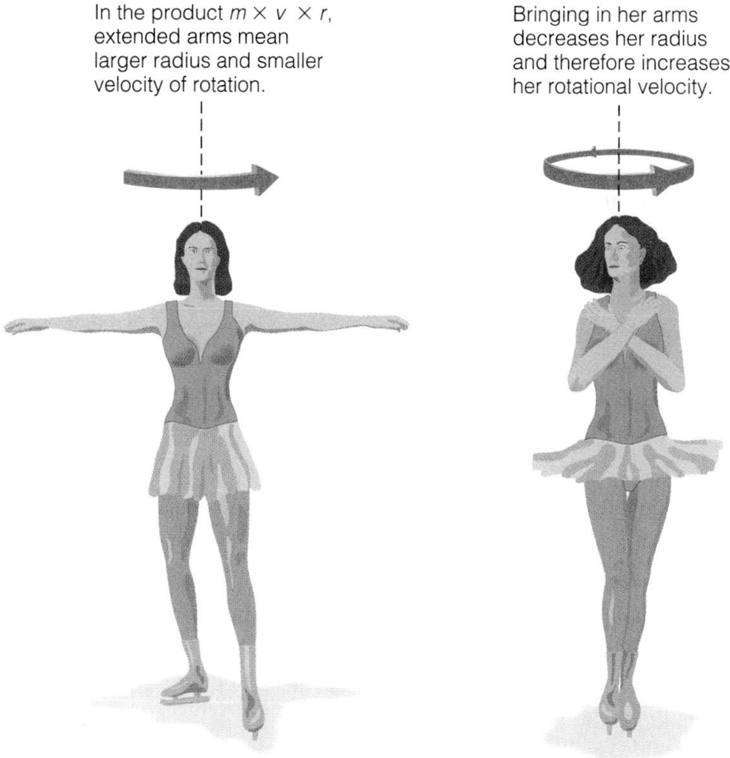

FIGURE 6.7 A spinning skater conserves angular momentum.

6.3 Planetary Motion: The Copernican Revolution

Now that we have described motion in general terms, we can turn to the important issue of planetary motion in our solar system. Although planetary motion is among the simplest motions found in nature, it remained mysterious for most of human history because of its *apparent* complexity in our sky [Section 1.5]. The prevailing view of European scholars at the dawn of the Renaissance remained that of Ptolemy, whose complex model had planets spinning on circles upon circles around the Earth [Section 4.3]. The sea change in human understanding of planetary motion is one of the most remarkable stories in the history of science. It began early in the sixteenth century, with the work of Nicholas Copernicus.

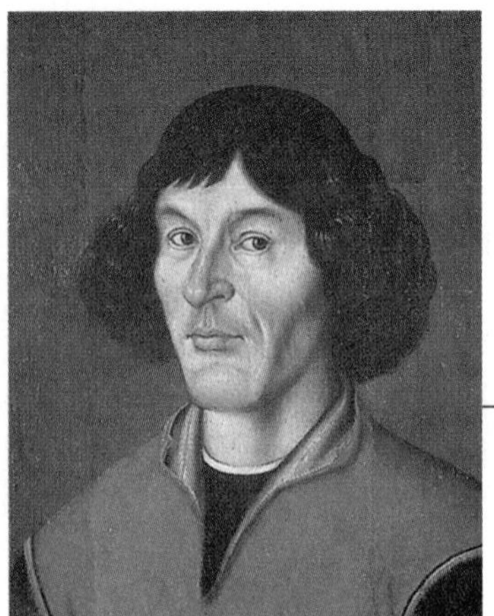
Copernicus

Nicholas Copernicus: The Revolution Begins

Copernicus was born in Toruń, Poland, on February 19, 1473. His family was wealthy, and he received a first-class education, studying painting, mathematics, medicine, and law. He began studying astronomy in his late 20s. By that time, tables of planetary motion based on the Ptolemaic model of the universe were noticeably inaccurate.[4] Copernicus concluded that planetary motion could be more simply explained in a Sun-centered solar system, as had been suggested by Aristarchus some 1,800 years earlier [Section 3.1], and he began developing a Sun-centered system for predicting planetary positions.

Copernicus was hesitant to publish his work for fear that his suggestion that the Earth moved would be considered heretical. Nevertheless, he discussed his system with other scholars and generated great interest in his work. At the urging of some of these scholars, including some high-ranking officials of the Church, he finally agreed to publish a book describing his system. The book, *De Revolutionibus Orbium Caelestium,* or "Concerning the Revolutions of the Heavenly Spheres," was published in 1543. Legend has it that Copernicus saw the first printed copy on the day he died—May 24, 1543.

In addition to its aesthetic advantages, the Sun-centered system of Copernicus allowed him to discover a mathematical relationship between a planet's true orbital period around the Sun and the time between successive appearances of the planet in its "full" phase (directly opposite the Sun in the sky).[5] He was also able to use geometrical techniques to estimate the distances of the planets from the Sun in terms of the Earth–Sun distance (i.e., distances in astronomical units [Section 3.1]). However, the actual model published by Copernicus did not predict planetary positions substantially more accurately than did the old Ptolemaic model. Moreover, it remained complex because Copernicus held to the ancient Greek belief that all heavenly motions must follow perfect circles. Because the true orbits of the planets are *not* circles, Copernicus found it necessary to add circles upon circles to his system, just as in the Ptolemaic system. As a result, the Copernican system won relatively few converts in the 50 years after it was published. After all, why throw out thousands of years of tradition for a new system that predicted planetary motion equally poorly?

Tycho Brahe: The Greatest Naked-Eye Observer of All Time

Part of the difficulty faced by astronomers who sought to improve either the Ptolemaic or the Copernican system was a lack of quality data. The telescope had not yet been invented, and existing naked-eye observations were not very accurate. In the late 1500s, a Danish nobleman named Tycho Brahe (1546–1601) set about correcting this problem.

When Tycho was a young boy, his family discouraged his interest in astronomy. He therefore hid his passion, learning the constellations from a miniature model of a celestial sphere that he kept hidden. In 1563, Tycho decided to observe a widely anticipated conjunction of Jupiter and Saturn in the sky. To his surprise, the conjunction occurred about a month later than had been predicted. He resolved to improve the state of astronomical prediction and set about compiling careful observations of stellar and planetary positions in the sky.

Tycho often was arrogant about both his noble birth and his learned abilities. At age 20, he fought a duel with another student over which of them was the better mathematician. Part of his nose was cut

---

[4]The best set of tables, known as the *Alfonsine Tables,* was compiled in 1252 under the guidance of the Spanish monarch Alfonso X (1221–1284). Commenting on the tedious nature of the calculations required by the Ptolemaic model, Alfonso X supposedly said that if he had been present at the creation, he would have recommended a simpler design for the universe.

[5]The true orbital period of a planet is called its *sidereal period* (because the orbit is fixed relative to the stars). A planet appears in its "full" phase when it is directly opposite the Sun in our sky and hence said to be at *opposition.* The time between successive appearances at opposition is the planet's *synodic period.*

Tycho's naked-eye observatory

Tycho Brahe

off, so he designed a replacement piece made of silver and gold. Tycho's fame grew after he observed what he called a *nova,* meaning "new star," in 1572 and proved that it was at a distance much farther away than the Moon.[6] Today, we know that Tycho saw a *supernova*—the explosion of a distant star. In 1577, Tycho observed a comet and proved that it, too, lay in the realm of the heavens; others, notably Aristotle, had argued that comets were phenomena of the Earth's atmosphere. King Frederick II of Denmark then decided to sponsor Tycho's ongoing work, providing him with money to build an unparalleled observatory for naked-eye observations.[7] Over a period of three decades, Tycho and his assistants compiled naked-eye observations accurate to within less than 1 arcminute. Because the telescope was invented shortly after his death, Tycho's data remain the best set of naked-eye observations ever made.

Despite the quality of his observations, Tycho never succeeded in coming up with a satisfying explanation for planetary motion. He did, however, succeed in finding someone who could: In 1600, he hired a young German astronomer named Johannes Kepler (1571–1630). Kepler and Tycho had a strained relationship while Tycho was living.[8] But in 1601, as Tycho lay on his deathbed, he bequeathed all his notebooks of observations to Kepler and begged him to find a system that would make sense of the observations so "that it may not appear I have lived in vain."

Kepler's Reformation: The Laws of Planetary Motion

Kepler was deeply religious and believed that understanding the geometry of the heavens would bring him closer to God. Kepler, like Copernicus, believed that Earth and the other planets traveled around the Sun in circular orbits, and he worked diligently to match circular motions to Tycho's data.

Kepler worked with particular intensity to find an orbit for Mars, which posed the greatest difficulties in matching the data to a circular orbit. After years of calculation, Kepler found a circular orbit that matched nearly all of Tycho's observations of Mars to within 2 arcminutes. In two cases, however, this orbit predicted a position for Mars that differed from Tycho's observations by the slightly greater margin of 8 arcminutes.

Kepler surely was tempted to ignore these two observations and attribute them to an error by Tycho.

[6]Tycho compared observations made by other astronomers at other locations on Earth to his own, proving that the nova had no observable parallax and must be more distant than the Moon.

[7]After Frederick II died in 1588, Tycho moved to Prague, where his work was supported by German emperor Rudolf II.

[8]A particularly moving version of the story of Tycho and Kepler can be found in episode 3 of Carl Sagan's *Cosmos* video series.

Kepler

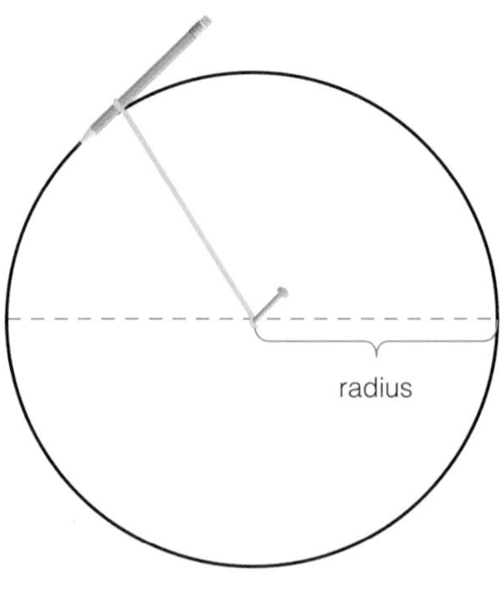

a Drawing a circle.

FIGURE 6.8

After all, 8 arcminutes is barely one-fourth the angular diameter of the full moon. But Kepler trusted Tycho's careful work, and the missed 8 arcminutes finally led him to abandon the idea of circular orbits—and to find the correct solution to the ancient riddle of planetary motion. About this event, Kepler wrote:

> *If I had believed that we could ignore these eight minutes [of arc], I would have patched up my hypothesis accordingly. But, since it was not permissible to ignore, those eight minutes pointed the road to a complete reformation in astronomy.*

TIME OUT TO THINK ***Some historians assert that Kepler's discovery represented the true birth of modern science because, for the first time, a scientist was willing to cast off long-held beliefs in a quest to match theory to observation. How did Kepler's work meet the criteria for the modern scientific method?***

Kepler summarized his discoveries with three simple laws that we now call **Kepler's laws of planetary motion.** He published the first two laws in 1610 and the third in 1618. (Be careful not to confuse Kepler's three laws of *planetary* motion with Newton's three laws that apply generally to *all* motion.)

Kepler's key discovery was that planetary orbits are not circles but instead are a special type of oval called an **ellipse.** You probably know how to draw a circle by putting a pencil on the end of a string, tacking the string to a board, and pulling the pencil around (Figure 6.8a). Drawing an ellipse is similar, except that you must stretch the string around *two* tacks (Figure 6.8b). The locations of the two tacks are called the **foci** (singular, **focus**) of the ellipse. By altering the distance between the two foci while keeping the same length of string, you can draw ellipses of varying **eccentricity,** a quantity that describes how much an ellipse deviates from a perfect circle (Figure 6.8c).

Kepler's first law states that *the orbit of each planet about the Sun is an ellipse with the Sun at one focus* (Figure 6.9). (There is nothing at the other focus.) This law tells us that a planet's distance from the Sun varies during its orbit: It is closest at the point called **perihelion,** and farthest at the point called **aphelion.**[9] The *average* of a planet's perihelion and aphelion distances is called its **semimajor axis;** we will refer to this simply as the planet's average distance from the Sun.

Kepler's second law states that, *as a planet moves around its orbit, it sweeps out equal areas in equal times.* As shown in Figure 6.10, this means that the planet moves a greater distance when it is near perihelion than it does in the same amount of time near aphelion; that is, the planet travels faster when it is nearer to the Sun and slower when it is farther from the Sun.

Kepler's third law describes how a planet's *period,* or the number of years it takes to complete one

[9] *Helios* is Greek for the Sun, the prefix *peri* means "near," and the prefix *ap* (or *apo*) means "away." Thus, *perihelion* means "near the Sun" and *aphelion* means "away from the Sun."

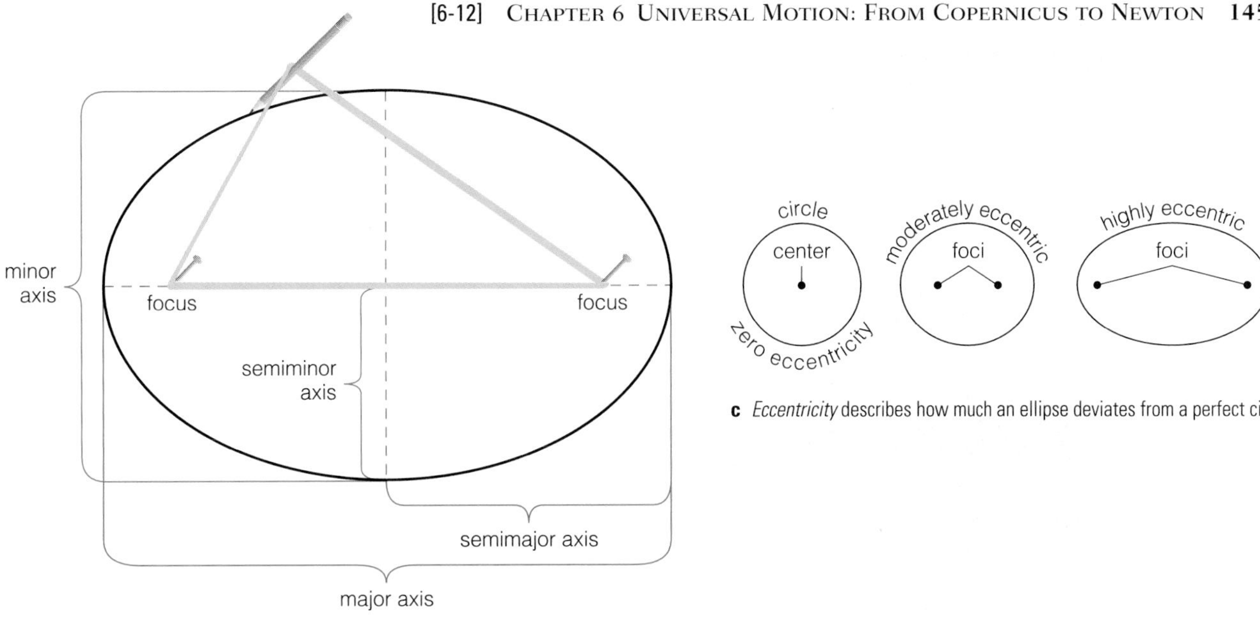

c *Eccentricity* describes how much an ellipse deviates from a perfect circle.

b Drawing an ellipse.

orbit of the Sun, is related to its average distance from the Sun in astronomical units (1 AU ≈ 150 million km). In particular, for any planet orbiting the Sun:

$$(\text{orbital period in years})^2 = (\text{average distance in AU})^3$$

This formula is often written more simply as $p^2 = a^3$, where it is understood that p is the orbital period measured in years and a is the average distance measured in AU (Figure 6.11a). Note that a planet's period does not depend on the eccentricity of its orbit: All orbits with the same semimajor axis have the same period. Nor does the period depend on the mass of the planet: Any object located an average of 1 AU from the Sun would orbit the Sun in a year (as long as the object's mass was small compared to the Sun's mass).

Kepler's third law shows that *more distant planets move at slower speeds* in their orbits about the Sun, which we can see by graphing planetary periods against average distances from the Sun (Figure 6.11b). For example, Saturn is slightly less than twice as far as Jupiter from the Sun but takes almost three times as long to orbit the Sun; thus, Saturn must be moving along its orbit at a slower average speed than Jupiter.

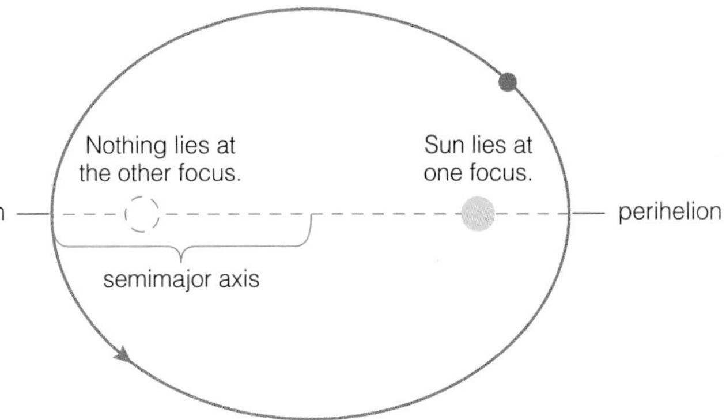

FIGURE 6.9 Kepler's first law: The orbit of each planet about the Sun is an ellipse with the Sun at one focus.

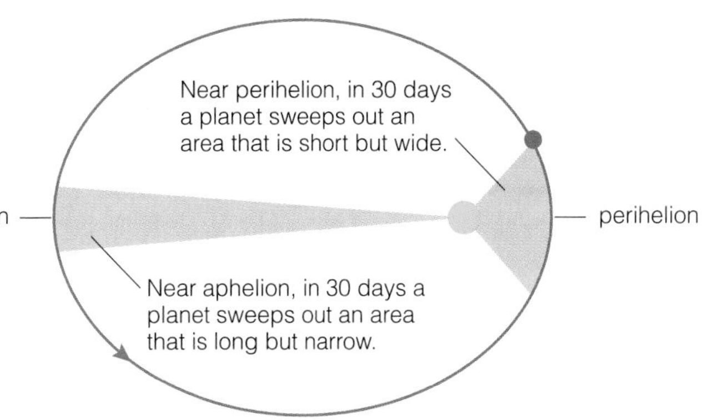

FIGURE 6.10 Kepler's second law: As a planet moves around its orbit, it sweeps out equal areas in equal times.

Galileo: The Death of the Earth-Centered Universe

The success of Kepler's laws in matching Tycho's data provided strong evidence in favor of Copernicus's placement of the Sun, rather than the Earth, at the center of the solar system. Nevertheless, many

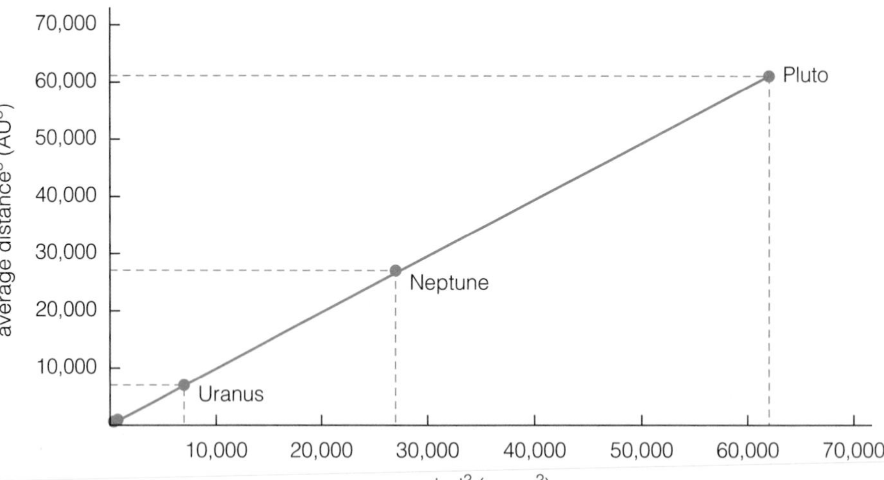

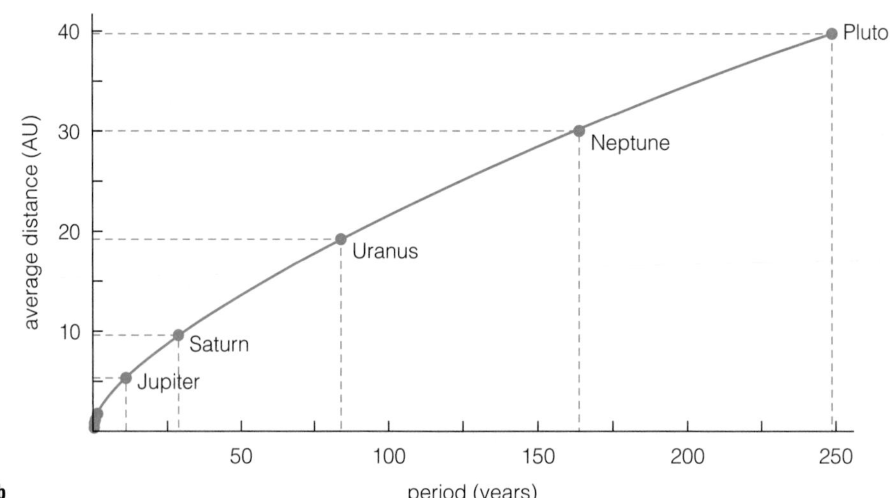

FIGURE 6.11 (**a**) Kepler's third law shows that the period of a planet squared is equal to its average orbital distance cubed. (**b**) A plot of average distance against orbital period. We can use this plot to find that planets nearer the Sun orbit faster than more distant planets.

scientists still voiced reasonable objections to the Copernican view. There were three basic objections, all rooted in the 2,000-year-old beliefs of Aristotle (384–322 B.C.) and other ancient Greeks. First, Aristotle had held that the Earth could not be moving because, if it was, objects such as birds, falling stones, and clouds would be left behind as the Earth moved along its way. Second, the idea of noncircular orbits contradicted the ancient Greek belief that the heavens—the realm of the Sun, Moon, planets, and stars—must be perfect and unchanging. The third objection was the ancient argument that stellar parallax ought to be detectable if the Earth orbits the Sun [Section 3.1]. Galileo (1564–1642), a contemporary and correspondent of Kepler, answered all three objections.

Galileo defused the first objection with experiments that almost single-handedly overturned the Aristotelian view of physics. His demonstration that gravity accelerates all objects by the same amount directly contradicted Aristotle's claim that heavier objects would fall to the ground faster. Aristotle also had taught that the natural tendency of any object was to come to rest. Galileo demonstrated that a moving object remains in motion *unless* a force acts to stop it, an idea now codified in Newton's first law of motion. Thus, he concluded that objects such as birds, stones, and clouds that are moving with the Earth should *stay* with the Earth unless some force knocks them away. This same idea explains why passengers in an airplane stay with the moving airplane even when they leave their seats.

Tycho's supernova and comet observations already had shown that the heavens could change, and Galileo shattered the idea of heavenly perfection after he built a telescope in late 1609.[10] Through his tele-

[10]Hans Lippershey invented the telescope in 1608. Galileo built his own telescope after hearing reports of Lippershey's invention, and he made many improvements to the design.

THINKING ABOUT . . .

Aristotle

Aristotle (384–322 B.C.) is among the best-known philosophers of the ancient world. Both his parents died while he was a child, and he was raised by a family friend. In his 20s and 30s, he studied under Plato (427–347 B.C.) at Plato's Academy. He later founded his own school, called the Lyceum, where he studied and lectured on virtually every subject. Historical records tell us that his lectures were collected and published in 150 volumes. About 50 of these volumes survive to the present day.

Many of Aristotle's scientific discoveries involved the nature of plants and animals. He studied more than 500 animal species in detail, including dissecting specimens of nearly 50 species, and came up with a strikingly modern classification system. For example, he was the first person to recognize that dolphins should be classified with land mammals rather than with fish. In mathematics, he is known for laying the foundations of mathematical logic. Unfortunately, he was far less successful in physics and astronomy, areas in which many of his claims turned out to be wrong.

Interestingly, Aristotle's philosophies were not particularly influential until many centuries after his death. His books were preserved and valued by Islamic scholars, but they were unknown in Europe until they were translated into Latin in the twelfth and thirteenth centuries. Aristotle achieved his near-reverential status only after St. Thomas Aquinas (1225–1274) integrated Aristotle's philosophy into Christian theology. In the ancient world, Aristotle's greatest influence came indirectly, through his role as the tutor of Alexander the Great [Section 4.3].

Galileo

FIGURE 6.12 The shadows cast by mountains and crater rims near the dividing line between the light and dark portions of the lunar face prove that the Moon's surface is not perfectly smooth.

scope, he saw sunspots on the Sun, which were considered "imperfections" at the time. He also used his telescope to prove that the Moon has mountains and valleys like the "imperfect" Earth by noticing the shadows cast near the dividing line between the light and dark portions of the lunar face (Figure 6.12). If the heavens were not perfect, then the idea of elliptical (rather than circular) orbits was not so objectionable.

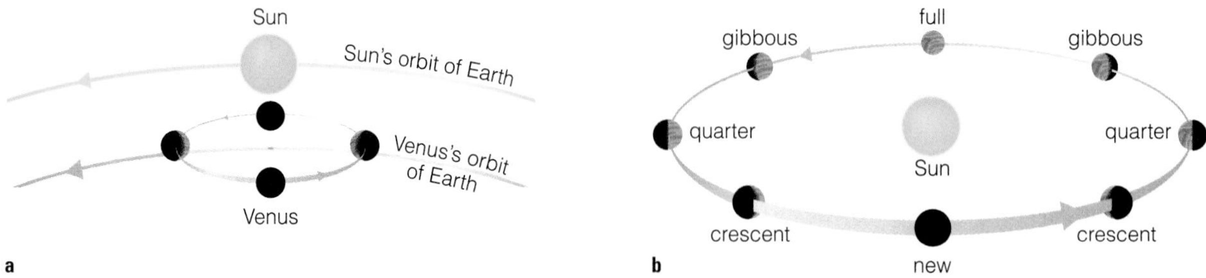

FIGURE 6.13 (**a**) In the Ptolemaic system, Venus follows a circle upon a circle that keeps it close to the Sun in our sky. Therefore, its phases would range only from new to crescent. (**b**) Galileo saw Venus go through a complete set of phases and therefore proved that it orbits the Sun.

The absence of observable stellar parallax had been of particular concern to Tycho. Based on his estimates of the distances of stars, Tycho concluded that his naked-eye observations were sufficiently precise to detect stellar parallax if the Earth orbited the Sun.[11] Refuting Tycho's argument required showing that the stars were more distant than Tycho had thought and therefore too distant for him to have observed stellar parallax. Although Galileo didn't actually prove this fact, he provided strong evidence in its favor. In particular, he saw with his telescope that the Milky Way resolved into countless individual stars, which helped him argue that the stars were far more numerous and more distant than Tycho had imagined.

The true death knell for an Earth-centered universe came with two of Galileo's earliest discoveries through the telescope. First, he observed four moons clearly orbiting Jupiter, *not* the Earth. Soon thereafter, he observed that Venus goes through phases like the Moon, proving that Venus must orbit the Sun and not the Earth (Figure 6.13).

At last the Copernican revolution was complete. Thanks to Galileo, by the mid-1600s the scientific community was near-unanimous in accepting Kepler's model of planetary motion. The debate turned to the question of *why* Kepler's laws hold true.

[11]Tycho based his estimates of stellar distances on what he *thought* were their angular sizes, but he was really measuring effects of atmospheric refraction. His planetary observations convinced him that the *planets* must orbit the Sun, so he advocated a model in which the Sun orbits the Earth while all other planets orbit the Sun. Few people took this model seriously, and Kepler's work soon made it a mere historical curiosity.

6.4 The Force of Gravity

The Aristotelian view of the world, held in Europe as near-gospel truth, was in tatters. The Earth was not the center of the universe, and the laws of physics were not what Aristotle had believed. But the final blow was still to come. Aristotle had maintained that the heavens were totally distinct from Earth and that the physical laws on Earth could not be applied to understanding heavenly motion.

In 1666, Isaac Newton saw an apple fall to the ground.[12] He suddenly realized that the force that brought the apple to the ground and the force that held the Moon in orbit were the same. That insight brought the Earth and the heavens together in one *universe*. It also heralded the birth of the modern science of *astrophysics,* in which physical laws discovered on Earth are applied to phenomena throughout the cosmos.

Newton and the Universal Law of Gravitation

Isaac Newton was born prematurely in Lincolnshire, England, on Christmas day in 1642.[13] His father, a farmer who had never learned to read or write, died 3 months before he was born. Newton had a difficult childhood and showed few signs of unusual talent. He attended Trinity College at Cambridge, where he earned his keep by performing menial labor, such as cleaning the boots and bathrooms of wealthier students and waiting on their tables.

Shortly after he graduated, the plague hit Cambridge, and Newton returned home. It was there that

[12]The story of the apple may or may not be true, but Newton himself told the story.

[13]England still was using the Julian calendar [Section S1.4] at the time; by the Gregorian calendar, Newton was born on January 4, 1643.

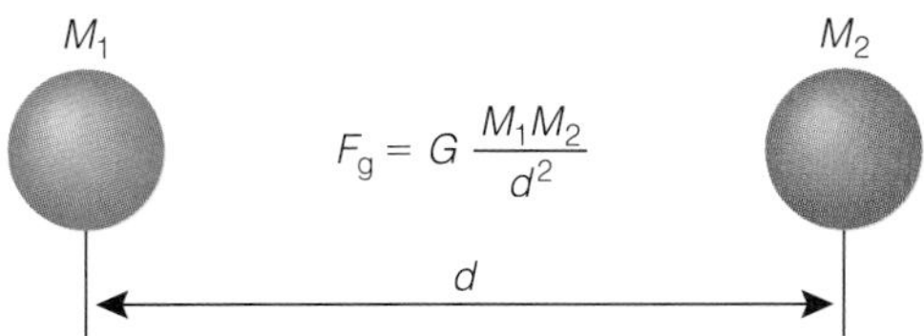

FIGURE 6.14 The law of universal gravitation.

Newton

he saw the apple fall and began his detailed investigations of gravity, light, optics, and mathematics. Over the next 20 years, Newton's work completely revolutionized mathematics and science. Besides his work on motion and gravity, he conducted crucial experiments regarding the nature of light, built the first reflecting telescopes, and invented the branch of mathematics called *calculus*. The compendium of Newton's discoveries is so tremendous that it would take a complete book just to describe them, and many more books to describe their influence on civilization. When Newton died in 1727, at age 84, the English poet Alexander Pope composed the following epitaph:

> *Nature, and Nature's laws lay hid in the Night.*
> *God said,* Let Newton be! *and all was Light.*

In 1687, Newton published *Philosophiae Naturalis Principia Mathematica* ("Mathematical Principles of Natural Philosophy"), usually referred to as *Principia*. In this book, Newton stated his three laws of motion and another law, called the **universal law of gravitation,** that describes the force of gravity.

Three simple statements summarize the universal law of gravitation:

- Every mass attracts every other mass through the force called *gravity*.
- The force of attraction between any two objects is *directly proportional* to the product of their masses. For example, doubling the mass of *one* object doubles the force of gravity between the two objects.
- The force of attraction decreases rapidly as the centers of the objects get farther apart. In particular, it decreases with the *square* of the distance between their centers; that is, the force follows an **inverse square law** with distance. For example, doubling the distance between two objects weakens the force of gravity by a factor of 2^2, or 4.

Mathematically, Newton's law of universal gravitation is written:

$$F_g = G\frac{M_1 M_2}{d^2} \quad \left(G = 6.67 \times 10^{-11} \frac{\text{m}^3}{\text{kg} \times \text{s}^2}\right)$$

where F_g is the force of gravitational attraction, M_1 and M_2 are the masses of the two objects, and d is the distance between their *centers* (Figure 6.14). The symbol G is a constant called the **gravitational constant;** its numerical value was not known to Newton but has since been measured by experiments.

TIME OUT TO THINK ***Consider the gravitational force between two objects with mass M_1 and M_2 separated by a distance d. How will the gravitational force change if the distance between them increases to $3 \times d$? How will the force change if the distance decreases to $0.1 \times d$? How will the force change if the mass of one of the objects magically triples? What if both objects triple in mass?***

The "Why" of Kepler's Laws, and More

For almost 70 years after Kepler published his first two laws in 1610, the outstanding unsolved problem in science was what caused Kepler's laws to hold true. Kepler himself speculated that his laws might be explained by a force holding the planets in their orbits about the Sun. He incorrectly guessed that this force might be related to magnetism, an idea shared by Galileo.[14] In *Principia,* Newton showed this force to be gravity.

Newton explained Kepler's laws by *solving* the law of universal gravitation together with the laws of motion. The solution is a bit like solving a pair of algebraic equations, but it requires the use of calculus. Newton found that the elliptical orbits with varying speeds described by Kepler's first two laws represent one possible solution. The easiest way to understand

[14]This idea was first suggested by William Gilbert (1544–1603), an early believer in the Copernican system.

the varying speeds is in terms of *conservation of angular momentum*. A planet's orbital angular momentum is the product $m \times v \times r$, where m is the mass of the planet, v is its orbital speed, and r is its distance ("radius") from the Sun. To keep this product constant, the planet's orbital speed (v) must go up when its distance from the Sun (r) goes down, and vice versa.

Newton's work showed that Kepler's first two laws apply not just to planets but to *any* object going around another object under the force of gravity.[15] The orbits of satellites around the Earth, of moons around planets, of asteroids around the Sun, and of binary stars around each other are all ellipses in which orbital speeds vary so that angular momentum is conserved. Moreover, Newton found that elliptical orbits are not the only possible solution to the equations involving the law of gravity. Orbits in the shape of *parabolas* or *hyperbolas* are also allowed (Figure 6.15). Elliptical orbits are **bound orbits** because gravity creates a bond that makes one object go around and around the other. In contrast, parabolas and hyperbolas are **unbound orbits.** A comet with an unbound orbit comes in toward the Sun just once, looping around the Sun and never returning.

[15]More technically, pairs of objects orbit each other in ellipses with their *center of mass* at one focus. If one object is much more massive than the other, the center of mass is very close to the massive object's center.

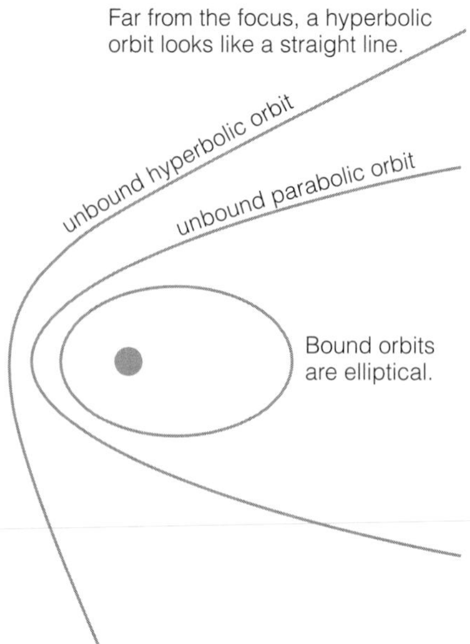

FIGURE 6.15 Orbits allowed by the law of gravity.

Kepler's third law also follows naturally from the law of gravity: The force is stronger nearer the Sun, so inner planets must move faster than outer planets to avoid falling toward the Sun. However, in solving the equations involving gravity, Newton showed that Kepler's statement $p^2 = a^3$ is only one special case of

Mathematical Insight 6.2 Using Newton's Version of Kepler's Third Law

Newton's version of Kepler's third law is remarkably powerful, as the following two examples show.

Example 1: Use the fact that the Earth orbits the Sun in 1 year at an average distance of 150 million km (1 AU) to calculate the mass of the Sun.

Solution: Because the Earth is much less massive than the Sun, the sum of their masses is approximately the mass of the Sun alone; that is, $M_{\text{Sun}} + M_{\text{Earth}} \approx M_{\text{Sun}}$. We can therefore use Newton's version of Kepler's third law in the following form:

$$(p_{\text{Earth}})^2 \approx \frac{4\pi^2}{G \times M_{\text{Sun}}} (a_{\text{Earth}})^3$$

We can solve this equation for the mass of the Sun by multiplying both sides by M_{Sun} and dividing both sides by $(p_{\text{Earth}})^2$:

$$M_{\text{Sun}} \approx \frac{4\pi^2}{G} \frac{(a_{\text{Earth}})^3}{(p_{\text{Earth}})^2}$$

The Earth's orbital period is $p_{\text{Earth}} = 1$ year, which you can confirm is the same as 3.15×10^7 seconds. Its average distance from the Sun is $a_{\text{Earth}} = 150$ million km, or 1.5×10^{11} m. Using these values and the experimentally measured value $G = 6.67 \times 10^{-11}\ \text{m}^3/(\text{kg} \times \text{s}^2)$, we find:

$$M_{\text{Sun}} \approx \frac{4\pi^2}{G} \frac{(a_{\text{Earth}})^3}{(p_{\text{Earth}})^2} = \frac{4\pi^2(1.5 \times 10^{11}\ \text{m})^3}{\left(6.67 \times 10^{-11} \dfrac{\text{m}^3}{\text{kg} \times \text{s}^2}\right)\left(3.15 \times 10^7\ \text{s}\right)^2} = 2 \times 10^{30}\ \text{kg}$$

a more general law. Newton's generalization of Kepler's third law is:

$$p^2 = \frac{4\pi^2}{G(M_1 + M_2)} a^3$$

In this form, the law applies to any pair of orbiting objects with bound orbits, and it is not necessary to measure the period (p) in years and the average distance (a) in astronomical units; you can use any units, as long as you properly match them to the units you use for the gravitational constant G. In addition, the law now includes the *masses* of the orbiting objects, which has important implications.

Suppose the two objects are the Sun and a planet. Because the Sun is much more massive than any planet, the sum $M_{\text{Sun}} + M_{\text{planet}}$ is pretty much just M_{Sun}. In that case, we can rewrite Newton's version of Kepler's third law as:

$$p^2_{\text{planet}} = \frac{4\pi^2}{G \times M_{\text{Sun}}} a^3_{\text{planet}}$$

We now find two remarkable results. First, note that the only characteristic of a planet that affects its orbital period is its average distance from the Sun. Generalizing this result, we find that the orbital period of any object orbiting a much more massive object depends only on its average distance. For example, the period of any satellite orbiting the Earth depends only on its average distance from the center of the Earth. All satellites orbiting 42,000 kilometers from the center of the Earth take 1 day to complete an orbit, and all satellites in the low-Earth orbit of the Space shuttle orbit the Earth in about 90 minutes. Thus, an astronaut on a space walk remains close to the Space Shuttle because the astronaut and the shuttle both orbit the Earth in the same amount of time.

Second, by knowing the period p and the average distance a of any planet, we can calculate the mass of the Sun. More generally, we can calculate the mass of any massive object by measuring the period p and the average distance a of something that orbits it. By measuring the period and average distance of any one of Jupiter's moons, we can calculate the mass of Jupiter. By measuring the period and distance of a small star orbiting a more massive star, we can calculate (approximately) the larger star's mass. In fact, Newton's version of Kepler's third law provides the primary means by which we determine masses throughout the universe.

6.5 Tides and Tidal Forces

You probably know that, at any location along a coast, the tide rises and falls twice each day. Newton's universal law of gravitation explains the origin of these

The mass of the Sun is about 2×10^{30} kg. Simply by knowing the Earth's orbital period and distance from the Sun, and the gravitational constant G, we have used Newton's version of Kepler's third law to "weigh" the Sun!

Example 2. A *geosynchronous satellite* orbits the Earth in the same amount of time that Earth rotates: 1 sidereal day. If a geosynchronous satellite is also in an equatorial orbit, it is said to be *geostationary* because it remains fixed in the sky (i.e., it maintains a constant altitude and direction) as seen from the ground (see problem 6). Calculate the orbital distance of a geosynchronous satellite.

Solution: A satellite is much less massive than the Earth, so $M_{\text{Earth}} + M_{\text{satellite}} \approx M_{\text{Earth}}$ and we can use Newton's version of Kepler's third law in the following form:

$$(p_{\text{satellite}})^2 \approx \frac{4\pi^2}{G \times M_{\text{Earth}}} (a_{\text{satellite}})^3$$

Because we want to know the satellite's distance, we solve for $a_{\text{satellite}}$ by multiplying both sides of the equation by ($G \times M_{\text{Earth}}$), dividing both sides by $4\pi^2$, and then taking the cube root of both sides:

$$a_{\text{satellite}} = \sqrt[3]{\frac{G \times M_{\text{Earth}}}{4\pi^2} (p_{\text{satellite}})^2}$$

We know that $p_{\text{satellite}} = 1$ sidereal day $\approx 86{,}164$ seconds. You should confirm that substituting this value and the mass of the Earth yields $a_{\text{satellite}} \approx 42{,}000$ km. Thus, a geosynchronous satellite orbits at a distance of 42,000 km above the *center* of the Earth.

tides. The Moon and the Earth attract each other gravitationally. Because the strength of this gravitational attraction declines with distance, the side of the Earth nearest the Moon feels a stronger attraction than the opposite side. The result is that, from the perspective of someone looking down on the Earth from above the North Pole, the oceans bulge in directions toward and away from the Moon (Figure 6.16). As the Earth rotates, the two daily high tides occur when your location passes through these two tidal bulges, and the two low tides occur when your location passes through the points halfway between.

Perhaps you are wondering why there are *two* tidal bulges. In a simple sense, the answer is that the oceans bulge toward the Moon because they are being pulled out from the Earth, and they bulge in a direction opposite the Moon because the Earth is being pulled out from under them. However, a better way to look at tides is to recognize that the attraction toward the Moon gets progressively weaker with distance *throughout* the Earth; that is, tides affect the solid Earth (and the atmosphere) as well as the oceans. This gradually decreasing attraction causes the Earth to stretch along the Earth–Moon line; the liquid oceans simply stretch more than the solid land. This "stretching force" is called a **tidal force.**

FIGURE 6.16 Tidal bulges face toward and away from the Moon. Arrows represent the strength and direction of the gravitational attraction toward the Moon.

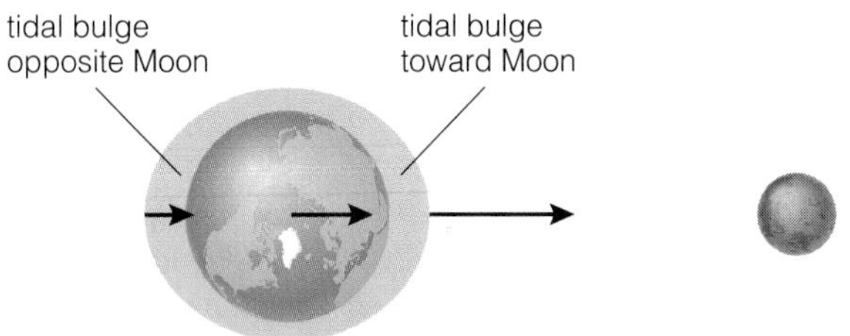

The Sun also exerts a tidal force on the Earth, causing the Earth to stretch along the Sun–Earth line. However, the greater distance to the Sun (than to the Moon) means that the *difference* in the Sun's pull on the near and far sides of the Earth is relatively small. (Of course, the *gravitational* force between the Earth and the Sun is greater than that between the Earth and the Moon; the Earth orbits the Sun, not the Moon.) In fact, the tidal force from the Sun is only about one-third as strong as the tidal force from the Moon. When the Sun and the Moon are both stretching the Earth along the same line, the tides are especially pronounced and are called *spring tides* (because the water tends to "spring up" from the Earth); this is the case at *both* new moon and full moon (Figure 6.17). Similarly, the tides are relatively small at first- and third-quarter moon, when the Sun's tidal force stretches the Earth along a line perpendicular to the stretch caused by the Moon; these tides are called *neap tides.*

FIGURE 6.17 This diagram represents the gravitational pulls from the Sun (yellow arrows) and the Moon (black arrows) at four phases of the Moon. Tides are enhanced at both new and full moon (spring tides) when the tidal forces from the Sun and the Moon stretch the Earth along the same line. Tides are diminished at first- and third-quarter moon (neap tides) when the tidal forces from the Sun and the Moon stretch the Earth along different (and perpendicular) lines.

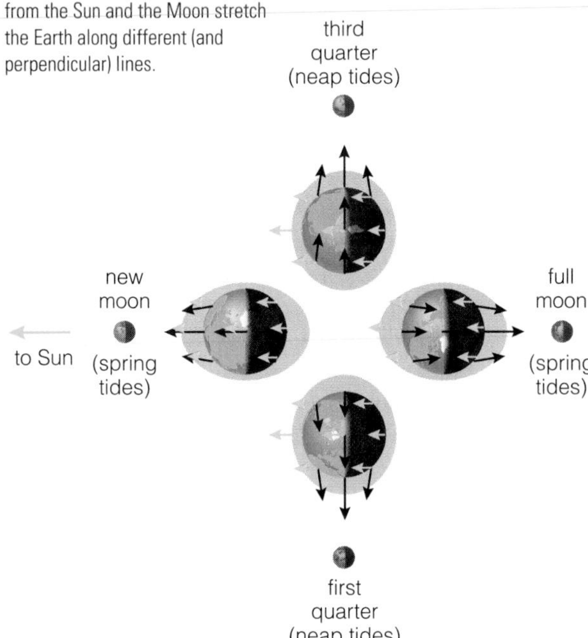

TIME OUT TO THINK ***The diagrams presented here show the oceans spread uniformly around the Earth. In reality, the tides are affected by the locations and shapes of land-ocean boundaries, with the result that tides are much more extreme in some places than in others. Where would you expect tides to be more extreme: along a long beach where the sand slopes gradually downward, or in a narrow, deep bay? Why?***

The tidal stretching of the Earth exerts internal friction, called **tidal friction,** that tends to fight against the Earth's rotation. As a result, the Earth's daily rotation is gradually slowing. Because angular momentum must be conserved, the angular momentum that the Earth loses as its rotation slows must reappear somewhere else. In fact, tidal forces transfer this angular momentum to the Moon, where it shows its presence by making the Moon move gradually farther from the Earth. Although these changes are barely noticeable on human time scales, they add up over time: A few billion years ago, a day may have been only 5 or 6 hours long, and the Moon may have been one-tenth or less of its current distance from the Earth. In later chapters, we will see many more examples of the important effects of tidal forces and tidal friction.

6.6 Orbital Energy and Escape Velocity

Consider a satellite in an elliptical orbit around the Earth. Its gravitational potential energy is greatest when it is farthest from the Earth, and smallest when it is nearest the Earth. Conversely, its kinetic energy is greatest when it is nearest the Earth and moving fastest in its orbit, and smallest when it is farthest from the Earth and moving slowest in its orbit. Throughout its orbit, its total *orbital energy*—the sum of its kinetic and gravitational potential energies—must be conserved.

Because any change in its orbit would mean a change in its total orbital energy, a satellite's orbit around the Earth cannot change if it is left completely undisturbed. If the satellite's orbit *does* change, it must somehow have gained or lost energy. For a satellite in low-Earth orbit, the Earth's thin upper atmosphere exerts a bit of drag that can cause it to lose energy and eventually plummet back to Earth. The satellite's lost orbital energy is converted to heat, which is why a falling satellite usually burns up in the atmosphere. Raising a satellite to a higher orbit requires that it gain energy by firing one of its rockets. The chemical potential energy of the rocket fuel is converted to gravitational potential energy as the satellite moves higher.

Generalizing from the satellite example shows that conservation of energy has a very important

Mathematical Insight 6.3 Using the Escape-Velocity Formula

The following examples demonstrate the use of the escape-velocity formula given on p. 155.

Example 1: Calculate the escape velocity from the Moon. Compare it to that from the Earth.

Solution: The mass and radius of the Moon are, respectively, $M = 7.4 \times 10^{22}$ kg and $R = 1.7 \times 10^6$ m. Plugging these numbers into the escape-velocity formula, we find:

$$v_{\text{escape}} = \sqrt{\frac{2 \times \left(6.67 \times 10^{-11} \frac{\text{m}^3}{\text{kg} \times \text{s}^2}\right) \times \left(7.4 \times 10^{24} \text{ kg}\right)}{1.7 \times 10^6 \text{ m}}} = 2{,}380 \text{ m/s} = 2.38 \text{ km/s}$$

The escape velocity from the Moon is 2.38 km/s, which is less than one-fourth the 11-km/s escape velocity from the Earth.

Example 2: Imagine that, in the future, a space station is orbiting the Earth in geosynchronous orbit, which is 42,000 km above the center of the Earth (see Mathematical Insight 6.2). Suppose we want to launch a spacecraft to Mars from this space station. At what velocity must the spacecraft be launched to escape the Earth's gravity? Would there be any advantage to launching from the space station instead of from Earth's surface?

Solution: We find the escape velocity from geosynchronous orbit by using the escape-velocity formula with the mass of the Earth ($M_{\text{Earth}} = 6.0 \times 10^{24}$ kg) and the distance of the orbit above the center of the Earth ($R = 42{,}000$ km $= 4.2 \times 10^7$ m):

$$v_{\text{escape}} = \sqrt{\frac{2 \times \left(6.67 \times 10^{-11} \frac{\text{m}^3}{\text{kg} \times \text{s}^2}\right) \times \left(6.0 \times 10^{24} \text{ kg}\right)}{4.2 \times 10^7 \text{ m}}} = 4{,}400 \text{ m/s} = 4.4 \text{ km/s}$$

The escape velocity from geosynchronous orbit is 4.4 km/s—considerably lower than the 11-km/s escape velocity from the Earth's surface. Thus, it would require substantially less fuel to launch the spacecraft toward Mars from the space station than from the Earth. Of course, this assumes that the space station is already in place and that the spacecraft is assembled at the space station.

implication for understanding motion throughout the cosmos: *Orbits cannot change spontaneously.* For example, an asteroid or a comet passing near a planet cannot spontaneously be "sucked in" to crash on the planet. It can hit the planet only if its current orbit already intersects the planet's surface or if it somehow gains or loses orbital energy so that its new orbit intersects the planet's surface. Of course, if the asteroid or comet *gains* energy, something else must *lose* exactly the same amount of energy, and vice versa.

One way that objects can exchange orbital energy is through **gravitational encounters,** in which they pass near enough so that each can feel the effects of the other's gravity. For example, Figure 6.18 shows a gravitational encounter between Jupiter and a comet headed toward the Sun on an unbound orbit. The comet's close passage by Jupiter allows the comet and Jupiter to exchange energy: The comet loses orbital energy and changes to a bound, elliptical orbit; Jupiter must gain the energy the comet loses. However, because Jupiter is so much more massive than the comet, the effect on Jupiter is unnoticeable. More generally, when two objects exchange orbital energy we expect one to lose energy and fall to a lower orbit while the other gains energy and is thrown to a higher orbit.

If an object gains enough energy, it may end up on an unbound orbit that allows it to *escape* from the gravitational influence of the object it is orbiting. For example, if we want to send a space probe to Mars, we must use a large rocket that gives the probe enough energy to achieve an unbound orbit and ultimately escape the Earth's gravitational influence. Although it would probably make more sense to talk of the probe's achieving "escape energy," we usually discuss the process of escape in terms of *escape velocity* (see Figure 6.3). For example, the escape velocity from the Earth is about 40,000 km/hr, or 11 km/s, meaning that this is the *minimum* velocity required to escape the Earth's gravity if you start from near the Earth's surface. Note that the escape velocity does not depend on the mass of the escaping object; *any* object must travel 11 km/s to escape from

FIGURE 6.18 Depiction of a comet in an unbound orbit of the Sun that happens to pass near Jupiter. The comet loses orbital energy to Jupiter, thereby changing to a bound orbit.

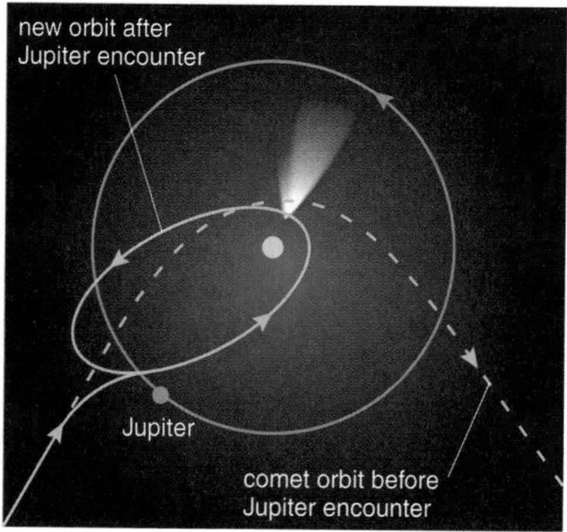

Mathematical Insight 6.4 The Acceleration of Gravity

The text on page 156 shows that the acceleration of a falling rock near the surface of the Earth is:

$$a_{\text{rock}} = G\frac{M_{\text{Earth}}}{(R_{\text{Earth}})^2}$$

Because this formula applies to *any* falling object, we call it the *acceleration of gravity* and abbreviate it with the letter g. Calculating g is easy. Simply look up the Earth's mass (6.0×10^{24} kg) and radius (6.4×10^6 m), and then "plug in":

$$g = G\frac{M_{\text{Earth}}}{(R_{\text{Earth}})^2} = \left(6.67 \times 10^{-11}\frac{\text{m}^3}{\text{kg} \times \text{s}^2}\right) \times \frac{6.0 \times 10^{24}\text{ kg}}{(6.4 \times 10^6\text{ m})^2} = 9.8\frac{\text{m}}{\text{s}^2}$$

Example 1: What is the acceleration of gravity on the surface of the Moon? (*Note:* $M_{\text{Moon}} = 7.4 \times 10^{22}$ kg and $R_{\text{Moon}} = 1.7 \times 10^6$ m.)

Solution: The formula for the acceleration of gravity on the Moon's surface is analogous to the one for Earth's surface. Thus, we find:

$$g_{\text{Moon}} = G\frac{M_{\text{Moon}}}{(R_{\text{Moon}})^2} = \left(6.67 \times 10^{-11}\frac{\text{m}^3}{\text{kg} \times \text{s}^2}\right) \times \frac{7.4 \times 10^{22}\text{ kg}}{(1.7 \times 10^6\text{ m})^2} = 1.7\frac{\text{m}}{\text{s}^2}$$

The acceleration of gravity on the Moon is 1.7 m/s^2, or about one-sixth that on the Earth. Thus, objects on the Moon weigh about one-sixth of what they would the Earth, whether it is an individual atom or molecule escaping from the atmosphere, a spacecraft being launched into deep space, or a rock blasted into the sky by a large impact.

We can calculate the escape velocity from the surface of any planet with the following simple formula:

$$v_{\text{escape}} = \sqrt{\frac{2 \times G \times M}{R}}$$

where M and R are the mass and radius of the planet, respectively, and G is the gravitational constant. As expected, the mass of the escaping object does not matter. (This formula can be derived from Newton's law of gravity, but we will not derive it here.)

Note that this formula gives the escape velocity from the planet's *surface.* The same formula also gives the escape velocity for an object that is already in orbit, but we must replace the radius of the planet with the object's *distance from the center* of the planet. The formula then applies to escape from the gravity of any massive object, including moons, stars, and galaxies.

Common Misconceptions: No Gravity in Space?

Most people are familiar with pictures of astronauts floating weightlessly in Earth orbit. Unfortunately, because we usually associate weight with gravity, many people assume that the astronauts' weightlessness implies a lack of gravity in space. Actually, there's plenty of gravity in space; even at the distance of the Moon, the Earth's gravity is strong enough to hold the Moon in orbit. In fact, in the low-Earth orbit of the Space Shuttle, the acceleration of gravity is scarcely less than it is on the Earth's surface. Why, then, are the astronauts weightless? Because, as we discussed earlier in the chapter, the Space Shuttle and all other orbiting objects are in a constant state of *free-fall,* and any time you are in free-fall, you are weightless (see Figure 6.2). Imagine what it feels like to be an astronaut: You'd have the sensation of free-fall—just as when falling from a diving board—the entire time you were in orbit. This constantly falling sensation makes most astronauts sick to their stomachs when they first experience weightlessness. Fortunately, they quickly get used to the sensation, which allows them to work hard and enjoy the view.

6.7 The Acceleration of Gravity

Throughout the remainder of the text, we will see many more applications of the universal law of gravitation. For the moment, let's look at just one more: Galileo's discovery that the acceleration of a falling object is independent of its mass.

If you drop a rock, the force acting on the rock is the force of gravity. The two masses involved are the masses of the Earth and the rock, denoted M_{Earth} and M_{rock}, respectively. The distance between their *centers* is the distance from the *center of the Earth* to

weigh on Earth. If you can lift a 50-kilogram barbell on Earth, you'll be able to lift a 300-kilogram barbell on the Moon.

Example 2: The Space Shuttle typically orbits at an altitude of 300 kilometers above the surface of the Earth. What is the acceleration of gravity at this altitude?

Solution: Because the Space Shuttle is significantly above the Earth's surface, we cannot use the approximation $d \approx R_{\text{Earth}}$ that we used in the text. Instead, we must go back to Newton's second law, set the gravitational force on the Space Shuttle equal to its mass times acceleration, and then solve for its acceleration:

$$G\frac{M_{\text{Earth}}M_{\text{shuttle}}}{d^2} = M_{\text{shuttle}}a_{\text{shuttle}} \quad \Rightarrow \quad a_{\text{shuttle}} = G\frac{M_{\text{Earth}}}{d^2}$$

In this case, the distance d is the 6,400-km radius of the Earth *plus* the 300-km altitude of the shuttle, or d = 6,700 km = 6.7×10^6 m. Thus, the gravitational acceleration of the shuttle when orbiting the Earth is:

$$a_{\text{shuttle}} = G\frac{M_{\text{Earth}}}{d^2} = \left(6.67 \times 10^{-11}\frac{\text{m}^3}{\text{kg} \times \text{s}^2}\right) \times \frac{6.0 \times 10^{24}\text{ kg}}{(6.7 \times 10^6\text{ m})^2} = 8.9\frac{\text{m}}{\text{s}^2}$$

The acceleration of gravity in low-Earth orbit is 8.9 m/s^2, or only slightly less than the 9.8 m/s^2 acceleration of gravity at the Earth's surface.

the center of the rock. If the rock isn't too far above the Earth's surface, this distance is approximately the radius of the Earth, R_{Earth} (about 6,400 km); that is, $d \approx R_{\text{Earth}}$. Thus, the force of gravity acting on the rock is:

$$F_g = G\frac{M_{\text{Earth}}M_{\text{rock}}}{d^2} \approx G\frac{M_{\text{Earth}}M_{\text{rock}}}{(R_{\text{Earth}})^2}$$

According to Newton's second law of motion, this force is equal to the product of the mass and the acceleration of the rock; that is,

$$G\frac{M_{\text{Earth}}\cancel{M_{\text{rock}}}}{(R_{\text{Earth}})^2} = \cancel{M_{\text{rock}}}a_{\text{rock}}$$

Note that M_{rock} "cancels" because it appears on both sides of the equation (as a multiplier), thereby giving Galileo's result that the acceleration of the rock does not depend on the mass of the rock. All falling objects near the Earth's surface, regardless of mass, fall with the same acceleration of gravity.

The fact that objects of different mass fall with the same acceleration struck Newton as an astounding coincidence, even though his own equations showed it to be so. For the next 240 years, this seemingly odd coincidence remained just that—a coincidence—in the minds of scientists. However, in 1915 Einstein discovered that it is not a coincidence at all. Rather, it reveals something deeper about the nature of gravity and of the universe; the new insights were described by Einstein in his *general theory of relativity* (the topic of Chapter S4).

If I have seen farther than others, it is because I have stood on the shoulders of giants.

ISAAC NEWTON

THE BIG PICTURE

We've covered a lot of ground in this chapter, from the scientific terminology of motion to the story of how universal motion finally was understood by Newton. Be sure you understand the following "big picture" ideas:

- Understanding the universe requires understanding motion. Although the terminology of the laws of motion may be new to you, the *concepts* are familiar from everyday experience. Think about these experiences so that you'll better understand the less familiar astronomical applications of the laws of motion.
- The Copernican revolution, which overthrew the ancient belief in an Earth-centered universe, did not occur instantaneously. It unfolded over a period of more than a century and involved careful observational, experimental, and theoretical work by many different people—especially Copernicus, Tycho Brahe, Kepler, and Galileo.
- Newton's discovery of the universal law of gravitation allowed him to explain *how* gravity holds planets in their orbits. Perhaps even more important, it showed that the same physical laws we observe on Earth apply throughout the universe. This universality of physics opens up the entire universe as a possible realm of human study.

Review Questions

1. How does *speed* differ from *velocity?* Give an example in which you can be traveling at constant speed, but not at constant velocity.
2. What do we mean by *acceleration?* Explain why the units of acceleration are m/s^2. How should we interpret these units?
3. What is the *acceleration of gravity* on Earth? If you drop a rock from very high, how fast will it be falling after 4 seconds (neglecting air resistance)?
4. What is the definition of *momentum?* How can momentum be affected by a *force?* What do we mean when we say that momentum will be changed only by a *net force?*

5. What is the difference between *mass* and *weight?*
6. What is *free-fall,* and why does it make you *weightless?* Have you ever been weightless? Explain.
7. Briefly describe why astronauts are weightless in the Space Shuttle. Why does the Space Shuttle require a high speed to achieve orbit? What would happen if it were launched with a speed greater than the Earth's *escape velocity?*
8. What is *angular momentum?* What is the formula for the angular momentum of an object moving in a circle? Explain the meaning of each term in the formula.
9. What kind of force can cause a change in angular momentum? How does this explain why it's easier to open a door by pushing near the doorknob than near the hinges?
10. State each of Newton's three laws of motion, including both ways of stating Newton's second law mathematically. For each law, give an example of its application.
11. What is the principle of *conservation of angular momentum?* How can a skater use this principle to vary his or her rate of spin?
12. Was Copernicus the first person to suggest a Sun-centered solar system? What advantages did his model have over the Ptolemaic model? In what ways did it fail to improve on the Ptolemaic model?
13. Why do we say that Tycho Brahe was the greatest naked-eye observer of all time? How were his observations important to the development of modern astronomy?
14. State Kepler's three laws of motion, and explain the meaning of each law. Note that both Kepler and Newton have sets of three laws named for them, which can sometime be confusing; find and describe a way to avoid becoming confused between Kepler's three laws and Newton's three laws.
15. Describe how Galileo helped spur acceptance of Kepler's Sun-centered model of the solar system. Be sure to describe the role of his experiments in physics and of his observations of the Sun and the Moon, Venus, Jupiter, and the Milky Way.
16. What is the universal law of gravitation? Summarize what this law says in words, and then state the law mathematically. Define each variable in the formula.
17. Describe how Newton's laws of motion and law of universal gravitation explain *why* each of Kepler's three laws is true.
18. What do we mean by *bound* and *unbound* orbits? Give a few examples of objects in each type of orbit.
19. How did Newton modify and extend Kepler's third law? Why is this modification so important to our understanding of the universe?
20. Explain how the Moon creates tides on the Earth and why there are *two* high and low tides each day. Also explain why the *tidal force* from the Sun is weaker than that from the Moon, despite the Earth's stronger gravitational attraction to the Sun. How do tides vary with the phase of the Moon? Why?
21. What is *tidal friction?* Briefly describe how tidal friction slows the Earth's rotation and why the Moon's distance from the Earth must increase at the same time.
22. Explain why orbits cannot change spontaneously. How can atmospheric drag cause an orbit to change? How can a *gravitational encounter* cause an orbit to change?
23. Define each of the variables in the escape-velocity formula. Explain how this formula is used.
24. Is there gravity in Earth orbit? Explain.

Discussion Questions

1. *Kepler's Choice.* Casting aside the idea that orbits must be perfect circles meant going against deeply entrenched beliefs, and Kepler said that it shook his deep religious faith. Given that only two of Tycho's observations disagreed with a perfectly circular orbit—and only by 8 arcminutes—do you think that most other people would have made the choice Kepler made to abandon perfect circles? Have you ever performed an experiment that disagreed with theory? Which did you question, the theory or your experiment? Why?
2. *Aristotle and Modern English.* Aristotle believed that the Earth was made from the four elements fire, water, earth, and air, while the heavens were made from *ether* (literally, "upper air"). The literal meaning of *quintessence* is "fifth element," and the literal meaning of *ethereal* is "made of ether." Look up these words in the dictionary. Discuss how their modern meanings are related to Aristotle's ancient beliefs.
3. *Tidal Complications.* The ocean tides on Earth are much more complicated than they might at first seem from the simple physics that underlies tides. Discuss some of the factors that make the real tides so complicated and how these factors affect the tides. Some factors to consider: the distribution of land and oceans; the Moon's varying distance from Earth in its orbit; the fact that the Moon's orbital plane is not perfectly aligned with the ecliptic and neither the Moon's orbit nor the ecliptic is aligned with the Earth's equator.

Problems

1. *Practice with Acceleration.*
   a. Some schools have an annual ritual that involves dropping a watermelon from a tall building. Suppose it takes 6 seconds for the watermelon to fall to the ground (which would mean it's about a 60-story building). If there were no air resistance so that the watermelon would fall with the acceleration of gravity, how fast would it be going when it hit the ground? Give your answer in m/s, km/hr, and mi/hr.
   b. As you sled down a steep, slick street, you accelerate at a rate of 4 m/s^2. How fast will you be going after 5 seconds? Give your answer in m/s, km/hr, and mi/hr.
   c. You are driving along the highway at a speed of 70 miles per hour when you slam on the brakes. If you decelerate at an average rate of -20 miles per hour per second, how long will it take to come to a stop?
2. *Gees.* Acceleration is sometimes measured in *gees*, or multiples of the acceleration of gravity: 1 gee means $1 \times g$, or 9.8 m/s^2; 2 gees means $2 \times g$, or 2×9.8 $m/s^2 = 19.6$ m/s^2; and so on. Suppose you experience 6 gees of acceleration in a rocket.
   a. What is your acceleration in meters per second squared?
   b. You will feel a compression force from the acceleration. How does this force compare to your normal weight?
   c. Do you think you could survive this acceleration for long? Explain.
3. *New Comet.* Imagine that a new comet is discovered and studies of its motion indicate that it orbits the Sun with a period of 1,000 years.
   a. What is the comet's average distance (semimajor axis) from the Sun? (*Hint:* Use Kepler's third law in its original form.)
   b. Suppose the comet's perihelion distance is 0.1 AU. What is its aphelion distance? (*Hint:* How is the average distance related to the perihelion and aphelion distances?)
4. *Eclipse Frequency in the Past.* Over billions of years, the Moon has been gradually moving farther from the Earth. Thus, the Moon used to be substantially nearer the Earth than it is today.
   a. How would the Moon's past angular size in our sky compare to its present angular size? Why?
   b. How would the length of a lunar month in the past compare to the length of a lunar month today? Why? (*Hint:* Think about Kepler's third law as it would apply to the Moon orbiting the Earth.)
   c. Based on your answers to parts (a) and (b), would eclipses (both solar and lunar) have been more or less common in the past? Why?
5. *The Gravitational Law.* Use the universal law of gravitation to answer each of the following questions.
   a. How does tripling the distance between two objects affect the gravitational force between them?
   b. Compare the gravitational force between the Earth and the Sun to that between Jupiter and the Sun. The mass of Jupiter is about 318 times the mass of the Earth.
   c. Suppose the Sun were magically replaced by a star with twice as much mass. What would happen to the gravitational force between the Earth and the Sun?
6. *Geostationary Orbit.* A satellite in geostationary orbit appears to remain stationary in the sky as seen from any particular location on Earth.
   a. Briefly explain why a geostationary satellite must orbit the Earth in 1 *sidereal* day, rather than 1 solar day.
   b. Communications satellites, such as those used for television broadcasts, are often placed in geostationary orbit. The transmissions from such satellites are received with satellite dishes, such as those that can be purchased for home use. In one or two paragraphs, explain why geostationary orbit is a convenient orbit for communications satellites.
7. *Measuring Masses.* Use Newton's version of Kepler's third law to answer each of the following questions.
   a. The Moon orbits the Earth in an average of 27.3 days at an average distance of 384,000 kilometers. Use these facts to determine the mass of the Earth. You may neglect the mass of the Moon in comparison to the mass of the Earth ($M_{\text{Moon}} \approx 1/80\ M_{\text{Earth}}$).
   b. Jupiter's moon Io orbits Jupiter every 42.5 hours at an average distance of 422,000 kilometers from the center of Jupiter. Europa orbits Jupiter every 85.2 hours at an average distance of 671,000 kilometers. Use either of these moons to calculate the mass of Jupiter. Does it matter which moon you use? Why or why not?
   c. Calculate the orbital period of the Space Shuttle in an orbit 300 kilometers above the Earth's surface.
   d. Pluto's moon Charon orbits Pluto every 6.4 days with a semimajor axis of 19,700 kilometers. Calculate the *combined* mass of Pluto and Charon. Compare this combined mass to the mass of the Earth.
8. *Measuring Masses in Other Star Systems.* Use Newton's version of Kepler's third law to answer the following questions. (*Hint:* The calculations for this problem are so simple that you will not need a calculator.)
   a. Imagine another solar system, with a star of the same mass as the Sun. Suppose there is a planet in that solar system with a mass twice that of

Earth orbiting at a distance of 1 AU from the star. What is the orbital period of this planet? Explain.

b. Suppose a solar system has a star that is four times as massive as our Sun. If that solar system has a planet the same size as Earth orbiting at a distance of 1 AU, what is the orbital period of the planet? Explain.

9. *Frequency of Tides.* As Figure 6.16 should make clear, high tide at any location on Earth would come every 12 hours—if the Moon never moved. However, the Moon orbits the Earth every 27.3 days, and it orbits in the same sense (counterclockwise as seen from above the North Pole) that the Earth rotates. Given this fact, what is the actual interval between high tides? Explain. (*Hint:* In Figure 6.16, draw the Moon's location 1 day later. How are the tidal bulges affected?)

10. *Head-to-Foot Tides.* You and the Earth attract each other gravitationally, so you should also be subject to a tidal force resulting from the difference between the gravitational attraction felt by your feet and that felt by your head (at least when standing). Explain why you can't feel this tidal force.

11. *Weights on Other Worlds.* Calculate the acceleration of gravity on the surface of each of the following worlds. How much would *you* weigh, in pounds, on each of these worlds?

    a. Mars (mass = 0.11 M_{Earth}, radius = 0.53 R_{Earth}).

    b. Venus (mass = 0.82 M_{Earth}, radius = 0.95 R_{Earth}).

    c. Jupiter (mass = 317.8 M_{Earth}, radius = 11.2 R_{Earth}). Bonus: Given that Jupiter has no solid surface, how could you weigh yourself on Jupiter?

    d. Jupiter's moon Europa (mass = 0.008 M_{Earth}, radius = 0.25 R_{Earth}).

    e. Mars's moon Phobos (mass = 1.1×10^{16} kg, radius = 12 km).

12. Calculate the escape velocity from each of the following. (Masses and radii are listed in problem 11.)

    a. The surface of Mars.

    b. The surface of Phobos.

    c. The cloudtops of Jupiter.

    d. Our solar system, starting from the Earth's orbit. (*Hint:* Most of the mass of our solar system is in the Sun; $M_{\text{Sun}} = 2.0 \times 10^{30}$ kg.)

    e. Our solar system, starting from Saturn's orbit.

13. *Research: Players in the Copernican Revolution.* A number of interesting personalities played important roles in the Copernican revolution besides Copernicus, Tycho, Kepler, and Galileo. Research one or more of these people and write a short biography of their scientific lives and their contributions to the Copernican revolution. Among the people you might consider: Nicholas of Cusa (1401–1464), who argued for a Sun-centered solar system and believed that other stars might be circled by their own planets; Leonardo da Vinci (1452–1519), who made many contributions to science, engineering, and art and also believed the Earth *not* to be the center of the universe; Rheticus (1514–1574), a student of Copernicus who persuaded him to publish his work; William Gilbert (1540–1603), who studied the Earth's magnetism and influenced the thinking of Kepler; Giordano Bruno (1548–1600), who was burned at the stake for his beliefs in the Copernican system and the atomism of Democritus; and Francis Bacon (1561–1626), whose writings contributed to the acceptance of experimental methods in science.

7
Light: the Cosmic Messenger

ANCIENT OBSERVERS COULD DISCERN ONLY THE MOST basic features—such as color and brightness—of the light that they saw. Over the past several hundred years, we have discovered that light contains far more information. Remarkably, analysis of light with proper instruments can reveal the chemical composition of distant objects, their temperature, how fast they rotate, and much more.

It is fortunate that light can convey so much information. Our present spacecraft reach only objects within our solar system, and except for an occasional meteorite falling from the sky the cosmos does not come to us. In contrast to the limited reach of spacecraft, light travels throughout the universe, carrying its treasury of information wherever it goes. Light, the cosmic messenger, brings the stories of distant objects to our home here on Earth.

FIGURE 7.1 A prism reveals that white light contains a spectrum of colors from red to violet.

7.1 Light in Everyday Life

Modern astronomy is largely the science of collecting, recording, and analyzing light from the universe. We collect light with telescopes, record it with film or electronic detectors, and then analyze it to decipher the messages it carries through the cosmos. Understanding light and its properties therefore is an important part of understanding astronomy. As usual, starting with everyday experiences is an excellent way to begin learning about something as complex as light.

Every moment we spend with our eyes open, we are treated to a spectacular light show. Of course, we are usually more concerned with the objects we see than with the light that enables us to see them. But let's turn the tables and concentrate more closely on the light itself.

Even without opening your eyes, it's clear that light is a form of energy, called *radiative energy* [Section 5.1]. Outside on a hot, sunny day, you can feel the radiative energy of sunlight being converted to thermal energy as it strikes your skin. On an economic level, you know that light is a form of energy because you have to pay for it. The rate at which a light bulb uses energy is usually printed on it—something like "100 watts." A watt is a unit of **power,** which describes the rate of energy use; each watt of power means that 1 joule of energy is used each second. That is:

$$1 \text{ watt} = 1 \text{ joule/s}$$

For every second that you leave a 100-watt light bulb turned on, you will have to pay the utility company for 100 joules of energy. Interestingly, the power requirement of an average human—about 10 million joules per day—is about the same as that of a 100-watt light bulb.

Another basic property of light is what our eyes perceive as *color.* You've probably seen a prism split light into a **spectrum** (plural, spectra), or rainbow of colors (Figure 7.1). You can also produce a spectrum with a **diffraction grating**—a piece of plastic or glass etched with many closely spaced lines.[1] The colors in a spectrum are unusually pure forms of the basic colors red, orange, yellow, green, blue, and violet. The wide variety of all possible colors comes from mixtures of these basic colors in varying proportions; *white* is simply what we see when the basic colors are mixed in roughly equal proportions. Your television takes advantage of this fact to simulate a huge range of colors by combining only three specific colors of red, green, and blue light.

TIME OUT TO THINK ***If you have a magnifying glass handy, hold it close to your TV set to see the individual red, blue, and green dots. If you don't have a magnifying glass, try splashing a few droplets of water onto your TV screen. What do you see? What are the drops of water doing?***

[1]You can purchase inexpensive (under $1) plastic diffraction gratings from many science supply stores, or a more sophisticated spectrometer (about $25) from the Astronomical Society of the Pacific (800-335-2624). Playing with a grating will help you understand light spectra.

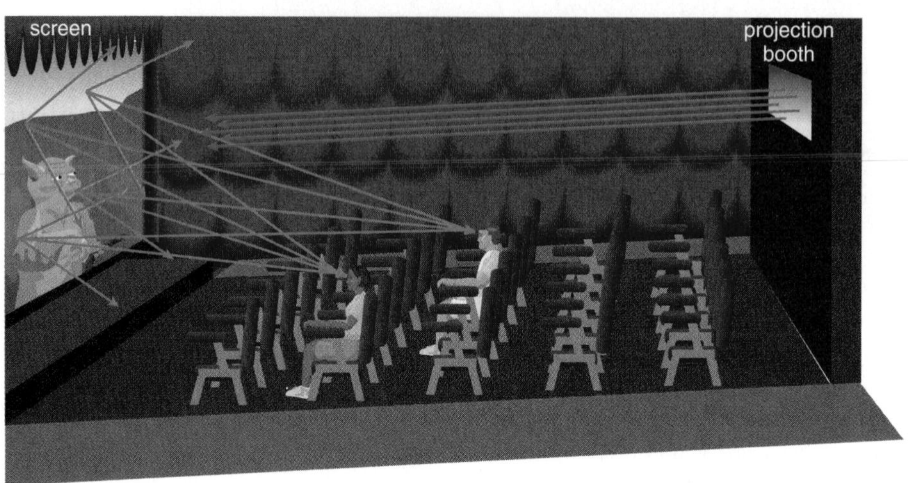

FIGURE 7.2 (**a**) A mirror reflects light along a path determined by the angle at which the light strikes the mirror. (**b**) A movie screen scatters light into an array of beams that reach every member of the audience.

Energy carried by light can interact with matter in four general ways.

- **Emission:** Matter can *emit* light. When you turn on a lamp, electricity flowing through the filament of the light bulb heats it to a point at which it emits visible light.
- **Absorption:** Matter can *absorb* light. If you place your hand near a lit light bulb, your hand absorbs some of the light, and this absorbed energy makes your hand warmer.
- **Transmission:** Some forms of matter, such as glass or air, *transmit* light; that is, they allow light to pass through them.
- **Reflection:** Matter can *reflect* light. A mirror reflects light in a very specific way, similar to the way a rubber ball bounces off a hard surface, so that the direction of a reflected beam of light depends on the direction of the incident (incoming) beam of light (Figure 7.2a). Sometimes, reflection is more random, so that an incident beam of light is **scattered** into many different directions. The screen in a movie theater provides a good example of scattering: The screen scatters a narrow beam of light from the projector into an array of beams that reach every member of the audience (Figure 7.2b).

Materials that transmit light are said to be **transparent,** and materials that absorb light are called **opaque.** Many materials are neither perfectly transparent nor perfectly opaque, and the technical term that describes *how much* light they transmit and absorb is **opacity.** For example, dark sunglasses have a higher opacity than clear glasses because they absorb more light.

Particular materials can affect different colors of light differently. A piece of red glass transmits red light but absorbs purple, blue, green, yellow, and orange light; that is, red glass is at least partially transparent to red light but opaque to all other colors of the spectrum. A healthy lawn looks green because it reflects green light but absorbs all other colors of the spectrum.

Now let's put all these ideas together and think about what happens when you walk into a dark room and turn on the light switch. The light bulb begins to emit white light, which is a mix of all the colors in the spectrum. Some of this light exits the room, transmitted through the windows. The rest of the light strikes the surfaces of objects inside the room, where each object's material properties determine the colors absorbed or reflected. The light coming from each object therefore carries an enormous amount of information about the object's location, shape and structure, and material makeup. You acquire this information when light enters your eyes, where it is absorbed by special cells (called *cones* and *rods*) that use the energy of the absorbed light to send signals to your brain. Your brain interprets the messages carried by the light, recognizing materials and objects in the room in the process we call *vision.*

In fact, the light carries even more information than your ordinary vision can recognize. Just as a microscope can reveal structure that is invisible to the naked eye, modern instruments can reveal otherwise invisible details in the spectrum of light. Learning to interpret these details is the key to unlocking the vast amount of information carried by light.

7.2 Properties of Light

Despite our familiarity with light, its nature remained a mystery for most of human history. The first real insights into the nature of light came with experiments performed by Isaac Newton in the 1660s. It was already well known that light passed through a prism separates into the rainbow of colors, but the most common belief held that the colors were a property of the prism rather than of the light itself. Newton dispelled this belief by placing a second prism in front of the light of just one color, such as red, from the first prism. He found that the color did not change any further, thereby proving that the colors were not a property of the prism but must be part of the white light itself.

Newton guessed that light, with all of its colors, is made up of countless tiny particles. However, later experiments by other scientists demonstrated that light behaves like waves. Thus began one of the most important debates in scientific history: Is light a wave or a particle? We must address this question if we hope to understand the messages conveyed by light; first let's consider the difference between a wave and a particle.

Particles and Waves

Marbles, baseballs, and individual atoms are all examples of *particles.* A particle of matter can sit still, or it can carry its matter from one place to another. If you throw a baseball at a wall, its matter moves from your hand to the wall. In contrast, imagine throwing a pebble into a pond (Figure 7.3). The ripples moving out from the place where the pebble lands are *waves,* consisting of *peaks,* where the water is higher than average, and *troughs,* where the water is lower than average. If you watch as the waves pass by a floating leaf, you'll see the leaf rise up with the peak and drop down with the trough, but the leaf does *not* move across the pond's surface with the wave. That is, the wave carries *energy* outward from the place where the pebble landed but does not carry matter along with it. In a sense, a particle is a *thing,* while a wave is a *pattern* revealed by its interaction with particles.

TIME OUT TO THINK *Hold a piece of rope with one end in each hand. By shaking one end up and down, make waves moving along the rope. Watch the motion of the peaks and troughs. As a peak moves along the rope, does any material move with it? Explain.*

FIGURE 7.3 Throwing a pebble into a pond generates waves traveling outward.

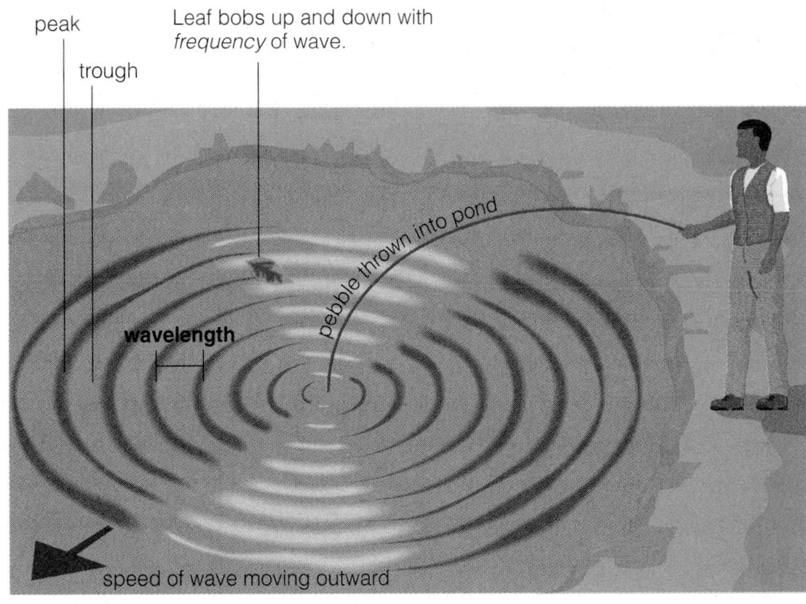

Three basic properties characterize the waves moving outward through the pond. Their **wavelength** is the distance between adjacent peaks. Their **frequency** is the number of peaks passing by any point each second. If a passing wave causes the leaf to bob up and down twice per second, then its frequency is 2 **cycles per second** (referring to the up and down "cycles" of the passing waves). Cycles per second often are called **hertz** (**Hz**), so we can also describe this frequency as 2 Hz. The third basic characteristic of the waves is the **speed** at which any peak travels across the pond.

The wavelength, frequency, and speed of a wave are related by a simple formula, which we can understand with the help of an example. Suppose a wave has a wavelength of 1 centimeter and a frequency of 2 hertz. The wavelength tells us that each time a peak passes by, the wave has traveled 1 centimeter. The frequency tells us that two peaks pass by each second. Thus, the speed of the wave must be 2 centimeters per second. If you try a few more similar examples, you'll find that the general rule is:

$$\text{wavelength} \times \text{frequency} = \text{speed}$$

Photons and Electromagnetic Waves

In our everyday lives, waves and particles appear to be very different. After all, no one would confuse the ripples on a pond with a baseball. However, light behaves as *both* a particle *and* a wave. Like particles, light comes in individual "pieces," called **photons,** that can hit a wall one at a time. Like ripples on a pond, each photon is characterized by its wavelength, frequency, and speed. We will discuss the implications of this "wave–particle duality" in Chapter S5. Here we need only discuss how we measure the wave and particle properties of light.

First, let's ignore the particle properties of light and look at its wave nature. Although waves don't carry material along with them, something must vibrate to transmit energy along a wave. For example, water waves are transmitted by the up and down vibrations of the water surface, and sound waves are transmitted by back-and-forth vibrations of the air as it responds to changing pressure. In the case of light, it is electric and magnetic *fields* that vibrate.

The concept of a **field** is a bit abstract. Fields associated with forces, such as electric and magnetic fields, describe how these forces affect a particle placed at any point in space. For example, we say that the Earth has a *gravitational field* because if you place an object above the Earth's surface, the force

Mathematical Insight 7.1 Wavelength, Frequency, and Energy

In the text, we found that the speed of any wave is the product of its wavelength and its frequency. Because all forms of light travel at the same speed, $c = 3 \times 10^8$ m/s, we can write:

$$\lambda \times f = c$$

where λ (the Greek letter lambda) stands for wavelength and f stands for frequency. Solving this formula allows us to find the wavelength of light if we know the frequency, or vice versa:

$$\lambda = \frac{c}{f} \quad \text{or} \quad f = \frac{c}{\lambda}$$

The formula for the radiative energy carried by a photon of light is:

$$E = h \times f \quad (h = 6.626 \times 10^{-34} \text{ joule} \times \text{s})$$

where h is a number called *Planck's constant.* Thus, the energy increases in proportion to the *frequency* of the photon. Because $f = c/\lambda$, we can also write this formula as:

$$E = \frac{hc}{\lambda}$$

showing that the energy *decreases* in proportion to the *wavelength* of the photon.

Example 1: The numbers on a radio dial for FM radio stations are their frequencies in megahertz (MHz), or millions of hertz. If your favorite radio station is "93.3 on your dial," it broadcasts radio waves with a frequency of 93.3 million cycles per second. What is the wavelength of these radio waves?

of gravity pulls the object to the ground. That is, any object placed in a gravitational field feels the force of gravity. In a similar way, a charged particle (such as an electron) placed in an *electric field* or a *magnetic field* feels electric or magnetic forces.

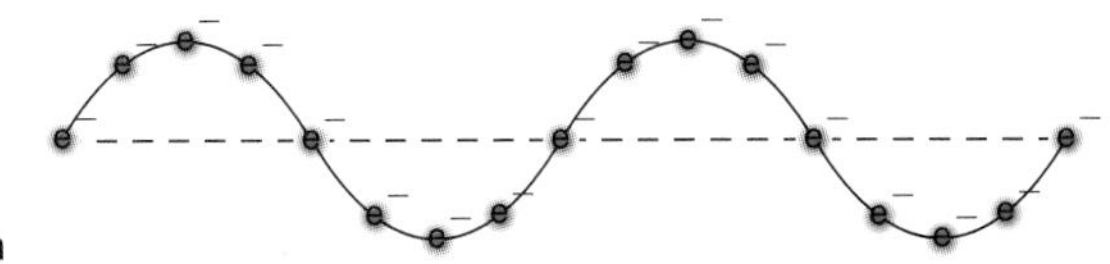

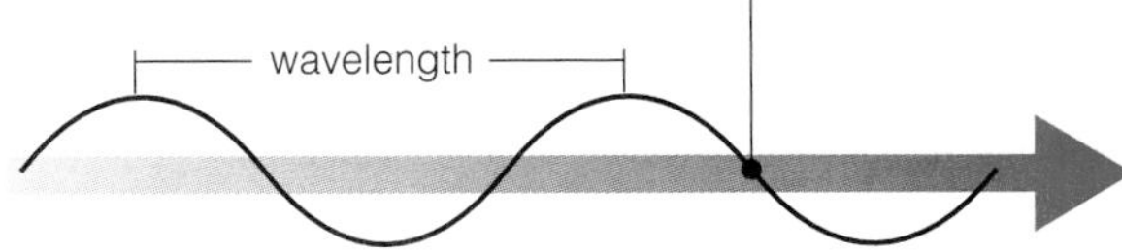

FIGURE 7.4 (**a**) A row of electrons would wriggle up and down as light passes by, showing that light carries a vibrating electric field. It also carries a magnetic field (not shown) that vibrates perpendicular to the direction of the electric field vibrations. (**b**) Characteristics of light waves. Because all light travels at the same speed, light of longer wavelength must have lower frequency.

Light is an **electromagnetic wave**—a wave in which electric and magnetic fields vibrate. Like a leaf on a rippling pond, an electron will bob up and down when an electromagnetic wave passes by. If you could set up a row of electrons, they would wriggle like a snake, revealing the wavelength and frequency of the passing wave (Figure 7.4a). The wavelength is the distance between adjacent peaks of the electric or magnetic field, and the frequency is the number of peaks that pass by any point each second (Figure 7.4b). All light travels at the same speed—about 300,000 kilometers per second, or 3×10^8 m/s—regardless of its wavelength or frequency. Therefore, light with a shorter wavelength must have a higher frequency, and vice versa.

Measuring the particle properties of light requires thinking of each photon as a distinct entity. Just as a baseball carries a specific amount of kinetic energy, each photon of light carries a specific amount of radiative energy. The shorter the wavelength of the light (or, equivalently, the higher its frequency), the higher the energy of the photons. For example, a photon with a wavelength of 100 nanometers (nm) has more energy than a photon with a 120-nm wavelength.[2]

[2]A nanometer (nm) is a billionth of a meter: 1 nm = 10^{-9} m. Many astronomers work with a unit called the Angstrom (Å): 1 nm = 10 Å.

Solution: We know the speed of light and the frequency, so the wavelength is:

$$\lambda = \frac{c}{f} = \frac{3 \times 10^8 \frac{\text{m}}{\text{s}}}{93.3 \times 10^6 \frac{1}{\text{s}}} = 3.2 \text{ m}$$

Note that, when working with frequency in equations, the "cycles" do not show up as a unit; that is, the units of frequency are simply 1/s, or "per second."

Example 2: The average wavelength of visible light is about 550 nanometers (1 nm = 10^{-9} m). What is the frequency of this light?

Solution: This time we know the wavelength, so the frequency of the light is:

$$f = \frac{c}{\lambda} = \frac{3 \times 10^8 \frac{\text{m}}{\text{s}}}{550 \times 10^{-9} \text{ m}} = 5.45 \times 10^{14} \frac{1}{\text{s}}$$

The frequency of visible light is about 5.5×10^{14} cycles per second, or about 550 trillion Hz.

Example 3: What is the energy of a visible light photon with wavelength 550 nm?

Solution: We know the wavelength, so the energy is:

$$E = \frac{hc}{\lambda} = \frac{(6.626 \times 10^{-34} \text{ joule} \times \text{s}) \times (3 \times 10^8 \frac{\text{m}}{\text{s}})}{550 \times 10^{-9} \text{ m}} = 3.6 \times 10^{-19} \text{ joule}$$

Note that this energy for a single photon is extremely small compared to, say, a 100-watt light bulb that uses 100 joules of energy per second.

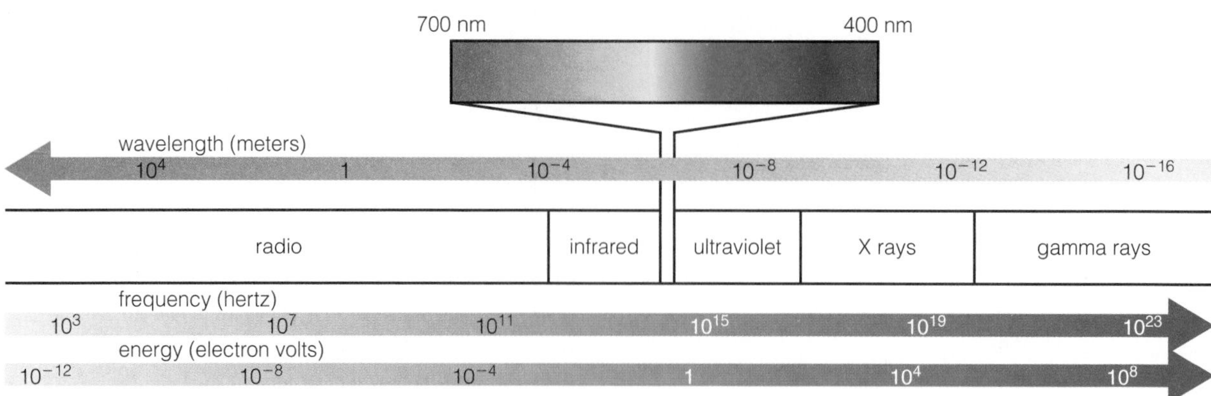

FIGURE 7.5 The electromagnetic spectrum.

7.3 The Many Forms of Light

Because light consists of electromagnetic waves, light is often called *electromagnetic radiation*[3] and the spectrum of light is called the **electromagnetic spectrum.** Photons of light can have *any* wavelength or frequency, so in principle the complete electromagnetic spectrum extends from a wavelength of zero to infinity. For convenience, we usually refer to different portions of the electromagnetic spectrum by different names (Figure 7.5).

The **visible light** that we see with our eyes has wavelengths ranging from about 400 nm at the blue end of the rainbow to about 700 nm at the red end.

[3]In this book, we use the terms *light* and *electromagnetic radiation* as synonyms, but some people use the term *light* to mean only *visible* light.

Common Misconceptions: Is Radiation Dangerous?

Many people associate the word *radiation* with *danger.* However, the word *radiate* simply means "to spread out from a center" (note the similarity between *radiation* and *radius* [of a circle]), and *radiation* is simply energy being carried through space. If energy is being carried by particles, such as protons or neutrons, we call it *particle radiation.* If energy is being carried by light, we call it *electromagnetic radiation.* High-energy forms of radiation are dangerous because they can penetrate body tissues and cause cell damage; these forms include particle radiation from radioactive substances such as uranium and plutonium and electromagnetic radiation such as ultraviolet, X rays, or gamma rays. Low-energy forms of radiation such as radio waves are usually harmless. And solar radiation, the light that comes from the Sun, is necessary to life on Earth. Thus, while some forms of radiation are dangerous, others are harmless or beneficial.

Light with wavelengths somewhat longer than red light is called **infrared,** because it lies beyond the red end of the rainbow. Light with very long wavelengths is called **radio** (or *radio waves*)—thus, radio is a form of light, *not* a form of sound.

On the other end of the spectrum, light with wavelengths somewhat shorter than blue light is called **ultraviolet,** because it lies beyond the blue (or violet) end of the rainbow. Light with even shorter wavelengths is called **X rays,** and the shortest-wavelength light is called **gamma rays.** Note that visible light is an extremely small part of the entire electromagnetic spectrum: The reddest red that our eyes can see has only about twice the wavelength of the bluest blue, but the radio waves from your favorite radio station are a billion times longer than the X rays used in a doctor's office.

As we move from the radio end toward the gamma-ray end of the spectrum, the wavelengths become shorter, and therefore the frequencies and energies of the photons increase. Visible photons happen to have enough energy to activate the molecular receptors in our eyes. Ultraviolet photons, with shorter wavelength than visible light, carry more energy—enough to harm our skin cells, causing sunburn or skin cancer. X-ray photons have enough energy to transmit easily through skin and muscle but not so easily through bones or teeth. That is why photographs taken with X-ray light allow doctors and dentists to see our underlying bone structures.

Interactions between light and matter depend on the types of light and matter involved. A brick wall is opaque to visible light but transmits radio waves, and glass that is transparent to visible light can be opaque to ultraviolet light. In general, certain types of matter tend to interact more strongly with certain types of light, so each type of light carries different information about distant objects in the universe. Astronomers therefore seek to observe light of all wavelengths, using telescopes adapted to detecting each different form of light, from radio waves to gamma rays.

Common Misconceptions: You Can't Hear Radio Waves and You Can't See X Rays

Most people associate the term *radio* with sound, but radio waves are a form of *light* with long wavelengths—too long for our eyes to see. Radio stations encode sounds (e.g., voices, music) as electrical signals, which they broadcast as radio waves. What we call "a radio" in daily life is an electronic device that receives these radio waves and decodes them to re-create the sounds played at the radio station. Television is also broadcast by encoding information (both sound and pictures) in the form of light called radio waves.

X rays are also a form of light, with wavelengths far too short for our eyes to see. In a doctor's or dentist's office, a special machine works rather like the flash on an ordinary camera but emits X rays instead of visible light. This machine flashes the X rays at you, and a piece of photographic film records the X rays that are transmitted through your body. Note that you never *see* the X-rays; you see only an image left on film by the transmitted X rays.

7.4 Light and Matter

Whenever matter and light interact, matter leaves its fingerprints. Examining the color of an object is a crude way of studying the clues left by the matter it contains. For example, a red shirt absorbs all visible photons except those in the red part of the spectrum, so we know that it must contain a dye with these special light-absorbing characteristics. If we take light and disperse it into a spectrum, we can see the spectral fingerprints in more detail.

Figure 7.6 shows a schematic spectrum of light from a celestial body such as a planet. The spectrum is a graph that shows the amount of radiation, or **intensity**,[4] at different wavelengths. At wavelengths where a lot of light is coming from the celestial body, the intensity is high; at wavelengths where there is little light, the intensity is low. Our goal is to see how the bumps and wiggles in this graph convey a wealth of information about the celestial body in question. Let's begin by going through a short list of ways that matter interacts with light and showing the spectra that result. We'll start with atoms—the simplest form of matter—and work our way up to more complex objects such as stars, rocks, and you.

Absorption and Emission by Atoms and Molecules

The universe is filled with a variety of wispy, tenuous gases, including clouds of gas between the stars and the thin atmospheres of stars and planets. You know that gases can absorb light; for example, ozone in the Earth's atmosphere absorbs ultraviolet light from space, preventing it from reaching the ground. You also know that gases can emit light, which is what makes interstellar clouds glow so beautifully. But *how* do gases absorb or emit light? Answering this question requires a close examination of the interactions between light and the atoms and molecules that make up a gas.

[4]Different kinds of instruments measure the "amount of radiation" in different ways, such as by the number of photons or by the total energy of all the photons at that wavelength. We use the generic term *intensity* to include various means of measuring the amount of radiation.

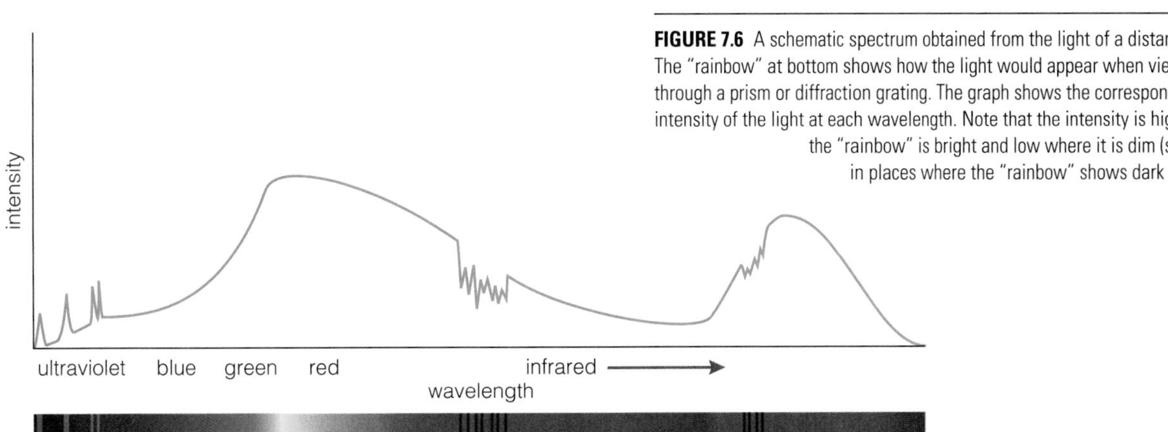

FIGURE 7.6 A schematic spectrum obtained from the light of a distant object. The "rainbow" at bottom shows how the light would appear when viewed through a prism or diffraction grating. The graph shows the corresponding intensity of the light at each wavelength. Note that the intensity is high where the "rainbow" is bright and low where it is dim (such as in places where the "rainbow" shows dark lines).

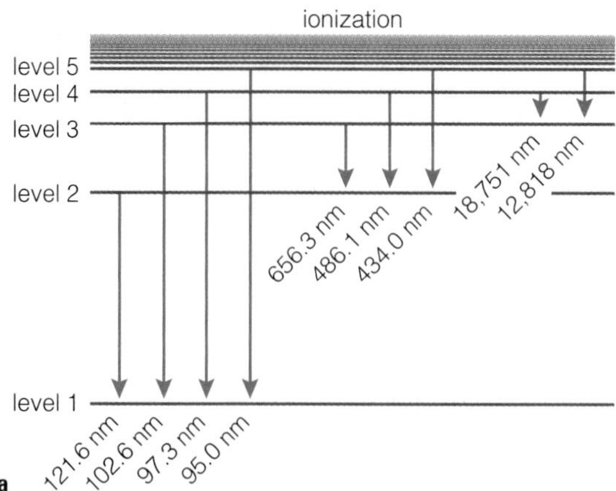

FIGURE 7.7 (**a**) Photons emitted by various energy level transitions in hydrogen. (**b**) The visible emission line spectrum from heated hydrogen gas. These lines come from transitions in which electrons fall from higher energy levels to level 2. (**c**) If we pass white light through a cloud of cool hydrogen gas, we get this absorption line spectrum. These lines come from transitions in which electrons jump from energy level 2 to higher levels.

Recall that the electrons in atoms can have only specific energies, somewhat like the specific heights of the rungs on a ladder [Section 5.4]. If an electron in an atom is bumped from a lower energy level to a higher one—by a collision with another atom, for example—it will eventually fall back to the lower level. The energy that the atom loses when the electron falls back down must go somewhere, and often it goes to *emitting* a photon of light. The emitted photon must have exactly the same amount of energy that the electron loses, which means that it has a specific wavelength (and frequency).

Figure 7.7a shows the allowed energy levels in hydrogen, along with the wavelengths of the photons emitted by various downward *transitions* of an electron from a higher energy level to a lower one. For example, transitions from level 2 to level 1 emit an ultraviolet photon of wavelength 121.6 nm, and transitions from level 3 to level 2 emit a red visible light photon of wavelength 656.3 nm.[5] If you heat some hydrogen gas so that collisions are continually bumping electrons to higher energy levels, you'll get an **emission line spectrum** consisting of the photons emitted as each electron falls back to lower levels (Figure 7.7b).

TIME OUT TO THINK *If nothing continues to heat the hydrogen gas, all the electrons eventually will end up in the lowest energy level (the ground state, or level 1). Use this fact to explain why we should* not *expect to see an emission line spectrum from a very cold cloud of hydrogen gas.*

Photons of light can also be absorbed, causing electrons to jump *up* in energy—but only if an in-

[5]Astronomers call transitions between level 1 and other levels the *Lyman* series of transitions. The transition between level 1 and level 2 is Lyman α, between level 1 and level 3 Lyman β, and so on. Similarly, transitions between level 2 and higher levels are called *Balmer* transitions. Other sets of transitions also have names, but they are less commonly used.

coming photon happens to have precisely the right amount of energy. For example, just as an electron moving downward from level 2 to level 1 in hydrogen emits a photon of wavelength 121.6 nm, absorbing a photon with this wavelength will cause an electron in level 1 to jump up to level 2.

Suppose a lamp emitting white light illuminates a cloud of hydrogen gas from behind. The cloud will absorb photons with the precise energies needed to bump electrons from a low energy level to a higher one, while all other photons pass right through the cloud. The result is an **absorption line spectrum** that looks like a rainbow with light missing at particular wavelengths (Figure 7.7c).

If you compare the bright emission lines in Figure 7.7b to the dark absorption lines in Figure 7.7c, you will see that the lines occur at the same wavelengths regardless of whether the hydrogen is absorbing or emitting light. Absorption lines simply correspond to upward jumps of the electron between energy levels, while emission lines correspond to downward jumps.

The energy levels of electrons in each chemical element are unique [Section 5.4]. As a result, each element produces its own distinct set of spectral lines, giving it a unique "spectral fingerprint." For example, Figure 7.8 shows emission line spectra for helium, sodium, and neon. Note that the spectra of atoms such as neon are more complex than the spectra of hydrogen and helium because they have many electrons and therefore many allowed energy levels.

TIME OUT TO THINK *Three common examples of objects with emission line spectra are storefront neon signs, fluorescent light bulbs, and the yellow sodium lights used in many cities at night. When you look at objects of different colors under such lights, do you see their normal colors? Why or why not?*

The fact that each chemical element has a unique spectral fingerprint makes spectral analysis extremely useful: When you see the fingerprint of a particular element, you immediately know that the gas producing the spectrum contains this element.

Movie Madness: Lois Lane's Underwear

In the 1978 movie *Superman,* the caped crusader claims to use his "X-ray vision" to determine that Lois Lane is wearing pink underwear. Sorry, but even if Superman really has "X-ray vision," his claim is impossible regardless of whether he sees X rays or whether his eyes emit them. First of all, there's no such thing as a *pink* X ray, because pink is a color in the visible part of the spectrum. Second, underwear neither emits nor reflects X rays. If it emitted X rays, then we'd all need to wear shielding to protect ourselves from its harmful effects. If it reflected X rays, then we could simply wear underwear instead of lead shields at the dentist's office. It's a good thing that Lois Lane was not aware of these problems with Superman's claim; otherwise, she might have suspected him of secretly rifling through her dresser!

For example, Figure 7.9 shows the spectral fingerprints of hydrogen, helium, oxygen, and neon in an emission line spectrum from the Orion Nebula. Not only does each chemical element produce a unique spectral fingerprint, but *ions* of a particular element (atoms that are missing one or more electrons) produce fingerprints different from those of neutral atoms [Section 5.3]. For example, the spectrum of doubly ionized neon (Ne^{++}) is different from that of singly ionized neon (Ne^{+}), which in turn is different from that of neutral neon (Ne). This fact can help us determine the temperature of a hot gas or plasma. At higher temperatures, more highly charged ions will be present, so we can estimate the temperature by identifying the ions that are creating spectral lines.

FIGURE 7.8 Emission line spectra for helium, sodium, and neon. The patterns and wavelengths of lines are different for each element, giving each a unique spectral fingerprint.

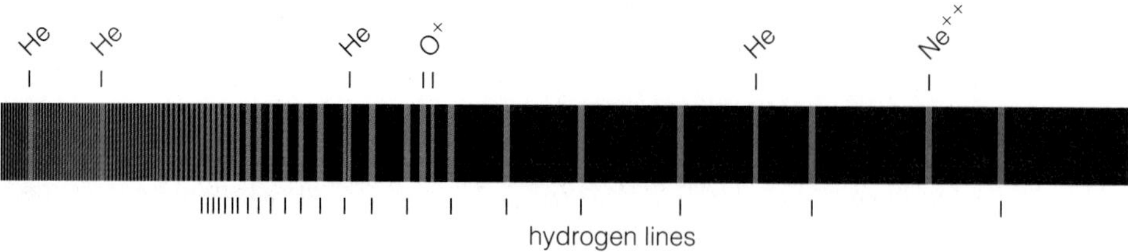

FIGURE 7.9 The emission line spectrum of the Orion Nebula. The lines are identified with the chemical elements or ions that produce them (He = helium; O = oxygen; Ne = neon).

Just as atoms and ions can absorb or emit light at particular wavelengths, so can *molecules.* Like electrons in atoms, the electrons in molecules can have only particular energies, and therefore molecules produce spectral lines when electrons change energy levels. However, because molecules are made of two or more atoms bound together, they can also have energy due to vibration or rotation (Figure 7.10a). It turns out that, just as its electrons can be in only specific energy levels, a molecule can rotate or vibrate only with particular amounts of energy. Thus, a molecule can absorb or emit a photon when it changes its rate of vibration or rotation. Because molecules can change energy in three different ways, their spectra look very different from the spectra of individual atoms. Molecules produce a spectrum with many sets of tightly bunched lines, called **molecular bands** (Figure 7.10b). The energy jumps in molecules are usually smaller than those in atoms—and therefore produce lower-energy photons—so most molecular bands lie in the infrared rather than in the visible or ultraviolet. That is one reason why infrared telescopes and instruments are so important to astronomers.

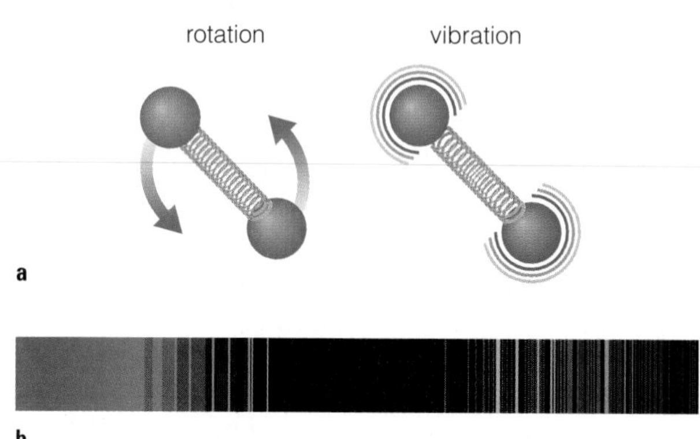

FIGURE 7.10 (**a**) We can think of a two-atom molecule as two balls connected by a spring. Although this model is overly simplistic, it illustrates how molecules can rotate and vibrate. (**b**) This spectrum of molecular hydrogen (H_2) shows that molecular spectra consist of lines bunched into broad *molecular bands.*

Thermal Radiation: Every Body Does It

In a low-density gas, individual atoms or molecules are essentially independent of one another [Section 5.3]. That is why thin, low-density clouds of gas produce relatively simple emission or absorption spectra, with lines (or bands) in locations determined by the energy levels of their constituent atoms (or molecules). But what happens in an opaque object, such as a star, a planet, or you? (Although saying that a star is opaque may sound strange, it is true because we cannot see *through* a star.) Because photons of light cannot pass through an opaque object, they bounce randomly around among the atoms or molecules inside it, constantly exchanging energy. Recall that the "bouncing around" of atoms and molecules tends to randomize their kinetic energies, giving them an average kinetic energy that characterizes the object's *temperature.* In a similar way, the "bouncing around" of the photons inside an opaque object randomizes their radiative energies in a way that also depends only on the object's temperature. These photons can escape only when they happen to reach the surface of the object (where no more material is "in the way" to block their escape). Thus, the radiation coming from an opaque object depends only on its temperature and hence is called **thermal radiation.**

An idealized, perfectly opaque object absorbs *all* radiation that strikes it and reemits the absorbed energy as thermal radiation; such an object is called a **thermal emitter.**[6] No real object is a perfect thermal emitter, but any sufficiently opaque object (roughly speaking, opaque enough so that we can't see through it) makes a close approximation. Thus, almost all familiar objects—including the Sun, the planets, and even you—glow with light that approximates thermal radiation. Two simple rules describe how the spectrum of a thermally emitting object depends on its temperature:

[6]A thermal emitter is also called a *blackbody,* and thermal radiation is often called *blackbody radiation.*

1. *Hotter objects emit more total radiation per unit surface area.* The radiated energy is proportional to the *fourth* power of the temperature expressed in Kelvin (*not* in Celsius or Fahrenheit). For example, a 600 K object has twice the temperature of a 300 K object and therefore radiates $2^4 = 16$ times as much total energy per unit surface area.
2. *Hotter objects emit photons with a higher average energy,* which means a shorter average wavelength.

You can see the first rule in action by playing with a light that has a dimmer switch: When you turn the switch up, the filament in the light bulb gets hotter and the light brightens; when you turn it down, the filament gets cooler and the light dims. (You can verify the changing temperature by placing your hand near the bulb.) You can see the second rule in action with a fireplace poker (Figure 7.11). When the poker is relatively cool, it emits only infrared radiation, which we cannot see. As it gets hot, it begins to glow red ("red hot"). If the poker continues to heat up, the average wavelength of the emitted photons gets shorter, moving toward the blue end of the visible spectrum. By the time it gets very hot, the mix of colors emitted by the poker looks white ("white hot").

Figure 7.12 shows how these two rules affect the spectra of idealized thermal emitters. The spectra of hotter objects show bigger "humps" because they emit more total radiation per unit area (rule 1). Hotter objects also have the peaks of their humps at shorter wavelengths because of the higher average energy of their photons (rule 2). Note that hotter objects emit more light at *all* wavelengths but the biggest difference appears at the shortest wavelengths. An object with a temperature of 310 K, which is about human body temperature, emits mostly in the

FIGURE 7.11 A fireplace poker gets brighter as it is heated, demonstrating rule 1 for thermal radiation (hotter objects emit more total radiation per unit surface area). In addition, its "color" moves from infrared to red to white as it is heated, demonstrating rule 2 (hotter objects emit photons with higher average energy).

Mathematical Insight 7.2 Laws of Thermal Radiation

The two rules of thermal radiation each have simple mathematical formulas. Rule 1, called the *Stefan–Boltzmann law* (named after its discoverers), is expressed as:

$$\text{emitted power per unit area} = \sigma T^4 \quad \left(\sigma = 5.7 \times 10^{-8} \frac{\text{watt}}{\text{m}^2 \times \text{Kelvin}^4}\right)$$

where σ (Greek letter *sigma*) is a constant.

Rule 2, called *Wien's law,* is expressed approximately as:

$$\lambda_{\text{max}} = \frac{2{,}900{,}000}{T\ (\text{Kelvin})}\ \text{nm}$$

where λ_{max} (*lambda-max*) is the wavelength of maximum intensity, which is the peak of the hump in a thermal radiation spectrum.

Example: Consider a perfect thermal emitter with a temperature of 15,000 K. How much power does it emit per unit area? What is its wavelength of peak intensity?

Solution: The emitted power per unit area from a 15,000 K thermal emitter is:

$$\sigma T^4 = 5.7 \times 10^{-8} \frac{\text{watt}}{\text{m}^2 \times \text{K}^4} \times (15{,}000\ \text{K})^4 = 2.9 \times 10^9 \frac{\text{watt}}{\text{m}^2}$$

Its wavelength of maximum intensity is:

$$\lambda_{\text{max}} = \frac{2{,}900{,}000}{15{,}000\ (\text{Kelvin})}\ \text{nm} = 190\ \text{nm}$$

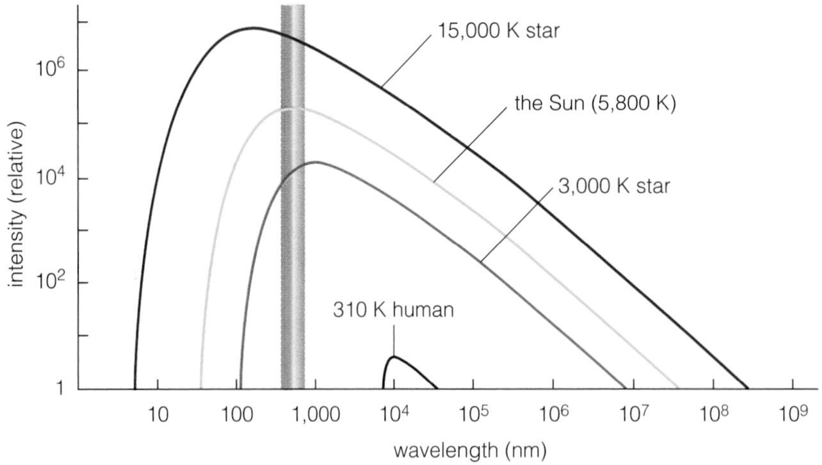

FIGURE 7.12 Graphs of idealized thermal radiation spectra. Note that, per unit surface area, hotter objects emit more radiation at every wavelength, demonstrating rule 1 for thermal radiation. The peaks of the spectra occur at shorter wavelengths (higher energies) for hotter objects, demonstrating rule 2 for thermal radiation.

infrared and emits no visible light at all—that explains why we don't glow in the dark! A relatively cool star, with a 3,000 K surface temperature, emits mostly red light; that is why some bright stars in our sky appear red, such as Betelgeuse (in Orion) and Antares (in Scorpio). The Sun's 5,800 K surface emits most strongly in green light (around 500 nm), but the Sun looks yellow or white because it also emits other colors throughout the visible spectrum. Hotter stars emit mostly in the ultraviolet, but because our eyes cannot see ultraviolet they appear blue or blue-white in color. If an object were heated to a temperature of millions of degrees, it would radiate mostly X rays. Some astronomical objects are indeed hot enough to emit X rays, such as disks of gas encircling exotic objects like neutron stars and black holes (see Chapter 17).

Summary of Spectral Formation

We can now summarize the circumstances under which objects produce thermal, absorption line, or emission line spectra.[7]

- Any opaque object produces thermal radiation over a broad range of wavelengths. If the object is hot enough to produce visible light, as is the case with the filament of a light bulb, we see a smooth, *continuous* rainbow when we disperse the light through a prism or a diffraction grating (Figure 7.13a). On a graph of intensity versus wavelength, the rainbow becomes the characteristic hump of a thermal radiation spectrum.
- When thermal radiation passes through a thin cloud of gas, the form of the fingerprints the cloud leaves depends on its temperature. If the background source of thermal emission is hotter than the cloud, the balance between emission and absorption in the cloud's spectral lines tips toward absorption. We then see absorption lines cutting into the thermal spectrum. This is the case when the light from the hot light bulb passes through a cool gas cloud (Figure 7.13b). On the graph of intensity versus wavelength, these absorption lines create dips in the thermal radiation spectrum; the width and depth of each dip depend on how much light is absorbed by the chemical responsible for the line.
- If the background source is colder than the cloud, or if there is no background source at all, the spectrum is dominated by bright emission lines produced by the cloud's atoms and molecules (Figure 7.13c). These lines create narrow peaks on a graph of intensity versus wavelength.

Reflected Light

We've now covered enough material to understand spectra emitted by objects generating their own light, such as stars and clouds of interstellar gas. But most of our daily experience involves *reflected* (or *scattered*) light. The source of the light is thermal radiation from the Sun or a lamp. After this light strikes the ground, clouds, people, or other objects, we see only the wavelengths of light that are reflected. For example, a red sweatshirt absorbs blue light and reflects red light, so its visible spectrum looks like the thermal radia-

[7]These three general rules determining the conditions for a thermal, absorption line, or emission line spectrum are often called *Kirchhoff's laws*.

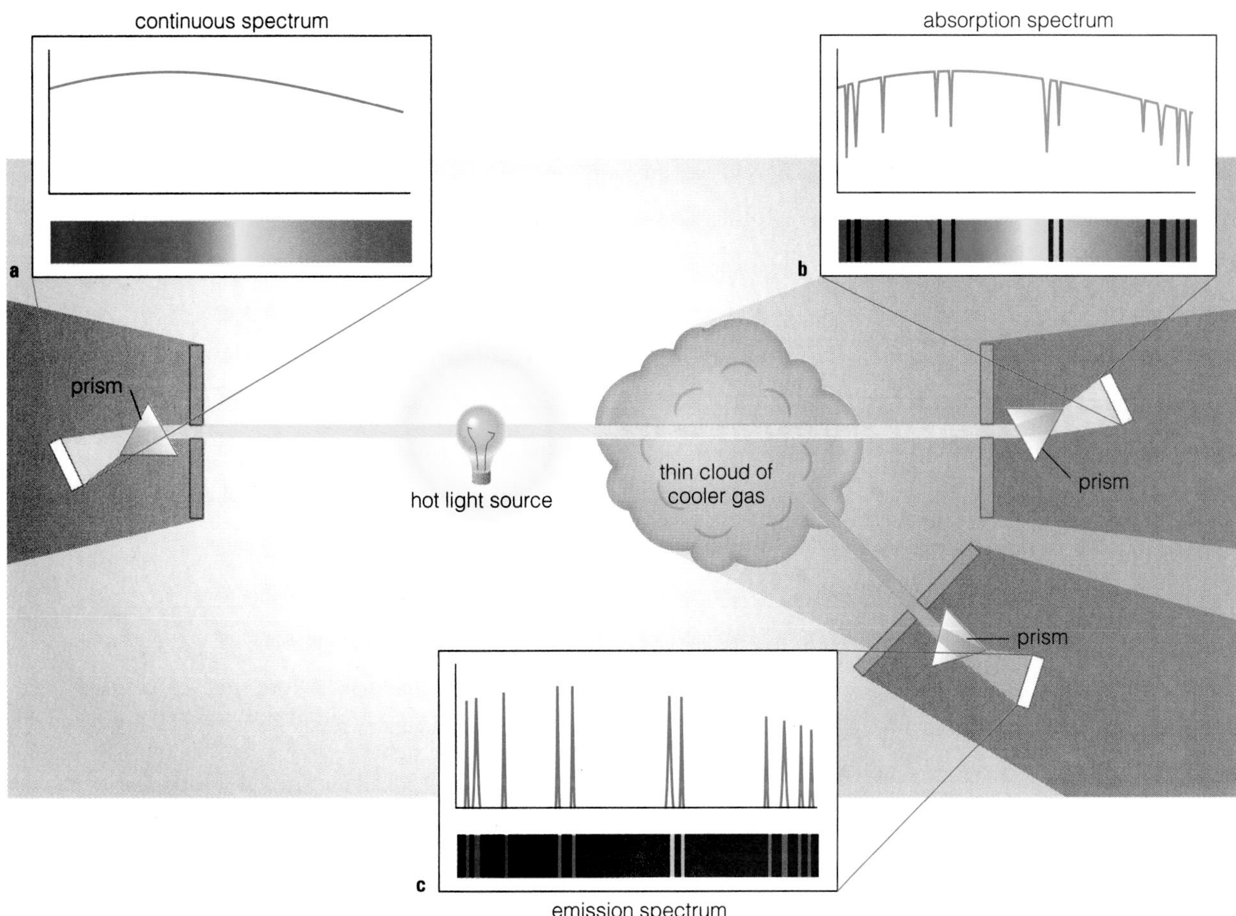

FIGURE 7.13 (**a**) An opaque object, such as a light bulb filament, produces a continuous spectrum of thermal radiation. (**b**) If thermal radiation passes through a thin gas that is cooler than the thermal emitter, dark absorption lines are superimposed on the continuous spectrum. (**c**) Viewed against a cold, dark background, the same gas produces an emission line spectrum.

tion spectrum of its light source—the Sun—but with blue light missing.

In the same way in which we distinguish lemons from limes, we can use color information in reflected light to learn about celestial objects. Different fruits, different rocks, and even different atmospheric gases reflect and absorb light at different wavelengths. Although the absorption features that show up in spectra of reflected light are not as distinct as the emission and absorption lines for thin gases, they still provide useful information. For example, the surface materials of a planet determine how much light of different colors is reflected or absorbed. The reflected light gives the planet its color, while the absorbed light heats the surface and helps determine its temperature.

Putting It All Together

Figure 7.14 again shows the complicated spectrum that we began with in Figure 7.6, but this time with labels indicating the processes responsible for its various features. What can we say about this object from its spectrum? The hump of thermal emission shows that this object has a surface temperature of about 225 K, well below the freezing point of water. The absorption bands in the infrared come mainly from carbon dioxide, which tells us that the object has a carbon dioxide atmosphere. The emission lines in the ultraviolet come from hot gas in a high, thin layer of the object's atmosphere. The reflected light looks like the Sun's 5,800 K thermal radiation except that the blue light is missing, so the object must be reflecting sunlight and must look red in color. Perhaps by now you have guessed that this figure represents the spectrum of the planet Mars.

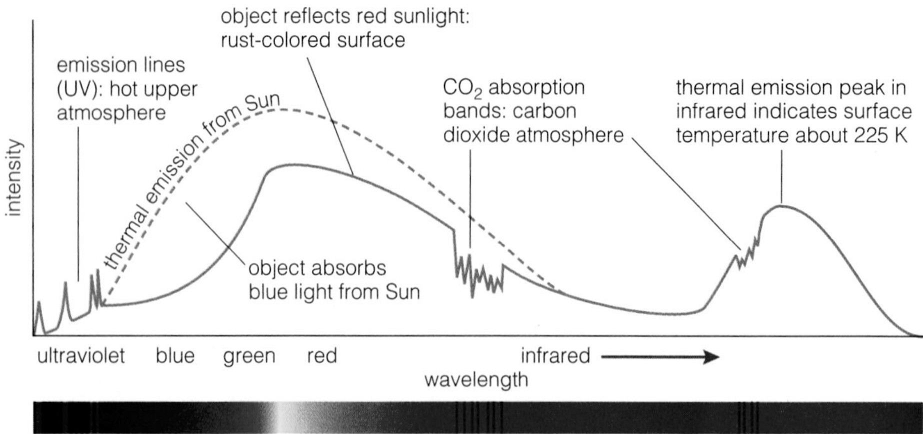

FIGURE 7.14 The spectrum of Figure 7.6, with interpretation. We can conclude that the object looks red in color because it absorbs more blue light than red light from the Sun. The absorption lines tell us that the object has a carbon dioxide atmosphere, and the emission lines tell us that its upper atmosphere is hot. The hump in the infrared (near the right of the diagram) tells us the object has a surface temperature of about 225 K. It is a spectrum of the planet Mars.

7.5 The Doppler Shift

The volume of information about temperature and composition that light contains is truly amazing, but we can learn even more once we identify the various lines in a spectrum. Among the most important pieces of information contained in light is information about motion. In particular, we can determine the radial motion (toward or away from us) of a distant object from changes in its spectrum caused by the **Doppler effect** [Section 3.4].

You've probably noticed the Doppler effect on *sound;* it is especially easy to notice when you stand near train tracks and listen to the whistle of a train (Figure 7.15). As the train approaches, its whistle is relatively high pitched; as it recedes, the sound is relatively low pitched. Just as the train passes by, you hear the dramatic change from high to low pitch—a sort of "weeeeeeee–oooooooooh" sound. You can visualize the Doppler effect by imagining that the train's sound waves are bunched up ahead of it, resulting in shorter wavelengths and thus the high pitch you hear as the train approaches. Behind the train, the sound waves are stretched out to longer wavelengths, resulting in the low pitch you hear as the train recedes.

The Doppler effect causes similar shifts in the wavelengths of light. If an object is moving toward us, then its entire spectrum is shifted to shorter wavelengths. Because shorter wavelengths are bluer when we are dealing with visible light, the Doppler shift of an object coming toward us is called a **blueshift.** If an object is moving away from us, its light is shifted to longer wavelengths; we call this a **redshift** because longer wavelengths are redder when we are dealing with visible light. Note that the terms *blueshift* and *redshift* are used even when we are not dealing with visible light.

Spectral lines provide the reference points we use to identify and measure Doppler shifts (Figure 7.16). For example, suppose we recognize the pattern of hydrogen lines in the spectrum of a distant object. We know the **rest wavelengths** of the hydrogen lines—that is, their wavelengths in stationary clouds of hydrogen gas—from laboratory experiments in which a tube of hydrogen gas is heated so that the wavelengths of the spectral lines can be measured. If the hydrogen lines from the object appear at longer wavelengths, then we know that they are redshifted and the object is moving away from us; the larger the shift, the faster the object is moving. If the lines appear at shorter wavelengths, then we know that they are blueshifted and the object is moving toward us.

TIME OUT TO THINK *Suppose the hydrogen emission line with a rest wavelength of 121.6 nm (the transition from level 2 to level 1) appears at a wavelength of 120.5 nm in the spectrum of a particular star. Given that these wavelengths are in the ultraviolet, is the shifted wavelength closer to or farther from blue visible light in the spectrum? Why, then, do we say that this spectral line is* blueshifted?

The Doppler effect not only tells us how fast a distant object is moving toward or away from us but also can reveal information about motion *within* the object itself. For example, suppose we look at spectral lines of a planet or star that happens to be rotat-

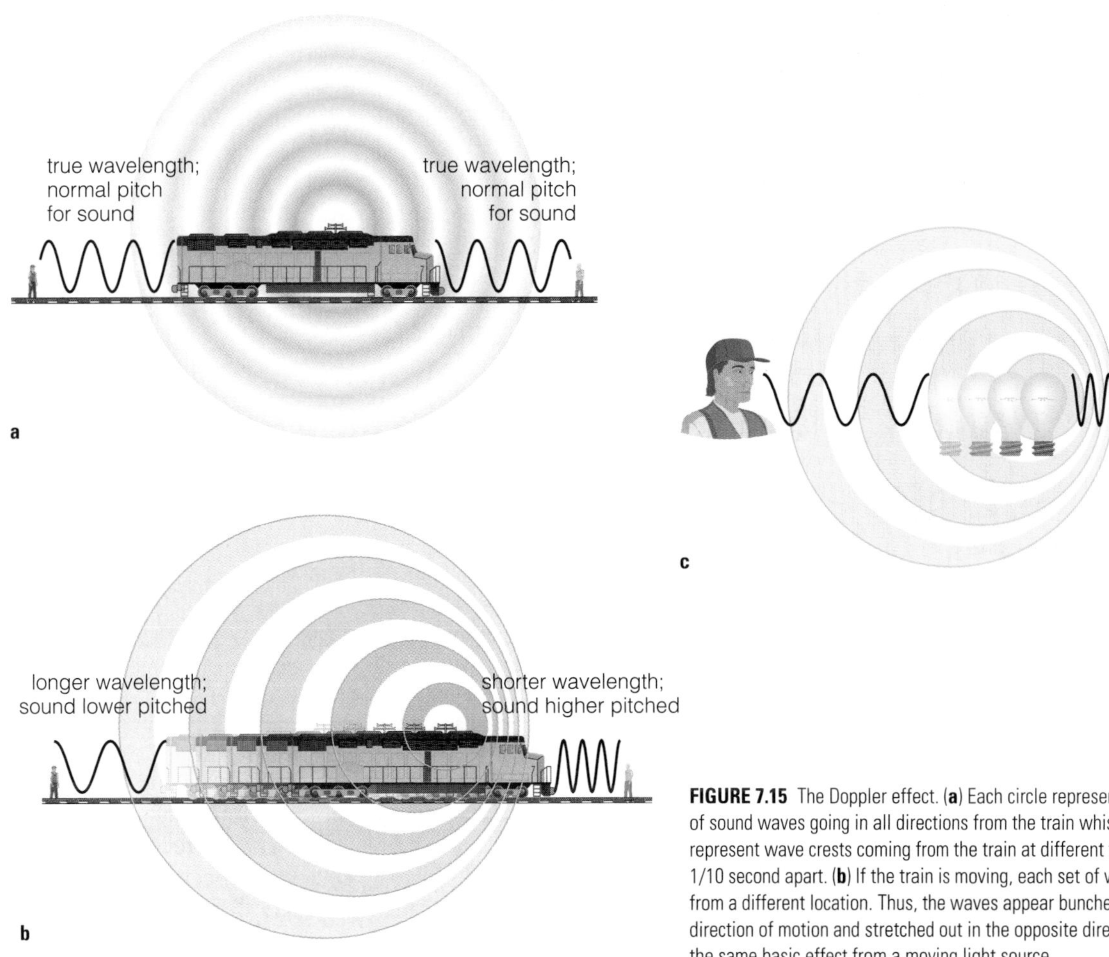

FIGURE 7.15 The Doppler effect. (**a**) Each circle represents the crests of sound waves going in all directions from the train whistle. The circles represent wave crests coming from the train at different times, say, 1/10 second apart. (**b**) If the train is moving, each set of waves comes from a different location. Thus, the waves appear bunched up in the direction of motion and stretched out in the opposite direction. (**c**) We get the same basic effect from a moving light source.

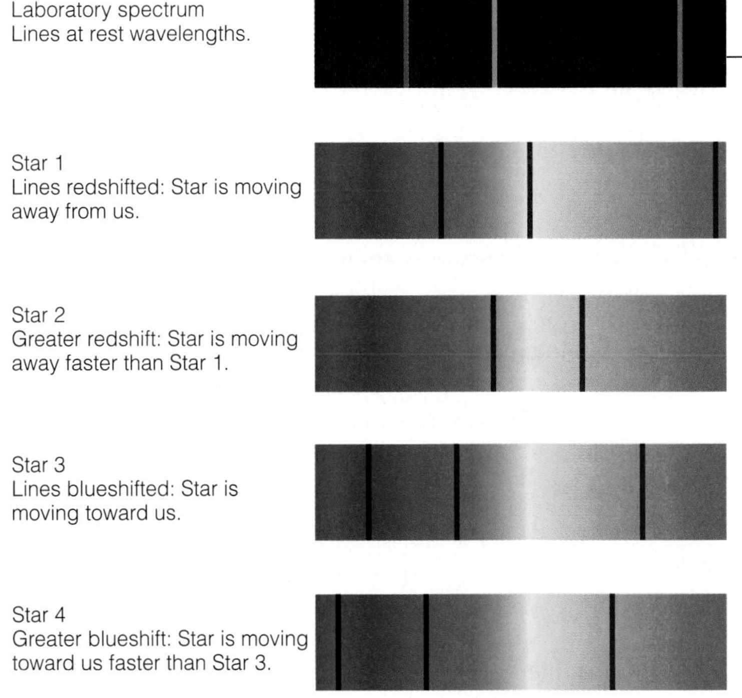

FIGURE 7.16 Spectral lines provide the crucial reference points for measuring Doppler shifts.

ing (Figure 7.17). As the object rotates, light from the part of it rotating toward us will be blueshifted, light from the part rotating away from us will be redshifted, and light from the center of the object won't be shifted at all. The net effect, if we look at the whole object at once, is to make each spectral line appear *wider* than it would if the object were not rotating. The faster the object is rotating, the broader in wavelength the spectral lines become. Thus, we can determine the rotation rate of distant objects by measuring the width of their spectral lines. As we will see later, this is only a small part of the information revealed through the Doppler effect on the spectra of celestial objects.

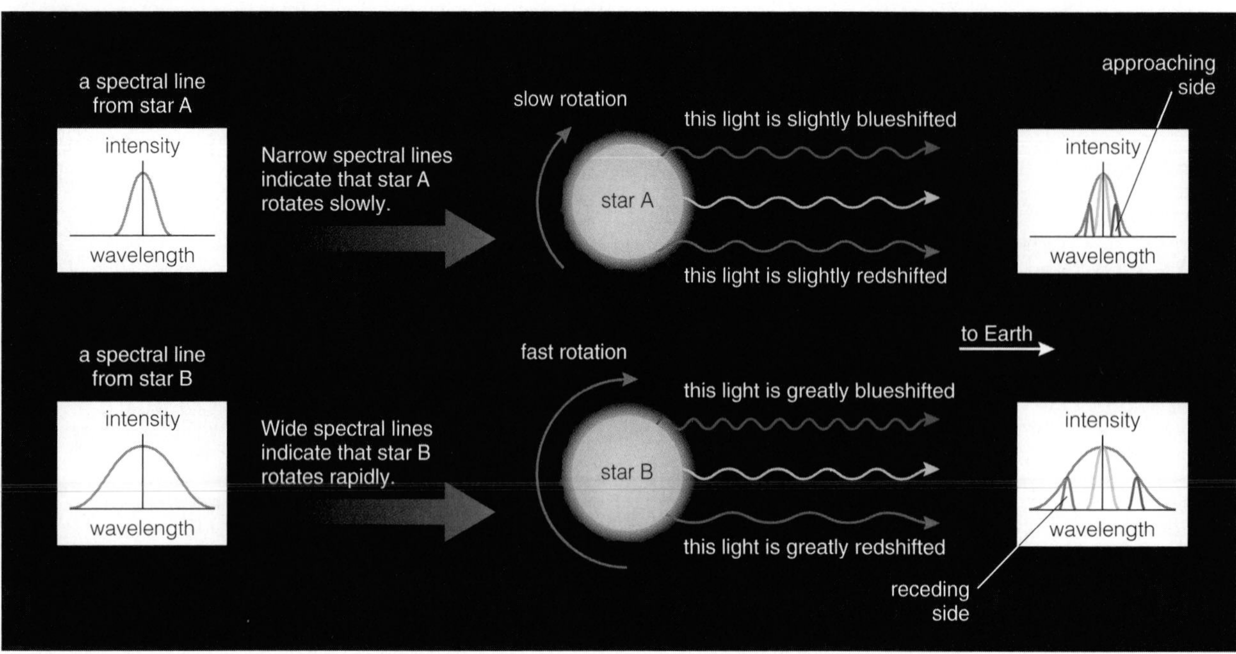

FIGURE 7.17 The Doppler effect broadens the widths of the spectral lines of rotating objects

Mathematical Insight 7.3 The Doppler Shift

As long as an object's radial speed is small compared to the speed of light (i.e., less than a few percent of c), a simple formula allows us to calculate the radial speed of an object from its Doppler shift:

$$\text{radial speed of object} = \frac{\text{shifted wavelength} - \text{rest wavelength}}{\text{rest wavelength}} \times \text{speed of light}$$

If the result is positive, the object has a redshift and is moving away from us; a negative result means the object has a blueshift and is moving toward us. The formula can also be written symbolically as:

$$\frac{v}{c} = \frac{\Delta\lambda}{\lambda_0}$$

where v is the object's radial speed, c is the speed of light, λ_0 is the rest wavelength of a particular spectral line, and $\Delta\lambda$ is its wavelength shift (positive for a redshift and negative for a blueshift).

Example: The rest wavelength of one of the visible lines of hydrogen is 656.285 nm. This line is easily identifiable in the spectrum of the bright star Vega (by its strength), but it appears at a wavelength of 656.255 nm. What is the radial speed of Vega?

Solution: The line's wavelength in Vega's spectrum is slightly shorter than its rest wavelength, so the line is blueshifted and Vega's radial motion is *toward* us. Vega's radial speed toward us is:

$$\frac{656.255 \text{ nm} - 656.285 \text{ nm}}{656.285 \text{ nm}} \times 300{,}000 \frac{\text{km}}{\text{s}} = (-4.57 \times 10^{-5}) \times 3 \times 10^5 \frac{\text{km}}{\text{s}}$$

$$= -13.7 \frac{\text{km}}{\text{s}}$$

The negative answer confirms that Vega is moving *toward* us at 13.7 km/s.

May the warp be the white light of morning,
May the weft be the red light of evening,
May the fringes be the falling rain,
May the border be the standing rainbow.
Thus weave for us a garment of brightness.

SONG OF THE SKY LOOM (NATIVE AMERICAN)

THE BIG PICTURE

If you look back, you'll see that this chapter has been devoted to one essential purpose: understanding how to read the messages contained in the spectra of distant objects. "Big picture" ideas that will help you keep your understanding in perspective include the following:

- There is far more to light than meets the eye. By dispersing light into a spectrum with a prism or a diffraction grating, we discover a wealth of information about the object from which the light has come. Most of what we know about the universe comes from information that we receive in the form of light.
- The visible light that our eyes can see is only a small portion of the complete electromagnetic spectrum. Different portions of the spectrum may contain different pieces of the story of a distant object, so it is important to study spectra at many wavelengths.
- The spectrum of any object is determined by the interactions of light and matter in that object. These interactions can produce *emission lines, absorption lines,* or *thermal radiation;* the spectra of most objects contain some degree of all three of these. Some spectra also include transmitted light (from a light source behind the object) and reflected light.
- By studying the spectra of a distant object, we can determine its composition, surface temperature, motion toward or away from us, rotation rate, and more.

Review Questions

1. What is the difference between *energy* and *power,* and what units do we use to measure each? If you have a 100-watt light bulb, how much energy does it use each second? Each minute? Each day?
2. What do we mean when we speak of a *spectrum* produced by a prism or *diffraction grating?*
3. Describe each of the four basic ways that light can interact with matter: *emission, absorption, transmission,* and *reflection.* How is *scattering* related to reflection?
4. What does it mean for a material to be *transparent?* To be *opaque?* What is *opacity?*
5. How does a particle differ from a wave? Define each of the following terms as they apply to waves: *wavelength, frequency, cycles per second, hertz, speed.* Explain why wavelength × frequency = speed.
6. What is a *photon?* In what way is a photon like a particle? In what way is it like a wave?
7. Briefly describe the concept of a *field,* and use it to explain why we say that light is an *electromagnetic wave.*
8. Explain why light with a shorter wavelength must have a higher frequency, and vice versa.
9. Describe the *electromagnetic spectrum.* In terms of frequency, wavelength, and energy, distinguish among *radio, infrared, visible light, ultraviolet, X rays,* and *gamma rays.*
10. Explain how transitions between energy levels can cause atoms to emit or absorb photons. Why will a hot gas produce an *emission line spectrum?* Why do we get an *absorption line spectrum* when white light passes through a cool cloud of gas?
11. Why do the lines of a particular element appear in the same place in both emission and absorption line spectra? Why do different chemical elements—and different ionization states of those elements—produce different spectra?
12. Explain how spectral lines can be used to determine the composition and temperature of their source.
13. What is *thermal radiation?* What is a *thermal emitter?* Give a few examples of objects that make close approximations to ideal thermal emitters.
14. Describe the two laws of thermal radiation and give a few examples of the use of each law. Explain how the effects of both laws can be seen in Figure 7.12. Why is "white hot" hotter than "red hot"?

15. Summarize the circumstances under which objects produce thermal, emission line, or absorption line spectra as shown in Figure 7.13.

16. How does reflection affect spectra? Study Figure 7.14 carefully, and explain each feature in the spectrum.

17. Describe the *Doppler effect* for light. Explain why light from objects moving toward us shows a *blueshift* while light from objects moving away from us shows a *redshift*. What can we say about an object with a large Doppler shift, compared to an object with a small Doppler shift?

18. Describe how the Doppler effect allows us to determine the rotation rates of distant stars.

Discussion Questions

1. *The Changing Limitations of Science.* In 1835, French philosopher Auguste Comte stated that the composition of stars could never be known by science. Although spectral lines had been seen in the Sun's spectrum by that time, not until the mid-1800s did scientists recognize that spectral lines give clear information about chemical composition (primarily through the work of Foucault and Kirchhoff). Why might our present knowledge have seemed unattainable in 1835? Discuss how new discoveries can change the apparent limitations of science. Today, other questions seem beyond the reach of science, such as the question of how life began on Earth. Do you think that such questions will ever be answerable by science? Defend your opinion.

2. *Your Microwave Oven. Microwaves* is a name sometimes given to light near the long-wavelength end of the infrared portion of the spectrum. A *microwave oven* emits microwaves that happen to have just the right wavelength needed to cause energy level jumps in water molecules. Use this fact to explain how a microwave oven cooks your food. Why doesn't a microwave oven make a plastic or ceramic dish get hot?

Problems

1. *Spectral Summary.* Clearly explain how studying an object's spectrum can allow us to determine each of the following properties of the object.
   a. The object's surface chemical composition.
   b. The object's surface temperature.
   c. Whether the object is a thin cloud of gas or something more substantial.
   d. Whether the object has a hot upper atmosphere.
   e. The speed at which the object is moving toward or away from us.
   f. The object's rotation rate.

2. *Planetary Spectrum.* Suppose you take a spectrum of light coming from a planet that looks blue to the eye. Do you expect to see any visible light in the planet's spectrum? Is the visible light emitted by the planet, reflected by the planet, or both? Which (if any) portions of the visible spectrum do you expect to find "missing" in the planet's spectrum? Explain your answers clearly.

3. *Human Wattage.* A typical adult uses about 2,500 Calories of energy each day.
   a. Using the fact that 1 Calorie is about 4,000 joules, convert the typical adult energy usage to units of joules per day.
   b. Use your answer from part (a) to calculate a typical adult's average *power* requirement, in watts. Compare this to that of a light bulb.

4. *Wavelength, Frequency, and Energy.*
   a. What is the frequency of a visible light photon with wavelength 550 nm?
   b. What is the wavelength of a radio photon from an "AM" radio station that broadcasts at 1,120 kilohertz? What is its energy?
   c. What is the energy (in joules) of an ultraviolet photon with wavelength 120 nm? What is its frequency?
   d. What is the wavelength of an X-ray photon with energy 10 keV (10,000 eV)? What is its frequency? (*Hint:* Recall that 1 eV = 1.60×10^{-19} joule.)

5. *How Many Photons?* Suppose that all the energy from a 100-watt light bulb came in the form of photons with wavelength 600 nm. (This is not quite realistic; see problem 11.)
   a. Calculate the energy of a *single* photon with wavelength 600 nm.
   b. How many 600-nm photons must be emitted each second to account for all the light from this 100-watt light bulb? Based on your answer, explain why we don't notice the particle nature of light in our everyday lives.

6. *Hotter Sun.* Suppose that the surface temperature of the Sun were about 12,000 K, rather than 6,000 K.
   a. How much more thermal radiation would the Sun emit?

b. How would the thermal radiation spectrum of the Sun be different?

c. Do you think it would still be possible to have life on Earth? Explain.

7. *Taking the Sun's Temperature.* The Sun radiates a total power of about 4×10^{26} watts into space. The Sun's radius is about 7×10^8 meters.

   a. Calculate the average power radiated by each square meter of the Sun's surface. (*Hint:* The formula for the surface area of a sphere is $A = 4\pi r^2$.)

   b. Using your answer from part (a) and the Stefan–Boltzmann law (see Mathematical Insight 7.2), calculate the average surface temperature of the Sun. (*Note:* The temperature calculated this way is called the Sun's *effective temperature.* It would be equal to the actual temperature only if the Sun were a perfect thermal emitter.)

8. *The Doppler Effect.* In hydrogen, the transition from level 2 to level 1 has a rest wavelength of 121.6 nm. Suppose you see this line at a wavelength of 120.5 nm in Star A, at 121.2 nm in Star B, at 121.9 nm in Star C, and at 122.9 nm in Star D. Which stars are coming toward us? Which are moving away? Which star is moving fastest relative to us (either toward or away)? Explain your answers without doing any calculations.

9. *Doppler Calculations.* Calculate the speeds of each of the stars described in problem 8. Be sure to state whether each star is moving toward or away from us.

10. *The Expanding Universe.* Recall from Chapter 3 that we know the universe is expanding because (1) all galaxies outside our Local Group are moving away from us and (2) more distant galaxies are moving faster. Explain how Doppler shift measurements allow us to know these two facts.

11. *Understanding Light Bulbs.* A standard (incandescent) light bulb uses a hot tungsten coil to produce a thermal radiation spectrum. The temperature of this coil is typically about 3,000 K.

   a. What is the wavelength of maximum intensity for a standard light bulb? Compare this to the 500-nm wavelength of maximum intensity for the Sun. Also explain why standard light bulbs must emit a substantial portion of their radiation as invisible, infrared light.

   b. Overall, do you expect the light from a standard bulb to be the same as, redder than, or bluer than light from the Sun? Why? Use your answer to explain why professional photographers use a different type of film for indoor photography than for outdoor photography.

   c. *Fluorescent* light bulbs primarily produce emission line spectra rather than thermal radiation spectra. Explain why, if the emission lines are in the visible part of the spectrum, a fluorescent bulb can emit more light than a standard bulb of the same wattage.

   d. Today, *compact fluorescent* light bulbs are designed to produce so many emission lines in the visible part of the spectrum that their light looks very similar to the light of ordinary bulbs. However, they are much more energy-efficient: A 15-watt compact fluorescent bulb typically emits as much visible light as a standard 75-watt bulb. Although compact fluorescent bulbs generally cost more than standard bulbs, is it possible that they could save you money? Besides initial cost and energy efficiency, what other factors must be considered?

S2

Telescopes and Spacecraft

We are in the midst of a great revolution in human understanding of the universe. Astonishing new discoveries about the early history of the universe, the lives of galaxies and stars, and the planets in our solar system are frequent features of the daily news.

The fuel for this astronomical revolution is the wealth of new data acquired with the aid of telescopes and spacecraft. New technologies have vastly improved the quality of data that can be obtained with telescopes on the ground, and telescopes lofted into space can overcome the observational difficulties posed by the Earth's atmosphere. Whereas past astronomers could study only the messages contained in visible light, today's astronomer can read the story of the universe contained in the entire spectrum, from radio to gamma rays.

In addition, some spacecraft travel to other planets for close-up study, and in a few cases, spacecraft have even sampled the surfaces or atmospheres of other worlds. In this chapter, we explore how telescopes and spacecraft are helping us to better understand our universe.

S2.1 Eyes and Cameras: Everyday Light Sensors

We observe the world around us with the five basic senses—touch, taste, smell, hearing, and sight. We learn about the world by using our brains to analyze and interpret the data recorded by these senses. The science of astronomy progresses similarly: We collect data about the universe, and then we analyze and interpret the data to develop theories about how the universe works. Within our solar system, we can analyze matter directly, collecting samples from the Earth's surface and from meteorites that fall to Earth, and occasionally sending spacecraft to sample the surfaces or atmospheres of other worlds. Nearly all other data about the universe come in the form of light.

Astronomers collect light with telescopes and record light with photographic film or electronic detectors. You are already familiar with many of the basic principles of both telescopes and detectors because of your everyday experience with eyes and cameras.

The Eye

The human eyes and brain make up a familiar and exquisite optical system. The eye is more complex than any human-made instrument, but its basic components are a *lens,* a *pupil,* and a *retina* (Figure S2.1). The retina contains light-sensitive cells (called *cones* and *rods*) that, when triggered by light, send signals to the brain via the optic nerve.

The lens of your eye is similar to a simple glass lens. Light rays that enter the lens farther from the center are bent more, and rays that pass directly through the center are not bent at all. In this way, parallel rays of light, such as the light from a distant star, converge to a point called the **focus** (Figure S2.2a). If you have perfect vision, the focus of your lens is on your retina.[1] That is why distant stars appear as *points* of light to our eyes or on photographs.

[1]If you are near-sighted, light focuses in front of the retina. If you are far-sighted, light focuses behind the retina. Corrective lenses adjust the angles of light rays before they enter the eye, thereby returning the focal plane to its proper place on the retina. The retina actually is curved, rather than a flat plane, but we are ignoring this detail.

FIGURE S2.1 A simplified diagram of the human eye.

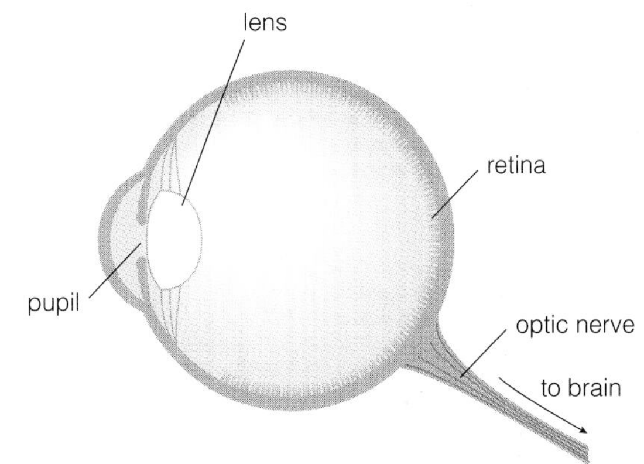

FIGURE S2.2 (**a**) A glass lens bends parallel rays of light to a point called the focus of the lens. In a healthy eye, the lens focuses light on the retina. (**b**) Light from different parts of an object focuses at different points to make an image of the object. Note that the image formed on the retina is upside-down. The world does not look upside-down because your brain flips the image back up.

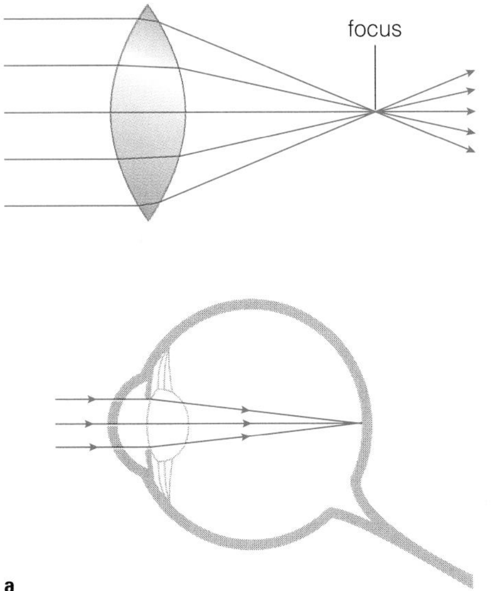

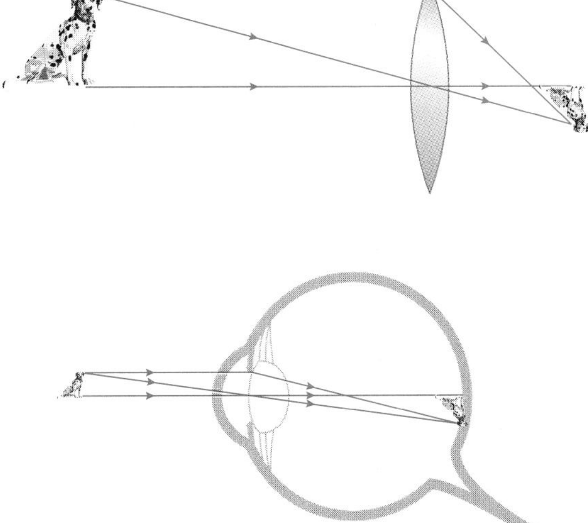

Light rays that come from different directions, such as from different parts of an object, converge at different points to form an **image** of the object (Figure S2.2b). The place where the image appears in focus is called the **focal plane** of the lens. Again, in a healthy eye, the focal plane is on your retina. Note that the image formed by a lens is upside down. It is flipped right side up by your brain, where the true miracle of vision occurs.

The pupil controls the amount of light that enters the eye by adjusting the size of its opening. The pupil dilates to a wider opening when light levels are low, allowing your eye to gather as many photons as possible. It constricts when light levels are high so that your eye is not overloaded with light.

TIME OUT TO THINK *Cover one eye and look toward a bright light. Then look immediately into a mirror and compare the openings of your two pupils. Which one is wider, and why? Why do eye doctors dilate your pupils when they want to examine the inside of your eyes?*

One way to quantify your visual acuity is to measure the smallest angle over which your eyes can tell that two dots—or two stars—are distinct. Imagine looking at a car heading toward you on a long, straight road. When the car is very far away, your eyes cannot distinguish the headlights individually, and they look like one light. But as the car comes closer, the angular separation of the two headlights gets bigger, and you eventually see two distinct headlights (Figure S2.3). The smallest angular separation at which you can tell that the headlights are distinct is the **angular resolution** of your eyes.

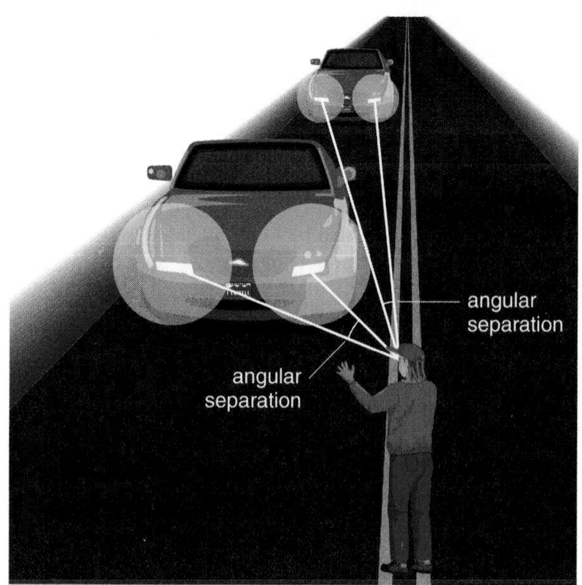

FIGURE S2.3 The angular separation of the headlights increases as the car comes closer.

Mathematical Insight S2.1 Angular Separation

The angular separation of two points is related to their actual separation and distance. Figure S2.4 shows two points with an actual separation s between them. The angle α is the angular separation between the two points when viewed from a distance d.

As long as d is much larger than s, we can think of s as a small portion of an imaginary circle with radius d. Recall that the circumference of a circle is $2 \times \pi \times$ radius, so the circumference of the dotted circle in the figure is $2\pi d$. Thus, the separation s represents a fraction $s/(2\pi d)$ of this circumference. We find the angle α by multiplying this fraction by the 360° in a full circle:

$$\alpha = \frac{s}{2\pi d} \times 360°$$

Example 1: Suppose the two headlights on a car are separated by 1.5 meters, and you are looking at the car from a distance of 500 meters. What is the angular separation of the headlights? Can your eyes resolve the two headlights?

Solution: The separation of the headlights is $s = 1.5$ m, and their distance is $d = 500$ m. Thus, their angular separation is:

$$\alpha = \frac{1.5 \text{ m}}{2\pi \times 500 \text{ m}} \times 360° = 0.17°$$

The angular resolution of the human eye is about $\frac{1}{60}° \approx 0.017°$, or 10 times smaller than the angular separation of the two headlights. Thus, your eyes can easily resolve the two headlights at a distance of 500 meters.

The human eye has an angular resolution of about 1 arcminute ($\frac{1}{60}°$), meaning that two stars will appear distinct if they lie greater than 1 arcminute apart in the sky. By extension, two stars separated by *less* than 1 arcminute appear to your naked eye as a single point of light. Most of the stars visible to the naked eye are in fact members of binary or multiple star systems, but because they are so far away the angular separation of the individual stars in the system is far below the angular resolution of our eyes.

Cameras and Film

The basic operation of a camera is quite similar to that of an eye (Figure S2.5). The camera lens plays the role of the lens of the eye, and film plays the role of the retina. The camera is "in focus" when the film lies in the focal plane of the lens. Fancier cameras even have an adjustable circular opening, called the camera's *aperture,* that controls the amount of light entering the camera just as the pupil controls the amount of light entering the eye.

The chemicals in photographic film darken or change color in response to light, thereby recording an image. Recording images on film offers at least two important advantages over simply looking at them or drawing them. First, an image on film is much more reliable and detailed than a drawing. Second, whereas the eye continually and automatically sends images to the brain, we can use a camera's *shutter* to control the amount of time, called the **exposure time,** over which light collects on a single

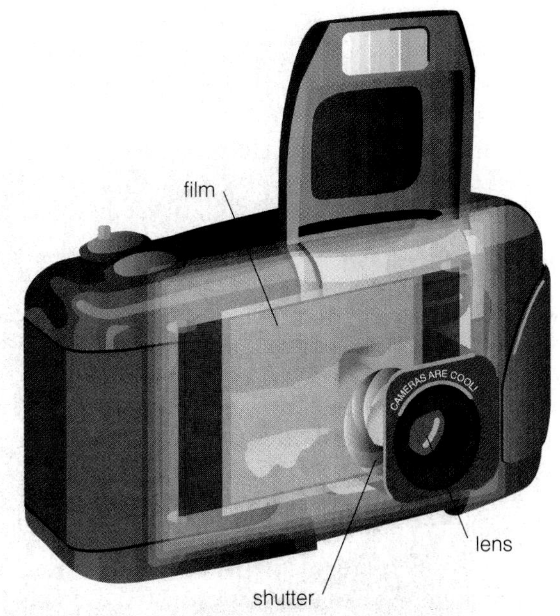

FIGURE S2.5 Basic characteristics of a camera.

Example 2: The angular diameter of the Moon is about 0.5°, and the Moon is about 380,000 km away. Estimate the diameter of the Moon. Compare your result to the actual value of 3,476 km.

Solution: We can think of the separation s as the diameter of the Moon. We are given $\alpha = 0.5°$ and $d = 380{,}000$. Thus, we must first solve the angular separation equation for s:

$$\alpha = \frac{s}{2\pi d} \times 360° \quad \Rightarrow \quad s = \frac{2\pi d}{360°} \times \alpha$$

Now we substitute the given values:

$$s = \frac{2\pi \times 380{,}000 \text{ km}}{360°} \times 0.5° = 3{,}316 \text{ km}$$

Our estimate of about 3,300 km is fairly close to the Moon's actual diameter (3,476 km), which we could find by using more precise values for the Moon's angular diameter and distance.

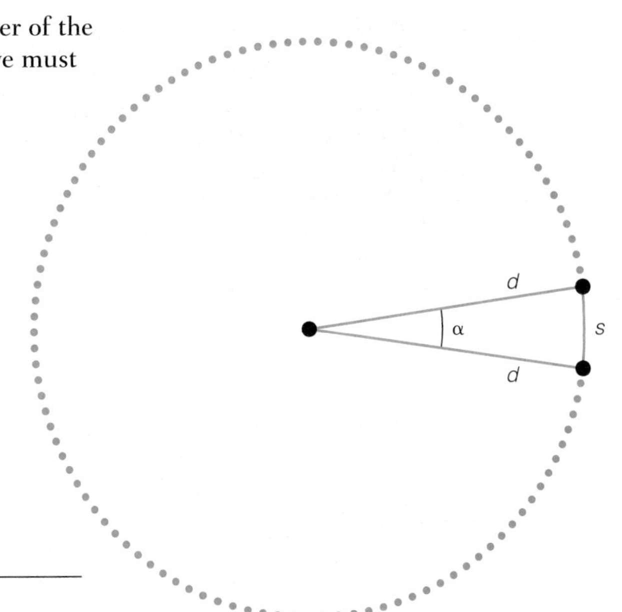

FIGURE S2.4 As long as the angle α is small, the length s and d are related to α by the formula $\alpha = \frac{s}{2\pi d} \times 360°$.

piece of film. A longer exposure time means that more photons reach the film, allowing more opportunities for the light-sensitive chemicals to change in response to light. As a result, long-exposure photographs can reveal images of objects far too faint for our eyes to see.

CCDs

Thanks to the advantages of film over the human eye, the advent of photography in the mid-1800s spurred a leap in astronomical data collection. However, photographic film is far from ideal for recording light. Most photons striking film have no effect at all: Fewer than 10% of the visible photons reaching the film cause a change in the light-sensitive chemicals.[2] Thus, your film may record nothing of use if too little light reaches it. At the other extreme, a long exposure time can *overexpose* (or *saturate*) your film, darkening so large a fraction of the light-sensitive chemicals that details of the image are lost.

Today, many cameras and camcorders come equipped to record images with electronic detectors called *charge-coupled devices,* or **CCDs,** that can accurately record up to 90% or more of the photons that strike them. A CCD is a chip of silicon carefully engineered to be extraordinarily sensitive to photons (Figure S2.6). The chip is physically divided into a grid of squares called *picture elements,* or **pixels** for short. When a photon strikes a pixel, it causes a bit of electric charge to accumulate. Each subsequent photon striking the same pixel adds to this accumulated electric charge. After an exposure is complete, a computer measures the total electric charge in each pixel, thus determining how many photons struck each pixel. The overall image is stored in the computer as an array of numbers representing the results from each of the pixels. At this point, the image can be manipulated through techniques of *image processing* to bring out details that otherwise might be missed in analyzing an image.

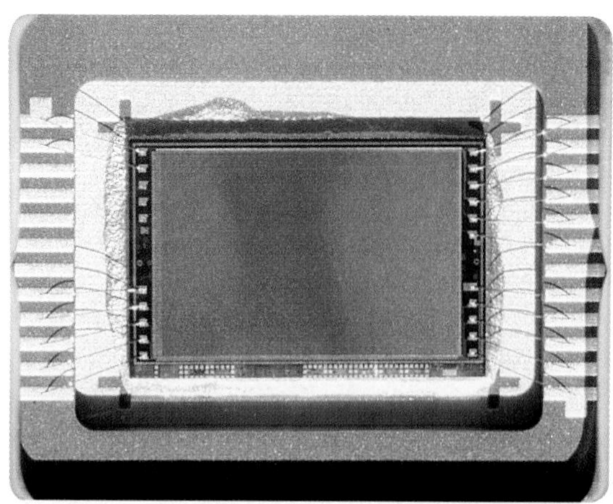

FIGURE S2.6 This photograph shows a magnified view of a CCD (charge-coupled device) that is only about 1 centimeter across. The blue and red colors are artifacts of the way the CCD reflects light.

S2.2 Telescopes: Giant Eyes

Our naked eyes allow us to admire the beauty of the night sky, but for most astronomical purposes they are completely inadequate. Their small size limits their angular resolution and the amount of light they can collect. They are sensitive only to visible light. And they are attached to bodies that require a pleasantly warm atmosphere for survival—an atmosphere that distorts light and prevents much of the electromagnetic spectrum from reaching the ground.

Telescopes solve all these problems. We can design telescopes to compensate for the distorting effects of our atmosphere or launch them into space aboard spacecraft. We can build telescopes and detectors that are sensitive to nonvisible light, such as radio waves or X rays. Most important, telescopes function as giant eyes, collecting far more light with far better angular resolution than our naked eyes.

Common Misconceptions: Magnification and Telescopes

You are probably familiar with how binoculars, magnifying glasses, and telephoto lenses for cameras make objects appear larger in size—the phenomenon we call magnification. Many people assume that astronomical telescopes are also characterized by their magnification. In fact, astronomers are more interested in a telescope's light-gathering power and resolution. The magnification of most telescopes is set to achieve the maximum amount of detail given the telescope's resolution—no more, no less. After all, there's little point in magnifying an image if it doesn't show any more detail.

Telescope Design

Telescopes come in two basic designs: *refracting* and *reflecting.* A **refracting telescope** operates much like an eye, using transparent glass lenses to focus the light from distant objects (Figure S2.7a). Note that the focus of the lens is *inside* the telescope, so

[2]The probability that an individual photon will be detected is called the *quantum efficiency.* It is about 1% for the human eye, up to about 10% for film, and 90% or more for CCDs.

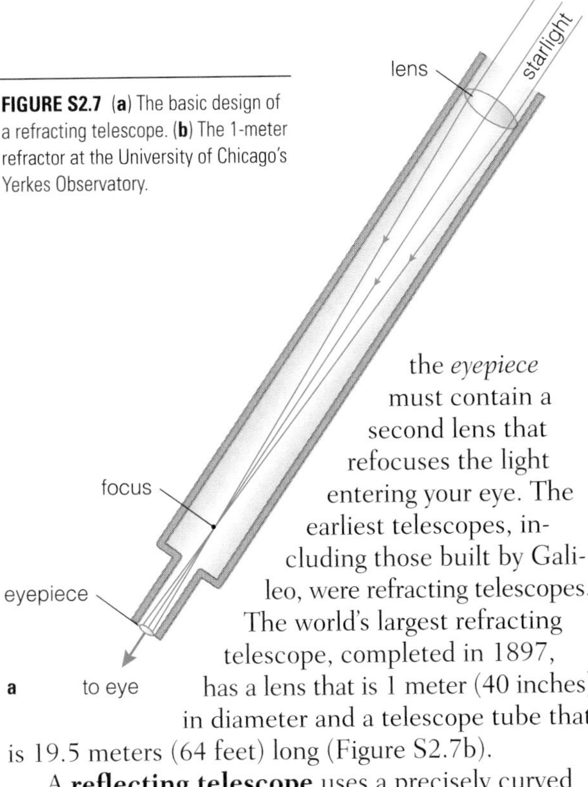

FIGURE S2.7 (**a**) The basic design of a refracting telescope. (**b**) The 1-meter refractor at the University of Chicago's Yerkes Observatory.

the *eyepiece* must contain a second lens that refocuses the light entering your eye. The earliest telescopes, including those built by Galileo, were refracting telescopes. The world's largest refracting telescope, completed in 1897, has a lens that is 1 meter (40 inches) in diameter and a telescope tube that is 19.5 meters (64 feet) long (Figure S2.7b).

A **reflecting telescope** uses a precisely curved **primary mirror** to gather and focus light (Figure S2.8a). Note that the focal plane, or *prime focus,*[3] of a reflecting telescope lies *in front of* the mirror. Thus, if we place a camera at the focal plane, it will block some of the light entering the telescope. This does not create a serious problem as long as the camera is small compared to the telescope mirror so that the amount of light blocked is small compared to the total amount of light entering. If the telescope is large enough, a "cage" in which an astronomer can work can be suspended over the focal plane (Figure S2.8b).

[3]Technically, the prime focus is the *point* at which rays that come straight into the telescope converge, which is at the center of the focal plane.

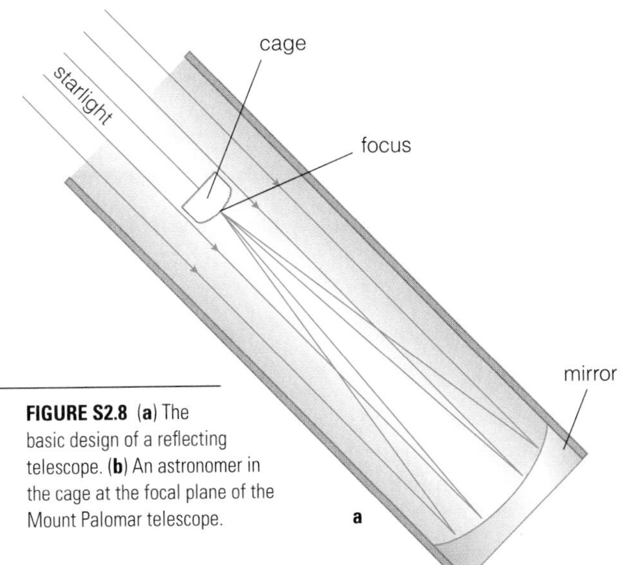

FIGURE S2.8 (**a**) The basic design of a reflecting telescope. (**b**) An astronomer in the cage at the focal plane of the Mount Palomar telescope.

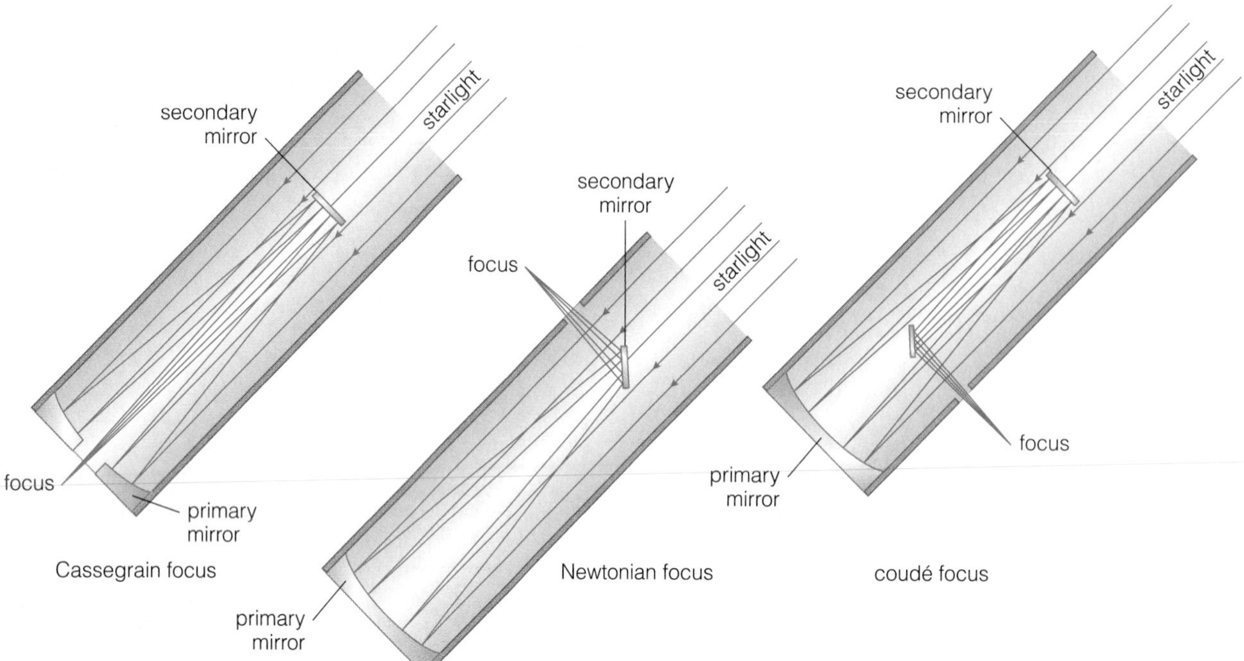

FIGURE S2.9 Alternative designs for reflecting telescopes.

An alternative and more common arrangement for reflecting telescopes is to place a **secondary mirror** just below the focal plane. The secondary mirror reflects the light to a location that is more convenient for viewing or for attaching instruments (Figure S2.9): through a hole in the primary mirror (a *Cassegrain focus*), to a hole in the side (a *Newtonian focus,* particularly common in telescopes used by amateurs), or to a third mirror that deflects light toward instruments that are not attached to the telescope (a *coudé focus*).

Obtaining clear, properly focused images requires high **optical quality** in either lenses or mirrors. The importance of optical quality is easy to appreciate: If your glasses are not very clear or not well made, you won't see very well. The glass lenses used in refracting telescopes pose several optical quality problems not shared by the mirrors of reflecting telescopes. Because light passes *through* a lens, proper focus requires a lens made of extremely high-quality glass polished to a very precise shape on both surfaces. In contrast, only the reflecting surface of a mirror must be precisely shaped, and the quality of the underlying glass is unimportant. Also, like prisms, lenses bend different wavelengths of light by different amounts, causing light of different colors to focus at different distances from the lens.[4] As a result, stellar images that ought to be sharp spread out slightly into a rainbow pattern. This problem does not occur with reflecting telescopes, because all colors of light reflect from a mirror in the same way.

Most of the optical quality problems of lenses can be minimized by using combinations of lenses and special optical coatings on the lenses to achieve a proper focus, but the sizes of lenses are limited by further obstacles. Making a large refracting telescope requires mounting a large glass lens by its edges at the top of a very long tube. The sheer weight of a large lens makes this mounting difficult, and because glass is not perfectly transparent, its thickness means that it absorbs some of the light passing through it. In contrast, the mirror of a reflecting telescope is mounted at the bottom, where its weight is a far less serious problem. Moreover, transparency is not an issue for mirrors. Given these factors, professional astronomers have turned to reflecting telescopes for most modern research.

Light-Collecting Area

The first of the two key properties of a telescope is its ability to collect light. The larger the lens or mirror of a telescope, the more light it can collect. The "size" of a telescope is usually described by the diameter of its primary lens or mirror. Thus, the **light-collecting area** is proportional to the square of its size:

$$\text{light-collecting area} = \pi \times (\text{radius})^2 = \pi \times \left(\frac{\text{diameter}}{2}\right)^2$$

For example, a 2-meter telescope has $2^2 = 4$ times the light-collecting area of a 1-meter telescope.

[4]The technical name for this problem is *chromatic aberration.*

For more than 40 years after its opening in 1947, the 5-meter Hale telescope on Mount Palomar outside San Diego was the world's largest.[5] Building larger telescopes posed a difficult technological challenge because large mirrors tend to sag under their own weight, degrading their optical quality. However, recent technological innovations have made it possible to build very large, low-weight mirrors, fueling a boom in large telescope construction (Table S2.1). For example, each of the twin 10-meter Keck telescopes in Hawaii has a primary mirror consisting of 36 smaller mirrors that function together (Figure S2.10). When you realize that each Keck telescope has more than a million times the light-collecting area of the human eye, it is no wonder that modern telescopes have so dramatically enhanced our ability to observe the universe.

[5]A 6-meter telescope built in Russia in 1976 is larger than the 5-meter Hale, but it suffers several problems of optical quality that make it far less useful.

Table S2.1 Largest Optical Telescopes

Size	Telescope Name	Sponsor	Location	Operational Date	Special Features
11.8 m	Large Binocular Telescope	U. Arizona, Arcetri Astrophysical Observatory (Italy)	Mt. Graham, AZ	2004*	Two 8.4-m mirrors on a common mount give light-collecting area of 11.8-m telescope.
10 m	Keck I	Cal Tech, U. California, NASA	Mauna Kea, HI	1993	Primary mirror consists of 36 1.8-m hexagonal segments.
10 m	Keck II	Cal Tech, U. California, NASA	Mauna Kea, HI	1996	Twin of Keck I; future plans for interferometry with Keck I.
9.2 m	Hobby–Eberly	U. Texas, Penn State, Stanford, Germany	Mt. Locke, Texas	1997	Consists of 91 1-m segments, for a total diameter of 11 m, but only 9.2 m can be used at a time; designed primarily for spectroscopy.
4 × 8 m	Very Large Telescope	European Southern Observatory	Cerro Paranal, Chile	2000*	Four separate 8-m telescopes can work individually or together as the equivalent of a 16-m telescope.
8.3 m	Subaru	National Observatory of Japan	Mauna Kea, HI	1999*	Japan's first large telescope project.
8 m	Gemini North	U.S., U.K., Canada, Chile, Brazil, Argentina	Mauna Kea, HI	1999*	Identical to Gemini South so similar observations can be made from each hemisphere.
8 m	Gemini South	U.S., U.K., Canada, Chile, Brazil, Argentina	Cerro Pachon, Chile	2000*	Identical to Gemini North.
6.5 m	Magellan	Carnegie Institute, Harvard, U. Michigan, MIT	Las Campanas, Chile	2000*	A second 6.5-m telescope is planned at a later date.
4.5 m/6.5 m	MMT	Smithsonian Institution, U. Arizona	Mt. Hopkins, AZ	1979/1999*	Original consisted of 6 1.8-m mirrors working together as a 4.5-m telescope; 1999 replacement with single 6.5-m mirror.
5 m	Hale	Cal Tech, JPL, Cornell	Mt. Palomar, CA	1947	World's largest functional telescope for more than 40 years.

*Scheduled completion date.

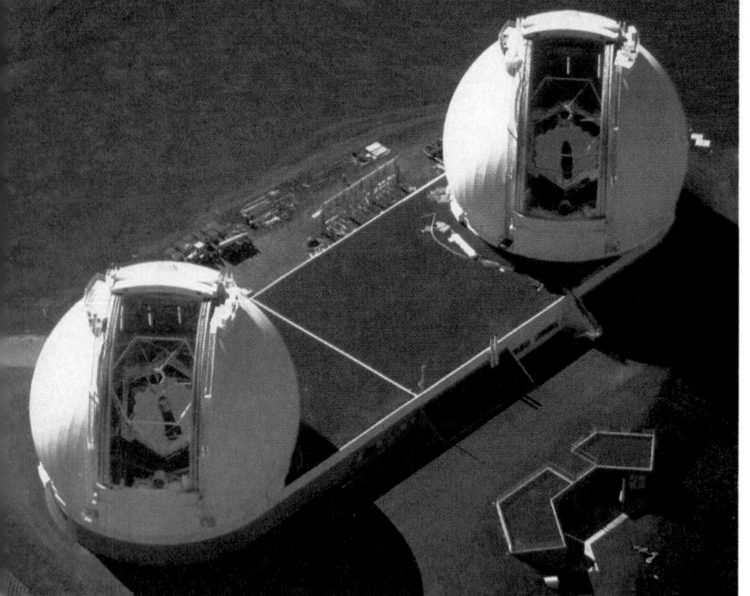

a

b

FIGURE S2.10 (**a**) The two Keck telescopes sit atop Mauna Kea in Hawaii. (**b**) The Keck telescopes use 36 hexagonal mirrors to function as a single 10-meter primary mirror.

FIGURE S2.11 This computer-generated image shows how overlapping sets of ripples on a pond interfere with one another. The effects of the two sets of ripples add in some places, making the water rise extra high or fall extra low, and cancel in other places, making the water surface flat. Light waves also exhibit interference. (The colors in this image are for visual effect only.)

TIME OUT TO THINK *What is the population of your home town? Suppose everyone in your home town looked at the dark sky at the same time. How would the total amount of light entering everyone's eyes compare with the light collected by a 10-meter telescope? Explain.*

Telescope Angular Resolution

The second key property of a telescope is its angular resolution. Many factors affect the angular resolution of a telescope, including optical quality and the distorting effects of the Earth's atmosphere, but the ultimate limit to a telescope's resolving power comes from the properties of light. Because light is an electromagnetic wave [Section 7.2], beams of light can interfere with one another like overlapping sets of ripples on a pond (Figure S2.11). This *interference* causes a blurring of images that limits a telescope's angular resolution even when all other conditions are perfect. That is why even a high-quality telescope in space, such as the Hubble Space Telescope, cannot have perfect angular resolution[6] (Figure S2.12).

The angular resolution that a telescope could achieve if it were limited only by the interference of light waves is called its **diffraction limit.**[7] It depends on both the diameter of the primary mirror and the wavelength of the light being observed.

- Larger telescopes have smaller diffraction limits for any particular wavelength of light. Thus, in the

[6]In fact, the Hubble Space Telescope's primary mirror was made with the wrong shape, which further limited its angular resolution until corrective optics were installed in 1993. The corrective optics consist of small mirrors that essentially undo the blurring created by the primary mirror.

[7]*Diffraction* is a technical term for the specific effects of interference that limit telescope resolution.

absence of atmospheric distortion, a high-quality 2-meter telescope can achieve better angular resolution than a comparable 1-meter telescope.

- Achieving a particular angular resolution requires a larger telescope for longer-wavelength light. For example, a radio telescope must be far larger than an optical telescope to achieve the same angular resolution because radio waves have a much longer wavelength than visible light.

Few ground-based optical telescopes actually achieve their diffraction-limited angular resolution, because the blurring caused by interference is usually much smaller than that caused by atmospheric distortion. Nevertheless, the angular resolution of large telescopes is quite impressive. Using a telescope limited by atmospheric distortion to an angular resolution of 1 arcsecond, you could read this book from halfway across a football field. The Hubble Space Telescope's 0.05-arcsecond resolution (for visible light) would allow you to read this book from a distance of about 800 meters ($\frac{1}{2}$ mile).

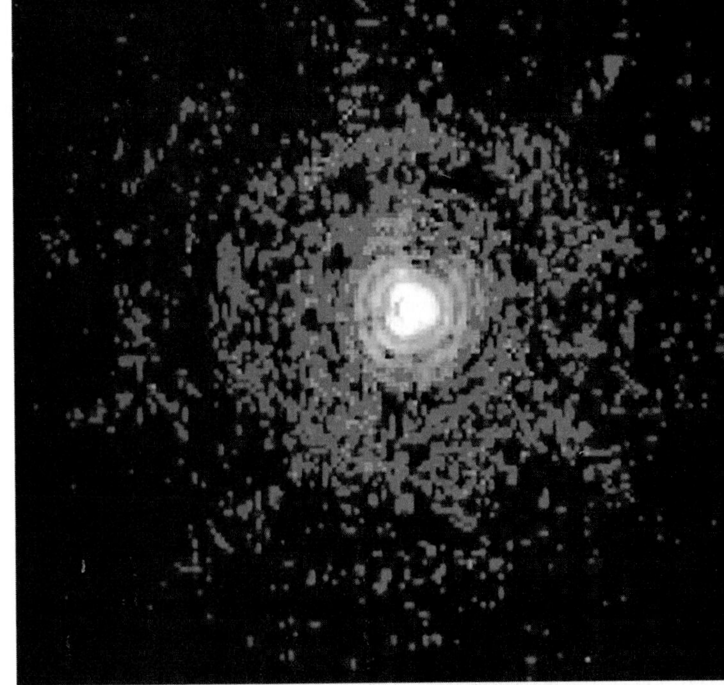

FIGURE S2.12 When examined in detail, a Hubble Space Telescope image of a star has rings resulting from the wave properties of light. With higher angular resolution, the rings would be smaller.

Mathematical Insight S2.2 The Diffraction Limit

A simple formula gives the diffraction-limited angular resolution of a telescope:

$$\text{diffraction limit (in arcseconds)} \approx 2.5 \times 10^5 \times \left(\frac{\text{wavelength of light}}{\text{diameter of telescope}}\right)$$

Note that both the wavelength of light and the diameter of the telescope must be in the same units.

Example 1: What is the diffraction-limited angular resolution of the 2.4-m Hubble Space Telescope for visible light with a wavelength of 500 nm?

Solution: For light with a wavelength of 500 nm (500×10^{-9} m), the diffraction limit of the Hubble Space Telescope is approximately:

$$2.5 \times 10^5 \times \left(\frac{\text{wavelength}}{\text{telescope diameter}}\right) = 2.5 \times 10^5 \times \frac{500 \times 10^{-9}\ \text{m}}{2.4\ \text{m}} = 0.05\ \text{arcsecond}$$

Example 2: Suppose you wanted to achieve a diffraction-limited angular resolution of 0.001 arcsecond for visible light of wavelength 500 nm. How large a telescope would you need?

Solution: In this case, we solve for the telescope diameter and then substitute the given values for the wavelength and diffraction limit:

$$\text{diffraction limit} \approx 2.5 \times 10^5 \times \left(\frac{\text{wavelength}}{\text{telescope diameter}}\right) \Rightarrow$$

$$\text{telescope diameter} \approx 2.5 \times 10^5 \times \left(\frac{\text{wavelength}}{\text{diffraction limit}}\right)$$

$$= 2.5 \times 10^5 \times \frac{500 \times 10^{-9}\ \text{m}}{0.001\ \text{(arcsecond)}} = 125\ \text{m}$$

An optical telescope would need a mirror diameter of 125 meters—longer than a football field—to achieve an angular resolution of 0.001 arcsecond.

S2.3 Uses of Telescopes

Astronomers use many different kinds of instruments and detectors to extract the information contained in the light collected by a telescope. Every astronomical observation is unique, and the best way to make a particular observation depends on the characteristics of the telescope being used, the types of instruments and detectors available, and what the astronomer wants to learn about the object(s) under study. However, most observations fall into one of three basic categories:

- **Imaging** yields pictures of astronomical objects.
- **Spectroscopy** involves dispersing light into a spectrum.
- **Timing** tracks how the light intensity hitting the detector varies with time.[8]

Major observatories typically have several different instruments capable of each of these tasks, and some instruments can perform all three basic tasks.

[8]Some astronomers include a fourth general category called *photometry,* which is the accurate measurement of light intensity from a particular object at a particular time. We do not call this out as a separate category because today's CCDs can do photometry at the same time that they are being used for imaging, spectroscopy, or timing.

FIGURE S2.13 In astronomy, color images are usually constructed by combining several images taken through different filters.

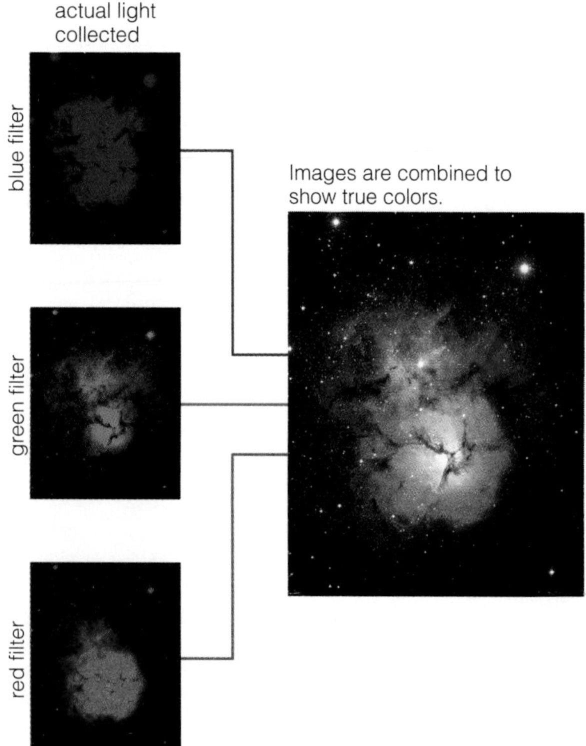

Imaging

At its most basic, an imaging instrument is simply a camera that puts a piece of film or a CCD at the focal plane of a telescope. Historically, most astronomical images were recorded with black-and-white film, because it is more sensitive than color film. Today, astronomers usually record images with CCDs. In either case, individual astronomical images are like black-and-white photographs, containing information about light intensity but not about color.

Astronomers usually obtain color information by placing a **filter** in front of the camera. For example, a red filter is a piece of glass that transmits only red light, so placing a red filter in front of a camera produces an image of the red light coming through the telescope. Similarly, placing a blue filter in front of the camera allows us to record an image of the blue light collected by the telescope. The richly hued astronomical pictures featured in this and other books generally combine several images recorded through different filters (Figure S2.13). Some filters transmit the light of only a very narrow range of wavelengths. For example, an *H-alpha filter* transmits light only from a specific transition in hydrogen atoms (the level 3 to level 2 transition with wavelength 656.3 nm).

Many telescopes collect nonvisible light, which can also be passed through filters and recorded with CCDs or other types of detectors. For example, an X-ray telescope can be used to make an X-ray image of the sky in its field of view. Even though the X-ray light that made the image is invisible to our eyes, we can display the results as a black-and-white photograph in which brighter intensity means more X rays. However, it is much easier for the human eye to perceive color differences than shades of gray. Therefore, we can see more detail by displaying the X-ray image in **false color,** in which, for example, the brightest parts of the image are shown as yellow or white and the faintest parts as deep blue or black (Figure S2.14). The colors are "false" because they are meaningless without a key to show how each color corresponds to the number of X-ray photons recorded.

TIME OUT TO THINK *Newspaper and television weather maps often use different colors to represent different temperatures. How is such a map similar to a false-color image? Can you think of other examples of the use of false color?*

Spectroscopy

Spectroscopy, the study of light spectra, reveals a wealth of information about an object, such as its chemical composition, temperature, and speed [Sec-

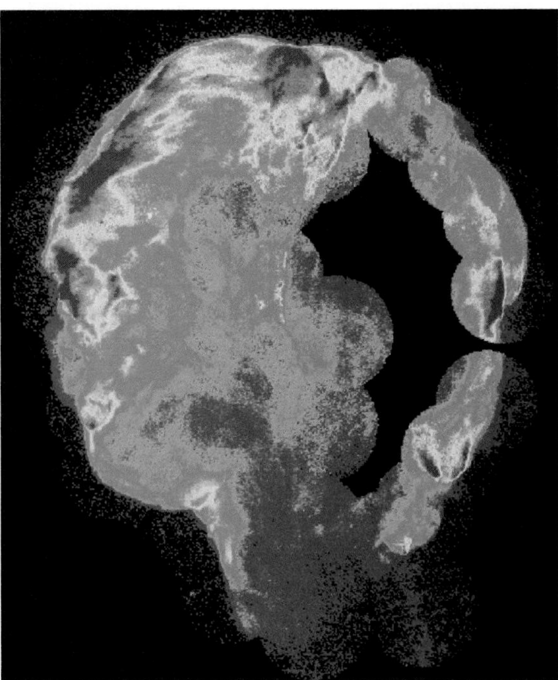

FIGURE S2.14 X rays are invisible, but we can color-code the information recorded by an X-ray detector to study how an object appears in X rays. This image shows the Cygnus Loop, a supernova remnant, as seen by the ROSAT X-ray telescope.

tion 7.4]. The instruments used for spectroscopy are called *spectrographs*. Light entering a simple spectrograph first passes through a narrow slit in a dark plate (Figure S2.15). A prism or diffraction grating then disperses the light into a spectrum, which is recorded with a detector such as a CCD. Each pixel along a horizontal line on the CCD records the intensity from a different part of the spectrum.

The **spectral resolution** of a spectrograph describes the degree of detail in the spectrum (Figure S2.16): The higher the spectral resolution, the more detail we can see. However, higher spectral resolution comes at a price. The greater the number of pixels over which the light is dispersed, the smaller the number of photons striking each pixel. Thus, for very faint objects, we may need a very large telescope and a long exposure time to obtain a high-resolution spectrum.

Timing

Many astronomical objects change over relatively short time scales. Weather patterns change on the planets. Some stars vary in brightness over periods of days or weeks. A few exotic objects vary dramatically in brightness in a fraction of a second. Timing experiments are observations designed to measure the variability of astronomical objects.

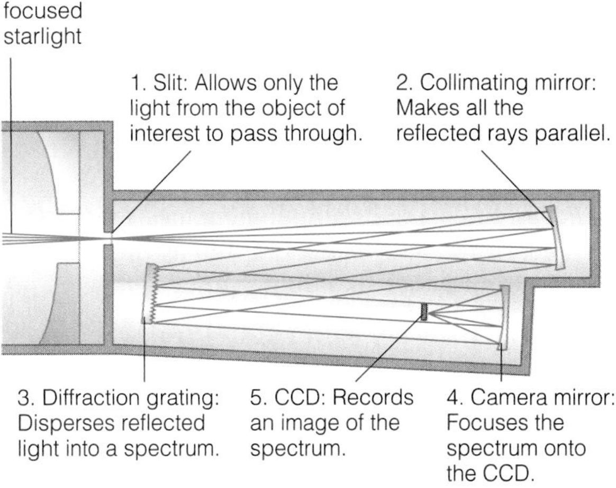

FIGURE S2.15 The basic design of a spectrograph.

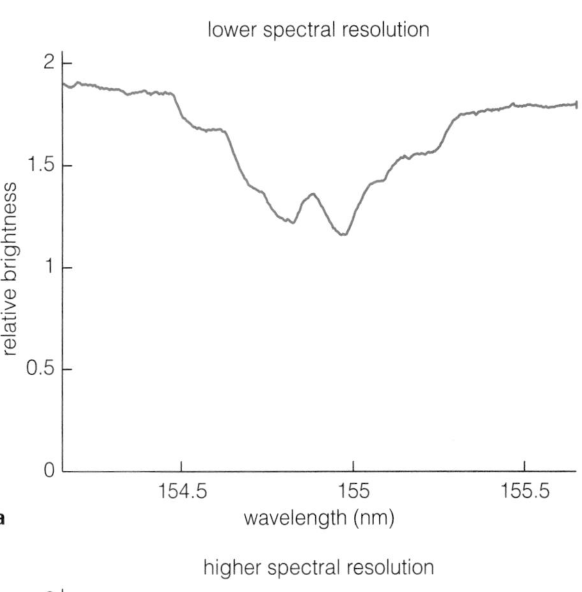

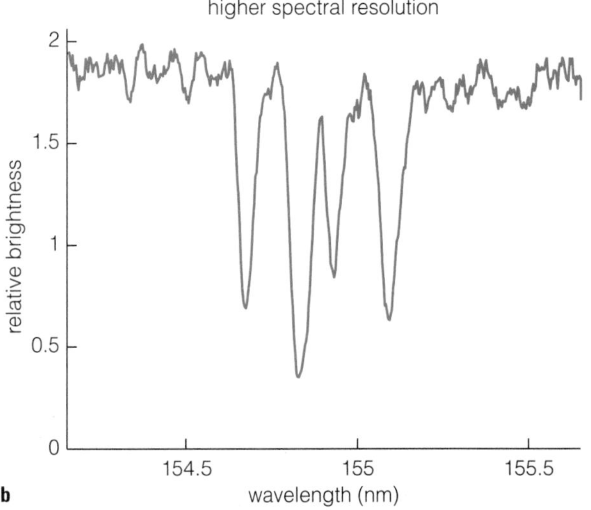

FIGURE S2.16 Higher spectral resolution means we can see more details in the spectrum. Both (**a**) and (**b**) show spectra of the same object for the same wavelength band, but the higher spectral resolution in (**b**) enables us to see individual spectral lines that appear merged together in (**a**).

If an object varies slowly enough, a timing experiment may be as simple as comparing a set of images or spectra obtained on different nights. Objects that vary rapidly require more sophisticated timing techniques. Some timing instruments can actually record the arrival time of every individual photon striking a detector. When a timing experiment measures how the intensity of light changes with time, the results are usually displayed as a **light curve,** a graph of intensity versus time (Figure S2.17).

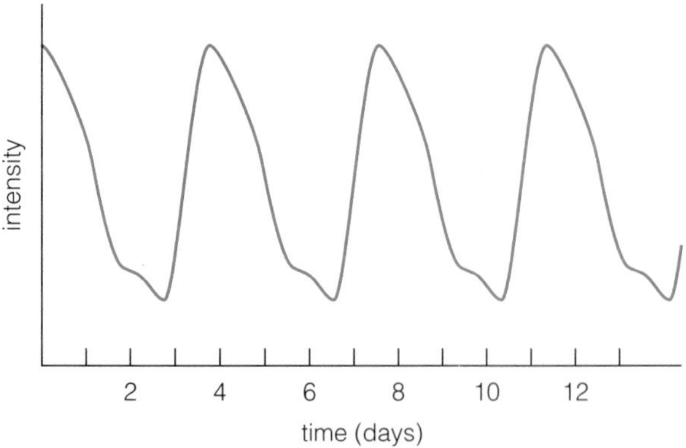

FIGURE S2.17 A light curve is a graph of how an object's intensity changes with time. This particular light curve represents a type of star called a Cepheid variable.

S2.4 Observing Sites

A great telescope with modern scientific instruments does not guarantee good astronomical observations. The *location* of the telescope is at least as important as its quality.

The primary problems to overcome in selecting a good location arise from properties of the Earth's atmosphere. The most obvious problem is weather—an optical telescope is useless under cloudy skies. Another problem is that our atmosphere scatters light so that the bright lights of cities cause **light pollution** that hinders astronomical observations. Somewhat less obvious is the fact that the atmosphere distorts light. Looking through our atmosphere is somewhat like looking up from the bottom of a swimming pool, trying to make out objects above the water. Air and water both bend light, and the constant motion of water in a swimming pool or of air in the atmosphere bends light in constantly shifting patterns. This ever-changing motion, or **turbulence,** of air in the atmosphere causes the familiar twinkling of stars. It also makes stars and other objects viewed through a telescope appear to dance about randomly in the focal plane, effectively limiting the angular resolution of the telescope.[9]

TIME OUT TO THINK *If you look down a long, paved street on a hot day, you'll notice the images of distant cars and buildings rippling and distorting. How are these distortions similar to the twinkling of stars? Why do you think these distortions are more noticeable on hot days than on cooler days?*

The most serious problem with the Earth's atmosphere is that it prevents most forms of light from reaching the ground at all. Figure S2.18 shows the depth to which different forms of light penetrate into the Earth's atmosphere. Only radio waves, visible light, parts of the infrared spectrum, and the longest wavelengths of ultraviolet light reach the ground. Gases such as water vapor and carbon dioxide, known as *greenhouse gases* [Section 10.2], absorb infrared light. Thus, telescopes on high mountaintops or in airplanes are better for infrared observations because they lie above most of the greenhouse gases. Most ultraviolet light is absorbed higher in the atmosphere, by ozone, so ultraviolet telescopes are useful only on high-flying balloons or in space. X rays and most gamma rays are absorbed very high in the atmosphere—which is

Common Misconceptions: Twinkle, Twinkle Little Star

Twinkling, or apparent variation in the brightness and color of stars, is *not* intrinsic to the stars. Instead, just as light is bent by water in a swimming pool, starlight is bent by the Earth's atmosphere. Air turbulence causes twinkling because it constantly changes how the starlight is bent. Hence, stars tend to twinkle more on windy nights and when they are near the horizon (and therefore viewed through a thicker layer of atmosphere). Planets also twinkle, but not nearly as much as stars: because planets have a measurable angular size, the effects of turbulence on any one ray of light are compensated for by the effects of turbulence on others, reducing the twinkling seen by the naked eye (but making planets shimmer noticeably in telescopes). While twinkling may be beautiful, it blurs telescopic images. Avoiding the effects of twinkling is one of the primary reasons for putting telescopes in space. There, above the atmosphere, astronauts and telescopes do not see any twinkling.

[9]The limitation created by atmospheric blurring is called the *seeing*. For example, seeing of 1 arcsecond means that objects less than 1 arcsecond apart cannot be resolved, even with a large and optically perfect telescope.

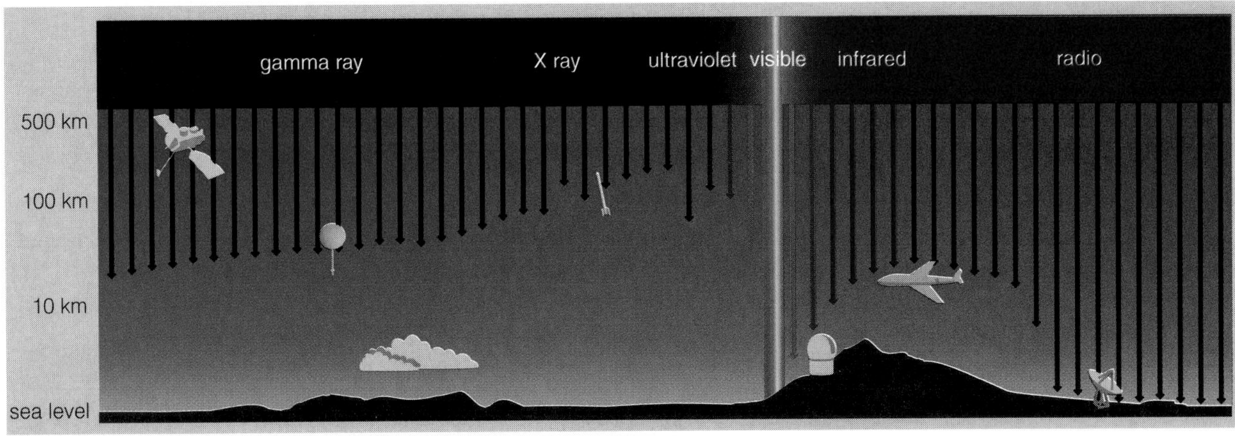

FIGURE S2.18 Different wavelengths of light penetrate to different depths in the Earth's atmosphere.

fortunate, because these high-energy forms of light would otherwise be lethal to most living organisms. X-ray and gamma-ray telescopes therefore are useful only in space.

Observing Sites on Earth

The key criteria for observing sites on the Earth's surface are that they be dark, dry, calm, and high. A *dark* site is one that is relatively unaffected by light pollution. A *dry* site has both few rainy days and low humidity. A *calm* site has relatively little air turbulence above it. Placing an observatory *high,* such as on a mountaintop, means that there is less atmosphere to absorb light.

The best combinations of dark, dry, calm, and high tend to be found on mountaintops situated near deserts or in the paths of the smooth, prevailing winds that blow eastward across the oceans. Islands often are ideal, and the 4,300-meter (14,000-foot) summit of Mauna Kea on the big island of Hawaii is home to many of the world's best observatories (Figure S2.19). Other sites for major observatories include the high coastal mountains of northern Chile and southern California that face the prevailing winds from the Pacific Ocean, the Canary islands off the northwest coast of Africa that face the prevailing winds of the Atlantic, and mountains in the Arizona desert.

Human development and population growth sometimes change the desirability of particular observing sites. The 2.5-meter telescope at Mount Wilson was the largest in the world when it was built in 1917 and would still be very useful today—if it weren't located so close to the lights of what was once the small town of Los Angeles. Similar but less serious light pollution hinders many other telescopes, including the Mount Palomar telescopes near San Diego and the telescopes of the National Optical

FIGURE S2.19 Observatories on the summit of Mauna Kea in Hawaii.

FIGURE S2.20 Artist's drawing of the SOFIA airborne observatory.

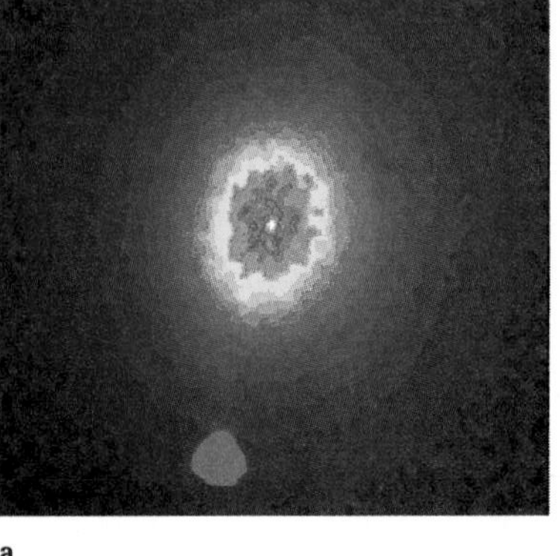

a

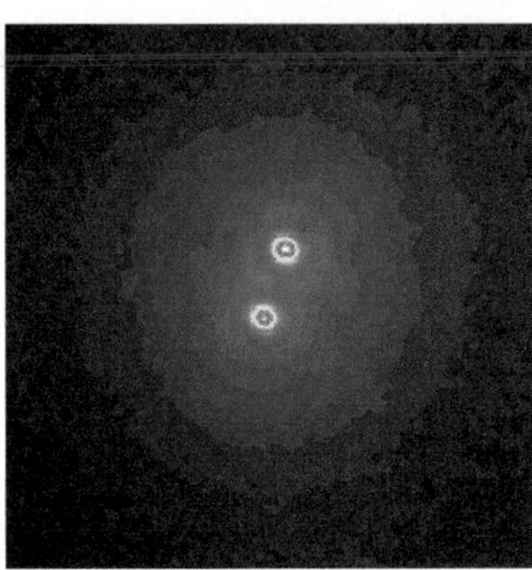

b

FIGURE S2.21 (a) Atmospheric distortion makes this ground-based image of a double star look like a single star. (b) Using the same telescope, but with adaptive optics, the two stars are clearly distinguishable.

Astronomy Observatory on Kitt Peak near Tucson.[10] Fortunately, communities such as San Diego and Tucson are working to reduce light pollution. Placing reflective covers on the tops of streetlights directs more light toward the ground rather than toward the sky. Using "low-pressure" (sodium) lights also helps. These lights shine most brightly in just a few wavelengths of light, so specially designed filters can absorb these wavelengths while transmitting most of the light from the astronomical objects under study. It's worth noting that both reflective covers and low-pressure lights also offer significant energy savings to communities that use them.

Modern technology provides solutions to some of the problems caused by the atmosphere. Putting a telescope on an airplane takes it above much of the atmosphere, allowing many infrared observations not possible from the ground. NASA is building an airborne observatory called SOFIA[11] that will have a 3-meter infrared telescope looking out through a large hole cut in the fuselage of a Boeing 747 airplane (Figure S2.20). Advanced technologies make it possible for the airplane to fly smoothly despite its large hole and to keep the telescope itself extremely stable even while the airplane is flying at speeds of hundreds of kilometers per hour.

Perhaps the most amazing new technology is **adaptive optics,** which can eliminate most atmospheric distortion. Recall that atmospheric turbulence causes a stellar image to dance around in the focal plane of a telescope. Adaptive optical systems essentially make the telescope's mirrors do an opposite dance, so that the distortions of the mirrors cancel out the atmospheric distortions. The mirror shape must change slightly many times each second to compensate for the rapidly changing atmospheric distortions. Moreover, each change in the mirror shape must shift the stellar image in a direction precisely opposite the shift caused by the atmosphere. Computers calculate the necessary changes in mirror shape by monitoring the distortions of a bright star near the object under study or of an *artificial star* made by shining a laser beam into the sky. The computer then instructs devices located behind the mirror to make the needed changes in the mirror shape. If everything works properly, the result is an image free of the usual effects of atmospheric distortion (Figure S2.21).

[10]Kitt Peak has an outstanding visitor's center and is well worth a visit if you are in the Tucson area.

[11]SOFIA is an acronym for *Stratospheric Observatory for Infrared Astronomy.*

Table S2.2 Selected Major Observatories in Space

Name	Launch Year	Key Capabilities
Infrared Space Observatory (ISO)	1995	Infrared imaging and spectroscopy
Hubble Space Telescope (HST)	1991	Optical, infrared, and ultraviolet imaging, spectroscopy, and timing
Extreme Ultraviolet Explorer (EUVE)	1992	Extreme ultraviolet imaging and spectroscopy
Far Ultraviolet Spectrographic Explorer (FUSE)	1999*	Ultraviolet spectroscopy
Roentgen Satellite (ROSAT)	1991	X-ray imaging, spectroscopy, and timing
Advanced Satellite for Cosmology and Astrophysics (ASCA)	1993	X-ray spectroscopy
Rossi X-Ray Timing Experiment (RXTE)	1995	X-ray timing and spectroscopy
Advanced X-Ray Astrophysical Facility (AXAF)	1999*	X-ray imaging and spectroscopy
X-Ray Multi-Mirror Mission (XMM)	1999*	X-ray spectroscopy
Compton Gamma Ray Observatory (CGRO)	1991	Gamma ray imaging and timing

*Scheduled launch year.

Telescopes in Space

The ultimate solution to the observing problems caused by the Earth's atmosphere is to put telescopes in space. Above the atmosphere, a telescope's angular resolution is limited only by its size and its optical quality. Although new technologies such as adaptive optics allow some ground-based telescopes to achieve angular resolutions comparable to those of telescopes in space, no technology can enable ground-based observations of radiation that doesn't reach the ground, such as ultraviolet and X-ray light.

The Hubble Space Telescope is the most famous observatory in space, but it is only one of many telescopes in orbit. Most of the other space telescopes observe only invisible wavelengths of light. Table S2.2 provides a partial list of space telescopes.

With the exception of a few telescopes operated by astronauts in the Space Shuttle's cargo bay or on the Russian space station *Mir*, telescopes in space are robotic and are operated by controllers on the ground. Scientists choose targets for astronomical observations, and telescope operators use radio waves to transmit computerized instructions to a telescope. The telescope then transmits the results of the observations back down to the ground by radio waves. Most space telescopes, including the Hubble Space Telescope (Figure S2.22), orbit the Earth.

Given the tremendous advantages of putting telescopes in space, you might wonder why astronomers continue to build and use telescopes on the ground. The answer comes down mostly to money. Three main factors make space telescopes extremely expensive compared to telescopes on the ground. First, the launch costs for even the smallest space telescopes are at least $20 million, and launch costs

Common Misconceptions: Closer to the Stars?

Many people mistakenly believe that space telescopes are advantageous because their locations above the Earth make them closer to the stars. You can quickly realize the error of this belief by thinking about scale. On the 1-to-10-billion scale discussed in Section 2.1, the Hubble Space Telescope is so close to the surface of the millimeter-diameter Earth that you would need a microscope to resolve its altitude. Thus, the distances to the stars are effectively the same no matter whether a telescope is on the ground or in space. The real advantages of space telescopes all arise from being above the many observational problems presented by the Earth's atmosphere.

a

b

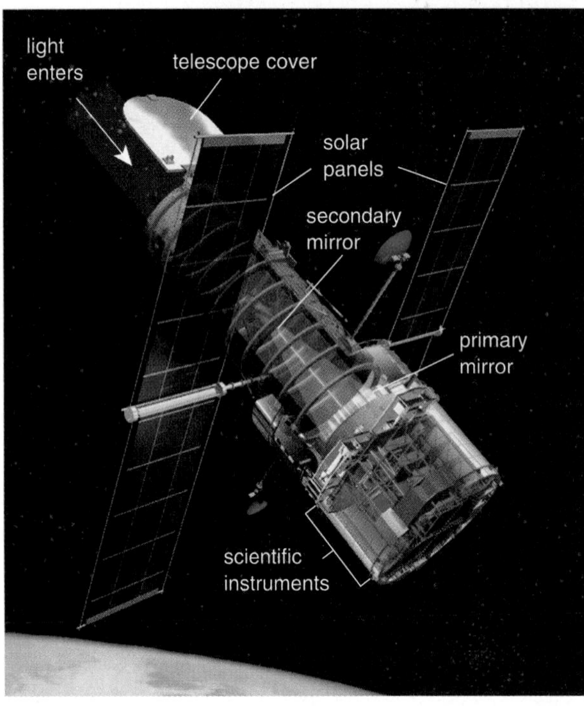

c

FIGURE S2.22 (**a**) The Hubble Space Telescope (HST) orbits the Earth. (**b**) Astronauts replacing an instrument on the HST in 1997. (**c**) Basic components of the HST.

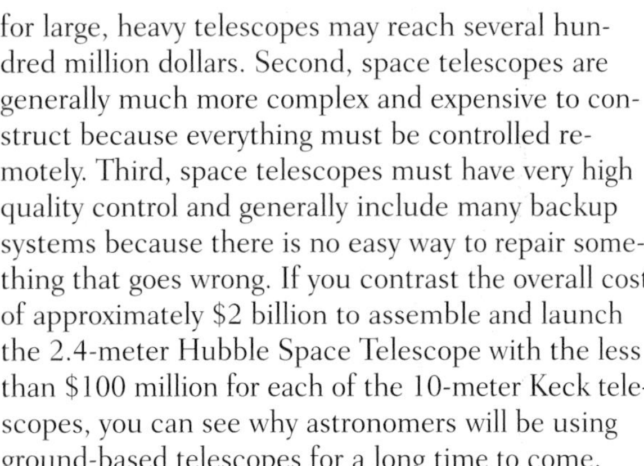

for large, heavy telescopes may reach several hundred million dollars. Second, space telescopes are generally much more complex and expensive to construct because everything must be controlled remotely. Third, space telescopes must have very high quality control and generally include many backup systems because there is no easy way to repair something that goes wrong. If you contrast the overall cost of approximately $2 billion to assemble and launch the 2.4-meter Hubble Space Telescope with the less than $100 million for each of the 10-meter Keck telescopes, you can see why astronomers will be using ground-based telescopes for a long time to come.

S2.5 Telescopes Across the Spectrum

If we studied only visible light, we'd be missing much of the picture. Planets are relatively cool and emit primarily infrared light, and their spectra feature ultraviolet emission lines from high in their atmospheres. The hot upper layers of stars like the Sun emit ultraviolet and X-ray light. Many objects emit radio waves, including some of the most exotic objects in the cosmos—objects that may contain *black holes* buried in their centers [Section 17.4]. Some violent events even produce gamma rays that travel through space to Earth. Indeed, most objects emit light over a broad range of wavelengths.

Today, astronomers study light from all regions of the spectrum. However, observing outside the visible-light region generally requires the use of somewhat different telescope or detector technologies.

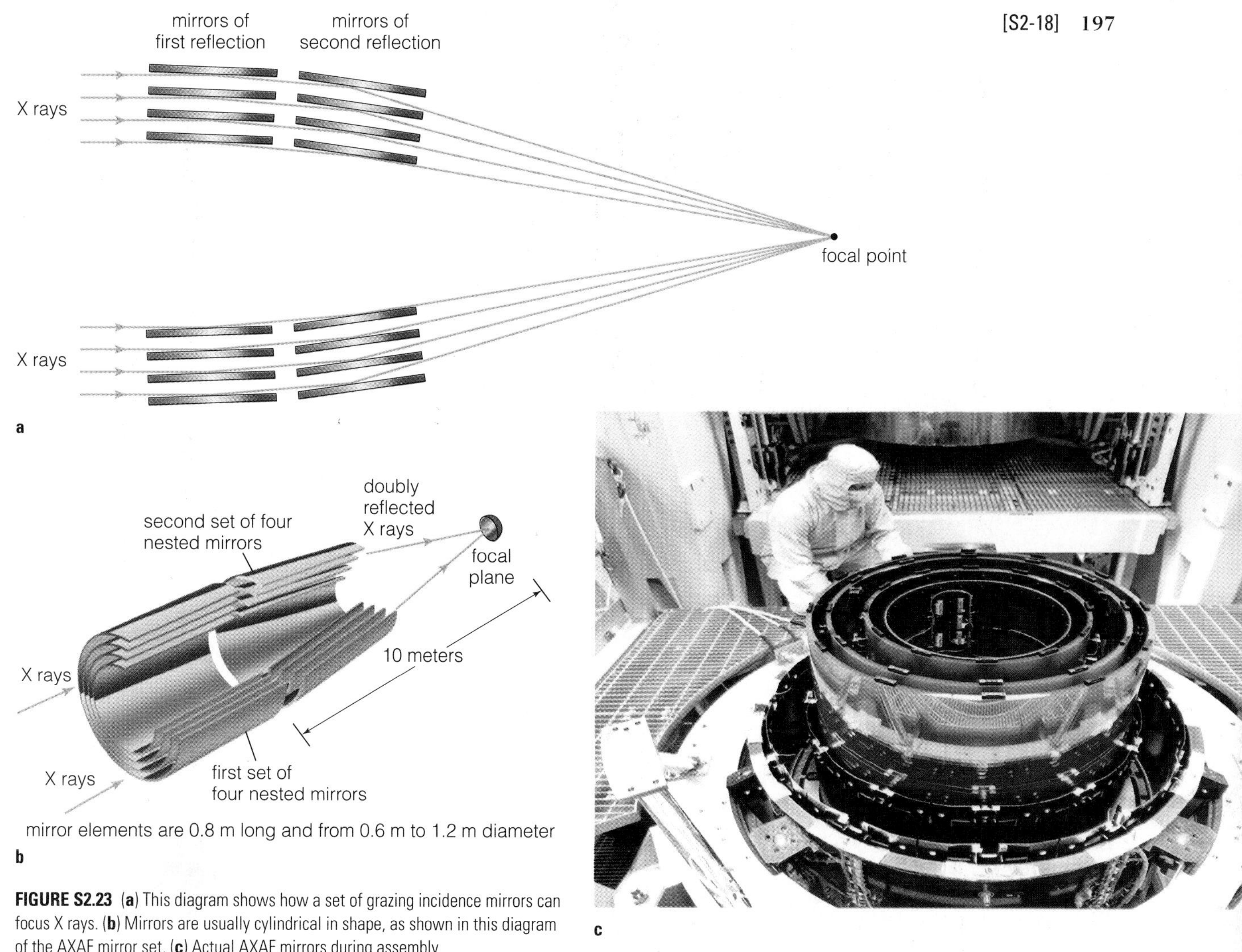

FIGURE S2.23 (**a**) This diagram shows how a set of grazing incidence mirrors can focus X rays. (**b**) Mirrors are usually cylindrical in shape, as shown in this diagram of the AXAF mirror set. (**c**) Actual AXAF mirrors during assembly.

Infrared and Ultraviolet Telescopes

Infrared and ultraviolet wavelengths begin just beyond the boundaries of visible light, so infrared or ultraviolet light is not very different from visible light in its basic behavior. Ordinary telescopes therefore can collect and focus most infrared and ultraviolet light. Special problems arise when we try to observe near the extreme ends of the infrared or ultraviolet. Extreme ultraviolet light (the shortest wavelengths of ultraviolet) behaves like X rays, which we'll discuss shortly. Extreme infrared light (the longest wavelengths of infrared) poses observing difficulties because all objects, including telescopes, emit thermal radiation [Section 7.4]. Ordinary telescopes are warm enough to emit significant amounts of long-wavelength infrared light, and this emission would interfere with any attempt to observe these wavelengths from the cosmos. One solution to this problem is to cool the telescope to a very low temperature so that it emits very little infrared radiation. The planned Space Infrared Telescope Facility (SIRTF) will be cooled with liquid helium to just a few degrees above absolute zero.

X-Ray and Gamma-Ray Telescopes

X rays have sufficient energy to penetrate many materials, including living tissue and ordinary mirrors.[12] While this property makes X rays very useful to medical doctors, it gives astronomers headaches. Trying to focus X rays is rather like trying to focus a stream of bullets. If bullets are fired directly at a metal sheet, they will puncture or damage the sheet. However, if the metal sheet is angled so that bullets barely graze its surface, it will slightly deflect the bullets. Specially designed mirrors can deflect X rays in much the same way. Such mirrors are called *grazing incidence* mirrors because X rays just graze their surfaces as they are deflected toward the focal plane. X-ray telescopes, such as the Advanced X-ray Astrophysics Facility (AXAF), generally consist of several nested grazing incidence mirrors (Figure S2.23).

[12]This discussion of X rays also applies to extreme ultraviolet wavelengths.

FIGURE S2.24 The Compton Gamma Ray Observatory being released by the Space Shuttle's robotic arm.

FIGURE S2.25 The Arecibo radio telescope in Puerto Rico.

TIME OUT TO THINK *If you look straight down at your desktop, you probably cannot see your reflection. But if you glance along the desktop surface (or another flat surface, such as that of a book), you should see reflections of objects in front of you. Explain how these reflections represent grazing incidence for visible light.*

Gamma rays (and very short wavelength X rays) can penetrate even grazing incidence mirrors and therefore cannot be focused in the traditional sense. Capturing such high-energy light at all requires detectors so massive that the photons cannot simply pass through them. The largest gamma-ray observatory is a 17-ton behemoth (the Compton Gamma Ray Observatory) that was launched into orbit by the Space Shuttle in 1991 (Figure S2.24).

Note that it is relatively easy to know *when* a detector captures a gamma ray (because gamma-ray photons carry so much energy), which makes gamma-ray timing experiments fairly easy. However, it is very difficult to determine the direction from which a gamma ray entered the detector. Thus, it is very difficult to make gamma-ray images, and those that have been made have relatively poor angular resolution. Gamma-ray spectroscopy is possible because detectors can measure the energies of individual gamma-ray photons.

Radio Telescopes and Interferometry

The Earth's atmosphere is transparent to most radio waves, so radio astronomy is possible from the ground. Because the atmosphere does not scatter radio waves the way it scatters visible light, radio telescopes work equally well day or night. In fact, the primary impediments to radio astronomy are radio and television broadcasts, cellular telephones, and other human-made sources of radio-wave emission. These emissions are essentially the radio-wave equivalent of light pollution; they interfere with attempts to observe the same wavelengths coming from the cosmos. Fortunately, governmental regulations and international treaties have kept parts of the radio spectrum free from human broadcasts so that astronomers can detect cosmic radio waves.

Radio telescopes use large metal dishes as "mirrors" to reflect radio waves. The dishes are cheap compared to optical mirrors because they don't have to be so smooth—the long wavelengths of radio waves make them much more forgiving about little bumps and imperfections in radio dishes; in fact, for longer-wavelength radio waves, a metal mesh works as well as a solid dish.

Radio telescopes operate somewhat differently from other kinds of telescopes, but the fundamental principles of light-collecting area and angular resolution remain the same. The larger the reflecting dish, the more radio waves it collects and the better its angular resolution for a given wavelength. However, the long wavelengths of radio waves mean that very large telescopes are necessary in order to achieve reasonable angular resolution. The largest radio dish in the world, the Arecibo telescope in Puerto Rico, stretches 305 meters (1,000 feet) across a natural valley (Figure S2.25). Despite its large size, Arecibo's angular resolution is only about 1 arcminute at commonly observed radio wavelengths (e.g., 21 cm)—a few hundred times worse than the visible light resolution of the Hubble Space Telescope.

In the 1950s, radio astronomers developed an ingenious technique for improving the angular resolution of radio telescopes: They learned to link together several separate radio dishes in an array that resolves details as if it were one enormous radio telescope (Figure S2.26). (The array does *not* have the light-collecting area of a single enormous telescope, but only the combined light-collecting area of the individual dishes.) This technique is called **interferometry** because it takes advantage of the wavelike properties of light that cause interference (see Figure S2.11). The procedure relies on the precise timing of when radio waves reach each dish and on supercomputers to analyze the resulting interference patterns.

FIGURE S2.26 Interferometry gives two (or more) small radio dishes the angular resolution of a much larger dish. However, their total light-collecting area is only the sum of the light-collecting areas of the individual dishes.

One famous array of radio telescopes, the Very Large Array (VLA) near Socorro, New Mexico, consists of 27 individual radio dishes that can be moved along railroad tracks laid down in the shape of a Y (Figure S2.27). Each arm of the railroad is 21 kilometers long, and each individual dish is 25 meters in diameter. The light-gathering capability of the VLA's 27 dishes is simply equal to their combined area, which is equivalent to that of a single telescope 130 meters across. But its angular resolution, achieved by spacing the individual dishes as widely as possible, is equal to that of a single radio telescope with a diameter of almost 40 kilometers.

Astronomers can achieve even higher angular resolution by linking radio telescopes around the world for individual observations (Figure S2.28). This technique, called *Very Long Baseline Interferometry* (VLBI), can provide an angular resolution equivalent to that of a radio telescope the size of the Earth. VLBI observations already have achieved angular resolution of less than 0.001 arcsecond, far superior to the angular resolution currently possible with visible light. VLBI can be further enhanced by linking signals from radio telescopes on Earth with signals from one or more radio telescopes in space. This technique has already been tested, and NASA hopes to launch additional radio telescopes for this purpose in the not too distant future.

FIGURE S2.27 The Very Large Array (VLA), New Mexico.

FIGURE S2.28 Interferometry with appropriately located radio telescopes can provide the angular resolution of a single radio telescope the size of the Earth.

In principle, interferometry can improve angular resolution not only for radio waves but also for any other form of light. In practice, interferometry becomes increasingly difficult for light with shorter wavelengths. Recent advances in technology have made interferometry possible in much of the infrared, and preliminary experiments in optical interferometry are quite promising. Indeed, one reason why *two* Keck telescopes were built close together on Mauna Kea was so they could be used for infrared interferometry, and someday for optical interferometry. The potential value of such interferometers is enormous. In the future, infrared interferometers may be able to obtain spectra from individual planets around other stars, allowing spectroscopy that could determine the composition of their atmospheres and help determine whether they harbor life.

S2.6 Spacecraft

The space age began in earnest with the launch of the Soviet Union's *Sputnik* satellite in 1957. Within the first decade after Sputnik, astronomers were already using spacecraft to carry telescopes into orbit and to make the first close-up studies of the Moon, Venus, and Mars. Today, orbiting spacecraft carry telescopes that allow us to collect light from nearly every region of the electromagnetic spectrum, while other spacecraft have visited all the planets in our solar system except Pluto. Spacecraft are even used to study the Earth: In fact, we can study many facets of our planet much more easily from space than from the ground. Because spacecraft are so important to astronomy, let's briefly investigate what they do and how they work.

Spacecraft Basics

With a few exceptions, such as the Space Shuttle and the Apollo spacecraft on missions to the Moon, nearly all spacecraft are *robotic,* meaning that no humans are aboard. Their operations are partly automatic and partly controlled by scientists who send instructions via radio signals from Earth.

Broadly speaking, spacecraft fall into four main categories according to their orbits:

1. **Earth-orbiters:** Simply being in space is enough to overcome the observational problems presented by Earth's atmosphere. Thus, most telescopes in space are on Earth-orbiting spacecraft. Earth orbit is also the obvious place to put spacecraft that study the Earth itself.
2. **Flybys:** Close-up study of other planets requires sending spacecraft to them. Flybys follow unbound orbits past their destinations, so they fly

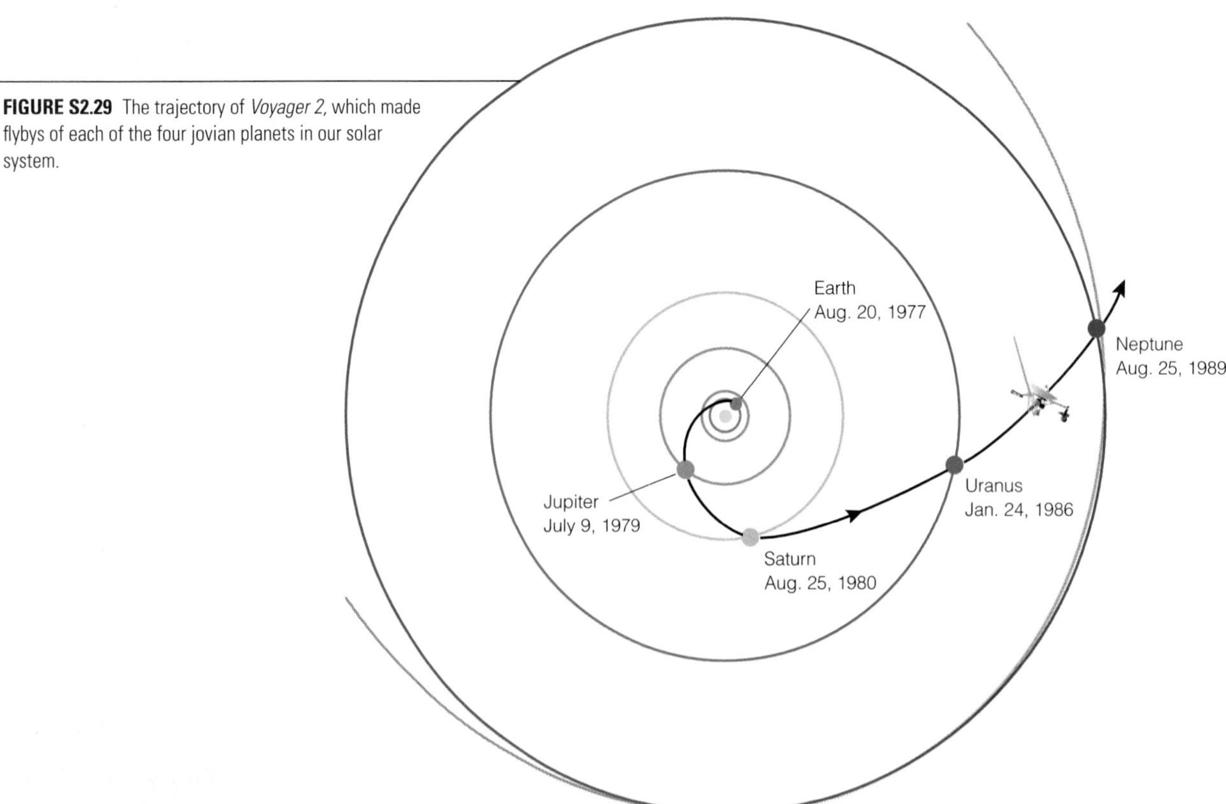

FIGURE S2.29 The trajectory of *Voyager 2,* which made flybys of each of the four jovian planets in our solar system.

past a world just once and then continue on their way. For example, *Voyager 2*'s orbit allowed it to make flybys of Jupiter, Saturn, Uranus, and Neptune (Figure S2.29).

3. **Orbiters** (of other worlds): A flyby can study a planet or moon close up—but only once. More detailed or longer-term studies require a spacecraft that goes into orbit around another world.
4. **Landers:** The most "up close and personal" studies of other worlds come from spacecraft that descend to the surface. We have placed landers on the Moon, Venus, and Mars, and one is scheduled to land on Saturn's moon Titan in 2004. Most landers never return to Earth; any samples they collect must be analyzed with onboard instruments, and the data radioed back to Earth. If a lander returns to Earth, it is called a *sample return mission*. As of the year 1999, samples have been returned only from the Moon. An *atmospheric probe* is a slight variation on a lander (Figure S2.30): In 1995, the *Galileo* orbiter dropped a probe into Jupiter's atmosphere (Jupiter has no solid surface to land on).

Table S2.3 lists a few of the most significant robotic missions to other worlds.

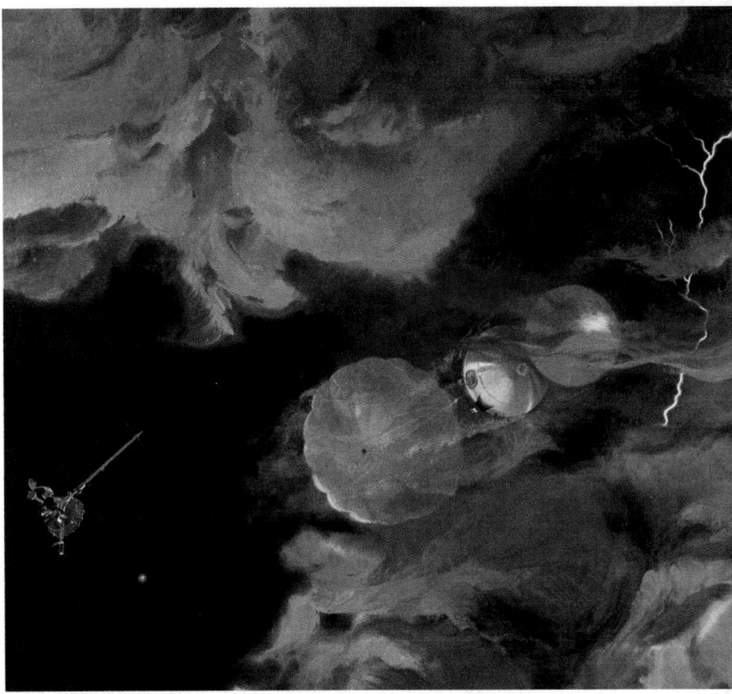

FIGURE S2.30 Artist's conception of the *Galileo* probe entering Jupiter's atmosphere. Its parachute is deployed to slow its descent. The *Galileo* orbiter appears to left.

Table S2.3 Selected Robotic Missions to Other Worlds

Name	Type	Destination	Arrival Date	Mission Highlights
Cassini	Orbiter	Saturn	2004*	Includes lander for Titan
Mars Surveyor 98	Orbiter/lander	Mars	1999*	Follow-up to Mars Global Surveyor and Pathfinder; landing site near south pole
Near Earth Asteroid Rendezvous (NEAR)	Orbiter	Eros (asteroid)	1999*	First spacecraft dedicated to in-depth study of an asteroid
Lunar Prospector	Orbiter	Moon	1998	Detailed mapping of lunar surface; search for ice in craters near lunar poles
Mars Global Surveyor	Orbiter	Mars	1997	Detailed imaging of surface from orbit
Mars Pathfinder	Lander	Mars	1997	Carried *Sojourner*, the first robotic rover on Mars
Galileo	Orbiter	Jupiter	1995	Dropped probe into Jupiter; close-up study of moons
Magellan	Orbiter	Venus	1990	Detailed radar mapping of surface of Venus
Voyager 1	Flyby	Jupiter, Saturn	1979, 1981	Unprecedented views of Jupiter and Saturn; continuing out of solar system
Voyager 2	Flyby	Jupiter, Saturn, Uranus, Neptune	1979, 1981, 1986, 1989	Only mission to Uranus and Neptune; continuing out of solar system
Viking 1 and 2	Orbiter and Lander	Mars	1976	First landers on Mars

*Scheduled arrival date.

Spacecraft Operations and Instruments

Robotic spacecraft essentially are fully automatic observatories and therefore must be equipped with a variety of components to carry out their duties. These components fall into just a few basic categories:

- *Scientific instruments* to achieve mission objectives.
- *On-board computers* to control spacecraft operations.
- *Radio antennas* for transmitting data to Earth and for receiving new computer instructions from Earth.
- *Propulsion systems* for executing orbital maneuvers and rotating the spacecraft. Orbital maneuvers are usually executed by firing small rockets, while *gyroscopes* can be used to rotate the spacecraft. (If you've never seen a gyroscope, you can find one at most science and nature stores.)
- *Power sources* to supply power for spacecraft operations.

Figure S2.31 shows the major components of a few important spacecraft. Note that scientific instrument packages differ from one spacecraft to another. Most Earth-orbiters carry telescopes and detectors to observe distant astronomical objects at a variety of wavelengths of light. Spacecraft that fly by or orbit other worlds often carry small telescopes and detectors, instruments for sampling gases and plasma in space, and instruments for measuring magnetic fields. A few orbiters also carry radar instruments that can map the surface of a planet in great detail. Landers may carry small weather stations that measure surface conditions, robot arms that collect samples, or rovers that can move about the surface.

Note also that most astronomical spacecraft use solar panels to generate the power needed to charge their on-board batteries. However, solar energy may be inadequate for spacecraft going to the outer solar system. For example, the Cassini spacecraft has a small canister of plutonium on board; the slow radioactive decay of the plutonium generates heat that

FIGURE S2.31 Major components of individual spacecraft.

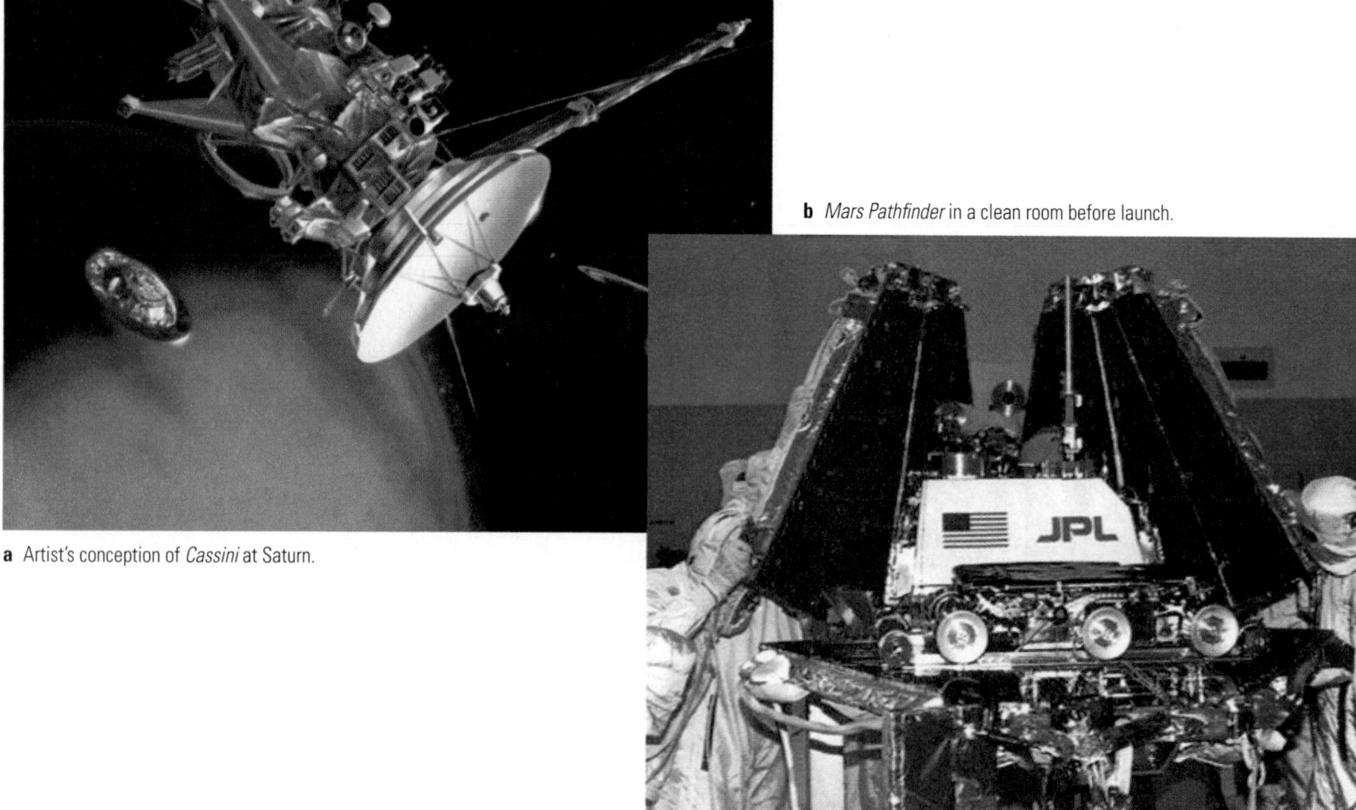

b *Mars Pathfinder* in a clean room before launch.

a Artist's conception of *Cassini* at Saturn.

is used to make electricity. Improved solar cell technology and spacecraft miniaturization may eventually make the use of plutonium batteries unnecessary.

Orbital Considerations and Cost

Because even the smallest spacecraft require very high speeds to leave the Earth, spacecraft must be launched atop large rockets. The heavier the spacecraft, the larger and more expensive the rocket required. The most obvious way to reduce the weight and launch cost of a spacecraft is to make it smaller. Just as powerful computers today are smaller than ever, new technologies are also reducing the size of many scientific instruments. The *Sojourner* rover that explored a small area of Mars in 1997 is not much bigger than a typical microwave oven (Figure S2.32).

Once a spacecraft is on its way, it requires no more fuel unless it needs to *change* its orbit [Section 6.6]. Thus, in general, a flyby is less expensive to launch than an orbiter: A spacecraft headed toward another world necessarily approaches on an unbound orbit

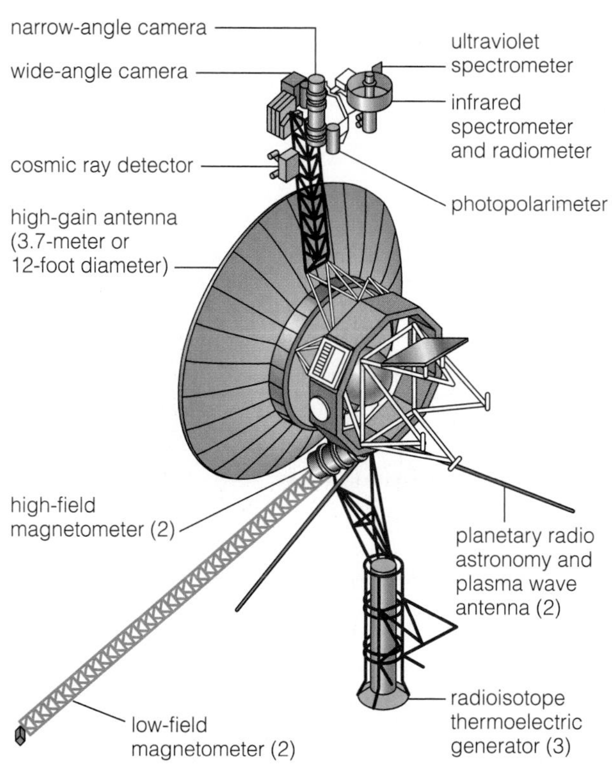

c Schematic illustration of Voyager (1 and 2).

FIGURE S2.32 The *Sojourner* rover. The rover is roughly the size of a standard microwave oven, making it look a bit like a child's toy.

and therefore is destined for a flyby unless it carries rocket engines and enough fuel to change its orbit to a bound one (elliptical or circular). The engines and fuel add weight to the spacecraft, adding to its launch costs. The spacecraft may also be heavier if it carries a lander, because the lander needs some way to slow its descent for a safe landing on the surface.

Two 1997 Mars missions used very imaginative techniques to reduce the amount of fuel they needed to carry. The *Pathfinder* lander (which carried *Sojourner*) used parachutes to slow its descent through the thin Martian atmosphere. Before the lander hit the surface at about 50 km/hr, it deployed air bags all around itself for protection (Figure S2.33). The air bags made the spacecraft bounce several times along the surface before it settled down at its landing site. The *Mars Global Surveyor*, an orbiter, saved on weight by carrying only enough rocket fuel to change its flyby trajectory to a highly eccentric orbit around Mars. To settle into a smaller and more circular orbit, it skimmed the Martian atmosphere at the low point

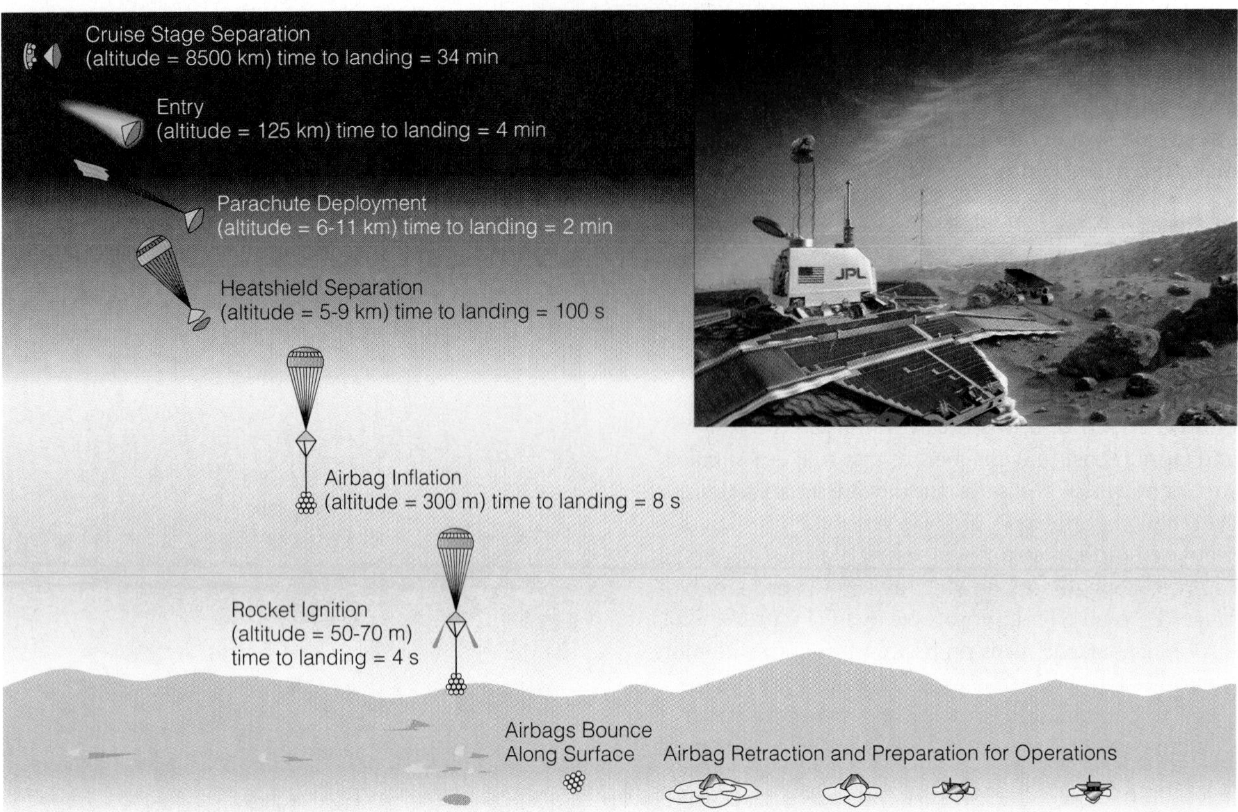

FIGURE S2.33 The *Pathfinder* lander used parachutes and air bags to land safely on Mars. The inset shows an artist's conception of the lander on Mars.

of each elliptical orbit. Atmospheric drag slowed the spacecraft, thereby removing its unwanted orbital energy.

Spacecraft headed to the outer solar system present another problem: The farther we wish to send a spacecraft from the Sun, the more energy it requires. So far, we've never launched a spacecraft to the outer solar system with enough energy to reach its final destination. Instead, we've given it only enough energy to make it to at least one planet and then have taken advantage of the planet's gravity to "slingshot" the spacecraft to higher speeds. In effect, a spacecraft gains energy by stealing a little bit of the planet's orbital energy around the Sun; because spacecraft are so small compared to planets, the effect on the planet's orbit is negligible. The *Cassini* mission uses this slingshot technique four times: twice at Venus, then with a close flyby of Earth, and finally with a flyby of Jupiter before its scheduled arrival at Saturn in 2004 (Figure S2.34).

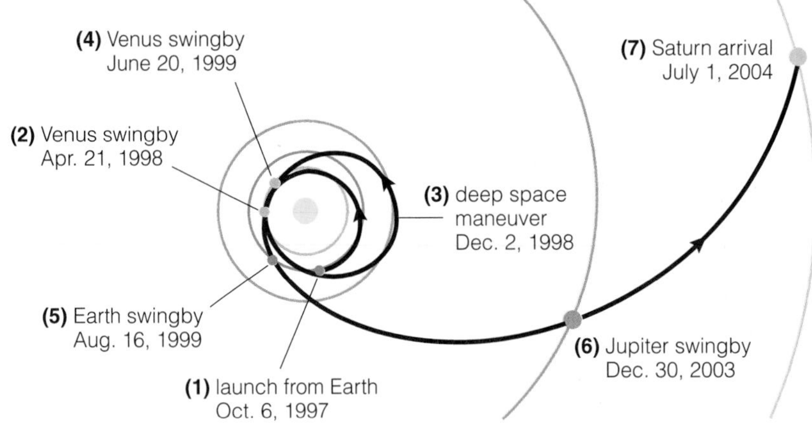

FIGURE S2.34 The trajectory of *Cassini* from Earth to Saturn.

The Future of Astronomy in Space

It will always be cheaper and easier to do astronomy from ground-based observatories, but space is likely to play an ever-increasing role in astronomy. NASA is considering future missions to every planet, as well as to some moons, asteroids, and comets. Mars is targeted for a particularly ambitious set of missions, with at least one mission scheduled about every 2 years (when Mars is located at the position in its

FIGURE S2.35 An artist's conception of a lunar observatory.

orbit making it easiest to reach from Earth). A sample return mission, bringing Mars rocks back to Earth, probably will occur in either 2002 or 2004. Some scientists hope to launch a human mission to Mars within the next couple of decades.

Telescopic astronomy in space also should advance spectacularly. The Hubble Space Telescope has given us unprecedented views of the cosmos, but NASA is already at work on even more impressive observatories to replace it. For the more distant future, many astronomers dream of an observatory on the far side of the Moon (Figure S2.35). Because the Moon has no atmosphere, it offers all the advantages of telescopes in space, in particular the ability to receive light of any wavelength without any atmospheric distortion. In addition, because the far side of the Moon never faces the Earth, the Moon itself would shield radio telescopes from interference from radio and television communication on Earth. Most important, the Moon has a solid surface like the Earth. Thus, telescopes on the Moon can be built and operated as simply as telescopes on the Earth. Of course, setting up a transportation system to and from a lunar observatory would be very expensive, but many scientists believe that once in place such an observatory would be fairly inexpensive to operate.

All of this has been discovered and observed these last days thanks to the telescope that I have [built], after having been enlightened by divine grace.

GALILEO

THE BIG PICTURE

In this chapter, we've focused on the technological side of astronomy: the telescopes and spacecraft that we use to learn about the universe. Keep in mind the following "big picture" ideas as you continue to learn about astronomy:

- Technology drives astronomical discovery. Every time we build a bigger telescope or a more sensitive detector, open up a new wavelength to study, or send a spacecraft to another world, we learn more about the universe than was possible before.
- Telescopes work much like giant eyes, enabling us to see the universe in great detail. New technologies for making larger telescopes, along with advances in adaptive optics and interferometry, are making ground-based telescopes more powerful than ever.
- For the ultimate in observing the universe, space is the place! Telescopes in space allow us to detect light from across the entire spectrum while also avoiding any distortion caused by Earth's atmosphere. Flybys, orbiters, and landers allow us to make detailed studies of other worlds in our solar system that are not possible from Earth.

Review Questions

1. Briefly describe how the eye collects light. What do we mean by a *focal plane?* What do we mean by *angular resolution?*
2. Briefly describe how a camera collects and records light. How do we control the *exposure time* with a camera?
3. What is a CCD? Briefly describe the advantages of using CCDs over using photographic film.
4. Briefly distinguish between a *refracting telescope* and a *reflecting telescope.* Why are most large telescopes reflectors? What do we mean by the *optical quality* of a telescope?
5. What are the two key functions of a telescope? How do we describe a telescope's *light-collecting area?*
6. What is the *diffraction limit* of a telescope, and how does it depend on the telescope's size and the wavelength of light being observed?
7. Define and differentiate between the three basic categories of astronomical observation: *imaging, spectroscopy,* and *timing.*
8. What is a *false-color* image? Why are false-color images useful?
9. What do we mean by *spectral resolution?* Explain why high spectral resolution is desirable but more difficult to achieve than lower spectral resolution.
10. Explain how *light pollution* and atmospheric *turbulence* affect astronomical observations made from the Earth's surface.
11. Study Figure S2.18 in detail, and describe how deeply each portion of the electromagnetic spectrum penetrates the Earth's atmosphere.
12. Describe the basic criteria for locating observing sites on Earth.
13. What is *adaptive optics?* What problems caused by the Earth's atmosphere can it solve? What problems caused by the Earth's atmosphere cannot be solved by any technology?
14. Briefly describe the advantages of putting telescopes in space. Are there any disadvantages? Explain.
15. Why is it useful to study radiation from across the spectrum? Briefly describe the characteristics of telescopes used to observe different portions of the spectrum.
16. Why must a radio telescope be larger than other telescopes to achieve the same angular resolution? Briefly describe how interferometry allows us to achieve high resolution with radio waves. What is the VLA? What is VLBI?
17. Describe each of the four main categories of spacecraft, and describe their basic components.
18. Briefly explain the trade-offs between orbital considerations and cost involved in planning a spacecraft mission. How were costs reduced in the Mars Pathfinder mission, the Mars Global Surveyor mission, and the Cassini mission?

Discussion Questions

1. *Science and Technology Funding.* Technological innovation clearly drives scientific discovery in astronomy, but the reverse is also true. For example, Newton's discoveries were made in part to explain the motions of the planets, but they have had far-reaching effects on our civilization. Congress often must make decisions between funding programs with purely scientific purposes ("basic research") and programs designed to develop new technologies. If you were a member of Congress, how would you try to allocate spending between basic research and technology? Why?
2. *A Lunar Observatory.* Do the potential benefits of building an astronomical observatory on the Moon justify its costs at the present time? If it were up to you, would you recommend that Congress begin funding such an observatory? Defend your opinions.

Problems

1. *Keck Versus the Eye.* Each Keck telescope has a 10-meter-diameter primary mirror. The opening of the human eye is about 0.8 cm (when the pupil is fully dilated). How many times greater is the light-collecting area of each Keck telescope than the light-collecting area of the human eye? (*Hint:* First calculate the light-collecting area of a Keck telescope and that of the human eye, and then divide your results.)
2. *Angular Separation Practice.*
   a. Suppose that two light bulbs are separated by 0.2 meter, and you are looking at the lights from a distance of 2 kilometers. What is the angular separation of the lights? Can your eyes resolve them, or will they look like a single light? Explain.
   b. The diameter of a dime is about 1.8 centimeters. What is its angular diameter if you view it from across a 100-meter-long football field?
3. *Calculating the Sun's Size.* The angular diameter of the Sun is about the same as that of the Moon (0.5°). Use this fact and the Sun's average distance of about 150 million km to estimate the diameter of the Sun. Compare your result to the Sun's actual diameter of 1.392 million km.
4. *Viewing a Dime with the HST.* The Hubble Space Telescope (HST) has an angular resolution of about 0.05 arcsecond. How far away would you have to place a dime (diameter = 1.8 cm) for its angular diameter to be 0.05 arcsecond? (*Hint:* Start by converting an angular diameter of 0.05 arcsecond into degrees.)
5. *Close Binary System.* Suppose that two stars in a binary star system are separated by a distance of 100 million km and are located at a distance of 100 light-years from Earth. What is the angular separation of the two stars? Give your answer in both degrees and arcseconds. Will the two stars appear as two distinct lights or as a single point of light to the Hubble Space Telescope?
6. *Diffraction Limit of the Eye.* Calculate the diffraction-limited resolution of the human eye, assuming a lens size of 0.8 cm, for visible light of 500-nm wavelength. How does this compare to the diffraction-limited resolution of a 10-meter telescope?
7. *The Size of Radio Telescopes.* What is the diffraction-limited resolution of a 100-meter radio telescope observing radio waves with a wavelength of 21 cm? Compare to the diffraction-limited resolution of the 2.4-meter Hubble Space Telescope for visible light. Use your results to explain why radio telescopes must be much larger than optical telescopes to be useful.
8. *Hubble's Field of View.* Large telescopes often have small fields of view. For example, the Hubble Space Telescope's (HST) "wide field" camera has a field of view that is roughly square and about 0.04° on a side.
   a. Calculate the angular area of the HST's field of view in square degrees.
   b. The angular area of the entire sky is about 41,250 square degrees. How many pictures would the HST have to take with its wide field camera to obtain a complete picture of the entire sky?
   c. Assuming that it requires an average of 1 hour to take each picture, how long would it take to acquire the number of pictures you calculated in part (b)? Use your answer to explain why astronomers would like to have more than one large telescope in space.
9. *Project: Twinkling Stars.* Using a star chart, identify 5 to 10 bright stars that should be visible in the early evening. On a clear night, observe each of these stars for a few minutes. Note the date and time, and for each star record the following information: approximate altitude and direction in your sky, brightness compared to other stars, color, how much the star twinkles compared to other stars. Study your record. Can you draw any conclusions about how brightness and position in your sky affect twinkling? Explain.
10. *Research: Planetary Mission.* Choose a current mission to a planet. Using the Internet, find as much information as you can about the mission. Write a two- to three-page summary of the mission, including discussion of its purpose, design, and cost.

PART III
LEARNING FROM OTHER WORLDS

8

Formation of the Solar System

How old is the Earth, and how did it come to be? Is it unique? Our ancestors could do little more than guess, but today we can answer with reasonable confidence: The Earth and the rest of our solar system formed from a great cloud of gas and dust about 4.6 billion years ago. Whereas ancient astronomers believed the Earth to be fundamentally different from objects in the heavens, we now see the Earth as just one of many worlds.

As we examine the evidence that helped shape the modern scientific view of the formation of our solar system, we'll highlight one common technique of scientific research: sifting through seemingly unrelated facts to find the most important trends, patterns, and characteristics—and then attempting to explain them. The explanation must be based on known physical laws, and the success or failure of a theory depends on how well it matches the facts. As we discover more and more planets around other stars, we hope to test our theories further and to modify them as necessary.

8.1 Comparative Planetology

Galileo's telescopic observations began a new era in astronomy in which the Sun, the Moon, and the planets could be studied for the first time as *worlds,* rather than merely as lights in the sky. During the early part of this new era, astronomers studied each world independently, but in the twentieth century we discovered that the similarities among worlds are often more profound than the differences.

Today, we know that the Sun, planets, moons, and other bodies in our solar system all formed at about the same time from the same cloud of interstellar gas and dust (*interstellar* means "between the stars") and in accord with the same physical laws. Thus, the differences between these worlds must be attributable to physical processes that we can study and understand by comparing the worlds to one another. This approach is called **comparative planetology;** the term *planetology* is used broadly to include moons, asteroids, and comets as well as planets. The idea is that we can learn more about the Earth, or any other world, by studying it in the context of other objects in our solar system—rather like learning about a person by studying his or her family, friends, and culture.

Before beginning the comparative study of worlds, we must be sure that the comparison will provide valid lessons. The basic premise of comparative planetology—that the similarities and differences among worlds in our solar system can be traced to common physical processes—assumes that the planets share a common origin. Thus, we begin our study of comparative planetology by seeking clues to our origins from the current state of our solar system; then we examine the modern theory of solar system formation that explains much of what we see.

8.2 The Origin of the Solar System: Four Challenges

Learning about the origin of the solar system is easiest if we begin by looking at our solar system with a "big picture" view, rather than focusing on individual worlds. Imagine that we have the perspective of an alien spacecraft making its first scientific survey of our solar system. What would we see?

First, as we learned from looking at a scale model of our solar system in Chapter 2, we'd find that our solar system is mostly empty—interplanetary space is a near-vacuum containing only very low density gas and tiny, solid grains of dust. But we'd soon focus on the widely spaced planets, mapping their orbits, measuring their sizes, compositions, and densities, and taking inventory of their moons and ring systems. Figure 8.1 shows the resulting schematic maps, and Table 8.1 summarizes the planetary properties. Note that asteroids and comets are also listed in the table because these objects orbit the Sun, though with orbital properties somewhat different from those of the planets.

Careful study of the data shown in the maps and the table reveals several general features of our solar system. Let's investigate by grouping our observations into four major challenges that must be met by any theory claiming to explain how our solar system developed.

Challenge 1: Patterns of Motion

If we look closely at the maps of solar system orbits in Figure 8.1, we'll see several striking patterns:

- All planets orbit the Sun in the same direction—counterclockwise when seen from high above the Earth's North Pole.
- All planetary orbits lie nearly in the same plane.
- Almost all planets travel on nearly circular orbits, and the spacing between planetary orbits increases with distance from the Sun according to a fairly regular trend.[1] The most notable exception to this trend is an extra-wide gap between Mars and Jupiter that is populated with asteroids.
- Most planets rotate in the same direction in which they orbit—counterclockwise when seen from above the Earth's North Pole—with fairly small axis tilts (i.e., less than about 25°).
- Almost all moons orbit their planet in the same direction as the planet's rotation and near the planet's equatorial plane.
- The Sun rotates in the same direction in which the planets orbit.

These patterns show that motion in the solar system is generally quite organized. If each planet had come into existence independently, we might expect planetary motions to be much more random. (Note that

[1]The spacing between planetary orbits is roughly described by a mathematical pattern, called *Bode's law,* discovered in the late 1700s by German astronomers Johann Titius and Johann Bode. Taken literally, this pattern "predicts" the existence of a planet at the location of the asteroid belt between Mars and Jupiter. But there is no known physical explanation for Bode's law, so it may be just a coincidence. Nevertheless, if the asteroid belt is counted as a planet, Bode's law closely predicts the locations of all the planets except Neptune.

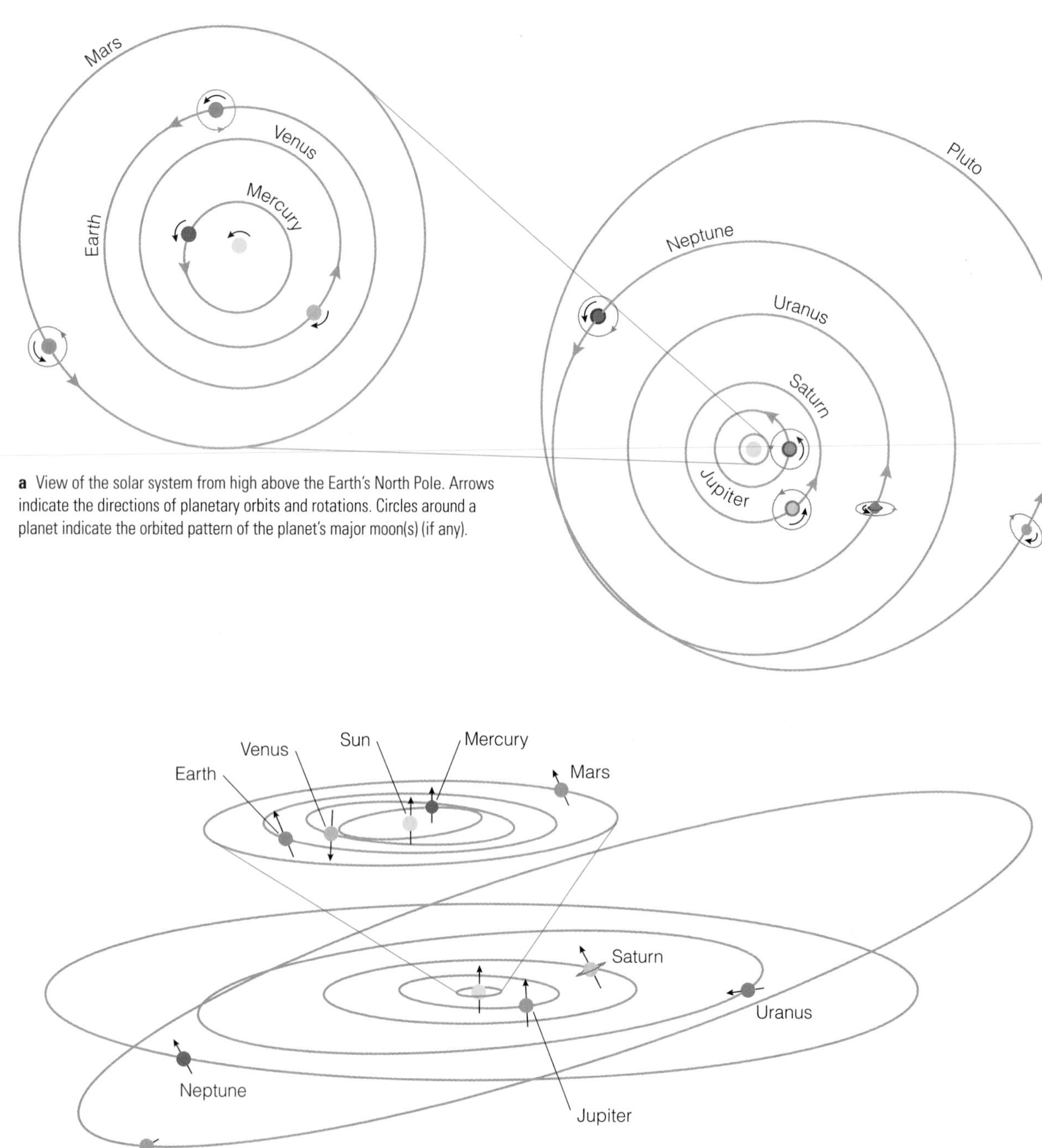

a View of the solar system from high above the Earth's North Pole. Arrows indicate the directions of planetary orbits and rotations. Circles around a planet indicate the orbited pattern of the planet's major moon(s) (if any).

b Side view of the solar system. Arrows indicate the orientation of the rotation axes of the planets. (Planetary tilts in this diagram are aligned in the same plane for easier comparison. Planets not to scale.)

FIGURE 8.1 The layout of the solar system.

the motions of asteroids are slightly more random, and those of comets even more so, as we will discuss shortly.)

The first challenge for any theory of solar system formation is to explain why motions in the solar system are generally so orderly.

Challenge 2: Categorizing Planets

Before reading any further, study the properties of the planets in Table 8.1. Can you categorize the planets into groups? How many categories do you need? What characteristics do planets within a category share?

Table 8.1 Planetary Facts*

Photo	Planet	Average Distance from Sun (AU)	Temperature†	Relative Size	Average Equatorial Radius (km)	Average Density (g/cm^3)	Composition	Moons	Rings?
	Mercury	0.387	700 K		2,440	5.43	Rocks, metals	0	No
	Venus	0.723	740 K		6,051	5.24	Rocks, metals	0	No
	Earth	1.00	290 K		6,378	5.52	Rocks, metals	1	No
	Mars	1.52	240 K		3,397	3.93	Rocks, metals	2 (tiny)	No
	Most asteroids	2–3	170 K		≤ 500	1.5–3	Rocks, metals	?	No
	Jupiter	5.20	125 K		71,492	1.33	H, He, hydrogen compounds‡	16	Yes
	Saturn	9.53	95 K		60,268	0.70	H, He, hydrogen compounds‡	18	Yes
	Uranus	19.2	60 K		25,559	1.32	H, He, hydrogen compounds‡	17	Yes
	Neptune	30.1	60 K		24,764	1.64	H, He, hydrogen compounds‡	8	Yes
	Pluto	39.5	40 K		1,160	2.0	Ices, rock	1	No
	Most comets	0–50,000	a few K§		a few km?	<1?	Ices, dust	?	No

*Appendix C gives a more complete list of planetary properties. †Surface temperatures for all objects except Jupiter, Saturn, Uranus, and Neptune, for which cloud-top temperatures are listed. ‡Includes water (H_2O), methane (CH_4), and ammonia (NH_3).
§Comets passing close to the Sun warm considerably, especially their outer layers.

Astronomers classify most of the planets in two distinct groups: the rocky **terrestrial planets** and the gas-rich **jovian planets.** The word *terrestrial* means "Earth-like" (*terra* is the Greek word for Earth), and the terrestrial planets are the four planets of the inner solar system—Mercury, Venus, Earth, and Mars. These are the worlds most like our own. They are relatively small, close to the Sun, and close together. They have solid, rocky surfaces and an abundance of metals deep in their interiors. They have few moons, if any, and none have rings. Note that many scientists count our Moon as a fifth terrestrial world because it shares these general characteristics, although it's not technically a planet.

The word *jovian* means "Jupiter-like," and the jovian planets are the four outer solar system planets—Jupiter, Saturn, Uranus, and Neptune. The jovian planets are much larger than the terrestrial planets, farther from the Sun, and widely separated from each other. These huge planets have little in common with Earth. They are made mostly of hydrogen, helium, and **hydrogen compounds** such as water, ammonia, and methane. They contain relatively small amounts of rocky material only deep in their cores. Jovian planets do not have solid surfaces; if you plunged into the atmosphere of a jovian planet, you would sink deeper and deeper until you were crushed by overwhelming pressure. The jovian planets are sometimes said to be made mostly of gas, but in fact the intense pressures and temperatures of the jovian planet interiors transform familiar gases (and solids) to phases unlike anything we see on Earth. Each jovian planet has rings and an extensive system of moons. These solid satellites are made mostly of low-density ices and rocks. Table 8.2 contrasts the general traits of the terrestrial and jovian planets.

The second challenge for any theory of solar system formation is to explain why the inner and outer solar system planets divide so neatly into two classes.

TIME OUT TO THINK *Are the distinctions in Table 8.2 clear-cut? Examine Table 8.1 to compare, for example, the radii of the largest terrestrial planet and the smallest jovian planet.*

Note that Pluto is left out in the cold, both literally and figuratively. On one hand, it is small and solid like the terrestrial planets. On the other hand, Pluto is far from the Sun, cold, and made of low-density ices. Pluto also has an unusual orbit, far more eccentric than the orbit of any other planet and substantially inclined to the plane in which the other planets orbit the Sun. For a long time, scientists considered Pluto to be a lone misfit. However, some scientists now classify Pluto with other icy bodies in the outer solar system: comets. Others propose the existence of a class of *Plutonian planets,* of which Pluto is the closest and biggest member.

Challenge 3: Asteroids and Comets

No formation theory is complete without an explanation of the most numerous objects in the solar system: asteroids and comets. **Asteroids** are small, rocky bodies that orbit the Sun primarily in the **asteroid belt** between the orbits of Mars and Jupiter (Figure 8.2).[2] Asteroids orbit the Sun in the same direction as the planets. Their orbits generally lie close to the plane of planetary orbits, although they are usually a bit tilted. Some asteroids have elliptical orbits that are quite eccentric compared to the nearly circular orbits of the planets. Almost 9,000 asteroids have been identified and cataloged, but these are probably only the largest among a much greater number of small asteroids. The largest asteroids are a few hundred kilometers in radius—much less than half of the Moon's radius.

Comets are small, icy bodies that spend most of their lives well beyond the orbit of Pluto; we generally recognize them only on the rare occasions when

[2]Another large population of asteroids—the Trojan asteroids—shares Jupiter's orbit, falling into two clumps leading and trailing Jupiter by 60°.

Table 8.2 Comparison of Terrestrial and Jovian Planets

Terrestrial Planets	Jovian Planets
Smaller size and mass	Larger size and mass
Higher density (rocks, metals)	Lower density (light gases, hydrogen compounds)
Solid surface	No solid surface
Closer to the Sun (and closer together)	Farther from the Sun (and farther apart)
Warmer	Cooler
Few (if any) moons and no rings	Rings and many moons

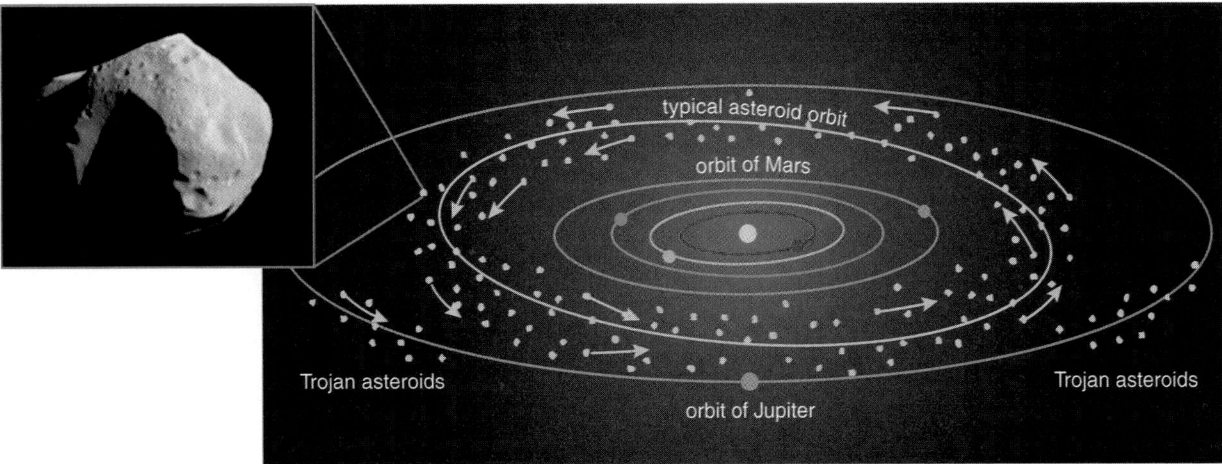

FIGURE 8.2 Most asteroids orbit the Sun between Mars and Jupiter. Their orbits are noticeably tilted and eccentric.

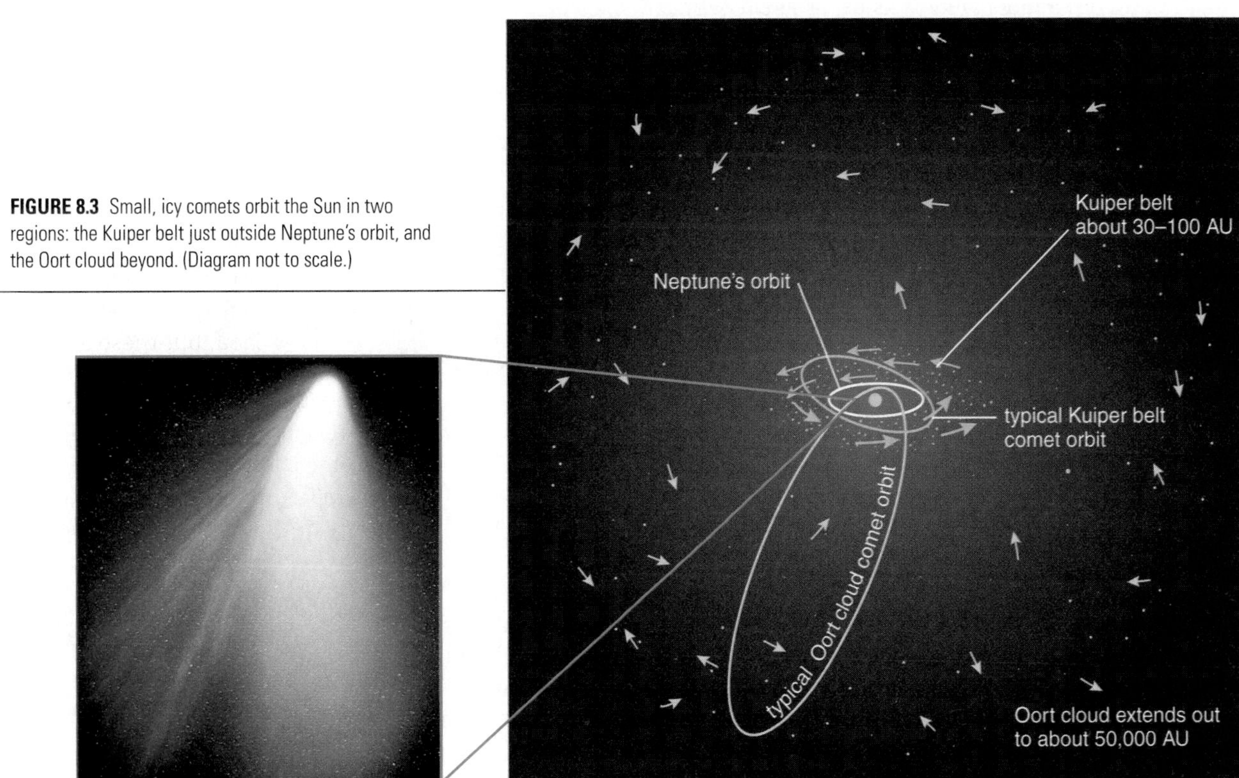

FIGURE 8.3 Small, icy comets orbit the Sun in two regions: the Kuiper belt just outside Neptune's orbit, and the Oort cloud beyond. (Diagram not to scale.)

one dives into the inner solar system and grows a spectacular tail. Based on the orbits of the relatively few comets that reach the inner solar system, astronomers have determined that many billions of comets must be orbiting the Sun, primarily in two broad regions (Figure 8.3). The first region, called the **Kuiper belt** (pronounced koy-per), begins somewhere in the vicinity of the orbit of Neptune (about 30 AU from the Sun) and extends to perhaps 100 AU from the Sun. Comets in the Kuiper belt have orbits that lie fairly close to the plane of planetary orbits, and they travel around the Sun in the same direction as the planets. The second region populated by comets, called the **Oort cloud,** is a huge, spherical region centered on the Sun and extending perhaps halfway to the nearest stars. The orbits of comets in the Oort cloud are completely random, with no pattern to their inclinations, orbital directions, or eccentricities.

The third challenge for any theory of solar system formation is to explain the existence and general properties of the large numbers of asteroids and comets.

Challenge 4: Exceptions to the Rules

The first three challenges involve explaining general patterns in our solar system. But some objects don't fit these patterns. For example, Mercury and Pluto have larger orbital eccentricities and inclinations than the other planets. The rotational axes of Uranus and Pluto are substantially tilted, and Venus rotates backward—clockwise, rather than counterclockwise, as viewed from high above Earth's North Pole. Unlike the other terrestrial planets, Earth has a large moon. Pluto has a moon almost as big as itself. While most moons of the jovian planets orbit with the same orientation as their planet's rotation, a few orbit in the opposite direction.

Allowing for these and other exceptions to the general patterns is the fourth challenge for any theory of solar system formation.

Summarizing the Challenges

Table 8.3 summarizes the major characteristics of our solar system that we have discussed as four challenges to be explained by our solar system formation theory. In the remainder of this chapter, we will examine the leading theory of solar system formation and put it to the test to see whether it can meet the four challenges. In this sense, we will be following the idealized scientific method by starting with the facts, putting forward a theory, and putting the theory to the test. In doing so, we have the benefit of hindsight: The historical path to our modern theory of the solar system was actually quite complex, with many competing theories being proposed and eventually discarded because they were unable to meet the challenges we have laid out.

The process of identifying the most important facts to explain is an essential first step in scientific analysis, followed by the development of a theory to explain these facts. We might tackle an environmental problem with a similar approach, for example, determining which pollutants are damaging an ecosystem and then developing a logical explanation for how these pollutants get there. A correct explanation of the facts is the first step in finding a solution.

Table 8.3 Four Major Characteristics of the Solar System

Large bodies in the solar system have orderly motions. All planets and most satellites have nearly circular orbits going in the same direction in nearly the same plane. The Sun and most of the planets rotate in this same direction as well.

Planets fall into two main categories: small, rocky terrestrial planets near the Sun and large, hydrogen-rich jovian planets farther out. The jovian planets have many moons and rings made of rock and ice.

Swarms of asteroids and comets populate the solar system. Asteroids are concentrated in the asteroid belt, and comets populate the regions known as the Kuiper belt and the Oort cloud.

Several notable exceptions to these general trends stand out, such as planets with unusual axis tilts or surprisingly large moons, and moons with unusual orbits.

8.3 The Nebular Theory of Solar System Formation

Over the past three centuries, several models have been put forth in attempts to meet the four challenges for a solar system formation theory. In the past few decades, a tremendous amount of evidence has accumulated in support of one model. This model, which we will call the **nebular theory,**[3] holds that our solar system formed from a giant, swirling **interstellar cloud** of gas and dust; such a cloud is also called a **nebula** (the Latin word for "cloud"). More generally, we now have evidence that all **star systems**—systems like our solar system that contain at least one star and, in some cases, planets or other orbiting material—form similarly from interstellar clouds.

Cosmic Recycling

We usually think of interstellar space as being empty, and it certainly is a vacuum by any earthly standards. A sugar cube–size volume of interstellar space contains, on average, only a single atom. The same volume of air on Earth contains 10^{20} atoms. Nevertheless, the Milky Way Galaxy is so vast that it contains a significant amount of interstellar matter—about 10% as much as is contained in all the stars of the Milky Way. Star systems are born within interstellar clouds in which the gas is somewhat cooler and denser than average (though still trillions of times less dense than air). Typical star-forming clouds contain enough material to form hundreds or thousands of stars (Figure 8.4).

[3]We use the term *nebular theory* to encompass the entire formation process described in this chapter; some texts use this term more narrowly.

Where does a star-forming cloud come from? Observations of many such clouds show that they represent just one step in a galactic recycling process (Figure 8.5) [Section 18.2]. Stars live for millions or billions of years (a star's lifetime depends on its mass, with massive stars having shorter lives), shining with the energy created by *nuclear fusion* [Section 14.3]. During their lives, stars return some of their material to interstellar space through **stellar winds** that blow outward from their surfaces. When their nuclear fuel is exhausted, stars die, sometimes in titanic explosions (called *supernovae*) that spew much of their material in all directions. This material mixes with other material in interstellar space, eventually forming new clouds that can collapse gravitationally into star systems. Then the cycle begins anew.

Our galaxy, like the entire universe, originally contained only hydrogen and helium (and trace amounts of lithium). All the heavier elements are produced in stars. Thus, the galactic recycling process gradually enriches the galaxy with heavier elements so that later generations of stars are born with a greater proportion of heavier elements than earlier generations. By the time our solar system formed, 4.6 billion years ago, about 2% of the original hydrogen and helium in the galaxy had been converted into heavier elements. Although 2% sounds like a small amount, it was more than enough to form the rocky terrestrial planets—and us. Nevertheless, hydrogen and helium remain the most abundant elements by far and make up the bulk of the Sun and the jovian planets.

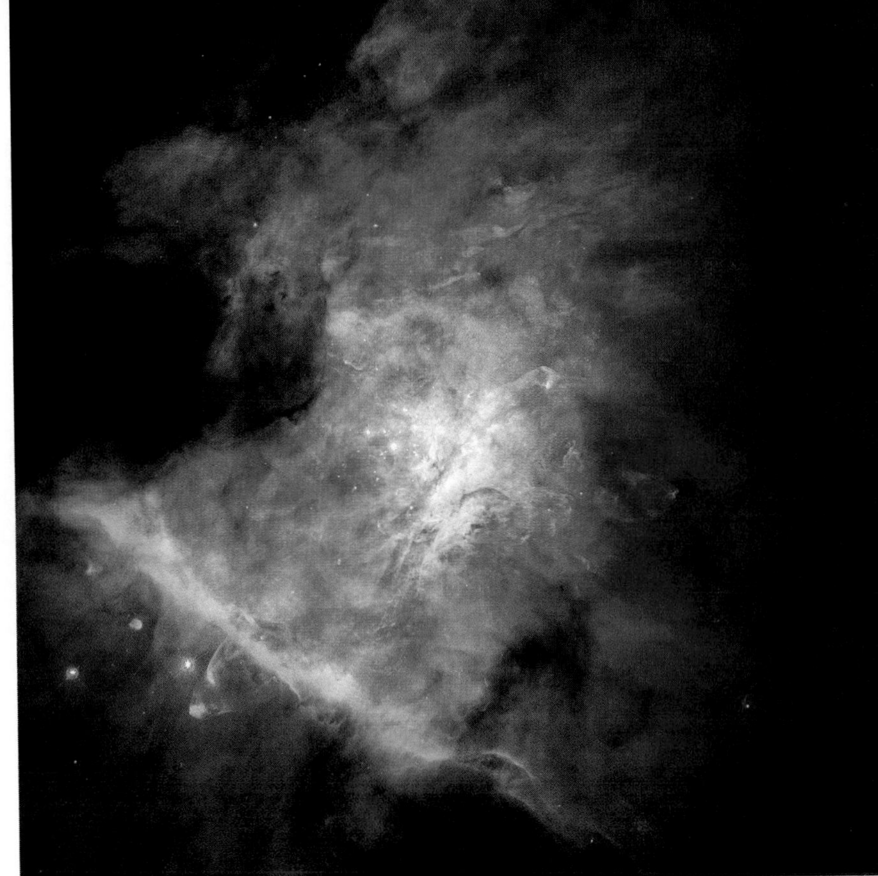

FIGURE 8.4 This photograph shows the central region of the Orion Nebula, an interstellar cloud in which star systems—possibly including planets—are forming. The photo is actually a composite of more than a dozen separate images taken with the Hubble Space Telescope. (See Figure 18.14 for a complete view of the Orion Nebula.)

TIME OUT TO THINK ***Could a solar system like ours have formed with the first generation of stars after the Big Bang? Explain.***

FIGURE 8.5 The galactic recycling process.

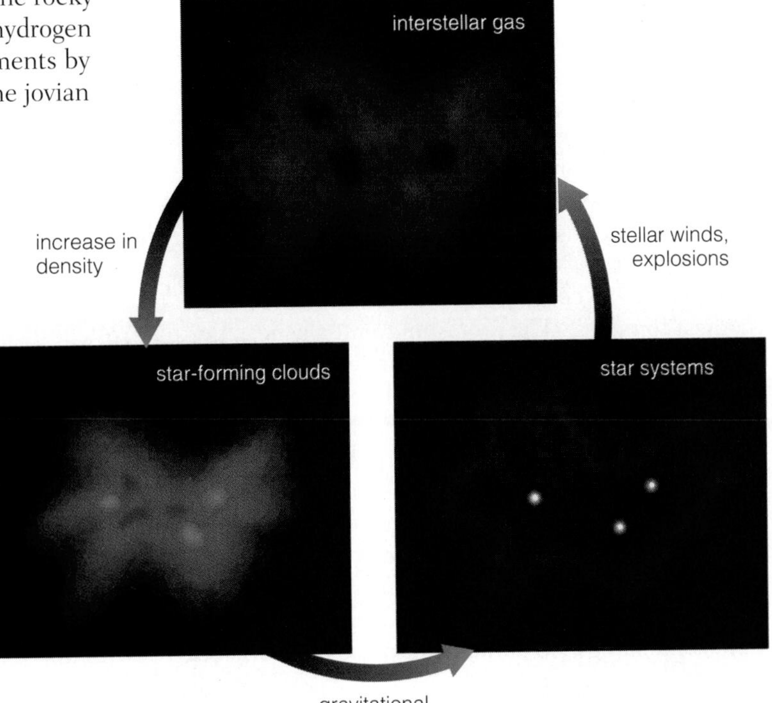

Collapse of the Solar Nebula

An individual star system forms from just a small part of a giant interstellar cloud. We refer to the collapsed piece of cloud that formed our own solar system as the **solar nebula.** The solar nebula collapsed under its own gravity; the

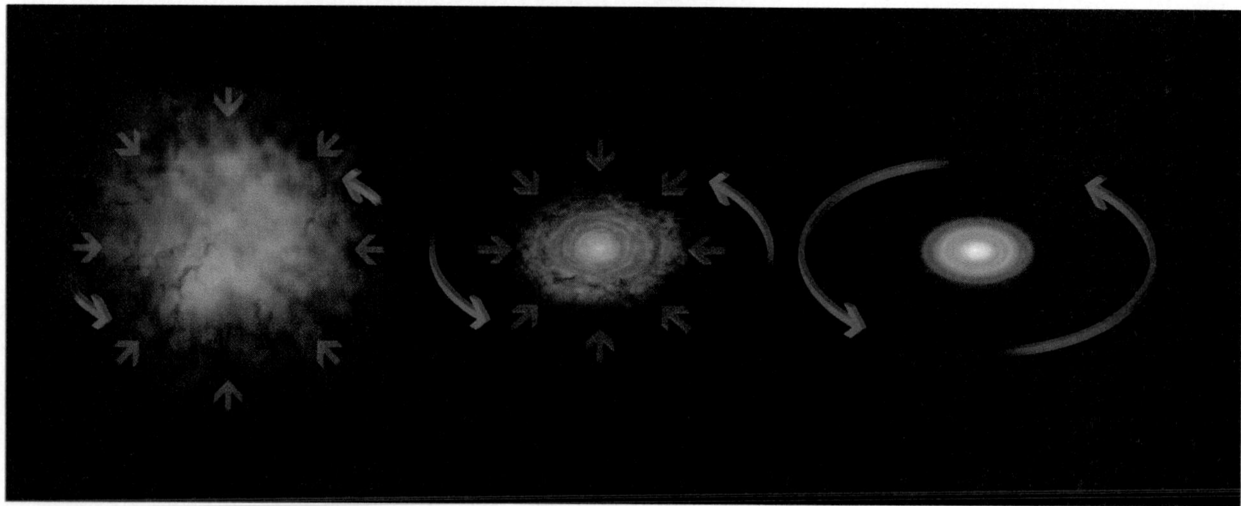

a The original cloud is large and diffuse, and its rotation is almost imperceptibly slow.

b The cloud heats up and spins faster as it contracts.

c The result is a spinning, flattened disk, with mass concentrated near the center.

FIGURE 8.6 This sequence of paintings shows the collapse of an interstellar cloud. In our solar nebula, the hot, dense central bulge became the Sun, and the planets formed in the disk.

may have been triggered by a cataclysmic event such as the impact of a shock wave from a nearby exploding star. Before the collapse, the low-density gas of the interstellar cloud may have been a few light-years in diameter. As it collapsed to a diameter of about 200 AU—roughly twice the present-day diameter of Pluto's orbit—three important processes gave form to our solar system (Figure 8.6).

First, the temperature of the solar nebula increased as it collapsed. Such heating represents energy conservation in action [Section 5.2]. As the cloud shrank, its gravitational potential energy was converted to the kinetic energy of individual gas particles falling inward. These particles crashed into one another, converting the kinetic energy of their inward fall into the random motions of thermal energy. Some of this energy was radiated away as thermal radiation. The solar nebula became hottest near its center, where much of the mass collected to form the **protosun** (the prefix *proto* comes from a Greek word meaning "earliest form of"). The protosun eventually became so hot that nuclear fusion ignited in its core—at which point our Sun became a full-fledged star.

Second, like an ice skater pulling in her arms as she spins, the solar nebula rotated faster and faster as it shrank in radius. This increase in rotation rate represents the conservation of angular momentum in action [Section 6.2].[4] The rotation helped ensure that not all of the material in the solar nebula collapsed onto the protosun: The greater the angular momentum of a rotating cloud, the more spread out it will be.

Third, the solar nebula flattened into a disk—the **protoplanetary disk** from which the planets eventually formed. This flattening is a natural consequence of collisions between particles, which explains why flat disks are so common in the universe (e.g., the disks of spiral galaxies like the Milky Way, ring systems around planets, and *accretion disks* around neutron stars or black holes [Section 17.3]). A cloud may start with any size or shape, and different clumps of gas within the cloud may be moving in random directions at random speeds. When the cloud collapses, these different clumps collide and merge, giving the new clumps the average of their differing velocities (Figure 8.7). The result is that the random motions of the clumps in the original cloud become more orderly as the cloud collapses, changing the cloud's original lumpy shape into a rotating, flattened disk. Similarly, collisions between clumps of material in highly elliptical orbits reduce their ellipticities, making their orbits more circular. You can see a similar effect if you shake some pepper into a bowl of water and quickly stir it in a random way. Because the water molecules are always colliding with one another, the motion of the pepper grains will settle down into a slow rotation representing the average of the original, random velocities.

TIME OUT TO THINK *Try the experiment with the pepper and bowl of water. Note that, no matter how you stir the water, it almost always settles down into a slow rotation in one direction or the other. How is this process similar to what took place in the solar nebula? How is it different?*

[4]The same thing happens if you sit in a swivel chair, set it spinning with your arms and legs extended, and then pull them in.

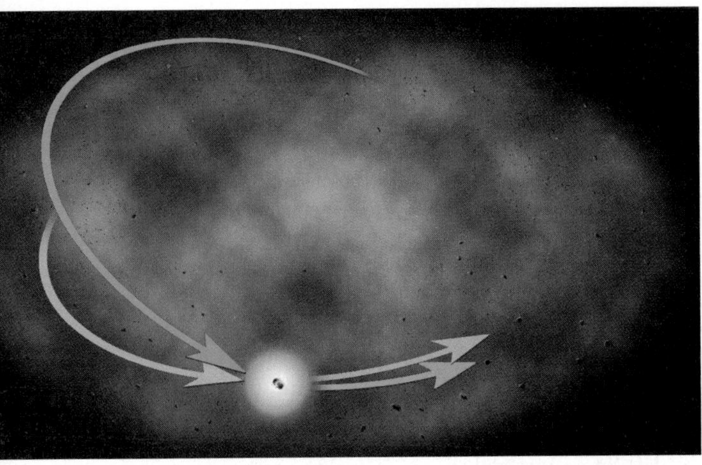

FIGURE 8.7 As shown in this painting, collisions between particles in the solar nebula average out their random motions and flatten the cloud into a disk. The green arrow represents the path of a particle that originally had a tilted orbit. After the collision, its orbit lies closer to the plane of the other particles. If the particles had started an eccentric orbit, collisions would have made its orbit more circular. (Particle sizes are highly exaggerated.)

These three processes—heating, spinning, and flattening—explain the tidy layout of our solar system. The flattening of the protoplanetary disk explains why all planets orbit in nearly the same plane. The spinning explains why all planets orbit in the same direction and also plays a role in making most planets rotate in this same direction. The fact that collisions in the protoplanetary disk tend to make highly elliptical orbits more circular explains why most planets in our solar system have nearly circular orbits today.

Evidence Concerning Nebular Collapse

The theory that our solar system formed from interstellar gas may sound reasonable, but we cannot accept it without hard evidence. Fortunately, we have strong observational evidence that the same processes are occurring elsewhere in our galaxy. A collapsing nebula emits thermal radiation [Section 7.4], primarily in the infrared. We've detected such infrared radiation from many other nebulae where star systems appear to be forming. We've even seen several structures around other stars that look similar to protoplanetary disks (Figure 8.8).

Other support for the nebular theory comes from sophisticated computer models that simulate the formation processes it describes. A simulation begins with a set of data representing the conditions that we observe in interstellar clouds. Then, with the aid of a computer, we apply the laws of physics to these data to simulate how the conditions in a real cloud would change with time. These computer simulations successfully reproduce most of the general characteristics of motion in our solar system, suggesting that the nebular theory is on the right track. However, the simulations do not yet successfully explain the observed patterns of spacing between planets; much work lies ahead before we will fully understand the early history of our solar system. All in all, though, the nebular theory meets the challenge of explaining most of the orderly motions of the planets and satellites. Now let's turn to our second challenge: explaining the strikingly different characteristics of the terrestrial and jovian planets.

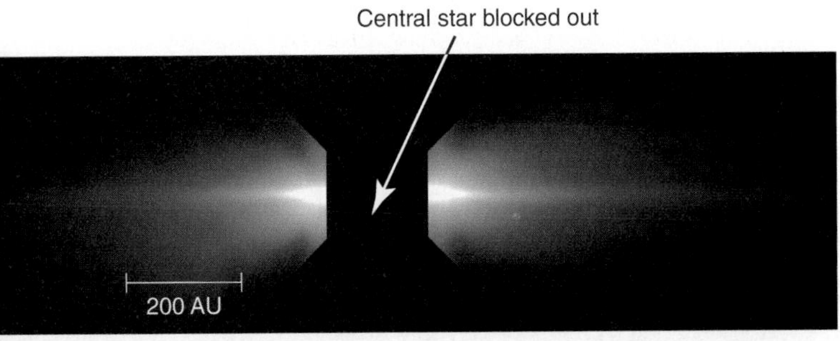

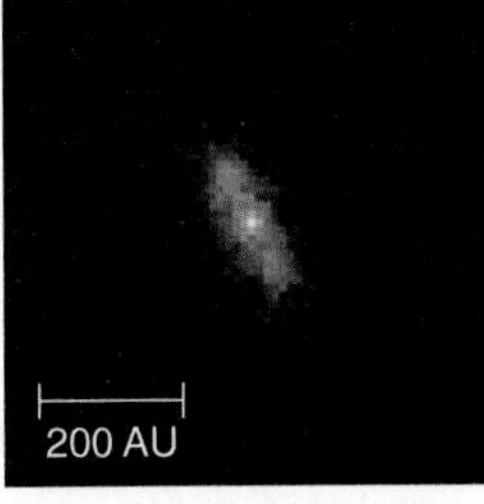

FIGURE 8.8 Evidence for disks around other stars.
a Dust disks similar to protoplanetary disks around the stars Beta Pictoris (left) and HR4796 (right). The central star is blocked out in the image of Beta Pictoris.

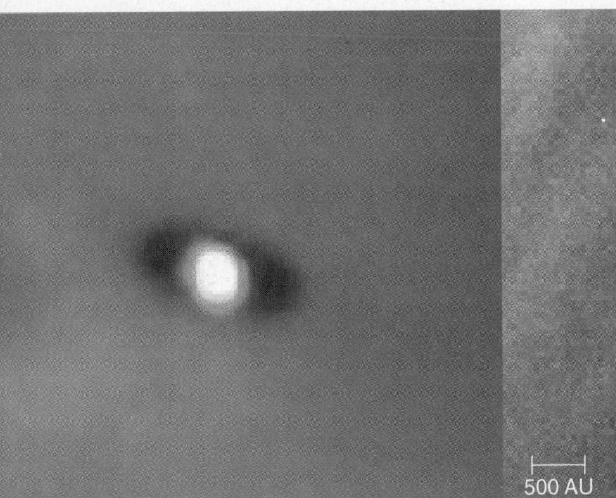

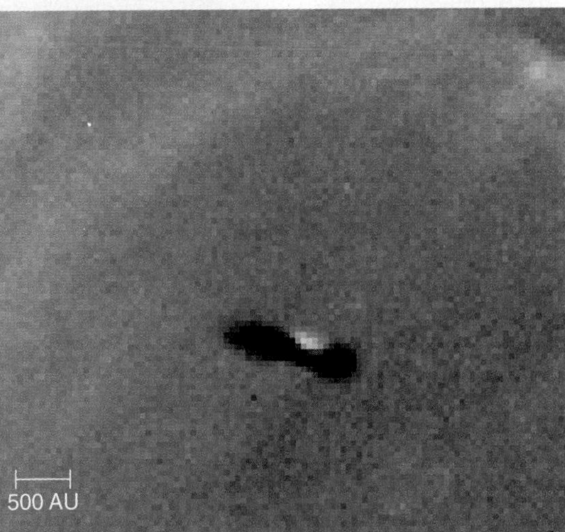

b Protoplanetary disks around stars in the Orion Nebula.

8.4 Building the Planets

The churning and mixing of the gas in the solar nebula ensured that its composition was about the same throughout: roughly 98% hydrogen and helium, and 2% heavier elements such as carbon, nitrogen, oxygen, silicon, and iron. How did the planets and other bodies in our solar system end up with such a wide variety of compositions when they came from such uniform material? To answer this question, we must investigate how the material in the protoplanetary disk came together to form the planets.

Condensation: Sowing the Seeds of Planets

In the center of the collapsing solar nebula, gravity drew much of the material together into the protosun. In the rest of the spinning protoplanetary disk, however, the gaseous material was so spread out that, at first, gravity was not strong enough to pull it together to form planets. The formation of planets therefore required "seeds"—solid chunks of matter that gravity could eventually draw together. Understanding these seeds is the key to explaining the differing compositions of the planets.

The vast majority of the material in the solar nebula was gaseous: High temperatures kept virtually all the ingredients of the solar nebula vaporized near the protosun. Farther out the nebula was still primarily gaseous because the hydrogen and helium that comprised 98% of the solar nebula remain gaseous even at extremely low temperatures. But the 2% of material consisting of heavier elements could form solid seeds where temperatures were low enough. The formation of solid or liquid particles from a cloud of gas is called **condensation.** Pressures in the solar nebula were so low that liquid droplets rarely formed, but solid particles could condense in the same way that snowflakes condense from water vapor in our atmosphere. We refer to such solid particles as **condensates.** The different kinds of planets and satellites formed out of the different kinds of condensates present at different locations in the solar nebula.

The ingredients of the solar nebula fell into four categories based on their condensation temperatures:[5]

- **Metals** include iron, nickel, aluminum, and other materials that are familiar on Earth but less common on the surface than ordinary rock. Some of these metals can remain solid at temperatures as high as 1,600 K. These materials made up less than 0.2% of the solar nebula's mass.
- **Rocks** are materials common on the surface of the Earth, primarily silicon-based minerals. Rocks are solid at temperatures and pressures found on Earth but typically melt or vaporize at temperatures of 500–1,300 K depending on their type. Rocky materials made up about 0.4% of the nebula by mass.
- **Hydrogen compounds** are molecules such as methane (CH_4), ammonia (NH_3), and water (H_2O) that solidify into **ices** below about 150 K. These compounds were significantly more abundant than rocks and metals, making up 1.4% of the nebula's mass.
- **Light gases** (hydrogen and helium) never condense under solar nebula conditions. These gases made up the remaining 98% of the nebula's mass.

The great temperature differences between the hot inner regions and the cool outer regions of the nebula determined what kinds of condensates were available to form planets (Figure 8.9). Let's examine these regions, beginning near the protosun and moving outward to cooler regions. Very near the protosun, where the nebula temperature was above 1,600 K, there were no condensates—everything remained gaseous. Farther out, where the temperature was slightly lower, metal flakes appeared. Near the distance of Mercury's orbit, flakes of rock joined the mix. Moving outward past the orbits of Venus and Earth, more varieties of rock minerals condensed. Near the location of the future asteroid belt, temperatures were low enough to allow minerals containing small amounts of water to condense as well. Dark, carbon-rich materials also condensed here.

FIGURE 8.9 Temperature differences in the solar nebula led to different kinds of condensed materials, sowing the seeds for two different kinds of planets.

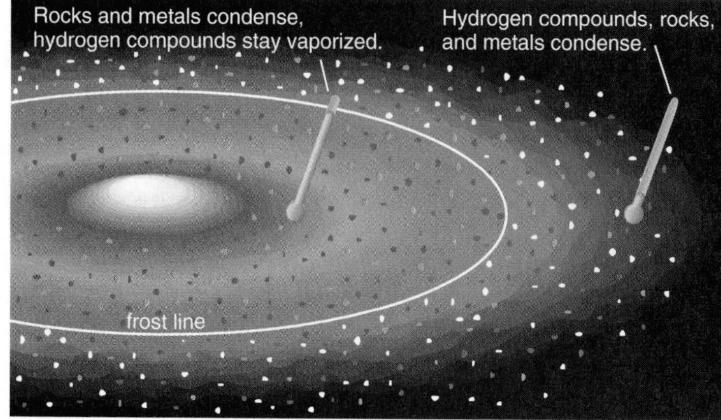

[5]Note that these categories, listed in order of declining condensation temperatures, also follow an order of declining atomic (or molecular) weights and densities in solid or liquid form.

Only beyond the **frost line,** which lay between the present-day orbits of Mars and Jupiter, were temperatures low enough (150 K ≈ −123°C) for hydrogen compounds to condense into ices. (Water ice did not form at the familiar 0°C because the pressures in the nebula were 10,000 times lower than on Earth.) Thus, the outer solar system contained condensates of all kinds: rocks, metals, and ices. However, ice flakes were nearly three times more abundant than flakes of rock and metal because of the greater abundance of hydrogen compounds in the solar nebula.

Common Misconceptions: Solar Gravity and the Density of Planets

You might think that the dense rocky and metallic materials were simply pulled to the inner part of the solar nebula by the Sun's gravity or that light gases simply escaped from the inner nebula because gravity couldn't hold them. But this is not the case; all the ingredients were orbiting the Sun together under the influence of the Sun's gravity. The orbit of a particle or a planet does *not* depend on its size or density, so the Sun's gravity is *not* the cause of the different kinds of planets. Rather, the different temperatures in the solar nebula are the cause.

Accretion: Assembling the Planetesimals

The first solid flakes that condensed from the solar nebula were microscopic in size. They orbited the protosun in the same orderly, circular paths as the gas from which they condensed. Individual flakes therefore moved with nearly the same speed as their neighboring flakes, allowing them to collide very gently. At this point, the flakes were far too small to attract one another by gravity, but they were able to stick together through electrostatic forces—the same "static electricity" that makes hair stick to a comb. Thus, the flakes grew slowly into larger particles. As the particles grew in mass, gravity began to aid the process of their sticking together, accelerating their growth. This process of growing by colliding and sticking is called **accretion.** The growing objects formed by accretion are called **planetesimals,** which essentially means "pieces of planets." Small planetesimals probably came in a variety of shapes, still reflected in many small asteroids today. Larger planetesimals (i.e., those several hundred kilometers across) became spherical due to the force of gravity pulling everything toward the center (Figure 8.10).

The growth of planetesimals was rapid at first: As planetesimals grew larger, they had both more surface area with which to collide and more gravity to attract other planetesimals. Some probably grew to hundreds of kilometers in size in only a few million years. This might sound like a long time, but it is only about 1/1,000 of the age of the solar system. However, once the planetesimals reached these relatively large sizes, further growth became more difficult. Gravitational encounters between planetesimals tended to alter their orbits, particularly those of the smaller planetesimals [Section 6.6]. With different orbits crossing each other, collisions between planetesimals tended to occur at higher velocities and hence were more destructive. Collisions started to produce *fragmentation* more often than accretion. Only the largest planetesimals avoided such shattering and continued to grow into full-fledged planets.

The sizes and compositions of the planetesimals depended on the temperature of the surrounding solar nebula. In the inner solar system, where only rocky and metallic flakes condensed, planetesimals were made of rock and metal. That is why the terrestrial planets ended up being composed of rocks and metals. Moreover, because rocky and metallic elements made up only about 0.6% of the material in the solar nebula, the planetesimals in the inner solar

FIGURE 8.10 Gravity is not strong enough to alter the irregular shapes of small objects. A larger object with the same shape will eventually be compressed to a sphere by the greater strength of gravity.

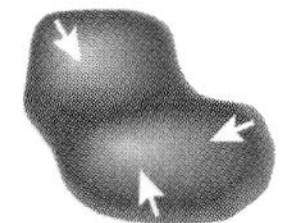

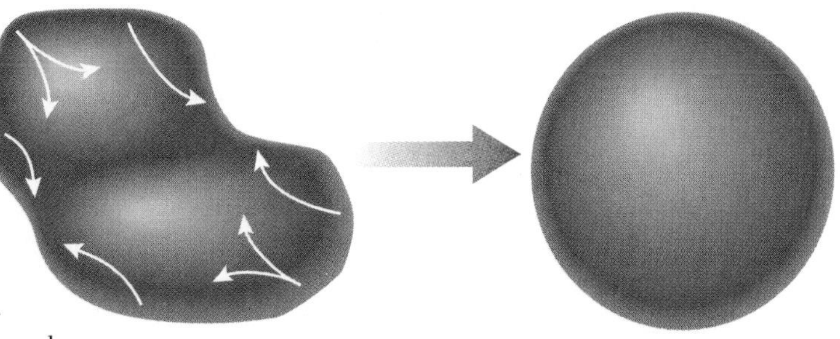

system could not grow very large, which explains why the terrestrial planets are relatively small.

Beyond the frost line between the orbits of Mars and Jupiter, where temperatures were cold enough for ices to condense, planetesimals were built from ice flakes in addition to flakes of rock and metal. Because ice flakes were much more abundant, these planetesimals were made mostly of ices and could grow to much larger sizes than could planetesimals in the inner solar system. The largest icy planetesimals of the outer solar system became the cores of the jovian planets.

If condensation and accretion really took place as we've discussed, the composition of solid objects in the solar system (i.e., not including the jovian planets) should gradually change with increasing distance from the Sun. Objects closest to the Sun should be rich in metals with some rocks, and the most distant objects should be rich in ice with a smaller proportion of rocks and metals.

We can test this prediction by examining planetary densities. However, we need to consider one subtlety when we compare the densities of the planets: The high pressure in planetary interiors crushes materials to a higher density than they would have on the surface. The stronger the gravity of the planet, the more the density is affected. Scientists take into account the effects of gravity by converting the actual average densities of the planets into **uncompressed densities**—the densities the materials would have on the surface. Mercury's uncompressed density is about 5.3 g/cm^3, only slightly lower than its actual average density (see Table 8.1). This value is intermediate between the densities of metal (7 g/cm^3) and rock (3 g/cm^3), suggesting that Mercury has a high metal content. (As we will discuss shortly, Mercury's metal content may have been further enhanced by a giant impact.) Venus and Earth have uncompressed densities of about 4.0 g/cm^3, suggesting proportionally less metal and more rock than Mercury.[6] Mars has an uncompressed density of about 3.7 g/cm^3, suggesting even less metal and more rock. Farther out, asteroids appear to include carbon-rich materials and water-rich minerals along with rocks and metals. And in the outer solar system, solid bodies such as moons and comets have low densities of 1–2 g/cm^3, suggesting a composition mostly of ices. These composition trends provide strong evidence that condensation and accretion really occurred in the manner we've discussed.[7]

[6]Gravity compresses Venus and Earth so that their actual densities are 30% higher than their uncompressed densities. If we could turn off gravity, Venus and Earth would grow by 30% in volume.

[7]We cannot include the densities of the jovian planets in this comparison because they are not solid, and hydrogen and helium gas have no natural "uncompressed density."

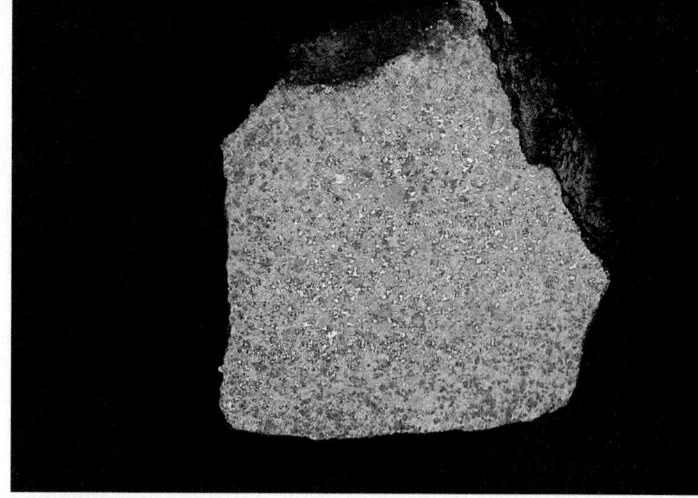

FIGURE 8.11 Shiny flakes of metal are clearly visible in this meteorite, mixed in among the rocky material. Such metallic flakes are just what we would expect to find if condensation really occurred in the solar nebula as described by the nebular theory.

Further evidence of accretion comes from study of *meteorites*—pieces of our solar system that have fallen to Earth. Many meteorites contain metallic grains embedded in a variety of rocky minerals (Figure 8.11). Meteorites thought to come from greater distances contain abundant carbon-rich materials, and some contain water [Section 12.3].

Nebular Capture: Making the Jovian Planets

Accretion proceeded rapidly in the outer solar system, since the presence of ices meant much more material in solid form. Some of the icy planetesimals of the outer solar system quickly grew to sizes many times larger than the Earth. At these large sizes, their gravity was strong enough to capture the far more abundant hydrogen and helium gas from the surrounding nebula. As they accumulated substantial amounts of gas, the gravity of these growing planets grew larger still—allowing them to capture even more gas. The process by which icy planetesimals act as seeds for capturing far larger amounts of hydrogen and helium gas, called **nebular capture,** led directly to the formation of the jovian worlds (Figure 8.12). It explains their huge sizes and the large abundance of hydrogen and helium reflected in their low average densities.

Nebular capture also explains the formation of the diverse satellite systems of the jovian planets. As the early jovian planets captured large amounts of gas from the solar nebula, the same processes that formed the protoplanetary disk—heating, spinning, and flattening—formed similar but smaller disks of material around these planets. Condensation (of metals, rocks, and a lot of ice) and accretion took place within these **jovian nebulae,** essentially creating a miniature solar system around each jovian planet. The spinning disks of the jovian nebulae

FIGURE 8.12 Large icy planetesimals in the cold outer regions of the solar nebula captured significant amounts of hydrogen and helium gas. This process of nebular capture created jovian nebulae, resembling the solar nebula in miniature, in which the jovian planets and satellites formed. This painting shows a jovian nebula (enlarged at left) located within the entire solar nebula.

explain why most of the jovian planet satellites orbit in nearly circular paths lying close to the equatorial plane of their parent planet and also why they orbit in the same direction in which their planet rotates. The composition of the jovian nebulae explains why the jovian planets possess systems of large, icy satellites.[8] Their densities of 1–3 g/cm^3 reflect their mixtures of icy and rocky condensates. Temperature differences within the jovian nebulae may have led to density differences between satellites analogous to the density differences between planets.

Our theory of condensation, accretion, and nebular capture meets the second challenge for our solar system formation theory: It explains the general differences between terrestrial and jovian planets. (Our theory can also explain the presence of planetary rings, but we postpone this interesting discussion until Chapter 11.) However, our theory does not yet explain why the spacing between the planets increases with distance from the Sun, although we can come up with some reasonable hypotheses. For example, the rapid accretion in the outer solar system may have allowed a few protoplanets to gobble up all their neighbors, leaving the jovian planets more widely spaced.

[8]Neptune lacks large, icy satellites in regular orbits, probably because the nebula was too thin to produce such satellites at Neptune's distance from the Sun. Another possible explanation involves Neptune's moon Triton [Section 11.5].

The Solar Wind: Clearing Away the Nebula

What happened to the remaining gas of the solar nebula? Apparently, this gas was blown into interstellar space by the **solar wind**, a flow of low-density plasma [Section 5.3] ejected by the Sun in all directions. Although the solar wind is fairly weak today, we have evidence that it was much stronger when the Sun was young—strong enough to have swept huge quantities of gas out of our solar system. The clearing of the gas interrupted the cooling process in the nebula. Had the cooling continued longer, ices might have condensed in the inner solar system. Instead, the solar wind swept the still-vaporized hydrogen compounds away from the inner solar system, along with the remaining hydrogen and helium throughout the solar system. When the gas cleared, the compositions of objects in the early solar system were essentially set.

We've seen that the nebular theory accounts quite well for the motions of the planets and moons and the compositional trends in the solar system. However, the nebular theory also makes one prediction that at first seems to contradict our observations:

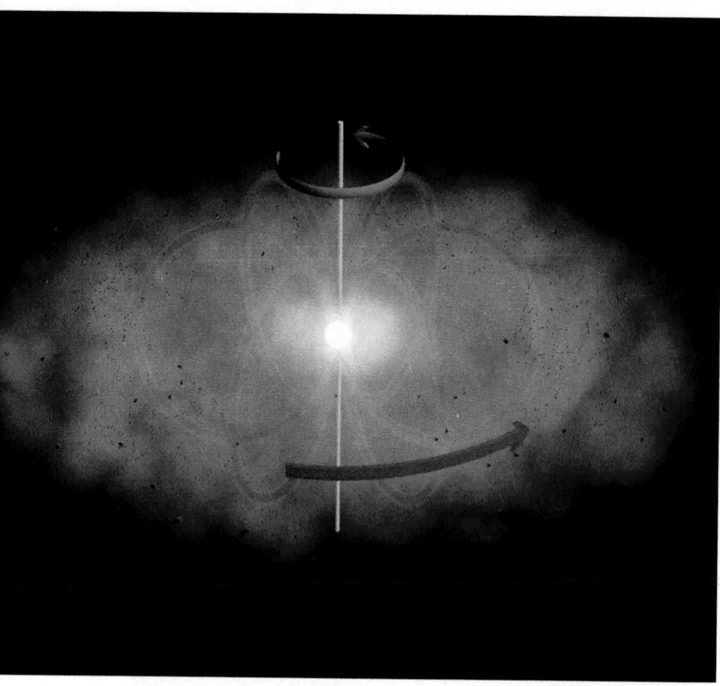

FIGURE 8.13 The magnetic braking process: Charged particles in the solar nebula tend to move with the Sun's magnetic field (represented by the purple loops). As the magnetic field rotates with the Sun, these charged particles are dragged through the disk. Friction between the charged particles and the rest of the disk slows the Sun's rotation. The Sun's relatively slow rotation probably resulted from this process. (Particle sizes are highly exaggerated.)

Conservation of angular momentum in the collapsing solar nebula means that the young Sun should have been rotating very fast, but the Sun actually rotates quite slowly, with each full rotation taking about a month.

Fortunately, this apparent contradiction between theory and observation has a simple resolution. Angular momentum cannot simply disappear, but it is possible to transfer angular momentum from one object to another—and then get rid of the second object. A spinning skater can slow her spin by grabbing her partner and then pushing him away. Beginning in the 1950s, scientists realized that the young Sun's rapid rotation would have generated a magnetic field far stronger than that of the Sun today. It made the Sun highly *active* (e.g., more and larger sunspots on its surface), a circumstance under which the Sun emits larger amounts of ultraviolet and X-ray light [Section 14.5]. This high-energy radiation ionized some of the remaining gas in the solar nebula, creating many charged particles.

As we will discuss in more detail in Chapter 14, charged particles and magnetic fields tend to stick together. As the sun rotated, its magnetic field dragged the charged particles along and added to their angular momentum. Because the particles were gaining angular momentum, the Sun was losing it. The strong solar wind then blew these particles into interstellar space, leaving the Sun with greatly diminished angular momentum and hence the much slower rotation that we see today. The process by which the Sun's magnetic field helped slow its rotation is called **magnetic braking,** because it effectively "applied the brakes" to the Sun's rapid rotation (Figure 8.13).

Although we cannot look into the past to see if magnetic braking really did slow the Sun's rotation, we can look for evidence of the magnetic braking process in other star systems. When we look at young stars that have recently formed in interstellar clouds, we find that nearly all of them rotate rapidly and have strong magnetic fields and strong stellar winds [Section 16.2]. In contrast, older stars almost invariably rotate slowly, just like our Sun. This fact suggests that nearly all stars have their original rapid rotations slowed by magnetic braking, just as we should expect from the nebular theory.

8.5 Leftover Planetesimals

We are now ready to turn to our third challenge: explaining the existence of asteroids and comets. The strong wind from the young Sun cleared excess gas from the solar nebula, but many planetesimals remained scattered between the newly formed planets. These "leftovers" became comets and asteroids. Like the planetesimals that formed the planets, they formed from condensation and accretion in the solar nebula. Thus, their compositions followed the pattern determined by condensation: planetesimals of rock and metal in the inner solar system, and icy planetesimals in the outer solar system.

Asteroids and comets originally must have had nearly circular orbits in the same plane as the orbits of the planets. But gravitational encounters with the newly formed planets soon made the orbits of the leftover planetesimals more random. Recall that two objects can exchange orbital energy when they pass near each other in a gravitational encounter [Section 6.6]. When two planetesimals of comparable size passed near each other, this exchange of energy altered both of their orbits. When small planetesimals passed near a large planet, the planet was hardly affected, but the planetesimals were flung off at high speed in random directions. The strong gravity of the jovian planets tugged and nudged the orbits of planetesimals even at great distances. The result was that the remaining planetesimals ended up with more highly elliptical orbits than the planets, and sometimes with large tilts relative to the plane of planetary orbits. Most of these asteroids and comets eventually either crashed into one of the planets or were flung out of the solar system, but many others still survive today.

Asteroids are the rocky, leftover planetesimals of the inner solar system. Some asteroids are scattered throughout the inner solar system, but most are concentrated in the "extra-wide" gap between Mars and Jupiter that contains the *asteroid belt.* This region probably once contained enough rocky planetesimals to form another terrestrial planet. However, the gravity of Jupiter (the largest and closest jovian planet) tended to nudge the orbits of these planetesimals, sending most of them on collision courses with the planets or with one another. The present-day asteroid belt contains the remaining planetesimals from this "frustrated planet formation." Although thousands of asteroids remain in the asteroid belt, their *combined* mass is less than 1/1,000 of Earth's mass. Jupiter's gravity continues to nudge these asteroids, changing their orbits and sometimes leading to violent, shattering collisions. Debris from these collisions often crashes to Earth in the form of meteorites.

Comets are the icy, leftover planetesimals of the outer solar system. Depending on where they formed, most comets ended up in one of two main groups. The icy planetesimals that cruised the space between Jupiter and Neptune couldn't grow to more than a few kilometers in size before suffering either a collision or a gravitational encounter with one of the jovian planets. Those that escaped being swallowed up by the jovian planets tended to be flung off at high speeds in random directions. Some may have been cast away at such high speed that they completely escaped the solar system and now drift through interstellar space. But billions of these small, icy planetesimals ended up in orbits with very large average distances from the Sun. The gravitational influences of neighboring stars continue to randomize the orbital directions and tilts relative to the plane of planetary orbits. These are the comets of the spherical *Oort cloud* (see Figure 8.3).

Beyond the orbit of Neptune, the icy planetesimals were much less likely to be destroyed by collisions or cast off by gravitational encounters. Instead, they remained in orbits going in the same direction as planetary orbits and concentrated near the plane of planetary orbits (but with somewhat more randomness than the orbits of the planets). They were also able to continue their accretion, and many may have grown to hundreds or even thousands of kilometers in diameter. These are the comets of the *Kuiper belt* (see Figure 8.3); Pluto is probably the largest member of this class.[9]

[9]Important discoveries often lead to confusion in nomenclature. The Kuiper belt is sometimes called the Kuiper disk, and its residents are sometimes called comets, Kuiper belt objects (KBOs), Kuiperoids, iceteroids, or Plutinos. To astronomers, they are technically just distant asteroids, even though they are icy instead of rocky.

Evidence that asteroids and comets really are leftover planetesimals comes from analysis of meteorites, spacecraft visits to comets and asteroids, and computer simulations of solar system formation. Note that the nebular theory, together with the gravitational encounters we've discussed, predicts the existence of comets in both the Oort cloud and the Kuiper belt—a prediction first made in the 1950s. Not until the early 1990s did astronomers verify the existence of objects orbiting within the Kuiper belt. The nebular theory has thus met our third challenge: explaining the existence of asteroids and comets. Moreover, it has suggested an explanation for the seemingly anomalous planet Pluto. Now we turn to our fourth and final challenge: explaining the exceptions to the general trends in our solar system.

8.6 The Early Bombardment: A Rain of Rock and Ice

The collision of a leftover planetesimal with a planet is called an **impact,** and the responsible planetesimal is called the **impactor.** On planets with solid surfaces, impacts leave the scars we call **impact craters.** Impacts were extremely common in the young solar system; in fact, impacts were part of the accretion process in the late stages of planetary formation. The vast majority of impacts occurred in the first few hundred million years of our solar system's history.

The heavy bombardment of planetary surfaces in the early solar system would have resembled a rain of rock and ice from space (Figure 8.14). The surface of the Earth was once scarred like the Moon's surface, but most impact craters on Earth were erased long ago by erosion and other geological processes. Only the outlines of a few large craters are recognizable on Earth. But vast numbers of craters remain on worlds that experience less erosion or other geological activity, such as the Moon and Mercury. Indeed, one way of estimating the age of a planetary surface (the time since the surface last changed in a substantial way) is to count the number of craters: If there are many craters, the surface must still look much as it did at the time of the early bombardment, about 4 billion years ago.

Although most impacts occurred long ago, asteroids and comets still occasionally crash into planets. On Earth, millions of particles the size of sand grains burn up in the atmosphere every day as *meteors.* Every month or so, an impactor a few meters across explodes high in the upper atmosphere, releasing as much energy as an atomic bomb. An impactor some 30 meters across is expected to strike the Earth every century or so, and an impactor about 10 kilometers in diameter appears to have been responsible for the

FIGURE 8.14 Around 4 billion years ago, Earth, its Moon, and the other planets were heavily bombarded by leftover planetesimals. This painting shows the young Earth and Moon glowing with the heat of accretion, and with an impact in progress on the Earth.

extinction of the dinosaurs and many other species about 65 million years ago [Section 12.6]. Today, we know of a few hundred **Earth-approaching asteroids** that may still crash into the Earth or Moon, and more are discovered every month.

The early rain of rock and ice did more than just scar planetary surfaces. It also brought the materials from which atmospheres, oceans, and polar caps eventually formed. Remember that the terrestrial planets were built from planetesimals of metal and rock. But Earth's oceans, the polar caps of Earth and Mars, and the atmospheres of Venus, Earth, and Mars are all made from hydrogen compounds that remained gaseous in the inner solar nebula. These materials must have arrived on the terrestrial planets after their initial formation, most likely brought by impacts of planetesimals formed farther out in the solar system. We don't yet know whether the impactors came from the asteroid belt, where rocky planetesimals contained small amounts of water and other hydrogen compounds that had condensed as ices, or whether they were comets containing huge amounts of ice. Either way, the water we drink and the air we breathe probably once were part of planetesimals floating beyond the orbit of Mars.

TIME OUT TO THINK *What was Jupiter's role in bringing water (and other materials that could not condense in the inner solar nebula) to Earth? How might Earth be different if Jupiter had never formed?*

Captured Moons

We can easily explain the orbits of most jovian planet satellites by their formation in a jovian nebula that swirled around the forming planet. But some moons have unusual orbits—orbits in the "wrong" direction (opposite the rotation of their planet) or with large inclinations to the planet's equator. These unusual moons are probably leftover planetesimals that were *captured* into orbit around a planet.

It's not easy for a planet to capture a moon. An object cannot switch from an unbound orbit (e.g., an asteroid whizzing by Jupiter) to a bound orbit (e.g., a moon orbiting Jupiter) unless it somehow loses orbital energy [Section 6.6]. Captures probably occurred when the capturing planet had a very extended atmosphere or, in the case of the jovian planets, its own miniature solar nebula. Passing planetesimals could be slowed by friction with the gas, just as artificial satellites are slowed by drag with the Earth's atmosphere. If friction reduced a planetesimal's orbital energy enough, it could have become an orbiting moon. Because of the random nature of the capture process, the captured moons would not necessarily orbit in the same direction as their planet or in its equatorial plane.

The two small moons of Mars—Phobos and Deimos—probably were asteroids captured by this process (Figure 8.15). They resemble asteroids seen in the asteroid belt and are much darker and lower in density than Mars. Jupiter probably also captured several of its moons. Two groups of four satellites circle Jupiter in unusual orbits: All are noticeably eccentric and tilted, and one of the two groups orbits Jupiter in the "wrong" direction. Astronomers speculate that two asteroids may have been captured by Jupiter and that each broke into several pieces in the process. Other jovian planets may also have captured moons, including one that is particularly large—Triton, the largest moon of Neptune. Triton is considerably larger than Pluto and orbits Neptune in a direction opposite to Neptune's rotation. It may be a captured "Plutonian planet" from the Kuiper belt [Section 11.5].

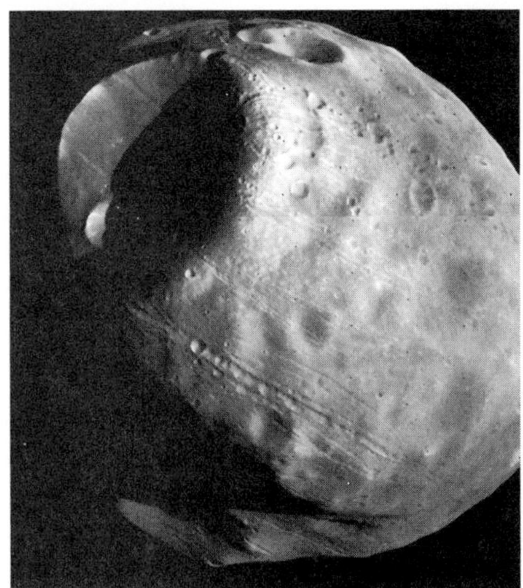

a Phobos

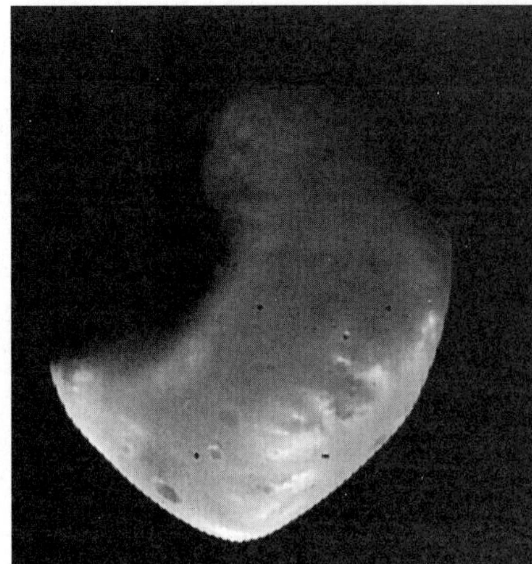

b Deimos

FIGURE 8.15 The two moons of Mars, shown here in photos taken by the Viking spacecraft, are probably captured asteroids. Phobos is only about 13 km across and Deimos is only about 8 km across—making each of these two moons small enough to fit within the boundaries of a typical large city.

The one unusual moon that cannot be explained by this process is our own: Our Moon is much too large to have been captured by Earth.

Giant Impacts and the Formation of Our Moon

The largest planetesimals remaining as the planets formed may have been huge—some may have been the size of Mars. When one of these planet-size planetesimals collided with a planet, the spectacle would have been awesome. Such a **giant impact** could have significantly altered the planet's fate.

What if an object the size of Mars had collided with Earth? Computer simulations address this interesting question. The Earth would have shattered from the impact, and material from the outer layers would have ended up in orbit around the Earth (Figure 8.16). There this material could have reaccreted to form a large satellite. Depending on exactly where and how fast the giant impactor struck the Earth, the blow might also have tilted the Earth's axis, changed its rotation rate, or completely torn it apart. Today, such a giant impact is the leading hypothesis for explaining the origin of our Moon. The Moon's composition is similar to that of the Earth's outer layers, exactly what we would predict for the kind of collision shown in Figure 8.16. The Moon is also depleted in easily vaporized ingredients, as would be expected from the tremendous heating that would have occurred during the collision. Moreover, we can rule out the idea of the Moon forming simultaneously with Earth. In that case, Earth and the Moon would have formed from the same material and therefore should have the same density, but the Moon's density is considerably lower than the Earth's.

In fact, many of the unusual properties of specific planets—properties that defy the general trends expected by the nebular theory—may be the results of giant impacts. Mercury may have lost much of its outer, rocky layer in a giant impact, leaving it with a huge metallic core. A giant impact might even have

FIGURE 8.16 Artist's conception of the impact of a Mars-size object with Earth, as may have occurred soon after Earth's formation. The ejected material comes mostly from the outer rocky layers and accretes to form the Moon, which is poor in metal.

contributed to the slow, backward rotation of Venus, which may have had a "normal" rotation until the arrival of the giant impactor. Giant impacts probably were also responsible for the axis tilts of many planets (including Earth) and for tipping Uranus on its side. Pluto's moon Charon may have formed in a giant-impact process similar to the one that formed our Moon.

Unfortunately, we can do little to test whether a particular giant impact really occurred billions of years ago. Even if we had telescopes powerful enough to witness similar events in other solar systems, they would be so rare that it would take great luck to observe one. The difficulty in proving the giant-impact hypothesis makes the idea controversial, and most planetary scientists didn't take such ideas seriously when they were first proposed. But no other idea so effectively explains the formation of our Moon and other "oddities" that we've discussed. Moreover, giant impacts certainly should have occurred, given the number of large, leftover planetesimals predicted by the nebular theory. Random giant impacts are the most promising explanation for the many exceptional circumstances noted in the fourth challenge.

Summary: Meeting the Challenges

The nebular theory explains the great majority of important facts contained in our four challenges. But you should not be left with the impression that competing theories were never put forth or that solar system formation is now a "solved problem." Theories have evolved hand-in-hand with the discovery of the nature of our solar system, and what we have presented here is the culmination of that great endeavor to date. Planetary scientists are still struggling with more quantitative aspects, seeking reasons for the exact sizes, locations, and compositions of the planets. Perhaps as we discover other planetary systems and examine their properties, we will be able to improve our understanding of solar system formation.

Assuming that the nebular theory is correct, it is interesting to ask whether our solar nebula was "destined" to form the solar system we see today. The first stages of planet formation were orderly and inevitable according to the nebular theory. Nebular collapse, condensation, and the first stages of accretion were relatively gradual processes that probably would happen all over again if we turned back the clock. But the final stages of accretion, and giant impacts in particular, are inherently random in nature and probably would not happen again in just the same way. A larger or smaller planet might form at Earth's location and might suffer from a larger giant impact or from none at all. We don't yet know whether these differences would fundamentally alter the solar system or simply change a few "minor" details—such as the possibility of life on Earth.

Mathematical Insight 8.1 Radioactive Decay

The amount of a radioactive substance decays by a factor of 2 with each half-life, so radioactivity is an example of *exponential decay.* We can express this decay process with a simple formula relating the current amount of a radioactive substance in a rock to the original amount:

$$\frac{\text{current amount}}{\text{original amount}} = \left(\frac{1}{2}\right)^{t/T_{\text{half}}}$$

where t is the time since the rock formed and T_{half} is the half-life of the radioactive material. This equation is graphed in Figure 8.17 for the decay of potassium-40 into argon-40. Note that the steady decline in the amount of potassium-40 with time is matched by a steady rise in the amount of argon-40. If you examine the graph over any 1.3-billion-year period (the half-life of potassium-40), you'll see that the amount of potassium declines by half.

Example: The famous Allende meteorite lit up the skies of Mexico as it shattered during its fall to Earth on February 8, 1969. Scientists collected pieces of the meteorite for study. To determine the age of the meteorite, they heated and chemically analyzed a small bit of meteorite. They found both potassium-40 and argon-40 present in a ratio of approximately 0.85 unit of potassium-40 atoms to 9.15 units of gaseous argon-40 atoms. (The units are unimportant, since only the relative amounts of the parent and daughter materials matter.) How old is the Allende meteorite?

8.7 The Age of the Solar System

The nebular theory accounts for the major physical properties of our solar system, supporting the idea that all the planets formed at about the same time from the same cloud of gas. But *when* did it all happen, and how do we know? The answer is that the solar system began forming about 4.6 billion years ago, a fact we learned by determining the age of the oldest rocks in the solar system. The solar system is therefore less than half as old as the universe: It is a middle-aged solar system in an old universe.

The concept of a rock's *age* is tricky: The rock's atoms were forged in stars and are therefore much older than the Earth. Atoms are not stamped with any date of manufacture, and old atoms are truly indistinguishable from young ones. By the age of a rock, we mean the time since those atoms became locked together, that is, the time *since the rock last solidified.*

Radioactive Dating

Most rocks contain minute amounts of **radioactive elements**—atoms whose nuclei have a tendency to break apart. (The term *radioactive* has nothing to do with radio waves; both terms come from the concept of particles or waves radiating away from a source.) A radioactive element starts with a certain number of protons and neutrons in its nucleus. When that nucleus breaks apart, or **decays,** it ejects some subatomic particles and leaves behind a **decay product**—a different element or isotope with a different number of protons and/or neutrons [Section 5.3]. Over time, the amount of the original radioactive material (the *parent*) in the rock decreases, and the amount of the decay product (the *daughter*) increases. Most nuclei are stable, but certain combinations of neutrons and protons in a nucleus are unstable and prone to break apart sooner or later. One kind of radioactive decay is called nuclear *fission.* The ejected particles are very energetic; this is the energy that drives nuclear power plants and that generates much of the Earth's internal heat.

The rate at which a particular radioactive substance decays is characterized by its **half-life**—the time it takes for half of the parent nuclei to decay. Each isotope of each element has its own half-life, ranging from a fraction of a second to billions of years. For example, potassium-40 (i.e., potassium with atomic weight 40) has a half-life of about 1.3 billion years. If a rock begins with a certain amount of potassium-40, half this amount will remain after 1.3 billion years, one-fourth the original amount will remain after 2.6 billion years, one-eighth the original amount will remain after 3.9 billion years, and so on (Figure 8.17). As the potassium-40 decays, it leaves behind its decay product, argon-40. Thus, by comparing the amounts of potassium-40 and argon-40 in

Solution: Since no argon gas should have been present in the meteorite when it formed, the 9.15 units of argon must originally have been potassium. Thus, the sample started with 0.85 + 9.15 = 10 units of potassium-40, of which 0.85 unit remains. Thus, the *current amount* of potassium-40 is 0.85 unit, and the *original amount* is 10 units. With these values, the radioactive decay equation becomes:

$$\left(\frac{1}{2}\right)^{t/T_{\rm half}} = \frac{\text{current amount}}{\text{original amount}} = 0.085$$

You can get a fairly accurate estimate of the age t by reading the graph in Figure 8.17 or by testing some guesses in the equation above using your calculator. The most accurate method is to solve the equation for the age t by using logarithms:

$$t = T_{\rm half} \times \frac{\log_{10}\left(\dfrac{\text{current amount}}{\text{original amount}}\right)}{\log_{10}\left(\dfrac{1}{2}\right)}$$

$$= 1.3 \text{ billion yr} \times \frac{\log_{10}(0.085)}{\log_{10}\left(\dfrac{1}{2}\right)} = 4.6 \text{ billion yr}$$

Thus, from the ratio of potassium-40 to argon-40, we conclude that the Allende meteorite solidified from the solar nebula about 4.6 billion years ago.

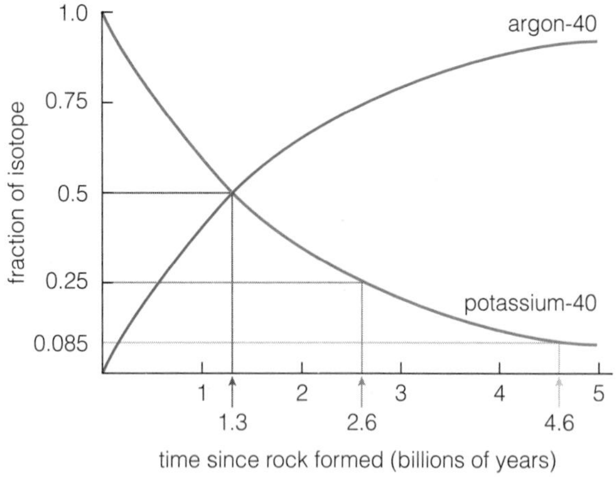

FIGURE 8.17 Radioactive decay of potassium-40 to argon-40. The red line shows the decreasing amount of potassium-40, and the blue line shows the increasing amount of argon-40. The half-life for this reaction is 1.3 billion years. At 2.6 billion years, only a quarter of the original potassium-40 remains. After 4.6 billion years, only 0.085 of the original amount remains.

the rock, we can determine its age. This process of determining the age of an object by studying the results of radioactive decay is called **radioactive dating.**

Radioactive dating is simple in principle, but in practice it requires careful laboratory work and a good understanding of "rock chemistry." For example, suppose you find a rock that contains equal numbers of atoms of potassium-40 and argon-40. If you assume that all the argon came from potassium decay, then the rock must be 1.3 billion years old (because half the original potassium atoms must have decayed to yield equal numbers of potassium and argon atoms). But this age is correct only if the assumption is valid; that is, the rock must not have contained any argon-40 when it formed. In this case, knowing a bit of chemistry helps. Potassium-40 is a natural ingredient of many minerals in rocks, but argon-40 is a gas that never combines with other elements and did not condense in the solar nebula. Therefore, if you find argon-40 gas trapped inside minerals, you can be sure that it came from the radioactive decay of potassium-40. Another assumption is that none of the argon-40 gas has escaped from the rock. But as long as the rock has not been significantly heated, even atoms of gases will remain trapped within it.

As another example, consider rocks from the ancient lunar highlands [Section 9.5]. These rocks contain minerals with a very small amount of uranium-238, which decays (in several steps) to lead-206 with a half-life of about 4.5 billion years. Lead and uranium have very different chemical behaviors, and some minerals start with virtually no lead. Laboratory analysis of such minerals in lunar rocks shows that they now contain almost equal proportions of uranium-238 and lead-206. We conclude that half the original uranium-238 has decayed, turning into the same number of lead-206 atoms. The lunar rock must be about one half-life old, or almost 4.5 billion years. (More precisely, the oldest lunar rocks are about 4.4 billion years old.)

TIME OUT TO THINK *If future scientists examine the lunar rocks 4.5 billion years from now, what proportions of uranium-238 and lead-206 will they find?*

Radioactive dating is possible with many other materials having a variety of different half-lives. In some cases it can be very difficult, particularly if some of the daughter material may have been present when the rock first formed. In that case, sophisticated chemistry may be needed to determine how much material is original and how much is the product of decay. Fortunately, rocks often contain several different radioactive materials, so scientists can date a rock by analyzing several different parent–daughter ratios. If the ages found from analyzing the different materials agree, we can be very confident that we know the true age of the rock.

Earth Rocks, Moon Rocks, and Meteorites

How on Earth can we measure the age of the solar system? In fact, nothing on Earth's surface remains from the formation era. Geological activity, such as plate tectonics and volcanoes, has ensured that virtually all the rocks present on the early Earth have since melted and resolidified. Because dating tells us only how long it has been since a rock last solidified, we cannot determine the age of our planet from its rocks, let alone the age of the solar system. Nevertheless, Earth rocks tell us that the solar system is old: The oldest rocks on the Earth date back about 4 billion years, although most of the Earth is covered in rocks only hundreds of millions of years old.

TIME OUT TO THINK *Suppose you do radioactive dating on a chunk of lava recently spewed out of Kilauea, an active volcano on the island of Hawaii. How old would it be? Explain.*

Determining the age of the solar system requires finding rocks from beyond the Earth—rocks that might not have melted or vaporized since the birth of the solar system. But how do we get rocks from space? Astronauts brought back numerous rocks from the

Moon, and the oldest of these date to about 4.4 billion years ago. But this still is not the age of the solar system, because the Moon's surface melted and resolidified as a result of the early bombardment. Our other main source of rocks from space is meteorites that have fallen to Earth.

Many meteorites appear to be unchanged since their formation and are therefore samples of the condensation and accretion processes in the early solar system. Careful analysis of radioactive elements in meteorites shows that the oldest ones formed about 4.6 billion years ago, marking the beginning of accretion from the solar nebula.[10] Calculations show that accretion probably lasted only about 0.1 billion years (100 million years), so Earth and the other planets formed about 4.5 billion years ago.

A Trigger for the Collapse?

The very existence of radioactive elements in meteorites is profoundly important, quite apart from the age information those elements provide. Radioactive elements are made only deep inside stars or in violent stellar explosions (supernovae). Thus, the presence of these elements in meteorites and on Earth underscores the fact that our solar system is made from the remnants of past generations of stars.

The discovery of daughter elements of short-lived radioactive elements has further implications. The rare isotope xenon-129 is found in some meteorites. Xenon is gaseous even at extremely low temperatures, so it could not have condensed and become trapped in planetesimals forming in the solar nebula. Thus, any xenon present in meteorites must be a product of radioactive decay. In fact, xenon-129 is a decay product of iodine-129. But iodine-129 has a half-life of just 17 million years, so this iodine must have traveled from the star in which it was produced to our solar nebula fairly quickly. (If the supernova had occurred a billion years earlier, the iodine would have decayed entirely into xenon, which would not have condensed or accreted.) This suggests that a nearby star exploded only a few million years (or less) before our solar system formed. The shock wave emanating from the exploding star may even have triggered the collapse of the solar nebula by giving gravity a little extra push. Once gravity got started, the rest of the collapse was inevitable.

---

[10]Museums and specialty catalogs sell meteorite samples. For a few dollars, you can buy a small, solidified piece of the solar nebula. It will be the oldest object you'll ever touch.

8.8 Planetary Systems Beyond Our Solar System

How common are planetary systems? What are the odds of another Earth-like planet? Are habitable planets as common in the real universe as they are in science fiction? We cannot definitively answer these questions—yet. But our observational and theoretical tools are helping us guess at possible answers.

Our theory of solar system formation suggests that other forming stars should also be surrounded by spinning, flattened disks in which planets might form, and numerous protoplanetary disks have been observed around stars. However, about half the "stars" in our sky are actually binary or multiple star systems. No one knows how planets might form in multiple star systems, or if their orbits are ever stable. However, as this book goes to press, astronomers have observed what appears to be a protoplanet ejected from a binary star system (Figure 8.18a).

We also need to think about what the term *planet* really means. Does a newly discovered object around another star have to be like one of our solar system's planets to qualify? Where do we draw the line between large planets and small stars? For the time being, let's be generous with our definition: Detect first and classify later.

At the beginning of the 1990s, we had no conclusive proof that planets existed around any star besides our Sun. By 1998, however, astronomers had detected about a dozen planetlike objects around other stars, providing the first definitive answer to the age-old question of whether other solar systems exist.

There are two basic ways to detect "extrasolar planets" (planets beyond our solar system). The most direct method involves detecting light emitted or reflected by a planet. The difficulty with this technique is that any light coming from a planet is likely to be overwhelmed by light coming from the star it orbits. For example, a Sun-like star would be a *billion times* brighter than the reflected light from an Earth-like planet. Because even the best telescopes blur the light from stars at least a little [Section S2.2], finding the small blip of planetary light amid the glare of scattered starlight would be very difficult. Direct detection should be easiest for large planets orbiting at relatively large distances around their stars. As of 1998, we have yet to directly detect a true planet orbiting a star, but we have found planetlike objects. For example, direct detection revealed an object called Gliese 229B that orbits about 40 AU from the star Gliese 229A (Figure 8.18b). With a mass between 20 and 50 times that of Jupiter, Gliese 229B is too big to be considered a planet but too small to have the nuclear fusion in its core that would qualify it as a star. Astronomers categorize Gliese 229B as a

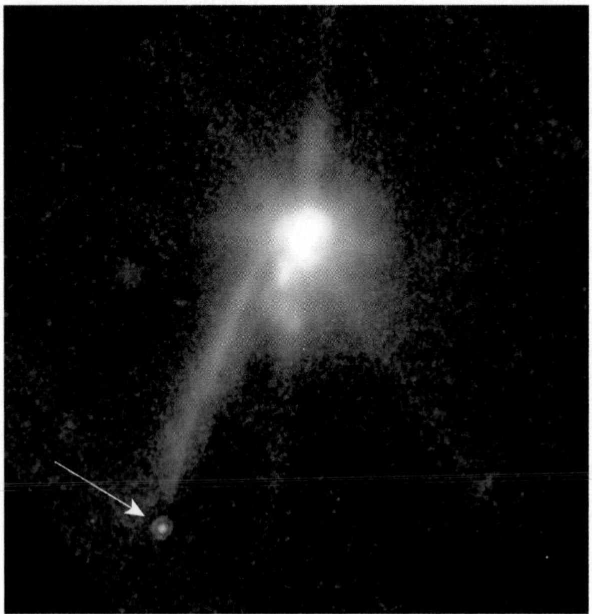

a This infrared image shows a possible protoplanet (indicated by arrow) ejected from a binary star system (bright spot in image), leaving behind a bright, curved trail where the protoplanet plowed through a dust cloud surrounding the stars.

b Gliese 229B (indicated by arrow) is a brown dwarf about 20–50 times more massive than Jupiter. Stray light from the star it orbits (Gliese 229A, out of picture to left) floods the image.

FIGURE 8.18 The Hubble Space Telescope obtained these images showing objects too faint to be stars.

type of "failed star" called a *brown dwarf* [Section 16.2]. Within a couple of decades, high-resolution interferometers [Section S2.5]—particularly ones that can detect the infrared thermal radiation from planets—may allow us to obtain images of true planets in other solar systems.

The second method is indirect: detecting a planet by virtue of the small gravitational "tug" it exerts on its star. This tug should cause the star's position to shift very slightly back and forth against the sky with each orbit of the planet. Such shifts are easiest to detect when they are caused by a large planet far from its star—though not too far, since observers must wait for one or more complete orbits to be sure of what they've seen. This technique has purportedly found two Jupiter-size planets around the nearby star Lalande 21185.[11] In some cases, it is easier to detect the gravitational effect of an orbiting planet by looking for Doppler shifts in a star's spectrum [Section 7.5]. An orbiting planet causes its star to alternately move slightly toward and away from us, which makes the star's spectral lines alternately shift toward the blue and toward the red (Figure 8.19a). This method works best for massive planets very near their stars and has been the most successful method so far. One early discovery was a 0.6-Jupiter-mass planet that tugs on the star 51 Pegasi (Figure 8.19b). This planet lies so close to its star that its year lasts only four of our days, and its temperature is probably over 1,000 K.

Many of the recently discovered planets seem quite different from those of our solar system (Figure 8.20). For example, several of the objects are much more massive than Jupiter yet lie quite close to their stars—a situation very different from that in our solar system, where all the jovian planets are found in the outer solar system. In at least one case, Earth-size objects have been detected around a *neutron star*—the bizarre, compact remains of a star that died in a titanic explosion [Section 17.3]. Such "planets" could not have survived the neutron star–forming explosion, so they probably formed during the explosion aftermath and not from a solar nebula like our own.

Do these unusual planets require us to throw out our theories of solar system formation? Not necessarily, because we may be seeing only the rare exceptions to general rules. Large planets in fast orbits are much easier to find than planets like Earth or even Jupiter. Systems like our own may be quite common

[11]Names of faint stars like Lalande 21185 and Gliese 229 refer to their entries in huge catalogs of star positions and properties. Brighter stars like 51 Pegasi are named for their constellations (in this case, Pegasus).

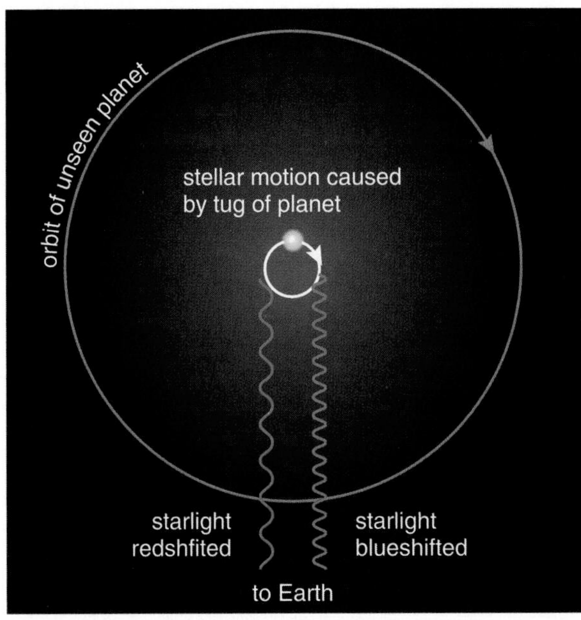

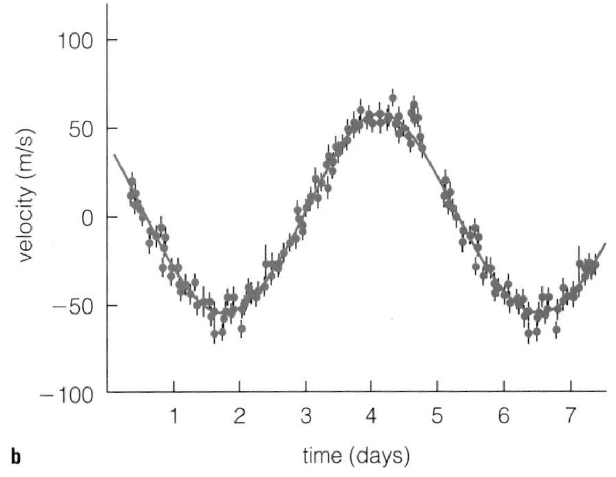

FIGURE 8.19 (**a**) Doppler shifts allow us to detect the slight motion of a star caused by an orbiting planet. (**b**) A periodic Doppler shift in the spectrum of 51 Pegasi shows the presence of a large planet with an orbital period of about 5 days. Dots are actual data points; bars through dots represent measurement uncertainty.

Planets Around Sun-Like Stars

inner solar system	Mercury · Venus · Earth · Mars
HD 187123	$0.52M_{Jup}$
HD 75289	$0.4M_{Jup}$
Tau Boo	$3.64M_{Jup}$
51 Peg	$0.44M_{Jup}$
Ups And	$0.63M_{Jup}$
HD 217107	$1.28M_{Jup}$
Gliese 86	$3.6M_{Jup}$
Rho1 55 Cancri	$0.85M_{Jup}$
HD 195019	$3.43M_{Jup}$
Gliese 876	$2.1M_{Jup}$
Rho Cr B	$1.1M_{Jup}$
HD 168443	$5.04M_{Jup}$
HD 114762	$11M_{Jup}$
70 Vir	$7.4M_{Jup}$
HD 210277	$1.36M_{Jup}$
16 Cyg B	$1.74M_{Jup}$
47 UMa	$2.42M_{Jup}$
14 Her	$4M_{Jup}$

0 1 2 3
orbital semimajor axis (AU)

FIGURE 8.20 Most of the planets found around other stars are closer to their stars and more massive than the planets in our solar system. (Planet sizes are not to scale with their distances shown.)

but hard to detect. Our early discoveries of extrasolar planets are like glimpsing animals in the rain forest: The jungle appears full of brightly colored parrots and frogs, but far more animals are present that don't catch our eye. However, if future surveys show that planets like Jupiter and Earth are rare, planetary scientists will have to go back to the drawing board.

The future looks bright for the expansion of planetary science to other solar systems. New technologies are rapidly improving our ability to find and study extrasolar planets, and we already know of more planets around other stars than around our own Sun. For the latest news in planet discoveries, check the book web site.

The evolution of the world may be compared to a display of fireworks that has just ended: some few red wisps, ashes and smoke. Standing on a cooled cinder, we see the slow fading of the suns, and we try to recall the vanished brilliance of the origin of the worlds.

G. LEMAÎTRE (1894–1966), Astronomer and Catholic priest

THE BIG PICTURE

We've seen that the nebular theory accounts for the major characteristics of our solar system. As you continue your study of the solar system, keep in mind the following "big picture" ideas.

- Close examination of the solar system unveils a wealth of patterns, trends, and groupings, leading us to the conclusion that all the planets formed from the same cloud of gas at about the same time.
- Chance events may have played a large role in determining how individual planets turned out. No one knows how different the solar system might be if we started over.
- Planet-forming processes are apparently universal. The discovery of protoplanetary disks and full-fledged planets around other stars brings planetary science to the brink of an exciting new era.

Review Questions

1. What is *comparative planetology?* What is its basic premise? Does comparative planetology apply only to the nine planets of our solar system? Explain.
2. Briefly summarize the observed patterns of motion in our solar system.
3. Summarize the differences between terrestrial planets and jovian planets. Why is the Moon grouped with the terrestrial planets? Where does Pluto fit in?
4. What are *asteroids?* Where are they found? What are *comets,* and how do those of the *Oort cloud* and *Kuiper belt* differ in terms of their orbits?
5. Summarize the four challenges that any theory of the solar system must explain. List a few of the key exceptions to the rules that must be explained.
6. What is the *nebular theory?* How does it get its name?
7. Briefly describe the cosmic recycling process that led to the formation of the *solar nebula.* What fraction of the material in the solar nebula consisted of elements other than hydrogen and helium?
8. Describe the processes that led the solar nebula to collapse into a spinning disk. Also describe the evidence supporting the idea that our solar system had a *protosun* and a *protoplanetary disk* early in its history.
9. Distinguish between *metals, rocks, hydrogen compounds,* and *light gases* in terms of condensation temperatures and relative abundance. Give a few examples of substances that fall into each category.
10. Explain how temperature differences in the solar nebula led to the condensation of different materials at different distances from the protosun. What do we mean by the *frost line* in the solar nebula?
11. What is *accretion?* What are *planetesimals?* Explain the role of electrostatic forces in getting accretion started and the role of gravity in accelerating accretion as planetesimals grow larger.
12. Briefly explain why accretion in the outer solar system was able to produce much larger planetesimals than in the inner solar system.
13. What do we mean by *uncompressed densities?* How do the uncompressed densities of the planets support the theory of accretion in the solar nebula?
14. Describe the process of *nebular capture* that allowed the jovian planets to grow to very large sizes (compared to terrestrial planets). How does this process explain why jovian planets have extensive satellite systems?

15. What is the *solar wind?* How did its strength in the past differ from its strength today? What role did it play in ending the growth of the planets?
16. Why does the Sun's slow rotation rate seem, at least at first, to contradict the nebular theory? Explain how the process of *magnetic braking* accounts for this apparent contradiction.
17. What evidence supports the idea that asteroids and comets are leftover planetesimals? Briefly describe how gravitational encounters led to the present distribution of asteroids and comets in our solar system.
18. Why were impacts so much more common in the past than today? How can we use this fact to estimate the age of a planetary surface by counting its impact craters? What are *Earth-approaching asteroids?*
19. What are the clues that a planetary satellite was captured instead of forming with the planet? Briefly describe the process by which we believe satellites were captured in the early solar system.
20. Describe the process by which a giant impact may have led to the formation of the Moon. What other oddities of our solar system might be explained by giant impacts?
21. Summarize how the nebular theory meets the four challenges laid out in this chapter. Describe a few of the remaining unanswered questions about the origin of our solar system.
22. Briefly describe the process of *radioactive dating* and how we use it to establish the age of our solar system. Why can't we determine the age of our solar system by dating Earth rocks? What kinds of rocks can we use?
23. How does the existence of radioactive elements in meteorites tell us that our solar system formed from the remains of past generations of stars? What evidence suggests that a nearby stellar explosion may have triggered the collapse of the solar nebula?
24. Why is direct detection of extrasolar planets so difficult? Describe how we can indirectly detect extrasolar planets today, and summarize the current evidence concerning planets in other solar systems.

Discussion Questions

1. *Theory and Observation.* Discuss the interplay between theory and observation that has led to our modern theory of the formation of the solar system. What role does technology play in allowing us to test this theory?
2. *Random Events in Solar System History.* According to our theory of solar system formation, numerous random events, such as giant impacts, had important consequences for the way our solar system turned out. Can you think of other random events that might have caused the planets to form very differently? If there had been a different set of random events, what important properties of our solar system would have turned out the same and what ones might be different? Discuss.
3. *Lucky to Be Here?* In considering the overall process of solar system formation, do you think it was very likely for a planet like Earth to have formed? Could random events in the early history of the solar system have prevented us from even being here today? Defend your opinion. Do your opinions on these questions have any implications for your belief in the possibility of Earth-like planets around other stars?

Problems

1. *Two Classes of Planets.* Explain in terms a friend or roommate would understand why the jovian planets are lower in density than the terrestrial planets even though they all formed from the same cloud.
2. *A Cold Solar Nebula.* Suppose the entire solar nebula had cooled to 50 K before the solar wind cleared it away. How would the composition and sizes of the terrestrial planets be different from what we see today? Explain your answer in a few sentences.
3. *No Nebular Capture.* Suppose the solar wind had cleared away the solar nebula before the process of nebular capture was completed in the outer solar system. How would the jovian planets be different? Would they still have satellites? Explain your answer in a few sentences.
4. *Angular Momentum.* Suppose our solar nebula had begun with much more angular momentum than it did. Do you think planets could still have formed? Why or why not? What if the solar nebula had started with zero angular momentum? Explain your answers in one or two paragraphs.
5. *Jupiter's Action.* Suppose that, for some reason, the planet Jupiter had never formed. How do you think the distribution of asteroids and comets in our solar system would be different? How would these differences have affected Earth? Explain your answer in a few sentences.
6. *Satellite Systems as Mini–Solar Systems.* The nebula surrounding the forming jovian planets was hotter near its center, so satellites forming from the nebula

could be built from different types of condensates, just as the planets were. Would satellite densities increase or decrease with increasing distance from the planet? According to Appendix C, which jovian planet satellite system best matches this prediction? Explain your answers in a few sentences.

7. *A Backward Planet.* Imagine a solar system in which all planets except one orbit the Sun in the same direction. Can the nebular theory explain the origin of this planet? How could you alter the theory of solar system formation to explain the "backward planet"? Explain your answer in a few sentences. (*Hint:* Think of "backward" planetary satellites.)

8. *Dating Lunar Rocks.* Suppose you are analyzing Moon rocks that contain small amounts of uranium-238, which decays into lead with a half-life of 4.5 billion years.
   a. In one rock from the "lunar highlands," you determine that 55% of the original uranium-238 remains; the other 45% has decayed into lead. How old is the rock?
   b. In a rock from the "lunar maria," you find that 63% of its original uranium-238 remains; the other 37% has decayed into lead. Is this rock older or younger than the highlands rock? By how much? (The significance of these ages will become clear in the next chapter.)

9. *Radioactive Dating with Carbon-14.* The half-life of carbon-14 is about 5,700 years.
   a. You find a piece of cloth painted with organic dyes. By analyzing the dye in the cloth, you find that only 77% of the carbon-14 originally in the dye remains. When was the cloth painted?
   b. A well-preserved piece of wood found at an archaeological site has 6.2% of the carbon-14 that it must have had when it was alive. Estimate when the wood was cut.
   c. Is carbon-14 useful for establishing the age of the Earth? Why or why not?

10. *Unusual Meteorites.* Some unusual meteorites thought to be chips from Mars [Section 12.3] contain small amounts of the radioactive element thorium-232 and its decay product lead-208. The half-life for this decay process is 14 billion years. A detailed analysis shows that 94% of the original thorium remains. How old are these meteorites? Compare your answer to the age of the solar system and comment briefly.

11. *51 Pegasi.* The star 51 Pegasi has about the same mass as our Sun, and the planet discovered around it has an orbital period of 4.23 days. The mass of the planet is estimated to be 0.6 times the mass of Jupiter.
    a. Use Kepler's third law to find the planet's average distance (semimajor axis) from its star. (*Hint:* Because the mass of 51 Pegasi is about the same as the mass of our Sun, you can use Kepler's third law in its original form, $p^2 = a^3$ [see Chapter 6]; be sure to convert the period into years.)
    b. Briefly explain why, according to our theory of solar system formation, it is surprising to find a planet the size of the 51 Pegasi planet orbiting at this distance.
    c. Hypothesize as to how the 51 Pegasi planet might have come to exist. Explain your hypothesis in a few sentences.

12. *Web Project: New Planets.* Starting from the book web site, search the web for up-to-date information about discoveries of planets in other solar systems. Create a personal "web journal," complete with pictures from the web, describing at least three recent discoveries of new planets. For each case, write one or two paragraphs in your journal describing the method used in the discovery, comparing the discovered planet to the planets of our own solar system, and summarizing whether the discovery poses any new challenges to our theory of solar system formation.

9 Geology of the Terrestrial Worlds

THINK BACK TO A TIME WHEN YOU WENT FOR A WALK OR A DRIVE through the open countryside. Did you see a valley? If so, the creek or river that carved it probably still flows at the bottom. Were the valley walls composed of lava, telling of an earlier time when molten rock gushed from volcanoes? Or were they made of neatly layered sedimentary rocks that formed when ancient seas covered the area? Were the rock layers tilted? That's a sign that movements of the Earth's crust have pushed the rocks around.

Since the beginning of the space age, we've been able to make similar observations and ask similar questions about geological features on other worlds. As a result, we can now make detailed geological comparisons between the different planets. We've learned that, even though all the terrestrial worlds are similar in composition and formed at about the same time, their geological histories have differed because of a few basic properties of each world. Thus, by comparing the geologies of the terrestrial worlds, we can learn much more about how the Earth works.

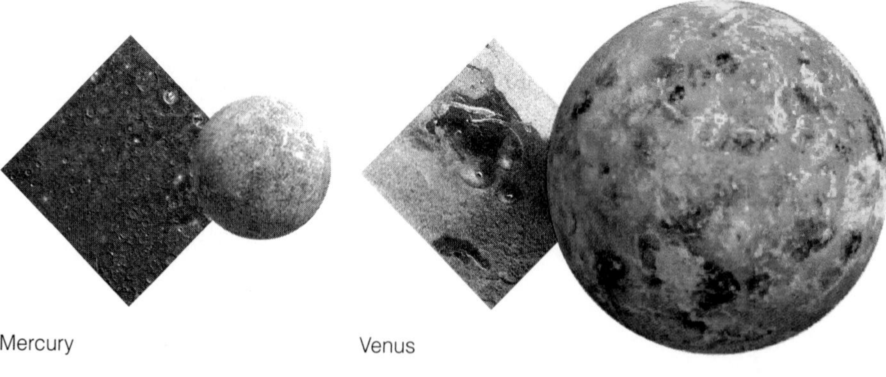

FIGURE 9.1 Global views of the terrestrial planets to scale and representative surface close-ups. The global view of Venus shows its surface without its atmosphere, based on radar data from the *Magellan* spacecraft; all other images are photos (or composite photos) taken from spacecraft.

9.1 Comparative Planetary Geology

The five terrestrial worlds—Mercury, Venus, Earth, the Moon, and Mars—share a common ancestry in their birth from the solar nebula, but their present-day surfaces show vast differences (Figure 9.1). Mercury and the Moon are battered worlds densely covered by craters except in areas that appear to be volcanic plains. The volcanic plains of Venus appear to have been twisted and torn by internal stresses, leaving bizarre bulges and odd volcanoes that dot the surface. Mars, despite its middling size, has the solar system's largest volcanoes and is the only planet other than Earth where running water played a major role in shaping the surface. Earth has surface features similar to all those on other terrestrial worlds and more—including a unique layer of living organisms that covers almost the entire surface of the planet. Our purpose in this chapter is to understand how the profound differences among the terrestrial surfaces came to be.

The study of surface features and the processes that create them is called **geology.** The root *geo* means "Earth," and *geology* originally referred only to the study of the Earth. Today, however, we speak of *planetary geology,* the extension of geology to include all the solid bodies in the solar system, whether rocky or icy.

Geology is relatively easy to study on Earth, where we can examine the surface in great detail. People have scrambled over much of the Earth's surface, identifying rock types and mapping geological features. We've pierced the Earth's surface with 10-kilometer-deep drill shafts and learned about the deeper interior by studying *seismic waves* generated by earthquakes. Thanks to these efforts, today we have a fairly good understanding of the history of our planet's surface and the processes that shaped it.

Studying the geology of other planets is much more challenging. The Moon is the only alien world from which we've collected rocks, some gathered by the Apollo astronauts and others by Russian robotic landers in the 1970s. We can also study the dozen or so meteorites from Mars that have landed on Earth. Aside from these few rocks, we have little more than images taken from spacecraft from which to decode the history of the other planets over the past 4.6 billion years. It's like studying people's facial expressions to understand what they're feeling inside—and what their childhood was like! Fortunately, this decoding is considerably easier for planets than for people.

Spacecraft have visited and photographed all the terrestrial worlds. Patterns of sunlight and shadow on photographs taken from orbit reveal features such as cliffs, craters, and mountains (Figure 9.2a). Besides ordinary, visible-light photographs, we also have spacecraft images taken with infrared and ultraviolet cameras, spectroscopic data, and in some cases three-dimensional data compiled with the aid of radar (Figure 9.2b). Finally, we have close-up photographs of selected locations on all the terrestrial worlds but Mercury, taken by spacecraft that have landed on their surfaces (Figure 9.3). The result is that we now understand the geology of the terrestrial worlds well enough to make detailed and meaningful comparisons among them.

Comparative planetary geology (one part of comparative planetology) hinges on the principle that a planet's surface features can be traced back to its fundamental properties. For example, we saw in Chapter 8 that a planet's composition depends primarily on how far from the Sun it formed. In the rest of this chapter, we will look for similar relationships between surface features and the fundamental properties and processes of the planets. But because surface geology depends largely on a planet's interior, we must first look inside the terrestrial worlds.

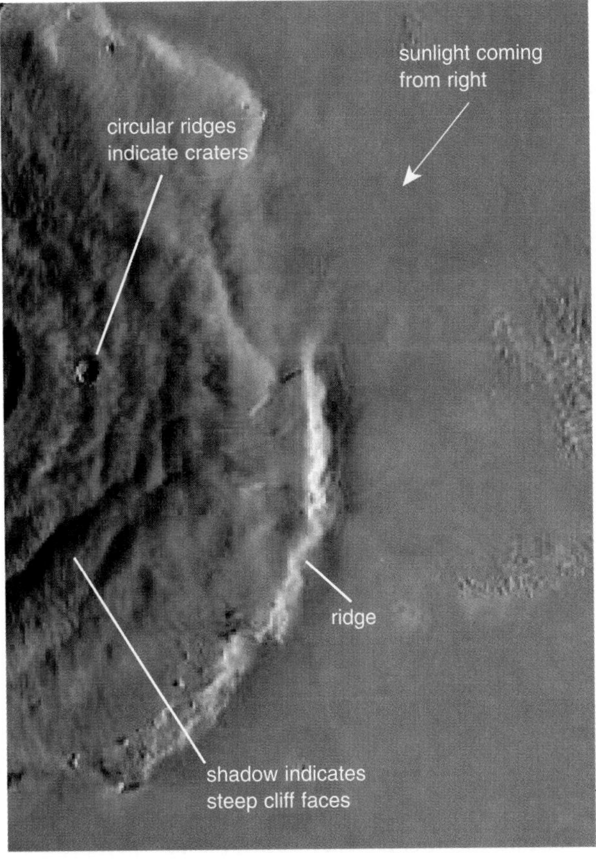

a

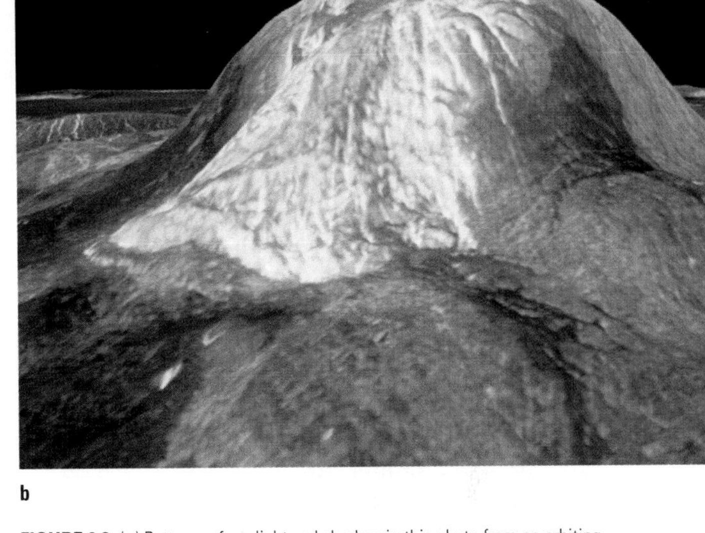

b

FIGURE 9.2 (**a**) Patterns of sunlight and shadow in this photo from an orbiting spacecraft reveal geological features on Mars. (**b**) The Magellan spacecraft used radar to penetrate the thick clouds of Venus. The radar data provided information on surface features and their altitudes, which were converted into this three-dimensional perspective with the aid of computers. The heights are magnified by a factor of more than 20 to make geological features easier to see.

9.2 Inside the Terrestrial Worlds

Planets are all approximately spherical[1] and quite smooth relative to their size. For example, compared to the Earth's radius of 6,378 km, the tallest mountains could be represented by grains of sand on a typical globe.

TIME OUT TO THINK *Find a globe of the Earth that shows mountains in raised relief (i.e., you can feel the mountains as bumps on the globe). Are the heights of the mountains correctly scaled relative to the Earth's size? Explain.*

It's easy to understand why the gaseous jovian planets ended up nearly spherical: Gravity always acts to pull material together, and a sphere is the most compact shape possible. It's a bit subtler when we deal with the rocky terrestrial worlds, because rock

[1]The Earth's equatorial diameter is actually about 40 km more than its diameter measured pole-to-pole, making its shape slightly *oblate* (flattened at the poles). This slight flattening is caused by the Earth's rotation; it's essentially the same process that makes a spinning ball of pizza dough flatten along its "equator." Faster-rotating planets, such as Jupiter, are even more oblate.

a Mercury

b Venus

FIGURE 9.3 Surface views of the terrestrial worlds. No spacecraft have landed on Mercury, so an artist's conception is shown; all other images are photos.

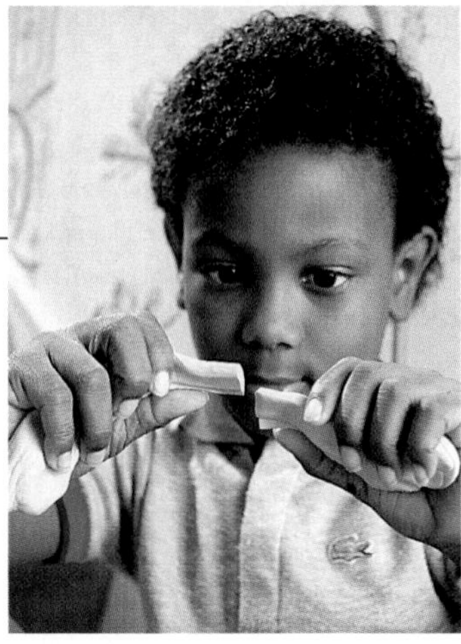

FIGURE 9.4 Silly Putty stretches when pulled slowly but breaks cleanly when pulled rapidly. Rock behaves just the same, but on a longer time scale.

can resist the pull of gravity. Indeed, many small moons and asteroids are "potato-shaped" precisely because their weak gravity is unable to overcome the rigidity of their rocky material. Before we can understand why the terrestrial worlds ended up smooth and spherical, and why they all have similar internal structures, we must first investigate the behavior of rock.

Solid as a Rock?

We often think of rocks as the very definition of strength, as in the expression "solid as a rock." But rocks are not always as solid as they may seem. Familiar rocks that you find outside are a hodgepodge of different minerals, each of which has a simple chemical formula. For example, *granite* is a common rock composed of crystals of quartz (SiO_2), feldspar ($NaAlSi_3O_8$ is one of several varieties), and other minerals. Rocks are solid because of electrical bonds between the mineral molecules. When subjected to sustained stress over millions or billions of years, these solid bonds can break and re-form, slowly allowing rocky material to deform and flow. In fact, the long-term behavior of rock is very much like that of Silly Putty™, which stretches when you pull it slowly but breaks if you pull it sharply (Figure 9.4).

The strength of a rock depends on its composition, its temperature, and the surrounding pressure.

c Earth

d Moon

e Mars

Higher temperatures also make rocks weaker. Just as Silly Putty becomes more pliable if you heat it, warm rocks are weaker and more deformable than cooler rocks of the same type. Also, some rocks contain traces of water, which can act as a lubricant to reduce the rock's strength. Finally, very high pressures like those found deep in planetary interiors can compress rocks so much that they stay solid even when temperatures are high enough to melt them under ordinary conditions.[2]

At high enough temperatures (usually over 1,000 K), some or all of the minerals in a rock may melt, allowing the rock to become *molten* (or partially molten, if only some minerals have melted). Just as different types of solid rock are more deformable than others, different types of rock behave differently when they are molten. Some molten rocks are runny like water, while others flow slowly like honey or molasses. We describe the "thickness" of a liquid with the technical term **viscosity.** Water has a low viscosity, while honey and molasses have much higher viscosities. As we'll soon see, the viscosities of different types of molten rock are very important to understanding volcanoes.

[2]Most substances are denser as solids than as liquids, so the compression caused by high pressure tends to make things solid. Water is a rare exception, becoming *less* dense when it freezes—which is why ice floats in water.

Terrestrial-World Layering

The rocky terrestrial worlds became spherical because of rock's ability to flow. When objects exceed about 500 km in diameter, gravity can overcome the strength of solid rock and make a world spherical in less than a billion years. If the object is molten, as was the case with the terrestrial worlds in their early histories, it can become spherical much more quickly. (For similar reasons, Earth's ocean surface is smoother and more spherical than its rocky surface.)

Gravity also gives the terrestrial worlds similar internal structures. You know that in a mixture of oil and water, the less dense oil rises to the top while the denser water sinks to the bottom. The process by which gravity separates materials according to density is called **differentiation** (because it results in layers made of *different* materials). The terrestrial worlds underwent differentiation early in their histories, at the time when they were molten throughout their interiors. The result was the formation of three layers of differing composition within each terrestrial world (Figure 9.5). Dense metals such as iron and nickel sank through molten rocky material to form a **core.** Rocky material composed of *silicates*—minerals that contain silicon and oxygen as well as other elements—came to rest above the core, forming a thick **mantle.** A low-density "scum" of rocks composed of the lightest silicates rose to form a **crust.**

The terms *core, mantle,* and *crust* are defined by the composition within each layer. But characterizing the layers by *rock strength* rather than composition turns out to be more useful for understanding geological activity. From this view, we speak of an outer layer of relatively rigid rock called the **lithosphere** (*lithos* means "stone" in Greek). The lithosphere generally encompasses the crust and the uppermost portion of the mantle. Beneath the lithosphere, the higher temperatures allow rock to deform and flow much more easily.[3] Thus, the lithosphere is essentially a layer of rigid rock that "floats" on the softer rock below. On the Earth, the lithosphere is broken into *plates* that move as the underlying rock flows gradually, creating the phenomenon called *continental drift* [Section 13.2].

The strength of the lithosphere determines how the flow of the "soft" rock in a planet's mantle affects the planet's surface. Interior heat and motion will lead to volcanoes and other geological activity on a planet with a relatively thin lithosphere. In contrast, geological activity is inhibited on a planet with a thick, strong lithosphere. The most important factor determining lithospheric thickness is internal temperature: Higher internal temperature makes rocks softer, leading to a thinner rigid lithosphere.

Figure 9.6 compares the interior structures of the terrestrial worlds. Note that the smaller worlds have thicker lithospheres, indicating that they have cooler interiors. The terrestrial worlds also differ in the relative sizes of their cores and mantles. Because cores are made from metals and mantles are made from rock, their relative sizes depend on the proportions

[3]The layer of softer rock beneath the lithosphere is sometimes called the *asthenosphere* (literally, the "weak-o-sphere"). Although the rock deforms more easily, it is not actually molten.

FIGURE 9.5 Interior structure of a generic terrestrial world.

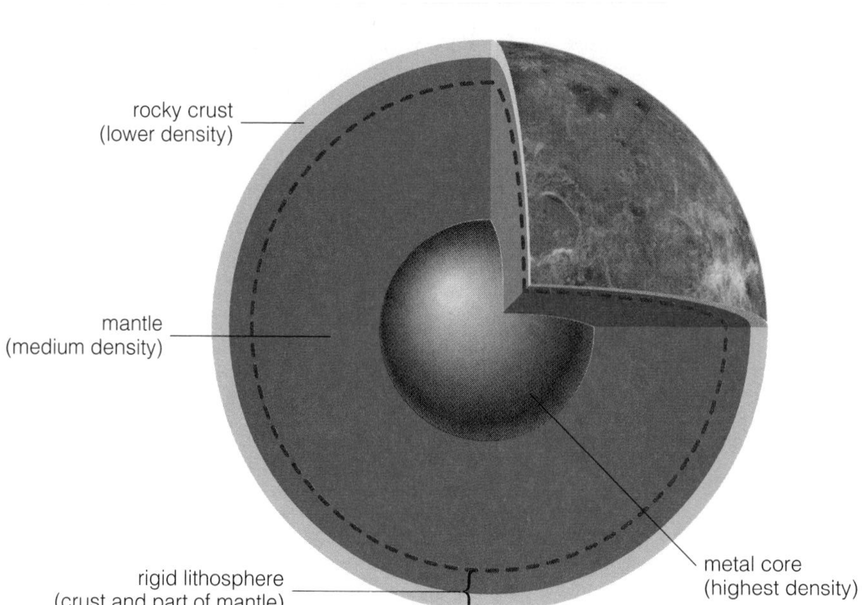

of metal and rock in a planet. These proportions resulted primarily from condensation in the solar nebula, but they may also have been affected by giant impacts [Section 8.6]. Planetary cores may be partially or completely molten, depending on the particular combination of internal temperature and pressure.

You may be wondering how we know what the interiors of the terrestrial worlds look like. After all, even on Earth our deepest drill shafts barely prick the lithosphere. Fortunately, several techniques allow us to study planetary interiors without actually having to go to them. First, we can measure a planet's size from telescopic or spacecraft images, and its mass by applying Newton's version of Kepler's third law to the orbital properties of a natural or artificial satellite [Section 6.4]. Together, these measurements give the average density of the planet. Second, the strength of gravity on a planet may differ slightly from place to place depending on the underlying interior structure, so we can learn about that structure through gravity measurements made from orbiting satellites. Third, a planet's magnetic field is generated in its interior, so measurements of the magnetic-field strength also provide information about interior structure. Fourth, lava brought up from deep inside a planet can tell us something about the interior composition. Finally, planetary vibrations (*seismic waves* from "planetquakes" [Section 13.2]) tell us about the interior structure in much the same way shaking a present offers a few hints on what's inside. These diverse types of measurements are combined with computer models based on known laws of physics to calculate a planet's interior structure. Our measurements are much more extensive for Earth than for any other world, so we know the most about Earth's interior. But scientists have applied some of the same techniques in more limited ways to other worlds, giving us confidence that the general pictures shown in Figure 9.6 are correct.

9.3 How Interiors Work

A planet's geological activity depends on its interior structure, and its interior structure depends on its internal temperature. In this section, we investigate how heat (thermal energy) is deposited in the planetary interior and how it leaks outward over time.

How Interiors Get Hot

It's often said that the Sun is the ultimate source of all our energy. On the surface, this statement is true: Sunlight bathes the Earth's surface with a power of roughly 1,000 watts per square meter (in the daytime), while the heat leaking outward from the Earth's interior contributes only about 0.05 watt per square meter. Thus, the Earth's surface receives more than 10,000 times as much energy from the Sun as from its own interior. In general, the energy a world receives from sunlight depends on its distance from the Sun. This energy, along with properties of the planet's atmosphere, determines the surface temperature [Section 10.2].

FIGURE 9.6 Interior structures of the terrestrial worlds, in order of size.

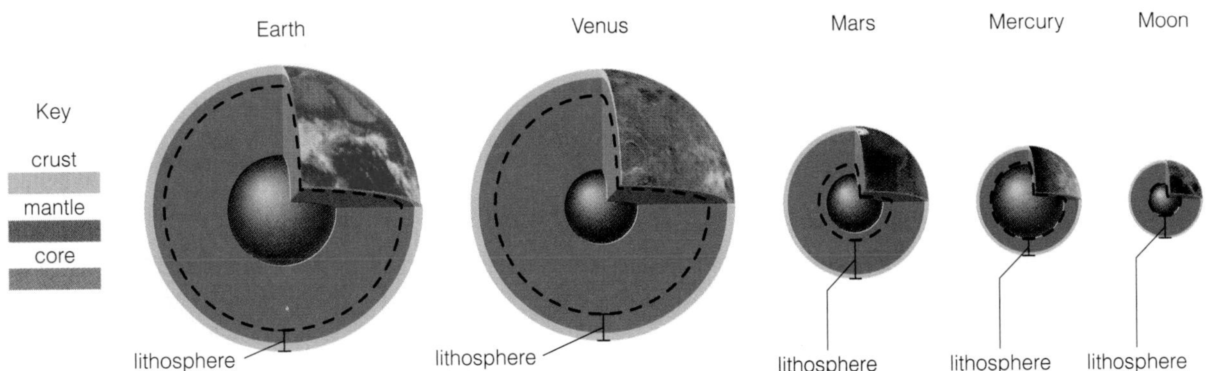

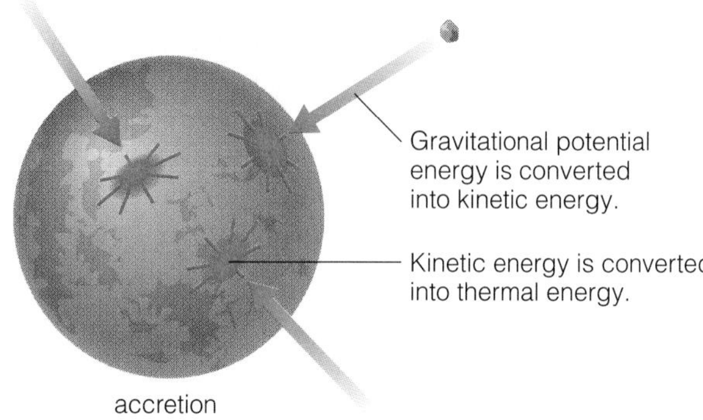

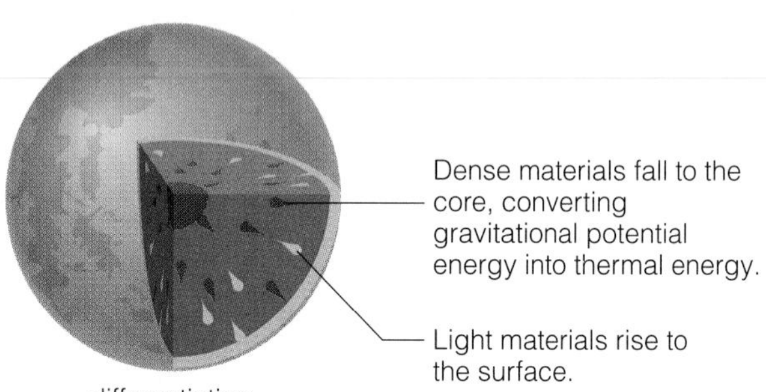

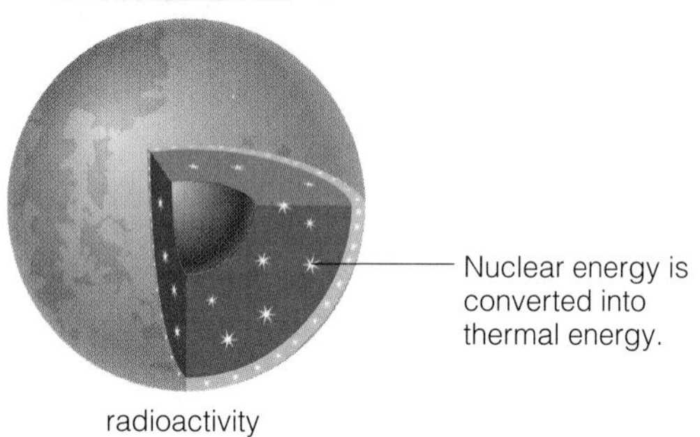

FIGURE 9.7 The three main internal energy sources in terrestrial planets. Only radioactivity is a major heat source today.

In contrast, the Sun contributes very little to the internal temperature of a planet. In fact, the Sun's only role is in helping to determine the starting point for the interior temperature structure: A terrestrial world is coolest at its surface and gets warmer as you go into its interior. Far more important sources of heat in planetary interiors are three principal **internal energy sources:** *accretion, differentiation,* and *radioactivity* (Figure 9.7). (A fourth process, called *tidal heating,* is not important for the terrestrial worlds but is for some of the jovian moons, particularly Io [Section 11.5].)

Accretion was the earliest major source of internal heat for the terrestrial worlds. The many violent impacts that occurred during the latter stages of accretion deposited so much energy that planet-wide melting occurred. Imagine the entire Earth engulfed in molten rock! This melting enabled the process of differentiation, in which the densest materials (metals) settled to the planet's core while lighter rocks rose to the surface.

Both accretion and differentiation yield heat by converting *gravitational potential energy* into thermal energy [Section 5.2]. An impactor that hits an accreting world starts with a lot of gravitational potential energy when it is far away. This gravitational potential energy is converted to kinetic energy as gravity makes the impactor accelerate toward the planet's surface. Upon impact, the kinetic energy is converted to sound and heat, thus adding to the thermal energy of the planet. Differentiation converts gravitational potential energy to thermal energy just like a brick falling to the bottom of a swimming pool. The brick loses gravitational potential energy as it falls, and this energy ultimately appears as thermal energy in the water.

The third internal heat source is the decay of radioactive elements such as uranium, potassium, thorium, and others. When radioactive nuclei decay, subatomic particles fly off at high speeds, colliding with neighboring atoms and heating them up. In essence, this transfers some of the mass-energy of the radioactive element ($E = mc^2$ [Section 5.2]) to the thermal energy of the planetary interior. Most radioactive elements are locked in minerals that make up the crust and mantle, but some end up in the core as well. Note that accretion and differentiation deposited heat in planetary interiors billions of years ago, but radioactivity continues to heat the planets to this very day.[4]

How Interiors Cool Off

Some of a planet's internal heat, or thermal energy, is always escaping. Heat flows outward from the hot interior toward the cooler surface through three main processes: *conduction, convection,* and *eruption* (Figure 9.8). **Conduction** is the process that makes heat flow from your hand to a glass full of ice water. It occurs because the microscopic jiggling of molecules is more energetic (faster) at higher temperatures. Some of the thermal energy of this motion is transferred from a warm rock to its cooler neighbors. Conduction is the main process by which heat flows upward through the lithosphere.

Some planetary interiors also lose heat through **convection,** in which hot material expands and rises

[4]Differentiation is still slowly occurring in the jovian planets.

while cooler material contracts and falls. Convection can occur any time a substance is strongly heated from underneath; you can see convection whenever you heat a pot of soup on the stove, and you're probably also familiar with it in the context of weather, in which warm air rises while cool air falls in our atmosphere. Even though it is solid, a planet's hot interior also convects, but much more slowly. The rising hot material transfers heat from deep in the planetary interior toward the surface. When the material fully cools, it begins to descend back down until the heating from below heats it enough to make it rise upward again. Thus, convection is an ongoing process. Each small region of rising and falling material is called a **convection cell.**

The third process by which heat sometimes escapes a planet's interior is volcanic **eruption.** An eruption directly transfers heat outward by depositing hot lava on the surface.

Regardless of whether heat reaches the surface through the slow leakage of convection and conduction or through the sudden burst of an eruption, this heat eventually radiates away into space. Recall that all objects emit *thermal radiation* characteristic of their temperatures [Section 7.4]. Because of their relatively low surface temperatures, planets radiate almost entirely in the infrared portion of the spectrum.

The interiors of the terrestrial planets are slowly cooling as their heat escapes. By now, 4.6 billion years after their formation, much of the original heat from accretion and differentiation has leaked away. Most of the heat flow today comes from radioactive decay, and even the rate of radioactive decay declines as the planets age (because a particular nucleus can decay only once). The interior cooling gradually changes the planet's structure, making the lithosphere thicker and leaving any regions of molten rock lying deeper inside.

Ultimately, the single most important factor in determining how long a planet stays hot is its size: Larger planets stay hot longer, just as larger baked potatoes stay hot longer than small ones. You can see why size is the critical factor by picturing a large planet as a smaller planet wrapped in extra layers of rock. The extra rock acts as insulation, so heat from the center takes longer to reach the surface on the larger planet than it would on the smaller one. The overall result is that the large terrestrial worlds—Earth and Venus—have scarcely cooled off over the life of the solar system. These worlds therefore have thin lithospheres and substantial geological activity. The smaller worlds—Mercury and the Moon—have mostly cooled off and therefore have very thick lithospheres that essentially make them geologically "dead." Mars, intermediate in size, has cooled significantly, resulting in an intermediate interior temperature.

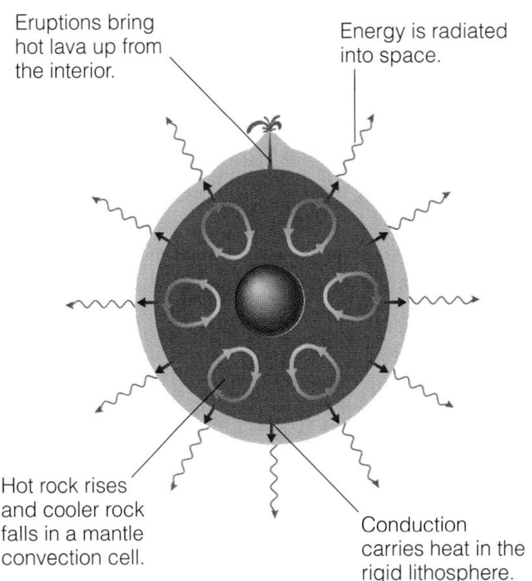

FIGURE 9.8 Heat escapes a planet's interior through conduction, convection, and eruption.

TIME OUT TO THINK *Can you think of another example in everyday life of a small object cooling off faster than a large one? What about a smaller object warming up more quickly than a large one? How do these examples relate to the issue of geological activity on the terrestrial worlds?*

Common Misconceptions: Pressure and Temperature

You might think that Earth's interior is hot just because the pressures are high. After all, if we compress a gas from low pressure to high pressure, it heats up. But the same is not necessarily true of rock. High pressure hardly compresses rock, so the compression causes little increase in temperature. Thus, while high pressures and temperatures sometimes go together in planets, they don't have to. In fact, after all the radioactive elements decay (billions of years from now), Earth's deep interior will become quite cool even though the pressure will be the same as it is today. The temperatures inside Earth and the other planets can remain high only if there is a source of heat, such as accretion, differentiation, or radioactivity.

Planetary Cores and Magnetic Fields

Interior structure also determines whether a planet has a **magnetic field** through which it can influence charged particles or magnetic materials. You are probably familiar with the general pattern of a bar magnet's magnetic field, which we can see by placing it among small iron filings (Figure 9.9a). A planet's magnetic field is generated by a process more similar to that of an *electromagnet,* in which the magnetic field arises as a battery forces charged particles (electrons) to move along a coiled wire (Figure 9.9b). Planets do not have batteries, but they do have charged particles in motion in their metallic cores (Figure 9.9c). Deep within these electrically conducting cores, molten metals rise and fall in convection cells. At the same time, the molten material spins with the planet's rotation. If this combination of convection and rotation is strong enough, it moves electrons in the same way they move in an electromagnet. The result is a planetary magnetic field. Similar combinations of convection and rotation of conducting materials (such as metallic hydrogen or ionized plasma) also give rise to magnetic fields in the jovian planets and in stars.

This simple analysis explains why Earth has the strongest magnetic field of the terrestrial worlds: It is the only one that has both a partially molten metallic core (which can convect) and reasonably rapid rotation. The Moon may lack a metallic core; even if it has one, it has certainly solidified and ceased convecting. In either case, it generates no magnetic field. Mars also has virtually no magnetic field, probably because of similar core solidification. In contrast, Venus probably has a molten metal layer in its core, but either its convection or its 243-day rotation pe-

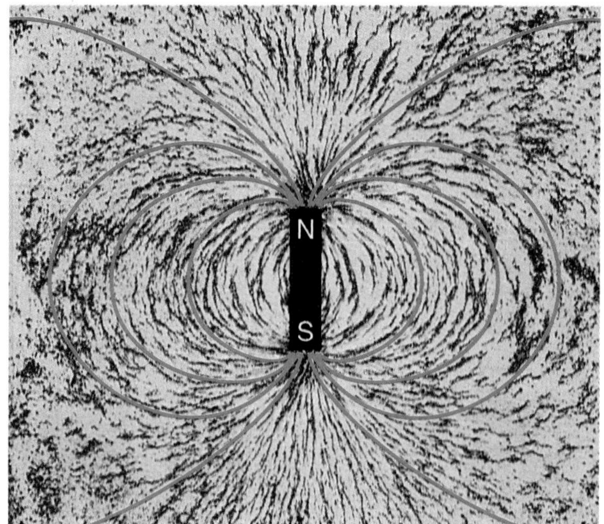

a This photo shows how a bar magnet influences iron filings (small black specks) around it. The *magnetic field lines* (red) represent this influence graphically.

FIGURE 9.9 Sources of magnetic fields.

Mathematical Insight 9.1 The Surface Area–to–Volume Ratio

We can see why large planets cool more slowly than smaller ones by thinking about their relative surface areas and volumes. Consider the Earth and the Moon. Both started out very hot inside but continually radiate some of this heat away from their surfaces to space. As heat escapes from the surface, more heat flows upward from the interior to replace it. This process will continue until the interior is no hotter than the surface. Because all the heat escapes from the surface, the key factor in a world's heat-loss rate is its *surface area.* You may recall that the surface area of a sphere is given by the formula $4\pi \times (\text{radius})^2$. Thus, the ratio of Earth's surface area to that of the Moon is:

$$\frac{\text{surface area of Earth}}{\text{surface area of Moon}} = \frac{4\pi \times (r_{\text{Earth}})^2}{4\pi \times (r_{\text{Moon}})^2} = \left(\frac{r_{\text{Earth}}}{r_{\text{Moon}}}\right)^2 = \left(\frac{6{,}378 \text{ km}}{1{,}738 \text{ km}}\right)^2 \approx 13$$

Thus, at a given temperature, Earth has about 13 times more area from which to radiate its heat away.

However, the total amount of heat deposited inside a world by radioactivity roughly depends on its mass or, if we assume that we're dealing with

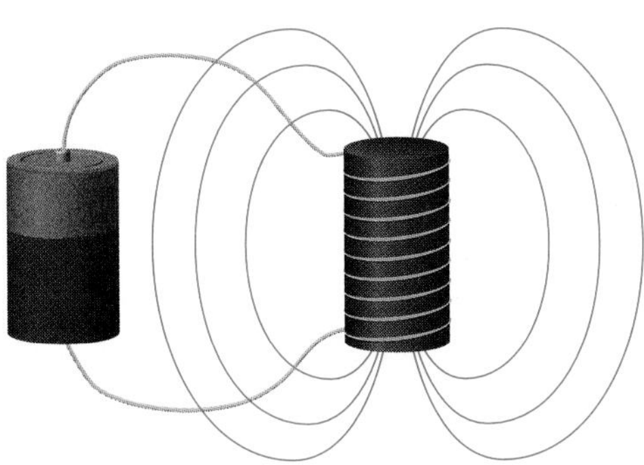

b A similar magnetic field is created by an electromagnet, which is essentially just a coiled wire attached to a battery. The field is created by the battery-forced motion of charged particles (electrons) along the wire.

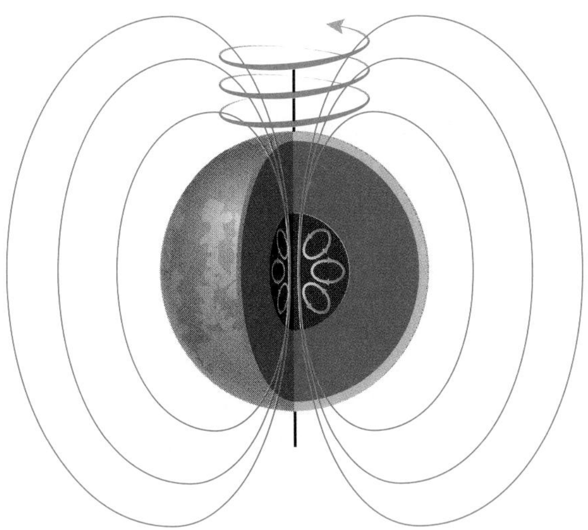

c A planet's magnetic field also arises from motion of charged particles. For a terrestrial planet, the charged particles are in a molten metallic core, and their motion arises from the planet's rotation and interior convection.

riod is too slow to generate a magnetic field. Mercury remains an enigma: It possesses a measurable magnetic field despite its small size and slow, 59-day rotation. The solution to this enigma may lie in Mercury's high density, which tells us that it has a huge metal core—a larger core than that of Mars, even though it is a smaller planet overall. Perhaps Mercury's core is still partly molten and convecting.

Although convection and rotation of electrically conducting material basically explain why planetary magnetic fields exist, many details remain mysterious. For example, we don't know why the Earth's magnetic axis intersects the surface in northern Canada, 11° away from the rotation axis at the North Pole. Nor do we know why Earth's magnetic field

worlds of similar density, its volume. The formula for the volume of a sphere is $\frac{4}{3}\pi \times (\text{radius})^3$, so the volume ratio for Earth and the Moon is:

$$\frac{\text{volume of Earth}}{\text{volume of Moon}} = \frac{\frac{4}{3}\pi \times (r_{\text{Earth}})^3}{\frac{4}{3}\pi \times (r_{\text{Moon}})^3} = \left(\frac{r_{\text{Earth}}}{r_{\text{Moon}}}\right)^3 = \left(\frac{6{,}378 \text{ km}}{1{,}738 \text{ km}}\right)^3 \approx 50$$

Thus, Earth's interior contains some 50 times more heat than the Moon's interior. This larger amount of internal heat offsets the larger radiating area. That is why the Earth is still hot inside while the Moon has mostly cooled off. Planetary scientists use sophisticated computer models of the different heat transport processes to calculate the cooling time of the planets more accurately.

The general idea that larger objects cool more slowly than smaller ones is described by the mathematical idea of the *surface area–to–volume ratio.* As the Earth–Moon example shows, larger objects have relatively smaller surface areas compared to their volumes. Thus, they lose internal heat more slowly and also take longer to be heated from the outside. This principle explains many everyday phenomena. For example, crushing a cube of ice into smaller pieces increases the total amount of surface area, while the total volume of ice remains the same. Thus, the crushed ice will cool a drink more quickly than would the original ice cube.

varies in strength over time. Even more mysterious is the fact, discovered from geological evidence, that Earth's entire magnetic field completely reverses its orientation (magnetic north becomes magnetic south, and vice versa) every half-million years or so.

Magnetic fields can be very useful. On Earth, they help us navigate. Their presence or absence and their strength on any planet provide important information about the planetary interior. But aside from this utility, magnetic fields have virtually no effect on the structure of a planet itself. They can, however, form a protective *magnetosphere* surrounding a planet, which in turn can have profound effects on a planet's atmosphere and any inhabitants [Section 10.2].

9.4 Shaping Planetary Surfaces

When we look around the Earth, we find an apparently endless variety of geological surface features. The diversity increases when we survey the other planets. But on closer examination, geologists have found that almost all the features observed result from just four major **geological processes** that affect planetary surfaces:

- **Impact cratering:** the excavation of bowl-shaped depressions *(impact craters)* by asteroids or comets striking a planet's surface.
- **Volcanism:** the eruption of molten rock, or *lava,* from a planet's interior onto its surface.
- **Tectonics:** the disruption of a planet's surface by internal stresses.
- **Erosion:** the wearing down or building up of geological features by wind, water, ice, and other phenomena of planetary weather.

Planetary geology once consisted largely of cataloging the number and kinds of geological features found on the planets. The field has advanced remarkably in recent decades, however, and it is now possible to describe how planets work in general and what kinds of geological features we expect to find on different planets. This progress has been made possible by an examination of the geological processes in detail and the determination of what factors affect and control them.

Understanding Geological Relationships

We can understand the most fundamental ideas of comparative geology by exploring just a few key cause-and-effect relationships. Some of the important relationships are fairly easy to see: A planet can have active volcanoes only if it has a sufficiently hot interior; water, a major agent of erosion, remains liquid only if the planet has a temperature in the proper range and an atmosphere with sufficient pressure.[5] Other relationships are subtler, such as that between the presence of an atmosphere and the planetary surface temperature. Moreover, the sheer number of relationships and interrelationships makes it difficult to keep track of them all. We will concentrate only on the most important geological relationships, keeping in mind that a deeper understanding requires the inclusion of much more detail.

Figure 9.10 shows a planetary flowchart (or "concept map") designed to help us visualize geological relationships. The first row lists four **formation properties** with which each planet was endowed at birth: *size* (i.e., mass and radius), *distance from the Sun, chemical composition,* and *rotation rate.* These properties generally remain unchanged throughout a planet's history, unless the planet suffers a giant impact. The second row lists four **geological controlling factors** that drive a planet's geological activity: *surface gravity, internal temperature, surface temperature,* and the characteristics of the planet's *atmosphere.*

In the following pages, we will extend the connections shown in Figure 9.10 to include the four geological processes that form surface features. But first, let's briefly discuss how to interpret this diagram and similar ones that follow. Notice that the diagram consists of blocks and arrows:

- Each block represents a particular property, controlling factor, or process.
- Each arrow represents a cause-and-effect relationship between one block and another. There are other connections besides those shown; the figure includes only those that have substantial effects on geology.

Arrows from the first row to the second row show how a planet's formation properties help determine the geological controlling factors. A single arrow shows that *surface gravity* (the strength of gravity at the surface) is determined solely by *size* (mass and radius), in accord with Newton's law of gravity. Two arrows point to *internal temperature,* indicating that the formation properties of both *size* and *composition* affect it: Size is the most important factor in determining how rapidly a planet loses its internal heat,

[5]If there is no atmosphere or if the atmospheric pressure is too low, ice goes directly from the solid phase to the gas phase through the process of sublimation [Section 5.3], and liquid water cannot exist. The same phenomenon occurs with dry ice (solid CO_2), which sublimes away under Earth's "low" pressure.

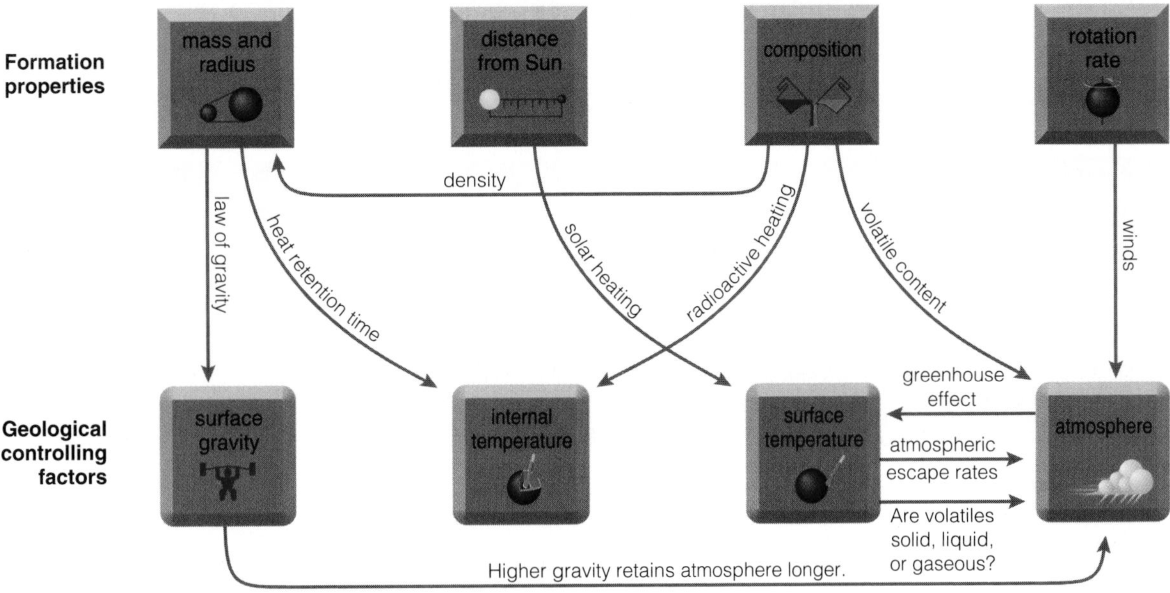

FIGURE 9.10 Major cause-and-effect relationships between a planet's formation properties and its geological controlling factors. Each block represents a particular property, controlling factor, or geological process. Each arrow represents a cause-and-effect relationship between one block and another. This diagram forms the foundation for several to follow.

and composition—particularly the amount of radioactive elements—determines how much new heat is deposited in the interior. Other arrows from the first to the second row show that a planet's *surface temperature* depends in part on its *distance from the Sun* and that its *atmosphere* is affected by both its *composition* and its *rotation rate.*

Arrows between the geological controlling factors show how they affect one another. Both *surface gravity* and *surface temperature* have arrows pointing to *atmosphere* because a planet's ability to hold atmospheric gases depends on both how fast the gas molecules move (which is related to temperature) and the strength of the planet's gravity. A second arrow from *surface temperature* to *atmosphere* shows that the temperature also influences what substances exist as gases (versus solid or liquid). The *atmosphere,* in turn, points back to *surface temperature* because a thick atmosphere can make a planet's surface warmer than it would be otherwise (through the *greenhouse effect* [Section 10.2]).

TIME OUT TO THINK *Before continuing, make sure you understand how to use the planetary flowchart by answering the following questions: What planetary formation properties determine whether a planet's interior is hot, and why? How does surface gravity affect a planet's atmosphere, and why? Do any of the relationships shown in Figure 9.10 surprise you? If so, why?*

The planetary flowcharts in this chapter contain all the information needed for a conceptual model of how planetary geology works—a model that allows us to answer questions about why different planets have different geological features. Figure 9.10 represents the beginning of this model. In the sections that follow, we will add each of the four geological processes to the set of relationships we have just covered, ending up with a fairly complete model for planetary geology. You do not need to memorize the model in this graphical form; instead, use it to help you understand planetary geology. You should also think about how you could adapt this technique of developing a conceptual model to any other problem that interests you.

Impact Cratering

The first of the four major geological processes we will add to our model is impact cratering, which occurs when a leftover planetesimal (such as a comet or an asteroid) crashes into the surface of a terrestrial world. Impacts can have devastating effects on planetary surfaces, which we can see both from the impact craters left behind and from laboratory experiments that reproduce the impact process (Figure 9.11). Impactors typically hit planets at speeds between 30,000 and 250,000 km/hr (10–70 km/s) and thus pack enough energy to vaporize solid rock and excavate a *crater* (the Greek word for "cup"). Craters

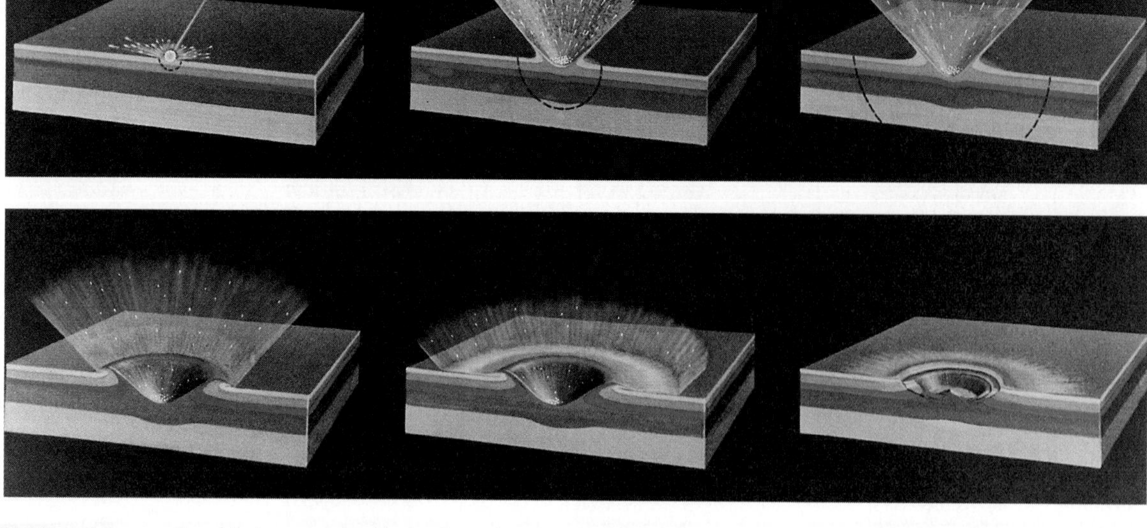

FIGURE 9.11 Artist's conception of the impact process. The last frame shows how, in a larger crater, the center can rebound just as water does after you drop a pebble in.

are generally circular because the impact blasts out material in all directions, no matter which direction the impactor came from. A typical crater is about 10 times wider than the impactor that created it, with a depth about 10–20% of the crater width. Thus, for example, a 1-kilometer-wide impactor creates a crater about 10 kilometers wide and 1–2 kilometers deep. Debris from the blast, called **ejecta,** shoots high into the atmosphere and then rains down over an area much larger than that of the impact crater itself. In large impacts, some of the atmosphere may be blasted away into space, and some of the rocky ejecta may completely escape from the planet.

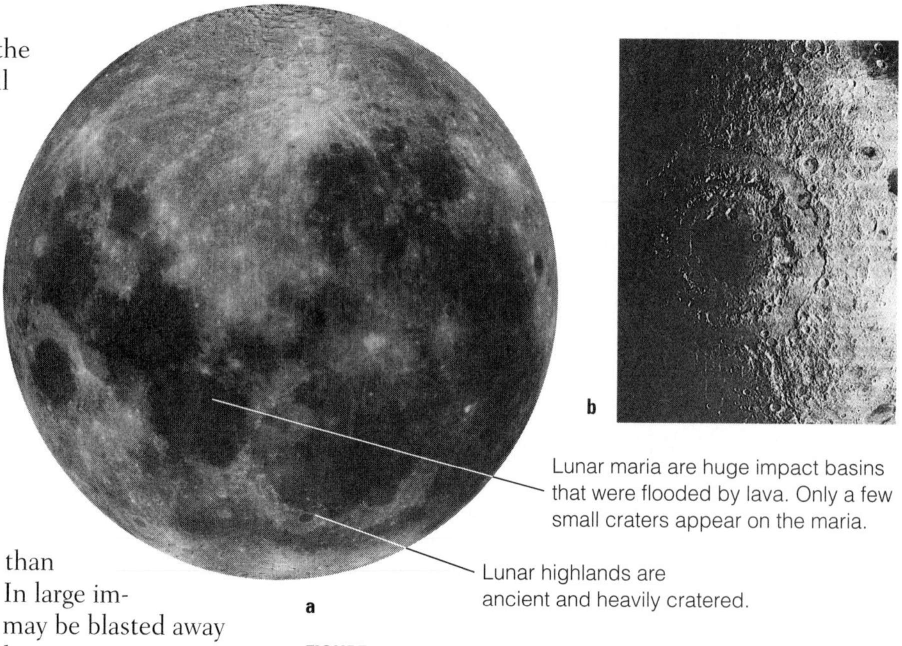

Craters come in all sizes and a few different shapes, but most are small and bowl-shaped. Small craters far outnumber large ones because there are far more small objects orbiting the Sun than large ones. When the largest impactors strike a planet, they form **impact basins.** Looking at the Moon with binoculars, you can see examples of lava-filled impact basins up to 1,100 kilometers across called **lunar maria** (Figure 9.12a). (*Maria* is the Latin word for "seas"; they got their name because their smooth appearance reminded early observers of oceans.) These large impacts violently fractured the Moon's lithosphere, making cracks through which lava escaped to flood the impact basins. The impacts that made the largest basins were so violent

b

Lunar maria are huge impact basins that were flooded by lava. Only a few small craters appear on the maria.

Lunar highlands are ancient and heavily cratered.

a

FIGURE 9.12 (**a**) The Moon's surface shows both heavily and lightly cratered areas. (**b**) This photo shows a multi-ring basin on the Moon known as Mare Orientale.

they sent out ripples that left tremendous *multi-ring basins* shaped like bull's-eyes (Figure 9.12b).

The present-day abundance of craters on a planet's surface tells us a great deal about its geological history. Even though impacts still occur today, the vast majority of craters formed during the "rain of rock and ice" that ended around 3.8 billion years ago [Section 8.6]. At that time, the terrestrial worlds were saturated (completely covered) with impact craters. A surface region that is still saturated with craters, such as the *lunar highlands*, must have remained essentially undisturbed for the last 3.8 billion years (Figure 9.12a). In contrast, the original craters must have

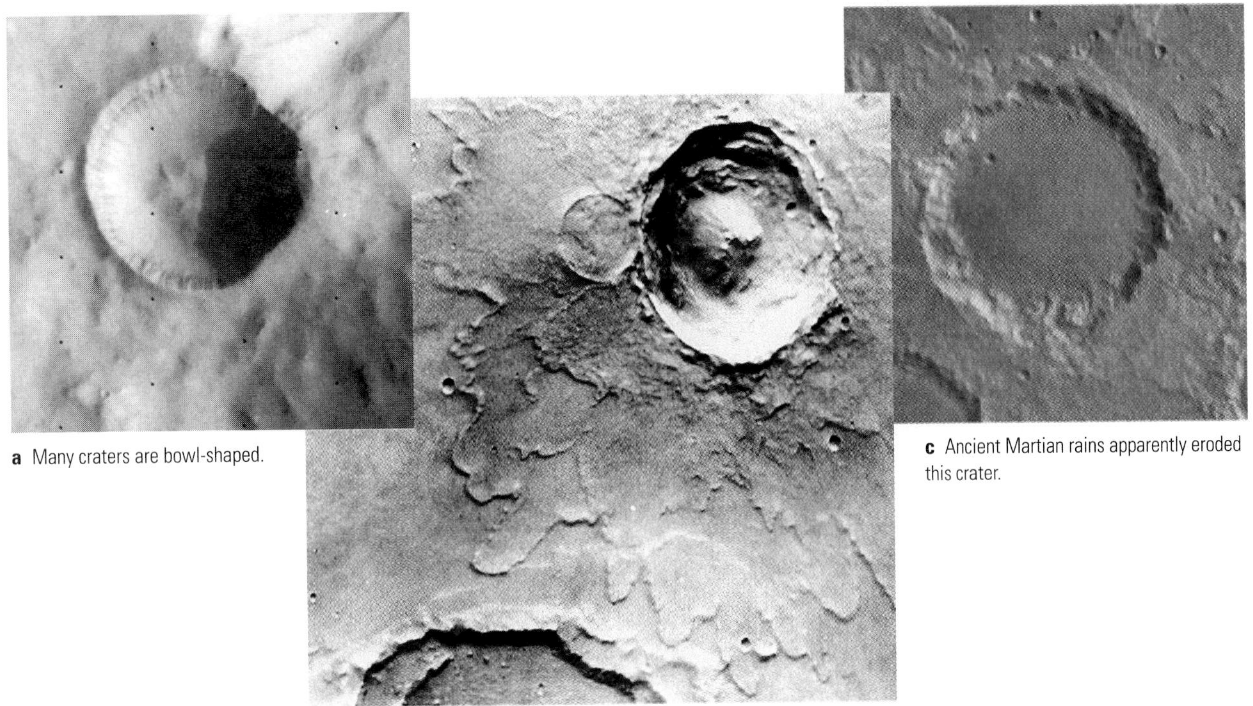

a Many craters are bowl-shaped.

b Impacts into icy ground may form muddy ejecta.

c Ancient Martian rains apparently eroded this crater.

FIGURE 9.13 Crater shapes on Mars tell us about Martian geology. These photos were taken from orbit by the Viking spacecraft.

been somehow "erased" in regions that now have few craters, such as the *lunar maria*. The flood of lava that formed the lunar maria covered any craters that had formed inside the impact basin, and the few craters that exist today within the maria must have formed from impacts occurring after the lava flows solidified. These craters tell us that the impact rate since the end of heavy bombardment has been quite small: The lunar maria have only 3% as many craters as the lunar highlands, but radioactive dating of moon rocks shows that the maria are still 3–3.5 billion years old.

TIME OUT TO THINK *Earth must also have been saturated with impact craters early in its history, but we see relatively few impact craters on Earth today. What processes erase impact craters on Earth?*

Craters with unusual shapes provide additional information about geological conditions on a planetary surface, as we can see by comparing a few craters on Mars. Craters that form in rocky surfaces usually have a simple bowl shape (Figure 9.13a). However, some Martian craters look as if they were formed in mud (Figure 9.13b), suggesting that underground water or ice vaporized upon impact and lubricated the flow of ejecta away from the crater. Other craters lack a sharp rim and bowl-shaped floor, suggesting that geological processes such as erosion have altered their shape over time (Figure 9.13c). Planetary geologists must be cautious in interpreting unusually shaped craters, because impacts are not the only process that can form craters. Fortunately, craters created by other processes, such as volcanic craters, tend to have distinctly different shapes than impact craters (see Figure 9.17b).

The most common impactors are sand-size particles called *micrometeorites* when they impact a surface. Such tiny particles burn up as meteors in the atmospheres of Venus, Earth, and Mars. But on worlds that lack significant atmospheres, such as Mercury and the Moon, the countless impacts of micrometeorites gradually pulverize the surface rock to create a layer of powdery "soil." On the Moon, the Apollo astronauts and their rovers left marks in this powdery surface (Figure 9.14). Because of the lack

FIGURE 9.14 Reminders of the Apollo missions to the Moon will last for millions of years.

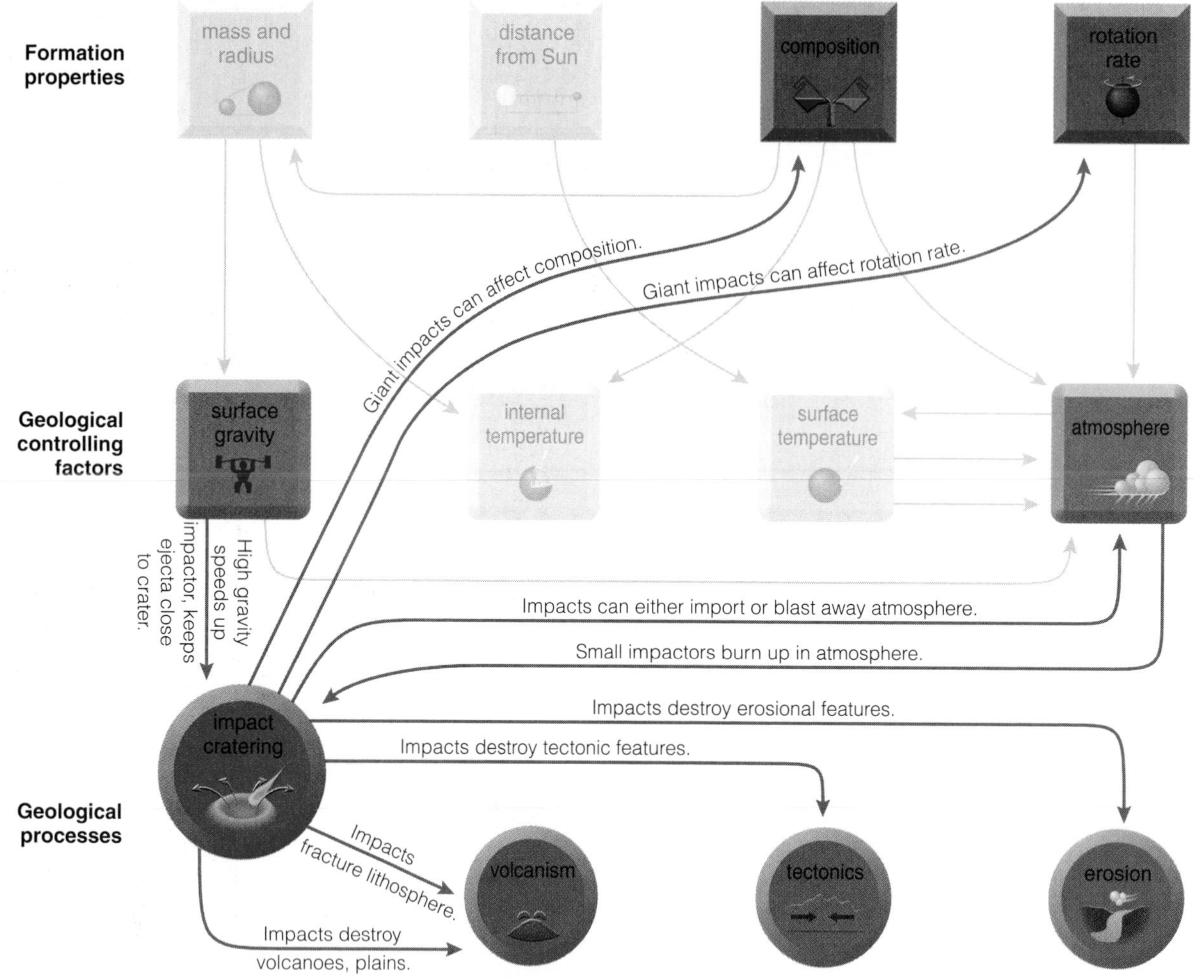

FIGURE 9.15 Major connections between impact cratering, other geological processes, controlling factors, and formation properties. This diagram includes all the blocks and arrows from Figure 9.10, but those that are not directly relevant to impact cratering are shaded lighter to clarify this illustration.

of wind and rain, these footprints and tire tracks will last millions of years, but they will eventually be erased by micrometeorite impacts. In fact, over millions and billions of years, these tiny impacts smooth out rough crater rims much as erosion processes do more rapidly on Earth.

Figure 9.15 summarizes the important cause-and-effect relationships linking impact cratering with other geological processes, controlling factors, and formation properties. Note that the planet's formation properties do not greatly affect whether impact craters form. Craters formed on all the terrestrial worlds, especially early in their histories. But giant impacts can affect formation properties by altering a planet's composition or rotation rate. Impacts certainly affect features made by other geological processes. For example, arrows pointing from impact cratering to tectonics, volcanism, and erosion show that impacts can obliterate volcanoes, cliffs, or rivers. A second arrow from impact cratering to volcanism shows that impacts can fracture the lithosphere, creating a path for volcanic lava to reach the surface. The relationships between impacts and the atmosphere are particularly interesting: Not only can an atmosphere burn up small impactors, but the impactors themselves can either bring in atmospheric ingredients or blast away some of the atmosphere. Before you read on, take a bit of time to analyze the remaining arrows and make sure all the connections make sense to you.

Volcanism

The second major geological process to be included in our model is volcanism. We find evidence for volcanoes and lava flows on all the terrestrial planets, as well as on a few of the moons of the outer solar

FIGURE 9.16 (**a**) Volcanism: Hot magma erupts to the surface. (**b**) Eruption of an active volcano on the flanks of Kilauea on the Big Island in Hawaii.

system.[6] Volcanic surface features differ quite a lot from one world to another, but some general principles apply.

Volcanoes erupt when underground molten rock, or **magma,** finds a path through the lithosphere to the surface (Figure 9.16). Magma rises for two main reasons: First, molten rock is generally less dense than solid rock, so it has a natural tendency to rise. Second, a *magma chamber* may be squeezed by tectonic forces, driving the magma upward under pressure. Any trapped gases expand as magma rises, sometimes leading to very dramatic eruptions. Volcanism is more likely on a planet with high internal temperatures and a thin lithosphere, because then magma lies relatively close to the surface. A thick lithosphere, on the other hand, means that magma is very deep and may not reach the surface easily.

The structure of volcanic flows depends on the viscosity of the lava that erupts onto the surface. The rock type known as **basalt** makes relatively low-viscosity lava when molten because it is made of relatively short molecular chains that don't tangle with one another. Temperature affects lava viscosity as well: The hotter the lava, the more easily the

[6]The moons may have "lavas" of water or other unusual molten materials. Water volcanism isn't quite as foreign as it sounds: Geysers erupt by processes similar to volcanic processes.

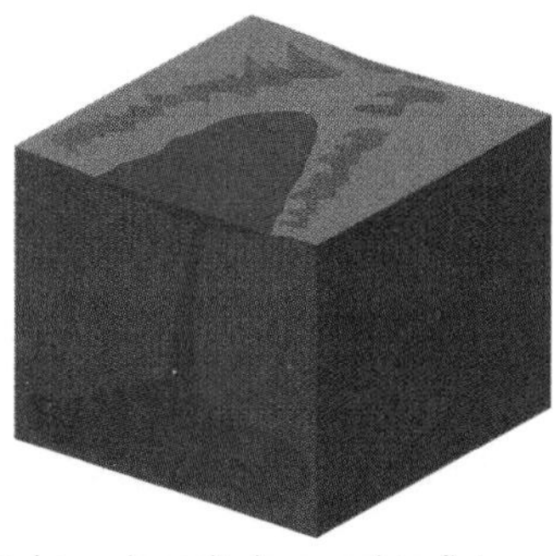

a Low-viscosity lava makes flat lava plains.

Lava plains (maria) on the Moon

b Medium-viscosity lava makes shallow-sloped shield volcanoes.

Olympus Mons (Mars)

c High-viscosity lava makes steep-sloped stratovolcanoes.

Mount St. Helens

FIGURE 9.17 Volcanoes produce different types of features depending primarily on the viscosity of the lava erupted.

chains can jiggle and slide along one another, and the lower the viscosity. The amount of water or gases trapped in the lava also affects viscosity. Such materials can act as a lubricant, decreasing the viscosity of subsurface magma. But when lavas erupt, materials such as water may form gas bubbles that *increase* the viscosity. (You can understand why bubbles increase viscosity by thinking about how much higher a "mountain" you can build with bubble-bath foam than you can with soapy water.)

The runniest basalt lavas flow far and flatten out before solidifying, creating vast *volcanic plains* such as the lunar maria (Figure 9.17a). Somewhat more viscous basalt lavas solidify before they can completely spread out, resulting in **shield volcanoes** (so-named because they are shield-shaped). Shield volcanoes can be very tall, but they are not very steep; most have slopes of only 5°–10° (Figure 9.17b). The mountains of the Hawaiian Islands are shield volcanoes; measured from the ocean floor to their summits, the Hawaiian mountains are the tallest (and widest) on Earth. Tall, steep **stratovolcanoes** such

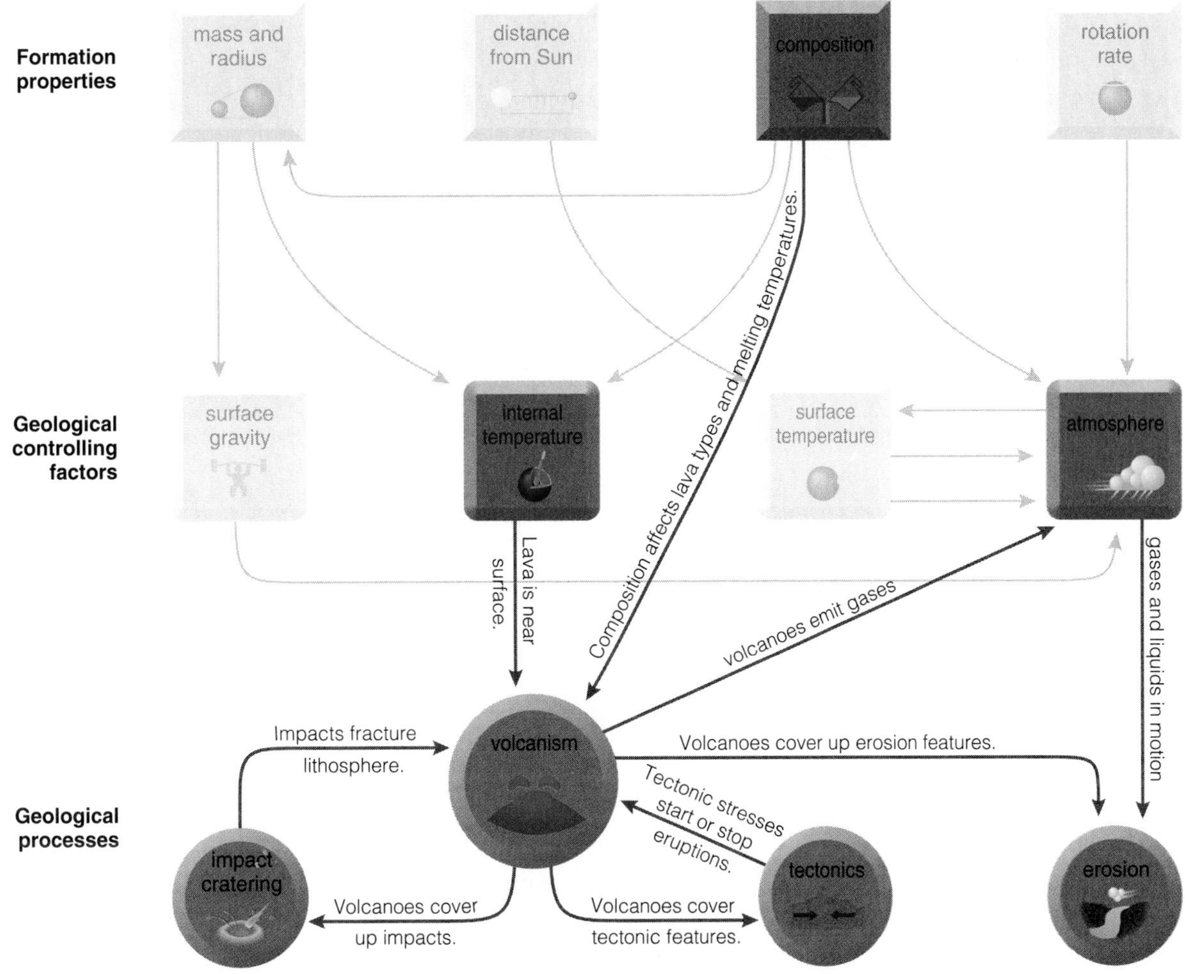

FIGURE 9.18 Major connections between volcanism, other geological processes, controlling factors, and formation properties. This diagram includes all the blocks and arrows from Figure 9.10, but those that are not directly relevant to volcanism are shaded lighter to clarify this illustration.

as Mount St. Helens are made from much more viscous lavas that can't flow very far before solidifying (Figure 9.17c).

Figure 9.18 summarizes the geological relationships that involve volcanism. The main point to keep in mind is that internal temperature exerts the greatest control over volcanism: A hotter interior means that lava lies closer to the surface and can erupt more easily. The formation property of composition is also very important, since it determines the melting temperatures and viscosities of lavas that erupt and therefore what types of volcanoes form. Volcanism is connected to other geological processes in several ways. First, lava flows can cover up or fill in previous geological features, which is why Venus has very few large impact craters. Second, tectonics and volcanism often go together: Tectonic stresses can either force lava to the surface or cut off eruptions. The final connection is one of the most important: Volcanism indirectly affects erosion, since volcanoes are the primary source of gases in the current terrestrial planet atmospheres [Section 10.4]. Before continuing, be sure you understand the other relationships shown in Figure 9.18.

Tectonics

We are now ready to discuss tectonics, the third major geological process in our model. The root of the word *tectonics* comes from Greek legend, in which Tecton was a carpenter. In geology, *tectonics* refers to the processes that do "carpentry" on planetary surfaces—that is, to the internal forces and stresses that act on the lithosphere to create surface features.

Several types of internal stress can drive tectonic activity. On planets with internal convection, the strongest type of stress often comes from the circulation of the convection cells themselves. The tops

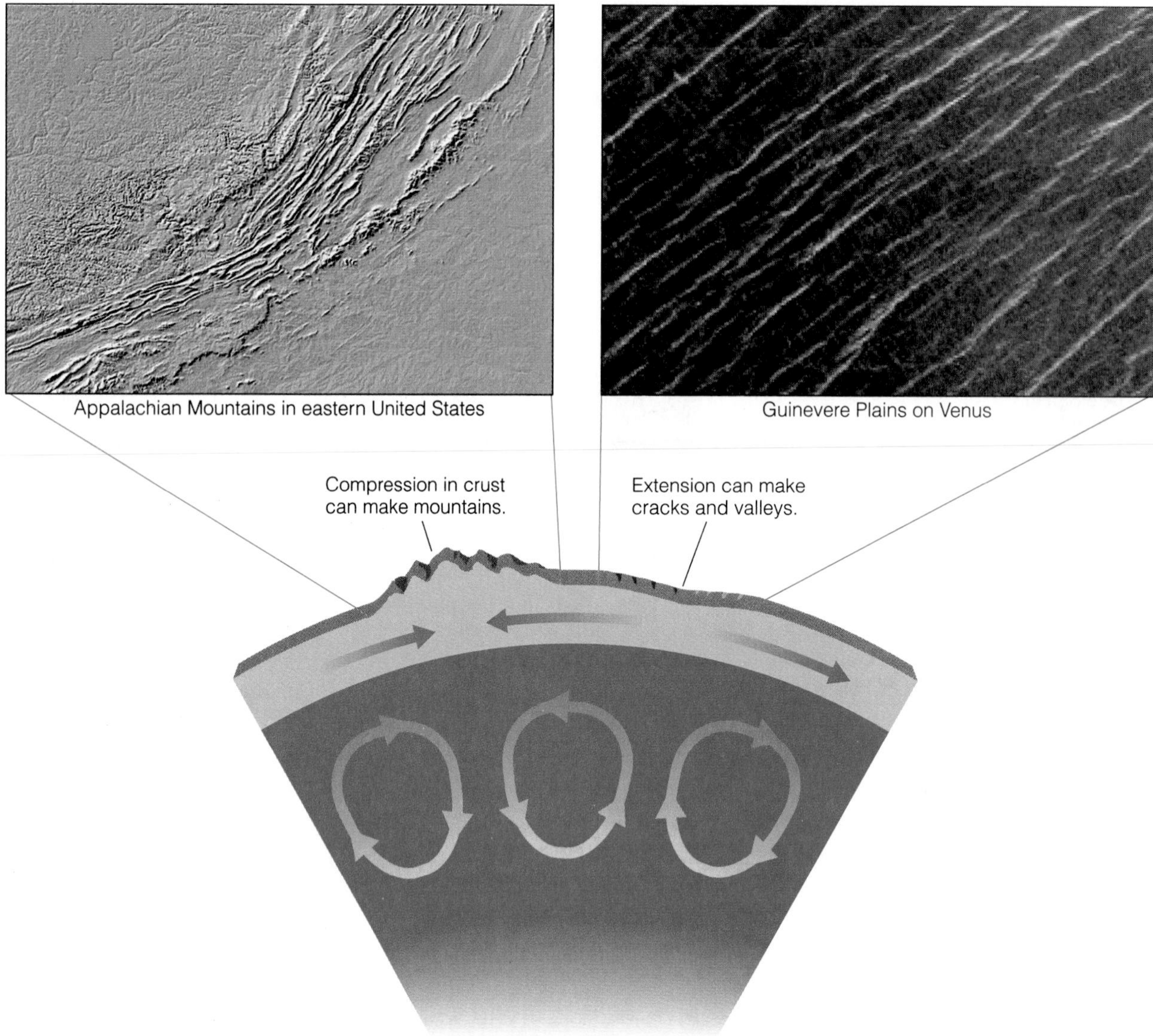

FIGURE 9.19 Tectonic forces can produce a wide variety of features. Mountains and fractured plains are among the most common.

of convection cells can drag against the lithosphere, sometimes forcing sections of the lithosphere together or apart. On Earth, these forces have broken the lithosphere into *plates* that move over, under, and around each other in what we call *plate tectonics* [Section 13.2]. Even if the crust does not break into plates, stresses from underlying convection cells may create vast mountain ranges, cliffs, and valleys. A second type of stress associated with convection comes from individual rising *plumes* of hot mantle material that push up on the lithosphere. Internal stress can also arise from temperature changes in the planetary interior. For example, the crust may be forced to expand and stretch if the planetary interior heats up from radioactive decay. Conversely, the mantle and lithosphere respond as a planetary core cools and contracts, leading to planet-wide compression forces. Tectonic stresses can also occur on a more local scale; for example, the weight of a newly formed volcano can bend or crack the lithosphere beneath it.

Tectonic features take an incredible variety of forms (Figure 9.19). Mountains may rise where the crust is compressed. Such crustal compression helped create the Appalachian Mountains of the eastern United States. Huge valleys and cliffs may result where the crust is pulled apart; examples include the Guinevere Plains on Venus and New Mexico's Rio

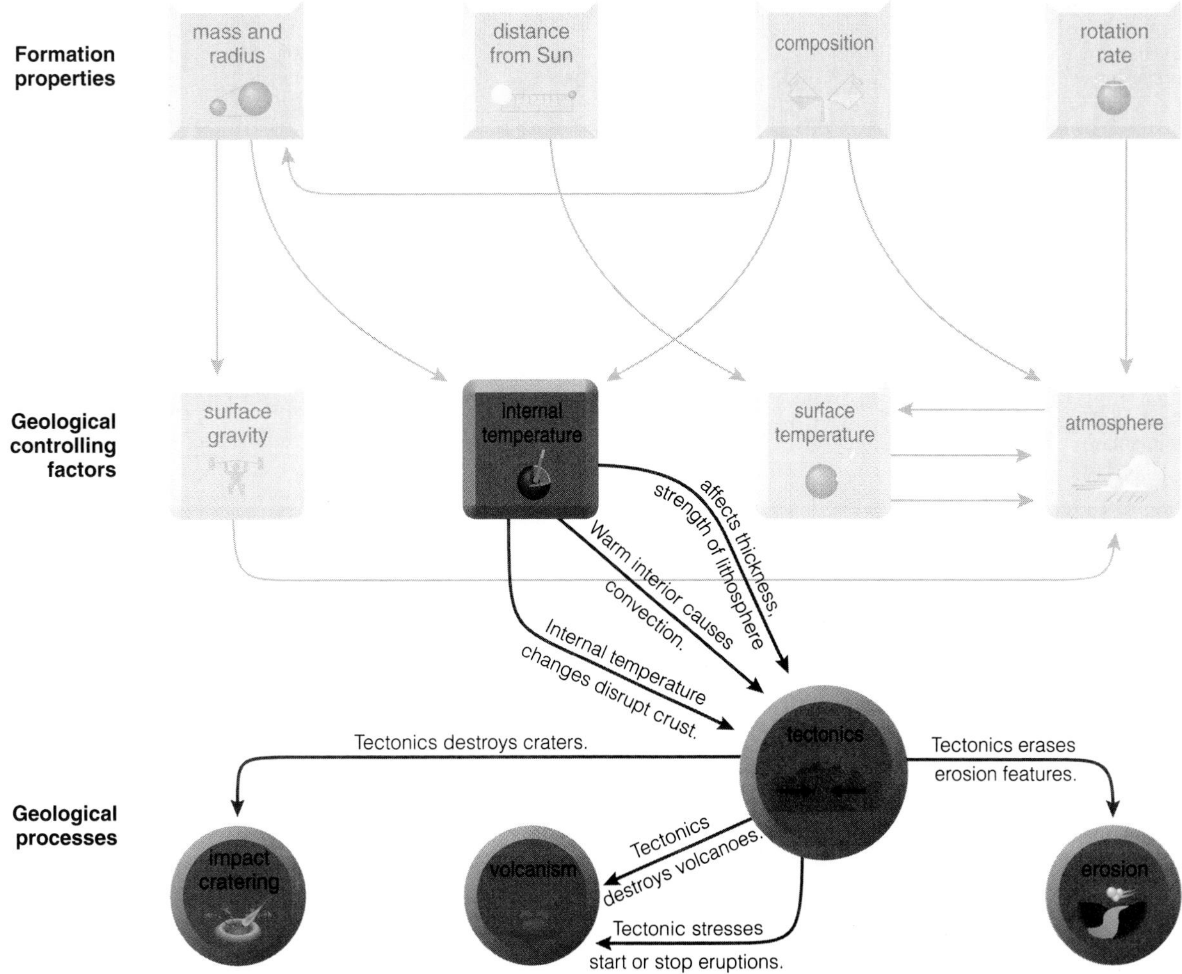

FIGURE 9.20 Connections between tectonics, other geological processes, controlling factors, and formation properties. This diagram includes all the blocks and arrows from Figure 9.10, but those that are not directly relevant to tectonics are shaded lighter to clarify this illustration.

Grande Valley here on Earth.[7] Tectonic forces may bend or break rocks, and tectonic activity on Earth is always accompanied by earthquakes. Other worlds undoubtedly experience "planet-quakes," and planetary geologists are eager to plant seismometers on them to probe their interiors.

Figure 9.20 summarizes the connections between tectonics and other geological processes and properties. Like volcanism, tectonics is most likely on larger worlds that remain hot inside. Tectonics may also have been important in the past on smaller worlds that underwent major shrinking or expansion.

[7]The Rio Grande came *after* the valley formed from tectonic processes.

Erosion

The last of the four major geological processes to be added to our model is erosion, which encompasses a variety of processes connected by a single theme: the breakdown and transport of rocks by **volatiles.** The term *volatile* means "evaporates easily" and refers to substances—such as water, carbon dioxide, and methane—that are usually found as gases, liquids, or surface ices on the terrestrial worlds. Wind, rain, rivers, flash floods, and glaciers are just a few examples of processes that contribute to erosion on Earth. Erosion not only breaks down existing geological features (wearing down mountains and forming gullies, riverbeds, and deep valleys), but also builds new ones (such as sand dunes, river deltas, and glacial deposits).

Virtually no erosion takes place on worlds without a significant atmosphere, such as Mercury and the

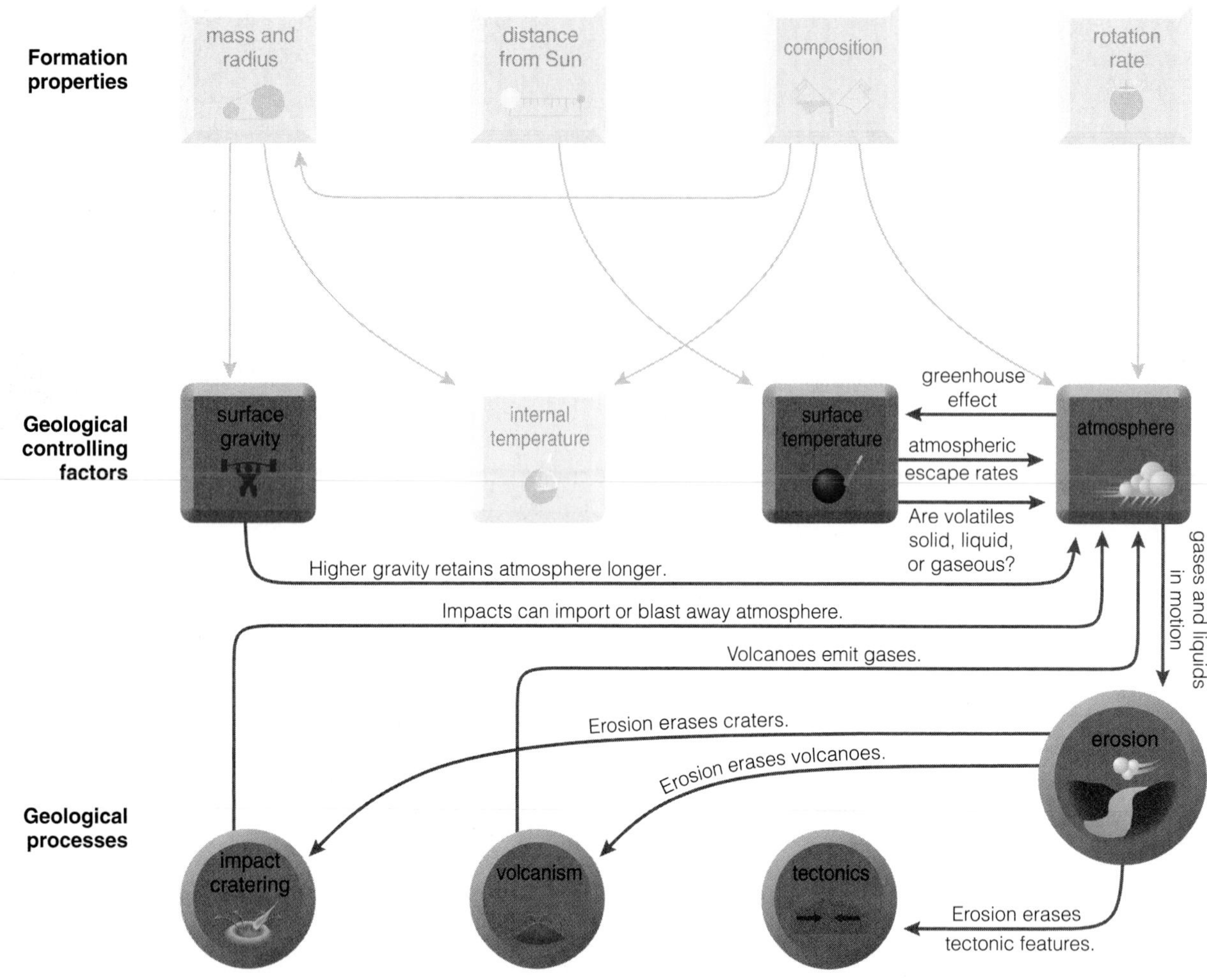

FIGURE 9.21 Connections between erosion, other geological processes, controlling factors, and formation properties. This diagram includes all the blocks and arrows from Figure 9.10, but those that are not directly relevant to erosion are shaded lighter to clarify this illustration.

Moon. But planets with atmospheres can have significant erosional activity. Larger planets are better able to generate atmospheres by releasing gases trapped in their interiors; their stronger gravity also tends to prevent their atmospheres from escaping to space [Section 10.4]. In general, a thick atmosphere is more capable of driving erosion than a thin atmosphere, but a planet's rotation rate is also important. Slowly rotating planets have correspondingly slow winds and therefore weak or nonexistent wind erosion. Surface temperature also matters: Erosion isn't very important if most of the volatiles stay frozen on the surface, but it can be very powerful when volatiles are able to evaporate and then recondense as rain or snow. Figure 9.21 summarizes the connections between erosion and other geological processes and properties; it also repeats the most important factors controlling the atmosphere.

9.5 A Geological Tour of the Terrestrial Worlds

Now that we've examined the processes and properties that shape the geology of the terrestrial worlds, we're ready to return to the issue that lies at the heart of this chapter: why the terrestrial worlds ended up so geologically different from one another. We'll take a brief "tour" of the terrestrial worlds. We could organize the tour by distance from the Sun, by density, or even by alphabetical order. But we'll choose the planetary property that has the strongest effect on geology: size, which controls a planet's internal heat. We'll start with our Moon, the smallest terrestrial world.

The Moon (1,731-km radius, 1.0 AU from Sun)

When you look at the Moon on a clear night, you can see much of its global geological history with your naked eye. The Moon is unique in this respect; other planets are too far away to see any surface details,

a Astronaut explores a small crater.

b Site of an ancient lava river.

c Lava flows filled impact basins to create maria like this one (Mare Imbrium). Wrinkles on the maria are evidence of minor tectonic forces.

FIGURE 9.22 Impact cratering and volcanism are the most important geological processes on the Moon.

and Earth is so close that we see only the local geological history. Twelve Apollo astronauts visited the lunar surface between 1969 and 1972, taking photographs, making measurements, and collecting rocks. Thanks to these visits and more recent observations of the Moon from other spacecraft, we know more about the Moon's geology than that of any other object, with the possible exception of the Earth.[8]

The Moon's composition differs somewhat from the Earth's. According to the leading theory, our Moon formed very early in the history of our solar system when a giant impact blasted away some of Earth's rocky outer layers without dredging up metallic core material [Section 8.6]. The Moon therefore has a lower metal content than Earth and a smaller (or possibly nonexistent) metal core. In addition, lunar rocks collected by astronauts contain a very low amount of volatiles compared to Earth rocks. Presumably, heat from the giant impact evaporated the volatiles and allowed them to escape into space before the Moon formed from the debris.

As the Moon accreted following the giant impact, the heat of accretion melted its outer layers. The lowest-density molten rock rose upward through the process of differentiation, forming what is sometimes called a *magma ocean*. This magma ocean cooled and solidified during the heavy bombardment that took place early in the history of the solar system, leaving the Moon's surface crowded with craters. We still see this ancient, heavily cratered landscape in the lunar highlands (Figure 9.22a). Lunar samples confirm that the highlands are composed of low-density rocks. As the bombardment tailed off, around 3.8 billion years ago, a few larger impactors struck the surface and formed impact basins.

Even as the heat of accretion leaked away, radioactive decay kept the Moon's interior molten long enough for volcanism to reshape parts of its surface. Between about 3 and 4 billion years ago, molten rock welled up through cracks in the deepest impact basins, forming the lunar maria. Because the lava that filled the maria rose up from the Moon's mantle, the maria contain dense, iron-rich rock that is darker in color than the rock of the lunar highlands. The contrasts

[8]Because much of Earth's surface is unexplored seafloor, some people claim that we know more about the Moon than about the Earth.

between the light-colored rock of the highlands and the dark rock of the maria make the "man-in-the-Moon" pattern that some people imagine when they look at the full moon.

The lunar lavas must have been among the least viscous (runniest) in the solar system, perhaps because the lack of volatiles meant a lack of bubbles (which increase viscosity) in the erupting lava. The lava plains of the maria cover a large fraction of the surface. Only a few small shield volcanoes exist on the Moon, and no steeper-sided stratovolcanoes have been found. Low-viscosity lavas also carved out long, winding channels (Figure 9.22b). These channels must once have been rivers of molten rock that helped fill the lunar maria.

Almost all the geological features of the Moon were formed by impacts and volcanism, although a few small-scale tectonic stresses wrinkled the surface as the lava of the maria cooled and contracted (Figure 9.22c). The Moon has virtually no erosion because of the lack of atmosphere, but the continual rain of micrometeorites has rounded crater rims. The heyday of lunar volcanism is long gone—3 billion years gone. Over time, the Moon's small size allowed its interior to cool, thickening the lithosphere. Today, the Moon's lithosphere probably extends to a depth of 1,000 km, making it far too thick to allow further volcanic or tectonic activity. The Moon has probably been in this geologically "dead" state for 3 billion years.

Despite its geological inactivity, the Moon is of prime interest to those hoping to build colonies in space. The *Lunar Prospector* arrived at the Moon in early 1998 and began mapping the surface in search of promising locations for a permanent human base. Although no concrete plans are yet in place as of 1998, several nations are exploring the possibility of building a human outpost on the Moon within the next couple of decades.

Mercury (2,439-km radius, 0.39 AU from Sun)

Mercury is the least studied of the terrestrial worlds. Its proximity to the Sun makes it difficult to study through telescopes, and it has been visited by only one spacecraft: *Mariner 10,* which collected data during three rapid flybys of Mercury in 1974–75. *Mariner 10* obtained images of only one hemisphere of Mercury (Figure 9.23b). The influence of Mercury's gravity on *Mariner 10*'s orbit allowed an accurate determination of Mercury's mass and high average density. Mercury is about 61% iron by mass, with a core that extends to perhaps 75% of its radius. Mercury's high metal content is due to its formation close to the Sun, possibly enhanced by a giant impact that blasted away its outer, rocky layers [Section 8.6].

Craters are visible almost everywhere on Mercury, indicating an ancient surface (Figure 9.23c). A

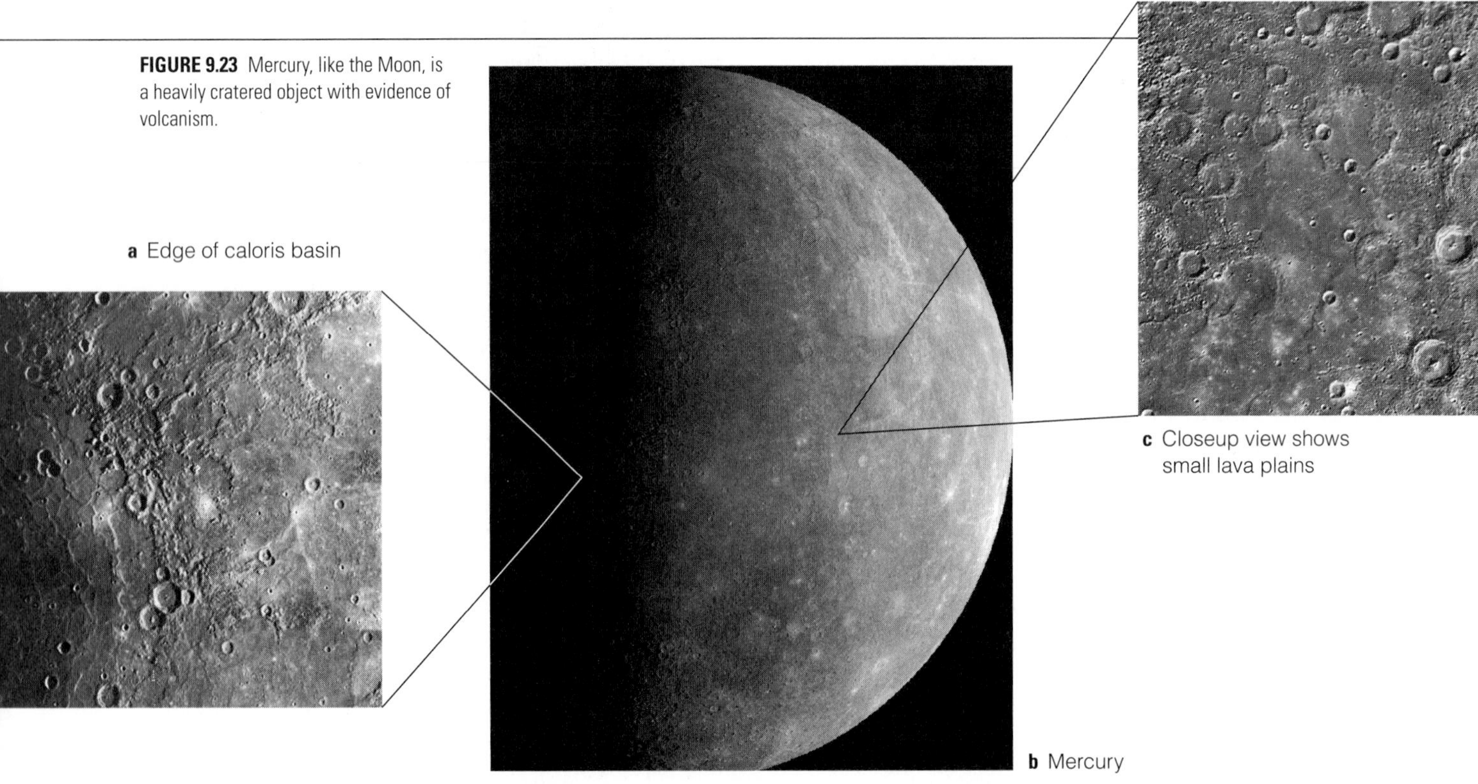

FIGURE 9.23 Mercury, like the Moon, is a heavily cratered object with evidence of volcanism.

a Edge of caloris basin

b Mercury

c Closeup view shows small lava plains

huge impact basin called the *Caloris Basin* (Figure 9.23a) covers a large portion of one hemisphere. Few craters have formed on top of the Caloris Basin, indicating that, like the large lunar basins, it must have been formed toward the end of the solar system's early period of heavy bombardment.

Despite the superficial similarity between the cratered surfaces of Mercury and the Moon, significant differences exist. Mercury has noticeably fewer craters than the lunar highlands (compare Figure 9.12a to Figure 9.23c). Smooth patches between many of the craters tell us that volcanic lava once flowed and covered up many of the smallest craters.

a

b

FIGURE 9.24 (**a**) This tall cliff on Mercury is several hundred kilometers long. (**b**) Cliffs formed from tectonic stresses (indicated by arrow) caused by Mercury's contraction.

Mercury has no large volcanic maria, but smaller lava plains are found almost everywhere, suggesting that Mercury went through at least as much volcanic activity as the Moon. Much like the Moon, Mercury passed through phases of accretion, differentiation, and heavy cratering. Radioactive heat accumulated enough to melt the interior again, leading to the volcanism that created the lava plains.

TIME OUT TO THINK *Geologists have discovered that the ejecta didn't travel quite as far away from craters on Mercury as it did on the Moon. Why not?* (Hint: *What planetary formation properties affect impact craters?*)

Tectonic processes played a more significant role on Mercury than on the Moon, as evidenced by tremendous cliffs. The cliff shown in Figure 9.24a is several hundred kilometers long and up to 3 kilometers high in places. The crater in the center of the image apparently crumpled during the formation of the cliff. This cliff, and many others on Mercury, probably formed when tectonic forces compressed the crust (Figure 9.24b). However, nowhere on Mercury do we find evidence of extension (stretching) to match this crustal compression. Can it be that the whole planet simply shrank? Apparently so. Early in its history, Mercury gained much more internal heat from accretion and differentiation than did the Moon, owing to its larger proportion of iron and stronger gravity. Then, as the core cooled, it contracted by perhaps as much as 20 kilometers in radius. This contraction crumpled and compressed the crust, forming the many compressional cliffs. The contraction probably also closed off volcanic vents, ending Mercury's volcanic activity.

Mercury lost its internal heat relatively quickly because of its small size. Its days of volcanism and tectonic shrinking probably ended within its first billion years. Mercury, like the Moon, has been geologically dead for most of its existence.

Mars (3,396-km radius, 1.52 AU from Sun)

The geological history of Mars is particularly interesting because it has long seemed the most promising place to look for life beyond Earth. Early telescopic observations of Mars revealed several uncanny resemblances to Earth. A Martian day is just over 24 hours, and the Martian rotation axis is tilted about the same amount as Earth's. Mars also has polar caps, which we now know to be composed primarily of frozen carbon dioxide, with smaller amounts of water ice. Telescopic observations also showed seasonal variations in surface coloration over the course of the Martian year (about 1.9 Earth years). All these discoveries led to the perception that Mars and Earth were at least cousins, if not twins. By the early 1900s, many astronomers—as well as the public—envisioned Mars as nearly Earth-like, possessing water, vegetation that changed with the seasons, and possibly intelligent life.

In 1877, Italian astronomer Giovanni Schiaparelli reported seeing linear features across the surface of

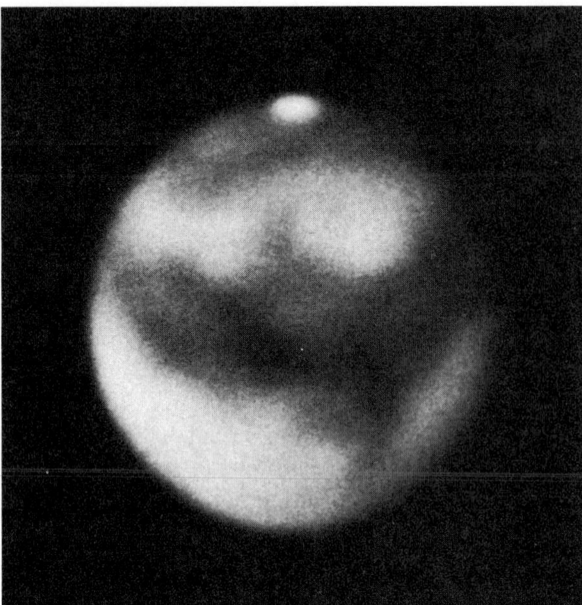

FIGURE 9.25 Can you see how the markings on Mars in the telescopic photo on the left might have resembled the geometrical features in the drawing by Percival Lowell on the right? Try blurring your eyes.

Mars through his telescope. He named these features *canali,* the Italian word for "channels."[9] This report inspired the American astronomer Percival Lowell to build an observatory in Flagstaff, Arizona, for the study of Mars. In 1895, Lowell claimed that canali were *canals* being used by intelligent Martians to route water around their planet. He suggested that Mars was falling victim to unfavorable climate changes and that water was a precious resource that had to be managed carefully. Lowell's studies drove rampant speculation about the nature of the Martians, fueling science fiction fantasies. The public mania drowned out the skepticism of astronomers who saw no canals through their telescopes or in their photographs (Figure 9.25).

The debate about Martian canals and cities was finally put to rest in 1965 with photos of the barren, cratered surface taken during the *Mariner 4* flyby. Schiaparelli's *canali* turned out to be nothing more than dark dust deposits redistributed by seasonal winds. Several other spacecraft studied Mars in the 1960s and 1970s, culminating with the impressive Viking missions in 1976. Each of the two Viking spacecraft deployed an orbiter that mapped the Martian surface from above and landers that returned surface images and regular weather reports for almost 5 years. More than 20 years passed before the next successful missions to Mars, the Mars Pathfinder mission and Mars Global Surveyor mission in 1997 [Section S2.6]. More missions to Mars are planned for every favorable launch window (roughly every 2 years) in the next decade and beyond.

The Viking and Mars Global Surveyor images show a much wider variety of geological features than seen on the Moon or Mercury. Some areas of the southern hemisphere are heavily cratered (Figure 9.26a), while the northern hemisphere is covered by huge volcanoes surrounded by extensive volcanic plains. Volcanism was expected on Mars—40% larger than Mercury, it should have retained a hot interior much longer. But no one knows why volcanism affected the northern hemisphere so much more than the southern hemisphere, or why the northern hemisphere on average is lower in altitude than the southern hemisphere.

Mars has a long, deep system of valleys called *Valles Marineris* running along its equator (Figure 9.26b). Named for the *Mariner 9* spacecraft that first imaged it, Valles Marineris is as long as the United States is wide, and almost four times deeper than the Grand Canyon. No one knows exactly how it formed; in fact, parts of the canyon are completely enclosed by high cliffs on all sides, so neither flowing lava nor water could have been responsible. Nevertheless, the linear "stretch marks" around the valley are evidence of tectonic stresses on a very large scale.

[9]It's not clear whether he really thought the channels contained water or was merely continuing the tradition, started with the Moon, of naming dark features after bodies of water.

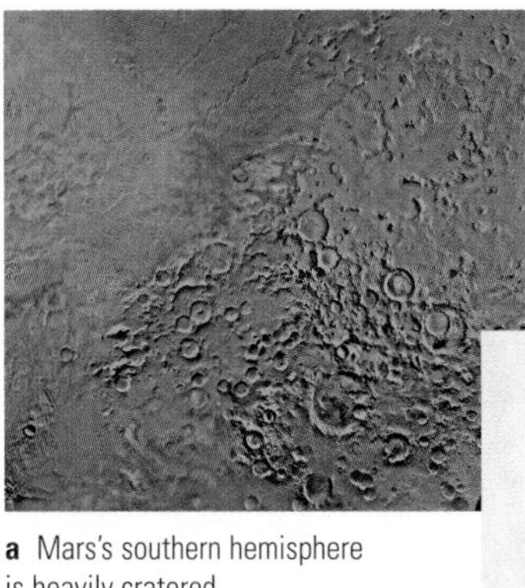

a Mars's southern hemisphere is heavily cratered.

b Valles Marineris is a huge valley created in part by tectonic stresses.

d Olympus Mons—the largest shield volcano in the solar system.

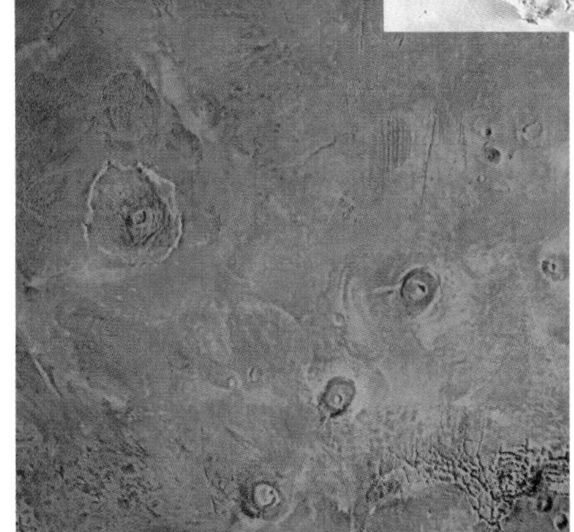

c A mantle plume was probably responsible for the volcanoes near the bottom of the image and the network of valleys near the top.

FIGURE 9.26 These photos of the Martian surface, taken from orbiting spacecraft, show that impact cratering has been an important process on Mars, but volcanism and tectonics have been even more important in shaping the current Martian surface.

More cracks are visible on the nearby *Tharsis Bulge,* a continent-size region that rises well above the surrounding Martian surface (Figure 9.26c). Tharsis was probably created by a long-lived plume of rising mantle material that bulged the surface upward while stretching and cracking the crust. The plume was also responsible for releasing vast amounts of basaltic lava that built up several gigantic shield volcanoes, including *Olympus Mons,* the largest shield volcano in the solar system (Figure 9.26d). Olympus Mons has roughly the same shallow slope as the volcanic island of Hawaii, but it is three times larger in every dimension. Its base is some 600 kilometers wide, and it stands 26 kilometers high—some three times higher than Mount Everest.

How old are the volcanoes? Are they still active? In the absence of red-hot lava (none has been seen so far), the abundance of impact craters provides the best answer. The Martian volcanoes are almost devoid of craters, but not quite. Planetary geologists therefore estimate that the volcanoes went dormant about a billion years ago. Geologically speaking, a billion years is not that long, and it remains possible that the Martian volcanoes will someday come back to life. However, the Martian interior is presumably cooling, and its lithosphere thickening. At best, it's only a matter of a few billion more years before Mars becomes as geologically dead as the Moon and Mercury.

Our Martian robotic explorers have also revealed geological features unlike any seen on the Moon or Mercury—features caused by erosion. Figure 9.27a no doubt reminds you of a dry riverbed on Earth seen from above. The eroded craters visible on some of the oldest terrain suggest that rain fell on the Martian surface billions of years ago (see Figure 9.13c). In some places, erosion has even acted below the surface, where underground water and ice have broken up or dissolved the rocks. When the water and ice disappeared, a wide variety of odd pits and troughs were left behind. This process may have contributed to the formation of parts of Valles Marineris.

Underground ice may be the key to one of the biggest floods in the history of the solar system. Figure 9.27b shows a landscape hundreds of kilometers across that was apparently scoured by rushing water. Geologists theorize that heat from a volcano melted great quantities of underground ice, unleashing a catastrophic flood that created winding channels and large sandbars along its way.

a This Viking photo (from orbit) shows ancient river beds that were probably created billions of years ago.

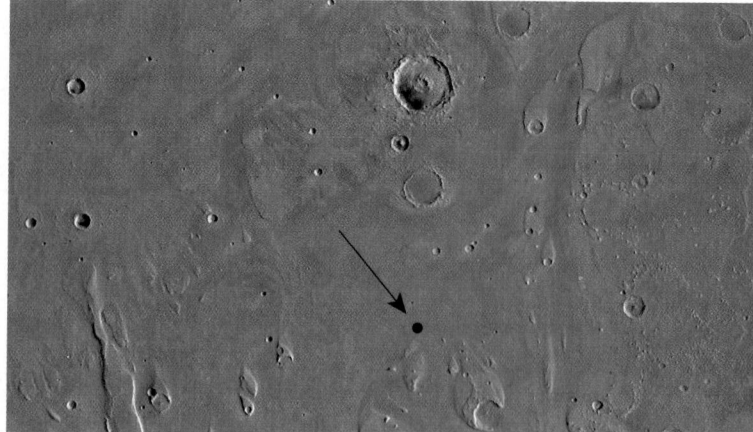

b This photo shows winding channels and sandbars that were probably created by catastrophic floods. The arrow shows the *Pathfinder* landing site.

c View from the floodplain (see **b** above) from the *Mars Pathfinder;* the *Pathfinder* landing site is now known as Carl Sagan Memorial Station.

FIGURE 9.27 Unlike the Moon and Mercury, Mars shows extensive evidence of erosion.

The Mars Pathfinder spacecraft went to this interesting region to see the flood damage close up. The lander released the *Sojourner* rover, named after Sojourner Truth, an African-American heroine of the Civil War era who traveled the nation advocating equal rights for women and blacks. The six-wheeled rover, no larger than a microwave oven, carried cameras and instruments to measure the chemical composition of nearby rocks. Together the lander and rover confirmed the flood hypothesis: Gentle, dry, winding channels were visible in stereo images; rocks of many different types had been jumbled together in the flood; and the departing waters had left rocks stacked against each other in just the same manner that floods do here on Earth (Figure 9.27c).

Despite evidence that water flowed in the past, Mars is dry today. The stark contrast between Mars's past and present is captured in a dramatic Mars Global Surveyor image (Figure 9.28) in which a channel that once carried water is now bone-dry and filled

FIGURE 9.28 Mars Global Surveyor image highlighting Mars's change from a wet planet to a dry one. The horizontal channel extending across the lower part of the image was probably once filled with water, but it is now filled with sand dunes.

with sand dunes. The strong Martian winds still stir up planet-wide dust storms, redistributing dust on the surface and slowly grinding away at ancient geological features. Thus, Mars is still geologically active, though only a shadow of its former self. Ironically, even though he lacked our modern geological evidence, Percival Lowell's supposition that the planet was drying up was basically correct! The reasons for Mars's climate change are fascinating and complex and will be an important focus of the next chapter.

The Mars Global Surveyor and Pathfinder missions are paving the way for more adventurous missions. *Mars Global Surveyor* is searching for optimal landing sites for future rovers—one will search for evidence of life past and present and another will gather rock samples for return to Earth. For the latest news on Mars missions, check the book web site.

Venus (6,051-km radius, 0.72 AU from Sun)

Venus is a large step up in size from Mars—it is almost as big as Earth. Thus, we would expect it to have a history of early impact cratering, followed by even more volcanism and tectonics than we have seen on Mars. Erosional activity is harder to predict, requiring a closer look at Venus's atmosphere.

Venus's thick cloud cover prevents us from seeing through to its surface, but we can study its geological features with radar. *Radar mapping* involves bouncing radio waves off the surface from a spacecraft and using the reflections to create three-dimensional images of the surface. From 1990 to 1993, the Magellan spacecraft used radar to map the surface of Venus, discerning features as small as 100 meters across. During its 4 years of observations, *Magellan* returned more data than all previous planetary missions combined. Like the planet itself, almost all the geological features are named for female goddesses and famous women in history.

Most of Venus is covered by relatively smooth, rolling plains with few mountain ranges. The surface has some impact craters (Figure 9.29a), but very few compared to the Moon or even Mars. Venus lacks very small craters, because small impactors burn up in its dense atmosphere. But even large craters are rare, indicating that they must have been erased by other geological processes. Volcanism is clearly at work on Venus, as evidenced by an abundance of volcanoes and lava flows. Some are shield volcanoes, indicating familiar basaltic eruptions (Figure 9.29b). Some volcanoes have steeper sides, probably indicating eruptions of a different kind of lava (Figure 9.29c).

The most remarkable features on Venus are tectonic in origin. Its crust is quite contorted; in some regions, the surface appears to be fractured in a regular pattern (Figure 9.29d; see also Figure 9.19). Many of the tectonic features are associated with volcanic features, suggesting a particularly strong linkage between volcanism and tectonics on Venus. One striking example of this linkage is a type of roughly circular feature called a *corona* (Latin for "crown") that probably resulted from a hot, rising plume in the mantle (Figure 9.29e). The plume pushed up on the crust, forming concentric tectonic stretch marks on the surface. The plume also forced lava to the surface, dotting the area with volcanoes.

Does Venus, like Earth, have plate tectonics that moves pieces of its lithosphere around? Convincing evidence for plate tectonics might include deep trenches and linear mountain ranges like those we see on Earth, but *Magellan* found no such evidence on Venus. The apparent lack of plate tectonics suggests that Venus's lithosphere is quite different from Earth's. Some geologists theorize that Venus's lithosphere is thicker and stronger, preventing its surface from fracturing into plates. Another possibility is that plate tectonics happens only occasionally on Venus. The uniform but low abundance of craters suggests that most of the surface formed around a billion years ago. Plate tectonics is one candidate for the process that wiped out older surface features and created a fresh surface at that time. For the last billion years, only volcanism, a bit of impact cratering, and small-scale tectonics have occurred.

One might expect Venus's thick atmosphere to drive strong erosional processes, but the view both from orbit and from the surface suggests that erosion has only a minimal effect. The Soviet Union landed two probes on the surface of Venus in 1975. Before the 700 K heat destroyed them, they returned images of a bleak, volcanic landscape with little evidence of erosional activity (see Figure 9.3). Venus apparently lacks the rains and winds that drive erosion, probably because of its high surface temperature and slow rotation rate.

The lack of strong erosion on Venus leaves the tectonic contortions of its surface exposed to view, even though some probably approach a billion years in age. Earth's terrain might look equally stark and rugged from space if not for the softening touches of wind, rain, and life. Like Earth, Venus probably has ongoing volcanic and tectonic activity, although we have no direct proof of it.

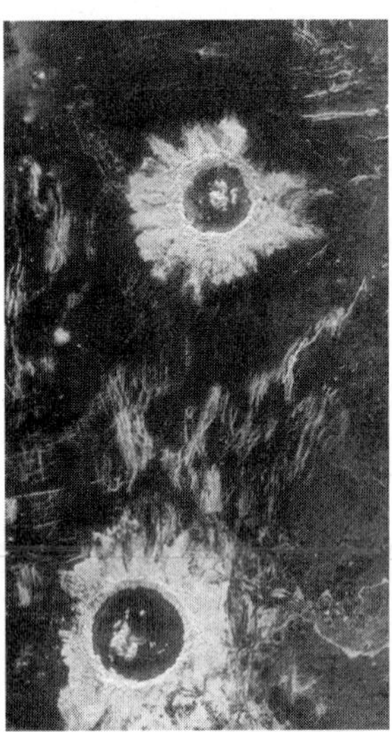

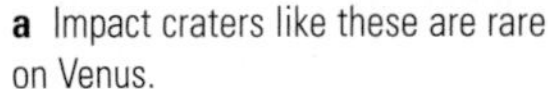

a Impact craters like these are rare on Venus.

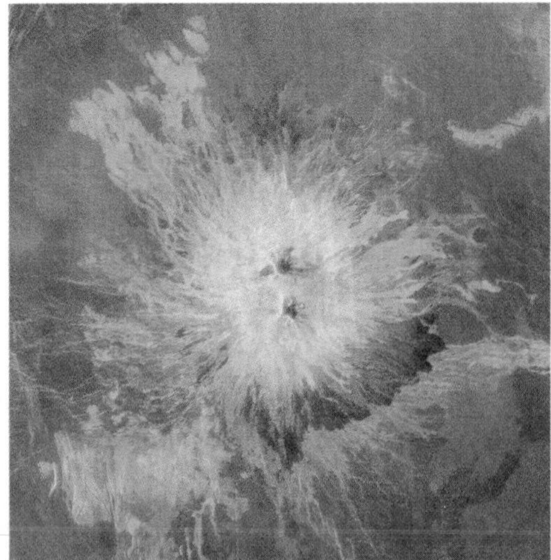

b Shield volcanoes like this one are common on Venus.

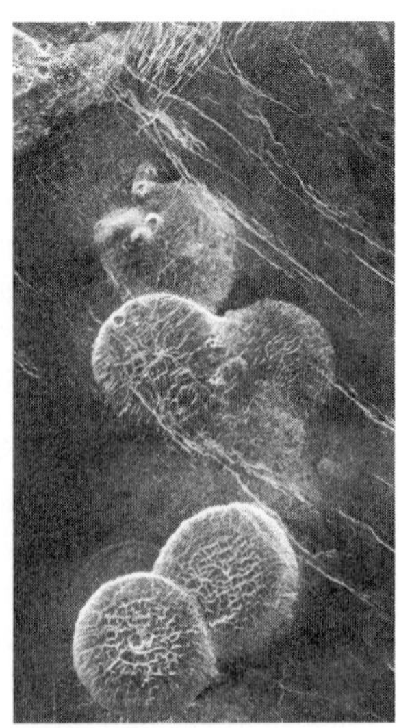

c These volcanoes were made from viscous lava.

d Tectonic forces have fractured and twisted the crust in this region.

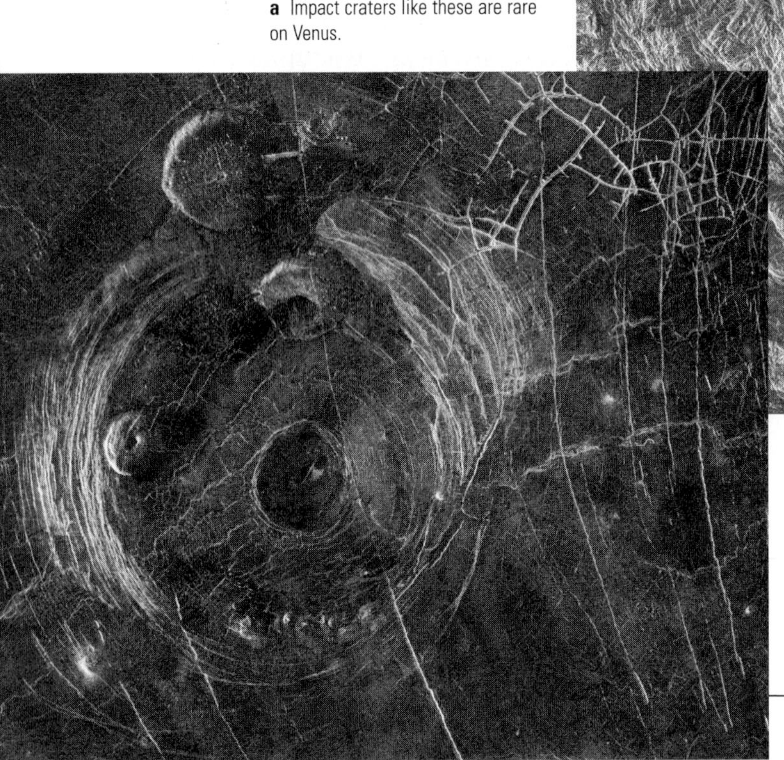

e A mantle plume probably created the round corona, which is surrounded by tectonic stress marks.

FIGURE 9.29 The surface of Venus is covered with abundant lava flows and tectonic features, along with a few large impact craters. Because these images were taken by the Magellan spacecraft radar, dark and light areas correspond to how well radio waves are reflected, not visible light. Nonetheless, geological features stand out well.

Looking Forward to Earth (6,378-km radius, 1.0 AU from Sun)

We've toured our neighboring terrestrial worlds and found a clear trend in the relationship of their volcanic and tectonic activity to size. Soon after the intense cratering ended about 3.8 billion years ago, the larger planets underwent extensive reworking of their surfaces through volcanic and tectonic activity, which removed the evidence of earlier craters. The largest terrestrial worlds may still have ongoing tectonic and volcanic activity.

What does our model of planetary geology predict for Earth? Earth is the largest of the terrestrial worlds, so we would expect to see all the activity of smaller worlds and more. In terms of volcanism and tectonics, Earth should be most like Venus because of their similar sizes. But in terms of erosion, Earth appears to be most like its smaller cousin Mars, with a surface sculpted by running water. Before we can fully understand the processes that shape the Earth, we need to look more deeply at the behavior of planetary atmospheres—the focus of the next chapter. We'll return to the subject of Earth's geology in Chapter 13, after we have completed our development of comparative planetology.

Nothing is rich but the inexhaustible wealth of nature. She shows us only surfaces, but she is a million fathoms deep.

RALPH WALDO EMERSON

THE BIG PICTURE

In this chapter, we have seen how the images returned from our robotic explorers have helped decipher the histories of the planets and have yielded clues to what their interiors must be like. As you continue your study of the solar system, keep the following "big picture" ideas in mind:

- Much of planetary geology can be distilled down to a few basic geological processes that depend on a handful of planetary properties.
- Every terrestrial world was once as heavily cratered as the Moon is today, but craters have been erased on the other worlds to varying degrees—mainly depending on each planet's size.
- Our model of planetary geology explains much of what we see on terrestrial worlds in our solar system and might help us learn what happens on planets in other solar systems.
- Understanding the differences and similarities between Venus, Earth, and Mars requires a closer look at planetary atmospheres and how they change.

Review Questions

1. Briefly explain how spacecraft allow us to study the geology of other worlds.
2. Briefly explain how the terrestrial planets ended up spherical in shape. Would you also expect small asteroids to be spherical? Why or why not?
3. What do we mean when we talk about rock that is *molten?* What is *viscosity?* Give examples of substances with low viscosity and substances with high viscosity.
4. How does *differentiation* occur, and under what circumstances? Explain how differentiation led to the *core–mantle–crust* structure of the terrestrial worlds.
5. What is the *lithosphere* of a planet? How does it correspond to the layering described in terms of core–mantle–crust? What determines the strength and thickness of a planet's lithosphere?
6. Briefly describe several ways by which we learn about planetary interiors.
7. Describe and distinguish among the three major processes by which planetary interiors get hot: *accretion, differentiation,* and *radioactivity.*
8. Describe and distinguish among the three major processes by which planetary interiors transfer heat to planetary surfaces: *conduction, convection,* and *eruption.* How does the heat escape from the planet's surface into space?
9. Describe the conditions under which convection occurs. What are *convection cells?*
10. What is the primary factor in determining how long a planetary interior remains hot, and why?
11. What is a *magnetic field?* Contrast the magnetic fields of terrestrial planets.
12. Briefly define each of the four major geological processes: *impact cratering, volcanism, tectonics,* and *erosion.*
13. Describe and distinguish between what we call *formation properties* and *geological controlling factors* in the context of planetary geology. Summarize in your own words the relationships shown in Figure 9.10.
14. Briefly describe how an impact affects a planetary surface. How large is the crater compared to the size of the impactor? What is the *ejecta* from an impact? What is an *impact basin?*
15. Explain how we can estimate the geological age of a planetary surface from its number of impact craters. How do we know that the *lunar maria* are younger than the *lunar highlands?*

16. What can we learn from the detailed shapes of craters? Give several examples from Mars.
17. What created the "soil" on the lunar surface? Explain why the footprints of the astronauts on the Moon will last a long time, but not forever.
18. Give two reasons why *magma* may rise to a planetary surface.
19. What is the difference between volcanic features made by low-, medium-, and high-viscosity lava? Explain how the viscosity of the lava leads to *volcanic plains, shield volcanoes,* or *stratovolcanoes.* Which of the terrestrial worlds have volcanic plains and shield volcanoes? Which have stratovolcanoes?
20. Describe several types of internal stresses that can drive tectonic activity on a planet. Under what conditions does a lithosphere break into *plates* and show *plate tectonics?* Briefly explain how interior temperature affects tectonics.
21. What are *volatiles,* and how are they important to erosion? Briefly explain why Mercury and the Moon have virtually no erosion.
22. Briefly summarize the geological history of the Moon. How and when did the maria form? Why is the Moon geologically "dead" today?
23. Briefly summarize the geological history of Mercury. What is the *Caloris Basin?* How did Mercury get its many huge cliffs? How and why did the geological history of Mercury differ from that of the Moon? How did we determine that Mercury has a huge iron core (relative to its size)?
24. Briefly discuss some of the superficial similarities between Mars and Earth that led many people in the early 1900s to believe that Mars must harbor life. What were the *canali* that some astronomers reported on Mars? Do they really exist?
25. Describe the mysteries of the differences in appearance between Mars's northern and southern hemispheres and of the origin of *Valles Marineris.* How does Valles Marineris compare in size to the Grand Canyon on Earth?
26. Describe the *Tharsis Bulge* and *Olympus Mons.* Briefly discuss the evidence concerning whether Mars remains volcanically active.
27. Briefly discuss the evidence for past water flow on Mars, including catastrophic floods and rainfall. Does Mars have any flowing water today?
28. Briefly describe the evidence for volcanic and tectonic activity on Venus. Does Venus have plate tectonics? Does it show signs of significant erosion?

Discussion Questions

1. *Making Models.* In this chapter, we developed a relatively simple conceptual model to describe the complex interactions that govern planetary geology. Use a similar approach to make a model for a complex problem of interest to you. The problem need not be astronomical; for example, you might consider global warming, changes in the stock market, or ways that class size could be reduced at your college or university. Your model need not use flowcharts, but it should include (1) some phenomenon you wish to understand, (2) the processes that affect it, and (3) the various factors those processes depend on. Discuss your model with others, and see if you can reach agreement on what factors are most important to understanding your chosen problem.
2. *Geological Connections.* Consider the four geological processes: impact cratering, volcanism, tectonics, and erosion. Which two do you think are most closely connected to each other? Explain your answer, giving several ways in which the processes are connected.
3. *Testing the Model.* Our theories of planetary geology seem to explain the terrestrial worlds fairly well, but do they work elsewhere in the solar system?
   a. According to the model based on the terrestrial worlds, which of the four geological processes would you predict to be most important on Jupiter's moon Io (1,815-km radius) and Uranus's moon Miranda (243-km radius)?
   b. Based on the photos of these moons (Figure 11.18 and Figure 11.25), are your predictions correct?
   c. In a strictly logical sense, what would you conclude about our model of planetary geology based on your answer to part (b)? Explain.

Problems

1. *Earth vs. Moon.* Why is the Moon heavily cratered, but not Earth? Explain in a paragraph or two, relating your answer to their *planetary formation properties.*
2. *Erosion.* Consider erosion on Mercury, Venus, the Moon, and Mars. Which of these four worlds has the greatest erosional activity, and why? Briefly explain why the other three worlds do not have significant erosion. In all your answers, relate erosional activity to the four *planetary formation properties.* Summarize your answers with a paragraph for each world.
3. *Miniature Mars.* Suppose Mars had turned out to be significantly smaller than its current size—say, the size of our Moon. How would this have affected the number of geological features due to each of the four major geological processes (impact cratering, volcan-

ism, tectonics, and erosion)? Do you think Mars would still be a good candidate for harboring extraterrestrial life? Why or why not? Summarize your answers in two or three paragraphs.

4. *Mystery Planet.* It's the year 2098, and you are designing a robotic mission to a newly discovered planet around a star that is nearly identical to our Sun. The planet is as large in radius as Venus, rotates with the same daily period as Mars, and lies 1.2 AU from its star. Your spacecraft will orbit but not land.
   a. Some of your colleagues believe that the planet has no metallic core. How could you support or refute their hypothesis? (*Hint:* Remember that metal is dense and electrically conducting.)
   b. Other colleagues believe that the planet formed with few if any radioactive elements. How could images of the surface support or refute their hypothesis?
   c. Others suspect it has no atmosphere, but the instruments designed to study the planet's atmosphere fail due to a software error. How could you use the spacecraft's photos of geological features to determine whether a significant atmosphere is (or was) present on this planet?

5. *Dating Planetary Surfaces.* We have discussed two basic techniques for determining the age of a planetary surface: studying the abundance of impact craters, and radioactive dating of surface rocks [Section 8.7].
   a. Which technique seems more reliable? Why?
   b. Which technique is easier to use for planets besides Earth, and why?

6. *Impact Energies.* A relatively small impact crater 10 km in radius could be made by a comet 1 km in radius traveling at 30 km/s.
   a. Assume the comet has a density of 1 g/cm^3 (1,000 kg/m^3), and calculate its total kinetic energy. *Hints:* First find the volume of the comet in m^3, then find its mass by multiplying its volume by its density, and finally calculate its kinetic energy. You'll need the following formulas; if you use mass in kg and velocity in m/s, the answer for kinetic energy will have units of joules:

   $$\text{volume of sphere} = \tfrac{4}{3}\pi \times (\text{radius})^3$$
   $$\text{kinetic energy} = \tfrac{1}{2} \times \text{mass} \times (\text{velocity})^2$$

   b. Convert your answer from part (a) to megatons of TNT, the unit used for nuclear bombs. Comment on the degree of devastation that the impact of such a comet could cause if it struck a populated region on Earth. (*Hint:* One megaton of TNT releases 4.2×10^{15} joules of energy.)

7. *Project: Geological Properties of Silly Putty.*
   a. Roll room-temperature Silly Putty into a ball, and measure its diameter. Place the ball on a table, and gently place one end of a heavy book on it. After 5 seconds, measure the height of the squashed ball.
   b. Warm the Silly Putty in hot water and repeat the experiment. Then chill the Silly Putty in ice water and repeat. How did the different temperatures affect the rate of "squashing"?
   c. Did heating and cooling have a small or large effect on the rate of "squashing"? How does this experiment relate to planetary geology?

8. *Project: Planetary Cooling in a Freezer.* To simulate the cooling of planetary bodies of different sizes, use a freezer and two small *plastic* containers of similar shape but different size. Fill them with cold water and place them in the freezer. Checking every hour or so, record the time and your estimate of the thickness of the "lithosphere" (the frozen layer) in the two tubs.
   a. Make a graph in which you plot the lithospheric thickness versus time for each tub. In which tub does the lithosphere thicken fastest?
   b. How long does it take each tub to freeze completely? What is the ratio of the two freezing times?
   c. Describe in a few sentences the relevance of your experiment to planetary geology.

9. *Project: Tectonics in a Bowl of Pudding.* Make a large bowl of pudding and let it set well. Eat half, making a vertical "cliff" halfway across the bowl.
   a. Watch and comment on how the shape of the cliff changes over the next several minutes.
   b. Observe any fractures that appear in the pudding crust. If none appear, weaken the material by jiggling the bowl or poking the crust. Note the shape and orientation of the fractures that appear.
   c. Does the pudding surface resemble any of the planetary images in this chapter? Comment.

10. *Web Project: Mars Exploration.* Follow the links from the book web site to find the latest results from current Mars missions.
    a. In a few sentences, describe the main scientific goals of one of the missions.
    b. Print out (or describe) an image of geological interest to you, and analyze it using the techniques developed in this chapter.
    c. Explain how the latest results have changed or improved our understanding of Mars.

11. *Amateur Astronomy: Observing the Moon.* Any amateur telescope is adequate to identify geological features on the Moon. The light highlands and dark maria should be evident, and the shading of topography will be seen in the region near the terminator (the line between night and day). Try to observe near first- or third-quarter phase. Sketch the Moon at low magnification, and then zoom in on a region of interest, noting its location on your sketch. Sketch your field of view, label its features, and identify the geological process that created them. Look for craters, volcanic plains, and tectonic features. Estimate the horizontal size of the features by comparing them to the size of the whole Moon (radius = 1,738 km).

10

Atmospheres of the Terrestrial Worlds

LIFE AS WE KNOW IT WOULD BE IMPOSSIBLE ON EARTH WITHOUT our atmosphere. In addition to supplying the oxygen we breathe, it also shields us from harmful ultraviolet and X-ray radiation coming from the Sun, protects us from continual bombardment by micrometeorites, generates rain-giving clouds, and traps just enough heat to keep Earth habitable. We couldn't have designed a better atmosphere if we had tried.

Despite forming under similar conditions, our neighboring planets ended up with atmospheres utterly inhospitable to us. No other planet in our solar system has air that we could breathe, or temperature and pressure conditions in which we could survive without a spacesuit. Mars, only slightly farther from the Sun than Earth, has an atmosphere that leaves its surface cold, dry, and barren. Venus, only slightly closer to the Sun, has an atmosphere that makes its surface conditions resemble a classical view of hell.

How did Earth end up with such fortunate conditions? To fully appreciate what happened on Earth, we must first understand what happened on the other terrestrial worlds.

10.1 Planetary Atmospheres

All the planets of our solar system have atmospheres to varying degrees, as do several of the solar system's larger moons. The jovian planets are essentially atmosphere throughout—they are largely made of gaseous material. In contrast, the atmospheres of solid bodies—terrestrial worlds, jovian moons with atmospheres, and Pluto—are relatively thin layers of gas that contain only a minuscule fraction of a world's mass. Although we will concentrate on understanding the atmospheres of the terrestrial worlds, similar principles apply to the atmospheres of all the planets and satellites.

The terrestrial atmospheres are even more varied than the terrestrial geologies. The Moon and Mercury have very little atmosphere at all—you would seem to be facing the blackness of space even when standing on their surfaces. What little atmosphere they possess probably comes from materials vaporized as the surface is bombarded by micrometeorites, the solar wind, or energetic solar photons. These materials consist mostly of individual atoms of potassium, sodium, and oxygen, among others. If you could condense the entire atmosphere of the Moon into solid form, you would have so little material that you could store it in a basement.

At the other extreme, Venus is shrouded by an atmosphere so thick that it prevents us from seeing the planetary surface. If you stood on the surface of Venus, you'd feel a crushing pressure 90 times greater than that on Earth and a searing temperature higher than that in the hottest ovens used for cooking. Moving through the air on Venus would feel something like a cross between swimming and flying, because its density at the surface is nearly 10% of the density of liquid water. You couldn't breathe the air, because it consists almost entirely of carbon dioxide and does not contain any of the molecular oxygen (O_2) on which we depend. The surface views are perpetually overcast as weak sunlight filters though the thick clouds above. Violent storms on Venus are virtually nonexistent, and corrosive chemicals that rain downward from high in the atmosphere evaporate before ever hitting the ground.

The atmosphere of Mars is also made mostly of carbon dioxide, but much less of it than on Venus. The result is very thin air with a pressure so low that your body tissues would bulge painfully if you stood on the surface without wearing a full spacesuit. In fact, the pressure and temperature are both so low that liquid water would rapidly disappear, some evaporating and some freezing into ice. Thus, the fact that we see dried-up riverbeds on Mars tells us that its atmosphere must have been very different in the past [Section 9.5]. Mars has seasons like Earth because of its similar axis tilt, and it is occasionally engulfed in planet-wide dust storms.

Earth's atmosphere of nitrogen and oxygen is the only one under which human life is possible, at least without the creation of an artificial environment. What makes Earth's atmosphere so special? The best way to answer this question is to compare planetary atmospheres. Each atmosphere is unique, but, as we saw with geology in Chapter 9, similar properties and processes determine the characteristics of all atmospheres. Table 10.1 summarizes the properties of the terrestrial atmospheres, and Figure 10.1 shows representative views from above the worlds and from the surfaces. We'll spend the rest of this chapter learning

Table 10.1 Atmospheres of the Terrestrial Worlds

World	Composition	Surface Pressure*	Winds, Weather Patterns	Clouds, Hazes
Mercury	Atoms of helium, sodium, oxygen, potassium	10^{-15} bar	None: too little atmosphere	None
Venus	96% CO_2, 3.5% N_2	90 bars	Slow winds, no violent storms	Sulfuric acid clouds
Earth	77% N_2, 21% O_2, 1% argon, H_2O (variable)	1 bar	Winds, hurricanes	H_2O clouds, pollution
Moon	Atoms of sodium, potassium	10^{-15} bar	None: too little atmosphere	None
Mars	95% CO_2, 2.7% N_2, 1.6% argon	0.007 bar	Winds, dust storms	H_2O and CO_2 clouds, dust

*1 bar ≈ the pressure at sea level on Earth.

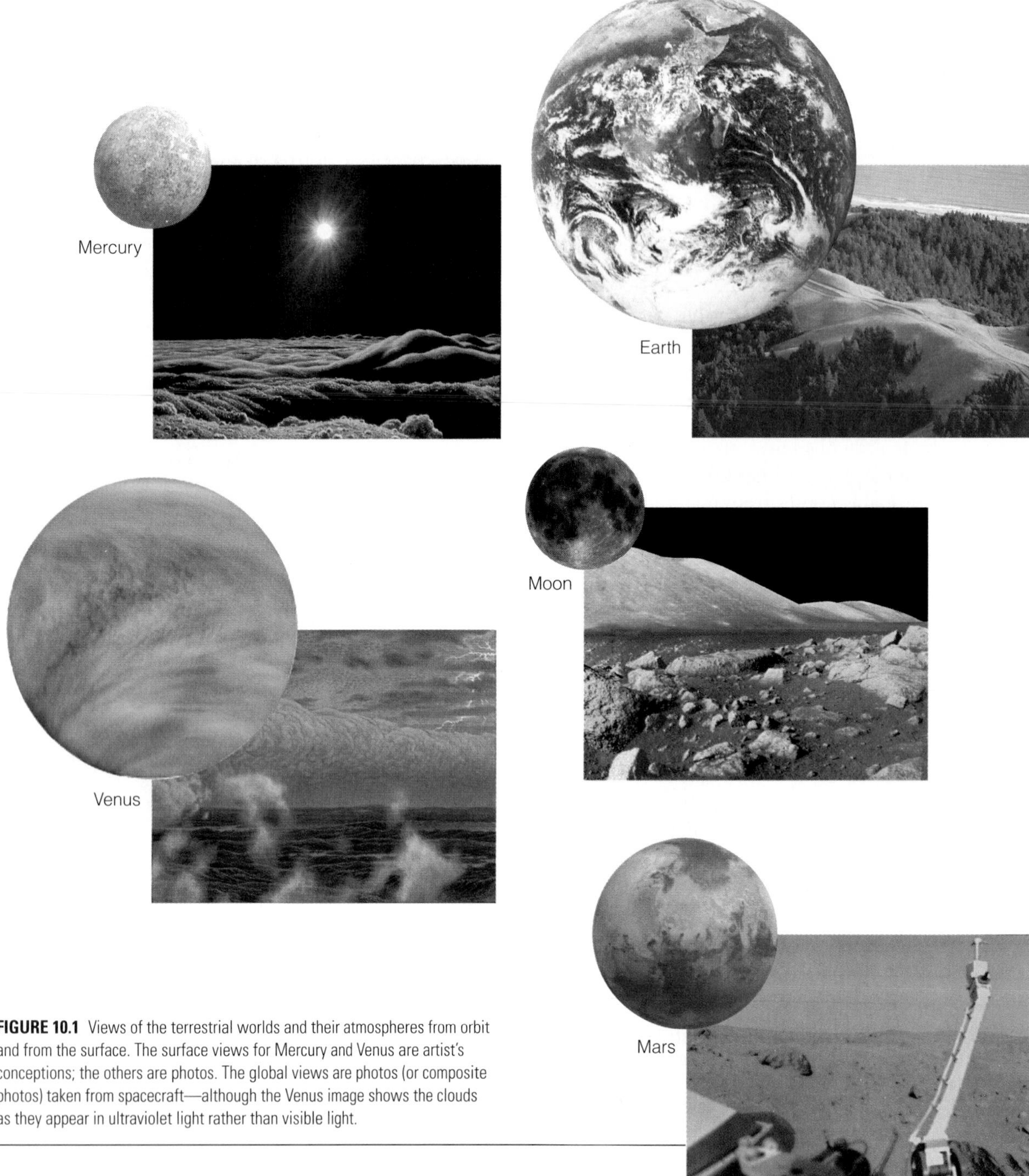

FIGURE 10.1 Views of the terrestrial worlds and their atmospheres from orbit and from the surface. The surface views for Mercury and Venus are artist's conceptions; the others are photos. The global views are photos (or composite photos) taken from spacecraft—although the Venus image shows the clouds as they appear in ultraviolet light rather than visible light.

how and why the terrestrial atmospheres came to differ so profoundly. Before we begin our comparative study, however, we first need to discuss a few basic properties of atmospheres. Let's begin with the most basic question: What is an atmosphere?

What Is an Atmosphere?

An *atmosphere* is a layer of gases that surrounds a world. The air that makes up any atmosphere is a mixture of many different gases that may consist either of individual atoms or of molecules. At the relatively low temperatures found throughout most planetary atmospheres, most atoms in gases combine to form molecules. For example, the air that we breathe consists of *molecular* nitrogen (N_2) and oxygen (O_2), as opposed to individual atoms (N or O). The wide variety of gases in planetary atmospheres falls into the category called *volatiles* [Section 9.4], materials that are usually gaseous or liquid at temperatures of a few hundred Kelvin. Common volatiles include nitrogen (N_2), oxygen (O_2), and water (H_2O) on Earth; carbon dioxide (CO_2) on Venus and Mars; and hydrogen (H_2), helium[1] (He), methane (CH_4), and ammonia (NH_3) on the jovian planets.

The individual atoms or molecules in air whiz around at high speeds and frequently collide with one another. On Earth, the nitrogen and oxygen molecules fly around at average speeds of about 500 meters per second—fast enough to cross your bedroom a hundred times in a second. Given that a single breath of air contains some 10^{22} molecules, you can imagine how frequently molecules collide. In fact, each molecule in the air around you will suffer a million collisions in the time it takes to read this paragraph.

All these collisions in a gas create the **gas pressure.** An easy way to understand pressure is to think about a balloon. The air molecules inside a balloon are constantly colliding with the balloon walls, exerting a pressure that tends to make the balloon expand. Because the particles in a gas move every which way, the pressure acts equally in all directions. At the same time, air molecules outside the balloon collide with the walls from the other side, tending to make the balloon contract. A sealed balloon remains stable when the inward pressure and the outward pressure are balanced. (We are neglecting any tension in the balloon material.) Now imagine that you blow more air into the balloon. The extra molecules inside mean more collisions with the balloon wall, momentarily making the pressure inside greater than the pressure outside. The balloon therefore expands until the inward and outward pressures are again in balance. If you heat the balloon, the gas molecules begin moving faster and therefore collide harder with the walls. Thus, heating a balloon also makes the inside pressure momentarily greater than the outside pressure, causing the balloon to expand. Conversely, cooling a balloon makes it contract, because the outside pressure momentarily exceeds the inside pressure.

We can understand *atmospheric pressure* using similar principles. Gas in an atmosphere is held down by gravity. The atmosphere over any area has some weight that presses downward, tending to compress the atmosphere beneath it. At the same time, gas pressure pushes in all directions, including upward, which tends to make the atmosphere expand. Planetary atmospheres exist in a perpetual balance between the downward weight of their gases and the upward push of their gas pressure.[2] The higher you go in an atmosphere, the less the weight of the gas above you. Thus, the pressure must also become less as you go upward, which explains why the pressure decreases as you climb a mountain or ascend in an airplane. You can visualize what happens by imagining the atmosphere as a very big stack of pillows. The pillows at the bottom are very compressed because of the weight of all the pillows above. As you go upward, the pillows are less and less compressed because less weight lies on top of them. The standard unit of pressure is the **bar** (as in *barometer*) and is roughly equal to Earth's atmospheric pressure at sea level.[3]

TIME OUT TO THINK *There is pressure under water just as there is pressure under air. Use this fact to explain why the pressure increases dramatically with depth in the oceans. How does this pressure affect deep-sea divers?*

Where Does an Atmosphere End?

The higher you go in an atmosphere, the lower the pressure becomes. Atmospheres therefore do not have clear upper boundaries but instead gradually fade away with increasing altitude. Nevertheless, there are a number of ways to characterize the thickness of an atmosphere surrounding a planet. One simple way is to ask when you would seem to have left the atmosphere and entered "space." Atmospheres scatter sunlight, which is what prevents us from seeing stars in the daytime. Thus, you might say that you had reached "space" when you had risen high enough that you could see stars in the daytime. The terrestrial atmospheres are remarkably thin when viewed in this way. For example, stars become visible in the daytime at an altitude of about 100 kilometers above the Earth—less than 2% of the Earth's radius (Figure 10.2a). But even the Space Shuttle, orbiting hundreds of kilometers above the Earth's surface, still runs into the upper vestiges of Earth's atmosphere (Figure 10.2b). So the question of where an atmosphere ends is more one of semantics than of science.

[1]Helium is *chemically inert* and never forms molecules.

[2]We call this balance *gravitational equilibrium* (or *hydrostatic equilibrium*), and it is also very important in stars [Section 14.1].

[3]1 bar $\approx$ 14.7 pounds/inch^2 ($\approx$ 100,000 newtons/m^2). That is, if you gathered up all the air directly above 1 square inch of the Earth's surface, it would weigh 14.7 pounds. You don't feel this weight on your shoulders because the pressure is also pushing up under your arms, inward against your sides, and so on in all directions.

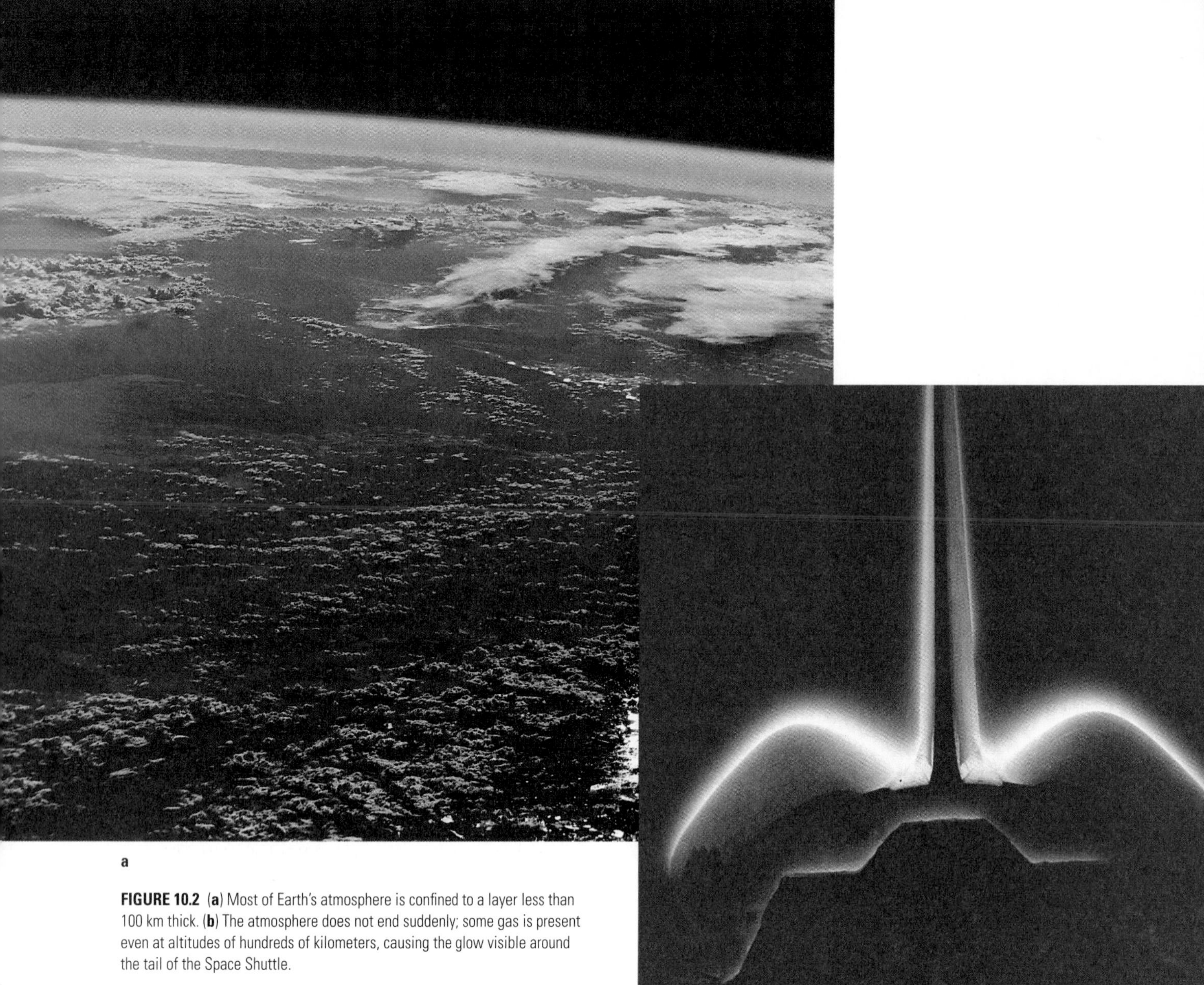

FIGURE 10.2 (**a**) Most of Earth's atmosphere is confined to a layer less than 100 km thick. (**b**) The atmosphere does not end suddenly; some gas is present even at altitudes of hundreds of kilometers, causing the glow visible around the tail of the Space Shuttle.

10.2 Atmospheric Structure

If you've ever driven up a mountain, you know that the atmosphere changes with altitude. Your popping ears and shortness of breath prove that the pressure is lower as you go higher, and the chill tells you that temperature also decreases with altitude. On airplane flights, you may have noticed that the ride generally becomes smoother at high altitudes, indicating that air becomes less turbulent up high. In this section, we discuss how atmospheric structure—that is, the variation in atmospheric properties with altitude—depends on the effects of sunlight on a planet's surface and on its atmospheric gases.

What If Planets Had No Atmospheres?

You may already know that an atmosphere can greatly affect a planet's temperature through the *greenhouse effect,* which we'll discuss in detail shortly. But first we need to look at how sunlight would heat a planet without an atmosphere—or at least without any of the gases that contribute to the greenhouse effect. In that case, the surface temperature of any planet would be determined by only three basic properties:

1. How far is the planet from the Sun? The closer to the Sun, the more energy the planet receives from sunlight on each square meter of its surface.
2. How much sunlight does the surface absorb, and how much does it reflect? The darker the surface, the more light it absorbs and the less it reflects. The term **albedo** describes the fraction of sunlight reflected by the planet. A higher albedo means more reflection and less absorption: albedo = 0 means no reflection at all (a perfectly black surface); albedo = 1 means all

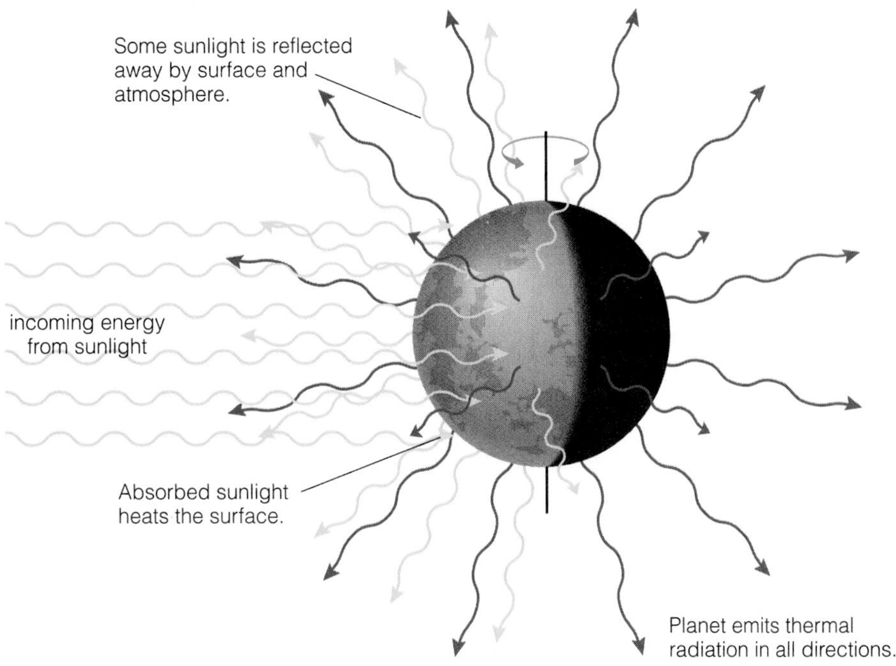

FIGURE 10.3 To maintain a steady temperature, a planet must emit precisely as much energy as it absorbs from sunlight.

light is reflected (a perfectly white surface). Clouds, snow, and ice have albedos of 0.7 or more, meaning that they reflect 70% or more of the light that hits them. Rocks have low albedos, usually between 0.1 and 0.25.

3. How fast does the planet rotate? A planet with a very long day will be very hot on the day side and very cold on the night side, but a planet with a short day will have similar temperatures both day and night. (A planet's atmosphere can also transport heat from dayside to nightside.)

TIME OUT TO THINK *On a hot summer day, you can gain a brief respite from walking barefoot across blacktop by stepping on a surface that is painted white. Use the idea of how albedo affects temperature to explain why.*

These three factors determine how much sunlight is actually absorbed by a planet and whether the energy is distributed over one or both sides of the planet. Any absorbed sunlight heats the planetary surface, which causes the planet to emit its own thermal radiation back into space (Figure 10.3). The energy absorbed from sunlight must be precisely balanced by the energy radiated away by the planet—otherwise the planet would rapidly heat up (if it absorbed more energy than it emitted) or cool down (if it emitted more energy than it absorbed).

We can use these simple ideas to calculate the temperature a planet would have in the absence of any additional warming by the atmosphere. Table 10.2 shows the results, giving both the "no atmosphere" temperatures and the actual surface temperatures for

Table 10.2 Temperatures of the Terrestrial Worlds

World	Distance to Sun (AU)	Albedo (0 = black, 1 = white)	Length of Day	"No Atmosphere" Temperature*	Observed Average Surface Temperature
Mercury	0.38	0.11	176 days	440 K	700 K (day), 100 K (night)
Venus	0.72	0.72	117 days	230 K	740 K
Earth	1.00	0.36	1 day	250 K	288 K
Moon	1.00	0.07	28 days	273 K	400 K (day), 100 K (night)
Mars	1.52	0.25	≈ 1 day	218 K	223 K

*Assumes rapid rotation, in which case day and night temperatures would be the same.

the terrestrial worlds. Note that the "no atmosphere" temperatures for Mercury and the Moon lie between their actual day and night temperatures—just as we should expect on worlds with very little atmosphere. The "no atmosphere" temperature for Mars is just 5 K lower than the actual value, but the "no atmosphere" temperature for Earth is almost 40 K lower than the actual surface temperature. If Earth were really this cold, our oceans would freeze (250 K = −23°C ≈ −9°F). The "no atmosphere" temperature for Venus is even farther from its actual temperature—about 500 K lower.

A Generic Planetary Atmosphere

Clearly, the presence of an atmosphere can make a planet much warmer than it would be otherwise. But how? The solution lies with how gases interact with the radiative energy of light.

Gases can *absorb, transmit,* or *scatter* light, and they can also *emit* their own radiation [Section 7.1]. Whether a particular gas absorbs, transmits, or scatters light depends on its composition and the wavelength of the light. Fortunately, we can make a few general observations about the common gases in atmospheres (Figure 10.4):

- Some gases absorb infrared light, which causes the gas molecules to begin "wiggling" (i.e., vibrating and rotating). Molecules that are particularly good at absorbing infrared light—such as carbon dioxide, methane, and water vapor—are called **greenhouse gases.**
- Atmospheric gases are generally transparent to visible light, but they can also *scatter* (change the direction of) some of the visible light.
- Ultraviolet photons can break molecules apart, particularly weakly bonded molecules such as **ozone** (O_3). This process of *molecular dissociation* [Section 5.3] produces atoms that may then react with other atoms or molecules or may even escape from the planet.
- X rays have enough energy to knock electrons free from atoms, thereby ionizing gases.[4]

[4]We are neglecting radio waves and gamma rays because the Sun emits very little of these forms of radiation.

Mathematical Insight 10.1 "No Atmosphere" Temperatures

The "no atmosphere" temperature of a planet—more accurately, the average temperature it would have if it did not have any greenhouse gases—depends on only three things: the planet's distance from the Sun, its albedo, and its rotation rate. As discussed in the text, the total energy the planet absorbs from sunlight must precisely balance the total energy the planet emits back into space. If we assume that a planet is rotating fast enough so that its dayside and nightside have the same temperature, we can derive a simple formula for the "no atmosphere" temperature by equating the total amounts of incoming and outgoing power (energy per unit time; e.g., watts):

$$\text{total power absorbed from sunlight} = \text{total power emitted by thermal radiation}$$

First, we need a formula for the total power absorbed. At Earth's distance from the Sun of 1 AU, the power per unit area contained in sunlight is 1,360 watts/m^2. The power per unit area of light follows an inverse square law with distance. For example, the power per unit area at a distance of 2 AU from the Sun is $(\frac{1}{2})^2 = \frac{1}{4}$ as great as at the distance of Earth, and the power per unit area at a distance of 0.5 AU is $(\frac{1}{0.5})^2 = 4$ times greater than at Earth [Section 15.2]. Thus, in general, the power per unit area contained in sunlight is:

$$1{,}360 \frac{\text{watt}}{\text{m}^2} \times \frac{1}{d^2}$$

where d is the distance from the Sun in AU. The *total* amount of power striking a planet is this power per unit area times the total area the planet has facing the Sun. As shown in Figure 10.5, this area is simply the area of a circle with the same radius as the planet, or $\pi \times (R_{\text{planet}})^2$. However, a planet does not absorb

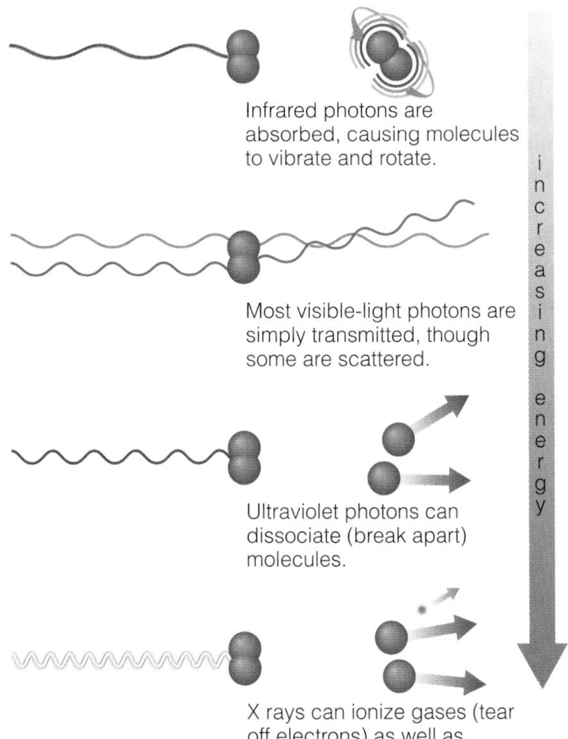

FIGURE 10.4 Light of different energies has different effects on atmospheric gases.

Different wavelengths of sunlight are absorbed in different layers of the atmosphere, causing the variation of temperature with altitude. Figure 10.6 shows the structure of a "generic" planetary atmosphere; it shows how and why the temperature varies with altitude. The uppermost portion of the atmosphere is called the **exosphere.** The exosphere is the region where the atmosphere "fades away" into space; it is hot because the gases absorb X rays from the Sun. The remaining X rays are absorbed in the next layer down, called the **thermosphere.** These X rays ionize atoms in the part of the thermosphere that we call the **ionosphere.** Ultraviolet photons are absorbed by certain gases in the **stratosphere.**[5] Visible photons penetrate the entire atmosphere, and some are absorbed by the ground. The ground radiates away this energy as infrared light, which is trapped in the lowest atmospheric layer, called the **troposphere.** In the rest of this section, we investigate in more detail the processes that make these basic atmospheric layers, from the ground up.

[5]Technically, the stratosphere is only the region where the temperature rises with altitude above the troposphere. The region where the temperature falls again is called the *mesosphere.* We will not make this distinction in this book.

all the power that reaches it from the Sun. It reflects the fraction that we call the *albedo* and therefore absorbs a fraction of *1 minus the albedo.* Putting these facts together, we can calculate the total amount of power that a planet absorbs from sunlight:

$$\text{total power absorbed} = \underbrace{\left(1{,}360\,\frac{\text{watt}}{\text{m}^2} \times \frac{1}{d^2}\right)}_{\substack{\text{power per unit area}\\ \text{at distance } d \text{ (AU)}}} \times \underbrace{\pi \times (R_{\text{planet}})^2}_{\substack{\text{total area of planet}\\ \text{facing the Sun}}} \times \underbrace{(1 - \text{albedo})}_{\substack{\text{fraction of sunlight}\\ \text{absorbed by surface}}}$$

FIGURE 10.5 A planet intercepts a circular area of sunlight.

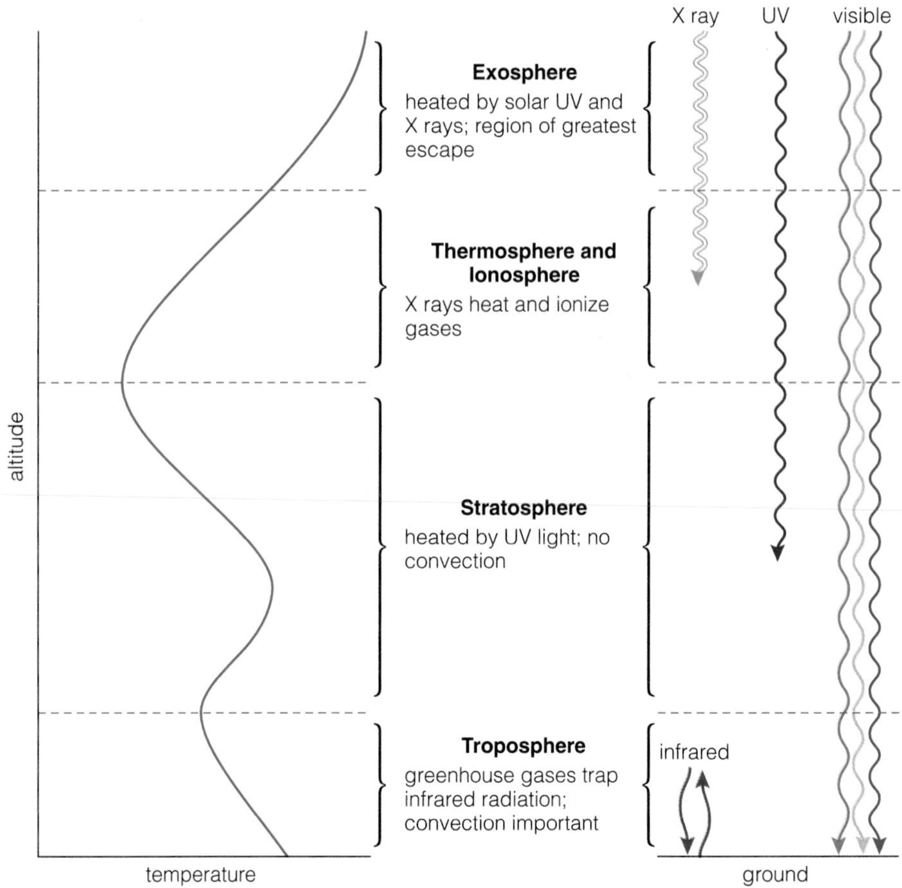

FIGURE 10.6 The structure of a generic planetary atmosphere: Solar X rays are absorbed in the thermosphere, ultraviolet light is absorbed in the stratosphere, and visible light reaches the ground. Planets that lack ultraviolet-absorbing molecules will lack a stratosphere, and planets with very little gas will have only an exosphere.

Next, we need a formula for the total power emitted by the planet. As discussed in Mathematical Insight 7.2, the power per unit area emitted by thermal radiation is σT^4, where T is the surface temperature of the planet and $\sigma = 5.7 \times 10^{-8} \frac{\text{watt}}{\text{m}^2 \times \text{Kelvin}^4}$. As shown in Figure 10.3, the planet radiates this power from its entire surface (assuming the temperature is the same everywhere). Because the surface area of a planet is $4\pi \times (R_{\text{planet}})^2$, the total amount of power that a planet emits by thermal radiation is:

$$\text{total power emitted} = \underbrace{\sigma T^4}_{\substack{\text{emitted power}\\ \text{per unit area}}} \times \underbrace{4\pi \times (R_{\text{planet}})^2}_{\substack{\text{total surface area}\\ \text{of planet}}}$$

We can now equate the formulas we've found for the total power absorbed and the total power emitted:

$$\underbrace{1{,}360 \frac{\text{watt}}{\text{m}^2} \times \frac{1}{d^2} \times \pi \times (R_{\text{planet}})^2 \times (1 - \text{albedo})}_{\text{total power absorbed}} = \underbrace{\sigma T^4 \times 4\pi \times (R_{\text{planet}})^2}_{\text{total power emitted}}$$

You should confirm that solving this equation for the temperature T yields:

$$T = \sqrt[4]{\frac{1{,}360 \frac{\text{watt}}{\text{m}^2} \times (1 - \text{albedo})}{4\sigma \times d^2}}$$

TIME OUT TO THINK *As we will discuss shortly, some planetary atmospheres lack ultraviolet-absorbing gases and therefore lack stratospheres. How would the structure of such a planetary atmosphere differ from that of the generic atmosphere shown in Figure 10.6?*

Visible Light: Warming the Surface and Coloring the Sky

The Sun emits most of its energy in the form of visible light, and atmospheres are generally transparent to visible light. Thus, it is primarily visible light from the Sun that reaches and warms a planetary surface.

Although most visible light passes directly through the atmosphere, some is scattered by molecules in the atmosphere. This scattering of sunlight prevents us from seeing stars in the daytime. If our atmosphere did not scatter light, you'd be able to block out the light of the Sun simply by holding your hand in front of it, and then you'd have no trouble seeing other stars. Scattering of light also prevents shadows from being pitch black. On the Moon, where there is no air to scatter light, you can see stars in the daytime, and shadowed regions are extremely dark.

The scattering of light by the atmosphere also explains why the sky is blue. It turns out that molecules scatter blue light (higher frequencies) much more effectively than red light (lower frequencies). Thus, although the Sun illuminates the atmosphere with all colors of light, effectively only the blue light gets scattered. When the Sun is overhead, this scattered blue light reaches your eyes from all directions, explaining why the sky appears blue (Figure 10.7). At sunset (or sunrise), the sunlight must pass through a greater amount of atmosphere on its way to you. Thus, most of the blue light is scattered away from you altogether (making someone else's sky blue), leaving only red light to color your sunset.

TIME OUT TO THINK *Suppose atmospheric molecules didn't scatter light at all. In that case, what color would the sky be? Explain.*

Infrared Light, the Greenhouse Effect, and the Troposphere

The small amount of sunlight emitted farther in the infrared is mostly absorbed in our generic atmosphere before it reaches the ground. But the planet itself emits significant amounts of infrared radiation: As we discussed earlier, a planet must radiate exactly as much energy as it absorbs from the Sun to maintain a constant temperature. Herein lies the key to how atmospheres make planetary surfaces warmer than they would be otherwise.

(Recall that the symbol $\sqrt[4]{\ }$ means the fourth root, or the square root of the square root.) You should also confirm that, by substituting the value $\sigma = 5.7 \times 10^{-8} \frac{\text{watt}}{\text{m}^2 \times \text{Kelvin}^4}$, this formula becomes:

$$T = 280\text{ K} \times \sqrt[4]{\frac{(1 - \text{albedo})}{d^2}}$$

This is our formula for the "no atmosphere" temperature of a planet, where d must be measured in AU. (You might pause for a moment to consider why the planet's size does not affect its "no atmosphere" temperature.)

Example: Calculate the "no atmosphere" temperature of Mercury, assuming that it rotates rapidly enough to make its temperature the same on its dayside and nightside.

Solution: Table 10.2 shows that Mercury's distance from the Sun is $d = 0.38$ AU and its albedo is 0.11. Plugging these numbers into our formula for the "no atmosphere" temperature, we find:

$$T = 280\text{ K} \times \sqrt[4]{\frac{(1 - \text{albedo})}{d^2}} = 280\text{ K} \times \sqrt[4]{\frac{(1 - 0.11)}{0.38^2}} = 440\text{ K}$$

Note that this value for Mercury's "no atmosphere" temperature agrees with the value shown in Table 10.2.

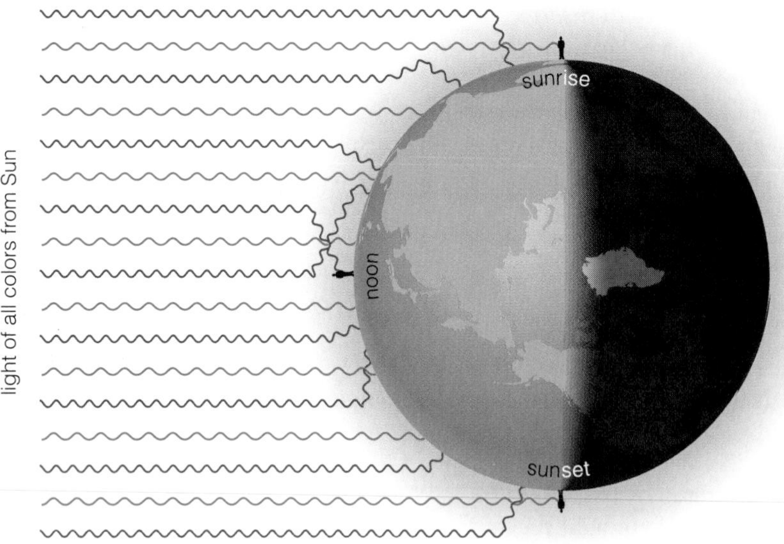

FIGURE 10.7 Atmospheric gases scatter blue light more than they scatter red light. During most of the day, you therefore see blue photons coming from most directions in the sky, making the sky look blue. But only the red photons reach your eyes at sunrise or sunset, when the light must travel a longer path through the atmosphere to reach you.

Figure 10.8 shows how the **greenhouse effect** works.[6] Carbon dioxide, water vapor, and other *greenhouse gases* in the atmosphere absorb some of the infrared radiation emitted upward from the planet's surface. These gases therefore warm up and emit infrared thermal radiation themselves—but in all directions. Some of this radiation is directed back down toward the surface, making the surface warmer than it would be from absorbing visible sunlight alone. The more greenhouse gases present, the greater the degree of greenhouse warming.

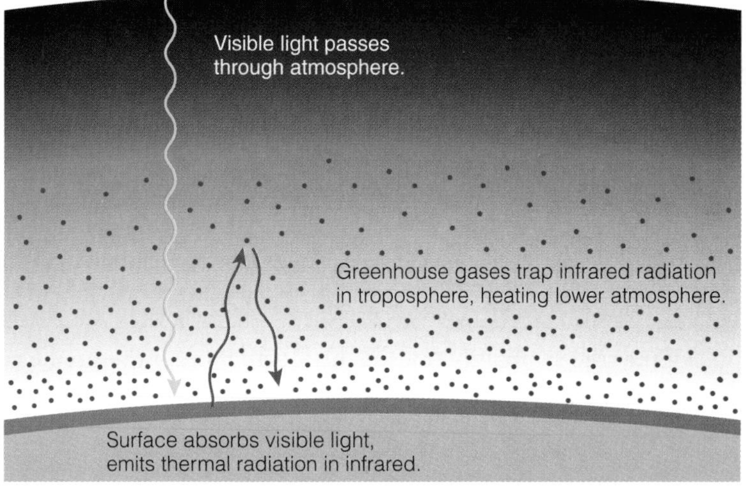

FIGURE 10.8 The greenhouse effect: The troposphere becomes warmer than it would be if it had no greenhouse gases.

TIME OUT TO THINK *Clouds on Earth are made of water, and H_2O is a very effective greenhouse gas—even when the water vapor condenses into droplets in clouds. Use this fact to explain why clear nights tend to be colder than cloudy nights.* (Hint: *Sunlight warms the surface only in the daytime, but the surface radiates its heat away both day and night.*)

Air is denser at lower altitudes, so the greenhouse effect primarily affects temperatures in the troposphere. Above the troposphere, there is so little air that greenhouse gases cannot effectively trap infrared radiation. The fact that the greenhouse effect makes air warmer near the ground drives *convection* in the troposphere [Section 9.3]. The warm air near the ground expands and rises, while cooler air near the top of the troposphere contracts and falls. The descending cooler air gets heated, and the cycle repeats. The

[6] The name *greenhouse effect* comes from botanical greenhouses, usually built mostly of glass, which keep the air inside them warmer than it would be otherwise. But it turns out that greenhouses trap heat simply by not letting hot air rise, a very different way of trapping heat than the infrared absorption that leads to the "greenhouse effect" in planetary atmospheres.

troposphere gets its name from this convection: *Tropos* is Greek for "turning." The process of convection carries heat upward and continually churns the air in the troposphere, causing the ever-changing conditions that we call the weather.

Local weather patterns called *inversions* occasionally make the air colder near the surface than higher up in the troposphere—the opposite of the usual condition, in which the troposphere is warmer at the bottom. Over cities on Earth, inversions often lead to smog problems by inhibiting the convection that normally carries pollution away to higher altitudes.

Ultraviolet Light and the Stratosphere

The primary source of heat in the troposphere—the absorption of infrared radiation from the ground—is not important in higher layers of the atmosphere. Once infrared radiation from the ground reaches the top of the troposphere, it travels essentially unhindered to space. Thus, we need to consider only the effects of sunlight in investigating atmospheric structure above the troposphere. As we move upward through the stratosphere, the primary source of atmospheric heating is absorption of ultraviolet light from the Sun. This heating tends to be stronger at higher altitudes because the ultraviolet light gets absorbed before it reaches lower altitudes.

As a result, the stratosphere gets warmer with increasing altitude—the opposite of the situation in the troposphere. Convection therefore cannot occur in the stratosphere: Heat cannot rise if the air is even hotter higher up.[7] The stratosphere gets its name because the lack of convection makes its air relatively stagnant and *stratified* (layered), rather like a sitting jar of oil and water. The lack of convection also means that the stratosphere essentially has no weather and no rain. That is why pollutants that reach the stratosphere—such as ozone-destroying chlorofluorocarbons (CFCs) [Section 13.5]—remain there for decades.

Note that a planet can have a stratosphere *only* if its atmosphere contains molecules that are particularly good at absorbing ultraviolet photons. Ozone plays this role on Earth. In fact, Earth is the only terrestrial planet with such an ultraviolet-absorbing layer and hence the only terrestrial planet with a stratosphere—at least in our solar system. (We'll discuss why Earth is unique in this way in Chapter 13.) The jovian planets also have stratospheres, as we'll discuss in the next chapter.

Common Misconceptions: The Greenhouse Effect Is Bad

The greenhouse effect is often in the news, usually in discussions about environmental problems. But the greenhouse effect itself is not a bad thing. In fact, we could not exist without it. Remember that the "no atmosphere" temperature of the Earth is well below freezing. Thus, the greenhouse effect is the only reason why our planet is not frozen over. Why, then, is the greenhouse effect discussed as an environmental problem? It is because human activity is adding more greenhouse gases to the atmosphere—which might change the Earth's climate. After all, while the greenhouse effect makes the Earth livable, it is also responsible for the searing 740 K temperature of Venus—proving that it's possible to have too much of a good thing.

X Rays and the Thermosphere and Ionosphere

Virtually all gases are good X-ray absorbers, because X rays have sufficient energy to ionize almost any atom and more than enough to dissociate almost any molecule. Solar X rays therefore are absorbed by the first gases they encounter as they enter the atmosphere. The result is that the upper atmosphere is strongly heated by solar X rays in the daytime, usually to much higher temperatures than those on the surface. That is why this upper region of the atmosphere is called the *thermo*sphere. Despite its high temperatures, the gas in the thermosphere would not feel hot to your skin because its density and pressure are so low [Section 5.2].

The thermosphere contains a small but important fraction of its gas as charged ions and free electrons—the result of ionization. The layer within the thermosphere that contains most of these ions is called the *ionosphere*. The ionosphere is very important for radio communications on Earth. Most radio broadcasts are completely reflected back to Earth's surface by the ionosphere, almost as though the Earth were wrapped in aluminum foil. Without this reflection by the ionosphere, radio communication would work only between locations in sight of each other.

[7] Convection lessens even in the upper troposphere, where temperature changes slowly with altitude. Thus, airplane rides become smoother at high altitudes because less convection means less turbulence.

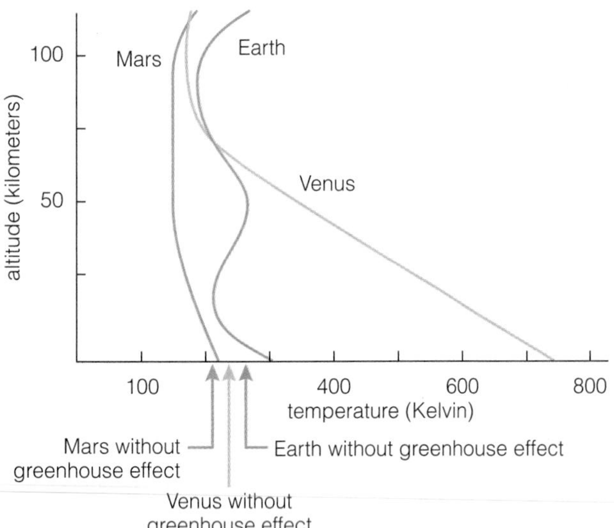

FIGURE 10.9 Temperature profiles for Venus, Earth, and Mars. Note that Venus and Earth are considerably warmer than they would be without the greenhouse effect. (The thermospheres and exospheres, not shown, are qualitatively alike.)

The Exosphere

The exosphere is the region that marks the fuzzy "boundary" between the atmosphere and space. (The prefix *exo* means "outermost" or "outside.") The gas density is so low in the exosphere that collisions between atoms or molecules are very rare. The high temperatures in the exosphere give the individual atoms or molecules high speeds, sometimes allowing them to escape from the planet. The Space Shuttle and many satellites actually orbit the Earth within its exosphere, and the small amount of gas in the exosphere can exert a bit of atmospheric drag. That is why satellites in low-Earth orbit eventually spiral deeper into the atmosphere and burn up [Section 6.4].

Comparative Structure of the Terrestrial Atmospheres

Now that we understand the structure of our generic atmosphere, let's see how it corresponds to actual terrestrial atmospheres. The atmospheres of the Moon and Mercury have so little gas that they essentially possess only the top layer, the exosphere. Venus, Earth, and Mars have denser atmospheres and are more comparable (Figure 10.9).

The similarities and differences among these three planetary atmospheres make sense in light of our study of the generic atmosphere. All three have a warm troposphere at the bottom, created by the greenhouse effect, and a warm thermosphere at the top, where solar X rays are absorbed. The greatest difference is the extra "bump" of Earth's stratosphere, where ultraviolet light is absorbed. Without the warming in our unique layer of stratospheric ozone, Earth's middle altitudes would be almost as cold as those on Mars.

Although the structures of the tropospheres are similar on Venus, Earth, and Mars, the tropospheric temperatures on the three planets are dramatically different. On Venus, the thick CO_2 atmosphere causes a greenhouse effect that raises the surface temperature about 500 K above what it would be without greenhouse gases. The thin CO_2 atmosphere of Mars creates only a weak greenhouse effect, making its surface temperature only 5 K warmer than its "no atmosphere" temperature. Earth has a moderate greenhouse effect that increases its temperature about 40 K, making it comfortable for life. The greenhouse effect on Earth is almost entirely due to gases that make up less than 2% of our atmosphere: mainly carbon dioxide, water vapor, and methane. The major constituents of Earth's atmosphere, N_2 and O_2, have no effect on infrared light and do not contribute to the greenhouse effect.[8] If they did, Earth might be too warm for life.

Common Misconceptions: Higher Altitudes Are Always Colder

Many people think that the low temperatures in the mountains are just the result of lower pressures, but Figure 10.9 shows that it's not that simple. The higher temperatures near sea level on Earth are a result of the greenhouse effect trapping more heat at lower altitudes. If Earth had no greenhouse gases, mountaintops wouldn't be so cold—or, more accurately, sea level wouldn't be so warm.

Magnetospheres and the Solar Wind

One more type of energy that we haven't yet considered comes from the Sun: the low-density breeze of charged particles that we call the *solar wind* [Section 8.4]. Although the solar wind does not significantly affect atmospheric structure, it can have other important influences. Among the terrestrial planets, only Earth has a strong enough

[8]Molecules with only two atoms—and especially molecules with two of the same kind of atom—are poor infrared absorbers because they have very few ways to vibrate and rotate.

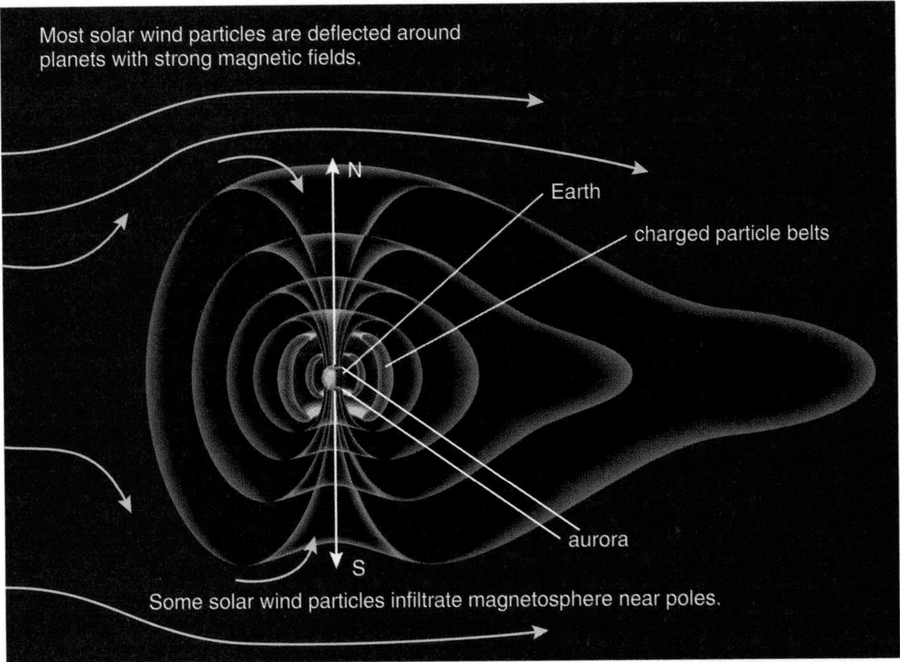

a Earth's magnetosphere and its interaction with the solar wind.

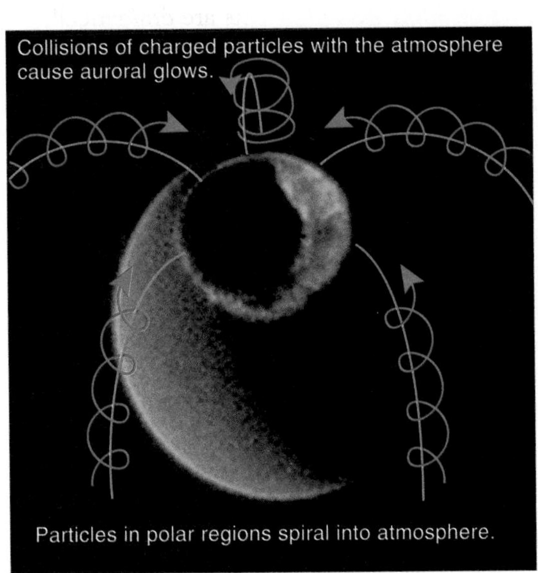

b Origin of the aurora.

c Earth's aurora as seen from the Space Shuttle.

FIGURE 10.10 A planet's magnetosphere acts like a protective bubble that shields the surface from charged particles coming from the solar wind. Among the terrestrial planets, only the Earth has a strong enough magnetic field to create a magnetosphere. The Earth's magnetosphere allows charged particles to strike the atmosphere only near the poles, thereby creating the phenomena of the aurora borealis and aurora australis.

magnetic field to divert the charged particles of the solar wind. Thus, solar wind particles can impact the exospheres of Venus and Mars—and the surfaces of Mercury and the Moon. In contrast, Earth's strong magnetic field creates a **magnetosphere** that acts like a protective bubble surrounding the planet (Figure 10.10). (Jovian planets also have magnetospheres [Section 11.4].) Charged particles cannot easily pass through a magnetic field, so a magnetosphere diverts most of the solar wind around a planet.

Some solar wind particles manage to infiltrate Earth's magnetosphere at its most vulnerable points, near the magnetic poles. Once inside, they have a difficult time escaping. Charged particles are restricted by magnetic forces to following magnetic

THINKING ABOUT . . .

Weather and Chaos

Scientists today have a very good understanding of the physical laws and mathematical equations that govern the behavior and motion of atoms in the air, oceans, and land. Why, then, do we have so much trouble predicting the weather? For a long time, most scientists assumed that the difficulty of weather prediction would go away once we had enough weather stations to collect data from around the world and sufficiently powerful computers to deal with all those data. However, we now know that weather is fundamentally unpredictable on time scales longer than a few weeks. To understand why, we must look at the nature of scientific prediction.

Suppose you want to predict the location of a car on a road 1 minute from now. You need two basic pieces of information: where the car is now, and how it is moving. If the car is now passing Smith Road and heading north at 1 mile per minute, it will be 1 mile north of Smith Road in 1 minute.

Now suppose you want to predict the weather. Again, you need two basic types of information: (1) the current weather and (2) how weather changes from one moment to the next.

You could attempt to predict the weather by creating a "model world." For example, you could overlay a globe of the Earth with graph paper and then specify the current temperature, pressure, cloud cover, and wind within each square. These are your starting points, or *initial conditions.* Next, you could input all the initial conditions into a computer, along with a set of equations (physical laws) that describe the processes that can change weather from one moment to the next.

Suppose the initial conditions represent the weather around the Earth *at this very moment* and you run your computer model to predict the weather for the next month in New York City. The model might tell you that tomorrow will be warm and sunny, with cooling during the next week and a major storm passing through a month from now. Now suppose you run the model again but make one minor change in the initial conditions—say, a small change in the wind speed somewhere over Brazil. For tomorrow's

field lines (Figure 10.10a). In many magnetospheres, the ions and electrons accumulate to make **charged particle belts** encircling the planet. (The charged particle belts around the Earth are called the *Van Allen belts* after their discoverer.) The high energies of the particles in charged particle belts can be very hazardous to spacecraft and astronauts passing through them.

Particles trapped in magnetospheres also cause the beautiful spectacles of light called *auroras* (see Figure 1.22). If a trapped charged particle has enough energy, it can follow the magnetic field all the way down to a planet's atmosphere. The charged particles collide with atmospheric atoms and molecules, causing them to radiate and produce auroras (Figure 10.10b,c). Because the charged particles follow the magnetic field, auroras are most common near the magnetic poles. On Earth, auroras therefore are best viewed from Canada, Alaska, and Russia in the Northern Hemisphere (the *aurora borealis,* or northern lights) and from Australia, Chile, and Argentina in the Southern Hemisphere (the *aurora australis,* or southern lights).

10.3 Weather and Climate

So far, we've mostly talked about a planet's atmosphere as if it were the same at all locations and at all times. The atmospheric structure profiles show only *average* temperatures at different altitudes in the atmosphere. But we know from experience on Earth that surface and atmospheric conditions constantly change. In any particular location, some days may be hotter or cooler than others, some may be clearer or cloudier, some may be calmer or stormier. This ever-varying combination of winds, clouds, temperature, and pressure is what we call **weather.**

Local weather can vary dramatically from one day to the next, or even from hour to hour. The weather can also vary greatly between places just short distances apart. For example, temperatures on a warm day may be several degrees hotter just a few kilometers inland from a coastal city. The complexity of weather makes it difficult to predict, although modern-day *meteorologists*[9] (people who study weather) often

[9]The word *meteor* comes from a Greek word meaning "a thing in the air," which explains why the very different concepts of *meteorology* and *meteors* have the same word origin.

weather, that this slightly different initial condition will not change the weather prediction for New York City. For next week's weather, the new model may yield a slightly different prediction. But for next month's weather, the two predictions may not agree at all!

The disagreement between the two predictions arises because the laws governing weather can cause very tiny changes in initial conditions to be greatly magnified over time. This extreme sensitivity to initial conditions is sometimes called the *butterfly effect:* If initial conditions change by as much as the flap of a butterfly's wings, the resulting prediction may be very different.

The butterfly effect is a hallmark of *chaotic systems.* Simple systems are described by *linear equations* in which, for example, increasing a cause produces a proportional increase in an effect. In contrast, chaotic systems are described by *nonlinear equations,* which allow for subtler and more intricate interactions. For example, the economy is nonlinear because a rise in interest rates does not automatically produce a corresponding change in consumer spending. Weather is nonlinear because a change in the wind speed in one location does not automatically produce a corresponding change in another location. Many (but not all) nonlinear systems exhibit chaotic behavior.

Despite their name, chaotic systems are not completely random. In fact, many chaotic systems have a kind of underlying order that explains the general features of their behavior, even while details at any particular moment remain unpredictable. In a sense, many chaotic systems are *predictably unpredictable.* Our understanding of chaotic systems is increasing at a tremendous rate, but much remains to be learned about them.

can predict weather quite accurately a few days in advance. On longer time scales, local weather is fundamentally unpredictable, at least on Earth.

Climate is the long-term average of weather and is generally stabler than weather. Deserts remain deserts and rain forests remain rain forests over periods of hundreds or thousands of years, while the day-to-day and even year-to-year weather may vary dramatically. Weather and climate can be hard to distinguish on a human time scale: Do a few hot summers in a row imply a change in climate, or are they merely coincidence? We'll return to this question when we study the issue of global warming on Earth in Chapter 13. For now, let's concentrate on understanding what drives planetary weather.

Seasonal Weather Patterns

The most obvious weather pattern on a planet is the change in the seasons. In general, a planet has seasons only if it has a significant axis tilt, so that over the course of the year first one hemisphere and then the other receives more direct sunlight [Section 3.1]. Otherwise, both hemispheres receive the same amount of sunlight year-round. For example, Venus has no seasons because its axis is nearly perpendicular to the ecliptic. Mars has seasonal changes like Earth because its 25° axis tilt is comparable to Earth's $23\frac{1}{2}°$ tilt. Thus, when one Martian hemisphere is in summer, the other is in winter, and vice versa. However, Martian seasons last almost twice as long as Earth seasons because a Martian year is almost twice as long as an Earth year.

A planet's changing distance from the Sun can also affect the severity of seasons. Earth's nearly circular orbit makes this effect unimportant. But Mars has a more elliptical orbit that puts it significantly closer to the Sun during southern hemisphere summer, making it warmer than summer in the northern hemisphere, and farther from the Sun during southern hemisphere winter, making it colder than winter in the northern hemisphere (Figure 10.11). Mars therefore has more extreme seasons in its southern hemisphere than in its northern hemisphere.

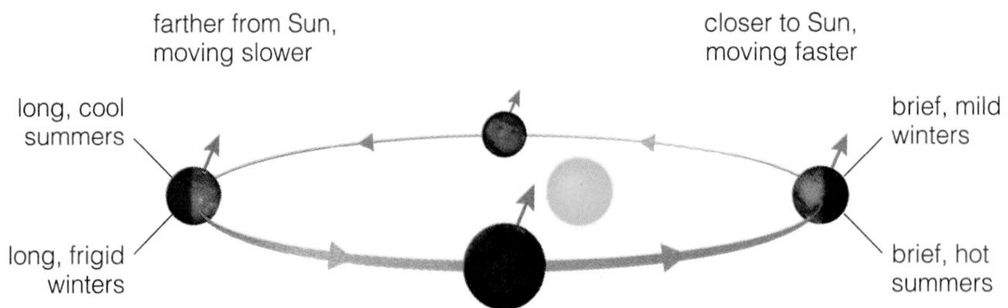

FIGURE 10.11 The ellipticity of Mars's orbit makes the severity of seasons different in the northern and southern hemispheres.

Global Wind Patterns

Winds are flows of air caused by pressure differences between different regions of a planet. On a local scale, winds vary as much as the weather, blowing in different directions and with different strengths at different times. But certain wind patterns are fixed on a global scale, creating what we call **global wind patterns** (or *global circulation*).

The first major factor affecting global wind patterns is atmospheric heating. In general, the equatorial regions of a planet receive more heat from the Sun than do polar regions. The excess heat makes the atmosphere expand above the equator, and air spills northward and southward at high altitudes (Figure 10.12). By itself, this process creates two huge **circulation cells** resembling convection cells but much larger. (They are often called *Hadley cells* after the person who first suggested their existence in 1735.) The circulation cells serve to transport heat both from lower to higher altitudes and from the equator to the poles. If this circulation is very efficient and if the atmosphere is dense enough to carry a lot of thermal energy, the resulting winds can equalize planetary temperatures from equator to pole. That is the case on Venus, where temperatures near the poles are essentially the same as those at the equator.

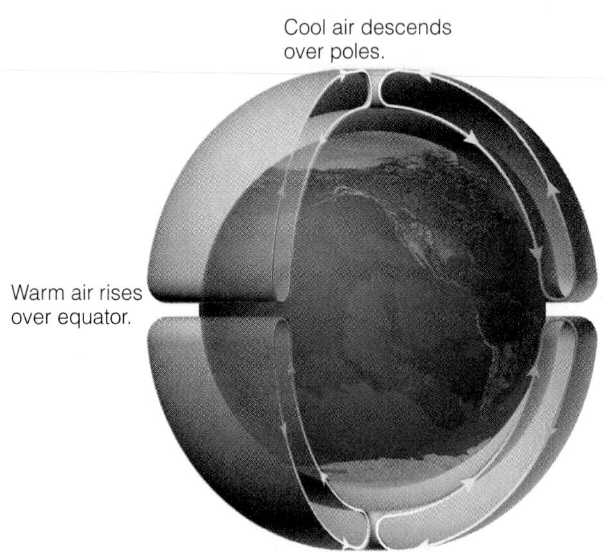

FIGURE 10.12 Circulation cells. Heat rises above the equator, setting up a flow of warm air toward the poles at high altitudes and a flow of cool air toward the equator near the surface. The planet's rotation is neglected for the moment.

TIME OUT TO THINK ***Suppose a planet rotated synchronously with its orbit around the Sun, so that it kept the same face toward the Sun at all times. What kind of circulation pattern would you expect on this planet? Explain.*** (Hint: ***Which regions of the planet receive the most sunlight, and which regions receive the least sunlight?***)

The second major factor affecting global wind patterns is the planet's rotation. It's easiest to understand this effect by analogy. The movement of air over a planet is like the motion of a ball rolled on a merry-go-round (Figure 10.13). Imagine that you are standing near the edge of a merry-go-round. Your speed around the axis will be higher than the speeds of the inner parts of the merry-go-round. If you roll a ball toward the center, the ball begins with your relatively high speed around the axis. As it rolls inward, this high speed around the axis makes it move ahead of the slower-moving inner regions from your perspective. If your merry-go-round rotates counterclockwise, the ball therefore deviates to the right instead of heading straight inward. The deviation caused by the merry-go-round's rotation is called the **Coriolis effect.** It also works in reverse: If you roll the ball outward from the center, the ball lags behind the faster-moving outer regions, again deviating to the right. If your merry-go-round instead rotates clockwise, the deviations are to the left.

The Coriolis effect occurs on rotating planets because equatorial regions travel faster around the rotation axis than do polar regions (see Figure 3.1).

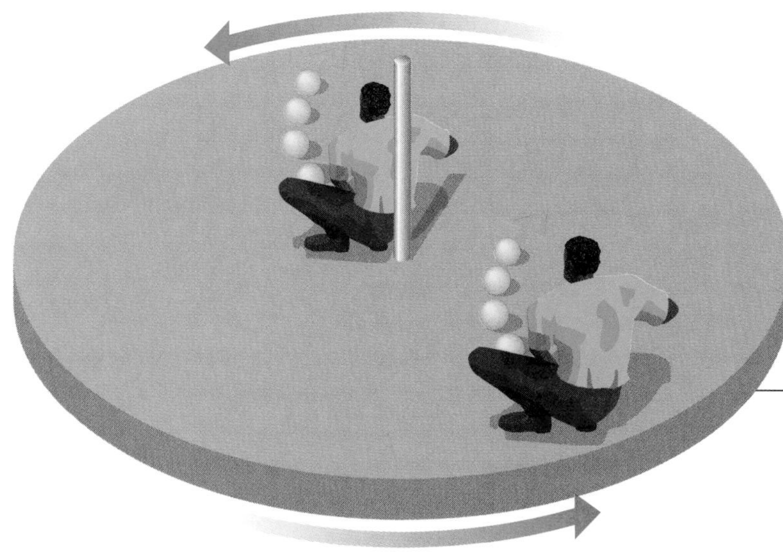

FIGURE 10.13 The Coriolis effect on a merry-go-round rotating counterclockwise: A ball rolled inward starts with a speed around the center faster than that of regions nearer the center; this "extra" speed makes the ball deviate to the right. A ball rolled outward starts with a speed around the center slower than that of regions farther out; this "lag" also makes the ball deviate to the right. If the merry-go-round were rotating clockwise, both balls would veer to the left.

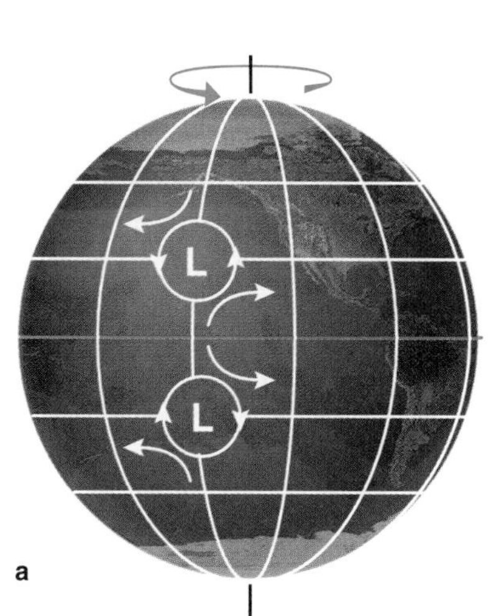

FIGURE 10.14 (**a**) The Coriolis effect causes air moving toward low-pressure regions to be diverted into a circular flow around the regions. The direction of flow is opposite in the two hemispheres. (**b**) You can see the direction of flow in a satellite photo of storms in the Pacific.

On Earth, air moving away from the equator deviates ahead of the Earth's rotation to the east, and air moving toward the equator deviates behind the Earth's rotation to the west (Figure 10.14a). Note that, either way, the deviation is to the *right* in the Northern Hemisphere and to the *left* in the Southern Hemisphere. One result is that air flowing toward low-pressure zones ends up circulating around them counterclockwise in the Northern Hemisphere and clockwise in the Southern Hemisphere (Figure 10.14b). Low-pressure regions ("L" on weather maps) are often associated with storms, including hurricanes. Because the circulating winds prevent air from flowing in to equalize the pressure, low-pressure regions tend to remain stable for many days, during which time

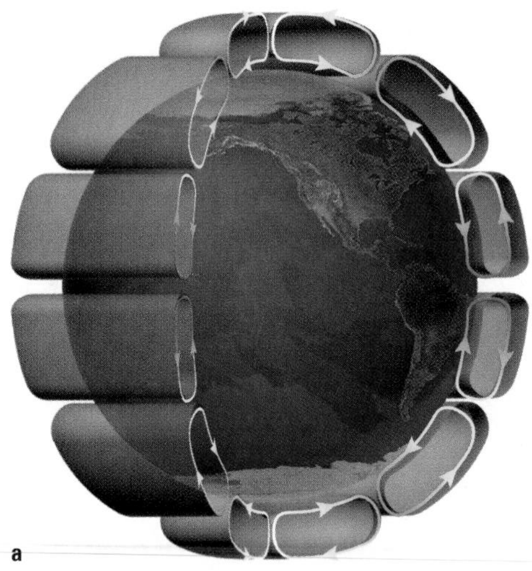

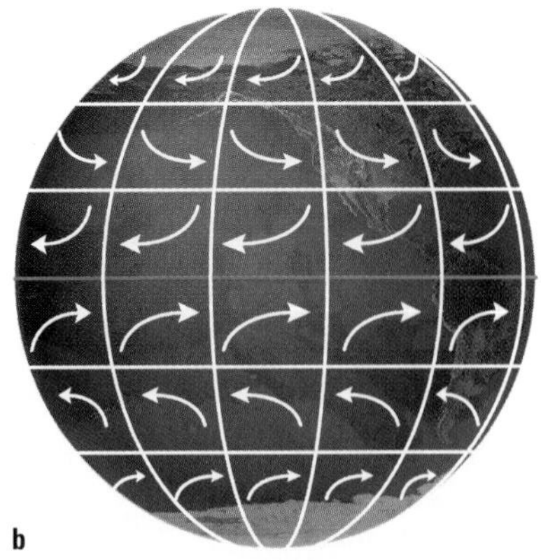

FIGURE 10.15 (**a**) On a rapidly rotating planet like Earth, the Coriolis effect causes each of the two large circulation cells of a slowly rotating planet (see Figure 10.12) to split into three cells. (**b**) The resulting surface wind patterns. Note that, at mid-latitudes such as over North America, the surface winds blow from the west.

prevailing winds can carry them thousands of kilometers across the planet. High-pressure regions ("H" on weather maps) can also last for days and travel great distances, but they are usually storm-free.

The Coriolis effect also diverts the global wind patterns. On Earth, this diversion causes each of the two huge circulation cells (shown in Figure 10.12) to split into three smaller cells (Figure 10.15a). Note that surface winds in the cells near the equator and near the poles flow toward the equator and hence are diverted westward by the Coriolis effect. In contrast, surface winds in the mid-latitude cells flow toward the poles and are diverted eastward by the Coriolis effect (Figure 10.15b). You've probably noticed that surface winds generally blow from west to east at mid-latitudes, which is why storms generally first hit the west coast of North America and then progress eastward.

TIME OUT TO THINK *Study Figure 10.15 a bit more. If you want to take maximum advantage of the prevailing winds, what course should you plot for a sailing trip from Europe to North America and back?*

Common Misconceptions: Do Toilets Flush Backward in the Southern Hemisphere?

A common myth holds that water circulates "backward" in the Southern Hemisphere; that is, toilets flush the opposite way, water spirals down sink drains the opposite way, and so on. If you visit an equatorial town, you may even find charlatans who, for a small fee, will demonstrate how to change the direction of water spiraling down a drain simply by stepping across the equator. This myth sounds similar to the Coriolis effect—which really does make hurricanes spiral in opposite directions in the two hemispheres—but it is completely untrue. The Coriolis effect is noticeable only when material moves significantly closer to or farther from the Earth's rotation axis—which means scales of hundreds of kilometers. It has no effect at all on small scales, such as the scales of toilets or drains. Apart from the shape of the basin, the only thing that affects the direction in which the water in a toilet or sink swirls is the initial angular momentum, which you can affect by swirling the water in one direction or the other. Conservation of angular momentum then dictates that the water will swirl faster and faster as it gets closer to the drain. You can find water in toilets and sinks swirling in either direction, and even tornadoes twisting in either direction, in both hemispheres.

The strength of the Coriolis effect depends on a planet's size and rotation rate: It is weakest on planets that are small and rotating slowly. Venus has a very weak Coriolis effect because of its slow rotation and hence has very slow surface winds; it never has hurricane-like storms. The top wind speeds measured by the Soviet Union's *Venera* landers were no more than about 6 kilometers per hour. Apart from faster winds near the top of the atmosphere (of unknown cause), Venus is a rather dull place in terms of weather: Its temperature is nearly the same everywhere, and the surface winds are calm.

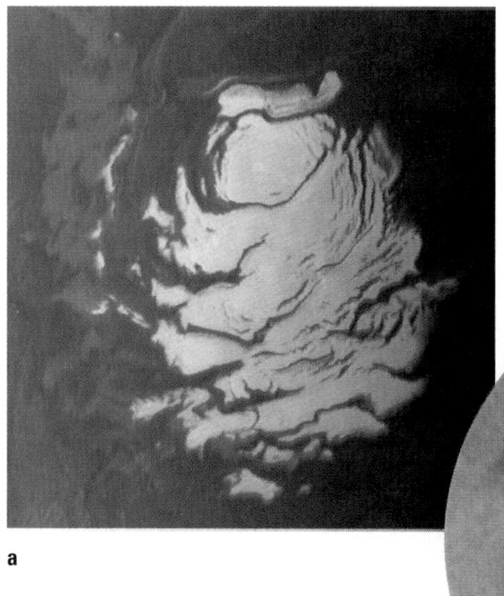

a

b

c

FIGURE 10.16 (**a**) This photo, taken by the *Viking Orbiter*, shows the ice cap at a Martian pole during the Martian winter. (**b**) Hubble Space Telescope image of a Martian dust storm (dark swirl over polar cap). (**c**) Close-up of the edge of a polar cap, showing alternating layers of ice and dust. Each layer is a few dozen meters thick and may take thousands of years to form.

Mars has a rotation period very close to that of Earth, but its smaller size means a weaker Coriolis effect. Its circulation cells therefore do not split like those on Earth. However, its thin atmosphere means that the circulation cells transport very little heat, so the poles remain much colder than the equator. Polar temperatures in the winter are so low (about 145 K, or −130°C) that carbon dioxide condenses into "dry ice" at the polar caps (Figure 10.16a). Because carbon dioxide is the major constituent of the Martian atmosphere, this condensation changes the atmospheric pressure by as much as 20% from season to season. The condensation occurring at the winter pole draws CO_2 out of the atmosphere, which is replenished by winds blowing from the opposite polar cap, where the dry ice is *subliming* [Section 5.3]. The combination of circulation cells, the Coriolis effect, and pole-to-pole circulation gives Mars very dynamic weather.

The direction of the pole-to-pole winds on Mars changes with the alternating seasons. Sometimes these winds initiate huge dust storms (Figure 10.16b), particularly when the more extreme summer in the southern hemisphere approaches. At times, the surface is so shrouded by airborne dust that surface markings are hidden from view and large dunes form and shift with strong winds. As the dust settles out onto the surface, it can change the color or albedo over vast areas—thereby creating the seasonal changes in appearance that fooled some astronomers in the late 1800s and early 1900s into thinking they saw changes in vegetation [Section 9.5]. Dust also settles onto the polar caps, making alternating layers of darker dust and brighter frost (Figure 10.16c). The dust storms leave Mars with a perpetually dusty sky, giving it a pale pink color.

Clouds and Precipitation

We may think of clouds as imperfections on a sunny day, but they have profound effects on a planet. Clouds fundamentally change the appearance of a planet, reflecting sunlight back to space and thus reducing the amount of sunlight that warms the surface. Clouds are also important to planetary geology because they are a prerequisite for rain, a major cause of erosion. Despite these significant effects, clouds on most planets form from very minor ingredients of the atmosphere (see Table 10.1).

Clouds form when one of the gases in the air condenses into liquid or solid form, a process usually resulting from convection (Figure 10.17). On Earth, water vapor is released into the atmosphere by evaporation of surface water (or sublimation of ice and snow). Convection carries the water vapor to high, cold regions of the troposphere, where it condenses into droplets or flakes of ice. Thus, clouds on Earth consist of countless tiny water droplets or ice flakes, a fact that you can feel if you walk through a cloud

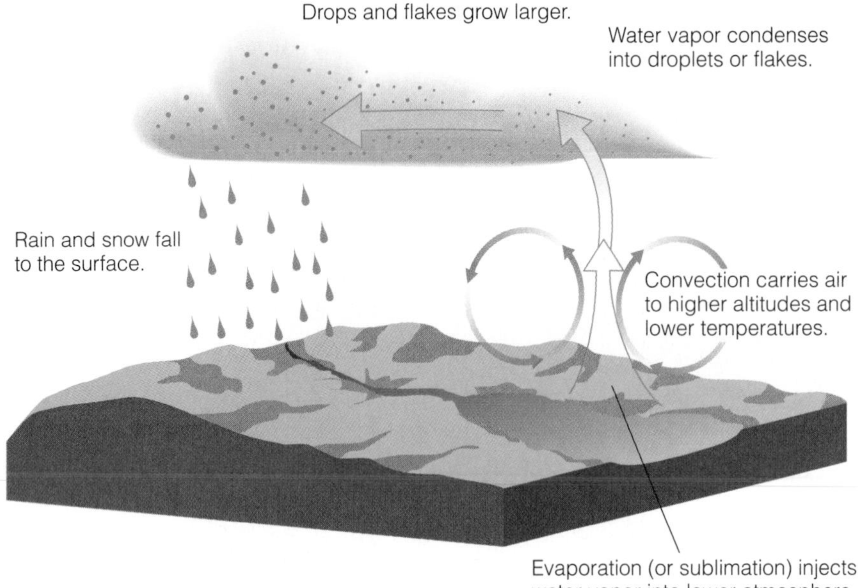

FIGURE 10.17 The cycle of water on Earth's surface and in the atmosphere. Other atmospheric ingredients on Earth and other planets have important cycles but may not be as apparent to the eye.

on a mountaintop. The droplets or flakes grow within the clouds until they become so large that the upward convection currents can't hold them aloft. They then begin to fall toward the surface as rain, snow, or hail.

Strong convection means more clouds and precipitation. Thunderstorms often form on late summer afternoons when the sunlight-warmed surface drives stronger convection. Earth's equatorial regions also experience more convection and rain than other regions. When you combine this fact with the global wind patterns (see Figure 10.15), you can understand why jungles lie at the equator and deserts lie at latitudes 20°–30° north and south of the equator. The equatorial rain depletes the moisture in the air as the winds carry it north or south, leaving little moisture to rain on the deserts.

Venus is covered with clouds because its high surface temperature drives strong convection and circulation (Figure 10.18). In contrast to Earth's water clouds, the clouds of Venus are made of sulfuric acid (H_2SO_4) dissolved in water droplets. These droplets form high in Venus's troposphere, where temperatures are 400 K cooler than on the surface. The droplets grow in the clouds until they form a sulfuric acid rain that falls toward the surface. But the droplets evaporate completely long before they reach the ground. Thus, it never rains on the surface of Venus, and the lower 30 kilometers of the atmosphere is very clear. The sulfuric acid is formed by chemical reactions involving sulfur dioxide (SO_2) and water, both of which come from volcanic eruptions. Because Venus lacks an ultraviolet-absorbing stratosphere, water molecules in its atmosphere are broken apart by ultraviolet light from the Sun. The atmosphere also loses sulfur dioxide through chemical reactions with surface rocks. The presence of sulfuric acid clouds therefore suggests that volcanoes on Venus must still be active (erupting at least occasionally) in order to replenish sulfur dioxide and water.

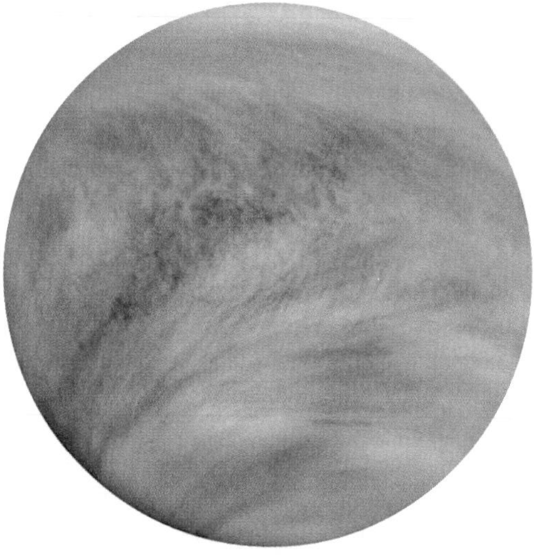

FIGURE 10.18 Clouds on Venus can be seen in this ultraviolet image taken by the *Pioneer Venus Orbiter.*

Clouds can also form when winds blowing over mountains carry air parcels to higher altitudes. Drop-

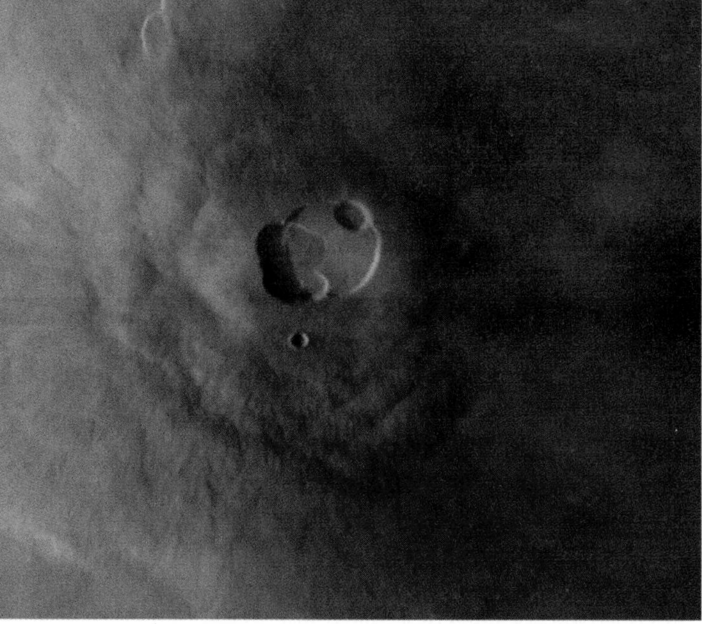

FIGURE 10.19 High-altitude clouds (the blue streaks in the upper right corner) over Olympus Mons, photographed by the *Mars Global Surveyor.*

lets form in an air parcel when it is pushed to high altitude over the mountain. These droplets then evaporate as the air parcel descends to the valley or plains below. The result is that droplets constantly form in the air over the mountain, creating a cloud that remains stationary even though strong winds are blowing. This process forms clouds over terrestrial mountain ranges like the Rockies, and also over tall Martian mountains like Olympus Mons (Figure 10.19). The clouds over the Martian mountains consist of water-ice crystals, and water-ice fog sometimes fills Martian canyons. But the total amount of water vapor in the Martian atmosphere is so small that it would make a frost layer less than 1/100 millimeter thick on the planet. Carbon-dioxide clouds can also form on Mars at much higher altitudes, but they are very rare.

Interestingly, Mars possesses a much larger amount of water than we ever see in its clouds. The polar caps contain large amounts of water ice in addition to frozen carbon dioxide. In fact, all the carbon dioxide at the northern polar cap sublimes during the northern summer, exposing a frozen-water polar cap. So much carbon dioxide is deposited on the southern polar cap during the long, deep winter that it does not all sublime during the relatively short summer. We therefore do not know how much water is hidden below. At high latitudes, water ice probably also permeates the Martian soil 1–2 meters below the surface, where not even the heat of a summer day can penetrate. This *ground ice* probably has had important geological effects. The Martian craters that look as if they were made in mud (see Figure 9.13b) probably formed when impacts melted the ground ice, and catastrophic flooding in the Martian past may have occurred when sudden volcanic heating melted the ground ice into liquid water.

Long-Term Climate Change

We know that long-term climate change can occur. For example, the Earth has been in and out of ice ages throughout its history. What can change the climate of an entire planet?

At least four factors can cause climate change. First, the Sun slowly grows in brightness as it ages [Section 14.3]. The Sun was about 30% fainter when the solar system first formed than it is today. Planets therefore may have been cooler early in the history of the solar system, or perhaps greenhouse gases were more abundant.

Second, changes in the tilt of a planet's axis can cause climate change. The axis tilt may slowly change over thousands or millions of years because of small gravitational tugs from moons, other planets, or the Sun. Although *average* planetary temperature may not be affected by the axis tilt, temperatures in different regions may change a great deal. A smaller axis tilt means year-round reduced sunlight in polar regions. Temperatures therefore drop in the polar regions, and perhaps even in mid-latitudes, causing *ice ages.* Earth and Mars probably both experience ice ages for this reason, and several of the icy satellites of the outer solar system undergo analogous cycles.

A third influence on climate change is changes in a planet's albedo. If a planet reflects more sunlight, it absorbs less—which can lead to planet-wide cooling. Ice ages may lead to further planetary cooling because the extra ice covering can increase the planet's albedo. Microscopic dust particles, or *aerosols,* released by volcanic eruptions can have the same effect. If the dust particles are blasted all the way into the stratosphere, they will remain suspended there for years, reflecting sunlight back to space. Human activity may be changing Earth's albedo. Smog particles can act like volcanic dust, reflecting sunlight before it reaches the ground. Deforestation can also increase Earth's albedo by removing sunlight-absorbing plants. But some human activity decreases albedo, such as paving a road with blacktop. Clearly, the overall human effect on Earth's albedo is not easy to gauge.

The fourth major cause of climate change is varying atmospheric conditions, particularly the abundance of greenhouse gases. If the abundance of greenhouse gases increases, the planet generally will warm. If the planet warms enough, increased evaporation and sublimation may add substantial amounts of gases to the planet's atmosphere, leading to an increase in atmospheric pressure. Conversely, if the abundance of greenhouse gases decreases, the planet generally will cool, and atmospheric pressure may decrease as gases freeze. We'll apply these concepts to the planets in the next section.

10.4 Atmospheric Origins and Evolution

Today's headlines are full of stories about the human effect on Earth's atmosphere, such as ozone depletion, global warming, and air pollution. But the atmospheres of Earth and the other terrestrial planets have seen much larger changes in their billions of years of history. What causes such large changes? To answer this question, we must examine both how a planet gets its atmospheric gases and how it can lose them.

How Atmospheres Are Created

The jovian planets obtained their atmospheres by capturing gas from the solar nebula, but the terrestrial worlds were too small to capture significant amounts of gas before the solar wind cleared away the nebula [Section 8.4]. Any gas that the terrestrial worlds did capture from the solar nebula consisted primarily of hydrogen and helium (just like the solar nebula as a whole). But these light gases escape easily from the terrestrial worlds and by now are long gone. The atmospheres of the terrestrial worlds therefore must have formed after the worlds themselves. Terrestrial atmospheres can get their gas from three different processes: *outgassing, evaporation/sublimation,* and *bombardment* (Figure 10.20).

The most important process is the release of gases by volcanic eruptions. This process is called **outgassing** because it releases gases from the planetary interior. Recall that these gases were originally brought to the terrestrial planets by comets and volatile-rich asteroids during the early bombardment that ended 3.8 billion years ago [Section 8.6]. The high pressures inside the planets trapped these gases in rocks in much the same way that a pressurized bottle traps the bubbles of carbonated beverages. When the rocks melt and erupt onto the surface as lava, the release of pressure expels the gases. The gases may be released with the lava itself or from nearby vents (Figure 10.21). The most common gases expelled are water (H_2O), carbon dioxide (CO_2), nitrogen (N_2), and sulfur-bearing gases (H_2S or SO_2).[10] Because outgassing depends on volcanism, it is most important on larger planets that retain their internal heat longer. Outgassing supplied most of the gas that became the atmospheres of Venus, Earth, and Mars.

Once outgassing creates an atmosphere, some of the gases may condense onto the surface. Earth's first volcanoes may have belched out water vapor over a dry, barren planet. As more and more water vapor filled the atmosphere, some of it began to condense and fall as rain—eventually filling our oceans. Similarly, Martian volcanoes outgassed large volumes of carbon dioxide, some of which is now frozen in the Martian polar caps. Surface liquids and ices continually exchange gas with the atmosphere. Sometimes the amount of gas released by *evaporation* (from liquids) or *sublimation* (from ices) exceeds the amount returning by condensation (from the atmosphere back into liquid or ice). These processes are considered the second most important way in which gases are supplied to an atmosphere. Even though the gases first entered the atmosphere through outgassing, sublimation and evaporation can resupply gases that the atmosphere has lost.

Sublimation is most important on relatively cold bodies where ices form. It adds gas to the Martian atmosphere with the changing Martian seasons, as the warmer southern summer releases vast amounts

[10]Aside from water, none of the gases existed in the solar nebula. Instead, they were created by chemical reactions that occurred *inside* the planets.

FIGURE 10.20 Three processes by which atmospheres gain gas.

of carbon dioxide from the south polar cap. It turns out to be even more important on some of the icy bodies of the outer solar system and in explaining the appearance of comets that happen to enter the inner solar system [Section 12.4].

Planets and satellites where outgassing and sublimation are negligible may still have thin atmospheres created by a third process: *bombardment* by micrometeorites, the solar wind, or energetic solar photons. Bombardment can make only a thin atmosphere (a denser atmosphere would protect the surface from the bombarding particles), but it is the primary source of the atmospheres of the Moon and Mercury. The atmosphere may be composed both of vaporized surface materials and of any gas released by the material doing the bombardment.

FIGURE 10.21 Volcanoes National Park, Hawaii. Volcanoes give off H_2O, CO_2, N_2, and sulfur-bearing gases. Some gas is given off from the lava itself, and some leaks out from nearby gas vents.

How Atmospheric Gases Are Lost

Just as several processes can add gas to a planetary atmosphere, an atmosphere can also lose gas in several ways: *thermal escape, bombardment, atmospheric cratering, condensation,* and *chemical reactions* (Figure 10.22). The terrestrial planets lost any original hydrogen and helium they once captured from the solar nebula through the process of **thermal escape,** in which an atom or molecule in the exosphere moves fast enough to escape the pull of gravity. Three factors determine whether an atmospheric gas can be lost to space by thermal escape:

1. The planet's *escape velocity,* or the speed at which a particle must travel to escape the pull of gravity—assuming it is traveling in the right direction and doesn't collide with anything else. The more massive the planet, the higher its escape velocity [Section 6.6].
2. The temperature. Individual particles in a gas move with a wide range of random speeds [Section 5.2], but the higher the temperature, the faster their *average* speed.

FIGURE 10.22 The five major processes by which atmospheres lose gas.

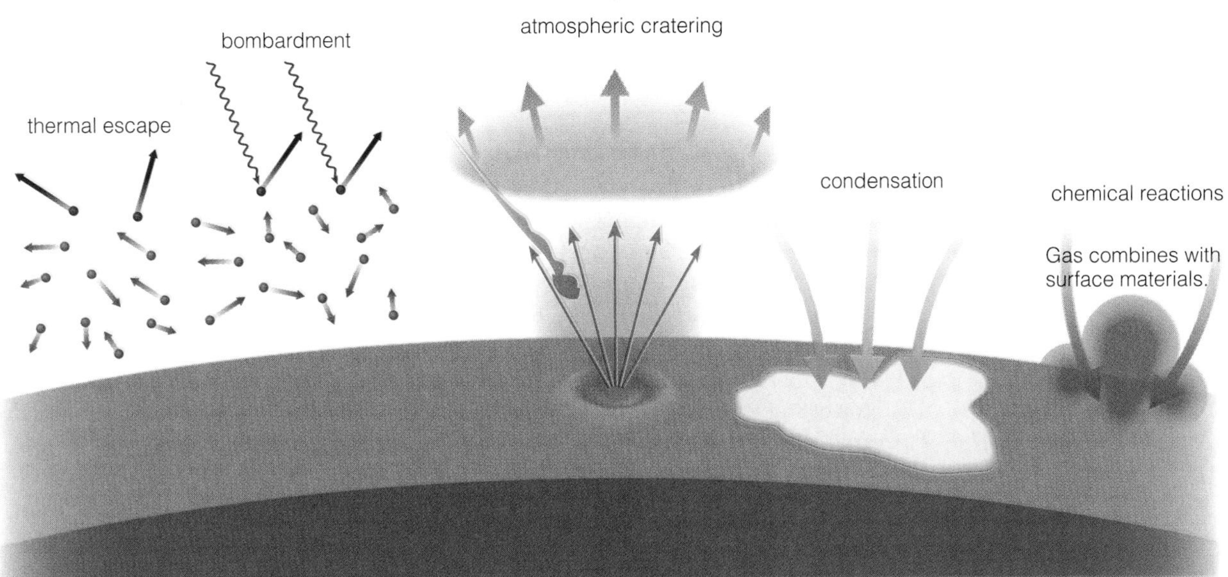

3. The mass of the gas particles. At a particular temperature, lighter particles (such as H or He) move around at faster average speeds than heavier ones (such as O_2 or CO_2) and therefore are more likely to achieve escape velocity.[11]

Mercury cannot hold much of an atmosphere because its small size and high temperature mean that nearly all gas particles eventually achieve escape velocity. Gases escape about as easily from the Moon because it is smaller than Mercury, even though it is also somewhat cooler. Light gases such as hydrogen and helium escape any planet much more easily than heavier gases. That is why Venus, Earth, and Mars retained heavier gases but lost light gases—including any hydrogen released when molecules such as H_2O were broken apart by solar ultraviolet radiation.

TIME OUT TO THINK *Suppose Mercury had formed at a much greater distance from the Sun, say, somewhere beyond the orbit of Mars. How would its lower surface temperature affect its ability to hold an atmosphere? Considering all you know about planetary formation and how atmospheric gases are created and lost, do you think Mercury would have a thicker atmosphere in this case? Explain.*

A second process that can cause atmospheric loss is *bombardment*—the same process that creates the atmospheres of the Moon and Mercury. Just as bombardment can eject atoms from a surface into a thin atmosphere, it can also eject atmospheric particles from a planet's exosphere altogether. Bombardment by the solar wind is particularly important because it can sweep ionized atoms from high in the atmosphere into space. Bombardment by energetic solar photons can also be important, because ultraviolet photons can break apart molecules and send fragments flying off at speeds faster than escape velocity. Note that bombardment can eject atoms that are too heavy for thermal escape. In general, planets with protective magnetic fields—such as Earth—are less susceptible to this type of atmospheric loss.

Larger impacts can also cause atmospheric escape. The same impacts that create craters on the surface may blast away a significant amount of the

[11]At a particular temperature, all gas particles have the same *average kinetic energy.* Because the kinetic energy is described by the formula $\frac{1}{2}mv^2$, low-mass particles must have higher average speeds (v) than high-mass particles of the same kinetic energy.

Mathematical Insight 10.2 Thermal Escape from an Atmosphere

Whether an atmosphere loses gas by thermal escape depends on how many of the gas particles (atoms or molecules) are moving faster than the escape velocity. In general, particles of a particular type in a gas move at a wide range of speeds that depend on the temperature. For example, Figure 10.23 shows the range of speeds of sodium atoms at the temperature of the Moon's extremely thin atmosphere. The peak in the figure represents the most common speed of the sodium atoms. This speed, called the *thermal velocity* of the sodium atoms, is given by the following formula:

$$v_{\text{thermal}} = \sqrt{\frac{2kT}{m}}$$

where m is the mass of a single atom, T is the temperature in Kelvin, and $k = 1.38 \times 10^{-23}$ joule/Kelvin is a constant called *Boltzmann's constant.* This formula shows that higher temperatures mean higher thermal velocities for the atoms or molecules in a gas. It also shows that, for a given temperature, particles of lighter gases (smaller m) move at faster speeds than particles of heavier gases.

If the thermal velocity of a particular type of gas is higher than the escape velocity, then most of the gas particles will be traveling fast enough to escape the atmosphere, and the planet will lose this atmospheric gas to space fairly quickly. However, even if the thermal velocity is lower then the escape velocity, the wide range of particle speeds in a gas means that some small fraction of the gas particles may be traveling faster than the escape velocity. These fast-moving particles will escape, as long as they're moving upward and don't hit another particle on their way out. Collisions among the remaining gas particles will continually ensure that a few always attain escape velocity, and thus the atmosphere can slowly atmosphere as well, a process sometimes called *atmospheric cratering*. For the most part, atmospheric cratering happened only during the early bombardment stage of the solar system. We expect it to have been most important on smaller worlds, where gravity's hold on the atmosphere is weaker. Mars probably lost a significant amount of atmospheric gas through atmospheric cratering.

Thermal escape, bombardment, and atmospheric cratering all cause a planet to lose gas permanently into space. Two other processes allow atmospheric gas to be absorbed back into the solid planet. Gases may simply *condense* into liquid or solid form if a planet cools down, especially in the cold polar regions. On Mars, carbon dioxide condenses into the polar caps during winter. Condensation probably also explains the presence of water ice in craters near the Moon's poles, discovered by the *Lunar Prospector* orbiter in 1998. The water probably came from comet impacts that once briefly created a thin lunar atmosphere. The water molecules from these impacts bounced around the surface at random until they condensed into ice in the polar craters. The bottoms of these craters lie in nearly perpetual shadow, keeping them so cold that the water remains perpetually frozen. Recent radar observations of Mercury suggest that it, too, may have water ice in its polar craters, probably for the same reason.

Finally, gases may become locked up in the surface through *chemical reactions* with rocks or liquids. On Earth, for example, carbon dioxide dissolves in the oceans, where it ultimately undergoes reactions that create *carbonate* rocks, such as limestone, on the ocean floor [Section 13.3]. In fact, volcanoes have outgassed huge amounts of carbon dioxide into Earth's atmosphere over its history, but nearly all this gas has been converted into limestone by chemical reactions.

Putting It All Together

We've seen that the processes by which atmospheric gases are created and lost are closely tied to planetary formation properties and other planetary processes. We now have the perspective to link what we have learned about planetary atmospheres with our understanding of planetary geology from the previous chapter. In particular, we are in a better position to appreciate how planetary size affects the amount of outgassing and therefore the abundance of atmospheric gases. A closer look at atmospheric change will allow us to understand the different erosional

leak away into space. In general, most of the gas particles of a particular type will escape over the age of the solar system if the thermal velocity is 20% or more of the escape speed.

(continued)

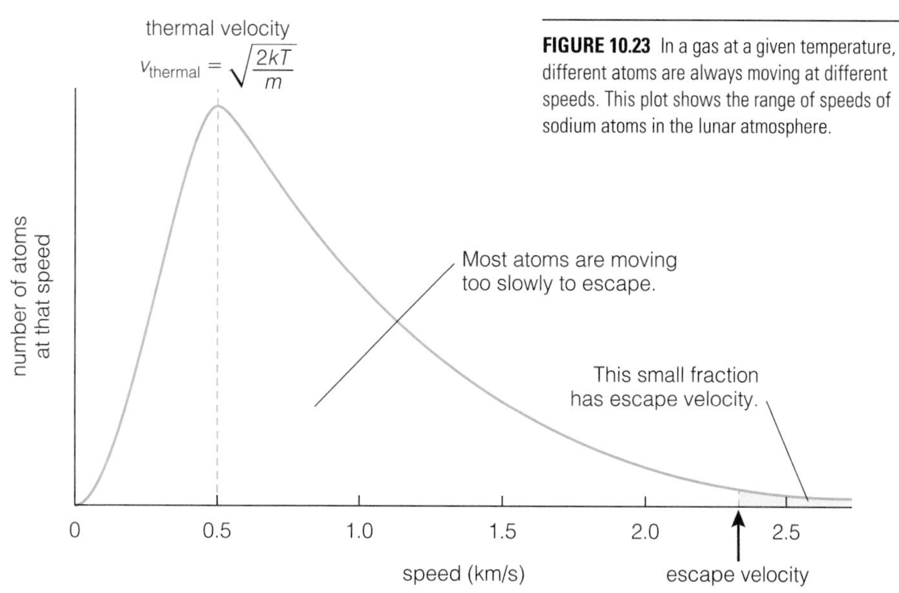

FIGURE 10.23 In a gas at a given temperature, different atoms are always moving at different speeds. This plot shows the range of speeds of sodium atoms in the lunar atmosphere.

histories of Earth, Mars, and Venus, which we left unresolved in the last chapter. Before you read further, you should refresh your memory of the "planetary flowcharts" in Chapter 9 and make your own analysis of atmospheric change on each of the terrestrial worlds.

10.5 History of the Terrestrial Atmospheres

We are now ready to return to the issue with which we began the chapter: how and why the terrestrial atmospheres came to differ so profoundly. Let's take a "historical tour" of the terrestrial atmospheres, in order of increasing planetary size.

The Moon and Mercury

The extremely thin atmospheres of the Moon and Mercury are the easiest to explain (Figure 10.24). Volcanic outgassing ceased long ago on these small worlds, and any early atmosphere was lost. Both Mercury and the Moon may have some ice in craters near their poles, but it remains perpetually frozen. Thus, the only source of atmospheric gas is bombardment. The relatively few particles blasted into the atmosphere by bombardment collide more often with the surface than with one another. They bounce around the surface like tiny rubber balls, arcing hundreds of kilometers into the sky before crashing back down. Sometimes these particles travel fast enough to escape the pull of gravity (thermal escape). Other times they are stripped away by solar wind bombardment. Thus, gas does not accumulate in these atmospheres, but a small amount is always present because bombardment occurs continuously. Both atmospheres are relatively hot and extend high above the surface.

Mars

Mars is only 40% larger in radius than Mercury, but its surface tells of a much more fascinating and complex atmospheric history. Pictures show clear evidence that water once flowed on the Martian surface. The abundant geological signs of water erosion on Mars present a major enigma to our understanding of the planet. The evidence includes not only obvious features such as dried-up streambeds (see Figure 9.27a), but also subtler features such as eroded crater rims and floors. Further evidence comes from the study of meteorites believed to have come to Earth from Mars [Section 12.3]. These *Martian meteorites* contain types of minerals (clays and carbonates) that can form only in the presence of liquid water. Some planetary geologists even believe they have

Example: The temperature on the dayside of the Moon is $T = 400$ K. Calculate the thermal velocities of both hydrogen atoms and sodium atoms. Then explain why the Moon (escape velocity $\approx$ 2.4 km/s) cannot retain a hydrogen atmosphere for very long but retains a thin exosphere of sodium. The mass of a hydrogen atom is 1.67×10^{-27} kg; the mass of a sodium atom is 23 times greater, or 3.84×10^{-26} kg.

Solution: From the formula above, the thermal velocity of the hydrogen atoms on the dayside of the Moon is:

$$\text{hydrogen: } v_{\text{thermal}} = \sqrt{\frac{2 \times (1.38 \times 10^{-23} \frac{\text{joule}}{\text{K}}) \times (400\ \text{K})}{1.67 \times 10^{-27}\ \text{kg}}} \approx 2{,}600\ \text{m/s} = 2.6\ \text{km/s}$$

Similarly, the thermal velocity of the sodium atoms is:

$$\text{sodium: } v_{\text{thermal}} = \sqrt{\frac{2 \times (1.38 \times 10^{-23} \frac{\text{joule}}{\text{K}}) \times (400\ \text{K})}{3.84 \times 10^{-26}\ \text{kg}}} \approx 540\ \text{m/s} = 0.54\ \text{km/s}$$

Note that the thermal velocity of the hydrogen atoms is slightly greater than the Moon's escape velocity of about 2.4 km/s. Thus, the Moon cannot retain a hydrogen atmosphere. However, the thermal velocity of sodium is only about one-fifth, or about 20%, of the escape velocity. Thus, the Moon cannot hold a sodium atmosphere indefinitely, but it can hold some of the sodium atoms that are continually entering its atmosphere through bombardment.

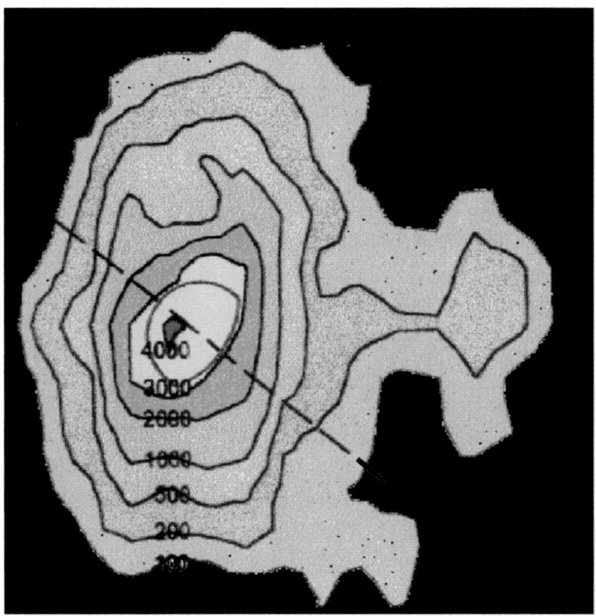

a Mercury's atmosphere. Color-coded contours represent the gas density from lowest (black) to highest (red). Mercury's partially illuminated disk is indicated as the gibbous-shaped blue contour near the center (just within the yellow region); the Sun lies to the left along the direction of the dashed line.

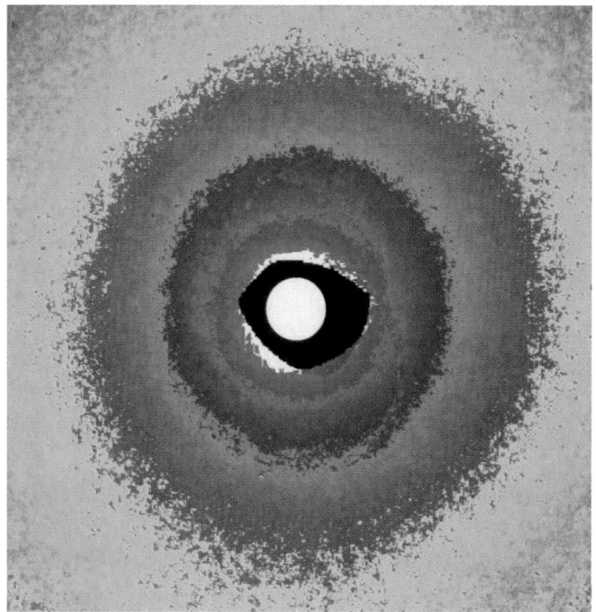

b The Moon's atmosphere. This composite image represents the gas density from lowest (green) to highest (red). The Moon itself was blocked to make this image, resulting in the central black area of no data. The Moon's size is shown schematically by the white circle.

FIGURE 10.24 The atmospheres of Mercury and the Moon—which are essentially exospheres only—viewed through instruments sensitive to emission lines from sodium atoms. Although these are extremely low-density atmospheres, they extend quite high.

found evidence in Viking images for ancient oceans and glaciers on Mars, although these conclusions are controversial.

The fact that liquid water once flowed on Mars means that its past must have been very different from its current "freeze-dried" state. The surface temperature must have been much warmer, and therefore Mars must have had a much stronger greenhouse effect and much higher atmospheric pressure. Computer simulations show that these requirements can be met by a carbon dioxide atmosphere about 400 times denser (i.e., having greater pressure) than the current Martian atmosphere—or about twice as thick as Earth's atmosphere.

The idea that Mars had a much denser atmosphere in the past is reasonable. Mars has plenty of ancient volcanoes, and calculations show that these volcanoes could have supplied all of the carbon dioxide needed to warm the planet. Even CO_2 ice clouds could have helped hold heat in the Martian troposphere. Moreover, if Martian volcanoes outgas carbon dioxide and water in the same proportions as volcanoes on Earth, Mars would have had enough water to fill oceans tens or even hundreds of meters deep.

The bigger question is not whether Mars once had a denser atmosphere, but where those atmospheric gases are today. Mars must somehow have lost most of its carbon dioxide gas. This loss reduced the strength of the greenhouse effect until the planet essentially froze over. The fate of the lost carbon dioxide is not completely clear. Some is locked up in the polar caps, and some in carbonate minerals like those identified in the Martian meteorites. Perhaps these carbonates formed in the same way as carbonates on Earth—through chemical reactions on seafloors after carbon dioxide dissolved in liquid oceans. Some of the gas in the early Martian atmosphere may also have been blasted away by large impacts (atmospheric cratering) and bombardment by the solar wind. We hope to learn more about the relative importance of these loss processes on Mars when future robotic spacecraft make better measurements of the distribution of carbonate rocks and the properties of Mars's current atmosphere.

The vast amounts of water once present on Mars are also gone. Some is tied up in the polar caps and underground ice, but most was probably lost forever. Mars lacks an ultraviolet-absorbing stratosphere, and water molecules are easily broken apart by ultraviolet photons. Once water molecules were broken apart, the hydrogen atoms would have been rapidly lost to space through thermal escape. Some of the remaining oxygen was probably lost by solar wind bombardment, and the rest was probably drawn out of the atmosphere through chemical reactions with surface rock. This oxygen literally rusted the Martian rocks, giving the "red planet" its distinctive tint.

The history of the Martian atmosphere holds important lessons for us on Earth. Mars was apparently once a world with pleasant temperatures and streams, rain, glaciers, lakes, and possibly oceans. It had all the necessities for "life as we know it." But the once hospitable planet turned into a frozen and barren desert at least 3 billion years ago, and it is unlikely that Mars will ever be warm enough for its frozen water to flow again.[12] If life once existed on Mars, it is either extinct or hidden away in a few choice locations, such as hot springs around not-quite-dormant volcanoes. As we think about the possibility of future climate change on Earth, Mars presents us with an ominous example of how much things can change.

Venus

Venus presents a stark contrast to Mars. Its thick carbon dioxide atmosphere creates an extreme greenhouse effect that bakes the surface to unbearable temperatures. Venus's larger size allowed it to retain more interior heat than Mars, leading to greater volcanic activity [Section 9.5]. The associated outgassing produced the vast quantities of carbon dioxide in Venus's atmosphere, as well as the much smaller amounts of nitrogen and sulfur dioxide. But the outgassing presents a quandary when we consider water. We expect similar gases to come out of volcanoes on Earth and Venus, since both planets have similar overall compositions. Water vapor is the most common gas from volcanic outgassing on Earth, but Venus is incredibly dry, with only minuscule amounts of water vapor in its atmosphere and not a drop on the surface. Did Venus once have huge amounts of water? If so, what happened to it?

The leading theory suggests that Venus did indeed outgas plenty of water. But Venus's proximity to the Sun kept surface temperatures too high for oceans to form as they did on Earth. Outgassed water vapor accumulated in the atmosphere along with carbon dioxide and other gases. Because both water and carbon dioxide are strong greenhouse gases, Venus rapidly grew warmer still. Before the water disappeared, the surface may have had double its current high temperature. Solar ultraviolet photons broke water molecules apart, and the high temperature allowed rapid escape of hydrogen (just as on Mars). The remaining oxygen was lost by a combination of chemical reactions with surface rocks and the bombardment by the solar wind. Thus, the water that once was on Venus is gone forever. In fact, Venus may have lost so much water that its crust and mantle also dried out.

It's difficult to prove that Venus lost an ocean's worth of water billions of years ago, but some evidence comes from the gases that didn't escape. Recall that most hydrogen nuclei contain just a single proton. But a tiny fraction of all hydrogen atoms in the universe contain a neutron in addition to the proton. This isotope of hydrogen is called *deuterium,* or sometimes "heavy hydrogen." Deuterium can be present in water molecules in place of one of the hydrogen atoms (making such molecules "HDO," or "heavy water," rather than H_2O). A deuterium atom is therefore released into the atmosphere when an ultraviolet photon breaks apart such a water molecule. Both hydrogen and deuterium escape, but less deuterium escapes due to its higher mass. If Venus once had huge amounts of water molecules that broke apart, the more rapid escape of ordinary hydrogen should have left its atmosphere enriched in deuterium. And, indeed, measurements show that deuterium is at least a hundred times more abundant (relative to ordinary hydrogen) on Venus than anywhere else in the solar system.[13]

The next time you see Venus shining brightly as the morning or evening "star," consider the radically different path it has taken from that of Earth—and thank your lucky star. If Earth had formed a bit closer to the Sun or if the Sun had been a shade hotter, our planet might have suffered the same greenhouse-baked fate.

Looking Ahead to Earth

Astronomers began their studies of Mars and Venus expecting these planets to be variants of Earth. Instead we've discovered that they are fundamentally different from Earth, particularly in the areas of atmospheric composition and climate. Rather than being a "role model" for understanding other planets, Earth looks more like the odd one out. The lesson of planetary exploration is that in order to understand Earth we must understand the solar system as a whole: its formation, geological processes, atmospheres, and even the jovian planets, asteroids, and comets. We are more than halfway through this list already, and after two more chapters we will come full circle back to Earth, ready to explore and explain our unique circumstances in Chapter 13.

---

[12]Although it's unlikely that Mars will become substantially warmer naturally, some people speculate that human intervention might somehow enhance the natural greenhouse effect.

[13]Some scientists remain skeptical that Venus lost huge amounts of water, instead suggesting that it was "born dry"—perhaps because it never received as much water as Earth or Mars. In that case, Venus's extraordinary abundance of deuterium relative to ordinary hydrogen requires an alternative explanation.

For the first time in my life I saw the horizon as a curved line. It was accentuated by a thin seam of dark blue light—our atmosphere. Obviously this was not the ocean of air I had been told it was so many times in my life. I was terrified by its fragile appearance.

ULF MERBOLD, ASTRONAUT (GERMANY)

THE BIG PICTURE

With what we have learned in this chapter and the previous one, we now have a complete "big picture" view of how the terrestrial planets started out so similar yet ended up so different. As you continue, keep the following important ideas in mind:

- Planetary atmospheres are collections of particles in search of balance. The vertical spread of an atmosphere is a balance between gravity and random thermal motions of molecules. Temperatures in different atmospheric regions are in balance with the amount of light absorbed there.
- Weather and climate change are the planets' vain attempts to equalize temperatures and pressures over their surfaces. It's a losing battle, because solar heating can never be uniform. But it's the very nature of gases to move toward equilibrium, even if they never attain it.
- The many levels of balance rely on an active, not static, atmosphere. From the unseen vibrations and rotations of speeding molecules to planet-wide winds and weather, atmospheres are in motion.
- The long-term evolution of planetary atmospheres reflects an imbalance between atmospheric gains and losses, which vary over billions of years. Complete atmospheric transformation over the age of the solar system appears to be the rule for large planets, not the exception.

Review Questions

1. In a few words, briefly summarize the basic characteristics of the atmospheres of each of the five terrestrial worlds.
2. What is an atmosphere? List a few of the most common atmospheric gases. Where does an atmosphere end?
3. Explain the microscopic origin of *gas pressure.* What is 1 *bar* of pressure?
4. Briefly describe the three factors that would determine planetary temperatures in the absence of greenhouse gases. How do the "no atmosphere" temperatures of the terrestrial planets compare to their actual temperatures? Why?
5. What do we mean by *albedo?* Give examples of materials with high and low albedos.
6. What types of gases absorb infrared light? Ultraviolet light? X rays? Give a few examples of *greenhouse gases.* What is *ozone,* and what does it do in our atmosphere?
7. Explain how to interpret the generic atmospheric structure profile shown in Figure 10.6. Define the *exosphere, thermosphere, ionosphere, stratosphere,* and *troposphere.*
8. Briefly explain why the sky is blue in the daytime and why sunrises and sunsets are red.
9. Briefly describe how the *greenhouse effect* makes a planetary surface warmer than it would be otherwise.
10. What is convection? Briefly explain why it occurs in the troposphere but not in the stratosphere.
11. Briefly explain why a planet can have a stratosphere only if it has ultraviolet-absorbing molecules in its atmosphere. What molecule plays this role on Earth?
12. Why are virtually all gases good X-ray absorbers? Explain how X-ray absorption in the thermosphere creates an ionosphere, and how the ionosphere affects radio communications.
13. Briefly describe the actual atmospheric structure of each of the five terrestrial worlds, and compare these actual structures to the generic structure shown in Figure 10.6.
14. What is a *magnetosphere?* What are *charged particle belts?* Briefly describe how the solar wind affects magnetospheres, and how auroras are produced.
15. Briefly distinguish between *weather* and *climate.*
16. Explain why Earth and Mars have seasons, but Venus does not. How do seasons on Mars differ from seasons on Earth? Why does atmospheric pressure on Mars change during the seasons, while atmospheric pressure on Earth remains steady year-round?
17. What do we mean by *global wind patterns?* Briefly explain why terrestrial planets tend to have huge *circulation cells* carrying warm air toward the poles and cool air toward the equator.

18. What is the origin of the *Coriolis effect?* How does it affect global wind patterns? Briefly explain why the Coriolis effect is stronger on Earth than on Mars and virtually nonexistent on Venus.
19. Briefly describe how clouds form and how they affect planetary weather. What causes rain? Why does it rain high in the atmosphere of Venus, but not on the surface?
20. Briefly describe the four factors that can lead to long-term climate change, and the effects of each factor.
21. Briefly describe the three processes that can add gas to a planetary atmosphere. What planetary properties determine the relative importance of each process?
22. Briefly describe the five processes by which a planetary atmosphere can lose gas.
23. Explain how a planet's escape velocity, temperature, and atmospheric composition determine the rate of thermal escape of gases from its atmosphere. Use this idea to explain why Mercury and the Moon retain no substantial atmospheres at all, and why the other terrestrial planets retain heavier gases but not light gases such as hydrogen and helium.
24. Summarize the processes that ensure that Mercury and the Moon have only thin exospheres. Why may there be ice in some polar craters on both Mercury and the Moon?
25. Briefly describe the atmospheric history of Mars. What evidence tells us that its atmospheric pressure and temperature were once much greater than they are today? Where did all the atmospheric gas go? Is it likely that Mars will ever again have a much thicker and warmer atmosphere? Why or why not?
26. Why should we expect Venus and Earth to have had very similar early atmospheres? How did the two planetary atmospheres become so different? In particular, what happened to the water on Venus? What happened to the carbon dioxide on Earth? How does the ratio of deuterium to ordinary hydrogen on Venus support the idea that it has lost enormous amounts of water?

Discussion Questions

1. *Charting Atmospheric Evolution.* How would you develop "planetary flowcharts" for the atmospheric source and loss processes analogous to those developed for geological processes in Chapter 9? Develop two charts, one for a source process and one for a loss process. Be sure your flowchart includes all the relevant connections described in this chapter.
2. *Mars: Past and Future.* Summarize the evolution of the Martian atmosphere since planetary formation, emphasizing the important source and loss processes. What is your prediction for the future of the Martian atmosphere?

Problems

1. *Cool Venus.* Table 10.2 shows that Venus's temperature in the absence of the greenhouse effect is lower than Earth's, even though it is closer to the Sun.
   a. Explain this unexpected result in a sentence or two.
   b. Now suppose that Venus has neither clouds nor greenhouse gases. What do you think would happen to the temperature of Venus? Why?
   c. How are clouds and volcanoes linked on Venus? What change in volcanism might result in the disappearance of clouds? Explain.
2. *Atmospheric Structure.* Study Figure 10.6, which shows the atmospheric structure for a generic terrestrial planet. For each of the following cases, make a sketch similar to Figure 10.6 showing how the atmospheric structure would be different, and explain the differences in words.
   a. Suppose the planet had no greenhouse gases.
   b. Suppose the Sun emitted no solar ultraviolet light.
   c. Suppose the Sun had a higher output of X rays.
3. *Inversions.* Consider a local inversion, in which the air is colder near the surface than higher up in the troposphere.
   a. Explain why convection is suppressed in an inversion.
   b. Cities that experience frequent inversions, such as Los Angeles and Denver, often have more serious problems with air pollution than other cities of similar size. Explain how an inversion can trap pollutants, keeping them close to the city in which they are generated.
4. *Coastal Winds.* During the daytime, heat from the Sun tends to make the air temperature warmer over land near the coast than over the water offshore. But at night, when land cools off faster than the sea, the temperatures tend to be cooler over land than over the sea. Use these facts to predict the directions in which winds generally blow during the day and at night in coastal regions; for example, do the winds blow out to

sea or in toward the land? (*Hint:* How are these conditions similar to those that cause planetary circulation cells?) Explain your reasoning in a few sentences. Diagrams may help.

5. *A Swiftly Rotating Venus.* Suppose Venus had rotated as rapidly as Earth throughout its history. Briefly explain how and why you would expect it to be different in terms of each of the following: geological processes, atmospheric circulation, magnetic field, and atmospheric evolution. Write a few sentences about each.

6. *Sources and Losses.* Choose one atmospheric source process and one atmospheric loss process. Describe for each the ways in which the process is related to the four planetary formation properties discussed in Chapter 9. (For example, how is the creation of atmospheric gas by bombardment related to a planet's size, distance from the Sun, composition, and rotation rate?) (*Hint:* A process may not depend on all four formation properties.)

7. *Terraforming Mars.* Some people have proposed that we might someday *terraform* Mars, making it more Earth-like so that we might live there more easily. One of the first steps in terraforming Mars would be to somehow warm it up so that its ice sublimes into the atmosphere and increases the atmospheric pressure. Suggest at least one way you could cause Mars to warm up if you had unlimited resources. Describe your suggestion(s) in one or two paragraphs, and also discuss whether you think your idea might ever be practical. (*Hint:* Consider the four factors that can cause climate change.)

8. *Habitable Planet Around 51 Peg?* A recently discovered planet orbits the star 51 Pegasi at a distance of only 0.051 AU. The star is approximately as bright as our Sun. The planet has a mass 0.6 times that of Jupiter, but no one knows if it is more similar to our jovian or to our terrestrial planets (if either).
   a. Suppose it is a terrestrial-type planet with an albedo of 0.15 (i.e., without clouds). Calculate its "no atmosphere" temperature, assuming it rotates fast enough to have the same temperatures on its dayside and nightside. How does the temperature compare to that of Earth?
   b. Repeat part (a), but this time assume that the planet is covered in very reflective clouds, giving it an albedo of 0.8.
   c. Based on your answers to parts (a) and (b), do you think it is likely that the conditions on this planet are conducive to life? Explain your answer in one or two paragraphs.

9. *Escape from Venus.*
   a. Calculate the escape velocity from Venus's exosphere (about 200 km above Venus's surface). (*Hint:* See Mathematical Insight 6.3.)
   b. Calculate and compare the thermal speeds of hydrogen and deuterium atoms at the exospheric temperature of 350 K. The mass of a hydrogen atom is 1.67×10^{-27} kg, and the mass of a deuterium atom is about twice as much.
   c. Comment in a few sentences on the relevance of the calculations in parts (a) and (b) to atmospheric evolution on Venus.

10. *Project: Atmospheric Science in the Kitchen.*
   a. Find an empty plastic bottle, such as a water bottle with a screwtop that makes a good seal. Warm the air inside by filling the bottle partway with hot water, then shaking and emptying the bottle. Seal the bottle and place it in the refrigerator or freezer. What happens after 15 minutes or so? Explain why this happens, imagining that you could see the individual air molecules.
   b. You may have noticed that loose ice cubes in the freezer gradually shrink away to nothing or that frost sometimes builds up in old freezers. What are the technical terms used in this chapter for these phenomena? On what terrestrial planet do these same processes play a major role in controlling the atmosphere?

11. *Web Project: Martian Weather.* Follow the links from the book web site to find the latest weather report for Mars from spacecraft and other satellites. What season is it in the northern hemisphere? When was the most recent dust storm? What surface temperature was most recently reported from Mars's surface, and where was the lander located? Summarize your findings by writing a 1-minute script for a television news update on Martian weather.

II
Jovian Planet Systems

In Roman mythology, the namesakes of the jovian planets are rulers among gods: Jupiter is the king of the gods, Saturn is Jupiter's father, Uranus is the lord of the sky, and Neptune rules the sea. But even the most imaginative of our ancestors did not foresee the true majesty of the four jovian planets. The smallest, Neptune, is still large enough to contain the volume of more than 50 Earths. The largest, Jupiter, has a volume some 1,400 times that of Earth. These worlds are totally unlike the terrestrial planets. They are essentially giant balls of gas, with no solid surface on which to stand.

Why should we care about a set of worlds so different from our own? Apart from satisfying natural curiosity about the diversity of the solar system, the jovian planet systems serve as a testing ground for our general theories of comparative planetology. Are our theories good enough to predict the nature of these planets and their elaborate satellite systems, or are different processes at work? We'll find out in this chapter as we investigate the nature of the jovian planets, along with their intriguing moons and beautifully complex rings.

11.1 The Jovian Worlds: A Different Kind of Planet

The jovian planets have long held an important place in many human cultures. Jupiter is the third-brightest object that appears regularly in our night skies, after the Moon and Venus, and ancient astronomers carefully charted its 13-year trek through the constellations of the zodiac. Saturn is fainter and slower-moving, but people of many cultures also mapped its 27-year circuit of the sky. Ancient skywatchers did not know of the planets Uranus and Neptune, but astronomers charted their orbits soon after their discoveries in 1781 and 1846, respectively [Section 2.1].

Scientific study of the jovian worlds dates back to Galileo [Section 6.3]; however, scientists did not fully appreciate the differences between terrestrial and jovian planets until they established the absolute scale of the solar system (that is, in units such as kilometers instead of AU). Once they had established distances, they could calculate the truly immense sizes of the jovian planets from their *angular* sizes (as measured through telescopes) [Section S2.1]. Knowing the distance scale also allowed them to measure the masses of the jovian planets; recall that Newton's version of Kepler's third law allows the calculation of a planet's mass from the orbital period and average distance of one of its moons [Section 6.4]. Together, the measurements of size and mass revealed the low densities of the jovian planets, proving that these worlds are very different from the Earth.[1]

The slow pace of discovery about the jovian planets gave way to revolutionary advances with the era of spacecraft exploration. The first spacecraft to the outer planets, *Pioneer 10* and *Pioneer 11,* flew past Jupiter and Saturn in the early 1970s. The Voyager missions followed less than a decade later. Both *Voyager 1* and *Voyager 2* visited Jupiter in 1979 and Saturn in 1981. *Voyager 2* continued on a "grand tour" of the jovian planets, flying past Uranus in 1986 and Neptune in 1989. The results were spectacular. Figure 11.1 shows a montage of the jovian planets compiled by the Voyager spacecraft. Exploration of the jovian planet systems continues with the Galileo spacecraft, orbiting Jupiter since 1995, and the Cassini mission, scheduled for arrival at Saturn in 2004.

Jupiter, Saturn, Uranus, and Neptune are so different from the terrestrial planets that we must create an entirely new mental image of the term *planet.*

[1]Isaac Newton proved that Jupiter was *relatively* less dense than the Earth before anyone could measure the *absolute* density of either.

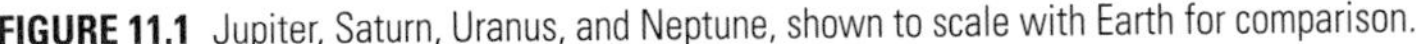

FIGURE 11.1 Jupiter, Saturn, Uranus, and Neptune, shown to scale with Earth for comparison.

Table 11.1. Comparison of Bulk Properties of the Jovian Planets

Planet	Average Distance from Sun (AU)	Mass (Earth masses)	Radius (Earth radii)	Average Density (g/cm³)	Bulk Composition
Jupiter	5.20	317	11.2	1.33	Mostly H, He
Saturn	9.53	90	9.4	0.70	Mostly H, He
Uranus	19.2	14	4.11	1.32	Hydrogen compounds and rocks, H and He
Neptune	30.1	17	3.92	1.64	Hydrogen compounds and rocks, H and He

Table 11.1 summarizes the bulk properties of these giant worlds. The most obvious difference between the jovian and terrestrial planets is size. Jupiter's mass in comparison to that of Earth is like the mass of a full-grown adult in comparison to that of a squirrel. By volume, Earth in comparison to Jupiter is like a pea in a cup of soup.[2]

The overall composition of the jovian planets—particularly Jupiter and Saturn—is more similar to that of the Sun than to any of the terrestrial worlds. Their primary chemical constituents are the light gases hydrogen and helium, although their atmospheres also contain many hydrogen compounds and their cores consist of a mixture of rocks, metals, and hydrogen compounds. Although these cores are small in proportion to the jovian planet volumes, each is still more massive than any terrestrial planet. Moreover, the pressures and temperatures near the cores are so extreme that the "surfaces" of the cores do not resemble terrestrial surfaces at all. In a sense, the general structures of the jovian planets are opposite those of the terrestrial planets: Whereas the terrestrial planets have thin atmospheres around rocky bodies, the jovian planets have proportionally small, rocky cores surrounded by massive layers of gas (Figure 11.2). A hypothetical spacecraft descending into Jupiter, for example, would have to travel *tens of thousands* of kilometers before reaching the core.

The jovian planets may be Sun-like in composition, but their masses are far too low to provide the interior temperatures and densities needed for nuclear fusion.[3] The jovian planets have undoubtedly lost internal heat during the more than 4 billion years since their formation and thus must have been much warmer in the distant past. This heat may have "puffed up" their atmospheres, making them larger and brighter in the past. If so, Jupiter would have been even more prominent in Earth's sky billions of years ago than it is today.

The jovian planets also rotate much more rapidly than any of the terrestrial worlds, but precisely defining their rotation rates can be difficult because they are not solid. We can measure a terrestrial rotation period simply by watching the apparent movement of a mountain or crater as a planet rotates, but on jovian planets we can observe only the movements of clouds. Cloud movements can be deceptive, because their apparent speeds may be affected by winds as well as by planetary rotation. Nevertheless, observations of clouds at different latitudes suggest that the jovian planets rotate faster near their equators than near their poles. (The Sun also exhibits this type of *differential rotation* [Section 14.5].) We can measure the rotation rates of the jovian interiors by tracking emissions from charged particles trapped in their

[2]Recall that volume is proportional to the cube of the radius. Thus, because Jupiter's radius is 11.2 times Earth's, its volume is $11.2^3 = 1{,}405$ times Earth's.

[3]Calculations show that nuclear fusion is possible only with a mass at least 80 times that of Jupiter. Some people have called Jupiter a "failed star" for this reason, but we prefer to think of it as a very successful planet.

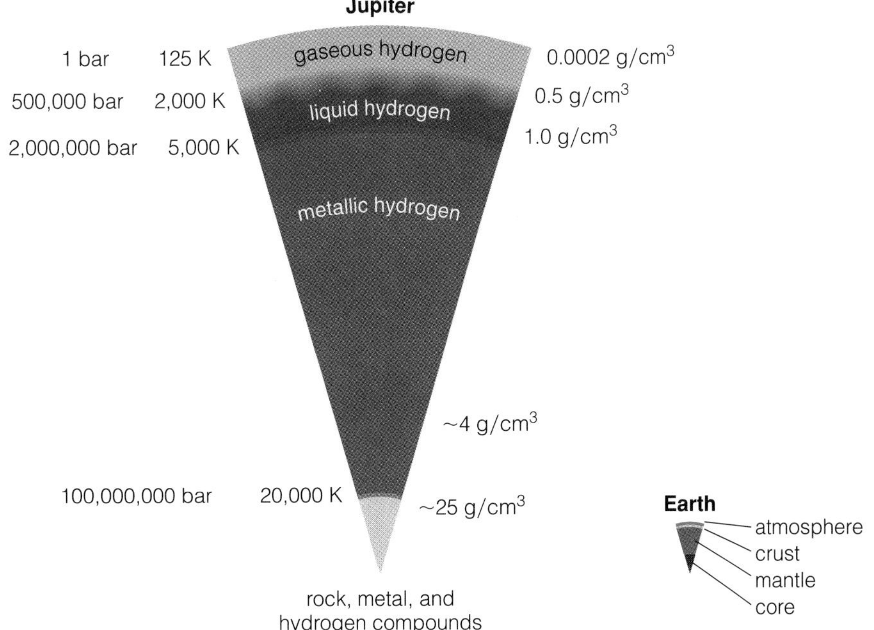

FIGURE 11.2 Jupiter's interior structure, labeled with the pressure, temperature, and density at various depths. Earth's interior structure is shown to scale for comparison. Note that Jupiter's core is only slightly larger than Earth but is about 10 times more massive.

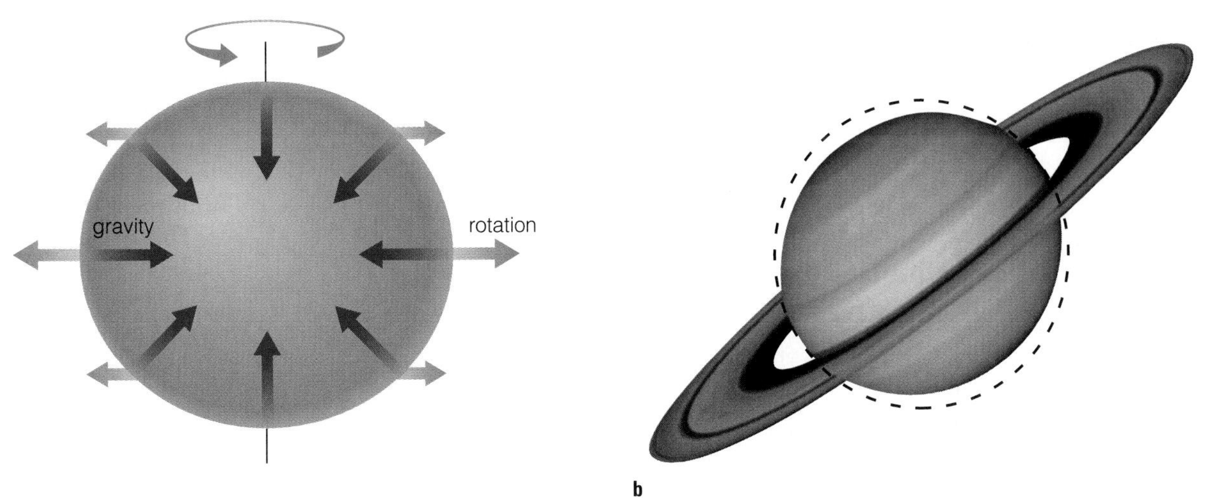

FIGURE 11.3 (**a**) Gravity alone makes a planet spherical, but rapid rotation flattens out the spherical shape by flinging material near the equator outward. (**b**) Saturn is clearly not spherical.

magnetospheres [Section 10.2]. This technique allows us to observe the rotation period of a magnetosphere, which should be the same as the rotation period deep in the interior, where the magnetic fields are generated. These measurements show that the jovian "days" range from only about 10 hours on Jupiter and Saturn to 16–17 hours on Uranus and Neptune.

TIME OUT TO THINK ***How would you measure the rotation period of Earth if you made telescopic observations from Mars? How might Earth's clouds mislead you if you focused on them?***

Even the shapes of the jovian planets are somewhat different from the shapes of the terrestrial planets. Gravity makes the jovian planets approximately spherical, but their rapid rotation rates make them slightly squashed (Figure 11.3). Material near the equator, where speeds around the rotation axis are highest, is flung outward in the same way that you feel yourself flung outward when you ride a merry-go-round. The result is that jovian planets bulge around the equator. The size of the equatorial bulge depends on the balance between the strength of gravity pulling the material inward and the rate of rotation pushing the material outward. The balance tips most strongly toward flattening on Saturn, with its rapid 10-hour rotation period and its relatively weak surface gravity. Saturn is about 10% wider at its equator than at its poles. In addition to altering a planet's appearance, the equatorial bulge itself exerts an extra gravitational pull that helps keep satellites and rings aligned with the equator.

We can trace almost all the major characteristics of the jovian planets back to their formation [Section 8.4]. The jovian planets formed far from the Sun where icy grains condensed in the solar nebula, yielding far more solid matter to accrete into planetesimals. As the massive ice/rock cores of the future jovian planets accreted, their strong gravity captured hydrogen and helium gas from the surrounding nebula. Some of this gas formed flattened "miniature solar nebulae" (or *jovian nebulae*) around the jovian planets. Just as solid grains condensed and combined into planets in the full solar nebula, solid grains in the nebulae surrounding the jovian planets condensed to form satellite systems. The deviations from the general patterns, such as the large 98° axis tilt of Uranus,[4] probably arose from giant impacts as the planets were forming. In the rest of this chapter, we will explore the features of the jovian planets and their satellites in greater depth, putting our theories of planetary development to the test.

11.2 Jovian Planet Interiors

As was the case with the terrestrial planets, developing a true understanding of the jovian planets requires looking deep inside them. Probing jovian planet interiors is even more difficult than probing terrestrial interiors, but several techniques allow us to learn what lies below the clouds. We've already discussed how Earth-based observations yielded the sizes and average densities of the jovian planets and how spacecraft measurements of magnetic and gravitational fields provide additional clues to their interior structure. Detailed observations of planetary shapes also provide information that we can use in computer models to learn about interior structure. For example, such calculations tell us that Saturn's core makes up a larger fraction of its mass than Jupiter's core.

[4]The 98° tilt is measured by assuming that Uranus was "supposed to" rotate counterclockwise as seen from above Earth's North Pole—just like most other planets. We could equivalently say that Uranus has an 82° tilt (180° − 98° = 82°) but rotates backward.

Spectroscopy from Earth and from spacecraft reveals the chemical compositions of the jovian upper atmospheres. Their deeper atmospheres are generally more difficult to probe, but two recent events provided rare opportunities. In July 1994, fragments of comet Shoemaker–Levy 9 slammed into Jupiter, blasting material from deeper in the atmosphere out into the open where astronomers could study it through telescopes [Section 12.6]. Then, in December 1995, NASA's Galileo spacecraft dropped a scientific probe into Jupiter's atmosphere. The Galileo probe survived to a depth of about 300 kilometers—a significant distance, but still much less than 1% of Jupiter's 70,000-kilometer radius—providing our first direct data from within a jovian world.

Beneath the depths at which we have compiled direct data, we learn about the jovian interiors by combining laboratory studies and theoretical models. These studies tell us how the ingredients of the jovian planets (especially hydrogen and helium) act under the tremendous temperatures and pressures expected deep below the cloud tops. Nowadays, elaborate computer models successfully match the observed sizes, densities, atmospheric compositions, and even shapes of the jovian planets. We therefore believe that we have a fairly clear understanding of their interiors. We'll begin our discussion of the jovian interiors by using Jupiter as a prototype; then we'll apply the principles of comparative planetology to understand the other jovian planet interiors.

Inside Jupiter

Imagine plunging head-on into Jupiter in a futuristic spacesuit that allows you to survive the incredible interior conditions. Near the cloud tops, you'll find the temperature to be a brisk 125 K (−148°C), the density to be a low 0.0002 g/cm^3, and the atmospheric pressure to be about 1 bar [Section 10.1]—the same as the pressure at sea level on Earth. The deeper you go, the higher the temperature, density, and pressure become (see Figure 11.2). By a depth of 7,000 km—about 10% of the planet's radius—you'll find that the temperature has increased to a scorching 2,000 K, the density has reached 0.5 g/cm^3 (half that of water), and the pressure is about 500,000 bars. Under these conditions, hydrogen acts more like a liquid than a gas. By the time you reach a depth of 14,000 km, the density is about the same as that of water and the temperature is near 5,000 K. The pressure of 2 million bars is so great that it forces hydrogen into an even more compact form: *metallic hydrogen*, which is so compact that all the molecules share electrons, just as happens in everyday metals.[5] The metallic and hence electrically conducting nature of this interior hydrogen is important in generating Jupiter's strong magnetic field. Continuing your descent, you'll reach Jupiter's core at a depth of 60,000 km, about 10,000 km from the center. The core is a mix of hydrogen compounds, rocks, and metals, but these materials bear little resemblance to familiar solids or liquids because of the extreme 20,000-K temperature and 100-million-bar pressure. The core materials probably remain mixed together, with a density of about 25 g/cm^3, rather than separating into layers of different composition as in the terrestrial planets. The total mass of Jupiter's core is about 10 times that of Earth, so the core alone would make an impressive planet.

Our study of the terrestrial planets taught us that internal heat sources can have a strong effect on the behavior of planets, so it is natural to ask if Jupiter and the other jovian planets are also affected by internal heat sources. Jupiter apparently has a tremendous amount of internal heat—so much that it emits almost twice as much energy as it receives from the Sun. (For comparison, recall that Earth's internal heat adds only 0.005% as much energy to the surface as does sunlight.) The magnitude of this internal heat source is much larger than can be explained by radioactive decay, so we must search for other explanations. The most probable explanation is that Jupiter is still slowly contracting, as if it has not quite finished forming from a jovian nebula. This contraction is so gradual that we cannot measure it directly, but it converts gravitational potential energy to thermal energy, keeping Jupiter hot inside.

[5]This phase change from nonmetal to metal does not occur in any everyday substances under normal conditions on Earth, and for a long time it was a theoretical prediction without experimental verification. Metallic hydrogen was finally produced under extreme pressure in laboratories in the early 1990s.

Comparing Jovian Planet Interiors

Now that we've discussed the interior of Jupiter, what can we say about the other jovian planets? Look back at the summary of the bulk properties of the jovian planets in Table 11.1. These properties may seem difficult to explain at first, because no clear correlation exists between size, density, and composition. But further examination of the behavior of gaseous materials removes the mystery.

Building a planet of hydrogen and helium is a bit like making one out of very fluffy pillows (Figure 11.4a). Imagine assembling a planet pillow by pillow. As each new pillow is added, those on the bottom are more compressed by those above. As the lower layers are forced closer together, their mutual gravitational attraction increases, compressing them even further. At first the stack grows substantially with each additional pillow, but eventually the growth slows until adding pillows hardly increases the height of the stack. This example explains why Jupiter is only slightly larger than Saturn in radius even though it is more than three times more massive. The extra mass of Jupiter compresses its interior to a much greater extent, making Jupiter's average density almost twice

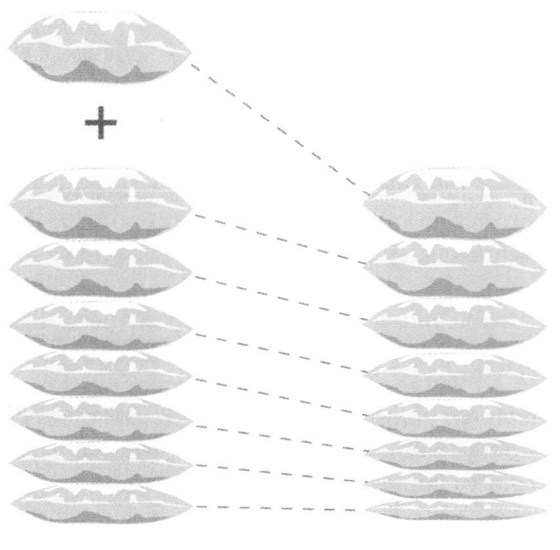

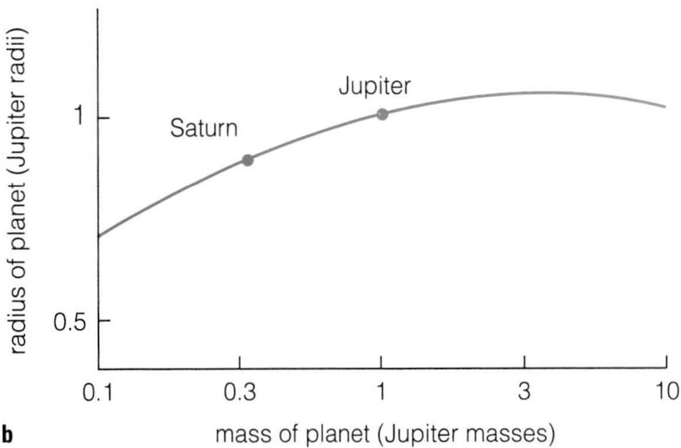

FIGURE 11.4 (**a**) Adding pillows to a stack may increase its height at first, but eventually it just compresses all the pillows in the stack. In a similar way, adding mass to a jovian planet would eventually just compress the planet to higher density without increasing the planet's radius. (**b**) This graph shows how the radius of a hydrogen/helium planet depends on the planet's mass. Jupiter's radius is only slightly larger than Saturn's, although it is three times more massive. For a planet much more massive than Jupiter, gravitational compression would actually make it smaller in size.

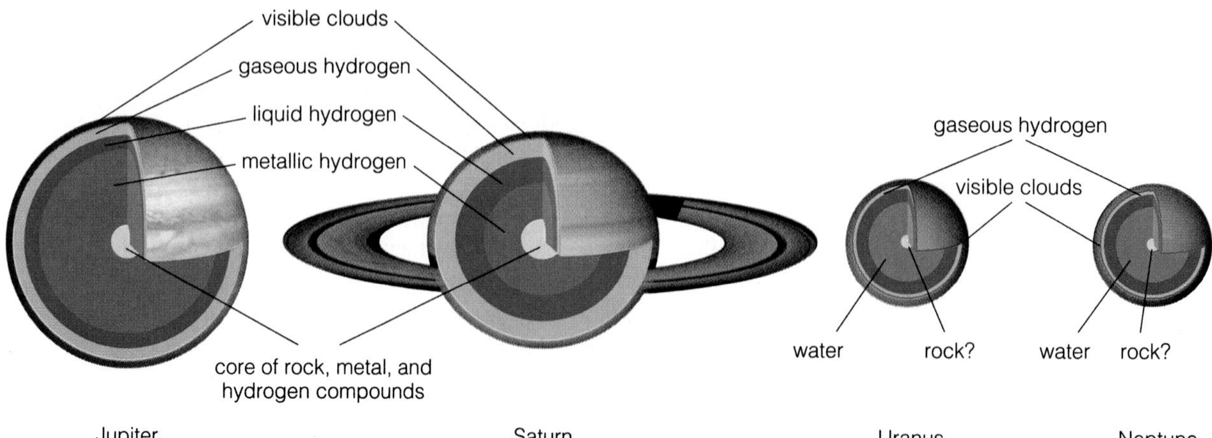

FIGURE 11.5 These diagrams compare the interior structures of the jovian planets (shown approximately to scale). All four planets have cores equal to about 10 Earth masses of rock, metal, and hydrogen compounds, and they differ primarily in the hydrogen/helium layers that surround the cores.

that of Saturn. In fact, Saturn's average density of 0.7 g/cm^3 is less than that of water. More precise calculations show that Jupiter's radius is almost the maximum possible for a jovian planet: If more hydrogen and helium were added to Jupiter, they would actually compress the interior enough to make the planet *smaller* instead of larger (Figure 11.4b). In fact, the smallest stars are significantly smaller in radius than Jupiter, even though they are 80 times more massive.

Using the same logic that explains why Saturn's density is lower than Jupiter's, we would predict that a hydrogen/helium planet smaller than Saturn should have an even lower density. The higher densities of Uranus and Neptune therefore tell us that they cannot have the same composition as Jupiter and Saturn. Instead, Uranus and Neptune must have a much larger fraction of higher-density material, such as hydrogen compounds and rocks, and a smaller fraction of "plain" hydrogen and helium.

Computer models of the jovian interiors based on these compositions yield the somewhat surprising result that the cores of all four jovian planets are quite similar—about 10 Earth masses of rock, metal, and hydrogen compounds. This suggests that all four jovian planets began with about the same size "seed" from accretion and that their differences stem from their capturing different amounts of additional gas from the solar nebula. Jupiter captured more than 300 Earth masses of gas from the solar nebula, Saturn captured about one-fourth as much gas as Jupiter, and Uranus and Neptune captured only a few Earth masses of solar-nebula gas. This general pattern makes sense, because icy planetesimals took longer to accrete in the outer solar system, where planetesimals were more spread out. Thus, the more distant jovian planets didn't have as much time as Jupiter to capture solar-nebula gas before the nebula was cleared by the solar wind [Section 8.4].

The similar cores also mean that the interior structures of the jovian planets differ mainly in the hydrogen/helium layers that surround the cores (Figure 11.5). Saturn is large enough for interior densities, temperatures, and pressures to eventually become high enough to force gaseous hydrogen into liquid and then metallic form, just as in Jupiter. But these conditions occur relatively deeper in Saturn than in Jupiter because of its smaller size. Uranus and Neptune have even smaller layers of hydrogen and helium, so pressures are not high enough to form liquid or metallic hydrogen. However, their cores of rock, metal, and hydrogen compounds may be liquid, making for very odd "oceans" buried deep inside them.

Like Jupiter, Saturn emits nearly twice as much energy as it receives from the Sun. Saturn's mass is too small for it to be generating all its excess heat by contracting like Jupiter. Instead, its lower temperatures probably allow helium to condense and "rain out" from higher regions in the interior. The gradual rain of dense helium droplets resembles the process of *differentiation* that once occurred in terrestrial planet interiors [Section 9.2]. Voyager observations confirmed that Saturn's atmosphere is somewhat depleted of helium (compared to Jupiter's atmosphere), just as we would expect if helium has been raining down into Saturn's interior for billions of years.

Neither Uranus nor Neptune has conditions that allow helium rain to form, and most of their original heat from accretion should have escaped long ago. This explains why Uranus emits virtually no excess internal energy. But the case of Neptune is very surprising: Like Jupiter and Saturn, it emits nearly twice as much energy as it receives from the Sun. The only reasonable explanation for Neptune's internal heat source is that the planet is somehow still contracting, rather like Jupiter, thereby converting potential energy into thermal energy. But no one knows why a planet of Neptune's size would still be contracting more than 4 billion years after its formation.

11.3 Jovian Planet Atmospheres

The jovian planets lack the geology of the terrestrial planets because they do not have solid surfaces. But whatever they lack in geology they more than make up for with their atmospheres. Fortunately, their atmospheric processes are quite similar to those we've already discussed for the terrestrial atmospheres, despite their much greater extent and very different compositions. Let's again begin by studying Jupiter as a prototype.

Jupiter's Atmosphere

Jupiter's atmosphere is mostly hydrogen and helium (about 75% hydrogen and 24% helium by mass), but it also contains trace amounts of an interesting and important assortment of other hydrogen compounds. Oxygen, carbon, and nitrogen are the most common elements besides hydrogen and helium, so the most common hydrogen compounds are methane (CH_4), ammonia (NH_3), and water (H_2O). Spectroscopy also reveals the presence of more complex compounds, including acetylene (C_2H_2), ethane (C_2H_6), propane (C_3H_8), and larger molecules that can act like haze particles. Although all these hydrogen compounds make up only a minuscule fraction of Jupiter's atmosphere, they are responsible for virtually all aspects of its appearance. Some of these compounds condense to form the clouds so prominent in telescope and spacecraft images, and others are responsible for Jupiter's great variety of colors. Without these compounds, Jupiter would be a uniform, colorless ball of gas.

TIME OUT TO THINK *Jupiter possesses substantial amounts of gases like methane, propane, and acetylene, highly flammable fuels used here on Earth. There is plenty of lightning on Jupiter to provide sparks, so why aren't we concerned about Jupiter exploding?* (Hint: *What's missing from Jupiter's atmosphere that's necessary for ordinary fire?*)

Jupiter is the only jovian planet that we've directly sampled. On December 7, 1995, following a 6-year trip from Earth, the Galileo spacecraft released a scientific probe (the size of a large suitcase) designed to study the atmosphere of Jupiter (Figure 11.6a). The probe collected temperature, pressure, composition, and radiation measurements for about an hour as it descended into Jupiter's atmosphere, until it was finally destroyed by the ever-increasing pressures and temperatures. The probe sent its data via radio signals back to the Galileo spacecraft in orbit around Jupiter, which relayed the data back to Earth.

The Galileo probe helped confirm that the thermal structure of Jupiter's atmosphere is very similar to that of terrestrial atmospheres (Figure 11.6b). Let's follow its fiery descent. The probe plunged into the atmosphere at a speed of over 200,000 km/hr, with a robust heat shield facing forward to protect the scientific instruments. Above the cloud tops, the probe found very low density gas that is heated to perhaps 1,000 K by solar X rays and by energetic particles from Jupiter's magnetosphere. This thin, hot upper region is Jupiter's *thermosphere.* Next, the probe encountered Jupiter's *stratosphere,* where solar ultraviolet photons are absorbed by a few minor ingredients in the atmosphere. This absorption gives the stratosphere a peak temperature of about 170 K. Chemical reactions driven by the solar ultraviolet photons also create a smoglike haze that masks the color and sharpness of the clouds below. Below the stratosphere lies Jupiter's *troposphere,* where the increased density greatly slowed the descent of the probe. At this point, the probe jettisoned its heat shield and released a parachute to further slow its descent. The temperature at the top of the troposphere is close to 125 K; it rises with depth because greenhouse gases trap both solar heat and Jupiter's own internal heat. The probe also found tremendous winds and turbulence in the troposphere, along with evidence of clouds. It may seem surprising that the atmospheric structure of Earth and Jupiter are so similar given that the two planets are so dissimilar in other ways. The basic similarities in how gases interact with light and gravity are evidently more important than differences in size, temperature, and composition.

As on the terrestrial planets, Jupiter's weather occurs mostly in its troposphere, where higher temperatures at lower altitudes drive vigorous convection. This convection is responsible for the thick clouds that enshroud Jupiter. As a parcel of gas rises upward through the troposphere, it encounters gradually lower temperatures. Soon the surrounding temperature is low enough for water vapor to condense into liquid droplets or flakes of ice, forming water

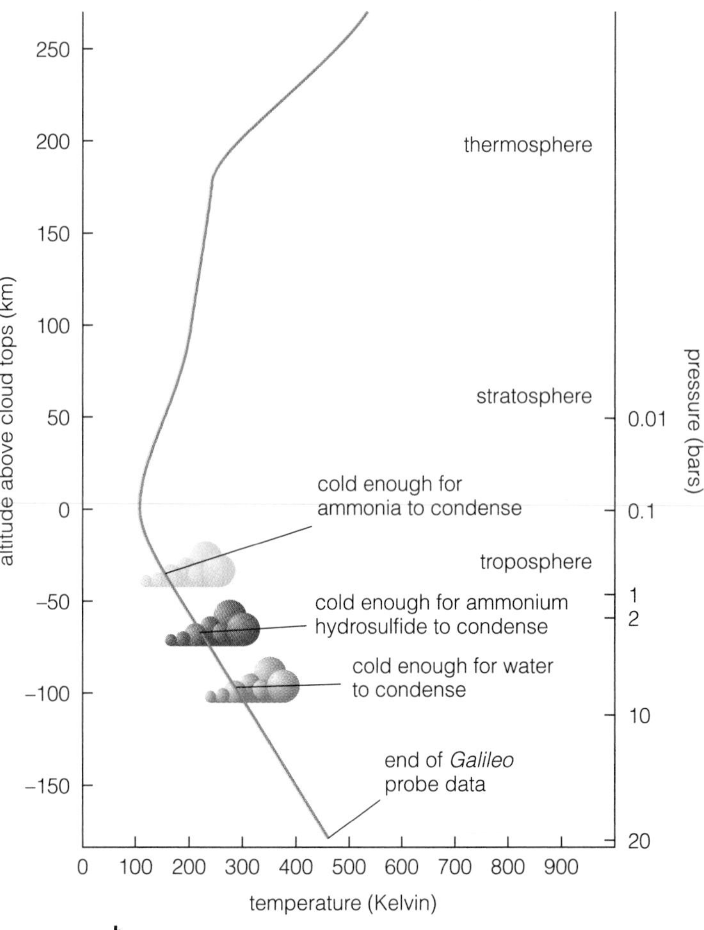

FIGURE 11.6 (**a**) Artist's conception of the Galileo probe entering Jupiter's atmosphere. This view, looking outward from beneath the clouds, shows the suitcase-sized probe falling with its parachute extended above it. (**b**) This graph shows the temperature structure of Jupiter's atmosphere (compare to Figure 10.6). Jupiter has at least three distinct cloud layers because different atmospheric gases condense at different temperatures and hence at different altitudes.

clouds similar to those on Earth. As the gas continues to rise, temperatures become low enough for another minor atmospheric ingredient, ammonium hydrosulfide (NH_4SH), to condense and form clouds. Still higher, the temperatures fall to the point at which ammonia (NH_3) condenses to form clouds of ammonia crystals—the prominent white clouds visible on Jupiter. Thus, Jupiter has several layers of clouds of different compositions (see Figure 11.6b). The relentless motion of Jupiter's atmosphere continuously regenerates the extensive clouds.

Jupiter also has planet-wide *circulation cells* similar to those on Earth, where solar heat causes equatorial air to expand and spill northward and southward toward the poles [Section 10.3]. However, whereas Earth's rotation splits its circulation cells into just three separate cells in each hemisphere, Jupiter's much more rapid rotation creates a strong Coriolis effect that causes its circulation cells to split into many huge bands encircling the entire planet at fixed latitudes (Figure 11.7a,b). The bands of rising air are called **zones,** and they appear white in color because ammonia clouds form as the air rises to high, cool altitudes. Ammonia "snowflakes" rain downward against the rising convective motions in the zones. The adjacent **belts** of falling air are depleted in the cloud-forming ingredients and do not contain any white ammonia clouds. Instead, they appear dark in color because we can see down to the red or tan ammonium-hydrosulfide clouds that form at lower altitudes. The distinction between Jupiter's belts and zones is analogous to the distinction on Earth between the cloudy, rainy equatorial zone (a region of generally rising air) and the clear desert skies found roughly 20°–30° north and south of the equator (regions of descending air). Infrared images confirm the temperature differences between Jupiter's warm belts and cool zones (Figure 11.7b,c).

Jupiter's global wind patterns are shaped by these alternating bands of rising air in zones and falling air in belts. The rising air in the zones indicates higher pressures than in the adjacent belts, so winds must flow from the high-pressure zones toward the low-pressure belts. Figure 11.7a shows the northward wind flowing from a zone to a belt in the northern hemisphere. The northward winds are quickly diverted into fast eastward-flowing winds by the very strong Coriolis effect. Similarly, the southward-flowing winds from zones to belts in the northern

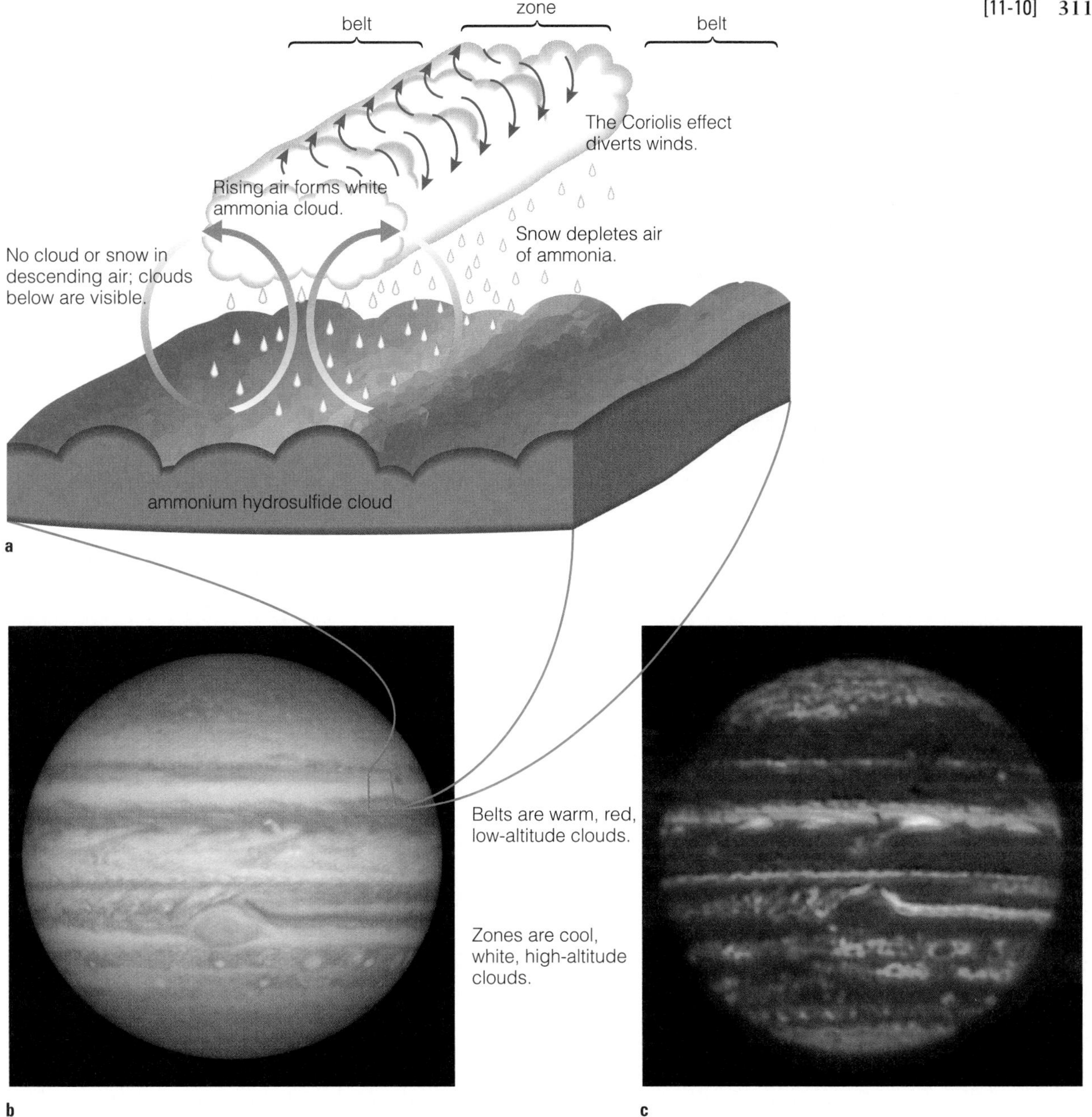

FIGURE 11.7 This figure explains the origin of Jupiter's banded appearance. (**a**) Belts and zones correspond to clouds of different composition at different altitudes. The Coriolis effect diverts motions in the circulation cells into strong easterly and westerly winds. (**b**) The color difference between belts and zones is evident in this Hubble Space Telescope image. (**c**) In this infrared image taken nearly simultaneously with (b), brightness indicates high temperatures.

hemisphere are diverted to the west. (Note that both these diversions of winds are to the right in the Northern Hemisphere; in the Southern Hemisphere, winds are diverted to the left [Section 10.3].) The winds are generally strongest at the equator and at the boundaries between belts and zones. Peak wind speeds exceed 400 km/hr.

Jupiter's **Great Red Spot** is perhaps the most dramatic weather pattern in the solar system (Figure 11.8). We know that it has been prominent in Jupiter's southern hemisphere for at least three centuries, because it has been seen ever since telescopes became powerful enough to detect it. The Great Red Spot is huge—more than twice as wide as the Earth. Could it be a tremendous low-pressure storm like a hurricane on Earth? Not quite. Hurricane winds circulate around low-pressure regions and therefore

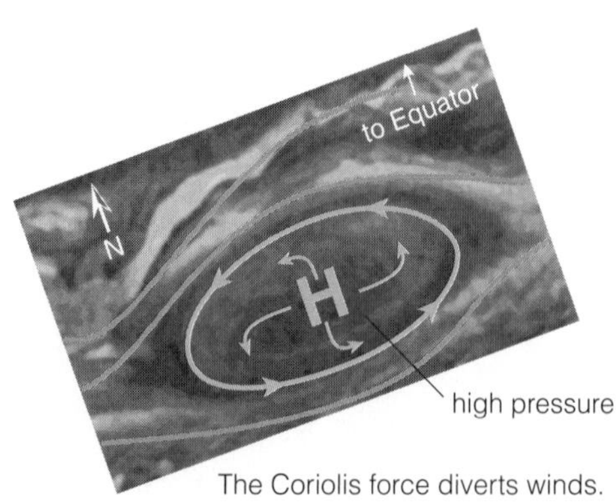

FIGURE 11.8 This photograph shows Jupiter's Great Red Spot, a huge high-pressure storm that is large enough to swallow two or three Earths. The smaller photo (right) of the Great Red Spot is overlaid with a weather map of the region.

circulate *clockwise* in the southern hemisphere [Section 10.3]. The winds in the Great Red Spot circulate counterclockwise, indicating that it is a *high*-pressure storm, possibly kept spinning by the influence of nearby belts and zones.

Other, smaller storms are always brewing in Jupiter's atmosphere—small only in comparison to the Great Red Spot. Brown ovals are low-pressure storms with their cloud tops deeper in Jupiter's atmosphere, and white ovals are high-pressure storms topped with ammonia clouds. No one knows what drives Jupiter's storms, why Jupiter has only one Great Red Spot, or why the Great Red Spot lasts so much longer than storms on Earth. Storms on Earth lose their strength when they pass over land, so perhaps Jupiter's biggest storms last for centuries because there is no solid surface below to sap their energy.

Jupiter's vibrant colors remain perplexing despite decades of intense study. Observations tell us that the high clouds of ammonia in the zones are usually white and that the lower clouds of ammonium hydrosulfide in the belts are brown or red. But pure ammonium-hydrosulfide crystals are as white as ammonia crystals, so the colors of the belt clouds must come from ingredients besides the ammonium-hydrosulfide crystals themselves. Sulfur compounds are a plausible suspect, because they form a variety of reds and tans (also seen on the surface of Jupiter's moon Io). Phosphorus compounds (including phosphine, PH_3) are another possibility. Whatever their origin, the red and tan compounds are found primarily at lower (and therefore warmer) altitudes. The chemical reactions that form them evidently require the extra energy of Jupiter's internal heat, or possibly lightning deep below the clouds. The bright red color of the very high altitude clouds in the Great Red Spot remains an even greater mystery. Perhaps chemical reactions caused by solar ultraviolet radiation produce the compounds responsible for these colors.

As far as we know, Jupiter's climate is steady and unchanging. Jupiter has no appreciable axis tilt and therefore has no seasons. In fact, Jupiter's polar temperatures are quite similar to its equatorial temperatures. Solar heating alone might leave the poles relatively cool, but Jupiter's internal heat source keeps the planet uniformly warm.

Comparing Jovian Planet Atmospheres

The most striking difference among the jovian atmospheres is their colors. Starting with Jupiter's distinct red colors, the jovian planets turn to Saturn's more subdued yellows, to Uranus's faint blue-green tinge, and finally to Neptune's pronounced blue hue (see Figure 11.1). What's responsible for this regular progression of colors? The primary constituents in all the jovian atmospheres (hydrogen and helium) are colorless, so the planetary colors must come from trace gases or chemical reactions that create colored compounds. In fact, we can understand the color differences by comparing the atmospheric structures and compositions of the four planets. All four are quite similar, except for the effects of the lower temperatures and lower gravities on the more distant planets (Figure 11.9).

Saturn's reds and tans almost certainly come from the same compounds that produce these colors on Jupiter—whatever those compounds may be. As on Jupiter, these compounds are probably created by chemical reactions beneath the cloud layers and carried upward by convection. However, because of its overall lower temperatures, Saturn's cloud layers lie deeper in its atmosphere than Jupiter's. As a result, a thicker layer of tan "smog" overlies the clouds and washes out Saturn's appearance. Saturn's lower gravity means that the atmosphere is less compressed, so cloud layers are more spread out vertically. The cloud layers are thicker, preventing us from seeing down to the lower, more richly hued cloud levels that we see on Jupiter.

The blue colors of Uranus and Neptune have a completely different explanation: methane gas, which is at least 20 times more abundant (by percentage) on these planets than on Jupiter or Saturn. The highest-altitude clouds of Uranus and Neptune are made from flakes of methane ice; Jupiter and Saturn lack such clouds because their warmer temperatures prevent methane ice from condensing. Methane gas above these clouds absorbs red light but transmits blue light (Figure 11.10). The clouds then reflect (scatter) the blue light upward, where it is again transmitted through the gas above. Thus, we see this reflected blue light when we look at the atmospheres of Uranus and Neptune. Uranus and Neptune may also have red and tan cloud layers like those of Jupiter and Saturn, but if so they are hidden deep below the methane gas and clouds.

The fainter blue of Uranus compared to Neptune indicates that less sunlight penetrates the atmosphere to the level of the clouds. Instead, the light is probably scattered by abundant smoglike haze. The extra haze on Uranus may arise because its extreme axis tilt keeps one hemisphere in bright sunlight for decades at a time during its 84-year orbit (see Figure 2.11), allowing more ultraviolet-driven chemical reactions. This continuous sunlight in one hemisphere may also explain why Uranus has a surprisingly hot thermosphere that extends thousands of kilometers above its cloud tops.

The Voyager spacecraft cameras monitored the meteorology of the four jovian planets with excellent resolution (Figure 11.11). Jupiter's weather phenomena (clouds, belts, zones, and other storms) are by far the strongest and most active. Saturn also possesses belts and zones (in more subdued colors), along with some small storms and an occasional large storm, but it lacks any weather feature as prominent as Jupiter's Great Red Spot. (For unknown reasons, Saturn's winds are even stronger than Jupiter's.) Neptune's atmosphere is also banded and has a high-pressure storm, called the Great Dark Spot, that acts much like Jupiter's Great Red Spot. The greatest surprise in the weather patterns on the jovian planets is the scarcity of clouds, belts and zones, and long-lived storms on Uranus. The weaker weather on Uranus

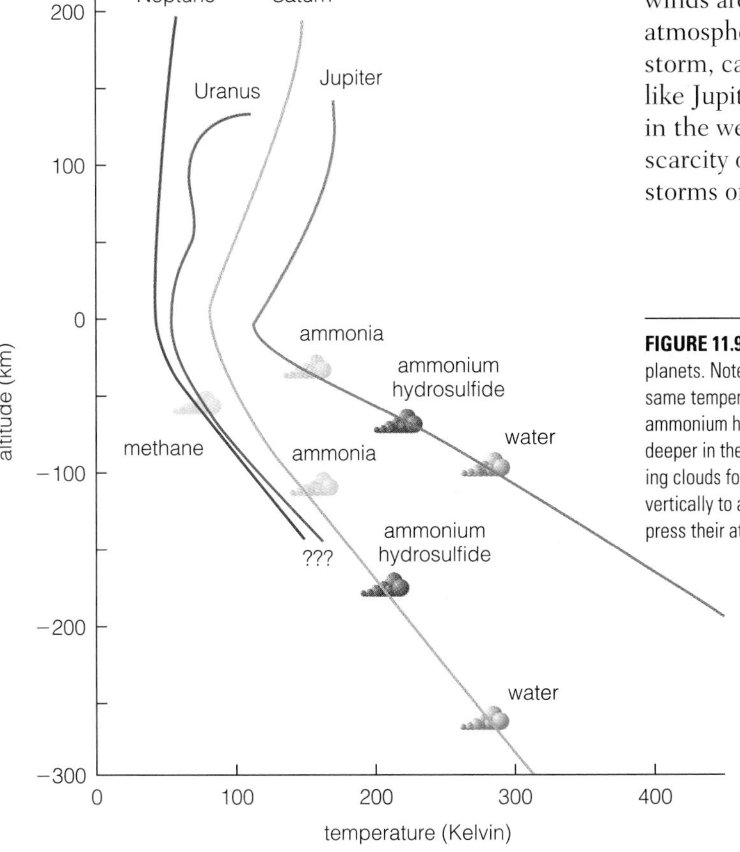

FIGURE 11.9 Temperature variation with altitude for each of the jovian planets. Note that clouds of a particular composition always form at about the same temperature: less than 100 K for methane, 150 K for ammonia, 200 K for ammonium hydrosulfide, and 270 K for water. These temperatures occur deeper in the atmospheres of planets farther from the Sun, so the corresponding clouds form deeper as well. Note also that the cloud layers are separated vertically to a greater extent on smaller planets with weaker gravity to compress their atmospheres. (The "???" indicates uncertainty beneath this level.)

FIGURE 11.10 Neptune and Uranus are blue because methane gas absorbs red light but transmits blue light. Clouds of methane ice flakes reflect the transmitted blue light back to space.

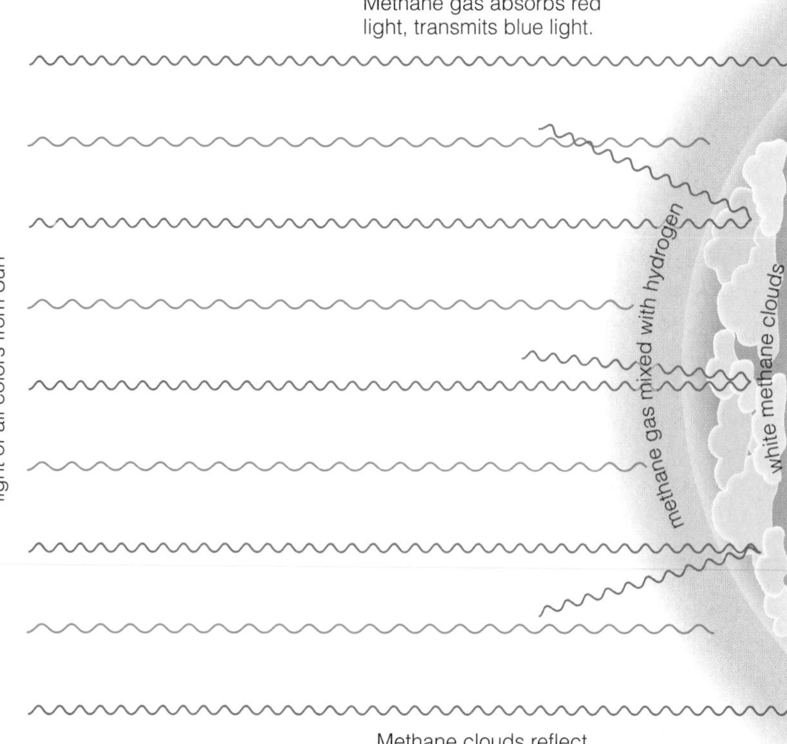

FIGURE 11.11 Close-ups of cloud patterns on the four jovian planets. Scientists combine series of images like these to make weather movies similar to those for Earth shown on the evening news. The Great Dark Spot is visible in the Neptune image.

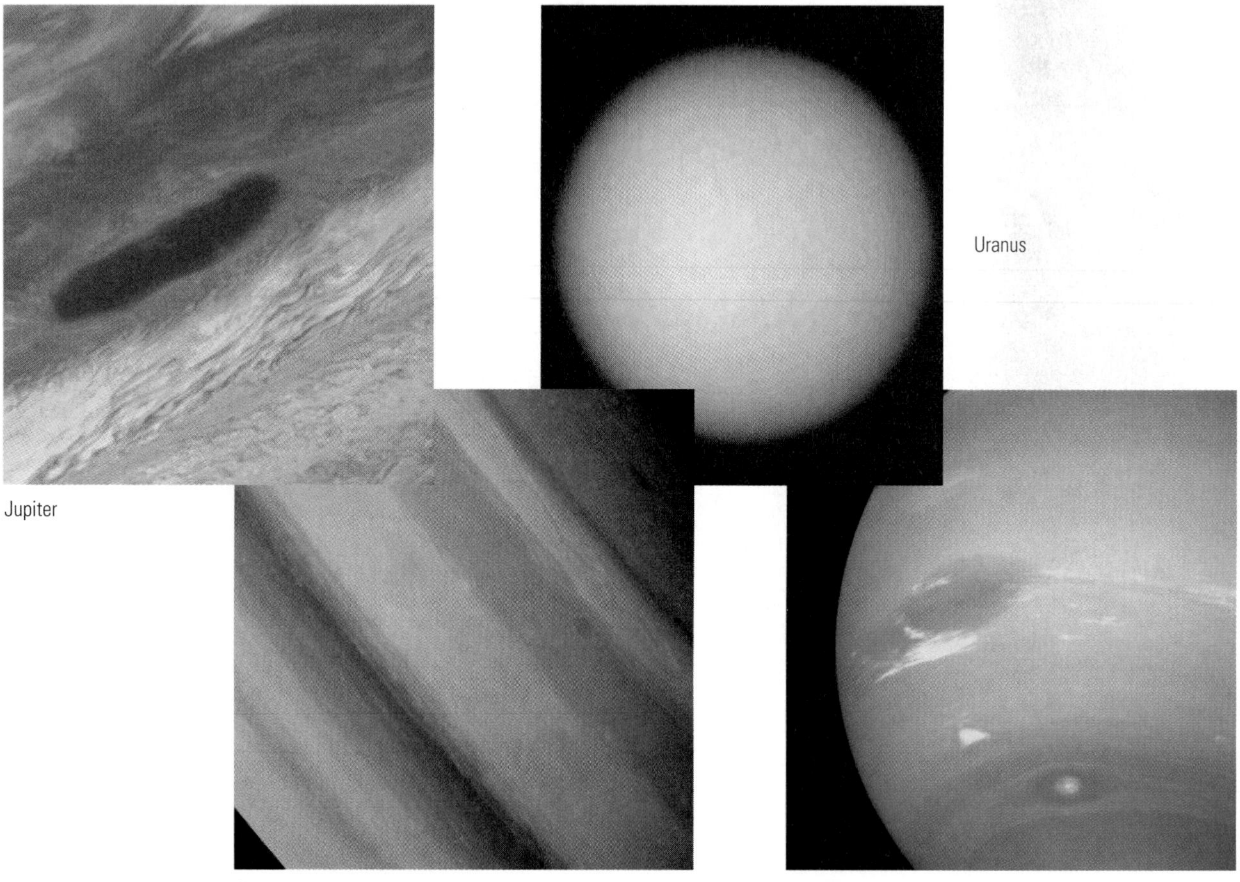

Jupiter

Uranus

Saturn

Neptune

probably comes from the fact that it is the only jovian planet without a lot of excess internal heat and therefore has weaker convection.

Seasonal changes play a relatively small role on the jovian planets. We might expect seasons on Saturn and Neptune because they have axis tilts similar to that of Earth, and some seasonal weather changes have been observed. But their internal heat sources maintain similar temperatures all over, just as Jupiter's internal heat keeps its polar and equatorial temperatures about the same. Uranus is the oddball. Its large axis tilt should produce extreme seasons, especially since Uranus appears to lack a strong internal heat source. Nevertheless, the winter pole of Uranus seems to have about the same temperature as the rest of the planet. The uniform temperature of Uranus remains a major mystery, and we still do not know whether it is affected by seasonal variations: Its "year" is so long that we've had little time to see what seasonal changes might occur.

11.4 Jovian Planet Magnetospheres

Each jovian planet is surrounded by a bubble-like magnetosphere consisting of the planet's magnetic field and the particles trapped within it. The jovian planets themselves dwarf the terrestrial planets, and their magnetospheres are even more impressive. The Voyager spacecraft were equipped with sophisticated instruments to measure planetary magnetic fields and observe the charged particle belts encircling the planets. We learned not only about the magnetospheres themselves, but also about their interactions with the planets beneath them and the satellite systems within them.

Jupiter's Magnetosphere

Jupiter's magnetic field is awesome—about 20,000 times stronger than Earth's. This strong magnetic field deflects the solar wind some 3 million km (about 40 Jupiter radii) before it even reaches Jupiter (Figure 11.12a). If our eyes could see this part of Jupiter's magnetosphere, it would be larger than the full

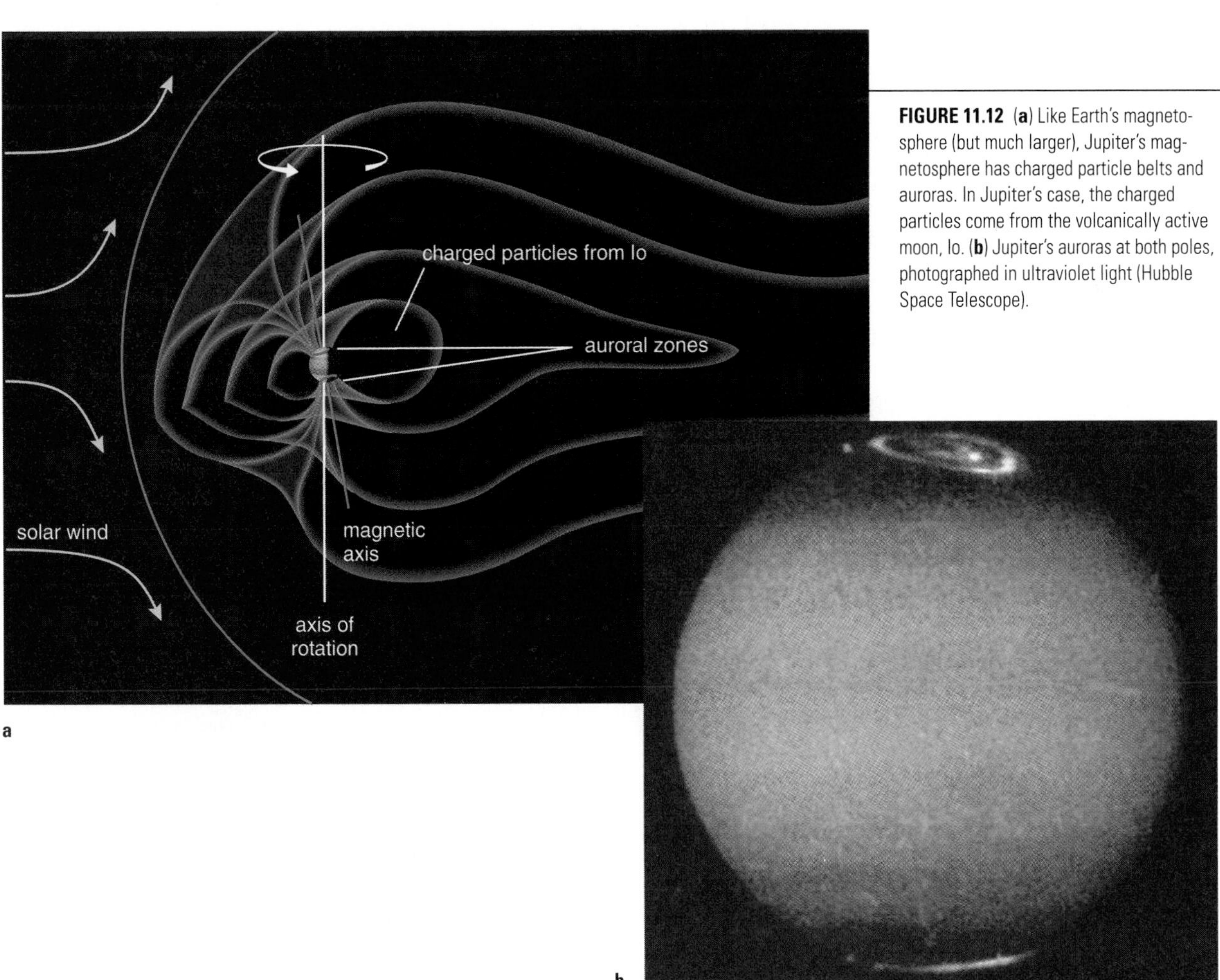

FIGURE 11.12 (**a**) Like Earth's magnetosphere (but much larger), Jupiter's magnetosphere has charged particle belts and auroras. In Jupiter's case, the charged particles come from the volcanically active moon, Io. (**b**) Jupiter's auroras at both poles, photographed in ultraviolet light (Hubble Space Telescope).

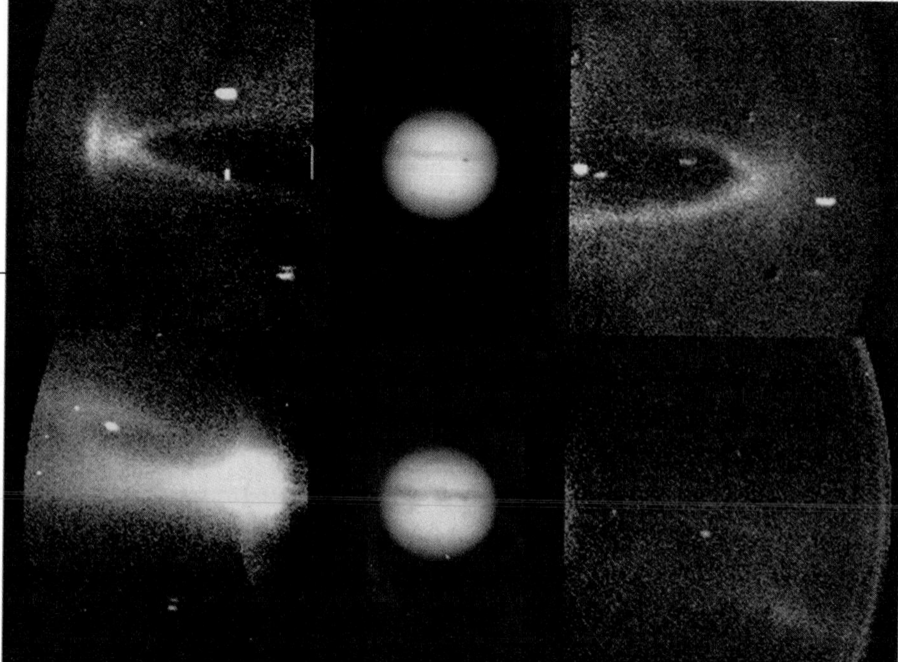

FIGURE 11.13 Top: These telescopic images (central Jupiter plus images to its left and right) show the Io torus as a thin white "donut" encircling Jupiter. Bottom: The large yellow blob is an overexposed image of Io; the yellow trail extending to the left is formed by escaping atoms, which are ionized and swept into the torus by Jupiter's strong magnetic field.

moon in our sky. The solar wind sweeps around Jupiter's magnetosphere, stretching it out on the far side of Jupiter all the way to Saturn's orbit. Just as on Earth, the magnetosphere traps charged particles and makes them spiral along the magnetic field lines. The more energetic particles cause *auroras* as they follow the magnetic field into Jupiter's upper atmosphere, colliding with atoms and molecules and causing them to radiate (Figure 11.12b). Jupiter's magnetosphere contains a great deal of plasma, some captured from the solar wind but most originating from Jupiter's volcanically active moon, Io.

Not only does Io help feed plasma into the magnetosphere, but the magnetosphere in turn has important effects on Io and other satellites of Jupiter. The charged particles bombard the surfaces of Jupiter's icy moons, with each particle blasting away a few atoms or molecules. This process alters the surface materials and can even generate thin bombardment atmospheres—just as bombardment creates the thin atmospheres of Mercury and the Moon [Section 10.4]. On Io, the bombardment leads to the escape of the atmospheric gases continuously released by volcanic outgassing. As a result, Io loses atmospheric gases faster than any other object in the solar system. The escaping gases (sulfur, oxygen, and a hint of sodium) are ionized and feed a donut-shaped charged particle belt, called the *Io torus,* that approximately traces Io's orbit (Figure 11.13).

Comparing Jovian Planet Magnetospheres

The strength of each jovian planet's magnetic field depends primarily on the size of the electrically conducting layer buried in its interior. Jupiter, with its huge metallic-hydrogen layer, has by far the strongest magnetic field. Saturn has a thinner layer of metallic hydrogen and hence a correspondingly weaker magnetic field. Uranus and Neptune have no metallic hydrogen at all. Their relatively weak magnetic fields must be generated in their core "oceans" of hydrogen compounds, rock, and metal. The actual size of a planet's magnetosphere depends on the pressure of the solar wind pushing on a planet's magnetic field, as well as on the magnetic-field strength itself. The pressure of the solar wind is weaker at greater distances from the Sun. Thus, the magnetospheric "bubbles" surrounding the outer planets, particularly Uranus and Neptune, are larger than they would be if these planets were closer to the Sun. This effect is not sufficient to offset their weak magnetic fields, however, so Uranus and Neptune have small magnetospheres (Figure 11.14).

While the trend of decreasing magnetic-field strengths makes sense, the magnetic fields still pose some unsolved problems. The magnetic fields of Jupiter and Saturn are fairly closely aligned with their rotation axes (10° tilt and 0° tilt, respectively), just as we might expect given that the magnetic fields are generated in their rotating interiors. But Voyager observations showed that the magnetic field of Uranus is tipped by a whopping 60° relative to its rotation axis (see Figure 11.14); its center is also significantly offset from the planet's center. Scientists briefly speculated that whatever giant impact caused the large tilt of Uranus's rotation axis also distorted the magnetic field. But this idea was discarded when Voyager discovered a similarly large magnetic field tilt (46°) for Neptune. Scientists still cannot explain the magnetic-field tilts of Uranus and Neptune.

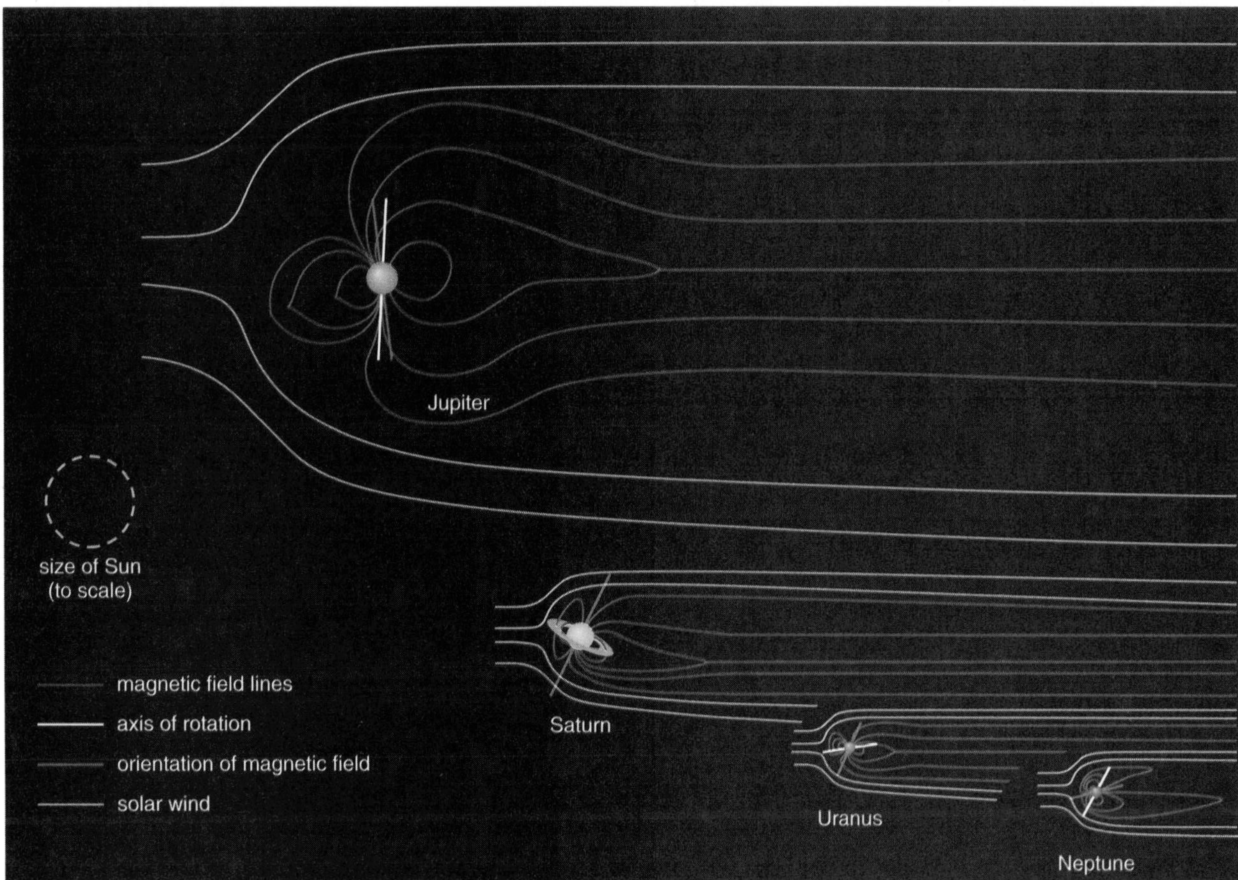

FIGURE 11.14 Comparison of jovian planet magnetospheres. The size of Jupiter's magnetosphere is particularly impressive in light of the greater pressure from the solar wind nearer the Sun. (Planets enlarged for clarity.)

No magnetosphere is as full of charged particles as Jupiter's, primarily because no other jovian planet has a satellite like Io. All magnetospheres also trap particles from the solar wind, but Jupiter again captures the most. Because trapped particles generate auroras, Jupiter has the brightest auroras, and the more distant jovian planets have progressively weaker auroras.

11.5 A Wealth of Worlds: Satellites of Ice and Rock

More than 50 known moons *(satellites)* orbit the jovian planets. Figure 11.15 shows the larger ones, and Appendix C lists fundamental data for all the known moons. It's worth taking a few minutes to study the figure and appendix, looking for patterns, before you read on.

Broadly speaking, the jovian moons can be divided into three groups: small moons less than about 300 km across, medium-size moons

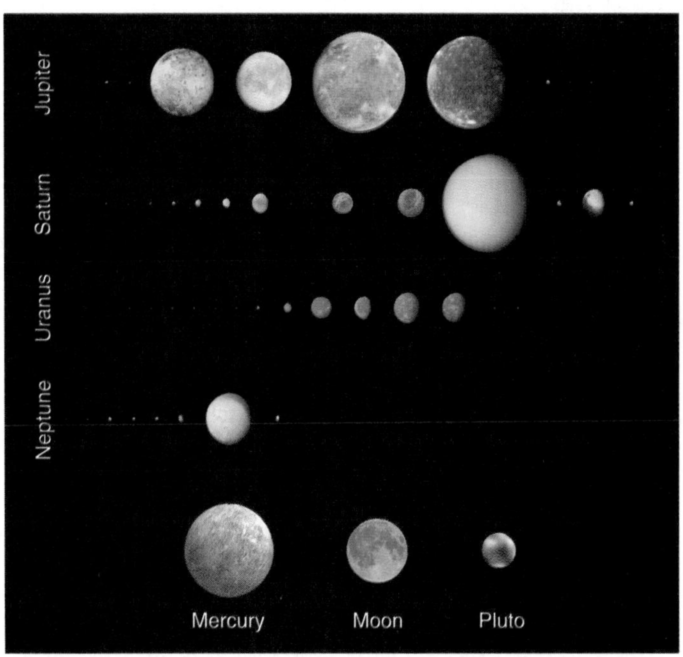

FIGURE 11.15 The larger satellites of the jovian planets, with sizes (but not distances) shown to scale. Mercury, the Moon, and Pluto are included for comparison.

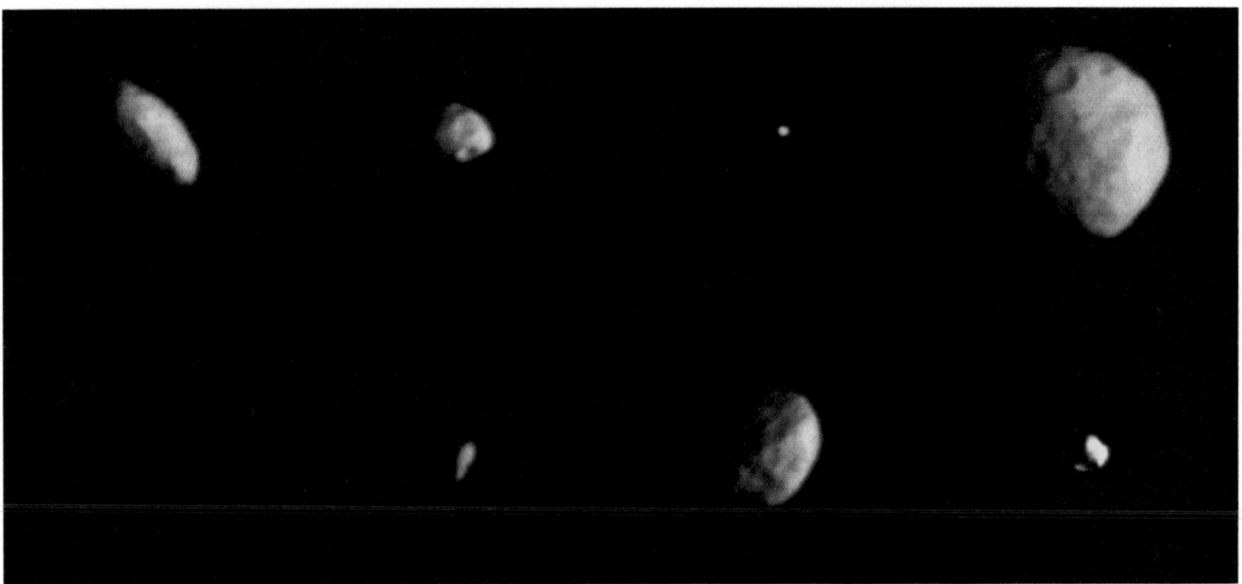

FIGURE 11.16 A montage of the small moons of Saturn, shown to scale. The small moons of the other jovian planets also are probably not spherical. These photographs were taken by the Voyager spacecraft.

ranging from about 300 to 1,500 km in diameter, and large moons more than 1,500 km in diameter. More than half of the known moons fall into the small category, and additional small moons may yet be discovered. The small moons generally look more like potatoes than spheres, because their gravities are too weak to force their rigid material into spheres (Figure 11.16) [Section 8.4]. Many of these moons (particularly those farther from their planets) also have unusual orbits—such as orbits that go backward relative to the planet's rotation or orbits that have large orbital tilts, significant eccentricities, or combinations of all these unusual features. Such orbits are signs that the moons are captured asteroids rather than moons that formed from the "miniature solar nebulae" that surrounded each jovian planet. In a few cases, several small moons share similar orbits. For example, four moons orbit Jupiter backward at distances between 21 and 24 million kilometers and probably once were part of a single larger captured asteroid that fragmented into several pieces.

Nearly all the medium-size and large moons follow the orbital patterns we expect from formation in "miniature solar nebulae": They have approximately circular orbits that lie close to the equatorial plane of their parent planet, and they orbit in the same direction that their planet rotates. These moons are planetlike in almost all ways. They are approximately spherical, each has a solid surface with its own unique geology, and some possess atmospheres, hot interiors, and even magnetic fields. A few are planetlike in size as well. The two largest moons—Jupiter's moon Ganymede and Saturn's moon Titan—are larger than the planet Mercury. Four others are larger than Pluto: Jupiter's moons Io, Europa, and Callisto, and Neptune's moon Triton.

Nearly all moons share an uncanny trait: They always keep the same face turned toward their planet, just as our Moon always shows the same face to Earth [Section 3.3]. This trait is called *synchronous rotation* because it means that a moon's rotation period and orbital period are synchronized, or the same (see Figure 3.13a).

Synchronous rotation arises from the effects of *tidal forces,* the same forces that make the Earth's oceans bulge outward in directions both toward and away from our Moon [Section 6.5]. The Earth also exerts a tidal force on the Moon that makes the Moon bulge along the line directed toward and away from Earth. In fact, the Earth exerts a greater tidal force on the Moon than vice versa because of its larger mass. The massive jovian planets exert even greater tidal forces on their nearby moons. In some cases, these tidal bulges make the moons several kilometers larger in one direction than in others.

We can understand how tidal bulging leads to synchronous rotation by imagining what would happen to a moon that did *not* rotate synchronously. A moon's tidal bulge is always aligned toward and away from a planet, so a moon with rotation faster or slower than synchronous rotation would be constantly reshaped as different parts of the moon rotated into

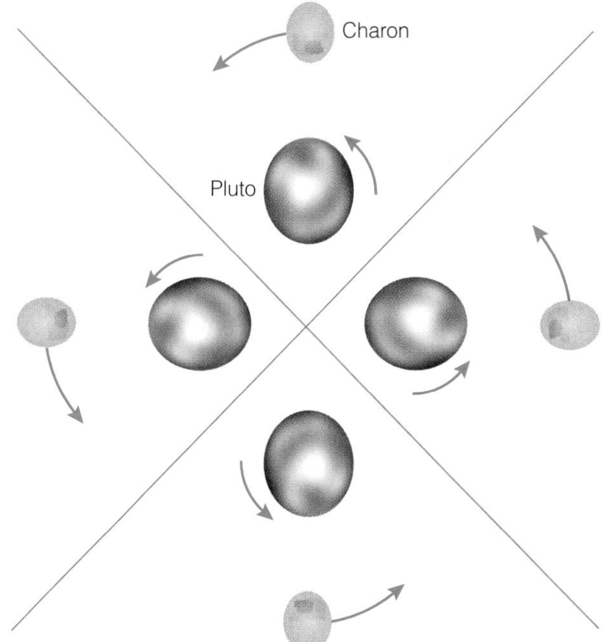

a Pluto and Charon rotate synchronously with each other, so that each always shows the same face to the other. If you stood on Pluto, Charon would remain stationary in your sky, always showing the same face (but going through phases like the phases of our Moon). Similarly, if you stood on Charon, Pluto would remain stationary in your sky, always showing the same face (and going through phases).

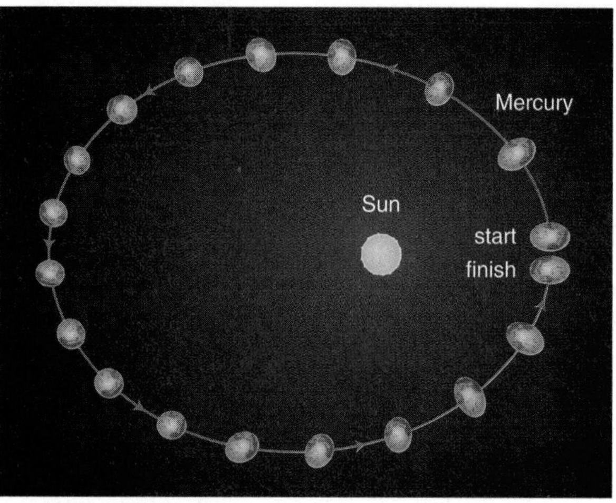

b These "snapshots" of Mercury, in which Mercury's exaggerated shape represents its tidal bulges, show that it rotates exactly one and a half times per orbit, or three times for every two orbits. This rotation pattern ensures that the tidal bulges are always aligned with the Sun at perihelion, when the tidal bulges are strongest. The orbital eccentricity is also exaggerated.

FIGURE 11.17 Synchronous rotation of planets

and out of the bulging region. This continual reshaping would create internal friction (*tidal friction*), heating the moon's interior. The energy for this heat must come from somewhere, and it would be drawn from the satellite's rotation. As a result, the rotation rate would gradually change until the moon no longer rotated through its tidal bulge—which means synchronous rotation. Thus, no matter how fast a moon is rotating when it forms, tidal forces can eventually bring it into synchronous rotation. Precise calculations suggest that almost all but the smallest and most distant moons have kept the same faces toward their planets for billions of years.

Incidentally, similar tidal processes can also affect planets. Pluto and its moon Charon are relatively close in size, so their mutual tidal forces have forced them to rotate synchronously with each other (Figure 11.17a). Like two dancers, each always keeps the same face toward the other. Mercury exhibits a variation on synchronous rotation caused by tidal forces from the Sun: It rotates exactly three times for every two orbits that it completes (Figure 11.17b). This unusual pattern ensures that Mercury's tidal bulge is always aligned with the Sun at perihelion (the part of its orbit nearest the Sun), where the Sun's tidal force is significantly stronger. Tidal forces from the Moon cause the Earth's rotation to gradually slow down [Section 6.5]. In a few hundred billion years (if the Earth and the Moon stay together that long), the Earth will always show the same face to the Moon.

Ice Geology

Prior to the Voyager missions, most scientists expected the jovian moons to be cold and geologically dead. After all, only two of the moons are even as large as Mercury, and Mercury has been geologically dead for a long time. Instead, Voyager provided many surprises, revealing worlds with spectacular past and present geological activity. We've since learned even more—about Jupiter's moons in particular—thanks primarily to data from the Galileo spacecraft.

How can moons have so much more geological activity than similar-size (or larger) planets? In a few cases, the answer involves unforeseen heat sources that we'll discuss shortly. But another important factor is the icy composition of most jovian moons. These moons accreted from solid particles that condensed in the "miniature solar nebulae" around the jovian planets. These nebulae were much cooler than the parts of the solar nebula nearer to the Sun and hence were rich with ice crystals of the various hydrogen compounds (mostly water, but also small amounts of ammonia and methane). As a result, the jovian planet satellites contain much higher proportions of ice relative to rock than do the terrestrial planets. We even find distinctions among the jovian satellite systems: Jupiter's satellites have the highest proportion of rock, while the satellites of more distant planets contain higher proportions of the more volatile methane and ammonia ices.

We can understand why icy moons have more geological activity than similar-size rocky moons by

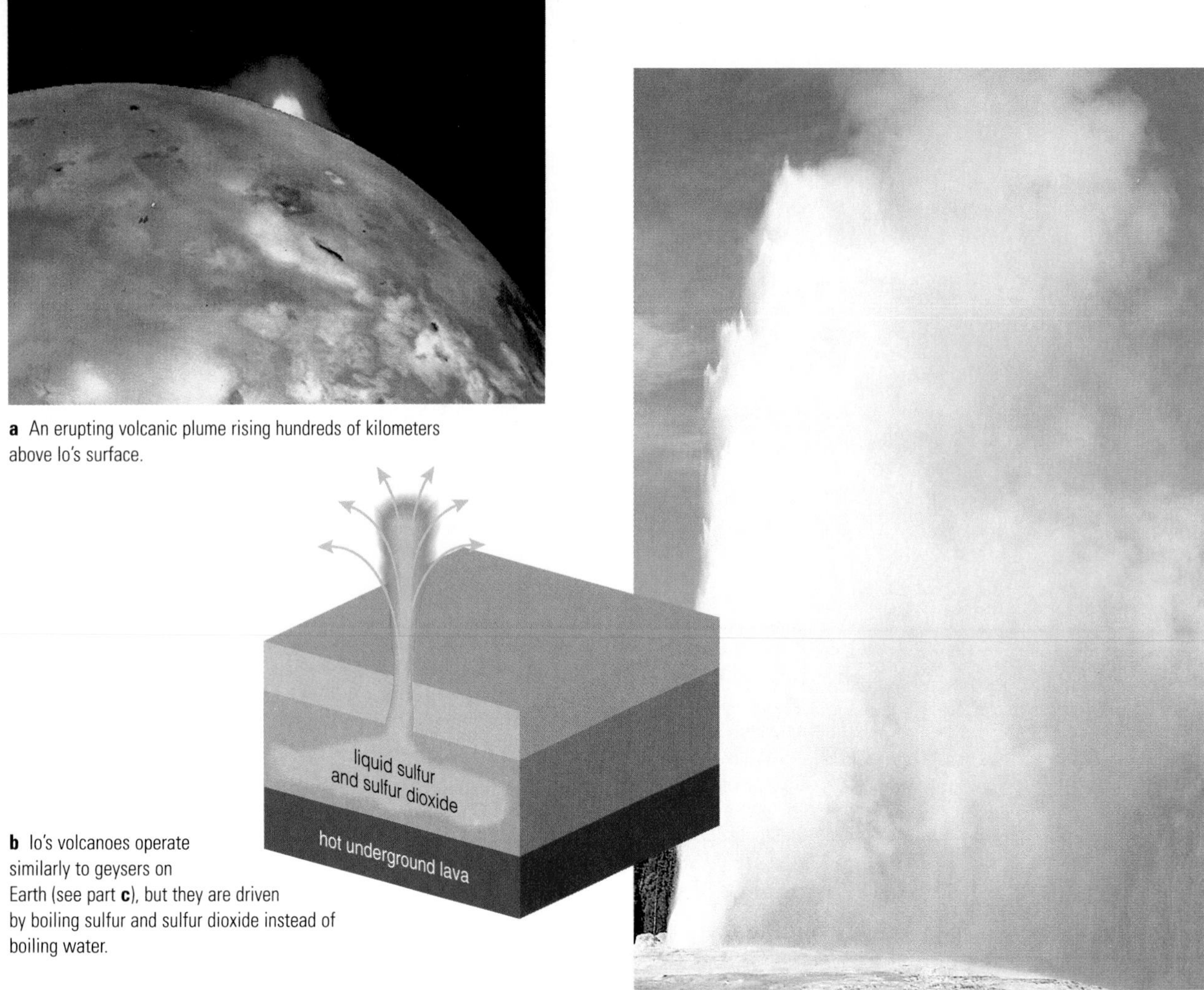

a An erupting volcanic plume rising hundreds of kilometers above Io's surface.

b Io's volcanoes operate similarly to geysers on Earth (see part **c**), but they are driven by boiling sulfur and sulfur dioxide instead of boiling water.

c Old Faithful geyser in Yellowstone National Park erupts when underground water reaches the boiling point and shoots up in plumes of hot water and steam.

FIGURE 11.18 Io is the most volcanically active body in the solar system.

investigating how ices compare to rocks in three geologically important properties: strength (or rigidity), radioactive content, and melting point. In terms of strength, the differences might be smaller than you would expect. Ice at a temperature of 100 K is almost as rigid as rock at a few hundred Kelvin. Nevertheless, the difference is great enough to cause ice mountains and ice cliffs on the jovian moons to sag and fade away more quickly than similar features made of rock on terrestrial planets. Most radioactive elements are found in rocks, not ices, so icy moons have relatively little internal heat generated by radioactive decay and hence lower internal temperatures. But ices also have much lower melting points than rocks, so the heat of accretion in many medium-size and large moons was enough to cause differentiation of their interiors. Thus, most of these moons have internal structures similar to those of the terrestrial planets, but with an additional, icy layer on the outside. Despite the icy composition, no satellite surface is "clean as the driven snow"; the ices are "dirtied" by rocky and carbon-rich compounds. Each satellite has its own proportions of rock and ice. Some are cleaner, some are dirtier, and some are clean on one side and dirty on the other.

All in all, the *ice geology* of the jovian moons bears many similarities to the rock geology of the terrestrial planets [Section 9.4]. *Impact cratering,* occurring mostly during the time of the early bombardment more than 4 billion years ago, probably left all the satellites with battered surfaces. *Volcanism* is certainly present on some jovian moons, and *tectonics* seems likely as well. Both of these processes are driven by internal heat and hence take place more easily on icy moons than on rocky planets because ice becomes deformable and melts at lower temperatures than rock. We know very little about *erosion* in the outer solar system, although we expect it to be relatively rare. Most satellites possess either no atmosphere or a very thin one, so in most cases wind or rain erosion is unlikely. Nevertheless, many of the jovian moons have interesting features that go beyond simple generalities. There are far too many moons to study all of them in depth here, but as we undertake our brief tour of the jovian moon systems, we will stop to investigate a few of the more intriguing moons in detail.

d The reddish color of the now-cooled lava flows extending from this volcano on Io (center black dot) suggests they were once molten sulfur.

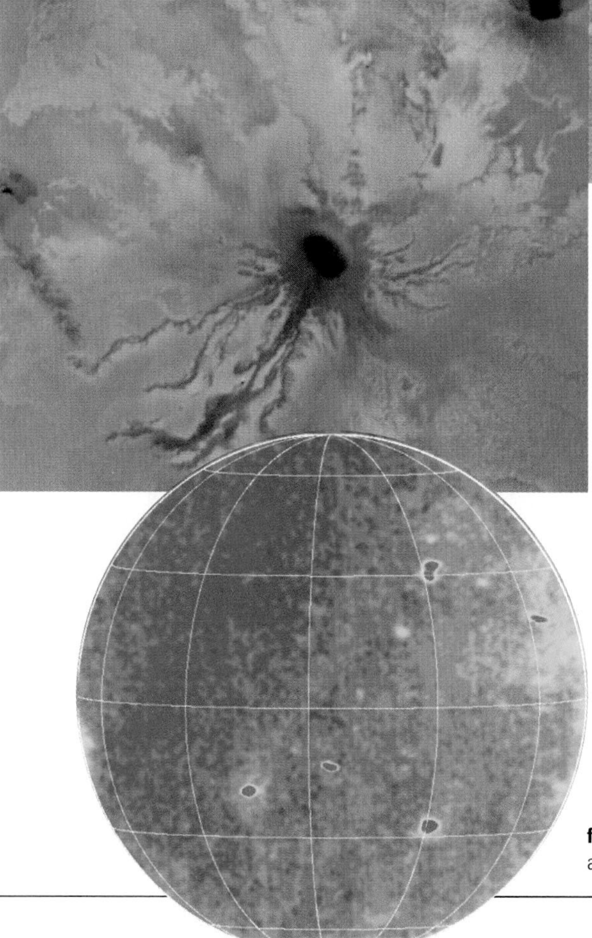

e This photo shows a shield volcano on Io that may be made of basaltic lava.

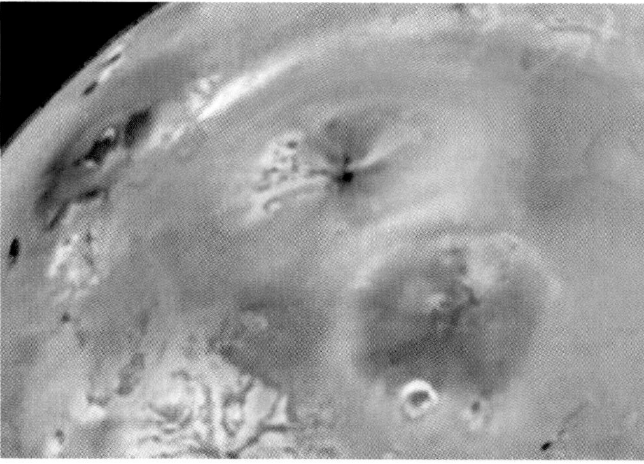

g This enhanced-color photo shows fallout (dark patch) from a volcanic plume on Io. The fallout region covers an area the size of Arizona. (The orange ring is the fallout from another volcano.)

f This false-color photo shows the glow of Io's volcanic vents (red) and atmosphere (green) when Io is in the darkness of Jupiter's shadow.

The Galilean Satellites of Jupiter: Io, Europa, Ganymede, and Callisto

The first stops on our tour are the Galilean satellites of Jupiter—the four moons that Galileo saw through his telescope and that disproved the idea that all celestial bodies circle the Earth [Section 6.3]. These four moons are large enough that they would count as planets if they orbited the Sun. They bear the names of four mythological lovers of the Roman god Jupiter: Io, Europa, Ganymede, and Callisto. Their densities decrease with distance from Jupiter, just as the densities of the planets decrease with distance from the Sun, suggesting that similar condensation and accretion patterns occurred within the "miniature solar nebula" around Jupiter and in the solar nebula. Io, the innermost Galilean satellite, formed from rocky and metallic condensates with little or no icy condensates. Europa is mostly rocky, but it formed with enough ice to give it an icy outer shell. Ganymede and Callisto, the outermost Galilean moons, formed from a mix of icy and rocky condensates.

Io Just a few months before the Voyager spacecraft reached Jupiter, a group of scientists made the astonishing prediction that we would find active volcanism on Io. This prediction went against the common belief that only much larger bodies could have substantial geological activity. But the prediction proved correct.[6] Direct proof came from the discovery of towering volcanic plumes reaching hundreds of kilometers above the surface and spreading fallout over almost a million square kilometers of Io's surface (Figure 11.18a). The indirect proof was just as mind-boggling: the complete lack of impact craters on the surface—not a single one, to the resolution of Voyager's cameras. Io's eruptions have buried *all* its impact craters.

Io's volcanoes take many forms. The plumes photographed by the Voyager and Galileo spacecraft are not eruptions of molten rock like those from volca-

[6]The heroes of this extraordinary discovery are theoreticians Pat Cassen, Stan Peale, and Ray Reynolds, and Linda Morabito, the Voyager navigation engineer who first spotted the plumes in calibration images.

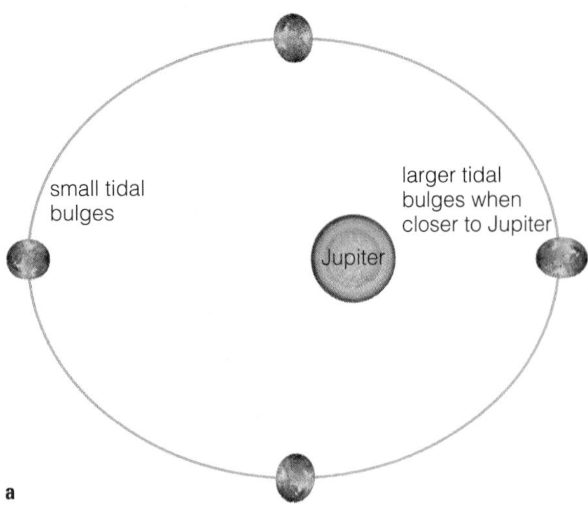

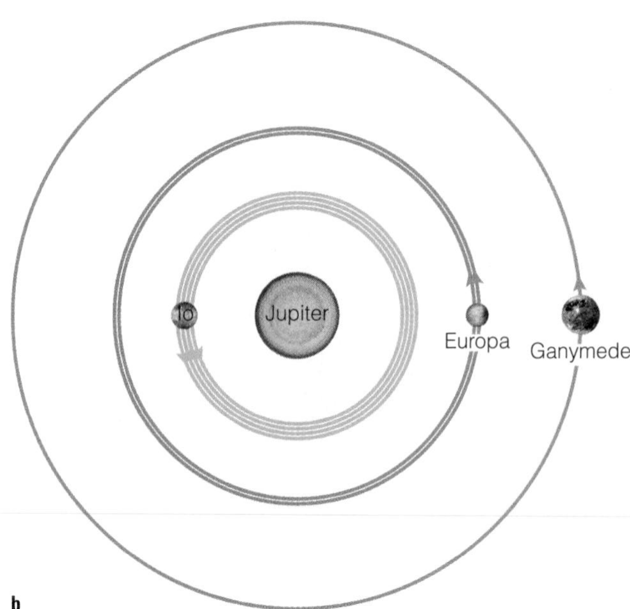

FIGURE 11.19 (**a**) Because Io's orbit is slightly elliptical, the strength and direction of Io's tidal bulges change. The bulges and orbital eccentricity are exaggerated. (**b**) About every seven Earth days (one Ganymede orbit, two Europa orbits, and four Io orbits), the three moons line up as shown. The small gravitational tugs repeat and make all three orbits slightly elliptical. (**c**) Tidal heating may give Europa a subsurface ocean beneath its icy crust. This artist's conception imagines a region where the crust has been disrupted by an undersea volcano.

noes on terrestrial planets. Instead, they resemble *geysers* on Earth, in which underground water is heated to the boiling point and shoots up in plumes of hot water and steam (Figure 11.18b,c). If Yellowstone's Old Faithful geyser erupted under Io's low gravity and thin atmosphere, its plume would shoot up about 30 kilometers. Io's plumes shoot up hundreds of kilometers because they are driven by boiling sulfur and sulfur dioxide, which are even more explosive when boiling than water. Not all volcanoes erupt so violently. Liquid sulfur flows out of some vents, creating distinctly red lava flows on the surface (Figure 11.18d). Some of the higher mountains found in some regions of Io were probably built by basaltic lava flows similar to those of terrestrial volcanoes (Figure 11.18e). Many of Io's volcanic vents glow red-hot (Figure 11.18f). Tectonic processes are probably also active on Io, but evidence for them is covered by the frequent lava flows and plume deposits everywhere on Io's surface. Galileo and Voyager images confirmed that fallout from volcanic plumes can cover areas the size of Arizona in a matter of months (Figure 11.18g).

Io's volcanoes also produce a very thin sulfur-dioxide atmosphere—about a billion times less dense than our own atmosphere. It is too thin to cause erosion, and Io's weak gravity allows a steady escape of the atmospheric gases. These volcanic gases supply the plasma in the Io torus (see Figure 11.13) and in Jupiter's magnetosphere.

Why is Io so geologically active? After all, it is about the size of our own geologically dead Moon, and it is not as icy in composition as most other jovian satellites. Io must have an additional internal heat source besides the usual combination of accretion, differentiation, and radioactivity that heats the interiors of the terrestrial worlds [Section 9.3]. This fourth internal heat source is **tidal heating,** and its theory was the basis for the pre-Voyager prediction of volcanic activity on Io.

How does tidal heating work? Tidal forces lock Io in synchronous rotation so that it keeps the same face toward Jupiter. But Io's orbit is slightly elliptical, so its orbital speed and distance from Jupiter vary. As a result, the strength and even the direction of the tidal force change very slightly over its orbit: Io is continuously flexed by Jupiter (Figure 11.19a). The flexing of Io's rocky crust and interior releases heat, just as flexing warms Silly Putty. In fact, Io's tidal heating releases more than 200 times more heat (per gram of mass) than the radioactive heat driving Earth's geology, which explains why Io is the most volcanically active body in the solar system.

But why is Io's orbit slightly elliptical, when almost all other large satellites' orbits are virtually circular? Io's neighbors are responsible: Gravitational nudges between Io, Europa, and Ganymede maintain **orbital resonances** between them (Figure 11.19b). During the time in which Ganymede completes one orbit of Jupiter, Europa completes exactly two orbits and Io completes exactly four orbits. The satellites therefore are periodically lined up, and the gravitational tugs these satellites exert on one another add up over time. The satellites are always tugged in exactly the same direction, and the satellite orbits

c

become very slightly elliptical as a result. (If the satellites didn't line up periodically, the tugs would be exerted in random directions at random times—as ineffective as pushing a child on a swing at random times.) The resulting tidal forces are strongest on Io because it is closest to Jupiter, but they are also important on Europa and Ganymede. The case of Europa is particularly interesting (Figure 11.19c).

Europa Europa's bizarre, fractured crust is proof enough that tidal heating has taken hold there (Figure 11.20a). The icy surface is nearly devoid of impact craters and may be only a few million years old. Jumbled icebergs (Figure 11.20b) suggest that an ocean of liquid water lies just a few kilometers below the surface, extending down to perhaps 100 km deep

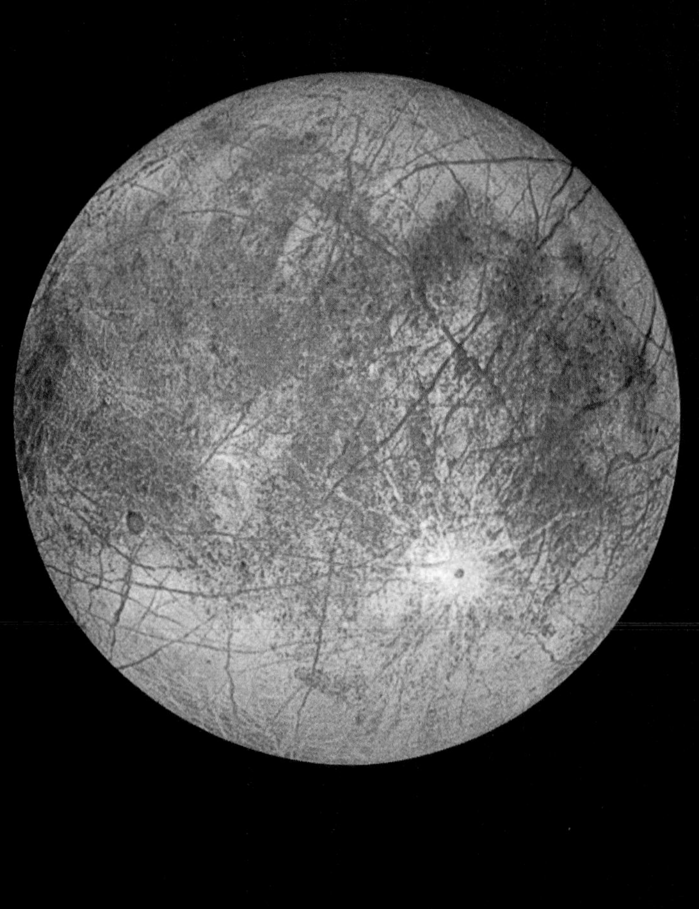

a Europa's icy crust is criss-crossed with cracks.

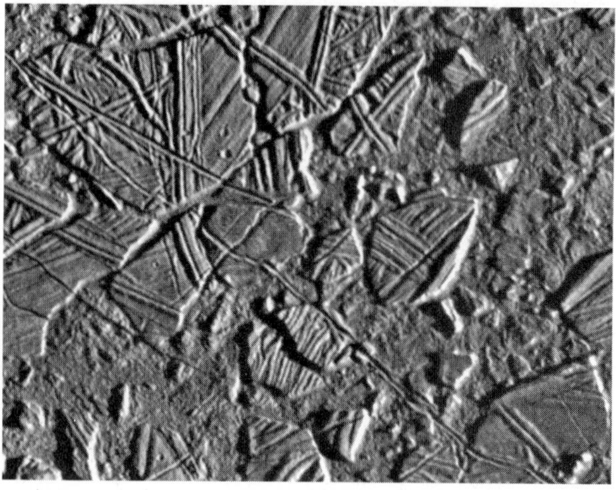

b Some regions show jumbled crust with icebergs, apparently frozen in slush.

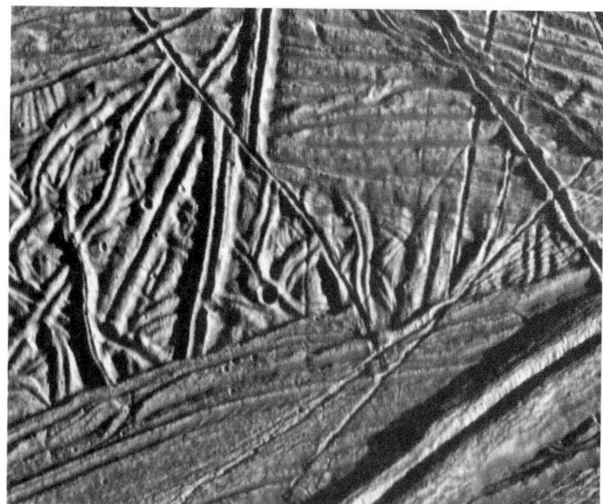

c Close-up photos show that surface cracks have a double-ridged pattern.

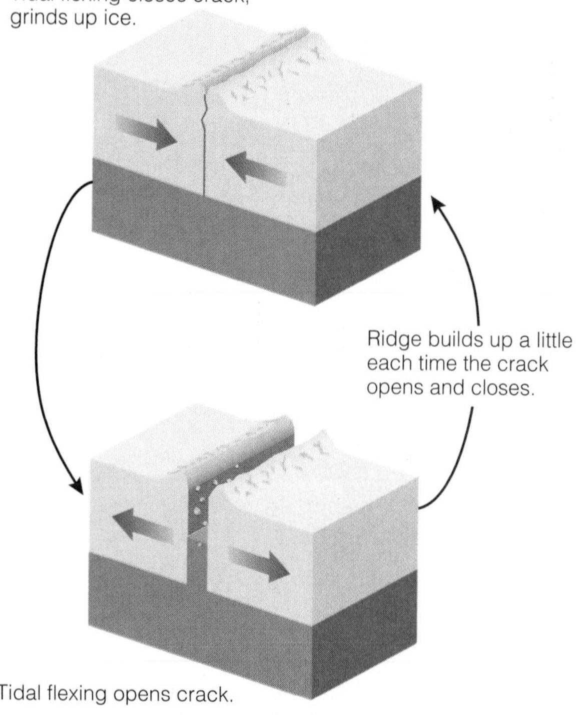

d A possible mechanism for making the double-ridged surface cracks.

FIGURE 11.20 Europa is one of the most intriguing moons in the solar system.

(Figure 11.20c). The icy shell overlying the ocean is repeatedly flexed by tidal forces, resulting in a global network of cracks. Close-up images of the cracks taken by the Galileo orbiter show that most have a remarkable double-ridged pattern, perhaps the result of debris piling up around a crack that is repeatedly opened and closed by tidal flexing (Figure 11.20c,d). On Earth, changing winds form similar double ridges in sea ice in the Arctic Ocean, though on much smaller scales. (We do not expect winds on Europa, although it does have a very thin atmosphere created by magnetospheric bombardment.)

In some ways, Europa resembles a rocky terrestrial planet wrapped in an icy crust.[7] The rocky object that makes up most of Europa is undoubtedly geologically active below the layer of ocean and ice. Perhaps basaltic lavas erupt on Europa's seafloors,

[7]Some scientists even suggest that the icy shell has not always rotated at exactly the same rate at the rocky center, leading to even greater cracking of the crust.

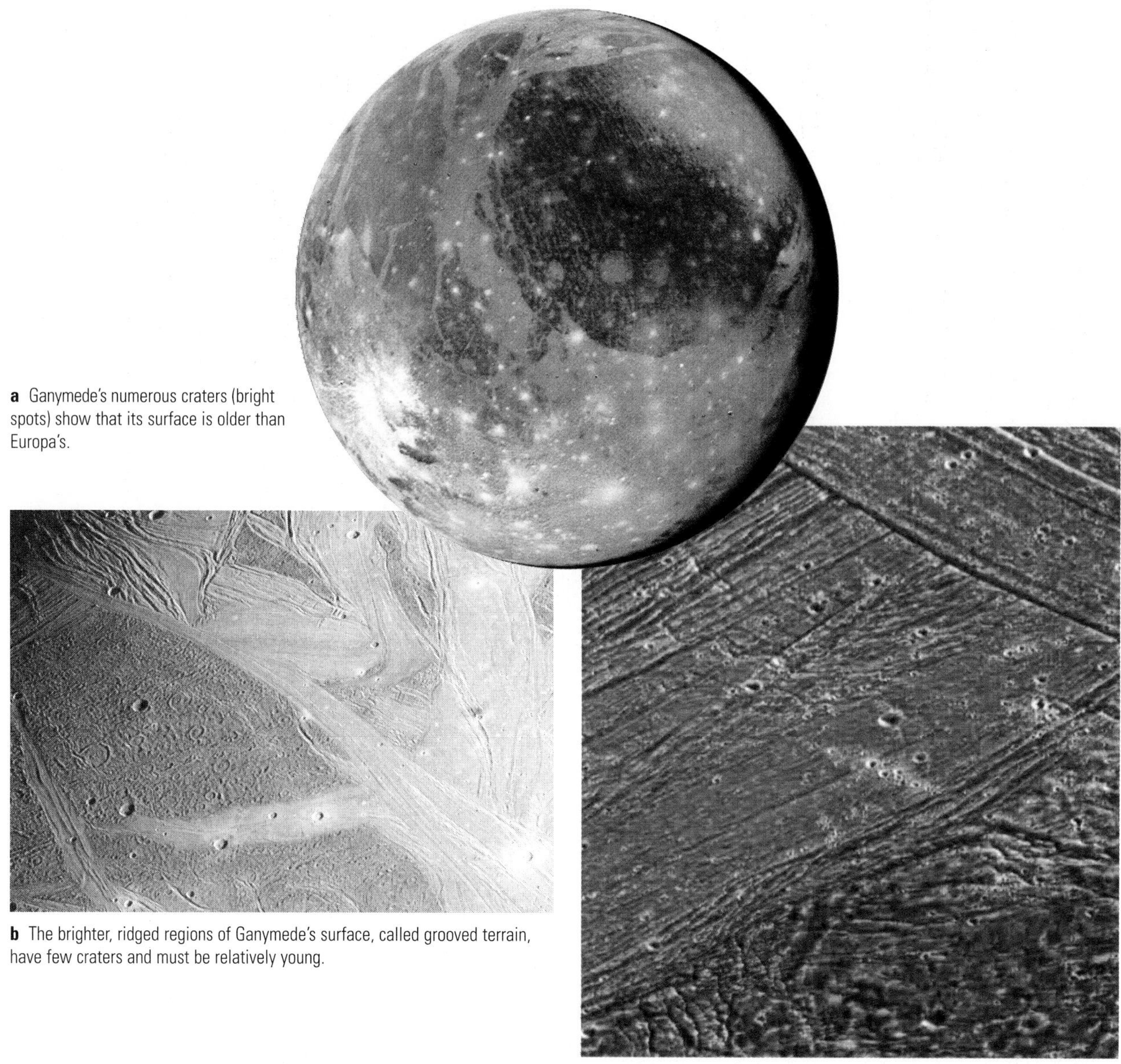

a Ganymede's numerous craters (bright spots) show that its surface is older than Europa's.

b The brighter, ridged regions of Ganymede's surface, called grooved terrain, have few craters and must be relatively young.

c A close-up photo of the grooved terrain.

FIGURE 11.21 Ganymede, the largest moon in the solar system.

sometimes violently enough to jumble up the icy crust above (see Figure 11.19c). And, just possibly, tidal heating makes Europa's oceans as hospitable for life as our own oceans [Section 13.6].

Ganymede Photographs of Ganymede tell the story of an interesting geological history. Parts of the surface have many impact craters, indicating that these regions are billions of years old (Figure 11.21a). But other areas show unusual "grooved terrain" unlike anything we've seen in terrestrial geology (Figure 11.21b,c). The grooves may form either through tectonic stresses or from water erupting along a crack in the surface; because ice expands when it solidifies (unlike rock), the grooves may form as the surface refreezes. A freezing surface patch on Ganymede therefore pushes outward, perhaps creating the grooves and ridges seen on its surface. Ganymede also shows some evidence of tectonics similar to plate tectonics on Earth—some chunks of the surface appear to have shifted relative to one another. Erosion is unimportant on Ganymede, since it has just a hint of an atmosphere generated by bombardment. The precise nature of Ganymede's internal heating remains mysterious, although it certainly is some combination of tidal heating and radioactive decay. Surprisingly, Ganymede has its own magnetic field—perhaps indicating a molten, convecting core.

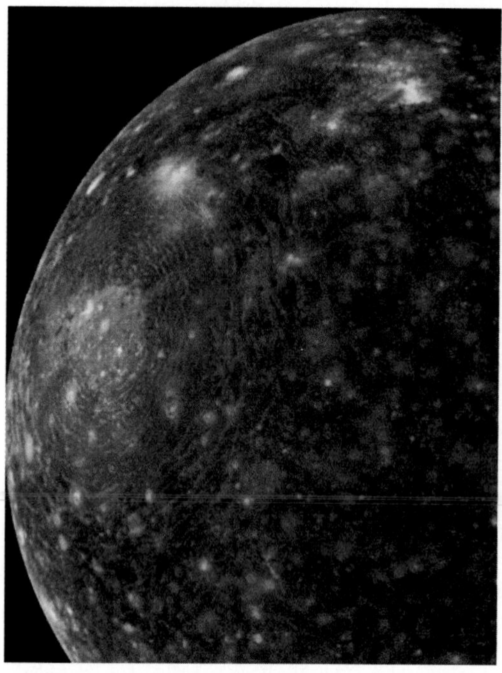

a Heavy cratering indicates an ancient surface.

b Close-up photos show a dark powder overlying the low areas of the surface.

FIGURE 11.22 Callisto shows no evidence of volcanic or tectonic activity.

Callisto The outermost Galilean satellite, Callisto, is about what we might expect for outer solar system satellites: a heavily cratered iceball (Figure 11.22a). The bright patches on its surface are caused by cratering: Large impactors dig up cleaner ice from deep down and spread it over the surface. The odd rings in the upper left of Callisto's image also come from impacts: The largest impactors shatter the icy sphere in a way that produces concentric rings around the impact site.

Despite its relatively large size (the third-largest moon in the solar system), Callisto lacks volcanic and tectonic features. Internal heat sources must be very small. In fact, gravity measurements by the Galileo spacecraft show that Callisto never even underwent differentiation: Dense rock and lighter ice are still thoroughly mixed throughout. Callisto has no tidal heating because it shares no orbital resonances with other satellites. Nonetheless, the surface holds some surprises. Close-up images show the surface to be covered by a dark, powdery substance concentrated in low-lying areas, leaving ridges and crests bright white (Figure 11.22b). No one knows the nature of this material or how it got there.

Titan and the Medium-Size Moons of Saturn

Leaving Jupiter's moons behind, our satellite tour takes us next to Saturn. Here we find an amazing moon shrouded in a thick atmosphere: Titan. It is Saturn's only large moon and is the second-largest moon in the solar system after Ganymede (see Figure 11.15).

Titan's hazy and cloudy atmosphere hides its surface (Figure 11.23a). The atmosphere is about 90% nitrogen—making it the only world besides Earth where nitrogen is the dominant atmospheric constituent. However, on Earth the rest of the atmosphere is mostly oxygen, while the rest of Titan's atmosphere consists of argon, methane, and other hydrogen compounds, including ethane (C_2H_6). Thus, we could not breathe the air on Titan and live. The methane and ethane actually give Titan an appreciable greenhouse effect, but the surface temperature is still a frigid 93 K (-180°C). The surface pressure on Titan is only somewhat higher than on Earth—about 1.5 bars at the surface, which would be fairly comfortable if not for the lack of oxygen and cold temperatures.

How did Titan end up with such an unusual atmosphere? Titan is composed mostly of ices, including methane and ammonia ice. Some of this ice sublimed long ago to form an atmosphere on Titan. Over billions of years, solar ultraviolet light broke apart the ammonia molecules (NH_3) into hydrogen and nitrogen. The light hydrogen escaped from the atmosphere (through thermal escape [Section 10.4]), but the heavier nitrogen molecules remained and accumulated.

FIGURE 11.23 Saturn's moon Titan.

a Titan is enshrouded by a hazy, cloudy atmosphere. The inset shows a close-up looking over the limb of Titan.

b Artist's conception of the surface of Titan, showing the possible ethane oceans.

We encounter a particularly intriguing idea when we examine what happened to the methane gas that sublimed into Titan's atmosphere long ago. Methane is no longer present in large quantities, so some process must have removed it from the atmosphere. The most likely process is chemical reactions, triggered by solar ultraviolet light, that transform methane into ethane. If this idea is correct, atmospheric ethane may form clouds and rain on Titan, perhaps creating oceans of liquid ethane on the surface (Figure 11.23b). According to some calculations, Titan may have produced enough ethane over its history to create ethane oceans a kilometer deep. Unfortunately, we don't yet know if such oceans really exist.

Because we cannot see its surface, we do not know whether Titan is geologically active. However, given its relatively large size and icy composition, it seems likely that Titan has a rich geological history. Our understanding of Titan should improve dramatically when the *Cassini* spacecraft reaches Saturn in 2004 [Section S2.6]. In addition to cameras and spectrographs, *Cassini* carries a radar mapper similar to the *Magellan* radar that mapped the surface of Venus [Section 9.5]. Thus, we will see the surface of

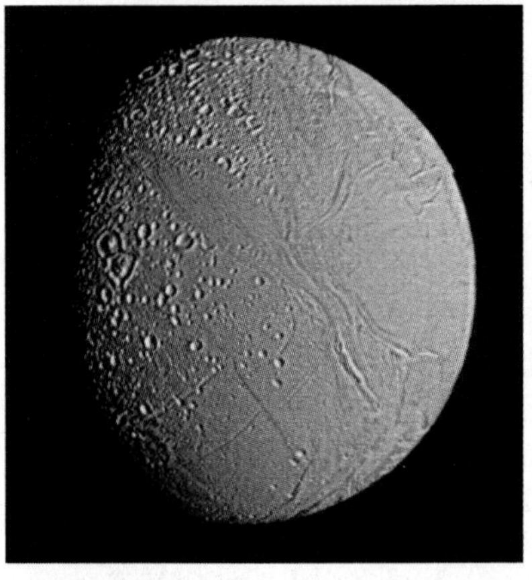
Enceladus

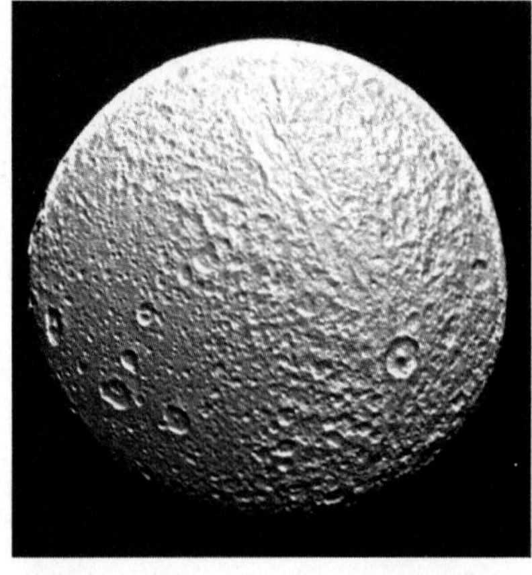
Tethys

Dione

FIGURE 11.24 Saturn's six medium-size moons, which range in diameter from 390 km (Mimas) to 1,530 km (Rhea). All but Mimas show some evidence of volcanism and/or tectonics.

Titan at high resolution and at last will learn about its geology—and whether the surface is covered by an ethane ocean. *Cassini* will also drop a probe (named *Huygens,* for a seventeenth-century pioneer in astronomy and optics [Section 1.1]) into Titan's atmosphere that will take pictures and measure atmospheric properties on the way down. Just in case, the probe is designed to float in liquid ethane.

TIME OUT TO THINK *Based on your understanding of the four geological processes on icy satellites, what kind of surface features do you think might be present on Titan?*

Given that Titan and Ganymede are similar in both size and composition, it's natural to wonder why Titan has so much atmosphere and Ganymede has so little. One possibility is that cooler temperatures in the solar nebula at Saturn's greater distance from the Sun allowed more methane and ammonia ice to condense. These ices are more volatile than the water ice that makes up Jupiter's satellites and hence are more likely to sublime and form an atmosphere. Alternatively, atmospheric cratering [Section 10.4] may have stripped away Ganymede's atmosphere but not Titan's. Atmospheric cratering was probably more important on Ganymede than on Titan because Jupiter's stronger gravity accelerates impactors to higher speeds. No one yet knows which of these two processes—condensation in the solar nebula or atmospheric cratering—bears more responsibility for the differences between Ganymede and Titan.

Saturn's medium-size moons (Rhea, Iapetus, Dione, Tethys, Enceladus, and Mimas) show evidence of substantial geological activity (Figure 11.24). While all show some cratered surfaces, Enceladus has grooved terrain similar to Ganymede's, and the others show evidence of flows of "ice lava." But these satellites are considerably smaller than Ganymede: Enceladus is barely 500 kilometers across—small enough to fit inside the boundaries of Colorado. How can such small moons support so much geological activity? The answer probably lies with "ice geology," in which icy combinations of water, ammonia, and methane can melt, deform, and flow at remarkably low temperatures. Thus, even small objects with relatively little internal heat can sustain fascinating geological activity. Iapetus is particularly bizarre, with some regions distinctly bright and others distinctly dark. The dark regions appear to be covered by a thin veneer of dark material, but we still don't know what it is or how it got there.

Mimas (radius = 195 km), the smallest of the "medium-size" moons of Saturn, is essentially a heavily cratered iceball. One huge crater is sometimes called "Darth Crater" because of Mimas's resemblance to the Death Star in the *Star Wars* movies. (The official name of the crater is Herschel.) The impact that created this crater probably came close to breaking Mimas apart.

The Medium-Size Moons of Uranus

Five medium-size moons orbit Uranus (Titania, Oberon, Ariel, Umbriel, and Miranda; see Figure 11.15). Prior to the Voyager visits, these five moons were the only moons known around Uranus, but Voyager images revealed 10 smaller moons, and astronomers recently discovered two more moons orbiting Uranus backward.

Rhea

Iapetus

Mimas

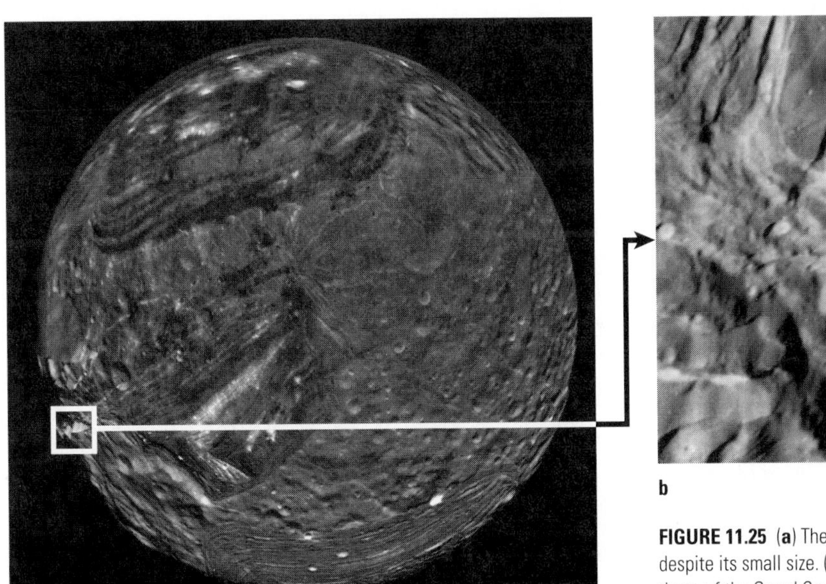

a

b

FIGURE 11.25 (**a**) The surface of Miranda shows astonishing tectonic activity despite its small size. (**b**) The cliff walls seen in this photo are higher than those of the Grand Canyon on Earth.

The medium-size moons pose several puzzles. For example, Ariel and Umbriel are virtual twins in size, yet Ariel shows evidence of volcanism and tectonics, while the heavily cratered surface of Umbriel suggests a lack of geological activity. Titania and Oberon also are twins in size, but Titania appears to have had much more geological activity than Oberon. No one knows why these two pairs of similar-size moons should vary so greatly in geological activity.

Miranda, the smallest of Uranus's medium-size moons (radius = 235 km), is the most surprising (Figure 11.25). Prior to the Voyager flyby, scientists expected Miranda to be a cratered iceball like Saturn's similar-size moon Mimas (see Figure 11.24). Instead, Voyager images of Miranda show tremendous tectonic features and relatively few craters. Why should Miranda be so much more geologically active than Mimas? Our best guess is that, whereas a large impact *almost* shattered Mimas, a large impact *did* shatter Miranda. The fragments orbiting Uranus eventually reaccreted into the moon we see today. This jumbled heap of odd-shaped fragments (possibly of different compositions and densities) deformed very slowly back into a roughly spherical shape, creating the odd tectonic features visible on Miranda's surface.

Triton, the Backward Moon of Neptune

Last stop on our satellite tour is Neptune, so distant that only two of its moons (Triton and Nereid) were known to exist before the Voyager visit. Triton is the only large moon in the Neptune system, and only one other moon (Proteus) even makes it into our "medium-size" category.

Triton may appear to be a typical satellite, but it is not. Its orbit is retrograde (it travels in a direction opposite to Neptune's rotation) and highly inclined to Neptune's equator. These are telltale signs of a captured satellite. But Triton is not small and potato-shaped like most captured asteroids. Instead it is large, spherical, and icy. In fact, it is larger than the planet Pluto. As such, Triton presents many challenges to our understanding of satellite captures. Nonetheless, it is almost certain that Triton once orbited the Sun instead of orbiting Neptune.

A quick study of the Triton photos shown in Figure 11.26 reveals a surface unlike any other in the solar system, and one that poses many unanswered questions. Are the flat surfaces like the lunar maria? Are the wrinkly ridges (nicknamed "cantaloupe terrain") tectonic in nature? Are the bright polar caps similar to those on Mars? Why is Triton's surface composition (a mixture of ices of nitrogen, methane, carbon dioxide, and carbon monoxide) unlike that of any other satellite? Clearly, Triton is more than a cratered iceball. The undisturbed craters suggest that major geological activity no longer continues, but Triton appears to have undergone some sort of icy volcanism in the past. Such geological activity is astonishing in light of Triton's extremely low (40 K) surface temperature, but it probably involved low-melting-point mixtures of water, methane, and ammonia ices. The internal heating source for Triton's geological activity is not known, but it may have involved tidal heating. Triton probably had a very elliptical orbit and a more rapid rotation when it was first captured by Neptune, but tidal forces would have circularized its orbit, slowed it to synchronous rotation, and heated its interior enough to cause its past geological activity.

To cap off Triton's odd geology, the sublimation of surface ices creates a thin atmosphere. Thin as it is, the atmosphere creates wind streaks on the surface, and unknown processes pump unusual plumes of gas and particles into the atmosphere. And, even at 30 AU from the Sun, the atmosphere has a small but measurable greenhouse effect, and solar ultraviolet radiation generates hazes. The combination of Triton's large orbital inclination and the substantial tilt of Neptune's rotation axis leads to extreme seasonal swings, and polar caps probably grow, shrink, and migrate from pole to pole. All of this takes place on a satellite that, because of its high albedo, is even colder

FIGURE 11.26 Neptune's moon Triton.

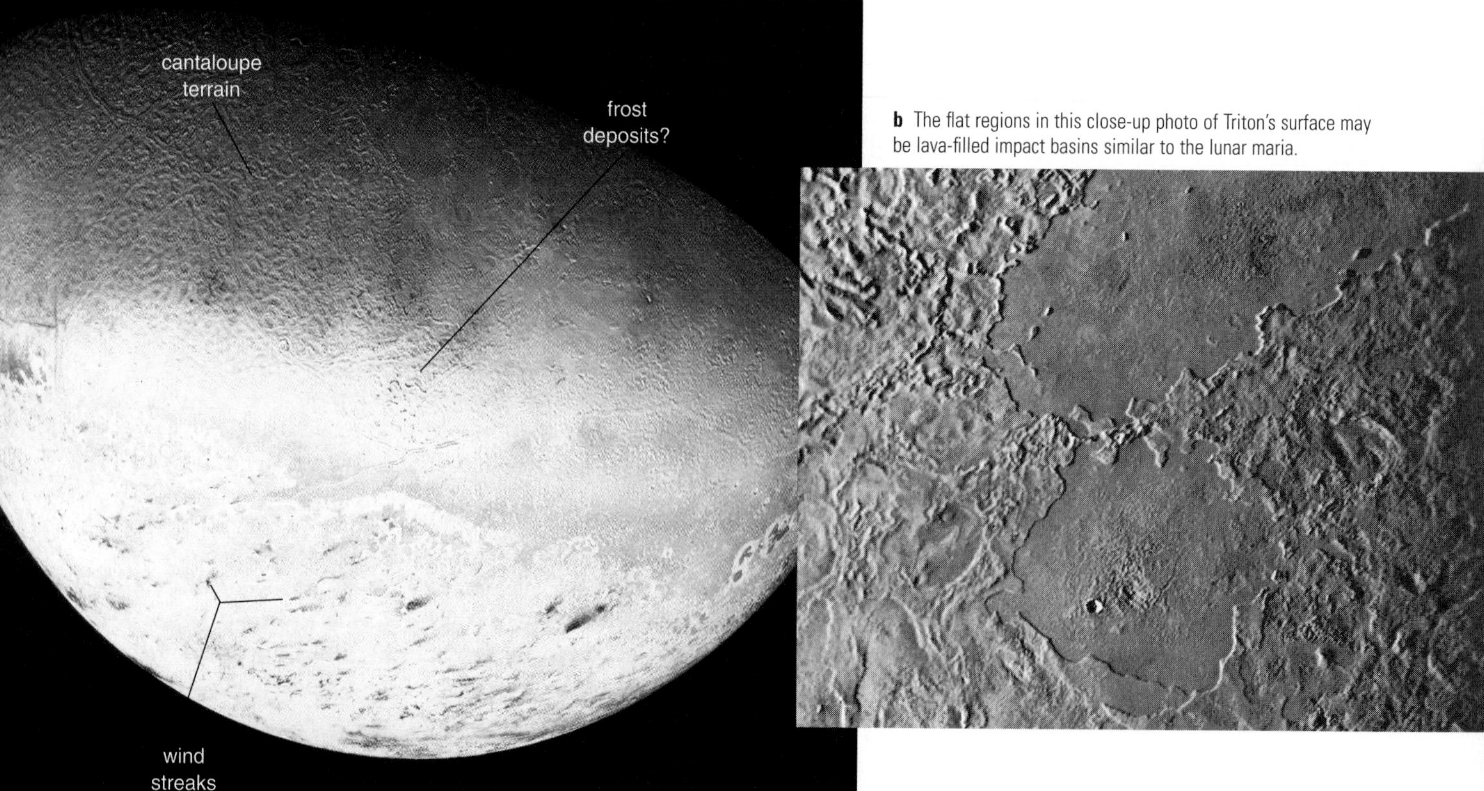

a Triton's southern hemisphere as seen by *Voyager 2*.

b The flat regions in this close-up photo of Triton's surface may be lava-filled impact basins similar to the lunar maria.

than Pluto. In fact, Triton is the coldest planet or satellite in the solar system.

If Triton really is a captured satellite, where did it come from? Could other objects like it still be orbiting the Sun? Intriguingly, we know of at least one object that appears to be quite similar to Triton: the planet Pluto. But that's a story for the next chapter.

Satellite Summary: The Active Outer Solar System

Our brief tour took us to most of the medium-size and large moons in the solar system. The major lesson of the tour is that icy-satellite geology is harder to predict than terrestrial-planet geology. Size isn't everything when it comes to geological activity. Ices form lavas more easily than rocks, so any icy object can have more geological activity than a rocky object of the same size. Satellites made of ices that include methane and ammonia may have even greater potential geological activity. In addition, we saw how tidal heating can supply added heat even when other internal heat sources are no longer important, as is certainly the case on Io and Europa and as occurs to lesser extents on Ganymede and Triton. Finally, we found at least one satellite (Miranda) where some geological activity may be caused not by internal heat but by reaccretion after shattering.

Satellite atmospheres also revealed surprises. We found examples of all three atmospheric sources (outgassing, sublimation, and bombardment [Section 10.4]). Io's atmosphere comes from volcanic outgassing, while Europa's and Ganymede's are attributed to bombardment. Titan's thick atmosphere probably came from sublimation, followed by extensive chemistry aided by solar ultraviolet light. Sublimation is also responsible for Triton's thin atmosphere, even out at the frigid edges of our solar system. The jovian planet satellites have turned out to be far more interesting and instructive than anyone could have imagined before the Voyager expeditions.

11.6 Jovian Planet Rings

We have completed our comparative study of the jovian planets themselves and of their major moons, but we have one more topic left to cover: their amazing rings. Saturn's rings have dazzled and puzzled astronomers since Galileo first saw them through his small telescope and suggested that they resembled "ears" on Saturn. For a long time, Saturn's rings were thought to be unique in the solar system, but we now know that all four jovian planets have rings. As we did for the planets themselves, we'll look at one ring system in detail and then examine the others for important similarities and differences. Saturn's rings are the clear choice as the standard for comparison.

Saturn's Rings

You can see Saturn's rings through a backyard telescope, but learning about their nature requires higher resolution. Even through large telescopes on Earth, the rings appear to be continuous, concentric sheets of material separated by gaps (Figure 11.27a). Spacecraft images reveal these "sheets" to be made of many more individual rings, each broken up into even smaller concentric *ringlets,* and different regions of a ring to range from transparent to opaque (Figure 11.27b). But even these appearances are somewhat deceiving. If we could wander into the rings, we'd see that they are made of countless individual particles orbiting Saturn together, each obeying Kepler's laws and occasionally colliding with nearby particles (Figure 11.27c). The particles range in size from large boulders to dust grains—far too small to be photographed, even when spacecraft like *Voyager* or *Cassini* pass nearby.

Spectroscopy reveals that Saturn's ring particles are made of relatively reflective (high-albedo) water ice. The rings look bright where there are enough particles to intercept most of the Sun's light and scatter it back toward us and appear more transparent where there are fewer particles. In regions where light cannot easily pass through, neither can another ring particle. In the densest parts of the rings, each particle collides with another every few hours.

TIME OUT TO THINK *Which ring particles travel faster: those at the inner edge of Saturn's rings, or those at the outer edge?* (Hint: *Think about Kepler's third law.*) *Can you think of a way to confirm your answer with telescopic observations?*

Saturn's rings lie in a thin plane directly above the planet's bulging equator, so they share Saturn's 27° axis tilt. Over the course of Saturn's year, the rings present a changing appearance to the Earth and Sun, appearing wide open at Saturn's solstices and edge-on at its equinoxes. The rings are perhaps the thinnest known astronomical structure: They are over 270,000 kilometers across but only a few tens of *meters* thick. Collisions help keep the ring thin. You can see why by imagining adding a new ring particle with an orbit tilted relative to the other particles. (The existence of many particles on tilted orbits would make the ring thicker.) The new particle

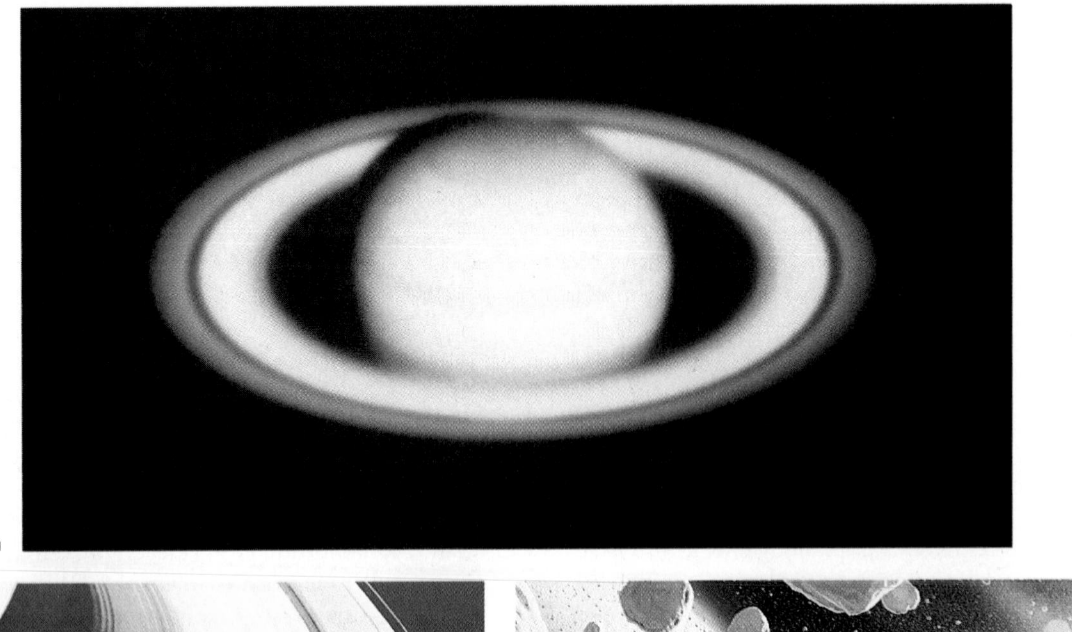

a

b

c

FIGURE 11.27 (**a**) Earth-based telescopic view of Saturn. (**b**) Voyager image of Saturn's rings against the disk. (**c**) Artist's conception of particles in a ring system. All the particles are moving slowly relative to one another and occasionally collide.

would collide with other particles each time its orbit intersected the ring plane, and its orbital tilt would be reduced with every collision. It wouldn't be long before these collisions would force the particle to conform to the orbital pattern of the other particles. Similarly, a new ring particle with a highly elliptical orbit would soon end up with a circular orbit.

Where do the rings come from? An important clue is that the rings lie close to the planet in a region where tidal forces are very influential. Within two to three planetary radii (of any planet), the tidal forces tugging an object apart become comparable to the gravitational forces holding it together. This region is called the **Roche zone.** Only relatively small objects held together by nongravitational forces (such as the electrostatic forces that hold solid rock—or spacecraft or human beings—together) can survive within the Roche zone. Thus, there are two basic scenarios for the origin of the rings. One possibility is that a wandering moon strayed too close to Saturn and was torn apart. A more likely scenario is that tidal forces prevented the material in Saturn's rings from accreting into a single large moon, instead forming many smaller moons.

Among the many mysteries of Saturn's rings are unusual dusty patches, called *spokes,* that can appear and change dramatically in a matter of hours (Figure 11.28). They are probably particles of microscopic dust that has been levitated out of the ring plane by forces associated with Saturn's magnetic field.

FIGURE 11.28 The dark patches in Saturn's rings are called spokes.

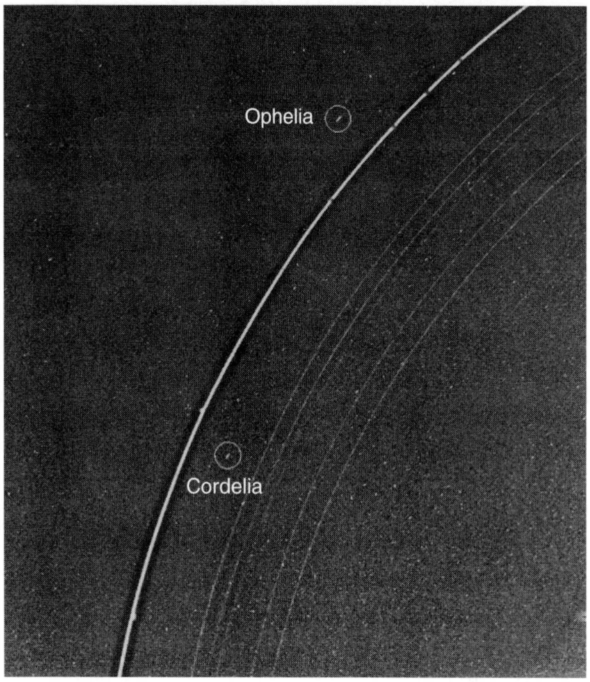

FIGURE 11.29 Voyager's best photograph of shepherd moons came during the *Voyager 2* Uranus flyby. Two satellites shepherd one of Uranus's rings. The satellite images are smeared out by their motion during the exposure.

Rings and Gaps

By the time the Voyager spacecraft were approaching Saturn, astronomers knew of at least six distinct rings around Saturn. Voyager scientists had hoped to find reasons for the six rings, but instead they found that the total number of individual rings and gaps may be as high as 100,000. Theorists are still struggling to explain all the rings and gaps, but some general ideas are now clear.

Rings and gaps are caused by particles bunching up at some orbital distances and being forced out at others. This bunching does not happen spontaneously but happens when gravity nudges the orbits of ring particles in some way. One source of nudging is tiny moons—called *gap moons*—located within the rings themselves. The gravity of a gap moon tugs gently on nearby particles, effectively nudging them farther away to other orbits.[8] This clears a gap in the rings around the moon's orbit. Voyager photographed a few gap moons, and there may be thousands more that are too small to see in the Voyager images. In some cases, Voyager images show two gap moons forcing particles trapped between them into line; the best such example comes from Uranus (Figure 11.29), but we find similar examples in Saturn's ring system. The gap moons are said to act as *shepherd moons* in such cases, because they shepherd the ring particles.

In some cases, ring particles orbiting Saturn may also be nudged by the gravity from Saturn's larger and more distant moons. For example, a ring particle orbiting about 120,000 kilometers from Saturn's center will circle the planet in exactly half the time it takes the moon Mimas to orbit. Every time Mimas returns to a certain location, the ring particle will also return to its original location and will experience the same gravitational nudge from Mimas. The nudges reinforce one another and eventually clear a gap in the rings (Figure 11.30a). This "Mimas 2:1 resonance" is essentially the same type of *orbital resonance* that affects the orbits of Io, Europa, and Ganymede. It is responsible for the large gap in Saturn's rings easily visible from Earth (called the *Cassini division*). Many similar resonances are responsible for other gaps, and resonances can even create beautiful ripples within the rings (Figure 11.30b).

Although we now know three causes of rings and gaps—gap moons, shepherd moons, and orbital resonances—we have not yet identified specific causes for the vast majority of the features within Saturn's rings. Perhaps the Cassini spacecraft will identify more gap moons and shepherd moons, or perhaps it will reveal even more puzzling structures.

[8]It may seem counterintuitive that a gap moon's gravitational attraction should nudge a ring particle *farther away*, but theoretical models show that this happens as a result of the combined gravitational interactions between the gap moon, the ring particles, and the planet.

a

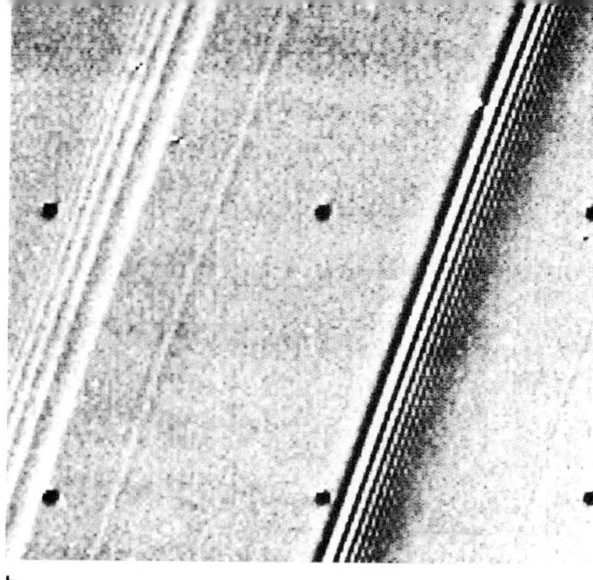

b

FIGURE 11.30 (**a**) The largest gap in Saturn's rings, called the Cassini division, is caused by an orbital resonance with the moon Mimas. (**b**) Another Mimas resonance creates remarkable ripples in Saturn's rings. The dark spots in the image are calibration marks for the camera.

Comparing Planetary Rings

The ring systems of Jupiter, Uranus, and Neptune are so much fainter than Saturn's ring system that it took almost four centuries longer to discover them. Ring particles in these three systems are far less numerous, generally smaller, and much darker. Saturn's rings were the only ones known until 1977, when the rings of Uranus were discovered during observations of a *stellar occultation*—a star passing behind Uranus as seen from the Earth. During the occultation, the star "blinked" on and off nine times before it disappeared behind Uranus, and nine more times as it emerged. Scientists concluded that these nine "blinks" were caused by nine thin rings encircling Uranus. Similar observations of stars passing behind Neptune gave more confounding results: Rings appeared to be present some of the time, but not at other times. Could Neptune's rings be incomplete or transient?

The Voyager spacecraft provided some definitive answers. Voyager cameras first discovered thin rings around Jupiter in 1979. Then, after providing incredible images of Saturn's rings, *Voyager 2* confirmed the existence of rings around Uranus as it flew past in 1986. In 1989, *Voyager 2* passed by Neptune and found that it does, in fact, have partial rings—at least when seen from Earth. The space between the ring segments is filled with dust that was not detectable from Earth.

How do Voyager images distinguish between boulders and dust? The secret is to look at how the ring particles scatter light (Figure 11.31a). Tiny particles like dust are not much bigger than the wavelengths of visible light, so they scatter sunlight only by very small angles from its original direction. Thus, rings that scatter light mostly forward must be made of tiny dust-size particles. Dust on your windshield scatters light in the same way, which is why it's so hard to see the road when you're driving into the Sun. Larger particles—marble- or boulder-size—mostly reflect light back toward the Sun. Thus, the Voyager spacecraft saw mostly the larger particles on their way toward each planet and saw mostly dust-size particles when they looked back toward the Sun after passing each planet (Figure 11.31b). In fact, *Voyager 2* detected vast dust sheets between the widely separated rings of Uranus and Neptune.

TIME OUT TO THINK *Turn on a flashlight or projector in a dark room. Set it down and walk a few steps away from it. Explain why you see so much airborne dust when you look back toward the projector but not when you look forward in the direction of the beam. How is this phenomenon similar to the scattering of light by ring particles?* (Note: *You can see this effect particularly well if you look back toward the projector in a movie theater.*)

A family portrait of the jovian ring systems shows many similarities (Figure 11.32). All rings lie in their planet's equatorial plane within the Roche zone. Particle orbits are fairly circular, with small orbital tilts relative to the equator. Gaps and ringlets are probably due to gap moons, shepherd moons, and orbital resonances. The differences also offer important insights: The larger size, higher reflectivity, and much greater number of particles in Saturn's rings compel us to wonder if different processes are at work there. The newly discovered rings brought challenges as well. The reasons for the slight tilt and eccentricity of Uranus's thin rings and the arcs within Neptune's rings are not yet understood.

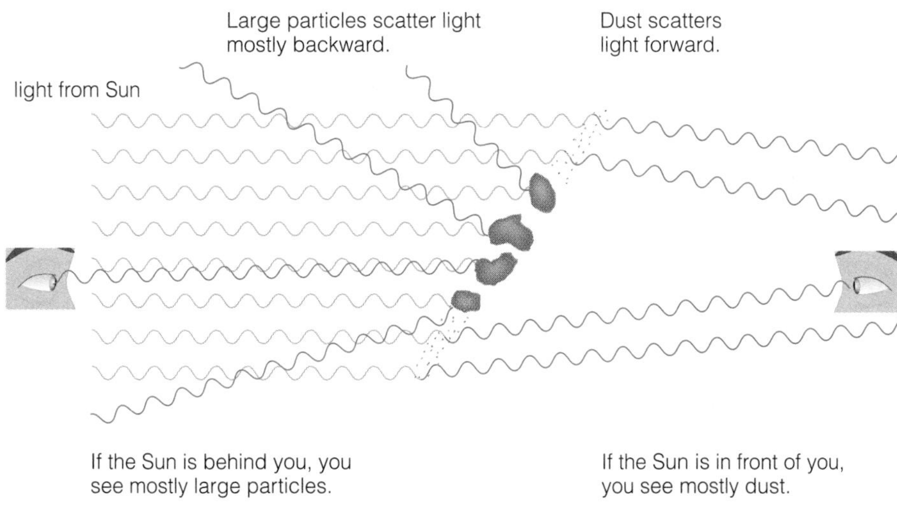

FIGURE 11.31 (**a**) You can tell whether ring particles are large or small even if you can't see an individual particle. (**b**) The rings of Uranus seen before the spacecraft arrived (top; Sun behind the spacecraft) and after it passed by (bottom; looking back toward the Sun). Only the larger particles that scatter light back toward the Sun are visible in the top image. Abundant dust particles become visible looking back toward the Sun (bottom).

Origin of the Rings

Where do rings come from? How old are they? Did they form with their planet out of the nebula, or are they a more recent phenomenon? These questions long puzzled astronomers when they had only one case (Saturn) to study, but the discovery that all jovian planets have rings has made the answers clearer.

Part of the answer lies in determining how long rings will last. Unlike planets and large satellites, ring particles may not last very long. Ring particles orbiting Saturn bump into one another every few hours, and even with low-velocity collisions the particles are chipped away over time. Basketball-size ring particles around Saturn are ground away to dust in a few million years. Other processes dismantle rings around the other jovian planets. The thermosphere of Uranus extends into its Roche zone, so atmospheric drag slowly causes ring particles to spiral into the planet. Dust orbiting Jupiter is slightly slowed by the flood of solar photons passing by, and the particles we now see in Jupiter's ring will fall into the planet in only a few million years. In summary, particles in the four ring systems cannot last as long as the 4.6-billion-year age of our solar system.

If ring particles are rapidly disappearing, some source of new particles must keep the rings supplied. The most likely source is collisions. As ring particles are ground away by collisions, small moons (perhaps the gap moons) occasionally collide and create more ring particles. Meteorites may also strike these moons, with the impacts releasing even more ring particles. We now believe that the "miniature solar nebulae" that surrounded the jovian planets formed many small satellites and that the gradual dismantling of these satellites creates the dramatic ring systems we see today. The appearance of any particular ring system probably changes dramatically over millions and billions of years. The spectacle of Saturn's brilliant rings may be a special treat of our epoch, one that could not have been seen a billion years ago and that may not last long on the time scale of our solar system.

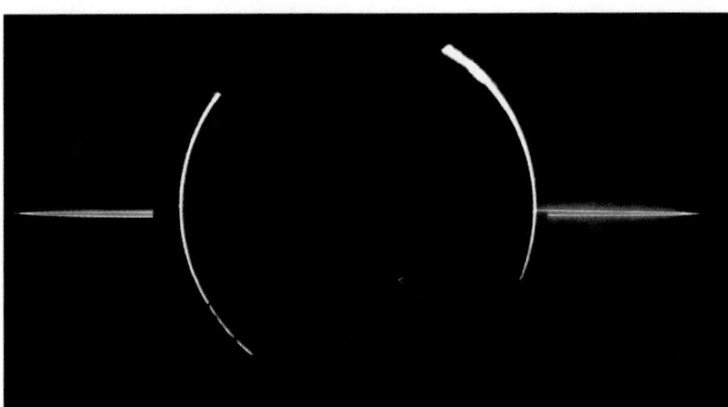
Jupiter

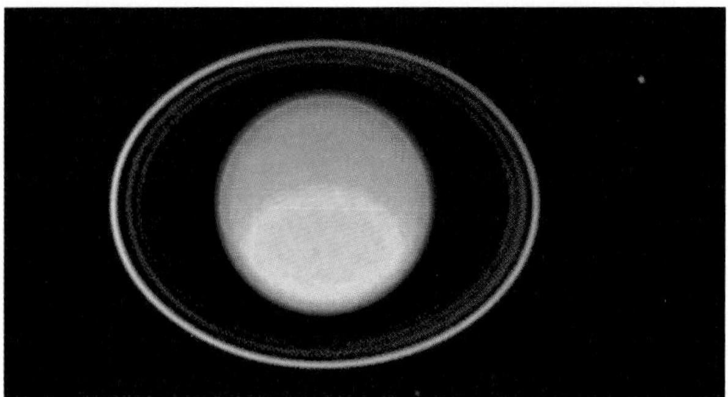
Uranus

Saturn

Neptune

FIGURE 11.32 Four ring systems. The planets are not shown to scale. Uranus's rings were photographed by the Hubble Space Telescope, the others by Voyager. The Neptune frame is made of two images taken on either side of the bright planet.

Do there exist many worlds, or is there but a single world? This is one of the most noble and exalted questions in the study of Nature.

ST. ALBERTUS MAGNUS (1206–1280)

THE BIG PICTURE

In this chapter, we saw that the jovian planets really are a "different kind of planet"—and, indeed, a different kind of planetary system. The jovian planets dwarf the terrestrial planets, and even some of their moons are as large as terrestrial worlds. As you continue your study of the solar system, keep the following "big picture" ideas in mind:

- The jovian planets may lack solid surfaces on which geology can work, but they are interesting and dynamic worlds with rapid winds, huge storms, strong magnetic fields, and interiors in which common materials behave in unfamiliar ways.
- Despite their relatively small sizes and frigid temperatures, many jovian satellites are geologically active by virtue of their icy compositions. Ironically, it was cold temperatures in the solar nebula that led to icy compositions and hence geological activity.
- Ring systems owe their existence to small satellites formed from the "miniature solar nebulae" that produced the jovian planets billions of years ago. The rings we see today are composed of particles liberated from those satellites surprisingly recently.
- Understanding the jovian planets forced us to modify many of our earlier ideas about the solar system, in particular by adding the concepts of ice geology, tidal heating, and orbital resonances. Each new set of circumstances offers further opportunities to learn how the universe works.

Review Questions

1. Briefly describe the past and present space exploration of the jovian worlds. Summarize what we've learned about the jovian planets in terms of composition, rotation rate, and shape. How did the jovian planets form?
2. Briefly summarize the techniques we use to study the interiors of the jovian planets.
3. Describe the interior structure of Jupiter. Contrast Jupiter's interior structure with the structures of the other jovian planets. Why is Jupiter denser than Saturn? Why is Neptune denser than Saturn?
4. How do we think Jupiter generates its internal heat? Does a similar process operate on any of the terrestrial planets? How do other jovian planets generate internal heat?
5. Describe the composition and temperature structure of Jupiter's atmosphere. In what ways is Jupiter's atmospheric structure similar to Earth's? In what ways is it different?
6. Why does Jupiter have several distinct layers of clouds? Why are clouds composed of a highly volatile gas such as ammonia found higher in the atmosphere than water clouds?
7. How are the circulation cells on Jupiter similar to and different from those on Earth? What are *belts* and *zones?* Why do the zones appear white, but not the belts? Why do the air currents of the circulation cells result in east-west winds?
8. What is the *Great Red Spot?* Is it the same type of storm as a hurricane on Earth? Explain.
9. Describe the possible origins of Jupiter's vibrant colors. Contrast these with the origins of the colors of the other jovian planets.
10. Does Jupiter have seasons? Why or why not? Do other jovian planets have seasons? Explain.
11. Why does Jupiter have auroras? Contrast Jupiter's magnetosphere with that of the Earth and of the other jovian planets.
12. What is the *Io torus?* Why don't Saturn, Uranus, or Neptune have a similar "torus"?
13. Briefly describe some of the general characteristics of the jovian moons.
14. What is *synchronous rotation?* What causes it? Explain why our Moon, most of the jovian moons, and Pluto and Charon rotate synchronously. How is Mercury's rotation similar to synchronous rotation?
15. How is the *ice geology* of the jovian moons similar to the rock geology of the terrestrial planets? How is it different?
16. Describe the volcanoes of Io and the mechanism of *tidal heating* that generates the internal heat needed for these volcanoes.
17. What are *orbital resonances?* Explain how the resonance between Io, Europa, and Ganymede makes their orbits slightly elliptical.
18. Describe the evidence suggesting that Europa may have a subsurface ocean of liquid water.
19. Describe the general surface characteristics of Ganymede and Callisto. What do these characteristics tell us about internal heating on these worlds?
20. Describe the atmosphere of Titan. What evidence suggests that Titan may have oceans of liquid ethane? Why do we expect to learn much more about Titan in 2004?
21. Describe a few general features of Saturn's medium-size moons. Why is it surprising to find evidence of geological activity on some of these moons, and how do we explain it?
22. Describe a few general features of Uranus's medium-size moons. Why is the geological activity on Ariel and Titania puzzling? How do we explain the odd appearance of Miranda?
23. How do we know that Triton is *not* a typical large jovian satellite? Describe Triton's general characteristics and give possible ideas of where Triton came from.
24. Describe the appearance and structure of Saturn's rings. Why are the rings so thin? Where do the rings lie in relation to Saturn's *Roche zone?*
25. Briefly describe how *gap moons* and orbital resonances can lead to gaps within ring systems.
26. Contrast the rings of the other jovian planets with Saturn's rings.
27. Describe two possible scenarios for the origin of planetary rings. What makes us think that ring systems must be continually replenished?

Discussion Questions

1. *A Miniature Solar System?* In what ways is the Jupiter system similar to a miniature solar system? In what ways is it different?
2. *Jovian Mission.* We can study terrestrial planets up close by landing on them, but jovian planets have no surfaces to land on. Suppose you are in charge of planning a long-term mission to "float" in the atmosphere of a jovian planet. Describe the technology you will use and how you will ensure survival for any people assigned to this mission.
3. *Extrasolar Planets.* Many of the newly discovered planets orbiting other stars are more massive than Jupiter and much closer to their stars. Assuming that they are in fact jovian (as opposed to terrestrial or something else), how would you expect these new planets to differ from the jovian planets of our solar system? How would any magnetospheres, satellites, or rings be different? Explain.

Problems

1. *The Importance of Rotation.* Suppose the material that formed Jupiter came together without any rotation, so that no "jovian nebula" formed and the planet today wasn't spinning. How else would the jovian system be different? Think of as many effects as you can, and explain each in a sentence.
2. *The Great Red Spot.* Based on the infrared and visible-wavelength images in Figure 11.7, is Jupiter's Great Red Spot warmer or cooler than nearby clouds? Does this mean it is higher or lower in altitude than the nearby clouds? (*Hint:* Use what you know about the belts and zones as you study these images.)
3. *Comparing Jovian Planets.* You can do comparative planetology armed only with telescopes and an understanding of gravity.
   a. The satellite Amalthea orbits Jupiter at just about the same distance in kilometers at which Mimas orbits Saturn. Yet Mimas takes almost twice as long to orbit. What can you deduce from this difference, qualitatively?
   b. Jupiter and Saturn are not very different in radius. When you combine this information with your answer to part (a), what can you conclude?
4. *Minor Ingredients Matter.* Suppose the jovian planet atmospheres were composed 100% of hydrogen and helium rather than 98% of hydrogen and helium. How would the atmospheres be different in terms of color and weather? Explain.
5. *Disappearing Satellite.* Io loses about a ton of sulfur dioxide per second to Jupiter's magnetosphere.
   a. At this rate, what fraction of its mass would Io lose in 5 billion years?
   b. Suppose sulfur dioxide currently makes up 1% of Io's mass. When will Io run out of this gas at the current loss rate?
6. *Ring Particle Collisions.* Each ring particle in the densest part of Saturn's rings collides with another about every 5 hours. If a ring particle survived for the age of the solar system, how many collisions would it undergo? Do you consider it likely that the ring particles we see in Saturn's rings today have been there since the formation of the solar system? Explain the ramifications of your answer for understanding the origin of the jovian planet rings.
7. *Project: Jupiter's Moons.* Using binoculars or a small telescope, view the moons of Jupiter. Make a sketch of what you see, or take a photograph. Repeat your observations several times (nightly, if possible) over a period of a couple of weeks. Can you determine which moon is which? Can you measure their orbital periods? Can you determine their approximate distances from Jupiter? Explain.
8. *Project: Saturn and Its Rings.* Using binoculars or a small telescope, view the rings of Saturn. Make a sketch of what you see, or take a photograph. What season is it in Saturn's northern hemisphere? How far do the rings extend above Saturn's atmosphere? Can you identify any gaps in the rings? Describe any other features you notice.
9. *Web Project: Galileo and Cassini Missions.* Follow the links from the book web site to find the latest news on the Galileo and Cassini missions.
   a. Print out (or describe) an image from the Galileo mission of interest to you, and interpret it using the techniques developed in this chapter.
   b. In a few sentences, describe the main scientific goals of the Cassini mission. What is the current health of the spacecraft, and how is the mission progressing?
10. *Web Project: Oceans of Europa.* The possibility of subsurface oceans on Europa holds great scientific interest and may even mean that life could exist on Europa. Use the links from the book web site to find some of the latest research concerning Europa. Write a one- to two-page summary of what we have learned about Europa since the time this book was published.

12 Remnants of Rock and Ice: Asteroids, Comets and Pluto

Asteroids and comets might seem insignificant in comparison to the planets and satellites we've been discussing. But there is strength in numbers, and the trillions of small bodies orbiting our Sun are far more important than their small size might suggest.

Few topics in astronomy have the potential to touch our lives more directly than asteroids and comets. The appearance of comets more than once altered the course of human history, when our ancestors acted upon superstitions related to comet sightings. More profoundly, asteroids or comets falling to the Earth have scarred our planet with impact craters and altered the course of biological evolution.

Asteroids and comets are also important for a very different reason: They are remnants from the birth of our solar system and therefore provide an important testing ground for our theories of how our solar system came to be. In this chapter, we will explore asteroids, comets, and those bodies that fall to Earth as meteorites. We will also examine the smallest planet, Pluto, and see that, while it may be a misfit among planets, it may be right at home among the smaller objects of our solar system.

12.1 Remnants from Birth

The objects we've studied so far—terrestrial planets, jovian planets, and large moons—have changed dramatically since their formation. Most planets have been transformed in almost every way: Their interiors have differentiated, their surfaces have been reshaped by geological processes, and their atmospheres have changed as some gases escape while others are added. But many small bodies remain virtually unchanged since their formation some 4.5 billion years ago. Thus, comets, asteroids, and meteorites carry the history of our solar system encoded in their compositions, locations, and numbers.

Using these small bodies to understand planetary formation is a bit like picking through the trash in a carpenter's shop to see how furniture is made. By studying the scraps, sawdust, paint chips, and glue, we can develop a rudimentary understanding of how the carpenter builds furniture. Similarly, the "scraps" left over from the formation of our solar system allow us to understand how the planets and larger moons came to exist. Indeed, much of our modern theory of solar system formation (presented in Chapter 8) was developed from studies of asteroids, comets, and meteorites.

Asteroids and comets are essentially leftovers from the process of accretion in the early solar system. As we discussed in Chapter 8, these leftovers fall into three fairly distinct groups, distinguished by their physical and orbital properties: *asteroids, Kuiper belt comets,* and *Oort cloud comets.* Asteroids are rocky or metallic in composition, and most orbit the Sun between the orbits of Mars and Jupiter. In contrast, comets are mostly icy in composition and spend most of their time well outside the orbits of the planets. This compositional difference is directly related to where the objects formed: Asteroids formed inside the *frost line* of the solar nebula, and comets formed outside [Section 8.4]. Kuiper belt comets have orbits like those of the planets, lying close to the ecliptic plane and orbiting the Sun counterclockwise as viewed from above Earth's North Pole. Their distance from the Sun ranges between about 30 AU (about the distance of Neptune) and 100 AU (more than twice the average distance of Pluto). Oort cloud comets have orbits that are randomly inclined to the ecliptic plane, and they generally lie at much greater distances from the Sun than the Kuiper belt comets—in some cases, perhaps halfway to the nearest stars.

In this chapter, we will investigate what we have learned from these remnants from the birth of our solar system. We'll also discuss the cosmic collisions that sometimes occur between leftover planetesimals and the existing planets. Before we continue, however, it's worth noting that the terminology used to describe these objects has a long history that can sometimes be confusing. The word *comet* comes from the Greek word for "hair," and comets get their name from the long, hairlike tails they display on the rare occasions when they come close enough to the Sun to be visible in our sky. *Asteroid* means "starlike," but asteroids are starlike only in their appearance through a telescope, not in any more fundamental properties. The term *meteor* refers to a bit of interplanetary dust burning up in our atmosphere; literally, it means "a thing in the air." Meteors are also sometimes called *shooting stars* or *falling stars* because some people once believed that they really were stars falling from the sky. Larger chunks of rock that survive the plunge through the atmosphere and hit the ground are called *fallen stars* or *meteorites,* which means "associated with meteors." To keep potential confusion to a minimum, we will use only the following terms and definitions in this book:

- *Asteroid:* a rocky leftover planetesimal orbiting the Sun.
- *Comet:* an icy leftover planetesimal orbiting the Sun—regardless of its size or whether or not it has a tail.
- *Meteor:* a flash of light in the sky caused by a particle entering the atmosphere, whether the particle comes from an asteroid or a comet.
- *Meteorite:* any piece of rock that falls from the sky, whether from an asteroid, a comet, or even another planet.

12.2 Asteroids

Asteroids are virtually undetectable to the naked eye and went unnoticed for almost two centuries after the invention of the telescope. The first asteroids were discovered about 200 years ago when astronomers were searching the "extra wide" gap between the orbits of Mars and Jupiter in hopes of discovering a previously unknown planet. It took 50 years to discover the first 10 asteroids (all large and bright), but today modern telescopes with advanced detectors discover that many on a typical night. Once an asteroid's orbit is calculated and verified, the asteroid is assigned a number (in order of verification), and the discoverer chooses a name (subject to the approval of the International Astronomical Union). The earliest discovered asteroids bear names of mythological figures; more recent discoveries carry names of scientists, cartoon heroes, pets, and rock stars.

The largest asteroid, Ceres, has a radius of about 500 kilometers—about half that of Pluto. About a dozen others are large enough that we would call them "medium-size" moons if they orbited a planet. Most asteroids are far smaller. There are probably more than 100,000 asteroids larger than 1 kilometer in diameter. However, even if they could all be put together, they'd make an object less than 1,000 kilometers in radius—far smaller than any terrestrial planet.

Painstaking observations have revealed the precise orbits of thousands of asteroids. The main **asteroid belt** lies between 2.2 and 3.3 AU from the Sun, though some interesting asteroids lie outside the belt (Figure 12.1). Asteroid orbits are distinctly elliptical and inclined relative to the ecliptic by as much as 20°–30°, but all orbit the Sun in the same direction as the planets. Science fiction movies often show the asteroid belt as a crowded and hazardous place, but the average distance between asteroids is millions of kilometers. The Galileo spacecraft had to be carefully targeted to fly close enough to an asteroid to photograph one.

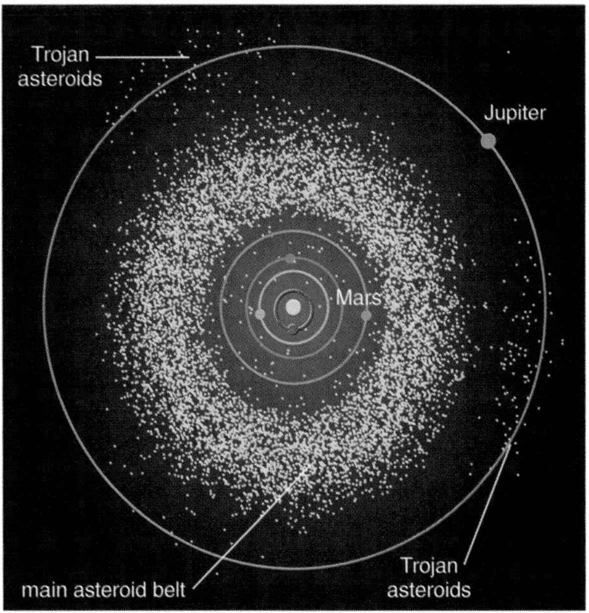

FIGURE 12.1 Positions of 8,777 asteroids for midnight, 1 January 2000. The asteroids themselves are much smaller than the dots on this scale.

Origin and Evolution of the Asteroid Belt

Why are asteroids concentrated in the asteroid belt, and why didn't a full-fledged planet form in this region? The answers probably lie with *orbital resonances.* Recall that an orbital resonance occurs whenever one object's orbital period is a simple ratio of another object's period, such as $\frac{1}{2}$, $\frac{1}{4}$, or $\frac{5}{3}$. In such cases, the two objects periodically line up with each other, and the extra gravitational attractions at these times can affect the objects' orbits. Consider, for example, the orbital resonances between Saturn's moons and particles in its rings: The moons are much larger than the ring particles and therefore force the ring particles out of resonant orbits, clearing *gaps* in the rings [Section 11.6]. In the asteroid belt, similar orbital resonances occur between asteroids and Jupiter, the most massive planet by far. For example, any asteroid in an orbit that takes 6 years to circle the Sun—half of Jupiter's 12-year orbital period—would receive the same gravitational nudge from Jupiter every 12 years and thus would soon be pushed out of this orbit. The same is true for asteroids with orbital periods of 4 years ($\frac{1}{3}$ of Jupiter's period) and 3 years ($\frac{1}{4}$ of Jupiter's period). (Music offers an elegant analogy: When a vocalist sings into an open piano, the strings of notes "in resonance" with the voice are "nudged" and begin to vibrate—not just the note being sung, but also notes with a half or a quarter of the note's frequency.)

Today, we see the results of these orbital resonances when we graph the number of asteroids with various orbital periods. Just as we would expect, there are virtually no asteroids with periods exactly $\frac{1}{2}$, $\frac{1}{3}$, or $\frac{1}{4}$ of Jupiter's. Thus, there are no asteroids with average distances from the Sun (semimajor axes) that correspond to these orbital periods. Figure 12.2 shows several other gaps in the asteroid belt, corresponding to orbital resonances that make other simple ratios with Jupiter's orbital period.[1] (By a *gap* in the asteroid belt we mean a gap in *average* distance; because asteroid orbits are elliptical, some asteroids pass through the gaps on parts of their orbits.)

The gravitational tugs from Jupiter that created the gaps in the asteroid belt probably also explain why no planet ever formed in this region. Early in the history of the solar nebula, the region of the asteroid belt probably contained enough rocky material to form another planet as large as Earth or Mars. Just as in other parts of the solar nebula, collisions among planetesimals frequently disrupted their orbits. However, the asteroid belt had an additional source of disruption: gravitational resonances with the forming planet Jupiter. The result was that planetesimals

[1] Once you determine the period for a particular orbital resonance, you can calculate the gap distance with the original version of Kepler's third law: $p^2 = a^3$, with p in years and a in AU. For example, if p = 6 yr ($\frac{1}{2}$ of Jupiter's period), then the gap distance is $a = 6^{2/3} = 3.3$ AU. The gaps in the asteroid belt are often called *Kirkwood gaps,* after their discoverer.

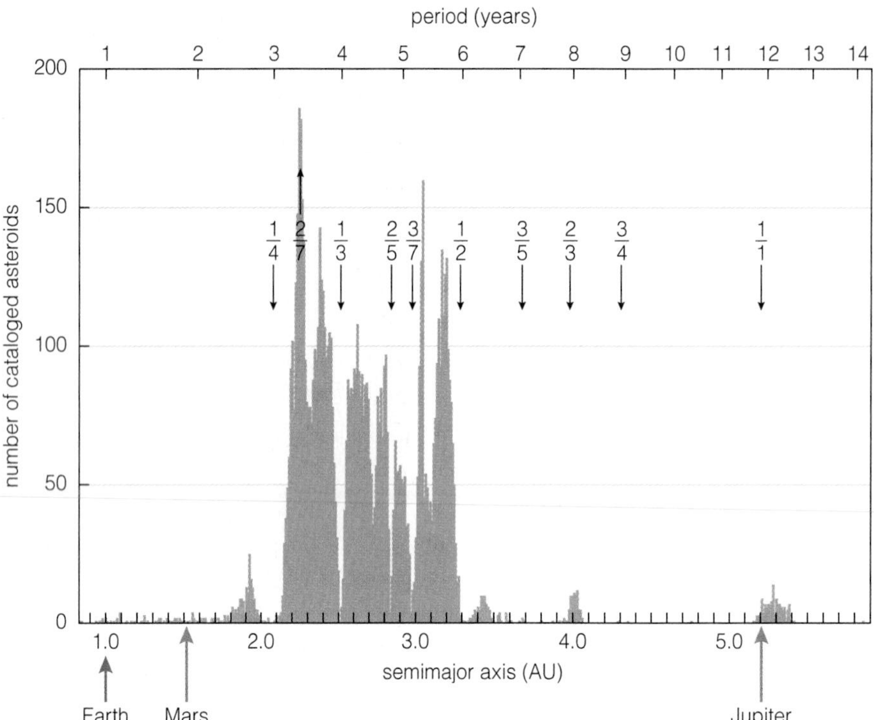

FIGURE 12.2 This graph shows the numbers of asteroids with different average distances (semimajor axes) from the Sun. Distances where we see few if any asteroids represent gaps. The ratio labeling each gap shows how the orbital period of objects at this distance compares to the 12-year orbital period of Jupiter. For example, the label $\frac{1}{3}$ means the orbital period of objects with this average distance is $\frac{1}{3}$ Jupiter's orbital period, or about 4 years.

could not collect together to form a planet. Over the past 4.5 billion years, the ongoing orbital disruptions have gradually kicked pieces of this "unformed planet" out of the asteroid belt altogether. Once booted from the main asteroid belt, these objects eventually either crash into a planet or a moon or are flung out of the solar system. (In extremely rare cases, they may become "captured" moons.) Thus, the asteroid belt slowly loses mass, explaining why the total mass of all its asteroids is now less than that of any terrestrial planet.

TIME OUT TO THINK *Gravitational forces prevented asteroids from accreting into a planet and prevented Saturn's ring particles from accreting into a single moon. How are the processes that affect asteroids and Saturn's rings similar? How are they different?*

The overlapping elliptical orbits of the remaining asteroids still lead to a major collision somewhere in the asteroid belt every 100,000 years or so. (By comparison, collisions occur somewhere within Saturn's rings many times per second, which has circularized the orbits and flattened the rings.) Thus, larger asteroids continue to be broken into smaller ones, with each collision also creating numerous dust-size particles. The asteroid belt has been grinding itself down in this way for over 4 billion years and will continue to do so for as long as the solar system exists.

Most of the asteroids found outside the asteroid belt—including the *Earth-approaching asteroids* that pass near Earth's orbit—are probably "impacts waiting to happen." But asteroids can safely congregate in two stable zones outside the main belt. These zones are found along Jupiter's orbit 60° ahead of and behind Jupiter (see Figure 12.1). The asteroids found in these two zones are called the *Trojan asteroids,* and the largest are named for the mythological Greek heroes of the Trojan War. The Trojan asteroids are stable because of a different type of orbital resonance with Jupiter. In this case, any asteroid that wanders away from one of theses zones is nudged back *into* the zone by Jupiter's gravity. The existence of such stable orbital resonances was first predicted by the French mathematician Joseph Lagrange more than 200 years ago—135 years before the discovery of the first asteroid in such an orbit.[2] It is possible that the population of Trojan asteroids is as large as that of main-belt asteroids, but the greater distance to the Trojan asteroids makes them more difficult to study or even count from Earth.

Learning About Asteroids

The general locations and orbits of asteroids tell us a lot about their origin, but we would like to know many other properties, such as masses, sizes, densities, shapes, and compositions. The first step in learning about an asteroid is determining its precise orbit. Asteroids are recognizable in telescopic images because they move noticeably relative to the stars in just a short time (Figure 12.3). Once we detect an asteroid, we can calculate its orbit by applying Kepler's laws [Section 6.3]—even if we've observed only a small part of one complete orbit.

Most asteroids appear as little more than points of light even to our largest Earth-based telescopes, so the best way to study an asteroid in depth is to send a spacecraft out to meet it. Spacecraft have visited only a handful of asteroids close-up in their native environment of the asteroid belt and a few other presumed asteroids "in captivity" orbiting other planets (e.g., Phobos and Deimos orbiting Mars, and the backward outer satellites of Jupiter). The Galileo spacecraft on its way to Jupiter photographed the asteroid Gaspra (Figure 12.4a) and an asteroid pair known as Ida and Dactyl (Figure 12.4b). The Near-Earth Asteroid Rendezvous mission (NEAR) photographed the asteroid Mathilde (Figure 12.4c). (NEAR is scheduled to match orbits with another asteroid, called Eros, for in-depth study beginning in 1999.) These spacecraft images reveal battered worlds that are probably fragments of larger asteroids shattered by collisions. For example, the abundance of craters on Gaspra's surface suggests that it was born in a violent collision a few hundred million years ago.[3]

[2]The locations of these stable zones are called *Lagrange points.* Similar Lagrange points lie 60° ahead of and behind the Moon in its orbit around the Earth. These locations have been suggested for space colonies because of their stability. Three additional Lagrange points lie along the line connecting any two massive objects (such as the Earth and the Moon). They are not quite stable but also are good places to "park" spacecraft because staying in place doesn't take much fuel.

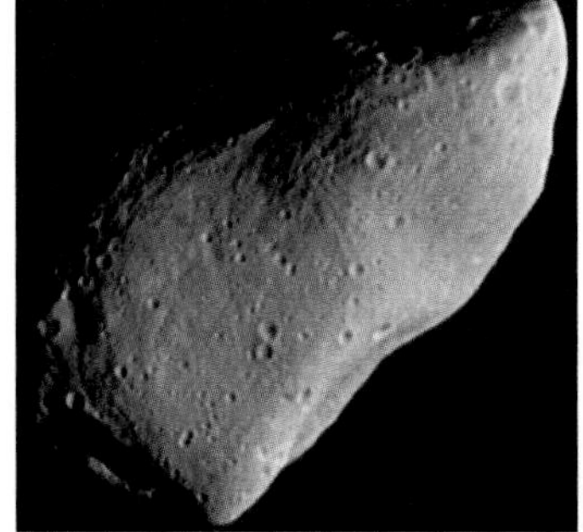

a Gaspra (16 km across)

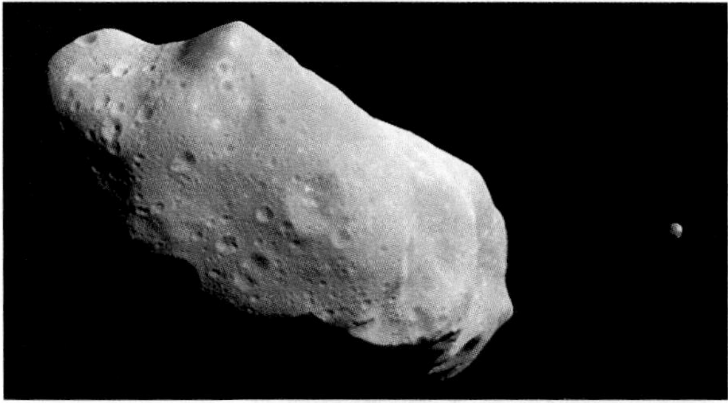

b Ida (53 km) and its tiny moon, Dactyl

c Mathilde (59 km)

FIGURE 12.4 Close-up views of three asteroids.

FIGURE 12.3 Because asteroids orbit the Sun, they move relative to the stars just as planets do. In this photograph, stars show up as white dots, and the motion of an asteroid makes it show up as a short streak.

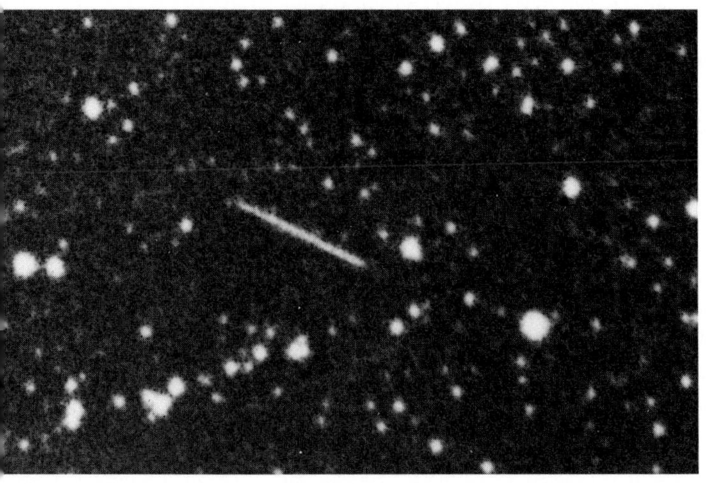

[3]We must carefully account for an object's "environment" when we determine ages from crater abundances. For example, far more impact craters are expected over a given area within a given amount of time for objects in the asteroid belt than on Mars or the Moon.

Ida appears to be somewhat older, apparently fragmented from a larger object about 1–2 billion years ago. Mathilde, with its greater abundance of craters, is even older.

We can directly measure the size of an asteroid only when we've seen it close up or in the rare cases when an asteroid is large enough to be resolved in a telescope. (We calculate physical size from the asteroid's angular size and distance [Section S2.1].) We can also determine asteroid sizes by analyzing their light. The total amount of sunlight that an asteroid reflects depends on its size and albedo [Section 10.2]. For example, if two asteroids have the same albedo, the one that reflects more light must be larger in size. We can determine an asteroid's albedo by taking advantage of the fact that asteroids not only reflect visible light from the Sun, but also emit their own infrared thermal radiation. Because the temperature of an asteroid depends primarily on how much sunlight it absorbs, comparing its infrared thermal emission to its visible light reflection tells us its albedo. For example, an asteroid that is very reflective (high albedo) will not absorb much solar energy, so it will be cold and will give off very little infrared thermal emission in comparison to its visible-light reflection (Figure 12.5). By using this technique, we have compiled good size estimates for more than a thousand asteroids.

FIGURE 12.5 By comparing measurements of reflected sunlight and thermal emission in the infrared, we can learn about asteroid albedos. Once we know an asteroid's albedo, we can determine its size.

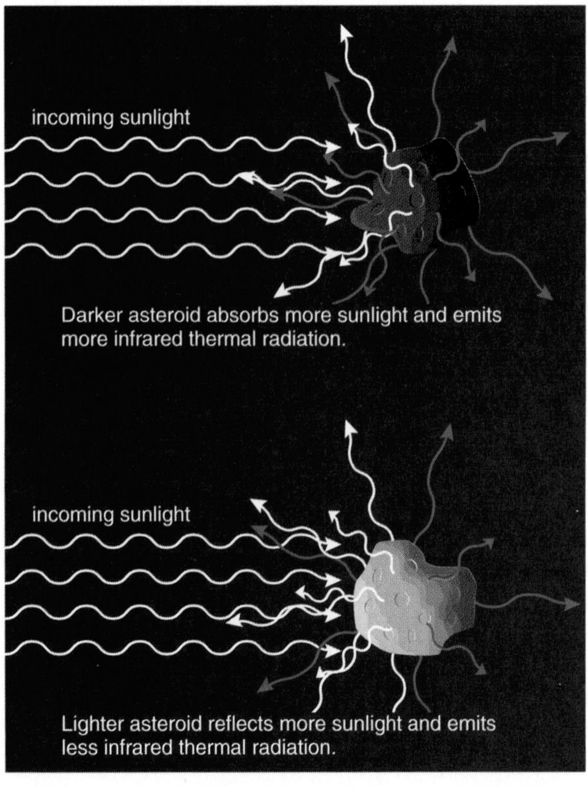

TIME OUT TO THINK *Imagine that you are observing a bright newly discovered asteroid. What additional observations would you need to determine whether it is very large, very close, or simply light-colored?*

Asteroid densities offer valuable insights into their origin and makeup but require difficult measurements of both size and mass. Ida is only about 50 kilometers across, but its weak gravity holds the tiny moon Dactyl in orbit—allowing us to apply Newton's version of Kepler's third law [Section 6.4] to find the mass of Ida. Using Ida's mass and volume (determined from photos), we find its average density to be about 2.5 g/cm^3. NEAR scientists calculated Mathilde's density using the small gravitational tug on the spacecraft itself, finding a value of around 1.5 g/cm^3, so low that the asteroid must contain substantial amounts of water or empty space (or both). We can estimate the masses of larger asteroids from their minute gravitational influences on planets; by combining these mass estimates with estimates of their volumes, we find that their densities range from 2 to 4 g/cm^3. The range of asteroid densities reveals substantial differences in composition, as expected from their formation in different parts of the solar nebula. Some asteroids have undergone significant changes since their formation, as we will discuss shortly. These changes can further alter their compositions.

Determining the shapes of asteroids is even more challenging. One technique involves looking for brightness variations as an asteroid rotates. A uniform spherical asteroid will not change its brightness as it rotates, but a potato-shaped asteroid will reflect more light when it presents its larger side toward the Sun and our telescope (Figure 12.6). Astronomers can also determine asteroid shapes by examining the shape of the shadows cast in starlight when an asteroid happens to pass right in front of a star. By combining the exact times of the star's disappearance and reappearance as clocked with telescopes in several different locations around the world, they can calculate the shape and size of the shadow and therefore the outline of the asteroid itself. Shape measurements show that only the largest asteroid (Ceres, 940-km diameter) is approximately spherical. The next two largest asteroids are somewhat oblong (Pallas, 540 km across, and Vesta, 510 km across). Smaller asteroids like Gaspra and Ida have still odder shapes. The strength of gravity on these smaller asteroids is evidently less than the strength of the rock.

In a few cases, astronomers have used radar to obtain even more detailed shape information about

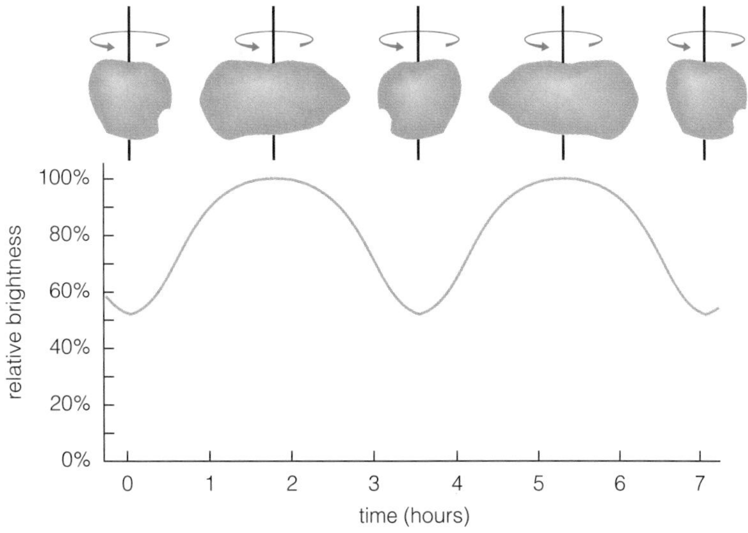

FIGURE 12.6 Gaspra's irregular shape leads to large brightness variations as it rotates. The diagrams at the top show its appearance from Earth at different times during its rotation, and the graph shows the corresponding brightness at those times. Note that Gaspra appears brighter when we are looking at a larger face because it reflects more sunlight.

asteroids that have passed close to the Earth. A radio dish sends out radio "pings" that bounce off the asteroid and reflect back to Earth.[4] By analyzing the reflected signals returned as the asteroid rotates, we can create a three-dimensional image of the asteroid (Figure 12.7). Surprisingly, radar images indicate that some asteroids are actually two objects held in contact with each other by a weak gravitational attraction. Apparently, this attraction is strong enough to hold the objects together, but not strong enough to force them to merge into a single body.

Finally, we can determine the compositions of asteroids through spectroscopy [Section S2.3]. Thousands of asteroids have been analyzed in this way, and they fall into three main categories: About 75% of asteroids are very dark and show absorption bands from carbon-rich materials. Another 15% are more reflective and show absorption bands characteristic of rocky materials. The remaining 10% include asteroids with spectral characteristics of metals such as iron. These differences provide important clues to their origins, as we will discuss shortly.

12.3 Meteorites

Spectroscopy is a pretty good method of determining the composition of an asteroid, but wouldn't it be nice to study an asteroid sample in a laboratory? No spacecraft has yet landed on an asteroid, but we have pieces of asteroids nonetheless—the rocks called *meteorites* that fall from the sky.

The reality of rocks falling from the sky wasn't always accepted. Stories of such events arose occasionally in human history, sometimes even influencing history. For example, the ancient Greek scientist Anaxagoras (500–428 B.C.) came up with his idea that planets and stars are flaming rocks in the heavens after hearing of such "fallen stars." But later scientists were skeptical of such stories. Thomas Jefferson (who worked in science as well as politics), upon hearing of a meteorite fall in Connecticut, reportedly said, "It is easier to believe that Yankee professors would lie than that stones would fall from heaven."

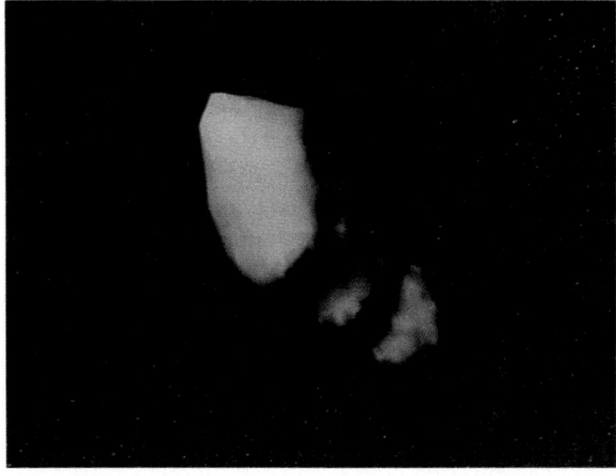

FIGURE 12.7 Radar image of Toutatis, a 5-km-long asteroid that passed only 3 million km from the Earth in 1992.

[4]The intense radar beacons beamed at these asteroids are probably the strongest human signals sent from Earth. Perhaps a distant alien civilization will someday pick up these signals and wonder what we were doing.

FIGURE 12.8 Large craters on Earth show that major impacts have played an important role in even relatively recent geological history. The Manicouagan Lakes crater in Canada formed about 200 million years ago and has been heavily eroded by glaciers since then.

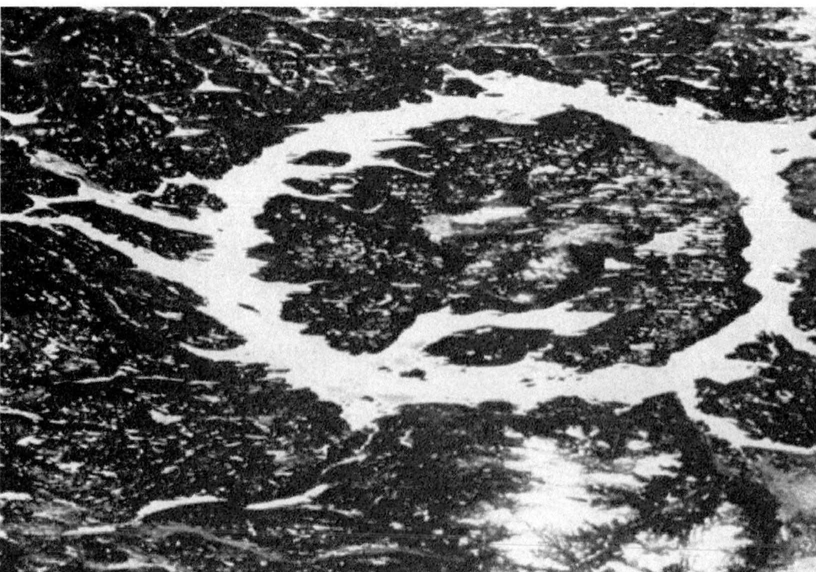

Today we know that rocks really do sometimes fall from the heavens. Some are large enough to do significant damage, blasting huge craters upon impact. Despite the forces of erosion, more than 100 impact craters can still be seen on Earth (Figure 12.8). Even more rarely, an impact may alter the entire history of our planet. It now seems likely that an impact was directly responsible for a chain of events that killed off the dinosaurs and paved an evolutionary path for humans.

The precise origin of meteorites was long a mystery, but today we know that most come from the asteroid belt. In recent decades, a few meteorite falls have been captured on film, allowing scientists to calculate the orbits that led them to crash to Earth. The results clearly show that the meteorites originated in the asteroid belt.

Rocks from space continue to rain down on Earth, and observers find a few meteorites every year by following the trajectories of very bright meteors called *fireballs*. Meteorites are often blasted apart in their fiery descent through our atmosphere, scattering fragments over an area several kilometers across. A direct hit on the head by a meteorite would be fatal, but there are no reliable accounts of human deaths from meteorites. However, meteorites have injured at least one human (by a ricochet, not a direct hit), damaged houses and cars, and killed animals

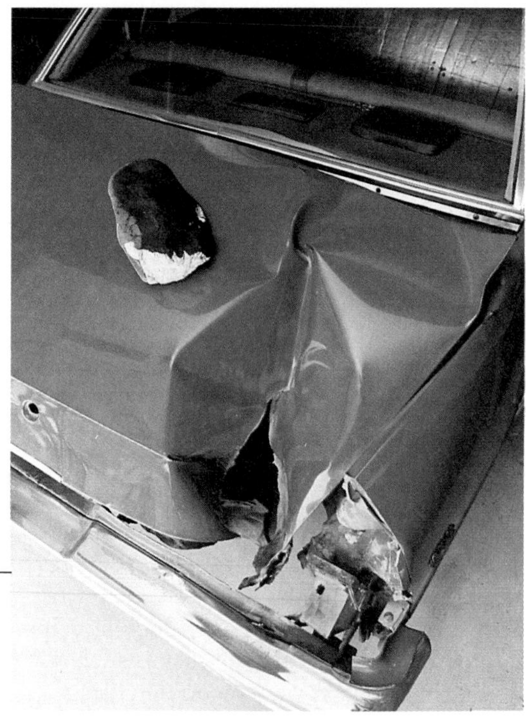

a

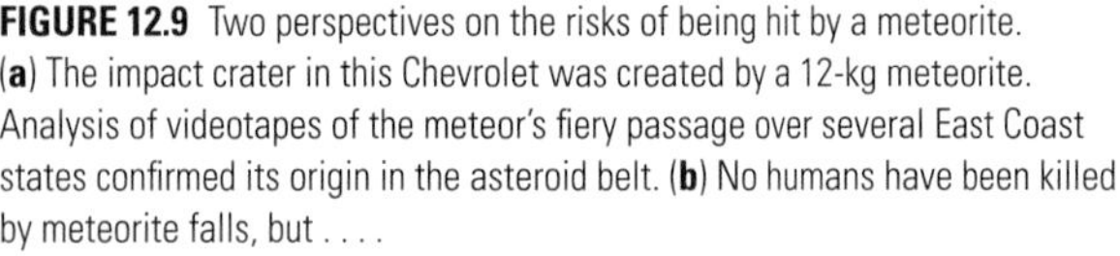

FIGURE 12.9 Two perspectives on the risks of being hit by a meteorite. (**a**) The impact crater in this Chevrolet was created by a 12-kg meteorite. Analysis of videotapes of the meteor's fiery passage over several East Coast states confirmed its origin in the asteroid belt. (**b**) No humans have been killed by meteorite falls, but . . . .

PEANUTS

b

a

Stony primitive meteorite: metal flakes intermixed with rocky material.

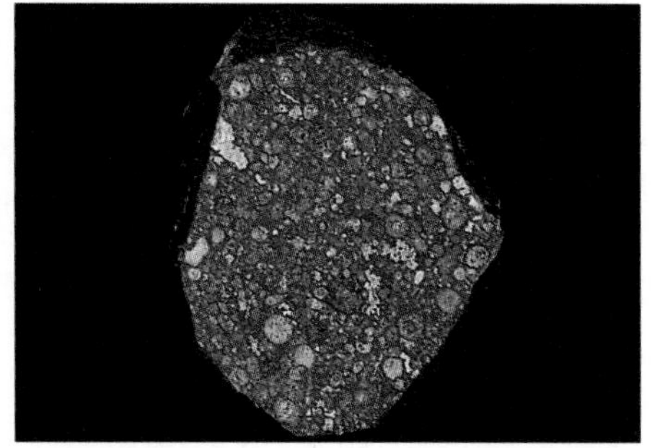

Carbon-rich primitive meteorite: similar to meteorite at left, but with carbon compounds added in.

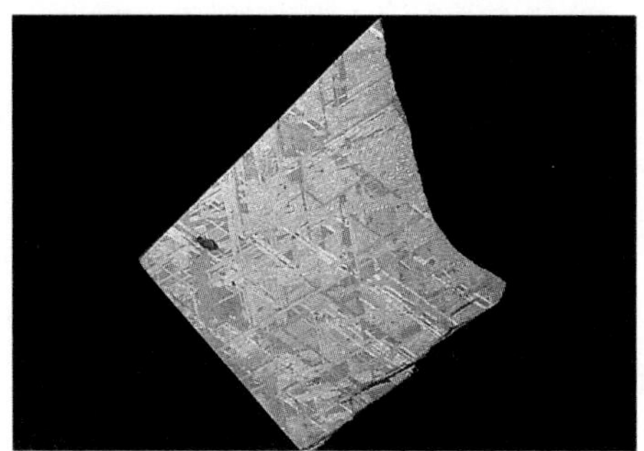

b

Differentiated iron meteorite: similar to planetary core.

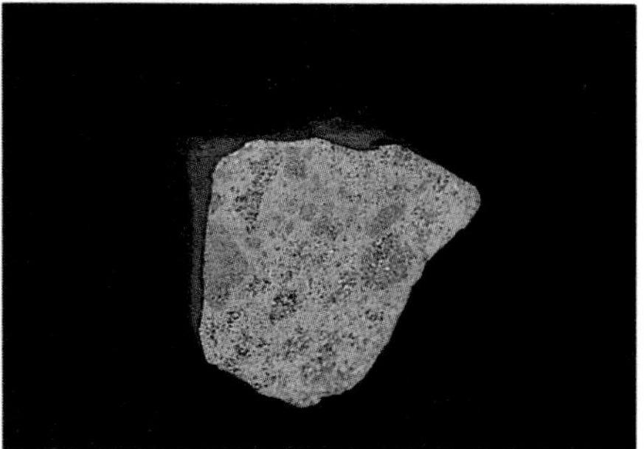

Differentiated stony meteorite: similar to volcanic rocks.

FIGURE 12.10 (**a**) Two examples of primitive meteorites. (**b**) Two examples of processed meteorites.

(Figure 12.9). Most meteorites fall into the oceans, which cover three-quarters of Earth's surface area.

Many more meteorites—ones that fell to Earth at some time in the past—are found by accident or by organized meteorite searches. Antarctica offers the greatest treasure-trove of meteorites, not because more fall there but because the icy surface makes meteorites easier to identify and the movement of glaciers carries rocky debris into a few small areas. The majority of meteorites in museums and laboratories come from Antarctica.

Comparing Meteorites

Most meteorites are difficult to distinguish from terrestrial rocks without a detailed scientific analysis, but a few clues can help. Meteorites are usually covered with a dark, pitted crust resulting from their fiery passage through the atmosphere. Some can be readily distinguished from terrestrial rocks by their high metal content.[5] The ultimate judge of extraterrestrial origin is laboratory analysis of the rock's composition. Terrestrial rocks may differ from one another in composition, but all contain isotopes of particular elements in roughly the same ratios. Thus, meteorites can be identified by their very different isotope ratios. The presence of certain rare elements such as iridium also indicates extraterrestrial origin; nearly all of Earth's iridium sank to the core long ago and hence is absent from surface rocks.

The tens of thousands of known meteorites fall into two basic categories. The vast majority of meteorites appear to be composed of a random mix of flakes from the solar nebula, and radioactive dating shows that they formed at the same time as the solar system itself—about 4.6 billion years ago [Section 8.7]. These **primitive meteorites** are our best source of information about conditions in the solar nebula (Figure 12.10a). Most primitive meteorites are composed of rocky minerals with an important difference from Earth rocks: A small but noticeable fraction of pure metallic flakes is mixed in. (Iron in Earth rocks is chemically bound in minerals.) Some primitive meteorites also contain substantial amounts of carbon compounds, and some even contain a small amount of water.

A much smaller group of meteorites appears to have undergone substantial change since the formation of the solar system, and radioactive dating shows some of these meteorites to be younger than the primitive meteorites. We call this group **processed meteorites** because they apparently were once part

[5]Most types of meteorites contain enough metal to attract a magnet hanging on a string. If you suspect that you have found a meteorite, most science museums will analyze a small chip free of charge.

of a larger object that "processed" the original material of the solar nebula into another form (Figure 12.10b).[6] Some processed meteorites resemble the Earth's core in composition: an iron/nickel mixture with trace amounts of other metals These meteorites have a density of about 7 g/cm^3 and played an important role in human history: As the most accessible form of metallic iron available on the Earth's surface, they helped humans make the transition from the bronze age to the iron age. Other processed meteorites have much lower densities and are made of rocks more similar to the Earth's crust or mantle. A few even have a composition remarkably close to that of the basalts erupted from terrestrial volcanoes [Section 9.4].

The Origin of Meteorites

We've seen that primitive meteorites may be either rocky or carbon-rich. Why are there two varieties? The answer lies in where they formed in the solar nebula. Carbon compounds condense only at the relatively low temperatures that were found in the solar nebula beyond about 3 AU from the Sun. Thus, we conclude that carbon-rich meteorites come from the outer portion of the asteroid belt lying beyond about 3 AU from the Sun. Laboratory spectra of these carbon-rich primitive meteorites confirm that they are a good match to the compositions of asteroids in the outer part of the asteroid belt. The remainder of the primitive meteorites lack carbon compounds and thus presumably formed closer to the Sun in what is now the inner part of the asteroid belt. Curiously, most of the meteorites collected on Earth are of the rocky variety, even though most asteroids are of the carbon-rich variety. Evidently, orbital resonances are more effective at pitching asteroids our way from the inner part of the asteroid belt. All the primitive meteorites apparently survived untouched since the beginning of the solar system—at least until they came crashing down to Earth.

The processed meteorites tell a more complex story. Their compositions appear similar to the cores, mantles, or crusts of the terrestrial worlds. Thus, they must be fragments of worlds that underwent *differentiation* just like the terrestrial worlds [Section 9.2]. That is, they are fragments of worlds that must have been heated to high enough temperatures to melt inside, allowing metals to sink to the center and rocks to rise to the surface. The processed meteorites with basaltic compositions must come from lava flows that occurred on the surfaces of some of the larger asteroids during a time when they had active volcanism. Perhaps as many as a dozen large asteroids were geologically active shortly after the formation of the solar system, but radioactive dating of processed meteorites suggests that these worlds soon became inactive.

Basaltic meteorites may simply have been chipped off the surface of a large asteroid by relatively small collisions. (By comparing meteorite and asteroid spectra, scientists have identified the asteroid Vesta as the likely source of some of these meteorites.) In contrast, the processed meteorites with core-like or mantle-like compositions must be fragments of worlds that completely shattered in collisions. These shattered worlds not only lost a chance to grow into planets, but they sent many fragments onto collision courses with other asteroids and other planets, including the Earth. Thus, processed meteorites essentially offer us an opportunity to study a "dissected planet."

If fragments such as the basaltic meteorites can be chipped off the surface of an asteroid, is it possible that they might also be chipped off a larger object such as a moon or a planet? In fact, a few processed meteorites have been found that don't appear to match the compositions of asteroids but instead appear to match the composition either of the Moon or of Mars. We now believe that some of these in fact were chipped off the Moon in impacts and that others were chipped off Mars. The analysis of these *lunar meteorites* and *Martian meteorites* is providing new insights into the conditions on the Moon and Mars. In at least one case, a Martian meteorite may be offering us clues about whether life once existed on Mars [Section 13.6].

12.4 Comets

Humans have watched comets in awe for millennia. The occasional presence of a bright comet in the night sky was hard to miss before the advent of big cities, inexpensive illumination, and television. In some cultures, these rare intrusions into the unchanging heavens foretold bad or good luck, and in most there was little attempt to interpret the event in astronomical terms. (Recall that comets were generally thought to be atmospheric phenomena until proved otherwise by Tycho Brahe [Section 6.3].) Nowadays, these leftover icy planetesimals can teach us about formation processes in the outer solar system, just as asteroids teach us about the inner solar system.

[6]The technical term for primitive meteorites is *chondrites,* because they contain seed-shaped "chondrules" that may be solidified droplets splashed out in the accretion process. Processed meteorites underwent major changes that destroyed the chondrules, so this category is called *achondrites.*

a

b

FIGURE 12.11 Brilliant comets can appear at almost any time, as demonstrated by the back-to-back appearances of (**a**) comet Hyakutake in 1996 and (**b**) comet Hale–Bopp in 1997, photographed at Mono Lake.

After Kepler developed his laws of planetary motion, other astronomers applied them to comets and found extremely eccentric orbits. Some comets visit the inner solar system only once before being ejected into interstellar space, others return on orbits spanning thousands of years, and a few come by more frequently. The most famous is Halley's comet, named for the English scientist Edmund Halley (1656–1742). Halley did not "discover" his comet but rather gained fame for recognizing that a comet seen in 1682 was the same one seen on a number of previous occasions. He used Newton's law of gravitation to calculate the comet's 76-year orbit and, in a book published in 1705, predicted that the comet would return in 1758. The comet was given his name when it reappeared as predicted, 16 years after his death. Halley's comet has returned on schedule ever since, including twice in the twentieth century—spectacularly in 1910 when it passed near the Earth, and unimpressively in 1986 when it passed by at a much greater distance. It will next visit in 2061.

Today, new comets carry a form of good luck. Thousands of amateur astronomers eagerly scan the skies with their telescopes and binoculars in hopes of discovering a new comet. The first discoverers (up to three) who report their findings to the International Astronomical Union have the comet named after them. Several comets are discovered every year, and most have never been seen before by human eyes. In 1996 and 1997, we were treated to back-to-back brilliant comets: comet Hyakutake and comet Hale–Bopp (Figure 12.11).

The Flashy Lives of Comets

Comets are icy planetesimals from the outer solar system. Their composition has been described as "dirty snowballs": ices mixed with rocky dust. Comets are completely frozen when they are far from the

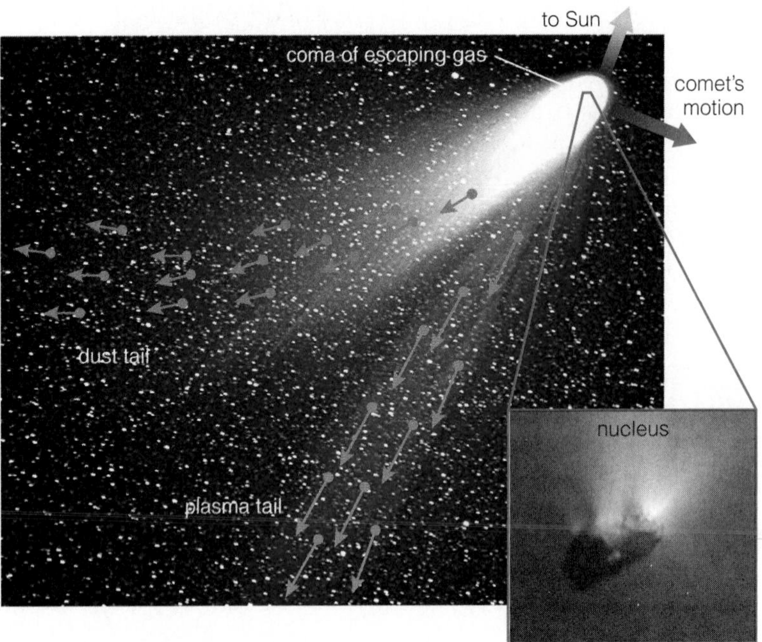

FIGURE 12.12 Anatomy of a comet. The inset photo is the nucleus of comet Halley photographed by the Giotto spacecraft; the coma and tails shown are those of comet Hale–Bopp.

Sun, and most are just a few kilometers across. But a comet takes on an entirely different appearance when it comes close to the Sun (Figure 12.12). The dirty snowball is the comet's **nucleus.** The sublimation of ices in the nucleus creates a rapidly escaping dusty atmosphere called the **coma,** along with a **tail** that points away from the Sun regardless of which way the comet is moving through its orbit. In most cases, we see two tails associated with comets: a **plasma tail** made of ionized gas, and a **dust tail** made of small solid particles. Although much of comet science focuses on the dramatic tail and coma, we will begin our closer look at comets with the humbler nucleus.

Comet nuclei are rarely observable with telescopes, as they are quite small and are shrouded by the dusty coma. The European Space Agency's Giotto spacecraft provided our best view of a nucleus when it passed within 600 kilometers of comet Halley's nucleus in 1986. (The spacecraft was named for the Italian painter Giotto, who apparently was inspired by a passage of Halley's comet to include a comet in a 1304 religious painting.) The flyby revealed a dark, lumpy, potato-shaped nucleus about 16 kilometers long and 8 kilometers in width and depth. Despite its icy composition, comet Halley's nucleus is darker than charcoal, reflecting only 3% of the light that falls on it. (It doesn't take much rocky soil or carbon-rich material to darken a comet.) The ice is not packed very solidly; some estimates of the density of Halley's nucleus are considerably less than 1 g/cm^3—suggesting that the nucleus is part ice and part empty space.

How can such a small nucleus put on such a spectacular show? Comets spend most of their lives in the frigid outer limits of our solar system. As a comet accelerates toward the Sun, its surface temperature increases and ices begin to sublime into gaseous form. By the time the comet comes within about 5 AU of the Sun, sublimation begins to form a noticeable atmosphere that easily escapes the comet's weak gravity. The escaping atmosphere drags away dust particles that were mixed with the ice and begins to create the dusty coma around the nucleus. Sublimation rates increase as the comet approaches the Sun, and gases jet away from patches on the nucleus at speeds of hundreds of meters per second (Figure 12.13). The jets can make faint pinwheel patterns within the coma, due to the slow rotation of comet nuclei. The rapid sublimation carries away so much dust that a spacecraft flyby is dangerous. The Giotto spacecraft, which approached Halley's nucleus at a relative speed of nearly 250,000 km/hr (70 km/sec), was sent reeling by dust impacts just minutes before closest approach, breaking radio contact with Earth.

As gas and dust move away from the comet, they are influenced by the Sun in different ways. The gases are ionized by ultraviolet photons from the Sun and then carried straight outward away from the Sun with the solar wind at speeds of hundreds of kilometers per second. These ionized gases form the *plasma tail,* which may extend hundreds of millions of kilometers away from the Sun. Dust-size particles experience a much weaker push caused by the pressure of sunlight itself (called *radiation pressure* [Section 16.4]), so the *dust tail* is swept away from the Sun in a slightly different direction (see Figure 12.12). The comet wheels around the Sun, always keeping its tails of dust and plasma pointed roughly away from the Sun. Comets also eject some larger, pebble-size particles that are not pushed away from the Sun. These form another, invisible, "tail" extending along the comet's orbit and are the particles responsible for meteor showers [Section 1.5].

After the comet loops around the Sun and begins to head back outward, sublimation declines, the coma dissipates, and some of the dust settles back to the surface. Nothing happens until the comet again comes sunward—in a century, a millennium, a million years, or perhaps never.

Active comets cannot last forever. In a close pass by the Sun, a comet may shed a layer 1 meter thick. Dust that is too heavy to escape accumulates on the surface. This thick dusty deposit helps make comets dark and may eventually block the escape of interior gas that makes comets so phenomenal. It is estimated that a comet loses about 0.1% of its volatiles on every pass around the Sun. No one is certain what happens after the volatiles stop escaping. Either the remaining dust disguises the dead comet as an asteroid

FIGURE 12.13 Comets exist as bare nuclei over most of their orbits and grow a coma and tails only when they approach the Sun. (Diagram not to scale.)

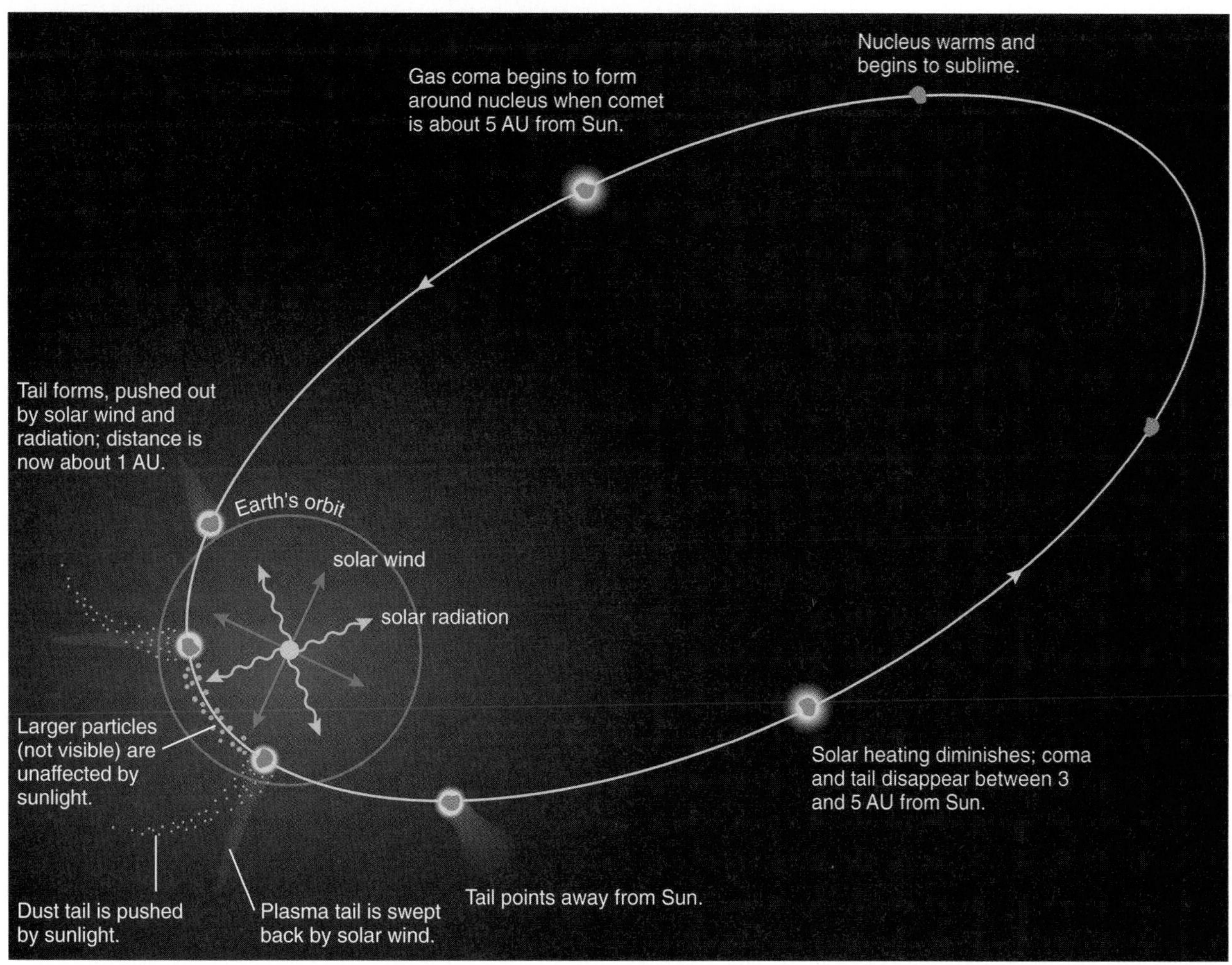

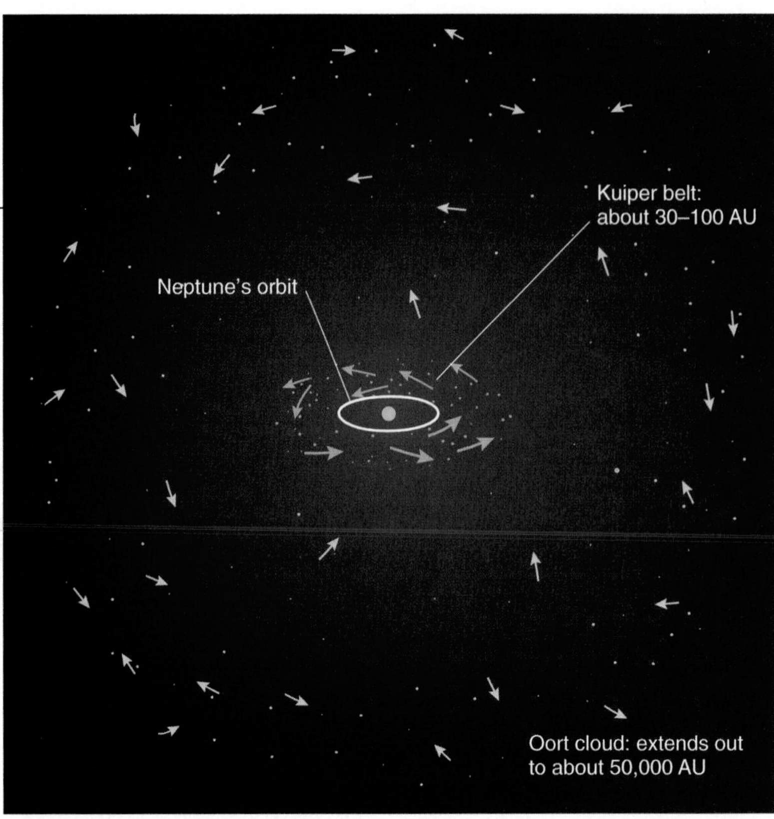

FIGURE 12.14 The Kuiper belt and the Oort cloud. Arrows indicate representative orbital motions of objects in the two regions. (Figure is not to scale.)

until it crashes into a planet or is ejected from the solar system, or the comet comes "unglued" without the binding effects of ices and simply disintegrates along its orbit.

Spectra of the coma and tail show emission lines and bands from water molecules and other hydrogen compounds that condensed in the outer regions of the solar nebula. They also show emission from carbon dioxide and carbon monoxide that condensed only in the coldest regions of the solar nebula, as well as a wide variety of more complex molecules. These emissions do not give a complete picture of a comet's ingredients, since sunlight and chemical reactions have altered the composition. Organic chemicals (containing carbon, hydrogen, oxygen, and nitrogen) may also be present, suggesting that comets carry the necessary building blocks for life. Comets may even contain some grains of *interstellar dust* that formed *before* our solar system. A complete understanding of the composition of comets still eludes us, but it may have critical implications for understanding their formation in the outer solar nebula and their importance to the planets.

The Origin of Comets

Comets are a rarity in the inner solar system, with only a handful coming within the orbit of Jupiter at any one time. However, because comets cannot last for very many passes by the Sun, comets must come from enormous reservoirs at much greater distances: the Kuiper belt and the Oort cloud (Figure 12.14). The existence of these reservoirs was first inferred by analysis of the orbits of comets that pass close to the Sun. Most comet orbits fit no pattern—they do not orbit the Sun in the same direction as the planets, and their elliptical orbits can be pointed in any direction. Comets Hyakutake and Hale–Bopp both fall into this class. It is believed that such comets are visitors from the Oort cloud. Other comets have a pattern to their orbits: They travel around the Sun in the same plane and direction as the planets, though on very elliptical orbits that take them beyond the orbit of Pluto. They also return much more frequently—typically every few decades or centuries—and are thought to be visitors from the Kuiper belt. Comets from either the Oort cloud or the Kuiper belt may occasionally pass near enough to a planet to have their orbits significantly altered, sometimes becoming trapped in the inner regions of the solar system. Halley's comet must have passed near a jovian planet in the distant past, causing it to end up on its current 76-year orbit of the Sun [Section 6.6]. Gravitational encounters with planets can also eject comets from the solar system altogether or send them crashing into the planets or the Sun.

Based on the number of comets that fall into the inner solar system, the Oort cloud must contain about a trillion (10^{12}) comets. However, Oort cloud comets normally orbit the Sun at such great distances—up to about 50,000 AU—that we have so far seen only the rare visitors to the inner solar system. The comets of the Oort cloud probably formed in the vicinity of the jovian planets and were flung into their large, random orbits by gravitational encounters with these planets. Oort cloud comets are so far from the Sun that the gravity of neighboring stars can alter their orbits, preventing some comets from ever returning near the planets and sending others plummeting toward the Sun.

In contrast, the comets of the Kuiper belt probably still lie in the same general region in which they formed, which is why their orbits match the pattern of the solar nebula as a whole. In the 1990s, astronomers began using the largest telescopes to search for Kuiper belt objects in their native environment rather than on their excursions into the inner solar system. The first discovery, given the name 1992QB1, orbits beyond Pluto with an orbital period of 296 years. It is a comet approximately 280 kilometers across. Subsequent searches have identified dozens of other objects that match the predicted orbital properties of Kuiper belt comets—fairly circular orbits in the same direction as the planets, with small orbital tilts. Gravitational nudges from the planets can cause Kuiper belt comets to fall in toward the Sun. The search for Kuiper belt comets is still in its infancy, and the 60 or so discovered so far are probably among the largest. Nevertheless, the work to date suggests that at least 100,000 comets more than 100 km across must populate the Kuiper belt, along with a billion objects around 10 km across (the typical size of comets that enter the inner solar system). Thus, the total mass of the Kuiper belt is far greater than that of the asteroid belt. Spectroscopy has provided some preliminary information on the composition of Kuiper belt objects. Although they are undoubtedly rich in ices, the Kuiper belt comets are covered with dark, carbon-rich compounds.

Planetary scientists would dearly love to compare samples of the ancient materials frozen inside the two types of comet. But comet samples falling to Earth burn up in the atmosphere and do not become meteorites, so we are limited to distant spectroscopic analysis of the gases sublimed off the nucleus. Astronomers cannot yet tell if Oort cloud comets show signs of having formed in closer, warmer, and denser regions of the solar nebula than their Kuiper belt counterparts. Spacecraft missions to these frigid regions of the solar system are prohibitive, so astronomers must be content to wait for comets to come to them and watch as the Sun dissects them meter by meter.

What are the limits on Kuiper belt objects? An odd object called *Chiron* orbits the Sun between Saturn and Uranus. About 170 km across, Chiron once was classified as a surprisingly distant asteroid. However, Chiron recently developed a coma, suggesting that it is actually a comet from the Kuiper belt. An even larger comet, designated 1996T066, may be 600 km across. If a comet can be 600 km across, could one be larger still? It's time to consider the odd planet Pluto.

12.5 Pluto: Lone Dog or Part of a Pack?

Pluto was discovered in 1930 by American astronomer Clyde Tombaugh, culminating a search that began not long after the discovery of Neptune in 1846 [Section 2.1]. Recall that Neptune's existence and location were predicted before its actual discovery by mathematical analysis of irregularities in the orbit of Uranus. Further analysis of Uranus's orbit suggested the existence of a "ninth planet," and Pluto was found only 6° from the predicted position of this planet. Initial estimates suggested that Pluto was much larger than Earth, but successively more accurate measurements derived ever-smaller sizes. We now know that Pluto has a radius of only 1,195 kilometers and a mass of just 0.0025 Earth mass—making it far too small to affect the orbit of Uranus. In fact, the discovery of Pluto near the predicted position of the "ninth planet" was coincidental; Uranus's supposed orbital irregularities were apparently just errors in measurement. Pluto might have escaped detection for decades without those errors.

Pluto has long seemed to be a misfit among the planets, fitting into neither the terrestrial nor the jovian category. Its 248-year orbit is also unusually elliptical and significantly tilted relative to the ecliptic (see Figure 2.13). Near perihelion, it actually comes closer to the Sun than Neptune—such was the case between 1979 and 1999. At aphelion, it is 50 AU from the Sun, putting it far beyond the realm of the jovian planets. Despite the fact that Pluto sometimes comes closer to the Sun than Neptune, there is no danger of the two planets colliding: For every two Pluto orbits, Neptune circles the Sun three times. Because of this stable *orbital resonance,* whenever Pluto is near Neptune's orbit, Neptune is always a safe distance away. The two will probably continue their dance of avoidance until the end of the solar system.

Pluto and Its Moon

Pluto's great distance and small size make it difficult to study, but the task became much easier with the discovery of Pluto's moon Charon in 1978. The original discovery was made when astronomers noticed a "bump" on Pluto's blurry image that moved from side

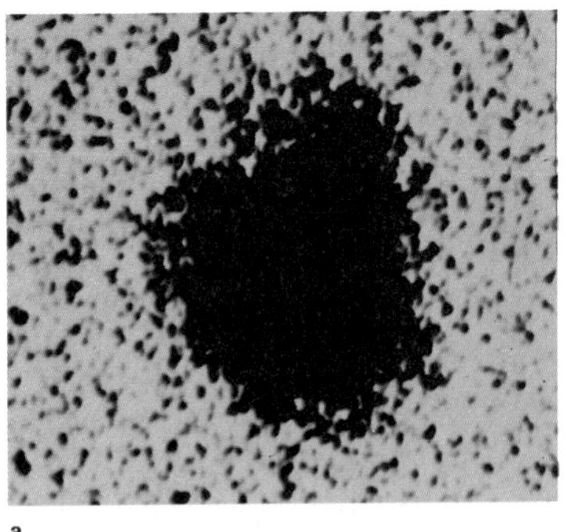
a

b

FIGURE 12.15 (**a**) The bump on the upper right of this fuzzy, black image of Pluto might not seem significant, but astronomers noticed that the bump moved from one side of Pluto to the other and back in six days. (**b**) The bump is actually Pluto's moon Charon, clearly resolved in this Hubble Space Telescope image. The object on the left is Pluto, with Charon on the right.

to side over a 6.4-day period (Figure 12.15a). Today we have much clearer pictures of Pluto and Charon from the Hubble Space Telescope (Figure 12.15b). The discovery of the moon enabled astronomers to accurately determine the mass of Pluto by applying Newton's version of Kepler's third law.[7] They also learned that Pluto's rotation axis is tipped 118° relative to its orbit, making it the third planet (in addition to Venus and Uranus) that rotates backward relative to the majority of planets.

Pluto's moon was discovered just in time, because astronomers soon learned that Charon's orbit was about to go edge-on as seen from the Earth—something that happens only every 124 years. From 1985 to 1990, astronomers carefully monitored the combined brightness of Pluto and Charon as they alternately eclipsed each other every few days. Detailed analysis of brightness variations allowed calculation of accurate sizes, masses, and densities for both Pluto and Charon, as well as the compilation of rough maps of their surface markings. Charon's diameter is more than half Pluto's, and its mass is about one-eighth the mass of Pluto. Furthermore, Charon orbits only 20,000 kilometers from Pluto. (For comparison, our Moon has a mass 1/80th that of Earth and orbits 240,000 kilometers away.) Some astronomers argue that Pluto and Charon qualify as a "double planet," but others argue that neither is large enough to qualify as a planet at all.

[7]If you look in an astronomy text published before 1978, you'll often see Pluto's mass listed with some estimate followed by "?" because of the uncertainty. After the discovery of Charon in 1978, this uncertainty disappeared, providing a clear example of the power of Newton's version of Kepler's third law.

Curiously, Pluto and Charon have slightly different densities and presumably slightly different compositions. Pluto's density of 2 g/cm^3 is a bit high for the expected ice/rock mix in the outer solar system, while Charon's density of 1.6 g/cm^3 is a bit low. The leading explanation is that Charon was created in a way very similar to how our Moon formed. Pluto may have suffered a giant impact with a large icy object in the Kuiper belt that blasted away its low-density outer layers, after which this material formed a ring around Pluto and eventually reaccreted into the low-density moon Charon.

Pluto currently has a thin atmosphere of nitrogen and other gases formed by sublimation of surface ices. However, the atmosphere is steadily thinning because Pluto's elliptical orbit is now carrying it farther from the Sun. (Pluto will not reach aphelion until 2113.) As Pluto recedes, its atmospheric gases are refreezing onto its surface. Maps of surface markings generated from observation of eclipses and Hubble Space Telescope images show that the surface has bright and dark patches, but we don't yet have enough information to understand why (Figure 12.16).

Despite the cold, the view from Pluto would be stunning. Charon would dominate the sky, appearing almost 10 times larger in angular size than our Moon. Pluto and Charon's mutual tidal pulls long ago made them rotate synchronously with each other (see Figure 11.17a). Thus, Charon is visible from only one side of Pluto and would hang motionless in the sky, always showing the same face. Moreover, the synchronous rotation means that Pluto's "day" is the same length as Charon's "month" (orbital period) of 6.4 Earth days. Charon would neither rise nor set

FIGURE 12.16 The surfaces of Pluto and Charon, as reconstructed from Hubble Space Telescope images. The pairs of spheres show views of opposite hemispheres.

in Pluto's skies but instead would cycle through its phases in place. The Sun would appear more than a thousand times fainter than it appears here on Earth, and it would be no larger in angular size than Jupiter is in our skies.

Pluto: Planet or Kuiper Belt Object?

Pluto is an object in search of a category. It is a misfit among the planets, but it seems somewhat less out of place when considered in the context of the Kuiper belt. After all, it orbits in the same vicinity as Kuiper belt comets, it has a cometlike composition of icy and rocky materials, and it has a cometlike atmosphere that grows as it comes nearer the Sun in its orbit and fades as it recedes. Could Pluto be simply the largest known member of the Kuiper belt?

A closer look at the orbits of Kuiper belt objects uncovers an uncanny resemblance to the orbit of Pluto. Like Pluto, many Kuiper belt objects lie in stable orbital resonances with Neptune, and more than a dozen Kuiper belt objects have the *same* period and semimajor axis as Pluto itself (and are nicknamed "Plutinos"). Aside from Pluto's large size compared to the known comets, the most obvious difference between Pluto and Kuiper belt comets is their surface brightness. Comet nuclei generally have surfaces that are quite dark from carbon compounds, while Pluto's high reflectivity indicates an icy surface. However, this difference is easily explained. Whenever either Pluto or a smaller Kuiper belt comet approaches the Sun, the most volatile ices sublimate into an atmosphere. These atmospheric gases escape completely from the smaller Kuiper belt objects, but Pluto's gravity holds them until they refreeze onto the surface. Over time, the smaller Kuiper belt objects have become depleted in the most volatile ices, leaving behind the dark, carbon-rich compounds on their surfaces. Pluto has retained its volatile ices, giving it a bright, icy surface.

All in all, Pluto stands apart from Kuiper belt objects only in its size. But Pluto may not have been so unusual in the early solar system. Recall that Neptune's moon Triton is quite similar to (and larger than) Pluto, and its orbit indicates that it must be a captured object. Thus, Triton was once a "planet" orbiting the Sun. Giant impacts of other large icy objects may have created Charon and might even have given Uranus its large tilt. Many Pluto-size objects probably roamed the Kuiper belt in its early days. Those that did not become trapped in stable resonances may have fallen inward to be accreted by the jovian planets or been thrown outward and ejected from the solar system. Pluto may be the largest surviving member of the Kuiper belt. However, it remains possible that we will someday discover other objects with sizes similar to that of Pluto, though probably farther away or considerably darker. As of 1998, much less than 1% of the sky had been thoroughly searched for large Kuiper belt objects.

TIME OUT TO THINK *What is your definition of a planet? Be as rigorous as possible. Exactly how many planets are there, according to your criteria? Does your list include Pluto, Charon, Ceres or other asteroids, or Triton (a captured cousin of Pluto)? How big would a new Kuiper belt object have to be to meet your definition? Try to convince a classmate of your definition. Is your definition useful?*

Planet or not, Pluto remains the largest known object in the solar system never visited by spacecraft. Close-up observations of its surface, atmosphere, and unusual moon are sure to teach us as much about icy bodies as past missions did about rocky bodies. NASA scientists and engineers are struggling with the difficult task of designing an affordable Pluto mission. Pluto presents real challenges: At typical spacecraft speeds, it may take a decade to reach Pluto. But the faster a probe goes, the less time it will have to study Pluto during its flyby. (A Pluto orbiter is simply too expensive.) Fortunately, recent advances in miniaturization are paying off. The current Pluto mission design fits on a coffee table but has instruments about as good as those on the Cassini spacecraft to Saturn, which is as large as a typical small classroom. (The Pluto probe design is also about 5–10 times cheaper than that of *Cassini.*) NASA's current goal is to send the mission to Pluto before its atmosphere completely refreezes onto the surface, but budget constraints may keep the mission from becoming a reality. (Check the text web site for updates on the status of a Pluto mission.)

12.6 Cosmic Collisions: Small Bodies Versus the Planets

The hordes of small bodies orbiting the solar system are slowly shrinking in number through collisions with the planets and ejection from the solar system. Many more must have roamed the solar system in the days of the early bombardment, when most impact craters were formed [Section 8.6]. But there are still plenty left, and cosmic collisions still occur on occasion, with important ramifications for Earth as well as for other planets.

Shoemaker–Levy 9 Impacts on Jupiter

We usually think about impacts in the context of solid bodies because such impacts leave long-lasting scars in the form of impact craters. But impacts must be even more common on the jovian planets than on the terrestrial planets because of their larger size and stronger gravity, although no impact craters can form on gaseous worlds. It's estimated that a major impact on Jupiter occurs about once every 1,000 years. We were privileged to witness one in 1994.

The husband-and-wife team of Gene and Carolyn Shoemaker, along with their colleague David Levy, are among the premier "comet hunters" of modern times.[8] During a routine search of the sky in March 1993, this team came across an unusual object that Carolyn Shoemaker described as a "squashed comet." The trio's ninth joint discovery, the comet was called *Shoemaker–Levy 9,* or *SL9* for short. On closer examination, the "squashed comet" turned out to be a string of comet nuclei all in a row (Figure 12.17a). The reason for SL9's odd appearance became clear when astronomers calculated its orbit backward in time. The calculations showed that it had passed very close to Jupiter only a few months earlier (in July 1992). Because it had passed well within Jupiter's Roche zone [Section 11.6], it apparently was ripped apart by tidal forces. Prior to that, SL9 was probably a single comet nucleus orbiting Jupiter, unnoticed by human observers. Similar breakups of comets near Jupiter have occurred before, as evidenced by a chain of craters on Callisto that must have formed when a string of comet nuclei crashed into its icy surface (Figure 12.17b).

The orbital calculations also showed that, thanks to gravitational nudges from the Sun, SL9 was on a collision course with Jupiter. Astronomers had more than a year to plan for observations of the impacts due in July 1994. The impacts of the 22 identified nuclei would take almost a week, and virtually every telescope on Earth and in space would be ready and watching. The only uncertainty was whether or not anything would happen. Even the best Hubble Space Telescope images were unable to determine whether the comet nuclei were massive kilometer-size chunks or merely insubstantial puffs of dust.

The answer came within minutes of the first impact: Infrared cameras recorded an intense fireball of hot gas rising thousands of kilometers above the impact site, which lay just barely on Jupiter's far side (Figure 12.18). As Jupiter's rapid rotation carried the scene into view, the collapsing plume smashed down on Jupiter's atmosphere, leading to another episode of heating and more infrared glow. News of each subsequent impact spread around the world rapidly via the Internet, and everyone waited anxiously to hear each observatory report. The Hubble Space Telescope caught impact plumes in action as they rose into sunlight and collapsed back down, leaving dust clouds high in Jupiter's stratosphere. The impact scars lingered for months, but Jupiter has recovered—for now.

It will be many years before all the data from the week of SL9 impacts are completely analyzed, but

[8] Gene Shoemaker died in 1997 in an automobile crash while on a trip to study impact craters in Australia. Carolyn Shoemaker and David Levy continue their joint projects.

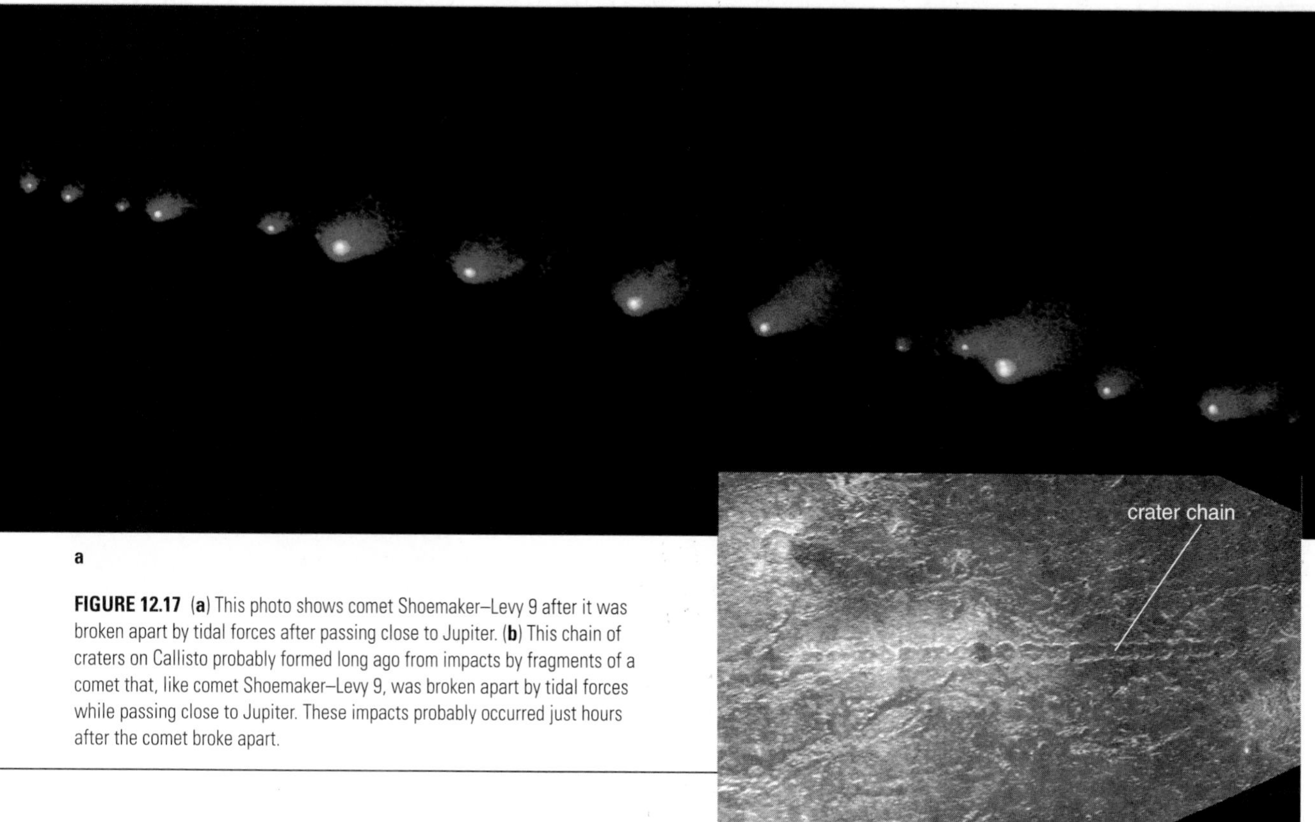

FIGURE 12.17 (**a**) This photo shows comet Shoemaker–Levy 9 after it was broken apart by tidal forces after passing close to Jupiter. (**b**) This chain of craters on Callisto probably formed long ago from impacts by fragments of a comet that, like comet Shoemaker–Levy 9, was broken apart by tidal forces while passing close to Jupiter. These impacts probably occurred just hours after the comet broke apart.

we have already learned a great deal. Material splashed out by the impacts provided us with a unique opportunity to study material from well beneath the layers of Jupiter that are ordinarily visible. More important, we learned much about the impact process itself. Many scientists correctly predicted the behavior of a rising plume of gas from an impact but did not consider the effects of the plume reimpacting on Jupiter. The reimpact of the plumes turned out to have even more significant global effects than the impact itself, heating an area 10,000 kilometers across and leaving dark, dusty clouds that eventually encircled the planet.

The SL9 impacts on Jupiter also provided two important sociological lessons. First, "Comet Crash Week" proved to be one of the best examples of international collaboration in history. With the aid of the Internet, scientists quickly and effectively shared data from observatories around the world. Second, extensive media coverage helped the event capture the public imagination, providing awareness that impacts are not relegated only to ancient geological history. Each individual fragment of SL9 crashing into Jupiter carried the equivalent force of a million hydrogen bombs. If such violent impacts can happen on other planets in our lifetime, could they also happen on Earth?

Comet Tails and Meteor Showers

Far smaller impacts happen on Earth all the time, lighting up the sky as *meteors*. If you watch the sky on a clear night, you'll typically see a few meteors each hour. Most meteors are created by single pieces of comet dust, each no larger than a pea, that enter our atmosphere at speeds of up to 250,000 km/hr (70 km/s). (A small fraction of meteors may be dust from asteroids rather than comets, and the brightest meteors come from larger particles.) The particles and the surrounding atmosphere are heated so much that we see a brief but brilliant flash. Meteor particles are completely vaporized at altitudes of 50–100 kilometers, so we never get complete cometary "meteorites" to study.[9] An estimated 25 million meteors occur worldwide every day, adding hundreds of tons of comet dust to the Earth daily.

Earth's orbit intersects many comet orbits over the course of the year, and each produces a *meteor shower* during which we may see dozens of meteors per hour (see Table 1.1). The most famous meteor shower, the *Perseids*, occurs every August when the Earth passes through the orbit of comet Swift–Tuttle.

[9]Microscopic comet dust *has* been collected by high-flying military aircraft.

a The Galileo spacecraft, on its way to Jupiter at the time, got a direct view of the impacts on Jupiter's night side.

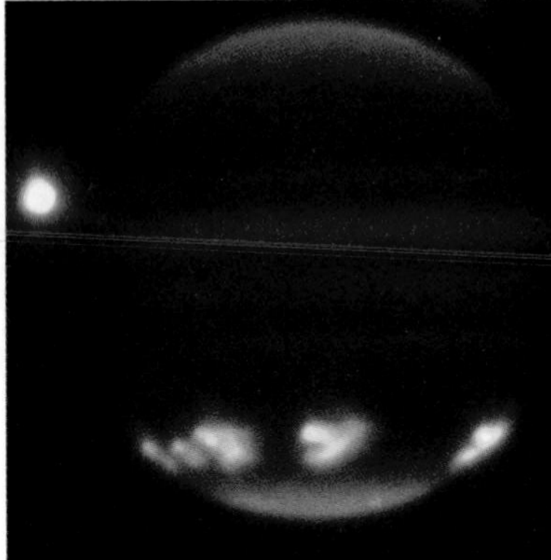

d This infrared photo shows high clouds created by several impacts. These clouds reflect infrared light from the Sun because they lie high above the methane gas that absorbs infrared light.

FIGURE 12.18 The SL9 impacts allowed astronomers their most direct view ever of cosmic collisions.

b In under 20 minutes, the Hubble Space Telescope observed an impact plume rise thousands of kilometers above Jupiter's clouds and then collapse back down.

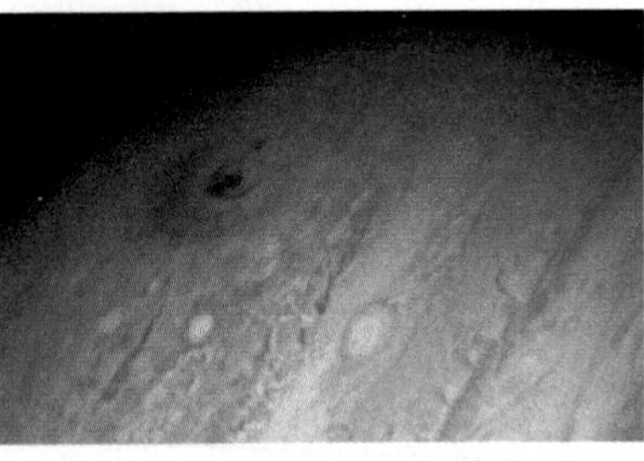

e In this visible-light photo from the Hubble Space Telescope, the high, dusty clouds left by the impacts appear as dark "scars" on Jupiter.

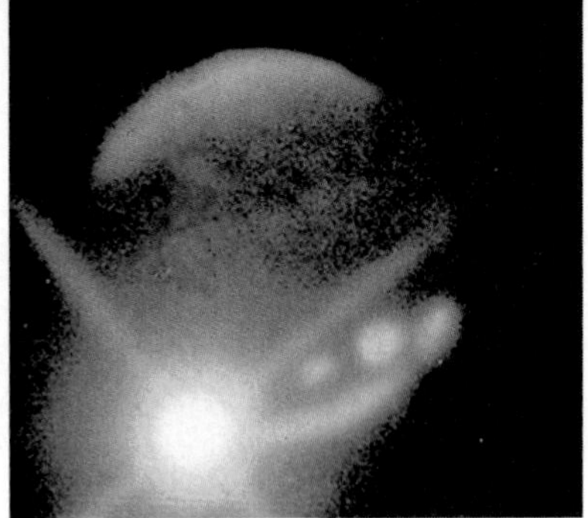

c This infrared photo shows the brilliant glow of a rising fireball from one of the impacts. Most of Jupiter appears dark because methane gas in its atmosphere absorbs infrared light.

f This Hubble Space Telescope photo shows how the impact scars became smeared due to atmospheric winds. The scars disappeared completely within a few months.

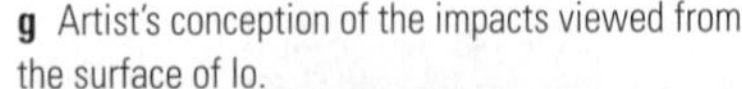

g Artist's conception of the impacts viewed from the surface of Io.

a Driving into a blizzard.

b Geometry of a meteor shower.

FIGURE 12.19 Snowflakes and meteor showers.

c Meteors photographed during the intense Leonid meteor shower in 1966. The constellation Leo lies to the left of the photo.

Although meteors all strike the Earth on parallel tracks, meteor showers appear to radiate from a particular direction in the sky. You may have experienced a similar illusion while driving into a blizzard at night: The snowflakes seems to diverge from a single direction that depends on the combination of your speed and the wind speed (Figure 12.19a,b).[10] The Perseids get their name because they appear to radiate from the constellation Perseus (Figure 12.19c). The Earth runs into more meteors on the side facing in the direction of Earth's orbital motion (just as more snow hits the front windshield while the car is moving), so the predawn sky is the best place to watch for meteors.

TIME OUT TO THINK *The associated "comet" for the Geminid meteor shower is an object called Phaethon, which is classified as an asteroid because we've never seen a coma or tail associated with it. How is it possible that Phaethon looks like an asteroid today but once shed the particles that create the Geminid meteor shower?*

Impacts and Mass Extinctions

Meteorites and impact craters bear witness to the fact that much larger impacts occasionally occur on Earth. Meteor Crater in Arizona (see Figure 2.8) formed about 50,000 years ago when a metallic impactor roughly 50 meters across crashed to Earth with the explosive power of a 20-megaton hydrogen bomb. Although the crater is only a bit more than 1 kilometer across, an area covering hundreds of square kilometers was probably battered by the blast and ejecta. Far bigger impacts have occurred, sometimes with catastrophic consequences for life on Earth.

Collecting geological samples in Italy in 1978, the father–son team of Luis and Walter Alvarez discovered a thin layer of dark sediments that had apparently been deposited 65 million years ago—about the same time that the dinosaurs and many other organisms suddenly became extinct. Subsequent studies found similar sediments deposited at the same

[10]This illusion occurs even before the air flow around the car deflects the motion of the snowflakes.

FIGURE 12.20 The arrow points to a layer of sediment laid down by the impact 65 million years ago. At the time, the rock layers above the arrow did not exist. Dust from the impact and soot from global wildfires settled down through the atmosphere onto the seafloor that once occupied this location in Colorado.

time at many sites around the world (Figure 12.20). Careful analysis showed this worldwide sediment layer to be rich in the element iridium, which is rare on Earth's surface. But iridium is common in primitive meteorites, which led the Alvarezes to a stunning conclusion: The extinction of the dinosaurs was caused by the impact of an asteroid or comet. This conclusion was not immediately accepted and still generates some controversy, but it now seems clear that a major impact coincided with the death of the dinosaurs. Moreover, while the dinosaurs were the most famous victims of this **mass extinction,** it seems that up to 99% of all living things were killed and that 75% of all *species* living on Earth were wiped out at that time.

How could an impact lead to mass extinction? The amount of iridium deposited worldwide suggests that the impactor must have been about 10 kilometers across. After a decade-long search, scientists identified what appears to be the impact crater from the event. Located off the coast of Mexico's Yucatán peninsula (Figure 12.21), it is 200 kilometers across, which is close to what we expect for a 10-kilometer impactor, and dates to 65 million years ago. Further evidence that the Yucatán crater is the right one comes from the distribution of small glassy spheres that formed when the molten impact ejecta solidified as it rained back to Earth. More of these glassy spheres are found in regions near the crater, and careful study of their distribution suggests that the impactor crashed to Earth at a slight angle. The impact almost immediately sent

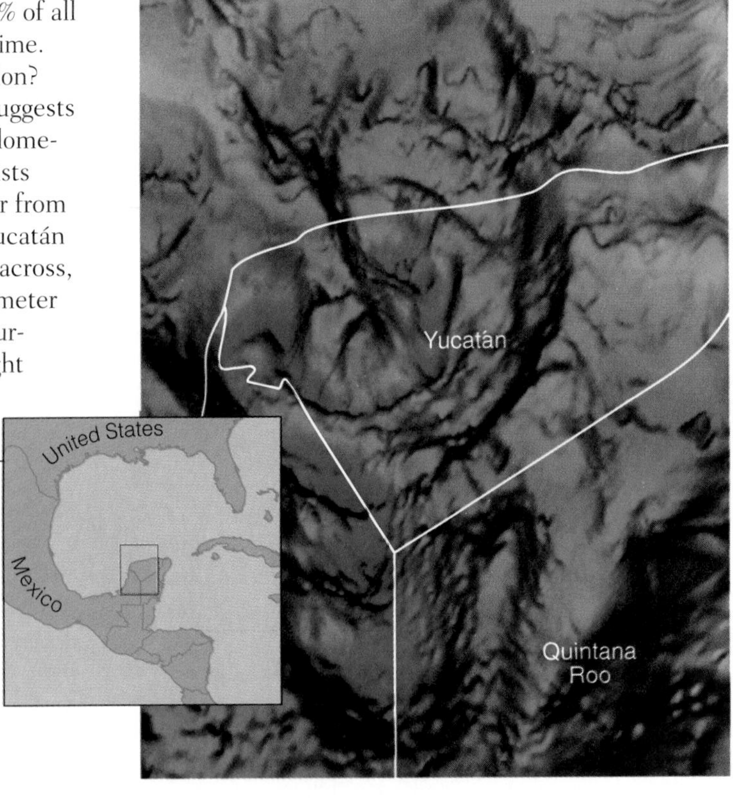

FIGURE 12.21 This false-color image, made with the aid of precise measurements of the strength of gravity, shows an impact crater with its center near the northwest coast of Yucatán. This crater, known as the Chicxulub crater, is thought to have been formed by the impact that killed the dinosaurs and most other species present 65 million years ago. The red box on the inset map shows the region covered by the image; the white lines on the image correspond to the coastlines and borders of Mexican states.

a A 10-km impactor, hurtling toward Earth at perhaps 100,000 km/hr or more, is about to enter the atmosphere.

FIGURE 12.22 Instant geology—and biology: Artist's conception (clockwise from upper left) of the impact that led to a mass extinction 65 million years ago.

b One second to impact.

c Impact.

d One month later, debris from the impact fills our atmosphere with dust.

e Many years later, after the dust has settled, we see the 200-km wide crater from the impact.

a shower of debris raining across much of North America and generated huge waves that may have sloshed more than 1,000 kilometers inland (Figure 12.22). Many North American species thus may have been wiped out shortly after impact. For the rest of the world, death may have come more slowly. Heat from the impact and returning ejecta probably ignited wildfires in forests around the world. Evidence of such wildfires is found in the large amount of soot that is also present in the iridium-rich sediments from 65 million years ago. The impact also sent huge quantities of dust high into the stratosphere, where it remained for several years, blocking out sunlight, cooling the surface, and affecting atmospheric chemistry. Plants died for lack of sunlight, and effects propagated through the food chain.

Perhaps the most astonishing fact is not that 75% of all species died, but that 25% survived. Among the survivors were a few small, rodentlike mammals. These mammals may have survived because they lived in underground burrows and managed to store enough food to outlast the long spell of cold, dark days. Small mammals had first arisen at about the same time as the dinosaurs, more than 100 million years earlier. But the sudden disappearance of the dominant dinosaurs made these mammals dominant. With an evolutionary path swept wide open, they rapidly evolved into a large assortment of much larger mammals—including us.

The event 65 million years ago was but one of at least a dozen mass extinctions in Earth's distant past, and several of these have also been linked to impacts. Could a similar event occur in the future, wiping out all that we have accomplished?

The Asteroid Threat: Real Danger or Media Hype?

The first step in analyzing the threat of future impacts is to examine the past evidence more clearly. Only one significant impact has clearly occurred in modern times. In 1908, an unusual explosion occurred in a sparsely inhabited region of Siberia. (The impact is known as the *Tunguska event.*) Entire forests were flattened and set on fire, and air blasts knocked over people, tents, and furniture up to 200 kilometers away (Figure 12.23). Seismic disturbances were recorded at distances up to 1,000 kilometers, and atmospheric pressure fluctuations were detected almost 4,000 kilometers away. Scientific studies were delayed for almost 20 years (due to the remoteness of the impact area and the politics of the time), but investigators eventually found meteoritic dust at the site—but no impact crater. These wide-reaching effects were apparently caused by a weak, stony meteorite only 30 meters across that exploded in the air before reaching the surface. Objects of this size probably strike our planet every century or so, usually over the ocean. But the death toll could be quite high if such an object struck a densely populated area.

Another way to gauge the threat of impacts is to look at asteroids that might strike Earth. The largest-known *Earth-approaching asteroid,* called Eros, is about 40 kilometers long. Dozens of smaller Earth-approaching asteroids are known, down to sizes of tens of meters across. Orbital calculations show that none of these known objects will impact Earth in the foreseeable future. However, the vast majority of Earth-approaching asteroids probably have not yet been detected. Astronomers estimate that some 2,000 undiscovered members of this class may be larger than 1 kilometer across, and another 300,000 may be larger than 100 meters across. Comets also pose a threat, particularly because their rapid plunge from the outer solar system gives little warning.

By combining observations of asteroids with information from past impact craters, we can estimate the frequency of impacts of various sizes (Figure 12.24). Larger impacts are obviously more devastating but thankfully are rare. Impactors a few meters across probably break up high in the Earth's atmosphere every few days, with each liberating the energy of an atomic bomb. However, the effects are dissipated before reaching the surface, so we do not notice them. (Military technology designed to detect nuclear bomb tests has recorded many such events.) Hundred-meter impactors that forge craters similar to Meteor Crater probably strike only about every 10,000 years or so. Impacts large enough to cause mass extinctions come tens of millions of years apart.

FIGURE 12.23 Damage from the 1908 impact over Tunguska, Siberia.

Thus, while it seems a virtual certainty that the Earth will be battered by many more large impacts, the chance of a major impact happening in our lifetime is quite small. Nevertheless, the chance is not zero, and until we know the orbits of *all* large Earth-approaching asteroids we cannot predict when the next impact might occur. Some people advocate a thorough search to find all potentially dangerous asteroids, a relatively cheap proposition. But others wonder whether the information would be useful: If you knew an asteroid would hit next year, could you do anything about it? Schemes to use nuclear weapons to save Earth by demolishing or diverting an asteroid abound, but no one knows whether current technology is up to the task. For the time being, we can only hope that our number does not come up soon.

While some people see danger in Earth-approaching asteroids, others see opportunity because Earth-approaching asteroids bring valuable resources tantalizingly close to Earth. Iron-rich asteroids are particularly enticing, since they probably contain many precious metals that mostly sank to the core on Earth. In the not-too-distant future, it may prove technically feasible and financially profitable to mine metals from asteroids and return these resources to Earth. It may also be possible to gather fuel and water from asteroids for use in missions to the outer solar system.

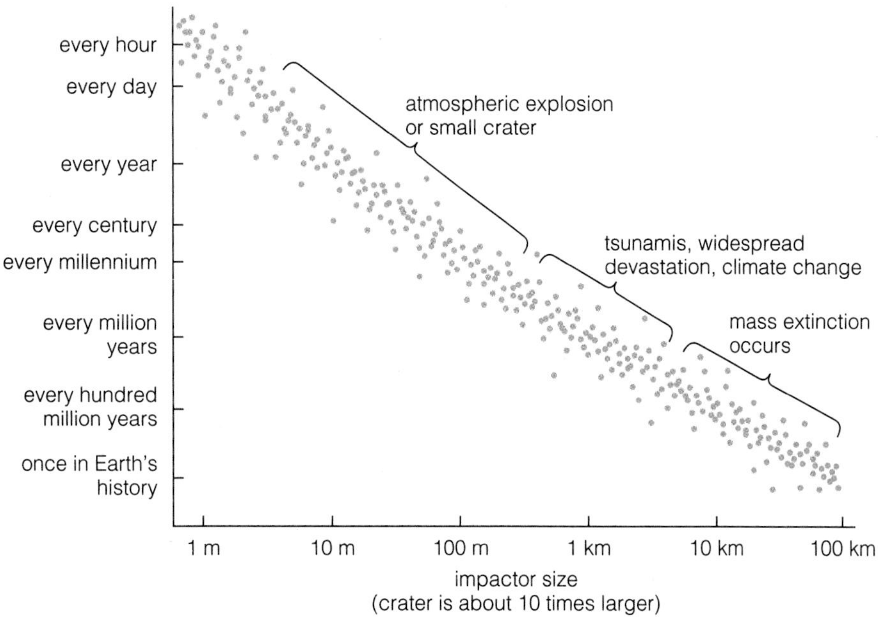

FIGURE 12.24 This graph shows how the frequency of impacts—and the magnitude of their effects—depends on the size of the impactor; note that smaller impacts are much more frequent than larger ones.

As we look out into the Universe and identify the many accidents of physics and astronomy that have worked to our benefit, it almost seems as if the Universe must in some sense have known that we were coming.

FREEMAN DYSON

THE BIG PICTURE

In this chapter, we focused on the solar system's smallest objects and found that they can have big consequences. Asteroids, comets, and meteorites teach us much about the evolution of the solar system. And, in the case of the impact 65 million years ago, one such object may have made our existence possible. Keep in mind the following "big picture" ideas as you continue your studies:

- The smallest bodies in the solar system—asteroids and comets—are our best evidence of how the solar system formed.
- The small bodies are subjected to the gravitational whims of the planets, particularly the jovian planets. The subtleties of resonances play a major role in sculpting the outer solar system.
- The interplay of large and small bodies brings us meteorites to teach us our origins, comets and meteor showers to light up the sky, and impacts that can alter or obliterate life as we know it.
- Pluto is called the ninth planet, but it bears much more similarity to the thousands of Kuiper belt comets than to the other eight planets.

Review Questions

1. Briefly explain why comets, asteroids, and meteorites are so useful in terms of helping us understand the formation and development of our solar system.
2. Define and distinguish among each of the following: *asteroid, comet, meteor, meteorite.*
3. How does the largest asteroid compare in size to the terrestrial worlds? How many asteroids are larger than 1 km in radius? How does the total mass of all the asteroids compare to the mass of a terrestrial world?
4. What is the *asteroid belt,* and where is it located?
5. Briefly describe how orbital resonances with Jupiter explain the numerous *gaps* in the asteroid belt. How did these resonances prevent a planet from forming in the vicinity of the asteroid belt?

6. What happens to asteroids that are kicked out of the asteroid belt by a gravitational encounter or a collision? In this context, describe the origin of the Earth-approaching asteroids.
7. What are the Trojan asteroids, and where are they found? What is the orbital period of the Trojan asteroids? (*Hint:* They share an orbit with an object whose orbital period you know.)
8. How do we detect asteroids with telescopic images? How do we determine the orbit of an asteroid?
9. Briefly describe how we can estimate an asteroid's size from its brightness and albedo. How can we estimate its albedo?
10. Explain how asteroid densities are determined. Does the range of observed densities show that asteroids have similar or dissimilar compositions?
11. Why do the largest asteroids have more spherical shapes than smaller asteroids? Briefly describe how we can determine asteroid shapes, even when we have not photographed them close up.
12. What do spectra tell us about the composition of asteroids? Describe the compositions of the three main groups of asteroids, as distinguished by spectroscopy.
13. How can we distinguish a meteorite from a terrestrial rock? Why do we find so many meteorites in Antarctica?
14. Distinguish between *primitive meteorites* and *processed meteorites* in terms of composition. How do the origins of these two groups of meteorites differ?
15. Why do primitive meteorites come in carbon-rich and rocky varieties? Where did each variety form in the solar nebula? Which type is more common among meteorites?
16. Explain how the existence of processed meteorites tells us that some asteroids once had active volcanism. How is the study of a processed meteorite an opportunity to look at a piece of a "dissected planet"?
17. What is the origin of processed meteorites with basaltic composition? What evidence suggests that some of these come from the Moon or Mars?
18. How did Halley's comet get its name? How are comets named today?
19. Describe the anatomy of a comet, including its *nucleus, coma, plasma tail,* and *dust tail.* Which of these components are present when a comet is close to the Sun? When it is far from the Sun? Explain.
20. Briefly describe the composition of the nucleus of Halley's comet. How does gas escape from the interior of the nucleus into space? Why do the tails point away from the Sun?
21. Why can't an active comet last forever? What happens to comets after many passes near the Sun?
22. Describe the Kuiper belt and the Oort cloud in terms of location and comet orbits. Explain how we infer the existence of these two comet reservoirs by studying the orbits of the rare comets that enter the inner solar system.
23. Where did comets that are now in the Oort cloud originally form? How did they end up in the Oort cloud? What can make an Oort cloud comet alter its orbit and plunge toward the inner solar system?
24. Describe the origin of the Kuiper belt comets. What do we know about the sizes of Kuiper belt comets?
25. Describe Pluto and Charon. How were they discovered? Why won't Pluto collide with Neptune even though their orbits cross? How did Charon probably form?
26. Briefly summarize the evidence suggesting that Pluto is a Kuiper belt comet. How does Neptune's moon Triton suggest that Pluto may not be unique in size among Kuiper belt comets? Is it possible that other Pluto-size objects remain to be discovered in our solar system? Explain.
27. Describe the impact of comet Shoemaker–Levy 9 on Jupiter. How was the impact studied? What did we learn from this impact?
28. Explain how meteor showers are linked to comets. Why do meteor showers seem to originate from a particular location in the sky?
29. Describe the evidence suggesting that the *mass extinction* that killed off the dinosaurs was caused by the impact of an asteroid or comet. How did the impact lead to the mass extinction?
30. How often should we expect impacts of various sizes on Earth? How could we alleviate the risk of major impacts?

Discussion Questions

1. *Impact Risk.* How can we compare the risk of death from an impact to the risk of death from other hazards? This question is very important if we are to assess how much money it is worth spending to search for potential impactors and ways of preventing them from hitting the Earth. One way of evaluating the risk involves multiplying the probability of an impact by the number of people it would kill. For example, the probability of a major impact that could kill half the world's population, or some 3 billion people, is estimated to be about 0.0000001 (10^{-7}) in any given year. If we multiply this low probability by the 3 billion people that would be killed, we find the "average" risk from an impact to be $10^{-7} \times (3 \times 10^{9}) = 300$ people killed per year—about the same as the average number of people killed in airline crashes. Do you think it is valid to say that the risk of being killed

by a major impact is about the same as the risk of being killed in an airplane crash? Should we pay to limit this risk, as we pay to limit airline crashes? Defend your opinions.

2. *Observational Bias.* Meteorites and long-tailed comets present "free samples" of more distant objects.
   a. Carbon-rich asteroids represent about 5% of meteorite falls on the Earth. How does this compare with the observed fraction of carbon-rich asteroids in the asteroid belt? The odds of a particular comet we observe originating from the Oort cloud versus the Kuiper belt are about 50–50. How does this compare with the relative number of objects in these two regions?
   b. Your answers to part (a) should show that meteorite samples and comet observations suffer from *bias*—that is, what we learn from their study may not accurately reflect the complete populations of the objects they represent. How might a similar bias play a role in more down-to-earth situations? For example, do you and your classmates form a representative sample of your city, nation, or planet? How might this affect the accuracy of a report on world events you might write together or the fairness of a decision you might vote on? In what ways can the effects of bias be removed?
3. *Rise of the Mammals.* Suppose the impact 65 million years ago had not occurred. How do you think our planet would be different? For example, do you think that mammals would still have eventually come to dominate the Earth? Would we be here? Defend your opinions.

Problems

1. *Orbital Resonances.* How would our solar system be different if orbital resonances had never been important? (For example, if Jupiter and asteroids continued to orbit the Sun but did not affect each other through resonances.) Describe at least five ways in which our solar system would be different. Which differences would you consider merely superficial, and which differences would be more profound? Be sure to consider the effects of orbital resonances discussed both in this chapter and in Chapter 11.
2. *Life Story of an Iron Atom.* Imagine that you are an iron atom in an iron meteorite recently fallen to Earth. Tell the story of how you got here, beginning from the time when you were in a gaseous state in the solar nebula 4.6 billion years ago. Include as much detail as possible. Your story should be scientifically accurate, but also creative and interesting.
3. *Asteroid Discovery.* You have discovered two new asteroids and have named them Barkley and Jordan. Both lie at the same distance from the Earth and have the same brightness when you look at them through your telescope. But Barkley is twice as bright as Jordan at infrared wavelengths. What can you deduce about the two asteroids' relative reflectivities and sizes? Which would make a better target for a mission to mine metal? Which would make a better target to obtain a sample of a carbon-rich planetesimal? Explain your answers in one or two paragraphs.
4. *The "Near Miss" of Toutatis.* The 5-km asteroid Toutatis passed a mere 3 million km from the Earth in 1992.
   a. In one or two paragraphs, describe what would have happened to the Earth if Toutatis had hit it.
   b. Suppose Toutatis was destined to pass *somewhere* within 3 million km of Earth. Calculate the probability that this "somewhere" would have meant that it slammed into Earth. Based on your result, do you think it is fair to say that the 1992 passage was a "near miss"? Explain. (*Hint:* You can calculate the probability by considering an imaginary dartboard of radius 3 million km on which the bullseye has the Earth's radius of 6,378 km.)
5. *Comet Temperatures.* Compare the strength of sunlight at 50,000 AU in the Oort cloud with its strength at 3 AU and 1 AU. Compare the "no atmosphere" temperatures for comets (see Mathematical Insight 10.1) at the three positions. At which location will the temperature be high enough for water ice to sublime (about 150 K)? Assume that the comets rotate rapidly and have an albedo of 3%.
6. *Project: Tracking a Meteor Shower.* Armed with an expendable star chart and a flashlight covered in red plastic, set yourself up comfortably to watch a meteor shower listed in Table 1.1. Each time you see a meteor, record its path on your star chart. Record at least a dozen, and try to determine the *radiant* of the shower—the point in the sky from which the meteors appear to radiate. On your chart, also include the direction of the Earth's motion, which you can find by determining which sign of the zodiac rises at midnight.
7. *Project: Dirty Snowballs.* If there is snow where you live or study, make a filthy snowball. (The ice chunks that form behind tires work well.) How much dirt does it take to darken snow? Find out by allowing your dirty snowball to melt and measuring the approximate proportions of water and dirt afterward.
8. *Web Project: NEAR and Rosetta.* Starting from the text web site, search for recent discoveries from the Near-Earth Asteroid Rendezvous mission (NEAR) *or* the European Space Agency's Rosetta mission. Write a one- to two-page summary of what you learn, giving special attention to anything that affects (or will affect) our understanding of phenomena or processes described in this chapter.

13

Planet Earth

PERHAPS YOU'VE HEARD THE EARTH DESCRIBED AS THE "THIRD rock from the Sun." At first glance, the perspective of comparative planetology may seem to support this dispassionate view of the Earth. However, a deeper comparison between Earth and its neighbors reveals a far more remarkable planet.

Earth is unique in the solar system in many ways. It is the only planet with surface oceans of liquid water and substantial amounts of oxygen in its atmosphere, and its geological activity is more diverse than that of any other terrestrial world. Most important, Earth is the only world known to harbor life, which has played a considerable role in shaping the Earth's surface and atmosphere.

So far, we have found that our theories of solar system formation and planetary evolution explain most of the general features of our solar system. In this chapter, we will apply and extend what we've learned in order to form a coherent picture of the combined geological, atmospheric, and biological evolution of our planet. We'll also discuss a few important lessons that the solar system teaches us about our future on Earth and that the Earth teaches us about the possibility of life elsewhere in the universe.

13.1 How Is Earth Different?

Take a look at Figure 13.1. The blue oceans and abundant white clouds show a planet unlike any other we have studied so far. It is, of course, our own planet Earth, and we are now ready to study it using the tools of comparative planetology we developed in the preceding five chapters.

We've already discussed many of the similarities and differences between Earth and other worlds. Earth's geology resembles that of the other terrestrial worlds, particularly Venus and Mars, in that it is characterized by the four processes of *impact cratering, volcanism, tectonics,* and *erosion* [Section 9.4]. But even here we find important differences. For example, Earth is the only planet on which the lithosphere is clearly broken into plates that move around in what we call *plate tectonics,* and erosion reshapes Earth's surface to a much greater extent than it does on any other world in the solar system.[1]

Earth's atmosphere is also unlike those of its nearest neighbors, despite a common origin in *outgassing.* Venus and Mars have atmospheres composed primarily of carbon dioxide (CO_2), but Earth's is composed primarily of nitrogen (N_2) with substantial amounts of oxygen (O_2) mixed in. While one other world—Titan—has an atmosphere composed mostly of nitrogen, no other world has substantial quantities of atmospheric oxygen. Furthermore, while the atmospheres of Venus, Earth, and Mars all possess warm tropospheres near the surface and hot thermospheres at high altitudes [Section 10.2], only Earth possesses an ultraviolet-absorbing *stratosphere.*

The greatest differences between Earth and other worlds lie in two features totally unique to Earth. First, the surface of the Earth is covered by huge amounts of water. Oceans cover nearly three-fourths of the Earth's surface, with an average depth of about 3 kilometers (1.8 miles). Water is also significant on land, where it flows through streams and rivers, fills lakes and underground water tables, and sometimes lies frozen in glaciers. Frozen water covers nearly the entire continent of Antarctica in the form of the southern ice cap. Another ice cap sits atop the Arctic Ocean in the north and also covers the large island of Greenland. Water plays such an important role on Earth that some scientists treat it as a distinct planetary layer, called the **hydrosphere,** between the lithosphere and the atmosphere.

The second totally unique feature of Earth is its diversity of life. We find life nearly everywhere on Earth's surface, throughout the oceans, and even underground. The layer of life on Earth is sometimes called the **biosphere.** As we will see shortly, the biosphere helps shape many of the Earth's physical characteristics. For example, the biosphere explains the presence of oxygen in Earth's atmosphere. Without life, Earth's atmosphere would be very different.

Comparative planetology compels us to explore these unique properties of Earth closely. If we find that the formation and evolution of our Earth violated the rules we've derived for other planets—if Earth is "irregular"—life might be unique to Earth. If Earth does follow the rules—if Earth is "regular" or even "standard"—then life may be common in the universe.

FIGURE 13.1 Dawn over the Atlantic Ocean, photographed from the Space Shuttle.

[1]Erosion may also be quite important on Titan, but we do not yet know [Section 11.5].

13.2 Our Unique Geology

It's not surprising that Earth shows much more geological activity than Mercury or the Moon, since those small worlds have cooled since their formation and are now geologically "dead." But why have Earth and Venus ended up so different, despite very similar sizes? And, although Earth is significantly larger than Mars, both planets probably once had similar surface conditions that allowed liquid water to flow. Why did Earth remain hospitable, while Mars is now dry and barren? These mysteries warrant a closer look.

Figure 13.2 shows shaded relief maps comparing the surfaces of Earth, Venus, and Mars. While the three planets have many superficial similarities, they also have important differences. We find fewer impact craters on Earth than on Venus or Mars. Most of Earth's volcanic mountains are steep-sided and therefore must have been made from a much more viscous lava than that found on the other planets. And Earth shows far more evidence of tectonic activity—surface reshaping driven by internal stresses. To our current knowledge, Earth is the only terrestrial planet on which the lithosphere is split into distinct *plates.* Earth also has a crust that comes in two very different varieties: a relatively thick, low-density crust underlying the *continents,* and a thinner, denser crust underlying the *seafloors* (Figure 13.3). Seafloor crust is made of basalt and is typically only 5–10 km thick, whereas continental crust is made of lower-density rock (such as granite) and is 20–70 km thick.

Why do the surfaces of Earth, Venus, and Mars look so different today, when all three planets share similar compositions and internal structures and all three have experienced at least some tectonic stresses and volcanic activity? To answer this question, we

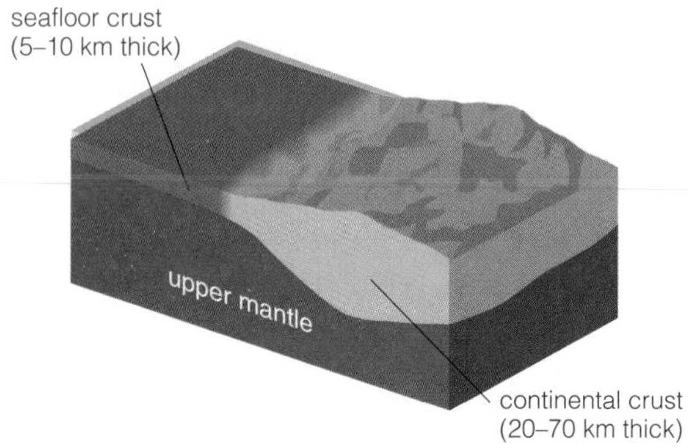

FIGURE 13.3 Earth has two distinct kinds of crust. Seafloor crust is denser, younger, and thinner than continental crust.

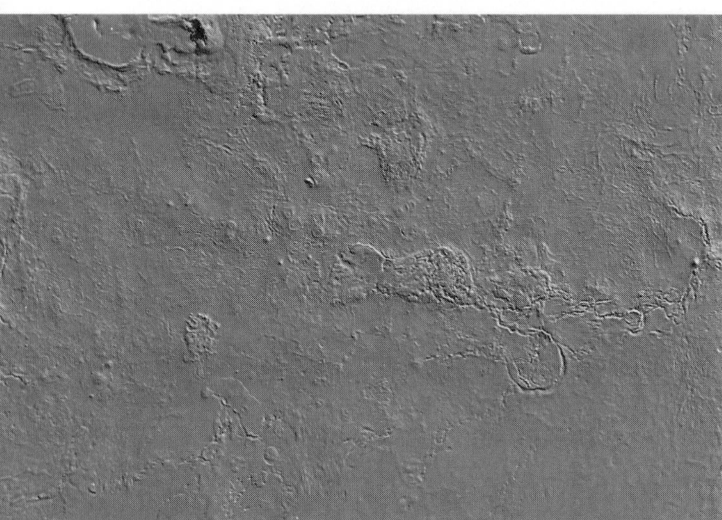

Venus

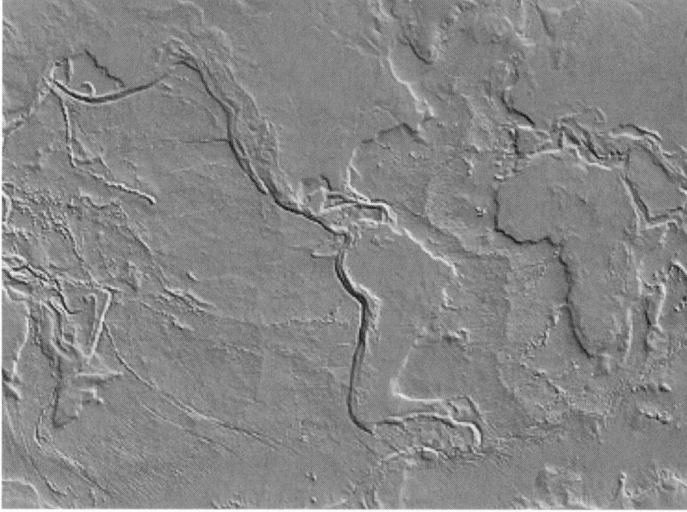

Earth

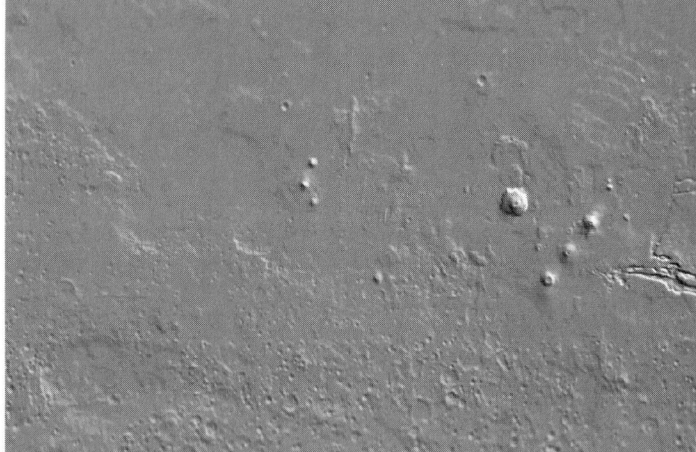

Mars

FIGURE 13.2 These shaded relief maps compare the surfaces of Venus, Earth, and Mars; the map for Earth shows the seafloor as it would appear if there were no oceans. Despite superficial similarities, there are important differences. For example, there are fewer impact craters on Earth or Venus than on Mars. Also note the unique distinction between the two kinds of surface (seafloor and continent) on Earth.

must compare the effects of the four geological processes on Earth to their effects on our neighbors.[2]

The Earth's Interior

As we discussed in Chapter 9, the first step in understanding geological processes is to learn as much as possible about a planet's interior structure. In the case of the Earth, geologists have created a three-dimensional "picture" of the interior by analyzing the propagation of seismic waves from earthquakes (Figure 13.4). The Earth's thin crust and the uppermost portion of the mantle make up a relatively cool and rigid *lithosphere* about 100 km thick [Section 9.2]. Below the lithosphere, the mantle is warm and partially molten in places; it supplies the magma for volcanic eruptions. Deeper in the mantle, higher pressures force the rock into the solid phase despite even higher temperatures. But even in solid form the mantle slowly flows, and convection continually carries heat up from below. The Earth's metallic core underlies the lower mantle. The outer region of the core is molten, but high pressures in the inner core force the metal into solid form.[3] Thus, the Earth has all the ingredients needed for exciting geology: a hot interior to supply energy for volcanism, convection to help generate tectonic stresses, and a relatively thin lithosphere that does not inhibit geological activity.

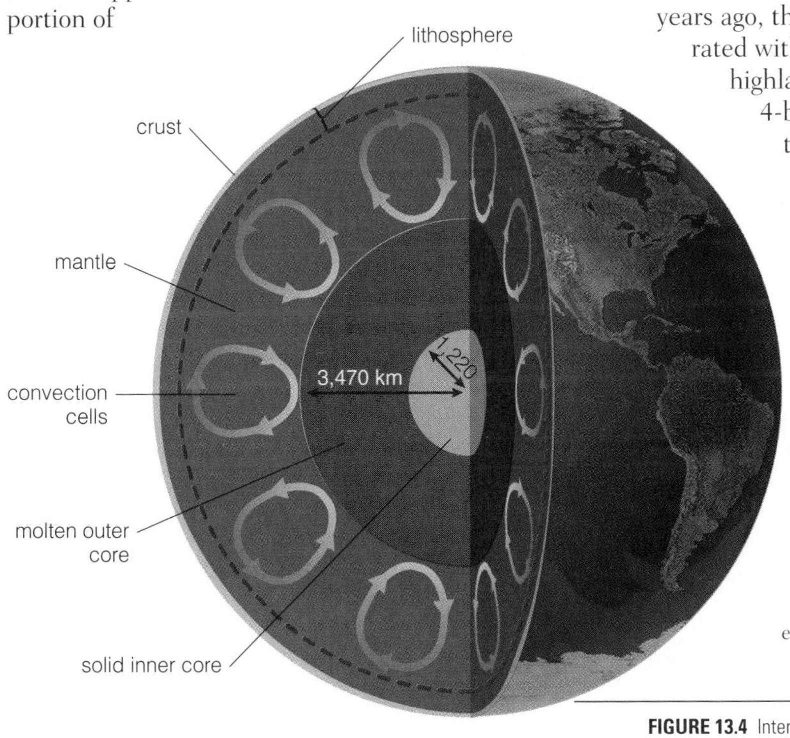

FIGURE 13.4 Internal structure of the Earth.

The Four Geological Processes on Earth

The first of the four geological processes, *impact cratering,* should have occurred at roughly similar rates on all the terrestrial worlds. All were subject to the intense early bombardment during the solar system's first few hundred million years. Four billion years ago, the Earth's surface may have been as saturated with craters as the densely cratered lunar highlands are today [Section 9.5]. We find no 4-billion-year-old craters on Earth now, so these ancient impacts must have been erased by other geological processes.

The impact rate fell substantially after the early bombardment, but occasional impacts still occur—sometimes with devastating consequences [Section 12.6]. The effects of erosion make impact craters surprisingly difficult to identify on Earth. Nevertheless, remnants of more than 100 impact craters have been found (Figure 13.5). Large impacts should

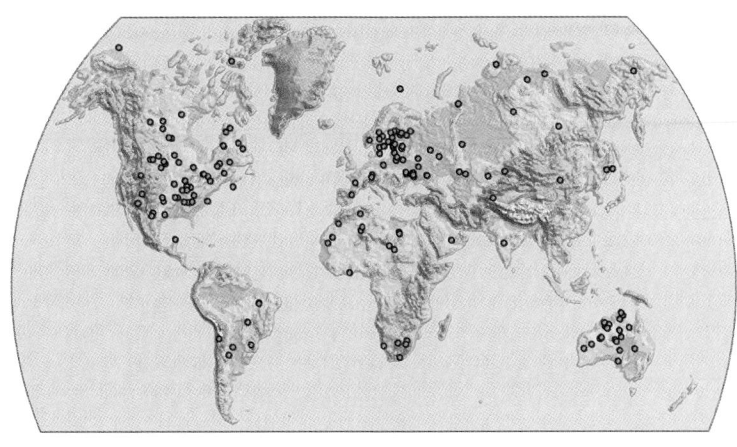

FIGURE 13.5 Locations of known impact craters on Earth.

[2]The presence of Earth's oceans might seem to invalidate geological comparisons with Mars and Venus. In fact, the mass of Earth's oceans is comparable to that of Venus's atmosphere, and both have relatively small direct effects on the four geological processes.

[3]Recent measurements suggest that the inner core is a huge single crystal that rotates at a slightly different rate than does the rest of the planet.

THINKING ABOUT . . .

Vibrations in the Earth

Earthquakes provide us with the most important probe of the Earth's interior—a silver lining to their terrifying destructive power. The frequent crustal motions caused by plate tectonics generate vibrations, or **seismic waves,** that propagate through the Earth. The word *seismic* comes from the Greek word for "shake," and seismic waves shake the Earth's surface when they arrive—even if they've traveled all the way through the Earth from the location where an earthquake originated. After an earthquake, *seismographs* located in many places around the world record the shaking of the surface as seismic waves arrive. By comparing the recordings from many different places, we can reconstruct the paths that the seismic waves took through the Earth's interior.

Seismic waves generally travel through the Earth at speeds of thousands of kilometers per hour, but the precise speeds and directions of the waves depend on the composition, density, pressure, temperature, and phase (solid or liquid) of the material they pass through. Thus, we can deduce all of these internal conditions through careful study of seismic waves.

Seismic waves come in two basic types, and the difference between the two types helps us study Earth's internal conditions. We can demonstrate the two wave types by imagining that the bonds holding rocks together are like Slinkys (Figure 13.6a). Pushing and pulling on one end of a Slinky generates a wave in which the Slinky is bunched up in some places and stretched out in others. Waves like

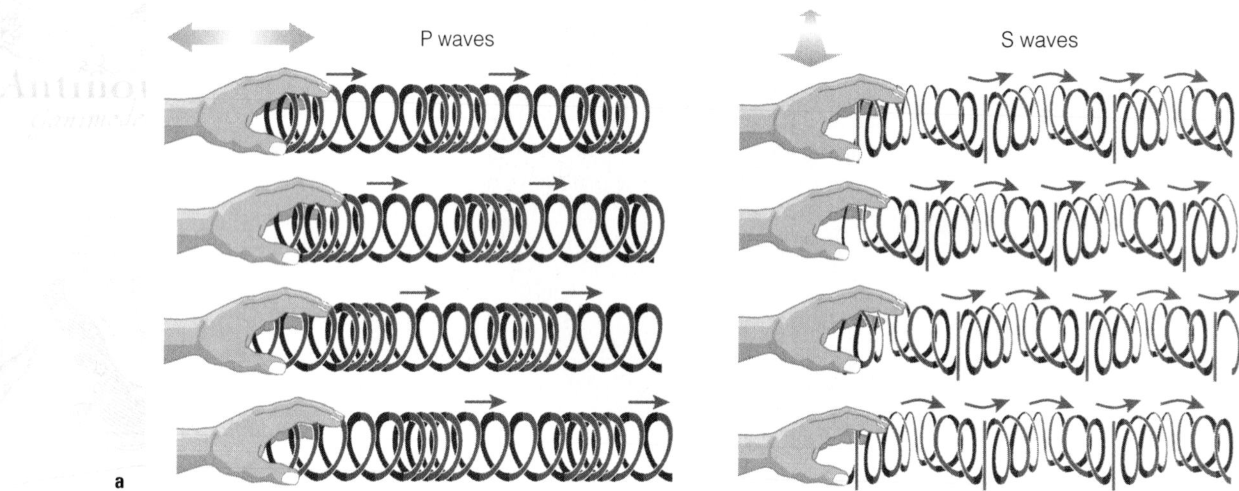

FIGURE 13.6 (**a**) Slinky examples demonstrating P and S waves.

occur more or less uniformly over the Earth's surface, including both continents and seafloors, because large impactors are not stopped by the oceans. Thus, the fact that we find fewer large craters on the seafloor than on the continents suggests that seafloor crust is much younger than is continental crust.

The rampant *erosion* on Earth arises primarily from processes involving water, but atmospheric winds also contribute. Wind tears away at geological features created by other processes, blows dust and sand across the globe, and builds features such as sand dunes. Note that wind erosion is more effective on Earth than on either Venus or Mars: Venus has very little surface wind because of its slow rotation, and wind on Mars does little damage because of the low atmospheric pressure.[4] On Earth, water contributes to erosion through a remarkable variety of processes (Figure 13.7). Perhaps the most obvious

[4]Note that high pressure alone is not enough to cause erosion. The lack of wind means little erosion on Venus despite high pressure. Similarly, erosion is relatively weak on Earth's seafloor because currents are slow despite the high pressure.

this in rock are called P waves. (The *P* stands for *primary,* because these waves travel fastest and therefore are the first to arrive, but is more easily remembered as *pressure* or *pushing.*) P waves are essentially a type of sound wave and therefore can travel through liquid or gas as well as through solid rock. In the second type of wave, shaking one end of the Slinky side-to-side or up and down sends a wave down its length. Such waves are called S waves in rock, and they travel through rock a bit more slowly than P waves. (The S stands for *secondary,* but it's easier to remember as *shear* or *shake.*) Unlike P waves, S waves cannot travel through liquids (or gases). Molecules in liquids are not connected by strong bonds like those that hold Slinkys together, so they cannot exert sideways forces on their neighbors. Thus, the fact that S waves do not reach the side of the world opposite an earthquake tells us that the Earth's interior must have a liquid layer in the outer region of the core (Figure 13.6b).

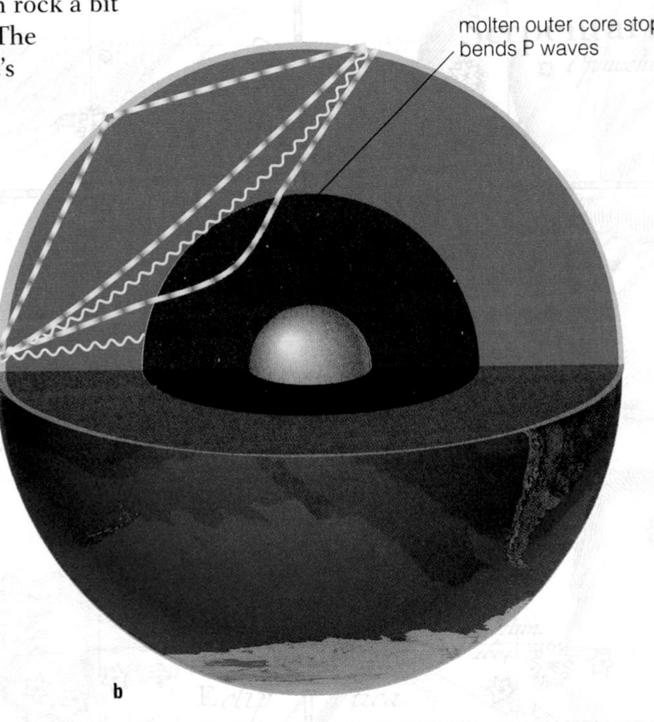

(**b**) Since S waves do not reach the side of the Earth opposite the earthquake, we infer that part of Earth's core is molten.

are rain and rivers breaking down mountains, carving canyons, and transporting sand and silt across continents to deposit them in the sea. But water erosion occurs on microscopic levels too: Water seeps into cracks and crevices in rocks and breaks them down from the inside, especially when this water freezes and expands. Further examples of water-erosion processes include the slow movement of glaciers, the flow of underground rivers, and the pounding of the ocean surf.

Volcanism and *tectonics* are also very important processes on Earth. The many active volcanoes prove that volcanism still reshapes the surface, and many small-scale cliffs and valleys offer evidence of tectonics similar to that found on other terrestrial worlds. But plate tectonics may be unique to Earth. So, to gain a deeper understanding of our planet's closely linked volcanism and tectonics, we must study plate tectonics in more detail.

Plate Tectonics

If you cut up a map and rearrange the continents, you'll find that they fit together in surprising ways. For example, the east coast of South America fits quite

FIGURE 13.7 A few examples of water erosion. (**a**) The Grand Canyon, carved by the flow of the Colorado River. (**b**) Yosemite Valley, carved when glaciers advanced during an ice age. (**c**) California houses falling into the sea as the pounding surf erodes the cliffs on which they were perched. (**d**) A river delta clogged with silt, an example of erosional processes building up a geological feature instead of breaking one down (straight channels created by humans).

nicely into the west coast of Africa (Figure 13.8). You might chalk this up to coincidence, but we also find similar types of distinct rocks and rare fossils in eastern South America and western Africa—suggesting that these two regions were once near each other. Based on such evidence, in 1912 German meteorologist and geologist Alfred Wegener proposed the idea of *continental drift:* that the continents gradually drift across the surface of the Earth, over time scales of tens of millions of years.

For decades after Wegener made his proposal, most geologists rejected the idea that continents could move, and many ridiculed the idea openly. However, the idea of continental drift gained favor in the 1950s as supporting evidence began to accumulate. Mapping of the seafloors revealed surprising structures (Figure 13.9): high *mid-ocean ridges* extending a total length of about 60,000 km along the ocean floors and *trenches* in which the ocean depth can reach more than 8 kilometers. Geologists soon recognized that these features represented boundaries between **plates**—pieces of the lithosphere that apparently float upon the denser mantle below. Convection cells in the upper mantle move the plates around the surface, forcing them together, apart, or sideways at their boundaries—with profound geological consequences. This discovery offered a model to explain how continents could drift about, and Wegener's idea finally gained acceptance under the new name **plate tectonics.** (Recall that tectonics is geological activity driven by internal stresses, so *plate tectonics* refers to the motion of plates driven by internal stresses.) Today we know that the Earth's lithosphere is broken into more than a dozen plates (Figure 13.10). Most major earthquakes and volcanic eruptions occur along plate boundaries.

TIME OUT TO THINK *Study the plate boundaries in Figure 13.10. Based on this diagram, explain why the West Coast states of California, Oregon, and Washington are more prone to earthquakes and volcanoes than other parts of the United States.*

Over millions of years, plate tectonics acts like a giant conveyor belt for the Earth's crust, carrying rock up from the mantle, transporting it across the seafloor, and then returning it down into the mantle (Figure 13.11). The mid-ocean ridges mark **spreading centers** between plates—places where hot mantle material rises upward and then spreads sideways, pushing the plates apart. New seafloor crust forms as the partially molten mantle material separates by density: Denser materials remain below in the mantle, while basalt rises up and emerges through a long string of underwater volcanoes. The newly formed basaltic crust cools and contracts as it spreads sideways from the central ridge, giving mid-ocean ridges their characteristic shape (see Figure 13.9). The new crust spreads throughout the area that opens as plates separate, forming only a relatively thin layer of crust over the mantle (see Figure 13.3). Worldwide along the mid-ocean ridges, new crust covers an area of about 2 km^2 every year—enough to replace the entire seafloor within a geologically short time of about 200 million years. Meanwhile, as new crust spreads over the surface, older crust must be returned

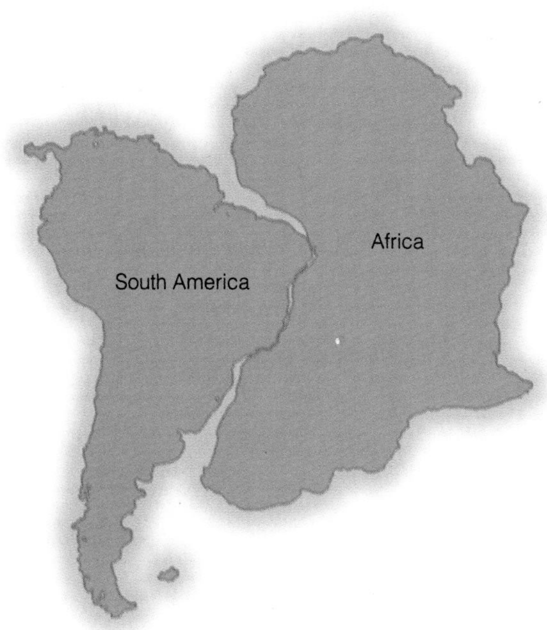

FIGURE 13.8 The "fit" of South America into Africa.

FIGURE 13.9 The discovery of mid-ocean ridges and deep trenches provided important evidence for the idea of continental drift. The relief map uses color to show elevation progressing from blue (lowest) to red (highest).

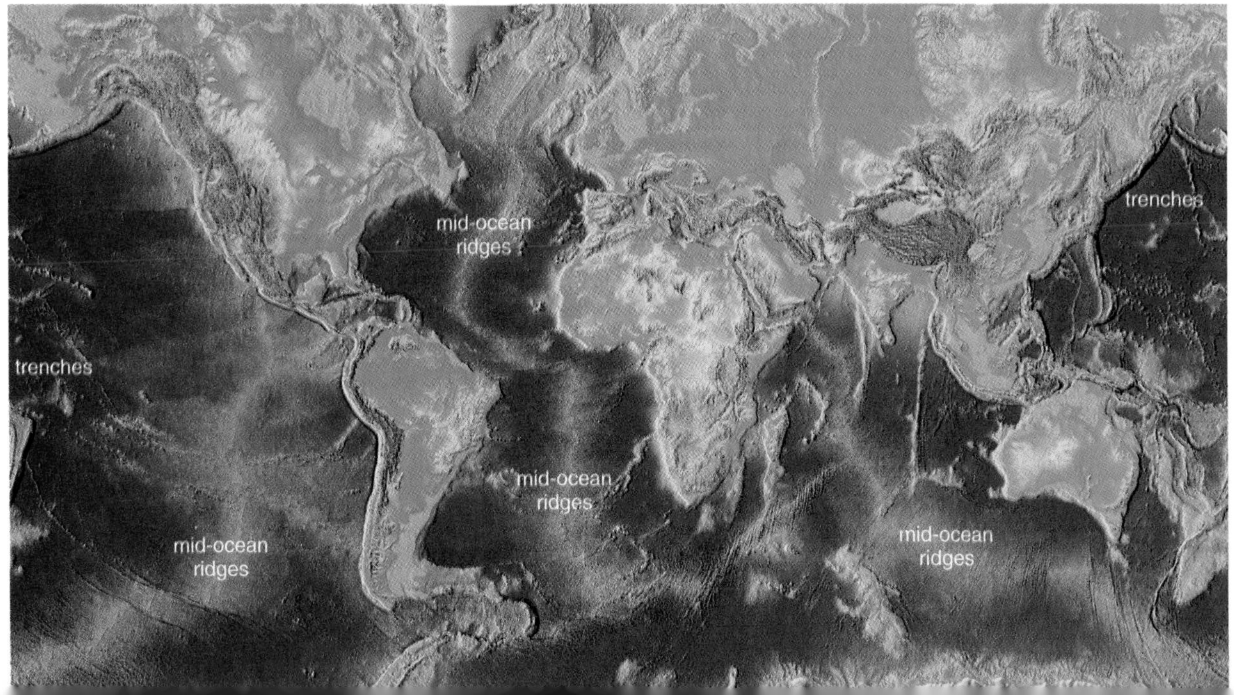

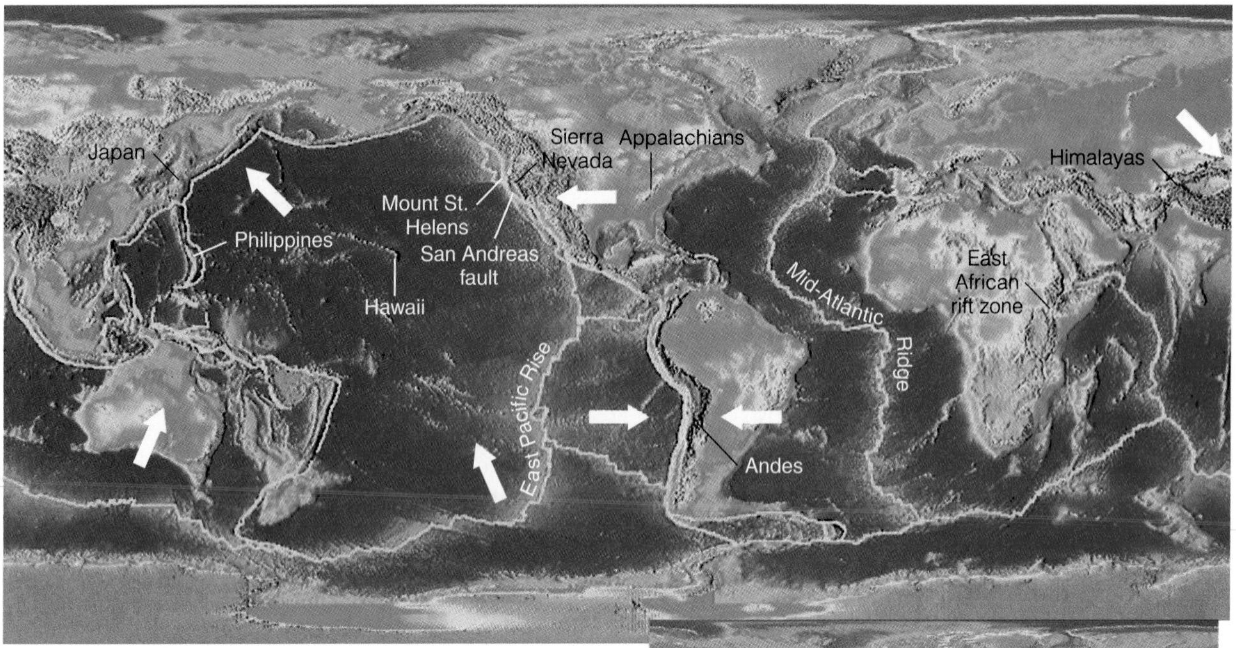

FIGURE 13.10 Solid lines show plate boundaries, and arrows represent directions in which the plates are moving. Important geological features discussed later in the text are identified. The small map shows that the locations of major earthquakes or volcanic eruptions (shown in red) during the past century are concentrated along plate boundaries.

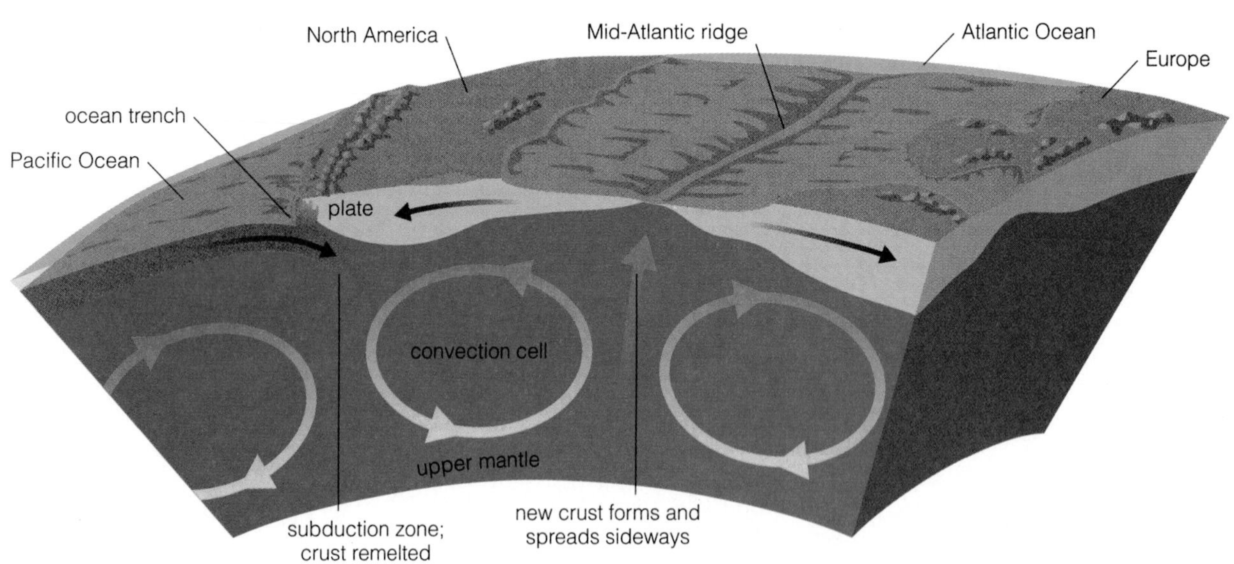

FIGURE 13.11 Plate tectonics acts like a giant conveyor belt, driven by mantle convection, that carries rock up from the mantle at mid-ocean ridges, transports it across the seafloor, and returns it down into the mantle at ocean trenches.

to the mantle. This return occurs at the locations of the deep ocean trenches, which mark places where one plate slides under another in a process called **subduction.** As we would expect from this model of seafloor crust formation, samples of seafloor crust are composed of basaltic material, and radioactive dating [Section 8.7] shows their age to be generally less than 200 million years.

The conveyor-like process of plate tectonics is undoubtedly driven by convection, although the precise nature of the convection remains uncertain. The crust may be simply dragged along by convection in the upper mantle. Alternatively, the tops of the convection cells may actually be the crust itself, with warm rock rising up under the ocean ridges and cooler, denser rock sinking into the interior in the subduction zones at the trenches. It's possible that both mechanisms play a role in driving plate tectonics.

Overall, the plates move across the Earth's surface at speeds of a few centimeters per year—about the same speed at which your fingernails grow—and we can use this motion to project the locations of the continents millions of years into the past or future. For example, at a speed of 2 centimeters per year, a plate will travel 2,000 kilometers in 100 million years. Over longer time scales, the plates may have altered their speeds and directions, but we can still project past locations of continents through careful comparisons of rock and fossil samples found in different places. For example, if 500-million-year-old rocks and fossils are very similar in two places, the two places were probably close together at that time—even if they are far apart now. Figure 13.12 shows several past arrangements of the continents, along with one future arrangement. Over the past billion

FIGURE 13.12 Past, present, and predicted future arrangements of the Earth's continents.

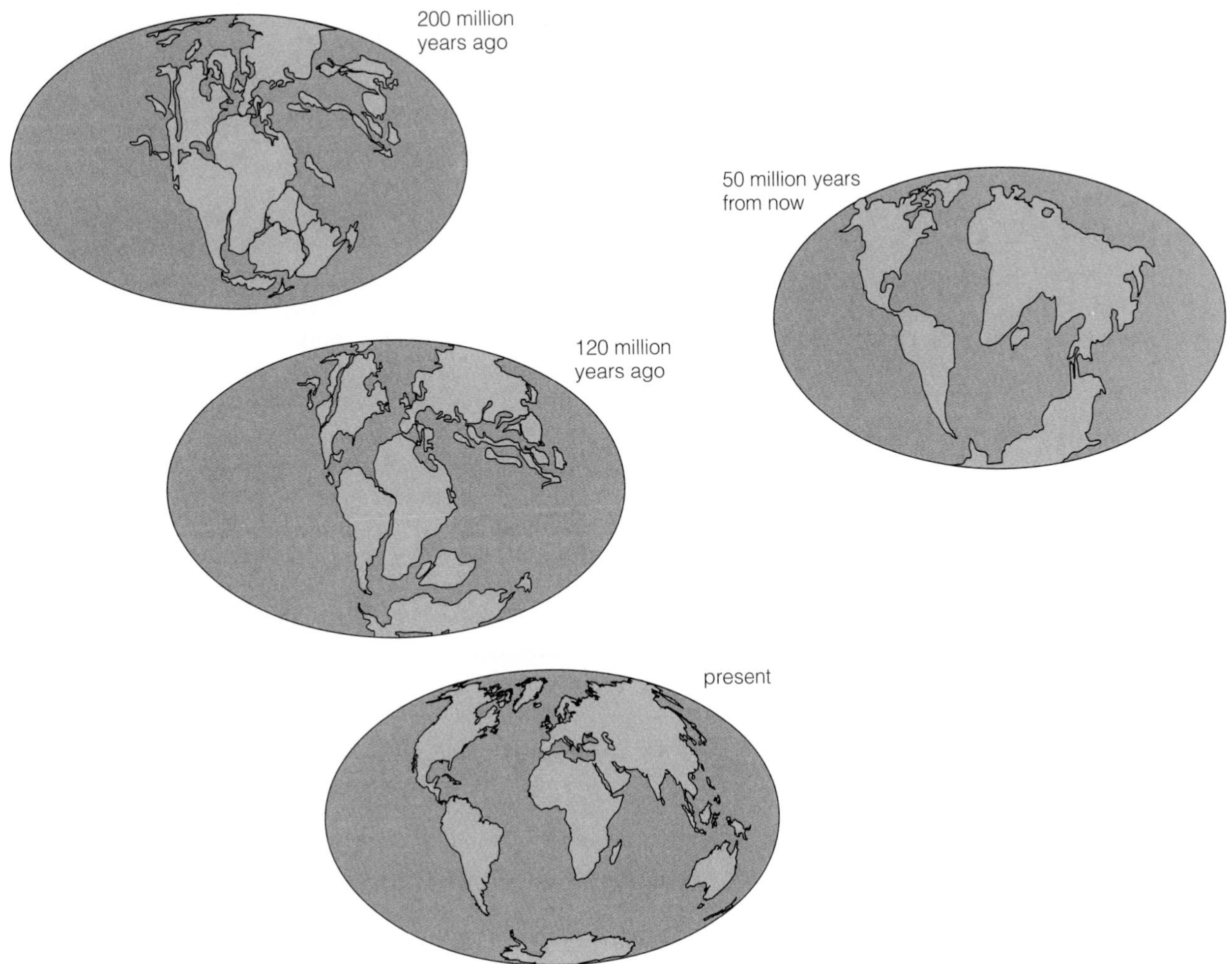

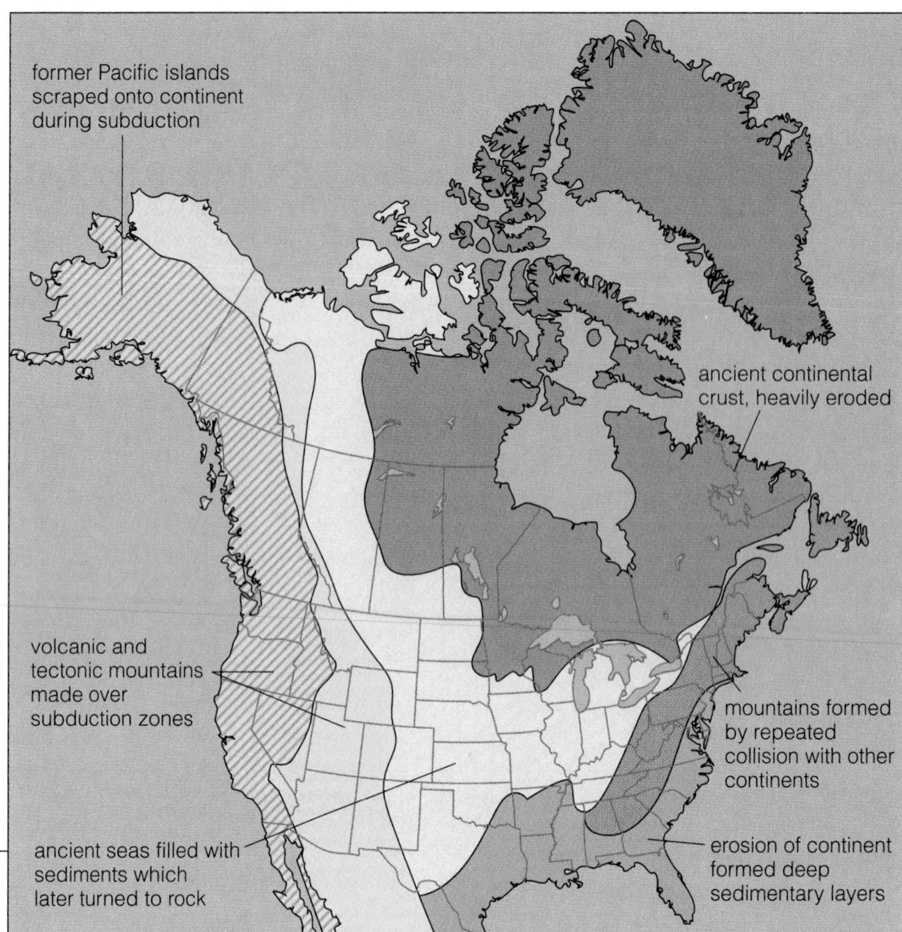

FIGURE 13.13 The major geological features of North America record the complex history of plate tectonics. Only the basic processes behind the largest features are shown here.

years, the continents have slammed together, pulled apart, spun around, and changed places on the globe. Central Africa once lay at Earth's South Pole, and Antarctica was once nearer the equator. At various times, such as was the case about 200 million years ago, the continents were all merged into a single giant continent, often called *Pangaea* (which means "all lands"). The current arrangement of the continents is no more permanent than any other.

Subduction and the Origin of Continents Where plates collide, subduction of seafloor crust creates continental crust. The basaltic seafloor crust remelts as subduction carries it back into the mantle, and the resulting magma separates by density. The lowest-density magma—which consists of rocks such as rhyolite, andesite, and granite—pushes upward to form continental crust. Thus, continental crust is made of lower-density material than seafloor crust. The present-day continents have been built up over billions of years; radioactive dating shows some rocks in continental crust to be as much as 4 billion years old. Despite being tens of kilometers thicker than seafloor crust, on average the continental crust rises only a few kilometers higher in altitude. The sheer weight of the continents presses down on the mantle, so most of their extra thickness goes deeper into the Earth. Thus, the continents poke up only a few kilometers higher than the seafloors (see Figure 13.3).

The rising magma along plate boundaries creates tremendous volcanic activity. Where one seafloor plate subducts under another, the resulting eruptions form a long string of volcanic islands. Alaska's Aleutian Islands exemplify this process at an early stage. As the process continues, the islands grow and merge; this has already happened with the islands of Japan and the Philippines, which once contained many smaller, individual islands that have merged to form the fewer islands we see today. As the islands continue to grow, they may eventually merge to create a new continent. In other cases, the plate carrying the islands may run up against an existing continental plate. The dense seafloor crust surrounding the islands subducts below the continents, but the islands resist subduction because they are made of low-density rock. The islands are essentially scraped off the seafloor and stuck onto the continent. Alaska, British Columbia, Washington, Oregon, and most of California began their existence as numerous Pacific islands that later attached to the growing North American continent (Figure 13.13). Seafloor continues to subduct below these states and provinces. Plate tectonics has gradually built up thick continental crust

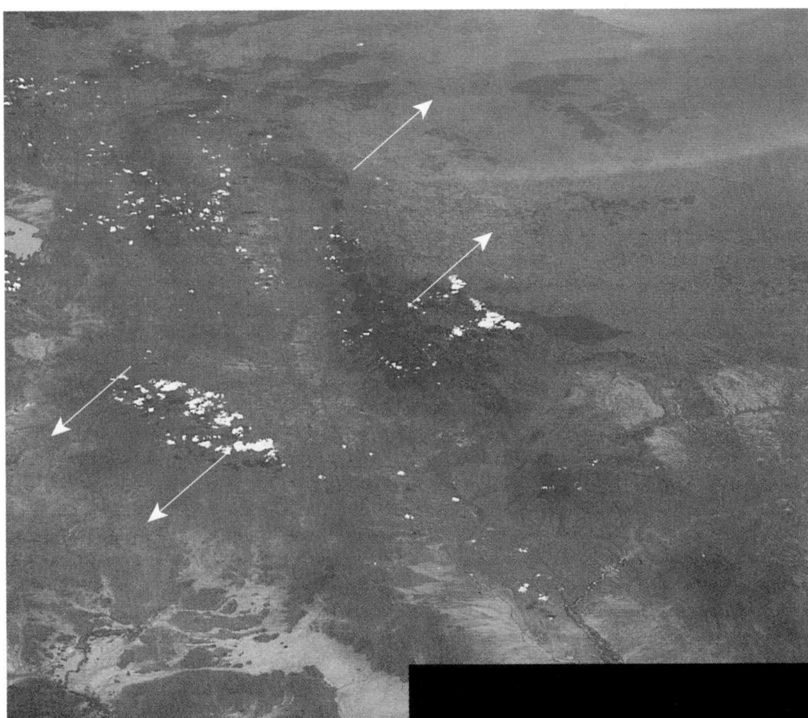

a

b

FIGURE 13.14 (**a**) Tectonic forces (arrows) are pulling apart eastern Africa, creating this rift valley photographed from space. (**b**) Tectonic forces have already torn the Arabian peninsula from Africa.

over about 45% of the Earth's surface,[5] and the fraction is still increasing. Billions of years ago, Earth was even more of an ocean planet than it is today.

The low-density magmas generated by subduction have relatively high viscosity, which affects the kind of volcanoes they construct. The rhyolite and andesite lavas that emerge from volcanoes along plate boundaries create steep-sided *stratovolcanoes* [Section 9.4] like Mount St. Helens in Washington and the mountains of the Andes range in South America.[6] Granite magma usually does not erupt onto the surface, but instead builds mountains through vast underground intrusions that make the rock layers above them bulge up. The upper rock layers may gradually erode away, leaving granite mountain ranges such as the Sierra Nevada in California.

Tectonic Stresses on Continents Continental crust is relatively thick, strong, and low-density, so it resists both spreading and subduction. Instead of simply separating like seafloor crust on either side of a mid-ocean ridge, continental crust thins and creates a *rift valley* when mantle convection tugs it apart. Figure 13.14a shows a rift valley in Africa. The rift will continue to grow, eventually tearing apart the continent and creating a seafloor spreading center. A similar process tore the Arabian peninsula from Africa in the past, creating the Red Sea (Figure 13.14b).

When two continent-bearing plates collide, neither plate can subduct (because of the low density of

[5]Some of this continental crust is underwater (e.g., continental shelves), so land areas represent less than 45% of the Earth's surface.

[6]In fact, *andesite* gets its name from the Andes range.

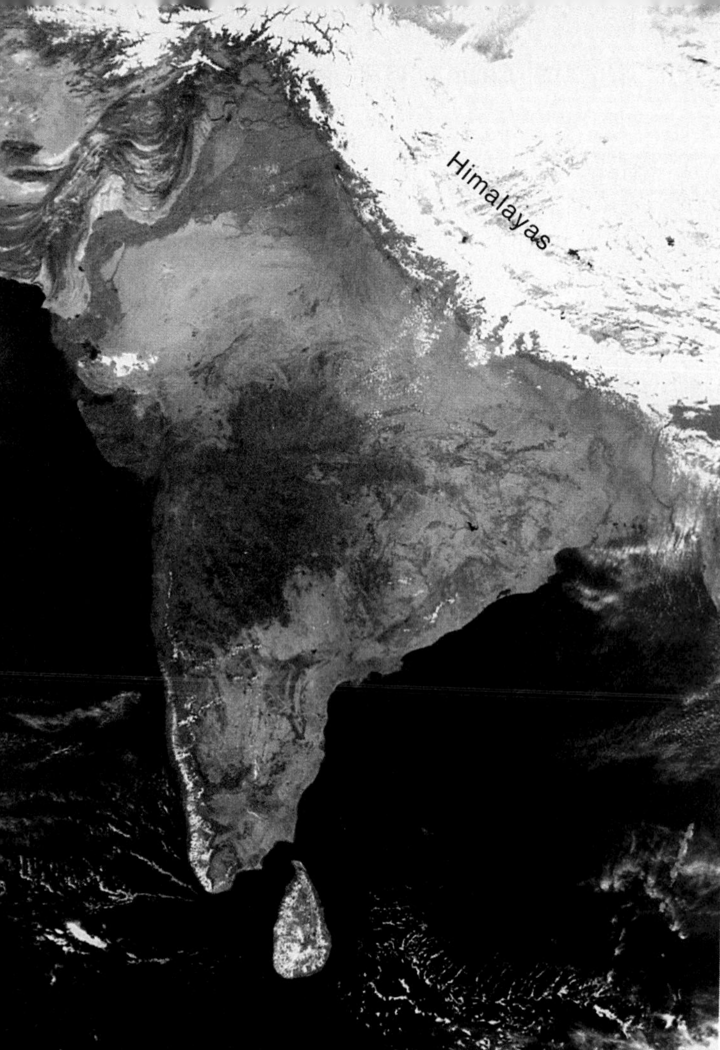

FIGURE 13.15 Space Shuttle photo of the Himalayas, which are still slowly growing as the Indian plate (carrying India) pushes into the Eurasian plate (carrying most of the rest of Asia).

the crust), and the resulting pressure creates tremendous mountain ranges. The Indian Plate is currently ramming into the Eurasian Plate, forming the Himalayas (Figure 13.15). These mountains, already the tallest on Earth, are still growing. The Appalachian range in the eastern United States formed by the same process, as North America collided twice with South America and then with western Africa over a period of hundreds of millions of years. The Appalachians were probably once as tall as the Himalayas are now, but erosion has gradually transformed them into the fairly modest mountain range we see today. Similar processes have contributed to the formation of the Rocky Mountains in the United States and Canada.

At other places on Earth, plates slip sideways relative to each other along a **fault,** or a fracture in the lithosphere. This is the case with the *San Andreas fault* in California, where plate motions are carrying Los Angeles on a 20-million-year trip to San Francisco (Figure 13.16a). Plates do not slip smoothly against each other. Their rough surfaces resist slippage until the tension grows too high; then everything slips along the fault in the violent motion that we call an *earthquake* (Figure 13.16b). The motions associated with earthquakes can raise mountains, level cities, and set the whole planet vibrating with seismic waves. In contrast to the usual motion of plates at the rate of a few centimeters per year, an earthquake can move plates by several *meters* in a few seconds. Although most earthquake faults lie along plate boundaries, a few more ancient faults are found elsewhere. As a result, devastating earthquakes occasionally occur in regions that we tend to think of as safe. In fact, one of the largest earthquakes in U.S. history occurred in Missouri in 1811.

Hot Spots and Mantle Plumes Not all volcanoes occur near plate boundaries. Sometimes, a plume of hot mantle material may rise in what we call a **hot spot** within a plate. The Hawaiian Islands are the result of a hot spot that has been erupting basaltic lava for tens of millions of years (Figure 13.17). The low viscosity of this lava produces broad *shield volcanoes* [Section 9.4]. Today, most of the lava erupts on the "Big Island" of Hawaii, giving much of this island a young, rocky surface. But about a million years ago, the mantle plume lay farther to the northwest—or, equivalently, the entire plate lay farther to the southeast. Back then, the eruptions built the island of Maui, parts of which are now heavily eroded and covered in lush vegetation. Before that, the mantle hot spot built the other islands of Hawaii, including Oahu (3 million years ago) and Kauai (5 million years ago); it also created Midway Island (27 million years ago), which has now almost completely eroded back to sea level. Thus, this single hot spot created a vast string of islands (and smaller underwater mountains) as the plate gradually moved over it. The process continues today: A new island named Loihi is currently forming beneath the ocean surface and will rise above sea level southeast of Hawaii in a million years or so. Hot spots can also appear beneath continental crust. The geysers and hot springs of Yellowstone National Park result from the heating of a mantle plume. This hot spot is currently migrating to the northeast (relative to the continental crust) because of the southwestern motion of the plate.

Summary of Earth's Geology

Two fundamental differences between the geology of Earth and that of other terrestrial worlds stand out:

1. Rampant erosion driven by the action of plentiful water.

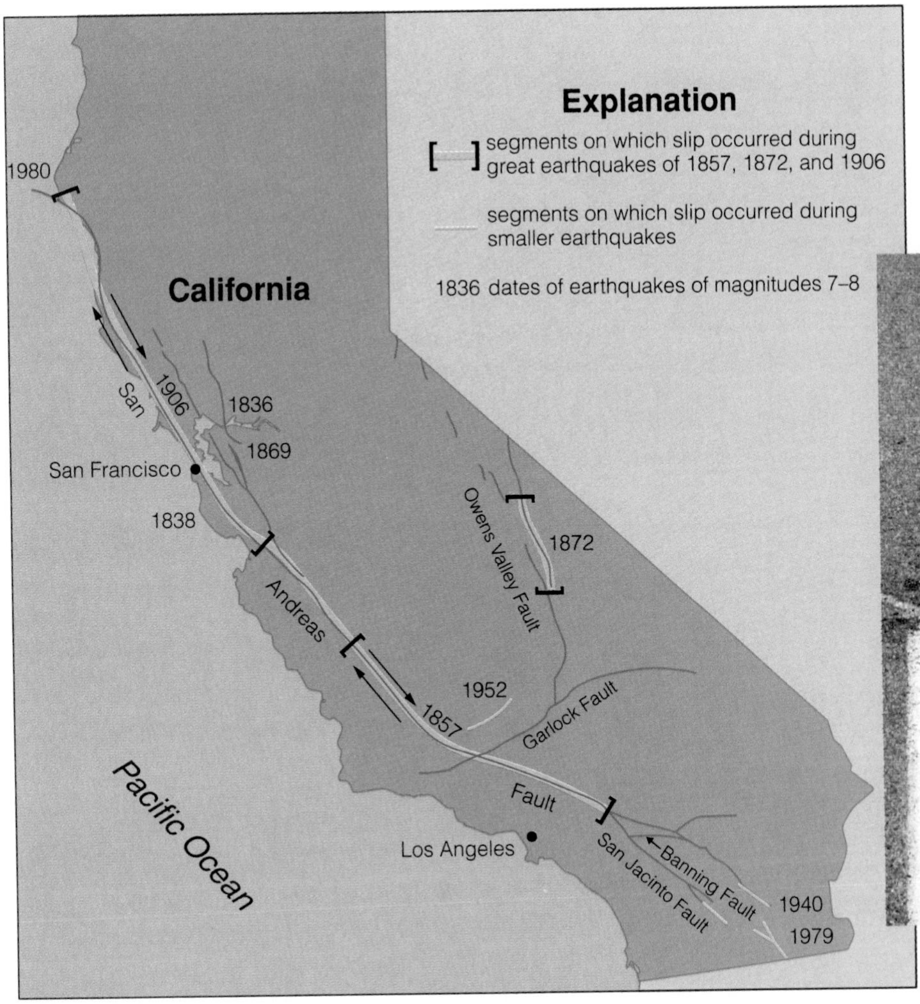

FIGURE 13.16 (**a**) This map shows how land on the west side of the San Andreas Fault is moving northward relative to land on the east side of the fault. The photo to the right of the map shows a place where a painted lane on a road is no longer straight because of this movement. (**b**) Earthquakes occur when plates slip violently along a fault. This photo shows damage from the 1995 earthquake in Kobe, Japan.

2. A different style of tectonics—plate tectonics—that recycles crust, drives high-viscosity volcanism to create continental crust, and causes frequent earthquakes.

Together, these two processes have reshaped the Earth's surface beyond recognition from its original state billions of years ago. They make the Earth's surface very young, perhaps the youngest in the solar system besides that of Io [Section 11.5].

But why does Earth's geology differ from that of our neighbors in these two ways? In the next section, we'll see how our unique atmosphere leads to the strong erosion on Earth. We'll also find that the interactions between our atmosphere and geology might help explain why Earth has plate tectonics but similar-size Venus does not.

FIGURE 13.17 The Hawaiian Islands are just the most recent of a very long string of islands made by a hot spot. The black and white image of Loihi was obtained by sonar, as it is currently entirely under water.

13.3 Our Unique Atmosphere

In Chapter 10, we learned that the superficial differences in the atmospheres of the terrestrial planets are fairly easy to understand. For example, the Moon and Mercury lack substantial atmospheres because of their small sizes, and a very strong greenhouse effect causes the high surface temperature on Venus. We also learned how solar heating, seasonal changes, and the Coriolis effect that comes from rotation lead to differing global circulation patterns on different worlds. But explaining how the compositions of the atmospheres came to be so different is more challenging, especially when we consider the cases of Venus, Earth, and Mars. Outgassing produced all three of these atmospheres, and the same volatiles—primarily water and carbon dioxide—were outgassed in all three cases. How, then, did Earth's atmosphere end up so different? In particular, any study of Earth's atmosphere must address four major questions:

1. Why did Earth retain most of its water—in the form of the oceans and other components of the hydrosphere—while Venus and Mars lost theirs?
2. Why does Earth have so little carbon dioxide (CO_2) in its atmosphere, when Earth should have outgassed about as much of it as Venus?
3. Why does Earth have so much more oxygen (O_2) than Venus or Mars? Where did it come from, and how does this highly reactive gas remain present in the atmosphere?
4. Why does Earth have an ultraviolet-absorbing stratosphere, while Venus and Mars do not?

To answer these questions, we must look into the history of Earth's atmosphere. The answers to all four questions turn out to be closely connected.

TIME OUT TO THINK *Recall that water ice could not condense in the region of the solar nebula where the terrestrial planets formed. How, then, did the water outgassed from volcanoes get here in the first place?* (Hint: *See Section* 8.6.)

Water and the Origin of the Hydrosphere

On a basic level, it's fairly easy to explain why the Earth retains so much water while Venus and Mars do not. As we learned in Chapter 10, Venus probably lost an "ocean-full" of water as solar ultraviolet photons split apart water molecules and the hydrogen atoms subsequently escaped to space. Mars probably lost some of its water in a similar way, and the rest is frozen at the polar caps and under the surface.

Let's look a little more closely at the sequence of events that led to these different histories. Picture a generic terrestrial planet billions of years ago: Erupting volcanoes were outgassing carbon dioxide, water vapor, and small amounts of nitrogen. As the water vapor increased, did this water vapor condense into oceans, freeze out as snow, or remain gaseous in the atmosphere? Even before Venus had a strong greenhouse effect, its proximity to the Sun made it warm enough to keep all its water gaseous in the atmosphere. The water vapor caused significant greenhouse warming and ensured that any additional water would also remain gaseous and warm the planet further. This tendency of the greenhouse effect to reinforce itself is called the **runaway greenhouse effect.** As volcanoes continued to belch out gases, Venus's atmosphere accumulated water vapor that later escaped into space. At the other extreme, Mars's temperature was low enough for the water vapor to freeze out of the atmosphere, resulting in thick polar caps.

On Earth, moderate temperatures allowed the water vapor to rain out of the atmosphere and liquid water to accumulate on the surface. A small amount of water vapor remained in the atmosphere and caused significant greenhouse warming, but additional water vapor released from volcanoes rained into the oceans and did not lead to a runaway greenhouse effect. The condensed water gradually formed the oceans and the other components of the hydrosphere. Thus, Venus lost its water because it was too hot, Mars lost its water because it was too cold, and Earth retained its water because conditions were "just right."

Where Is All the CO_2?

Measurements of outgassing by active volcanoes and estimates of past volcanic activity suggest that Earth must have outgassed nearly as much carbon dioxide throughout its history as Venus. But Venus today has an atmosphere with a CO_2 concentration of 96%, a surface pressure 90 times greater than that on Earth, and about 500°C of greenhouse warming. In contrast, the proportion of carbon dioxide in Earth's atmosphere is much less than 1%; in fact, we usually measure CO_2 concentration on Earth in units of *parts per million.* (For example, 300 parts per million means $300 \div 1{,}000{,}000 = 0.0003 = 0.03\%$.) Clearly, some atmospheric loss process must have removed nearly all of the outgassed CO_2 on Earth.

The secret of Earth's CO_2 is wrapped up in the history of our oceans. The primary process that removes atmospheric CO_2 on Earth involves liquid water. Carbon dioxide can dissolve in water, and the oceans actually contain about 60 times more carbon dioxide than the atmosphere (still a very small amount compared to the total amount that must have been outgassed). Most of the carbon dioxide is locked up in rocks on the seafloor through chemical reactions. Rainfall erodes silicate rocks on the Earth's surface and carries the eroded minerals to the oceans. There the minerals react with dissolved carbon dioxide to form *carbonate* minerals [Section 10.4], which fall to the ocean floor, building up thick layers of carbonate rock such as *limestone.* This is part of the **carbonate–silicate cycle** (Figure 13.18). Because the process requires liquid water, it operates only on Earth, where it has removed about as much CO_2 from Earth's atmosphere as now remains in Venus's atmosphere. (If Venus's CO_2 could somehow be removed, the remaining atmosphere would have about as much nitrogen as Earth's.) Had Earth been slightly closer to the Sun, its warmer temperature might have evaporated the oceans, leaving the CO_2 in the atmosphere and causing a runaway greenhouse effect as on Venus. Conditions might not have been so conducive to "life as we know it."

The interplay of atmosphere and geology may also explain why Earth has plate tectonics while Venus does not. The apparent lack of plate tectonics on Venus suggests that its lithosphere is stronger and thicker than Earth's. How can this be, given that both planets probably have similar internal temperatures? The answer may lie in the history of Venus's atmosphere. As Venus lost its oceans, even the small amounts of water dissolved in the crust and mantle were baked out, removing the lubricating and softening effects of volatiles trapped in rocks. So Venus's atmospheric evolution may have thickened the lithosphere to the point where Earth-like plate tectonics cannot occur. This theory is still controversial,[7] but it underscores the surprising possibility that a planet's atmosphere can affect its interior, not just the other way around.

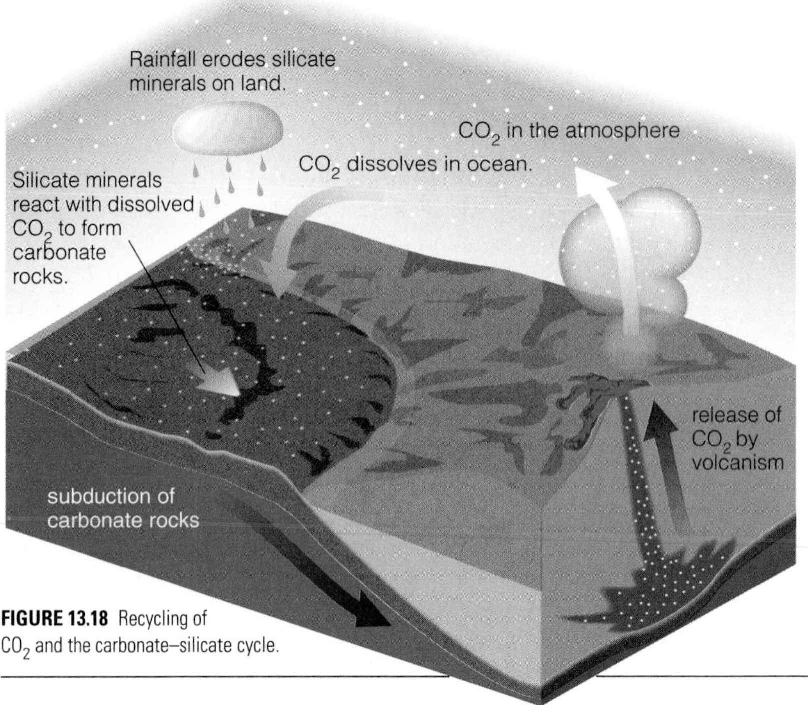

FIGURE 13.18 Recycling of CO_2 and the carbonate–silicate cycle.

The Origin of Oxygen, Ozone, and the Stratosphere

While both water and carbon dioxide are products of outgassing, molecular oxygen (O_2) is not. In fact, no geological process can explain the great abundance of oxygen (about 20%) in Earth's atmosphere. Moreover, oxygen is a highly reactive chemical that would rapidly disappear from the atmosphere if it were not continuously resupplied. Fire, rust, and the discoloration of freshly cut fruits and vegetables are everyday examples of **oxidation**—chemical reactions that remove oxygen from the atmosphere. Similar reactions between oxygen and surface materials (especially iron-bearing minerals) give rise to the reddish appearance of many of Earth's rock layers, such as those in Arizona's Grand Canyon. Thus, we must explain not only how oxygen got into the Earth's atmosphere in the first place, but also how the amount of oxygen remains relatively steady even while oxidation reactions tend to bind oxygen into rocks at a rapid rate.

The answer to the oxygen mystery is *life*. The process that supplies oxygen to the atmosphere is *photosynthesis*, which converts CO_2 to O_2. The carbon becomes incorporated into amino acids, proteins, and other components of living organisms. Today, plants and single-celled photosynthetic organisms return oxygen to the atmosphere in approximate balance with the rate at which animals and oxidation reactions consume oxygen, so the oxygen content of the atmosphere stays relatively steady. Earth originally developed its oxygen atmosphere when photosynthesis added oxygen at a rate greater than these processes could remove it from the atmosphere.

TIME OUT TO THINK *Mars also has red rocks. Is this evidence of life or of something else?* (Hint: *See Section 9.5.*)

Life and oxygen also explain the presence of Earth's ultraviolet-absorbing stratosphere. In the upper atmosphere, chemical reactions involving solar ultraviolet light transform some of the O_2 into molecules of O_3, or *ozone*. The O_3 molecule is more weakly bound than O_2, which allows it to absorb solar ultraviolet energy even better, giving rise to the warm stratosphere and preventing harmful ultraviolet radiation from reaching the surface. Mars and Venus lack photosynthetic life and therefore have too little O_2 and too little ozone to form a stratosphere.

[7]Some planetary geologists think that Venus's lithosphere is actually thinner than Earth's.

Feedback Processes and Carbon Dioxide Balance

The Earth today is just warm enough for liquid water and "life as we know it" thanks to a moderate greenhouse effect caused by the presence of water vapor and carbon dioxide in our atmosphere. Our good fortune might seem to be based on atmospheric properties alone, but in fact our luck lies deeper—with Earth's geological processes. Much of the CO_2 is injected into the atmosphere by volcanoes near subduction zones: Carbonate rocks carried beneath the seafloor by subduction melt and release their CO_2 (see Figure 13.18). The balance between the rate at which carbonate rocks form in the oceans and the rate at which carbonate rocks melt in subduction zones largely determines the amount of CO_2 in Earth's atmosphere. If plate tectonics—especially subduction—stopped on Earth, or if it occurred at a substantially different rate, the amount of atmospheric CO_2 would undoubtedly be different. The eventual effect on Earth's temperature might prove fatal to us.

Life also plays a role in the CO_2 balance, because the photosynthetic cycle consumes carbon dioxide as it produces oxygen.[8] Moreover, just as the carbon dioxide locked up in carbonate rocks can later be released, so can the carbon dioxide consumed by living organisms. When organic materials burn, their carbon reacts with atmospheric oxygen to produce carbon dioxide. Fossil fuels—oil, coal, and natural gas—are the remains of living organisms that died long ago. When we burn these fuels, we add carbon dioxide to the atmosphere, changing the CO_2 balance.

The full story of the Earth's carbon dioxide balance is even more complex because of **feedback relationships**—relationships in which a change in one property amplifies (*positive feedback*) or counteracts (*negative feedback*) the behavior of the rest of the system. You are undoubtedly familiar with audio feedback: If you bring a microphone too close to a loudspeaker, it picks up and amplifies small sounds from the speaker, and these amplified sounds are again picked up by the microphone and further amplified, causing a loud screech. This is an example of positive feedback. The screech usually leads to a form of negative feedback: The embarrassed person holding the microphone moves away from the loudspeaker, thereby stopping the positive audio feedback.

A planet's carbon dioxide balance, and hence its greenhouse-induced temperature, is subject to both positive and negative feedback. The hotter a planet's surface, the more of its greenhouse gases will be in the atmosphere instead of in the oceans and crust. But this increase in greenhouse gases further increases the temperature—an example of positive feedback (which ultimately led to the runaway greenhouse effect on Venus). However, in some cases, warming temperatures can also lead to negative feedback, at least on planets with liquid water. The warmer temperature leads to more evaporation and cloud formation, and the higher albedo of clouds causes cooling as more sunlight is reflected back to space. Furthermore, higher temperatures increase the rate of carbonate rock formation, pulling more CO_2 from the atmosphere and cooling the planet. The complexity of these relationships makes predicting planetary climates difficult.

Common Misconceptions: Ozone—Good or Bad?

Ozone often generates confusion, because in human terms it is sometimes good and sometimes bad. In the stratosphere, ozone acts as a protective shield from the Sun's ultraviolet radiation. However, ozone is poisonous to most living creatures and therefore is a bad thing when it is found near the Earth's surface. In fact, ozone is one of the main ingredients in urban air pollution, produced as a byproduct of automobiles and industry. Some people wonder whether we might be able to transport this ozone to the stratosphere and thereby alleviate the effects of ozone depletion. Unfortunately, this plan won't work. Even if we could find a way to transport this ozone, all the ozone ever produced in urban pollution would barely make a dent in the amount lost from the stratosphere.

TIME OUT TO THINK *The Earth's oceans have an albedo of about 5%, but ice has an albedo closer to 90%. Suppose it became just cold enough to freeze the oceans. How would the change in their albedo further affect the Earth's temperature? Is this an example of positive or negative feedback?*

Although we do not yet know precisely how these feedback relationships work in detail, they have clearly played a major role throughout the terrestrial planet histories. Careful studies of ancient rocks show that liquid water was plentiful throughout Earth's history, so Earth's temperature has apparently been quite stable for billions of years. However, stars like the

[8] Life in the oceans plays an additional role: One mechanism by which carbonate rocks form in the ocean involves seashells that sink to the ocean floor.

Sun increase in brightness as they age, and billions of years ago our Sun was probably about 30% fainter than it is today [Section 14.3]. With Earth's current abundance of greenhouse gases, our planet would have been frozen over. Thus, greenhouse gases must have been more abundant early in Earth's history. Venus, on the other hand, might have experienced hospitable conditions if it possessed "only" its current inventory of greenhouse gases—but it probably also had more greenhouse gases in its early history, leading to its runaway greenhouse effect. On Mars, extra greenhouse gases once provided the warm temperatures needed for liquid water to flow, but Mars lost so much of this gas that it became colder even as the Sun grew brighter.

Summary of Earth's Atmosphere

Venus, Earth, and Mars all began on similar paths, releasing similar gases into their atmospheres by outgassing. But only Earth had conditions "just right" to support liquid oceans throughout its history. The oceans helped remove carbon dioxide from the atmosphere, ensuring that the greenhouse effect on Earth remained just strong enough to keep conditions hospitable for life, but not so strong as to create a runaway greenhouse effect like that on Venus.

But the physical connections between the Earth's geology and its atmosphere cannot fully explain the conditions on Earth today. In particular, the presence of oxygen and ozone can be explained only as products of life. Thus, to complete our understanding of the unique place of Earth in the solar system, we must now turn our attention to the role of life on Earth.

13.4 Life

Fossils, the petrified remains of living organisms, tell the story of life on Earth (Figure 13.19). Most fossils formed when the dead organisms fell to the bottom of a body of water, where they were gradually buried by layers of sediment and eventually compressed into rocks. In some areas, layers of sediments have been deposited one on top of the others for billions of years, burying a detailed record of life in the process (Figure 13.20). Erosion or tectonic activity later exposes the fossils. Some fossils are remarkably well preserved, and we find fossils of large animals, plants, and even microscopic, single-celled organisms.

FIGURE 13.19 Dinosaur fossils.

FIGURE 13.20 The rock layers of the Grand Canyon record 2 billion years of Earth history.

The key to reconstructing the history of life is to determine the dates at which fossil organisms lived. The *relative* ages of fossils found in different layers can be determined easily: Deeper layers formed earlier and contain more ancient fossils. Radioactive dating confirms these relative ages and gives us fairly accurate absolute ages for fossils [Section 8.7].

For relatively recent geological history, such as the age of the dinosaurs that began about 250 million years ago and ended abruptly 65 million years ago, the numerous fossil skeletons of extinct animals prove that life on Earth has undergone dramatic changes. Life in earlier epochs is more difficult to characterize. Primitive life-forms without skeletons leave fewer fossils, erosion erases much old fossil evidence, and subduction destroys other fossil evidence deep beneath the Earth's surface. Nevertheless, geologists and biologists have developed a fairly clear picture of the history of life on Earth, at least in broad outline. As we trace this history through time, you should ask yourself three key questions:

1. How did Earth's physical conditions affect the development of life?

2. How has life influenced the physical characteristics of Earth?
3. What can we learn about the prospects for finding life elsewhere in the solar system or universe?

The Origin of Life (4.4–3.5 billion years ago)

Life arose on Earth at least 3.5 billion years ago, and possibly much earlier than that. We've found recognizable fossils in rocks as old as 3.5 billion years—microscopic fossils of single-celled bacteria and larger fossils of bacterial "colonies" called *stromatolites* (Figure 13.21). But indirect evidence suggests that life originated at least 350 million years earlier (3.85 billion years ago), even though such ancient rock samples are too contorted for fossils to remain recognizable. This evidence comes from analysis of different isotopes of carbon [Section 5.3]. Most carbon atoms are atoms of carbon-12 (6 protons and 6 neutrons), but a small proportion of carbon atoms are instead carbon-13 (6 protons and 7 neutrons).[9] Living organisms incorporate carbon-12 slightly more easily than carbon-13, and as a result the fraction of carbon-13 is always a bit lower in fossils than in rock samples that lack fossils. In fact, all life and all fossils tested to date show the same characteristic ratio of the two carbon isotopes. Rocks as old as 3.85 billion years show the same ratio, strongly suggesting that these rocks contain remnants of life. Beyond 3.85 billion years, we cannot yet determine whether life existed, since rocks more ancient than this no longer exist on Earth's surface. It is conceivable that life began as early as 4.4 billion years ago.

TIME OUT TO THINK ***You may have noticed that, while we've been talking about life, we haven't actually defined the term. In fact, it's surprisingly difficult to draw a clear boundary between life and nonlife. How would you define life? Explain your reasoning.***

If life did originate much before 3.85 billion years ago, it may have been in for a rough time. The transition from intolerable conditions during planetary formation to more hospitable conditions was a gradual one. The early bombardment that continued after the end of accretion gradually died away over a period of a few hundred million years. But large impacts capable of sterilizing the planet by temporarily boiling all the oceans may have occurred as late as 4.0–3.8 billion years ago. Earthquakes and volcanic eruptions were probably more frequent and more violent as well, due to greater radioactive heating in Earth's interior. Plate tectonics may not yet have begun, and oceans may have covered the entire surface. Indeed, some biologists speculate that life may have arisen several times in the Earth's early history, only to be extinguished by the hostile conditions. If so, life might have turned out very different if one less—or one more—sterilizing impact had occurred.

An even deeper question concerns not just *when* life arose, but *how*. No one knows the precise answer

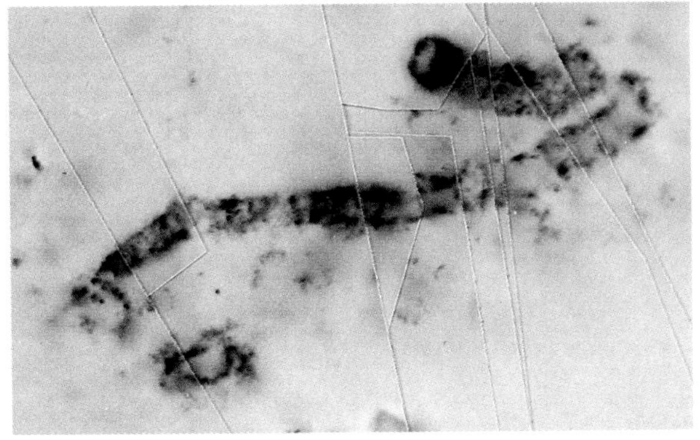

FIGURE 13.21 (**a**) Microscopic fossil bacteria. (**b**) Stromatolite fossils (up to 3.5 billion years old). (**c**) Living stromatolites today. The biological origin of fossil stromatolites was questioned until living examples were found.

[9]Recall that carbon also comes in a third form, carbon-14 (6 protons and 8 neutrons), but this form is radioactive with a half-life of 5,700 years. Thus, no carbon-14 survives in rocks from billions of years ago.

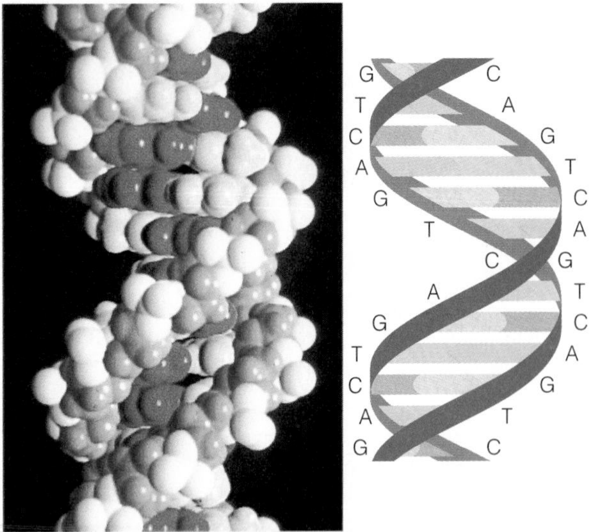

FIGURE 13.22 Left: A model of a small piece of a DNA molecule. Right: This diagram shows that a DNA molecule is made from two intertwined strands, with "ladder steps" connecting pairs of chemical bases. Note that A always connects to T and G always connects to C. The genetic code is a language in which triplets (along a single strand) of the chemical bases make "words" that represent particular amino acids.

to this question, but we are confident of one thing: Whether life arose once or multiple times, one particular type of organism came to dominate the entire Earth. That is, every organism living today apparently developed from a single ancestor.

The idea of a common ancestor comes from several lines of evidence based on the fact that all known life-forms share uncanny chemical resemblances to one another—similarities that are not expected on simple chemical grounds and that are far too improbable and numerous to be considered coincidences. First, all known organisms use virtually the same limited set of chemical building blocks. For example, proteins are made from building blocks called **amino acids,** and all living organisms use the same set of 20 different amino acids—even though more than 70 amino acids exist. Similarly, all living organisms use the same basic molecule (called ATP) to store energy within cells, and all use molecules of DNA to transmit their genes from one generation to the next. (Some viruses use RNA, rather than DNA, as their genetic material.)

A second line of evidence for a common ancestor comes from the fact that all living organisms share nearly the same **genetic code**—the "language" that living cells use to read the instructions chemically encoded in DNA. You are probably familiar with the structure of DNA, which consists of two long strands that look somewhat like train tracks, wound together in the shape of a double helix (Figure 13.22). Each letter shown along the DNA strands represents one of four *chemical bases,* denoted by A, G, T, and C (for the first letters of their chemical names). The instructions for assembling the cell are written in the precise arrangement of these four chemical bases: Genetic "words" composed of three consecutive bases represent particular amino acids. (For example, the "word" CAG represents one amino acid, and TCA represents a completely different amino acid. Chemical reactions between the DNA and its surroundings in effect read the codes.) There is no known reason why living organisms should follow a particular genetic code, and biochemists believe that DNA could use different sequences of these bases—or possibly even different bases altogether—to encode the same genetic information. Thus, the fact that all living organisms share a common genetic code implies a common ancestry.

A third line of evidence for a common ancestor comes from detailed analysis of the sequence of chemical bases in the DNA of different organisms. In particular, biologists have mapped out the sequence in many different organisms that holds the instructions for making a molecule called rRNA (ribosomal RNA). They've discovered that this sequence includes "unused" segments that apparently do not affect the structure or function of rRNA. Biologists can trace the process of evolution by comparing these unused segments. For example, suppose the unused segments look the same in two organisms but a little bit different in a third organism. Then we can conclude that the first two organisms are more closely related to each other than to the third. By doing many such comparisons, biologists have developed an evolutionary "tree of life" (Figure 13.23). Although many details in the structure of this tree remain uncertain, it clearly shows that living organisms share a common ancestry.

Comparison of DNA sequences also allows at least reasonable guesses as to which living organisms most resemble the common ancestor of all life. Surprisingly, the answer appears to be organisms living in the deep oceans around seafloor volcanic vents called *black smokers* (after the dark, mineral-rich water that flows out of them) and in hot springs in places like Yellowstone (Figure 13.24). These organisms thrive in temperatures as high as 125°C.[10] Unlike most life at the Earth's surface, which depends on sunlight, the ultimate energy source for these organisms is chemical reactions in water volcanically heated by the internal heat of the Earth itself.

Of course, knowing that all life shares a common ancestor still does not tell us how that ancestor first

[10]The high pressures at the seafloor prevent the water from boiling until it reaches 450°C.

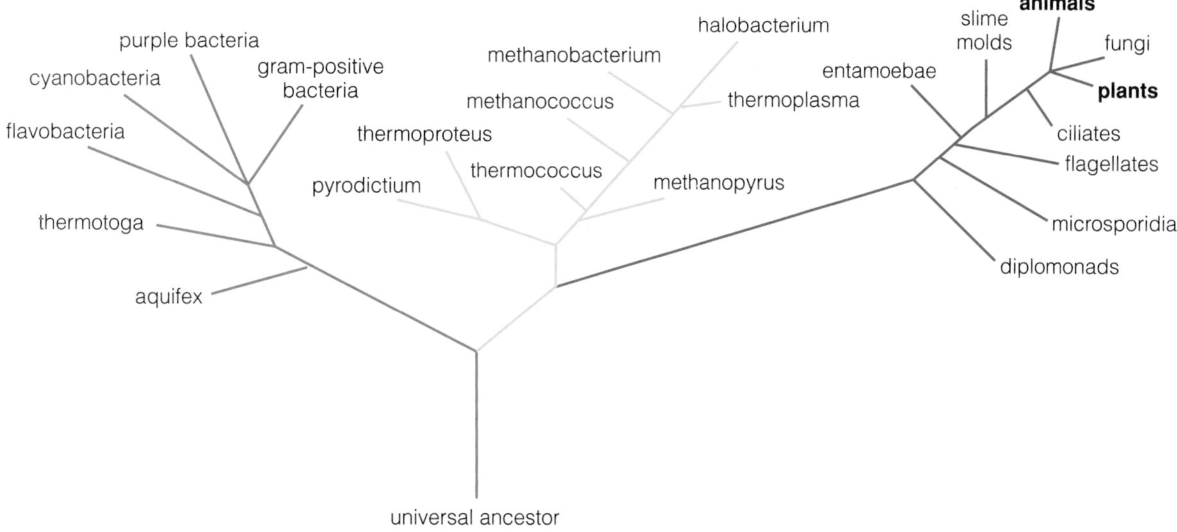

FIGURE 13.23 The tree of life, showing evolutionary relationships according to modern biology as of early 1998. Note that just two small branches represent *all* plant and animal species.

FIGURE 13.24 (**a**) Life around a black smoker deep beneath the ocean surface. (**b**) This aerial photo shows a hot spring in Yellowstone National Park that is filled with colorful bacterial life.

arose. The step from simple chemical building blocks to our common DNA-bearing ancestor (from non-life to life) is huge, to say the least. No fossils record the transition, nor has it ever been duplicated in the laboratory.

The bare necessities for "life as we know it"—chemicals, energy, and water—were undoubtedly available on the early Earth. The building blocks of life, including amino acids and nucleic acids, are composed primarily of carbon, oxygen, nitrogen, and hydrogen—chemicals that were readily available. Energy to fuel chemical reactions was present on the surface in the form of lightning and ultraviolet light from the Sun and was also present in the oceans in the heated water near undersea volcanoes. We know that oceans were present, because we find ancient carbonate rocks that must have formed in water.

Laboratory experiments demonstrate that the chemical building blocks of life could have formed spontaneously and rapidly in the conditions that prevailed early in Earth's history. In these experiments, researchers replicate the chemical conditions of the early Earth by, for example, including the types of ingredients outgassed by volcanoes and "sparking" the mixture with electricity to simulate lightning or other energy sources. Within just a few days, the mixture spontaneously becomes rich with amino acids and nucleic acids. Such chemical reactions were undoubtedly an important source of organic molecules on Earth, but they may not have been the only source—some meteorites contain complex organic molecules.

With the building blocks for life plentiful, it may only have been a matter of time before chemical reactions created a molecule that could make a copy of itself. Such a *self-replicating molecule* might have spread quickly through regions with hospitable conditions. **Mutations**—errors in the copying process—could have created variations on the original molecule, some with increased complexity. If life really began in this way, then some ancient self-replicating molecule was the ancestor of modern DNA—and of all life on Earth.

Early Evolution in the Oceans (3.5–2.0 billion years ago)

Regardless of its origin, life soon thrived in the oceans in the form of single-celled organisms, tapping a variety of energy sources including sunlight (i.e., through photosynthesis). Individual organisms that survived and reproduced passed on copies of their DNA to the next generation. However, the transmission of DNA from one generation to the next is not always perfect. Mutations can change the arrangement of chemical bases in a strand of DNA, making the genetic information in a new cell slightly different from that of its parent. Mutations can be caused by many factors, including high-energy *cosmic rays* from space [Section 18.2], ultraviolet light from the Sun, particles emitted by the decay of radioactive elements in the Earth, and various toxic chemicals. Most mutations are lethal, killing the cell in which the mutation occurs. However, some mutations may make a cell better able to survive in its surroundings, and the cell then passes on this improvement to its offspring.

The process by which mutations that make an organism better able to survive get passed on to future generations is called **natural selection.** The idea of natural selection was proposed by Charles Darwin (1809–82) as a way of explaining evolution (which simply means "change with time") in animal species. Today, we believe that natural selection is the primary mechanism by which evolution proceeds: Over time, natural selection may help individuals of a species become better able to compete for scarce resources and may also lead to the development of entirely new species from old ones. The fossil record provides strong evidence that evolution *has* occurred, while natural selection explains *how* it occurs.

Fossils show that evolution progressed remarkably slowly for most of Earth's history. For at least a billion years after life first arose, the most complex life-forms were still single-celled. Some 2 billion years ago, the land was still inhospitable because of the lack of a protective ozone layer. Continents much like those today were surrounded by oceans teeming with life, but the land itself was probably as barren as Mars is today, despite pleasant temperatures and plentiful rainfall.

Altering the Atmosphere (beginning about 2 billion years ago)

The process of photosynthesis appears to have developed quite early in the history of life. Over billions of years, the abundant single-celled organisms in the ocean pulled carbon dioxide from the atmosphere and put back oxygen. For a long time, nearly all the highly reactive oxygen gas was pulled back out of the atmosphere by reactions with surface rocks. However, by about 2 billion years ago, the oxidation of surface rocks was fairly complete, and oxygen began to accumulate in the atmosphere. Today it constitutes about 20% of the atmosphere, but this fraction may vary over periods of millions of years.

The buildup of oxygen had two far-reaching effects for life on Earth. First, it made possible the development of oxygen-breathing *animals* (Figure 13.25). No one knows precisely how or when the first oxygen-breathing organism appeared, but for that animal the world was filled with food. You can imagine how quickly these creatures must have spread around the world, and the fact that other organisms could now be eaten changed the "rules of the game" for evolution.

The second major effect of the oxygen buildup was the formation of the ozone layer, which made it safe for life to move onto the land. Although it took more than a billion years, natural selection eventually led to plants that could survive and thrive on land, and animals followed soon after, taking advantage of this new food source.

FIGURE 13.25 Thanks to the buildup of oxygen in the atmosphere (and small amounts dissolved in the oceans), animals like these trilobites became possible. Trilobites were among the most complex animals alive a few hundred million years ago. The largest specimens reached 75 cm (30 in.) in length.

This epoch highlights the ability of life to fundamentally alter a planet. Not only did life change the Earth's atmosphere, but in doing so it made two other kinds of life possible: oxygen-breathing animals and land organisms.

An Explosion of Diversity (beginning about 0.54 billion years ago)

As recently as 540 million years ago, most life-forms were still single-celled and tiny. Note that life had already been present on Earth for at least 3 billion years by that time, demonstrating the remarkable slowness with which evolution progressed for most of Earth's history. But the fossil record reveals a dramatic diversification of life beginning about 540 million years ago. Although this change occurred over a period of about 40 million years, it was so dramatic in comparison to the events of the previous 3 billion years that it is often called the *Cambrian explosion*. (Cambrian is the name geologists give the period from about 540 million to 500 million years ago.) We can trace the origin of most of today's plant and animal species—from insects to trees to vertebrates—back to this period. Apparently, once life began to diversify, the increased competition among the many more complex species accelerated the process of natural selection and the evolution of new species.

Some species have been more successful and more adaptable than others. Dinosaurs dominated the landscape of the Earth for more than 100 million years. But their sudden demise 65 million years ago paved the way for the evolution of large mammals—including us. The earliest humans appeared on the scene only a few million years ago (after 99.9% of Earth's history), and our few centuries of industry and technology have come after 99.99999% of Earth's history. Despite our recent arrival (in geological terms), modern humans are by some measures the most successful species ever to inhabit the Earth. Humans survive and prosper in virtually every type of land environment. With proper equipment, we can survive underwater, and a few people have lived under the sea for extended periods of time in experimental habitats. We are even developing the technology to survive away from our planet, in the inhospitable environment of space. And our population has been growing exponentially for the past few centuries, roughly doubling in just the past 40 years (Figure 13.26).

The tremendous growth of the human population suggests a chilling analogy: Exponential growth is also the mark of a cancerous tumor—which may seem very successful at surviving while it is growing but ultimately dies when it kills its host. Is it possible that our tremendous success is killing our host, the Earth? We can gain some perspective on this question by looking at a few lessons that the solar system teaches us about our relationship with the Earth.

TIME OUT TO THINK *The pace of change in human existence has accelerated since the advent of modern humans a few million years ago: Civilization arose about 10 thousand years ago, the industrial revolution began just a couple of centuries ago, and the computer era started just a few decades ago. Do you think the pace of change can continue to accelerate? What do you think the next stage will be?*

FIGURE 13.26 This graph shows human population over the past 12,000 years. Note the tremendous population growth that has occurred in just the past few centuries.

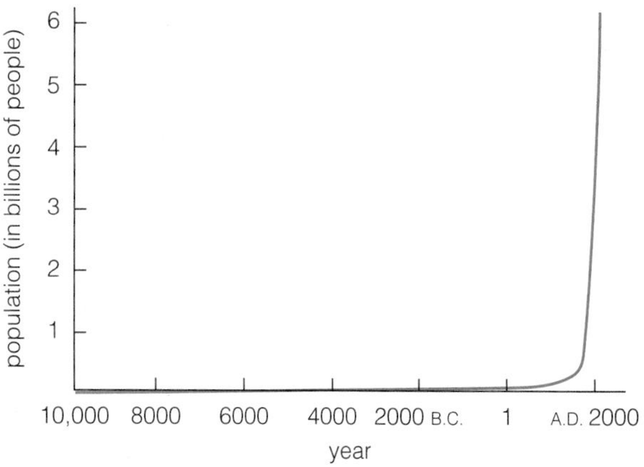

13.5 Lessons from the Solar System

The United States and other nations have carried out the exploration of the solar system for many reasons, including scientific curiosity, national pride, and technological advancement. Perhaps the most important return on this investment was one not expected: an improved understanding of our own planet and an enhanced ability to understand some very important global issues. In this section, we explore three particularly important issues about which comparative planetology offers important insights: global warming, ozone depletion, and mass extinctions.

Global Warming

Venus stands as a searing example of the effects of large amounts of atmospheric carbon dioxide. Earth remains habitable only because natural processes have locked up most of its carbon dioxide on the seafloor. However, humans are now tinkering with this difference between the two planets by adding carbon dioxide and other greenhouse gases to the Earth's atmosphere. The primary way we release carbon dioxide is through the burning of fossil fuels (coal, natural gas, and oil)—the carbon-rich remains of plants buried millions of years ago. But other human actions, such as deforestation and the subsequent burning of trees, also affect the carbon dioxide balance of our atmosphere. The inescapable result is that the amount of atmospheric carbon dioxide has risen steadily since the dawn of the industrial age—and continues to rise (Figure 13.27).

Although it may seem logical to conclude that an increase in the concentration of greenhouse gases should warm the Earth, the Earth is a very complex system. As a result, the debate over whether human activity is changing the Earth's climate is fierce. Measuring even something as seemingly simple as the average temperature of the entire planet can be surprisingly difficult. Until the advent of satellite temperature measurements, temperature data came only from ground-based weather stations and were subject to bias. For example, most weather stations are restricted to land and are located near cities (which tend to be warmer than surrounding areas). Despite measurement difficulties, however, there is growing scientific consensus that the average temperature of the Earth has warmed slightly (less than about 1°C) over the past 50 years—although there is much less agreement about why. It's likely that the cause is the human addition of greenhouse gases to the atmosphere, but it's also possible that the change is natural and would have occurred even without human activity. Some scientists even suggest that the Sun itself may be warming and thus contributing to the global temperature rise.

The question of whether human activity is inducing global warming is not merely academic: The consequences of a significant warming could be widespread and disastrous. The primary way scientists try to predict the effects of global warming is by making sophisticated computer models of the climate. Different models give different results, but some predict a warming of more than 5°C during the next century. Past ice ages occurred when the average temperature dropped by about 5°C; a similar increase could melt polar ice and flood the Earth's densely populated coastal regions. Such warming would also increase

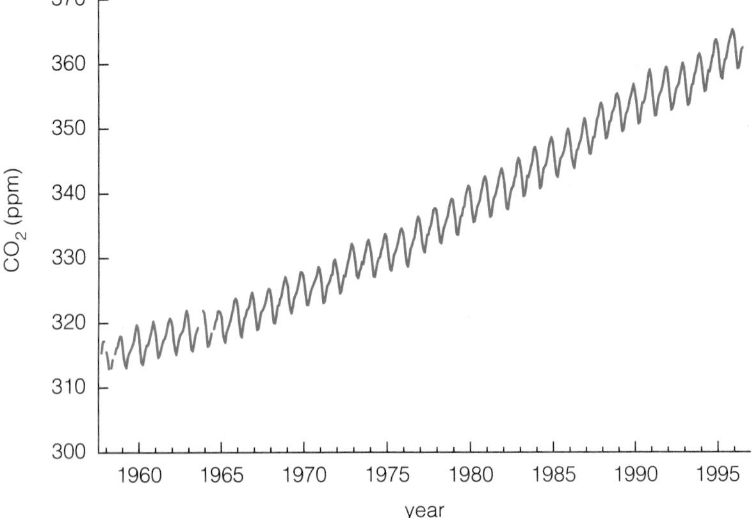

FIGURE 13.27 These data, collected for many years on Mauna Loa (Hawaii), show the increase in atmospheric carbon dioxide concentration. Yearly wiggles represent seasonal variations in the concentration, but the long-term trend is clearly upward. The concentration is measured in parts per million (ppm).

evaporation from the oceans, tending to make storms of all types both more frequent and more destructive: Coastal regions would be hit by more hurricanes, intense thunderstorms and associated tornadoes would strike more frequently, and even winter blizzards would be more severe and more damaging. Computer models also show that global warming would not affect all regions in the same way. Some regions would warm much more than the average, while others might even cool. Rainfall patterns would also shift, and much of the world's current cropland might become unusable. The most ominous possibility is the collapse of entire ecosystems. If a regional climate changes more rapidly than local species can adapt or migrate, these species might become extinct. Given the complex interactions of the Earth's biosphere, hydrosphere, and atmosphere, the consequences of such ecosystem changes are impossible to foresee.

The uncertainty in our understanding of global warming is underscored by the fact that most current computer models predict that the Earth should already have warmed to a greater degree than we've observed. This suggests that some process may be counteracting the effects of the added greenhouse gases. One hypothesis is that increased evaporation from the oceans might lead to more clouds that prevent sunlight from reaching the surface. Another is that recent volcanic eruptions, such as the 1991 eruption of Mount Pinatubo in the Philippines, have injected particles into the stratosphere that reflect sunlight, effectively increasing the Earth's albedo. The most frightening possibility is that reflective particles are indeed being injected into the atmosphere, but their source is sulfate particles from coal-burning industries. At first, this idea might seem to suggest that the pollutants from coal burning counter the effects of the added carbon dioxide, but in reality this can only postpone and worsen the problem: Sulfate particles fall out of the atmosphere in a matter of years, while carbon dioxide may remain for half a million years.

Given the current uncertainties, it is impossible for anyone to know for sure whether global warming will even be a serious problem in the next century let alone make specific predictions of its severity. Our studies of the planets show that surface temperatures depend on many factors and that positive and negative feedback processes can alter climates in surprising ways. Nevertheless, the fact that we cannot explain the current climate conditions on Earth suggests that we should be very careful about tampering with them.

TIME OUT TO THINK *The most obvious way to prevent global warming is to stop burning the fossil fuels that add greenhouse gases to the atmosphere. But much of the world economy depends on energy produced from fossil fuels. Given the uncertainties about the effects of global warming in the next century, what if any changes do you think are justified at present? What specific proposals would you make if you were a world leader?*

Ozone Depletion

For 2 billion years, the Earth's ozone layer has shielded the surface from hazardous ultraviolet radiation, but in the past two decades humans have begun to destroy the shield. The idea that human activity might damage the ozone was first suggested in the early 1970s, but the magnitude of the problem was not recognized until the discovery of an **ozone hole** over Antarctica in the mid-1980s (Figure 13.28). The "hole" is a place where the concentration of ozone in the stratosphere is dramatically lower than it would naturally be. The ozone hole appears for a few months during each Antarctic spring and has gradually worsened over the years. Since its discovery, satellite measurements have suggested that **ozone depletion** has caused ozone levels to fall by as much as a few percent worldwide. If ozone levels continue to decline, more solar ultraviolet radiation will reach the surface. Humans will face substantially greater risk from skin cancer, and plants and animals will suffer more genetic damage from mutations induced by ultraviolet radiation—with unknown consequences for the biosphere as a whole.

The apparent cause of the ozone depletion is human-made chemicals known as CFCs (chlorofluorocarbons). CFCs were invented in the 1930s and have been widely used in air conditioners and refrigerators, in the manufacture of packaging foam, as propellants in spray cans, in industrial solvents used in the computer industry, and for many other purposes. Indeed, CFCs once seemed like an almost ideal chemical: They are useful in many industries, cheap and easy to produce, and chemically inert (i.e., they do not burn, break down, or react with anything on the Earth's surface). Ironically, their inertness is also their downfall; because CFCs are gases and are not destroyed by chemical reactions in the lower atmosphere, they eventually rise intact into the stratosphere. There they are broken down by the Sun's ultraviolet light, and one of the by-products is chlorine—a very reactive gas that catalytically destroys

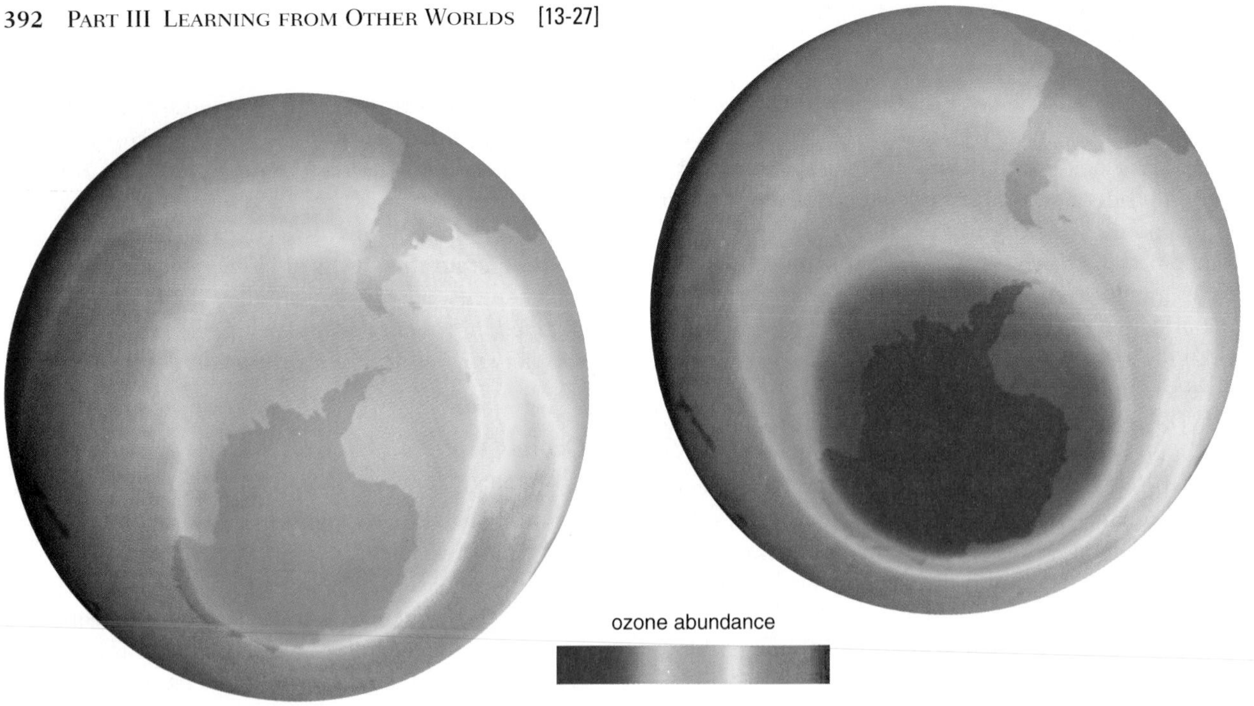

FIGURE 13.28 The extent of the ozone hole over Antarctica in 1979 (left) and 1998 (right). The ozone hole worsened between 1979 and 1998. Red indicates normal ozone concentration, and blue indicates severe depletion.

ozone without being consumed in the process. On average, a single chlorine atom in the stratosphere can destroy 100,000 ozone molecules before it is consumed itself in some other chemical reaction. Chlorine's ability to attack ozone is enhanced at low temperatures, which explains why ozone depletion was first observed over Antarctica.

Mars and Venus each played an important role in our understanding of ozone and chlorine. Mars lacks an ozone layer, so ultraviolet radiation reaches the surface. As a result, the surface is sterile: Terrestrial life could not survive on the surface, and the building blocks of life would be torn apart when exposed to the Sun. But it was the study of Venus that first alerted scientists to the dangers on Earth. Chlorine is a minor but important ingredient of Venus's atmosphere, and models of its chemical cycles showed that chlorine could rapidly destroy weak molecules such as ozone. When the models were altered to apply to Earth's atmosphere, CFCs and ozone depletion were connected.

This tale of three planets may have a happy ending, thanks to the awareness it brought of the dangers posed by ozone destruction. Today, international treaties ban the production of CFCs, although previously existing CFCs may still be used and recycled. If these treaties are obeyed, ozone destruction probably will not be a serious problem in the future, but the problem will remain at some level for decades to come. Most of the CFCs ever produced are still in intact air conditioners, refrigerators, or other products, and most of these CFCs will eventually escape into the atmosphere. Moreover, because the stratosphere lacks the weather of the troposphere [Section 10.2], CFCs (and their breakdown products) remain in the stratosphere for decades once they arrive. But the greater danger lies in the possibility of treaty failures. Some of the replacements for CFCs may be only marginally less damaging than the CFCs themselves. In addition, the replacements are generally much more expensive to produce than CFCs, and a large global black market now exists for CFCs produced in violation of the treaties. Like most global issues, the final outcome of the ozone problem remains in doubt.

TIME OUT TO THINK ***You've just learned that your car air conditioner is broken and will cost $200 to fix. Then you hear about a shop that can fix it for only $100—undoubtedly by violating the laws concerning CFCs. Is the small benefit to our ozone layer worth the $100 difference?***

Mass Extinctions

When geology was a young science, its practitioners thought that entire mountain ranges and huge valleys were formed suddenly by cataclysmic events. Later geologists denounced this *catastrophism* and replaced it with *uniformitarianism,* which holds that geological change occurs gradually by processes acting over very long periods. Uniformitarianism certainly holds for three of our four geological processes, but impact cratering represents the "instant geology" of catastrophism. Many geologists initially rejected the importance of impact cratering as a geological process, but they were eventually convinced by the unambiguous evidence of impacts on the Moon, by astronomical studies of comets and asteroids, and by careful analysis of the few impact craters still present on Earth. Geology today is a combination of uniformitarianism and catastrophism.

Scientific ideas about evolution are undergoing a similar transformation. Biologists once assumed that evolution always proceeded gradually, with new species arising slowly and others occasionally becoming extinct. It now appears that the vast majority of all species in Earth's history died out in sudden *mass extinctions*—for which impacts are the prime suspect. The mass extinction that killed off the dinosaurs is only the most famous of these events [Section 12.6]; the fossil record shows evidence of at least a dozen other mass extinctions, some with even greater species loss than in the dinosaur event. While some species died out from direct effects of the impact, most probably died due to the disappearance from their ecosystem of species on which they depended. Furthermore, the survival or extinction of species appears to be random during mass extinctions: Surviving species show no discernible advantage over those that perish. In a sense, mass extinctions clear the blackboard, allowing evolution to get a fresh start with a new set of species. If we re-formed the Earth from scratch, life (and even intelligent life) might well arise again, but there is no scientific basis for thinking that evolution would have followed the same course.

The story of mass extinctions teaches us about the danger of large impacts, but it also teaches a potentially more important lesson: Even without an impact, the Earth may be undergoing a mass extinction right now. In the normal course of evolution, the rate of species extinction is fairly low—perhaps one species lost per century. According to some estimates, human activity is now driving species to extinction so rapidly that half of today's species may be lost by the end of the twenty-first century. It is not just that humans are hunting and fishing many species to extinction. Far more are lost due to habitat destruction and pollution. And when a few species are lost by direct human influence, entire ecosystems that depend on those species may collapse. If global warming alters the climate significantly or if ozone depletion leads to more genetic damage in plants and animals, the rate of species extinction might increase even more. In any case, the "event" of losing half the world's species in just a few hundred years certainly qualifies as a mass extinction on a geological time scale, and the biological effects could be as dramatic as those of any impact. Are we unwittingly clearing the way for a new set of dominant species?

13.6 Lessons from Earth: Life in the Solar System and Beyond

Just as comparative planetology teaches us much about the Earth, the study of the Earth can teach us much about other worlds. In particular, it teaches us about the conditions for life and may help us answer what is surely one of the deepest philosophical questions of all time: Are we alone in the universe? It is fitting to close this portion of the book by extending our understanding of biology to the planets and beyond. The study of the possibility of life elsewhere in the universe, called *exobiology* (or astrobiology), is a logical extension of comparative planetology.

In the time between the Copernican revolution [Section 6.3] and the space age, many people expected the other planets in our solar system to be Earth-like and to harbor intelligent life. In fact, a reward was supposedly once offered for the first evidence of intelligent life on another planet *other than Mars.* Venus, a bit closer to the Sun than Earth, was often pictured as a tropical paradise. Such expectations were dashed by the bleak images of Mars returned by spacecraft and the discovery of the runaway greenhouse effect on Venus. Many scientists began to believe that only Earth has the right conditions for life, intelligent or otherwise. Recently, the pendulum has begun to swing the other way, spurred primarily by two developments. First, biologists are learning that life thrives under a much wider range of conditions than once imagined. Second, planetary scientists are developing a much better understanding of conditions on other worlds. It now seems quite likely that conditions in at least some places on other worlds might be conducive to life.

The Hardiness, Diversity, and Probability of Life

Even on Earth, biologists long assumed that many environments were uninhabitable. But recent discoveries have found life surviving in a remarkable range of conditions. The teeming life surviving at temperatures as high as 125°C near underwater volcanic vents and in hot springs is only one of many surprises. Biologists have found microorganisms living deep inside rocks in the frozen deserts of Antarctica and inside basaltic rocks buried more than a kilometer underground. Some bacteria can even survive radiation levels once thought lethal—apparently, they have evolved cellular machinery that repairs mutations as fast as they occur. The newly discovered diversity of microscopic life has forced scientists to redraw the "tree of life" (see Figure 13.23), crowding familiar plants and animals into one corner. The majority of these microorganisms need neither sunlight, oxygen, nor "food" in the form of other organisms. Instead, they tap a variety of chemical reactions for their survival.

The new view of terrestrial biology forces us to rethink the possibility of life elsewhere. First, life on Earth thrives at extremes of temperature, pressure, and atmospheric conditions that overlap conditions found on other worlds. The Antarctic valleys, for example, are as dry and cold as certain parts of Mars, and the conditions found in terrestrial hot springs may have been duplicated on a number of planets and moons at certain times in the past. Second, life harnesses energy sources readily available on other planets. The basalt-dwelling bacteria, for example, would probably survive if they were transplanted to Mars. Third, life on Earth has evolved from a common ancestor into every imaginable ecological niche (and some unimaginable ones as well). The diversification of life on Earth has basically tested the limits of our planet, and there is no reason to doubt that it would do so on other planets. Thus, if life ever had a foothold on any other planet in the past, some organisms might still survive in surprising ecological niches today even if the planet has undergone substantial changes. The only real question is whether life ever got started elsewhere in the first place.

What is the probability of life arising from nonliving ingredients? This is probably the greatest unknown in exobiology. The fact that we exist and are asking the question does not tell us the probability; it merely tells us that it happened once. But the rapidity with which life arose on Earth may provide a clue. As we've discussed, we find fossil evidence for life dating almost all the way back to the end of the period of early bombardment in the solar system, suggesting that life arises easily and perhaps inevitably under the right conditions. Some people even speculate that primitive life arose many times during the heavy bombardment, only to be extinguished by violent impacts just as many times. If life indeed arises easily given the right conditions, we must search the solar system for those conditions, past or present.

TIME OUT TO THINK *The preceding discussion implies that the rapid appearance of life on Earth means that life is highly probable. Do you agree with this logic? What alternative conclusions could you reach?*

Life in the Solar System

Speculation about life in the solar system usually begins with Mars, for good reason. Before it dried out billions of years ago, its early atmosphere gave the surface hospitable conditions that rivaled those on Earth, with ample running water, the necessary raw chemical ingredients for life, and a variety of familiar energy sources. Many of Earth's organisms would have thrived under early Martian conditions, and some could even survive in places in today's Martian environment. Our first attempt to search for life on Mars came with the Viking missions to Mars in the 1970s, which included two landers equipped to search for the chemical signs of life [Section 9.5]. No life was found. But the landers sampled only two locations on the planet and tested soils only very near the surface. If life once existed on Mars, it either has become extinct or is hiding in other locations.

Today, a renewed debate about Martian life is under way, thanks in part to the study of a Martian meteorite found in Antarctica in 1984. The meteorite apparently landed in Antarctica 13,000 years ago, following a 16-million-year journey through space after being blasted from Mars by an impact. The rock itself dates to 4.5 billion years ago, indicating that it solidified shortly after Mars formed and therefore was present during the time when Mars was warmer and wetter. Painstaking analysis of the meteorite reveals indirect evidence of past life on Mars, including layered carbonate minerals and complex molecules (called polycyclic aromatic hydrocarbons), both of which are associated with life when they are found in Earth rocks. Even more intriguing, highly magnified images of the meteorite reveal eerily lifelike forms (Figure 13.29). These forms bear a superficial resemblance to terrestrial bacteria, although they are about a hundred times smaller—about the same size as recently discovered terrestrial "nanobacteria" and viruses. Nevertheless, many scientists dispute the conclusion that these features suggest the past existence of life on Mars, claiming that nonbiological causes can also explain many of the meteorite's unusual features.

Future missions to Mars will search for life using more sophisticated techniques than those used by the Viking missions. The best place to look for fossil remains of extinct life is probably in ancient valley bottoms or dried-up lake beds. If any hot springs surround Mars's not-quite-dormant volcanoes, they may be a good place to look for surviving life. A thorough search for Martian life will probably require the return of rock samples to Earth or human exploration of the planet.

Martian meteorites also remind us that the planets may occasionally exchange rocks dislodged by major impacts. The harsh conditions under which some life on Earth exists suggest that living organisms might survive such impacts and even survive the journey from one planet to another. Earth's basalt-dwelling bacteria, for example, could probably survive an impact-cratering event, the ensuing millions of years in space, and a violent impact on Mars.

If a meteorite from Earth once landed in hospitable conditions on Mars, it might have introduced life to Mars or wiped out life already present. In a sense, Earth, Venus, and Mars have been "sneezing" on one another for billions of years. Life could conceivably have originated on any of these three planets and been transported to the others.[11]

TIME OUT TO THINK *Suppose we someday discover living organisms on Mars. How will we be able to tell whether these organisms share a common ancestor with living organisms on Earth?*

Besides Mars, the best places to search for life probably are the satellites of the jovian planets. Several of them may meet the requirements of having liquid water, appropriate chemicals, and energy sources for life. In particular, Europa appears to have a planet-wide ocean beneath its icy crust [Section 11.5], the ice and rock from which Europa formed undoubtedly included the necessary chemicals, and its internal heating might lead to undersea volcanic vents. Thus, the real question about Europa may be whether it is possible that its ocean has existed for billions of years *without* developing life. NASA is already considering missions to search for signs of life on Europa. The most ambitious involves landing a robotic spacecraft on the surface that will melt its way through the icy crust to reach the ocean below.

Another enticing place to look for life past or present is Saturn's moon Titan. We've already found evidence of complex chemical reactions on Titan, many involving the same elements used by life on Earth. Liquid water and energy are in shorter supply, but both might have been supplied by impacts that heated and melted the icy surface early in Titan's history. Life might have arisen in a slushy pond during the brief period that it remained liquid and might have survived after the pond froze. Unlike the case on Earth, where early impacts probably sterilized the planet, impacts on Titan may have made life possible.

Despite our new awareness of the diversity of life on Earth, we may still be underestimating the range of conditions in which life can exist. Could life arise in liquids other than water, or possibly in an atmosphere? Some people have speculated that life could develop in the clouds of Jupiter or other jovian planets. Might life be based on different elements than those used on Earth? Are there other energy sources that we have not considered? Is our definition of life too narrow? Clearly, it will be a long time before we know all the places where life might exist even within our solar system.

Life Around Other Stars

Only a few places in our solar system seem hospitable to life, but many more hospitable worlds may be orbiting some of the hundred billion other stars in the Milky Way Galaxy or stars in some of the billions of other galaxies in the universe. Might some of the stars be orbited by planets that are as hospitable as our own Earth?

Interstellar clouds throughout the galaxy contain the basic chemical ingredients of life—carbon, oxygen, nitrogen, and hydrogen, as well as other elements. All stars are born from such interstellar clouds, so it seems likely that other solar nebulae should have given rise to planetary systems similar to our own [Section 8.8]. Terrestrial planets anywhere will receive light from their parent star, and if they are large enough, like Venus or Earth, they will have plenty of internal heat. Thus, the chemicals and energy

[11]Some scientists even suggest that the life on Earth may have originated beyond our own solar system and been brought here by interstellar dust or meteorites.

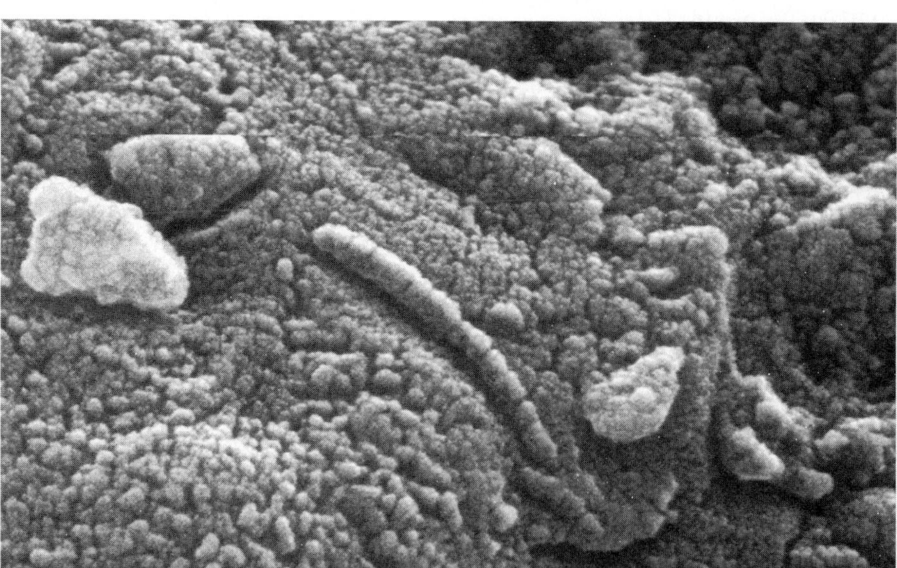

FIGURE 13.29 Microscopic view of seemingly lifelike structures in a Martian meteorite.

sources for life should be present on many planets throughout the universe.

Planets with liquid water on their surfaces may be somewhat more rare, particularly if we consider only planets where oceans can endure for billions of years. Even if a planet is large enough to outgas substantial quantities of water and retain its atmosphere, the stories of Venus and Mars tell us that oceans are not guaranteed. To keep oceans for billions of years, the planet must lie within a range of distances from its star, sometimes called the **habitable zone,** that has temperatures just right for liquid water. How big is the habitable zone in our own solar system? We know that Venus is too close to the Sun, so this zone must begin outside the orbit of Venus. Mars is a borderline case: If it had been large enough to retain its atmosphere and sustain a stronger greenhouse effect, Mars might still be habitable today. Overall, the habitable zone around our star probably ranges from 0.8 to 1.5 AU. This zone may be broader around brighter stars and narrower around fainter stars. Computer models of solar system formation suggest that one or more terrestrial planets will usually form within a star's habitable zone, as long as the star is not part of a binary star system.

The bottom line is that, according to our theories of solar system formation, planets with all the necessities for life should be quite common in the universe. The only major question is whether these ingredients combine to form life. The fact that life arose very early in Earth's history suggests that it may be very easy to produce life under Earth-like conditions, but we will not know for sure unless and until we find other life-bearing planets. NASA is currently developing plans for orbiting telescopes that may be able to detect ozone in the spectra of planets around other stars—and, at least in our solar system, substantial ozone implies life. In addition, radio astronomers are searching the skies in hopes of receiving a signal from some extraterrestrial civilization. Perhaps, in a decade or two, we will discover unmistakable evidence of life. On that day, if it comes, we will know that we are not alone.

TIME OUT TO THINK *Consider the following statements: (1) We are the only intelligent life in the entire universe. (2) Earth is one of many planets inhabited by intelligent life. Which do you think is true? Do you find either philosophically troubling?*

> *Looking outward to the blackness of space, sprinkled with the glory of a universe of lights, I saw majesty—but no welcome. Below was a welcoming planet. There, contained in the thin, moving, incredibly fragile shell of the biosphere is everything that is dear to you, all the human drama and comedy. That's where life is; that's where all the good stuff is.*
>
> LOREN ACTON, U.S. ASTRONAUT

THE BIG PICTURE

Our ancestors have observed the Sun, the Moon, and the planets for thousands of years, but only recently did we learn that these other worlds have much to teach us about our own Earth. Through our study of solar system formation and comparative planetology, we have learned to look at Earth from a very new perspective. Keep in mind the following "big picture" ideas:

- Earth has been shaped by the same geological and atmospheric processes that shaped the other terrestrial worlds. Earth is not a special case from a planetary point of view but rather a place where natural processes led to conditions conducive to life.
- Most of Earth's unique features can be traced to the fact that abundant water has remained liquid throughout our planet's history—thanks to our distance from the Sun and the size of our planet.
- Life arose early on Earth and played a crucial role in shaping our planet's history. The abundant oxygen in our atmosphere is just one of many phenomena that demonstrate how life can transform a planet.
- Humans are ideally adapted to the Earth today, but there is no guarantee that the Earth will remain as hospitable in the future. The study of our solar system teaches us how planets can change.
- The conditions necessary for "life as we know it" are probably common in the universe and may even be found in our own solar system. So far, we have no proof that life exists elsewhere.

Review Questions

1. Briefly summarize how Earth is different from the other terrestrial planets. In particular, define *hydrosphere* and *biosphere.*
2. Briefly describe the interior structure of the Earth, including its molten outer core and its lithosphere. How does seafloor crust differ from continental crust?
3. Do we find evidence of impact cratering on Earth? Why do we find fewer craters on the seafloor than on the continents?
4. Why is erosion more important on Earth than on Venus or Mars? Describe how erosion affects the Earth's geology.
5. What do we mean by *plates* and *plate tectonics*? How fast do plates move? How has plate movement changed the arrangement of the continents over long time scales?
6. Briefly describe the conveyor-like process of plate tectonics, with emphasis on the creation of new crust at *spreading centers* and the destruction of old crust at *subduction zones.*
7. Describe how subduction has created the continents over billions of years. Give examples of how this process has affected Japan, the Philippines, and the western United States. How does it explain the presence of stratovolcanoes and granite mountain ranges?
8. Briefly describe how tectonic stresses can create rift valleys, tall mountain ranges, and faults. Give examples of places that have been affected in each of these ways.
9. What is a *hot spot?* Describe how hot spots affect Hawaii and Yellowstone.
10. From the standpoint of comparative planetology, list four major questions about the Earth's atmosphere. Briefly answer each of them.
11. What is a *runaway greenhouse effect?* Why did it occur on Venus but not on Earth?
12. Describe how the *carbonate–silicate cycle* helps maintain the relatively small CO_2 content of our atmosphere.
13. What are *oxidation* reactions? If there were no life on Earth, would Earth's atmosphere still contain significant amounts of oxygen or ozone?
14. Give examples of positive and negative *feedback relationships.* How does feedback affect the CO_2 balance in our atmosphere? How does the fact that the Sun has grown in brightness tell us that this balance must have been different in the past?
15. Approximately when did life arise on Earth? How do we know?
16. What evidence tells us that all life today shares a common ancestor? How do biologists compare DNA sequences to establish the "tree of life"? Where do plants and animals fall on this diagram?
17. What are *mutations?* How do mutations drive the process of *natural selection?*
18. Briefly describe how life gradually altered the Earth's atmosphere until it reached its current state.
19. Summarize the slow evolution of life on the early Earth. What was the Cambrian explosion?
20. What is global warming? Briefly describe some of the potential dangers of global warming and why so many uncertainties are involved in knowing whether global warming represents a real threat.
21. Briefly describe the phenomena and causes of the Antarctic *ozone hole* and worldwide *ozone depletion.*
22. Is the Earth currently undergoing a mass extinction? Explain.
23. Summarize why we now think that life can survive in a much broader range of conditions than we did just a few decades ago. Why is this important to the prospect of finding life elsewhere?
24. Describe the evidence from Martian meteorites suggesting that life may once have existed on Mars. Explain how life might have originated on one terrestrial planet and been transferred to others.
25. Why are Europa and Titan considered prospects for harboring life?
26. What is a *habitable zone?* Using this idea, briefly describe the prospect of finding life on planets around other stars.

Discussion Questions

1. *Evidence of Our Civilization.* Imagine a future archaeologist, say 10,000 years from now, trying to piece together a picture of human civilization at our time (i.e., around the year 2000). What types of evidence of our civilization are most likely to survive for 10,000 years? (Will our buildings survive? Our infrastructure, such as highways and water pipes? Information in the form of books or computer data?) What geological processes are likely to destroy evidence over the next 10,000 years? Next, discuss the evidence that will remain, and why, for an archaeologist living 100 million years from now. Be sure to consider the effects of all four geological processes, as well as continental drift.
2. *Cancer of the Earth?* In the text, we discussed how the spread of humans over the Earth resembles, at least in some ways, the spread of cancer in a human body. Cancers end up killing themselves because of the damage they do to their hosts. Do you think we are in

danger of killing ourselves through our actions on the Earth? If so, what should we do to alleviate this danger? Overall, do you think the cancer analogy is valid or invalid? Defend your opinions.

3. *Contact.* Suppose we discover microbial life on another planet in our solar system. Would this discovery alter your view of our place in the universe? If so, how? What if we made contact with an intelligent species from another world? Do you think it is likely that either kind of life exists elsewhere in the universe? Do you think either kind will be discovered in your lifetime? Explain.

Problems

1. *Change in Formation Properties.* Consider Earth's four formation properties of size, distance, composition, and rotation rate. Choose one property, and suppose it had been different (e.g., smaller size, greater distance). Describe how this change might have affected Earth's subsequent history and the possibility of life on Earth.

2. *Growing Population.* Since about 1950, human population has grown with a doubling time of about 40 years; that is, the population doubles every 40 years. The current population (in 1999) is about 6 billion. If population continues to grow with a doubling time of 40 years, what will it be in 40 years? In 80 years? Do you think this will actually happen? Why or why not?

3. *Feedback Processes in the Atmosphere.*
   a. Give an everyday example (not used in the book) of a positive or negative feedback process.
   b. If the Sun were much fainter and the oceans froze, the formation of carbonate rocks on the seafloor would stop. Since volcanoes would still erupt and outgas, what would happen to the Earth's atmosphere and surface temperature? Is this an example of positive or negative feedback? Explain your answers in one or two paragraphs.

4. *Defining Life.* Write a definition of life and explain the basis of your definition in a few sentences. Then evaluate whether each of the following three cases meets your definition. Explain why or why not in a paragraph for each case. (i) The first self-replicating molecule on Earth lies near the boundary of life and nonlife. Does it meet your definition of life? (ii) Modern-day viruses are essentially packets of DNA encased in a microscopic shell of protein. Viruses cannot replicate themselves; instead, they reproduce by infecting living cells and "hijacking" the cell's reproduction machinery to make copies of the viral DNA and proteins. Are viruses alive? (iii) Imagine that humans someday travel to other stars and discover a planet populated by what appear to be robots programmed to mine metal, refine it, and assemble copies of themselves. Examination of fossil "robots" shows that they have improved, perhaps because cosmic rays have caused errors in their programs. Is this race of robots alive? Would your answer depend on whether another race had built the first robots?

5. *Ozone Signature.* Suppose a powerful future telescope is able to take a spectrum of a terrestrial planet around another star. The spectrum reveals the presence of significant amounts of ozone. Why would this discovery strongly suggest the presence of life on this planet? Would it tell us whether the life is microscopic or more advanced? Summarize your answers in one or two paragraphs.

6. *Essay: Explaining Ourselves to the Aliens.* Imagine that someday we make contact with intelligent aliens and even learn to communicate. Even so, many facets of life we take for granted may be utterly incomprehensible to them: music (perhaps they have no sense of hearing), money (perhaps there has never been a need), love (perhaps they don't feel this emotion), meals (perhaps they photosynthesize). Write a page attempting to explain one of these concepts, or another of your choosing. Remember that even the words in your explanation require definition; start at the lowest possible level.

7. *Research: Volcanoes and Earthquakes.* Write a one- to two-page report about one major earthquake or volcanic eruption of the past 200 years. Report on the geological circumstances that led to the event, as well as on its geological and biological consequences.

8. *Research: Local Geology.* Write a one- to two-page report about the geology of an area you know well—perhaps the location of your campus or your hometown. Which of the four geological processes has played the most important role, in your opinion? Has plate tectonics played a direct role in shaping the area?

9. *Research: Human Threats to the Earth.* Write an in-depth research report, three to five pages in length, about current understanding and controversy regarding one of the following issues: global warming, ozone depletion, or the loss of species on Earth due to human activity. Be sure to address both the latest knowledge about the issue and proposals for alleviating any dangers associated with it. End your report by making your own recommendations about what, if anything, needs to be done to prevent damage to the Earth.

10. *Web Project: Life on Mars.* Follow the links from the text web site to find the latest information regarding the controversy over evidence for life in Martian meteorites. Write a one- to two-page report summarizing the current state of the controversy and your own opinion as to whether life once existed on Mars.

PART IV
A DEEPER LOOK AT NATURE

S3 Space and Time

THE UNIVERSE CONSISTS OF MATTER AND ENERGY MOVING through *space* with the passage of *time.* Up to this point in the book, we have discussed the concepts of space and time as though they are absolute and distinct—just as they appear in everyday life. But what if this appearance is deceiving?

Almost 100 years ago, Albert Einstein discovered that space and time are not what they appear to be. Instead, space and time are intertwined in a remarkable manner described by Einstein's *theory of relativity.* Because space and time are such fundamental concepts, understanding relativity is important to understanding the universe.

The theory of relativity is *not* difficult to understand, despite popular myths to the contrary. It does, however, require us to think in new ways. That is our task in this chapter and the next. By the time we are finished, you will see that Einstein brought about a revolution in human thinking with many important ramifications for understanding our place in the universe.

S3.1 Einstein's Revolution

Imagine that, with the aid of a long tape measure, you carefully measure the distance you walk from home to work to be 5.0 kilometers. You wouldn't expect any argument about this distance. For example, if a friend drives her car along the same route and measures the distance with her car's odometer, she ought to get the same measurement of 5.0 kilometers—as long as her odometer is working properly.

Likewise, you would expect agreement about the time it takes you to walk from home to work. Suppose your friend continues driving and you call her on a cellular phone just as you leave your house at 8:00 A.M. and again just as you arrive at work at 8:45 A.M. You'd certainly be surprised if she argued that your walk took an amount of time other than 45 minutes.

Distances and times appear absolute and distinct in our daily lives. We expect everyone to agree on the distance between two *points,* such as the *locations* of home and work. We also expect agreement about the time between two *events,* such as *leaving* home and *arriving* at work. Thus, it came as a huge surprise to everyone when, in 1905, Albert Einstein showed that these expectations are not strictly correct.

With extremely precise measurements, the distance you measure between home and work will be *different* from the distance measured by a friend in a car, and you and your friend will also disagree about the time it takes you to walk to work. At ordinary speeds, the differences will be so small as to be unnoticeable. But if your friend could drive at a speed close to the speed of light, the differences would be substantial.

Disagreements about distances and times are only the beginning of the astonishing ideas contained in Einstein's **theory of relativity.** Einstein developed this theory in two parts. His **special theory of relativity,** published in 1905, showed how space and time are intertwined but did not deal with the effects of gravity. His **general theory of relativity,** published in 1915, offered a surprising new view of gravity—a view that we will use to help us understand topics such as the expansion and fate of the universe and the strange objects known as *black holes.*

In this chapter, we will focus on the new view of space and time in Einstein's special theory of relativity. In particular, we will see how this theory supports each of the following ideas:

- Nothing can travel faster than the speed of light, and no material object can even reach the speed of light.
- If you carefully observe anyone or anything moving by you at a speed close to the speed of light, time will run more slowly for the moving object. That is, a person moving by you ages more slowly than you, a clock moving by you ticks more slowly than your clock, a computer moving by you runs more slowly than your similar computer, and so on.
- If you observe two events to occur simultaneously, such as flashes of light in two different places at the same time, a person moving by you at a speed close to the speed of light will not agree that the two events were simultaneous.
- If you carefully measure the size of something moving by you at a speed close to the speed of light, you will find that its length (in the direction of its motion) is shorter than it would be if the object were not moving.
- If you could measure the mass of something moving by you at a speed close to the speed of light, you would find its mass to be greater than the mass it would have if it were stationary. From this fact you can conclude, as did Einstein, that $E = mc^2$.

Although the consequences of relativity may sound like science fiction or fantasy, their reality is supported by a vast body of observational and experimental evidence. They also follow logically from a few simple ideas. If you keep an open mind and think deeply as you read, you'll soon be confident and conversant in the ideas of relativity.

What Is Relativity?

Suppose a supersonic airplane is flying at a speed of 1,650 km/hr from Nairobi, Kenya, to Quito, Ecuador. How fast is the plane going? At first, this question sounds trivial—we have just said that the plane is going 1,650 km/hr.

But wait. . . . Nairobi and Quito are both nearly on the Earth's equator, and the equatorial speed of the Earth's rotation is the same 1,650 km/hr that the plane is flying [Section 3.1]. Moreover, the east-to-west flight from Nairobi to Quito is opposite the direction of the Earth's rotation (Figure S3.1). Thus, if you could observe the plane from far off in space, it would appear to stay put *while the Earth rotated beneath it.* When the flight began, you would see the plane lift straight off the ground in Nairobi. The plane would then remain stationary while the Earth's rotation carried Nairobi away from it and Quito toward it. When Quito finally reached the plane's position, the plane would drop straight down to the ground.

We have two alternative viewpoints about the plane's flight. People on Earth would say that the plane is traveling westward across the surface of the Earth.

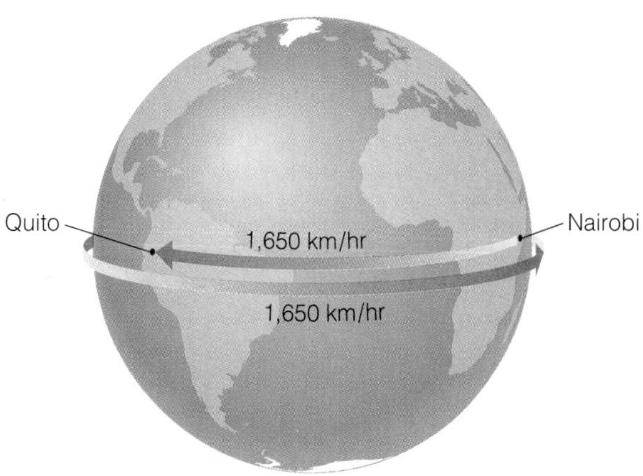

FIGURE S3.1 A plane flying at 1,650 km/hr from Nairobi to Quito travels precisely opposite the Earth's rotation. Viewed from afar, the plane would remain stationary while the Earth rotated underneath it.

Observers in space would say that the plane is stationary while the Earth rotates eastward beneath it. Both viewpoints are equally valid. In fact, there are many other equally valid viewpoints about the plane's flight. Observers looking at the solar system as a whole would see the plane moving at a speed of more than 100,000 km/hr—the Earth's speed in its orbit around the Sun. Observers living in a distant galaxy would see the plane moving away from them at a very high speed, carried along with the Earth and the Milky Way by the expansion of the universe. The only thing all these observers would agree on is that the plane is traveling at 1,650 km/hr *relative to* the surface of the Earth.

This example shows that questions like "Who is really moving?" and "How fast are you going?" have no absolute answers. Einstein's *theory of relativity* gets its name from the fact that it tells us that measurements of motion, as well as measurements of time and space, make sense only when we describe whom or what they are being measured relative to.

TIME OUT TO THINK

Suppose you are running on a treadmill and the readout says you are going 8 miles per hour. What is the 8 miles per hour measured relative to? How fast are you going relative to the ground? How fast would an observer on the Moon see you going? Describe a few other possible viewpoints on your speed.

Note that the theory of relativity does *not* say that *everything* is relative. In particular, the theory claims that two things in the universe are absolute:

1. The laws of nature are the same for everyone.
2. The speed of light is the same for everyone.

As we'll see shortly, all the astounding consequences of relativity follow directly from these two seemingly innocuous statements.

Making Sense of Relativity

One reason why relativity has a reputation for being difficult to grasp, despite its underlying simplicity, is that most of its ideas and consequences are not obvious in everyday life. They become obvious only when we deal with speeds close to the speed of light or with gravitational fields far stronger than that of the Earth. Because we don't commonly experience such extreme conditions, we have no *common* sense about them. Thus, to say that relativity violates common sense is not really accurate. The theory of relativity is perfectly consistent with everything we have come to expect in daily life.

Making sense of relativity really requires only that you learn to view your everyday experiences from a new, broader perspective. Fortunately, you have learned to broaden your perspective in a similar way before. At a very young age, you learned "common sense" meanings for *up* and *down: Up* is above your head, *down* is toward your feet, and things tend to fall down. One day, however, you learned that the Earth is round. When you looked at a globe with the Northern Hemisphere on the top, you were immediately confronted with a **paradox**—a situation that *seems* to violate common sense or to contradict itself. Your common sense told you that Australians should fall off the Earth (Figure S3.2a), but you knew that

FIGURE S3.2 Learning that the Earth is round helps children revise their "common sense" understanding of *up* and *down.*

early-childhood common sense

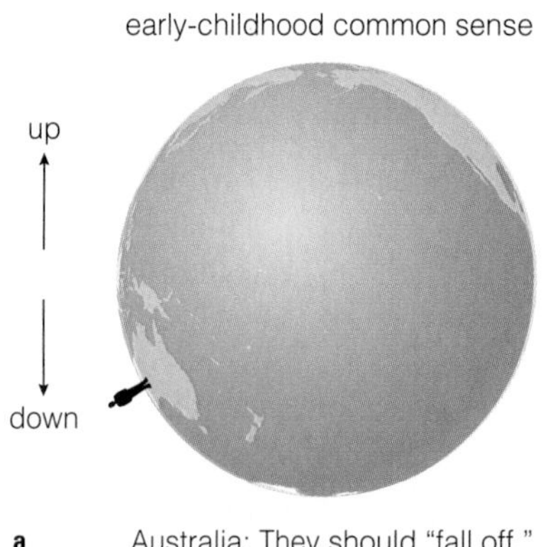

a Australia: They should "fall off."

revised common sense

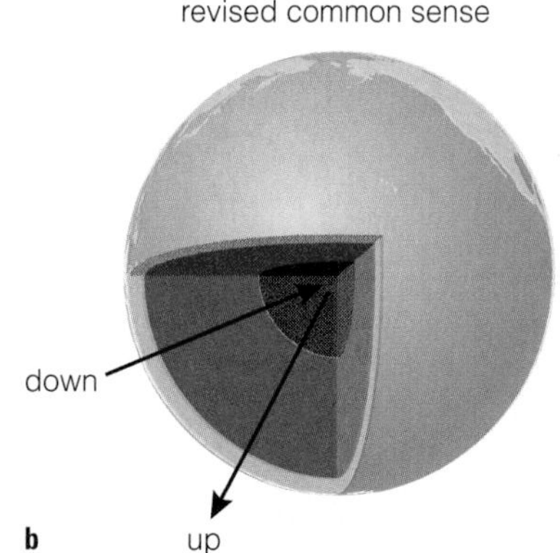

b

they don't. To resolve this paradox, you were forced to accept that your "common sense" understanding of *up* and *down* was incorrect. You therefore revised your common sense to accept that *up* and *down* are determined relative to the center of the Earth (Figure S3.2b).

TIME OUT TO THINK *Why do you suppose most maps and globes show the Northern Hemisphere on the top and the Southern Hemisphere on the bottom? If you hung your map upside down and rewrote the words so they read right-side up, would the map be equally valid?*

As a teenager, Einstein wondered what the world would look like if he could travel at or beyond the speed of light. But he inevitably encountered paradoxes when he thought about this question. Ultimately, he resolved the paradoxes only when he recognized that our common sense ideas about space and time must change if we are to extend them to the realm of very high speeds or very strong gravitational fields. Just as we all once learned a new common sense about up and down, we now must learn a new common sense about space and time.

S3.2 Relative Motion

We will study relativity with the aid of *thought experiments* in which we create imaginary situations and logically follow them through to their conclusions. A starting point for our thought experiments will be the assumption that the two absolutes of relativity are true.

The first of the two absolutes, that the laws of nature are the same for everyone, is probably not surprising. If you're on an airplane with the shades drawn during a very smooth flight, you won't feel any sensation of motion. Thus, you should expect to get the same results from any experiments you perform on the airplane that someone else would get performing the same experiments on the ground. These equivalent results prove that the laws of nature are the same in the airplane and on the ground. In the language of relativity, the ground and the airplane represent different **frames of reference** (or *reference frames*) because they are moving relative to each other. The idea that the laws of nature are the same for everyone means that they do not depend on your frame of reference.

The second absolute of relativity, that the speed of light is the same for everyone, is far more surprising. In general, we expect people in different reference frames to give different answers for the speed of the same moving object. For example, suppose you roll a ball down the aisle of an airplane. The ball will be rolling slowly in your airplane reference frame, but it will also have the speed of the airplane in the reference frame of a person on the ground (Figure S3.3a). However, if you turn on a flashlight and measure the speed of the emitted light, a person on the ground will find exactly the same speed for the light beam (Figure S3.3b). That is, people in different reference frames can disagree about the speeds of material objects, but everyone always agrees on the speed of light—regardless of where the light comes from.

How can we know that everyone always measures the same speed of light? Einstein reached this conclusion because it was the only way he could resolve the paradoxes he encountered when he thought about traveling at the speed of light. But observations and experiments are the ultimate judge of any scientific theory. Many observations and experiments, some of which we'll study later in the chapter, have verified that the speed of light really is an absolute in nature. *The absoluteness of the speed of light is an experimentally verified fact.*

FIGURE S3.3 (**a**) Observers in different reference frames measure different speeds for material objects. (**b**) Everyone measures the same speed of light.

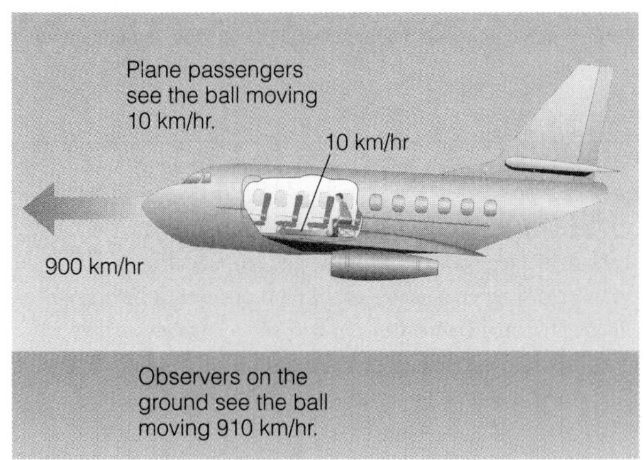

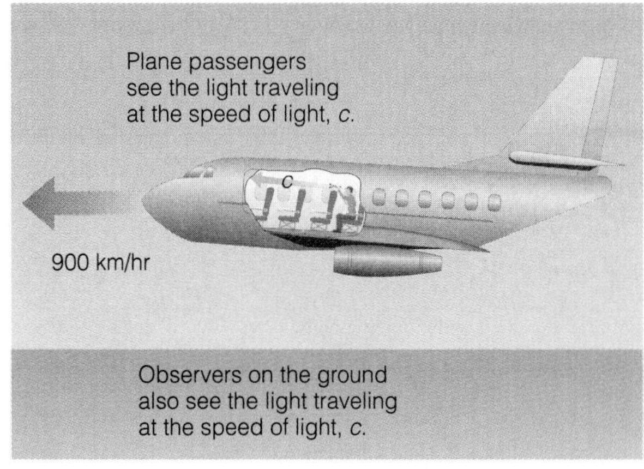

This statement may not seem important, but it will force you to let go of many of your intuitive beliefs about how the universe works. We will investigate why this statement forces such a change by constructing a series of thought experiments about relative motion viewed from different reference frames. If you follow the logic of the thought experiments carefully, you will come to understand the strange predictions of relativity. Later, we will investigate the experimental evidence that shows these predictions to be accurate. Because special relativity does not deal with the effects of gravity, the thought experiments are easiest to visualize by imagining that they take place in deep space, far from any gravitational fields. We will use spaceships floating freely without engine power. Because everything in and around these spaceships is weightless and floats freely, we call the reference frames of these spaceships **free-float frames.**[1]

Thought Experiments at Ordinary Speeds

First, to make sure that you're comfortable with ideas of relative motion in everyday life, let's analyze a few thought experiments involving ordinary objects and ordinary speeds.

Thought Experiment 1 (Figure S3.4) Imagine that you are floating freely in a spaceship. Because you feel no sensation of motion, you perceive yourself to be at rest, or traveling at zero speed. As you look out your window, you see your friend Jackie in her own spaceship, moving away at a constant speed of 90 km/hr. How does the situation appear to Jackie?

We can answer the question by logically analyzing the experimental situation. Because Jackie is moving at *constant velocity* relative to you, she must also be in a free-float frame. She therefore perceives *herself* to be at rest, and she would say that *you* are moving away from her at 90 km/hr.

Note that both points of view—yours and Jackie's—are equally valid. You both would find the same results for any experiments performed in your own spaceship, and your research would lead you to exactly the same laws of nature. You could argue endlessly about who is really moving, but your argument would be pointless because all motion is relative.

Thought Experiment 2 (Figure S3.5) We begin with the same situation as in Thought Experiment 1, but this time you put on your spacesuit and strap yourself to the outside of your spaceship. You happen to have a baseball, which you throw in Jackie's direction at a speed of 100 km/hr. How is the ball moving relative to Jackie?

From your point of view, Jackie and the ball are both going in the same direction. Jackie is going 90 km/hr and the ball is going 100 km/hr, so the ball is going 10 km/hr faster than Jackie. The ball will therefore overtake and pass her.

[1]A more traditional name for a free-float frame is an *inertial reference frame.*

FIGURE S3.4 Thought Experiment 1.

you

Jackie

90 km/hr

stationary

your point of view

FIGURE S3.5 Thought Experiment 2.

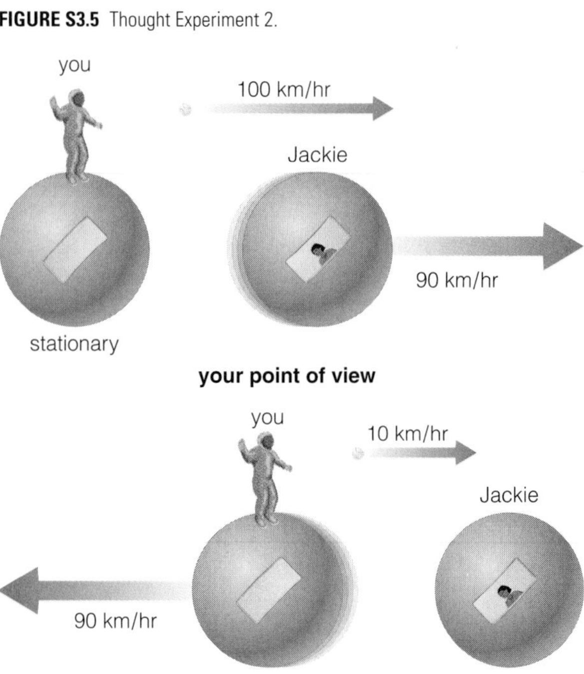

From Jackie's point of view, *she* is stationary and *you* are moving away from her at 90 km/hr. Thus, she sees the ball moving toward her at 10 km/hr—the ball's speed of 100 km/hr relative to you *minus* the 90 km/hr at which you are moving relative to her. Note that both you and Jackie agree that the ball will pass her at a *relative* speed of 10 km/hr.

Thought Experiment 3 (Figure S3.6) This time you throw a baseball in Jackie's direction at 90 km/hr.

FIGURE S3.6 Thought Experiment 3.

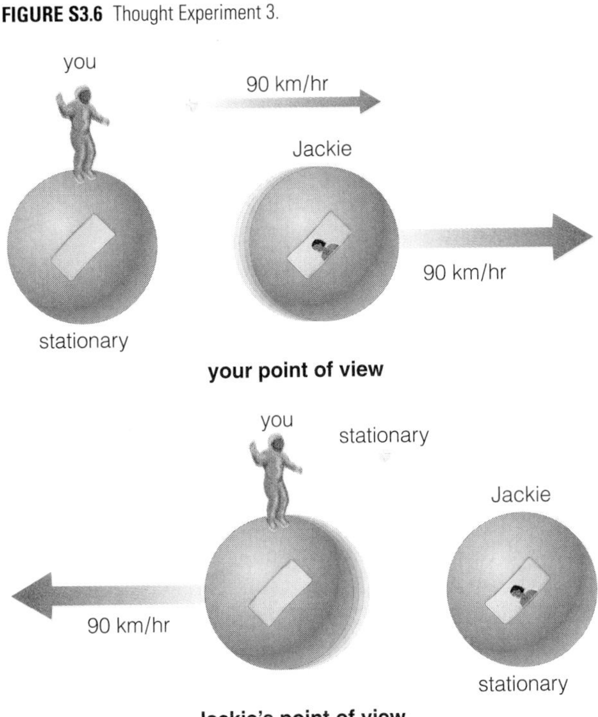

Thus, from your point of view, the ball is traveling at exactly the same speed as Jackie. Therefore, you'll see the ball forever chasing her through space, neither catching up nor falling behind.

From Jackie's point of view, the 90 km/hr at which you threw the ball exactly matches your 90 km/hr speed away from her. Therefore, the ball is *stationary* in her reference frame. Think about this for a moment: Before you throw the baseball, Jackie sees it moving away from her at 90 km/hr because it is in your hand. At the moment you release the baseball, it suddenly becomes stationary in Jackie's reference frame, floating in space at a fixed distance from her spaceship. Many hours later, after you have traveled far away, Jackie will still see the ball floating in the same place. If she wishes, she can put on her spacesuit and go out to retrieve it. Or she can just leave it there—from her point of view, it's not going anywhere, and neither is she.

TIME OUT TO THINK *In Thought Experiment 3, suppose instead that you throw the ball in Jackie's direction at 80 km/hr. In that case, what will Jackie see the ball doing? Next, suppose that Jackie is moving* toward *you rather than away from you. What would she see the ball doing in that case?*

Thought Experiments at High Speeds

The absoluteness of the speed of light did not come into play in our first three thought experiments because the speeds were so small compared to the speed of light. For example, 100 km/hr is less than *1 ten-millionth* of the speed of light. Now let's raise the speeds way up and explore the strange consequences of the absoluteness of the speed of light.

Thought Experiment 4 (Figure S3.7) Imagine that Jackie is moving away from you at 90% of the speed of light, or $0.9c$. (Recall that c is the symbol for the speed of light, which is about 300,000 km/s.) How does the situation appear to Jackie?

Other than the much higher speed, this situation is just like that in Thought Experiment 1. Jackie perceives *herself* to be at rest and sees *you* moving away from her at $0.9c$.

FIGURE S3.7 Thought Experiment 4.

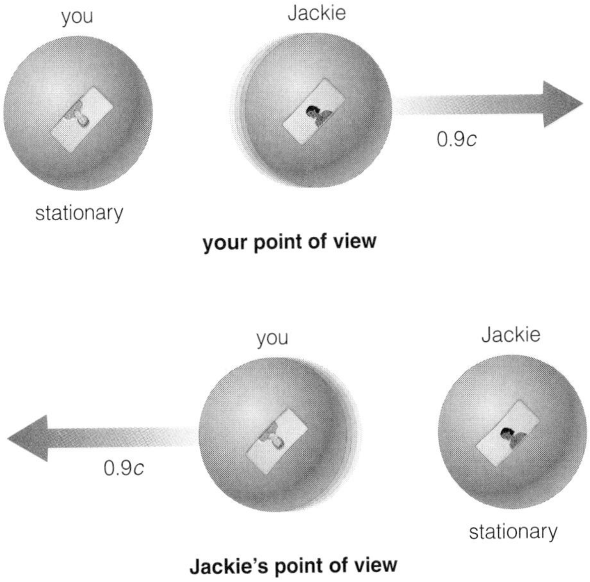

you
Jackie
0.9*c*
stationary
your point of view
you
Jackie
0.9*c*
stationary
Jackie's point of view

Thought Experiment 5 (Figure S3.8) Now, instead of throwing a baseball, you climb out of your spaceship and point a flashlight in Jackie's direction. How is the beam of light moving relative to Jackie?

From your point of view, Jackie and the flashlight beam are both going in the same direction. Jackie is going at 90% of the speed of light, or 0.9*c*, and the light beam is going at the full speed of light, or *c*. Thus, you see the light beam going 0.1*c* faster than Jackie.[2] Nothing should be surprising so far.

From Jackie's point of view, *she* is stationary and *you* are moving away from her at 0.9*c*. Following the "old common sense" used in our earlier thought experiments, we would expect Jackie to see the light moving toward her at 0.1*c*—the light's speed of *c* minus your speed of 0.9*c*. *But this answer is wrong!* Relativity tells us that the speed of light is always the same for everyone. Therefore, Jackie must see the beam of light coming toward her at *c*, *not* at 0.1*c*.

FIGURE S3.8 Thought Experiment 5.

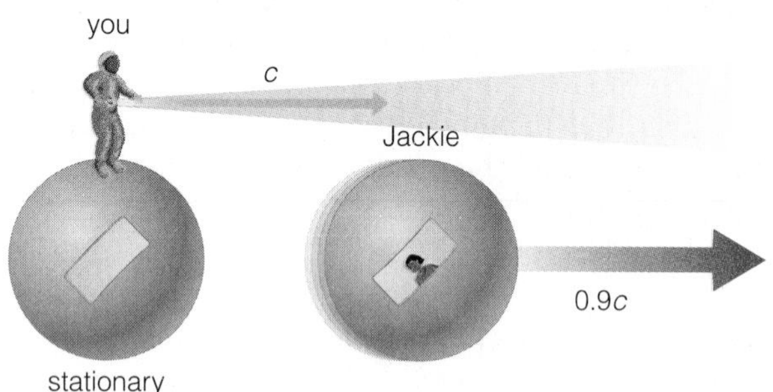

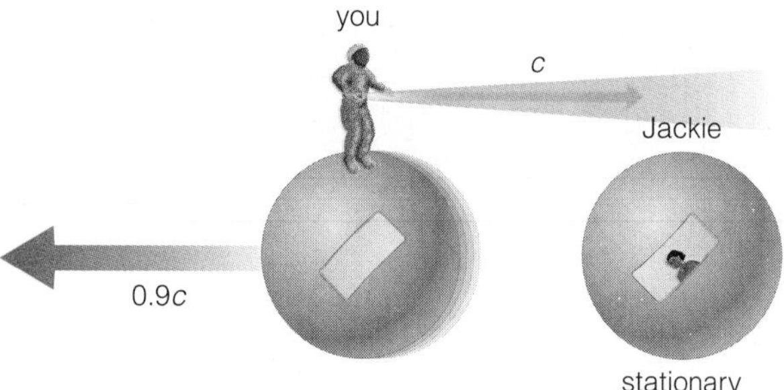

Note that you and Jackie no longer agree about her speed *relative* to the speed of the light beam: She'll see the light beam pass by her at the speed of light, *c*, but you'll see it going only 0.1*c* faster than she is going. By our old common sense, this result sounds preposterous. But we found it by using simple logic, starting with the assumption that the speed of light is the same for everyone. As long as this assumption is true—and, remember, the absoluteness of the speed of light is an experimentally verified fact—our conclusions follow logically.

[2]You can't actually *see* a light beam moving forward. When we say that you see the beam moving at the speed of light, we really mean that you would find it to be moving at this speed if you made a careful measurement with instruments at rest in your reference frame.

TIME OUT TO THINK ***Suppose Jackie is moving away from you at a speed of 0.99999c, which is short of the speed of light by only 0.00001c, or 3 km/s. How much faster than Jackie will you see the light going? How fast will Jackie see the light going as it passes her?***

You Can't Reach the Speed of Light

You might be wondering what Jackie would see if she were moving away from you at the speed of light or faster—which is essentially the same question Einstein asked when he wondered how the world would look if he could travel at or beyond the speed of light. However, once we accept the absoluteness of the speed of light, it's easy to prove that neither she nor you nor any other material object can ever reach the speed of light, let alone exceed it.

Thought Experiment 6 You have just built the most incredible rocket imaginable, and you are taking it on a test ride. Soon you are going faster than anyone had ever imagined possible—and then you put the rocket into second gear! You keep going faster and faster and faster. Here is the key question: Are you ever traveling faster than the speed of light?

Before we answer this question, the fact that all motion is relative forces us to answer another question: In what reference frame is your speed being measured? Let's begin with *your* reference frame. Imagine that you turn on your rocket's headlights. Because the speed of light is absolute, you must see the headlight beams travel away from you at $c =$ 300,000 km/s. Note that this is true no matter how long you have been firing your rocket engines; in your own reference frame, you cannot possibly outrace your own headlight beams.

Could someone in a different reference frame say that you are traveling faster than the speed of light? Observers in different reference frames will measure your speed differently, but *all* observers will agree on two key points: (1) Your headlight beams are moving out ahead of you, and (2) these light beams are traveling at $c =$ 300,000 km/s. Clearly, if you are being outraced by your headlights and if the headlights are

THINKING ABOUT . . .

What If Light Can't Catch You?

If you're like most students learning about relativity for the first time, you're probably already looking for loopholes in the logic of our thought experiments. For example, confronted with Thought Experiment 6, you might be tempted to ask, "What happens if you're traveling away from some planet faster than light, so the light from the planet can't catch you?" While it's surely true that light couldn't catch you if you were going faster than the speed of light, it's also irrelevant: If you can't see light from the planet, there is no way for you to know that the planet even exists.

In fact, what relativity really tells us about the speed of light is that it is a limit on the speed at which *information* can be transmitted. There are numerous circumstances in the universe in which, on philosophical grounds, an object may *seem* to be exceeding the speed of light. However, these circumstances do not provide any means of sending information or objects at speeds faster than the speed of light.

As an example, consider the implications of the fact that the universe is expanding [Section 3.5]. Recall that the more distant a galaxy, the faster the expansion of the universe is carrying it away from us. In principle, somewhere far away, there could be a point beyond which the expansion is carrying galaxies away from us faster than the speed of light. However, we cannot observe such galaxies because their light cannot reach us. Scientifically, we say that such galaxies are beyond the bounds of our observable universe, and the observable universe is the only "universe" that we can study.

Other examples of things that *seem* to move faster than the speed of light arise frequently in the strange world of quantum mechanics (see Chapter S5). According to quantum principles, measuring a particle in one place can (in certain specific circumstances) affect a particle in another place *instantaneously*—even if the particle is many light-years away. In fact, this process has been observed in laboratories over short distances. This instantaneous effect of one particle on another may at first seem to violate relativity, but it does not. The built-in randomness of quantum mechanics prevents this technique from being used to transmit information to a distant point. Moreover, if we wish to confirm that the second particle really was affected, we will have to receive a signal carrying information about the particle—and that signal can travel no faster than the speed of light.

traveling at the speed of light, you must be traveling *slower* than the speed of light. It does not matter who is measuring your speed; it can be you, someone on Earth, or anyone else in any other reference frame. No one can ever observe you to be traveling as fast as a light beam.

In case you are still not convinced, let's turn the situation around. Imagine that, as you race by some planet, a person on the planet turns on a light beam. Because the speed of light is absolute, the light beam will race past you at $c = 300{,}000$ km/s. The person on the planet will also see the light traveling at $c = 300{,}000$ km/s and will see the light outrace you. Thus, again, everyone will agree that you are traveling slower than the speed of light.

The same argument applies to any moving object. As long as the speed of light is absolute, no material object can reach or exceed it. Building a spaceship to travel at the speed of light is not a mere technological challenge—it simply cannot be done.

S3.3 The Reality of Space and Time

In the beginning of this chapter, we listed five key predictions of the theory of relativity. So far, we have shown only how the first one—that no material object can reach or exceed the speed of light—follows from the two absolutes of relativity. Let's continue on and see why our old conceptions of time and space must be revised.

Time Differs in Different Reference Frames

In preparation for our next thought experiment, imagine that you are on a moving train tossing a ball straight up so that it bounces straight back down from the ceiling. How do the path and speed of the ball appear to an observer along the tracks outside the train? As shown in Figure S3.9, the outside observer sees the ball going forward with the train at the same time that it is going up and down, so the ball's path always slants forward. Because the outside observer sees the ball moving with this forward speed

Inside the train, the ball goes up and down.

Outside the train, the ball appears to be going faster: It has the same up-and-down speed, plus the forward speed of the train.

The faster the train is moving, the faster the ball appears to be going to the outside observer.

FIGURE S3.9 A ball tossed straight up and down on a moving train follows a slanted path according to an outside observer.

Because Jackie, the laser, and the mirror are all moving from your point of view, the light's path looks slanted as it goes from floor to ceiling and back. Thus, from your point of view, the light travels a *longer* path in going from the floor to the ceiling than it does from Jackie's point of view, just as the ball tossed in a train took a longer path from the outside observer's point of view.

By our old common sense, this would be no big deal. You and Jackie would agree on how long it takes the light to go from the floor to the ceiling and back, just as you and an outside observer would agree on how long it takes you to toss a ball up and down on a train. You would explain the light's longer path by saying that the light is moving faster relative to you than it is relative to Jackie, just as the outside observer sees the tossed ball moving faster because of the forward motion of the train. But, because we are dealing with *light,* you and Jackie must both see its speed to be the *same*—300,000 km/s—even though you see its path slanted forward with the movement of the spaceship.

So what? Both you and Jackie see the light traveling at the same speed, but *you* see it traveling a longer distance—and, at a given speed, it takes more time to travel a longer distance. Thus, *your clock will record more time* than Jackie's as the light travels from floor to ceiling and back. That is, if you watch Jackie's clock, you will see it running slower than your own. It doesn't matter what you and Jackie use to measure the time for the light trip. *Anything* that

in addition to its up-and-down speed while you see the ball going only up and down, he would measure a *faster* overall speed for the ball than you would. If the train is moving slowly, the outside observer sees the ball's path slant forward only slightly and would say that the ball's overall speed is only slightly faster than you would report. If the train is moving rapidly, the ball's path leans much farther forward, and the ball appears to be going considerably faster to the outside observer than to you.

Thought Experiment 7 (Figure S3.10) Inside her spaceship, Jackie has a laser on her floor that is pointed up to a mirror on her ceiling. She momentarily flashes the laser light and uses a very accurate clock to time how long it takes the light to travel from the floor to the ceiling and back. As Jackie zips by you at a speed close to the speed of light, you observe her experiment through a window in her spaceship. Using your own very accurate clock, you also time the laser light's trip from Jackie's floor to her ceiling and back.

FIGURE S3.10 Thought Experiment 7.

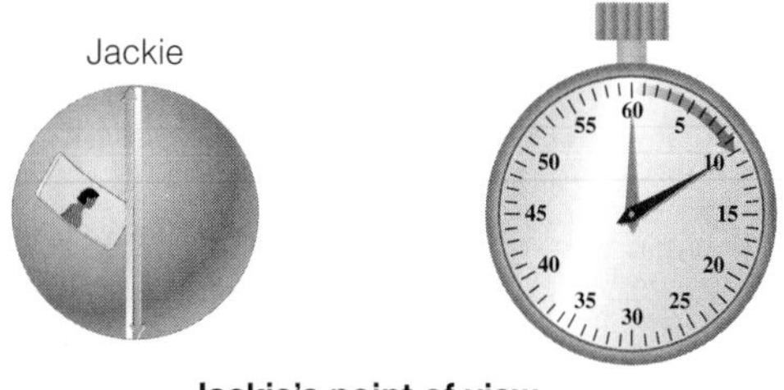

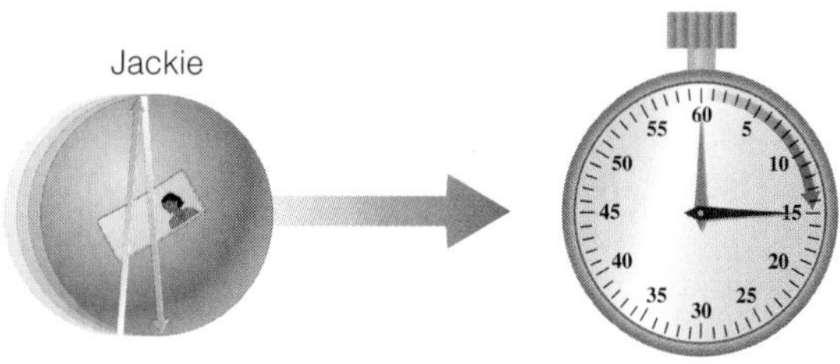

can measure time will be going slower in Jackie's reference frame than in yours, including mechanical clocks, electrical clocks, heartbeats, and biochemical reactions. Our astonishing conclusion: From your point of view, *time itself* is running slower for Jackie.

How much slower is time running for Jackie? It depends on her speed relative to you. If she is moving slowly compared to the speed of light, you will scarcely be able to detect the slant, and your clock and Jackie's clock will tick at nearly the same rate. The faster she is moving, the more slanted the light path you see and the greater the difference between the rate of her clock and that of yours. Generalizing, we reach the following conclusion.

From your point of view, time runs slower in the reference frame of anyone moving relative to you.

The faster the other reference frame is moving, the slower time passes within it.

This effect is called **time dilation** because it tells us that time is *dilated*, or expanded, in a moving reference frame.

The Relativity of Simultaneity

Our old common sense tells us that everyone must agree when two events happen at the same time or when one event happens before another. If you see two apples—one red and one green—fall from two different trees and hit the ground at the same time, you expect everyone else to agree that they landed at the same time. If you see the green apple land

Mathematical Insight S3.1 The Time Dilation Formula

Look at the different paths that you and Jackie see the light take from the floor to the ceiling in Figure S3.10. We can form a right triangle from these paths, which we can use to find an exact formula for time dilation. Recall that *distance = speed × time,* and note how the sides are constructed in Figure S3.11a:

- *You* see the light take a slanted path from Jackie's floor to ceiling because of her motion relative to you. Let's use t to represent the time that *you* measure as the light travels this path. Because the light travels at the speed of light, c, the length of this path is $c \times t$.
- The distance that *you* see Jackie travel while the light goes from her floor to her ceiling is $v \times t$, where v is her speed relative to you.
- We know that Jackie measures time differently than you, so we'll use t' (read "t-prime") to represent the time *she* measures as the light travels from her floor to her ceiling. The distance she sees the light travel therefore is $c \times t'$.

Because we have a right triangle, we can solve for the time on Jackie's clock (t') in terms of the time on your clock (t) by using the Pythagorean theorem and

(continued)

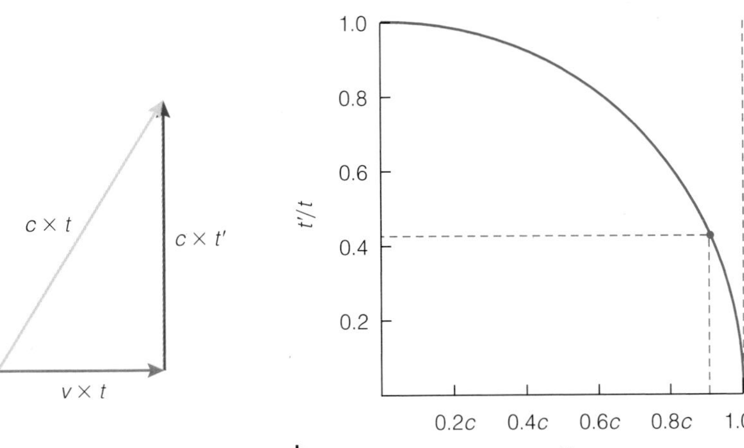

FIGURE S3.11 (**a**) The setup for finding the time dilation formula. (**b**) The time dilation factor graphed against speed.

before the red apple, you are very surprised if someone else says that the red apple hit the ground first. Prepare yourself to be surprised, because our next thought experiment will show that observers in different reference frames will not necessarily agree about the order or simultaneity of events that occur in different places.

Before we go on, note that observers in different reference frames *must* agree about the order of events that occur in the *same* place. For example, suppose you grab a cookie and eat it. In your reference frame, both events (picking up the cookie and eating it) occur in the same place. Thus, it could not possibly be the case that someone else would see you eat the cookie before you pick it up.

Thought Experiment 8 Jackie is coming toward you in a brand new, extra-long spaceship. She is in the center of her spaceship, which is totally dark. At the instant that Jackie happens to be next to you, two lights flash at either end of her spaceship. After a brief moment, the two flashes reach you at *exactly the same time:* a flash of green light from the front of her spaceship, and a flash of red light from the rear of her spaceship (Figure S3.12a). However, during the very short time that the light flashes are traveling toward Jackie, her speed carries her toward the point where you saw the green flash occur and away from the point where you saw the red flash occur (Figure S3.12b). Therefore, the green flash will reach her before the red flash: You'll see her illuminated first by green light and then by red light. Jackie must also see herself first turn green and then turn red, agreeing with you that the green light reached her before the red light.

So far, nothing is surprising. But remember that Jackie considers herself to be in the center of a *sta-*

Mathematical Insight S3.1 (continued)

a bit of algebra. (Recall that the Pythagorean theorem states that $a^2 + b^2 = c^2$ for a right triangle with side lengths a, b, and c, where c is the hypotenuse.)

Start with the Pythagorean theorem: $(ct')^2 + (vt)^2 = (ct)^2$

Expand the squares: $c^2t'^2 + v^2t^2 = c^2t^2$

Subtract v^2t^2 from both sides: $c^2t'^2 = c^2t^2 - v^2t^2$

Divide both sides by c^2t^2: $\dfrac{t'^2}{t^2} = \dfrac{c^2 - v^2}{c^2}$

Simplify: $\dfrac{t'^2}{t^2} = \dfrac{c^2}{c^2} - \dfrac{v^2}{c^2} = 1 - \left(\dfrac{v}{c}\right)^2$

Take the square root of both sides: $\dfrac{t'}{t} = \sqrt{1 - \left(\dfrac{v}{c}\right)^2}$ or $t' = t\sqrt{1 - \left(\dfrac{v}{c}\right)^2}$

The final result, called the *time dilation formula*, tells us the *ratio* of time in a moving reference frame to time in a reference frame at rest. This ratio is graphed against speed in Figure S3.11b. Note that $t'/t \approx 1$ at speeds that are small compared to the speed of light, meaning that clocks in both reference frames tick at about the same rate. But as v approaches c, the amount of time passing in the moving reference frame gets smaller and smaller compared to the time passing in the reference frame at rest.

Example: Suppose that Jackie is moving past you at a speed of 0.9c. While 1 hour passes for you, how much time passes for Jackie?

Solution: Jackie's speed of $v = 0.9c$ means $v/c = 0.9$. The variable t represents *your* time, so $t = 1$ hour. Substituting into the time dilation formula yields:

$$t' = t\sqrt{1 - \left(\frac{v}{c}\right)^2} = (1\text{ hr})\sqrt{1 - (0.9)^2} = (1\text{ hr})\sqrt{1 - 0.81} = (1\text{ hr})\sqrt{0.19} \approx 0.44\text{ hr}$$

which is about 26 minutes. Thus, while 1 hour passes for you, only 26 minutes pass for Jackie. Note that you can also find this answer from the graph in Figure S3.11b: At a speed of $v = 0.9c$, the graph shows that the ratio of Jackie's time to your time is 0.44.

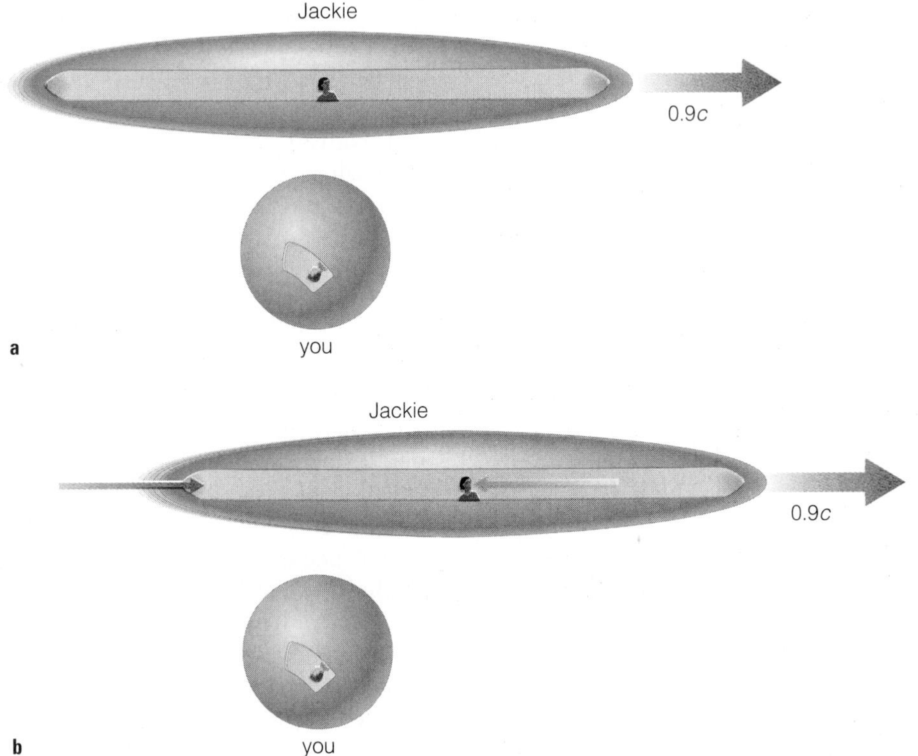

FIGURE S3.12 (**a**) In your reference frame, the green and red flashes occur simultaneously at the instant when Jackie is immediately adjacent to you. (**b**) Because her forward motion carries her toward the point at which the green flash occurred, she is illuminated by the green light before the red light. She must agree that the green light reaches her first and therefore concludes that the green flash happened *before* the red flash.

tionary spaceship. From her point of view, the flashes from the front and the rear both travel toward her at the *same speed* (the speed of light), and both have the same distance to travel. Therefore, because the green flash from the front of the spaceship reaches her *before* the red flash from the rear, she must conclude that the green flash happened first and the red flash happened some time later. Thus, while you say that the two flashes were simultaneous, Jackie disagrees: She says that the green flash occurred before the red flash.

Effects on Length and Mass

The fact that time is different in different reference frames means that lengths (or distances) and masses must also be affected, although the explanations are a bit subtler. The following two thought experiments use the idea of time dilation to help us understand the effects on length and mass.

Thought Experiment 9 Jackie is back in her original spaceship, coming toward you at high speed. As usual, both you and she agree on your *relative* speed; you disagree only about who is stationary and who is moving. Now imagine that Jackie tries to measure the length of your spaceship as she passes by you. In your reference frame, you'll see Jackie's time running slower. Because her clocks record less time than yours as she passes from one end of your spaceship to the other, she must measure the length of your spaceship to be shorter than you measure it to be. (Recall that distance is speed × time.) Similarly, you will measure Jackie's spaceship as having a shorter length than it would have if it were at rest in your reference frame, an effect called **length contraction** (Figure S3.13). Note that lengths are affected

FIGURE S3.13 People and objects moving by you are contracted in their direction of motion.

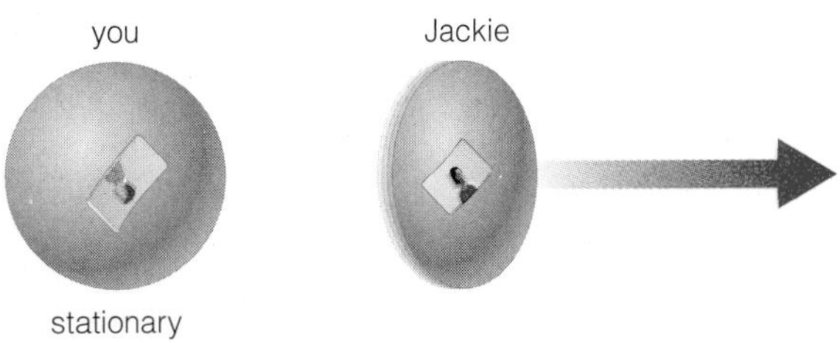

only in the direction of motion; her spaceship is shorter from your point of view, but its height and depth are unaffected. Generalizing, we reach the following conclusion:

From your point of view, lengths of objects moving by you (or distances between objects moving by you) are shorter in their direction of motion than they would be if the objects were at rest. The faster the objects are moving, the shorter the lengths.

Thought Experiment 10 Imagine that Jackie has an identical twin sister with an identical spaceship, and suppose that her sister is at rest in your reference frame while Jackie is moving by you at high speed. At the instant Jackie passes by, you give both Jackie and her sister identical pushes (Figure S3.14). If Jackie and her sister are truly identical, the force of your push should have the same effect on each of them; for example, it might cause each of them to gain 1 km/s of speed relative to you. However, because you'll see Jackie's time running slower than yours and her sister's, you will conclude that Jackie feels the force of your push for a *shorter* time than her sister feels your push. (For example, if you and Jackie's sister measure the duration of the push to be 1 microsecond, Jackie's clock will show the duration to be less than 1 microsecond.) Because Jackie feels the force of your push for a shorter time than her sister, the push must have a *smaller* effect on Jackie's velocity. In other words, you'll find that your push has less effect on Jackie than on her sister, despite the fact that you gave them identical pushes. According to Newton's laws of motion [Section 6.2], the only way the same push can have a smaller effect on Jackie's velocity is if her mass is *greater* than her sister's mass. This effect is sometimes called **mass increase:**

From your point of view, objects moving by you have greater mass than they have at rest. The faster an object is moving, the greater the increase in its mass.

Mass increase provides another way of understanding why no material object can reach the speed of light. The faster an object is moving relative to you, the greater mass you'll find it to have. Thus, at higher speeds, the same force will have less effect on an object's velocity. In fact, as an object approaches the speed of light, you will find its mass to be heading toward infinity. No force can accelerate an infinite mass, so the object can never gain that last little bit of speed needed to push it to the speed of light.

Velocity Addition

We have just one more important effect to discuss: As the next thought experiment shows, you and

FIGURE S3.14 Thought Experiment 10.

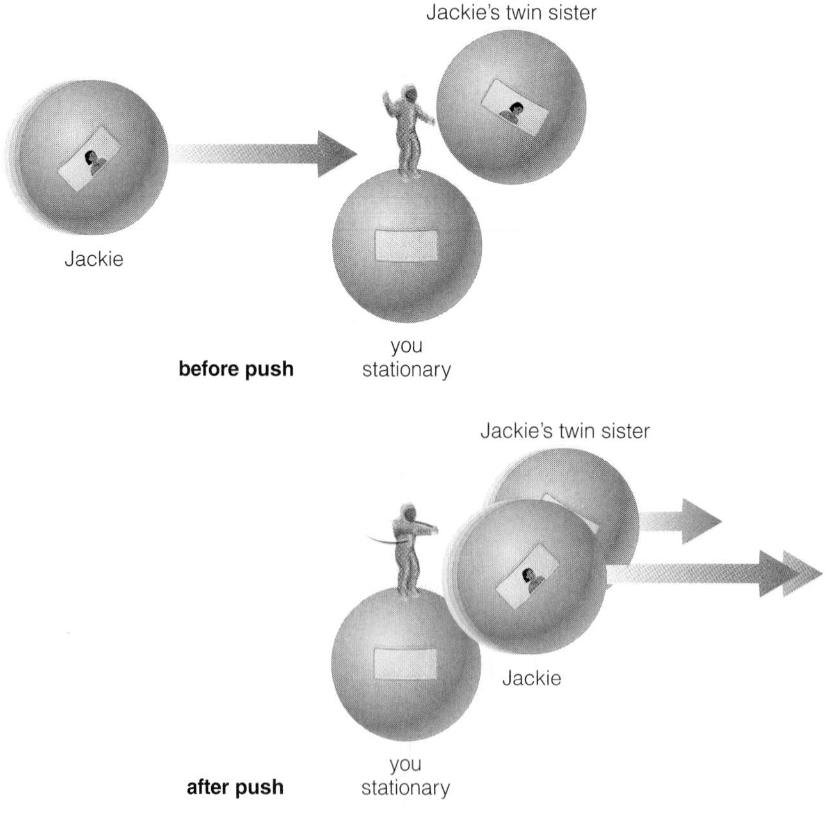

Jackie will disagree about the speed of a material object moving relative to both of you. The only speed you will agree on is the speed of light.

Thought Experiment 11 (Figure S3.15) Jackie is moving toward you at $0.9c$. Your friend Bob jumps into his spaceship and starts heading in Jackie's direction at $0.8c$ (from your point of view). How fast will Jackie see him approaching her?

By our old common sense, Jackie should see Bob coming toward her at $0.9c + 0.8c = 1.7c$. But we know this answer is wrong, because $1.7c$ is faster than the speed of light. Jackie must see Bob coming toward her at a speed less than c, but he will be coming faster than the speed of $0.9c$ at which she sees you coming. Thus, she'll conclude that Bob's speed is somewhere between $0.9c$ and c. (In this case, it turns out that she'll see him coming at $0.988c$; see Mathematical Insight S3.2.)

FIGURE S3.15 Thought Experiment 11.

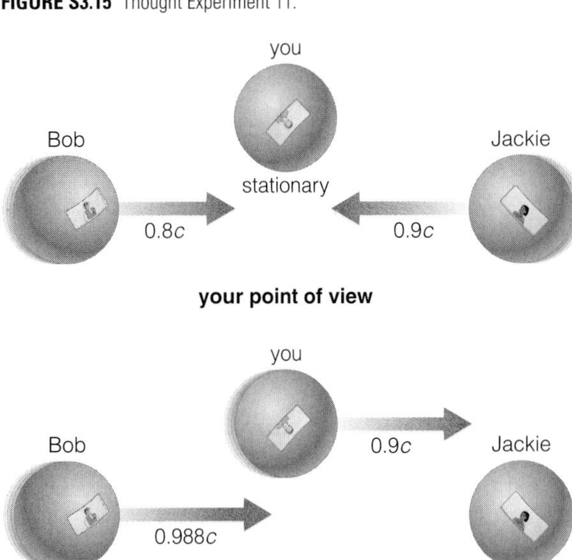

Mathematical Insight S3.2 Formulas of Special Relativity

We found a formula for time dilation in Mathematical Insight S3.1. Although we will not go through the derivations, it is possible to find similar formulas for length contraction and mass increase. In summary, the three formulas are:

$$\text{time in moving reference frame} = (\text{time in rest frame}) \times \sqrt{1 - \left(\frac{v}{c}\right)^2}$$

$$\text{length in moving reference frame} = (\text{rest length}) \times \sqrt{1 - \left(\frac{v}{c}\right)^2}$$

$$\text{moving mass} = \frac{(\text{rest mass})}{\sqrt{1 - \left(\frac{v}{c}\right)^2}}$$

There's also a simple formula for velocity addition. Suppose you see Jackie moving at speed v_1 and Jackie sees a second object moving relative to her at speed v_2. By our old common sense, you would see the second object moving at speed $v_1 + v_2$. But the speed that you actually see is:

$$\text{speed of second object} = \frac{v_1 + v_2}{1 + \left(\frac{v_1}{c} \times \frac{v_2}{c}\right)}$$

Example 1: Length Contraction. Suppose Jackie is moving by you at $0.99c$. Because her spaceship is the same model as yours, you know that it is 100 meters long when it is at rest. How long is her spaceship as it moves by you?

Solution: We use the length contraction formula to calculate the length of the moving spaceship:

$$\begin{aligned}\text{length in moving frame} &= (\text{rest length}) \times \sqrt{1 - \left(\frac{v}{c}\right)^2} \\ &= (100\text{ m}) \times \sqrt{1 - (0.99)^2} = 14\text{ m}\end{aligned}$$

(continued)

S3.4 Is It True?

We've shown how all the major predictions of special relativity follow directly from the absoluteness of the speed of light and from the fact that the laws of nature are the same for everyone. But despite the clear logic of our thought experiments, our conclusions remain tentative until they pass observational or experimental tests. Are the bizarre predictions of relativity true?

The Absoluteness of the Speed of Light

The first thing we might wish to test is the surprising premise of relativity: the absoluteness of the speed of light. In principle, we can test this premise by measuring the speed of light coming from many different objects and going in many different directions, and verifying that the speed is always the same.

The speed of light was first measured in the late 1600s, but experimental evidence for the *absoluteness* of the speed of light did not come until 1887, when A. A. Michelson and E. W. Morley performed their now-famous *Michelson–Morley experiment.* This experiment showed that the speed of light is not affected by the motion of the Earth around the Sun. Countless subsequent experiments verified and extended the results of the Michelson–Morley experiment: The speed of light is always the same.

Experimental Tests of Special Relativity

Although *we* cannot yet travel at speeds at which the effects of relativity should be obvious, tiny subatomic particles *can* reach such speeds. In machines called *particle accelerators,* physicists accelerate subatomic particles to speeds near the speed of light and study what happens when the particles collide. The collisions involve large amounts of kinetic energy, some of which is converted into mass-energy that emerges as a shower of newly produced particles [Section 5.2]. Many of these particles have very short lifetimes, at

Mathematical Insight S3.2 (continued)

You would measure her spaceship as only 14 meters long, instead of its rest length of 100 meters.

Example 2: Mass Increase. A fly has a mass of 1 gram at rest. It is an unusual fly, however, in that it can travel at 0.9999c. What is the mass of the fly at that speed?

Solution: We use the mass increase formula to calculate the mass of the moving fly:

$$\text{moving mass} = \frac{(\text{rest mass})}{\sqrt{1 - \left(\frac{v}{c}\right)^2}} = \frac{1\text{ g}}{\sqrt{1 - (0.9999)^2}} = 70.7\text{ g}$$

At 0.9999*c*, the mass of the fly is almost 71 grams, or more than 70 times its rest mass.

Example 3: Velocity Addition. Jackie is moving toward you at 0.9*c*. Your friend Bob jumps into his spaceship and starts heading in Jackie's direction at 0.8*c* (from your point of view). How fast will Jackie see him approaching her? (See Figure S3.15.)

Solution: According to Jackie, your speed is $v_1 = 0.9c$. Bob's speed *relative to you* is $v_2 = 0.8c$ in the same direction. Thus, she sees him coming at:

$$\underset{\text{(relative to Jackie)}}{\text{Bob's speed}} = \frac{v_1 + v_2}{1 + \left(\frac{v_1}{c} \times \frac{v_2}{c}\right)} = \frac{0.9c + 0.8c}{1 + (0.9 \times 0.8)} = \frac{1.7c}{1.72} = 0.988c$$

She sees him moving at almost 99% of the speed of light.

THINKING ABOUT . . .

Measuring the Speed of Light

The only reason it is difficult to measure the speed of light is that light travels so fast. If you stand a short distance from a mirror and turn on a light, the reflection seems to appear instantaneously. Such observations led Aristotle (384–322 B.C.) to conclude that light travels at infinite speed, a view that was still held by many scientists as recently as the late 1600s.

One way to make the measurement easier is to place a mirror at increasingly great distances. If the speed of light truly were infinite, the reflection would always appear instantaneously. However, if it takes time for the light to travel to and from the mirror, you should eventually find a delay between the time you turn on the light and the time you see the reflection. Galileo tried a version of this experiment using the distance between two tall hills, but he was unable to detect any delay (instead of using a mirror, he stationed an assistant on the distant hill to signal back when he saw the light). He concluded that the speed of light, if not infinite, was too fast to be measured between hills on Earth with the technology of his day.

A delay in seeing reflected light was first detected in 1675 by the Danish astronomer Olaus Roemer, who used the four largest moons of Jupiter as his "mirrors." By that time, the orbital periods of the moons were well known, so it was possible to predict the precise moments at which each moon would be eclipsed by Jupiter. To his surprise, Roemer found that the eclipses occurred progressively earlier than expected during those times of year when Earth was moving toward Jupiter in its orbit and progressively later when Earth was moving away from Jupiter. He realized that the eclipses actually were occurring at the proper times, but the light was taking longer to reach us when we were farther from Jupiter. Roemer's observations proved that the speed of light is finite and led to the correct value of 300,000 km/s for its speed.[3]

As technology advanced, it became possible to measure light travel time between mirrors at much closer distances. In 1849 and 1850, the French physicists Fizeau and Foucault (also famous for the Foucault pendulum) performed a series of experiments using rotating mirrors to measure the speed of light much more precisely. Modern devices for measuring the speed of light take advantage of the wave properties of light, particularly the fact that light waves can interfere with one another. Such devices, called *interferometers,* were refined by A. A. Michelson in the early 1880s and used in the Michelson–Morley experiment. Details of how these experiments work can be found in many physics texts.

[3]The measurement depends on knowing the Sun–Earth and Sun–Jupiter distances, which were not well known in 1675. As a result, Roemer calculated the speed of light to be 227,000 km/s. Redoing his calculations using the modern values of these distances yields 300,000 km/s.

the end of which they decay (change) into other particles. For example, a particle called the π^+ ("pi plus") meson has a lifetime of about 18 nanoseconds (billionths of a second) when produced at rest. But π^+ mesons produced at speeds close to the speed of light in particle accelerators last much longer than 18 nanoseconds—just as predicted by the time dilation formula.

Particle accelerators also offer experimental evidence that nothing can reach the speed of light. It is relatively easy to get particles traveling at 99% of the speed of light in particle accelerators. But no matter how much more energy is put into the accelerators, the particle speeds get only fractionally closer to the speed of light. Some particles have been accelerated to speeds within 0.00001% of the speed of light, but none have ever reached the speed of light.

Although the effects of relativity are obvious only at very high speeds, modern techniques of measuring time are so precise that effects can be measured even at ordinary speeds. For example, a 1975 experiment compared the amount of time that passed on an airplane flying in circles to the time that passed on the ground. Over 15 hours, the airborne clocks lost a bit under 6 nanoseconds to the ground clocks, matching the result expected from relativity.

A Great Conspiracy?

Perhaps you're thinking, "I still don't believe it." After all, how can you know that the scientists who report the experimental evidence are telling the truth? Perhaps physicists are making up the whole thing as part of a great conspiracy designed to confuse everyone else so they can take over the world!

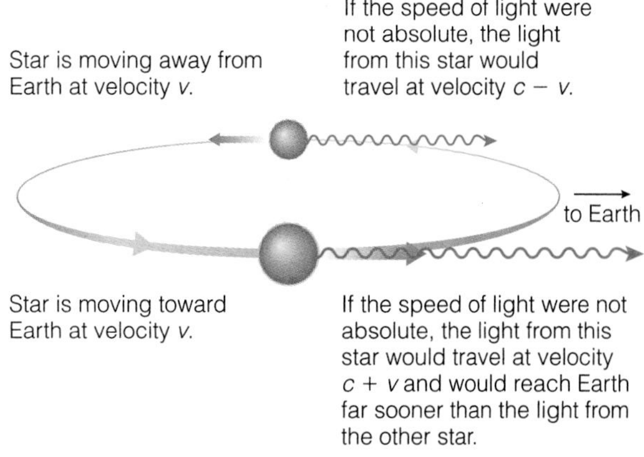

FIGURE S3.16 If the speed of light were *not* absolute, the speed at which light from each star in a binary system would come toward Earth would depend on its velocity toward us in the binary orbit.

What you need is evidence that you can see for yourself. What about nuclear energy? Einstein's famous formula $E = mc^2$, which explains the energy release in nuclear reactions, is a direct consequence of the special theory of relativity—and one that you can derive for yourself with a bit of algebra (see Mathematical Insight S3.3). Every time you see film of an atomic bomb, use electrical power from a nuclear power plant, or feel the energy of sunlight that was generated in the Sun by nuclear fusion, you are gaining experimental evidence of relativity.

Another test you can do yourself is to look through a telescope at a binary star system. If the speed of light were *not* absolute, the speed at which light from each star comes toward Earth would depend on its velocity toward us in the binary orbit (Figure S3.16). Imagine, for example, that one star is currently moving directly away from us in its orbit. Light from the star at this point would approach us at speed $c - v$. Some time later, when the same star is moving toward us in its orbit, its light would approach us at speed $c + v$. This light would therefore tend to catch up with the light that the star emitted from the other side of its orbit. If the orbital speed and distance were just right, we might see the same star on both sides of the orbit at once! More generally, because the light from each star would come toward us at a different speed from each point in its orbit, we would see each star in multiple positions in its orbit simultaneously. Thus, each star would appear as a short line of light rather than as a point—if the speed of light were not absolute. The fact that we always see

Mathematical Insight S3.3 Deriving $E = mc^2$

We can derive $E = mc^2$ from the mass increase formula, calling the moving mass m and the rest mass m_0:

$$\text{moving mass} = \frac{\text{(rest mass)}}{\sqrt{1 - \left(\frac{v}{c}\right)^2}} \quad \text{or} \quad m = m_0\left(1 - \frac{v^2}{c^2}\right)^{-\frac{1}{2}}$$

We also need a special mathematical approximation: If x is small compared to 1, then:

$$(1 + x)^{-\frac{1}{2}} \approx 1 - \frac{1}{2}x \quad \text{(for } x \text{ small compared to 1)}$$

You can verify this approximation by using your calculator to check that it holds for a few small values of x, such as $x = 0.05$ or $x = 0.001$.

Note what happens if we substitute $x = -v^2/c^2$ in this approximation:

Start with the approximation: $(1 + x)^{-\frac{1}{2}} \approx 1 - \frac{1}{2}x$

Substitute $x = -v^2/c^2$: $\left(1 + -\frac{v^2}{c^2}\right)^{-\frac{1}{2}} \approx 1 - \frac{1}{2}\left(-\frac{v^2}{c^2}\right)$

Simplify: $\left(1 - \frac{v^2}{c^2}\right)^{-\frac{1}{2}} \approx 1 + \frac{1}{2}\frac{v^2}{c^2}$

We can use this new form of the approximation to rewrite the mass increase formula. The condition that x must be small compared to 1 now means that distinct stars in binary systems therefore demonstrates that the speed of light *is* absolute.[4]

Finally, you can explore the paradoxes that would occur if the speed of light were not absolute. For example, imagine that two cars, both going about 100 km/hr, collide at an intersection (Figure S3.17). You witness the collision from far down one street. If the speed of light were *not* absolute, the light from the car that was coming toward you would have a speed of $c + 100$ km/hr, while the light from the other car would approach you only at c. You therefore would see the car coming toward you reach the intersection slightly *before* the other car and thus would see events unfold differently than the passengers in the car or eyewitnesses in other locations. This difference would be scarcely noticeable, because 100 km/hr is only about *one-millionth* the speed of light, but the difference would increase if you could watch from *very* far away. For example, if you had a super telescope and watched such a collision on a planet in a galaxy 1 million light-years from Earth, you'd see the first car reach the intersection a *year* before the second car. This is a serious paradox: From the viewpoint of the passengers in the cars, they have collided, yet you saw one car reach the collision point long before the other car had even started its journey!

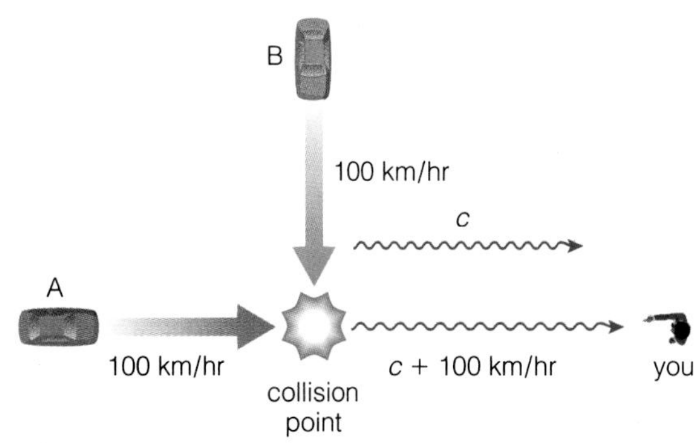

FIGURE S3.17 If the speed of light were *not* absolute, the light from a car coming toward you would approach you faster than the light from a car going across your line of sight.

[4]This conclusion would not follow if light waves were carried by a medium in the same way that sound waves are carried by air. Scientists in the 1800s believed that such a medium, which they called the *ether*, permeated all of space. The Michelson–Morley experiment ruled out the existence of such a medium, so we are left with the conclusion that c is absolute.

$-v^2/c^2$ must be small; that is, the speed v must be small compared to the speed of light c.

Start with the mass increase formula: $$m = m_0\left(1 - \frac{v^2}{c^2}\right)^{-\frac{1}{2}}$$

Substitute $\left(1 - \frac{v^2}{c^2}\right)^{-\frac{1}{2}} \approx 1 + \frac{1}{2}\frac{v^2}{c^2}$: $$m \approx m_0\left(1 + \frac{1}{2}\frac{v^2}{c^2}\right)$$

Expand right side: $$m \approx m_0 + \frac{1}{2}\frac{m_0 v^2}{c^2}$$

Multiply both sides by c^2: $$mc^2 \approx m_0c^2 + \frac{1}{2}m_0v^2$$

You may recognize the last term on the right as the ***kinetic energy*** [Section 5.2] of an object with mass m_0. Because the other two terms also have units of mass multiplied by speed squared, they also must represent some kind of energy. Einstein recognized that the term on the left represents the *total* energy of a moving object. He then noticed that, even if the speed is *zero* ($v = 0$) so that there is *no* kinetic energy, the equation states that the total energy is *not* zero; instead, it is m_0c^2. In other words, when an object is not moving at all, it still contains energy by virtue of its mass. Thus, $E = mc^2$ is a direct consequence of Einstein's theory of relativity.

TIME OUT TO THINK *The preceding paradox presents us with a choice. If the speed of light* is not *absolute, different people can witness the same events in very different ways. If the speed of light* is *absolute, measurements of time and space are relative. Einstein preferred the latter choice. Do you? Explain.*

Of course, like any scientific theory, the theory of relativity can never be *proved* beyond all doubt. But it is supported by a tremendous body of evidence, some of which you can see for yourself. This evidence is real and cannot be made to disappear. If anyone ever comes up with an alternative theory, the new theory will still have to explain the many experimental results that seem to support relativity so well.

S3.5 Toward a New Common Sense

We've used thought experiments to show that our old common sense doesn't work, and we've discussed how actual experiments verify the ideas of our thought experiments. But we haven't yet figured out what new common sense should replace the old. Surprisingly, another thought experiment that may at first make everything seem even more bizarre will help us understand what is really going on.

Thought Experiment 12 Suppose Jackie is moving by you at a speed close to the speed of light. From our earlier thought experiments, we know that you'll measure her time as running slow, her length as having contracted, and her mass as having increased. But what would *she* say?

From Jackie's point of view, she's not going anywhere—you are moving by *her* at high speed. Because the laws of nature are the same for everyone, she must reach exactly the same conclusions from her point of view that you reach from your point of view. That is, she'll say that *your* time is running slow, *your* length is contracted, and *your* mass is increased!

Now we have a severe argument on our hands. Imagine that you are looking into Jackie's spaceship with a super telescope. You can clearly see that her time is running slow because everything she does is in slow motion. You send her a radio message saying "Hi Jackie! Why are you doing everything in slow motion?" Because the radio message travels at the absolute speed of light, Jackie has no trouble receiving your message, and she responds with her own radio message back to you.

As you listen to her response, you'll hear it in slow motion—"Hheeeellllloooo tthhheeerrreee"—thus verifying that her time is running slow. But once you record the entire message and use your computer to speed up her voice so that it sounds normal, you'll hear Jackie say, "I'm not moving in slow motion, *you are!*"

You can argue back and forth all you want, but it will get neither of you anywhere. Then you come up with a brilliant idea. You hook up a video camera to your telescope and record a film showing that Jackie's clock is moving slower than yours. You put your videotape in a very fast rocket and shoot it off toward her. When the video arrives and she watches it, she'll see that you are right and that *she* is in slow motion.

Before you declare victory in the argument, one slight problem remains. Jackie had the same brilliant idea, and a rocket from Jackie arrives with a videotape that she made. You put it into the videotape player and watch what appears to be clear proof that Jackie is right—her tape shows that *you* are in slow motion!

How can it be that you see Jackie's time running slow while she sees your time running slow? Think back to our earlier discussion of up and down, and imagine that an American child and an Australian child are talking on the telephone. The Australian says, "Isn't the Moon beautiful up in the sky right now?" The American replies, "What are you talking about? The Moon isn't up right now!" According to childhood common sense about up and down, the two children appear to be contradicting each other and could argue endlessly. However, once they realize that up and down are measured *relative* to the center of the Earth, they realize that the argument stems only from incorrect definitions of up and down (Figure S3.18).

In a similar way, the argument between you and Jackie arises because you are using the old common sense in which we think of space and time as absolutes and expect the speed of light to be relative. The theory of relativity tells us that we have it backward. The speed of light is the absolute, and time and space are relative. By the new common sense,

FIGURE S3.18 The Moon is up for the Australian observer but down for the American observer.

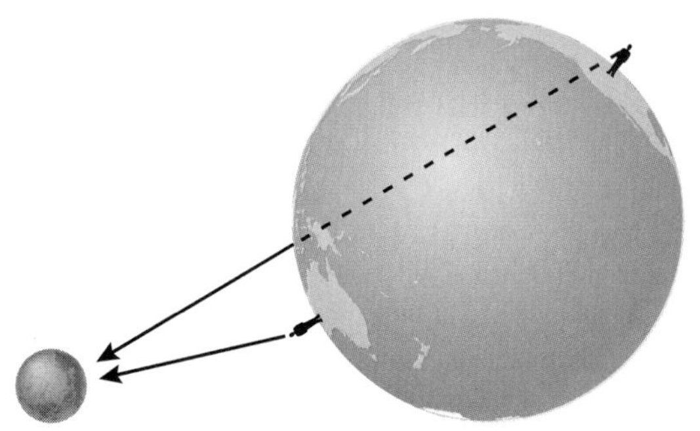

the fact that you and Jackie disagree about whose time is running slow is no more surprising than the fact that the two children disagree about whether the Moon is up or down. The disagreement is meaningless because it involves inadequate definitions of time and space. What counts are *results,* and every experiment that you perform will agree with every experiment that Jackie performs. You are both experiencing the same laws of nature, albeit in ways different from those our old common sense would have suggested.

S3.6 Ticket to the Stars

The fact that we cannot exceed the speed of light might at first make distant stars seem forever out of reach. But time dilation and length contraction actually offer a ticket to the stars, if we can ever build spaceships capable of traveling at speeds close to the speed of light.

Suppose you want to take a trip to the star Vega, about 25 light-years away. Further, suppose you have a ship that can travel at very close to the speed of light—say, at 0.999*c*. The trip will take you about 25 years from our point of view on Earth (since you are going at nearly the speed of light over a distance of 25 light-years), and the return trip will take another 25 years. If you leave in the year 2025, you will arrive at Vega in the year 2050 and return to Earth in 2075.

However, from your point of view, you remain stationary while Earth rushes away from you and Vega rushes toward you at 0.999*c*. You'll therefore find the length from Earth to Vega contracted from its rest length of 25 light-years; with the length contraction formula, the contracted distance turns out to be just over 1 light-year. Because Vega is coming toward you at 0.999*c* and has only 1 light-year to travel from your point of view, you'll be at Vega in only about 1 year. Your return trip to Earth will also take about 1 year, so the round-trip time is only about 2 years from your point of view (Figure S3.19). If you leave at age 40, you'll return as a 42-year-old.

Although it sounds contradictory by our old common sense, both points of view are correct.[5] If you leave in 2025 at age 40, you'll return to Earth at age 42—but in the year 2075. That is, while you will have aged only 2 years, all your surviving friends and family will be 50 years older than when you left.

You could make even longer trips within your lifetime with a sufficiently fast spaceship. For example, the Andromeda Galaxy is about 2.5 million light-years away, so the round-trip to any star in the Andromeda Galaxy would take at least 5 million years

[5]Given that you see the Earth–Vega distance contracted, shouldn't you also claim that it is *Earth's* time running slow, rather than yours? This question underlies the so-called *twin paradox,* in which you make the trip while your twin sister stays home on Earth (see Chapter S4). The resolution comes from the fact that, because you must turn around at Vega, you effectively change reference frames relative to Earth at least once during your trip. A careful analysis of the changing reference frames is beyond the scope of this book, but it turns out that you do indeed measure less total time than your stay-at-home twin.

FIGURE S3.19 At high speed, a traveler to a distant star will age less than people back home on Earth.

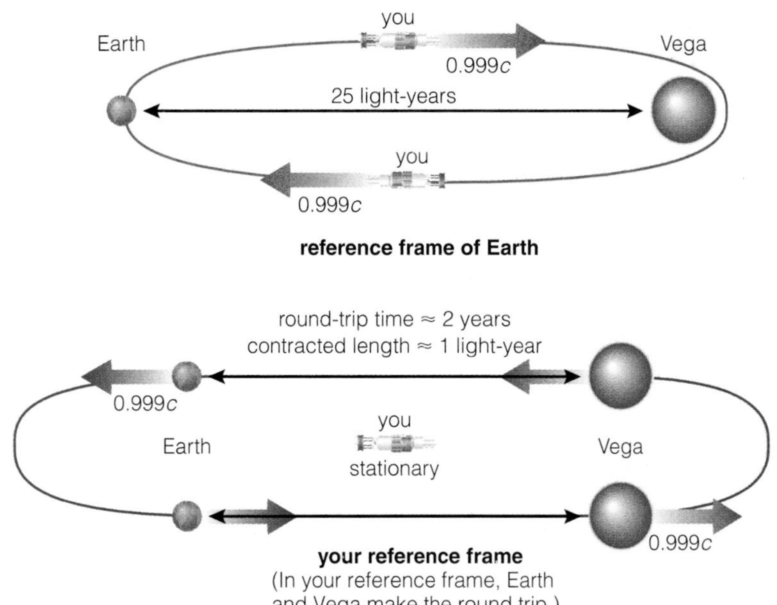

from the point of view of observers on Earth. However, if you could travel at a speed within 50 parts in a trillion of the speed of light (i.e., $c - 5 \times 10^{-11}c$), the trip would take only about 50 years from your point of view. You could leave Earth at age 30 and return at age 80—but you would return to an Earth on which your friends, your family, and everything you knew had been gone for 5 million years.

In terms of time, relativity offers only a one-way ticket to the stars. You can go a long distance and return to the *place* that you left, but you cannot return to the epoch from which you left.

Henceforth space by itself, and time by itself, are doomed to fade away into mere shadows, and only a kind of union of the two will preserve an independent reality.

HERMANN MINKOWSKI, 1908

THE BIG PICTURE

In this chapter, we have studied Einstein's special theory of relativity, learning that space and time are intertwined in remarkable ways that are quite different from what we might have expected from everyday experience. As you look back on our new viewpoint about space and time, keep in mind the following "big picture" ideas.

- The ideas of relativity all derive from two simple facts: The laws of nature are the same for everyone, and the speed of light is absolute. The thought experiments in this chapter simply derived the consequences of these facts.
- Thought experiments are useful, but the ultimate judge of any theory is observation or experiment. The theory of relativity has been extensively tested and verified.
- Although the ideas of relativity may sound strange at first, you can understand them simply and logically if you allow yourself to develop a "new common sense" that incorporates them.
- Space and time are properties of the universe itself, and the new understanding of them gained through the theory of relativity will enable you to better appreciate how the universe works.

Review Questions

1. What is the *theory of relativity?* How does *special relativity* differ from *general relativity?*
2. List five major predictions of the special theory of relativity.
3. Suppose you are riding on a stationary bike and the speedometer says you are going 30 km/hr. What does this number mean? What does it tell you about the idea of relative motion?
4. According to the theory of relativity, what are the two absolutes in the universe? Which one is more surprising, and why?
5. What is a *paradox?* How can a paradox lead us to a deeper understanding of an issue?
6. What do we mean by a *frame of reference?* What is a *free-float frame?*
7. Suppose you see a friend moving by you at some constant speed. Explain why your friend can equally well say that he is stationary and you are moving by him.
8. In your own words, restate Thought Experiment 6 to prove that no material object can be observed to travel at or above the speed of light.
9. What is *time dilation?* Explain how and why your measurements of time will differ from those of someone moving by you.
10. Explain why observers in different reference frames will not necessarily agree about the order of two events that occur in different places.
11. What is *length contraction?* How will your measurements of the size of a spaceship differ for a spaceship moving by you and the same spaceship at rest in your reference frame?
12. What is *mass increase?* How does the mass of an object moving by you compare to its rest mass?
13. Construct your own variation on Thought Experiment 11 to show why velocities must add up differently than we would expect according to our "old common sense."

14. Briefly describe several experimental tests that support special relativity.

15. Why do tests of $E = mc^2$ also test special relativity? Describe several tests of $E = mc^2$ that you can see for yourself.

16. If you watch a friend moving by you, you'll say that her time is running slow, her length is contracted, and her mass is greater than her rest mass. How will she perceive her own time, length, and mass? Why? How will she perceive *your* time, length, and mass?

17. Suppose you could take a trip to a distant star at a speed very close to the speed of light. Explain how relativity makes it possible for you to make this trip in a reasonably short amount of time. What will you find when you return home to Earth?

Discussion Questions

1. *Common Sense.* Discuss the meaning of the term *common sense.* How do we develop common sense? Can you think of other examples, besides the example of the meaning of *up* and *down,* in which you've had to change your common sense? Do you think that the theory of relativity contradicts common sense? Why or why not?

2. *Photon Philosophy.* Extend the ideas of time dilation and length contraction to think about how the universe would look if you were a photon traveling at the speed of light. Do you think there's any point to thinking about how a photon "perceives" the universe? If so, discuss any resulting philosophical implications. If not, explain why not.

3. *Ticket to the Stars.* Suppose that we someday acquire the technology to travel among the stars at speeds near the speed of light. Imagine that many people make journeys to many places. Discuss some of the complications that would arise from people aging at different rates depending on their travels.

Problems

1. *Relative Motion Practice.* In all of the following, assume that you and your friends are in free-float reference frames.
   a. Bob is coming toward you at a speed of 75 km/hr. You throw a baseball in his direction at 75 km/hr. What does he see the ball doing?
   b. Shawn is traveling away from you at a speed of 120 km/hr. He throws a baseball that, according to him, is going 100 km/hr in your direction. What do you see the ball doing?
   c. Carol is going away from you at 75 km/hr, and Sam is going away from you in the opposite direction at 90 km/hr. According to Carol, how fast is Sam going?
   d. Consider again the situation in part (c). Suppose you throw a baseball in Sam's direction at a speed of 120 km/hr. What does Sam see the ball doing? What does Carol see the ball doing?

2. *Moving Spaceship.* Suppose you are watching a spaceship go past at a speed close to the speed of light.
   a. How do clocks on the spaceship run, compared to your own clocks?
   b. If you could measure the length, width, and height of the spaceship as it passed by, how would these measurements compare to the spaceship's size if it were stationary?
   c. If you could measure the mass of the spaceship, how would it compare to its rest mass?
   d. How would a passenger on the spaceship view your time, size, and mass?

3. *Relativity of Simultaneity.* Consider the situation in Thought Experiment 8, about the green and red flashes of light at opposite ends of Jackie's spaceship. Suppose your friend Bob is traveling in a spaceship in the opposite direction from Jackie. Further imagine that he is also precisely aligned with you and Jackie at the instant the two flashes of light occur (in your reference frame).
   a. According to Bob, is Jackie illuminated first by the green flash or the red flash? Explain.
   b. According to you, which flash illuminates Bob first? Why?
   c. According to Bob, which flash occurred first? Explain. How does Bob's view of the order of the flashes compare to your view and to Jackie's view?

4. *Time Dilation 1.* A clever student, after learning about the theory of relativity, decides to apply his knowledge in order to prolong his life. He decides to spend the rest of his life in a car, traveling around the freeways at 55 miles per hour (89 km/hr). Suppose he drives for a period of time during which 70 years pass on Earth; how much time will pass in the car? (*Hint:* If you are unable to find a difference, be sure to explain why.)

5. *Time Dilation 2.* An even more clever student, upon realizing the folly of the student in the previous problem, decides on a better approach for prolonging her life. She decides to spend time cruising around the local solar neighborhood at a speed of $0.95c$ (95% of the speed of light). How much time will pass on her spacecraft during a period in which 70 years pass on Earth?

6. *Time Dilation with Subatomic Particles.* Recall that a π^+ meson produced at rest has a lifetime of 18 nanoseconds (1.8×10^{-8} s). Thus, in its own reference frame, a π^+ meson will always "think" it is at rest and therefore will decay after 18 nanoseconds. Suppose a π^+ meson is produced in a particle accelerator at a speed of $0.998c$.
   a. Use the time dilation formula to calculate how long scientists will see the particle last before it decays.
   b. Briefly explain how an experiment like this helps to verify the special theory of relativity.
7. *Ticket to the Stars.* Suppose you stay home on Earth while your twin sister takes a trip to a distant star and back in a spaceship that travels at 99% of the speed of light. If both of you are 25 years old when she leaves and you are 45 years old when she returns, how old is your sister when she gets back?
8. *The Betelgeuse Red Stars.* Like the fans in Boston, the fans of interstellar baseball at Betelgeuse (in the constellation Orion) have been deprived of a championship team for a long time. In fact, the Betelgeuse Red Stars have not won the Universe Series for nearly 200,000 years. (They did, however, come very close to winning 75,000 years ago; their defeat was sealed only when a routine ground ball went through the legs of the infamous Zargon Buckner.) Realizing that time may be running out for their team—Betelgeuse is expected to explode as a supernova within the next 100,000 years—the Red Stars management has decided to break some league rules (hopefully without getting caught) in hopes of winning the series. They have therefore extended a lucrative offer to Hideo Nomo, of planet Earth, if he will leave his home and join the Betelgeuse Red Stars as their new pitcher. Although he was reluctant to leave friends and family behind, Nomo finally was swayed to join the Red Stars. Interestingly, in an interview with the *Intergalactic Press,* Nomo said it was the travel opportunity, rather than the money, that lured him to Betelgeuse. Nomo was given a ticket to travel to Betelgeuse on an express spaceship at 95% of the speed of light. During the trip, he decided to try some pitching in the ship's on-board stadium. He found that, with the replacement body parts provided by the Red Stars management, his fastball was considerably improved: He was now able to throw a pitch at 80% of the speed of light. Assuming that he throws a pitch in the same direction the spacecraft is traveling, use the velocity transformation to calculate how fast we would see the ball moving from Earth.
9. *Racing a Light Beam, Part 1.* A long time ago, in a galaxy far away, there was a civilization that hosted an Olympic competition every 4 of their years. One year, during an Olympiad held in the city of Sole in the nation of Kira, a sprinter by the name of Jo shattered their world record for the 100-meter dash. Alas, Jo was disqualified for having ingested illegal substances, and his record was eliminated from the books. Rather than sulk, however, Jo decided that human competition was too easy anyway. He announced that he would instead race a beam of light. Sponsors lined up, crowds gathered, and the event was sold to a pay-per-view audience. Everything was set. The starting gun went off. Jo raced out of the starting blocks and shattered the old world record, running the 100 meters in 8.7 seconds. The light beam, represented by a flashlight turned on at precisely the right moment, of course emerged from its "blocks" at the speed of light. How long did it take the light beam to cover the 100-meter distance? What can you say about the outcome of this race?
10. *Racing a Light Beam, Part 2.* Following his humiliation in the race against the light beam, Jo went into hiding for the next 2 years. By that time, most people had forgotten about both him and the money they had wasted on the pay-per-view event. However, Jo was secretly in training during his hiding. He worked out hard and tested new performance-enhancing substances. One day, he emerged from hiding and called a press conference. "I'm ready for a rematch," he announced. Sponsors were few this time and spectators scarce in the huge Olympic stadium where Jo and the flashlight lined up at the starting line. But those who were there will never forget what they saw, although it all happened very quickly! Jo blasted out of the starting blocks at 99.9% of the speed of light. The light beam, emitted from the flashlight, took off at the speed of light. The light beam won again—but barely! After the race, TV commentators searched for Jo, but he seemed to be hiding again. Finally, they found him in a corner of the locker room, sulking under a towel. "What's wrong? You did great!" said the commentators. Jo looked back sadly, saying, "Two years of training and experiments, for nothing!" Let's investigate what happened.
    a. As seen by spectators in the grandstand, how much faster than Jo is the light beam?
    b. As seen by Jo, how much faster is the light beam than he is? Explain your answer clearly.
    c. Using your results from parts (a) and (b), explain why Jo can say that he was beaten just as badly as before, while the spectators can think he gave the light beam a good race.
    d. Although Jo was disappointed by his performance against the light beam, he did find one pleasant surprise: The 100-meter course seemed short to him. In Jo's reference frame during the race, how long was the 100-meter course?

S4 Spacetime and Gravity

WHAT IS GRAVITY? NEWTON CONSIDERED GRAVITY TO BE A mysterious force that somehow reached across vast distances of space to hold the Moon in orbit around the Earth and the planets in orbit around the Sun. His law of gravity explained the actions and consequences of this mysterious force but said nothing about how the force was transmitted through space.

Einstein removed the mystery of how gravity acts at a distance. As he extended his theory of relativity, Einstein found that he could explain gravity in terms of the structure of space and time. In Einstein's view, the orbits of the Moon and the planets are as natural as motion in a straight line.

As we investigate Einstein's revolutionary view of gravity, we will see that the consequences of his discoveries abound in astronomy and explain phenomena ranging from the peculiar orbit of Mercury to the expansion of the universe. We will also see how space and time merge together into a four-dimensional *spacetime* and that our universe is a strange world of curved space and altered time containing *black holes* through which material can leave the universe, never to return.

S4.1 Einstein's Second Revolution

Imagine that you and everyone around you believe the Earth to be flat. As a wealthy patron of the sciences, you decide to sponsor an expedition to the far reaches of the world. You select two fearless explorers and give them careful instructions. Each is to journey along a perfectly straight path, but they are to travel in opposite directions. You provide each with a caravan for land-based travel and boats for water crossings, and you tell each to turn back only after discovering "something extraordinary."

Some time later, the two explorers return. You ask, "Did you discover something extraordinary?" To your surprise, they answer in unison, "Yes, but we both discovered the same thing: We ran into each other, despite having traveled in opposite directions along perfectly straight paths."

Although this outcome would be extraordinarily surprising if you truly believed the Earth to be flat, we are not surprised because we know that the Earth is round (Figure S4.1). In a sense, the explorers followed the *straightest possible paths*, but these "straight" lines follow the curved surface of the Earth.

FIGURE S4.1 Travelers going in opposite directions along paths that are as straight as possible will meet as they go around the Earth.

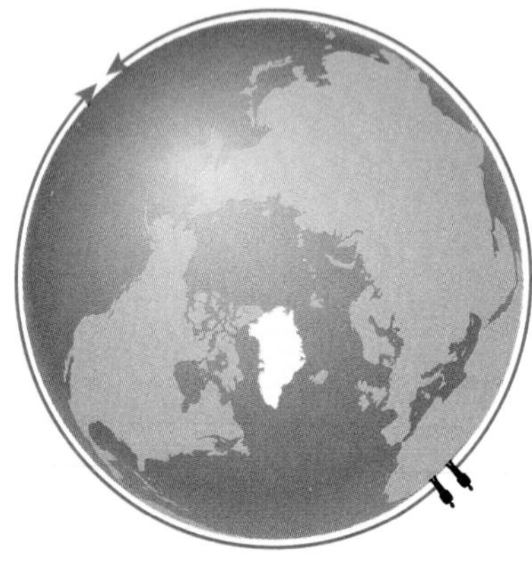

Now let's consider a somewhat more modern scenario. You are floating freely in a spaceship somewhere out in space. Hoping to learn more about space in your vicinity, you launch two small probes along straight paths in opposite directions. Each probe is equipped with a camera that transmits pictures back to your spaceship. Imagine that, to your astonishment, the probes one day transmit pictures of each other! That is, although you launched them in opposite directions and neither has ever fired its engines, the probes have somehow met. In fact, this situation arises quite naturally with any orbiting objects in space. If you launch two probes in opposite directions from a space station, they will meet as they orbit the Earth.

Since the time of Newton, we've generally explained the curved paths of the two probes as an effect caused by the force of gravity. However, by analogy with the explorers journeying in opposite directions on the Earth, might we instead conclude that the probes meet because *space* is somehow curved? The idea that space could be curved certainly sounds strange at first. While it's easy to visualize a surface curving *through* space, our mind cannot visualize three-dimensional space as being curved. The idea that space can be curved lies at the heart of Einstein's second revolution—a revolutionary view of gravity contained in his *general theory of relativity*, published in 1915.

From the special theory of relativity, we already know that space and time are inextricably linked. In fact, the three dimensions of space and the one dimension of time together form an inseparable, *four-*dimensional combination called **spacetime.** General relativity tells us that matter shapes the "fabric" of spacetime in a manner analogous to the way heavy weights distort a taut rubber sheet or trampoline (Figure S4.2). Of course, we cannot place weights "upon" spacetime because all matter exists *within* spacetime, and we cannot visualize distortions of spacetime. But, using a rubber sheet as an analogy, we can begin to appreciate the principles of general relativity that govern the structure of space and time.

It is difficult to overstate the significance of general relativity to our understanding of the universe. For example, all the following ideas come directly from Einstein's general theory of relativity:

- Gravity arises from distortions of spacetime. It is *not* a mysterious force that acts at a distance. The presence of mass causes the distortions, and the resulting distortions determine how other objects move through spacetime.
- Time runs slow in gravitational fields. The stronger the gravity, the slower time runs.
- *Black holes* can exist in spacetime, and falling into a black hole means leaving the observable universe.
- The universe has no boundaries and no center, yet it might still have a finite volume.
- Large masses that undergo rapid changes in motion or structure emit *gravitational waves* that travel at the speed of light.

FIGURE S4.2 A rubber-sheet analogy to spacetime: Matter distorts the "fabric" of four-dimensional spacetime in a manner analogous to the way heavy weights distort a taut, two-dimensional rubber sheet. The greater the mass, the greater the distortion of spacetime.

S4.2 The Equivalence Principle

Special relativity shows that there is no single, absolute answer to the question "Who is moving?" when two people pass each other at a constant velocity. Each individual can claim to be at rest, and each claim is equally valid. However, the situation seems quite different when accelerations are involved.

Imagine that you and your friend Jackie from Chapter S3 are both floating freely in space when your rocket engine fires (Figure S4.3). Jackie sees you accelerating away, with your speed growing ever faster, so she sends you a radio message saying, "Good-bye, have a nice trip!"

TIME OUT TO THINK *Suppose you start from rest in Jackie's reference frame and she sees you accelerate at 1g ($\approx$ 10 m/s^2). Approximately how fast will Jackie see you going after 1 second? After 10 seconds? After a minute?* (Hint: *See Section 6.1.*)

From your perspective, it is Jackie who is receding into the distance at ever-faster speeds, so you reply: "Thanks, but I'm not going anywhere. You're the one accelerating into the distance." On closer examination, however, your reply seems to have a flaw. If you had been moving at constant velocity relative to Jackie, you both would have been floating freely, making it impossible to determine who was "really" moving. However, your acceleration makes you feel a force. Instead of floating weightlessly like Jackie, you are held to the floor of your spaceship.

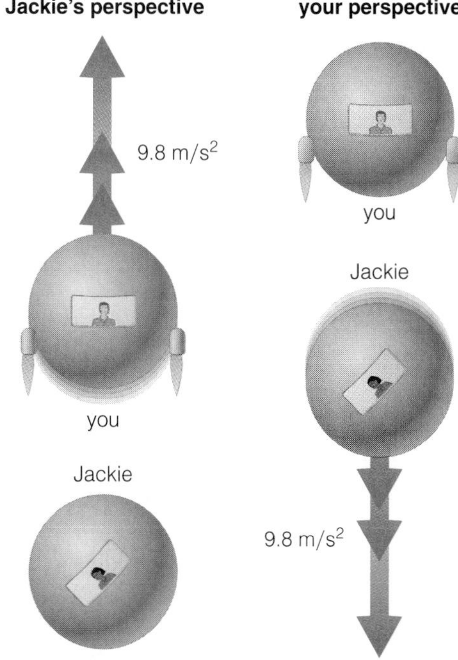

FIGURE S4.3 Jackie is floating freely in her spaceship. Your engines are firing, and you feel a force allowing you to stand on the floor of your spaceship.

If you happen to be accelerating at 1g, or 9.8 m/s^2, you'll feel just as you feel when you are stationary on Earth, with your normal Earth weight [Section 6.1]. Thus, Jackie can respond: "Oh yeah? If you're not going anywhere, why are you stuck to the floor of your spaceship, and why do you have your engines turned on? Furthermore, if I'm accelerating, why am I not feeling any forces?"

You must admit that Jackie is asking very good questions. It certainly *looks* as if you really are the one who is moving and that you *cannot* legitimately claim to be stationary. Einstein didn't like the appearance of situations like this one, because he believed that *all* motion should be relative.

In 1907, Einstein hit upon what he later called "the happiest thought of my life." His revelation consisted of the idea that, whenever you feel weight (as opposed to weightlessness), you can equally well attribute it to effects of either acceleration or gravity. This idea is called the **equivalence principle.** Stated more precisely, it says:

The effects of gravity are exactly equivalent to the effects of acceleration.[1]

To clarify the meaning of the equivalence principle, imagine that you are sitting inside with doors closed and window shades pulled down when your room is magically removed from the Earth and sent hurtling through space with an acceleration of 1*g* (Figure S4.4). According to the equivalence principle, you have no way of knowing that you've left the Earth. Any experiment you performed, such as dropping balls of different weights, would yield the same results you'd get on the Earth.

Now back to Jackie's questions. She claims that *you* must be accelerating because you feel weight as you stand in your spaceship. The equivalence principle tells us that, with equal validity, you can claim to feel weight because of gravity. From this perspective, space is filled with a gravitational field pointing "downward" toward the back of your spaceship. You are stationary only because your rocket engine prevents you from falling, and you feel weight just as a person hovering in a helicopter over the Earth does. Jackie, because she is not using her engines, is falling through the gravitational field. As you may recall, anyone in *free-fall* feels weightless [Section 6.1]. In summary, you can claim that the situation is much as it would be if you were hovering over a cliff while Jackie had fallen over the edge (Figure S4.5). Thus, you can respond: "Sorry, Jackie, but I still say that you have it backward. I'm using my engines to prevent my spaceship from falling, and I feel weight because of *gravity*. You're weightless because you're in free-fall. I hope you won't be hurt by hitting whatever lies at the bottom of this gravitational field!"

The equivalence principle allows us to claim that *all* motion is relative; it is the starting point for general relativity. Just as we derived the strange conse-

[1] Technically, this equivalence holds only within small regions of space. Over larger regions, we can detect *tidal forces* that can arise from gravity but not from acceleration.

FIGURE S4.4 The equivalence principle states that the effects of gravity are exactly equivalent to the effects of acceleration. Thus, according to the equivalence principle, you cannot tell the difference between being in a closed room on Earth and being in a closed room accelerating through space at 1*g*.

THINKING ABOUT . . .

Einstein's Leap

Given that the similarities in the effects of gravity and acceleration were well known to scientists as far back in time as Newton, you may be wondering why the equivalence principle is so surprising. The answer is that the similarities were generally attributed to coincidence—although a very puzzling coincidence. It is as if other scientists imagined that nature was showing them two boxes, one labeled "effects of gravity" and the other labeled "effects of acceleration." They shook, weighed, and kicked the boxes but could never find any obvious differences between them. They concluded: "What a strange coincidence! The boxes seem the same from the outside even though they contain different things." Einstein's revelation was, in essence, to look at the boxes and say that it is not a coincidence at all. The boxes appear the same from the outside because they contain the same thing.

In many ways, Einstein's assertion of the equivalence principle represented a leap of faith—although it was a faith he would willingly test through scientific experiment. He made the assertion because he thought the universe would make more sense if it were true, not because of any compelling observational or experimental evidence for it at the time. This leap of faith sent him on a path far ahead of his scientific colleagues.

From a historical viewpoint, special relativity was a "theory waiting to happen" because it was needed to explain two significant problems left over from the nineteenth century: the perplexing constancy of the speed of light, demonstrated in the Michelson–Morley experiment [Section S3.4], and some seeming peculiarities of the laws of electromagnetism. Indeed, several other scientists were very close to discovering the ideas of special relativity when it was published by Einstein in 1905, and *someone* was bound to come up with special relativity around that time. In contrast, general relativity was a tour de force by Einstein. He recognized that unsolved problems remained after completing the theory of special relativity, and he alone took the leap of faith required to accept the equivalence principle. Without Einstein, general relativity probably would have remained undiscovered for several decades beyond 1915, when he completed and published the theory.

quences of special relativity from the idea that the speed of light is absolute, the astounding predictions of general relativity follow from the equivalence principle. And, just as special relativity led us to recognize some underlying truths about nature, such as that space and time are different for observers in different reference frames, general relativity also leads us to a new and deeper understanding of the universe.

S4.3 Understanding Spacetime

It's easy to say that you can equally well attribute your weight to effects of gravity or acceleration, but the two effects tend to *look* very different. A person standing on the surface of the Earth appears to be motionless, while an astronaut accelerating through space continually gains speed. How can gravity and acceleration produce such similar effects when they look so different? The theory of general relativity answers that we're not seeing the whole picture: Instead of looking just at the three dimensions of space, we must learn to "look" at the *four* dimensions of spacetime.

FIGURE S4.5 If you hover over a cliff while Jackie falls, you feel weight, you need your engines to keep you from falling, and Jackie is weightless because she is in free-fall.

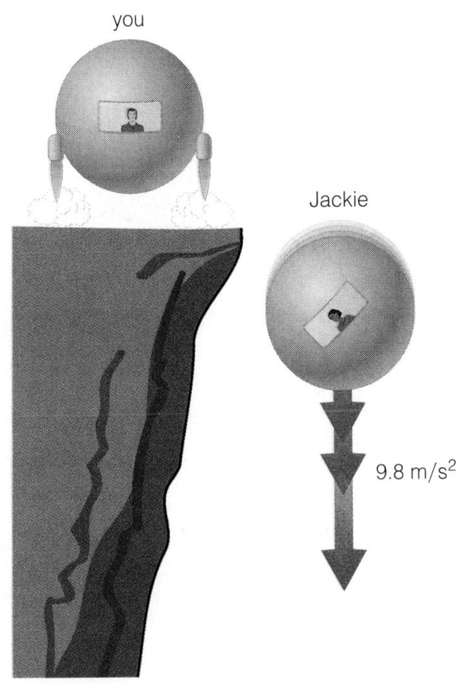

Four Dimensions

The concept of **dimension** describes the number of independent directions in which movement is possible. A **point** has zero dimensions. If you were a geometric prisoner confined to a point, you'd have no place to go. Sweeping a point back and forth along one direction generates a **line** (Figure S4.6). The line is one-dimensional because only one direction of motion is possible (going backward is considered the same as going forward by a negative distance). Sweeping a line back and forth generates a two-dimensional **plane.** The two directions of possible motion are horizontal and vertical; any other direction is just a combination of these two. If we sweep a plane back and forth, it fills three-dimensional **space,** with the three independent directions of length, width, and depth.

We live in three-dimensional space, so we cannot visualize any direction that is distinct from length, width, and depth (and combinations thereof). However, just because we cannot see "other" directions doesn't mean they don't exist. Thus, we can imagine sweeping space back and forth in some "other" direction to generate a **four-dimensional space.** Although we have no hope of visualizing a four-dimensional space, we can easily describe it mathematically. In algebra, we do one-dimensional problems with the single variable x, two-dimensional problems with the variables x and y, and three-dimensional problems with the variables x, y, and z. A four-dimensional problem simply requires adding a fourth variable, as in x, y, z, and w. We could continue to five dimensions, six dimensions, and many more. Any space with more than three dimensions is called a **hyperspace,** which means "beyond space."

Spacetime

Spacetime is a four-dimensional space in which the four directions of possible motion are length, width, depth, and *time.* Note that time is not "the" fourth dimension; it is simply one of the four. (However, time differs in an important way from the other three dimensions; see Mathematical Insight S4.1.) We cannot picture all four dimensions of spacetime at once, but we can imagine what things would look like if we could. In addition to the three spatial dimensions of spacetime that we ordinarily see, every object would be stretched out through time. Objects that we see as three-dimensional in our ordinary lives would appear as "solid" four-dimensional objects in spacetime. If we could see in four dimensions, we could look through time just as easily as we look to our left or right. If we looked at a person, we could see every event in the person's life. If we wondered what really happened during some historical event, we'd simply look to find the answer.

TIME OUT TO THINK ***Try to imagine how you would look in four dimensions. How would your body, stretched through time, appear? Imagine that you bumped into someone on the bus yesterday. What would this event look like in spacetime?***

FIGURE S4.6 We can generate one-, two-, and three-dimensional spaces by sweeping back-and-forth a point, a line, and a plane, respectively.

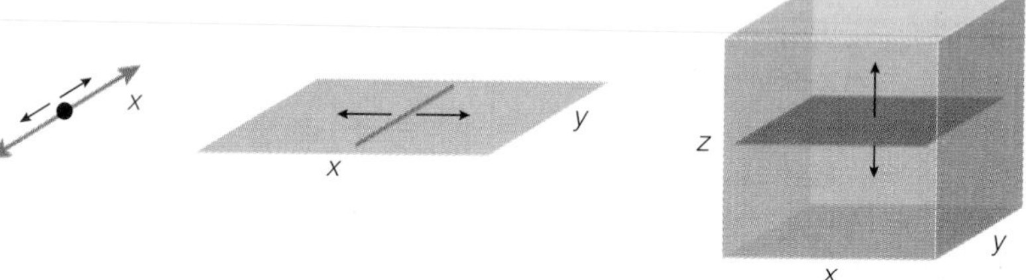

This spacetime view of objects provides a new way of understanding why different observers can disagree about measurements of time and distance. Because we can't visualize four dimensions, let's use a three-dimensional analogy. Suppose you give the same book to many different people and ask each person to measure the book's dimensions. Everyone will get the same results, agreeing on the three-dimensional structure of the book (Figure S4.7a). But suppose instead that you show each person only a two-dimensional picture of the book rather than the book itself. The pictures can look very different, even though they show the same book in all cases (Figure S4.7b). If the people believed that the two-dimensional pictures reflected reality, they might argue endlessly about what the pictured object really looks like.

In our ordinary lives, we perceive only three dimensions, and we assume that this perception reflects reality. But spacetime is actually four-dimensional. Just as different people can see different two-dimensional pictures of the same three-dimensional book, different observers can see different three-dimensional "pictures" of the same spacetime reality. These different "pictures" are the differing perceptions of time and space of observers in different reference frames. Thus, different observers will get different results when they measure time, length,

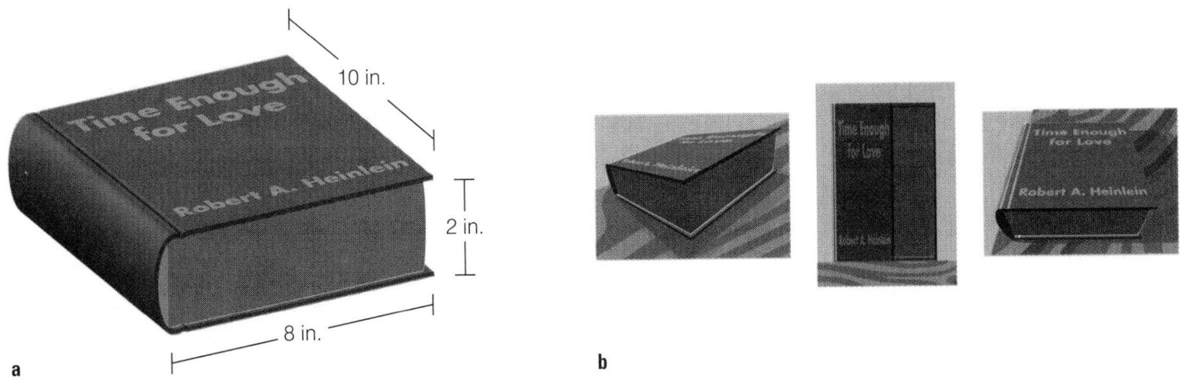

FIGURE S4.7 (**a**) A book has an unambiguous three-dimensional shape. (**b**) Two-dimensional pictures of the book can look very different.

or mass, even though all are actually looking at the same spacetime reality. In the words of the authors of a famous textbook on relativity:

> *Space is different for different observers.*
> *Time is different for different observers.*
> *Spacetime is the same for everyone.*[2]

[2]From E. F. Taylor and J. A. Wheeler, *Spacetime Physics*, 2d ed., Freeman, 1992.

Spacetime Diagrams

Suppose you drive your car along a straight road from home to work as shown in Figure S4.9a. At 8:00 A.M., you leave your house and accelerate to 60 km/hr. You maintain this speed until you come to a red light, where you decelerate to a stop. After the light turns green, you accelerate again to 60 km/hr, which you maintain until you slow to a stop when you reach work at 8:10. What does your trip look like in spacetime?

Mathematical Insight S4.1 Spacetime Geometry

As you read this chapter, you hopefully will not find the basic ideas of four-dimensional geometry and spacetime difficult to grasp. However, working with spacetime geometry is mathematically complex because the four dimensions are not all equivalent. In particular, time enters the equations of spacetime geometry in a different way than the three dimensions of space.

To gain insight into the nature of spacetime geometry, consider two points in a plane separated by amounts $x = 3$ along the horizontal axis and $y = 4$ along the vertical axis (Figure S4.8a). You may recall from geometry that the *distance* between the two points is $\sqrt{x^2 + y^2}$. Now consider the same two points viewed from a coordinate system that happens to be rotated so that both points lie along the x-axis (Figure S4.8b). In this coordinate system, the horizontal separation is $x = 5$ and the vertical separation is $y = 0$. However, the distance $\sqrt{x^2 + y^2}$

(continued)

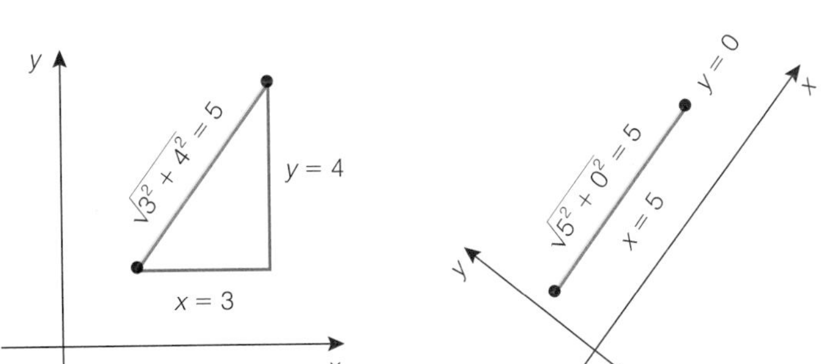

FIGURE S4.8 The distance between two points in a plane is the same regardless of how we set up a coordinate system.

If we could see all four dimensions of spacetime, we'd see all three dimensions of your car and your trip stretched out through the 10 minutes of time taken for your trip. We can't visualize all four dimensions at once, but in this case we have a special situation: Your trip progressed along only one dimension of space because you took a straight road. Therefore, we can represent your trip in spacetime by drawing a graph showing your path through one dimension of space on the horizontal axis and your path through time on the vertical axis (Figure S4.9b). This type of graph is called a **spacetime diagram.**

The car's path through four-dimensional spacetime is called its **worldline.** Any particular point along a worldline represents a particular **event.** That is, an event is a specific place and time. For example, the lowest point on the worldline in Figure S4.9b represents the event of your leaving your house: The place is 0 km from home, and the time is 8:00 A.M. You can see three very important properties of any worldline in Figure S4.9b:

1. The worldline of an object at rest is vertical (i.e., parallel to the *time* axis). The object is going nowhere in space, but it still moves through time.
2. The worldline of an object moving at constant velocity is straight but slanted. The more slanted the worldline, the faster the object is moving.
3. The worldline of an accelerating object is curved. If the object's speed is increasing, its worldline curves toward the horizontal. If its speed is decreasing, its worldline gradually becomes more vertical.

In Figure S4.9b, we used units of minutes for time and kilometers for distance. In relativity it is easiest to work with spacetime diagrams in which we use units related by the speed of light, such as *seconds* for time and *light-seconds* for distance. In this case, light itself follows 45° lines on the spacetime diagram because light travels 1 light-second of distance with each second of time. For example, suppose you are sitting still in your chair, so that your worldline is vertical (Figure S4.10a). If at some particular time you flash a laser beam pointed to your right, the worldline of the light goes diagonally to the right. If you flash the laser to the left a few seconds later, its worldline goes diagonally to the left. Worldlines for several other objects are shown in Figure S4.10b.

TIME OUT TO THINK *Explain why, in Figure S4.10, the worldlines of objects we see in our everyday life would be nearly vertical.*

We can use spacetime diagrams to clarify the relativity of time and space. Suppose you see Jackie

Mathematical Insight S4.1 (continued)

between the points is still the same. This fact should not be surprising: *Distance* is a real, physical quantity, while the x and y separations are artifacts of a chosen coordinate system. The same idea holds if we add a z-axis, perpendicular to both x and y (you can represent the z-axis with a pencil that sticks straight up out of the page), to make a three-dimensional coordinate system: Different stationary observers using different coordinate systems can disagree about the x, y, and z separations, but they will always agree on the distance $\sqrt{x^2 + y^2 + z^2}$.

We can think of spacetime as having a fourth axis, which we will call the t-axis for time. We might expect that, just as different observers always agree on the three-dimensional distance between two points, they will also agree on some kind of four-dimensional "distance" that has the formula $\sqrt{x^2 + y^2 + z^2 + t^2}$. However, it turns out that different observers will instead agree on the value of the quantity $\sqrt{x^2 + y^2 + z^2 - t^2}$, which is called the *interval.* (Technically, the interval formula should use ct rather than t so that all the terms have units of length.) That is, different observers can disagree about the values of x, y, z, and t separating two events, but all will agree on the interval between the two events.

It is the *minus sign* that goes with the time dimension in the interval formula that makes the geometry of spacetime surprisingly complex. For example, the three-dimensional distance between two points can be zero only if the two points are in the same place. But the interval between two events can be zero even if they are in different places in spacetime, as long as $x^2 + y^2 + z^2 = t^2$. (For example, the interval is zero between any two events connected by a light path on a spacetime diagram.) If you study general relativity further, you will see many more examples of how this strange geometry comes into play.

FIGURE S4.9 (**a**) A trip from home to work on a straight road. (**b**) Spacetime diagram for the trip.

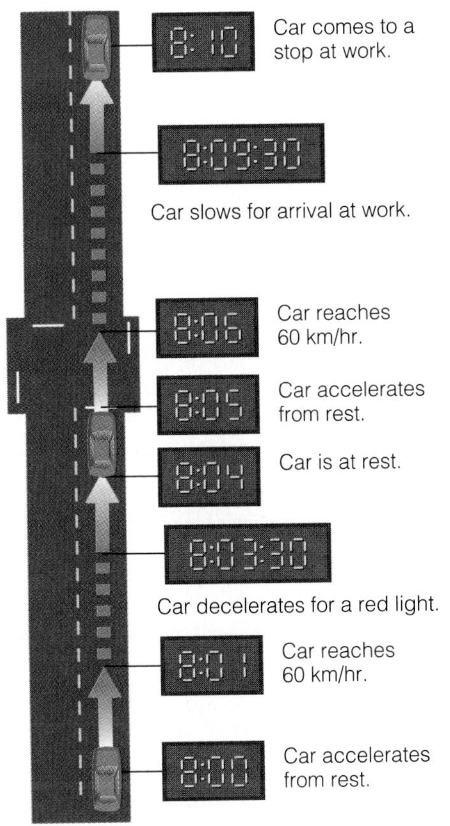

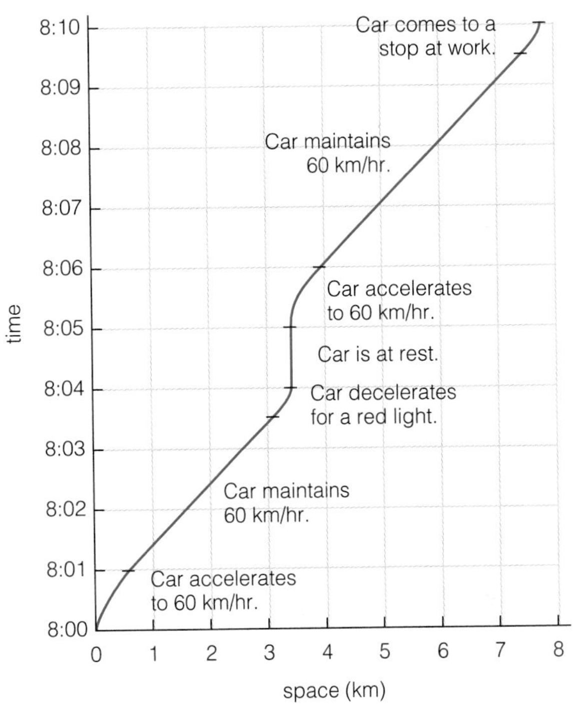

FIGURE S4.10 (**a**) Light follows 45° lines on a spacetime diagram that uses units of seconds for time and light-seconds for space. (**b**) This spacetime diagram shows several sample worldlines. Note that objects at rest have vertical worldlines, objects moving at constant velocity have straight but slanted worldlines, and accelerating objects have curved worldlines.

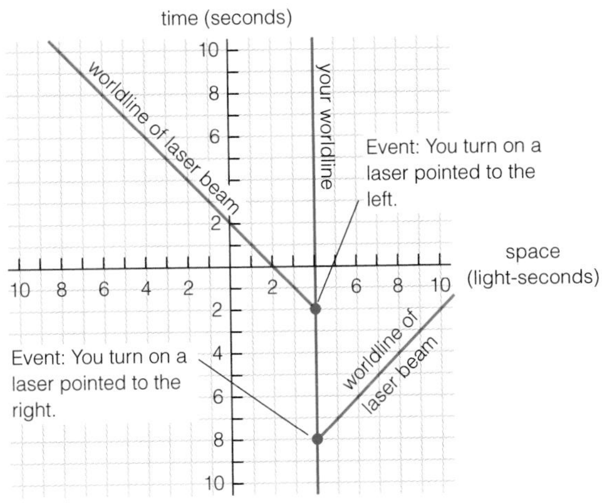

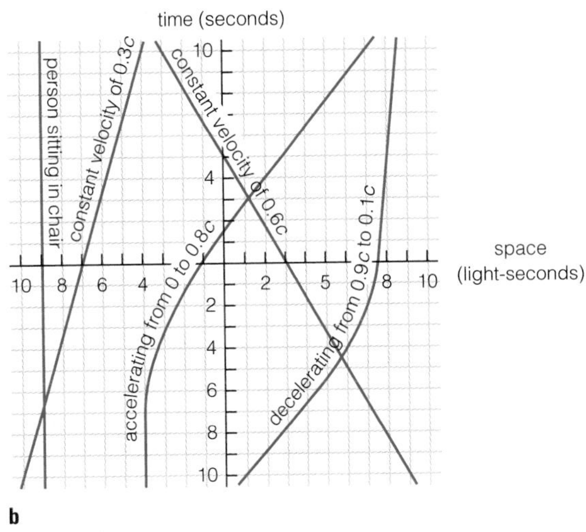

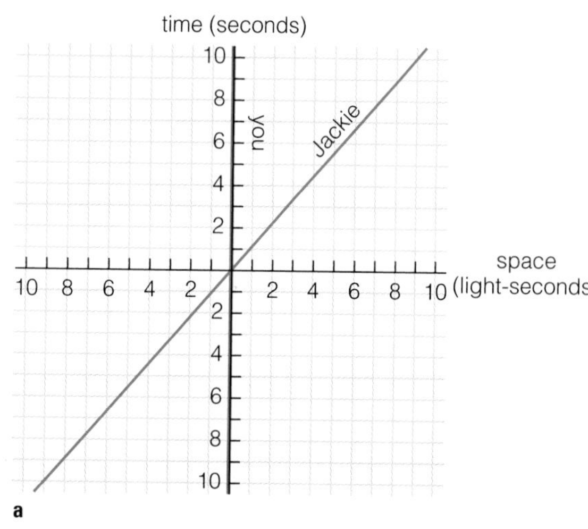

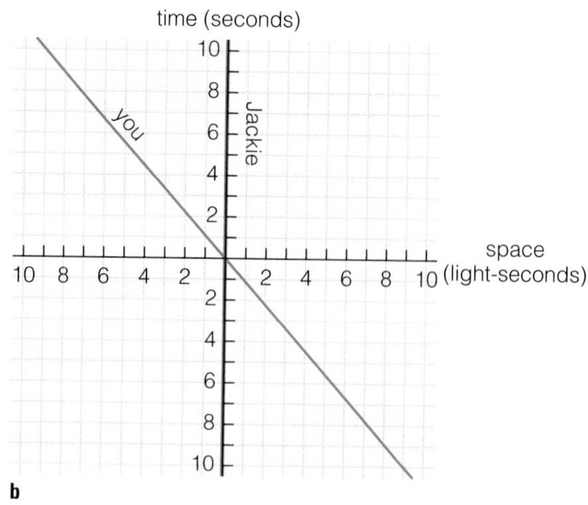

FIGURE S4.11 Jackie is moving by you at 0.9*c*. (**a**) The spacetime diagram from your point of view. (**b**) The spacetime diagram from Jackie's point of view.

moving past you in a spaceship at 0.9*c*. Figure S4.11a shows the spacetime diagram from your point of view: You are at rest and therefore have a vertical worldline; Jackie is moving and has a slanted worldline. Of course, Jackie claims that *you* are moving by her and therefore would draw the spacetime diagram shown in Figure S4.11b, in which her worldline is vertical and yours is slanted. Special relativity tells us that you would see Jackie's time running slow, while she sees *your* time running slow (along with effects on length and mass). We know there is no contradiction here, just a problem with our old common sense about space and time. From a four-dimensional perspective, the problem is that the large angle between your worldline and Jackie's means that neither of you is looking at the other "straight-on" in spacetime. Thus, you and Jackie are both looking at the same four-dimensional reality but from different three-dimensional perspectives. Like two people looking at each other cross-eyed, it's not surprising that, if you see Jackie's time running slow, she sees the same thing when she looks at you.

Spacetime Curvature

So far, we've been viewing spacetime diagrams drawn on the flat pages of this book. But, as we discussed in the beginning of this chapter, spacetime can be curved. Unfortunately, while it's easy to visualize the curvature of a two-dimensional surface, we have no hope of visualizing the curvature of space or spacetime. What, then, do we mean when we talk about space or spacetime being curved?

The answer lies in the rules of geometry, which are easiest to study on two-dimensional surfaces. Consider the curved surface of the Earth (Figure S4.12a). Note that there really is no such thing as a "straight" line on the Earth. Instead, the shortest and *straightest possible* path between two points on the Earth's surface is a piece of a **great circle**—a circle whose center is at the center of the Earth. For example, the equator is a great circle, and any "line" of longitude is part of a great circle. However, circles of latitude (besides the equator) are *not* great circles because their centers are *not* at the center of the Earth. If you are seeking the shortest and straightest route between two cities, you must follow a *great-circle route*. For example, Philadelphia and Beijing are both at about 40°N latitude, but the shortest route between them does *not* follow the circle of 40°N latitude; instead, it follows a great-circle route that extends far to the north (Figure S4.12b).

TIME OUT TO THINK *Find a globe and locate New Orleans and Katmandu (Nepal). Explain why the shortest route between these two cities goes almost directly over the North Pole. Why do you think that airplanes generally follow great-circle routes?*

Now let's contrast the rules of geometry on the surface of the Earth or any other sphere with the rules of geometry in a flat plane. You are probably familiar with the rules of geometry illustrated in Figure S4.13a, such as that two parallel lines never meet and that the circumference of a circle is $2\pi r$. However, as shown in Figure S4.13b, these rules do *not* hold if we draw lines and shapes on the surface of a sphere. If we draw two great circles that look parallel in one place, they do *not* obey the "flat plane" rule that parallel lines never meet; instead, they eventually converge. Similarly, the circumference

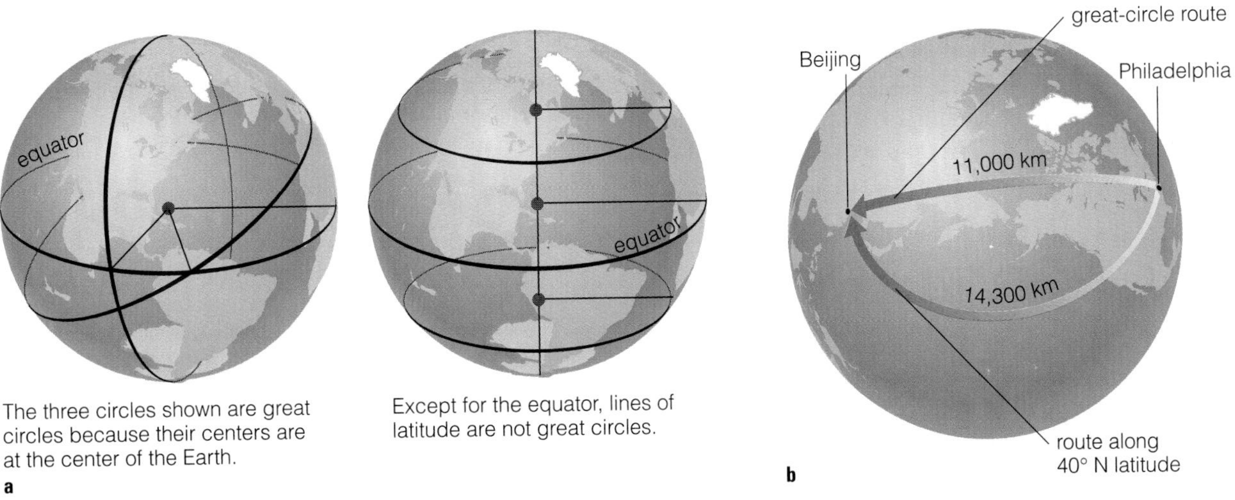

FIGURE S4.12 (**a**) A great circle is any circle on the Earth that has its center at the center of the Earth. (**b**) The shortest and straightest possible path between two points on Earth is always a piece of a great circle.

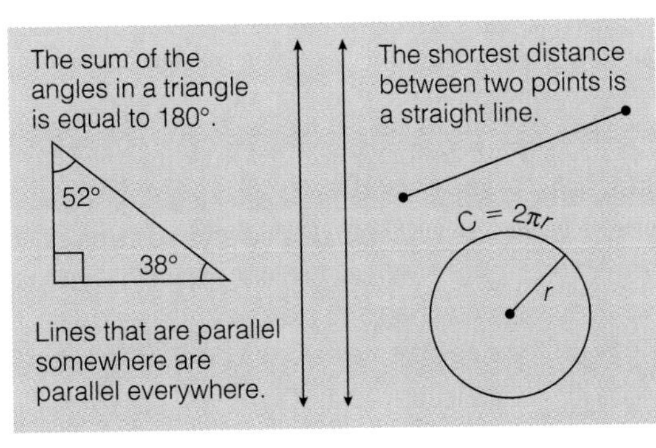

of a circle on the surface of the sphere is *not* $2\pi r$. Instead, it is *less* than $2\pi r$.

Generalizing these ideas to more than two dimensions, we say that space, or spacetime, has a **flat geometry** if the rules of geometry for a flat plane hold. For example, if the circumference of a circle in space really *is* $2\pi r$, then space has a flat geometry. (Flat geometry is also known as *Euclidean geometry*, after the Greek mathematician Euclid [c. 325–270 B.C.].) However, if the circumference of a circle in space turns out to be less than $2\pi r$, we say that space has a **spherical geometry** because the rules are those that hold on the surface of a sphere. Flat and spherical geometries are two of three general types of geometry. The third general type of geometry is called **saddle-shaped geometry** (also called *hyperbolic geometry*) because its rules are most easily visualized on a two-dimensional surface shaped like a saddle (Figure S4.13c). In this case, the circumference of a circle is *greater* than $2\pi r$.

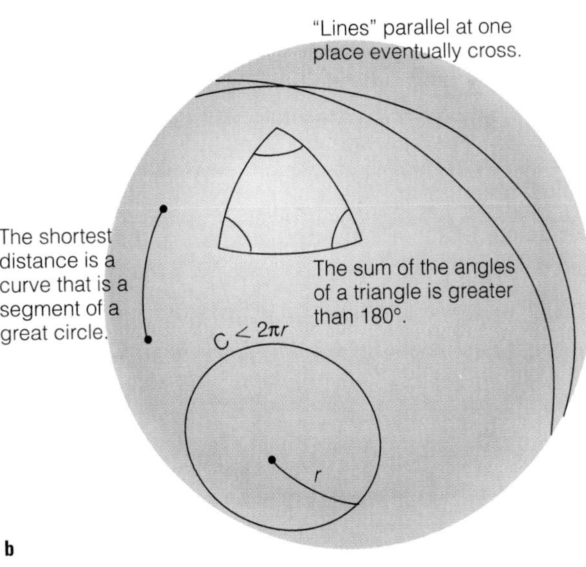

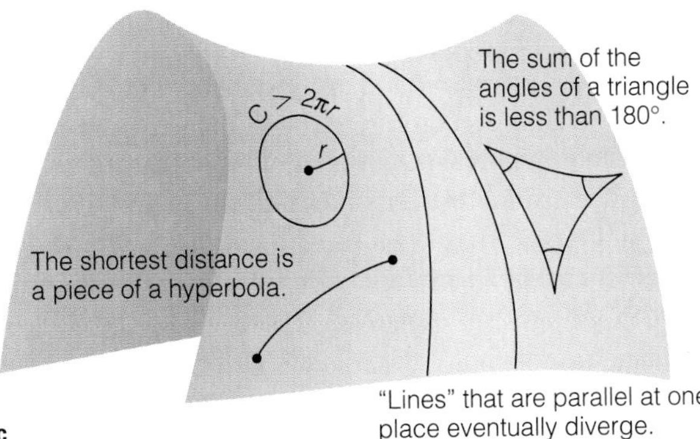

FIGURE S4.13 (**a**) Rules of flat geometry. (**b**) Rules of spherical geometry. (**c**) Rules of saddle-shaped geometry.

The actual geometry of spacetime turns out to be a mixture of all three general types. That is, just as a two-dimensional surface can look flat in some places, like a piece of a sphere in others, and like a piece of a saddle in still others, different regions of spacetime obey different sets of geometrical rules.

"Straight" Lines in Curved Spacetime

One of the keys to understanding spacetime is being able to tell whether an object is following the *straightest possible path* between two points in spacetime. However, given that we can visualize neither the time part of spacetime nor the curvature of spacetime, how do we know whether an object is traveling on the straightest possible path through spacetime?

Einstein used the equivalence principle to provide the answer. According to the equivalence principle, we can attribute a feeling of weight either to experiencing a force generated by acceleration or to being in a gravitational field. Similarly, any time we feel weightless, we may attribute it either to being in free-fall or to traveling at constant velocity far from any gravitational fields. Because traveling at constant velocity means traveling in a straight line, Einstein reasoned that objects experiencing weightlessness for *any* reason must be traveling in a "straight" line—that is, along a line that is the straightest possible path between two points in spacetime.

If you are floating freely, then your worldline is following the straightest possible path through spacetime. If you feel weight, then you are not on the straightest possible path.

This conclusion provides us with a remarkable way to examine the geometry of spacetime. Recall that any orbit is a free-fall trajectory. The Space Shuttle is always free-falling toward the Earth, but its forward velocity always moves it ahead just enough to "miss" hitting the ground. The Earth is constantly free-falling toward the Sun, but its orbital speed keeps us going around and around instead of ever hitting the Sun. According to the equivalence principle, all orbits represent paths of objects that are following the straightest possible path through spacetime. Thus, the shapes and speeds of orbits reveal the geometry of spacetime, which leads us to an entirely new view of gravity.

TIME OUT TO THINK *Suppose you are standing on a scale in your bathroom. Is your worldline following the straightest possible path through spacetime? Explain.*

S4.4 A New View of Gravity

Newton's law of gravity claims that every mass exerts a gravitational attraction on every other mass, no matter how far away it is. However, on close examination, this idea of "action at a distance" is rather mysterious. For example, how does the Earth feel the Sun's attraction and know to orbit it? Newton himself was troubled by this idea. A few years after publishing his law of gravity in 1687, Newton wrote:

> *That one body may act upon another at a distance through a vacuum, . . . and force may be conveyed from one to another, is to me so great an absurdity, that I believe no man, who has . . . a competent faculty in thinking, can ever fall into it.*[3]

Einstein's general theory of relativity removes the idea of "action at a distance" by stating that the Earth feels *no* forces—it simply follows the straightest possible path through spacetime. Thus, the fact that the Earth goes around the Sun tells us that spacetime itself is curved. In other words,

What we perceive as gravity arises from the curvature of spacetime.

We cannot picture curvature of four-dimensional spacetime, but a two-dimensional analogy illustrates many of the basic ideas. We use a stretched rubber sheet to represent spacetime and assume also that there is no friction on the rubber sheet just as there is no friction in space. We represent the Sun by placing a heavy mass on the rubber sheet, which causes the sheet to curve and form a bowllike depression (Figure S4.14). Freely moving objects follow paths that are as straight as possible given the curvature of the rubber sheet; the central mass is not grabbing them, communicating with them, or doing anything else to influence their motion. The path of an object coming from far away bends as it passes by the central mass—this path is the type we called an *unbound orbit* in Chapter 6. Other objects are "trapped" in the bowl and therefore follow circular or elliptical orbits around the central mass. By analogy, Einstein tells us that planets orbit the Sun because the Sun's mass curves spacetime in such a way that no other motion could be more natural. In summary,

Mass causes spacetime to curve, and the curvature of spacetime determines the paths of freely moving masses.

---

[3]Letter from Newton, 1692–93, as quoted in J. A. Wheeler, *A Journey into Gravity and Spacetime,* Scientific American Library, 1990, page 2.

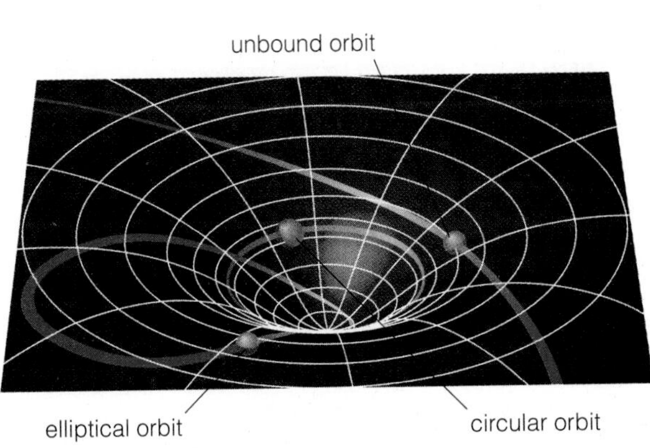

FIGURE S4.14 On a rubber sheet, marbles "orbit" a central mass by following the straightest paths that are possible.

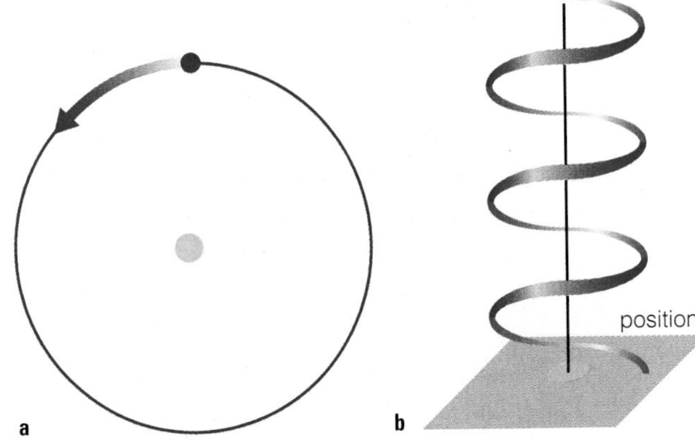

FIGURE S4.15 (**a**) If we ignore time, the Earth appears to return to the same point with each orbit of the Sun. (**b**) If we include a time axis, we see that the Earth never returns to the same point in spacetime because it always moves forward in time.

The rubber-sheet analogy to spacetime is useful, but we should keep three very important subtleties in mind:

- The rubber sheet is supposed to represent the universe, and it makes no sense to think of placing a mass "upon" the universe; instead, we should think of the masses as being *within* the rubber sheet.
- The rubber sheet allows us to picture orbits in only two dimensions. For example, it allows us to show that different planets orbit at different distances from the Sun and that some have more highly elliptical orbits than others, but it does not allow us to represent the fact that not all planets orbit the Sun in the same plane.
- The rubber-sheet analogy does not show the *time* part of spacetime at all. Bound orbits on the sheet or in space appear to return to the same point with each circuit of the Sun. However, objects cannot return to the same point in spacetime because they always move forward through time. For example, with each orbit of the Sun, the Earth returns to the same place in space (relative to the Sun) but to a time that is a year later (Figure S4.15).

The Strength of Gravity

The more spacetime curves, the stronger gravity becomes. The rubber-sheet analogy shows that there are two basic ways to increase the curvature of spacetime. First, a larger mass results in greater curvature at particular distances away from it. For example, the Sun curves spacetime more than any planet, and the Earth curves spacetime more than the Moon. Second, for an object of a given mass, spacetime curvature is greater near the object's surface if the object is denser. For example, suppose we could compress the Sun into a type of "dead" star called a *white dwarf* [Section 17.2]. Because its total mass is still the same, there is no effect on the curvature of spacetime far from the Sun, but spacetime is much more curved near the compressed Sun's surface (Figure S4.16a). Thus, gravity feels much stronger on a white dwarf star than on the Sun.

If we continued to compress the Sun to smaller and smaller size, spacetime would curve more and more at the Sun's surface. Eventually, we could create a bottomless pit in spacetime, which is what we call a **black hole** (Figure S4.16b). Nothing can escape from within a black hole, and we can never again detect or observe an object that falls into a black hole. The boundary that marks the "point of no return" is called the **event horizon,** because events that occur within this boundary can have no influence on our observable universe. Thus, a black hole is truly a hole in the observable universe. (We'll discuss black holes further in Chapter 17.)

Gravitational Time Dilation

Given that gravity arises from curvature of spacetime, you should not be surprised to learn that gravity affects time as well as space. We can learn about the effects of gravity on time by considering the effects of accelerated motion and then invoking the equivalence principle.

Imagine that you and Jackie are floating weightlessly at opposite ends of a spaceship. You both have watches that flash brightly each second and that you have synchronized beforehand. Because you are both floating freely with no relative motion between you, you are both in the same reference frame. Therefore, you will see each other's watches flashing at the same rate.

spacetime around the Sun today

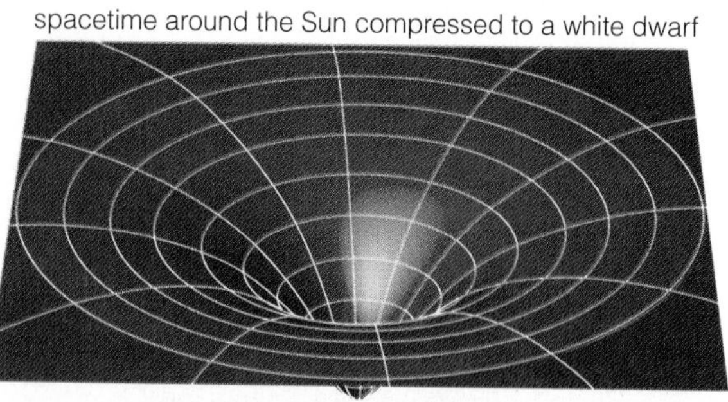

a The left diagram represents the curvature of space around the Sun today; the yellow circle in the center represents the Sun itself. The right diagram represents the curvature that would result if the Sun were compressed to the size of a white dwarf (about the size of the Earth). There is no change beyond the old boundaries of the Sun, but the curvature becomes much greater within the region that was formerly occupied by the Sun.

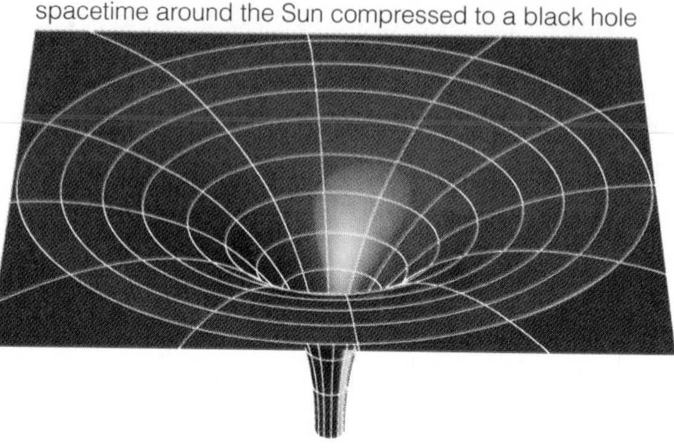

b If we kept compressing the Sun, the curvature within the region where the Sun used to be would be greater and greater, eventually creating a *black hole* in the universe

FIGURE S4.16 These rubber sheet diagrams represent the curvature of space around an object with the mass of our Sun. Remember that these diagrams have limitations, as described in the three bullets on page 435.

But suppose you fire the spaceship engines so that the spaceship begins to accelerate, with you at the front and Jackie at the back. When the ship begins accelerating, you and Jackie will no longer be weightless. But the acceleration introduces an even more important change into the situation, which we can understand by imagining the view of someone floating weightlessly outside the spaceship. Remember that observers moving at different relative speeds are in different reference frames. When the spaceship is accelerating, its speed is constantly increasing relative to the outside observer, which means that both you and Jackie are constantly changing reference frames. Moreover, the flashes from your watches take a bit of time to travel the length of the spaceship. Thus, by the time a particular flash from Jackie's watch reaches you (or a flash from your watch reaches Jackie), both of your reference frames are different from what they were at the time the flash was emitted.

Because you are in the *front* of the accelerating spaceship, your changing reference frames are always carrying you *away* from the point at which each of Jackie's flashes is emitted. Thus, the light from each of her flashes will take a little *longer* to reach you than it would have if the ship were not accelerating. As a result, instead of seeing Jackie's flashes 1 second apart, you'll see them coming a little *more* than 1 second apart. That is, you'll see Jackie's watch flashing slower than yours (Figure S4.17a). You will therefore conclude that time is running slow at the back end of the spaceship.

From Jackie's point of view at the *back* of the accelerating spaceship, her changing reference frames are always carrying her *toward* the point at which each of your flashes is emitted. Thus, the light from each of your flashes will take a little *less* time to reach her than it would have if the ship were not accelerating, so she'll see them coming a little *less* than 1 second apart. She will see your watch flashing *faster* than hers and conclude that time is running *fast* at the front end of the spaceship. Note that you and Jackie agree: Time is running slower at the back end of the spaceship and faster at the front end. The greater the acceleration of the spaceship, the greater the difference in the rate at which time passes at the two ends of the spaceship.

Now we apply the equivalence principle, which tells us that we should get the same results for a spaceship at rest in a gravitational field as we do for

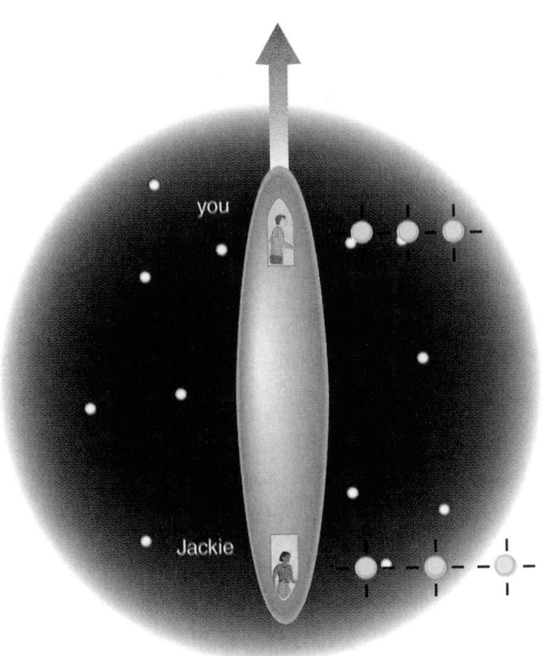

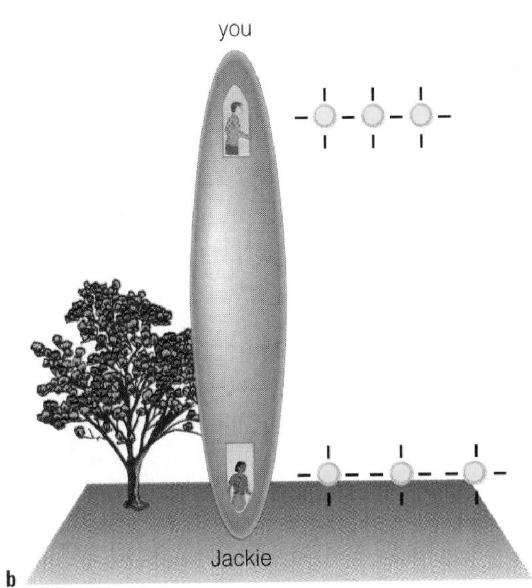

FIGURE S4.17 (**a**) A thought experiment with flashing watches shows that time runs slower at the back of an accelerating spaceship. The yellow dots represent the flashes from the watches, and the spacing between the dots represents the time between the flashes. (**b**) By the equivalence principle, time must also run slower at lower altitudes in a gravitational field.

a spaceship accelerating through space. Thus, if the spaceship were at rest on a planet, time would also have to be running slower at the bottom of the spaceship than at the top (Figure S4.17b). That is, time must run slower at lower altitudes than at higher altitudes in a gravitational field. This effect is known as **gravitational time dilation.**

The stronger the gravity—and hence the greater the curvature of spacetime—the larger the factor by which time runs slow. Time runs slower on the surface of the Sun than on the Earth, and slower on the surface of a white dwarf star than on the Sun. Perhaps you've already guessed that the extreme case is a black hole: As seen by anyone watching from a distance, time comes to a stop at the event horizon. If you could observe clocks placed at varying distances from the black hole, you'd see that clocks nearer the event horizon run slower and clocks *at* the event horizon show time to be frozen.

TIME OUT TO THINK *Where would you age more slowly, on the Earth or on the Moon? Explain.*

The Geometry of the Universe

On a small scale, the surface of the Earth appears flat in some places and curved by hills and valleys in others. However, when we expand our view to the entire Earth, it's clear that the overall geometry of the Earth's surface is like the surface of a sphere. In a similar way, but in two more dimensions, the four-dimensional spacetime of our universe presumably has some overall geometry determined by the masses within it. This overall geometry must be one of the three general types of geometry discussed earlier: flat, spherical, or saddle-shaped (see Figure S4.13).

In geometry, a plane is infinite in extent. An idealized, saddle-shaped (hyperbolic) surface is also infinite in extent. In the same way, spacetime would be infinite in extent if the universe were flat or saddle-shaped overall, and therefore the universe would have no center and no edges. If the overall geometry of the universe were spherical, spacetime would be finite like the surface of the Earth, but it would still have no center and no edges. Just as you can sail or fly around the Earth's surface endlessly, you could fly through the universe forever and never encounter an edge. And just as the *surface* of the Earth has no center—New York is no more "central" than Beijing or any other place on the Earth's surface—there is no center to the universe.[4] (In Chapter 21, we will see how the fate of the universe is related to the overall geometry of spacetime.)

[4]The three-dimensional Earth *does* have a center, but this center is not part of the two-dimensional surface of the Earth. Similarly, if four-dimensional spacetime is spherical, any "center" will be visible only by looking in at least five dimensions. Such a center has no meaning, because it is not part of the four spacetime dimensions of our universe.

THINKING ABOUT . . .

The Twin Paradox

Imagine two twins, one of whom stays on Earth while the other takes a high-speed trip to a distant star and back. In Chapter S3, we said that the twin who takes the trip will age less than the twin who stays home on Earth. But shouldn't the traveling twin be allowed to claim that she stayed stationary while Earth made a trip away from her and back? And in that case, shouldn't the twin on Earth be the one who ages less? This question underlies the so-called *twin paradox*. It can be analyzed in several different ways, but we will take an approach that offers some insights into the nature of spacetime.

Suppose you and Jackie are floating weightlessly next to each other with synchronized watches. While you remain weightless, Jackie uses her engines to accelerate a short distance away from you, decelerate to a stop a bit farther away, and then turn around and return. From your point of view, Jackie's motion means that you'll see her clocks ticking slower than yours. Thus, upon her return, you expect to find that less time has passed for her than for you. But how does Jackie view the situation?

The two of you can argue endlessly about who is really moving, but one fact is obvious to both of you: During the trip, you remained weightless while Jackie felt *weight* holding her to the floor of her spaceship. Jackie can account for her weight in either of two ways. First, she can agree with you that she was the one who accelerated. But, because we know that time runs slow in an accelerating spaceship, she'll therefore agree that her clocks ran slower than yours. Alternatively, she can claim that she felt weight because her engines counteracted a gravitational field in which she was stationary while you fell freely. But we also know that time runs slow in gravitational fields, and therefore she'll still agree that her clocks ran slower than yours. Thus, no matter how you or Jackie looks at it, the result is the same: Less time passes for Jackie.

Figure S4.18a shows a spacetime diagram for this experiment. You and Jackie both moved between the same two events in spacetime: the start and end points of Jackie's trip. However, note that your path between the two events is shorter than Jackie's. Because we have already concluded that less time passes for Jackie, we are led to a remarkable insight about the passage of time:

Between any two events in spacetime, more time passes on the shorter (and hence straighter) path.

FIGURE S4.18 (**a**) A person floating weightlessly must be following the straightest possible path through spacetime. Because this is not the case in Jackie's diagram (**b**), her diagram must be distorted.

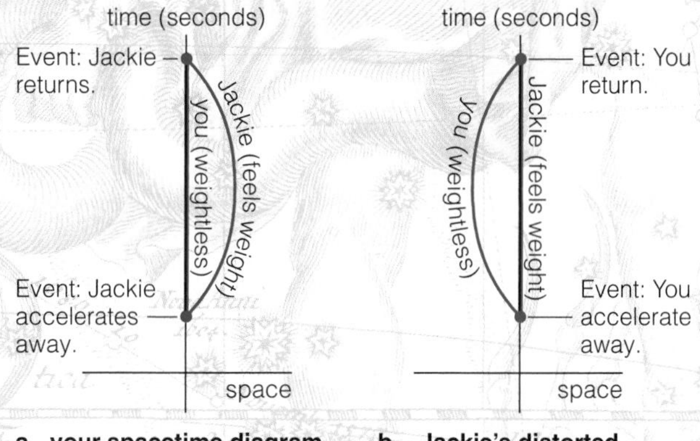

S4.5 Is It True?

Starting from the principle of equivalence, we've used logic and analogies to develop the ideas of general relativity. However, as always, we should not accept these theoretical conclusions unless they withstand observational and experimental tests. Like the predictions of special relativity, those of general relativity have faced many tests and have passed with flying colors.

Mercury's Peculiar Orbit

The first observational test passed by the theory of general relativity involved observations of the orbit of Mercury. Newton's law of gravity predicts that Mercury's orbit should precess slowly around the Sun because of the gravitational influences of other planets (Figure S4.20). Careful observations of Mercury's orbit during the 1800s showed that it does indeed precess, with each precession cycle taking more than 20,000 years. However, careful calculations made with Newton's law of gravity could not completely account for the observed precession. Although the

The maximum amount of time you can record between two events in spacetime occurs if you follow the straightest possible path—that is, the path on which you are weightless.

The subtlety arises if Jackie chooses to claim that she is at rest and attributes her weight to gravity. In that case, she might be tempted to draw the spacetime diagram in Figure S4.18b, on which you appear to have the longer path through spacetime. The rule that time runs slower on longer paths would then seem to imply that *your* clocks should have recorded less time, contradicting our earlier claims. But the contradiction is an illusion. If Jackie wishes to assert that she felt gravity, she must also claim that the gravity she felt implies that spacetime is curved in her vicinity, and therefore she should *not* have drawn a spacetime diagram on a flat piece of paper.

Jackie's problem is analogous to that of a pilot who plans a trip from Philadelphia to Beijing on a flat map of the Earth (Figure S4.19). On the flat map, it appears that the pilot has plotted the straightest possible path. However, this appearance is an illusion: The shortest and straightest path really is the great-circle route that appears curved on the flat map of the Earth (see also Figure S4.12b). A flat map of the Earth distorts reality because the actual geometry of the Earth's surface is spherical. Just as the distortions in a map of the world do not change the actual distances between cities, the way we choose to draw a spacetime diagram does not alter the reality of spacetime. The solution to the twin paradox is that the two twins do not share identical situations: The twin who turns around at the distant star must have a more strongly curved worldline than the stay-at-home twin. Thus, more time must pass for the stay-at-home twin, and the traveler does indeed age less during the journey.

FIGURE S4.19 On a flat map of the Earth, what looks like a straight line is not really as straight as possible.

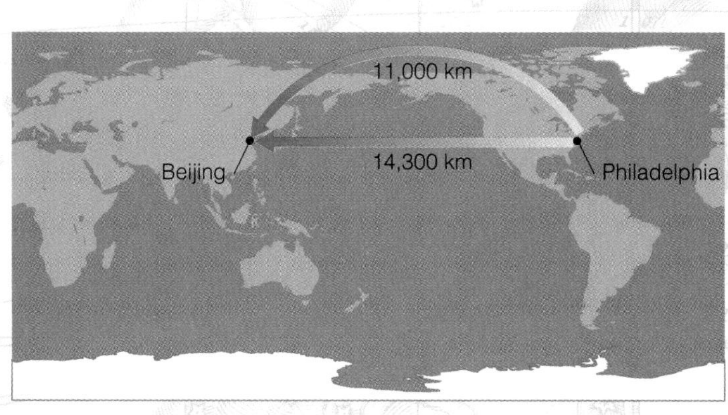

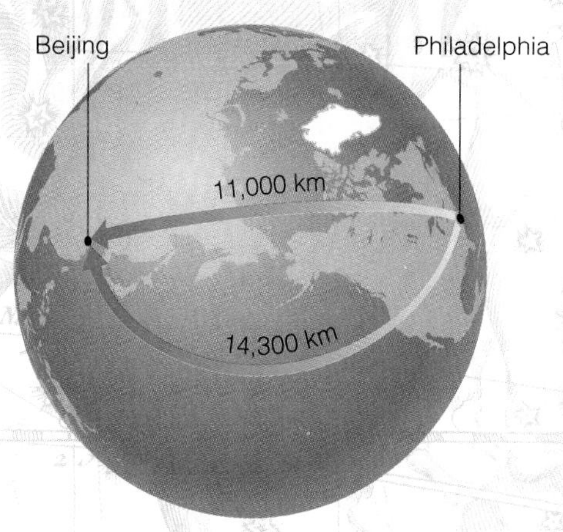

discrepancy was small, further observations verified that it was real.

Einstein was aware of this discrepancy and had hoped that he would be able to explain it from the time he first thought of the equivalence principle in 1907. When he finally succeeded in November 1915, he was so excited that he was unable to work for the next 3 days. He later called the moment of this success the high point of his scientific life.

In essence, Einstein showed that the discrepancy arose because Newton's law of gravity assumes that time is absolute and space is flat. In reality, time runs slower and space is more curved on the part of Mercury's orbit that is nearer the Sun. The equations of general relativity take this distortion of spacetime into account, providing a predicted orbit for Mercury that precisely matches its observed orbit.

TIME OUT TO THINK *Suppose the perihelion of Mercury's orbit were even closer to the Sun than it actually is. Would you expect the discrepancy between the actual orbit and the orbit predicted by Newton's laws to be greater than or less than it actually is? Explain.*

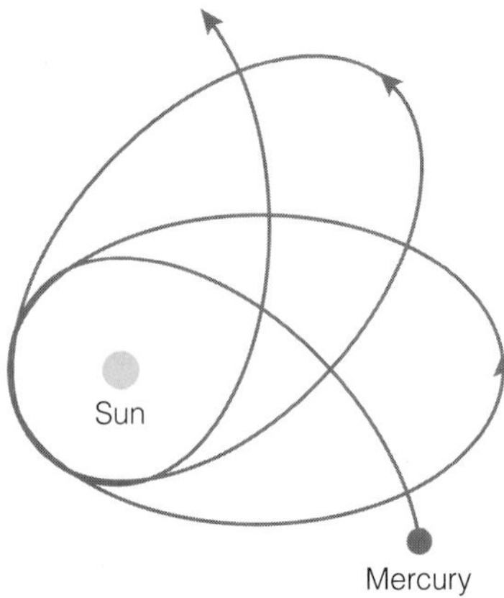

FIGURE S4.20 Mercury's orbit slowly precesses around the Sun. Note: The amount of precession with each orbit is highly exaggerated in this picture.

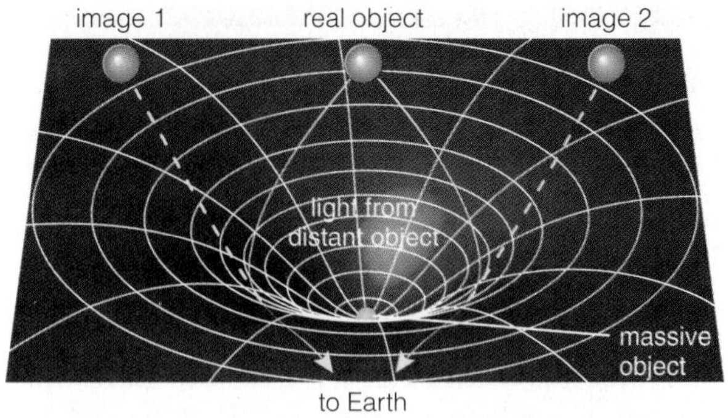

FIGURE S4.22 Gravitational lensing can create distorted or multiple images of a distant object whose light passes by a massive object on its way to Earth.

Gravitational Lensing

We can also test Einstein's claim that space is curved by observing the trajectories of light rays moving through the universe. Because light always travels at the same speed, never accelerating or decelerating, light must always follow the straightest possible path. If space itself is curved, then light paths will appear curved as well.

Suppose we could carefully measure the angular separation between two stars during the daytime, just when the light from one of the stars passes near the Sun. The curvature of space near the Sun forces the light beam passing closer to the Sun to curve more than the light beam from the other star (Figure S4.21). Therefore, the angular separation of the two stars will appear smaller than their true angular separation (which we would know from nighttime measurements). This effect was first observed during a total eclipse in 1919. This second spectacular success of general relativity brought Einstein worldwide fame.

Even more dramatic effects occur when a distant star or galaxy, as seen from Earth, lies directly behind another object with a strong gravitational field (Figure S4.22). The mass of the intervening object curves spacetime in its vicinity, altering the trajectories of light beams passing nearby. Different light paths can curve so much that they end up converging at Earth, grossly distorting the appearance of the star or galaxy. Depending on the precise four-dimensional geometry of spacetime between us and the observed star or galaxy, the image we see may be magnified or distorted into arcs, rings, or multiple images of the same object (Figure S4.23). This type of distortion is called **gravitational lensing,** analogous to the bending of light by a glass lens.

FIGURE S4.21 Light from Star A passes through a more highly curved region of spacetime than light from Star B, making the angular separation of the two stars appear smaller than their true angular separation.

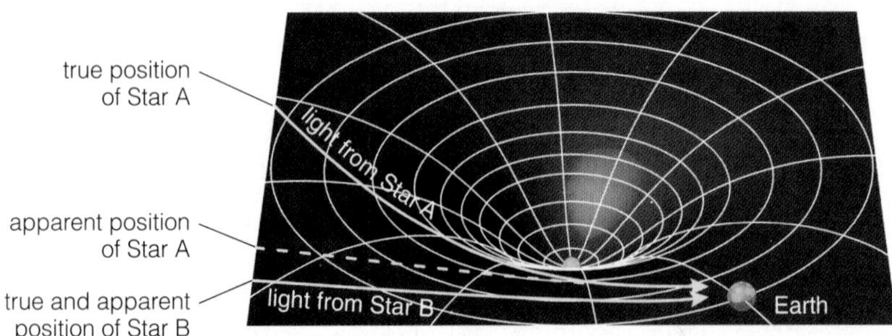

Effects on Time

The prediction of gravitational time dilation can be tested by comparing clocks located in places with different gravitational field strengths. Even in the Earth's weak gravity, experiments have demonstrated that clocks at low altitude tick more slowly than identical clocks at higher altitude. The effect is extremely small—it would add up to only a few billionths of a second over a human lifetime—but the differences agree precisely with the predictions of general relativity.

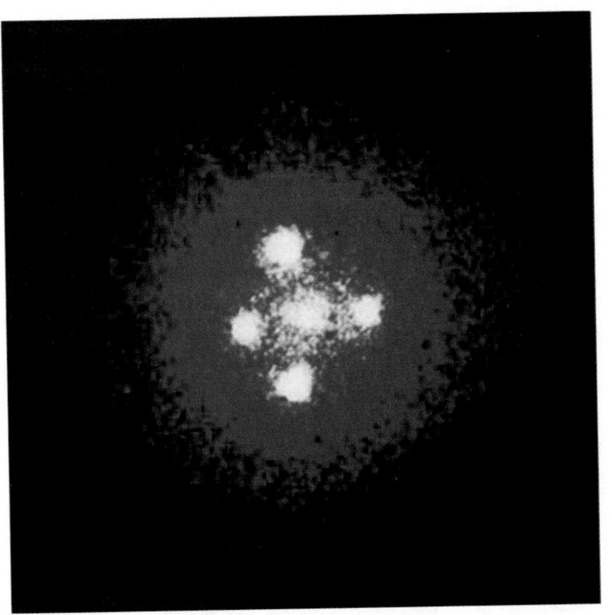

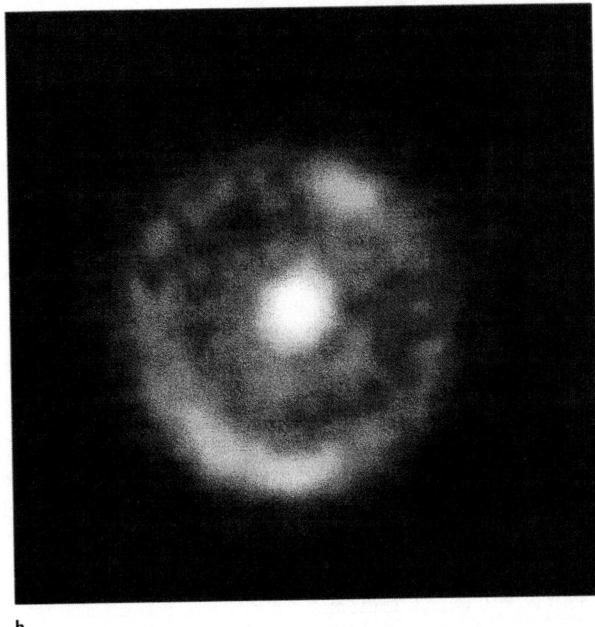

a

b

FIGURE S4.23 (**a**) In this case of gravitational lensing, called Einstein's Cross, the gravity of a foreground galaxy (center) bends light from a single bright background object so that it reaches us along four different paths—creating four distinct images of a single object. (**b**) When one galaxy lies directly behind another, the foreground galaxy can bend light on all sides so that the light converges on the Earth, forming an Einstein Ring like that pictured here.

Surprisingly, it's even easier to compare the passage of time on Earth with the passage of time on the surface of the Sun and other stars. Because stellar gases emit and absorb *spectral lines* with particular frequencies [Section 7.4], they serve as natural atomic clocks. Suppose that, in a laboratory on Earth, we find that a particular type of gas emits a spectral line with a frequency of 500 trillion cycles per second. If this same gas is present on the Sun, it will also emit a spectral line with a frequency of 500 trillion cycles per second. General relativity claims that time should be running very slightly slower on the Sun than on Earth; that is, 1 second on the Sun lasts *longer* than 1 second on Earth, or, equivalently, a second on Earth is shorter than a second on the Sun. Thus, during 1 second on Earth, we will see *fewer* than 500 trillion cycles from the gas on the Sun. Because lower frequency means longer or redder wavelength, the spectral lines from the Sun ought to be *redshifted.* Note that this redshift has nothing to do with the Doppler shifts that we see from moving objects [Section 7.5]. Instead, it is a **gravitational redshift,** caused by the fact that time runs slow in gravitational fields. Gravitational redshifts have been measured for spectral lines on the Sun and on many other stars. The results agree with the predictions of general relativity.

The Search for Gravitational Waves

If the curvature of space suddenly changes somewhere, it can have effects on distant parts of the universe. For example, the effect of a star suddenly imploding or exploding is rather like the effect of dropping a rock into a pond, generating ripples of curvature that propagate outward through space. Similarly, two massive stars orbiting each other closely and rapidly generate ripples of curvature in space rather like those of a blade turning in water. Einstein called these ripples **gravitational waves.** Similar in character to light waves but far weaker, gravitational waves have no mass and travel at the speed of light.

The distortions of space carried by gravitational waves should compress and expand objects they pass by. In principle, we could detect gravitational waves by looking for such waves of compression and expansion, but these effects are expected to be extremely weak. No one has yet succeeded in detecting gravitational waves, although several new approaches to detecting them are being developed and tested.

Despite the lack of direct detection, we are quite certain that gravitational waves exist because of a special set of observations carried out over the past quarter-century. In 1974, astronomers Russell Hulse and Joseph Taylor discovered an unusual binary star

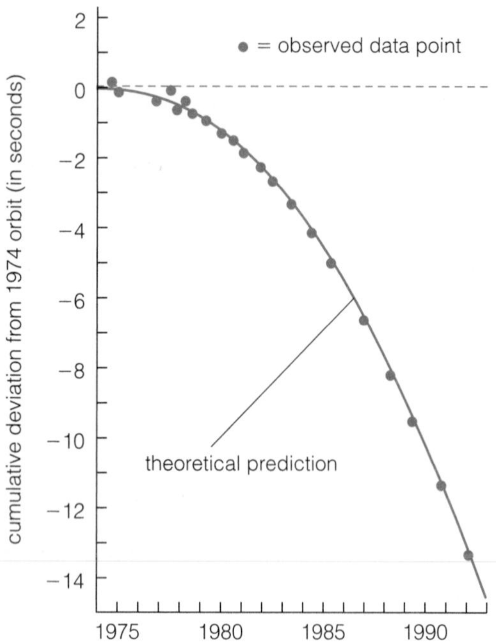

FIGURE S4.24 The decrease in the orbital period of the Hulse–Taylor binary star system matches what we expect if the system is emitting gravitational waves.

system in which both stars are highly compressed,[5] allowing them to orbit each other extremely closely and rapidly. General relativity predicts that this system should be emitting a substantial amount of energy in gravitational waves. If the system is losing energy to these waves, the orbits of the two stars should be steadily decaying. Observations show that the rate at which the orbital period is decreasing matches the prediction of general relativity, a strong suggestion that the system really is losing energy by emitting gravitational waves (Figure S4.24).

S4.6 Hyperspace, Wormholes, and Warp Drive

If you're a fan of *Star Trek, Star Wars,* or other science fiction, you've seen spaceships bounding about the galaxy with seemingly little regard for Einstein's prohibition on traveling faster than the speed of light. In fact, these stories do not necessarily have to violate the precepts of relativity as long as they exploit potential "loopholes" in the known laws of nature.

[5]The two stars are *neutron stars* (see Chapter 17). One is also a *pulsar,* a neutron star that spins rapidly, producing pulses of radiation that we can detect on Earth. Precision timing of the arrival of pulses allows scientists to measure the decrease in the orbital period of the system.

Let's begin with an analogy. Suppose you want to take a trip between Hawaii and South Africa, which happen to lie diametrically opposite on the Earth (Figure S4.25). Ordinarily, we are restricted to traveling along the Earth's surface by car, boat, or plane, and the most direct trip covers about 20,000 kilometers. But suppose you could somehow drill a hole through the center of the Earth and fly through the hole from Hawaii to South Africa. In that case, the trip would be only about 12,000 kilometers.

FIGURE S4.25 If you could take a shortcut *through* the Earth, the trip from Hawaii to South Africa would be shorter than is possible on the *surface* of the Earth.

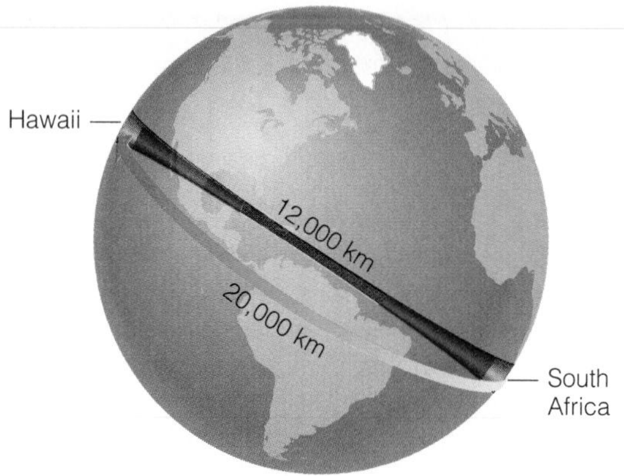

Now consider a trip from Earth to the star Vega, about 25 light-years away. From the point of view of someone who stays home on Earth, this trip must take at least 25 years in each direction. But suppose space happens to be curved in such a way that Earth and Vega are much closer together as viewed from a multidimensional *hyperspace,* just as Hawaii and South Africa are closer together if we can go through the Earth than if we must stay on its surface. Further, suppose there is a tunnel through hyperspace, often called a **wormhole,** through which we can travel (Figure S4.26). If the tunnel is short—perhaps only a few kilometers in length—then a spaceship needs to travel only a few kilometers through the wormhole to go from Earth to Vega. The trip might take only a few minutes in each direction! Note that there is no violation of relativity, because the spaceship has not exceeded the speed of light. It has simply taken a shortcut through hyperspace.

If there is no wormhole, perhaps we might discover a way to "jump" through hyperspace and return to the universe anywhere we please. Such hyperspace jumps are the fictional devices used for space travel in the *Star Wars* movies. Or perhaps we might

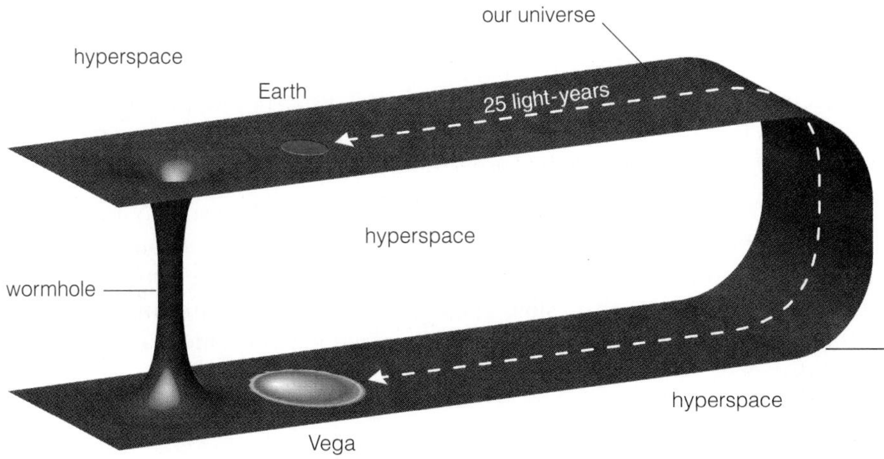

FIGURE S4.26 The curved sheet represents our universe, in which a trip from Earth to Vega covers a distance of 25 light-years. This trip could be much shorter if there were a wormhole that created a shortcut through hyperspace.

discover a way to warp spacetime to our own specifications, thereby allowing us to make widely separated points in space momentarily touch in hyperspace—this fictional device is the basic premise behind *warp drive* in the *Star Trek* series.

Do wormholes really exist? If so, could we really travel through them? Is it possible that we might someday discover a way to jump into hyperspace or create a warp drive? Our current understanding of physics is insufficient to answer these questions definitively. For the time being, the known laws of physics do not prohibit any of these exotic forms of travel. These loopholes are therefore ideal for science fiction writers, because they might allow rapid travel among distant parts of the universe without violating the established laws of relativity.

However, many scientists believe we will eventually find that these exotic forms of travel are *not* possible. Their primary objection is that wormholes seem to make time travel possible. If you could jump through hyperspace to another place in our universe, couldn't you also jump back to another *time?* If you used a trip through hyperspace to travel into the past, could you prevent your parents from ever meeting?

The paradoxes we encounter when we think about time travel are severe and seem to have no resolution. Most scientists therefore believe that time travel will prove to be impossible, even though we don't yet know any laws of physics that prohibit it. In the words of physicist Stephen Hawking, time travel should be prohibited "to keep the world safe for historians."

If time travel is not possible, it is much more difficult to see how shortcuts through hyperspace could be allowed. Nevertheless, neither time travel nor travel through hyperspace can yet be categorically ruled out in the same way that we can rule out the possibility of exceeding the speed of light. Until we learn otherwise, the world remains safe for science fiction writers who choose their fictional space travel techniques with care, avoiding any conflicts with relativity and other known laws of nature.

S4.7 The Last Word

We now know that space and time are intertwined in ways that would have been difficult to imagine before Einstein's work. So, for the last word in our study of relativity, let's turn to Einstein himself. The quote below is what he said about a month before his death on April 18, 1955.[6]

[6]This quotation was found with the aid of Alice Calaprice, author of *The Quotable Einstein*, Princeton University Press, 1996.

[Death] signifies nothing. . . . the distinction between past, present, and future is only a stubbornly persistent illusion.

ALBERT EINSTEIN

THE BIG PICTURE

Just as the Copernican revolution overthrew the ancient belief in an Earth-centered universe, Einstein's revolution overthrew the ancient belief that space and time are distinct and absolute. Now that we have explored Einstein's revolution in some detail in the past two chapters, keep in mind the following "big picture" ideas:

- We live in four-dimensional spacetime. Disagreements among different observers about measurements of time and space occur because the different observers are looking at this four-dimensional reality from different three-dimensional perspectives.

- Gravity arises from curvature of spacetime. Once we recognize this fact, the orbits of planets, moons, and all other objects are perfectly natural consequences of the curvature, rather than results of a mysterious "force" acting over great distances.
- Although the predictions of relativity may seem quite bizarre, they have been verified many times by observations and experiments.
- Some questions remain well beyond our current understanding. In particular, we do not yet know whether travel through hyperspace might be possible, thereby allowing some of the imaginative ideas of science fiction to become reality.

Review Questions

1. Explain what we mean by the *straightest possible path* on the Earth's surface.
2. What do we mean by *spacetime?*
3. List five major ideas that come directly from the general theory of relativity.
4. What is the *equivalence principle?* Give an example that clarifies its meaning.
5. What do we mean by *dimension?* Describe a *point,* a *line,* a *plane,* a *three-dimensional space,* and a *four-dimensional space.* What does *hyperspace* mean?
6. Explain the meaning of the following statement: *"Space is different for different observers. Time is different for different observers. Spacetime is the same for everyone."*
7. What is a *spacetime diagram?* What is an *event?* What is a *worldline?*
8. Consider a spacetime diagram that shows worldlines for three objects: One worldline is vertical, one is straight but slanted, and one is curved. In the reference frame of the person who drew the diagram, how is each object moving?
9. Briefly describe how the rules of geometry are different depending on whether the geometry is *flat, spherical,* or *saddle-shaped.*
10. How can you tell whether you are following the straightest possible path through spacetime?
11. According to general relativity, what is gravity? In this view, why does the Earth orbit the Sun?
12. Briefly describe the idea behind the "rubber sheet" analogy to spacetime. List three ways in which this analogy is limited.
13. Explain why the curvature of spacetime near a star depends on both the star's mass and its size.
14. What is a *black hole?* What do we mean by the *event horizon* of a black hole?
15. What is *gravitational time dilation?* What determines how much time is slowed in a gravitational field?
16. Explain how the universe can be finite yet have no center and no edges.
17. Briefly describe several observational tests that support general relativity.
18. What is *gravitational lensing?* According to general relativity, why does it occur?
19. Why do we see a *gravitational redshift* when we look at spectral lines from the Sun?
20. What are *gravitational waves?* What evidence supports their existence?
21. What is the current evidence regarding the possibility of travel through hyperspace, wormholes, or warp drive? Explain how a *wormhole* might allow us to travel between two places that are 10 light-years apart in less than 10 years.

Discussion Questions

1. *Relativity and Fate.* In principle, if we could see all four dimensions of spacetime, we could see future events as well as past events. In his novel *Slaughterhouse Five,* writer Kurt Vonnegut used this idea to argue that our futures are predetermined and that there is no such thing as free will. Do you agree with this argument? Why or why not?
2. *Philosophical Implications of Relativity.* According to our description of spacetime, you exist in spacetime as a "solid" object stretching through time. In that sense, you cannot erase anything you've ever said or done from spacetime; if we could see in four dimensions, we would be able to see everything you've ever said or done. Do you think these ideas have any important philosophical implications? Discuss.
3. *Wormholes and Causality.* Suppose it turns out that travel through wormholes *is* possible and that it is possible to travel into the past. Discuss some of the paradoxes that would occur. In light of these paradoxes, do you believe that travel through wormholes will turn out to be prohibited? Why or why not?

Problems

1. *Worldlines at Low Speed.* Make a spacetime diagram and draw worldlines for each of the following situations. Explain your drawings. j
   a. A person sitting still in a chair.
   b. A person driving by at a constant velocity of 50 km/hr.
   c. A person driving by at a constant velocity of 100 km/hr.
   d. A person accelerating from a stop sign to a speed of 50 km/hr.
   e. A person decelerating from 50 km/hr to a stop.
2. *Worldlines at High Speed.* Make a spacetime diagram on which the time axis is marked in seconds and the space axis is marked in light-seconds. Assume you are floating weightlessly and therefore consider yourself at rest. You see Sebastian moving to your right at $0.5c$ and Michaela moving to your left at $0.7c$. Sebastian passed your location 2 seconds ago, and Michaela passed your location 4 seconds ago. Draw worldlines for Sebastian, Michaela, and yourself. Explain.
3. *Galileo and the Equivalence Principle.* Galileo demonstrated that all objects near the Earth's surface should fall with the same acceleration, regardless of their mass. According to general relativity, why shouldn't the mass of a falling object affect its rate of fall? Explain in one or two paragraphs.
4. *Long Trips at Constant Acceleration.* On a realistic trip to the stars, we could not suddenly jump to a speed near the speed of light without being killed by the forces associated with the sudden acceleration. Thus, a more realistic trip would have us accelerate at a comfortable rate, such as $1g$, until we are halfway to our destination and then decelerate at the same rate until we reach our destination. Explain why we would be comfortable with this acceleration. By our own reckoning, would we notice anything unusual about lengths, masses, or the passage of time on our spaceship? Why or why not?
5. *Long Trips at Constant Acceleration: Earth Time.* Suppose you stay on Earth and watch a spaceship leave on a long trip at a constant acceleration of $1g$.
   a. At an acceleration of $1g$, approximately how long will it take before you see the spaceship traveling away from Earth at *half* the speed of light? Explain. (Use $g = 9.8\ \text{m/s}^2$.)
   b. Describe how you will see its speed change as it continues to accelerate. Will it keep gaining speed at a rate of 9.8 m/s each second? Why or why not?
   c. Suppose the ship travels to a star that is 500 light-years away. According to you back on Earth, *approximately* how long will this trip take? Explain.
6. *Long Trips at Constant Acceleration: Spaceship Time.* Consider again the spaceship on a long trip with a constant acceleration of $1g$. Although the derivation is beyond the scope of this book, it is possible to show that, as long as the ship is gone from Earth for many years, the amount of time that passes on the spaceship during the trip is approximately:

$$T_{\text{ship}} = \frac{2c}{g} \ln\left(\frac{g \times D}{c^2}\right)$$

   In this formula, D is the distance to the destination and "ln" stands for the natural logarithm. (Your calculator probably has a key for taking natural logarithms [usually labeled "ln"], so you can use this formula even if you are not familiar with them.) If you use D in meters, $g = 9.8\ \text{m/s}^2$, and $c = 3 \times 10^8$ m/s, the answer will be in units of *seconds*.
   a. Suppose the ship travels to a star that is 500 light-years away. How much time will pass on the ship? Compare this to the amount of time that passes on Earth (see problem 5c). (*Hint:* Be sure you convert the distance from light-years to meters, and convert your answer from seconds to years.)
   b. Suppose the ship travels to the center of the Milky Way Galaxy, about 28,000 light-years away. How much time will pass on the ship? Compare this to the amount of time that passes on Earth.
   c. The Andromeda Galaxy is about 2.5 million light-years away. Suppose you had a spaceship that could constantly accelerate at $1g$. Could you go to the Andromeda Galaxy and back within your lifetime? Explain. What would you find when you returned to Earth?
7. *Project: Movie Science Fiction.* Choose a popular science fiction movie that involves interstellar travel and study it closely as you watch it. What aspects of the movie are consistent with relativity and other laws of physics? What aspects of the movie violate the laws of relativity or other laws of physics? Write a two- to three-page summary of your findings.
8. *Research: The Eötvös Experiment.* Galileo's result that all objects fall to the Earth with the same acceleration (neglecting air resistance) is very important to general relativity: If it were not true, general relativity would be in serious trouble. Describe the experiments of Baron Roland von Eötvös, who tested Galileo's conclusions in the late 1800s. How did the results of these experiments influence Einstein as he worked on general relativity? Write a one- to two-page summary of your findings.
9. *Research: Gravitational Wave Detection.* Research the current status of an experiment designed to detect gravitational waves directly. Write a one- to two-page summary of your findings.
10. *Research: Wormholes.* Some scientists have thought seriously about wormholes and their consequences. Find and read a popular article or book about wormholes. Write a short summary of the article, and discuss your opinion of the implications of wormholes according to the article.

S5
Building Blocks of the Universe

THE MICROSCOPIC REALM OF ATOMS AND NUCLEI SEEMS FAR removed from the vast realm of planets, stars, and galaxies. Nevertheless, much of what we know about the cosmos today would have remained mysterious without a thorough understanding of these tiny particles. They are the building blocks from which all else is made, and the behavior of very large objects frequently depends on the laws that govern their tiniest pieces.

We've already seen that matter and energy behave in some strange ways when we break them down into small units. For example, we know that electrons in atoms can have only specific energies and that photons act sometimes like particles and sometimes like waves. These are only a few of the strange realities in the realm of the very small, where the laws of nature can seem utterly bizarre to even the most practiced scientists.

In this chapter, we will examine the laws of nature that underlie the structure of matter. We will look more deeply into the building blocks of nature, investigating current knowledge of the fundamental particles and forces that make up the universe. We will see that the strange laws of the microscopic world play a fundamental role in diverse processes such as nuclear fusion in the Sun and the collapse of a star into a black hole.

S5.1 The Quantum Revolution

Around the same time that Einstein was discovering the principles of relativity, he and others were also investigating the behavior of matter and energy. Their discoveries in this area were no less astonishing. In 1905, the same year he published the special theory of relativity, Einstein showed that light behaves like particles (photons) in addition to behaving like waves [Section 7.2]. In 1911, British physicist Ernest Rutherford (1871–1937) discovered that atoms consist mostly of empty space, begging the question of how matter can ever feel solid. In 1913, Danish physicist Niels Bohr (1885–1962) discovered that electrons in atoms can have only particular energies; that is, electron energies are *quantized* [Section 5.4]. Thus, the realm of the very small is often called the *quantum* realm, and the science of the quantum realm is called **quantum mechanics.**

Other scientists soon built upon the work of Einstein, Rutherford, and Bohr. By the mid-1920s, our ideas about the structure and nature of atoms and subatomic particles were undergoing a total revolution. The repercussions of this *quantum revolution* continue to reverberate today. They have forced us to reexamine our "common sense" about the fundamental nature of matter and energy. They have also driven a technological revolution, because the laws of quantum mechanics make modern electronics possible. Most important, at least from an astronomical point of view, the combination of new ideas and new technology is enabling us to look ever deeper into the heart of matter and energy—the ultimate building blocks of the universe.

This chapter discusses key ideas of the quantum revolution that are important to the study of astronomy. The following are among these surprising new ideas:

- The protons, neutrons, and electrons that we usually consider the building blocks of atoms are not truly fundamental. Instead, the building blocks of ordinary matter are *quarks* and *leptons.* Quarks and leptons, in turn, belong to a category called *fermions,* while photons belong to an entirely distinct category of particles called *bosons.*
- *Antimatter* is real and is readily produced in the laboratory. When a particle and its antiparticle meet, the result is mutual annihilation and release of energy.
- Just four forces govern all interactions between particles: gravity, electromagnetism, and the strong and weak nuclear forces. In fact, these four forces are themselves manifestations of a smaller number of truly fundamental forces. Perhaps a single unified force rules all of nature.
- Our everyday common sense tells us that particles and waves are different, but the quantum laws show that *all* tiny particles exhibit the same *wave–particle duality* that Einstein demonstrated for photons.
- The quantum laws have important astronomical consequences. For example, a strange quantum effect called *degeneracy pressure* can prevent the core of a dying star from collapsing. *Quantum tunneling* helps make nuclear fusion possible in the Sun. And phantomlike *virtual particles* may be important to the ultimate fate of black holes and the universe itself.

No matter how bizarre the quantum laws may seem, they lead to concrete predictions that can be tested experimentally and observationally. Some of these tests require sophisticated technological equipment found only in advanced physics laboratories. Others require billion-dollar particle accelerators. But some are performed every day, right before your eyes. Every time you see a ray of sunlight, you are seeing the product of nuclear reactions that are made possible by the quantum laws. And every time you turn on a computer, a calculator, a television, or any other "high-tech" electronic device, the strange laws of quantum mechanics are being put to work for your benefit.

S5.2 Fundamental Particles and Forces

More than 2,400 years ago, Democritus proposed that matter is made from building blocks that he called *atoms* [Section 5.3]. He believed that atoms were **fundamental particles,** the most basic units of matter, impossible to divide any further. By this definition, the particles we now call atoms are not truly fundamental. By the 1930s, we had learned that atoms themselves are made of protons, neutrons, and electrons. Following this realization, scientists had briefly hoped that these three particles were the true fundamental building blocks of the universe. However, under more extreme conditions, matter starts to display greater variety, and strange new particles begin to appear.

We can generate many unusual particles with the aid of **particle accelerators** (sometimes called *atom smashers*), such as Fermilab near Chicago and CERN on the border between Switzerland and France (Figure S5.1). The large magnets inside particle accelerators accelerate familiar particles such as electrons or protons to very high speeds—often extremely close to the speed of light. When these particles collide with one another or with a stationary target, they

FIGURE S5.1 The Fermilab particle accelerator in Illinois.

release a substantial amount of energy within a very small space. Some of this energy spontaneously turns into mass, producing a shower of particles. (Recall that $E = mc^2$ tells us not only that mass can turn into energy, but also that energy can turn into mass.) Scientists recognize particles of different types by their differing behavior. Whenever scientists observe a particle that behaves in previously unseen ways, it is cataloged as a new type of particle and given a name.

Hundreds of different particles had been discovered by the early 1960s, and scientists began to wonder whether any of them were truly fundamental. At about that time, physicist Murray Gell-Mann proposed a scheme in which all these particles could be built from just a few fundamental components. Gell-Mann's scheme has since blossomed into what physicists call the *standard model* for the structure of matter. The standard model has proved very successful and has even predicted the existence of new particles later discovered in particle accelerators. In this section, we briefly describe the fundamental particles according to the standard model, along with their special relationships with the forces of nature.

Properties of Particles

Each particular type of subatomic particle, such as an electron, proton, or neutron, has its own peculiar behavior that is determined by just a few basic properties. The most important of these basic properties are *mass, charge,* and *spin.* Mass is already very familiar to you, and the effects of charge, such as static electricity or lightning, are also part of your everyday experience. Spin, on the other hand, is a property evident only in the quantum realm of the very small.

The word *spin* indicates that this property is related to angular momentum. Recall that a spinning ice skater or a spinning baseball has rotational angular momentum [Section 6.1]. By analogy, we say that a subatomic particle, such as an electron, has **spin angular momentum**—or **spin,** for short—as it "spins" on its axis. However, because an electron is not a particle in the same sense as a baseball, it does not actually spin in the same sense as a baseball. *Spin* is simply a term used to describe the angular momentum that belongs to the electron. Just as all electrons have exactly the same mass and electric charge, all electrons have exactly the same amount of spin. Similar considerations hold for all other types of subatomic particle: Every particle of a particular type has a particular amount of mass, charge, and spin.

In fact, spin is a very important property for subatomic particles, since all particles turn out to fall into one of two broad classes based on their spin: the

fermions, named for Enrico Fermi (1901–1954), and the **bosons,** named for Satyendra Bose (1894–1974).[1] The most familiar fermions are electrons, neutrons, and protons. The most familiar bosons are photons. In this book, we will not deal with the distinction between fermions and bosons except as a matter of classification. However, one more important aspect of spin applies to fermions such as electrons, protons, and neutrons: The spin can be oriented in two ways, usually called *spin up* and *spin down.* These two orientations, which correspond to the two opposite senses of rotation, clockwise and counterclockwise, are often represented by arrows (Figure S5.2). Be sure to remember that denoting a particle's spin by a small dot with an arrow is a representation of convenience; subatomic particles are *not* tiny spinning balls.

FIGURE S5.2 The two possible states of an electron's, neutron's, or proton's spin are spin up and spin down, represented by arrows.

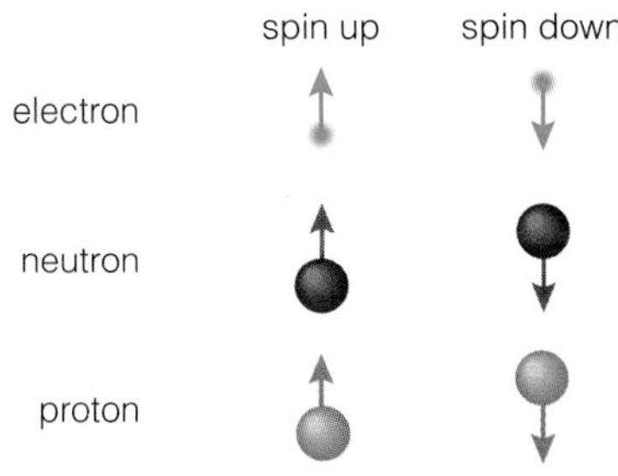

Quarks and Leptons: The Building Blocks of Matter

While electrons appear to be truly fundamental, protons and neutrons are made from even smaller particles, called **quarks.** Protons and neutrons each contain two different types of quarks: the **up quark,** which has an electric charge of $+\frac{2}{3}$, and the **down quark,** which has an electric charge of $-\frac{1}{3}$.[2] Two up quarks and one down quark form a proton, giving it an overall charge of $+\frac{2}{3}+\frac{2}{3}-\frac{1}{3}=+1$. One up quark and two down quarks form a neutron, making it neutral: $+\frac{2}{3}-\frac{1}{3}-\frac{1}{3}=0$ (Figure S5.3).

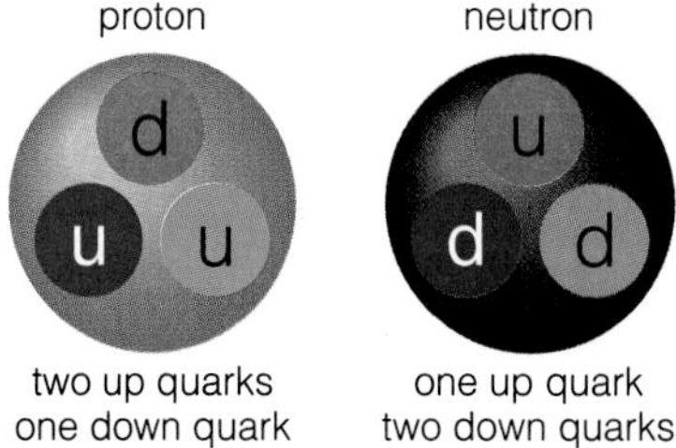

FIGURE S5.3 The quark composition of protons and neutrons.

To sum up so far, the fundamental building blocks of atoms are the up quark, the down quark, and the electron. But what about the hundreds of other particles discovered by scientists? The standard model organizes all these particles into a relatively simple hierarchy. First, as we discussed earlier, particles are classified as either fermions or bosons, depending on their *spin.* Quarks, electrons, protons, and neutrons are all fermions. Next, the fermions fall into two groups: those made from quarks and those not made from quarks. Fermions not made from quarks, such as electrons, are called **leptons.**[3]

TIME OUT TO THINK *In most high school science classes, students are told that the fundamental building blocks of atoms are protons, neutrons, and electrons. However, based on the preceding discussion, explain why our current understanding really holds that the fundamental building blocks of atoms are* **quarks** *and* **leptons.**

Particles made from quarks can consist of either two or three quarks. Interestingly, experiments in particle accelerators indicate that a single quark cannot exist in isolation; it must always live either in a pair or in a threesome with other quarks.

We can determine what types of quarks a particle is made of by analyzing the particle's behavior. In the standard model, a total of six different types of quarks[4] are needed to explain the characteristics of all particles observed to date. We have already met the up and down quarks that make up protons and neutrons; the other four quarks have the rather exotic names *strange, charmed, top,* and *bottom.* Six different types of leptons go with the six types of quarks: the electron and the *electron neutrino,* the *muon* and the *mu neutrino,* and the *tauon* and the *tau neutrino.*

[1]Physicists measure the angular momentum of subatomic particles in units of Planck's constant divided by 2π. In these special units, *fermions* have half-integer spins (e.g., $\frac{1}{2}, \frac{3}{2}, \frac{5}{2}$), and *bosons* have integer spins (e.g., 0, 1, 2).

[2]All charge values are given relative to the charge of protons (+1) and electrons (−1).

[3]There is also a name for the particles made from quarks: They are called *hadrons.* However, in this book we simply refer to them as "particles made from quarks."

[4]Physicists say that there are six different *flavors* of quarks, rather than *types* of quarks.

Table S5.1 Fundamental Fermions

The Quarks	The Leptons
Up	Electron
Down	Electron neutrino
Strange	Muon
Charmed	Mu neutrino
Top	Tauon
Bottom	Tau neutrino

All the quarks and leptons are listed in Table S5.1. All six quarks and all six leptons have been detected in experiments. The detection of the top quark was a particularly impressive success of the standard model: It was first predicted to exist in the 1970s and was verified experimentally in 1995.

At this point, you may be wondering what all these bizarrely named particles have to do with astronomy. In part, the answer is simply that these particles are the fundamental building blocks of everything, from atoms to people, planets, stars, and galaxies. In addition, in Chapter 22, we'll see that the events that unfolded during the first fraction of a second after the Big Bang depended critically on the various types of fundamental particles. We'll also see that *neutrinos,* in particular, are important in several astronomical processes, including nuclear fusion, the explosion of stars, and the fate of the universe.

Incidentally, neutrinos get their name, which means "little neutral ones," from the fact that they are electrically neutral and extremely lightweight—far less massive than even electrons. As of 1998, no one had yet succeeded in measuring the precise masses of neutrinos. Yet neutrinos turn out to be extremely common—they outnumber protons, neutrons, and electrons combined by a factor of roughly a billion. As a result, their total mass could conceivably be large enough to determine whether the universe continues to expand forever or whether the expansion will someday halt and the universe begin to collapse [Section 21.6].

Antimatter

Anyone who has watched *Star Trek* has heard of **antimatter.** Although *Star Trek* is science fiction, antimatter is not. It really exists. In fact, every quark and every lepton has a corresponding *antiquark* and *antilepton.* The antiparticle is like an exact opposite of its corresponding ordinary particle. For example, an *antielectron* (also called a *positron*) is identical to an ordinary electron except that it has a positive charge instead of a negative charge.

When a particle and its corresponding antiparticle meet, the result is mutual **annihilation** (Figure S5.4a). The combined mass of the particle and antiparticle turns completely into energy in accord with $E = mc^2$. Because our universe is made predominantly of ordinary matter, antimatter generally does not last very long. Whenever an antiparticle is produced, it quickly meets an ordinary particle, and the two annihilate each other to make energy.

This process also works in reverse. When conditions are right, pure energy can turn into a particle–antiparticle pair. For example, whenever an electron "pops" into existence, an antielectron also pops into existence with it (Figure S5.4b). This process of **pair production** happens routinely in particle accelerators here on Earth and on a much grander scale in outer space. In fact, during the first few moments after the Big Bang, the universe's energy fields were so intense that particle–antiparticle pairs popped rapidly in and out of existence at virtually every point in space.

FIGURE S5.4 (**a**) In this representation, an electron and a positron (antielectron) annihilate each other to make energy in the form of a pair of photons. (**b**) Here, a concentration of energy leads to pair production of an electron and a positron.

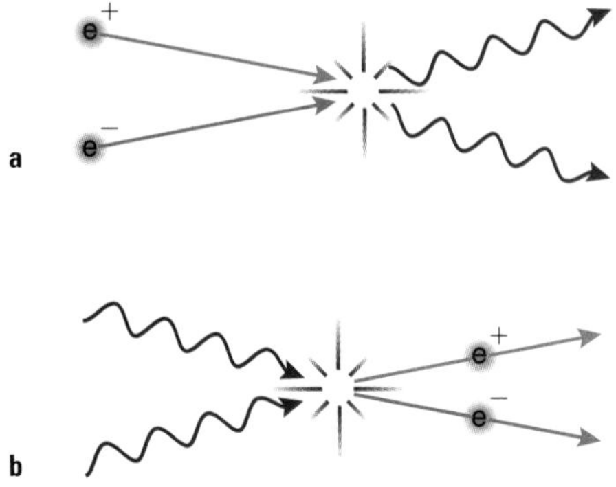

When we include antiparticles, the total number of quarks and leptons really is twice as high as shown in Table S5.1. That is, there are really 12 quarks: the six quarks listed and their six corresponding antiquarks (e.g., up quark and anti–up quark). Similarly, there are 12 leptons: the six listed and their six corresponding antileptons. The net total of 24 different fundamental particles is quite complex.[5] As we'll see

[5]Moreover, each of the six quarks and six antiquarks is believed to come in three distinct varieties, called *colors.* (The term *color* is not meant to be literal; rather, it describes a property of quarks that we cannot visualize.) For example, the up quark comes in three colors, often called red, green, and blue.

shortly, this complexity leads many scientists to believe that additional simplifying principles of particle physics still await discovery.

Forces

Without forces, the universe would be infinitely boring, a uniform sea of fundamental particles drifting aimlessly about. Forces supply the means through which particles communicate and exchange momentum, attracting or repelling one another depending on their properties. For example, particles with mass interact with other massive particles via the force of *gravity,* and particles with charge interact with other charged particles via the force of *electromagnetism.* Electromagnetism rules the processes of chemistry and biology, grouping electrons and protons into atoms, atoms into molecules, and molecules into living cells. Gravity drives the action on larger scales, holding people on planets, planets in solar systems, and stars in galaxies.

Besides gravity and electromagnetism, we know of two other fundamental forces in the universe, called the **strong force** and the **weak force.** Both act only on extremely short-distance scales—so short that these forces can be felt only *within* atomic nuclei. You can see why the strong force must exist by remembering that the nuclei of all elements except hydrogen (which has just a single proton) contain more than one proton. Protons are positively charged, so the electromagnetic force pushes them apart. If the electromagnetic force were unopposed, atomic nuclei would always fly apart. The strong force, so named because it is strong enough to overcome electromagnetic repulsion, is what holds nuclei together. The weak force, also very important in nuclear reactions, is a bit more subtle. All particles made from quarks respond to the strong force, but neutrinos, for example, feel only the weak force.

According to the standard model, each force is transmitted by an **exchange particle** that transfers momentum between two interacting particles. Photons are the exchange particle for the electromagnetic force. That is, photons carry electromagnetic force through the universe. Motions of electrons in a star create photons. The photons then cross light-years of space to our eyes, where they generate an electromagnetic disturbance that we see as starlight. Similarly, *gravitons* carry the gravitational force through the universe.[6] *Gluons,* which get their name because they act like glue to bind nuclei together, carry the strong force. The weak force is carried by particles sometimes called *weak bosons.* (All of the exchange particles are bosons.) The four forces and their exchange particles are listed in Table S5.2.

Table S5.2 also has columns describing the relative strengths of the four forces. For example, note that within an atomic nucleus the strong force is about 100 times stronger than the electromagnetic force, enabling the strong force to hold nuclei together. Outside atomic nuclei, the strong and weak forces vanish, leaving only gravity and electromagnetism to be felt in our daily lives.

Given the weakness of gravity in comparison to the other forces, you might wonder why it is important at all. The answer is that the other forces are out of the picture when we deal with large masses. The strong and weak forces vanish beyond atomic nuclei. The electromagnetic force cancels itself out in large objects, because such objects contain virtually equal numbers of protons and electrons, making them electrically neutral and unresponsive to the electromagnetic force. In contrast, all the matter we know about has positive mass, so gravity never cancels itself out. Thus, gravity is the only force left to act between massive enough objects. That is why gravity dominates the universe on large scales.

The Quest for Simplicity

The standard model involves four forces that mediate interactions between particles built from six types of

[6]*Gravitons* is the name for the particles that correspond to gravitational waves in general relativity, analogous to the correspondence between photons and electromagnetic waves. Gravitations have not yet been deducted.

Table S5.2 The Four Forces

Force	Relative Strength Within Nucleus*	Relative Strength Beyond Nucleus	Exchange Particles	Major Role
Strong	100	0	Gluons	Holding nuclei together
Electromagnetic	1	1	Photons	Chemistry and biology
Weak	10^{-5}	0	Weak bosons	Nuclear reactions
Gravity	10^{-43}	10^{-43}	Gravitons	Large-scale structure

*The force laws for the strong and weak forces are more complex than the inverse square laws for the electromagnetic force and gravity; hence, the numbers given for the strong and weak forces are very rough.

THINKING ABOUT . . .

Does God Play Dice?

Suppose we knew all the laws of nature and, at some particular moment in time, what every single particle in the universe was doing. Could we then predict the future of the universe for all time?

Until the twentieth century, nearly every philosopher would have answered yes. The idea was so pervasive that many philosophers concluded that God was like a watchmaker: God simply started up the universe, and the future was forever after determined. This idea that everything in the universe is predictable from its initial state is called *determinism;* a universe that runs predictably, like a watch, is called a *deterministic universe.*

The discovery of the uncertainty principle shattered the idea of a deterministic universe by telling us that, at best, we can only make statements about the *probability* of the precise future location of a subatomic particle. Because everything is made of subatomic particles, the uncertainty principle implies a degree of built-in randomness to the universe.

The idea that nature is governed by probability rather than certainty unsettled many people, including Einstein. Although he was well aware that the theories of quantum physics had survived many experimental tests, Einstein maintained a belief that the theories were incomplete. He believed that scientists would one day discover a deeper level of nature at which uncertainty would be removed. To summarize his philosophical objections to uncertainty, Einstein said, "God does not play dice."

Einstein did more than simply object on philosophical grounds. He also proposed a number of thought experiments in which he showed that the uncertainty principle implied paradoxical results. Claiming that such paradoxes made no sense, he argued that the uncertainty principle must not be correct. In the years since Einstein's death in 1955, advances in technology have made performing some of Einstein's thought experiments possible. The results have proved to be just as paradoxical as Einstein claimed they would be. That is, the experiments have confirmed the uncertainty principle at the same time that they have posed apparent logical paradoxes.

What can we make of an idea, such as the uncertainty principle, that seems to violate common sense at the same time that it survives every experimental test? Under the tenets of science, experiment is the ultimate judge of theory, and we must accept the results despite philosophical objections. In a sense, Einstein's objection that "God does not play dice" reflected his beliefs about how the universe *should* behave. Niels Bohr argued instead that nature need not fit our preconceptions with his famous reply to Einstein: "Stop telling God what to do."

quarks and six types of leptons, plus their corresponding antiparticles. It explains many experimental and observational results quite successfully; however, some scientists think that this model is still too complicated. They seek an even more basic theory of matter that reduces the number of forces and fundamental particles.

Theoretical work in the 1970s, verified experimentally in the 1980s, showed that the electromagnetic and weak forces are really just two different aspects of a single force, called the *electroweak force.* Many scientists hope that future discoveries will show three or even all four forces to be simply different aspects of a single, unified force governing *all* interactions in nature. As we will see in Chapter 22, these *unified theories* might be necessary to understanding the goings-on during the first fraction of a second after the Big Bang.

S5.3 The Uncertainty Principle

So far, we've used the word *particle* as if we were talking about tiny balls of matter. However, subatomic particles frequently act like waves, too. Unfortunately, our brains seem incapable of visualizing this dual wave–particle nature of subatomic particles. Nevertheless, we managed to invent the science of quantum mechanics, which enables us to determine with great accuracy the properties of matter.

Two fundamental laws that lie at the heart of quantum mechanics lead to most of its bizarre predictions. The first of these two laws is the **uncertainty principle,** discovered by Werner Heisenberg (1901–1976) in 1927. Here is one way of stating it:

The more we know about where a particle is located, the less we can know about its momentum, and the more we know about its momentum, the less we can know about its position.

FIGURE S5.5 A photograph of a ball taken with a blinking strobe light allows us to determine both where the ball was and where it was going at each moment in time.

the electron's momentum. Conversely, measuring the electron's momentum requires hitting it with a low-energy photon that will not disturb it much. Because low-energy light has long wavelengths, we'll no longer have a very good idea of where the electron is located.

TIME OUT TO THINK *Colloquially, we often express the uncertainty principle by saying that we can't know both where a particle is and where it is going. How does this statement relate to the more precise statement that we can't know both the particle's location and its momentum? In what ways is the colloquial statement accurate, and in what ways is it an oversimplification?*

We can illuminate the meaning of the uncertainty principle by considering how we might measure the trajectories of a baseball and an electron. In the case of a baseball, we could photograph it with a blinking strobe light. The resulting photograph shows us both where the ball was and where it was going at each moment in time (Figure S5.5). In scientific terms, knowing the path of the baseball means that we are measuring both its *location* and its *velocity* at each instant or, equivalently, its location and its *momentum*. (Recall that momentum is mass times velocity.)

Now imagine trying to observe an electron in the same way. We will detect the electron only if it manages to scatter some of the photons streaming by it. But whereas the photons are tiny particles of light compared to a baseball, they are quite large compared to the minuscule electron. The precision with which we can pinpoint the electron's location depends on the wavelengths of the photons. If we use visible light with a wavelength of 500 nanometers, we can measure the electron's location only to within 500 nanometers—which is about *5,000 times* the size of a typical atom. That is, if we see a flash from a row of 5,000 atoms, we do not even know which atom contains the electron that caused the flash! To locate the electron more precisely, we must use shorter-wavelength light, such as ultraviolet or X rays. But now we encounter our next problem: To determine the electron's path, we must observe the flashes from its interactions with one photon after another. Yet, each photon's energy delivers a "kick" that disturbs the electron, thereby changing the momentum that we are trying to measure. The higher the energy of the photon—which means the shorter its wavelength—the more it alters the electron's momentum.

It is almost as if nature is playing a perverse trick on us. Locating the electron precisely requires hitting it with a short-wavelength photon, but the high energy of this photon prevents us from determining

The uncertainty principle applies to *all* particles, not just to electrons. In fact, it applies even to large "particles" such as baseballs, but it is unnoticeable at this level. Consider what happens when we look at a baseball with visible light of 500-nanometer wavelength. Just as with the electron, we can locate any part of the baseball only to within 500 nanometers, but an uncertainty of 500 nanometers (about 0.00002 inch) is negligible in comparison to the size of the baseball. Moreover, the energy of visible light is so small compared to the energy of the baseball (including its mass-energy) that it has no noticeable effect on the baseball's momentum. That is why Newton's laws work perfectly well when we deal with the motion of baseballs, cars, planets, or other objects in the macroscopic[7] world. But Newton's laws fail us in the microscopic quantum realm, where we must deal with the implications of the uncertainty principle.

Wave–Particle Duality

Our thought experiment suggested that a particle such as an electron somehow "hides" its precise path from us. However, the uncertainty principle runs deeper than this—it implies that the electron *does not even have* a precise path. From this point of view, the concepts of location and momentum do not exist

[7]The prefix *macro* comes from the Greek word for "large," and the term *macroscopic* is used to contrast the large world of objects visible to the naked eye with the microscopic world visible only through microscopes.

independently for electrons in the way that they appear to exist for objects in everyday life. Instead of imagining the electron as following some complex but hidden path, we need to think of the electron as being "smeared out" over some volume of space [Section 5.4]. This "smearing out" of electrons and other particles holds the essence of the idea of wave–particle duality.

If we choose to regard the electron as a particle, we are imagining that we can locate it precisely by hitting it with a short-wavelength photon. In that case, we know the precise location of the electron at each instant but can never predict where it will be at the next instant. We see the electron in one place at one moment and in another place at another moment, but we have no idea how it passed through the regions in between. A mathematical description of quantum mechanics allows us to calculate the *probability* that we'll find the electron in any particular place at any particular time. For example, Figure S5.6 shows the probability patterns for the electron in several energy levels of hydrogen; the brighter the region on the diagram, the higher the probability of finding the electron at any particular instant. From the point of view of the electron as a particle, the "smeared out" electron cloud around the hydrogen nucleus represents those places where we are most likely to find the electron at any instant.

Alternatively, we can choose to regard the electron as a wave. We can measure the momentum of a wave, such as that of a ripple moving across a pond, but we cannot say that a wave has a single, precise location (Figure S5.7). Instead, the wave is spread out over some region of the water in the same way that an electron is "smeared out" over some volume of space.

The fact that electrons exhibit both particle and wave properties demonstrates that our common sense from the macroscopic world does not translate well to the quantum world. We say that electrons, like photons of light, exhibit wave–particle duality. In fact, all subatomic "particles" exhibit wave–particle duality; we call them *particles* only for convenience. Like a photon of light, each particle has a wavelength.[8] When the wavelength of the particle is small, we can locate the particle fairly precisely, but the particle's momentum is highly uncertain. When the wavelength of the particle is large, its momentum becomes well defined, but its location grows fuzzy.

Quantifying the Uncertainty Principle

We can quantify the uncertainty principle with a simple mathematical statement:

$$\begin{pmatrix}\text{uncertainty} \\ \text{in location}\end{pmatrix} \times \begin{pmatrix}\text{uncertainty} \\ \text{in momentum}\end{pmatrix} \approx \text{Planck's constant}$$

[8]Electrons usually have very small wavelengths and therefore can be used to locate other particles with high precision. This is the principle behind *electron microscopes,* in which short-wavelength electrons are used to study microscopic objects. Whereas the resolution of visible-light microscopes is limited to the roughly 500-nanometer wavelength of visible light, electron microscopes can achieve resolutions of less than 0.1 nanometer.

FIGURE S5.6 Probability patterns for the electron in several energy levels of hydrogen.

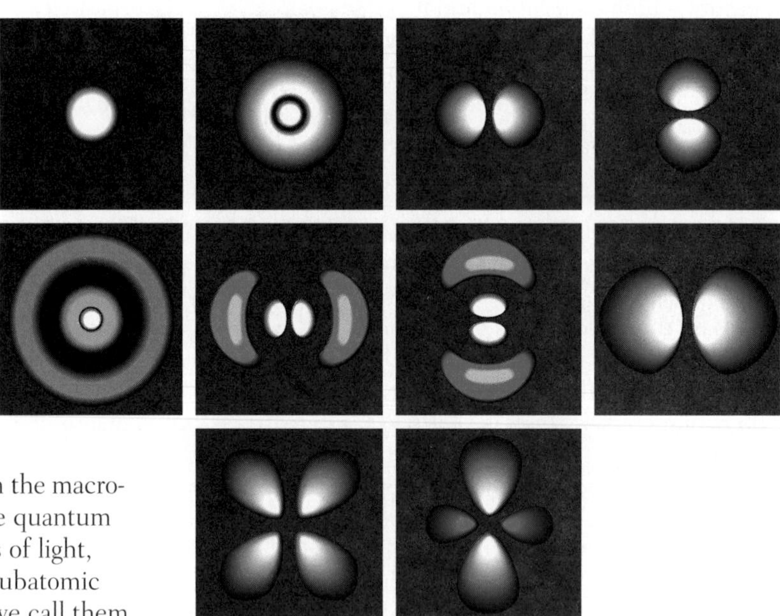

FIGURE S5.7 A wave has a well-defined momentum (represented by the arrows), but not a single, precise location.

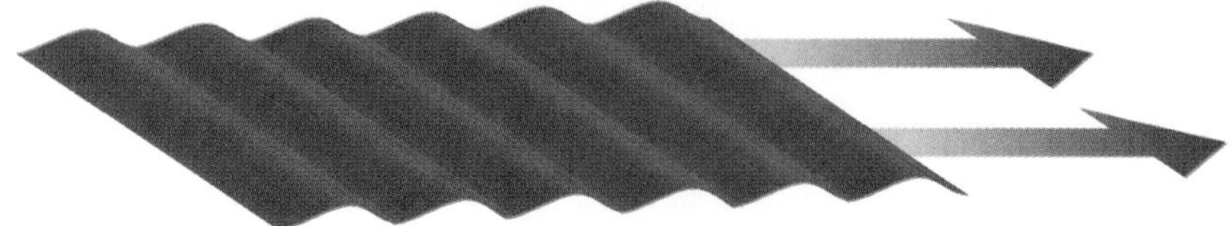

Planck's constant is a fundamental constant in nature rather like the gravitational constant (G) in Newton's law of gravity and the speed of light; its numerical value is 6.626×10^{-34} joule $\times$ s. (Recall that Planck's constant appears in the formula for the energy of a photon [Section 7.2].)

This formula quantifies what we have already learned: Because the product of the uncertainties is roughly constant, when one uncertainty (either location or momentum) goes down, the other must go up. For example, if we determine the location of an electron with a particular amount of uncertainty, such as to within 500 nanometers, we can use the formula to calculate the amount of uncertainty in the electron's momentum. Note that the numerical value of Planck's constant is quite small, which explains why uncertainties are scarcely noticeable for macroscopic objects.

A second way of writing the uncertainty principle is mathematically equivalent but leads to additional insights. Instead of expressing the uncertainty principle in terms of location and momentum, this alternative version expresses it in terms of the amount of *energy* that a particle has and *when* it has this energy. This version reads:

$$\begin{matrix}\text{uncertainty} \\ \text{in energy}\end{matrix} \times \begin{matrix}\text{uncertainty} \\ \text{in time}\end{matrix} \approx \text{Planck's constant}$$

Mathematical Insight S5.1 Electron Waves in Atoms

In the text, we've said that an electron in an atom is "smeared out" over some volume of space. In fact, the physics is much more precise than this vague statement implies. If we choose to view the electron as a wave, an electron in an atom can be regarded as a *standing wave*. You are probably familiar with standing waves on a string that is anchored in place at its two ends, such as a violin string (Figure S5.8a). Such waves are called standing waves because each point on the string vibrates up and down, but the wave does not appear to move along the length of the string. Moreover, because the string is anchored at both ends, only wave patterns with a half-integer (e.g., $0, \frac{1}{2}, 1, \frac{3}{2}, 2$) number of wavelengths along the string are possible. Other patterns, such as having three-fourths of a wavelength along the string (Figure S5.8b), are not possible without breaking one end of the string away from its anchor point.

An electron viewed as a standing wave is anchored by the electromagnetic force holding it in the atom. Like the waves on a string, only particular wave patterns are possible. However, these patterns are more complex than waves on a string because they are three-dimensional wave patterns. The allowed wave patterns for an electron in an atom can be calculated with the famous *Schrödinger equation*, developed by Erwin Schrödinger in 1926. These allowed wave patterns correspond directly to the allowed energies of the electron in the atom, and thus the Schrödinger equation enables scientists to predict what energies should be allowed in different atoms. The fact that the Schrödinger equation successfully predicts the energy levels that are measured in the laboratory (by analysis of spectral lines) is one of the great triumphs of quantum mechanics.

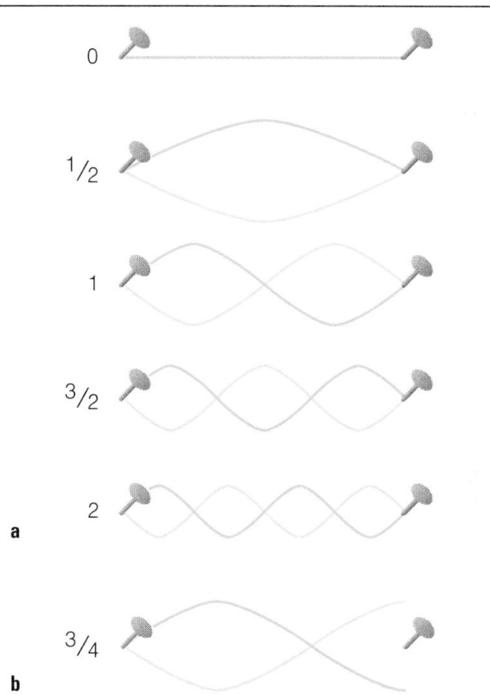

FIGURE S5.8 (**a**) Standing waves on a string must have a half-integer number of wavelengths. (**b**) Any other number of wavelengths is not possible without breaking the string away from one of its anchors.

An amusing way to gain further appreciation for the uncertainty principle in both its forms is to imagine a game of "quantum baseball." Suppose a quantum pitcher is pitching an electron that you try to hit with your quantum bat. You'll find it extremely difficult! With the first version of the uncertainty principle, the problem is that you'll never know both where the electron is and where it is headed next. You might see it right in front of you, but because you don't know which direction it's going, you can't know whether to swing straight, up, down, or sideways. With the second version of the uncertainty principle, your problem is that you might know the electron's energy, which tells you how hard you need to swing, but you'll never know *when* to swing. Either way, your chances of hitting the electron are completely random, governed by probabilities that can be calculated with quantum mechanics.

TIME OUT TO THINK *Explain why real baseball players don't have the same problems that would arise with quantum baseball.* (Hint: *How do the quantum uncertainties compare to the size of a real baseball and a real bat?*)

S5.4 The Exclusion Principle

The uncertainty principle is the first of the two fundamental laws that lie at the heart of quantum mechanics. The second fundamental law, called the **exclusion principle,** was discovered by Wolfgang Pauli (1900–1958) in 1925. In its simplest sense, the exclusion principle says that two particles cannot be in the same place at the same time. (It applies only to fermions, not to bosons.) A more complete understanding of the exclusion principle requires investigating the properties of particles a little more deeply.

The Quantum State of a Particle

Scientists use the term *state* to describe the current conditions of an object. For example, if you are relaxing in a chair, a scientist might say that you are in a "state of rest." A more precise description of your state in the chair might be something like "Your current state is a velocity of zero (at rest), a heart rate of 65 beats per minute, a breathing rate of 12 breaths per minute, a body temperature of 37°C, a metabolic rate of 200 Calories per hour," and so on.

TIME OUT TO THINK *Suppose you have the following information about the current state of a friend: velocity of 5 km/hr, heart rate of 160 beats per minute, metabolic rate of 1,200 Calories per hour. Which of the following is your friend most likely doing: walking slowly as she reads a book, driving in her car, riding a bicycle down a hill, or swimming at a hard pace? How do you know?*

Completely describing a person's state is quite complicated. Fortunately, describing the state of a subatomic particle such as an electron, proton, or neutron is much easier. In general, a complete description of the state of a subatomic particle—called its **quantum state**—specifies only its location, momentum, orbital angular momentum, and spin to the extent allowed by the uncertainty principle. Like the energy of an electron in an atom, each property that describes a particle's quantum state is *quantized,* meaning that it can take on only particular values and not other values in between.

Statement and Meaning of the Exclusion Principle

Earlier we described the exclusion principle in simple terms by saying that two particles cannot be in the same place at the same time. Now let us state the exclusion principle more precisely:

Two fermions of the same type cannot occupy the same quantum state at the same time.

This principle has many important implications. One of the most important is in chemistry, where it dictates how electrons occupy their various states in atoms. For example, an electron occupying the lowest energy level in an atom necessarily has a particular amount of orbital angular momentum and a restricted range of locations. The electron's energy level fully determines its quantum state, except for its spin. Because electrons have only two possible spin states, up and down, only two electrons can occupy the lowest energy level (Figure S5.9). If you tried to put a third electron into the lowest energy level, it would have the same spin—and hence the same quantum state—as one of the two electrons already there. The exclusion principle won't allow that, so the third electron must go into a higher energy level. If you take a course in chemistry, you'll learn how a similar analysis of higher energy levels explains the chemical properties of all the elements, including their arrangement in the periodic table of the elements (see Appendix 2).

The uncertainty principle and the exclusion principle together determine the sizes of atoms and of everything made of atoms, including your own body. The uncertainty principle ensures that electrons cannot be packed into infinitesimally tiny spaces. If

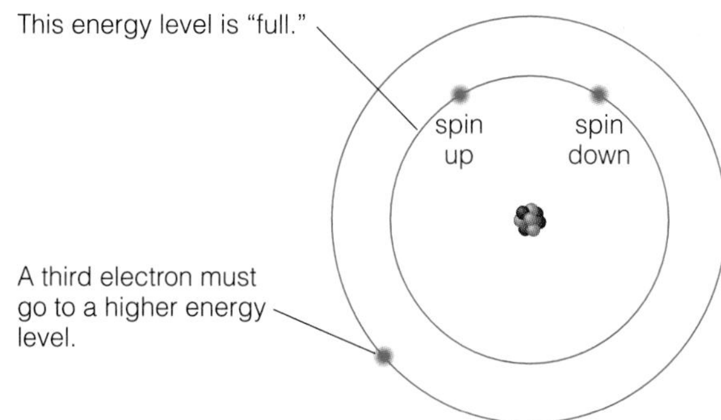

FIGURE S5.9 Only two electrons, one with spin up and the other with spin down, can share a single energy level in an atom.

you tried to confine an electron in too small a space in an atom, its momentum would become so large that the electromagnetic attraction of the nucleus could no longer retain it. The exclusion principle ensures that each electron has to have its own space. Thus, despite the fact that atoms are almost entirely empty, the two fundamental laws of quantum mechanics explain the solidity of matter. In the words of physicist Richard Feynman (1918–1988), "It is the fact that electrons cannot all get on top of each other that makes tables and everything else solid."

The exclusion and uncertainty principles also govern the behavior of tightly grouped protons, neutrons, or any other kinds of fermions. Just as these principles give atoms their size, they also determine the size of nuclei, because they limit how closely protons and neutrons can pack together. As we will see later, these quantum effects can even influence the lives of stars.

Before we go on to discuss some astronomical implications of the quantum laws, it's worth noting that photons and other bosons do *not* obey the exclusion principle. For example, laser beams are so intense because many photons *can* be in the same state at the same time. In addition, it is possible under special conditions for two or more fermions to act together like a single boson. Such conditions lead to some amazing behavior, including *superconductivity,* in which electricity flows without any resistance, and *superfluidity,* in which extremely cold liquids can "creep" up the walls of a container. Many popular books on quantum laws explain these phenomena in more detail.

S5.5 Key Quantum Effects in Astronomy

The uncertainty principle and the exclusion principle lead to some truly strange consequences in the subatomic realm. Amazingly, this odd microscopic behavior produces important effects on much larger scales. In fact, we cannot fully understand how stars are born, shine brightly throughout their lives, and die without understanding the implications of the quantum laws. In this section, we investigate three quantum effects of great importance in astronomy.

Degeneracy Pressure

Under ordinary conditions in gases, pressure and temperature are closely related. For example, suppose we inflate a balloon, filling it with air molecules. The individual molecules zip around inside the balloon, continually bouncing off its walls (Figure S5.10). The force of these molecules striking the walls of the balloon creates **thermal pressure,** which keeps the balloon inflated.[9] If we cool the balloon by, say, putting it in a freezer, the molecules slow down. Slowing the

[9]Pressure is defined as the *force per unit area* that pushes on the walls of the balloon. Thermal pressure is the type of pressure directly related to the temperature of a gas.

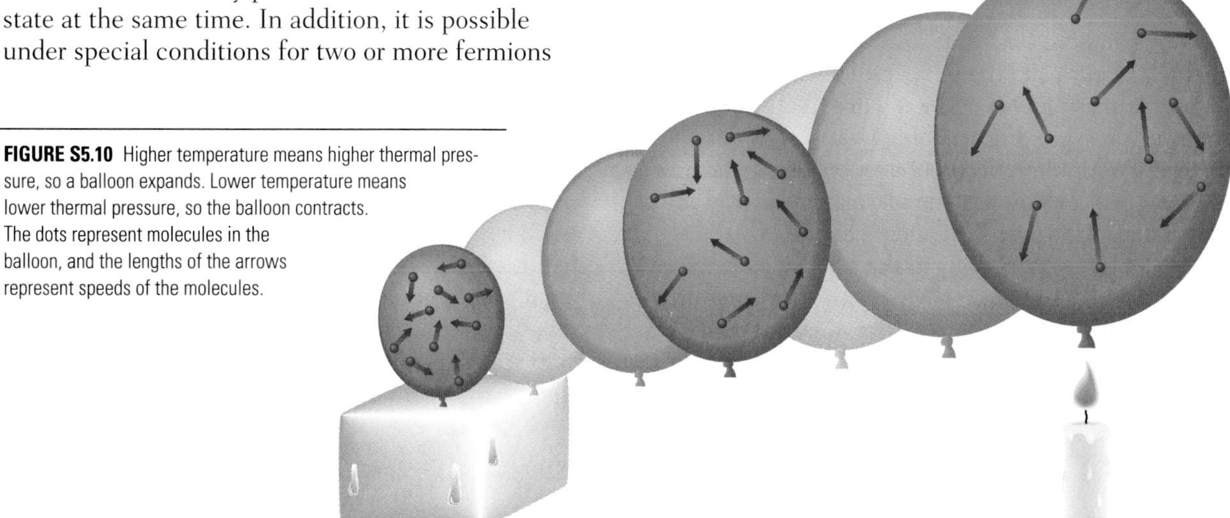

FIGURE S5.10 Higher temperature means higher thermal pressure, so a balloon expands. Lower temperature means lower thermal pressure, so the balloon contracts. The dots represent molecules in the balloon, and the lengths of the arrows represent speeds of the molecules.

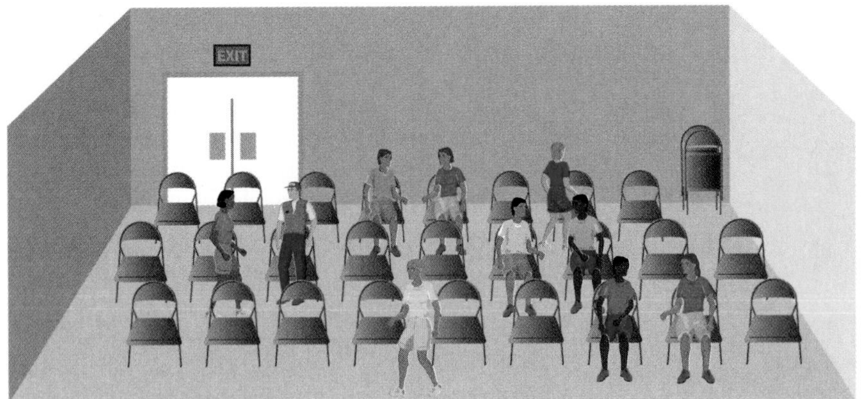

a

FIGURE S5.11 The auditorium analogy to degeneracy pressure. Chairs represent available quantum states, and people moving from chair to chair represent electrons. (**a**) When there are many more available quantum states (chairs) than electrons (people), an electron is unlikely to try to enter the same state as another electron. The only pressure comes from the temperature-related motion of the electrons, which is the thermal pressure. (**b**) When the number of electrons (people) approaches the number of available quantum states (chairs), finding an available state requires that the electrons move faster than they would otherwise. This extra motion creates degeneracy pressure.

b

molecules reduces the force with which they strike the walls, reducing the thermal pressure and shrinking the balloon. Heating the balloon speeds up the molecules, raising the thermal pressure and causing the balloon to expand.

Thermal pressure is the dominant type of pressure at low to moderate densities. However, quantum effects produce an entirely different type of pressure under conditions of extremely high density, one that does not depend on temperature at all. Consider what happens when we compress a *plasma* of positively charged ions and free electrons. At first, the energy we expend in crushing the plasma makes the electrons and protons move faster and faster, increasing the pressure and the temperature. But suppose we let the plasma cool off for a while. As the plasma cools, its pressure drops, enabling us to compress it further. Continuing this process of cooling and compression, we can squeeze the plasma down to a very dense state. However, we cannot continue this process indefinitely. According to the exclusion principle, no two electrons can have exactly the same position, momentum, and spin. Just as in an atom, all the electrons can't get on top of one another at once. The compression must stop at some point, no matter how cold the plasma. This resistance to compression, stemming from the exclusion principle, is what we call **degeneracy pressure.**

The following analogy might help you visualize how degeneracy pressure works. Imagine a small number of people in an auditorium filled with folding chairs. Each person can move freely about and sit in any empty chair. The chairs represent available quantum states, the people represent electrons darting from place to place, and their motions represent thermal pressure as the electrons move from one quantum state to the next. The exclusion principle corresponds to the rule that two people cannot sit in the same chair at the same time. As long as the chairs greatly outnumber the people, two people will rarely fight over the same chair (Figure S5.11a).

Now suppose that ushers begin removing chairs from the front of the auditorium (compression), gradually forcing people to move to the back. Soon everyone has to crowd toward the back of the auditorium, where the number of remaining chairs is just slightly larger than the number of people. Now when a person moves to a particular chair, there's a good chance that it's already occupied. Ultimately, when the number of people equals the number of chairs, people can still move from place to place, but only if they swap seats with somebody else (Figure S5.11b). The ushers can't take away any more chairs, and the compression must stop. In other words, all the available states are filled.

The uncertainty principle also influences degeneracy pressure, though in a way that does not perfectly fit this analogy. In a highly compressed plasma, the available space for each electron is very small, so the electrons' positions must be precisely defined. According to the uncertainty principle, their momentum must then be extremely uncertain, which means that the electrons must be moving very fast on average. This requirement that highly compressed electrons have to move quickly holds *even if the object is cold.*[10] This quantum-mechanical trade-off between position and momentum is at the root of degeneracy pressure. If you want to compress lots of electrons into a tiny space, you need to exert an enormous force to rein in their momentum.

Degeneracy pressure caused by the crowding of electrons, or **electron degeneracy pressure,** affects the lives of stars in several different ways. In some cases, it can prevent a collapsing cloud of gas from becoming a star in the first place, creating what is called a *brown dwarf* [Section 16.2]. In stars like the Sun, it determines how they begin burning helium near the end of their lives [Section 16.3]. When stars die, most leave behind an extremely dense stellar corpse called a *white dwarf,* which is also supported by electron degeneracy pressure [Section 17.2].

Not all stars meet this fate, because electron degeneracy pressure cannot grow infinitely strong. Under extreme compression, the average speed of the electrons begins to approach the speed of light. Nothing can go faster than the speed of light, so we eventually reach a limit to how much degeneracy pressure the electrons can exert. Once a dying star reaches that limit, electron degeneracy pressure cannot prevent it from shrinking further. The star then collapses until it becomes a ball of neutrons, called a *neutron star* [Section 17.3]. Neutron stars support themselves through **neutron degeneracy pressure,** which is just like electron degeneracy pressure except that it is caused by neutrons and occurs at much higher densities.[11]

Neutron degeneracy pressure cannot grow infinitely strong either. It begins to fail when the speed of the neutrons approaches the speed of light. According to our present understanding, nothing can stop the collapse of an object once its gravity overcomes neutron degeneracy pressure. Such an object collapses indefinitely, becoming the mysterious type of structure in spacetime that we call a *black hole.*

Quantum Tunneling

The next quantum effect we'll investigate arises from the uncertainty principle and has important implications not only in astronomy, but also in modern technology. Imagine that, as an unfortunate result of a case of mistaken identity, you're sitting on a bench in a locked jail cell (Figure S5.12a). Another bench is on the other side of the cell wall. If you could magically transport yourself from the bench on the inside to the bench on the outside, you'd be free. Alas, no such magic ever occurs for humans.

But what if you were an electron? In that case, the uncertainty principle would prevent us from predicting your precise location. At best, we could state only the probability of your being in various locations. While the probability that you remain in your cell might be greatest, there is always some small probability that you will be found outside your cell. With a bit of luck, you might suddenly find yourself free, thanks to the uncertainty principle (Figure S5.12b). This process, in which an electron or any

FIGURE S5.12 (**a**) A person is confined to the bench inside the jail cell, even though it would take no more energy to sit on the outside bench. (**b**) If you were an electron, you could "magically" become free through the process of quantum tunneling.

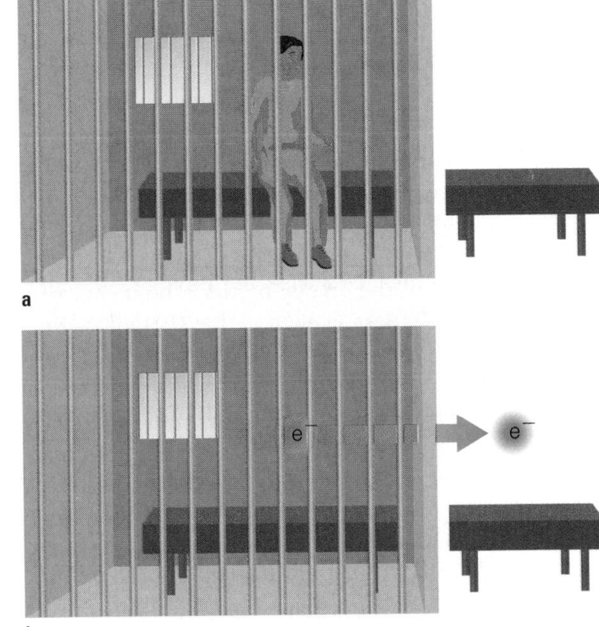

[10] In this sense, an object is *cold* if there's no way to get heat from it. Consider a plasma in which all the available momentum states are filled up to a certain level. To extract heat from the plasma, you'd have to slow down some of its particles. But that means moving them to lower-momentum states, which are already taken. The exclusion principle thus prevents any energy from escaping, so the plasma is cold even though the electrons may be moving at high speeds.

[11] The reason why neutron degeneracy pressure comes into play only at much higher densities than electron degeneracy pressure is that neutrons have much greater mass than electrons. A neutron near the speed of light possesses over 1,800 times more momentum than an electron at the same speed. Thus, the positions of neutrons can be over 1,800 times more precise, enabling them to occupy a much smaller volume of space.

other subatomic particle "magically" goes through a wall-like barrier, is called **quantum tunneling.**

We can gain a deeper understanding of quantum tunneling by thinking in terms of the *energy* needed to cross a barrier. If you are sitting in a jail cell, the barrier is the wall of the cell. The reason you cannot escape the cell is that you don't have enough energy to crash through the wall. Just as the cell wall keeps you imprisoned, a barrier of electromagnetic repulsion can imprison an electron that does not have enough energy to crash through it. However, recall that we can write the uncertainty principle in the following form:

$$\begin{matrix}\text{uncertainty} \\ \text{in energy}\end{matrix} \times \begin{matrix}\text{uncertainty} \\ \text{in time}\end{matrix} \approx \text{Planck's constant}$$

Because of the uncertainty inherent in energy, there is always *some* chance that the electron will have more energy than we think at a particular moment, allowing it to cross the barrier anyway. From this point of view, quantum tunneling comes about because of uncertainty in energy rather than uncertainty in location.

Both points of view on quantum tunneling are equivalent, but the latter viewpoint illustrates a rather remarkable "loophole" in the law of conservation of energy. To cross the barrier, the particle must briefly gain some excess energy. Thanks to the uncertainty principle, this "stolen" energy need not come from anywhere as long as the particle returns it within a time period *shorter* than the uncertainty in time. In that case, we cannot be certain that any energy was ever missing! It's like stealing a dollar and putting it right back before anyone notices, so that no harm is done—except that the particle uses the stolen energy to cross the barrier before returning it.

This tale of phantom quantum energy thefts may sound utterly ridiculous at first, but the process of quantum tunneling is readily observed, and it is extremely important. In fact, the "microchips" used in all modern computers and many other modern electronic devices work because of quantum tunneling by electrons. We can control the rate of quantum tunneling, and hence the electric current, by adjusting the "height" of the energy barrier. The higher the energy barrier, the less likely it is that particles will tunnel through it.

Even more amazingly, our universe would look much different were it not for quantum tunneling. The nuclear fusion reactions that power stars occur when atomic nuclei smash together so hard that they stick. However, nuclei tend to repel each other because they are positively charged (they contain only positive protons and neutral neutrons) and like charges repel. This repulsion creates an electromagnetic barrier that prevents nuclear fusion under most conditions. Inside the electromagnetic barrier, the attraction of the strong force takes over. But, even at the high temperatures inside stars, atomic nuclei don't have enough energy to crash through the electromagnetic barrier. Instead, they rely on quantum tunneling to sneak through to the region where the strong force dominates. In other words, quantum tunneling is what makes nuclear fusion possible in stars like our Sun.[12]

Virtual Particles

The same "loophole" in the law of conservation of energy that allows particles to tunnel through otherwise impenetrable barriers permits even bigger crimes. Entire particles can "pop" into existence from nowhere—their mass made from stolen energy—as long as they "pop" back out of existence before anyone can verify that they ever existed. A somewhat fanciful analogy will be helpful to understanding this bizarre concept.

Imagine that the law of conservation of energy is enforced by a "great cosmic accountant." (In reality, of course, the law is enforced naturally.) Further, imagine that the cosmic accountant keeps a storehouse of energy in a large bank vault and ensures that, anytime something borrows some energy, it returns the energy in a precisely equal amount. A particle that pops into existence is like a bank robber who steals some energy from the vault. If the particle is caught by the cosmic accountant, someone will have to pay for the stolen energy. However, the particle won't be caught as long as it pops back out of existence quickly enough, returning the stolen energy. The length of time the particle is allowed to exist is so short that the uncertainty principle prevents anyone from knowing that energy is missing.

Particles that pop in and out of existence before anyone can possibly detect them are called **virtual particles.** Although these particles are undetectable themselves, quantum theories predict that virtual particles should exert real, measurable effects on those particles we can detect. Remarkably, many of these predictions have been verified experimentally. In fact, modern theories of the universe propose that empty space—what we call a *vacuum*—actually "bubbles" with virtual particles that pop rapidly in and out of existence.

TIME OUT TO THINK ***Imagine that you write a check for $100, but you have no money in your checking account. Your check is not necessarily***

[12]Quantum tunneling is not as crucial to fusion in stars with core temperatures much higher than the Sun's.

doomed to bounce—as long as you deposit the needed $100 before the check clears. Explain how this $100 of "virtual money" is similar in concept to virtual particles.

Astronomically speaking, the most important process involving virtual particles concerns black holes. We can understand the process by extending our analogy to the case of a virtual electron popping into existence near a black hole. The virtual electron cannot pop into existence alone; it must be accompanied by a virtual positron (an antielectron) so that electric charge is conserved. The virtual electron–positron pair must return the stolen energy before being caught by our imaginary "great cosmic accountant," and they do so by annihilating each other (Figure S5.13a). But suppose a virtual electron–positron pair appears very close to, but outside of, the event horizon of a black hole. The electron and positron are supposed to annihilate each other quickly, but a terrible problem comes up: One of the particles crosses the event horizon during its brief, virtual existence (Figure S5.13b). From the perspective of our cosmic accountant, this virtual particle was never accounted for in the first place, so there's no problem with the fact that it will never be seen again. But the other particle is suddenly caught like a deer in the headlights. Without its virtual mate, it has no way to annihilate itself, and it is caught red-handed by the great cosmic accountant.

If we strip away the fanciful imagery of a cosmic accountant, the end result is the creation of *real* particles, not virtual ones, just outside the event horizon of a black hole. Note that nothing escapes from inside the black hole. Rather, these real particles are created from the gravitational potential energy of the black hole. Around the black hole, these real electrons and positrons annihilate each other, producing real photons that are radiated into space. To an outside observer, the black hole would appear to be radiating, even though nothing ever escapes from inside it. This effect was first predicted by Stephen Hawking in the 1970s and is therefore called **Hawking radiation.** As we've just seen, the ultimate source of Hawking radiation is the gravitational potential energy of the black hole. The continual emission of Hawking radiation causes the black hole to shrink slowly in mass, or *evaporate,* over very long periods of time.

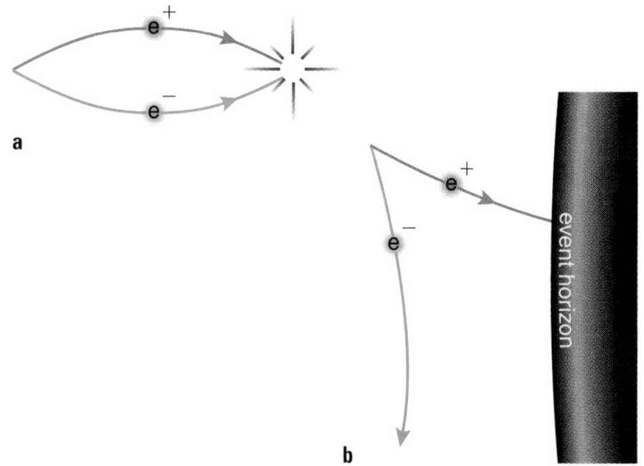

FIGURE S5.13 (**a**) Pairs of virtual electrons and positrons can pop into existence, as long as they annihilate each other before they can be detected. (**b**) Near the event horizon of a black hole, one member of a virtual pair crosses the event horizon; the other particle is left with no partner and cannot annihilate itself.

The idea that black holes can evaporate remains untested, but if it is true it may have profound implications for both the origin and the fate of the universe. Some scientists speculate that black holes of all sizes might have been created during the Big Bang. If so, some of the smaller ones should be evaporating today. The fact that we have not yet detected any such evaporation sets limits on the number of small black holes that might have formed in the Big Bang. At the other end of time, if the universe continues to expand forever, black holes may be the last large masses left after all the stars have died. In that case, the slow evaporation caused by Hawking radiation will mean that even black holes cannot last forever, and the universe eventually will contain nothing but a fog of photons and subatomic particles, separated from one another by incredible distances as the universe continues to grow in size.

There is a theory which states that if ever anyone discovers exactly what the Universe is for and why it is here, it will instantly disappear and be replaced by something even more bizarre and inexplicable.

There is another which states that this has already happened.

DOUGLAS ADAMS, from *The Restaurant at the End of the Universe*

THE BIG PICTURE

In this chapter, we have studied the quantum revolution and its astronomical consequences. As you look back, keep in mind the following "big picture" ideas:

- The quantum revolution is arguably the third great revolution in our understanding of the universe. The first was the Copernican revolution, which demolished the ancient belief in an Earth-centered universe. The second was relativity, which radically revised our ideas about space, time, and gravity. The quantum revolution has changed our ideas about the fundamental nature of matter and energy.
- Strange as the laws of quantum mechanics may seem, they can be readily tested and confirmed through observation and experiment. The quantum laws, like relativity, now stand on very solid ground. So far, they have passed every experimental test yet devised for them.
- The tiny quantum realm may seem remote from the large scales we are accustomed to in astronomy, but it is exceedingly important. The laws of quantum mechanics are necessary for understanding many astronomical processes, including nuclear fusion in the Sun, the degeneracy pressure that supports stellar corpses, and the possible evaporation of black holes.

Review Questions

1. What do we mean by the quantum realm? What is *quantum mechanics?*
2. List five major ideas that come directly from the laws of quantum mechanics.
3. What do we mean by *fundamental particles?* How do we investigate fundamental particles in *particle accelerators?*
4. What is *spin?* What are the two basic categories of particles based on spin?
5. What are *quarks* and *leptons?* Explain why we say that quarks and leptons are subsets of the fermions.
6. List the six quarks and six leptons in the standard model. Describe the quark composition of a proton and of a neutron.
7. What are *neutrinos?* Why might they turn out to be important to the fate of the universe?
8. What is *antimatter?* How does a positron differ from an electron? What happens when a particle and its antiparticle meet?
9. How is antimatter produced, and why is it always produced along with matter in *pair production?*
10. List the four basic forces in nature and the *exchange particles* for each.
11. The strong force and the weak force are sometimes called nuclear forces. Why?
12. Why do many scientists believe that the standard model will eventually be replaced by a simpler model of nature?
13. What is the *uncertainty principle?* How is it related to the idea of wave–particle duality?
14. Describe two ways of quantifying the uncertainty principle, and give an example showing the meaning of each.
15. What do we mean by the *quantum state* of a particle?
16. What is the *exclusion principle?* What types of particles obey it? Briefly explain how the exclusion principle determines how electrons fill energy levels in atoms.
17. What is *degeneracy pressure?* How does it differ from *thermal pressure?* Describe the auditorium analogy for degeneracy pressure.
18. Explain why the uncertainty principle implies that the particles in a highly compressed plasma must move at very high speeds. How does this fact explain why there is a limit to the strength of degeneracy pressure? Why is this limit higher for *neutron degeneracy pressure* than for *electron degeneracy pressure?*
19. What is *quantum tunneling?* How is it important to modern electronics? How is it important to nuclear fusion in the Sun?
20. What do we mean by *virtual particles?* Briefly explain how the concept of virtual particles suggests that black holes may be able to evaporate through *Hawking radiation.*

Discussion Questions

1. *Common Sense Versus Experiment.* Even the most highly trained physicists find the results of quantum mechanics to be strange and counter to our everyday common sense. Yet the predictions of quantum mechanics have passed every experimental test yet posed to them. Discuss whether it is important to reconcile

our common sense with experimental results and, if so, how we can do it.

2. *The Meaning of the Uncertainty Principle.* When first hearing about it, many people assume that the uncertainty principle means that we cannot *measure* the position and momentum of a particle precisely. But, according to current understanding, it really tells us that the particle *does not have* a precise position and momentum in the sense that we would expect from everyday life. How do these two viewpoints differ? Discuss the different philosophical consequences of these two viewpoints.

3. *Antimatter Engines.* In the *Star Trek* series, starships are powered by matter–antimatter annihilation. Explain why matter–antimatter annihilation is the most efficient possible source of power. What practical problems would we face in developing matter–antimatter engines?

Problems

1. *The Strong Force.* The strong nuclear force is the force that holds the protons and neutrons in the nucleus together. Based on the fact that atomic nuclei can be stable, briefly explain how you can conclude that the strong force must be even stronger than the electromagnetic force, at least over very short distances.

2. *Chemistry and Biology Are Electromagnetic.* All chemical and biological reactions involve the creation and breaking of chemical bonds, which are bonds between the electrons of one atom and the electrons of others. Given this fact, explain why the electromagnetic force governs all chemical and biological reactions. Also explain why the strong force, the weak force, and gravity play no role in these reactions.

3. *Why Does Gravity Dominate on Large Scales?* As shown in Table S5.2, the electromagnetic force between two charged particles is much greater than the strength of gravity between them, no matter how far apart they are. Nevertheless, it is *gravity,* rather than the electromagnetic force, that dominates the universe on large scales. Briefly explain why.

4. *Quantum Tunneling and Life.* In one or two paragraphs, explain the role of quantum tunneling in creating the elements from which we are made. (*Hint:* Recall that we are *star stuff* in the sense that the elements of our bodies were produced by nuclear fusion inside stars.)

5. *Comparing Gravity and the EM Force.* In this problem, we compare the strength of gravity to the strength of the electromagnetic (EM) force for two interacting electrons. Since both electrons are negatively charged, they will want to *repel* each other according to the EM force; since electrons have mass, they will want to *attract* each other according to gravity. Let's see which effect will dominate. You will need the following information for this problem:

   - The force law for gravitation is:

   $$F_g = G\frac{M_1 M_2}{d^2} \quad \left(G = 6.67 \times 10^{-11}\,\frac{\text{N} \times \text{m}^2}{\text{kg}^2}\right)$$

   where M_1 and M_2 are the masses of the two objects, d is the distance between them, and G is the gravitational constant. ("N" is the abbreviation for *newtons,* the metric unit of force.)

   - The force law for electromagnetism is:

   $$F_{EM} = k\frac{q_1 q_2}{d^2} \quad \left(k = 9.0 \times 10^{9}\,\frac{\text{N} \times \text{m}^2}{\text{Coul}^2}\right)$$

   where q_1 and q_2 are the *charges* of the two objects (in *Coulombs,* the standard unit of charge), d is the distance between them, and k is a constant. ("Coul" is an abbreviation for *Coulomb.*)

   - The *mass* of an electron is 9.10×10^{-31} kg.
   - The *charge* of an electron is -1.6×10^{-19} Coul.

   a. Calculate the gravitational force, in *newtons,* that attracts the two electrons if they are separated by a distance of 10^{-10} meter (about the size of an atom).

   b. Calculate the electromagnetic force, in *newtons,* that repels the two electrons.

   c. How many times stronger is the electromagnetic repulsion than the gravitational attraction for the two electrons?

6. *Evaporation of Black Holes.* The time it takes for a black hole to evaporate through the process of Hawking radiation can be calculated using the following formula, in which M is the mass of the black hole in kilograms and t is the lifetime of the black hole in seconds:

   $$t = 10{,}240\,\pi^2 \frac{G^2 M^2}{hc^4} \quad \left(h = 6.63 \times 10^{-34}\,\frac{\text{kg} \times \text{m}^2}{\text{s}};\; G = 6.67 \times 10^{-11}\,\frac{\text{m}^3}{\text{kg} \times \text{s}^2}\right)$$

   a. Without doing any calculations, explain how this formula implies that lower-mass black holes have much shorter lifetimes than more massive ones and that the evaporation process accelerates as a black hole loses mass.

   b. What is the lifetime of a black hole with the mass of the Sun ($M_{Sun} = 2.0 \times 10^{30}$ kg)? How does this compare to the current age of the universe?

   c. In Chapter 22, we will see that some scientists speculate that the universe will eventually consist only of gigantic black holes and scattered subatomic particles. The largest black holes that conceivably might form would have a mass of about a trillion (10^{12}) Suns. Calculate the lifetime of such a giant black hole. (*Hint:* Your calculator probably will be unable to handle the large numbers involved in this problem; you will need to rearrange the numbers so that you can calculate the powers of 10 without your calculator.)

7. *Mini–Black Holes.* Some scientists speculate that black holes of many different masses might have been formed during the early moments of the Big Bang. Some of these black holes might be mini–black holes, much smaller in mass than those that can be formed by the crush of gravity in today's universe.
   a. Calculate the lifetime of a mini–black hole with the mass of the Earth (about 6×10^{24} kg). Compare this to the current age of the universe.
   b. Calculate the mass of a black hole that might have formed in the Big Bang and be completing the evaporation today. Compare this to the mass of the Earth. For this calculation, assume that the universe is 12 billion years old.

PART V
STELLAR ALCHEMY

14
Our Star

TODAY, ASTRONOMY INVOLVES THE STUDY OF THE ENtire universe, but the root of the word *astronomy* originally came from the Greek word for "star." Although we have learned a lot about the universe up to this point in the book, only now do we turn our attention to the study of the stars, the namesakes of our subject.

When we think of stars, we usually think of the beautiful points of light visible on a clear night. But the nearest and most easily studied star is visible only in the daytime—our Sun. Of course, the Sun is important to us in many more ways than as an object for astronomical study. The Sun is the source of virtually all light, heat, and energy reaching the Earth, and life on Earth's surface could not survive without it.

In this chapter, we will study the Sun in some depth. We will learn how the Sun makes life possible on Earth. Equally important, we will study our Sun as a star, so that in subsequent chapters we can more easily understand stars throughout the universe.

14.1 Why Does the Sun Shine?

Ancient peoples recognized the vital role of the Sun in their lives. Some worshiped the Sun as a god, and others created elaborate mythologies to explain its daily rise and set. Only recently, however, have we learned how the Sun provides us with light and heat.

Most ancient thinkers viewed the Sun as some type of fire, perhaps a lump of burning coal or wood. The Greek philosopher Anaxagoras (c. 500–428 B.C.) imagined the Sun to be a very hot, glowing rock about the size of the Greek peninsula of Peloponnesus (comparable in size to Massachusetts). His belief was probably influenced by his learning of a stony meteorite that fell on the shore of the Aegean Sea in 468 B.C. Anaxagoras concluded that the planets and stars must be flaming rocks in the sky, making him one of the first people in history to believe that the heavens and the Earth are made from the same types of materials.

By the mid-1800s, the size and distance of the Sun were reasonably well known, and scientists seriously began to address the question of how the Sun shines. Two early ideas held either that the Sun was a cooling ember that had once been much hotter or that the Sun generated energy from some type of chemical burning similar to the burning of coal or wood. Although simple calculations showed that a cooling or chemically burning Sun could shine for a few thousand years—an age that squared well with biblically based estimates of the age of the Earth that were popular at the time—these ideas suffered from fatal flaws. If the Sun were a cooling ember, it would have been much hotter just a few hundred years earlier, making it too hot for civilization to have existed. Chemical burning was ruled out because, when scientists examined the idea in detail, they found that such burning could not generate enough energy to account for the rate of radiation observed from the Sun's surface.

A more plausible hypothesis of the late 1800s suggested that the Sun generates energy by contracting in size, a process called **gravitational contraction.** If the Sun were shrinking, it would constantly be losing gravitational potential energy that could be converted into thermal energy, thereby keeping the Sun hot. Because of its large mass, the Sun would need to contract only very slightly each year to maintain its temperature—so slightly that the contraction would have been impossible to measure with nineteenth-century technology. Unfortunately for this model, calculations showed that the Sun could last no more than about 25 million years generating energy by gravitational contraction, even if it had started from an infinitely large size. Geologists of the late 1800s had already established the age of the Earth to be far older than 25 million years, leaving astronomers in the embarrassing position of being unable to explain how the Sun could shine for so long.

Only after Einstein published his special theory of relativity, which included his discovery of $E = mc^2$, did the true energy-generation mechanism of the Sun become clear. Although it took several decades to work out the details, we now know that the Sun generates energy by *nuclear fusion,* a source so efficient that the Sun can shine for 10 billion years. Because the Sun is only 4.6 billion years old today [Section 8.7], we expect it to keep shining for some 5 billion years to come.

Our current model of solar-energy generation by nuclear fusion means that the Sun's size is generally stable, maintained by a balance between the competing forces of gravity pulling inward and pressure pushing outward. This balance is called **gravitational equilibrium** (also referred to as *hydrostatic equilibrium*). It means that, at any point within the Sun, the weight of overlying material is supported by the underlying pressure. A stack of acrobats provides a simple example of this balance (Figure 14.1). The bottom person supports the weight of everybody above him, so the pressure on his body is very large.

FIGURE 14.1 An acrobat stack is in gravitational equilibrium: The lowest person supports the most weight and feels the greatest pressure, and the overlying weight and underlying pressure decrease for those higher up.

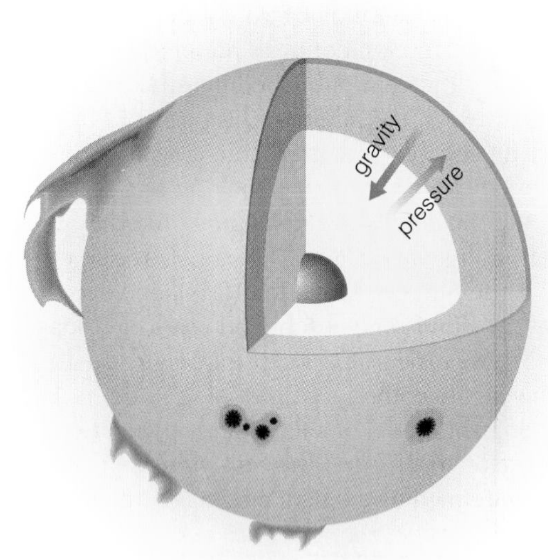

FIGURE 14.2 Gravitational equilibrium in the Sun: At each point inside, the pressure pushing up balances the weight of the overlying layers.

At each higher level, the overlying weight is less, so the pressure decreases. Gravitational equilibrium in the Sun means that the pressure increases with depth, making the Sun extremely hot and dense in its central core (Figure 14.2).

TIME OUT TO THINK ***The Earth's atmosphere is also in gravitational equilibrium, with the weight of upper layers supported by the pressure in lower layers. Use this idea to explain why the air gets thinner at higher altitudes.***

Interestingly, while gravitational contraction is not an important energy-generation mechanism in the Sun today, it was important in the distant past and will be important again in the distant future. Our Sun was born from a collapsing cloud of interstellar gas. The contraction of the cloud released gravitational potential energy, raising the interior temperature higher and higher. When the central temperature and density reached the values necessary to sustain nuclear fusion, the newly released energy provided enough pressure to halt the collapse. That is, the onset of fusion brought the Sun into gravitational equilibrium. About 5 billion years from now, when the Sun finally exhausts its nuclear fuel, the internal pressure will drop, and gravitational contraction will begin once again. As we will see later, some of the most important and spectacular processes in astronomy hinge on this ongoing "battle" between the crush of gravity and a star's internal sources of pressure.

Thus, the answer to the question "Why does the Sun shine?" is that about 4.6 billion years ago *gravitational contraction* made the Sun hot enough to sustain nuclear fusion in its core. Ever since, energy liberated by fusion has maintained the Sun's *gravitational equilibrium* and kept the Sun shining steadily, supplying the light and heat that sustain life on Earth.

14.2 Plunging to the Center of the Sun: An Imaginary Journey

In the rest of this chapter, we will discuss in detail how the Sun produces energy and how that energy travels to Earth. But first, to get a "big picture" view of the Sun, let's imagine you have a spaceship that can somehow withstand the immense heat and pressure of the solar interior, and take an imaginary journey from Earth to the center of the Sun.

As you begin your voyage from Earth, the Sun appears as a whitish ball of glowing gas. With spectroscopy [Section S2.3], you verify that the Sun is made of 70% hydrogen and 28% helium. Heavier elements make up the remaining 2%. The total power output of the Sun, called its **luminosity,** is an incredible 3.8×10^{26} watts. That is, every second, the Sun radiates a total of 3.8×10^{26} joules of energy into space (recall that 1 watt = 1 joule/s). If we could somehow capture and store just 1 second's worth of the Sun's luminosity, it would be enough to meet current human energy demands for roughly the next 500,000 years! Of course, only a tiny fraction of the Sun's total energy output reaches Earth, with the rest dispersing in all directions into space. Most of this energy is radiated in the form of visible light, but once you leave the protective blanket of the Earth's atmosphere you'll encounter significant amounts of other types of solar radiation, including dangerous ultraviolet and X rays. You'll need substantial shielding on your spaceship to prevent serious radiation burns from these high-energy forms of light.

Through a telescope, you can see that the Sun seethes with churning gases, and at most times you'll see at least a few **sunspots** blotching its surface (Figure 14.3). If you focus your telescope solely on a sunspot, you'll find that it is blindingly bright; sunspots appear dark only in contrast to the even brighter solar surface that surrounds them. A typical sunspot is large enough to swallow the entire Earth, dramatically illustrating that the Sun is immense by any earthly standard. The Sun's radius is nearly 700,000 kilometers, and its mass is 2×10^{30} kilograms—it is about 300,000 times more massive than the Earth. You can determine that the Sun is rotating by watching the sunspots day after day. If you watch very carefully, you may notice that sunspots near the solar equator circle the Sun faster than those at higher solar latitudes. This observation reveals that, unlike a spinning ball, the entire Sun does *not* rotate at the same rate. Instead, the solar equator completes one rotation in about 27 days, and the rotation period increases with latitude to about 31 days near the solar poles. Table 14.1 summarizes some of the basic properties of the Sun.

TIME OUT TO THINK *As a brief review, describe how we measure the mass of the Sun using Newton's version of Kepler's third law.* (Hint: *Look back at Chapter 6.*)

As you and your spaceship continue to fall toward the Sun, you notice an increasingly powerful headwind exerting a bit of drag on your descent. This headwind, called the **solar wind,** is created by the energy of photons and subatomic particles flowing outward

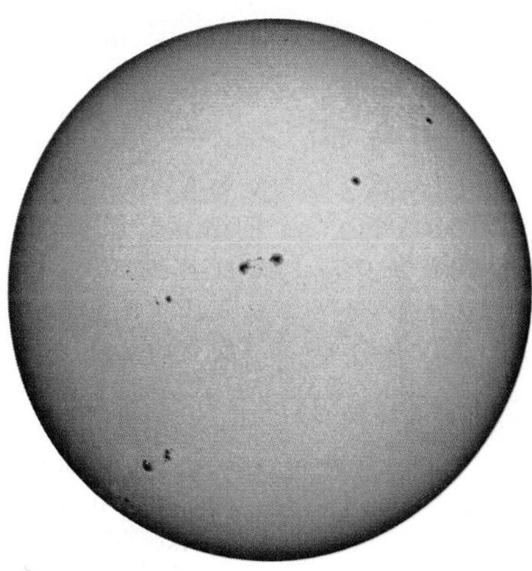

FIGURE 14.3 This photo of the visible surface of the Sun shows several dark sunspots.

from the solar surface. The solar wind helps shape the magnetospheres of planets and blows off the material that forms the tails of comets [Section 12.4].

A few million kilometers above the solar surface, you enter the solar **corona,** the tenuous uppermost layer of the Sun's atmosphere (Figure 14.4). Here you find the temperature to be astonishingly high—about 1 million Kelvin. This region emits most of the Sun's X rays. However, the density here is so low that your spaceship feels relatively little heat despite the million-degree temperature [Section 5.2]. Nearer the surface, the temperature suddenly drops to about 10,000 K in the **chromosphere,** the primary source of the Sun's ultraviolet radiation. At last, you plunge

Table 14.1 Basic Properties of the Sun

Radius (R_{Sun})	696,000 km (about 109 times the radius of the Earth)
Mass (M_{Sun})	2×10^{30} kg (about 300,000 times the mass of the Earth)
Luminosity (L_{Sun})	3.8×10^{26} watts
Composition (by percentage of mass)	70% hydrogen, 28% helium, 2% heavier elements
Rotation rate	27 days (equator) to 31 days (poles)
Surface temperature	5,800 K (average); 4,000 K (sunspots)
Core temperature	15 million K

through the visible surface of the Sun, called the **photosphere,** where the temperature averages just under 6,000 K. Although the photosphere looks like a well-defined surface from Earth, it consists of gas far less dense than the Earth's atmosphere.

Throughout the solar atmosphere, you notice that the Sun has its own version of weather, in which conditions at a particular altitude differ from one region to another. Some regions of the chromosphere and corona are particularly hot and bright, while other regions are cooler and less dense. In the photosphere, sunspots are cooler than the surrounding surface, though they are still quite hot and bright by earthly standards. In addition, your compass goes crazy as you descend through the solar atmosphere, indicating that solar weather is shaped by intense magnetic fields. Occasionally, huge magnetic storms occur, shooting hot gases high into space.

Up to this point in your journey, you have seen the Earth and the stars behind you, but as you slip beneath the photosphere, blazing light engulfs you. You are inside the Sun, and your spacecraft is tossed about by incredible turbulence. If you can hold steady long enough to see what is going on around you, you'll notice spouts of hot gas rising upward, surrounded by cooler gas cascading down from above. You are in the **convection zone,** where energy generated in the solar core travels upward, transported by the rising of hot gas and falling of cool gas called *convection* [Section 9.3]. With some quick thinking, you may realize that the photosphere above you is the top of the convection zone and that convection is the cause of the Sun's seething, churning appearance.

As you descend through the convection zone, the surrounding density and pressure increase substantially, along with the temperature. Soon you reach depths at which the Sun is far denser than water; nevertheless, it is still a *gas*—or, more specifically, a *plasma* of positively charged ions and free electrons—

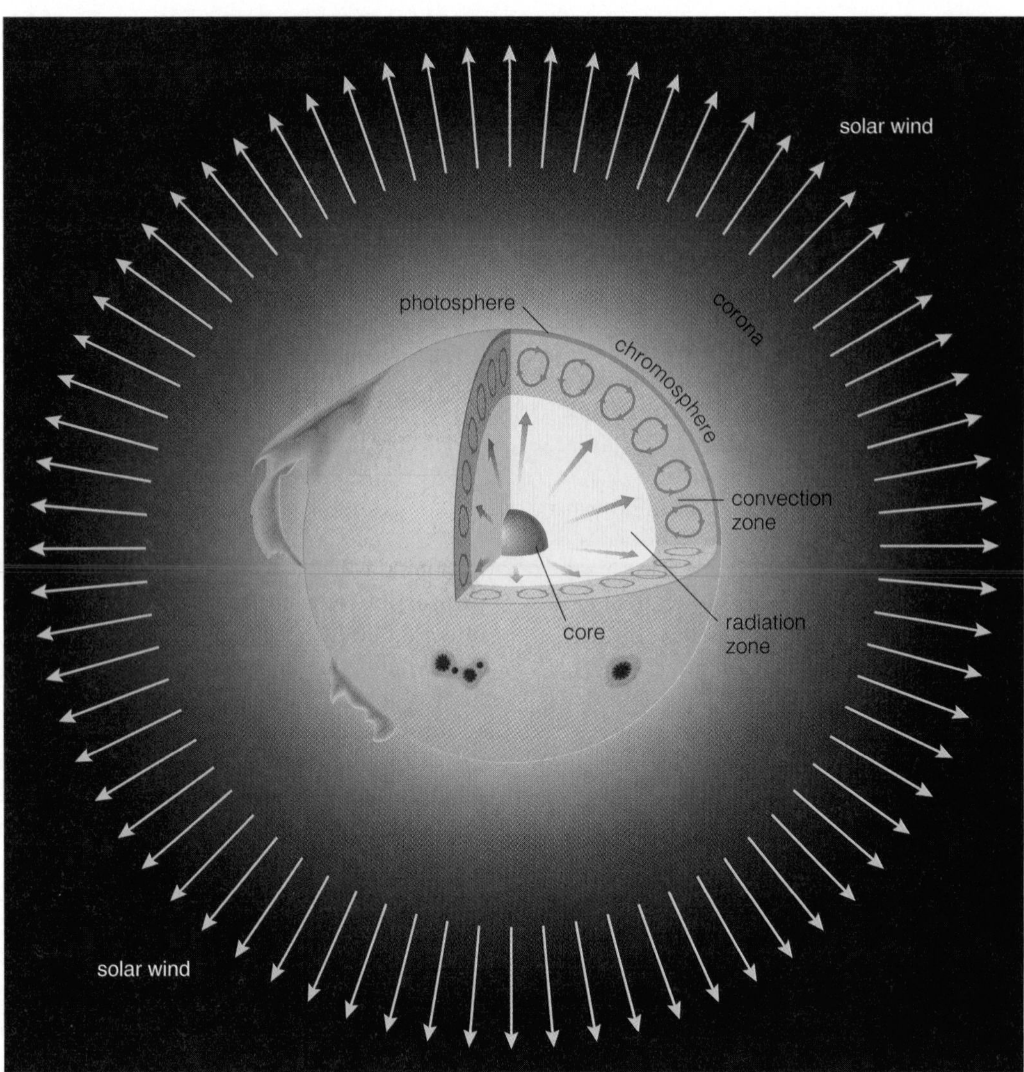

FIGURE 14.4 The basic structure of the Sun.

because each particle moves independently of its neighbors [Section 5.3].

About a third of the way down to the center, the turbulence of the convection zone gives way to the stabler plasma of the **radiation zone,** where energy is carried outward primarily by photons of light. The temperature is now almost 10 million K, and your spacecraft is bathed in X rays trillions of times more intense than the visible light at the solar surface. No real spacecraft could survive, but your imaginary one keeps plunging straight down to the solar **core.** There you finally find the source of the Sun's energy: nuclear fusion transforming hydrogen into helium. At the Sun's center, the temperature is about 15 million K, the density is more than 100 times that of water, and the pressure is 200 billion times that on the surface of the Earth.

With your journey complete, it's time to turn around and head back home. We'll continue this chapter by studying fusion in the solar core and then tracing the flow of the energy generated by fusion as it moves outward through the Sun.

14.3 The Cosmic Crucible

The prospect of turning common metals like lead into gold enthralled medieval alchemists. Sometimes they tried primitive scientific approaches, such as melting various ores together in a vessel called a crucible. Other times they tried magic. But their get-rich-quick schemes never managed to work. Today we know that there is no easy way to turn other elements into gold, but it *is* possible to transmute one element or isotope into another. If a nucleus gains or

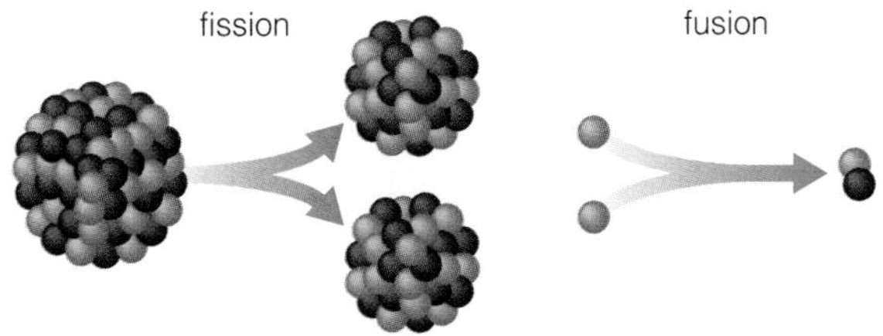

FIGURE 14.5 Nuclear fission splits a nucleus into smaller pieces, while nuclear fusion combines smaller nuclei into a larger nucleus.

loses protons, its atomic number changes and it becomes a different element. If it gains or loses neutrons, its atomic weight changes and it becomes a different isotope [Section 5.3]. The process of splitting a nucleus so that it loses protons or neutrons is called **nuclear fission.** The process of combining nuclei to make a nucleus with a greater number of protons or neutrons is called **nuclear fusion** (Figure 14.5). Human-built nuclear power plants rely on nuclear fission of uranium or plutonium. The nuclear power plant at the center of the Sun relies on nuclear fusion, turning hydrogen into helium.

Nuclear Fusion

The 15-million-K plasma in the solar core is like a "soup" of hot gas, with bare, positively charged atomic nuclei (and individual negatively charged electrons) whizzing about at extremely high speeds. At any one time, some of these nuclei are on high-speed collision courses with each other. In most cases, electromagnetic forces deflect the nuclei, preventing actual collisions, because positive charges repel one another. However, if nuclei collide with sufficient energy, they can stick together to form a heavier nucleus (Figure 14.6).

Sticking positively charged nuclei together is not easy. The **strong force,** which binds protons and neutrons together in atomic nuclei, is the only force in nature that can overcome the electromagnetic repulsion between two positively charged nuclei [Section S5.2]. In contrast to gravitational and electromagnetic forces, which drop off gradually as the distances between particles increase (by an inverse square law), the strong force is more like glue or Velcro: It overpowers the electromagnetic force over very small distances but is insignificant when the distances between particles exceed the typical sizes of atomic nuclei. The trick to nuclear fusion therefore is to push the positively charged nuclei close enough together for the strong force to outmuscle electromagnetic repulsion.

The high pressures and temperatures in the solar core are just right for fusion of hydrogen nuclei into helium nuclei. The high temperature is important because the nuclei must collide at very high speeds if they are to come close enough together to fuse. (Quantum tunneling is also important to this process [Section S5.5].) The higher the temperature, the harder the collisions, making fusion reactions more likely at higher temperatures. The high pressure of the overlying layers is necessary because, without it, the hot plasma of the solar core would simply explode into space, shutting off the nuclear reactions. In the Sun, the pressure is high and steady, allowing some 600 million tons of hydrogen to fuse into helium every second.

Incidentally, this tendency of nuclear fusion to turn itself off unless it occurs under very high and steady pressure is what makes nuclear fusion reactions so difficult to sustain on Earth. Scientists have tried for decades to overcome this difficulty so that we might use fusion as a source of energy. In many ways, fusion would be an ideal energy source. The fuel, hydrogen, is readily available in water, and the primary by-product, helium, is a harmless gas.[1] Unfortunately, laboratory fusion reactions always turn themselves off quickly, making it difficult to generate a commercially useful amount of energy in a controlled fashion. Even in a hydrogen bomb, fusion

[1] Fusion reactors will produce some nuclear waste, because high-energy neutrons generated in the reactions can create radioactive isotopes in the reactor walls.

FIGURE 14.6 Positively charged nuclei can fuse only if a high-speed collision brings them close enough for the strong nuclear force to come into play.

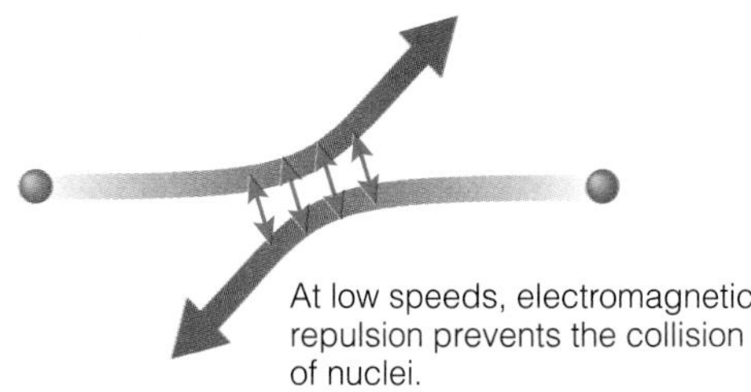

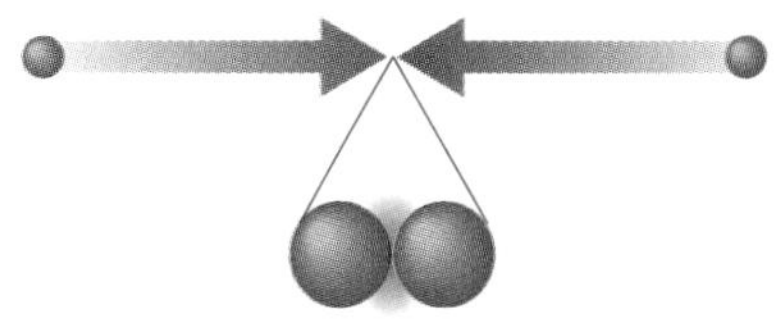

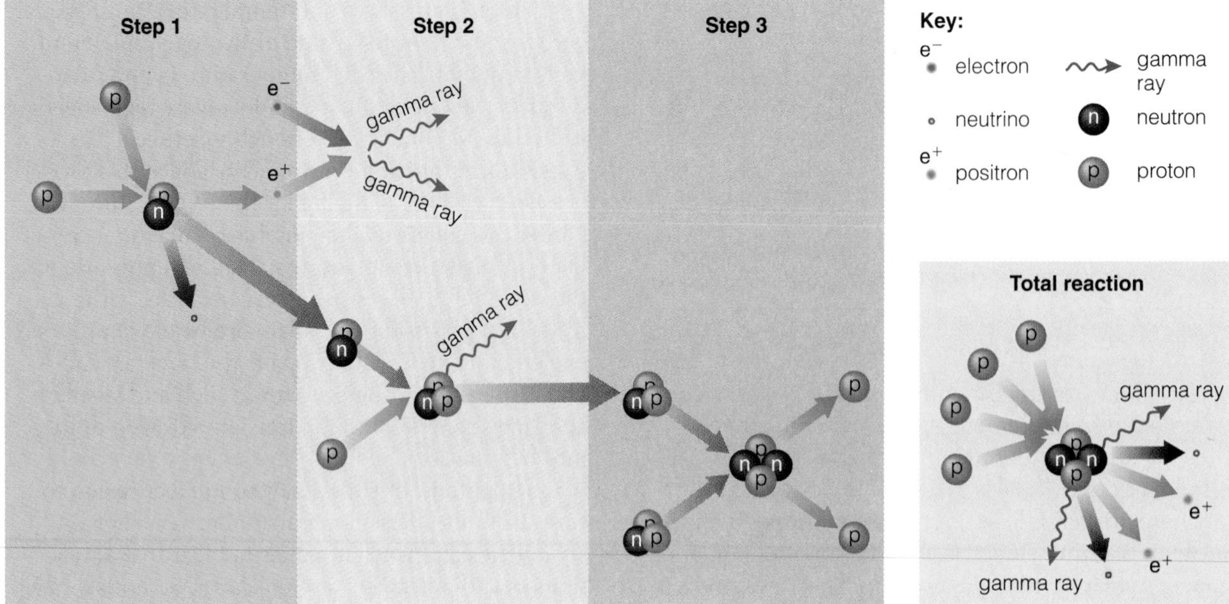

FIGURE 14.7 Hydrogen fuses into helium in the Sun by way of the proton–proton chain. In Step 1, two protons fuse to create a deuterium nucleus consisting of a proton and a neutron. In Step 2, the deuterium nucleus and a proton fuse to form helium-3, a rare form of helium. In Step 3 , two helium-3 nuclei fuse to form helium-4, the common form of helium.

happens only briefly. When nuclear burning ignites, the pressure inside the bomb builds rapidly. Because this pressure cannot be contained, the bomb blows apart with incredibly destructive force.

Hydrogen Fusion in the Sun: The Proton–Proton Chain

Recall that hydrogen nuclei are nothing more than individual protons, while the most common form of helium consists of two protons and two neutrons. Thus, the overall hydrogen fusion reaction in the Sun is:

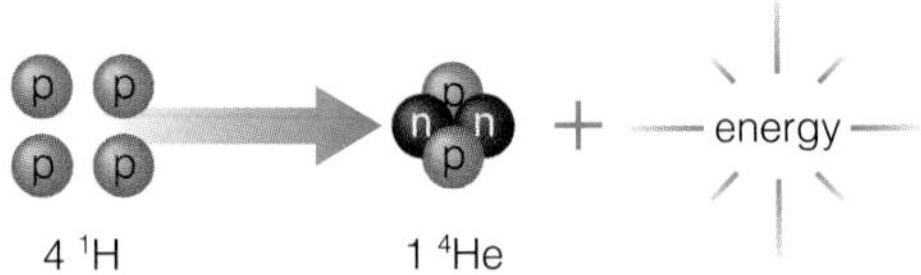

However, collisions between two nuclei are far more common than three- or four-way collisions, so this overall reaction proceeds through steps that involve just two nuclei at a time. The sequence of steps that occurs in the Sun is called the **proton–proton chain** because it begins with collisions between individual protons (hydrogen nuclei); it is illustrated in Figure 14.7.

Step 1. Two protons fuse to form a nucleus consisting of one proton and one neutron, which is the isotope of hydrogen known as *deuterium.* Note that this step converts a proton into a neutron, reducing the total nuclear charge from $+2$ for the two fusing protons to $+1$ for the resulting deuterium nucleus. The lost positive charge is carried off by a *positron,* the antimatter version of an electron with a positive rather than negative charge [Section S5.2]. A *neutrino*—a minuscule particle with a very tiny mass—is also produced in this step.[2] The positron won't last long; it soon meets up with an ordinary electron, resulting in the creation of two gamma-ray photons through matter–antimatter annihilation.

Step 2. A fair number of deuterium nuclei are always present along with the protons and other nuclei in the solar core, since step 1 occurs so frequently in the Sun (about 10^{38} times per second). Step 2 occurs when one of these deuterium nuclei collides and fuses with a proton. The result is a nucleus of helium-3, a rare form of he-

[2]Producing a neutrino is necessary because of a law called *conservation of lepton number:* The number of leptons (e.g., electrons or neutrinos [see Chapter S5]) must be the same before and after the reaction. The lepton number is zero before the reaction because there are no leptons. Among the reaction products, the positron (antielectron) has lepton number -1 because it is antimatter, and the neutrino has lepton number $+1$. Thus, the total lepton number remains zero.

lium with two protons and one neutron. This reaction also produces a gamma-ray photon.

Step 3. The third and final step of the proton–proton chain requires the addition of another neutron to the helium-3, thereby making normal helium-4. This final step can proceed in several different ways, but the most common route involves a collision of two helium-3 nuclei. Each of these helium-3 nuclei resulted from a prior, separate occurrence of step 2 somewhere in the solar core. The final result is a normal helium-4 nucleus and two protons.

Total reaction. Note that, somewhere in the solar core, steps 1 and 2 must each occur twice to make step 3 possible. Six protons go into each complete cycle of the proton–proton chain, but two come back out. Thus, the overall proton–proton chain converts four protons (hydrogen nuclei) into a helium-4 nucleus, two positrons, two neutrinos, and two gamma rays.

Each resulting helium-4 nucleus has a mass that is slightly less (by about 0.7%) than the combined mass of the four protons that created it. Overall, fusion in the Sun converts about 600 million tons of hydrogen into 596 million tons of helium every second; the "missing" 4 million tons of matter becomes energy in accord with Einstein's formula $E = mc^2$. About 2% of this energy is carried off by the neutrinos. Neutrinos rarely interact with matter (because they respond only to the weak force [Section S5.2]), so most of the neutrinos created by the proton–proton chain pass unscathed from the solar core, through the solar surface, and out into space. The rest of the energy emerges as kinetic energy of the nuclei and radiative energy of the gamma rays. As we will see shortly, this energy slowly percolates to the solar surface, eventually emerging as the sunlight that bathes the Earth.

The Solar Thermostat

The rate of nuclear fusion in the solar core, which determines the energy output of the Sun, is very sensitive to temperature. A slight increase in temperature would mean a much higher fusion rate, and a slight decrease in temperature would mean a much lower fusion rate. If the Sun's rate of fusion varied erratically, the effects on the Earth might be devastating. Fortunately, the Sun's central temperature is steady, thanks to gravitational equilibrium—the balance between the pull of gravity and the push of internal pressure.

Suppose that, for some reason, the core temperature rose very slightly. The rate of nuclear fusion would soar, generating excess energy that would increase the pressure. The push of this pressure would temporarily exceed the pull of gravity, causing the core to expand and cool. With this cooling, the fusion rate would drop back down. The expansion and cooling would continue until gravitational equilibrium was restored, at which point the fusion rate would be back at its original value. An opposite process restores the normal fusion rate if the core temperature drops. A decrease in core temperature would lead to decreased nuclear burning, a drop in the central pressure, and contraction of the core. As the core shrank, its temperature would rise until the burning rate returned to normal.

The response of the core pressure to changes in the nuclear fusion rate is essentially a *thermostat* that keeps the Sun's central temperature steady. Any random change in the core temperature is automatically corrected by the change in the fusion rate and the accompanying change in pressure.

While the processes involved in gravitational equilibrium prevent erratic changes in the fusion rate, they also ensure that the fusion rate gradually rises over billions of years. Because each fusion reaction converts *four* hydrogen nuclei into *one* helium nucleus, the total number of *independent particles* in the solar core is gradually falling. This gradual reduction in the number of particles causes the solar core to shrink in the same way that a balloon deflates when you let out air. The slow shrinking of the solar core means that it must generate energy more rapidly to counteract the stronger compression of gravity, so the solar core gradually gets hotter as it shrinks. Theoretical models indicate that the Sun's core temperature should have increased enough to raise its fusion rate and the solar luminosity by about 30% since the Sun was born 4.6 billion years ago.

The gradual increase in the solar luminosity poses a bit of a puzzle, because geological evidence shows that the Earth's temperature has remained fairly steady since the Earth finished forming over 4 billion years ago. How has the Earth maintained its temperature while the Sun's energy output has increased by some 30%? Apparently, the Earth has its own thermostat that keeps the surface temperature steady even as the Sun increases in brightness. This "Earth thermostat" is probably related to the greenhouse effect [Section 13.3], which may have been much stronger in the distant past than it is today. Studying how the Sun changes with time and how the Earth responds to these changes may teach us much about how the Earth's physical and biological systems regulate the greenhouse effect.

"Observing" the Solar Interior

We cannot see inside the Sun, so you may be wondering how we can know so much about what goes on underneath its surface. The primary way we learn about the interior of the Sun and other stars is by creating *mathematical models* that use the laws of physics to predict the internal conditions. A basic model uses the Sun's observed composition and mass as inputs to equations that describe gravitational equilibrium and the solar thermostat. With the aid of a computer, we can use the model to calculate the Sun's temperature, pressure, and density at any depth. We can then predict the rate of nuclear fusion in the solar core by combining these calculations with knowledge about nuclear fusion gathered in laboratories here on Earth. Remarkably, such models correctly "predict" the radius, surface temperature, luminosity, age, and many other properties of the Sun. However, current models do not predict *everything* about the Sun correctly, so scientists are constantly working to discover what is missing from them. The fact that the models successfully predict so many observed characteristics of the Sun gives us confidence that they are on the right track and that we really do understand what is going on inside the Sun.

A second way to learn about the inside of the Sun is to observe "sun quakes"—vibrations of the Sun that are similar to the vibrations of the Earth caused by earthquakes, although they are generated very differently. Earthquakes occur when the Earth's crust suddenly shifts, generating *seismic waves* that propagate through the Earth's interior [Section 13.2]. We can learn about the Earth's interior by recording seismic waves on the Earth's surface with seismographs. Sun quakes result from waves of pressure (sound waves) that propagate deep within the Sun at all times. These waves cause the solar surface to vibrate when they reach it. Although we cannot set up seismographs on the Sun, we can detect the vibrations of the surface by measuring Doppler shifts [Section 7.5]: Light from portions of the surface that are rising toward us is slightly blueshifted, while light from portions that are falling away from us is slightly redshifted. The vibrations are relatively small but measurable (Figure 14.8). In principle, we can deduce a great deal about the solar interior by carefully analyzing these vibrations. (By analogy to seismology on Earth, this type of study of the Sun is called *helioseismology* [recall that *helios* means "sun"].) Results to date confirm that mathematical models of the solar interior are on the right track and at the same time provide data that can be used to improve the models further.

The Solar Neutrino Problem

There is one other way to study the solar interior at present, and that is to observe the neutrinos coming from fusion reactions in the core. Don't panic, but as you read this sentence about a thousand trillion solar neutrinos will zip through your body. Fortunately, they won't do any damage, because neutrinos rarely interact with anything. Neutrinos created by fusion in the solar core fly quickly through the Sun as if passing through empty space. In fact, while an inch

Mathematical Insight 14.1 Mass–Energy Conversion in the Sun

We can calculate how much mass the Sun loses through nuclear fusion by comparing the input and output masses of the proton–proton chain. A single proton has a mass of 1.6726×10^{-27} kg, so four protons have a mass of 6.690×10^{-27} kg.

A helium-4 nucleus has a mass of only 6.643×10^{-27} kg, slightly less than the mass of the four protons. The difference is:

$$6.690 \times 10^{-27}\text{ kg} - 6.643 \times 10^{-27}\text{ kg} = 4.7 \times 10^{-29}\text{ kg}$$

which is 0.7%, or 0.007, of the original mass. Thus, for example, when 1 kilogram of hydrogen fuses, the resulting helium weighs only 993 grams, while 7 grams of mass turn into energy.

To calculate the *total* amount of mass converted to energy in the Sun each second, we use Einstein's equation $E = mc^2$. The total energy produced by the Sun each second is 3.8×10^{26} joules, so we can solve for the total mass converted to energy each second:

$$E = mc^2 \Rightarrow m = \frac{E}{c^2} = \frac{3.8 \times 10^{26}\text{ joules}}{\left(3.0 \times 10^8 \frac{\text{m}}{\text{s}}\right)^2} = 4.2 \times 10^9\text{ kg}$$

FIGURE 14.8 Vibrations on the surface of the Sun can be detected by Doppler shifts. In this schematic representation, red indicates falling gas, and blue indicates rising gas.

of lead will stop an X ray, stopping an average neutrino would require a slab of lead more than a light-year thick! Clearly, counting neutrinos is dauntingly difficult, because they stream right through any detector built to capture them.

TIME OUT TO THINK *Is the number of solar neutrinos zipping through our bodies significantly lower at night?* (Hint: *How does the thickness of the Earth compare with the thickness of a slab of lead needed to stop an average neutrino?*)

Nevertheless, neutrinos *do* occasionally interact with matter, and it is possible to capture a few solar neutrinos with a large enough detector. Neutrino detectors are usually placed deep inside mines so that the overlying layers of rock block all other kinds of particles coming from outer space—but not neutrinos, which pass through rock easily. The first major solar neutrino detector was built in the 1960s and was located 1,500 meters underground in the Homestake gold mine in South Dakota (Figure 14.9). The detector for this "Homestake experiment" consisted of a 400,000-liter vat of chlorine-containing dry-cleaning fluid. It turns out that, on very rare occasions, a chlorine nucleus can capture a neutrino and change into a nucleus of radioactive argon. By looking for radioactive argon in the tank of cleaning fluid, experimenters could count the number of neutrinos captured in the detector.

Remarkably, from the many trillions of solar neutrinos that passed through the tank of cleaning fluid each second, experimenters expected to capture an average of just one neutrino per day. This predicted capture rate was based on measured properties of chlorine nuclei and models of nuclear fusion in the Sun. However, over a period of more than two decades, neutrinos were captured only about once every 3 days on average. That is, the Homestake experiment detected only about one-third of the predicted

The Sun loses about 4 billion kilograms of mass every second, which is roughly equivalent to the combined mass of nearly 100 million people.

Example: How much hydrogen is converted to helium each second in the Sun?

Solution: We have already calculated that the Sun loses 4.2×10^9 kg of mass each second and that this is only 0.7% of the mass of hydrogen that is fused:

$$4.2 \times 10^9 \text{ kg} = 0.007 \times (\text{mass of hydrogen fused})$$

We now solve for the mass of hydrogen fused:

$$\begin{aligned} \text{mass of hydrogen fused} &= \frac{4.2 \times 10^9 \text{ kg}}{0.007} \\ &= 6.0 \times 10^{11} \text{ kg} \times \frac{1 \text{ metric ton}}{10^3 \text{ kg}} \\ &= 6.0 \times 10^8 \text{ metric tons} \end{aligned}$$

The Sun fuses 600 million tons of hydrogen each second, of which about 4 million tons becomes energy, and the remaining 596 million tons becomes helium.

FIGURE 14.9 This tank of dry-cleaning fluid (visible underneath the catwalk), located deep within South Dakota's Homestake mine, was a solar neutrino detector. The chlorine nuclei in the cleaning fluid turned into argon nuclei when they captured neutrinos from the Sun.

number of neutrinos. This disagreement between model predictions and actual observations is called the **solar neutrino problem.**

TIME OUT TO THINK *Although the observed number of neutrinos falls short of theoretical predictions, experiments like Homestake prove that at least some neutrinos are coming from the Sun. Explain why this provides direct evidence that nuclear fusion really is taking place in the Sun right now.* (Hint: *See Figure 14.7.*)

The shortfall of neutrinos found with the Homestake experiment has led to many more recent attempts to detect solar neutrinos using more sophisticated detectors (Figure 14.10). The chlorine nuclei in the Homestake experiment could capture only high-energy neutrinos that are produced by one of the rare pathways of step 3 in the proton–proton chain. More recent experiments can detect lower-energy neutrinos, including those produced by step 1 of the proton–proton chain, and therefore offer a better probe of fusion in the Sun. To date, all these experiments have still found fewer neutrinos than current models of the Sun predict. This discrepancy between model and experiment probably means one of two things: Either something is wrong with our models of the Sun, or something is missing in our understanding of how neutrinos behave.

FIGURE 14.10 The Super-Kamiokande neutrino detector in Japan.

For the moment, many physicists and astronomers are betting that we understand the Sun just fine and that the discrepancy has to do with the neutrinos themselves. One intriguing idea arises from the fact that neutrinos come in three types: electron neutrinos, muon neutrinos, and tau neutrinos [Section S5.2]. Fusion reactions in the Sun produce only electron neutrinos, and most solar neutrino detectors can detect only electron neutrinos. In the standard model of particle physics, neutrinos cannot change type. However, a variant of this model predicts that some electron neutrinos can change into muon and tau neutrinos as they fly out through the solar plasma. In that case, our detectors would count fewer than the expected number of electron neutrinos because some of them have turned into muon or tau neutrinos on their journey from the solar core to the Earth. We won't know the verdict until further experiments are conducted, at which point either we will learn how neutrinos really behave, or we will have to revise our models of the Sun's core. Our understanding of nuclear fusion in the Sun will remain incomplete until we solve the puzzling mystery of the solar neutrinos.

14.4 From Core to Corona

Energy liberated by nuclear fusion in the Sun's core must eventually reach the solar surface, where it can be radiated into space. The path that the energy takes to the surface is long and complex. In this section, we follow the long path that energy travels after its production in the core.

The Path Through the Solar Interior

In Chapter 7, we discussed how atoms can absorb or emit photons. In fact, photons can also interact with any charged particle, and a photon that "collides" with an electron can be deflected into a completely new direction. Deep in the solar interior, the plasma is so dense that the gamma-ray photons resulting from fusion travel only a fraction of a millimeter before colliding with an electron. Because each collision sends the photon in a random new direction, the photon bounces around the core in a haphazard way, sometimes called a *random walk* (Figure 14.11). With each random bounce, the photon drifts farther and farther, on average, from its point of origin. As a result, photons from the solar core gradually work their way outward. The technical term for this slow, outward migration of photons is **radiative diffusion** (to *diffuse* means to "spread out"; *radiative* refers to the photons of light or radiation). Along the way, the photons exchange energy with their surroundings; because the surrounding temperature declines as the photons move outward through the Sun, they are gradually transformed from gamma rays to photons of lower energy. By the time the energy of fusion reaches the surface, the photons are primarily visible light. On average, the energy released in a fusion reaction takes about a million years to reach the solar surface.

TIME OUT TO THINK *Radiative diffusion is just one type of diffusion. Another is the diffusion of dye through a glass of water. If you place a concentrated spot of dye at one point in the water, each individual dye molecule begins a random walk as it bounces among the water molecules. The result is that the dye gradually spreads through the entire glass. Can you think of any other examples of diffusion in the world around you?*

Radiative diffusion is the primary way by which energy moves outward through the *radiation zone*, which stretches from the core to about two-thirds of the Sun's radius (see Figure 14.4). Above this point, where the temperature has dropped to about 2 million K, the solar plasma absorbs photons more readily (rather than just bouncing them around). This is the beginning of the solar *convection zone*, where the build-up of heat resulting from photon absorption causes bubbles of hot plasma to rise upward in the process known as **convection** [Section 9.3]. Convection occurs because hot gas is less dense than cool gas. Like a hot-air balloon, a hot bubble of solar plasma rises upward through the cooler plasma above it. Meanwhile, cooler plasma from above slides around the rising bubble and sinks to lower layers, where it is heated. The rising of hot plasma and sinking of cool plasma form a cycle that transports energy outward from the top of the radiation zone to the solar surface (Figure 14.12).

FIGURE 14.11 A photon in the solar interior bounces randomly among electrons, slowly working its way outward in a process called radiative diffusion.

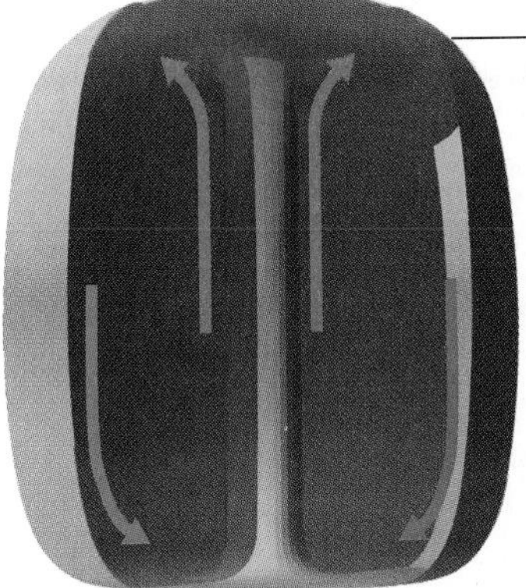

FIGURE 14.12 Convection transports energy outward in the Sun's convection zone. This schematic diagram shows how a column of hot gas rises while cooler gas descends around it.

The Solar Surface

The Earth has a solid crust, so its surface is well defined. In contrast, the Sun is made entirely of gaseous plasma. Defining where the surface of the Sun begins is therefore something like defining the surface of a cloud: From a distance it looks quite distinct, but up close the surface is fuzzy, not sharp. We generally define the solar surface as the layer that appears distinct from a distance. This is the layer we identified as the *photosphere* when we took our imaginary journey into the Sun. More technically, the photosphere is the layer of the Sun from which photons finally escape into space after the million-year journey of solar energy outward from the core.

Most of the energy produced by fusion in the solar core ultimately leaves the photosphere as thermal radiation [Section 7.4]. The average temperature of the photosphere is about 5,800 K, corresponding to a thermal radiation spectrum that peaks in the green portion of the visible spectrum. In our sky, the Sun appears somewhat more yellow—and even red at sunset—because the Earth's atmosphere scatters blue light. It is this scattered light from the Sun that makes our skies blue [Section 10.2].

Although the average temperature of the photosphere is 5,800 K, actual temperatures vary significantly from place to place. The photosphere is marked throughout by the bubbling pattern of **granulation** produced by the underlying convection (Figure 14.13). Each *granule* appears bright in the center, where hot gas bubbles upward, and dark around the edges, where cool gas descends. If we made a movie of the granulation, we'd see it bubbling rather like a pot of boiling water. Just as bubbles in a pot of boiling water burst on the surface and are replaced by new bubbles, each granule lasts only a few minutes before being replaced by other granules bubbling upward.

FIGURE 14.13 Granulation of the photosphere is evident in this photo of two sunspots. Each white granule is the top of a rising column of hot gas. At the darker lines between the granules, cooler gas is descending below the photosphere.

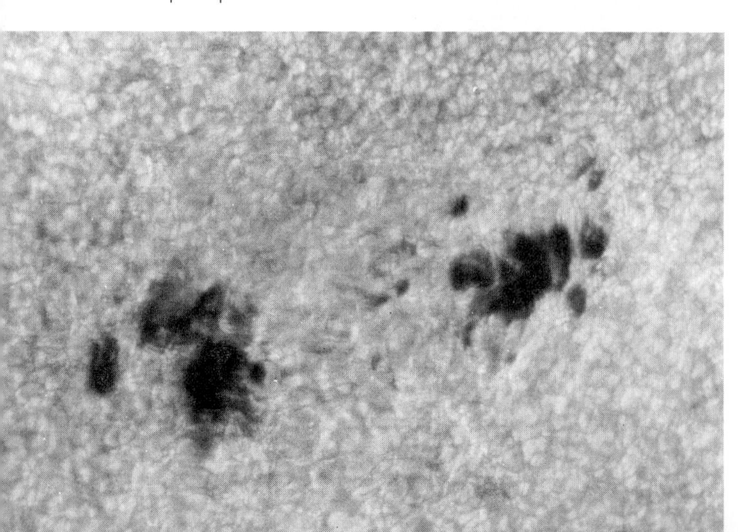

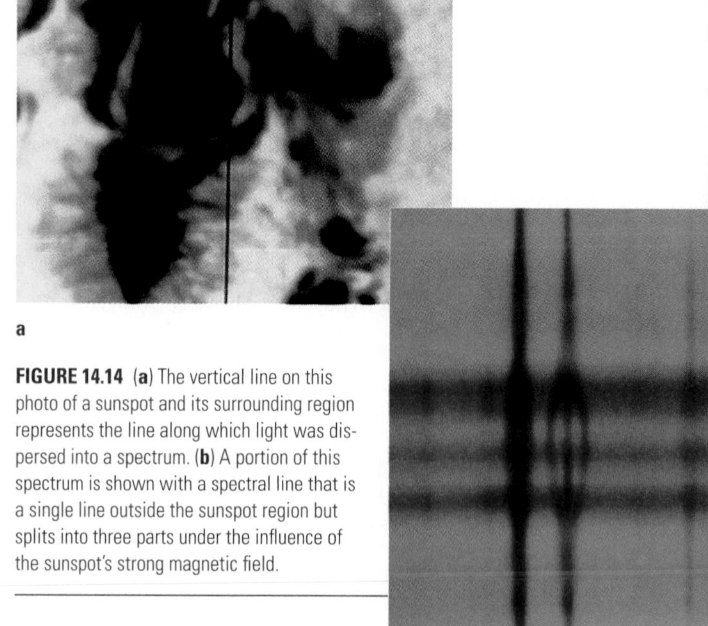

FIGURE 14.14 (**a**) The vertical line on this photo of a sunspot and its surrounding region represents the line along which light was dispersed into a spectrum. (**b**) A portion of this spectrum is shown with a spectral line that is a single line outside the sunspot region but splits into three parts under the influence of the sunspot's strong magnetic field.

Sunspots and Magnetic Fields

Sunspots are the most striking features on the solar surface. The temperature of the plasma in sunspots is about 4,000 K, which is significantly cooler than the 5,800-K plasma of the surrounding photosphere. If you think about this for a moment, you may wonder how sunspots can be so much cooler than their surroundings. What keeps the surrounding hot plasma from heating up the sunspots?

Something must be preventing hot plasma from entering the sunspots, and that "something" turns out to be magnetic fields. Strong magnetic fields can alter the energy levels in atoms and ions and therefore can alter the spectral lines they produce. More specifically, magnetic fields cause some spectral lines to split into two or more closely spaced lines[3] (Figure 14.14). Thus, scientists can map magnetic fields on the Sun by studying the spectral lines in light from different parts of the solar surface. Such maps reveal sunspots to be regions of intense magnetic fields.

Magnetic fields are invisible, but in principle we could visualize a magnetic field by laying out many compasses; each compass needle would point to local magnetic north. We can represent the magnetic field by drawing a series of lines, called **magnetic field lines,** connecting the needles of these imaginary compasses (Figure 14.15a). The strength of the magnetic field is indicated by the spacing of the lines: Closer lines mean a stronger field (Figure 14.15b).

[3]This splitting of spectral lines due to magnetic fields is called the *Zeeman effect.*

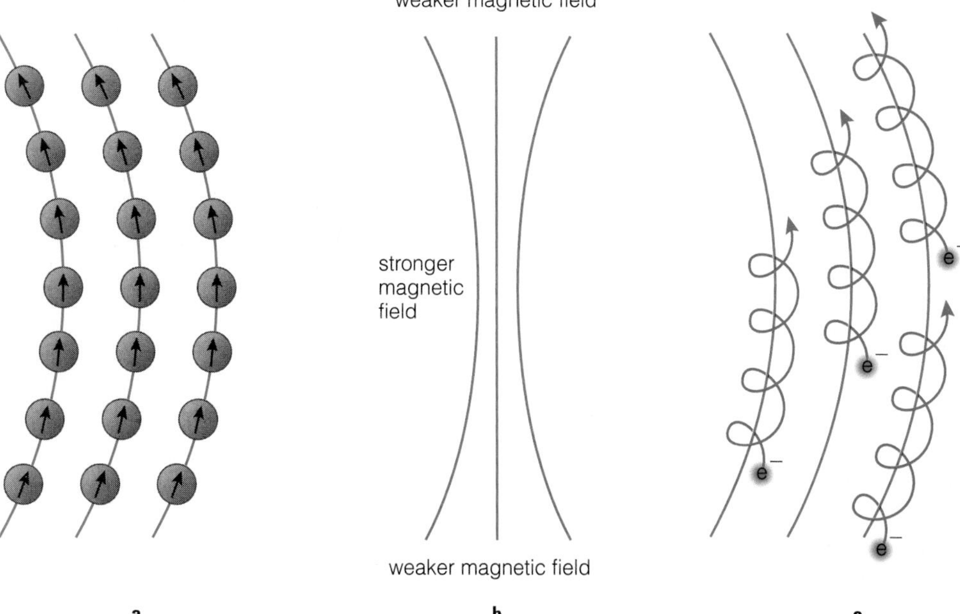

FIGURE 14.15 (**a**) Magnetic field lines follow the directions that compass needles would point. (**b**) Lines closer together indicate a stronger field. (**c**) Charged particles follow paths that spiral along magnetic field lines

Because these imaginary field lines are so much easier to visualize than the magnetic field itself, we usually discuss magnetic fields by talking about how the field lines would appear. Charged particles, such as the ions and electrons in the solar plasma, follow paths that spiral along the magnetic field lines (Figure 14.15c). Thus, the solar plasma can move freely *along* magnetic field lines but cannot easily move perpendicular to them.

The magnetic field lines act somewhat like elastic bands, twisted into contortions and knots by turbulent motions in the solar atmosphere. Sunspots occur where the most taut and tightly wound magnetic fields poke nearly straight out from the solar interior (Figure 14.16). Note that sunspots tend to occur in pairs connected by a loop of magnetic field lines. These tight magnetic field lines suppress convection within each sunspot and prevent surrounding plasma from sliding sideways into the sunspot. With hot plasma unable to enter the region, the sunspot plasma becomes cooler than that of the rest of the photosphere.

The magnetic field lines connecting two sunspots sometimes soar high above the photosphere, through the chromosphere, and into the corona. These vaulted loops of magnetic field sometimes appear as **solar prominences,** in which the field traps gas that may glow

FIGURE 14.16 Pairs of sunspots are connected by tightly wound magnetic field lines.

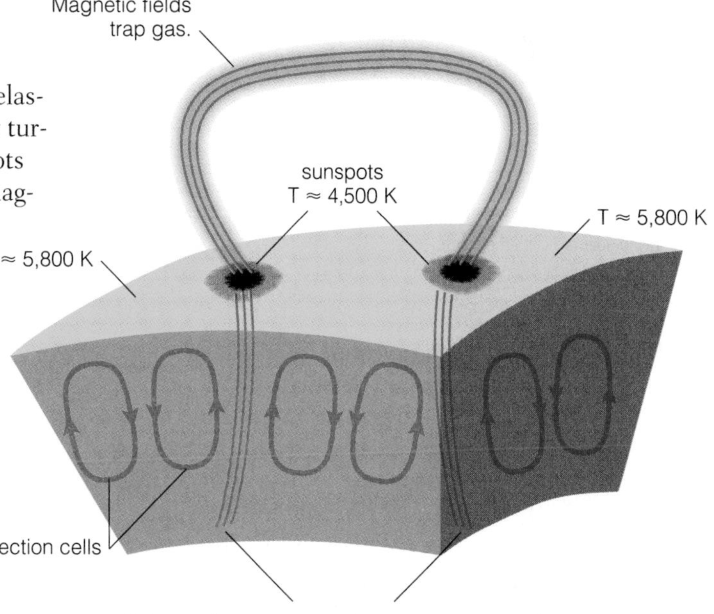

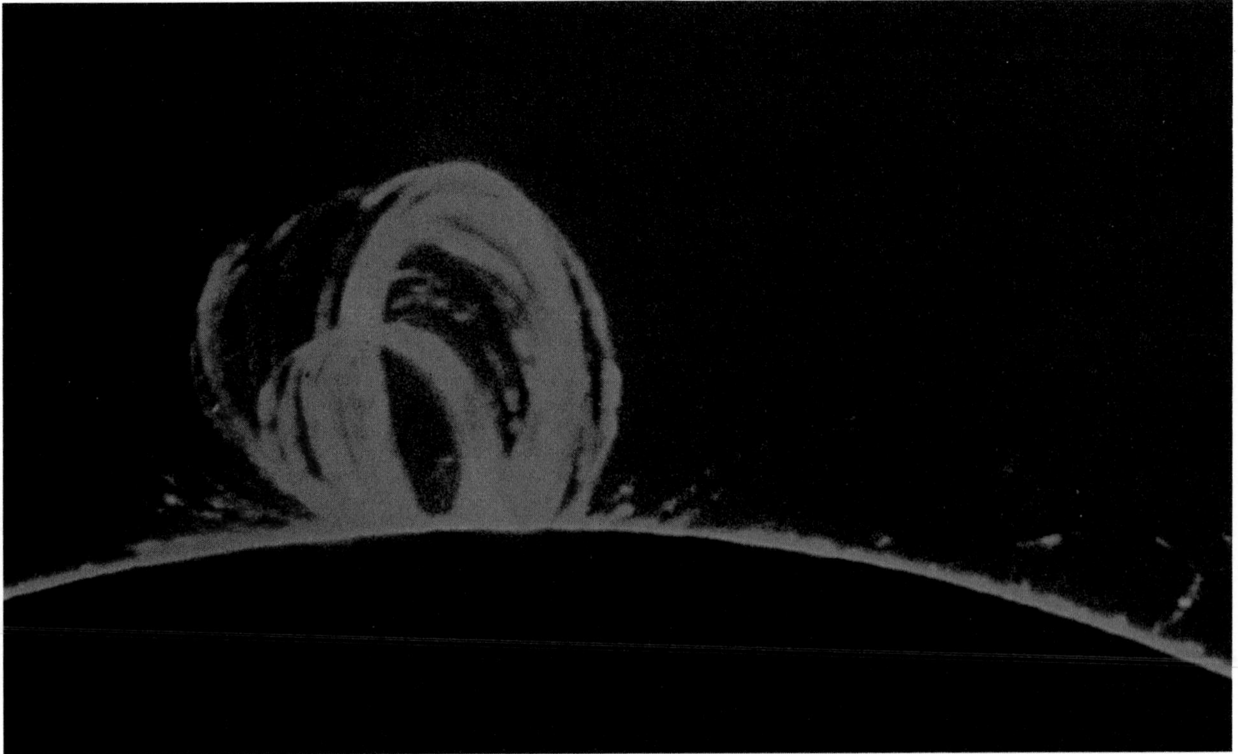

FIGURE 14.17 This photo shows a large solar prominence, many times the size of Earth, which consists of glowing gas trapped by magnetic field lines arching high above the surface of the Sun.

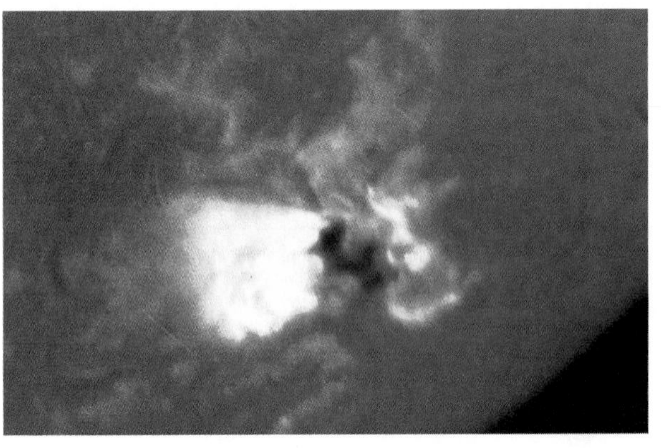

a

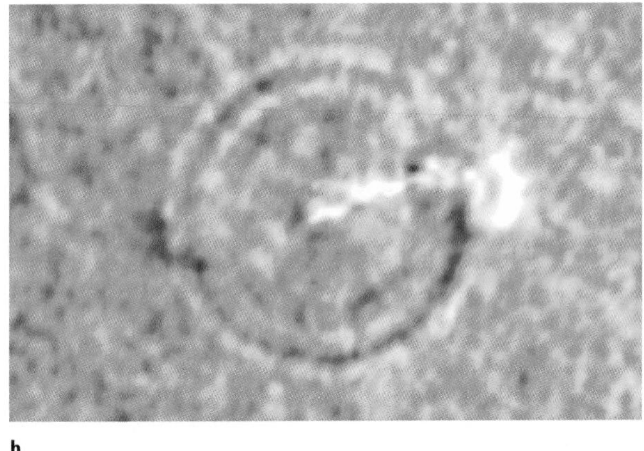

b

FIGURE 14.18 (**a**) Hot gas erupting from the Sun's surface during a solar flare. (Photo taken with a filter that shows only light from a particular transition in hydrogen atoms.) (**b**) This close-up photo of the Sun's surface shows a wave (dark ring) moving outward from the location of a solar flare.

for days or weeks as it cools or until the field collapses. Some prominences rise to heights of more than 100,000 kilometers above the Sun's surface (Figure 14.17).

The most dramatic events on the solar surface are **flares,** which emit bursts of X rays and fast-moving charged particles into space (Figure 14.18). Flares generally occur in the vicinity of sunspots, leading us to suspect that they occur when the magnetic field lines become so twisted and knotted that they can bear the tension no longer. The magnetic field lines suddenly snap like tangled elastic bands twisted beyond their limits, releasing a huge amount of energy. This energy heats the nearby plasma to 100 million K over the next few minutes to few hours, generating X rays and accelerating some of the charged particles to nearly the speed of light.

The Chromosphere and Corona

The high temperatures of the chromosphere and corona perplexed scientists for decades. After all, temperatures gradually decline as we move outward from the core to the top of the photosphere. Why should this decline suddenly reverse? Some aspects of this atmospheric heating remain a mystery today, but we have at least a general explanation: The Sun's strong magnetic fields carry energy upward from the churning solar surface to the chromosphere and corona. More specifically, the rising and falling of gas in the convection zone probably shakes magnetic field lines beneath the solar surface. This shaking generates waves along the magnetic field lines that carry energy upward to the solar atmosphere. Precisely how the waves deposit their energy in the chromosphere and corona is not known, but the waves agitate the low-density plasma of these layers, somehow heating them to high temperatures.

Note that, according to this model of solar heating, the same magnetic fields that keep sunspots cool make the overlying plasma of the chromosphere and corona hot. We can test this idea observationally. The gas of the chromosphere and corona is so tenuous that we cannot see it with our eyes except during a total eclipse, when we can see the faint visible light scattered by electrons in the corona [Section 3.3]. However, the roughly 10,000-K plasma of the chromosphere emits strongly in the ultraviolet, and the million-K plasma of the corona is the source of virtually all X rays coming from the Sun. Figure 14.19 shows an X-ray image of the Sun: As the solar heating model predicts, the brightest regions of the corona tend to be directly above sunspot groups.

Some regions of the corona, called **coronal holes,** barely show up in X-ray images. More detailed analyses show that the magnetic field lines in coronal holes project out into space like broken rubber bands, allowing particles spiraling along them to escape the Sun altogether. These particles streaming outward from the corona constitute the *solar wind,* which blows through the solar system at an average speed of about 500 kilometers per second and has important effects on planetary surfaces, atmospheres, and magnetospheres [Section 10.2]. Somewhere beyond the orbit of Neptune, the pressure of interstellar gas must eventually halt the solar wind. The Pioneer and Voyager spacecraft that visited the outer planets in the 1970s and 1980s are still traveling outward from our solar system and may soon encounter this "boundary" (called the *heliopause*) of the realm of the Sun. It's also worth noting that, in the same way that meteorites provide us with samples of asteroids we've never visited, solar wind particles captured by satellites provide us with a sample of material from the Sun. Analysis of these solar particles has reassuringly verified that the Sun is made mostly of hydrogen, just as we conclude from studying the Sun's spectrum.

14.5 Solar Weather and Climate

Individual sunspots, prominences, and flares are short-lived phenomena, somewhat like storms on the Earth, and they constitute what we call *solar weather* or **solar activity.** You know from personal experience that the Earth's weather is notoriously unpredictable, and the same is true for the Sun: We cannot predict precisely when or where a particular sunspot or flare will appear. The Earth's *climate,* on the other hand, is quite regular from season to season. So it is with the Sun, where despite day-to-day variations the general nature and intensity of solar activity follow a predictable cycle.

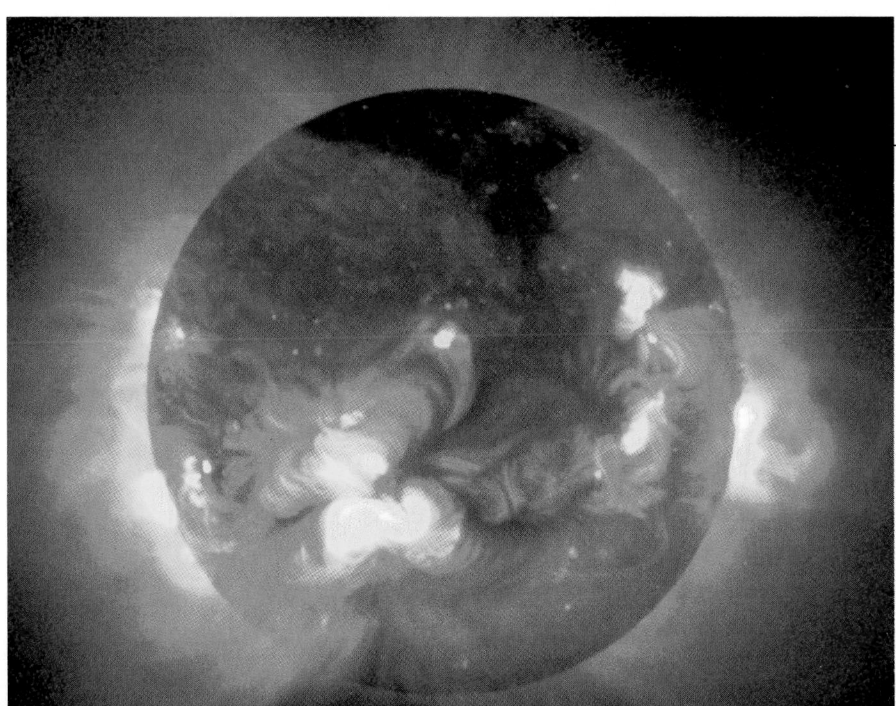

FIGURE 14.19 An X-ray image of the Sun reveals the corona: Brighter regions of this image correspond to regions of stronger X-ray emission.

The Sunspot Cycle

Long before we realized that sunspots were magnetic disturbances, astronomers had recognized patterns in sunspot activity. The most notable pattern is the number of sunspots visible on the Sun at any particular time. Thanks to telescopic observations of the Sun recorded by astronomers since the 1600s, we know that the number of sunspots gradually rises and falls in a **sunspot cycle** with a period of about 11 years (Figure 14.20a). At the time of **solar maximum,** when sunspots are most numerous, we may see dozens of sunspots on the Sun at one time. In contrast, we see few if any sunspots at the time of **solar minimum.** Note that the sunspot cycle is not perfectly predictable: The interval between solar maxima is sometimes as long as 15 years or as short as 7 years. More dramatically, between the years 1645 and 1715, sunspot activity virtually ceased.[4] The frequency of prominences and flares also follows the sunspot cycle, with these events being most common at solar maximum and least common at solar minimum.

[4]This period from 1645 to 1715 is called the *Maunder minimum,* after E. W. Maunder, who discovered it while sifting through historical records of sunspot activity.

Another feature of the sunspot cycle is a gradual change in the latitudes at which individual sunspots form and dissolve (Figure 14.20b). As a cycle begins at solar minimum, sunspots form primarily at mid-latitudes (30° to 40°) on the Sun. The sunspots tend to form at lower latitudes as the cycle progresses, appearing very close to the solar equator as the next solar minimum approaches.

A less obvious feature of the sunspot cycle is that, at each solar minimum, something peculiar happens to the Sun's magnetic field. The field lines connecting all pairs of sunspots (see Figure 14.16) tend to point in the same direction throughout an 11-year solar cycle (within each hemisphere); for example, all compass needles might point from the easternmost sunspot to the westernmost sunspot in a pair. However, as the cycle ends at solar minimum, the magnetic field reverses: In the subsequent solar cycle, the field lines connecting pairs of sunspots point in the opposite direction. Apparently, the entire magnetic field of the Sun flip-flops every 11 years. These magnetic reversals hint that the sunspot cycle is related to the generation of magnetic fields on the Sun. They also tell us that the *complete* cycle really averages 22 years, since it takes two 11-year cycles before the magnetic field is back the way it started.

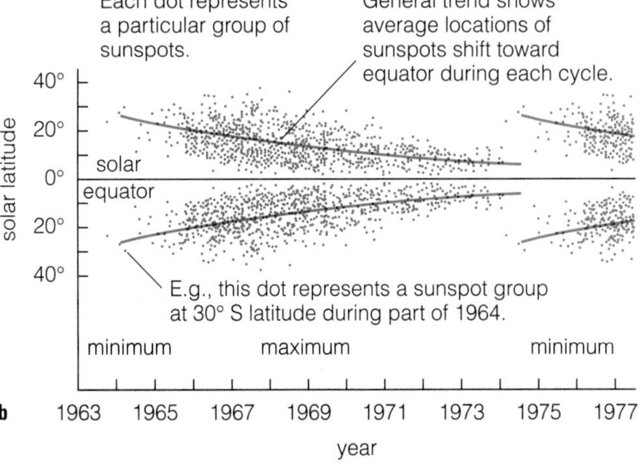

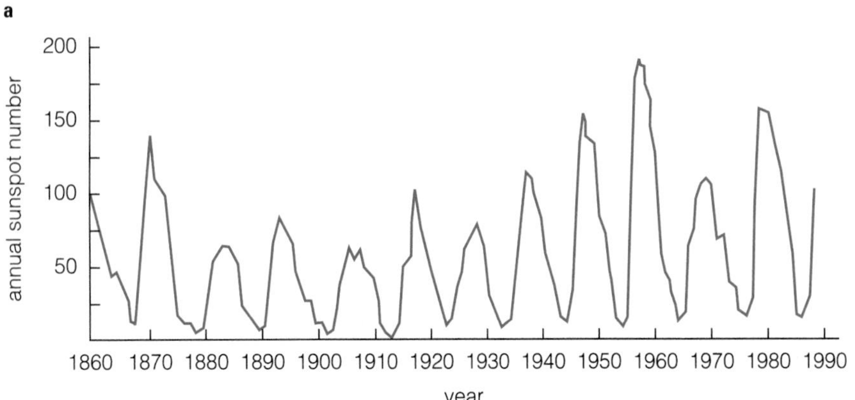

FIGURE 14.20 (**a**) This graph shows how the number of sunspots on the Sun changes with time. Note the approximately 11-year cycle. (**b**) This graph shows how the latitudes at which sunspot groups appear tend to shift during a single sunspot cycle.

What Causes the Sunspot Cycle?

The causes of the Sun's magnetic fields and the sunspot cycle are not well understood, but we believe we know the general nature of the processes involved. Convection is thought to dredge up weak magnetic fields generated in the solar interior, amplifying them as they rise. The Sun's **differential rotation**—faster at its equator than near its poles—then stretches and shapes these fields.

Imagine what happens to a magnetic field line that originally runs along the Sun's surface directly from the north pole to the south pole. At the equator, the field line circles the Sun in 27 days, but at higher latitudes the field line lags behind. Gradually, differential rotation winds the field line more and more tightly around the Sun (Figure 14.21). This process, operating at all times over the entire Sun, produces the contorted field lines that generate sunspots and other solar activity.

Making more specific statements about how the Sun's magnetic field develops and changes in time requires the aid of sophisticated computer models. Scientists are working hard on such models, but the behavior of these fields is so complex that approximations are necessary, even with the best supercomputers. Using these computer models, scientists have successfully replicated such features of the sunspot cycle as changes in the number and latitude of sunspots and the magnetic field reversals that occur about every 11 years. However, much still remains mysterious, including why the period of the sunspot cycle varies and why solar activity is different from one cycle to the next.

Solar Activity and the Earth

During solar maximum, solar flares and other forms of solar activity feed large numbers of highly energetic charged particles (protons and electrons) into the solar wind. Do these forms of solar weather affect the Earth? In at least some ways, the answer is a definitive yes.

The solar wind continually carries charged particles from the solar corona to the Earth. These particles then stream along the Earth's magnetic field lines, piling up near the poles (Figure 14.22). Most of these charged particles collide with atoms in the Earth's upper atmosphere, causing electrons in the atoms to jump to higher energy levels [Section 5.4]. These excited atoms subsequently emit visible-light photons as they drop to lower energy levels, creating the shimmering light of the *aurora* (see Figure 1.22).

Particles streaming from the Sun after the occurrence of solar flares or other major solar storms can also have practical impacts on society. For example, these particles can hamper radio communications, disrupt electrical power delivery, and damage the electronic components in orbiting satellites. During a particularly powerful magnetic storm on the Sun in March 1989, the U.S. Air Force temporarily lost

FIGURE 14.21 The Sun rotates more quickly at its equator than it does near its poles, a behavior known as differential rotation. Because gas circles the Sun faster at the solar equator, it drags the Sun's north-south magnetic field lines into a more twisted configuration. The magnetic field lines linking pairs of sunspots, depicted here as green and black blobs, trace out the directions of these stretched and distorted field lines.

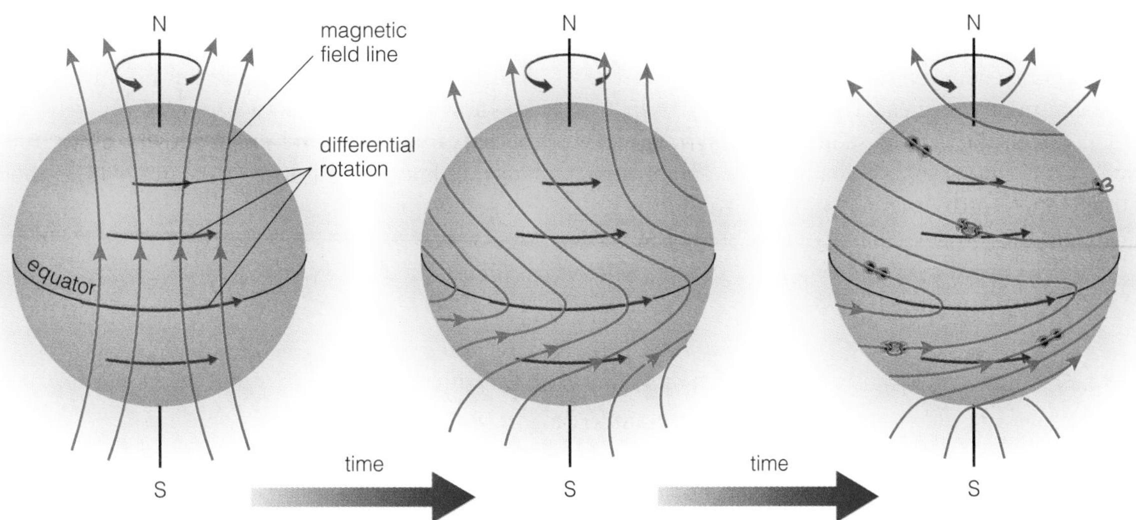

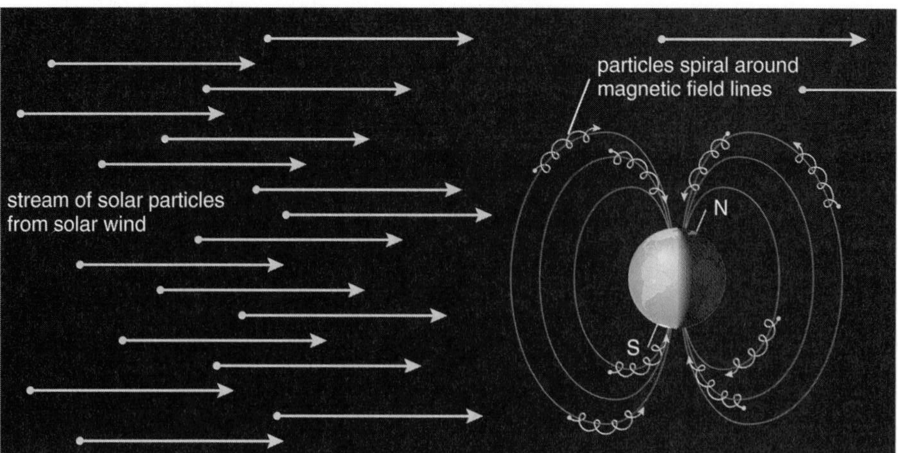

FIGURE 14.22 The Earth's magnetic field captures charged particles from the solar wind and directs them toward the poles; these particles cause auroras.

track of over 2,000 satellites, and powerful currents induced in the ground circuits of the Quebec hydroelectric system caused it to collapse for more than 8 hours. The combined cost of the power lost in the United States and Canada exceeded $100 million. In January 1997, AT&T lost contact with a $200-million communications satellite, probably because of damage caused by particles coming from another powerful solar storm.

Satellites in low-Earth orbit are particularly vulnerable during solar maximum, because the increase in solar X rays and energetic particles heats the Earth's upper atmosphere, causing it to expand. The density of the gas surrounding low-flying satellites therefore rises, exerting drag that saps their energy and angular momentum. If this drag proceeds unchecked, the satellites ultimately plummet back to Earth. Satellites in low orbits, including the Hubble Space Telescope, therefore require occasional boosts to prevent them from falling out of the sky. (Hubble is scheduled to get boosts from the Space Shuttle in 2000 and 2003.)

Connections between solar activity and the Earth's climate are much less clear. The period from 1645 to 1715, when solar activity seems to have virtually ceased, was a time of exceptionally low temperatures in Europe and North America known as the *Little Ice Age*. Did the low solar activity cause these low temperatures, or was their occurrence a coincidence? No one knows for sure. Some researchers have claimed that certain weather phenomena, such as drought cycles or frequencies of storms, are correlated with the 11- or 22-year cycles of solar activity. However, the data supporting these correlations are weak in many cases, and even real correlations may be coincidental.

Part of the difficulty in linking solar activity with climate is that no one understands how the linkage might work. Although emissions of ultraviolet light, X rays, and high-energy particles increase substantially from solar minimum to solar maximum, the total luminosity of the Sun barely changes at all. (The Sun becomes slightly brighter during solar maximum, but only by about 0.01%.) Thus, if solar activity really is affecting Earth's climate, it must be through some very subtle mechanism. For example, perhaps the expansion of the Earth's upper atmosphere that occurs with solar maximum somehow causes changes in weather.

The question of how solar activity is linked to the Earth's climate is very important, because we need to know whether global warming is affected by solar activity in addition to human activity. Unfortunately, for the time being at least, we can say little about this question.

I say Live, Live, because of the Sun,
The dream, the excitable gift.

ANNE SEXTON (1928–74)

THINKING ABOUT . . .

Long-Term Change in Solar Activity

Figure 14.20 shows that the sunspot cycle varies in length and intensity, and it sometimes seems to disappear altogether. With these facts as background, many scientists are searching for longer-term patterns in solar activity. Unfortunately, the search for longer-term variations is difficult because telescopic observations of sunspots cover a period of only about 400 years. Some naked-eye observations of sunspots recorded by Chinese astronomers go back almost 2,000 years, but these records are sparse, and naked-eye observations may not be very reliable. We can also guess at past solar activity from descriptions of solar eclipses recorded around the world: When the Sun is more active, the corona tends to have longer and brighter "streamers" visible to the naked eye.

Another way to gauge past solar activity is to study the amount of carbon-14 in tree rings. High-energy *cosmic rays* [Section 18.2] coming from beyond our own solar system produce radioactive carbon-14 in the Earth's atmosphere. During periods of high solar activity, the solar wind tends to grow stronger, shielding the Earth from some of these cosmic rays. Thus, production of carbon-14 drops when the Sun is more active. All the while, trees steadily breathe in atmospheric carbon, in the form of carbon dioxide, and incorporate it year by year into each ring. We can therefore estimate the level of solar activity in any given year by measuring the level of carbon-14 in the corresponding ring.

No clear evidence has yet been found of longer-term cycles of solar activity, but the search goes on. Moreover, theoretical models predict a long-term trend. The Sun must have rotated much faster when it was young [Section 8.4]. Because a combination of convection and differential rotation generates solar activity, a faster rotation rate should have meant much more activity. Observations of other stars that are similar to the Sun but rotate faster confirm that these stars are much more active. We find evidence for many more "starspots" on these stars than on the Sun, and their relatively bright ultraviolet and X-ray emissions suggest that they have brighter chromospheres and coronas—just as we would expect if they are more active than the Sun.

THE BIG PICTURE

In this chapter, we have examined our Sun, the nearest star. When you look back at this chapter, make sure you understand these "big picture" ideas:

- The ancient riddle of why the Sun shines is now solved. The Sun shines with energy generated by fusion of hydrogen into helium in the Sun's core. After a million-year journey through the solar interior and an 8-minute journey through space, a small fraction of this energy reaches the Earth and supplies sunlight and heat.
- Gravitational equilibrium, the balance between pressure and gravity, determines the Sun's interior structure and maintains its steady nuclear burning rate. If the Sun were not so steady, life on Earth might not have been possible.
- The Sun's atmosphere displays its own version of weather and climate, governed by solar magnetic fields. Solar weather has important influences on the Earth.
- The Sun is important not only as our source of light and heat, but also because it is the only star near enough for us to study in great detail. In the coming chapters, we will use what we've learned about the Sun to help us understand other stars.

Review Questions

1. Briefly describe how the mechanism of *gravitational contraction* generates energy. Although this mechanism does not generate energy for the Sun today, explain how it was important in the Sun's early history.
2. What two forces are balanced in what we call *gravitational equilibrium?* Briefly describe how the idea of gravitational equilibrium implies that pressure must increase with depth in any planet or star.
3. What is the Sun made of? How do we know?
4. Briefly describe the Sun's luminosity, mass, radius, and average surface temperature. Put each property in perspective by comparing it to some familiar quantity related to Earth.

5. Briefly describe the distinguishing features of each of the Sun's major layers shown in Figure 14.4.
6. What are sunspots? Why do they appear dark in pictures of the Sun?
7. Briefly distinguish between *nuclear fission* and *nuclear fusion.* Which one is used in nuclear power plants? Which one is used by the Sun?
8. Why does fusion require high temperatures and pressures? Explain why nuclei normally want to repel each other, and describe the conditions under which the *strong force* can bind nuclei together. Why is fusion-based nuclear power difficult to achieve on Earth?
9. What is the overall nuclear fusion reaction in the Sun? Briefly describe each of the steps of the *proton–proton chain* that lead to this overall reaction.
10. How does the idea of gravitational equilibrium explain why the Sun's core temperature and fusion rate are self-regulating, like a thermostat? Explain why the same idea ensures that the Sun's rate of fusion must gradually rise over billions of years.
11. Explain how *mathematical models* allow us to predict conditions inside the Sun. How can we be confident that the models are on the right track?
12. How are "sun quakes" similar to earthquakes? How are they different? Describe how we can observe them and how they help us learn about the solar interior.
13. What is the *solar neutrino problem?* Describe a possible solution to this problem.
14. Why does the energy produced by fusion in the solar core take so long to reach the solar surface? Describe the process of *radiative diffusion* that transports energy through the Sun's radiation zone.
15. What is *convection?* Describe how convection transports energy through the Sun's convection zone.
16. Describe the appearance and temperature of the Sun's photosphere. What is *granulation?* How would granulation appear in a movie?
17. Briefly describe how *magnetic field lines* are related to what we would see if we placed compasses in a magnetic field.
18. How are magnetic fields related to sunspots? Describe how magnetic fields explain each of the following: the fact that sunspot plasma is cool, the existence and appearance of *prominences,* and the dramatic explosions of *solar flares.*
19. Why is the chromosphere best viewed with ultraviolet telescopes? Why is the corona best viewed with X-ray telescopes? Why must such viewing be done from space, rather than from the ground on Earth?
20. Describe the current theory of how the chromosphere and corona are heated. What evidence supports it?
21. What are *coronal holes?* How are they related to the solar wind?
22. What do we mean by *solar activity?* Explain how it is similar to weather on short time scales and to climate on longer time scales.
23. Describe the *sunspot cycle.* What do we mean by *solar maximum* and *solar minimum?* Explain what happens to the Sun's magnetic field with each new sunspot cycle and why the complete cycle is really 22 years rather than 11 years.
24. Describe a few known ways solar activity affects the Earth. Why is it so difficult to establish links between solar activity and the Earth's climate?

Discussion Questions

1. *The Solar Neutrino Problem.* Discuss the solar neutrino problem and its potential solutions. How serious do you consider this problem? Do you think current theoretical models of the Sun could be wrong in any fundamental way? Why or why not?
2. *The Sun and Global Warming.* One of the most pressing environmental issues on Earth concerns the extent to which human emissions of greenhouse gases are warming our planet. Some people claim that part or all of the observed warming over the past century may be due to changes on the Sun, rather than to anything that humans have done. Discuss how a better understanding of the Sun might help us understand the threat posed by greenhouse gas emissions. Why is it so difficult to develop a clear understanding of how the Sun affects the Earth's climate?

Problems

1. *Flying Around the Sun.* A commercial jet airplane traveling at about 800 km/hr could fly all the way around the Earth in only a little over 2 days. Suppose we could somehow adapt a plane so that it could fly through the Sun's photosphere at the same speed without burning up. How long would it take to fly all the way around the Sun at 800 km/hr? Use your answer to comment on the size of the Sun in comparison to the size of the Earth. (*Hint:* The circumference of the Sun is $2 \times \pi \times$ radius of Sun.)
2. *Differential Rotation.* The Sun and Jupiter exhibit differential rotation, but Earth does not.

a. Explain clearly what we mean by *differential rotation.*

b. Briefly explain why it is possible for jovian planets to rotate differentially, but it is not possible for terrestrial planets.

c. Would you expect differential rotation to be common among other stars? Why or why not?

3. *Solar Luminosity and Weather.* Although ultraviolet and X-ray emissions may vary much more dramatically, the Sun's total luminosity varies up to only about 0.01% with solar activity. In principle, this variation could slightly affect solar heating of the Earth. However, the Earth also receives varying amounts of heat from the Sun due to its elliptical orbit, and the two hemispheres receive varying amounts of solar heat with the seasons. Contrast the degree of variation in solar heating of the Earth due to changes in solar luminosity, Earth's elliptical orbit, and the seasons. (You may answer qualitatively, without making any calculations.) Overall, do you think the changes in the Sun's total luminosity could affect the weather? Summarize your conclusions in a few sentences.

4. *Number of Fusion Reactions in the Sun.* Use the fact that each cycle of the proton–proton chain converts 4.7×10^{-29} kg of mass into energy (see Mathematical Insight 14.1), along with the fact that the Sun loses a total of about 4.2×10^{9} kg of mass each second, to calculate the total number of times the proton–proton chain occurs each second in the Sun.

5. *The Lifetime of the Sun.* The total mass of the Sun is about 2×10^{30} kg, of which about 75% was hydrogen when the Sun formed. However, only about 13% of this hydrogen ever becomes available for fusion in the core; the rest remains in layers of the Sun where the temperature is too low for fusion.

a. Based on the given information, calculate the total mass of hydrogen that is available for fusion over the lifetime of the Sun.

b. Combine your results from part (a) and the fact that the Sun fuses about 600 billion kg of hydrogen each second to calculate how long the Sun's initial supply of hydrogen can last. Give your answer in both seconds and years.

c. Given that our solar system is now about 4.6 billion years old, when will we need to worry about the Sun running out of hydrogen for fusion?

6. *Solar Power Collectors.* This problem leads you through the calculation and discussion of how much solar power can be collected by solar cells on Earth.

a. Imagine a giant sphere surrounding the Sun with a radius of 1 AU. What is the surface area of this sphere, in square meters? (*Hint:* The formula for the surface area of a sphere is $4\pi r^2$.)

b. Because this imaginary giant sphere surrounds the Sun, the Sun's entire luminosity of 3.8×10^{26} watts must pass through it. Calculate the power passing through each square meter of this imaginary sphere in *watts per square meter.* Explain why this number represents the maximum power per square meter that can be collected by a solar collector in Earth orbit.

c. List several reasons why the average power per square meter collected by a solar collector on the ground will always be less than what you found in part (b).

d. Suppose you want to put a solar collector on your roof. If you want to optimize the amount of power you can collect, how should you orient the collector? (*Hint:* The optimum orientation depends on both your latitude and the time of year and day.)

7. *Solar Power for the United States.* The total annual U.S. energy consumption is about 2×10^{20} joules.

a. What is the average *power* requirement for the United States, in watts? (*Hint:* 1 watt = 1 joule/s.)

b. With current technologies and solar collectors on the ground, the best we can hope is that solar cells will generate an average (day and night) power of about 200 watts/m^2. (You might compare this to the maximum power per square meter you found in problem 6b.) What total area would we need to cover with solar cells to supply all the power needed for the United States? Give your answer in both square meters and square kilometers.

c. The total surface area of the United States is about 2×10^{7} km^2. What fraction of the U.S. area would have to be covered by solar collectors to generate all of the U.S. power needs? In one page or less, describe potential environmental impacts of covering so much area with solar collectors. Also discuss whether you think these environmental impacts would be greater or lesser than the impacts of using current energy sources such as coal, oil, nuclear power, and hydroelectric power.

8. *The Color of the Sun.* The Sun's average surface temperature is about 5,800 K. Use Wien's law (see Mathematical Insight 7.2) to calculate the wavelength of peak thermal emission from the Sun. What color does this wavelength correspond to in the visible light spectrum? In light of your answer, why do you think the Sun appears white or yellow to our eyes?

9. *Research: Space-Based Solar Collectors.* Some people have suggested placing solar collectors in Earth orbit as a way of generating power for us here on the ground. Research some of these suggestions. Write a one- to two-page essay summarizing the advantages, disadvantages, and challenges of erecting a space-based array of solar collectors.

10. *Web Project: Current Solar Activity.* Find web sites describing current activity on the Sun. Where are we in the sunspot cycle right now? When is the next solar maximum or minimum expected? Have there been any major solar storms in the past few months? If so, did they have any significant effects on the Earth? Summarize your findings in a one- to two-page report.

15

Stars

On a clear, dark night, a few thousand stars are visible to the naked eye. Many more become visible through binoculars, and with a powerful telescope we can see so many stars that we could never hope to count them. Like individual people, each individual star is unique. Like the human family, all stars share much in common.

Today, we know that stars are born from clouds of interstellar gas, shine brilliantly by nuclear fusion for millions or billions of years, and then die, sometimes in dramatic ways. This chapter outlines how we study and categorize stars and how we have come to realize that stars, like people, change over their lifetime.

15.1 Snapshot of the Heavens

Imagine that an alien spaceship flies by Earth on a simple but short mission: The visitors have just 1 minute to learn everything they can about the human race. In 60 seconds, they will see next to nothing of each individual person's life. Instead, they will obtain a collective "snapshot" of humanity that shows people from all stages of life engaged in their daily activities. From this snapshot alone, they must piece together their entire understanding of human beings and their lives, from birth to death.

We face a similar problem when we look at the stars. Compared with stellar lifetimes of millions or billions of years, the few hundred years humans have spent studying stars with telescopes is rather like the aliens' 1-minute glimpse of humanity. We see only a brief moment in any star's life, and our collective snapshot of the heavens consists of such frozen moments for billions of stars. From this snapshot, we try to reconstruct the life cycles of stars while also analyzing what makes one star different from another.

Thanks to the efforts of hundreds of astronomers studying this snapshot of the heavens, including a notable group of women in the early twentieth century, stars are no longer mysterious points of light in the sky. We now know that all stars form in great clouds of gas and dust. All stars begin their life with roughly the same chemical composition: About three-quarters of the star's mass at birth is hydrogen, and about one-quarter is helium, with no more than about 2% consisting of elements heavier than helium. During most of any star's life, the rate at which it generates energy depends on the same type of balance between the inward pull of gravity and the outward push of internal pressure that governs the rate of fusion in our Sun. Despite these similarities, stars appear different from one another for two primary reasons: They differ in mass, and we see different stars at different stages of their lives.

The key that finally unlocked these secrets of stars was an appropriate classification system. Before the twentieth century, humans classified stars primarily by their brightness and location in our sky. The brightest stars within each constellation still bear Greek letters designating their place in order of brightness. For example, the brightest star in the constellation Centaurus is Alpha Centauri, the second brightest is Beta Centauri, the third brightest is Gamma Centauri, and so on. However, a star's brightness and membership in a constellation tell us little about its true nature. A star that appears bright could be either extremely powerful or unusually close, and two stars that appear right next to each other in our sky might not be true neighbors if they lie at significantly different distances from Earth.

Today, astronomers classify a star primarily according to its *luminosity* and *surface temperature*. Our task in this chapter is to learn how this extraordinarily effective classification system reveals the true natures of stars and their life cycles. We begin by investigating how to determine a star's luminosity, surface temperature, and mass.

15.2 Stellar Luminosity

A star's **luminosity** is the total amount of power it radiates into space, which can be stated in *watts* [Section 14.2]. For example, the Sun's luminosity is 3.8×10^{26} watts. We cannot measure a star's luminosity directly, because its brightness in our sky depends on its distance as well as its true luminosity. For example, our Sun and Alpha Centauri A (the brightest of the three stars in the Alpha Centauri system [Section 2.2]) are similar in luminosity. But Alpha Centauri A is a feeble point of light in the night sky, while our Sun provides enough light and heat to sustain life on Earth. The difference in brightness arises because Alpha Centauri A is about 270,000 times farther from Earth than is the Sun.

More precisely, we define the **apparent brightness** of any star in our sky as the amount of light reaching us *per unit area*. (A more technical term for apparent brightness is *flux*.) The apparent brightness of any light source obeys an *inverse square law* with distance, similar to the inverse square law that describes the force of gravity [Section 6.4]. If we viewed the Sun from twice the Earth's distance, it would appear dimmer by a factor of $2^2 = 4$. If we viewed it from 10 times the Earth's distance, it would appear $10^2 = 100$ times dimmer. From 270,000 times the Earth's distance, it would look like Alpha Centauri A—dimmer by a factor of $270{,}000^2$, or about 70 billion times.

We can see why apparent brightness follows an inverse square law by imagining that we could surround a star with three giant transparent spheres at distances of 1, 2, and 3 AU (Figure 15.1). The light all comes from the star, so the same total amount of light must pass through each sphere. If we focus our attention on the light passing through a small square on the sphere located at 1 AU, we see that the same amount of light must pass through *four* squares of the same size on the sphere located at 2 AU. Thus, each square on the sphere at 2 AU receives only $\frac{1}{2^2} = \frac{1}{4}$ as much light as the square on the sphere at 1 AU. Similarly, the same amount of light passes through *nine* squares of the same size on the sphere

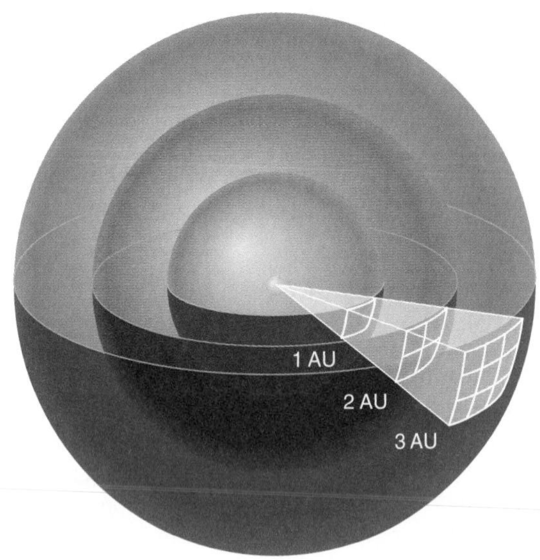

FIGURE 15.1 The inverse square law for light. At greater distances from a star, the same amount of light passes through an area that gets larger with the square of the distance. The amount of light per unit area therefore declines with the square of the distance.

located at 3 AU. Thus, each of these squares receives only $\frac{1}{3^2} = \frac{1}{9}$ as much light as the square on the sphere at 1 AU. Generalizing, we see that the amount of light received per unit area decreases with increasing distance by the square of the distance—an inverse square law.

This inverse square law leads to a very simple and important formula relating the apparent brightness, luminosity, and distance of any light source. We will call it the **luminosity–distance formula:**

$$\text{apparent brightness} = \frac{\text{luminosity}}{4\pi \times (\text{distance})^2}$$

Note that, because the standard units of luminosity are watts, the units of apparent brightness are *watts per square meter.* Since we can always measure the apparent brightness of a star, this formula provides a way to calculate a star's luminosity if we can first measure its distance, or to calculate a star's distance if we somehow know its luminosity.[1]

Although watts are the standard units for luminosity, it's often more meaningful to describe stellar luminosities in comparison to the Sun by using units of **solar luminosity:** $L_{\text{Sun}} = 3.8 \times 10^{26}$ watts. For example, Proxima Centauri, the nearest of the three stars in the Alpha Centauri system and hence the nearest star besides our Sun, is only about 0.00006 times as luminous as the Sun, or $0.00006L_{\text{Sun}}$. Betelgeuse, the bright left-shoulder star of Orion, has a luminosity of $52{,}000L_{\text{Sun}}$, meaning that it is 52,000 times more luminous than the Sun.

Measuring the Apparent Brightness

We can measure a star's apparent brightness by using a detector, such as a CCD, that records how much energy strikes its light-sensitive surface each second. For example, such a detector would record an apparent brightness of 2.7×10^{-8} watt per square meter from Alpha Centauri A. The only difficulties involved in measuring apparent brightness are making sure the detector is properly calibrated and, for ground-based telescopes, taking into account the absorption of light by the Earth's atmosphere.

No detector can record light of all wavelengths, so we necessarily measure apparent brightness in only some small range of the complete spectrum. For example, the human eye is sensitive to visible light but does not respond to ultraviolet or infrared photons. Thus, when we perceive a star's brightness, our eyes are measuring the apparent brightness only in the visible region of the spectrum. Clearly, when we measure the apparent brightness in visible light, we can calculate only the star's *visible-light luminosity.* Similarly, when we observe a star with a spaceborne X-ray telescope, we measure only the apparent brightness in X rays and can calculate only the star's *X-ray luminosity.* We will use the terms **total luminosity** and **total apparent brightness** to describe the luminosity and apparent brightness we would measure *if* we could detect photons across the entire electromagnetic spectrum.[2] If we do not specify a particular region of the spectrum, we are referring to the apparent brightness or luminosity in visible light.

Measuring Distance Through Stellar Parallax

Once we have measured a star's apparent brightness, the next step in determining its luminosity is to measure its distance. The most direct way to measure the distances to stars is with *stellar parallax,* the small annual shifts in a star's apparent position caused by the Earth's motion around the Sun [Section 3.1]. Recall that you can observe parallax of your finger by holding it at arm's length and looking at it alternately with one eye closed and then the other. Astronomers measure stellar parallax by comparing observations of a

[1]The luminosity–distance formula is strictly correct only if interstellar dust does not absorb or scatter the starlight along its path to Earth. In practice, astronomers must take into account the effects of interstellar dust before applying the luminosity–distance formula.

[2]Astronomers refer to the total luminosity or flux (apparent brightness) as the *bolometric* luminosity or flux.

nearby star made 6 months apart (Figure 15.2). The nearby star appears to shift against the background of more distant stars because we are observing it from two opposite points of the Earth's orbit. The star's **parallax angle** is defined as *half* the star's annual back-and-forth shift.

Measuring stellar parallax is difficult because stars are so far away, making their parallax angles very small.[3] Even the nearest star, Proxima Centauri, has a parallax angle of only 0.77 arcsecond. For increasingly distant stars, the parallax angles quickly become too small to measure even with our highest-resolution telescopes. Current technology allows us to measure parallax only for stars within a few hundred light-years—not much farther than what we call our *local solar neighborhood* in the vast, 100,000-light-year-diameter Milky Way Galaxy [Section 2.3].

[3]Two other problems make parallax measurements difficult, but both can be alleviated with careful work. (1) We must be sure that the "background" objects against which we observe the shifts of a nearby star truly are so distant that they have no detectable parallax of their own. (2) Besides the *apparent* shift of parallax, stars also have *real* motions relative to us. During the 6 months it takes to notice a parallax shift, these real motions can alter the positions of nearby stars by small but measurable amounts. We must account for a star's actual motion before we can determine the true parallax angle.

By definition, the distance to an object with a parallax angle of 1 arcsecond is one **parsec,** abbreviated **pc.** (The word *parsec* comes from the words

FIGURE 15.2 Parallax makes the apparent position of a nearby star shift back and forth with respect to distant stars over the course of each year. If we measure the parallax angle p in arcseconds, the distance d to the star in parsecs is $\frac{1}{p}$. The angle in this figure is greatly exaggerated: All stars have parallax angles of less than 1 arcsecond.

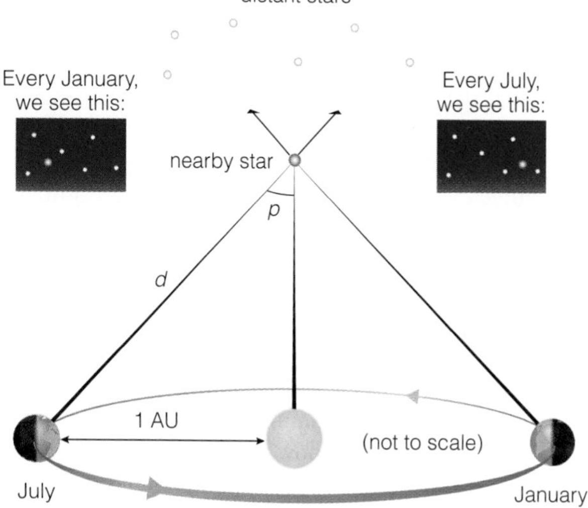

Mathematical Insight 15.1 The Luminosity–Distance Formula

We can derive the luminosity–distance formula by extending the idea illustrated in Figure 15.1. Suppose we are located a distance d from a star with luminosity L. The apparent brightness of the star is the power per unit area that we receive at our distance d. We can find this apparent brightness by imagining that we are part of a giant sphere with radius d, similar to any one of the three spheres in Figure 15.1. The surface area of this giant sphere is $4\pi \times d^2$, and the star's entire luminosity L must pass through this surface area. (Recall that the surface area of any sphere is $4\pi \times \text{radius}^2$.) Thus, the apparent brightness at distance d is the power per unit area passing through the sphere:

$$\text{apparent brightness} = \frac{\text{star's luminosity}}{\text{surface area of imaginary sphere}} = \frac{L}{4\pi \times d^2}$$

This is our luminosity–distance formula.

Example: What is the apparent brightness of sunlight as seen from the Earth?

Solution: The Sun's luminosity is $L_{\text{Sun}} = 3.8 \times 10^{26}$ watts, and the Earth's distance from the Sun is $d = 1.5 \times 10^{11}$ meters. Thus, the apparent brightness of sunlight at the Earth's surface is:

$$\frac{L}{4\pi \times d^2} = \frac{3.8 \times 10^{26}\ \text{watts}}{4\pi \times (1.5 \times 10^{11}\ \text{m})^2} = 1.3 \times 10^3\ \text{watts/m}^2$$

This apparent brightness of about 1,300 watts per square meter represents the maximum power that could be collected by a detector on Earth that directly faces the Sun, such as a solar power (or *photovoltaic*) cell. In reality, the apparent brightness is usually lower because the Earth's atmosphere absorbs some sunlight, particularly when it is cloudy.

parallax and *arcsecond*.) With a little geometry and Figure 15.2, it is possible to show that:

$$1 \text{ pc} = 3.26 \text{ light-years} = 3.09 \times 10^{13} \text{ km}$$

If we use units of arcseconds for the parallax angle, a simple formula allows us to calculate distances in parsecs:

$$d \text{ (in parsecs)} = \frac{1}{p \text{ (in arcseconds)}}$$

For example, the distance to a star with a parallax angle of 1/2 arcsecond is 2 parsecs, the distance to a star with a parallax angle of 1/10 arcsecond is 10 parsecs, and the distance to a star with a parallax angle of 1/100 arcsecond is 100 parsecs. Astronomers often express distances in parsecs or light-years interchangeably, so you should learn to convert quickly between them by remembering that 1 pc = 3.26 light-years. Thus, 10 parsecs is about 32.6 light-years; 1,000 parsecs, or 1 kiloparsec (1 kpc), is about 3,260 light-years; and 1 million parsecs, or 1 megaparsec (1 Mpc), is about 3.26 million light-years.

Enough stars have measurable parallax to give us a fairly good sample of the many different types of stars. For example, we know of over 300 stars within about 33 light-years (10 parsecs) of the Sun. About half are members of binary or multiple star systems. Most are tiny, dim red stars such as Proxima Centauri—so dim that we cannot see them with the naked eye, despite the fact that they are relatively close. A few nearby stars, such as Sirius (2.7 parsecs), Vega (8 parsecs), Altair (5 parsecs), and Fomalhaut (7 parsecs), are white in color and bright in our sky, but many more of the brightest stars in the sky lie farther away. The intrinsic luminosities of stars must therefore span a wide range, since so many nearby stars appear dim while many more distant stars appear bright.

The Magnitude System

The easiest way to describe apparent brightness is in watts per square meter, and luminosities are easily described in either watts or solar luminosities (L_{Sun}). However, many amateur and professional astronomers describe stellar brightness using the ancient **magnitude system** devised by the Greek astronomer Hipparchus (c. 190–120 B.C.). The magnitude system originally classified stars according to how bright they look to our eyes—the only instruments available in ancient times. The brightest stars received the designation "first magnitude," the next brightest "second magnitude," and so on. The faintest visible stars were magnitude 6. We call these descriptions **apparent magnitudes** because they compare how bright different stars *appear* in the sky. Star charts often use dots of different sizes to represent the apparent magnitudes of stars.

Mathematical Insight 15.2 The Parallax Formula

There are several ways to derive the formula relating a star's distance and parallax angle: here is one of the easiest. Figure 15.2 shows that the parallax angle p is part of a right triangle in which the length of the side opposite p is the Earth–Sun distance of 1 AU and the length of the hypotenuse is the distance d to the object. Using the definition that the *sine* of an angle in a right triangle is the length of its opposite side divided by the length of the hypotenuse, we find:

$$\sin p = \frac{\text{length of opposite side}}{\text{length of hypotenuse}} = \frac{1 \text{ AU}}{d}$$

We generally use this formula to calculate d after we have measured p, so it is more useful to solve it for d:

$$d = \frac{1 \text{ AU}}{\sin p}$$

By definition, 1 parsec is the distance to an object with a parallax angle of 1 arcsecond (1″), or 1/3,600 degree. (Recall that 1° = 60′ and 1′ = 60″.) Substituting these numbers in the parallax formula and using a calculator to find that $\sin 1'' = 4.84814 \times 10^{-6}$, we get:

$$1 \text{ pc} = \frac{1 \text{ AU}}{\sin 1''} = \frac{1 \text{ AU}}{4.84814 \times 10^{-6}} = 206{,}265 \text{ AU}$$

In modern times, the magnitude system has been extended and more precisely defined. As a result, several bright stars have apparent magnitudes of 0, which is *brighter* than magnitude 1, and the brightest star in the night sky, Sirius, has an apparent magnitude of −1. The modern magnitude system also defines **absolute magnitudes** as a way of describing stellar luminosities. A star's absolute magnitude is the apparent magnitude it would have *if* it were at a distance of 10 parsecs from Earth. For example, the Sun's absolute magnitude is about 4.8, meaning that the Sun would have an apparent magnitude between 4 and 5 if it were located at a distance of 10 parsecs—bright enough to be visible, but not conspicuous, on a dark night.

You might want to acquaint yourself with the magnitude system, because it is still used by many astronomers, both amateurs and professionals. However, this ancient system can be difficult to use: The brighter the star, the smaller its magnitude, and the mathematical relationship between magnitudes is complicated. We generally avoid using the magnitude system in this book.

15.3 Stellar Surface Temperature

The other basic property of stars (besides luminosity) needed for modern stellar classification is surface temperature. Measuring a star's surface temperature is somewhat easier than measuring its luminosity because the measurement is not affected by the star's distance. Instead, we determine surface temperature directly from the star's color or spectrum. One note of caution before we continue: We can measure only a star's *surface temperature,* not its interior temperature. (Interior temperatures are calculated with theoretical models [Section 14.3].) When astronomers speak of the "temperature" of a star, they usually mean the surface temperature unless they say otherwise.

A star's surface temperature determines the color of light it emits. A red star is cooler than a yellow star, which in turn is cooler than a blue star. The naked eye can distinguish colors only for the brightest stars, but colors become more evident when we view stars through binoculars or a telescope (Figure 15.3). Astronomers can determine the "color" of a star more precisely by comparing its apparent brightness as viewed through two different filters [Section S2.3]. For example, a cool star such as Betelgeuse, with a surface temperature of about 3,000 K, emits more red light than blue light and therefore looks much brighter when viewed through a red filter than when viewed through a blue filter. In contrast, a hotter star such as Sirius, with a surface temperature of about 10,000 K, emits more blue light than red light and thus looks much brighter through a blue filter than through a red filter.

By using the fact that 1 AU = 149.6 million km, we can then show that 1 parsec is also equivalent to 3.09×10^{13} km and to 3.26 light-years.

We need one more fact from geometry to derive the parallax formula given in the text. As long as the parallax angle, p, is small, $\sin p$ is proportional to p. For example, $\sin 2''$ is twice as large as $\sin 1''$, and $\sin \frac{1}{2}''$ is half as large as $\sin 1''$. (You can verify these examples with your calculator.) Thus, if we use $\frac{1}{2}''$ instead of $1''$ for the parallax angle in the formula above, we get a distance of 2 pc instead of 1 pc. Similarly, if we use a parallax angle of $\frac{1}{10}''$, we get a distance of 10 pc. Generalizing, we get the simple parallax formula given in the text:

$$d \text{ (in parsecs)} = \frac{1}{p \text{ (in arcseconds)}}$$

Example: Sirius, the brightest star in our night sky, has a measured parallax angle of $0.379''$. How far away is it in parsecs? In light-years?

Solution: From the formula, the distance to Sirius in parsecs is

$$d \text{ (in pc)} = \frac{1}{0.379} = 2.64 \text{ pc}$$

Since 1 pc = 3.26 light-years, this distance is equivalent to:

$$2.64 \text{ pc} \times 3.26 \frac{\text{light-years}}{\text{pc}} = 8.60 \text{ light-years}$$

Spectral Type

The emission and absorption lines in a star's spectrum provide an independent and more accurate way to measure its surface temperature. Stars displaying spectral lines of highly ionized elements must be fairly hot, while stars displaying spectral lines of molecules must be relatively cool [Section 7.4]. Astronomers thus classify stars according to surface temperature by assigning a **spectral type** determined from

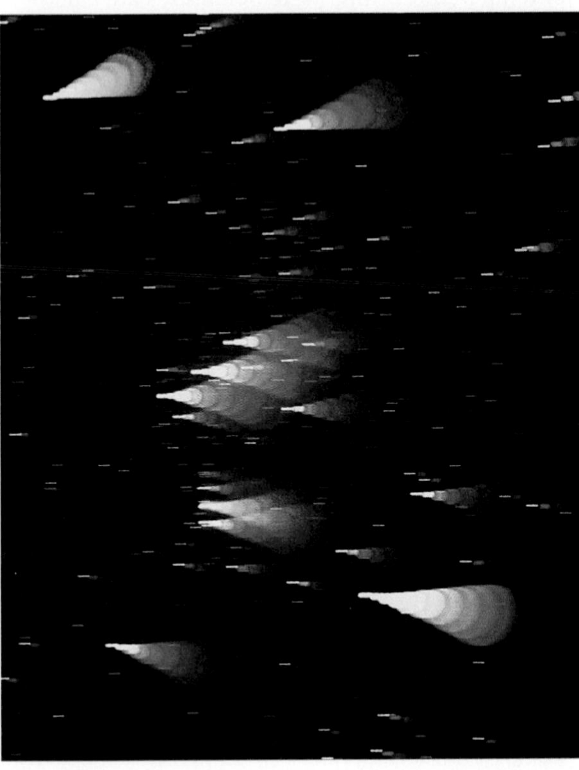

FIGURE 15.3 This specialized photograph shows the true colors of stars in Orion. Spreading each star's light into an extended cone reduces the effects of overexposure.

Table 15.1 The Spectral Sequence

Spectral Type	Example(s)	Temperature Range	Key Absorption Line Features
O	Stars of Orion's Belt	>30,000 K	Lines of ionized helium, weak hydrogen lines
B	Rigel	30,000 K–10,000 K	Lines of neutral helium, moderate hydrogen lines
A	Sirius	10,000 K–7,500 K	Very strong hydrogen lines
F	Polaris	7,500 K–6,000 K	Moderate hydrogen lines, moderate lines of ionized calcium
G	Sun, Alpha Centauri A	6,000 K–5,000 K	Weak hydrogen lines, strong lines of ionized calcium
K	Arcturus	5,000 K–3,500 K	Lines of neutral and singly ionized metals, some molecules
M	Betelgeuse, Proxima Centauri	<3,500 K	Molecular lines strong

Mathematical Insight 15.3 The Modern Magnitude Scale

The modern magnitude system is defined so that each difference of 5 magnitudes corresponds to a factor of exactly 100 in brightness. For example, a magnitude 1 star is 100 times brighter than a magnitude 6 star, and a magnitude 3 star is 100 times brighter than a magnitude 8 star. Because 5 magnitudes corresponds to a factor of 100 in brightness, a single magnitude corresponds to a factor of $(100)^{1/5} \approx 2.512$. The following formula summarizes the relationship between stars of different magnitudes:

$$\frac{\text{apparent brightness of Star 1}}{\text{apparent brightness of Star 2}} = (100^{1/5})^{m_2 - m_1}$$

where m_1 and m_2 are the apparent magnitudes of Stars 1 and 2, respectively. If we replace the apparent magnitudes with absolute magnitudes (designated M instead of m), the same formula applies to stellar luminosities:

$$\frac{\text{luminosity of Star 1}}{\text{luminosity of Star 2}} = (100^{1/5})^{M_2 - M_1}$$

Example 1: On a clear night, stars dimmer than fifth magnitude are quite difficult to see. Today, sensitive instruments on large telescopes can detect objects as the types of spectral lines present in a star's spectrum. The hottest stars, with the bluest colors, are spectral type O, followed in order of declining surface temperature by spectral types B, A, F, G, K, and M. The time-honored mnemonic for remembering this sequence, OBAFGKM, is "Oh Be A Fine Girl/Guy, Kiss Me!" Table 15.1 summarizes the characteristics of each spectral type.

Each spectral letter classification is subdivided into numbered subcategories (e.g., B0, B1, . . . , B9,

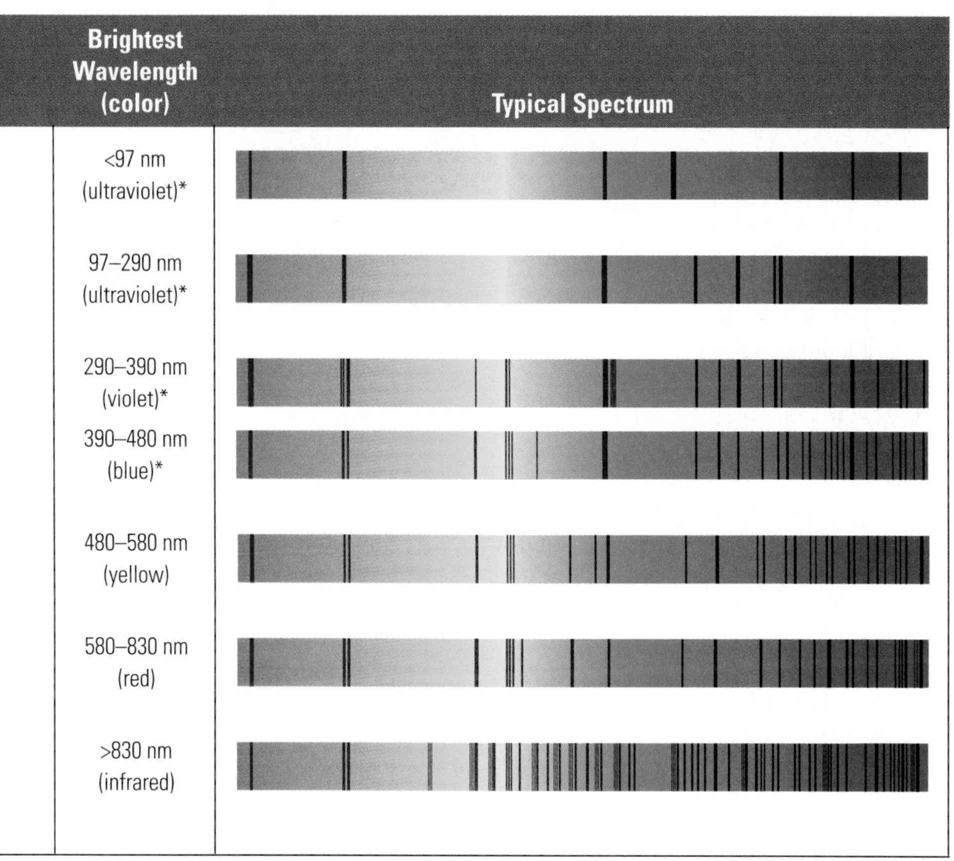

Brightest Wavelength (color)	Typical Spectrum
<97 nm (ultraviolet)*	
97–290 nm (ultraviolet)*	
290–390 nm (violet)*	
390–480 nm (blue)*	
480–580 nm (yellow)	
580–830 nm (red)	
>830 nm (infrared)	

*All stars above 6,000 K look more or less white to the human eye because they emit plenty of radiation at all visible wavelengths.

faint as thirtieth magnitude. How much more sensitive are such telescopes than the human eye?

Solution: We imagine that our eye sees "Star 1" with magnitude 5 and the telescope detects "Star 2" with magnitude 30; then we compare:

$$\frac{\text{apparent brightness of Star 1}}{\text{apparent brightness of Star 2}} = (100^{1/5})^{30-5} = (100^{1/5})^{25} = 100^5 = 10^{10}$$

The magnitude 5 star is 10^{10} = 10 billion times brighter than the magnitude 30 star, so the telescope is 10 billion times more sensitive than the human eye.

Example 2: The Sun has an absolute magnitude of about 4.8. The star Capella has an absolute magnitude of −0.7. How much more luminous is Capella than the Sun?

Solution: We use Capella as Star 1 and the Sun as Star 2:

$$\frac{\text{luminosity of Capella}}{\text{luminosity of Sun}} = (100^{1/5})^{4.8-(-0.7)} = (100^{1/5})^{5.5} = 100^{1.1} \approx 160$$

Capella is about 160 times more luminous than the Sun.

A0, A1 . . .). The larger the number, the cooler the star. For example, the Sun is designated spectral type G2, which means it is slightly hotter than a G3 star but cooler than a G1 star.

TIME OUT TO THINK *Invent your own mnemonic for the OBAFGKM sequence. To help get you thinking, here are two examples: (1) Only Bungling Astronomers Forget Generally Known Mnemonics; and (2) Only Business Acts For Good, Karl Marx.*

History of the Spectral Sequence

You may wonder why the spectral types follow such a peculiar order—OBAFGKM. The answer lies in the history of stellar spectroscopy. Astronomical research never paid well, and many astronomers of the 1800s were able to do research only because of family wealth. One such astronomer was Henry Draper (1837–1882), an early pioneer of stellar spectroscopy. After Draper died in 1882, his widow presented a series of large gifts to Harvard College Observatory for the purpose of building upon his work. The observatory director, Edward Pickering (1846–1919), used the gifts to improve the facilities and to hire numerous assistants, whom he called "computers"; Pickering added money of his own, as did other wealthy donors.

Most of Pickering's hired computers were women who had studied physics or astronomy at women's colleges such as Wellesley and Radcliffe. Women had few opportunities to advance in science at the time; Harvard, for example, did not allow women as either students or faculty. But Pickering's project of studying and classifying stellar spectra provided plenty of work and opportunity for his computers, and many of the Harvard Observatory women ended up among the most prominent astronomers of the late 1800s and early 1900s. One of the first computers was Williamina Fleming (1857–1911), who followed Pickering's suggestion and classified stellar spectra according to the strength of their hydrogen lines: Type A had the strongest hydrogen lines, type B slightly weaker ones, and so on to type O, which had the weakest hydrogen lines. Pickering published Fleming's classifications of more than 10,000 stars in 1890.

As more stellar spectra were obtained and as spectra were studied in greater detail, it became clear that the classification scheme based solely on hydrogen lines was inadequate. Ultimately, the task of finding a better classification scheme fell to Annie Jump Cannon (1863–1941), who joined Pickering's team in 1896 (Figure 15.4). Building upon the work of Fleming and another of Pickering's computers, Antonia Maury (1866–1952), Cannon soon realized that the spectral classes fell into a natural order—but it was not Pickering's original ABC . . . order determined by hydrogen lines alone. Moreover, she found that some of the original classes overlapped others and could therefore be eliminated. As it turned out, the natural sequence consisted of just a few of Pickering's original classes in the order OBAFGKM. Cannon also added the subdivisions by number and became so adept that she could properly classify a stellar spectrum with little more than a momentary glance. During her lifetime, she personally classified over 400,000 stars. She became the first woman ever awarded an honorary degree by Oxford University, and in 1929 the League of Women Voters named her one of the 12 greatest living American women.

FIGURE 15.4 Women astronomers pose with Edward Pickering at Harvard College Observatory in 1913. Annie Jump Cannon is fifth from the left in the back row.

The astronomical community adopted Cannon's system of stellar classification in 1910. However, no one at that time knew *why* spectra followed the OBAFGKM sequence. Many astronomers guessed, incorrectly, that the different sets of spectral lines reflected different compositions for the stars. The correct answer—that all stars are made primarily of hydrogen and helium and that a star's surface temperature determines the strength of its spectral lines—was discovered by Cecilia Payne-Gaposchkin (1900–1979), another woman working at Harvard Observatory (Figure 15.5). Relying on insights from what was then the newly developing science of quantum mechanics, Payne-Gaposchkin showed that the differences in spectral lines from star to star merely reflected changes in the ionization level of the emitting atoms. For example, O stars have weak hy-

FIGURE 15.5 Cecilia Payne-Gaposchkin.

drogen lines because, at their high surface temperatures, nearly all their hydrogen is ionized. Without an electron to "jump" between energy levels, ionized hydrogen can neither emit nor absorb its usual specific wavelengths of light. At the other end of the spectral sequence, M stars are cool enough for molecules to form, thus explaining their strong molecular absorption lines. Payne-Gaposchkin described her work and her conclusions in a dissertation published in 1925. A later review of twentieth-century astronomy called her work "undoubtedly the most brilliant Ph.D. thesis ever written in astronomy."[4]

15.4 Stellar Masses

The most important property of a star is its mass, but stellar masses are much harder to measure than luminosities or surface temperatures. The most dependable method for "weighing" a star relies on Newton's version of Kepler's third law:

$$p^2 = \frac{4\pi^2}{G(M_1 + M_2)}a^3 \quad \left(G = 6.67 \times 10^{-11} \frac{\text{m}^3}{\text{kg} \times \text{s}^2}\right)$$

where M_1 and M_2 are the masses of two orbiting objects, and p and a are the orbital period and average distance (semimajor axis) of either of the two objects [Section 6.4]. Note that this law can be applied only when we can measure both p and a. Thus, unless we have identified a planet orbiting a star, we can measure stellar masses only in binary star systems in which we can determine the orbital properties of the two stars.

[4] O. Struve and V. Zebergs, *Astronomy of the 20th Century* (New York: Macmillan, 1962).

About half of all stars orbit a companion star of some kind. In some cases, particularly in binary star systems that are relatively close to us, we can resolve both stars through a telescope. Mizar, the second star in the handle of the Big Dipper, is an example of such a **visual binary** (Figure 15.6). (The star Alcor appears very close to Mizar to the naked eye but does *not* orbit it.) The two stars in the visual binary system, Mizar A and Mizar B, gradually change positions, indicating that they are orbiting each other. We can observe only a portion of their orbit in a lifetime, because their orbital period is thousands of years.

Sometimes we observe a star slowly shifting position in the sky as if it were a member of a visual binary, but its companion is too dim to be seen. In such cases, we can use the wobbling of the star's position to infer the existence and mass of its unseen companion. Early twentieth-century astronomers used this method to discover that Sirius has a dim companion (Figure 15.7). With modern telescopes, we can see both Sirius A (the brightest star in our sky) and Sirius B (the companion, which is a white dwarf) as a visual binary.

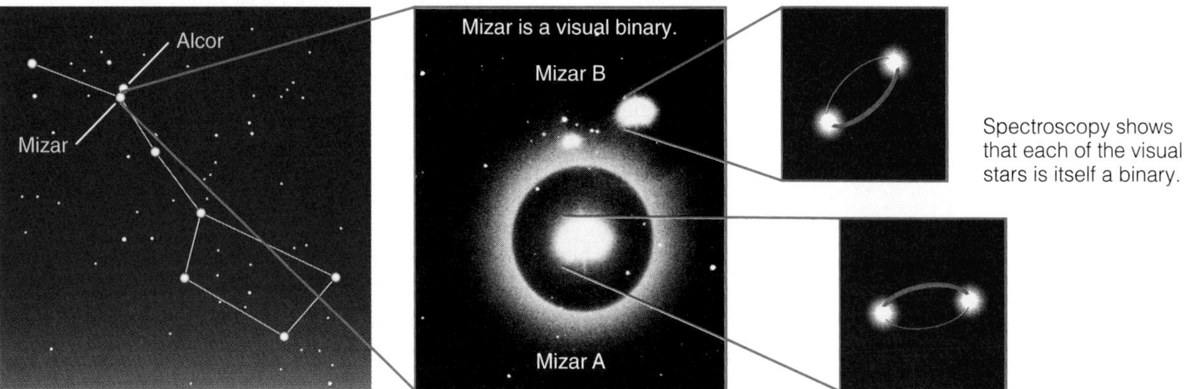

FIGURE 15.6 Mizar looks like one star to the naked eye but is actually a system of four stars. A telescope shows a visual binary, but each of the visual stars is actually a spectroscopic binary.

Visual binaries are relatively rare. More often, the two stars of a binary system are sufficiently close to appear as one even through the largest telescopes. If the two stars happen to be orbiting in the plane of our line of sight, we call the system an **eclipsing binary** because each star will periodically eclipse, or pass in front of, the other (Figure 15.8). When neither star is eclipsed, we see the combined light of both stars. But when one star eclipses the other, the apparent brightness of the system drops because

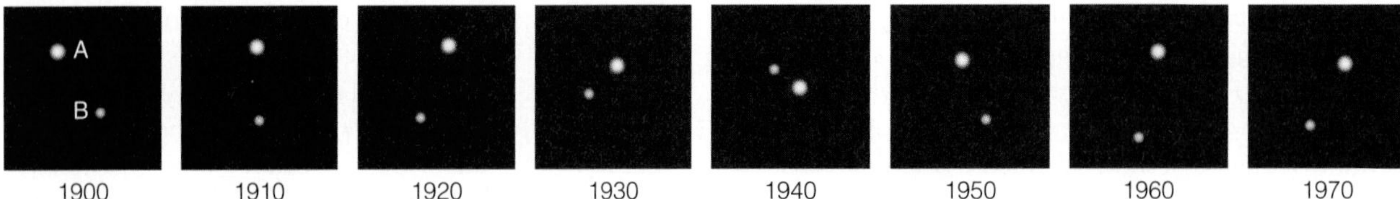

FIGURE 15.7 Each frame represents the relative positions of Sirius A and Sirius B at 10-year intervals from 1900 to 1970. The back-and-forth "wobble" of Sirius A allowed astronomers to infer the existence of Sirius B even before the two stars could be resolved in telescopic photos.

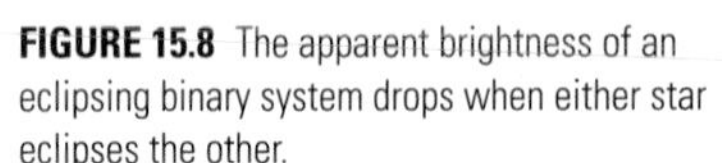

FIGURE 15.8 The apparent brightness of an eclipsing binary system drops when either star eclipses the other.

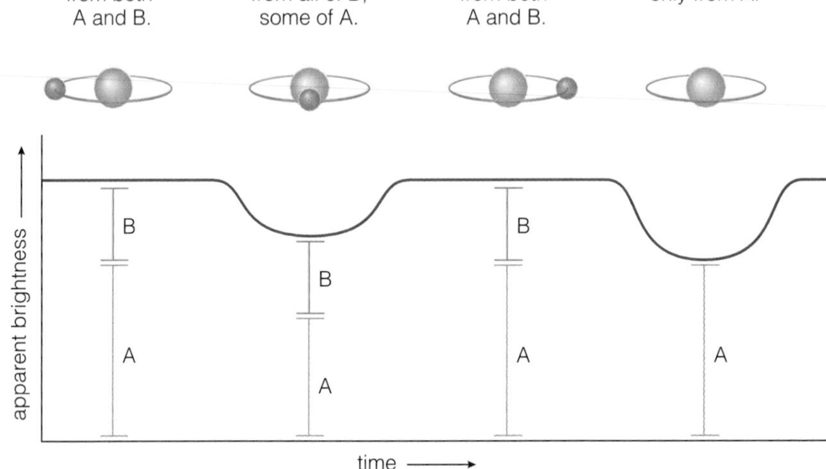

Mathematical Insight 15.4 Orbital Separation and Newton's Version of Kepler's Third Law

Measurements of stellar masses rely on Newton's version of Kepler's third law:

$$p^2 = \frac{4\pi^2}{G(M_1 + M_2)}a^3 \quad \left(G = 6.67 \times 10^{-11} \frac{\text{m}^3}{\text{kg} \times \text{s}^2}\right)$$

As described in the text, the orbital period p is generally easy to measure for binary star systems. We can rarely measure the semimajor axis a directly, but we can calculate it in cases in which we can measure the orbital velocity v. If we assume the orbit is a circle of radius a, the circumference of the orbit is $2\pi a$. Because the star makes one circuit of this circumference in one orbital period p, the star's velocity is:

$$v = \frac{\text{distance traveled in one orbit}}{\text{period of one orbit}} = \frac{2\pi a}{p}$$

Solving for a, we find:

$$a = \frac{pv}{2\pi}$$

Once we know both p and a, we can use Newton's version of Kepler's third law to calculate the *sum* of the masses of the two stars ($M_1 + M_2$). We can then calculate the individual masses of the two stars by taking advantage of the fact that the *relative* velocities of the two stars around the center of mass are inversely proportional to their relative masses.

some of the light is blocked from view. A *light curve,* or graph of apparent brightness against time, reveals the pattern of the eclipses [Section S2.3]. The most famous example of an eclipsing binary is Algol, the "demon star" in the constellation Perseus (*algol* is Arabic for "the ghoul"), which becomes significantly dimmer for a few hours about every 3 days as the brighter of its two stars is eclipsed by its dimmer companion.

If a binary system is neither visual nor eclipsing, we may be able to detect its binary nature by observing Doppler shifts in its spectral lines [Section 7.5]. If one star is orbiting another, it periodically moves toward us and away from us in its orbit, and its spectral lines show blueshifts and redshifts as a result of this motion (Figure 15.9). Because we detect the binary nature of such star systems by studying their spectra, they are called **spectroscopic binary** systems. Sometimes we see two sets of lines shifting back and forth—one set from each of the two stars in the system (a *double-line* spectroscopic binary). Other times we see a set of shifting lines from only one star because its companion is too dim to be detected (a *single-line* spectroscopic binary). Interestingly, both Mizar A and Mizar B are spectroscopic binaries. Thus, Mizar, which appears as a single star to the naked eye, is actually an amalgam of four stars, all bound together by gravity.

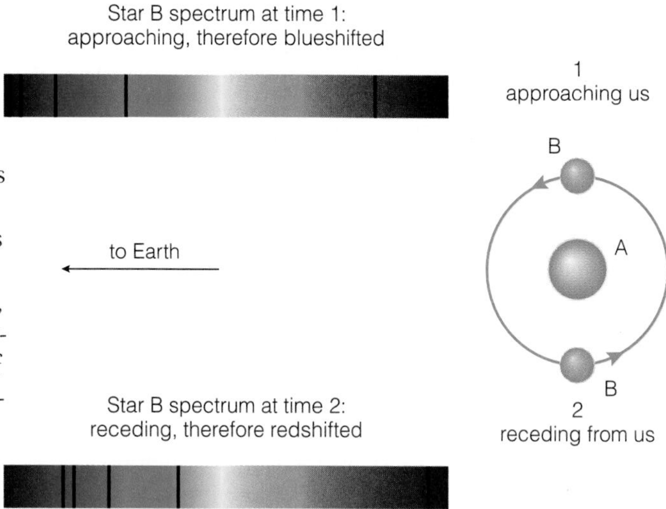

FIGURE 15.9 The light from a star in a binary system is alternately blueshifted as it comes toward us in its orbit and redshifted as it moves away from us.

Even with a binary star system, we can apply Newton's version of Kepler's third law only if we can measure both the orbital period and the separation of the two stars. Measuring orbital period is fairly easy: In a visual binary, we simply observe how long each orbit takes (or extrapolate from part of an orbit); in

Example: The spectral lines of two stars in a particular eclipsing binary system shift back and forth with a period of 2 years ($p = 6.2 \times 10^7$ seconds). The lines of one star (Star 1) shift twice as far as the lines of the other (Star 2); the amount of the Doppler shift indicates an orbital speed $v = 100{,}000$ m/s for Star 1. What are the masses of the two stars? Assume that each of the two stars traces a circular orbit around their center of mass.

Solution: We first compute the semimajor axis a of Star 1 from its orbital velocity v:

$$a = \frac{pv}{2\pi} = \frac{(6.2 \times 10^7 \text{ s}) \times (100{,}000 \text{ m/s})}{2\pi} = 9.9 \times 10^{11} \text{ m}$$

Now we use Newton's version of Kepler's third law to calculate the sum of the stellar masses:

$$(M_1 + M_2) = \frac{4\pi^2 \times a^3}{G \times p^2} = \frac{4\pi^2 \times (9.9 \times 10^{11} \text{ m})^3}{\left(6.67 \times 10^{-11} \dfrac{\text{m}^3}{\text{kg} \times \text{s}^2}\right)(6.2 \times 10^7 \text{ s})^2}$$
$$= 1.5 \times 10^{32} \text{ kg}$$

The fact that the lines of Star 1 shift twice as far as those of Star 2 tells us that Star 1 moves twice as fast as Star 2, and hence that Star 1 is half as massive as Star 2. In other words, Star 2 is twice as massive as Star 1. Using this fact and their combined mass of 1.5×10^{32} kg, we conclude that the mass of Star 2 is 1.0×10^{32} kg and the mass of Star 1 is 0.5×10^{32} kg.

an eclipsing binary, we measure the time between eclipses; and in a spectroscopic binary, we measure the time it takes the spectral lines to shift back and forth. In contrast, determining the average separation of the stars in a binary system is usually much more difficult: Except in rare cases, like visual binaries, in which we can measure the separation directly, we can calculate the separation only if we know the actual orbital speeds of the stars. The primary technique for measuring stellar speeds relies on the Doppler effect, but Doppler shifts tell us only the portion of a star's velocity that is directly toward us or away from us. Orbiting stars generally do not move directly along our line of sight, so their actual velocities can be significantly greater than those we measure through the Doppler effect. However, the orbits of eclipsing binary stars take them directly toward us and directly away from us once each orbit, enabling us to measure their true orbital velocities.[5] Eclipsing binaries are therefore particularly important to the study of stellar masses. As an added bonus, eclipsing binaries allow us to measure stellar radii directly: Because we know how fast the stars are moving across our line of sight as one eclipses the other, we can determine their radii by timing how long each eclipse lasts.

TIME OUT TO THINK ***Suppose two orbiting stars are moving in a plane perpendicular to our line of sight. Would the spectral features of these stars appear shifted in any way? Explain.***

15.5 The Hertzsprung–Russell Diagram

During the first decade of the twentieth century, a similar thought occurred independently to astronomers Ejnar Hertzsprung, working in Denmark, and Henry Norris Russell, working in the United States at Princeton University. Each decided to make a graph plotting stellar luminosities on one axis and spectral types on the other. Such graphs, from which we can immediately read a star's luminosity and surface temperature, are now known as **Hertzsprung–Russell (H–R) diagrams.** (Mass is not shown on an H–R diagram, but later we will see how the H–R diagram helps us understand the importance of mass.) Soon after they began making their graphs, Hertzsprung and Russell uncovered some previously unsuspected patterns in the properties of stars. As we will see shortly, understanding these patterns and the H–R diagram is central to the study of stars.

A Basic H–R Diagram

Figure 15.10 displays an example of an H–R diagram. Note the following key features of the diagram's construction:

- The horizontal axis represents stellar surface temperature or, equivalently, spectral type.
- Temperature increases *from right to left* because Hertzsprung and Russell based their diagrams on the spectral sequence OBAFGKM.
- The vertical axis represents stellar luminosity, in units of the Sun's luminosity (L_{Sun}).
- Stellar luminosities span a wide range, so we keep the graph compact by making each tick mark represent a luminosity 10 times larger than the prior tick mark.
- Each dot on the diagram represents the spectral type and luminosity of a single star. For example, the dot representing the Sun in Figure 15.10 corresponds to the Sun's spectral type of G2 and its luminosity of $1L_{Sun}$.

Because luminosity increases upward on the diagram and surface temperature increases leftward, stars near the upper left are hot and luminous. Similarly, stars near the upper right are cool and luminous, stars near the lower right are cool and dim, and stars near the lower left are hot and dim.

TIME OUT TO THINK ***Explain how the colors of the dots in Figure 15.10 help indicate stellar surface temperature. Do these colors tell us anything about interior temperatures? Why or why not?***

The H–R diagram also provides direct information about stellar radii, because a star's luminosity depends on both its surface temperature and its surface area or radius. Recall that surface temperature determines the amount of power emitted by the star *per unit area:* Higher temperature means greater power output per unit area [Section 7.4]. Thus, if two stars have the same surface temperature, one can be more luminous than the other only if it is larger in size. Stellar radii therefore must increase as we go from the high-temperature, low-luminosity corner on the lower left of the H–R diagram to the low-temperature, high-luminosity corner on the upper right, as shown in Figure 15.11.

[5]In other binaries, we can calculate an actual orbital velocity from the velocity obtained by the Doppler effect if we also know the system's orbital inclination. Astronomers have developed techniques for determining orbital inclination in a relatively small number of cases.

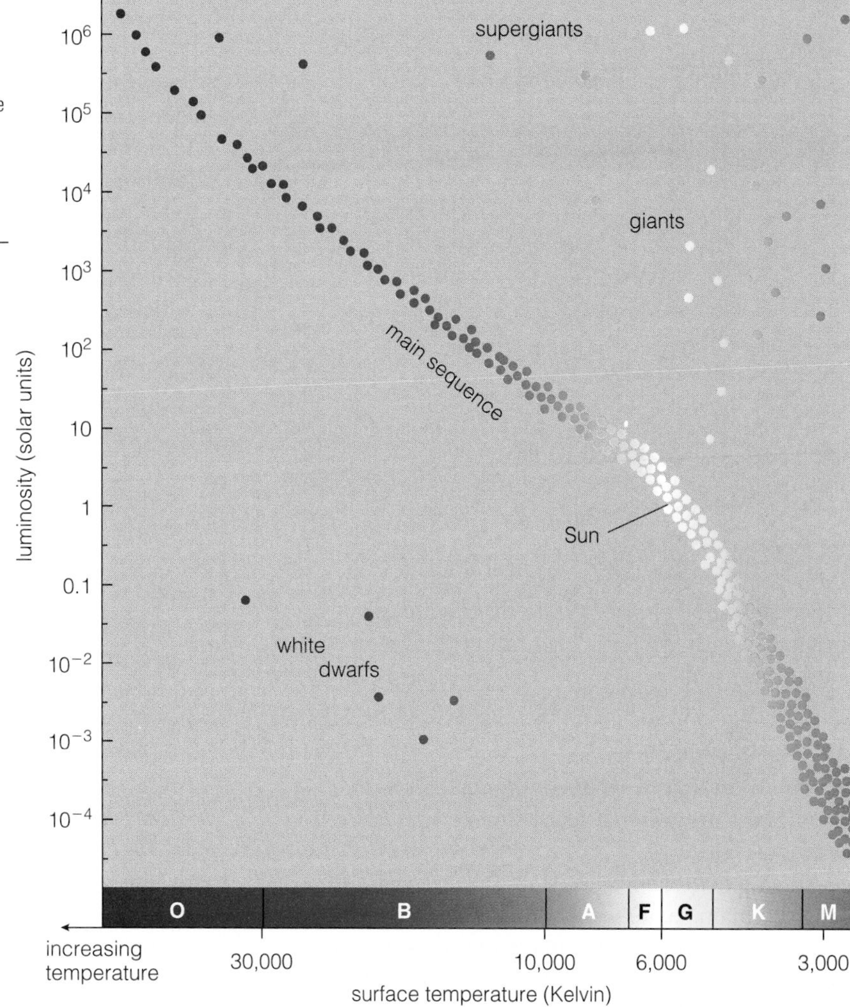

FIGURE 15.10 An H–R diagram, one of astronomy's most important tools, shows how the surface temperatures of stars, plotted along the horizontal axis, relate to their luminosities, plotted along the vertical axis. (See page 500 for a more complete description.)

Patterns in the H–R Diagram

Figure 15.10 also shows that stars do not fall randomly throughout the H–R diagram but instead fall into several distinct groups:

- Most stars fall somewhere along the **main sequence,** the prominent streak running from the upper left to the lower right on the H–R diagram. Note that our Sun is a main-sequence star.
- The stars in the extreme upper right are called **supergiants** because they are very large in addition to being very bright.
- Just below the supergiants are the **giants,** which are somewhat smaller in radius and lower in luminosity.
- The stars in the lower left are small in radius and appear white in color because of their high temperature; we therefore call these stars the **white dwarfs.**

When classifying a star, astronomers generally report both the star's spectral type and a **luminosity class** that describes the region of the H–R diagram in which the star falls. Table 15.2 summarizes the luminosity classes. Note that luminosity class I represents supergiants,[6] luminosity class III represents giants, and luminosity class V represents main-sequence stars, while luminosity classes II and IV are intermediate to the others. For example, the complete spectral classification of our Sun is G2 V: The

[6]Astronomers subdivide luminosity class I into Ia and Ib, with Ia the larger and brighter of the two.

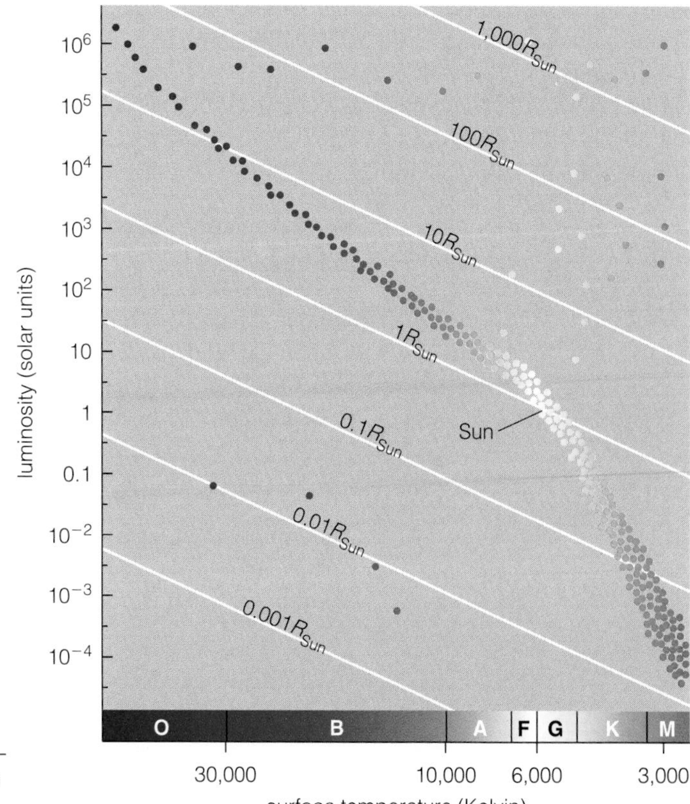

FIGURE 15.11 Stellar radii increase as we go diagonally up and right on the H–R diagram.

Table 15.2 Stellar Luminosity Classes

Class	Description
I	Supergiants
II	Bright giants
III	Giants
IV	Subgiants
V	Main sequence

G2 spectral type means it is yellow in color, and the luminosity class V means it is a main-sequence star. Betelgeuse is M2 I, making it a *red supergiant*. Proxima Centauri is M5 V—similar in color and surface temperature to Betelgeuse, but far dimmer because of its much smaller size. White dwarfs are usually designated with the letters *wd* rather than with a Roman numeral.

The Main Sequence

The common trait of main-sequence stars is that, like our Sun, they are fusing hydrogen into helium in their cores. Because stars spend the majority of their lives fusing hydrogen, most stars fall somewhere along the main sequence of the H–R diagram.

But why do main-sequence stars span such a wide range of luminosities and surface temperatures? By measuring the masses of stars in binary systems, astronomers discovered that stellar masses increase upward along the main sequence (Figure 15.12). At the upper end of the main sequence, the hot, luminous O stars can have masses as high as 100 times that of the Sun ($100M_{\text{Sun}}$). On the lower end, cool, dim M stars may have as little as 0.1 times the mass of the Sun ($0.1M_{\text{Sun}}$). Far more stars fall on the lower end of the main sequence than on the upper end, which tells us that low-mass stars are far more common than high-mass stars.

The orderly arrangement of stellar masses along the main sequence tells us that *mass* is the most important attribute of a hydrogen-burning star. Luminosity depends directly on mass because the weight of a star's outer layers determines the nuclear burning rate in its core [Section 14.3]. In fact, the nuclear burning rate, and hence the luminosity, is very sensitive to mass. For example, a $10M_{\text{Sun}}$ star on the main sequence is more than 1,000 times more luminous than the Sun.

The relationship between mass and surface temperature is a little more subtle. In general, a very luminous star must either be very large or have a very high surface temperature, or some combination of both. Stars on the upper end of the main sequence are thousands of times more luminous than the Sun but only about 10 times larger than the Sun in radius. Thus, their surfaces must be significantly hotter than the Sun's surface to account for their high luminosities. Main-sequence stars more massive than the Sun therefore have higher surface temperatures, and those less massive than the Sun have lower surface temperatures. That is why the main sequence slices diagonally from the lower right to the upper left on the H–R diagram.

Main-Sequence Lifetimes

A star has a limited supply of core hydrogen and therefore can remain as a hydrogen-fusing main-sequence star for only a limited time—the star's **main-sequence lifetime** (or *hydrogen-burning lifetime*).

Mathematical Insight 15.5 Calculating Stellar Radii

Almost all stars are too distant for us to measure their radii directly. However, we can calculate a star's radius from its luminosity with the aid of the thermal radiation laws. As given in Mathematical Insight 7.2, the amount of thermal radiation emitted by a star of surface temperature T is:

$$\text{emitted power per unit area} = \sigma T^4 \quad \left(\sigma = 5.7 \times 10^{-8} \frac{\text{watt}}{\text{m}^2 \times \text{Kelvin}^4}\right)$$

The luminosity L of a star is this power per unit area multiplied by the total surface area of the star. If the star has radius r, its surface area is given by the formula $4\pi r^2$. Thus:

$$L = 4\pi r^2 \times \sigma T^4$$

With a bit of algebra, we can solve this formula for the star's radius r:

$$r = \sqrt{\frac{L}{4\pi\sigma T^4}}$$

Example: Betelgeuse has a luminosity of $52{,}000L_{\text{Sun}}$ and a surface temperature of about 3,000 K. What is its radius?

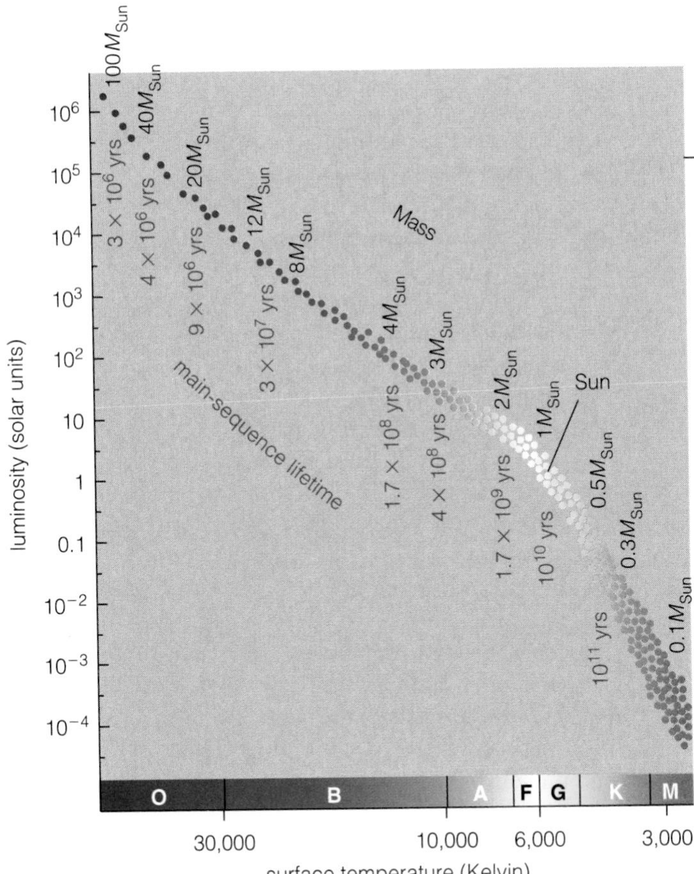

FIGURE 15.12 Along the main sequence, more massive stars are brighter and hotter but have shorter lifetimes. (Stellar masses are given in units of solar masses: 1 $M_{Sun} = 2 \times 10^{30}$ kg.)

Because stars spend the vast majority of their lives fusing hydrogen into helium, we sometimes refer to the main-sequence lifetime as simply the "lifetime." Like masses, stellar lifetimes vary in an orderly way as we move up the main sequence: Massive stars near the upper end of the main sequence have *shorter* lives than less massive stars near the lower end (see Figure 15.12).

Why do more massive stars live shorter lives? A star's lifetime depends on both its mass and its luminosity. Its mass determines how much hydrogen fuel the star initially contains in its core. Its luminosity determines how rapidly the star uses up its fuel. Massive stars live shorter lives because, even though they start their lives with a larger supply of hydrogen, they consume their hydrogen at a prodigious rate.

The main-sequence lifetime of our Sun is about 10 billion years [Section 14.1]. A 30-solar-mass star has roughly 30 times more hydrogen fuel than the Sun but burns it with a luminosity some 300,000 times greater. Consequently, its lifetime is roughly 30/300,000 = 1/10,000 as long as the Sun's—corresponding to a lifetime of only a few million years. Cosmically speaking, a few million years is a remarkably short time, which is one reason why massive stars are so rare: Most of the massive stars that have ever been born are long since dead. (A second reason is that lower-mass stars form in larger numbers than higher-mass stars [Section 16.2].) The fact that any massive stars exist tells us that stars form continuously in our galaxy. The massive, bright O stars we see today formed approximately at the time when our distant ancestors began to walk on two feet, long after the dinosaurs became extinct, and these stars will die long before they have a chance to complete even one orbit around the center of the galaxy.

Solution: First, we must make our units consistent by converting the luminosity of Betelgeuse into watts. Remembering that $L_{Sun} = 3.8 \times 10^{26}$ watts, we find:

$$L_{Bet} = 52{,}000 \times L_{Sun} = 52{,}000 \times 3.8 \times 10^{26}\ \text{watts} = 2.0 \times 10^{31}\ \text{watts}$$

Now we can use the formula derived above to calculate the radius of Betelgeuse:

$$r = \sqrt{\frac{L}{4\pi\sigma T^4}} = \sqrt{\frac{2.0 \times 10^{31}\ \text{watts}}{4\pi \times \left(5.7 \times 10^{-8} \dfrac{\text{watt}}{\text{m}^2 \times \text{K}^4}\right) \times (3{,}000\ \text{K})^4}}$$

$$= \sqrt{\frac{2.0 \times 10^{31}\ \text{watts}}{5.8 \times 10^{7} \dfrac{\text{watts}}{\text{m}^2}}} = 5.8 \times 10^{11}\ \text{m}$$

The radius of Betelgeuse is about 580 billion meters or, equivalently, 580 million kilometers. Note that this is nearly four times the Earth–Sun distance of 150 million kilometers.

TIME OUT TO THINK *Would you expect to find life on planets orbiting massive O stars? Why or why not?* (Hint: *Compare the lifetime of an O star to the amount of time that passed from the formation of our solar system to the origin of life on Earth.*)

On the other end of the scale, a 0.3-solar-mass star emits a luminosity just 0.01 times that of the Sun and consequently lives roughly 0.3/0.01 = 30 times longer than the Sun. In a universe that is now between 10 and 16 billion years old, even the most ancient of these small, dim M stars still survive and will continue to shine faintly for hundreds of billions of years to come.

Giants, Supergiants, and White Dwarfs

Giants and supergiants are stars nearing the ends of their lives because they have already exhausted their core hydrogen. Surprisingly, stars grow more luminous when they begin to run out of fuel. As we will discuss in the next chapter, a star generates energy furiously during the last stages of its life as it tries to stave off the inevitable crushing force of gravity. As ever-greater amounts of power well up from the core, the outer layers of the star expand, making it into a giant or supergiant. The largest of these celestial behemoths have radii more than 1,000 times that of the Sun. Because they are so bright, giants and supergiants can be seen even when they are not especially close to us. Many of the brightest stars visible to the naked eye are giants or supergiants; they are often identifiable by their reddish color. Nevertheless, giants and supergiants are rarer than main-sequence stars because, in our snapshot of the heavens, we catch most stars in the act of hydrogen burning and relatively few in a later stage of life.

Giants and supergiants eventually run out of fuel entirely. A giant with a mass similar to that of our Sun ultimately ejects its outer layers, leaving behind a "dead" core in which all nuclear fusion has ceased. White dwarfs are these remaining embers of former giants. They are hot because they are essentially exposed stellar cores, but they are dim because they lack an energy source and radiate only their leftover heat into space. A typical white dwarf is no larger in size than the Earth, although it may have a mass as great as that of our Sun. (Giants and supergiants with masses much larger than that of the Sun ultimately explode, leaving behind neutron stars or black holes as corpses [Section 16.4].)

Pulsating Variable Stars

Not all stars shine steadily like our Sun. Any star that significantly varies in brightness with time is called a *variable star.* A particularly important class of variable stars includes stars that have a peculiar problem with achieving the proper balance between the power welling up from the stellar core and the power being radiated from the stellar surface. Sometimes the photosphere of such a star is too opaque; energy and pressure build up beneath it, and the star expands in size. However, this expansion puffs the photosphere outward, making it too transparent. So much energy then escapes that the underlying pressure drops, and the star contracts again. In a futile quest for a steady equilibrium, the atmospheres of these **pulsating variable stars** alternately expand and contract, causing the star to rise and fall in luminosity. Figure 15.13 shows a typical light curve for a pulsating variable star, with the star's brightness graphed against time. Any pulsating variable star has its own particular period between peaks in luminosity, which we can discover easily from its light curve. These periods can range from as short as several hours to as long as several years.

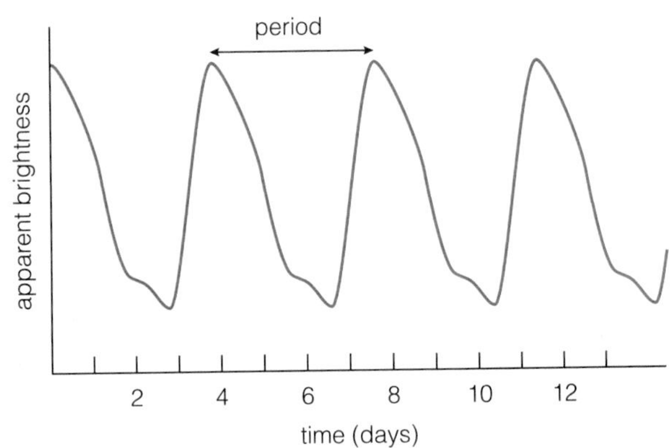

FIGURE 15.13 A typical light curve for a Cepheid variable star. Cepheids are giant, whitish stars whose luminosities regularly pulsate over periods of a few days to about a hundred days. The pulsation period of this Cepheid is about four days.

Most pulsating variable stars inhabit a strip (called the *instability strip*) on the H–R diagram that lies between the main sequence and the red giants (Figure 15.14). A special category of very luminous pulsating variables lies at the top of this strip: the **Cepheid variables,** or **Cepheids** (so named because the first identified star of this type was the star Delta Cephei).

Cepheids fluctuate in luminosity with periods of a few days to a few months. In 1912, another woman astronomer at Harvard, Henrietta Leavitt, discovered that the periods of these stars are very closely related to their luminosities: The longer the period, the more luminous the star (Figure 15.15).[7] This **period–luminosity relation** holds because larger (and hence more luminous) Cepheids take longer to pulsate in and out in size.

Once we measure the period of a Cepheid variable, we can use the period–luminosity relation to determine its luminosity. We can then calculate its distance with the luminosity–distance formula. In fact, as we'll discuss in Chapter 19, Cepheids provide our primary means of measuring distances to other galaxies and thus teach us the true scale of the cosmos. The next time you look at the North Star, Polaris, gaze upon it with renewed appreciation. Not only has it guided generations of navigators in the Northern Hemisphere, but it is also one of these special Cepheid variable stars.

[7]Cepheids actually come in two types: Type I Cepheids are found in young clusters of stars, and Type II Cepheids are usually found in old clusters of stars. Figure 15.15 shows the period–luminosity relation only for Type I Cepheids; the slope of the relation is similar for Type II Cepheids, but these stars are not as bright.

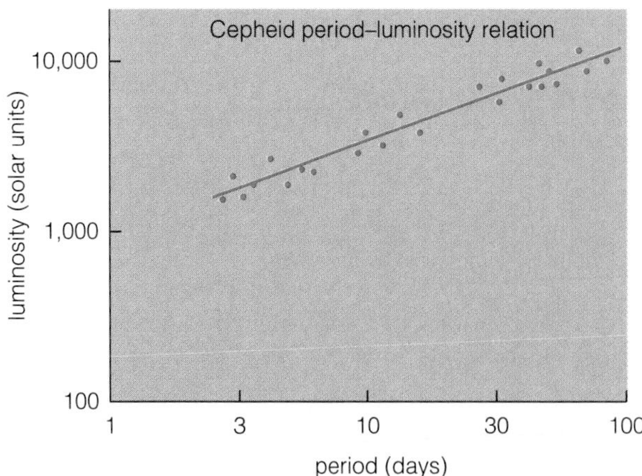

FIGURE 15.15 The period–luminosity relation for Cepheids. The pulsation period of a Cepheid is related directly to its luminosity—the more luminous a Cepheid is, the slower it pulsates.

15.6 Star Clusters

All stars are born from giant clouds of gas. Because a single interstellar cloud can contain enough material to form many stars, stars almost inevitably form in groups. In our snapshots of the heavens, many stars still congregate in the groups in which they formed. These groups are of two basic types: modest-size **open clusters** and densely packed **globular clusters.**

Open clusters of stars are always found in the disk of the galaxy. They can contain up to several thousand stars and typically span about 30 light-years (10 parsecs). The most famous open cluster is the *Pleiades,* a prominent clump of stars in the constellation Taurus (Figure 15.16). The Pleiades are often called the *Seven Sisters,* although only six of the cluster's several thousand stars are easily visible to the naked eye. Other cultures have other names for this beautiful group of stars; in Japanese it is *Subaru,* which is why the logo for Subaru automobiles is a diagram of the Pleiades.

Globular clusters are found in both the halo and the disk of our galaxy. A globular cluster can contain more than a million stars concentrated in a ball typically 60 to 150 light-years across (20 to 50 parsecs). Its innermost region can have 10,000 stars packed within just a few light-years (Figure 15.17). The view from a planet in a globular cluster would be marvelous, with thousands of stars lying closer than Alpha Centauri is to the Sun. Because a globular cluster's stars nestle so closely, they engage in an intricate

FIGURE 15.14 An H–R diagram with the instability strip highlighted.

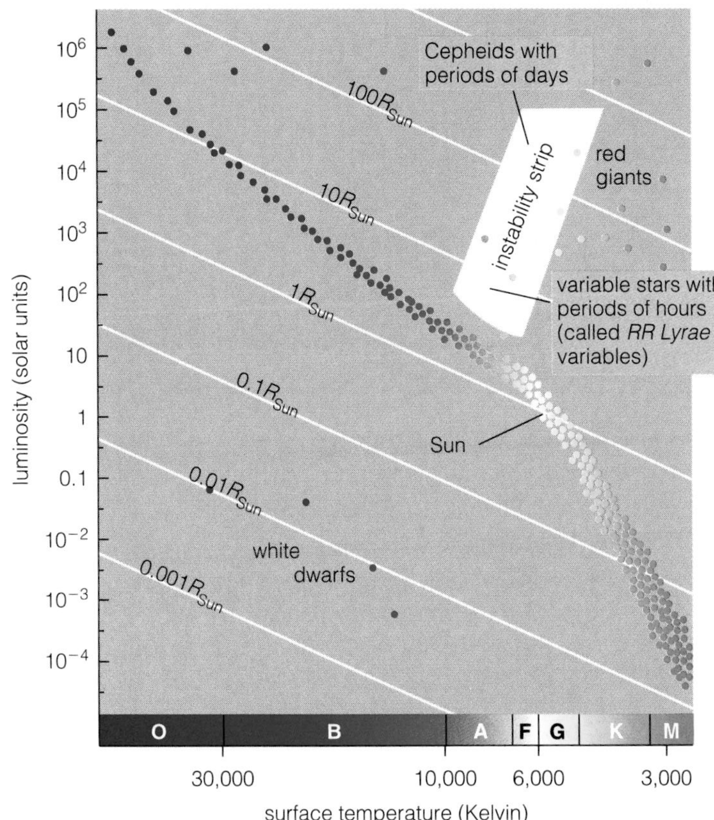

FIGURE 15.16 A photo of the Pleiades, a nearby open cluster of stars. The most prominent stars in this open cluster are of spectral type B, indicating that the Pleiades are less than 100 million years old, relatively young for a star cluster.

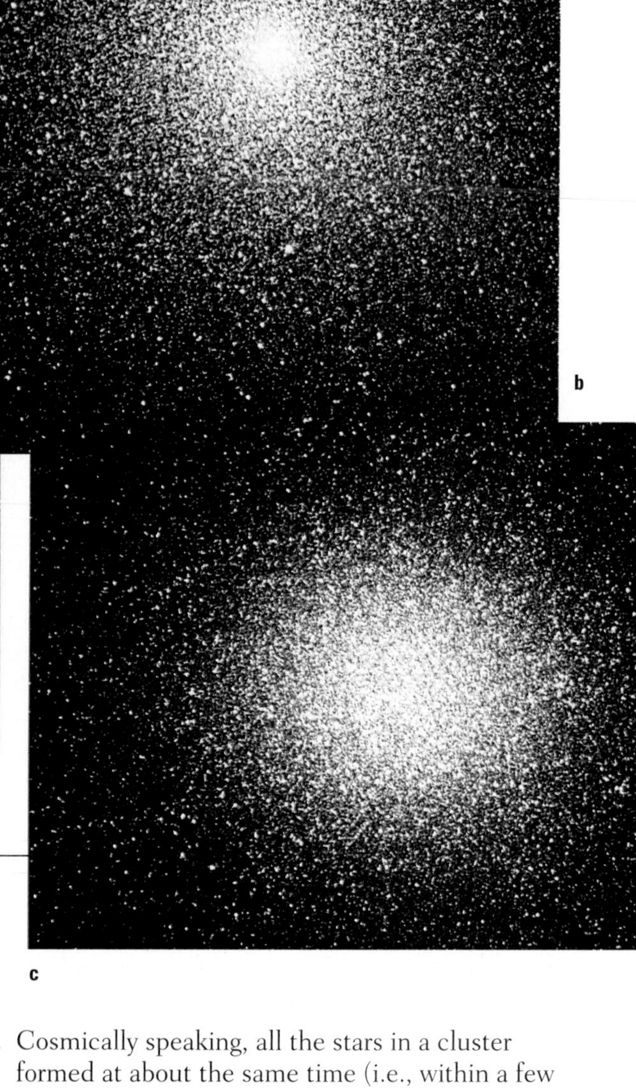

FIGURE 15.17 These photos of globular clusters show tight groups of 10 billion years old stars, bound together by gravity. (**a**) M5 in Serpens (**b**) 47 Tucanae (**c**) Omega Centauri.

and complex dance choreographed by gravity. Some stars zoom from the cluster's core to its outskirts and back again at speeds approaching the escape velocity from the cluster, while others orbit the dense core more closely. When two stars pass especially close to each other, the gravitational pull between them deflects their trajectories, altering their speeds and sending them careening off in new directions. Occasionally, a close encounter boosts one star's velocity enough to eject it from the cluster. Through such ejections, globular clusters gradually lose stars and grow more compact.

Star clusters are extremely useful to astronomers for two key reasons:

1. All the stars in a cluster lie at about the same distance from Earth.
2. Cosmically speaking, all the stars in a cluster formed at about the same time (i.e., within a few million years of one another).

Astronomers can therefore use star clusters as laboratories for comparing the properties of stars, as yardsticks for measuring distances in the universe, and as timepieces for measuring the age of our galaxy.

Cluster Distances from Main-Sequence Fitting

The fact that stars in a cluster all lie at approximately the same distance allows us to determine the distance

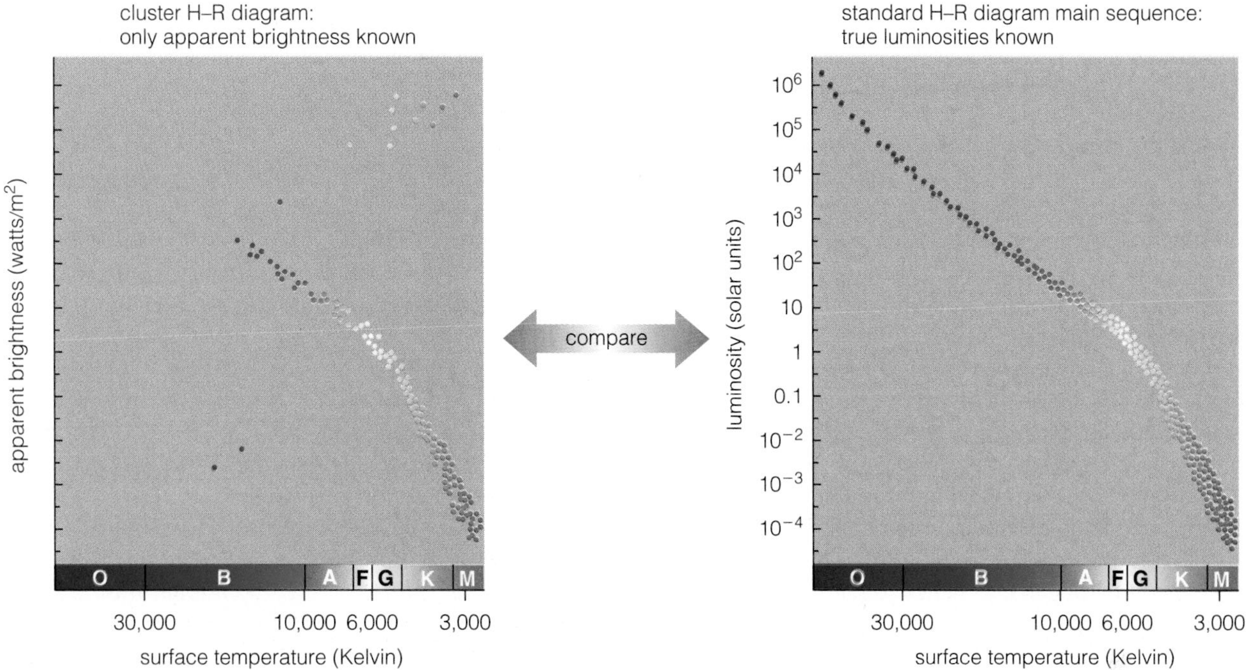

FIGURE 15.18 Main-sequence fitting allows us to determine the distance to a cluster by "fitting" the cluster main sequence, for which we measured only apparent brightness, to the standard main sequence. Once we know both apparent brightness and luminosity for cluster stars, we can calculate their distances with the luminosity–distance formula.

by making an H–R diagram of the cluster stars. However, we construct the diagram by plotting apparent brightnesses rather than true luminosities along the vertical axis. Once we plot a sufficient number of stars, the cluster's main sequence appears prominently. Because we already know the true luminosities of main-sequence stars from the standard H–R diagram, we can calculate the distance to the cluster with the luminosity–distance formula. This method of obtaining the cluster distance is called **main-sequence fitting** because it relies on "fitting" the main sequence on the cluster diagram to the standard main sequence (Figure 15.18). We'll discuss this technique further in Chapter 19.

Cluster Ages from Main-Sequence Turnoff

We can use the fact that all cluster stars form at about the same time to determine the age of a star cluster. Again, we begin by making an H–R diagram of the cluster stars. For example, Figure 15.19 shows an H–R diagram for the Pleiades. Note that most of the stars in the Pleiades fall along the standard main sequence, with one important exception: The Pleiades' stars trail away

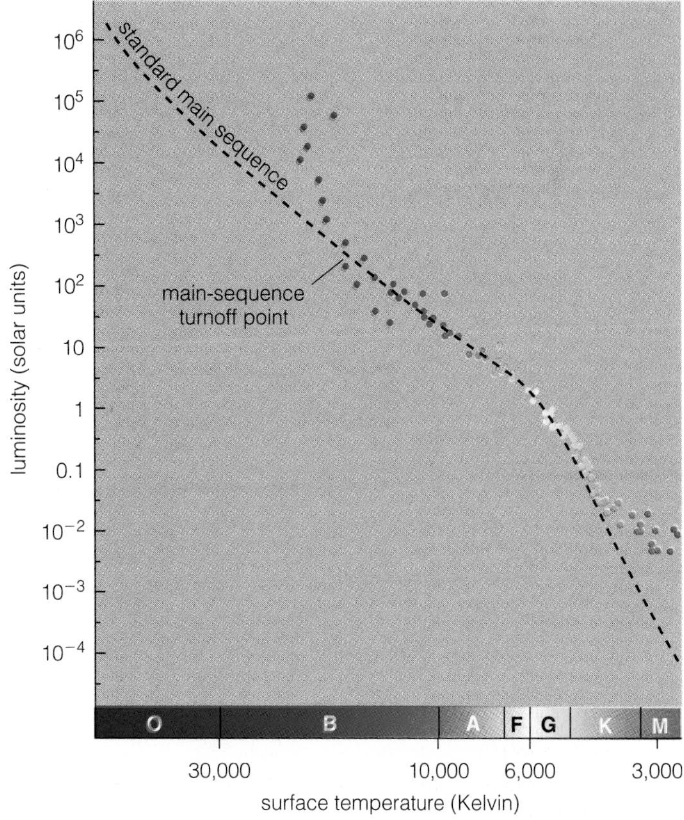

FIGURE 15.19 An H–R diagram for the stars of the Pleiades. Note that the Pleiades is missing its upper main-sequence stars.

to the right of the main sequence at the upper end. That is, the Pleiades are missing the hot, short-lived O main-sequence stars. Apparently, the Pleiades are old enough for its O stars to have already ended their hydrogen-burning lives. At the same time, they are young enough for its B stars to still survive on the main sequence.

The precise point on the H–R diagram at which the Pleiades' main sequence diverges from the standard main sequence is called the **main-sequence turnoff** point. In this cluster, it occurs around spectral type B6. The main-sequence lifetime of a B6 star is roughly 60 million years, so this must be the age of the Pleiades. Any star in the Pleiades that was born with a main-sequence spectral type hotter than B6 had a lifetime shorter than 60 million years and hence is no longer found on the main sequence. Stars with lifetimes longer than 60 million years are still fusing hydrogen and hence remain as main-sequence stars.

Over the next few billion years, the B stars in the Pleiades will die out, followed by the A stars and the F stars. Thus, if we could make H–R diagrams for the Pleiades every few million years, we would find that the main sequence gradually grows shorter. Comparing the H–R diagrams of other open clusters makes this effect more apparent (Figure 15.20). In each case, we can determine the cluster's age from the lifetimes of the stars at its main-sequence turnoff point:

age of the cluster = lifetime of stars at main-sequence turnoff point

Thus, stars in a particular cluster that once resided above the turnoff point on the main sequence have already exhausted their core supply of hydrogen, while stars below the turnoff point remain on the main sequence.

TIME OUT TO THINK *Suppose a star cluster is precisely 10 billion years old. On an H–R diagram, where would you expect to find its main-sequence turnoff point? Would you expect this cluster to have any main-sequence stars of spectral type F? Would you expect it to have main-sequence stars of spectral type K? Explain.* (Hint: *Remember that the lifetime of our Sun is 10 billion years.*)

The technique of identifying main-sequence turnoff points is our most powerful tool for evaluating the ages of star clusters. We've learned, for example, that most open clusters are relatively young and that none are older than about 5 billion years. In contrast, the stars at the main-sequence turnoff points in globular clusters are usually less massive than our Sun (Figure 15.21). Because stars like our Sun have a lifetime of about 10 billion years and these stars have already died in globular clusters, we can conclude that globular-cluster stars are older than 10 billion years. More precise studies of the turnoff points in globular clusters, coupled with theoretical calculations of stellar lifetimes, place the ages of these clusters at between 12 and 16 billion years, making them the oldest known objects in the galaxy.[8] In fact, globular clusters place a constraint on the possible age of the universe: If

FIGURE 15.20 Each curve on this diagram represents the H–R diagram for a different cluster of stars. Each cluster has a unique main-sequence turnoff point. The older the cluster, the more of its upper-main-sequence stars are missing from its H–R diagram and the farther down the main sequence its turnoff point appears.

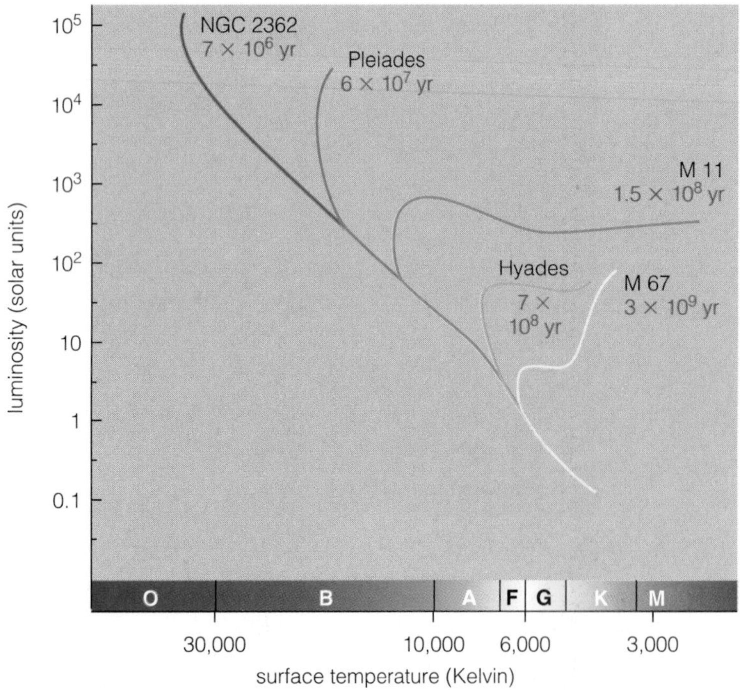

[8]Finding ages of globular clusters from their main-sequence turnoff points can be tricky because their stars contain fewer heavy elements than the younger stars in the galactic disk [Section 18.3]. Because heavy elements are more effective than hydrogen and helium at absorbing blue light, stars like the Sun appear slightly redder than globular-cluster stars of the same mass and surface temperature.

stars in globular clusters are 12 billion years old, then the universe must be at least this old.

More Clues to Stellar Evolution

Take another look at the H–R diagrams of open and globular star clusters (Figures 15.20 and 15.21). The youngest star clusters have almost no red giants at all, while the oldest ones are full of them. The H–R diagrams of old clusters show a diagonal "bridge" of stars connecting the main-sequence turnoff point to the red giants and sometimes include a *horizontal branch* of giant stars. In the youngest clusters, in contrast, stars are found scattered across the top of the H–R diagram, extending horizontally to the right from the turnoff point. These features of star clusters offer important clues to how stars change as they age. When stars like the Sun run out of core hydrogen, they leave the main sequence, growing in radius and luminosity until they become red giants. Massive stars at the top of the main sequence evolve somewhat differently, expanding in radius but maintaining a nearly constant luminosity. In the next chapter, we will learn just what happens inside aging stars that causes them to behave in these curious ways.

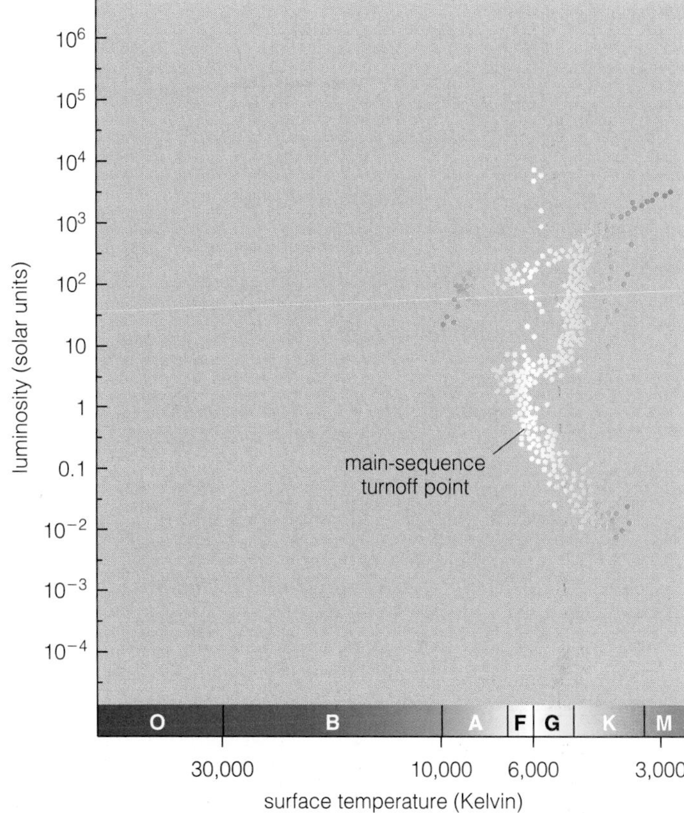

FIGURE 15.21 The H–R diagram of globular cluster Omega Centauri shows a turnoff point below the luminosity of our Sun, indicating an age of more than 10 billion years.

> *"All men have the stars," he answered, "but they are not the same things for different people. For some, who are travelers, the stars are guides. For others they are no more than little lights in the sky. For others, who are scholars, they are problems. For my businessman they were wealth. But all these stars are silent. You—you alone—will have the stars as no one else has them."*
>
> ANTOINE DE SAINT-EXUPERY, from *The Little Prince*

THE BIG PICTURE

We have classified the diverse families of stars visible in the night sky. Much of what we know about stars, galaxies, and the universe itself is based on the fundamental properties of stars introduced in this chapter, so make sure you understand the following "big picture" ideas:

- All stars are made primarily of hydrogen and helium, at least at the time they form. The differences between stars come about primarily because of differences in mass and age.
- Much of what we know about stars comes from studying the patterns that appear when we plot stellar surface temperatures and luminosities in an H–R diagram. Thus, the H–R diagram is one of the most important tools of astronomers.
- Stars spend most of their lives as main-sequence stars, fusing hydrogen into helium in their cores. The most massive stars live only a few million years, while the least massive stars will survive until the universe is many times its present age.
- Much of what we know about the universe comes from studies of star clusters. Here again, H–R diagrams play a vital role. For example, H–R diagrams of star clusters allow us to determine their distances and ages.

Review Questions

1. Briefly explain how we can learn about the lives of stars, despite the fact that stellar lives are far longer than human lives.
2. What basic composition are all stars born with? Given that all stars begin their lives with this same basic composition, why do stars differ from one another?
3. What do we mean by a star's *luminosity?* What are the standard units for luminosity? Explain what we mean when we measure luminosity in units of L_{Sun}.
4. What do we mean by a star's *apparent brightness?* What are the standard units for apparent brightness? Explain how a star's apparent brightness depends on its distance from us.
5. State the *luminosity–distance formula,* and explain why it is so important.
6. What do we mean by a star's *visible-light luminosity* or *visible-light apparent brightness?* By its *X-ray luminosity* or *X-ray apparent brightness?* Explain why we sometimes talk about such wavelength-specific measures, rather than *total luminosity* and *total apparent brightness.*
7. Briefly explain how we calculate a star's distance in parsecs by measuring its *parallax angle* in arcseconds.
8. What is the *magnitude system?* What are its origins? Briefly explain what we mean by the *apparent magnitude* and *absolute magnitude* of a star. Which one is equivalent to luminosity? Which one is equivalent to apparent brightness?
9. Briefly describe how a star's *spectral type* is related to its surface temperature. List the seven basic spectral types in order of decreasing temperature. Describe how the basic spectral types are subdivided by having numbers appended to them (e.g., A3, G5).
10. Briefly summarize the roles of Annie Jump Cannon and Cecilia Payne-Gaposchkin in discovering the spectral sequence and its meaning. Explain.
11. Describe and distinguish between *visual binary* systems, *eclipsing binary* systems, and *spectroscopic binary* systems.
12. Explain why we can directly measure stellar masses only in binary star systems and why eclipsing binaries are particularly important to the measurement of stellar masses.
13. Draw a sketch of a basic *Hertzsprung–Russell diagram (H–R diagram).* Label the *main sequence, giants, supergiants,* and *white dwarfs.* Where on this diagram do we find stars that are cool and dim? Cool and luminous? Hot and dim? Hot and bright?
14. What do we mean by a star's *luminosity class?* On your sketch of the H–R diagram, identify the regions for luminosity classes I, III, and V.
15. Explain how mass determines the luminosity and surface temperature of a main-sequence star. How do masses differ as we look along the main sequence on an H–R diagram?
16. What do we mean by a star's *main-sequence lifetime?* How do lifetimes differ as we look along the main sequence on an H–R diagram? Explain why more massive stars live shorter lives.
17. True or false? All giants, supergiants, and white dwarfs were once main-sequence stars. Explain.
18. What are *pulsating variable stars?* Why do they vary periodically in brightness?
19. What are *Cepheid variables?* Describe the *period–luminosity relation* for Cepheids, and explain how it enables us to determine distances to Cepheids that are much too far away for making parallax measurements.
20. The two basic types of star clusters are *open clusters* and *globular clusters.* Describe in general terms how these types differ in their number of stars, the age of their stars, and their location in the galaxy.
21. State two key reasons why star clusters are so important to astronomers.
22. Describe the technique of *main-sequence fitting,* and explain how it enables us to determine the distance to a star cluster.
23. Explain why H–R diagrams look different for star clusters of different ages. What is a *main-sequence turnoff* point? How does the location of the main-sequence turnoff point tell us the age of the star cluster?

Discussion Question

1. *Classification.* Edward Pickering's team of female "computers" at Harvard University made many important contributions to astronomy, particularly in the area of systematic stellar classification. Why do you think rapid advances in our understanding of stars followed so quickly on the heels of their efforts? Can you think of other areas in science where huge advances in understanding followed directly from improved systems of classification?

Problems

1. *The Inverse Square Law for Light.* The Earth is about 150 million km from the Sun, and the apparent brightness of the Sun in our sky is about 1,300 watts/m^2. Using these two facts and the inverse square law for light, determine the apparent brightness we would measure for the Sun *if* we were located at the following positions.
   a. Half the Earth's distance from the Sun.
   b. Twice the Earth's distance from the Sun.
   c. Five times the Earth's distance from the Sun.

2. *The Luminosity of Alpha Centauri A.* Alpha Centauri A lies at a distance of 4.4 light-years and has an apparent brightness in our night sky of 2.7×10^{-8} watt/m^2. Recall that 1 light-year = 9.5×10^{12} km = 9.5×10^{15} m.
   a. Use the luminosity–distance formula to calculate the luminosity of Alpha Centauri A.
   b. Suppose you have a light bulb that emits 100 watts of visible light. (*Note:* This is *not* the case for a standard 100-watt light bulb, in which most of its 100 watts goes to heat and only about 10–15 watts is emitted as visible light.) How far away would you have to put the light bulb for it to have the same apparent brightness as Alpha Centauri A in our sky? (*Hint:* Use the 100 watts as L in the luminosity–distance formula, and use the apparent brightness you found in part (a). Then solve for the distance.)

3. *More Practice with the Luminosity–Distance Formula.* Use the luminosity–distance formula to answer each of the following questions.
   a. Suppose a star has the same luminosity as our Sun (3.8×10^{26} watts) but is located at a distance of 10 light-years. What is its apparent brightness?
   b. Suppose a star has the same apparent brightness as Alpha Centauri A (2.7×10^{-8} watt/m^2) but is located at a distance of 200 light-years. What is its luminosity?
   c. Suppose a star has a luminosity of 8×10^{26} watts and an apparent brightness of 3.5×10^{-12} watt/m^2. How far away is it? Give your answer in both kilometers and light-years.
   d. Suppose a star has a luminosity of 5×10^{29} watts and an apparent brightness of 9×10^{-15} watt/m^2. How far away is it? Give your answer in both kilometers and light-years.

4. *Parallax and Distance.* Use the parallax formula to calculate the distance to each of the following stars. Give your answers in both parsecs and light-years.
   a. Alpha Centauri: parallax angle of 0.742″.
   b. Procyon: parallax angle of 0.286″.

5. *The Magnitude System.* Use the definitions of the magnitude system to answer each of the following questions.
   a. Which is brighter in our sky, a star with apparent magnitude 2 or a star with apparent magnitude 7? By how much?
   b. Which has a greater luminosity, a star with absolute magnitude +4 or a star with absolute magnitude −6? By how much?

6. *Stellar Data.* Consider the following data table for several bright stars. Note that M_v is absolute magnitude and m_v is apparent magnitude.

Star	M_v	m_v	Spectral Type	Luminosity Class
Aldebaran	−0.2	+0.9	K5	III
Alpha Centauri A	+4.4	0.0	G2	V
Antares	−4.5	+0.9	M1	I
Canopus	−3.1	−0.7	F0	II
Fomalhaut	+2.0	+1.2	A3	V
Regulus	−0.6	+1.4	B7	V
Sirius	+1.4	−1.4	A1	V
Spica	−3.6	+0.9	B1	V

   Answer each of the following questions, including a brief explanation with each answer.
   a. Which star appears brightest in our sky?
   b. Which star appears faintest in our sky?
   c. Which star has the greatest luminosity?
   d. Which star has the smallest luminosity?
   e. Which star has the highest surface temperature?
   f. Which star has the lowest surface temperature?
   g. Which star is most similar to the Sun?
   h. Which star is a red supergiant?
   i. Which star has the largest radius?
   j. Which stars have finished burning hydrogen in their cores?
   k. Among the main-sequence stars listed, which one is the most massive?
   l. Among the main-sequence stars listed, which one has the longest lifetime?

7. *Measuring Stellar Mass.* The spectral lines of two stars in a particular eclipsing binary system shift back and forth with a period of 6 months. The lines of both stars shift by equal amounts, and the amount of the Doppler shift indicates that each star has an orbital speed of 80,000 m/s. What are the masses of the two stars? Assume that each of the two stars traces a circular orbit around their center of mass. (*Hint:* See Mathematical Insight 15.4.)

8. *Ages of Star Clusters.* Consider the following sketches of H–R diagrams for star clusters. Based on these sketches, which one is the youngest? Which one is the oldest? Explain.

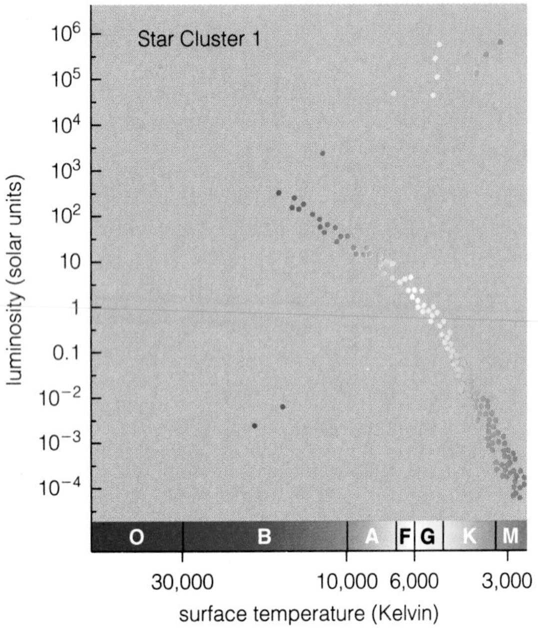

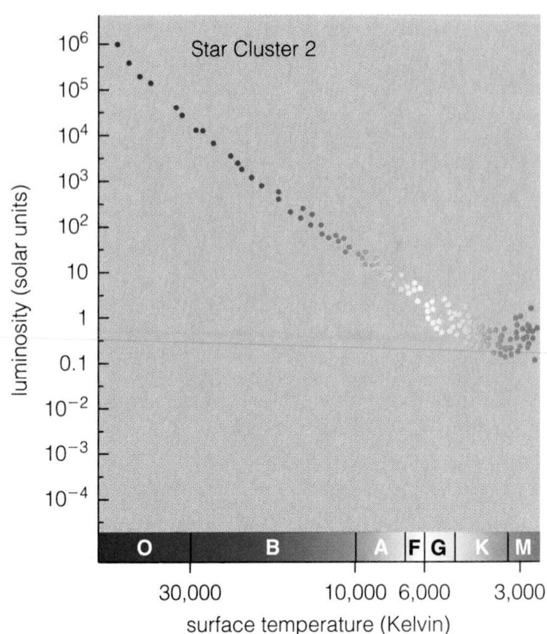

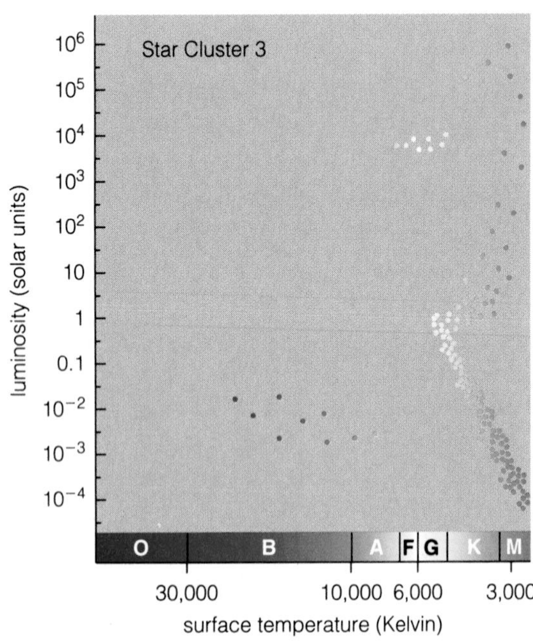

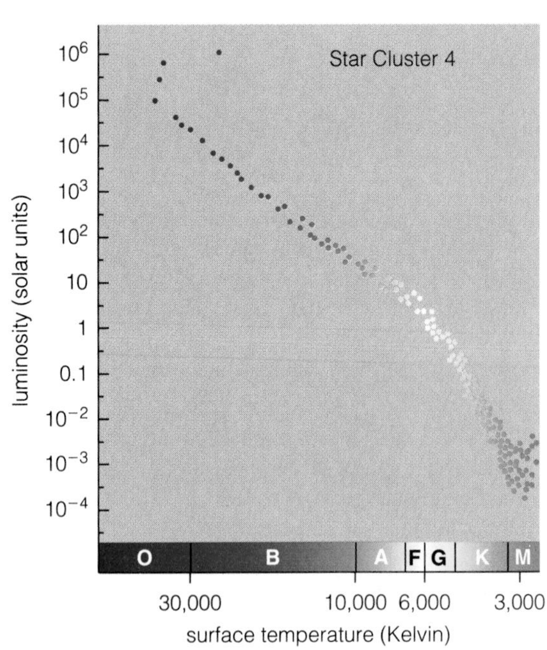

9. *Calculating Stellar Radii.* Sirius A has a luminosity of $26L_{Sun}$ and a surface temperature of about 9,000 K. What is its radius? (*Hint:* See Mathematical Insight 15.5.)

10. *Data tables:* Study the spectral types listed in Appendix D for the 20 brightest stars and for the stars with 12 light-years. Why do you think the two lists are so different?

11. *Research: Women in Astronomy.* Until fairly recently, men greatly outnumbered women in professional astronomy. Nevertheless, many women made crucial discoveries in astronomy throughout history. Do some research about the life and discoveries of a woman astronomer from any time period, and write a two- to three-page scientific biography.

16

Star Stuff

We inhale oxygen with every breath. Iron-bearing hemoglobin carries this oxygen through the bloodstream. Chains of carbon and nitrogen form the backbone of the proteins, fats, and carbohydrates in our cells. Calcium strengthens our bones, while sodium and potassium ions moderate communications of the nervous system. What does all this have to do with astronomy? The profound answer, recognized only in the latter half of the twentieth century, is that these elements were created by stars.

We've already discussed in general terms how the elements in our bodies came to exist: Hydrogen and helium were produced in the Big Bang, and heavier elements were created later by stars and scattered into space by stellar explosions. There, in the spaces between the stars, these elements mixed with interstellar gas and became incorporated into subsequent generations of stars.

In this chapter, we will discuss the origins of the elements in greater detail by delving into the lives of stars. As you read, keep in mind that no matter how far the stars may seem to be removed from our everyday lives, they actually are connected to us in the most intimate way possible: Without the births, lives, and deaths of stars, none of us would be here.

16.1 Lives in the Balance

The story of a star's life is in many ways the story of an extended battle between two opposing forces: gravity and pressure. The most common type of pressure in stars is **thermal pressure**—the familiar type of pressure that keeps a balloon inflated and that increases when the temperature or thermal energy increases.

A star can maintain its internal thermal pressure only if it continually generates new thermal energy to replace the energy it radiates into space. This energy can come from two sources: *nuclear fusion* of light elements into heavier ones and the process of *gravitational contraction,* which converts gravitational potential energy into thermal energy [Section 14.1]. These energy-production processes operate only temporarily, although in this case "temporarily" means millions or billions of years. In contrast, gravity acts eternally, and any time gravity succeeds in shrinking a star's core the strength of gravity grows. (Recall that the force of gravity inside an object grows stronger if it either gains mass or shrinks in radius [Section 6.4].) Because a star cannot rely on thermal energy to resist gravity forever, its ultimate fate depends on whether something other than thermal pressure manages to halt the unceasing crush of gravity.

The outcome of a star's struggle between gravity and pressure depends almost entirely on its birth mass. All stars are born from spinning clumps of gas, but newborn stars can have masses ranging from less than 10% of the mass of our Sun to about 100 times that of our Sun. The most massive stars live fast and die young, proceeding from birth to explosive death in just a few million years. The lowest-mass stars consume hydrogen so slowly that they will continue to shine until the universe is many times older than it is today.

Because stars come in such a great variety of masses, we can simplify our discussion of stellar lives by dividing stars into three basic groups: **low-mass stars** are born with less than about two times the mass of our Sun, or less than 2 *solar masses* ($2M_{\text{Sun}}$) of material; **intermediate-mass stars** have birth weights between about 2 and 8 solar masses; and **high-mass stars** include all stars born with masses greater than about 8 solar masses. Both low-mass and intermediate-mass stars swell into red giants near the ends of their lives and ultimately become white dwarfs. High-mass stars also become red and large in their latter days, but their lives end much more violently. We will focus most of our discussion in this chapter on the dramatic differences between the lives of low- and high-mass stars. Because the life stages of intermediate-mass stars are quite similar to the corresponding stages of high-mass stars until the very ends of their lives, we will include them in our discussion of high-mass stars.

Given the brevity of human history compared to the life of any star, you might wonder how we can know so much about stellar life cycles. As with any scientific inquiry, we study stellar lives by comparing theory and observation. On the theoretical side, we use mathematical models based on the known laws of physics to predict the life cycles of stars. On the observational side, we study stars of different mass but the same age by looking in star clusters whose ages we have determined by main-sequence turnoff [Section 15.6]. Occasionally, we even catch a star in its death throes. Theoretical predictions of the life cycles of stars now agree quite well with these observations. In the remainder of this chapter, we will examine in detail our modern understanding of the life stories of stars and how they manufacture the variety of elements—the *star stuff*—that make our lives possible.

16.2 Star Birth

A star's life begins in an interstellar cloud (Figure 16.1). Star-forming clouds tend to be quite cold—typically only 10–30 K. (Recall that 0 K is absolute zero, and temperatures on Earth are around 300 K.) They also tend to be quite dense compared to the rest of the gas between the stars, although they would qualify as a superb vacuum by earthly standards. Like the galaxy as a whole, star-forming clouds are made almost entirely of hydrogen and helium.

The clouds that form stars are sometimes called **molecular clouds,** because their low temperatures allow hydrogen atoms to pair up to form hydrogen molecules (H_2). The relatively rare atoms of elements heavier than helium can also form molecules, such as carbon monoxide or water, or tiny, solid grains of dust. More important, the cold temperatures and relatively high densities allow gravity to overcome thermal pressure more readily in molecular clouds than elsewhere in interstellar space, initiating the gravitational collapse that gives birth to new stars.

From Cloud to Protostar

As a piece of a molecular cloud starts to collapse, gravitational contraction increases the cloud's thermal energy. However, the cloud quickly radiates this thermal energy away, preventing the pressure from building enough to resist gravity. During this early phase of collapse, the temperature remains below 100 K, and the cloud glows in long-wavelength infrared light (Figure 16.2).

FIGURE 16.1 A photo of two star-forming clouds in the constellation Scorpio.

The ongoing collapse increases the cloud's density, making it increasingly difficult for radiation to escape. Eventually, its central regions grow completely opaque, trapping much of the thermal energy produced by gravitational contraction. Because thermal energy can no longer escape easily, the cloud's internal temperature and pressure rise dramatically. This rising pressure begins to fight back against the crush of gravity, and the dense cloud fragment becomes a **protostar**—the seed from which a star will grow. A protostar looks starlike, but its core is not yet hot enough for fusion.

The law of conservation of angular momentum implies that a **protostellar disk** should encircle a forming protostar, similar to the *protoplanetary disk* from which the planets of our solar system formed [Section 8.3]. A collapsing cloud fragment necessarily starts with some overall angular momentum (the sum total of the angular momenta of every constituent gas particle), although it might be unmeasurable when the fragment is large. But conservation of angular momentum ensures that the fragment spins faster and faster as it collapses (Figure 16.3). The rotating cloud also flattens to form the protostellar disk as it shrinks. Protostellar disks sometimes coalesce into true planetary systems, but we do not yet know how commonly this occurs.

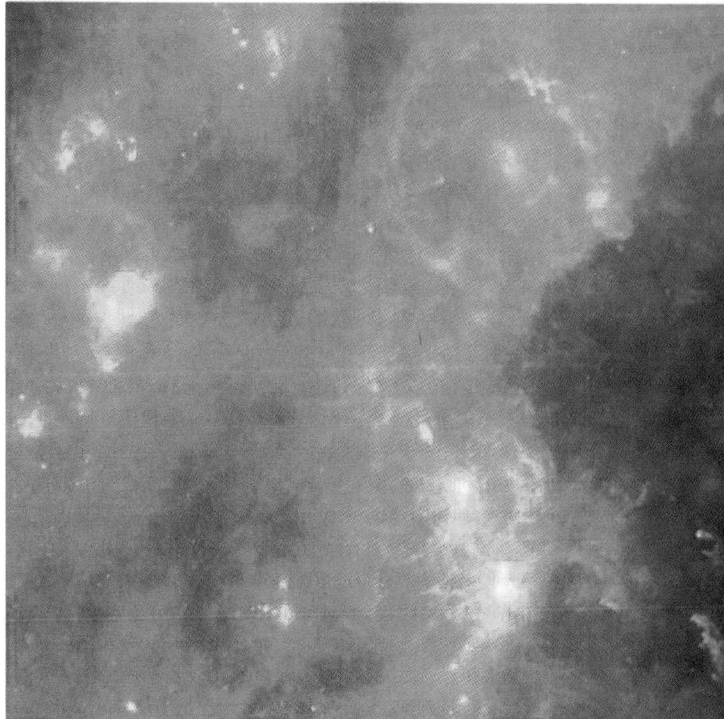

FIGURE 16.2 This false-color picture shows infrared radiation from a region within the constellation Orion. The colors correspond to the temperature of the emitting gas: Red is cooler, and blue is hotter. Many of the white dots in the picture are protostars forming within molecular clouds. (Infrared Astronomical Satellite.)

FIGURE 16.3 Artist's conception of star birth. A cloud fragment (top left) necessarily has some angular momentum, which causes it to spin faster and flatten into a protostellar disk as it collapses (center). In the late stages of collapse (bottom right), the central protostar has a strong protostellar wind and may fire jets of high-speed gas outward along its rotation axis.

TIME OUT TO THINK *The term* protostellar disk *refers to any disk of material surrounding a protostar. The term* protoplanetary disk *refers to a disk of material that later produces full-fledged planets. Do you think that all protostellar disks are also protoplanetary disks? If so, why? If not, what do you think might prevent planets from forming in some protostellar disks?*

The protostellar disk probably plays a large role in eventually slowing the rotation of the protostar. The protostar's rapid rotation generates a strong magnetic field. As the magnetic field lines sweep through the protostellar disk, they transfer some of the protostar's angular momentum outward, slowing its rotation [Section 8.4] to slow. The strong magnetic field also helps generate a strong **protostellar wind**—an outward flow of particles similar to the *solar wind* [Section 14.2]—that may carry additional angular momentum from the protostar to interstellar space.

Rotation is likely to be responsible for the formation of some binary star systems. Protostars that are unable to rid themselves of enough angular momentum spin too fast to become a stable single star and tend to split in two. These two fragments can each go on to form a separate star. If the stars are particularly close together, the resulting pair is called a **close binary** system, in which two stars coexist in close proximity, rapidly orbiting each other.

Observations show that the latter stages of a star's formation can be surprisingly violent. Besides the strong wind, many young stars also fire high-speed streams, or **jets,** of gas into interstellar space (Figure 16.4). No one knows exactly how protostars generate these jets, but two high-speed streams generally flow out along the rotation axis of the protostar, shooting in opposite directions. We also sometimes see blobs of material along the jets (called *Herbig–Haro objects* after their discoverers). These blobs appear to be collections of gas swept up as the jet plows into the surrounding interstellar material. Together, winds and jets play an important role in clearing away the cocoon of gas that surrounds a forming star, eventually allowing the star to emerge into view.

A Star Is Born

A full-fledged star is born when the core temperature of the protostar exceeds 10 million K—hot enough for hydrogen fusion to operate efficiently by the *proton–proton chain* [Section 14.3]. The ignition of fusion halts the protostar's gravitational collapse. The new star's interior structure stabilizes, with the thermal energy generated by fusion maintaining the balance between gravity and pressure that we call *gravitational equilibrium* [Section 14.1]. The star is now a hydrogen-burning, *main-sequence* star [Section 15.5].

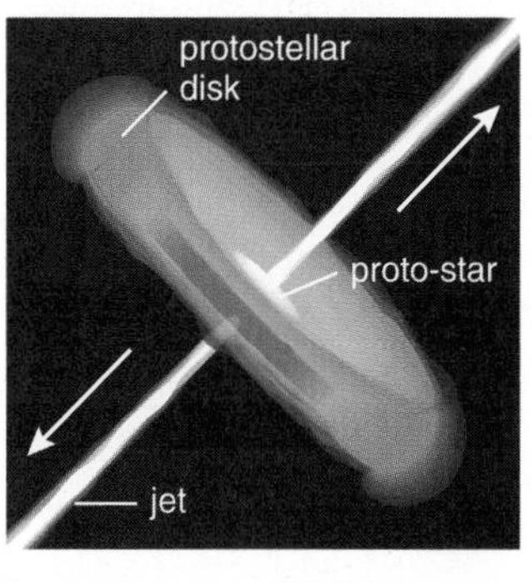

a

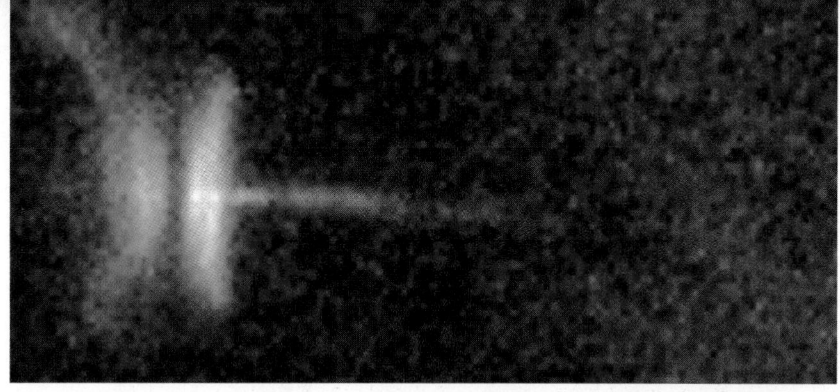
b

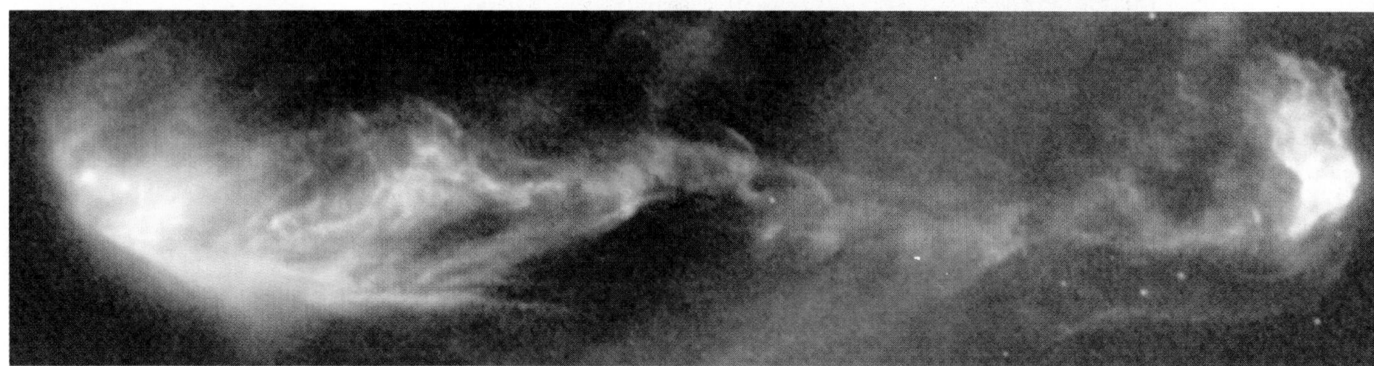
c

FIGURE 16.4 Some protostars can be seen shooting jets of matter into interstellar space. (**a**) Schematic illustration of protostellar disk-jet structure. (**b**) Close-up view of an actual protostellar disk and jet. We are seeing the disk edge-on, as in (**a**). The disk's top and bottom surfaces, illuminated by the protostar, are shown in green, and the jets emerging along the disk's axis are shown in red. The dark central layers of the disk block our view of the protostar itself. (**c**) A wider angle observation of two oppositely directed jets emanating from a protostar and ramming into surrounding interstellar gas.

The length of time from the formation of a protostar to the birth of a main-sequence star depends on the star's mass. Remember that massive stars do everything faster. The collapse of a high-mass protostar may take only a million years or less. Collapse into a star like our Sun takes about 50 million years, and collapse into a small star of spectral type M may take more than a hundred million years. Thus, the most massive stars in a young star cluster may live and die before the smallest stars finish their prenatal years.

We can summarize the transitions that occur during star birth with a special type of H–R diagram. Recall that a standard H–R diagram shows luminosities and surface temperatures for many different stars [Section 15.5]. In contrast, this special H–R diagram shows part of a **life track** (also called an *evolutionary track*) for a single star; for reference, the diagram also shows the standard main sequence. Each point along a star's life track represents its surface temperature and luminosity at some moment during its life. Figure 16.5 shows a life track for the prenatal period in the life of a $1M_{\text{Sun}}$ star like our Sun. Note that this period includes four distinct stages:

Stage 1. When the protostar first assembles from a cloud of gas, its surface temperature is quite low, putting it far to the right on the H–R diagram. However, its surface area is quite large at this time, so its luminosity is also quite large—perhaps 100 times the luminosity it will have when it becomes a Sun-like main-sequence star.

Stage 2. Because of its large luminosity, the young protostar rapidly loses the energy it generates through gravitational contraction, and its collapse continues at a relatively rapid rate. Its surface temperature warms only slightly during the next few million years, but its shrinking size reduces its luminosity dramatically. The life track therefore progresses almost straight downward on the H–R diagram.

Stage 3. Once the core temperature reaches a few million K, hydrogen nuclei begin to fuse into helium. However, the rate of fusion is still not high enough to halt the collapse of the star, although the rate of shrinkage slows considerably. As the star shrinks, its surface temperature rises. The net result of this shrinkage and heating is a slight increase in luminosity during the next 10 million years. The life track therefore progresses leftward and slightly upward on the H–R diagram.

Stage 4. The core temperature and rate of fusion continue to increase gradually for the next few

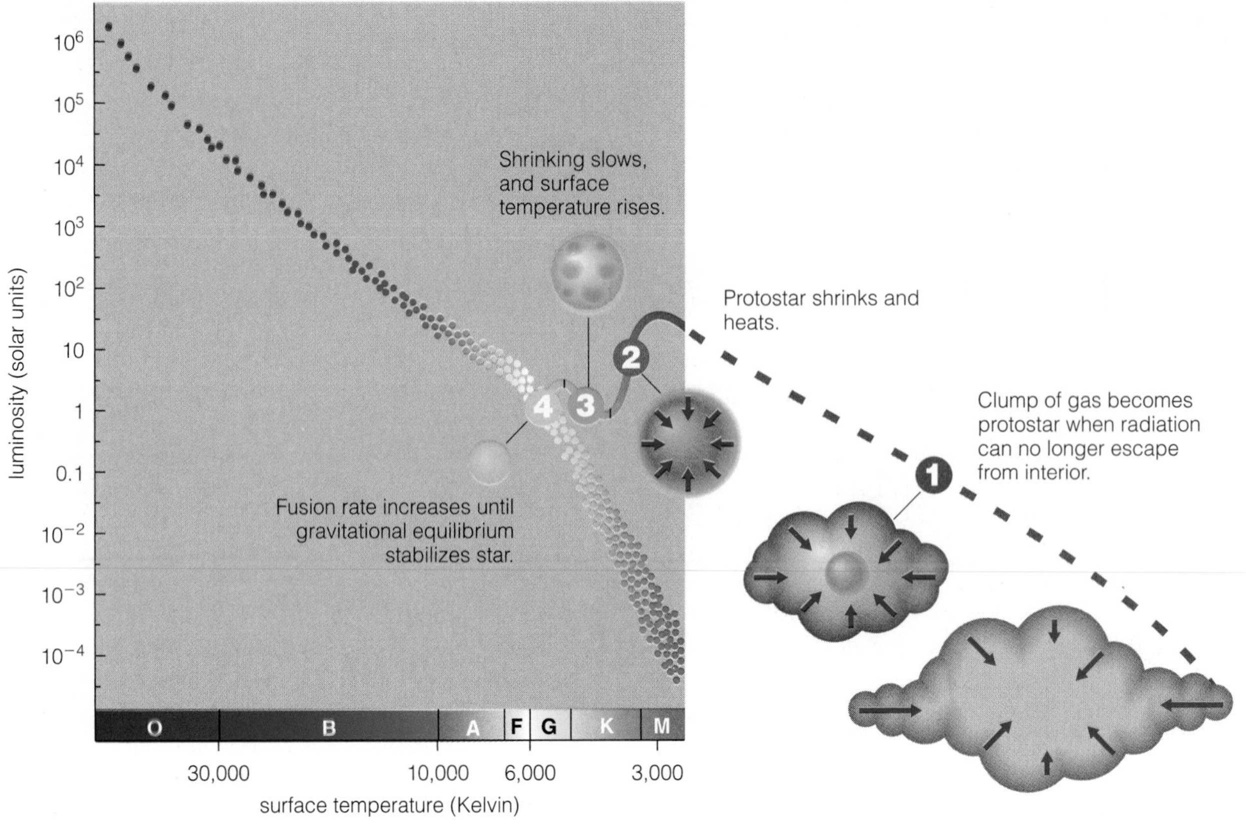

FIGURE 16.5 The life track of a $1M_{Sun}$ star from protostar to main-sequence star.

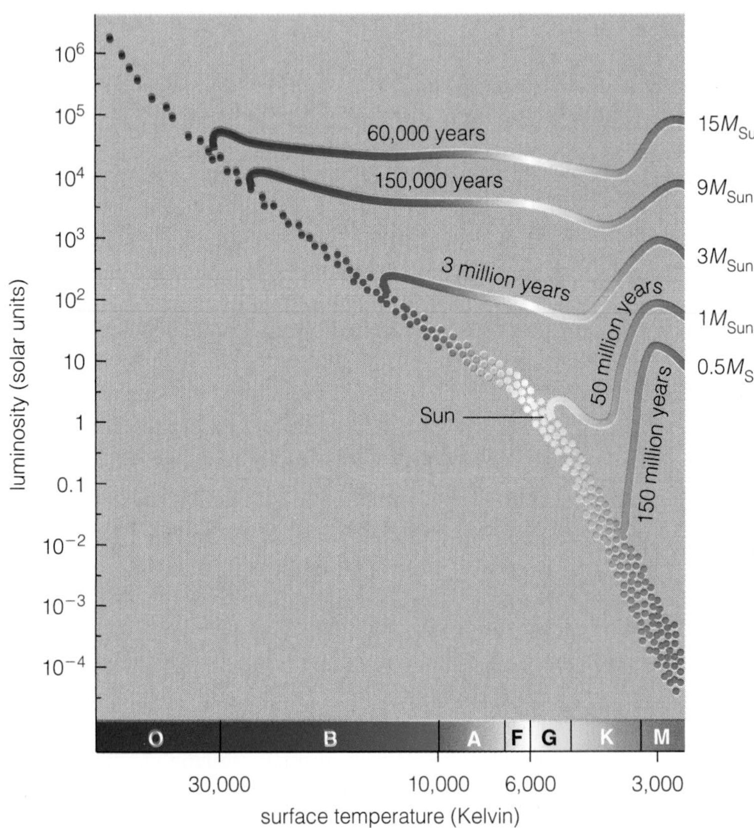

FIGURE 16.6 Life tracks from protostar to the main sequence for stars of different masses.

tens of millions of years. Finally, the rate of fusion becomes high enough to establish gravitational equilibrium. At this point, fusion becomes self-sustaining and the star settles into its hydrogen-burning, main-sequence life.

Stars of different masses follow a similar set of stages during their prenatal periods. Figure 16.6 illustrates life tracks on an H–R diagram for the prenatal stages of several stars of different masses.

TIME OUT TO THINK *Explain in your own words what we mean by a* life track *for a star. Why do we say that Figures 16.5 and 16.6 show only* pre–main-sequence *life tracks? In gen-*

eral terms, predict the appearance on Figure 16.6 of pre–main-sequence life tracks for a $25M_{Sun}$ star and for a $0.1M_{Sun}$ star.

Stellar Birth Weights

A single group of molecular clouds can contain thousands of solar masses of gas, which is why stars generally are born in clusters. We do not yet fully understand the processes that govern the clumping and fragmentation of these clouds into protostars with a wide array of masses. However, we can observe the results. In a newly formed star cluster, stars with low masses greatly outnumber stars with high masses. For every star with a mass between 10 and 100 solar masses, there are typically 10 stars with masses between 2 and 10 solar masses, 50 stars with masses between 0.5 and 2 solar masses, and a few hundred stars with masses below $0.5M_{Sun}$. Thus, although the Sun lies nearly in the middle of the overall *range* of stellar masses, most stars in a new star cluster are less massive than the Sun. With the passing of time, the balance tilts even more in favor of the low-mass stars as the high-mass stars die away.

The masses of stars have limits. Theoretical models indicate that stars above about $100M_{Sun}$ generate power so furiously that gravity cannot contain their internal pressure. Such stars effectively tear themselves apart and drive their outer layers into space. Observations confirm the absence of stars much larger than $100M_{Sun}$—if any stars this massive existed nearby, they would be so luminous that we would easily detect them.

On the other end of the scale, calculations show that the central temperature of a protostar with less than $0.08M_{Sun}$ never climbs above the 10-million-K threshold needed for efficient hydrogen fusion. Instead, a strange effect called **degeneracy pressure** halts the gravitational contraction of the core before hydrogen burning can begin. The quantum mechanical origins of degeneracy pressure are discussed in Chapter S5; you can visualize the origin of degeneracy pressure with an analogy to an auditorium in which the chairs represent all possible places that subatomic particles can be located and people represent the particles. Protostars with masses above $0.08M_{Sun}$ are like auditoriums with many more available chairs than people, so the people (particles) can easily squeeze into a smaller section of the auditorium. But the cores of protostars with masses below $0.08M_{Sun}$ are like auditoriums with so few chairs that the people (particles) fill nearly all of them. Because there are no extra chairs available, the people (particles) cannot squeeze into a smaller section of the auditorium. This resistance to squeezing is why degeneracy pressure halts gravitational contraction. Note that the degeneracy pressure arises *only* because particles have no place else to go. Thus, unlike thermal pressure, degeneracy pressure has nothing to do with temperature.

Because degeneracy pressure halts the collapse of a protostellar core with less than $0.08M_{Sun}$ before fusion becomes self-sustaining, the result is a "failed star" that slowly radiates away its internal thermal energy, gradually cooling with time. Such objects, called **brown dwarfs,** occupy a fuzzy gap between what we call a planet and what we call a star. (Note that $0.08M_{Sun}$ is about 80 times the mass of Jupiter.) Because degeneracy pressure does *not* rise and fall with temperature, the gradual cooling of a brown dwarf's interior does not weaken its degeneracy pressure. In the constant battle of any "star" to resist the crush of gravity, brown dwarfs are winners, albeit dim ones: Their degeneracy pressure will not diminish with time, so gravity will never gain the upper hand.

Brown dwarfs radiate primarily in the infrared and actually look deep red in color rather than brown. But they are far dimmer than normal stars and therefore are extremely difficult to detect, even if they are quite nearby. The first bona fide brown dwarf was discovered in 1995—a $0.05M_{Sun}$ object (called Gliese 229B) in orbit around a much brighter star (Gliese 229A) [Section 8.8].

Astronomers are actively searching for more brown dwarfs, because they may be far more important than their dimness and small size might at first suggest. If the trend that makes small stars far more common than massive stars continues to masses below $0.08M_{Sun}$, brown dwarfs might outnumber ordinary stars by a huge margin. These barely detectable balls of gas might be the most common form of ordinary matter in the universe.[1]

16.3 Life as a Low-Mass Star

In the grand hierarchy of stars, our Sun ranks as rather mediocre. We should be thankful for this mediocrity. If the Sun had been a high-mass star, it would have lasted only a few million years, dying before life could have arisen on Earth. Instead, the Sun has shone steadily for nearly 5 billion years and will continue to do so for about 5 billion more. In this section, we investigate the lives of low-mass stars like our Sun.

[1] By "ordinary" we mean matter composed of atoms. The mysterious *dark matter* may not be "ordinary" by this definition (see Chapter 21).

Slow and Steady

Low-mass stars spend their main-sequence lives fusing hydrogen into helium in their cores slowly and steadily via the *proton–proton chain* [Section 14.3]. As in the Sun, the energy released by nuclear fusion in the core may take a million years to reach the surface, where it finally escapes into space as the star's luminosity. The energy moves outward from the core through a combination of *radiative diffusion* and *convection* [Section 14.4]. Recall that radiative diffusion transports energy through the random bounces of photons from one electron to another and that convection transports energy by the rising of hot plasma and the sinking of cool plasma.

Radiative diffusion is more effective at transporting energy upward in the deeper, hotter plasma near a star's core. In higher layers, where the temperature is cooler, some of the ions in the plasma retain electrons. These ions can absorb photons and thereby tend to prevent photons from continuing upward by radiative diffusion. When the energy welling up from fusion in the core reaches the point at which radiative diffusion is inhibited, convection must take over as the means of transporting energy upward. This point represents the beginning of a star's *convection zone* (Figure 16.7). In the Sun, the temperature is cool enough to allow convection in the outer one-third of its interior. More massive stars have hotter interiors and hence shallower convection zones. Lower-mass stars have cooler interiors and deeper convection zones. In very low mass stars, the convection zone extends all the way down to the core. The highest-mass stars have no convection zone at all near their surface, but they have a convective core because they produce energy so furiously.

The depth of the convection zone plays a major role in determining whether a star has *activity* similar to that of the sunspot cycle on our Sun [Section 14.5]. Recall that the Sun's activity arises from the twisting and stretching of its magnetic fields by convection and rotation. The most active stars are low-mass M stars that happen to have fast rotation rates in addition to their deep convection zones. The churning interiors of these stars are in a constant state of turmoil, twisting and knotting their magnetic field lines. When these field lines suddenly snap, the result can be a spectacular flare that makes the most powerful solar flares look puny by comparison. For a few minutes or hours, the flare can produce more radiation in X rays than the total amount of light coming from the star in infrared and visible light. Life on a planet near one of these **flare stars** might be quite difficult. Proxima Centauri, the nearest star to the Sun, is a flare star.

A low-mass star gradually consumes its core hydrogen, converting it into helium over a period of billions of years. In the process, the declining number of independent particles in the core (four independent protons fuse into just one independent helium nucleus) causes the core to shrink and heat very gradually, pushing the luminosity of the main-sequence star slowly upward as it ages [Section 14.3].

FIGURE 16.7 Among main-sequence stars, convection zones extend deeper in lower-mass stars. High-mass stars have convective cores but have no convection zones near their surfaces.

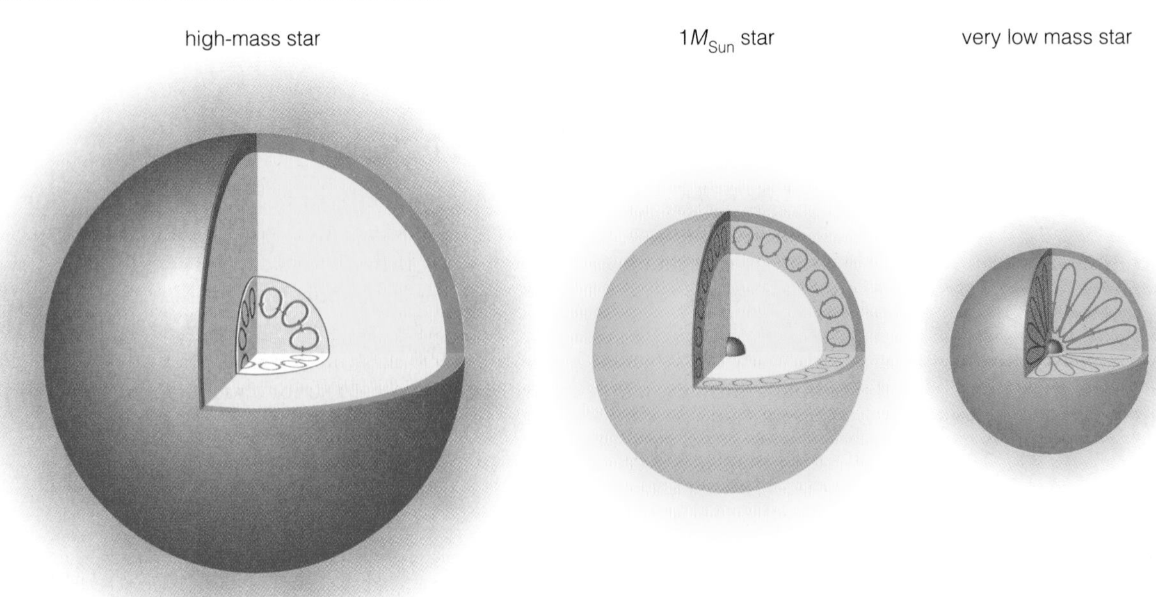

But the most dramatic changes occur when nuclear fusion exhausts the hydrogen in the star's core.

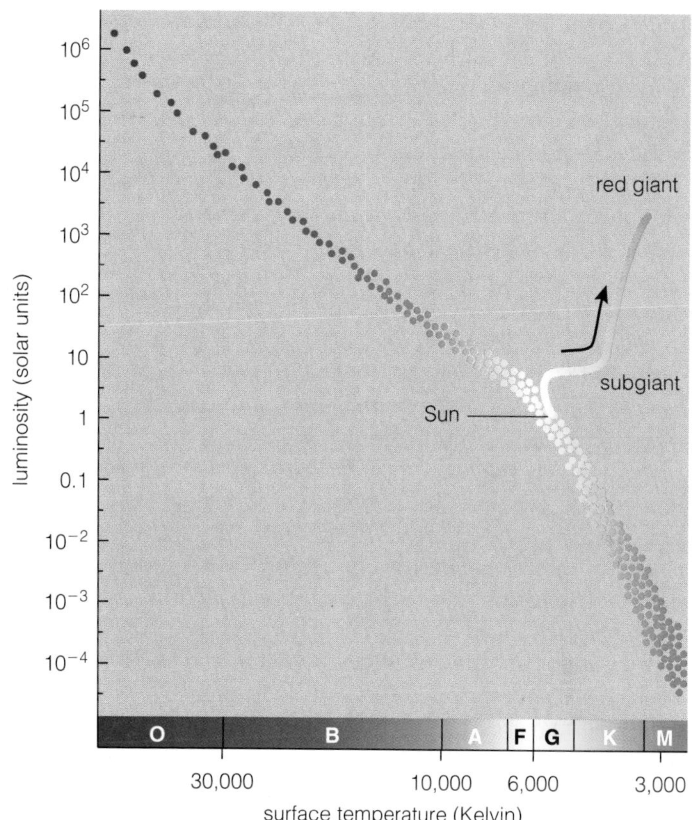

FIGURE 16.8 The life track of a $1M_{Sun}$ star on an H–R diagram from the end of its main-sequence life until it becomes a red giant.

Red Giant Stage

The energy released by hydrogen fusion during a star's main-sequence life maintains the thermal pressure that holds gravity at bay. But when the core hydrogen is finally depleted, nuclear fusion ceases. With no fusion to supply thermal energy, the core pressure can no longer resist the crush of gravity, and the core begins to shrink more rapidly.

Surprisingly, the star's outer layers expand outward while the core is shrinking. On an H–R diagram, the star's life track moves almost horizontally to the right as the star grows in size to become a **subgiant** (Figure 16.8). Then, as the expansion of the outer layers continues, the star's luminosity begins to increase substantially, and the star slowly becomes a **red giant.** For a $1M_{Sun}$ star, this process takes about a billion years, during which the star's radius increases about 100-fold and its luminosity grows by an even greater factor. (Like all phases of stellar lives, the process occurs faster for more massive stars and slower for less massive stars.) This process may at first seem paradoxical: Why does the star grow bigger and more luminous at the same time that its core is shrinking?

We can find the answer by considering the interior structure of the star. The core is now made of helium—the "ash" left behind by hydrogen fusion—but the surrounding layers still contain plenty of fresh hydrogen. Gravity shrinks both the *inert* (nonburning) helium core and the surrounding *shell* of hydrogen, and the shell soon becomes hot enough to sustain hydrogen fusion (Figure 16.9). In fact, the shell becomes so hot that this **hydrogen shell burning** proceeds at a higher rate than core hydrogen fusion did during the star's main-sequence life. The result is that the star becomes more luminous than ever before. The thermal pressure generated by the hydrogen shell burning overcomes gravity in the

FIGURE 16.9 After a star ends its main-sequence life, its inert helium core contracts while hydrogen shell burning begins. The high rate of fusion in the hydrogen shell forces the star's upper layers to expand outward.

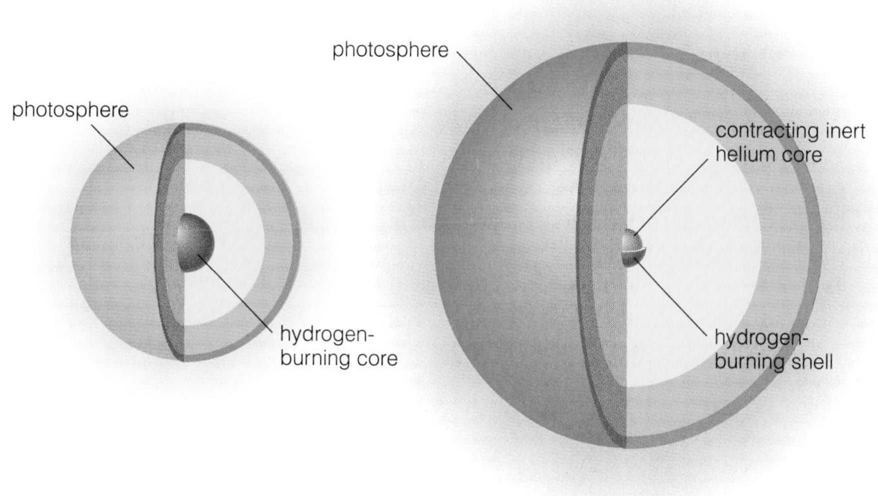

star's upper layers, pushing them to expand outward. What was once a fairly dim main-sequence star balloons into a luminous red giant. Note that the red giant is large on the outside, but most of its mass is buried deep in a shrunken stellar core.

The situation grows more extreme as long as the helium core remains inert. Recall that, in a main-sequence star like the Sun, a rise in the fusion rate causes the core to inflate and cool until the fusion rate drops back down; in Chapter 14, we referred to this self-correcting process as the solar *thermostat* [Section 14.3]. But thermal energy generated in the hydrogen-burning shell of a red giant cannot do anything to inflate the inert core that lies underneath. Instead, newly produced helium keeps adding to the mass of the helium core, amplifying its gravitational pull and shrinking it further. The hydrogen-burning shell shrinks along with the core, growing hotter and denser. The fusion rate in the shell consequently rises, feeding even more helium ash to the core. The star is caught in a vicious circle. The core and shell continue to shrink in size and grow in luminosity, while thermal pressure continues to push the star's upper layers outward. This cycle breaks down only when the inert helium core reaches a temperature of about 100 million K, at which point helium nuclei can fuse together.[2] Throughout the expansion phase, **red-giant winds** carry away much more matter than the winds from main-sequence stars, but at much slower speeds.

TIME OUT TO THINK ***Before you read on, briefly summarize why a star grows larger and brighter after it exhausts its core hydrogen. When does the growth of a red giant finally halt, and why? Would a star's red giant stage be different if the temperature required for helium fusion were around 200 million K, rather than 100 million K? Why?***

Helium Burning

Recall that fusion occurs only when two nuclei come close enough together for the attractive *strong force* to overcome electromagnetic repulsion [Section 14.3]. Helium nuclei have two protons (and two neutrons) and hence a greater positive charge than the single proton of a hydrogen nucleus. The greater charge means that helium nuclei repel one another more strongly than hydrogen nuclei. **Helium fusion** therefore occurs only when nuclei slam into one another at much higher speeds than needed for hydrogen fusion. Therefore, helium fusion requires much higher temperatures. The helium fusion process[3] converts three helium nuclei into one carbon nucleus:

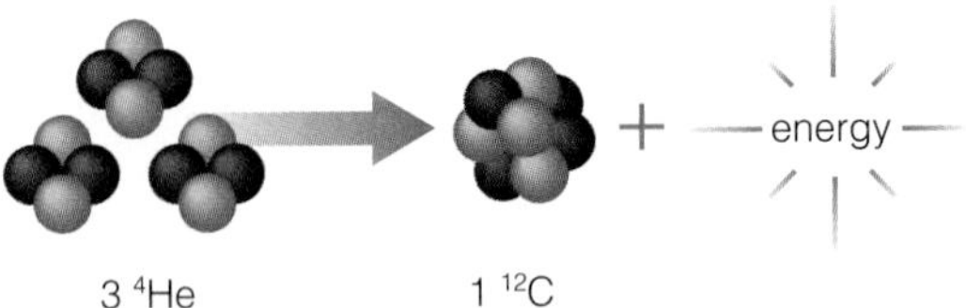

Energy is released because the carbon-12 nucleus has a slightly lower mass than the three helium-4 nuclei, and the lost mass becomes energy in accord with $E = mc^2$.

There's one subtlety in the ignition of helium burning in low-mass stars. Theoretical models show that, in the inert helium core of a low-mass star, the thermal pressure is too low to counteract gravity. Instead, the models show that the pressure fighting against gravity is *degeneracy pressure*—the same strange type of pressure that supports brown dwarfs. Because degeneracy pressure does *not* increase with temperature, the onset of helium fusion heats the core rapidly *without* causing it to inflate. The rising temperature causes the helium fusion rate to rocket upward in what is called a **helium flash.**

The helium flash dumps enormous amounts of new thermal energy into the core. In a matter of seconds, the rapidly rising thermal pressure becomes the dominant pressure pushing back against gravity; the core is no longer degenerate and begins to expand. This core expansion pushes the hydrogen-burning shell outward, lowering its temperature and its burning rate. The result is that, even though the star now has core helium fusion *and* hydrogen shell burning taking place simultaneously, the total energy production falls from its peak during the red giant phase. The reduced total energy output of the star reduces its luminosity, thereby allowing its outer layers to contract from their peak size during the red giant phase (Figure 16.10). As the outer layers contract, the star's surface temperature also increases somewhat.

Because the helium-burning star is now smaller and hotter than it was as a red giant, its life track on the H–R diagram drops downward and to the left (Figure 16.11a). Note that the helium cores of all

[2]In a very low mass star, the inert helium core may never become hot enough to fuse helium. The core collapse will instead be halted by degeneracy pressure, ultimately leaving the star a *helium white dwarf.*

[3]The helium fusion process is also called the *triple-alpha reaction* (helium nuclei are sometimes called *alpha particles*). This reaction actually proceeds in two steps: (i) ^{4}He + ^{4}He ⇒ ^{8}Be (*Be* is beryllium); (ii) ^{4}He + ^{8}Be ⇒ ^{12}C. However, ^{8}Be nuclei are extremely unstable; left alone, they decay back into two ^{4}He nuclei in a mere 2.6×10^{-16} second. Thus, the helium fusion reaction is completed only when a third ^{4}He nucleus crashes into the ^{8}Be before it breaks apart.

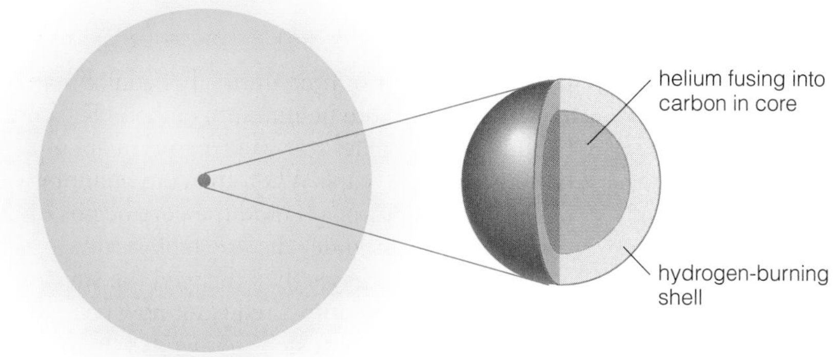

a

b

FIGURE 16.10 (**a**) Core structure of a helium-burning star. (**b**) Approximate relative sizes of a low-mass star as a main-sequence star, a red giant, and a helium-burning star. The scale is not precise; in particular, the size of the main-sequence star is even smaller compared to the others.

FIGURE 16.11 (below) (**a**) After the helium flash, a star's surface shrinks and heats, so the star's life track moves downward and to the left on the H–R diagram. (**b**) This H–R diagram plots the luminosity and surface temperature of individual stars in a cluster (i.e., it does *not* show life tracks).

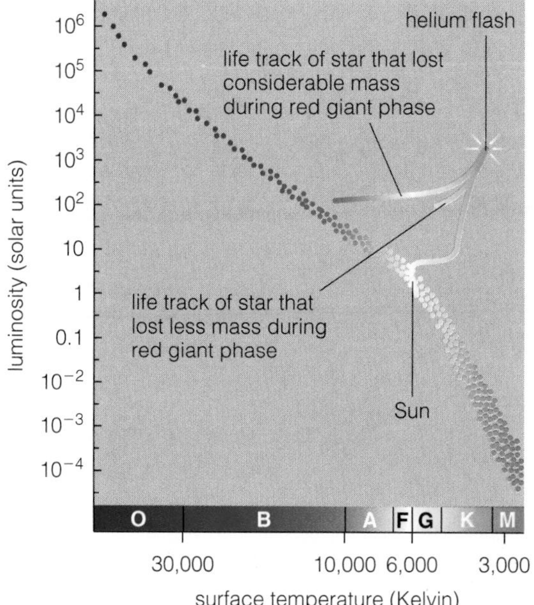

a

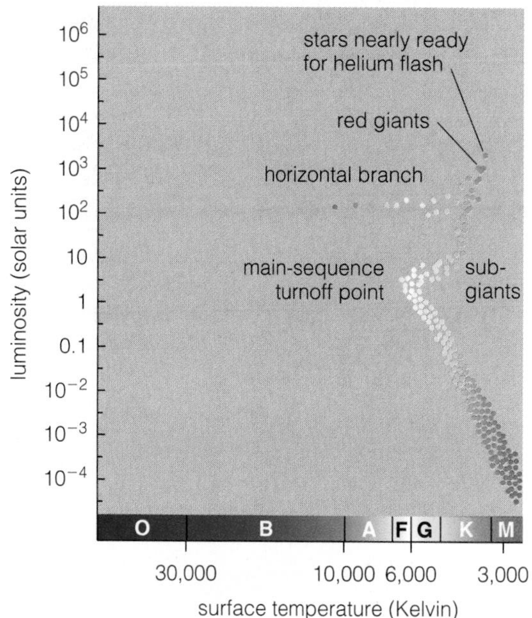

b

low-mass stars fuse helium into carbon at about the same rate, so these stars all have about the same luminosity. However, the outer layers of these stars can have different masses, depending on how much mass they lost through their red-giant winds. Stars that lost more mass end up with smaller radii and higher surface temperatures and hence are farther to the left on the H–R diagram. In a cluster of stars, those stars that are currently in their helium-burning phase therefore all have about the same luminosity but differ in surface temperature. Thus, on an H–R diagram for a cluster of stars, the helium-burning stars are arranged along a **horizontal branch**[4] (Figure 16.11b). Note that the cluster H–R diagram clearly shows the main-sequence turnoff point. Stars that have recently left the main sequence are subgiants on their way to becoming red giants. In the upper-right corner of the red giant region are stars that are almost ready for the helium flash. The stars that have already become helium-burning, horizontal-branch stars are slightly dimmer and hotter.

[4]The leftward extent of the horizontal branch on the H–R diagram differs for different clusters; in some clusters, all the "horizontal-branch" stars are bunched in a stubby *clump* on the right end of where the horizontal branch would otherwise be.

Last Gasps

It is only a matter of time until a horizontal-branch star fuses all its core helium into carbon. The core helium in a low-mass star will run out in about a hundred million years. When the core helium is exhausted, fusion turns off, and the core begins to shrink once again under the crush of gravity.

The basic processes that changed the star from a main-sequence star into a red giant now resume, but this time it is helium fusion that ignites in a shell around an inert carbon core. Meanwhile, the hydrogen shell still burns atop the helium layer. Both shells contract along with the inert core, driving their temperatures and fusion rates much higher. The luminosity of this double-shelled star grows greater than ever, and its outer layers swell to an even huger size. On the H–R diagram, the star's life track once again turns upward (Figure 16.12). Theoretical models show that helium burning inside such a star never finds a happy equilibrium but instead proceeds in a series of **thermal pulses** during which the fusion rate spikes upward every few thousand years.

The furious burning in the helium and hydrogen shells cannot last long—maybe a few million years or less for a $1M_{\text{Sun}}$ star. The star's only hope of extending its life lies with the carbon core, but this is a false

FIGURE 16.12 The life track of a $1M_{\text{Sun}}$ star from main-sequence star to white dwarf.

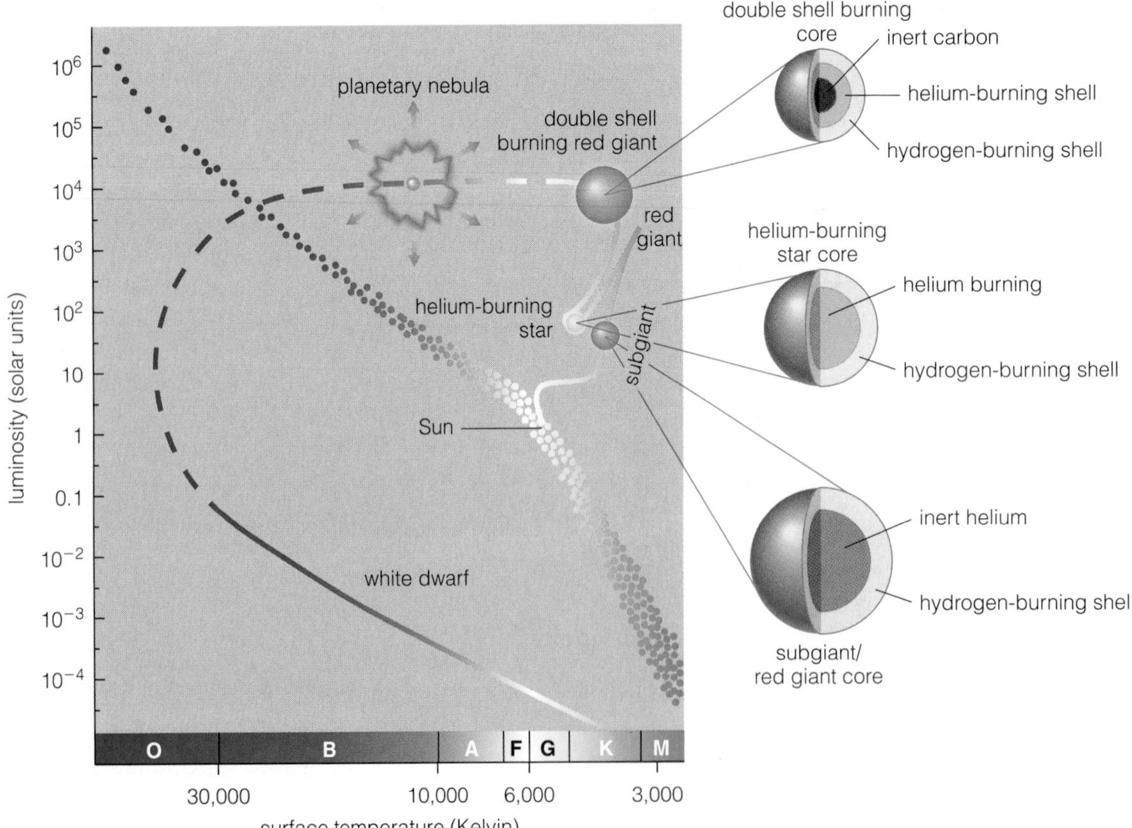

hope in the case of low-mass stars. Carbon fusion is possible only at temperatures above about 600 million K. Before the core of a low-mass star ever reaches such a lofty temperature, degeneracy pressure halts its gravitational collapse.

For a low-mass star with a carbon core, the end is near. The huge size of the dying star means that it has a very weak grip on its outer layers. As the star's luminosity and radius keep rising, matter flows from its surface at increasingly high rates. Meanwhile, during each thermal pulse, strong convection dredges up carbon from the core, enriching the surface of the star with carbon. Red giants whose atmospheres become especially carbon-rich in this way are called **carbon stars.** Carbon stars have cool, low-speed stellar winds, and the temperature of the gas in these winds drops with distance from the stellar surface. At the point at which the temperature has dropped to 1,000–2,000 K, some of the gas atoms in these slow-moving winds begin to stick together in microscopic clusters, forming small, solid particles of dust. These dust particles continue their slow drift with the stellar wind into interstellar space, where they become **interstellar dust grains.** The process of particulate formation is very similar to the formation of smoke particles in a fire. Thus, in a sense, carbon stars are the most voluminous polluters in the universe. However, this "carbon smog" is essential to life: Most of the carbon in your body (and in all life on Earth) was manufactured in carbon stars and blown into space by their stellar winds.

TIME OUT TO THINK *Suppose the universe contained only low-mass stars. Would elements heavier than carbon exist? Why or why not?*

Before a low-mass star dies, it treats us to one last spectacle. Through winds and other processes, the star ejects its outer layers into space. The result is a huge shell of gas expanding away from the inert, degenerate carbon core. The exposed core is still very hot and therefore emits intense ultraviolet radiation that ionizes the gas in the expanding shell, causing it to glow brightly as what we call a **planetary nebula.** Despite their name, planetary nebulae have nothing to do with planets. The name comes from the fact that nearby nebulae look disk-shaped (like planets) through a small telescope. Through a larger telescope, more detail is visible. The famous Ring Nebula in the constellation Lyra is a planetary nebula, as is the Helix Nebula in the southern sky (Figure 16.13).

The glow of the planetary nebula fades as the exposed core cools and the ejected gas disperses into space. The nebula will disappear within a million years, leaving behind the cooling carbon core. On the H–R diagram, the life track now represents this "dead" core (see Figure 16.12). At first the life track heads to the left, because the core is initially quite hot and luminous. But the life track soon veers downward and to the right as the remaining ember cools and fades. You already know these naked, inert cores by another name: *white dwarfs* [Section 15.5].

In the ongoing battle between gravity and a star's internal pressure, white dwarfs are a sort of stalemate. As long as no mass is added to the white dwarf from some other source (such as a companion star in a binary system), neither the strength of gravity nor the strength of the degeneracy pressure that holds gravity at bay will ever change. Thus, a white dwarf is little more than a decaying corpse that will cool for the indefinite future, eventually disappearing from view as a *black dwarf.*

The Fate of Life on Earth

The evolutionary stages of low-mass stars are immensely important to the Earth, because we orbit a low-mass star—the Sun. Fortunately, the Sun should cause few problems for Earth over the next 5 billion years. It will gradually brighten during its remaining time as a main-sequence star, but it has already been doing so for the past 4 billion years with no serious consequences for life on Earth. (Apparently, Earth's climate self-adjusts to maintain stable temperatures as the Sun heats over long time scales.) However, somewhere around the year A.D. 5,000,000,000, the hydrogen in the solar core will run out. The Sun will then expand to a subgiant about three times as large and twice as bright as it is today. This increased luminosity will probably begin to evaporate the oceans, perhaps leading to a runaway greenhouse effect that could raise Earth's average temperature to something similar to that of Venus and probably hotter [Section 13.3]. If any life survives, it will have to be very tolerant of heat.

Things will only get worse as the Sun grows into a red giant over the next several hundred million years. Just before helium flash, the Sun will be about 100 times larger in radius and over 1,000 times more luminous than it is today. Earth's surface temperature will exceed 1,000 K, and the oceans will long since have boiled away. By this time, any surviving humans will need to have found a new home. Saturn's moon Titan [Section 11.5] might not be a bad choice: Its surface temperature will have risen from well below freezing today to about the present temperature of Earth. The Sun will shrink somewhat after helium burning begins, providing a temporary lull in incineration while the Sun spends 100 million years as a helium-burning star.

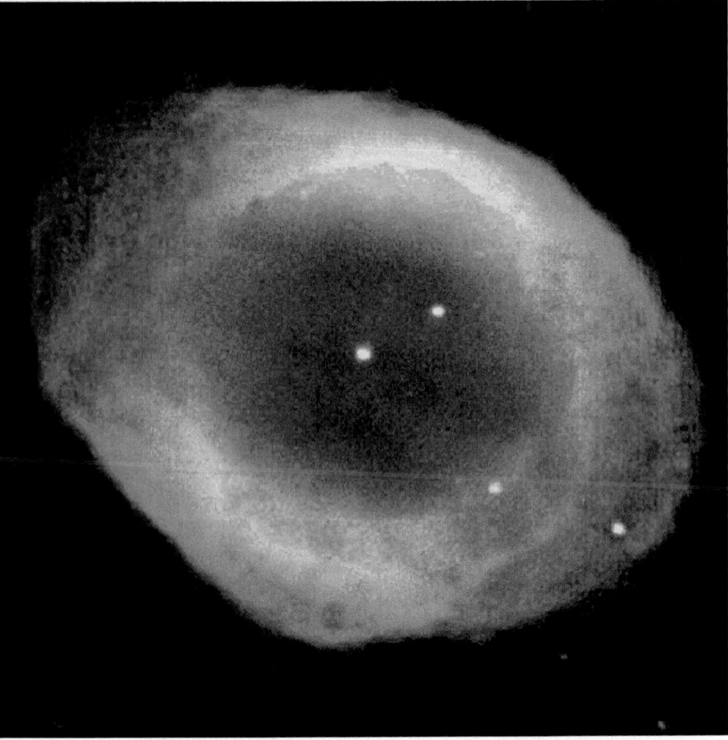

a The Ring Nebula

b The Helix Nebula

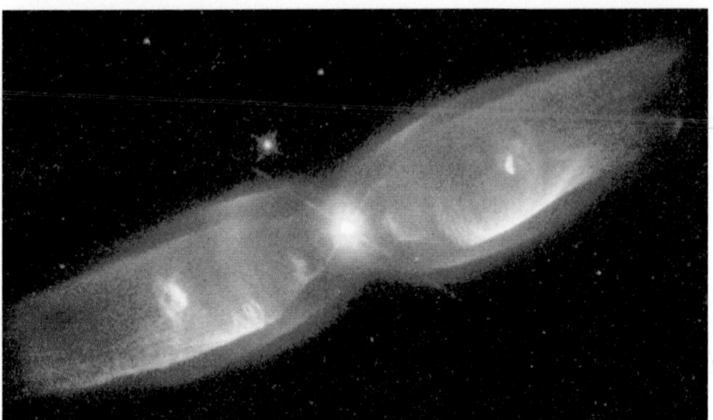

d The Twin Jet Nebula

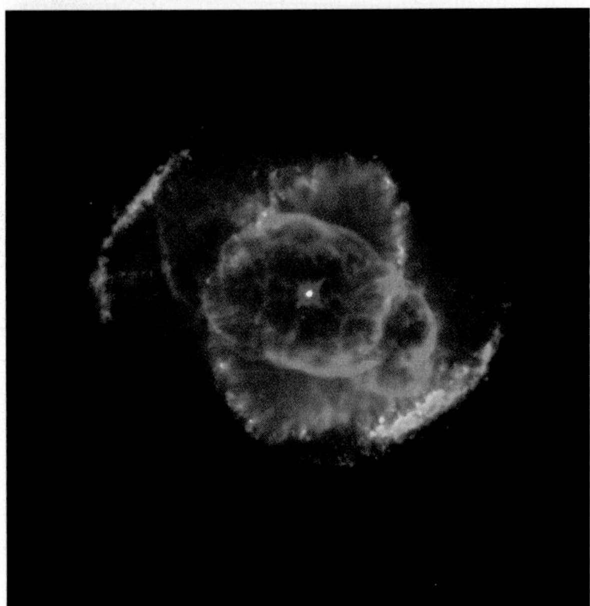

c The Cat's Eye Nebula

FIGURE 16.13 Planetary nebulae occur when low-mass stars in their final death throes cast off their outer layers of gas, as seen in these photos. The hot core that remains ionizes and energizes the richly complex envelope of gas surrounding it.

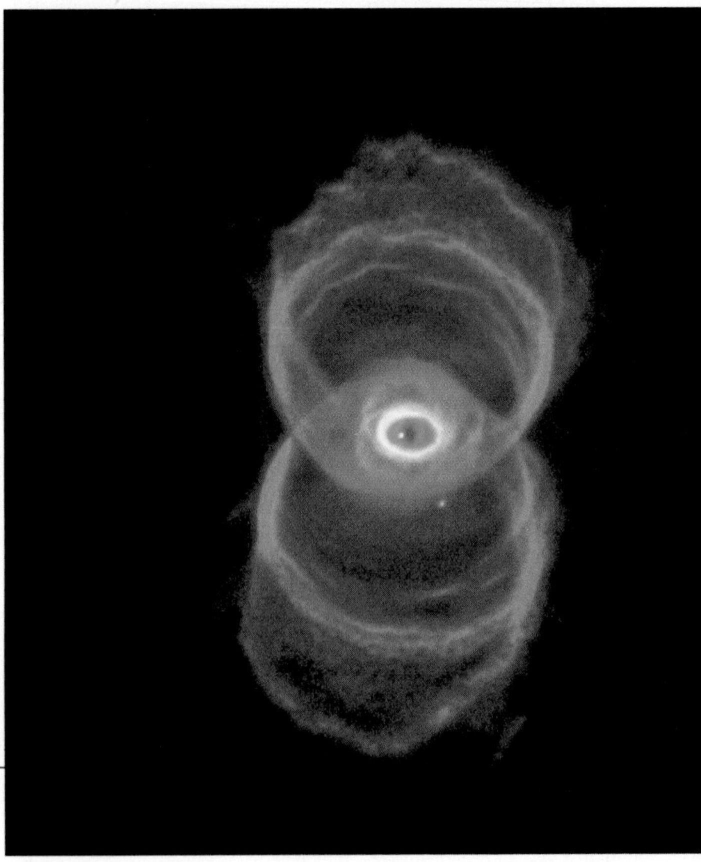

e The Hourglass Nebula

THINKING ABOUT . . .

Five Billion Years

The Sun's demise in about 5 billion years might at first seem worrisome, but 5 billion years is a very long time. It is longer than Earth has yet existed, and human time scales pale by comparison. A single human lifetime, if we take it to be about 100 years, is only 2×10^{-8}, or two hundred-millionths, of 5 billion years. Because 2×10^{-8} of a human lifetime is about 1 minute, we can say that a human lifetime compared to the life expectancy of the Sun is roughly the same as 60 heart beats in comparison to a human lifetime.

What about human creations? The Egyptian pyramids have often been described as "eternal." But they are slowly eroding due to wind, rain, air pollution, and the impact of tourists, and all traces of them will have vanished within a few hundred thousand years. While a few hundred thousand years may seem like a long time, the Sun's remaining lifetime is more than 1,000 times longer.

On a more somber note, we can gain perspective on 5 billion years by considering evolutionary time scales. During the past century, our species has acquired sufficient technology and power to destroy human life totally, if we so choose. However, even if we make that unfortunate choice, some species (including many insects) are likely to survive. Would another intelligent species ever emerge on Earth? There is no way to know, but we can look to the past for guidance. Many species of dinosaurs were biologically quite advanced, if not truly intelligent, when they were suddenly wiped out about 65 million years ago. Some small rodentlike mammals survived, and here we are 65 million years later. We therefore might guess that another intelligent species could evolve some 65 million years after a human extinction. If these beings also destroyed themselves, another species could evolve 65 million years after that, and so on. But even at 65 million years per shot, the Earth would have *nearly* 80 more chances for an intelligent species to evolve in 5 billion years (5 billion $\div$ 65 million $\approx$ 77). Perhaps one of those species will not destroy itself, and descendants of Earth might move on to other star systems by the time the Sun finally dies. Perhaps this species will be our own.

Anyone who survives the Sun's helium-burning phase will need to prepare for one final disaster. After exhausting its core helium, the Sun will expand again during its last million years. Its luminosity will soar to thousands of times what it is today, and it will grow so large that solar flames might lap at the Earth's surface. Then it will eject its outer layers as a planetary nebula that will engulf Jupiter and Saturn and drift on past Pluto into interstellar space. If the Earth is not destroyed, its charred surface will be cold and dark in the faint, fading light of the white dwarf that the Sun has become. From then on, the Earth will be little more than a chunk of rock circling the corpse of our once-brilliant star.[5]

16.4 Life as a High-Mass Star

Human life would be impossible without both low-mass stars and high-mass stars. The long lives of low-mass stars allow evolution to proceed for billions of years, but only high-mass stars produce the full array of elements on which life depends. Fusion of elements heavier than helium to produce elements heavier than carbon requires extremely high temperatures in order to overcome the larger electromagnetic repulsion of more highly charged nuclei. Reaching such temperatures requires an extremely strong crush of gravity on a star's core—a crush that occurs only under the immense weight of the overlying gas in high-mass stars.

In the final stages of their lives, the highest-mass stars proceed to fuse increasingly heavy elements until they have exhausted all possible fusion sources. When fusion ceases, gravity drives the core to implode suddenly, which, as we will soon see, causes the star to self-destruct in the titanic explosion we call a *supernova*. The fast-paced life and cataclysmic death of a high-mass star—surely one of the great dramas of the universe—are the topics of this section.

Brief but Brilliant

During its main-sequence life, the strong gravity of a high-mass star compresses its hydrogen core to higher temperatures than we find in lower-mass stars. You already know that the rate of fusion via the proton–proton chain increases substantially at higher temperatures, but the even hotter core temperatures of

[5]The Earth might still retain interior heat for billions of years more, so it is conceivable that a few members of an advanced civilization could eke out an existence by tapping this geothermal energy.

high-mass stars enable protons to slam into carbon, oxygen, or nitrogen nuclei as well as into other protons. Although carbon, nitrogen, and oxygen make up less than 2% of the material from which stars form in interstellar space, this 2% is more than enough to be useful in a stellar core. The carbon, nitrogen, and oxygen act as *catalysts* for hydrogen fusion, making it proceed at a far higher rate than would be possible by the proton–proton chain alone. (A *catalyst* is something that aids the progress of a reaction without being consumed in the reaction.) The lives of high-mass stars are truly brief but brilliant.

The chain of reactions that leads to hydrogen fusion in high-mass stars is called the **CNO cycle;** the letters *CNO* stand for carbon, nitrogen, and oxygen, respectively. The six steps of the CNO cycle are shown in Figure 16.14. Note that, just as in the proton–proton chain [Section 14.3], four hydrogen nuclei go in while one helium-4 nucleus comes out. The amount of energy generated in each reaction cycle therefore is the same as in the proton–proton chain: It is equal to the difference in mass between the four hydrogen nuclei and the one helium nucleus, multiplied by c^2. The CNO cycle is simply another, faster way to accomplish hydrogen fusion.

The escalated fusion rates in high-mass stars generate remarkable amounts of power. Many more photons stream from the photospheres of high-mass stars than from the Sun, and many more photons are bouncing around inside. Although photons have no mass, they act like particles and carry *momentum* [Section 6.1], which they can transfer to anything they hit, imparting a very slight jolt. The combined jolts from the huge number of photons streaming outward through a high-mass star apply a type of pressure called **radiation pressure.**

TIME OUT TO THINK *We have now discussed three types of pressure: thermal pressure, degeneracy pressure, and radiation pressure. Before continuing, briefly review each type of pressure and describe the conditions under which each is important.*

Radiation pressure can have dramatic effects on high-mass stars. In the most massive stars, radiation pressure is more important than thermal pressure in keeping gravity at bay. Near the photosphere, the radiation pressure can drive strong, fast-moving stellar winds. The wind from a massive star can expel as much as 10^{-5} solar mass of gas per year at speeds greater than 1,000 km/s. This wind would cross the United States in about 5 seconds and send a mass equivalent to that of our Sun hurtling into space in only 100,000 years. Such a wind cannot last long because it would blow away all the mass of even a very massive star in just a few million years.

FIGURE 16.14 The CNO cycle. Hydrogen fuses into helium via the CNO cycle in the cores of massive stars. This diagram illustrates the six steps of this cycle. Each cycle consumes four hydrogen nuclei (Steps 1, 3, 4, and 6) and produces one helium nucleus (Step 6), giving it the same result as the proton-proton chain. The carbon, nitrogen, and oxygen nuclei help the cycle proceed, but overall these nuclei are neither consumed nor created in the cycle.

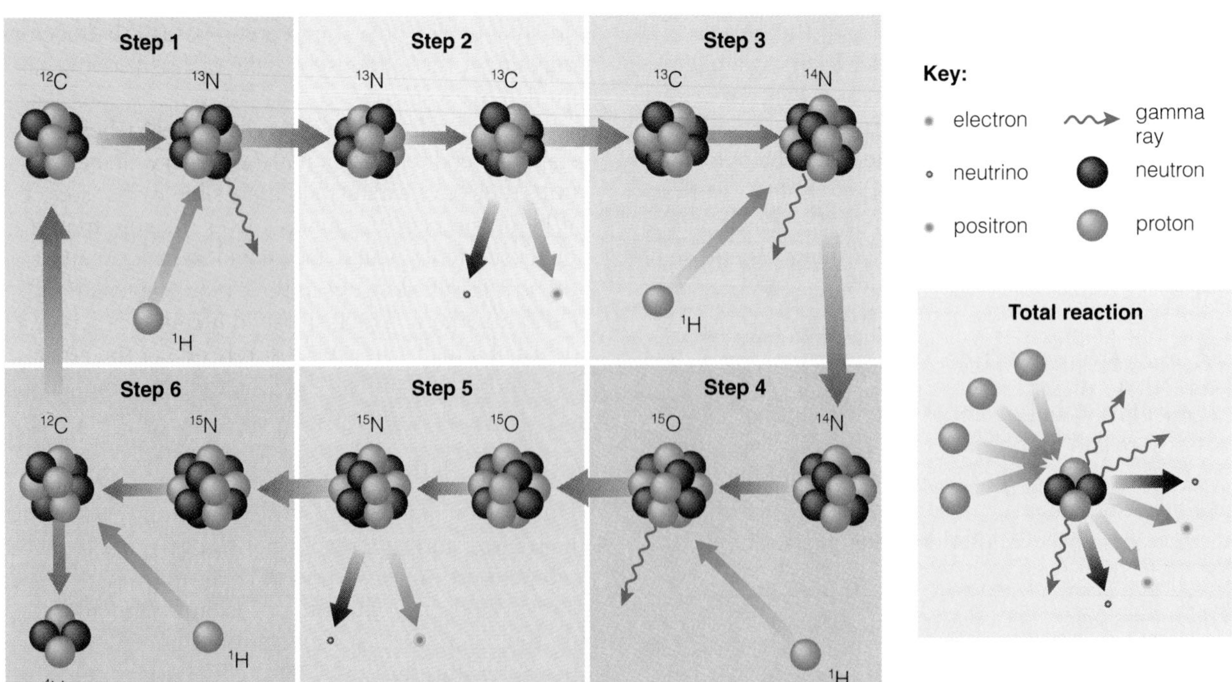

Advanced Nuclear Burning

The exhaustion of core hydrogen in a high-mass star sets in motion the same processes that turn a low-mass star into a red giant, but the transformation proceeds much more quickly. The star develops a hydrogen-burning shell, and its outer layers begin to expand outward. At the same time, the core contracts, and this gravitational contraction releases energy that raises the core temperature until it becomes hot enough to fuse helium into carbon. However, there is no helium flash in stars of more than 2 solar masses. The high core temperatures induced by core contraction keep the thermal pressure high, preventing degeneracy pressure from being a factor. Helium burning therefore ignites gradually, just as hydrogen burning did at the beginning of the star's main-sequence life.

A high-mass star fuses helium into carbon so rapidly that it is left with an inert carbon core after just a few hundred thousand years or less. Once again, the absence of fusion leaves the core without a thermal energy source to fight off the crush of gravity. The inert carbon core shrinks, the crush of gravity intensifies, and the core pressure, temperature, and density all rise. Meanwhile, a helium-burning shell forms between the inert core and the hydrogen-burning shell. The star's outer layers swell again.

Up to this point, the life stories of intermediate-mass stars (2–$8M_{\text{Sun}}$) and high-mass stars ($>8M_{\text{Sun}}$) are very similar. However, degeneracy pressure prevents the cores of intermediate-mass stars from reaching the temperatures required to burn carbon or oxygen into anything heavier.[6] These stars eventually blow away their upper layers and finish their lives as white dwarfs. The rest of a high-mass star's life, on the other hand, is unlike anything that a low-mass star or an intermediate-mass star ever experiences.

The crush of gravity in a high-mass star is so overwhelming that degeneracy pressure never comes into play in the collapsing carbon core. The gravitational contraction of the core continues, and the core temperature soon reaches the 600 million K required to fuse carbon into heavier elements. Carbon fusion provides the core with a new source of energy that restores the balance versus gravity, but only temporarily. In the highest-mass stars, carbon burning may last only a few hundred years. When the core carbon is depleted, the core again begins to collapse, shrinking and heating until it can fuse a still-heavier element. The star is engaged in the final phases of a desperate battle against the ever-strengthening crush of gravity. Each successive stage of core nuclear burning proceeds more rapidly than prior stages. In a field in which we usually think in terms of millions or billions of years, we are suddenly dealing with time scales that are short even for humans.

The nuclear reactions in the star's final stages of life become quite complex, and many different reactions may take place simultaneously (Figure 16.15). The simplest sequence of fusion stages involves **helium capture**—the fusing of helium nuclei into progressively heavier elements. (Some helium nuclei still remain in the core, but not enough to continue helium fusion efficiently.) Helium capture can fuse carbon into oxygen, oxygen into neon, neon into magnesium, and so on. At high enough temperatures, a star's core plasma can fuse heavy nuclei to one another. For example, fusing carbon to oxygen creates silicon, fusing two oxygen nuclei creates sulfur, and fusing two silicon nuclei generates iron.[7] Some of these heavy-element reactions release free neutrons, which may fuse with heavy nuclei to make still rarer elements. The star is forging the variety of elements that, in our solar system at least, became the stuff of life.

Each time the core depletes the elements it is fusing, it shrinks and heats until it becomes hot enough for other fusion reactions. Meanwhile, a new type of shell burning ignites between the core and the overlying shells of fusion. Near the end, the star's central region resembles the inside of an onion, with layer upon layer of shells burning different elements (Figure 16.16). During the star's final few days, iron begins to pile up in the silicon-burning core.

Despite the dramatic events taking place in its interior, the high-mass star's outer appearance changes only slowly. As each stage of core fusion ceases, the surrounding shell burning intensifies and further inflates the star's outer layers. Each time the core flares up again, the outer layers may contract a bit. The result is that the star's life track zigzags across the top of the H–R diagram (Figure 16.17). In very massive stars, the core changes happen so quickly that the outer layers don't have time to respond, and the star progresses steadily toward becoming a red supergiant. Betelgeuse, the upper-left shoulder star of Orion, is the best-known red supergiant star. Its radius is over 800 solar radii, or about four times the distance from the Sun to Earth. We have no way of knowing what stage of nuclear burning is now taking place in Betelgeuse's core. Betelgeuse may have a

[6]Some of the more massive intermediate-mass stars may undergo further nuclear reactions, but these reactions cease before the point at which iron forms.

[7]Fusion of two silicon nuclei and some other processes that lead to iron actually first produce nickel-56 (28 protons and 28 neutrons), but this decays rapidly to cobalt-56 (27 protons and 29 neutrons) and then to iron-56 (26 protons and 30 neutrons).

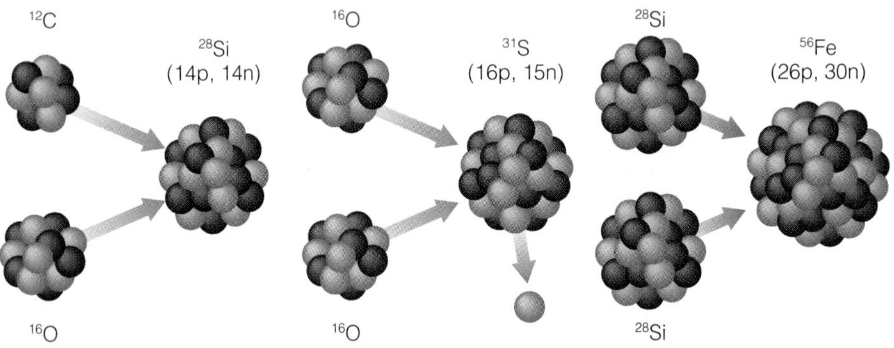

FIGURE 16.15 A few of the many nuclear reactions that occur in the final stages of a high-mass star's life.

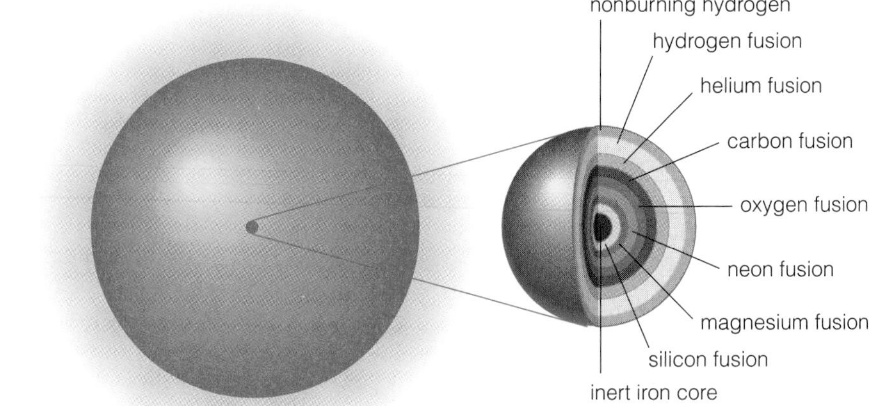

FIGURE 16.16 The multiple layers of nuclear burning in the core of a high-mass star during the final days of its life.

few thousand years of nuclear burning still ahead, or it may be piling up iron in its core as you read. If the latter is the case, then sometime in the next few days[8] we will witness the most dramatic event that ever occurs in the universe.

[8]This statement assumes that we've taken into account the light-travel time from Betelgeuse. Betelgeuse is about 500 light-years away, so an iron pileup would have had to occur about 500 years ago for us to find out about it today.

Iron: Bad News for the Stellar Core

As a high-mass star develops an inert core of iron, the core continues shrinking and heating while iron continues to pile up from nuclear burning in the surrounding shells. If iron were like the other elements in prior stages of nuclear burning, this core contraction would stop when iron fusion ignited. However, iron is unique among the elements in a very important

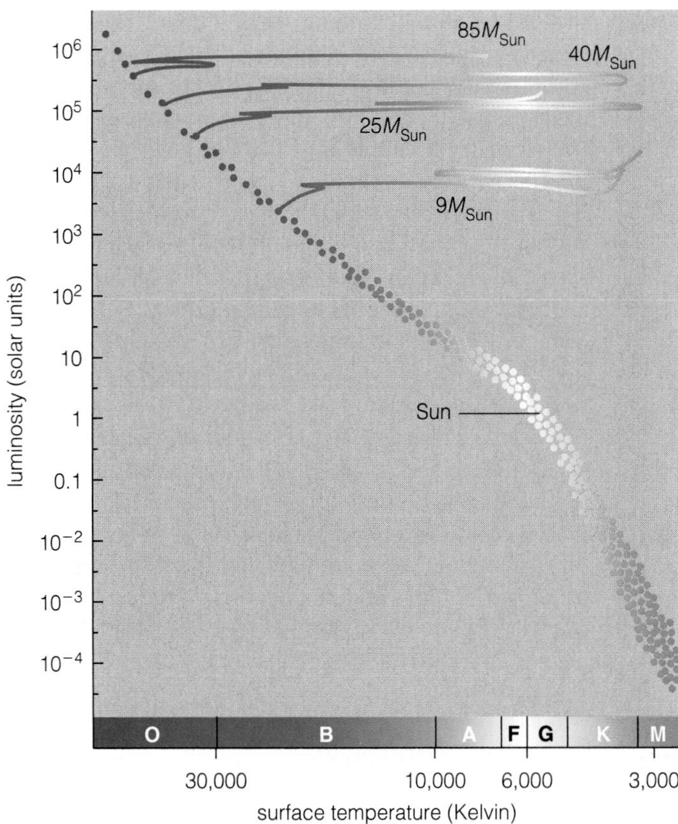

FIGURE 16.17 Life tracks on the H–R diagram from main-sequence star to red supergiant for selected high-mass stars. (Based on models from A. Maeder and G. Meynet.)

way: It is the one element from which it is *not* possible to generate any kind of nuclear energy.

To understand why iron is unique, remember that only two basic processes can release nuclear energy: *fusion* of light elements into heavier ones, and *fission* of very heavy elements into not-so-heavy ones. Recall that hydrogen fusion converts four protons (hydrogen nuclei) into a helium nucleus that consists of two protons and two neutrons. Thus, the total number of *nuclear particles* (protons and neutrons combined) does not change. However, this fusion reaction generates energy (in accord with $E = mc^2$) because the *mass* of the helium nucleus is less than the combined mass of the four hydrogen nuclei that fused to create it—despite the fact that the *number* of nuclear particles is unchanged.

In other words, fusing hydrogen into helium generates energy because helium has a lower *mass per nuclear particle* than hydrogen. Similarly, fusing three helium-4 nuclei into one carbon-12 nucleus generates energy because carbon has a lower mass per nuclear particle than helium—which means that some mass disappears and becomes energy in this fusion reaction. In fact, the decrease in mass per nuclear particle from hydrogen to helium to carbon is part of a general trend shown in Figure 16.18: The mass per nuclear particle tends to decrease as we go from light elements to iron, which means that fusion of light nuclei into heavier nuclei generates energy. Note that this trend reverses beyond iron: The mass per nuclear particle tends to increase as we look to still heavier elements. As a result, the only way to generate nuclear energy from elements heavier than iron is through fission into lighter elements. For example, uranium has a greater mass per nuclear particle than lead, so some mass is converted into energy when uranium nuclei split apart and ultimately leave behind lead as a by-product.

Iron itself has the lowest mass per nuclear particle of all nuclei and therefore cannot release energy by either fusion or fission. Thus, once the matter in a stellar core turns to iron, it can burn no further. Unable to generate thermal energy or pressure by nuclear burning, the iron core's only hope of resisting the crush of gravity lies with degeneracy pressure. But iron keeps piling up in the core until degeneracy pressure can no longer support it. What ensues is the ultimate nuclear-waste catastrophe.

TIME OUT TO THINK *How would the universe be different if hydrogen, rather than iron, had the lowest mass per nuclear particle? Why?*

FIGURE 16.18 Overall, the average mass per nuclear particle declines from hydrogen to iron and then increases. Selected nuclei are labeled to provide reference points. (This graph shows the most general trends only; a more detailed graph would show numerous up-and-down bumps superimposed on the general trend.)

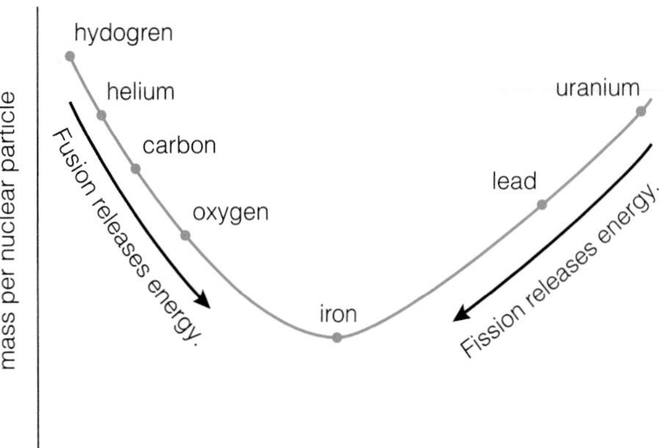

Supernova

The degeneracy pressure that briefly supports the inert iron core arises because the laws of quantum mechanics prohibit electrons from getting too close together [Section S5.5]. But once gravity pushes the electrons past the quantum mechanical limit, they can no longer exist freely. The electrons disappear by combining with protons to form neutrons, releasing neutrinos in the process (Figure 16.19). The electron degeneracy pressure suddenly vanishes, and gravity has free rein.

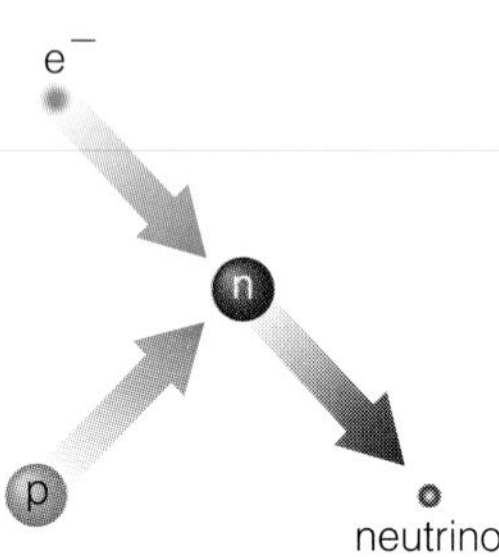

FIGURE 16.19 During the final, catastrophic collapse of a high-mass stellar core, electrons and protons combine to form neutrons, accompanied by the release of neutrinos.

In a fraction of a second, an iron core with a mass comparable to that of our Sun and a size larger than that of Earth collapses into a ball of neutrons just a few kilometers across. The collapse halts only because the neutrons have a degeneracy pressure of their own. The entire core then resembles a giant atomic nucleus. If you recall that ordinary atoms are made almost entirely of empty space [Section 5.3] and that almost all their mass is in their nuclei, you'll realize that a giant atomic nucleus must have an astoundingly high density.

The gravitational collapse of the core releases an enormous amount of energy—more than a hundred times what the Sun will radiate over its entire 10-billion-year lifetime! Where does this energy go? It drives the outer layers off into space in a titanic explosion—a **supernova.** The ball of neutrons left behind is called a **neutron star.** In some cases, the remaining mass may be so large that gravity also overcomes neutron degeneracy pressure, and the core continues to collapse until it becomes a *black hole.*

Theoretical models of supernovae successfully reproduce the observed energy outputs of real supernovae, but the precise mechanism of the explosion is not yet clear. Two general processes could contribute to the explosion. In the first process, neutron degeneracy pressure halts the gravitational collapse, causing the core to rebound slightly and ram into overlying material that is still falling inward. Until recently, most astronomers thought that this *core-bounce* process ejected the star's outer layers. But current models of supernovae suggest that the more important process involves the neutrinos that are formed when electrons and protons combine to make neutrons. Although these ghostly particles rarely interact with anything [Section 14.3], so many are produced when the core implodes that they drive a shock wave that propels the star's upper layers into space.

The shock wave sends the star's former surface zooming outward at a speed of 10,000 km/s, heating it so that it shines with dazzling brilliance. For about a week, a supernova blazes as powerfully as 10 billion Suns, rivaling the luminosity of a moderate-size galaxy. The ejected gases slowly cool and fade in brightness over the next several months, but they continue to expand outward until they eventually mix with other gas in interstellar space. The scattered debris from the supernova carries with it the variety of elements produced in the star's nuclear furnace, as well as additional elements created when some of the neutrons produced during the core collapse slam into other nuclei. Millions or billions of years later, this debris may be incorporated into a new generation of stars.

The Origin of Elements

Before we leave the subject of massive star life cycles, it's useful to consider the evidence that indicates we actually understand the origin of the elements. We cannot see inside stars, so we cannot directly observe elements being created in the ways we've discussed. However, the signature of nuclear reactions in massive stars is written in the patterns of elemental abundances across the universe.

For example, if massive stars really produce heavy elements (that is, elements heavier than hydrogen and helium) and scatter these elements into space when they die, the total amount of these heavy elements in interstellar gas should gradually increase with time (because additional massive stars have died). Thus, we should expect stars born recently to contain a greater proportion of heavy elements than stars born in the distant past, because they formed from interstellar gas that contained more heavy elements. Stellar spectra confirm this prediction: Older stars do indeed contain smaller amounts of heavy elements than younger stars. For very old stars in globular clusters, elements besides hydrogen and helium typically make up as little as 0.1% of the total mass. In contrast, about 2–3% of the mass of young stars that formed in the recent past is in the form of heavy elements.

We gain even more confidence in our model of elemental creation when we compare the abundances of different elements. For example, because helium-capture reactions add two protons (and two neutrons) at a time, we expect nuclei with even numbers of protons to outnumber those with odd numbers of protons that fall between them. Sure enough, even-numbered nuclei such as carbon, oxygen, and neon are relatively abundant (Figure 16.20). Similarly, because certain elements heavier than iron are made only by rare fusion reactions shortly before and during a supernova, we expect these elements to be extremely rare. Again, this prediction made by our model of nuclear creation is verified by observations.

Supernova Observations

The study of supernovae owes a great debt to astronomers of many different epochs and cultures. Careful scrutiny of the night skies allowed the ancients to identify several supernovae whose remains still adorn the heavens. The most famous example concerns the Crab Nebula in the constellation Taurus. The Crab Nebula is a **supernova remnant**—an expanding cloud of debris from a supernova explosion (Figure 16.21). A spinning neutron star lies at the center of the Crab Nebula, providing evidence that supernovae really do create neutron stars. Photographs taken years apart show that the nebula is growing larger at a rate of several thousand kilometers per second. Calculating backward from its present size, we can trace the nebula's birth to somewhere near A.D. 1100.

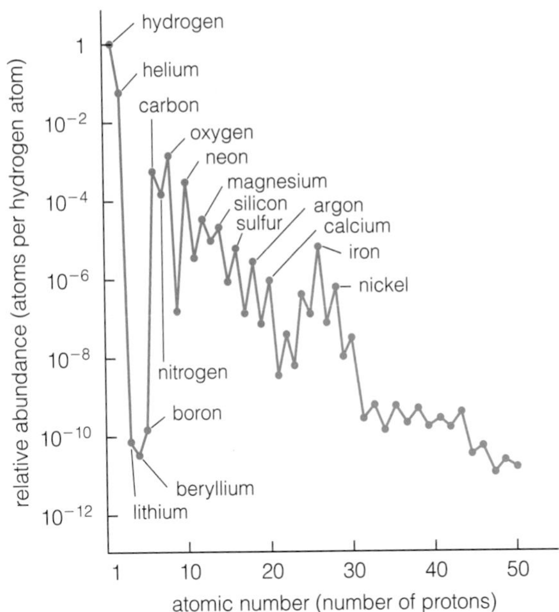

FIGURE 16.20 This graph shows the observed relative abundances of elements in the galaxy in comparison to the abundance of hydrogen. For example, the abundance of nitrogen is about 10^{-4}, which means that there are about $10^{-4} = 0.0001$ times as many nitrogen atoms in the galaxy as hydrogen atoms.

FIGURE 16.21 The Crab Nebula is the remnant of the supernova observed in A.D. 1054.

before

Thanks to observations made by ancient astronomers, we can be even more precise.

The official history of the Sung Dynasty in China contains a record of a remarkable celestial event:

> *In the first year of the period Chih-ho, the fifth moon, the day chi-ch'ou, a guest star appeared approximately several [degrees] southeast of Thien-kuan. After more than a year it gradually became invisible.*[9]

This description of the sudden appearance and gradual dimming of a "guest star" matches what we expect for a supernova, and the location "southeast of Thien-kuan" corresponds to the Crab Nebula's location in Taurus. Moreover, the Chinese date described in the excerpt corresponds to July 4, 1054, telling us precisely when the Crab supernova became visible on Earth. Descriptions of this particular supernova also appear in Japanese astronomical writings, in an Arabic medical textbook, and possibly in Native American paintings in the southwestern United States.[10] Curiously, European records do not mention this supernova, even though it would have been clearly visible.

Historical records of supernovae allow us to age-date the remnants we see today to determine the kinds of supernovae that produced them and to assess how frequently stars explode in our region of the Milky Way Galaxy. At least four supernovae have been observed during the last thousand years, appearing as brilliant new stars for a few months in the years 1006, 1054, 1572, and 1604. The supernova of 1006, the brightest of these four, could be seen during the daytime and cast shadows at night. Supernovae may even have influenced human history. The Chinese were meticulous in recording their observations because they believed that celestial events foretold the future, and they may have acted in accord with such fortune-telling. The 1572 supernova was witnessed by Tycho Brahe, and it helped convince him and others that the heavens were not as perfect and unchanging as Aristotle had imagined [Section 6.3]. Kepler saw the 1604 supernova at a time when he was struggling to make planetary orbits fit perfect circles; perhaps this "imperfection" of the heavens helped push him to test elliptical orbits instead.

No supernova has been seen in our own galaxy since 1604, but today astronomers routinely discover supernovae in other galaxies. The nearest of these extragalactic supernovae, and the only one near enough to be visible to the naked eye, burst into view in 1987. Because it was the first supernova detected that year, it was given the name **Supernova 1987A.** Supernova 1987A was the explosion of a star in the *Large Magellanic Cloud,* a small galaxy that orbits the Milky Way and is visible only from southern latitudes. The Large Magellanic Cloud is about 150,000 light-years away, so the star really exploded some 150,000 years ago.

As the nearest supernova witnessed in four centuries, Supernova 1987A provided a unique opportunity to study a supernova and its debris in detail. Astronomers from all over the planet traveled to the Southern Hemisphere to observe it, and several orbiting spacecraft added observations in many different wavelengths of light. Older photographs of the Large Magellanic Cloud allowed astronomers to determine precisely which star exploded (Figure 16.22). It turned out to be a blue star, not the red supergiant expected when core fusion has ceased. The most likely explanation is that the star's outer layers were unusually thin and warm near the end of its life (perhaps because some of its matter spilled onto a close

[9]From Murdin and Murdin, *Supernovae* (Cambridge: Cambridge University Press, 1985, p. 4).

[10]A petroglyph in Chaco Canyon, New Mexico, made at about the right time appears to show a star that may be the supernova in about the right place relative to the Moon and Venus. However, there is some controversy surrounding this interpretation of the painting.

after

FIGURE 16.22 Before and after photos of the location of Supernova 1987A. In the before picture, the arrow indicates the star that exploded. Note that the supernova actually appeared as a bright *point* of light; it appears larger than a point in the photograph only because of overexposure.

binary companion), changing its appearance from that of a red supergiant to a blue one. The surprising color of the pre-explosion star demonstrates that we still have much to learn about supernovae. Reassuringly, most other theoretical predictions of stellar life cycles were well matched by observations of Supernova 1987A.

One of the most remarkable findings from Supernova 1987A was that bursts of neutrinos coming from its direction were recorded by two neutrino detectors, one in Japan and one in Ohio. These data confirmed that the explosion released most of its energy in the form of neutrinos, suggesting that we are correct in believing that the stellar core undergoes sudden collapse to a ball of neutrons. The capture of neutrinos from Supernova 1987A has spurred scientific interest in building more purposeful "neutrino telescopes." Perhaps these neutrino telescopes will open new fields of astronomical research in the coming decades.

TIME OUT TO THINK ***When Betelgeuse explodes as a supernova, it will be more than 10 times brighter than the full moon in our sky. If Betelgeuse had exploded a few hundred or a few thousand years ago, do you think it could have had any effect on human history? How do you think our modern society would react if we saw Betelgeuse explode tomorrow?***

Summary of Stellar Lives

We have seen that the primary factor determining how a star lives its life is its mass. Low-mass stars live long lives and die in planetary nebulae, leaving behind white dwarfs. High-mass stars live short lives and die in supernovae, leaving behind neutron stars and black holes. Both types of stars are crucial to life. Near the ends of their lives, low-mass stars can become carbon stars, which are the source of most of the carbon in our bodies. High-mass stars produce the vast array of other chemical elements on which life depends. Figure 16.23 summarizes the life cycles of stars of different masses.

16.5 The Lives of Close Binary Stars

For the most part, stars in binary systems proceed from birth to death as if they were isolated and alone. The exceptions are close binary stars. Algol, the "demon star" in the constellation Perseus, consists of two stars that orbit each other closely: a $3.7M_{\text{Sun}}$ main-sequence star and a $0.8M_{\text{Sun}}$ subgiant. A moment's thought reveals that something quite strange is going on. The stars of a binary system are born at the same time and therefore must both be the same age. We know that more massive stars live shorter lives, and therefore the more massive star must exhaust its core hydrogen and become a subgiant before the less massive star does. How, then, can Algol's less massive star be a subgiant while the more massive star is still burning hydrogen as a main-sequence star?

This so-called *Algol paradox* reveals some of the complications in ordinary stellar life cycles that can arise in close binary systems. The two stars in close binaries are near enough to exert significant tidal forces on each other [Section 6.5]. The gravity of each star attracts the near side of the other star more strongly than it attracts the far side. The stars therefore stretch into football-like shapes rather than remaining spherical. In addition, the stars become *tidally locked* so that they always show the same face to each other, just as the Moon always shows the same face to Earth [Sections 3.3, 11.5].

During the time that both stars are main-sequence stars, the tidal forces have little effect on their lives. But when the more massive star (which exhausts its core hydrogen sooner) begins to expand into a red giant, gas from its outer layers can spill over onto its companion. This **mass exchange** occurs when the

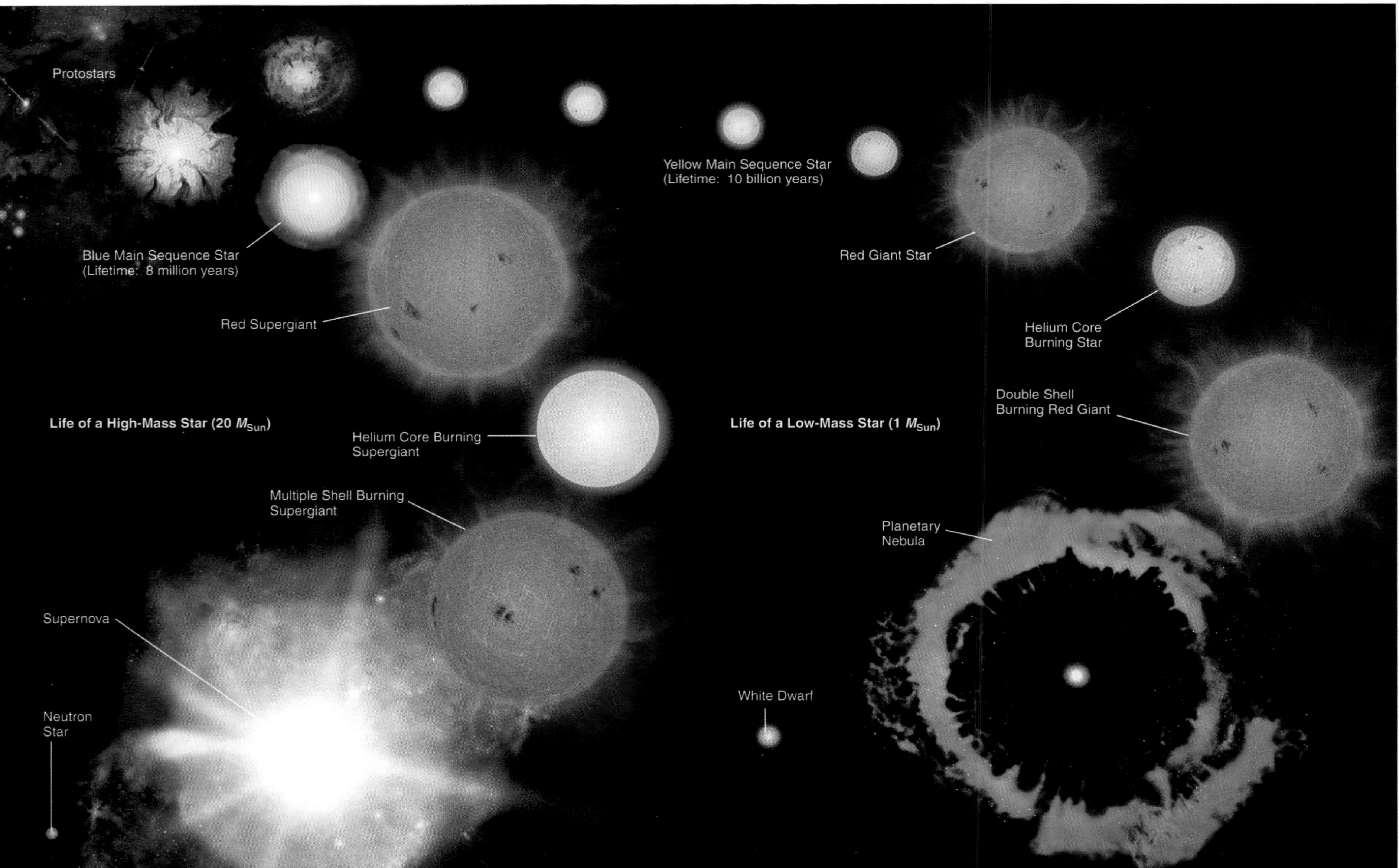

FIGURE 16.23 Summary of stellar lives. The life stages of a high-mass star (on the left) and low-mass star (on the right) are depicted in clockwise sequences beginning with the protostellar stage in the upper left corner.

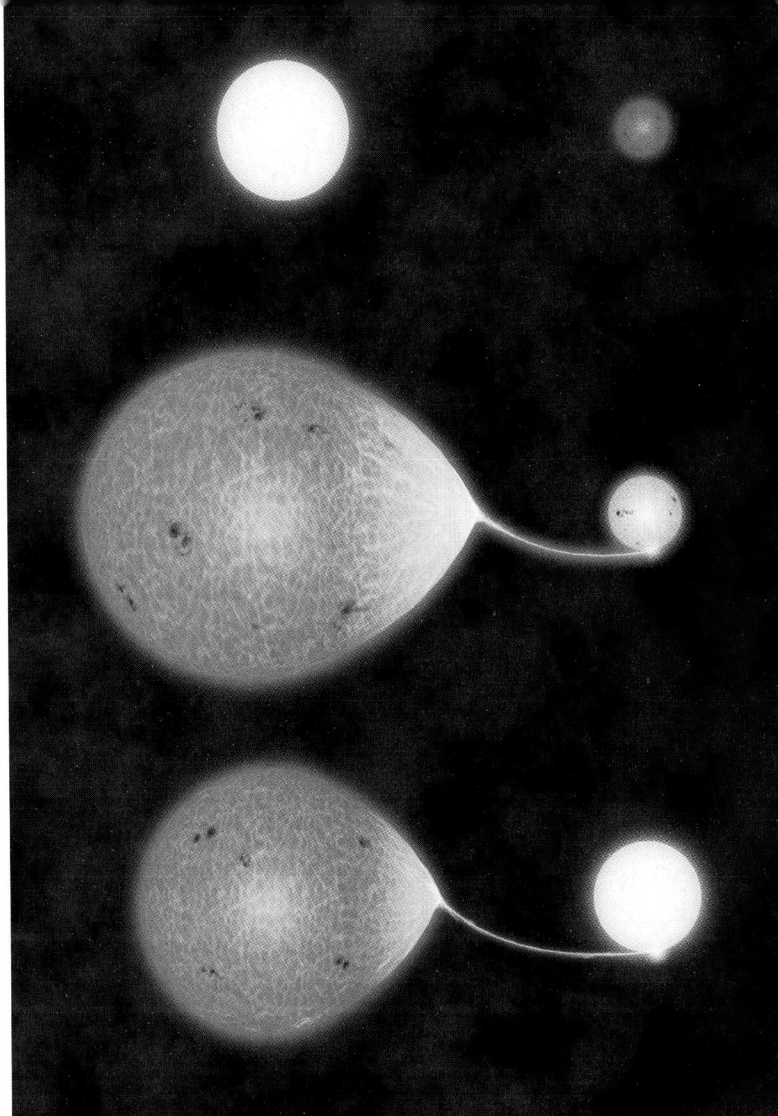

Algol, shortly after its birth. The higher mass main-sequence star (left) evolved more quickly than its lower mass companion (right).

Algol, onset of mass transfer. When the more massive star expanded into a red giant, it began losing some of its mass to its main-sequence companion.

Algol today. As a result of the mass transfer, the red giant has shrunk to a subgiant, and the main-sequence star on the right is now the more massive of the two stars.

FIGURE 16.24 Artist's conception of the development of the Algol close binary system.

giant grows so large that its tidally distorted outer layers succumb to the gravitational attraction of the smaller companion star. The companion then begins to gain mass at the expense of the giant.

The solution to the Algol paradox should now be clear (Figure 16.24). The $0.8M_{Sun}$ subgiant *used to be* much more massive. As the more massive star, it was the first to begin expanding into a red giant. But, as it expanded, so much of its matter spilled over onto its companion that it is now the less massive star.

The future may hold even more interesting events for Algol. The $3.7M_{Sun}$ star is still gaining mass from its subgiant companion. Thus, its life cycle is actually accelerating as its increasing gravity raises its core hydrogen fusion rate. Millions of years from now, it will exhaust its hydrogen and begin to expand into a red giant itself. At that point, it can begin to transfer mass *back* to its companion. Even stranger things can happen in other mass-exchange systems, particularly when one of the stars is a white dwarf or a neutron star. But that is a topic for the next chapter.

> *I can hear the sizzle of newborn stars, and know anything of meaning, of the fierce magic emerging here. I am witness to flexible eternity, the evolving past, and know I will live forever, as dust or breath in the face of stars, in the shifting pattern of winds.*
>
> JOY HARJO, *Secrets from the Center of the World*

THE BIG PICTURE

In this chapter, we answered the question of the origin of elements that we first discussed in Chapter 1. As you look back over this chapter, remember these "big picture" ideas:

- Virtually all elements besides hydrogen and helium were forged in the nuclear furnaces of stars. Carbon can be released from low-mass stars near the ends of their lives (carbon stars), and many other elements are released into space by massive stars in supernova explosions.

- The tug-of-war between gravity and pressure determines how stars behave from the time of their birth in a cloud of molecular gas to their sometimes violent death.
- Low-mass stars like our Sun live long lives and die with the ejection of a planetary nebula, leaving behind a white dwarf.
- High-mass stars live fast and die young, exploding dramatically as supernovae.
- Close binary stars can exchange mass, altering the usual course of stellar evolution.

Review Questions

1. What is *thermal pressure?* What are the two energy sources that can help a star maintain its internal thermal pressure?
2. Distinguish between the mass ranges for low-mass stars, intermediate-mass stars, and high-mass stars. Why is it useful to group stars into these mass ranges when we discuss the lives of stars?
3. What is a *molecular cloud?* Briefly describe the process by which a *protostar* forms from gas in a molecular cloud. Why are protostars surrounded by *protostellar disks?*
4. What is a *close binary?* Under what conditions does a close binary form?
5. Describe some of the activity seen in protostars, such as strong *protostellar winds* and *jets.*
6. What do we mean by a star's *life track* on an H–R diagram? How does an H–R diagram that shows life tracks differ from a standard H–R diagram?
7. Summarize the four stages that mark a star's life from its first appearance as a protostar to its becoming a main-sequence star. Identify the four stages on the star's life track. How do these stages differ for stars of different masses?
8. Describe the observed trends in the number of stars that are born with different masses. What types of stars are most common? What types are least common?
9. What is *degeneracy pressure?* How does it differ from thermal pressure? Explain why degeneracy pressure can support a stellar core against gravity even when the core becomes very cold.
10. What is a *brown dwarf?* Explain why brown dwarfs may be the most common form of ordinary matter in the universe, even though they are difficult to detect.
11. What are the two basic ways by which energy can move outward from a star's core to its surface? Explain why stars of different masses have convection zones of different depths, and briefly describe the interior structure of stars of different masses.
12. What is a *flare star?* What two properties of flare stars make them so active?
13. What happens to the core of a star when it exhausts its hydrogen supply? Why does *hydrogen shell burning* begin around the inert core?
14. Explain why a star expands into a *subgiant* and then a *red giant,* even while its core is contracting. What is a *red-giant wind?*
15. Why does *helium fusion* require much higher temperatures than hydrogen fusion? Briefly describe the overall reaction by which helium fuses into carbon.
16. Why does helium burning begin with a *helium flash* in low-mass stars? How do the core structure and rate of fusion change after the helium flash?
17. Explain why a star's overall radius shrinks from its peak size as a red giant after helium fusion begins. Why do helium-burning stars in a star cluster all fall on a *horizontal branch* on an H–R diagram?
18. Describe what happens to a low-mass star after it exhausts its core helium supply. What are *thermal pulses?* What makes a star become a *carbon star?* How is the existence of carbon stars important to life on Earth?
19. What is a *planetary nebula?* What happens to the core of a star after a planetary nebula occurs?
20. Briefly describe how the Sun will change, and how Earth will be affected by these changes, over the next several billion years.
21. What is the *CNO cycle?* Why are high-mass stars able to use the CNO cycle, while low-mass stars do not?
22. What is *radiation pressure?* Why is it more important in high-mass stars than in low-mass stars? What effects does radiation pressure have on the structure of a high-mass star?
23. Describe some of the nuclear reactions that can occur in high-mass stars after they exhaust their core helium. Why does this continued nuclear burning occur in high-mass stars but not in low-mass stars?
24. What special feature of iron nuclei determines how massive stars end their lives?
25. Describe the mechanism of a *supernova.* What is left behind after a supernova?
26. Summarize some of the observational evidence supporting our ideas about how the elements formed and showing that supernovae really occur.
27. What is the *Algol paradox?* What is its resolution? Use your answer to summarize how the lives of stars in close binaries can differ from the lives of single stars.

Discussion Questions

1. *Connections to the Stars.* In ancient times, many people believed that our lives were somehow influenced by the patterns of the stars in the sky, a belief that survives to this day in astrology. Modern science has not found any evidence to support this belief but instead has found that we have a connection to the stars on a much deeper level: In the words of Carl Sagan, we are "star stuff." Discuss in some detail our real connections to the stars as established by modern astronomy. Do you think these connections have any philosophical implications in terms of how we view our lives and our civilization? Explain.

2. *Humanity in A.D. 5,000,000,000.* Do you think it is likely that humanity will survive until the Sun begins to expand into a red giant 5 billion years from now? Why or why not? If the human race does survive, how do you think people in A.D. 5,000,000,000 will differ from people today? What do you think they will do when faced with the impending death of the Sun? Debate these questions, and see if you and your friends can come to any agreement on possible answers.

Problems

1. *Homes to Civilizations.* We do not yet know how many stars have Earth-like planets, nor do we know the likelihood that such planets might harbor advanced civilizations like our own. However, some stars can probably be ruled out as candidates for advanced civilizations. For example, given that it took a few billion years for humans to evolve on Earth, it seems unlikely that advanced life would have had time to evolve around a star that is only a few million years old. For each of the following types of star, decide whether you think it is possible that it could harbor an advanced civilization, and explain your reasoning in one or two paragraphs.
   a. A $10M_{\text{Sun}}$ main-sequence star.
   b. A flare star.
   c. A carbon star.
   d. A $1.5M_{\text{Sun}}$ red giant.
   e. A $1M_{\text{Sun}}$ horizontal branch star.
   f. A red supergiant.

2. *Close Binary Disks.* We believe that the primary way protostars shed angular momentum is by transferring it from the central star to the protostellar disk. We also believe that some binary stars form when the protostar is unable to shed enough angular momentum. If these beliefs are both correct, how might protostellar disks in binary systems differ from those in single-star systems? Could there be any implications regarding the formation of planets in binary systems? Explain.

3. *Rare Elements.* Lithium, beryllium, and boron are elements with atomic number 3, 4, and 5, respectively. But despite their being three of the five simplest elements, Figure 16.20 shows that they are rare compared to many heavier elements. Suggest a reason for their rarity. (*Hint:* Consider the process by which helium fuses into carbon.)

4. *Future Skies.* As a red giant, the Sun's angular size in the Earth's sky will be about 30°. What will sunset and sunrise be like? About how long will they take? Do you think the color of the sky will be different from what it is today? Explain.

5. *Massive Star Instability.* If you study the life track of a high-mass star on the H–R diagram, you will see that it passes through the *instability strip* discussed in Chapter 15. What happens to the star during the time it resides in the instability strip? What would we call this type of star during this time? Explain. Based on the fact that stars of this type are massive, explain why they are so relatively rare.

6. *Research: Historical Supernovae.* As discussed in the text, historical accounts exist for supernovae in the years 1006, 1054, 1572, and 1604. Choose one of these supernovae and learn more about historical records of the event. Did the supernova influence human history in any way? Write a two- to three-page summary of your research findings.

7. *Web Project: Coming Fireworks in Supernova 1987A.* Astronomers believe that the show from Supernova 1987A is not yet over. In particular, sometime between now and about 2010, the expanding cloud of gas from the supernova is expected to ram into surrounding material, and the heat generated by the impact is expected to create a new light show. Using the web, learn more about how Supernova 1987A is changing and what we might expect to see from it in the future. Summarize your findings in a one- to two-page report.

8. *Web Project: Picturing Star Birth and Death.* Photographs of stellar birthplaces (i.e., molecular clouds) and death places (e.g., planetary nebulae and supernova remnants) can be strikingly beautiful, but only a few such photographs are included in this chapter. Search the web for additional photographs of these types; look not only for photos taken in visible light, but also for false-color photographs made from observations in other wavelengths of light. Put each photograph you find into a personal on-line journal, along with a one-paragraph description of what the photograph shows. Try to compile a journal of at least 20 such photographs.

17
The Bizarre Stellar Graveyard

WELCOME TO THE AFTERWORLD OF STARS, THE FASCINATING domain of white dwarfs, neutron stars, and black holes. To scientists, these dead stars are ideal laboratories for testing the most extreme predictions of general relativity and quantum theory. To most other people, the eccentric behavior of stellar corpses demonstrates that the universe is stranger than they had ever imagined.

Dead stars behave in unusual and unexpected ways that challenge our minds and stretch the boundaries of what we believe is possible. Stars that have finished nuclear burning have only one hope of staving off the crushing power of gravity: the strange quantum mechanical effect of degeneracy pressure. But even this strange pressure cannot save the most massive stellar cores, which collapse into oblivion as black holes. Prepare to be amazed by the eerie inhabitants of the stellar graveyard!

17.1 A Star's Final Battle

In the previous chapter, we saw that an ongoing "battle" between the inward crush of gravity and the outward push of pressure governs a star's life from the time when it first begins to form in an interstellar cloud to the time when it finally exhausts its nuclear fuel. Throughout most of this time, the pressure that holds gravity at bay is *thermal pressure,* which results from the heat produced as the star fuses light elements into heavier ones in its core. But the nuclear fuel eventually runs out. In the end, after the star dies in a planetary nebula or supernova, the fate of the stellar corpse lies in the outcome of a final battle between gravity and pressure—but this time the source of the pressure is the quantum mechanical effect called *degeneracy pressure.*

We have already discussed how degeneracy pressure successfully resists the crush of gravity in the stellar corpses known as white dwarfs and neutron stars, as well as in several other cases (e.g., brown dwarfs and inert stellar cores). Because these objects are supported by degeneracy pressure, they are known collectively as *degenerate objects,* and the matter within them is called *degenerate matter.* Recall that degeneracy pressure arises when subatomic particles are packed as closely together as the laws of quantum mechanics allow [Section S5.5]. More specifically, white dwarfs are supported against the crush of gravity by **electron degeneracy pressure,** in which the pressure arises from densely packed electrons. In neutron stars, it is *neutrons* that are packed tightly together, thereby generating **neutron degeneracy pressure.**

White dwarfs and neutron stars would be strange enough if the story ended here, but it does not. Sometimes, gravity wins the battle with degeneracy pressure, and the stellar corpse collapses without end, crushing itself out of existence and forming a *black hole.* A black hole is truly a hole in the universe. If you enter a black hole, you leave our observable universe and can never return. In the rest of this chapter, we will study the bizarre properties and occasional catastrophes of the stellar corpses known as white dwarfs, neutron stars, and black holes.

17.2 White Dwarfs

A **white dwarf** is the inert core left over after a star has ceased nuclear burning, so its composition reflects the products of the star's final nuclear-burning stage. The white dwarf left behind by a $1M_{\text{Sun}}$ star like our Sun will be made mostly of carbon, the product of the star's final helium-burning stage. The cores of very low mass stars never become hot enough to fuse helium and thus end up as helium white dwarfs. Some intermediate-mass stars progress to carbon burning but do not create any iron (and hence do not explode as supernovae). These stars leave behind white dwarfs containing large amounts of oxygen or even heavier elements.

Despite the ordinary-sounding compositions of white dwarfs, their degenerate matter is unlike anything ever seen on Earth. Gravity and electron degeneracy pressure have battled to a draw in white dwarfs. Electron degeneracy pressure finally halts the collapse of a $1M_{\text{Sun}}$ stellar corpse when it shrinks to about the size of the Earth. If you recall that our Earth is smaller than a typical sunspot, it should be clear that it's no small feat to pack the entire mass of the Sun into the volume of the Earth. The density of such a white dwarf is so high that a pair of standard dice made from its material would weigh about 5 tons (Figure 17.1).

FIGURE 17.1 Two dice made from white dwarf material would weigh 5 tons—about as much as three cars.

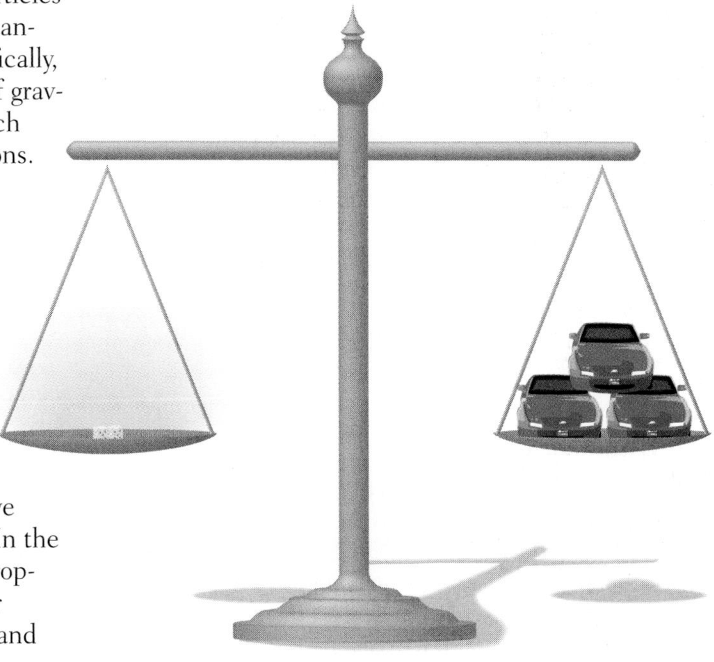

Surprisingly, more massive white dwarfs are smaller and denser than less massive ones. For example, a $1.3M_{\text{Sun}}$ white dwarf is half the diameter and about 10 times denser than a $1.0M_{\text{Sun}}$ white

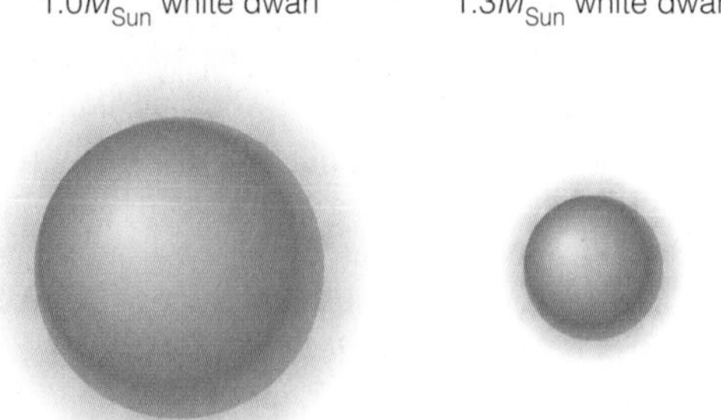

FIGURE 17.2 Contrary to what you might expect, more massive white dwarfs are actually *smaller* (and thus denser) than less massive white dwarfs.

dwarf (Figure 17.2). The more massive white dwarf is smaller and denser—even though it contains more matter—because its greater gravity can compress this matter into a smaller volume. As a white dwarf becomes denser, its electrons must move faster, and degeneracy pressure becomes stronger [Section S5.5]. The most massive white dwarfs are therefore the smallest, because they must be extremely dense for degeneracy pressure to balance their greater gravity.

The fact that adding mass to a white dwarf makes it shrink also explains why red giants become more luminous as they age [Section 16.3]. Recall that degeneracy pressure supports the inert helium core of a low-mass red giant, so this core is essentially a white dwarf buried inside a star. As the hydrogen-burning shell deposits more helium ash onto the degenerate core, the mass of the core continually increases. Therefore the core continually contracts, and the surrounding shell of hydrogen shrinks along with it. Because the shell gets hotter as it shrinks, its hydrogen fusion rate increases, steadily raising the luminosity of the red giant.

The White Dwarf Limit

Theoretical calculations show that the mass of a white dwarf cannot exceed a **white dwarf limit** of about $1.4M_{Sun}$. (The white dwarf limit is more commonly called the *Chandrasekhar limit* after its discoverer.) The limit comes about because the electron speeds are higher in more massive white dwarfs, and these speeds approach the speed of light in white dwarfs with masses approaching $1.4M_{Sun}$. Neither electrons nor anything else can travel faster than the speed of light, so electrons can do nothing to halt the crush of gravity in a stellar corpse with a mass above the white dwarf limit. Such an object must inevitably collapse into a compact ball of neutrons, at which point *neutron* degeneracy pressure can stop the crush of gravity.

Strong observational evidence supports this theoretical limit on the mass of a white dwarf. Many known white dwarfs are members of binary systems, and hence their masses can be measured [Section 15.4]. In every observed case, the white dwarfs have masses below $1.4M_{Sun}$, just as we would expect from theory.

White Dwarfs in Close Binary Systems

Left to itself, a white dwarf will never succumb to the crush of gravity because its electron degeneracy pressure does not lessen even as the white dwarf cools into a cold black dwarf. However, white dwarfs in close binary systems do not necessarily rest in peace.

A white dwarf in a close binary system can gain substantial quantities of mass if its companion is a main-sequence or giant star. When a clump of mass first spills over from the companion to the white dwarf, it has some small orbital velocity. The law of conservation of angular momentum dictates that the clump must orbit faster and faster as it falls toward the white dwarf's surface. The infalling matter therefore forms a whirlpool-like disk around the white dwarf (Figure 17.3). The process in which material falls onto another body is called *accretion* [Section 8.4], so this rapidly rotating disk is called an **accretion disk.**

TIME OUT TO THINK *Explain why the infalling matter forms a* disk *around the white dwarf (as opposed to making, say, a spherical distribution of matter).* (Hint: *See Sections 8.3 and 16.2.*) *How is an accretion disk similar to a protoplanetary disk? How is it different?*

Accreting gas gradually spirals inward through the accretion disk and eventually falls onto the surface of the white dwarf. This happens because the individual gas particles in the accretion disk, like anything else that orbits a massive body, obey Kepler's laws [Section 6.3]. Gas in the inner parts of the accretion disk moves faster than gas in the outer parts; because of these differences in orbital speed, gas in any particular part of the accretion disk "rubs" against slower-moving gas just outside of it. This "rubbing" generates friction and heat in the same way that rubbing your palms together makes them warm. The friction slowly causes the orbits of individual gas particles to decay, making the orbits smaller and smaller until they fall onto the white dwarf surface. Meanwhile, additional gas spilling over from the companion constantly replenishes the accretion disk.

Accretion can provide the "dead" white dwarf with a new energy source—making it a stellar zombie. Theory predicts that the heat generated by friction

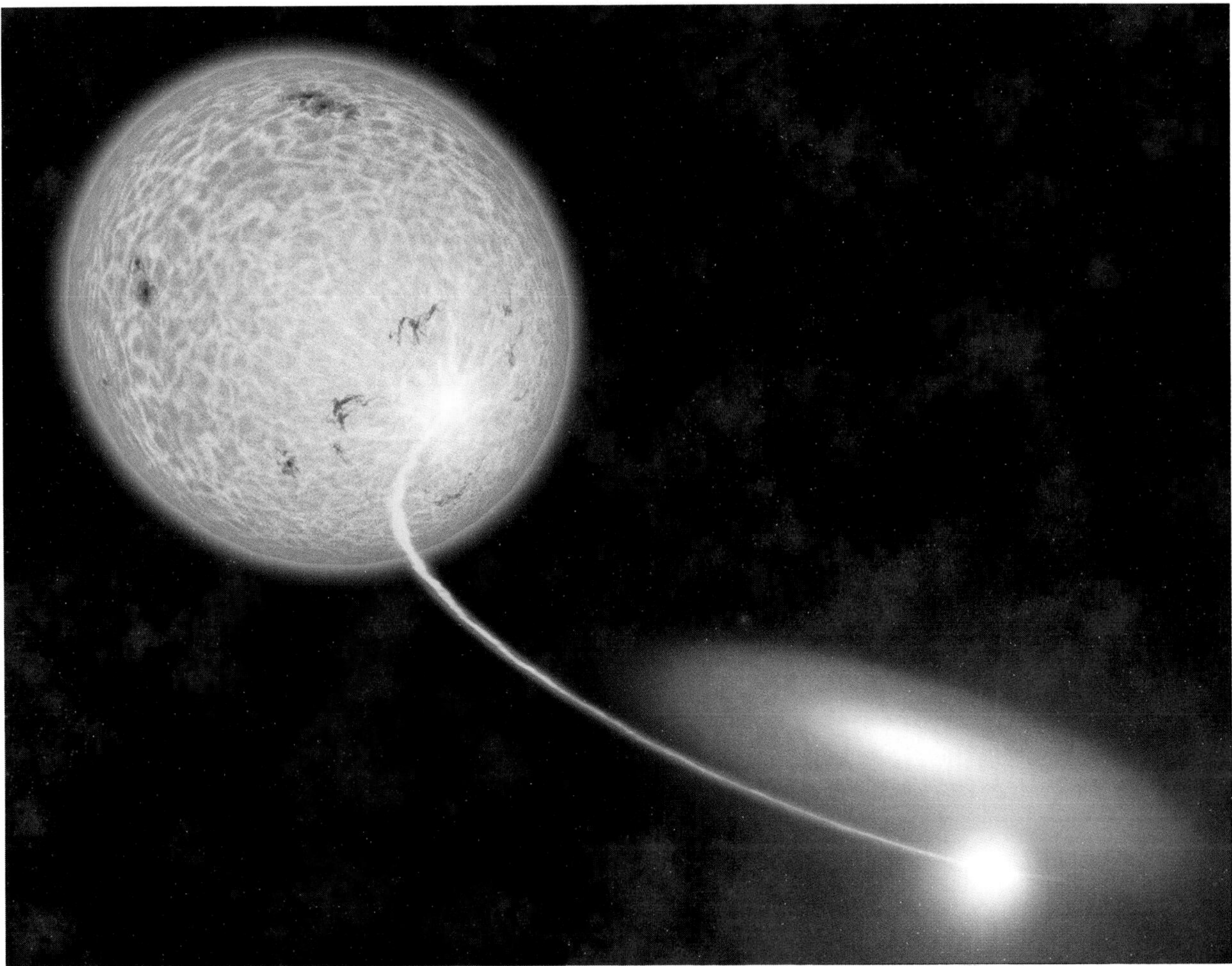

FIGURE 17.3 This artist's conception shows how mass spilling from a companion star (left) toward a white dwarf (right) forms an accretion disk around the white dwarf. The white dwarf itself is in the center of the accretion disk—too small to see on this scale. Matter streaming onto the disk creates a hot spot at the point of impact.

should make the accretion disk hot enough to radiate optical and ultraviolet light, and sometimes even X rays. Thus, although accretion disks are far too small and too far away to be seen directly with telescopes, we should be able to detect their intense ultraviolet or X-ray radiation. Searches for this radiation have turned up strong evidence for such accretion disks around many white dwarfs. In some cases, the brightness of these systems is highly variable—sudden increases in brightness by a factor of 10 or more may persist for a few days and then fade away, only to repeat a few weeks or months later. Such brightening probably arises when instabilities in the accretion disk cause some of the matter to fall suddenly onto the white dwarf surface, with an accompanying release of gravitational potential energy.[1]

[1]This type of brightening is sometimes called a *dwarf nova.*

Accreting white dwarfs occasionally flare up even more dramatically. Remember that a white dwarf is "dead" because it has no hydrogen left to fuse. This situation changes in an accreting white dwarf. The gas spilling onto an accreting white dwarf comes from the upper layers of its companion star and thus is composed mostly of hydrogen. A thin shell of fresh hydrogen builds up on the surface of the white dwarf as more and more material rains down from the accretion disk. The pressure and temperature at the bottom of this shell increase as the shell grows, and hydrogen fusion ignites when the temperature exceeds 10 million K. The white dwarf, no longer just a zombie, suddenly blazes back to life as its hydrogen shell burns. This thermonuclear flash causes the binary system to shine for a few glorious weeks as a **nova,** which can radiate as brightly as 100,000 Suns (Figure 17.4a). The nova generates heat and pressure, ejecting most of the material that has accreted onto

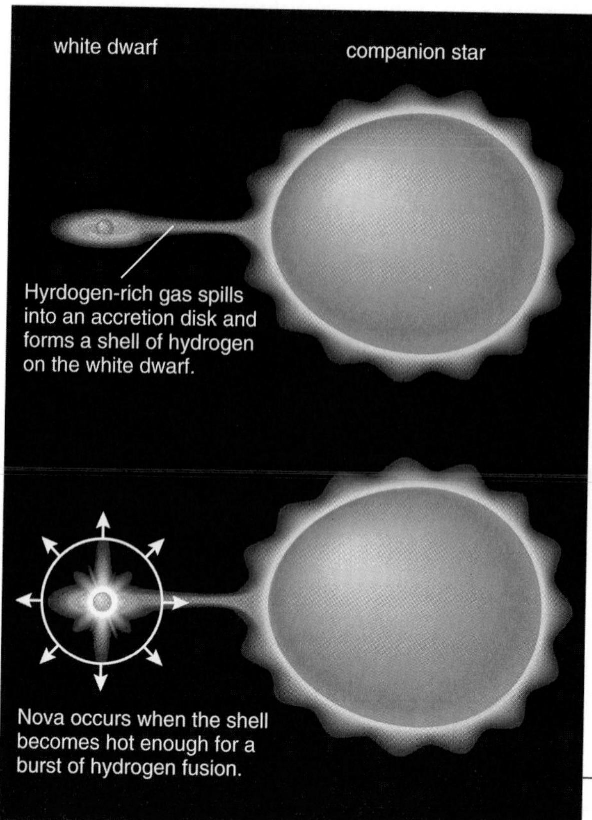

a

b

FIGURE 17.4 (**a**) Diagram of the nova process. (**b**) Hubble Space Telescope image showing blobs of gas ejected from the nova T Pyxidis. The bright spot at the center of the blobs is the binary star system that generated the nova.

the white dwarf. This material expands outward, creating a *nova remnant* that sometimes remains visible years after the nova explosion (Figure 17.4b).

Accretion resumes after a nova explosion subsides, so the entire process can repeat itself. The time between successive novae in a particular system depends on the rate at which hydrogen accretes on the white dwarf surface and on how highly compressed this hydrogen becomes. The compression of hydrogen is greatest for the most massive white dwarfs, which have the strongest surface gravities. In some cases, novae have been observed to repeat after just a few decades. More commonly, accreting white dwarfs may have to wait 10,000 years between nova outbursts.

Note that, according to our modern definitions, a nova and a supernova are quite different events: A nova is a relatively minor detonation of hydrogen fusion on the surface of a white dwarf in a close binary, while a supernova is the total explosion of a star. However, the word *nova* simply means "new," and historically a nova referred to any star that appeared to the naked eye where none was visible before. Because supernovae generate far more power than novae—the light of 10 billion Suns in a supernova versus 100,000 Suns in a nova—a very distant supernova can appear as bright in our sky as a nova that is relatively close. Thus, we could not distinguish between novae and supernovae prior to modern times.

White Dwarf Supernovae

Each time a nova occurs, the white dwarf ejects some of its mass. Each time a nova subsides, the white dwarf begins to accrete matter again. Theoretical models cannot yet tell us whether the net result is a gradual increase or decrease in the white dwarf's mass. Nevertheless, in at least some cases, accreting white dwarfs in binary systems continue to gain mass as time passes. If such a white dwarf gains enough mass, it can one day reach the $1.4M_{\text{Sun}}$ white dwarf limit.[2] This day is the white dwarf's last.

The electron degeneracy pressure that has been supporting the white dwarf gives out as soon as it reaches the $1.4M_{\text{Sun}}$ limit. Gravity is suddenly unopposed, and the white dwarf begins to collapse. The gravitational potential energy released in the collapse

[2]Another way in which white dwarfs might exceed the $1.4M_{\text{Sun}}$ white dwarf limit is through the merger of two white dwarfs in a close binary.

quickly heats the white dwarf, which soon reaches the temperature needed for carbon fusion. Because the white dwarf material is degenerate, carbon fusion ignites almost instantly throughout the star. This "carbon bomb" detonation is similar to the helium flash that occurs in low-mass red giants but releases far more energy [Section 16.3]. The white dwarf explodes completely in what we will call a **white dwarf supernova.**

TIME OUT TO THINK *Why do we expect that most collapsing white dwarfs undergo* carbon *fusion, as opposed to fusion of hydrogen, helium, or some other element?*

A white dwarf supernova shines as brilliantly as a supernova that occurs at the end of a high-mass star's life [Section 16.4]. To distinguish the two types, we'll refer to the latter as a **massive star supernova.** Both white dwarf supernovae and massive star supernovae result in the destruction of a star. A massive star supernova is thought inevitably to leave behind either a neutron star or a black hole, but nothing remains after the "carbon bomb" detonation of a white dwarf supernova.

Astronomers can distinguish between white dwarf and massive star supernovae by studying their light.[3] Because white dwarfs contain almost no hydrogen, the spectra of white dwarf supernovae show no spectral lines of hydrogen. In contrast, massive stars usually have plenty of hydrogen in their outer layers at the time they explode, so hydrogen lines are prominent in the spectra of massive star supernovae. A second way to distinguish the two types of supernovae is to plot *light curves* that show how their luminosity fades with time [Section S2.3]. Figure 17.5 contrasts typical light curves for white dwarf and massive star supernovae. Note that both types of supernovae reach a peak luminosity of about 10 billion Suns ($10^{10}L_{Sun}$), but white dwarf supernovae fade steadily while massive star supernovae fade in two distinct stages.

Not only are white dwarf supernovae dramatic, but they also provide one of the primary means by which we measure large distances in the universe. Massive star supernovae differ in intrinsic brightness, but white dwarf supernovae always occur in white dwarfs that have just reached the $1.4M_{Sun}$ limit. The light curves of all white dwarf supernovae therefore look amazingly similar, and their maximum luminosities are nearly identical. This fact is extremely useful: Once we know the true luminosity of one white dwarf supernova, we essentially know the luminosities of them all. Thus, whenever we discover a white dwarf supernova in a distant galaxy, we can determine the galaxy's distance by using the luminosity–distance formula [Section 15.2].

[3]Observationally, astronomers classify supernovae as *Type II* if their spectra show hydrogen lines, and *Type I* otherwise. All Type II supernovae are assumed to be massive star supernovae. However, a Type I supernova can be either a white dwarf supernova or a massive star supernova in which the star blew away all its hydrogen before exploding. Type I supernovae appear in two classes whose light curves differ, called *Type Ia* and *Type Ib*. The former are thought to be white dwarf supernovae.

17.3 Neutron Stars

White dwarfs with densities of 5 tons per teaspoon may seem incredible, but neutron stars are stranger still. A **neutron star** is the ball of neutrons created by the collapse of the iron core in a massive star supernova. Typically just 10 kilometers across yet more massive than the Sun, neutron stars are essentially giant atomic nuclei, with two important differences: (1) They are made almost entirely of neutrons, and (2) gravity, not the strong force, is what binds them together.

The force of gravity at the surface of a neutron star is truly awe-inspiring. Escape velocity is one-half the speed of light. The strong gravity causes photons to emerge with a *gravitational redshift* that increases their wavelengths to about 15% longer than normal,

FIGURE 17.5 Typical light curves for white dwarf and massive star supernovae.

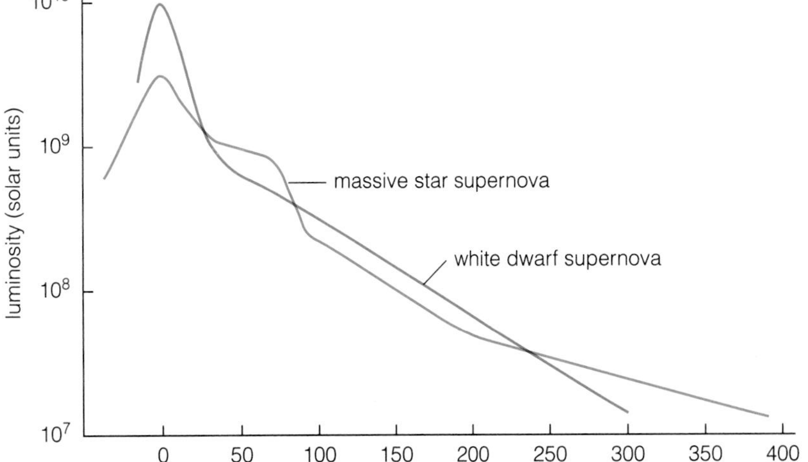

FIGURE 17.6 A paper clip made from neutron star material would outweigh Mount Everest.

just as predicted by Einstein's general theory of relativity [Section S4.5]. If you foolishly chose to visit a neutron star's surface, your body would be squashed immediately into a microscopically thin puddle of subatomic particles.

Things would be only slightly less troubling if a bit of neutron star came to visit you. A paper clip made from neutron star material would outweigh Mount Everest (Figure 17.6). If such a paper clip magically appeared in your hand, you could not prevent it from falling. Down it would plunge, passing through the Earth like a rock falling through space. It would gain speed until it reached the center of the Earth, and its momentum would carry it onward until it slowed to a stop on the other side of our planet. Then it would fall again. If it came in from space, each plunge of the neutron star material would drill a different hole through the rotating Earth. In the words of Carl Sagan, the inside of the Earth would "look briefly like Swiss cheese" (until the melted rock flowed to fill in the holes) by the time friction finally brought the piece of neutron star to rest at the center of the Earth.

In the unfortunate event that an *entire* neutron star came to visit you, it would not fall at all. Because it would be only about 10 kilometers across, the neutron star would probably fit in your hometown, but it would be 300,000 times more massive than the Earth. The neutron star's immense surface gravity would quickly destroy your hometown and the rest of civilization. By the time the dust settled, the former Earth would be a shell no thicker than your thumb on the surface of the neutron star.

Pulsars

Theorists first speculated about neutron stars in the 1930s. But until observational proof of their existence arrived in 1967, most astronomers assumed that neutron stars were too strange to exist. The proof came largely through the efforts of Jocelyn Bell, then a 24-year-old graduate student at Cambridge University. Bell had helped her adviser, Anthony Hewish, build a radio telescope ideal for discovering fluctuating sources of radio waves. She was busily trying to interpret the flood of data pouring out of this instrument in October 1967 when she noticed a peculiar signal. After ruling out other possibilities, she concluded that *pulses* of radio waves were arriving from somewhere near the direction of the constellation Cygnus at precise 1.337301-second intervals (Figure 17.7). No known astronomical object pulsated so regularly. In fact, the pulsations came at such precise intervals that they were nearly as reliable for measuring time as the most precise human-made clocks. For a while, the mysterious source of the radio waves was dubbed "LGM" for Little Green Men, only half-jokingly. Today we refer to such rapidly pulsing radio sources as **pulsars.**

The mystery of pulsars was soon resolved. By the end of 1968, astronomers had found two smoking guns: Pulsars sat at the centers of both the Crab Nebula and the Vela Nebula, the gaseous remnants of supernovae (Figure 17.8). The pulsars are neutron stars left behind by the supernova explosions.

The pulsations arise because the neutron star is spinning rapidly as a result of the conservation of angular momentum: As an iron core collapses into a neutron star, its rotation rate must increase as it shrinks in size. The collapse also bunches the magnetic field lines running through the core far more tightly, greatly amplifying the strength of the magnetic field. Shortly after the supernova event, the magnetic field of the remaining neutron star is a trillion times stronger than the Earth's. These intense magnetic fields somehow direct beams of radiation out along

FIGURE 17.7 About 20 seconds of data from the first pulsar discovered by Jocelyn Bell in 1967. Arrows mark the pulses, which come precisely 1.337301 seconds apart.

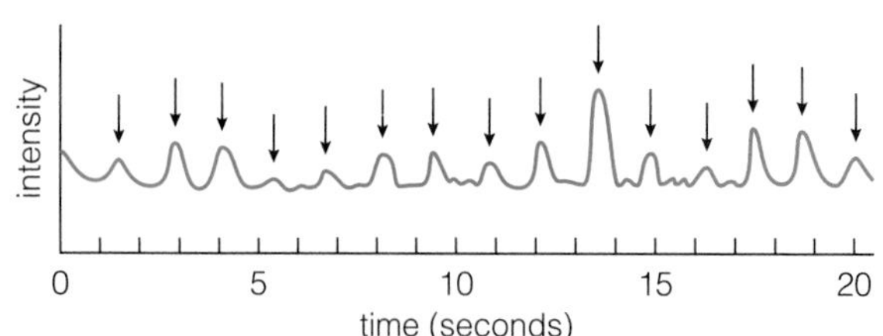

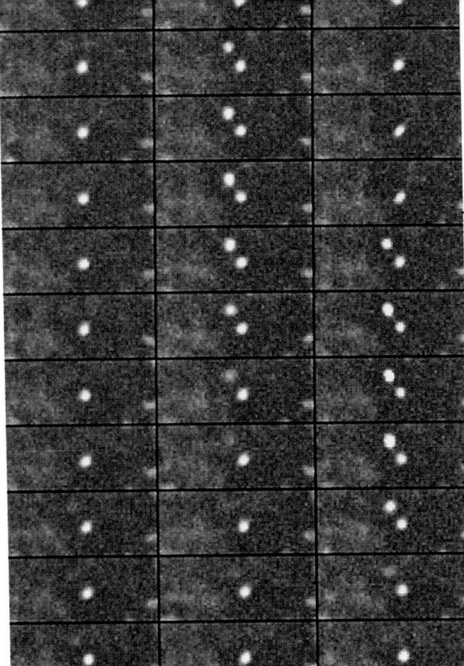

FIGURE 17.8 These time-lapse photos of the pulsar in the Crab Nebula, which lies to the upper left of the unrelated star at center, show how the pulsar flashes on and off about 30 times a second.

the magnetic poles, although we do not yet know exactly how the radiation is generated. If a neutron star's magnetic poles do not align with its rotation axis (just as the Earth's magnetic poles do not coincide with its geographic poles), the beams of radiation sweep round and round (Figure 17.9). Like lighthouses, these neutron stars actually emit a fairly steady beam of light, but we see pulses of light with each sweep of the beam past our location.

Pulsars are not quite perfect clocks, because each revolution of a pulsar takes ever so slightly longer. This gradual slowing of the neutron star's spin occurs because the continual twirling of the magnetic field generates electromagnetic radiation that carries away energy and angular momentum. The pulsar in the Crab Nebula, for example, currently spins about 30 times per second. Two thousand years from now, it will spin less than half as fast. Eventually, a pulsar's spin slows so much and its magnetic field weakens so much that we can no longer detect it. In addition, some spinning neutron stars may be oriented so that their beams do not sweep past our location. Thus, we have the following rule: All pulsars are neutron stars, but not all neutron stars are pulsars.

TIME OUT TO THINK ***Suppose we do*** not ***see pulses from a particular neutron star and hence do not call it a pulsar. Is it possible that a civilization living in some other star system would see this neutron star as a pulsar? Explain.***

b

FIGURE 17.9 **(a)** A pulsar is a rapidly rotating neutron star that beams radiation along its magnetic axis. **(b)** This artwork likens a pulsar (top) to a lighthouse (bottom). If a pulsar's radiation beams are not aligned with its rotation axis, they will sweep through space. Each time one of these beams sweeps across Earth, we see a pulse of radiation.

We know that pulsars must be neutron stars, because no other massive object could spin so fast. A white dwarf, for example, can spin no faster than about once per second; an increase in spin would tear it apart because its surface would be rotating faster than the escape velocity. Yet pulsars have been discovered that rotate as fast as 625 times per second. Only an object as extremely small and dense as a neutron star could spin so fast without breaking apart.

Interestingly, the first planets discovered outside our solar system orbit a pulsar. In 1992, the pulsar known as PSR B1257+12 was found to have a pulsation period that periodically becomes very slightly shorter or longer. A detailed analysis of the pulsation periods suggests that they can be explained by gravitational tugs of planetlike bodies on the pulsar. In fact, at least two planets appear to be orbiting PSR B1257+12, with orbital periods of about 67 and 98 days, respectively. This discovery came as an immense surprise to astronomers, since these planets are orbiting the remains of a star that exploded. Why weren't the planets destroyed during their star's red supergiant stage? Why didn't they fly off into space when the supernova ejected most of the star's mass? Why didn't the supernova explosion kick the pulsar away from its planets? The planets probably formed *after* the explosion. It's likely that the pulsar once had a stellar companion that came too close. Tidal forces ripped the companion star apart, and the debris formed a protoplanetary disk that then coalesced into planets.

Neutron Stars in Close Binary Systems

Like their white dwarf brethren, neutron stars in close binary systems can come back to life as brilliant stellar zombies. As is the case with white dwarfs, gas overflowing from a companion star can create a hot, swirling accretion disk around the neutron star. However, in the neutron star's mighty gravitational field, infalling matter releases an amazing amount of gravitational potential energy. Dropping a brick onto a neutron star would liberate as much energy as an atomic bomb. Because the gravitational energies of accretion disks around neutron stars are so tremendous, they are much hotter and more luminous than those around white dwarfs. The high temperatures in the inner regions of the accretion disk make it radiate powerfully in X rays. Some close binaries with neutron stars emit 100,000 times more energy in X rays than our Sun emits in all wavelengths of light combined. Due to this intense X-ray emission, close binaries that contain accreting neutron stars are often called **X-ray binaries.** Their existence confirms many of the strange properties of neutron stars.

Today we know of hundreds of X-ray binaries in the Milky Way. We can make an **all-sky map** (Figure 17.10) showing the distribution of X-ray binaries by taking the celestial sphere and "flattening" it in

FIGURE 17.10 This all-sky map shows the distribution of X-ray binaries in all directions in the sky. Notice that most X-ray binaries, represented by green dots, are found in the disk of the Milky Way Galaxy, which runs horizontally across the center of this map. To visualize how the entire map corresponds to what we see in the sky, recall that the disk of the Milky Way forms a band of light that wraps a full 360° around our sky [Section 1.5]. You could reconstruct the circular band we see in the sky by cutting out this picture and attaching the left end to the right end to form a circular strip. The points directly at the top and bottom of this map correspond to the two opposite directions that are each 90° away from the band of the Milky Way in our sky—just as the points at the top and bottom of a flat map of the Earth represent the north and south poles, which are each 90° away from the Earth's equator.

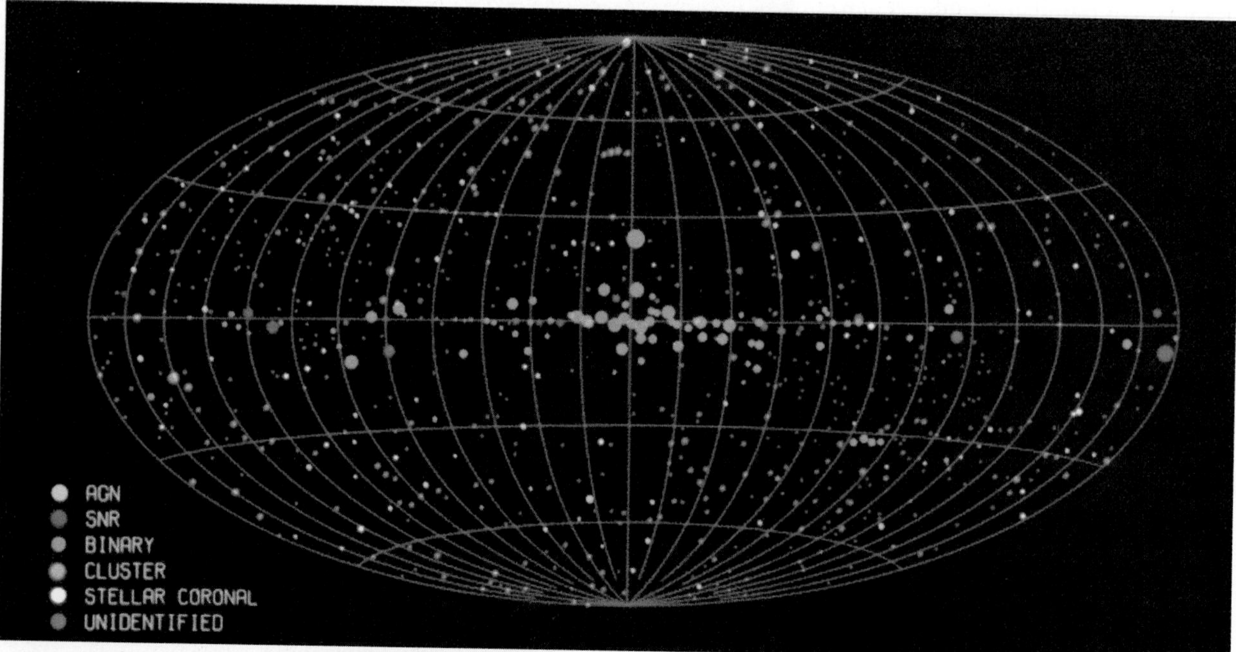

the same way we make flat maps of the Earth. We orient the map so that the disk of the Milky Way runs horizontally. Note that X-ray binaries are concentrated in the disk, just like most of our galaxy's stars, gas, and dust. X-ray sources are usually named by their constellation and order of discovery. For example, Sco X-1 was the first X-ray source discovered in the direction of the constellation Scorpio. Sco X-1 is now known to be an X-ray binary in which a low-mass star orbits a neutron star about every 19 hours.

The emission from most X-ray binaries pulsates rapidly, confirming the identification of the X-ray source as a neutron star. In contrast to radio pulsars, whose spins slow with time, the pulsation rates of X-ray binaries tend to accelerate. Matter accreting onto the neutron star adds angular momentum, speeding up the neutron star's rotation (Figure 17.11). By the time the companion star finally stops overflowing, the neutron star may be rotating hundreds of times per second. Because such pulsars spin every few thousandths of a second, they are sometimes called **millisecond pulsars.**

One of the strangest millisecond pulsars is known as the "black widow" because it appears to be destroying its companion. The companion star must once have had a reasonable mass, but almost all of it has spilled over onto the black widow pulsar. The result is that the pulsar now spins every 1.6 milliseconds, while the companion orbits it every 9 hours. The companion also eclipses the pulsar for about an hour of each orbit. Such a long eclipse means that the companion must still have an extended atmosphere, despite the fact that its mass is a mere $0.02M_{\text{Sun}}$. Evidently, the pulsar's energy output gradually evaporates matter from the surface of its companion. In a few million years, the companion star will have vanished entirely, leaving behind a solitary millisecond pulsar.

Like accreting white dwarfs that occasionally erupt into novae, accreting neutron stars sporadically erupt with enormous luminosities. Hydrogen-rich material from the companion star builds up on the surface of the neutron star. Pressures at the bottom of this accreted hydrogen shell, only a meter thick, maintain steady fusion. Helium accumulates beneath the hydrogen-burning shell, and helium fusion suddenly ignites when the temperature builds to 100 million K. The helium burns rapidly to carbon and heavier elements, generating a burst of energy that flows from the neutron star in the form of X rays. These **X-ray bursters** typically flare every few hours to every few days. Each burst lasts only a few seconds, but during that short time the system radiates power equivalent to 100,000 Suns, all in X rays. Within a minute after a burst, the X-ray burster cools back down and resumes accreting.

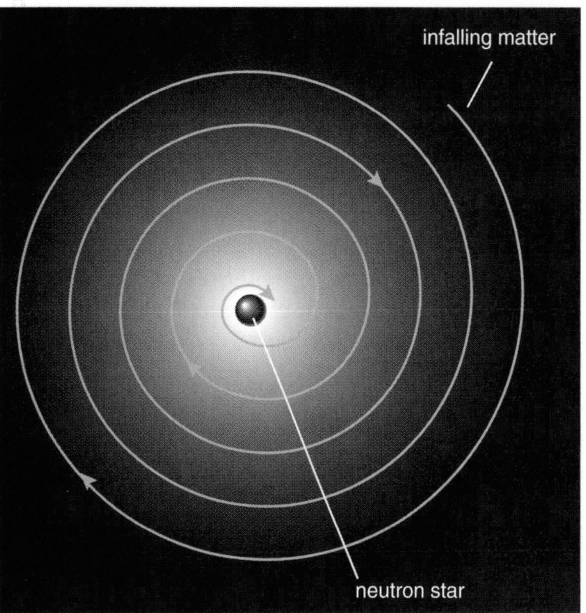

FIGURE 17.11 Matter accreting onto a neutron star adds angular momentum, increasing the neutron star's rate of spin.

17.4 Black Holes: Gravity's Ultimate Victory

We know that white dwarfs cannot exceed $1.4M_{\text{Sun}}$ because gravity overcomes electron degeneracy pressure above that mass. The mass of a neutron star has a similar limit. The precise *neutron star limit* is not known, but it certainly lies below $3M_{\text{Sun}}$. A collapsing stellar core that weighs more than $3M_{\text{Sun}}$ faces the ultimate oblivion: becoming a **black hole.**

A black hole originates in the collapse of the iron core that forms just prior to the supernova of a very high mass star. Any star born with more than about $10M_{\text{Sun}}$ dies in a supernova, but most of the star's mass is blown into space by the explosion. As a result, the collapsed cores left behind by most supernovae are neutron stars. However, theoretical models show that the most massive stars might not succeed in blowing away all their upper layers. If enough matter falls back onto the core, raising its mass above the neutron star limit, then neutron degeneracy pressure will not be able to fend off gravity.

A core whose mass exceeds the neutron star limit will continue to collapse catastrophically. Then, when the core has only a fraction of a second left, gravity plays its cruelest trick. Usually, the gravitational potential energy released as a star collapses boosts its temperature and pressure enough to fight off gravity. However, in a star destined to become a black hole, the enhanced temperature and pressure just make gravity stronger. According to Einstein's theory of relativity, energy is equivalent to mass ($E = mc^2$) and thus must exert some gravitational attraction. The gravity of pure energy usually is negligible, but not in a stellar core collapsing beyond the neutron star limit. Here the energy associated with the temperature and pressure concentrated in the tiny core acts like additional mass, hastening the collapse. To the best of our current understanding, *nothing* can halt the crush of gravity. The core collapses without end, forming a black hole. Gravity has achieved its ultimate triumph.

A Hole in the Universe

The idea of a black hole was first suggested in the late 1700s by British philosopher John Mitchell and French physicist Pierre Laplace. It was already known from Newton's laws that the escape velocity from any object depends only on its mass and size: More compact objects of a particular mass have higher escape velocities [Section 6.6]. Mitchell and Laplace speculated about objects so compact that their escape velocity exceeded the speed of light. Because they worked in a time before it was known that light always travels at the same speed, they assumed that light emitted from such an object would behave like a rock thrown upward, slowing to a stop and falling back down.

Einstein's work showed that black holes are considerably more bizarre. He found that space and time are not distinct, as we usually think of them, but instead are bound up together as four-dimensional **spacetime** [Section S4.3]. Moreover, in his general theory of relativity, Einstein showed that what we perceive as gravity arises from *curvature of spacetime.* The concept "curvature of spacetime" is quite strange, because we cannot even visualize four dimensions, let alone visualize their curvature. However, as discussed in more detail in Chapter S4, we can draw an analogy to spacetime curvature with a "rubber sheet" diagram in which we show how different masses would affect a stretched rubber sheet. In this analogy, a black hole is a region in which spacetime is stretched so far that it becomes a bottomless pit (Figure 17.12). Keep in mind that a black hole is not really shaped like a funnel; the illustration is only a two-dimensional analogy. Black holes are actually spherical, and they are black because not even light can escape from them.

A black hole really is a *hole* in the observable universe, and the boundary between the inside of the black hole and the universe outside is called the **event horizon.** Within the event horizon, the escape velocity exceeds the speed of light, so nothing—not even light—can ever get out. The event horizon gets its name because information can never reach us from events that occur within it.

A black hole has an event horizon because, according to general relativity, light always follows the *straightest possible path* through spacetime. If space happens to be curved, as it is near a black hole (or any other massive object), then the path of a light beam will also be curved. As the rubber-sheet analogy in Figure 17.12b shows, it is possible to draw a series of concentric circles around a black hole. However, it is not really possible to measure a *radius* for these circles. Because of the extreme curvature of

Mathematical Insight 17.1 The Schwarzschild Radius

The Schwarzschild radius (R_S) of a black hole is given by a simple formula:

$$\text{Schwarzschild radius} = R_S = \frac{2GM}{c^2}$$

where M is the black hole's mass, $G = 6.67 \times 10^{-11}$ m^3/(kg $\times$ s^2) is the gravitational constant, and $c = 3 \times 10^8$ m/s is the speed of light. With a bit of calculation, this formula can also be written as:

$$\text{Schwarzschild radius} = R_S = 3.0 \times \frac{M}{M_{\text{Sun}}} \text{ km}$$

Example 1: What is the Schwarzschild radius of a black hole with a mass of $10M_{\text{Sun}}$?

Solution: The latter version of the formula is easier to use in this case. We set $M = 10M_{\text{Sun}}$ to find:

$$R_S = 3 \times \frac{10M_{\text{Sun}}}{M_{\text{Sun}}} = 30 \text{ km}$$

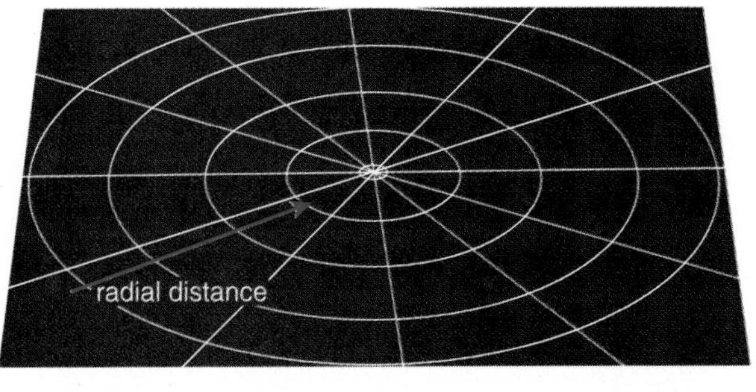

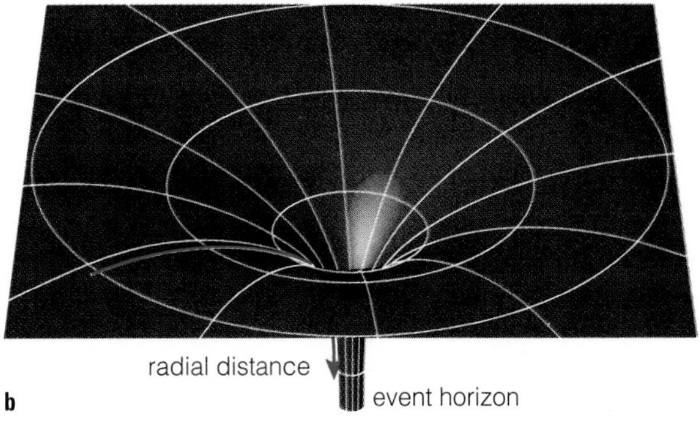

FIGURE 17.12 (**a**) A two-dimensional representation of "flat" spacetime. The circumference of each circle is 2π times its radius. (**b**) A two-dimensional representation of the "curved" spacetime around a black hole. The black hole's mass distorts spacetime, making the radial distance between two circles larger than it would be in "flat" spacetime.

spacetime near a black hole, their centers are within the event horizon and hence are not part of our observable universe. Thus, we define the radius of a circle around a black hole as the radius it *would* have if geometry were flat (Euclidean) as shown in Figure 17.12a. (That is, radius = circumference ÷ 2π.)

The radius of the event horizon is known as the **Schwarzschild radius,** named for Karl Schwarzschild (1873–1916), who computed it from Einstein's general theory of relativity in 1916. The Schwarzschild radius of a black hole depends only on its mass: The larger the mass, the larger the Schwarzschild radius. Schwarzschild computed his famous radius only a month after Einstein published his theory; he did the work while serving in the German Army on the Russian front during World War I. Sadly, he died less than a year later of an illness contracted during the war.

There is no way to tell what has fallen into a black hole in the past. A black hole that forms from the collapse of a stellar core has the mass of that core, but no recognizable material remains. Because stellar cores normally rotate, conservation of angular momentum dictates that black holes should be rotating rapidly when they form. Aside from mass and angular momentum, the only other measurable property of a black hole is its electrical charge, and most black holes are probably electrically neutral.[4] Any other information carried by objects that plunge into a black hole is irrevocably lost from the universe.

[4]Physicists whimsically say that "black holes have no hair" and refer to the idea that the only measurable properties of a black hole are mass, angular momentum, and charge as the "no-hair theorem."

The Schwarzschild radius of a $10M_{\text{Sun}}$ black hole is about 30 km.

Example 2: As far as we know, black holes in the present-day universe can be formed only when an object weighs more than the roughly $3M_{\text{Sun}}$ neutron star limit. However, Stephen Hawking and others have speculated that much less massive *mini–black holes* might have formed during the Big Bang. Suppose a mini–black hole has the mass of the Earth (about 6×10^{24} kg). What is its Schwarzschild radius?

Solution: In this case, the first version of the formula is more convenient:

$$R_S = \frac{2 \times \left(6.67 \times 10^{-11} \dfrac{\text{m}^3}{\text{kg} \times \text{s}^2}\right) \times (6 \times 10^{24}\ \text{kg})}{\left(3 \times 10^8\ \dfrac{\text{m}}{\text{s}}\right)^2} \approx 0.009\ \text{m}$$

The mini–black hole would have a Schwarzschild radius of only 9 millimeters, making it small enough to fit on the tip of your finger. But don't try to hold it—it weighs as much as the entire Earth!

Voyage to a Black Hole

Imagine that you are a pioneer of the future, making the first visit to a black hole. You've selected a black hole with a mass of $10M_{Sun}$ and a Schwarzschild radius of 30 km. As your spaceship approaches the black hole, you fire its engines to put the ship on a circular orbit a few thousand kilometers above the event horizon. Note that this orbit will be perfectly stable—there is no need to worry about getting "sucked in."

Your first task is to test Einstein's general theory of relativity. As discussed in Chapter S4, this theory predicts that time should run more slowly as the force of gravity grows stronger. It also predicts that light coming out of a strong gravitational field should show a redshift, called a *gravitational redshift,* that is due to gravity rather than to the Doppler effect. You test these predictions with the aid of two identical clocks whose numerals glow with blue light. You keep one clock aboard the ship and push the other one, with a small rocket attached, directly toward the black hole (Figure 17.13). The small rocket automatically fires its engines just enough so that the clock falls only gradually toward the event horizon. Sure enough, the clock on the rocket ticks more slowly as it heads toward the black hole, and its light becomes increasingly redshifted. When the clock reaches a distance of about 10 km above the event horizon, you see it ticking only half as fast as the clock on your spaceship, and its numerals are red instead of blue.

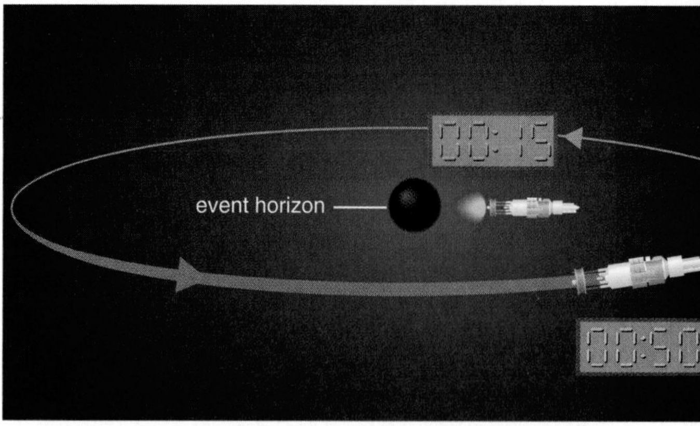

FIGURE 17.13 Time runs more slowly on the clock nearer to the black hole, and gravitational redshift makes its glowing blue numerals appear red from your spaceship.

The rocket has to expend fuel rapidly to keep the clock hovering in the strong gravitational field, and it soon runs out of fuel. The clock plunges toward the black hole. From your safe vantage point inside the spaceship, you see the clock ticking more and more slowly as it falls. However, you soon need a radio telescope to "see" it, as the light from the clock face shifts from the red part of the visible spectrum, through the infrared, and on into the radio. Finally, its light is so far redshifted that no conceivable telescope could detect it. Just as the clock vanishes from view, you see that the time on its face has frozen to a stop.

Curiosity overwhelms the better judgment of one of your colleagues. He hurriedly climbs into a spacesuit, grabs the other clock, resets it, and jumps out of the airlock on a trajectory aimed straight for the black hole. Down he falls, clock in hand. He watches the clock, but because he and the clock are traveling together, its time seems to run normally and its numerals stay blue. From his point of view, time seems to neither speed up nor slow down. In fact, he'd say that *you* were the one with the strange time, as he would see your time running increasingly fast and your light becoming increasingly blueshifted. When the clock reads, say, 00:30, he and his clock pass through the event horizon. There is no barrier, no wall, no hard surface. The event horizon is a mathematical boundary, not a physical one. From his point of view, the clock keeps ticking. He is inside the event horizon, the first human being ever to leave our universe.

Common Misconceptions: Black Holes Don't Suck

What would happen if our Sun suddenly became a black hole? For some reason, it has become part of our popular culture for most people to believe that Earth and the other planets would be inevitably "sucked in" by the black hole. This is not true. Although the sudden disappearance of light and heat from the Sun would be bad news for life, the Earth's orbit would not change. Newton's law of gravity tells us that the allowed orbits in a gravitational field are ellipses, hyperbolas, and parabolas [Section 6.4]; note that "sucking" is not on the list! A spaceship would get into trouble only if it came so close to a black hole—within about three times its Schwarzschild radius—that gravity would deviate significantly from Newton's law. Otherwise, a spaceship passing near a black hole would simply swing around it on an ordinary orbit (ellipse, parabola, or hyperbola). In fact, because most black holes are so small—typical Schwarzschild radii are smaller than any star or planet, and smaller even than most asteroids—a black hole is actually one of the most difficult things in the universe to fall into by accident.

Back on the spaceship, you watch in horror as your overly curious friend plunges to his death. Yet, from your point of view, he will *never* cross the event horizon. You'll see time come to a stop for him and his clock just as he vanishes from view due to the huge gravitational redshift of light. When you return home, you can play a videotape for the judges at your trial, proving that your friend is still a part of our universe. Strange as it may seem, all of this is true according to Einstein's theory. From your point of view, your friend takes *forever* to cross the event horizon; from his point of view, it is but a moment's plunge before he leaves the universe.

The truly sad part of this story is that your friend did not live to experience the crossing of the event horizon. The force of gravity he felt as he approached the black hole grew so quickly that it actually pulled much harder on his feet than on his head, simultaneously squeezing him from side to side (Figure 17.14). In essence, your friend was stretched in the same way the oceans are stretched by the tides, except that the *tidal force* near the black hole is trillions of times stronger than the tidal force of the Moon on the Earth [Section 6.5]. No human could survive it.

If he had thought ahead, your friend might have waited to make his jump until you visited a much larger black hole, like one of the *supermassive black holes* thought to reside in the center of many galaxies [Section 20.5]. A 1 billion M_{Sun} black hole has a Schwarzschild radius of 3 billion kilometers—about the distance from our Sun to Uranus. Although the gravitational forces at the event horizon of all black holes are equally great, the larger size of the supermassive black hole makes its tidal forces much weaker and hence nonlethal. Your friend could safely plunge through the event horizon. Again, from your point of view, the crossing would take forever, and you would see time come to a stop for him just as he vanished from sight because of the gravitational redshift. Again, he would experience time running normally, and he would see time in the outside universe running increasingly fast as he approached the event horizon. Unfortunately, anything he learned as he watched the future of the universe unfold would do him little good as he plunged to oblivion inside the black hole.

Singularity and the Limits to Knowledge

The center of a black hole is thought to be a place where gravity crushes all matter to an infinitely tiny and infinitely dense point called a **singularity.** This singularity is the point at which all the mass that created the black hole resides.

We can never know what really happens inside a black hole, because no information can ever emerge from within the event horizon. Nevertheless, we can use Einstein's theory of relativity to predict conditions inside the black hole, as long as we don't try to describe conditions too close to the singularity. The singularity itself is more puzzling, because the equations of modern physics yield conflicting predictions when we try to apply them to an infinitely compressed mass. General relativity predicts that spacetime should grow infinitely curved as it enters the pointlike singularity. Quantum physics predicts that, as a consequence of the uncertainty principle [Section S5.3], spacetime should fluctuate chaotically in regions smaller than 10^{-35} meters across. No current theory adequately accommodates these divergent claims. We will not fully understand how a singularity behaves until we have developed a quantum theory of gravity that encompasses both general relativity and quantum mechanics. This uncertainty in our current knowledge is a gold mine for science fiction writers, who speculate about using black holes for exotic forms of travel through spacetime [Section S4.6].

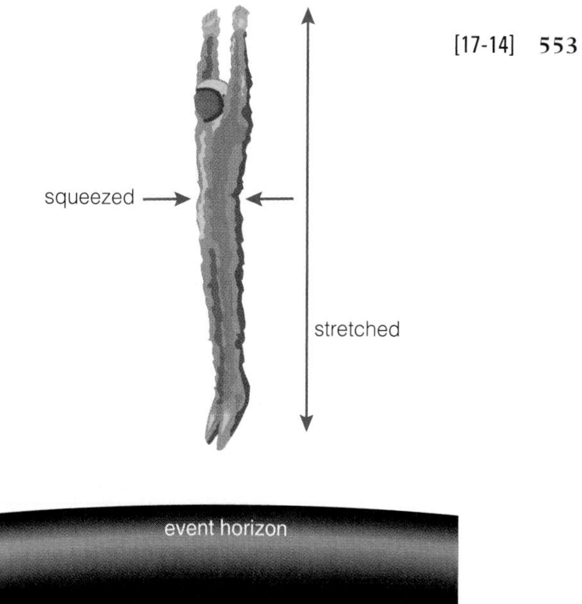

FIGURE 17.14 Tidal forces would be lethal near a black hole formed by the collapse of a star. The black hole would pull more strongly on the astronaut's feet than on his head, stretching him lengthwise and squeezing him from side to side.

Evidence for Black Holes

Have we discovered any black holes yet? We think so, but the evidence is indirect. Black holes emit no light. They are impossible to see, so we must look for their effects on surrounding matter. Black holes in close binaries should be among the easiest to identify.[5] Gas overflowing a black hole's stellar companion will form a hot, X ray–emitting accretion disk

[5]Supermassive black holes in the center of galaxies should also be relatively easy to identify; we will discuss the evidence that these exist in Chapter 20.

a

FIGURE 17.15 (**a**) The location of Cygnus X-1 in the sky. (**b**) Artist's conception of the Cygnus X-1 system. The X rays come from the high-temperature gas in the accretion disk surrounding the black hole.

similar to the disks that circle accreting neutron stars. The X rays can escape because the disk emits them from well outside the event horizon.

We strongly suspect that a few X-ray binaries contain black holes rather than neutron stars. The trick is to tell the difference, since the accretion disk is likely to be just as hot and to emit about as many X rays whether it circles a neutron star or a black hole. However, we can distinguish between the two possibilities if we can measure the mass of the accreting object. One of the most promising *black hole candidates* is in an X-ray binary called Cygnus X-1 (Figure 17.15). This system contains an extremely bright star with an estimated mass of $18M_{Sun}$; based on Doppler shifts of its spectral lines, this star orbits an unseen companion with a mass of about $10M_{Sun}$. Although there is some uncertainty in these mass estimates, the mass of the invisible accreting object clearly exceeds the $3M_{Sun}$ neutron star limit. Moreover, careful studies of variations in the system's X-ray emission indicate that this massive accreting object must be very small in size—far too small to be an ordinary star [Section 20.5]. Thus, based on our current knowledge, the accreting object in Cygnus X-1 cannot be anything other than a black hole.

TIME OUT TO THINK *As discussed earlier, some X-ray binaries that contain neutron stars emit frequent X-ray bursts and are called X-ray bursters. Could an X-ray binary that contains a black hole exhibit the same type of X-ray bursts? Why or why not?* (Hint: *What is the source of the X-ray bursts from an X-ray binary with a neutron star, and where is it located? Does a similar location exist for a system containing a black hole?*)

Of course, confirming that black holes are real with 100% certainty is very difficult. But our current theories successfully explain neutron stars, and the general theory of relativity that leads to the idea of black holes is also on solid ground. Unless something is dramatically wrong in our current theories about the mass limit of neutron stars or some other unknown type of compact object can have a huge mass, black holes must be real.

17.5 The Mystery of Gamma-Ray Bursts

In the early 1960s, the United States began launching a series of top-secret satellites designed to look for gamma rays emitted by nuclear bomb tests. The satellites soon began detecting occasional bursts of gamma rays, typically lasting a few seconds (Figure 17.16). It took several years for military scientists to become convinced that these **gamma-ray bursts** were coming from space, not from some sinister human activity. They publicized the discovery in 1973.

An increasingly large armada of satellites launched over the next two decades detected hundreds of other gamma-ray bursts. However, not only was their origin unknown, but it was very difficult to tell what direction the bursts came from. The problem is that gamma

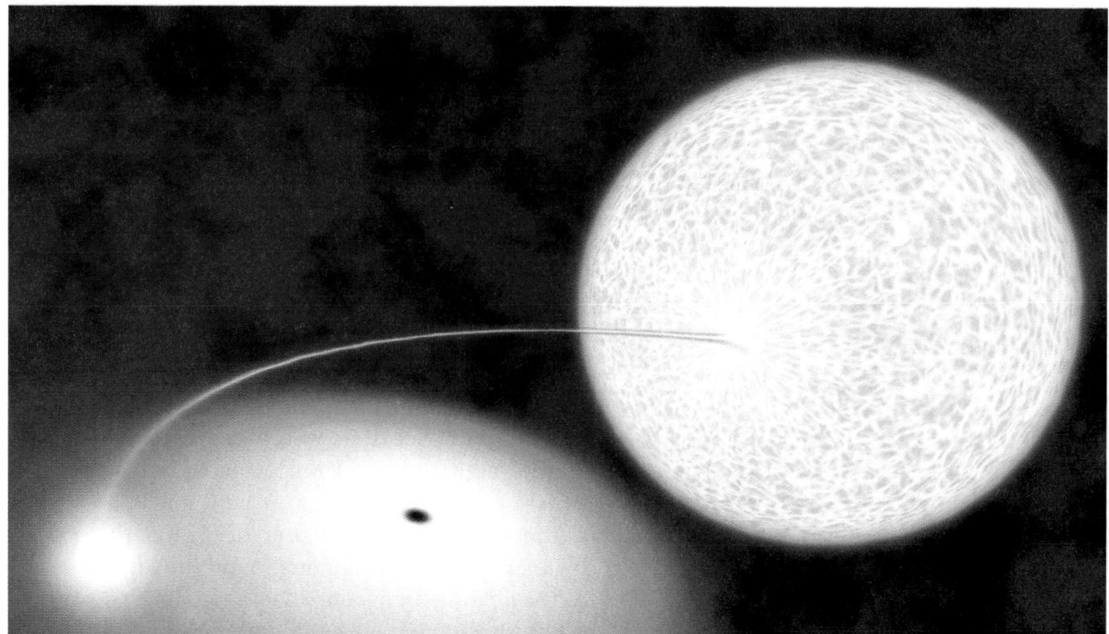

b

FIGURE 17.16 A typical gamma-ray burst light curve, showing dramatic fluctuations in gamma-ray intensity over a period of two minutes.

rays are very difficult to focus. A detector can record that it has been hit by a gamma ray but can provide little information on the direction of the gamma ray. With no specific proof, most astronomers assumed that gamma-ray bursts, like X-ray bursts, came from explosive events associated with neutron stars in X-ray binaries.

In 1991, NASA launched the *Compton Gamma Ray Observatory,* or *Compton* for short, which carried an array of eight detectors designed expressly to study gamma-ray bursts. By comparing the data recorded by all eight detectors, scientists could determine the direction of a gamma-ray burst within about 1°. The results were stunning. Compton detected gamma-ray bursts at a rate of about one per day and compiled a catalog of more than a thousand gamma-ray bursts within a few years. The results, plotted on an all-sky map, are shown in Figure 17.17. A quick comparison with the all-sky map of X-ray binaries in Figure 17.10 makes one thing very clear: The gamma-ray bursts are *not* concentrated in the disk of the Milky Way as the X-ray binaries are and thus must not be associated with neutron stars in the disk of our galaxy.

So where do gamma-ray bursts come from? Careful analysis of Figure 17.17 confirms what your eyes probably tell you: The gamma-ray bursts come from completely random directions in the sky. This fact seems to rule out the possibility that they come from anywhere in the Milky Way Galaxy. If the bursts came from objects distributed spherically about the Milky Way Galaxy, we would see a concentration of them in the direction of the galactic center. (Remember that we are located more than halfway out from the center of our galaxy.) The most reasonable conclusion is that gamma-ray bursts originate far outside our own galaxy. Additional evidence for this conclusion came in 1997, when astronomers first observed

THINKING ABOUT . . .

Too Strange to Be True?

Theoretical calculations predicted the existence of neutron stars and black holes long before their observational discovery, but many astronomers considered these theoretical results too strange to be true.

The story begins with Subrahmanyan Chandrasekhar, an astrophysicist from India. Chandrasekhar was only 19 when he completed the calculations showing that there is a white dwarf limit of $1.4M_{\text{Sun}}$, and he boldly predicted that a more massive white dwarf would collapse under the force of gravity. He did this work in 1931 while traveling by ship to England, where he hoped to impress the eminent British astrophysicist Sir Arthur Stanley Eddington. However, Eddington ridiculed Chandrasekhar for believing that white dwarfs could collapse. Neutrons had not yet been discovered, little was known about fusion, and no one had any idea what supernovae were. The idea of gravity achieving an ultimate victory seemed nonsensical to Eddington, who speculated that some type of force must prevent gravity from crushing any object.

Subrahmanyan Chandrasekhar

Sir Arthur Stanley Eddington

A few more radical thinkers took collapsing stars more seriously. A Russian physicist, Lev Davidovich Landau, independently computed the white dwarf limit in 1932. Neutrons were discovered just a few months later, and Landau speculated that stellar corpses above the white dwarf limit might collapse until neutron degeneracy pressure halted the crush of gravity. Most astronomers found the idea of neutron stars to be unacceptably weird. But two European scientists who had emigrated to California, Fritz Zwicky and Walter Baade, were not so skeptical. Without knowing of Landau's ideas, they also concluded that neutron stars were possible. In 1934, they suggested that a supernova might result when a stellar core collapses and forms a neutron star—an extraordinarily insightful guess. By 1938, physicist Robert Oppenheimer, working at Berkeley, was con-

the afterglow of a gamma-ray burst in other wavelengths. The higher resolution possible in these other wavelengths allowed astronomers to pinpoint the burst's origin in a distant galaxy.

If gamma-ray bursts really are coming from great distances, then we can estimate their true luminosities by using their apparent brightnesses and assuming that they occur in galaxies roughly halfway across the observable universe. But these estimates open up an even greater mystery, because they imply that gamma-ray bursts are *by far* the most powerful bursts of en-

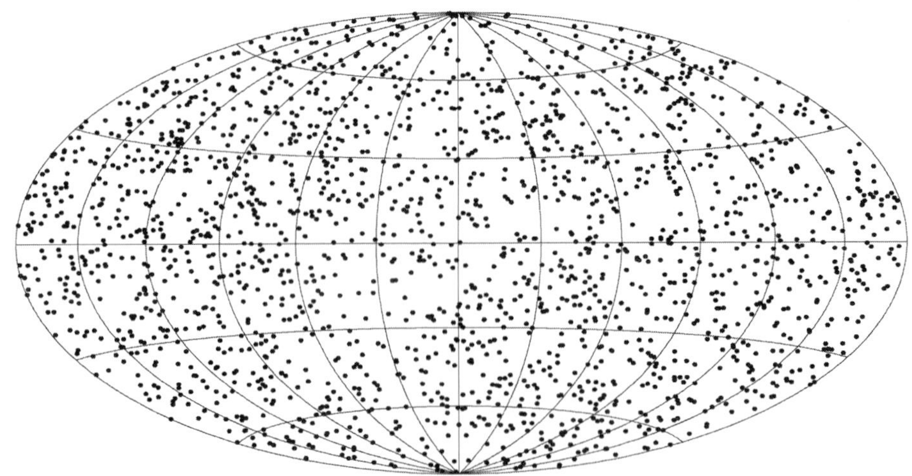

FIGURE 17.17 An all-sky diagram showing the distribution of gamma-ray bursts. Notice that the distribution appears random.

templating whether neutron stars had a limiting mass of their own. He and his coworkers concluded that the answer was yes and that neutron degeneracy pressure could not resist the crush of gravity when the mass rose above a few solar masses. Because no known force could keep such a star from collapsing indefinitely, Oppenheimer speculated that gravity would achieve ultimate victory, crushing the star into a black hole.

Lev Davidovich Landau

Astronomers gradually came to accept Chandrasekhar's $1.4M_{\text{Sun}}$ white dwarf limit, because observations found no white dwarfs more massive than this. However, most astronomers held to a belief that high-mass stars would inevitably shed enough mass late in life to prevent the formation of a more massive collapsed object. Jocelyn Bell's 1967 discovery of pulsars shattered this belief. Within a few months, Thomas Gold of Cornell University correctly suggested that the pulsars were spinning neutron stars. These discoveries forced astronomers to admit that nature was far stranger than they had expected. The verification that neutron stars really did exist made the prospect of the still-stranger black holes much less difficult to accept.

Jocelyn Bell

Robert Oppenheimer

Chandrasekhar, who had long since moved to the University of Chicago, was awarded a Nobel Prize in 1984 for his lifelong contributions to astronomy. Landau won a Nobel Prize in 1962 for his work on condensed states of matter. Oppenheimer went on to lead the *Manhattan Project* that developed the atomic bomb in 1945. Eddington died in 1944, still convinced that white dwarf stars could not collapse.

ergy that ever occur in the universe. For a few seconds, a gamma-ray burst releases as much total energy as a *hundred quadrillion* (10^{17}) Suns, or some 10,000 times the total luminosity of the brightest galaxy.

Astronomers still do not know what causes such massive outbursts of energy. One hypothesis suggests that the bursts come from the collision of *two* neutron stars in a binary system. As discussed in Chapter S4, the neutron stars in such a system gradually spiral in toward each other (because they lose energy to *gravitational waves* [Section S4.5]) and must eventually be destroyed in some type of catastrophic collision. Rough calculations suggest that such neutron star collisions should happen *somewhere* in the universe about once each day, which is consistent with the one-per-day frequency of detected gamma-ray bursts. However, why a neutron star collision would produce the peculiar spectra and light curves of gamma-ray bursts is not clear, and many scientists have raised theoretical objections to this model.

Thus, we are left with one of the greatest mysteries in all of science: Gamma-ray bursts are the most powerful events in the universe, but we have virtually no idea what causes them.

Now, my suspicion is that the universe is not only queerer than we suppose, but queerer than we can suppose.

J. B. S. HALDANE, *Possible Worlds*, 1927

THE BIG PICTURE

We have now seen what happens to stars after they die. What a mind-bending experience! Nevertheless, try to keep these "big picture" ideas straight in your head:

- Despite the strange nature of stellar corpses, clear evidence exists for white dwarfs and neutron stars, and the case for black holes is very strong.
- White dwarfs, neutron stars, and black holes can all have close stellar companions from which they accrete matter. These binary systems produce some of the most spectacular events in the universe, including novae, white dwarf supernovae, and X-ray bursters.
- Black holes are truly holes in the observable universe that strongly warp space and time around them. The nature of black hole singularities remains beyond the frontier of current scientific understanding.
- Gamma-ray bursts were once thought to be related to neutron stars in our galaxy, but recent evidence indicates that this idea is wrong. Their origin is one of the greatest mysteries in the universe.

Review Questions

1. What do we mean by a *degenerate object?* What do we mean by *degenerate matter?*
2. What is the difference between *electron degeneracy pressure* and *neutron degeneracy pressure?* Which type supports a white dwarf? Why are neutron stars so much smaller in size than white dwarfs?
3. With degeneracy pressure, how are the speeds of electrons (or neutrons) related to the mass of the degenerate object? Explain why this relationship between speed and mass implies upper limits on the possible masses of white dwarfs and neutron stars. What is the *white dwarf limit?* What is the *neutron star limit?*
4. Contrast the densities of white dwarf material with the densities of ordinary objects on Earth.
5. Explain why more massive white dwarfs are smaller in size. How does this idea explain why red giants become more luminous as they age?
6. What is an *accretion disk?* Under what conditions does an accretion disk form? Explain how the accretion disk provides a white dwarf with a new source of energy that we can detect from Earth.
7. Describe the process of a *nova.* Why do novae only occur in close binary star systems? Can the same star system undergo a nova event more than once? Explain.
8. Contrast the process of a *white dwarf supernova* with that of a *massive star supernova.* Observationally, how can we distinguish between these two types of supernovae?
9. What is a *gravitational redshift,* and how does it affect the light we see from neutron stars? Describe the incredible density of neutron star material.
10. What is a *pulsar?* Briefly describe how they were discovered and how they get their name. How do we know that pulsars must be neutron stars?
11. Explain why neutron stars in close binary systems are often called *X-ray binaries.*
12. Briefly explain why the spin rates of isolated pulsars (i.e., those not in binary systems) tend to slow with time while the spin rates of pulsars in X-ray binaries tend to increase with time. Explain how the latter can lead to a *millisecond pulsar.*
13. What is an *X-ray burster?* How is the process of an X-ray burst similar to that of a nova? How do X-ray bursts differ from novae observationally?
14. Briefly explain the process by which the core of a very high mass star can collapse to form a *black hole.*
15. How do black holes affect the structure of *spacetime?* Use these ideas to explain why light beams near a black hole do not follow perfectly straight lines.
16. What is the *event horizon* of a black hole? How does it get its name? How is it related to the *Schwarzschild radius?*
17. Suppose you are orbiting a black hole. Describe how you will see the passage of time on an object approaching the black hole. What will you notice about light from the object? Explain why an object falling toward the black hole will eventually fade from view yet never cross the event horizon from your point of view.
18. Suppose you are falling into a black hole. How will you perceive the passage of your own time? How will you perceive the passage of time in the universe around you? Briefly explain why your trip will be lethal if the black hole is relatively small in mass, and why you may survive to cross the event horizon of a supermassive black hole.
19. What do we mean by the *singularity* of a black hole? How do we know that our current theories are inadequate to explain what happens at the singularity?
20. Briefly describe the observational evidence supporting the idea that Cygnus X-1 contains a black hole.
21. What are gamma-ray bursts? Why are they so mysterious?

Discussion Questions

1. *Too Strange to Be True?* Despite strong theoretical arguments for their existence, many scientists rejected the possibility that neutron stars or black holes could really exist until they were confronted with very strong observational evidence. Some people claim that this type of scientific skepticism demonstrates that scientists are too unwilling to give up their deeply held scientific beliefs. Others claim that this type of skepticism is necessary for scientific advancement. What do you think? Defend your opinion.

2. *Black Holes in Popular Culture.* Phrases such as "it disappeared into a black hole" are now common in popular culture. Give a few examples in which the term *black hole* is used in popular culture but is not meant to be taken literally. In what ways are these uses correct in their analogies to real black holes? In what ways are they incorrect? Why do you think such an esoteric scientific idea as that of a black hole has so captured the public imagination?

Problems

1. *Life Stories of Stars.* Write a one- to two-page short story for each of the following scenarios. Each story should be detailed and scientifically correct, but also creative. That is, it should be entertaining while at the same time proving that you understand stellar evolution. Be sure to state whether "you" are a member of a binary system.
   a. You are a white dwarf of $0.8M_{\text{Sun}}$. Tell your life story.
   b. You are a neutron star of $1.5M_{\text{Sun}}$. Tell your life story.

2. *White Dwarf Density.* A typical white dwarf has a mass about that of the Sun ($M_{\text{Sun}} = 2 \times 10^{30}$ kg) and a radius about that of the Earth ($R_{\text{Earth}} = 6.4 \times 10^3$ km $= 6.4 \times 10^8$ cm).
   a. Calculate the average density of a white dwarf, in units of *kilograms per cubic centimeter.* (*Hint:* The average density of an object is its mass divided by its volume; the volume of a sphere is $4/3 \times \pi \times \text{radius}^3$.)
   b. Compare the mass of 1 cm^3 of white dwarf material with that of some familiar object of your choice.

3. *Neutron Star Density.* A typical neutron star has a mass of about $1.5M_{\text{Sun}}$ and a radius of 10 km.
   a. Calculate the average density of a neutron star, in *kilograms per cubic centimeter.*
   b. Compare the mass of 1 cm^3 of neutron star material to the mass of Mount Everest ($\approx 5 \times 10^{10}$ kg).

4. *A Black Hole?* You've just discovered a new X-ray binary, which we will call *Hyp-X1* ("Hyp" for hypothetical). The system Hyp-X1 contains a bright, B2 main-sequence star orbiting an unseen companion. The separation of the stars is estimated to be 20 million km, and the orbital period of the visible star is 4 days.
   a. Use Newton's version of Kepler's third law to calculate the sum of the masses of the two stars in the system. (*Hint:* See Mathematical Insight 15.4.) Give your answer in both kilograms and solar masses ($M_{\text{Sun}} = 2.0 \times 10^{30}$ kg).
   b. Determine the mass of the unseen companion. Is it a neutron star or a black hole? Explain. (*Hint:* A main-sequence star with spectral type B2 has a mass of about $10M_{\text{Sun}}$.)

5. *Schwarzschild radii.* Calculate the Schwarzschild radius (in km) for each of the following.
   a. A $10^8 M_{\text{Sun}}$ black hole in the center of a quasar.
   b. A $5M_{\text{Sun}}$ black hole that formed in the supernova of a massive star.
   c. A mini–black hole with the mass of the Moon.
   d. A mini–black hole formed when a super-advanced civilization decides to punish you (unfairly) by squeezing you until you become so small that you disappear inside your own event horizon.

6. *Extreme Escape Velocities.* Calculate the escape velocity from each of the following. Give each answer in three ways: in units of km/s, in units of km/hr, and as a fraction of the speed of light. (*Hint:* Use the escape velocity formula given in Section 6.6.)
   a. The surface of a white dwarf with a mass about that of the Sun and a radius about that of the Earth ($R_{\text{Earth}} = 6.4 \times 10^6$ meters).
   b. The surface of a neutron star with a mass of about $1.5M_{\text{Sun}}$ and a radius of 10 km.

7. *Challenge Problem: A Neutron Star Comes to Town.* Suppose a neutron star were suddenly to appear in your hometown. How thick a layer would the Earth form as it wraps around the neutron star's surface? To make the problem easier, you may assume that the layer formed by the Earth has the same average density as the neutron star. (*Hint:* Consider the mass of the Earth to be distributed in a spherical *shell* over the surface of the neutron star and then calculate the thickness of such a shell with the same mass as the Earth. The volume of a spherical shell is approximately its surface area times its thickness: $V_{\text{shell}} = 4\pi r^2 \times$ thickness. Because the shell will be thin, you can assume that its radius is the radius of the neutron star.)

8. *Web Project: Gamma-Ray Bursts.* Go to the text web site and seek out the latest information about gamma-ray bursts. Write a one- to two-page essay on recent discoveries and how they may shed light on the mystery of gamma-ray bursts.

PART VI

GALAXIES AND BEYOND

18 Our Galaxy

In previous chapters, we saw how stars forge new elements and expel them into space. We also studied how interstellar gas clouds enriched with these stellar by-products form new stars and planetary systems. These processes do not occur in isolation. Instead, they are part of a dynamic system that acts throughout our Milky Way Galaxy.

You are probably familiar with the idea that all living species on Earth interact with one another and with the land, water, and air to form a large, interconnected ecosystem. In a similar way, but on a much larger scale, our galaxy is a self-contained system that cycles matter from stars into interstellar space and back into stars again. The birth of our solar system and the evolution of life on Earth would not have been possible without this "galactic ecosystem."

In this chapter, we will study our Milky Way Galaxy. We will investigate the galactic processes that maintain an ongoing cycle of stellar life and death, examine the structure and motion of the galaxy, and explore the mysteries of the galactic center. Through it all, we will see that we are not only "star stuff" but "galaxy stuff"—the product of eons of complex recycling and reprocessing of matter and energy in the Milky Way Galaxy.

18.1 The Milky Way Revealed

On a dark night, you can see a faint band of light slicing across the sky through several constellations, including Sagittarius, Cygnus, Perseus, and Orion. This band of light looked like a flowing ribbon of milk to the ancient Greeks, and we now call it the *Milky Way.* In the early 1600s, Galileo used his telescope to prove that the light of the Milky Way comes from a myriad of individual stars. Together these stars make up the kind of stellar system we call a *galaxy,* echoing the Greek word for "milk," *galactos.*

Today we know that our Milky Way Galaxy holds over 100 billion stars and is just one among tens of billions of galaxies in the universe. If we could stand outside our galaxy, we would see it as a flat **disk** of stars with a bright central **bulge,** spectacular **spiral arms,** and a dimmer, rounder **halo** surrounding everything (Figure 18.1). A few hundred **globular clusters** of stars [Section 15.6] circle our galaxy's center in orbits extending tens of thousands of light-years into the halo. Two entire galaxies, known as the Large and Small Magellanic Clouds, orbit at even greater distances of some 150,000 light-years (50 kpc).[1] Although the Magellanic Clouds are relatively small for galaxies (a few billion stars each), they are far larger than globular clusters (which typically contain a few hundred thousand stars).

FIGURE 18.1 Artist's conception of the Milky Way viewed from its outskirts.

Our knowledge of the Milky Way's true size and shape was long in coming. Clouds of interstellar gas and dust known collectively as the **interstellar medium** fill the galactic disk, obscuring our view when we try to peer directly through it. The smoggy nature of the interstellar medium hides most of our galaxy from us and long fooled astronomers into believing that we lived near our galaxy's center. Astronomer Harlow Shapley finally proved otherwise in the 1920s, when he demonstrated that the Milky Way's globular clusters orbit a point tens of thousands of light-years from our Sun. He concluded that this point, not our Sun, must be the center of the galaxy. We now know that our Sun lies near the outskirts of the galactic disk, about 28,000 light-years (8.5 kpc) from its center.

Our own galaxy can be difficult to study not only because it is dusty, but also because we see it from the inside. The Milky Way's dustiness is no longer such a hindrance, because technologies developed in the past few decades allow us to observe the Milky Way's radio and infrared light. These wavelengths penetrate the enshrouding interstellar medium, enabling us to see into regions of the galaxy previously obscured from view. Still, determining our galaxy's structure from our location within the galactic disk is somewhat like trying to draw a picture of your house without ever leaving your bedroom. Just as it is easier to draw pictures of other houses that you can see out the window, it is easier to measure the sizes and shapes of other galaxies. Nevertheless, the Milky Way is still the only galaxy whose inner workings we can examine up close.

We once thought of the Milky Way as a band of light, and later we saw it as a collection of stars. But now we see it revealed as a dynamic and complex system. In the rest of this chapter, we will see that our galaxy in many ways is like a forest of stars whose ecology is shaped by the cycle of stellar life and death and whose structure is determined by gravity and the majestic rotation of the galactic disk.

18.2 The Star–Gas–Star Cycle

Stars have formed, fused atomic nuclei, and exploded throughout the history of our galaxy. Generations of stars continually recycle the same galactic matter through their cores, gradually raising the overall abundance of elements made by fusion. Elements

[1]Recall that 1 parsec (pc) ≈ 3.26 light-years, so 1 kiloparsec (kpc) ≈ 3,260 light-years and 1 megaparsec (Mpc) ≈ 3.26 million light-years.

THINKING ABOUT . . .

Discovering the Milky Way

The river of light in our sky that we call the Milky Way appears indistinct to our eyes. Galileo, looking through his telescope in 1610, was the first to realize that the Milky Way's light comes from innumerable faint stars. But the question of the size and shape of the Milky Way remained unanswered. In the late 1700s, British astronomers William and Caroline Herschel (brother and sister) tried to determine the shape of the Milky Way more accurately by counting how many stars lay in each direction. Their approach wrongly indicated that the Milky Way's width was five times its thickness. More than a century later, in the early 1900s, Dutch astronomer Jacobus Kapteyn and his colleagues used a more sophisticated star-counting method to gauge the size and shape of the Milky Way. Their results seemed to confirm the general picture found by the Herschels and suggested that the Sun lay very near the center of the galaxy.

Kapteyn's results made astronomers with a sense of history slightly nervous. Only four centuries earlier, before Copernicus challenged the Ptolemaic system, astronomers had erroneously believed that the Sun and planets orbited the Earth. Now Kapteyn was claiming that the Sun was nearly at the center of the Milky Way. Kapteyn was aware that obscuring material could deceive us by hiding the rest of the galaxy like some kind of interstellar fog. Kapteyn tried to find evidence for stellar obscuration, found none, and concluded that his picture was correct.

While Kapteyn was counting stars, American astronomer Harlow Shapley was studying globular clusters. He found that these clusters appeared to be centered around a point tens of thousands of light-years from the Sun. (His original estimate was that the center of the galaxy was 45,000 light-years from the Sun; today's accepted distance is 28,000 light-years.) Shapley concluded that this point marked the true center of our galaxy and that Kapteyn must be wrong.

Today we know that Shapley was right. The Milky Way's interstellar medium is the "fog" that misled Kapteyn. Robert Trumpler, working at California's Lick Observatory in the 1920s, established the existence of this dusty gas through his work on open clusters of stars. By assuming that all open clusters had about the same diameter, he estimated the distances to these clusters from their apparent sizes in the sky, as you might estimate the distances of cars at night from the separation of their headlights. However, according to these distance estimates, the stars in distant clusters were too dim, just as a car's headlights might be in foggy weather. They seemed to have lower luminosities than similar stars nearby. Trumpler concluded that light-absorbing material fills the spaces between the stars, partially obscuring the distant clusters and making them appear fainter than they would appear otherwise. Thus, we learned only in the past century that the stars visible in the night sky occupy a minuscule portion of the observable universe.

heavier than helium, usually called **heavy elements**[2] by astronomers, now constitute about 2% of the galaxy's gaseous content; the overall composition of the galaxy is about 70% hydrogen, 28% helium, and 2% heavy elements. The process of adding to the abundance of heavy elements is called **chemical enrichment,** and it is an inevitable by-product of continual star formation. However, the synthesis of elements is only one part of the galactic ecocycle. If the galaxy did not reincorporate these elements into new stars and their accompanying planetary systems, the mineral riches fused in the bellies of stars would be wasted.

TIME OUT TO THINK *Based on the idea of chemical enrichment, which types of stars must contain a higher proportion of heavy elements: stars in globular clusters or stars in open clusters? (Hint: Recall from Chapter 15 that stars in globular clusters are all very old, while stars in open clusters are relatively young.)*

Capturing the newly made elements released by stars is not an easy task. When a star explodes as a supernova, the ejected matter flies out at speeds of several thousand kilometers per second—far exceeding the escape velocity for the galaxy. Were it not for

[2]Astronomers sometimes refer to elements heavier than helium as *metals*—a very different use of the term than that used in daily life, where *metal* refers to substances like copper, silver, and gold that shine when polished and often are good electrical conductors.

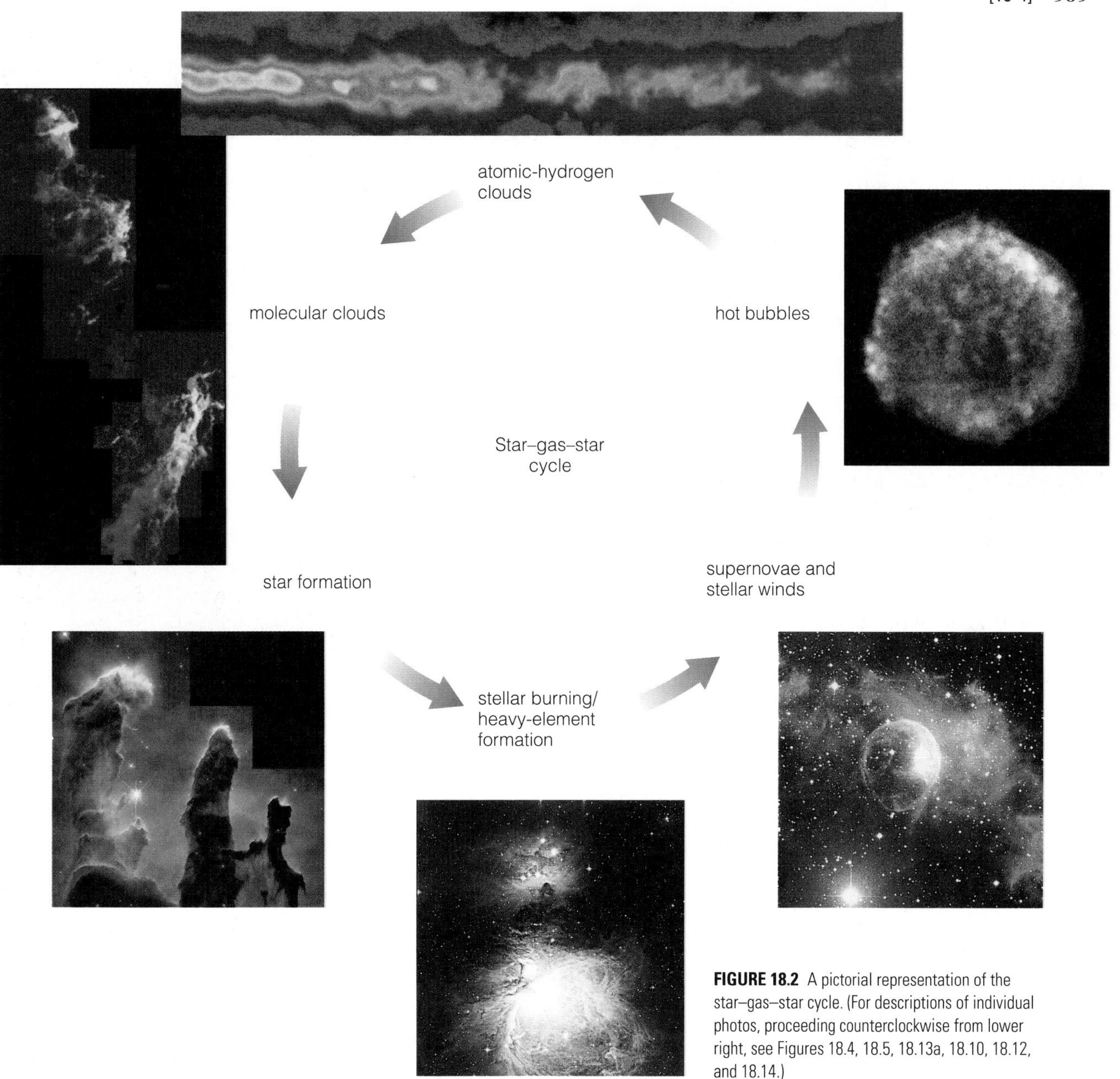

FIGURE 18.2 A pictorial representation of the star–gas–star cycle. (For descriptions of individual photos, proceeding counterclockwise from lower right, see Figures 18.4, 18.5, 18.13a, 18.10, 18.12, and 18.14.)

the interstellar medium, the new heavy elements released in the supernova would fly straight out of the Milky Way into intergalactic space. Instead, the blobs of matter expelled from the supernova collide with the interstellar medium, slow down, and eventually stop.

Supernovae of one star after another stir and heat the interstellar medium while feeding it new heavy elements. These elements eventually blend with the older, less chemically enriched hydrogen gas in the vicinity. However, before new stars can form, this gas must cool and form clouds. The hot gas cools first into clouds of atomic hydrogen (that is, neutral hydrogen atoms as opposed to ionized hydrogen or hydrogen molecules) and then into clouds of molecular hydrogen (H_2). The cooling of interstellar gas into clouds of molecular hydrogen takes millions of years. These clouds subsequently give birth to new stars more highly enriched in heavy elements, thus completing the star–gas–star cycle (Figure 18.2). Let's look at the stages of this cycle more closely.

Gas from Stars

All stars return much of their original mass to interstellar space in two basic ways: through stellar winds that blow throughout their lives, and through "death events" of planetary nebulae (for low-mass stars) or supernovae (for high-mass stars). Low-mass stars generally have weak stellar winds while they are on the main sequence, but their winds grow stronger

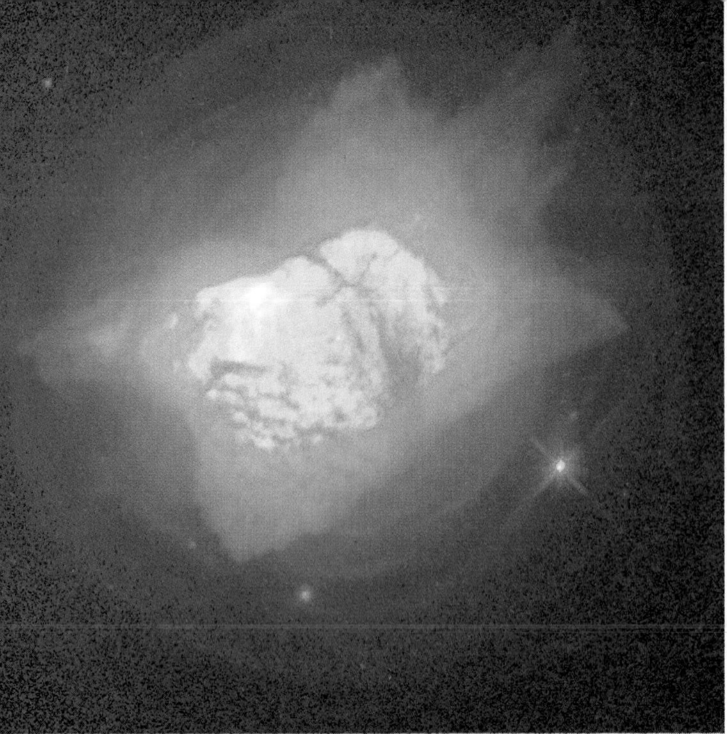

FIGURE 18.3 A dying low-mass star returns gas to the interstellar medium in a planetary nebula, as viewed by the Hubble Space Telescope.

FIGURE 18.4 This photo shows a bubble of hot, ionized gas blown by a wind from the hot star at its center.

and carry more material into space when they become red giants. By the time a low-mass star like the Sun ends its life with the ejection of a planetary nebula [Section 16.3], it has returned almost half its original mass to the interstellar medium (Figure 18.3).

High-mass stars lose mass much more dynamically and explosively. The powerful winds from supergiants and massive O and B stars recycle large amounts of matter into the galaxy. At the ends of their lives, these stars explode as supernovae. The high-speed gas ejected into space by these winds and supernovae sweeps up surrounding interstellar material, excavating a **bubble** of hot, ionized gas around the exploding star. Although the bubble in Figure 18.4 looks much like a soap bubble, it is actually an expanding shell of hot gas. The glowing surface is the edge of the expanding bubble, where gas piles up as the bubble sweeps outward through the interstellar medium. These hot, tenuous bubbles are quite common, filling roughly 20–50% of the Milky Way's disk. However, they are not always easy to detect. While some are hot enough to emit profuse amounts of X rays and others emit strongly in visible light, many others are evident only through radio emission from the shells of atomic hydrogen gas that surround them.

Shock Waves and Supernova Remnants

You might be wondering why gas tends to pile up on the expanding edge (surface) of a hot bubble. The answer requires thinking about the nature of sound waves and closely related shock waves. A **sound wave** is a wave of alternately rising and falling pressure. The sounds that we hear most commonly are caused by waves of pressure traveling through air. When these pressure waves reach your eardrum, they make it vibrate back and forth. Your brain interprets the vibrations of your eardrum as sound. Sound waves can travel through almost any material, and the speed of sound depends on the properties of the material through which it travels. For example, sound travels faster through water than through air, and faster still through solids. Sound waves can even travel through the near-vacuum of space, although the low density of space means that these sound waves would not be audible to the human ear.

The speed of sound in interstellar space is much higher than the speed of sound in air, but stellar winds and supernovae can force gas to travel through space at even higher speeds. The result is a **shock wave**—a wave of pressure generated by gas moving faster than the speed of sound. Surrounding material essentially receives no warning before a shock wave slams into it. A shock wave therefore sweeps up surrounding gas as it travels, creating a thickening "wall" of fast-moving gas on its leading edge. If you have

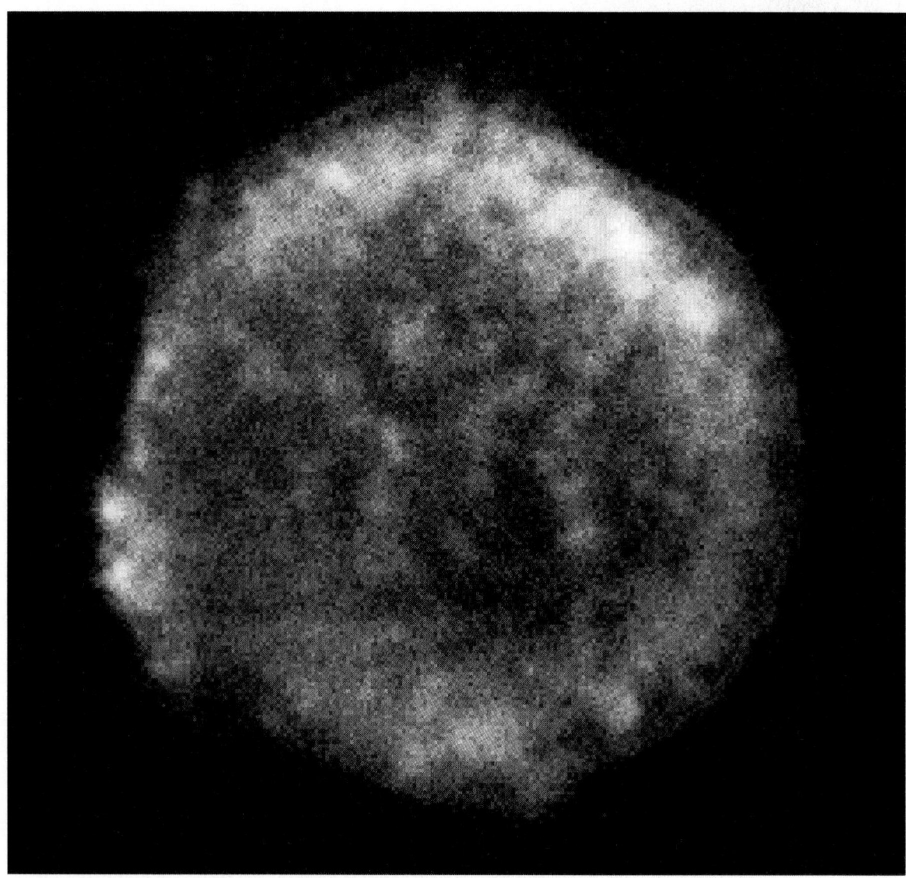

FIGURE 18.5 This false-color image shows X-ray emission from hot electrons in a young supernova remnant—the remnant from the supernova observed by Tycho Brahe in 1572.

ever heard a sonic boom from a supersonic jet airplane, you have directly experienced a shock wave and the abruptness with which it arrives. In interstellar space, a shock wave compresses, heats, ionizes, and accelerates all the gas in its path.

TIME OUT TO THINK *The familiar sound produced by the crack of a whip is also a shock wave; it occurs when the end of the whip exceeds the speed of sound through air. Using what you now know about how a shock wave produces a "wall" of fast-moving gas, explain why the sound from the crack of a whip is so sharp and loud.*

Movie Madness: The Sound of Space

In many science fiction movies, a thunderous sound accompanies the demolition of a spaceship. If the movie makers wanted to be more realistic, they would silence the explosion. You perceive sound when gas atoms moving with sound waves collide with your eardrums and cause them to vibrate, and it takes many trillions of atoms to move your eardrums noticeably. Because the density of gas in interstellar space is so low, the chances are extremely small that even one atom per second from an interstellar sound wave, let alone trillions, would enter your ears and collide with your eardrums. To the human ear, the sound of space is silence.

Supernovae generate shock waves that travel faster than shock waves generated by stellar winds. Thus, when a massive star explodes in a supernova, the resulting shock wave rapidly overtakes any shock waves that the star has previously generated by stellar winds. When we observe a *supernova remnant,* we are seeing the aftermath of its shock wave [Section 16.4]. Figure 18.5 shows a young supernova remnant whose shocked gas is hot enough to emit X rays. In contrast, older supernova remnants are cooler because their shock waves have swept up more material and must share their energy among more particles (Figure 18.6). Eventually, the shocked gas radiates away most of its original energy, and the expanding wall of gas slows to subsonic speeds. As the energy dissipates and the gas cools, the supernova's cargo of new elements merges with the surrounding interstellar medium.

In addition to their role as the movers and shakers of the interstellar medium, shock waves from supernovae can act as subatomic particle accelerators. Some of the electrons in supernova remnants accelerate to nearly the

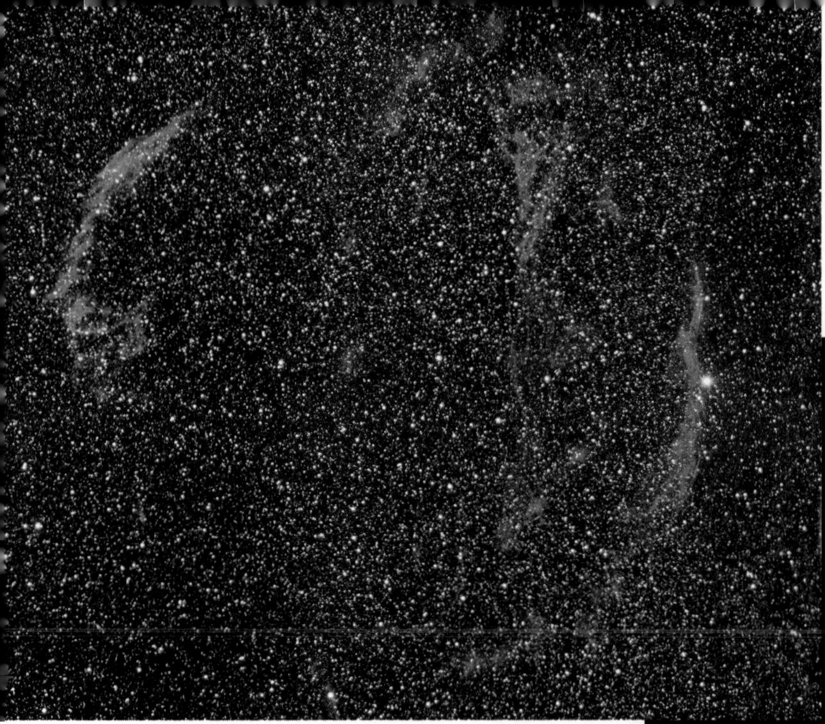

a

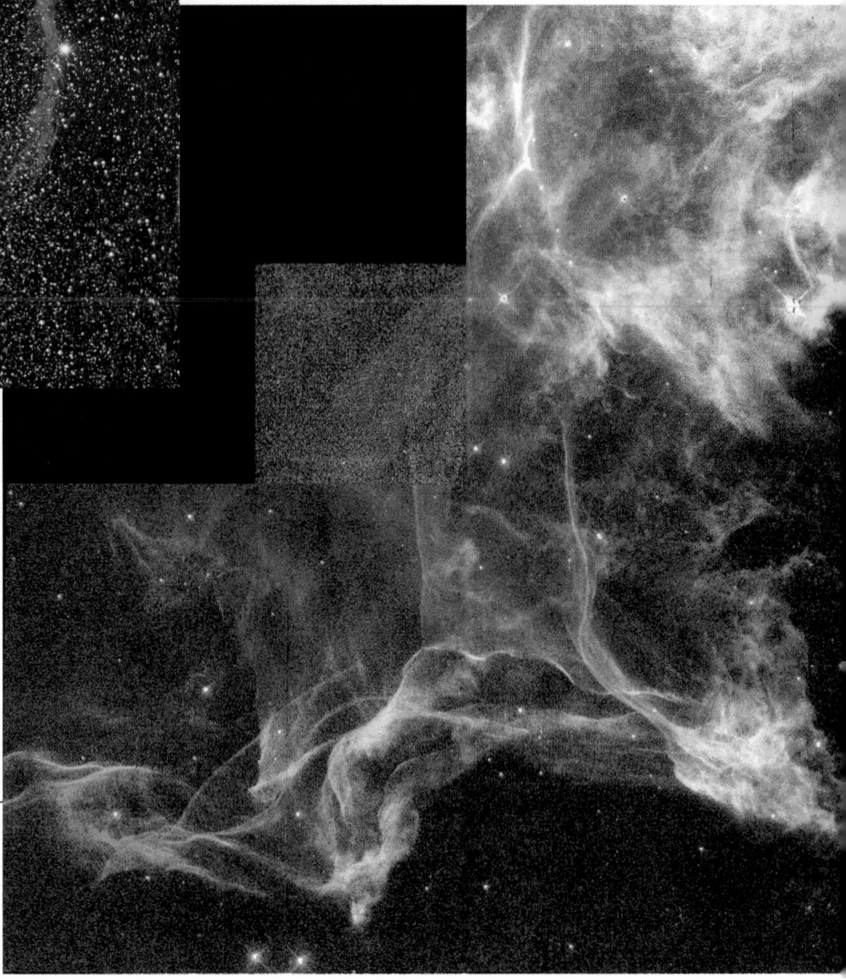

FIGURE 18.6 The optical emission from a supernova remnant, the Cygnus Loop. (**a**) This large-scale view shows the entire remnant glowing in optical light. (The angular size of this remnant in our sky is six times that of the Moon.) (**b**) The close-up view, from the Hubble Space Telescope, displays the fine filamentary structure of the remnant. (**c**) An optical spectrum from the Cygnus Loop shows the strong emission lines that account for the distinct colors in the Hubble Space Telescope photo. Blue represents emission from singly ionized oxygen, green represents emission from atomic hydrogen, and red represents emission from singly ionized sulfur.

b

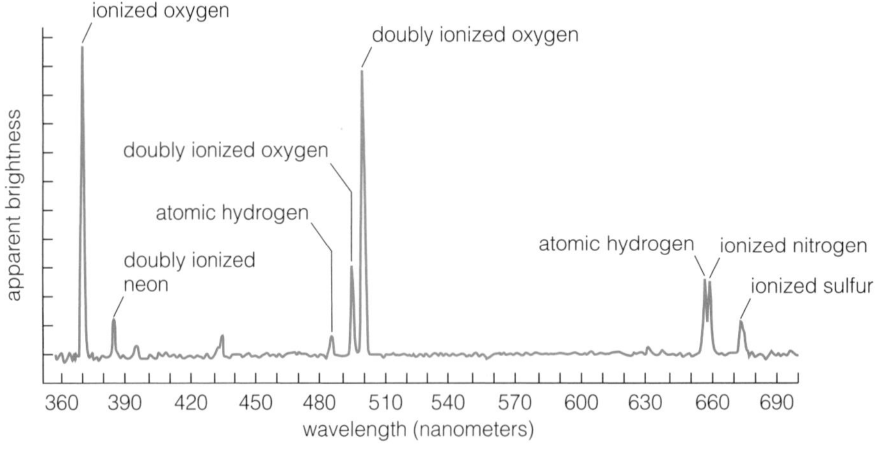

c

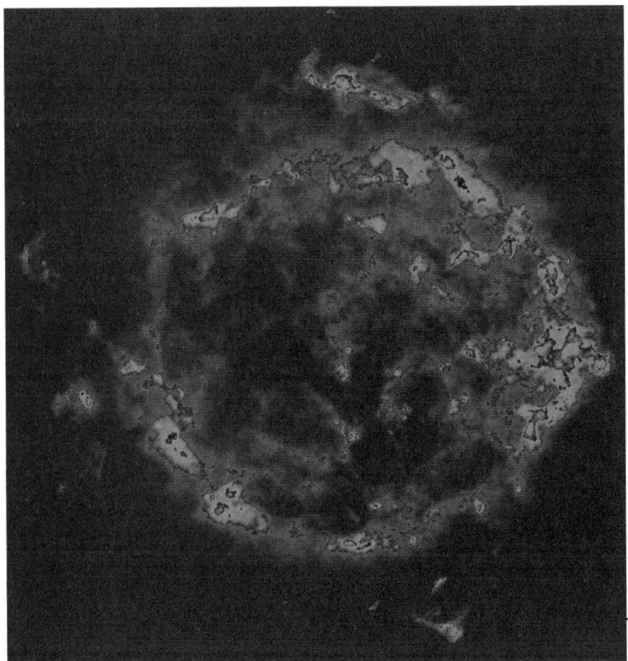

a

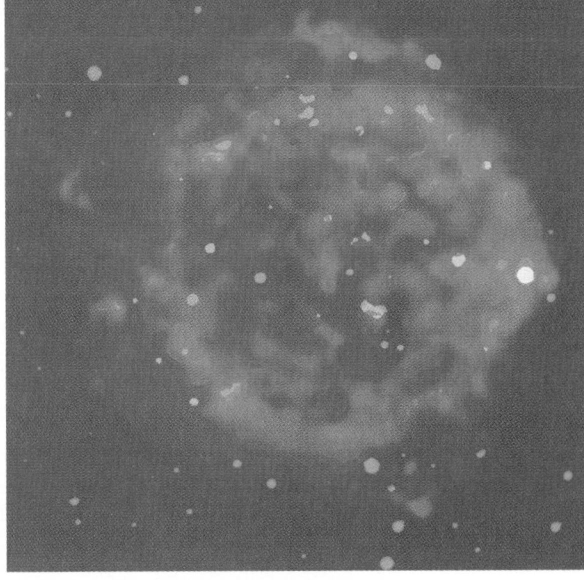

b

FIGURE 18.7 (**a**) Radio emission caused by electrons spiraling around magnetic field lines in the young supernova remnant Cassiopeia A. (**b**) A composite image made from radio (red), X-ray (blue), and optical (green) images of this remnant. Note the similarity of the structures in each band of the spectrum. (The green dots are unrelated stars.)

speed of light as they interact with the shock wave. These fast electrons emit radio waves as they spiral around magnetic field lines threading the supernova remnant[3] (Figure 18.7).

Supernova remnants probably also generate the **cosmic rays** that permeate the interstellar medium and bombard the Earth's atmosphere. Cosmic rays are made of electrons, protons, and atomic nuclei that zip through interstellar space at close to the speed of light. Some cosmic rays penetrate the Earth's atmosphere and reach the Earth's surface, and an average of at least one cosmic-ray particle strikes your body each second. At the altitudes at which jet planes fly, high above most of the Earth's protective atmosphere, the cosmic-ray bombardment rate is 100 times higher. Even more cosmic rays funnel along magnetic field lines to the Earth's magnetic poles, so for safety reasons airlines restrict how often their flight crews cross the Earth's polar regions.

Superbubbles and Fountains

The bubble associated with a single supernova remnant can grow to a size of about a hundred light-years (30 pc), but in some areas of the Milky Way we see cavities of hot gas over a thousand light-years (300 pc) wide. These huge cavities arise because stars tend to form in clusters. The hottest, most massive stars in a cluster can end their lives and explode within a few hundred thousand years of one another. The shock waves from these individual supernovae soon overlap, combining their energy into one very powerful shock wave. This extra-large shock wave forms an enormous **superbubble** in the interstellar medium (Figure 18.8). Subsequent supernovae from the cluster explode inside the superbubble, adding even more energy.

In many places in our galactic disk, we see what appear to be elongated bubbles extending from young clusters of stars to distances of 3,000 light-years (1 kpc) or more above the disk. These probably are places where superbubbles have grown so large that they cannot be contained within the disk of the Milky Way. Once the superbubble breaks out of the disk, where nearly all of the Milky Way's gas resides, nothing remains to slow its expansion except gravity. Such a *blowout* is in some ways similar to a volcanic eruption, but on a galactic scale: Hot plasma erupts from the disk and shoots high into the galactic halo.

In fact, there is growing evidence that processes like blowouts continually cycle gas between the Milky Way's disk and the halo, an idea summarized in a model known as the **galactic fountain.** According to this model, fountains of hot, ionized gas rise from the disk into the halo through the elongated bubbles carved by blowouts. The gravity of the galactic disk slows the rise of the gas, eventually pulling it back down. Near the top of its trajectory, the ejected gas starts to cool and form clouds of atomic hydrogen. These clouds cool further as they plunge back

[3]The radio emission from electrons moving close to the speed of light around magnetic field lines is sometimes called *synchrotron radiation*.

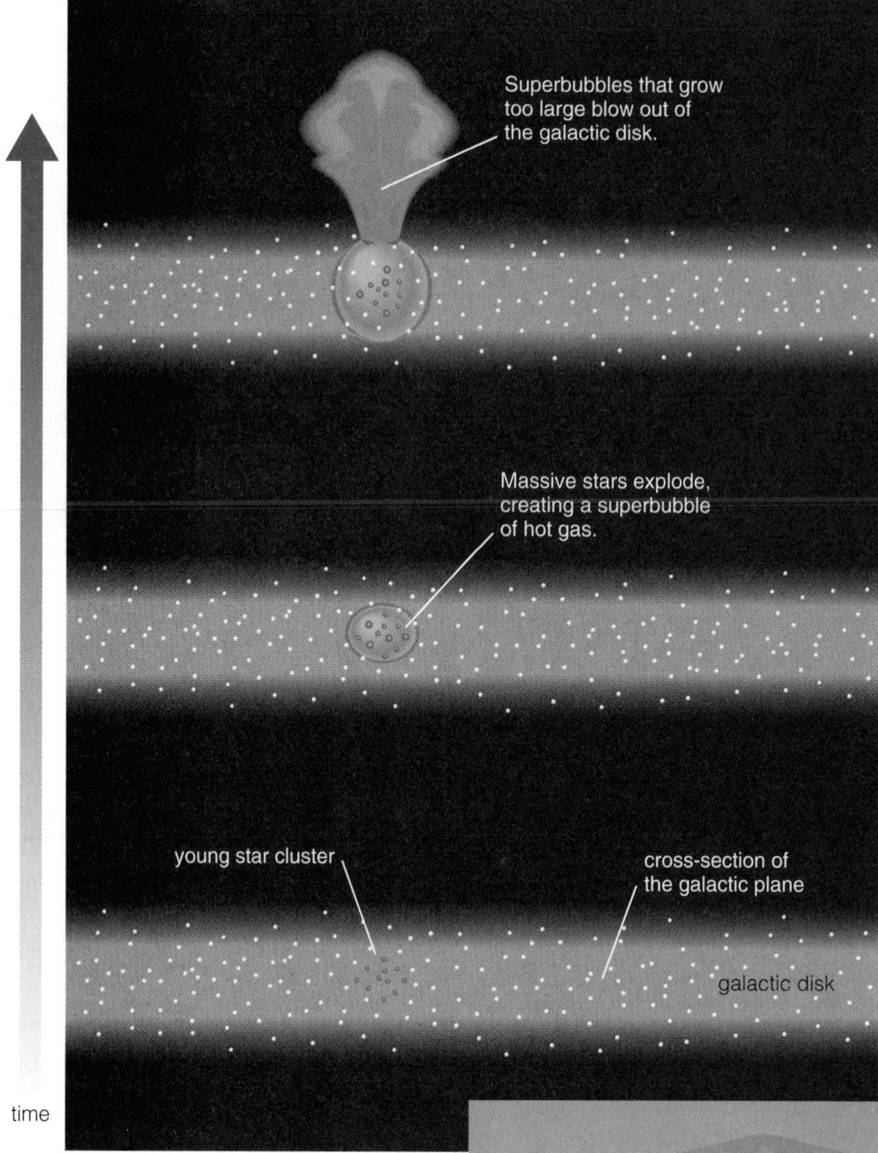

a

FIGURE 18.8 (**a**) Schematic sequence (bottom to top) showing the development of a superbubble. (**b**) Image from computer simulation of gas temperatures in a superbubble blowout (blue = coldest gas; red = hottest gas).

down, ultimately rejoining the layer of atomic hydrogen gas in the disk (Figure 18.9).

The galactic fountain model is plausible but difficult to verify. We do indeed see some hot gas high above the galaxy's disk. We also see cooler clouds that appear to be raining down from the halo. However, we can only see this "rain" directly above and below us, making it difficult to demonstrate beyond a doubt that galactic fountains circulate the products of supernovae throughout the Milky Way.

b

Atomic Hydrogen Gas

The hot, ionized gas in bubbles, superbubbles, and fountains is dynamic and widespread, but it is a relatively small fraction of the gas in the Milky Way. Atomic hydrogen gas is much more common, and we can map its distribution in the Milky Way with radio observations. Atomic hydrogen emits a spectral line with a wavelength of 21 cm in the radio portion of the electromagnetic spectrum. We see the radio emission from this **21-cm line** coming from all directions, telling us that atomic hydrogen gas is distributed throughout the galactic disk. Based on the overall strength of the 21-cm emission, the amount of atomic hydrogen gas in our galaxy must be about 5 billion solar masses, which is a few percent of the galaxy's total mass.

Atomic hydrogen gas tends to be found in two distinct forms: large, tenuous clouds of warm (10,000 K) atomic hydrogen and smaller, denser clouds of cool (100 K) atomic hydrogen. If you were to take an interstellar voyage across the Milky Way, you would spend the majority of your time cruising through regions of warm atomic hydrogen interspersed with bubbles of hot, ionized gas. Every few hundred parsecs, you would encounter a cooler, denser cloud of atomic hydrogen. In the warm regions, you would detect about one atom per cubic centimeter and a weak magnetic field (weaker than Earth's magnetic field by a factor of about 100,000). In the cooler clouds, you would find a density of about 100 atoms per cubic centimeter and a stronger magnetic field.

Matter remains in the warm atomic hydrogen stage of the star–gas–star cycle for millions of years. Gravity slowly draws blobs of this gas together into tighter clumps, which radiate energy more efficiently as they grow denser. The blobs therefore cool and contract, forming the smaller clouds of cool atomic hydrogen. This process of shrinkage and coagulation takes a much longer time than the other steps in the journey from star death to star birth. The slowness of the transition from warm gas to cool clouds accounts for the large amount of gaseous matter in the atomic hydrogen stage of the star–gas–star cycle.

Remember that, although we speak of clouds of *hydrogen,* all interstellar material actually has a composition of about 70% hydrogen, 28% helium, and 2% heavy elements. Some of the heavy elements in regions of atomic hydrogen are in the form of tiny, solid **dust grains:** flecks of carbon and silicon minerals that resemble particles of smoke and form in the winds of red-giant stars [Section 16.3]. Once formed, dust grains remain in the interstellar medium unless they are heated and destroyed by a passing shock wave or incorporated into a protostar. Although dust grains make up only about 1% of the mass of the atomic hydrogen clouds, they are responsible for the absorption of visible light that prevents us from seeing through the disk of the galaxy.

Molecular Clouds

As the temperature drops further in the center of a cool cloud of atomic hydrogen, hydrogen atoms combine into molecules, making a **molecular cloud.** Molecular clouds are the coldest, densest collections of gas in the interstellar medium. The total mass of molecular clouds in the Milky Way is somewhat uncertain, but it is probably about the same as the total mass of atomic hydrogen gas—about 5 billion solar masses. Throughout much of this molecular gas, temperatures hover only a few degrees above absolute zero, and gas densities are a few hundred molecules per cubic centimeter.

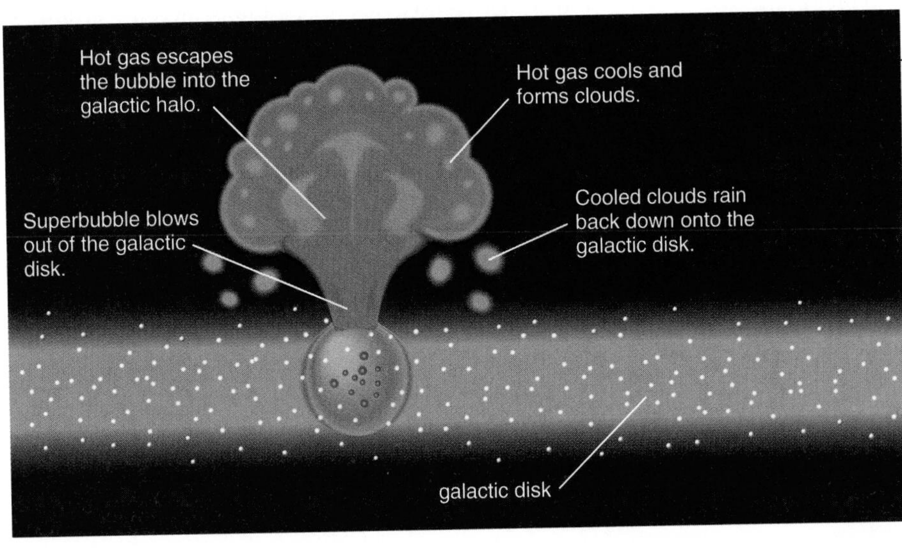

FIGURE 18.9 Schematic picture of a galactic fountain. In the galactic fountain model for our interstellar medium, multiple supernovae create hot superbubbles that blow hot gas out into the galactic halo. The hot gas in the halo then cools into clouds that rain back down on the disk.

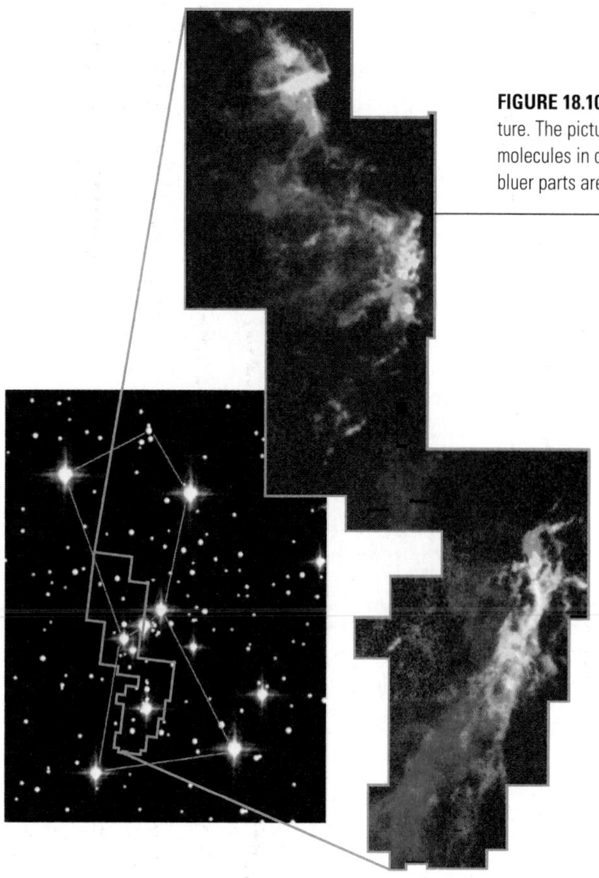

FIGURE 18.10 Image of a molecular cloud in the constellation Orion, showing its complex structure. The picture was made by measuring Doppler shifts of emission lines from carbon monoxide molecules in different locations. The colors indicate gas motions: relative to the cloud as a whole, bluer parts are moving toward us and redder parts are moving away from us.

Molecular hydrogen (H_2) is by far the most abundant molecule in molecular clouds, but it is difficult to detect because temperatures are usually too cold to excite the emission lines of H_2. As a result, most of what we know about molecular clouds comes from observing spectral lines of molecules that make up only a tiny fraction of a cloud's mass. The most abundant of these molecules is carbon monoxide (CO, also a common ingredient in car exhaust), which produces strong emission lines in the radio portion of the spectrum at the 10–30 K temperatures of molecular clouds. Many other molecules also produce radio emission lines, and astronomers have used these lines to identify more than 80 different kinds of molecules in molecular clouds. Among the more familiar are water (H_2O), ammonia (NH_3), and alcohol (CH_3OH).

Molecular clouds are heavy and dense compared to the rest of the interstellar gas and therefore tend to settle toward the central layers of the Milky Way's disk. This tendency creates a phenomenon you can see with your own eyes: the dark fissures running through the luminous band of light in our sky that we call the Milky Way [Section 1.5].

Molecular clouds are far from uniform in density and temperature. Molecular hydrogen gas tends to congregate into *giant molecular clouds* that hold up to a million solar masses of gas. However, these giant clouds contain many clumps of denser material, and within each dense clump are several smaller clumps of even denser gas (Figure 18.10). Gravity acts more powerfully within these dense clumps, gathering molecules into the compact *cores* that eventually become protostars.

The final stages of star formation can be quite disruptive to a molecular cloud. Recall that protostars often have violent jets of gas spurting outward [Section 16.2]. The turbulence of these jets stirs up nearby regions of the molecular cloud, probably preventing other stars from forming in the vicinity of the growing protostar, at least for a while (Figure 18.11).

Completion of the Cycle

Once a few stars form in a cluster, their radiation begins to erode the surrounding gas in the molecular cloud. Ultraviolet photons from high-mass stars heat and ionize the gas, and winds and radiation pressure push the ionized gas away. This kind of feedback prevents much of the gas in a molecular cloud from turning into stars.

The process of molecular cloud erosion is sometimes spectacular. Figure 18.12 shows the *Eagle Nebula,* a complex of clouds where new stars are currently forming. The dark, lumpy columns are molecular clouds. Off to the side (outside the picture), newly formed massive stars glow with ultraviolet radiation. This radiation sears the surface of the molecular clouds, destroying molecules and stripping electrons from atoms. As a result, matter "evaporates" from the molecular clouds and joins the hotter ionized gas encircling them. Only the densest knots of gas resist evaporation. Stars are forming in some of these dense knots, which remain compact while the rest of the cloud erodes. These star-forming knots are the tips of the dark "whiskers" protruding from the columns of molecular gas in Figure 18.12.

We have arrived back where we started in the star–gas–star cycle. The most massive stars now forming in the Eagle Nebula will explode within a few million years, filling the region with hot gas and newly formed heavy elements. Farther in the future, this gas will once again cool and coalesce into molecular clouds, forming new stars, new planets, and maybe even new civilizations.

FIGURE 18.11 Jets from protostars disrupt nearby regions of molecular clouds. This photo shows a jet emanating from the protostar on the left. All along the jet in the center and on the far right of the photo, gas shocked by the jet is glowing in optical light.

FIGURE 18.12 A portion of the Eagle Nebula as seen by the Hubble Space Telescope.

Table 18.1 Typical States of Gas in the Interstellar Medium

State of Gas	Primary Constituent	Approximate Temperature	Approximate Density (atoms per cm^3)	Description
Hot bubbles	Ionized hydrogen	1,000,000 K	0.01	Pockets of gas heated by supernova shock waves
Warm atomic gas	Atomic hydrogen	10,000 K	1	Fills much of galactic disk
Cool atomic clouds	Atomic hydrogen	100 K	100	Intermediate stage of star–gas–star cycle
Molecular clouds	Molecular hydrogen	30 K	300	Regions of star formation
Molecular cloud cores	Molecular hydrogen	60 K	10,000	Star-forming clouds

Despite the recycling of matter from one generation of stars to the next, the star–gas–star cycle cannot go on forever. With each new generation, some of the galaxy's gas becomes permanently locked away in brown dwarfs that never return material to space and in stellar corpses left behind when stars die (white dwarfs, neutron stars, and black holes). The interstellar medium therefore is slowly running out of gas, and the rate of star formation will gradually taper off over the next 50 billion years or so. Eventually, star formation will cease.

Putting It All Together: The Distribution of Gas in the Milky Way

As we look at different regions of the galaxy, we see various stages of the star–gas–star cycle playing themselves out. Because the cycle proceeds so slowly compared to a human lifetime, each stage appears to us as a snapshot. We therefore see the interstellar medium in a wide variety of manifestations, ranging from the tenuous million-degree gas of bubbles to the cold, dense gas of molecular clouds. Table 18.1 summarizes the different states in which we see interstellar gas in the galactic disk.

Figure 18.13 shows seven views of the disk of the Milky Way Galaxy. Each view represents a panorama made by photographing the Milky Way's disk in every direction from Earth. You can visualize how one of these views corresponds to the sky by imagining cutting it out, bringing its ends together to form a circular band, and then lining up the band with the Milky Way on a model of the celestial sphere. Note that each view shows the Milky Way as it appears in a different set of wavelengths of light and thus reveals different features of the galactic disk.

- Figure 18.13a shows variations in the intensity of radio emission from the 21-cm line of atomic hydrogen. Thus, it maps the distribution of atomic hydrogen gas, demonstrating that this gas fills much of the galactic disk.
- Figure 18.13b shows variations in the intensity of radio emission lines from carbon monoxide (CO) and therefore maps the distribution of molecular clouds. Note that these cold, dense clouds are concentrated in a narrow layer near the midplane of the galactic disk.
- Figure 18.13c shows variations in the intensity of infrared emission (at wavelengths of 60–100 micrometers) from interstellar dust grains. Note that the regions of strongest emission from dust correspond to the locations of molecular clouds in Figure 18.13b.
- Figure 18.13d shows infrared light from stars at wavelengths that penetrate clouds of gas and dust (1–4 micrometers). Thus, this image shows how our galaxy would look if there were no dust blocking our view. The galactic bulge is clearly evident at the center.
- Figure 18.13e shows the galactic disk in visible light, just as it appears in the night sky. (Of course, only part of the Milky Way is above the horizon at any one time.) Because visible light cannot penetrate interstellar dust, the dark blotches correspond closely to the bright patches of molecular radio emission and infrared dust emission in Figure 18.13b and c.
- Figure 18.13f shows the distribution of X-ray light from the galactic disk. The pointlike blotches in this view are mostly X-ray binaries [Section 17.3],

a 21-cm radio emission from atomic hydrogen gas.

b Radio emission from carbon monoxide reveals molecular clouds.

c Infrared (60–100 μm) emission from interstellar dust.

d Infrared (1–4 μm) emission from stars that penetrates most interstellar material.

e Visible light emitted by stars is scattered and absorbed by dust.

f X-ray emission from hot gas bubbles (diffuse blobs) and X-ray binaries (pointlike sources).

g Gamma-ray emission from collisions of cosmic rays with atomic nuclei in interstellar clouds.

FIGURE 18.13 Panoramic views of the Milky Way in different bands of the spectrum. The center of the galaxy, which lies in the direction of the constellation Sagittarius, is in the center of each strip. The rest of each strip shows all other directions in the Milky Way disk as seen from Earth. (Imagine attaching the left and right ends of each strip to form a circular band that corresponds to the 360° band of the Milky Way in our sky.)

but the rest of the X-ray emission comes primarily from hot gas bubbles. Note that the hot gas is less concentrated toward the midplane than the atomic and molecular gas, because it tends to rise into the halo. (The prominent yellow blob on the lower right is the Vela supernova remnant.)

- Figure 18.13g shows gamma-ray emission from the Milky Way. Most of the gamma-ray emission is produced by collisions between cosmic-ray particles and atomic nuclei in interstellar clouds. Such collisions happen most frequently where gas densities are highest, so the gamma-ray emission corresponds closely to the locations of molecular and atomic gas. (Gamma rays from the pulsar at the center of the Vela supernova remnant are also prominent on the lower right.)

TIME OUT TO THINK *Carefully compare and contrast the different views of the Milky Way's disk in Figure 18.13. Why do regions that appear dark in some views appear bright in others? What kinds of general patterns do you notice?*

18.3 Galactic Environments

The star–gas–star cycle has operated continuously since the Milky Way's birth, yet new stars are not spread evenly across the galaxy. Some regions seem much more fertile than others. Galactic environments rich in molecular clouds tend to spawn new stars easily, while gas-poor environments do not. A quick tour of some characteristic galactic environments will help you spot where the action is.

Out in the Halo

A census of stars in the Milky Way's *halo* would turn up many senior citizens and very few newborns. Most of the halo stars are old, red, and dim and much smaller in mass than our Sun. Halo stars also contain far fewer heavy elements than our Sun, sometimes having heavy-element proportions as low as 0.02% (in contrast to about 2% in the Sun). The relative lack of chemical enrichment in the halo indicates that its stars formed early in the galaxy's history—before many supernovae had exploded, adding heavy elements to star-forming clouds.

FIGURE 18.14 A photo of the Orion Nebula, an ionization nebula energized by ultraviolet photons from hot stars.

The halo's gas content corroborates this view. The halo is virtually gas-free compared to the disk, with very few detectable molecular clouds. Apparently, the bulk of the Milky Way's gas settled into the disk long ago. The halo environment is a place where lack of gas caused star formation to cease long ago. Now only very old stars still survive, and new stars are rarely born.

TIME OUT TO THINK *How does the halo of our galaxy resemble the distant future fate of the galactic disk? Why?*

Our Neighborhood

Our own stellar neighborhood is more active than the halo and typifies much of the galactic disk. Within about 33 light-years (10 pc), we know of over 300 stars. Most are dim, red, spectral type M stars. A few, including Sirius, Vega, Altair, and Fomalhaut, are bright, white stars younger than our Sun. While stars of many different ages and different proportions of heavy elements are in our neighborhood, no very massive, short-lived stars (i.e., spectral type O or B) are present. We therefore infer that stars form periodically in our neighborhood but that no star clusters have formed here recently.

Our quiet suburb of the galaxy has not always been as calm as it is today. X-ray telescopes in space reveal hot, X ray–emitting gas coming from nearby in every direction. Surrounding this hot gas, at distances ranging up to a few hundred light-years (100 pc), lies a region of much cooler gas. Apparently, we and all our stellar neighbors live inside a hot bubble. The existence of this *Local Bubble* means that a number

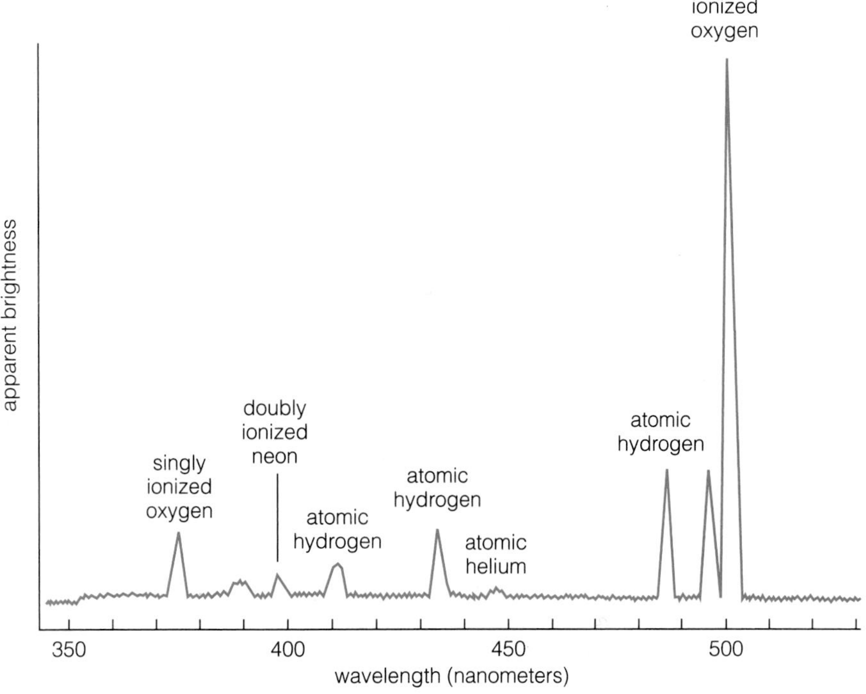

FIGURE 18.15 Optical spectrum of an ionization nebula. The prominent emission lines in the spectrum reveal the atoms and ions that emit most of the light. Through careful modeling of these lines, we can determine the nebula's chemical composition.

of supernovae must have detonated within our stellar neighborhood over the past several million years.

Hot-Star Hangouts

The hot spots in our galaxy are the neighborhoods of high-mass stars. Because hot, massive stars live fast and die young, they never get a chance to move very far from their birthmates. Thus, we find them in star clusters close to the molecular clouds from which they formed. These environments are highly active and extraordinarily picturesque.

We find colorful, wispy blobs of glowing gas known as **ionization nebulae**[4] throughout the galactic disk, particularly in the spiral arms. The energy that powers ionization nebulae comes from neighboring hot stars that irradiate them with ultraviolet photons. These photons ionize and excite the atoms in the nebulae, causing them to emit light. The Orion Nebula, about 1,400 light-years away in the "sword" of the constellation Orion, is among the most famous. Few astronomical objects can match its spectacular beauty (Figure 18.14).

Most of the striking colors in an ionization nebula come from particular spectral lines produced by particular atomic transitions. For example, the transition in which an electron falls from energy level 3 to energy level 2 in a hydrogen atom generates a red photon with a wavelength of 656 nanometers [Section 7.4]. Ionization nebulae appear predominantly red in photographs because of all the red photons released by this particular transition. (Jumps from level 2 to level 1 are even more common, but they produce ultraviolet photons that can be studied only with ultraviolet telescopes in space.) Transitions in other elements produce other spectral lines of different colors (Figure 18.15).

[4]Ionization nebulae are sometimes called *emission nebulae* or *H II regions;* H II is an abbreviation astronomers use for ionized hydrogen (H I means neutral hydrogen).

Common Misconceptions: What Is a Nebula?

The term *nebula* means *cloud,* but in astronomy it can refer to many different kinds of objects—a state of affairs that sometimes leads to misconceptions. Many astronomical objects look "cloudy" through small telescopes, and in past centuries astronomers called any such object a nebula as long as they were sure it wasn't a comet. For example, galaxies were called nebulae because they looked like either fuzzy round blobs or fuzzy spiral blobs. Using the term *nebula* to refer to a galaxy now sounds somewhat dated, given the enormous differences between these distant star systems and the much smaller clouds of gas that populate the interstellar medium. Nevertheless, some people still refer to spiral galaxies as "spiral nebulae." Today, we generally use the term *nebula* to refer to true interstellar clouds, but be aware that the term is still sometimes used in other ways.

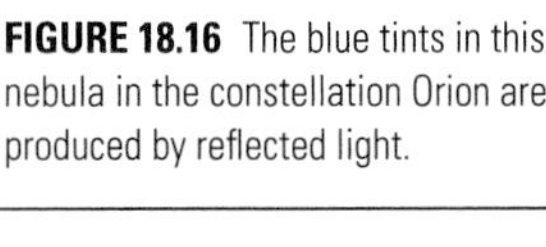

FIGURE 18.16 The blue tints in this nebula in the constellation Orion are produced by reflected light.

FIGURE 18.17 A photo of the Horsehead Nebula and its surroundings.

The blue and black tints in some nebulae have a different origin. Starlight reflected from dust grains produces the blue colors, because interstellar dust grains scatter blue light much more readily than red light (Figure 18.16).[5] (This effect is similar to the scattering of sunlight in our atmosphere that makes the sky blue [Section 10.2].) The black regions of nebulae are dark, dusty gas clouds that block our view of the stars beyond them. Figure 18.17 shows a multicolored nebula characteristic of a hot-star neighborhood.

TIME OUT TO THINK ***In Figure 18.17, identify the red ionized regions, the blue reflecting regions, and the dark obscuring regions. Briefly explain the origin of the colors in each region.***

[5]Nebulae that shine by reflected light alone are called *reflection nebulae.* They are always bluer in color than the stars supplying the light.

18.4 The Milky Way in Motion

The internal motions of the Milky Way range from the chaos of stars soaring on randomly oriented orbits through the halo to the stately rotation of the galactic disk. These motions, combined with the law of gravity, can tell us how mass is distributed within the Milky Way. When we decipher these movements, we find that the matter binding stars and gas to our galaxy extends far beyond its visible reaches and that the galaxy's spiral arms are propagating waves of new star formation. Let's investigate the motion of the Milky Way in more detail.

Orbits in the Disk and Halo

If you could stand outside the Milky Way and watch it for a few billion years, the disk would resemble a huge merry-go-round. All the stars in the disk, including our Sun, orbit the center in the same direction. Moreover, like horses on a merry-go-round, individual stars bob up and down through the disk as they orbit. The general orbit of a star around the galaxy arises from its gravitational attraction toward the galactic center, while the bobbing arises from the localized pull of gravity within the disk itself (Figure 18.18): A star that is "too far" above the disk is pulled back into the disk by gravity; because the density of interstellar gas is too low to slow the star, it flies through the disk until it is "too far" below the disk on the other side. Gravity then pulls it back in the other direction. This ongoing process produces the bobbing of the stars. These up-and-down motions spread the disk stars over a thickness of about 1,000 light-years—which is quite thin in comparison to the 100,000-light-year diameter of the disk. Each orbit takes over 200 million years in the vicinity of our Sun and each up-and-down "bob" takes a few tens of millions of years.

The orbits of stars in the halo and bulge are much less organized than the orbits of stars in the disk (see Figure 18.18). Individual bulge and halo stars travel around the galactic center on more or less elliptical paths, but the orientations of these paths are relatively random. Neighboring halo stars can circle the galactic center in opposite directions. They swoop from high above the disk to far below it and back again, plunging through the disk at velocities so high that the disk's gravity hardly alters their trajectories. Near the Sun, we see several fast-moving halo stars in the midst of hasty excursions through the local region of the disk. One such star is Arcturus, the fourth-brightest star in the night sky.

FIGURE 18.18 Characteristic orbits of disk stars and halo stars around the galaxy. The up-and-down motion of the disk star orbit is exaggerated compared to the radius of the orbit.

TIME OUT TO THINK *Is there much danger that the Sun or Earth will someday be hit by a halo star swooping through the disk of the galaxy? Why or why not?* (Hint: *Think about the typical distances between stars, as illustrated by use of the 1-to-10-billion scale in Chapter 2.*)

The Sun's orbital path around the galaxy is called the **solar circle,** and its radius is our 28,000-light-year distance from the galactic center. By measuring the speeds of globular clusters relative to the Sun, we've determined that the Sun and its neighbors orbit the center of the Milky Way at a speed of about 220 km/s. Even at this speed, one circuit of the Sun around the solar circle takes 230 million years. Early dinosaurs ruled the Earth when it last visited this side of the galaxy.

Orbits and Galactic Mass

Recall that Newton's law of gravity determines how quickly objects orbit one another. This fact, embodied in Newton's version of Kepler's third law, allows us to determine the mass of a relatively large object when we know the period and average distance of a much smaller object in orbit around it [Section 6.4]. A closely related law, which we will call the *orbital velocity law,* allows us to "weigh" the galaxy using the Sun's orbital velocity and its distance from the galactic center. If we call the Sun's orbital velocity v and its distance (radius) from the galactic center r, the orbital velocity law tells us that the mass of the galaxy within the Sun's orbit (M_r) is:

$$M_r = \frac{r \times v^2}{G}$$

Note that the orbital velocity law tells us only how much mass lies *within* the Sun's orbit. It does not tell us the mass of the entire galaxy, because matter lying outside the solar circle has very little effect on the Sun's orbital velocity. Every part of the galaxy exerts gravitational forces on the Sun as it orbits, but the net force from matter outside the solar circle is relatively small because the pulls from opposite sides of the galaxy virtually cancel one another. In contrast, the net gravitational forces from mass

Mathematical Insight 18.1 The Orbital Velocity Law

As we discussed in the text, the orbital velocity law allows us to use the orbital characteristics of a star or gas cloud to calculate the mass of the galaxy *within* the object's orbital circle. Because this law comes from the same law of gravity that Newton used to derive his version of Kepler's third law, we can derive it by starting from the form of Kepler's third law that we found in Mathematical Insight 6.2:

$$p^2 = \frac{4\pi^2}{G \times M} \times a^3 \quad \left(G = 6.67 \times 10^{-11} \frac{\text{m}^3}{\text{kg} \times \text{s}^2}\right)$$

where M is the mass of the massive object and p and a are the orbital period and semimajor axis of a smaller orbiting object, respectively. Solving for M, we find:

$$M = \frac{4\pi^2 \times a^3}{G \times p^2}$$

In Mathematical Insight 15.4, we showed that, if the object has a circular orbit, its orbital speed is:

$$v = \frac{\text{distance traveled in one orbit}}{\text{period of one orbit}} = \frac{2\pi a}{p}$$

Solving for p, we find:

$$p = \frac{2\pi a}{v}$$

Substituting this expression for p into Kepler's third law solved for the mass M, we find:

$$M = \frac{4\pi^2 \times a^3}{G \times p^2} = \frac{4\pi^2 \times a^3}{G \times \left(\frac{2\pi a}{v}\right)^2} = \frac{\cancel{4\pi^2} \times a^{\cancel{3}}}{G \times \frac{\cancel{4\pi^2 a^2}}{v^2}} = \frac{a \times v^2}{G}$$

within the solar circle all pull the Sun in the same direction—toward the galactic center. Thus, the Sun's orbital velocity responds almost exclusively to the gravitational pull of matter inside its orbit. Substituting the Sun's 28,000-light-year distance and 220-km/s orbital velocity into the orbital velocity law, we find that the total amount of mass within the solar circle is about 2×10^{41} kilograms, or about 100 billion solar masses.

The Distribution of Mass in the Milky Way

Just as we can use the orbit of the Sun to determine the mass of the galaxy within the solar circle, we can use the orbital motion of any other star to measure the mass of the Milky Way within the star's own orbital circle. In principle, we could determine the complete distribution of mass in the Milky Way by applying the orbital velocity law to the orbits of stars at every different distance from the galactic center. In practice, interstellar dust obscures our view of disk stars beyond a few thousand light-years, making it very difficult to measure stellar velocities. However, radio waves penetrate this dust, so we can see the 21-cm line from atomic hydrogen gas no matter where the gas is located in the galaxy. Such studies allow us to build a map of atomic hydrogen clouds. It is such maps that reveal our galaxy's large-scale spiral structure and also its overall pattern of rotation.

We can get a sense of the distribution of mass in the Milky Way by making a diagram called a **rotation curve,** which plots *rotational velocity* against *distance from the center.* As a simple example of the concept, let's construct a rotation curve for a merry-go-round. Every object on a merry-go-round goes around the center with the same rotational period, but objects farther from the center move in larger circles. Thus, objects farther from the center move at faster speeds, and the rotation curve for a merry-go-round is a straight line that rises steadily outward (Figure 18.19a). In contrast, the rotation curve for our solar system drops off with distance from the Sun because inner planets orbit at faster speeds than outer planets (Figure 18.19b). This drop-off in speed with distance occurs because virtually all the mass

---

Because we are dealing with a circular orbit, we can replace a with the radius r and write this equation as:

$$M_r = \frac{r \times v^2}{G}$$

The subscript r of M_r reminds us that we have calculated the mass only within a distance r of the galactic center.

Example: Calculate the mass of the Milky Way Galaxy within the solar circle.

Solution: The orbital velocity of the Sun around the center of the galaxy is $v = 220$ km/s $= 2.2 \times 10^5$ m/s. The radius of the solar circle is the Sun's 28,000-light-year distance from the galactic center. Using the fact that a light-year is about 9.46×10^{15} meters, we find this radius to be equivalent to $r = 2.6 \times 10^{20}$ m. We can now find the mass within the solar circle by substituting these values for v and r into the orbital velocity law:

$$M_r = \frac{r \times v^2}{G} = \frac{(2.6 \times 10^{20}\ \text{m}) \times \left(2.2 \times 10^5\ \dfrac{\text{m}}{\text{s}}\right)^2}{6.67 \times 10^{-11}\ \dfrac{\text{m}^3}{\text{kg} \times \text{s}^2}} = 1.9 \times 10^{41}\ \text{kg}$$

The mass of the Milky Way Galaxy within the solar circle is about 2×10^{41} kg. If you recall that the mass of the Sun is about 2×10^{30} kg, then it is clear that the mass within the solar circle is equivalent to about 10^{11}, or one hundred billion, solar masses.

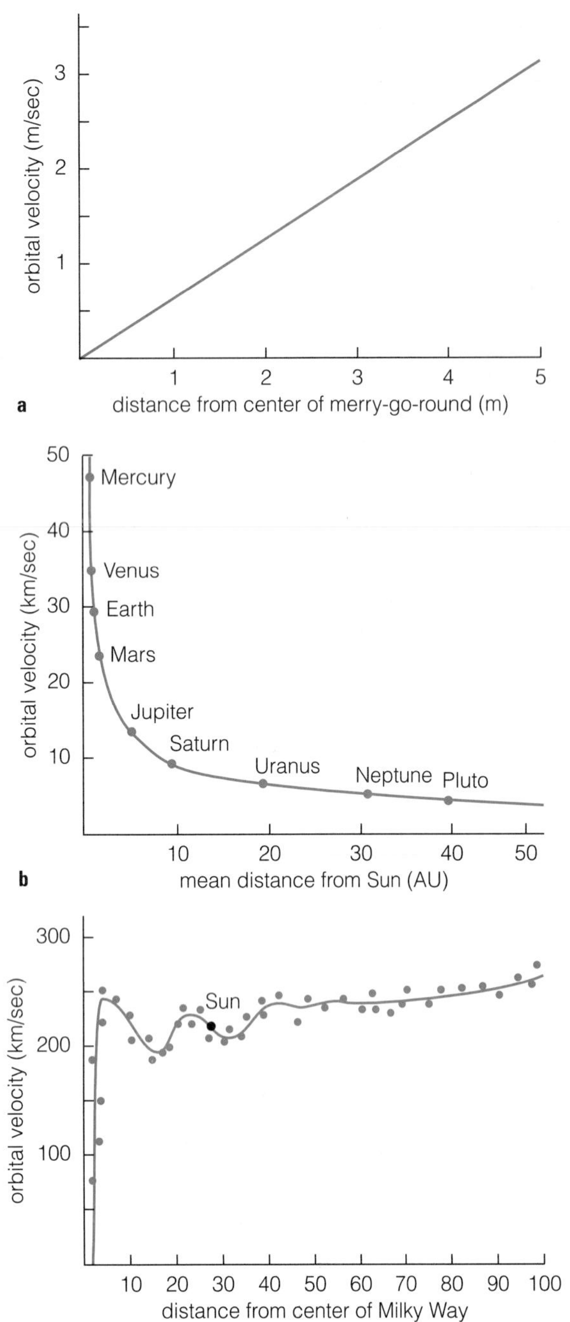

FIGURE 18.19 (**a**) A rotation curve for a merry-go-round. (**b**) The rotation curve for the planets in our solar system. (**c**) The rotation curve for the Milky Way Galaxy.

of the solar system is concentrated in the Sun. The gravitational force holding a planet in its orbit decreases with distance from the Sun, and a smaller force means a lower orbital speed. The rotation curve of any astronomical system whose mass is concentrated toward the center therefore drops steeply.

The rotation curve for the Milky Way Galaxy is shown in Figure 18.19c. Each individual dot represents the distance from the galactic center and the orbital speed of a particular star or cloud of atomic hydrogen. The curve running through the dots represents a "best fit" to the data. Note that, beyond the inner few thousand light-years, the orbital velocities remain approximately flat. This behavior contrasts sharply with the steeply declining rotation curve of the solar system. Thus, unlike the solar system, most of the mass of the Milky Way must *not* be concentrated at its center. Instead, the orbits of progressively more distant hydrogen clouds must encircle more and more mass. The solar circle encompasses about 100 billion solar masses, but a circle twice as large surrounds twice as much mass, and a larger circle surrounds even more mass. Because of the difficulty involved in finding clouds to measure on the outskirts of the galaxy, we have not yet found the "edge" of this mass distribution.

The flatness of the Milky Way's rotation curve came as an immense surprise to astronomers because it implies that most of our galaxy's mass must lie well beyond our Sun, tens of thousands of light-years from the galactic center. In fact, a more detailed analysis suggests that most of this mass is distributed in the galactic halo and that the halo might outweigh all the disk stars *combined* by a factor of 10. However, aside from the light of the relatively small number of halo stars, we have detected very little other radiation coming from this enormous amount of mass. Because we see so little light coming from outside the Sun's orbit, most of the halo's huge mass cannot be in the form of orbiting stars. The nature of this mass remains unknown—which means that we do not yet know the nature of the vast majority of matter in our own galaxy [Section 3.4]. We call this mysterious mass **dark matter** because it does not emit any light that we have yet detected. The nature of this dark matter is one of the greatest mysteries in astronomy today. We will investigate it in much more depth in Chapter 21.

Spiral Arms

The disks of spiral galaxies like the Milky Way display sweeping spiral arms that can stretch tens of thousands of light-years from their bulges. At first glance, spiral arms look as if they ought to rotate with the stars, like the fins of a giant pinwheel in space. However, because stars near the center of the galaxy take less time to complete an orbit than more distant stars, the arms would gradually wind themselves up if they really did rotate with the stars in the disk. By the time a spiral galaxy was a few billion years old, the spiral arms would look like a tightly

FIGURE 18.20 NGC 2997, a spiral galaxy with two prominent spiral arms. Note how blue the spiral arms are compared with the yellower tones of the bulge and the regions of the disk between the arms. Because blue stars only live for a few million years, the relative blueness of these spiral arms tells us that stars must be forming more actively here than elsewhere in the galaxy.

wound coil. Because we do not see such tightly coiled spiral arms, the explanation for spiral structure must be more complex.

Detailed images show the spiral arms to be home to most of the young, bright, blue stars in a galaxy (Figure 18.20). Clusters of these hot, short-lived massive stars populate the spiral arms. Between the arms, we find many fewer hot stars. We also see enhanced amounts of molecular and atomic gas in the spiral arms, and streaks of interstellar dust often obscure the inner sides of the arms themselves (Figure 18.21). All these clues suggest that spiral arms result from waves of star formation that propagate through the disks of spiral galaxies.

Theoretical models suggest that disturbances called **spiral density waves** are probably the source of our galaxy's spiral arms. To visualize how these waves operate, think about waves in the ocean. If an earthquake in Alaska initiates a tidal wave (tsunami) that slams into the northern shore of Hawaii, the water pummeling the Hawaiian coastline does not travel all the way from Alaska. The water in any given location merely sloshes back and forth, but the wave carries energy across the ocean. The enhanced water pressure beneath the crest of a wave momentarily causes the water level to rise ahead of the wave. As the crest passes, the water level falls again, and the wave moves farther along.

Spiral density waves are like waves in the ocean in two respects: They need some sort of disturbance to generate them, and they do not carry stars or interstellar gas with them as they move through the disk.

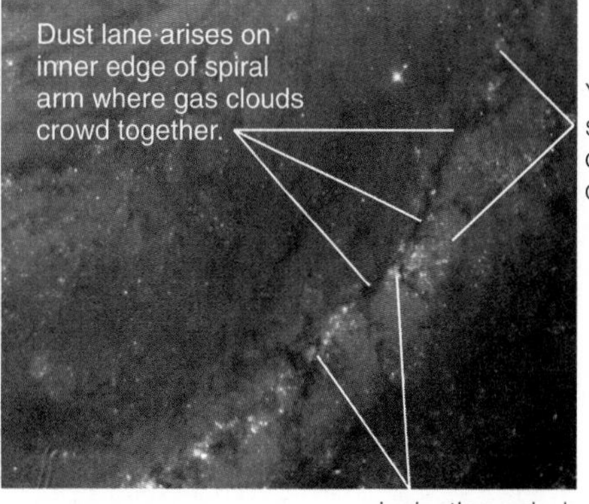

FIGURE 18.21 The relationship of dust, gas, and new star clusters in a spiral arm.

Tidal interactions with other galaxies passing nearby are one possible source of disturbance in a galaxy's disk. Once such a disturbance starts a spiral density wave in motion, the wave will continue to propagate through the galaxy's disk, perhaps for billions of years.

Gravitational interactions between stars keep spiral density waves going. The spiral arms themselves are waves of enhanced density, so stars and gas at the location of any wave crest are packed slightly more densely than normal. Thus, the wave crest is also a location of enhanced gravitational attraction within the disk. This gravitational attraction pulls the stars ahead of the crest more closely together; after the crest passes, the stars move apart into looser groups.

Interstellar clouds react even more strongly to the gravitational attraction of the wave crest. Gas clouds collide with one another most frequently in the densest part of the wave, collecting into groups (Figure 18.22). Because gas clouds crowd so thickly into the crest of the wave, stars form more profusely in spiral arms than elsewhere in the disk. A froth of young stars and ionization nebulae thus marks the crests of spiral density waves.

TIME OUT TO THINK *Perhaps you have noticed what happens when the scene of an accident blocks one side of a highway. Curious drivers on the other side of the highway slow down to check out the accident, and traffic backs up behind them. After the drivers have passed by, they speed back up, and traffic thins out. How is this disturbance in the flow of traffic similar to a spiral density wave?*

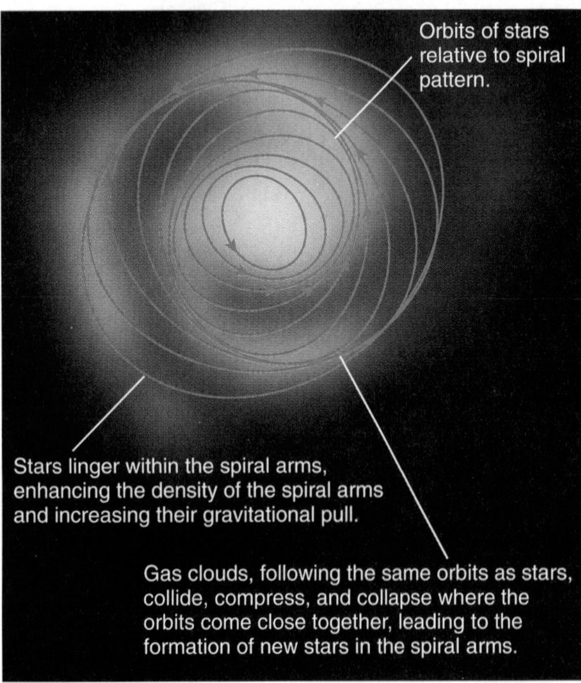

FIGURE 18.22 Generation of spiral density waves in a two-armed spiral galaxy. Once spiral arms appear, they can strengthen themselves because their enhanced gravity pulls orbiting stars into elliptical orbits with greater elongation.

The bluish hue of the spiral arms comes from their prolific star formation. Massive, short-lived, blue stars die out quickly as the wave crest passes by, long before the next spiral wave sweeps through. These luminous hot stars are therefore found close to the spiral arms in which they formed. Longer-lived yellow and red stars live through the passage of many spiral density waves and therefore are distributed relatively evenly throughout the galactic disk.

18.5 The Mysterious Galactic Center

The center of the Milky Way Galaxy lies in the direction of the constellation Sagittarius. This region of the sky does not look particularly special to our unaided eyes; however, if we could remove the interstellar dust that obscures our view, the elliptical shape of the galaxy's central bulge would fill Sagittarius, and its brilliance would be one of the night sky's most spectacular sights.

Because the Milky Way is relatively transparent to long-wavelength radiation, we can use radio and infrared telescopes to peer into its heart (Figure 18.23). Deep in the galactic center, we find swirling clouds of gas and a cluster of several million stars. Bright radio emission traces out the magnetic fields that thread this turbulent region. At the core of it all sits a source of bright radio emission named Sagittarius A* (pronounced "Sagittarius A-star"), or Sgr A* for short, that is quite unlike any other radio source in our galaxy.

The motions of the gas and stars in Sgr A* indicate that it contains a few million solar masses within a region no larger than about 3 light-years (1 pc) across (Figure 18.24). The star cluster in Sgr A* cannot account for all that mass. Many astronomers therefore suspect that Sgr A* contains a black hole weighing around a million solar masses.

Other astronomers are skeptical. The most plausible black-hole candidates in our galaxy—those in binary systems like Cygnus X-1 [Section 17.5]—accrete matter from their companions. The resulting accretion disks radiate brightly in X rays. If Sgr A* truly is a giant black hole, we might expect it to be accreting some of the surrounding gas from the galactic center. Because of its large mass, it would have a large accretion disk and would shine brightly in X rays. (Many

FIGURE 18.23 The galactic center at various wavelengths.

a In the optical band, dusty gas obscures our view of the galactic center, marked by the 0.65° green square.

b In this infrared image, roughly the size of the green square in (**a**), we can see through the dust to the star cluster at the center.

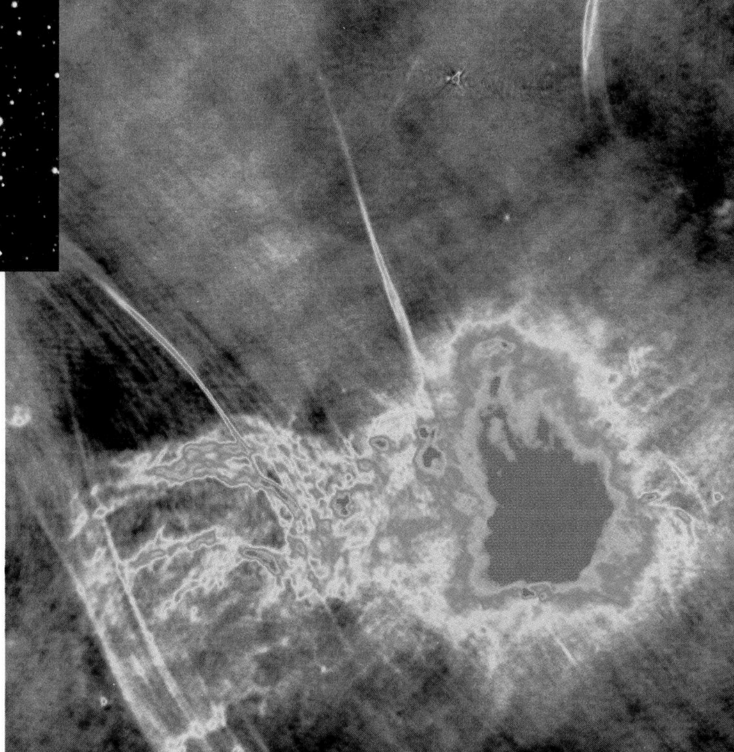

c This close-up radio image of the galactic center shows the strange environment of Sgr A*; the bright filaments trace magnetic field lines full of charged particles presumably energized by the central black hole.

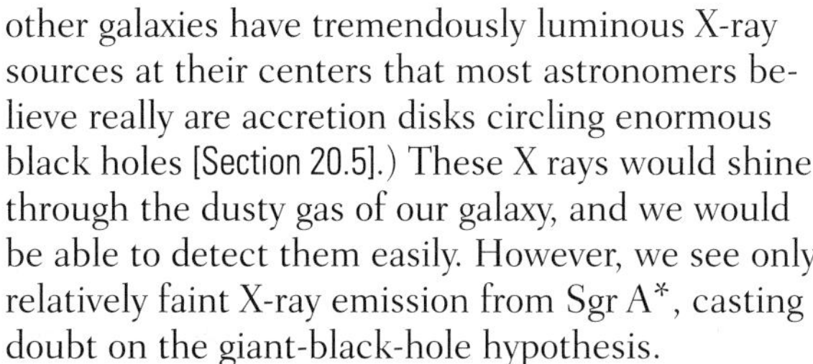

other galaxies have tremendously luminous X-ray sources at their centers that most astronomers believe really are accretion disks circling enormous black holes [Section 20.5].) These X rays would shine through the dusty gas of our galaxy, and we would be able to detect them easily. However, we see only relatively faint X-ray emission from Sgr A*, casting doubt on the giant-black-hole hypothesis.

It remains possible that Sgr A* contains a huge black hole that simply has run out of gas to accrete. Even more intriguingly, the black hole might be consuming the high-energy radiation from the accreting matter before this radiation can escape. Nevertheless, Sgr A* does not behave like other black holes whose accretion disks are more apparent. Because the black hole at the center of our galaxy is so well hidden, the nature of Sgr A* remains mysterious.

FIGURE 18.24 Star velocities near the galactic center.

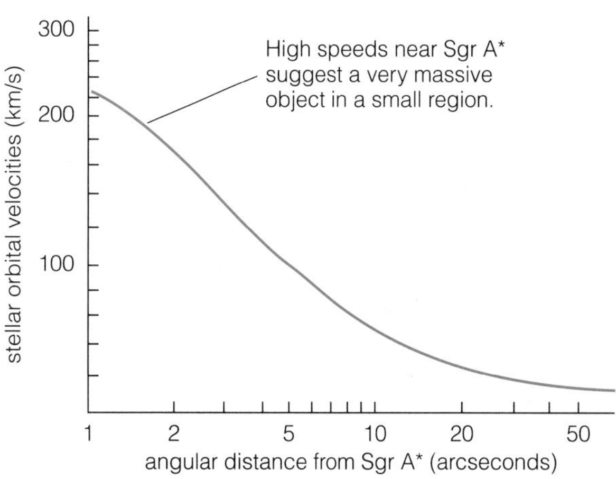

The infinitude of creation is great enough to make a world, or a Milky Way of worlds, look in comparison with it what a flower or an insect does in comparison with the Earth.

IMMANUEL KANT

THE BIG PICTURE

In this chapter, we have explored our galaxy, focusing on the recycling of gas that continually produces new stars and on the motions of the Milky Way that reveal its mass. When you review this chapter, pay attention to these "big picture" ideas:

- The inability of visible light to penetrate deeply through interstellar gas and dust concealed the true nature of our galaxy until recent times. Modern astronomical instruments reveal the Milky Way as a dynamic system of stars and gas that continually gives birth to new stars and planetary systems.
- Stellar winds and explosions make interstellar space a violent place. Hot gas tears through the pervasive atomic hydrogen gas in the galactic disk, leaving expanding pockets of hot plasma and fast-moving clouds in its wake. All this violence might seem quite dangerous, but it performs the great service of mixing essential new elements throughout the Milky Way.
- We can use the orbital speeds of stars and gas clouds around the center of the galaxy to determine the distribution of mass in the Milky Way. To our great surprise, we've discovered that most of the mass of the galaxy lies in the halo, not in the galactic disk where most of the stars are located. The nature of this *dark matter* is one of the greatest mysteries in astronomy today.
- Although the elements from which we are made were forged in stars, we could not exist if stars were not organized into galaxies. The Milky Way Galaxy acts as a giant recycling plant, converting gas expelled from each generation of stars into the next and allowing some of the heavy elements to solidify into planets like our own.

Review Questions

1. Draw simple sketches of our galaxy as it would appear face-on and edge-on, identifying the *disk, bulge, halo,* and *spiral arms.* What kind of star clusters are found in the halo?
2. What do we mean by the *interstellar medium?* How does it obscure our view of most of the galaxy? How can we see through it?
3. Briefly explain how Shapley concluded that the Sun lies quite far from the center of the Milky Way. By modern measurements, about how far is the Sun from the galactic center?
4. What do we mean by *heavy elements?* How much of the Milky Way's gas is in the form of heavy elements? Where are heavy elements made?
5. What is the *star–gas–star cycle,* and how does it lead to *chemical enrichment?* Use the idea of chemical enrichment to explain why stars that formed early in the history of the galaxy contain a smaller proportion of heavy elements than stars that formed more recently.
6. Distinguish between the following forms of gas: ionized hydrogen, atomic hydrogen, and molecular hydrogen.
7. What are *bubbles?* How are they made? What kinds of light do they emit?
8. What is a *shock wave?* What does a shock wave do as it sweeps through the interstellar medium? Why does material in a shock wave from a supernova cool as it expands? Why does it eventually slow down and dissipate?
9. What process produces the radio emission we see from supernova remnants?
10. What are *cosmic rays?* Where do we think they come from?
11. What is a *superbubble?* How is it made? Describe what happens in a blowout in which a superbubble breaks out of the galactic disk.
12. Describe the *galactic fountain* model. What evidence supports this model?
13. What is the most common form of gas in the interstellar medium? What is the wavelength of the radio emission line characteristic of this gas?
14. What are *dust grains?* Where do they form?
15. How do *molecular clouds* form? How do we observe them? Why do molecular clouds tend to settle toward the central layers of the Milky Way's disk?
16. Describe the general structure of a giant molecular cloud. What happens in molecular cloud *cores?*

17. Will the star–gas–star cycle continue forever? Why or why not?
18. Describe how the panoramic strips in Figure 18.13 correspond to the Milky Way that we see in our night sky. Briefly summarize what we learn from these panoramas in different wavelengths.
19. Describe the stars we find in the *halo* in terms of mass, luminosity, color, and proportion of heavy elements. Why do halo stars have these characteristics?
20. Describe the stars in the Sun's neighborhood. What evidence suggests that we live inside a hot *Local Bubble?*
21. What is an *ionization nebula?* Why are such nebulae found near massive young stars? What produces the striking red color of these nebulae? What produces the blue and black tints in some nebulae?
22. Contrast the general patterns of the orbits of stars in the disk and the orbits of stars in the halo.
23. At what speed does the Sun travel around the *solar circle?* About how long does one orbit take? Briefly explain how we can use the characteristics of the Sun's orbit to determine the mass of the Milky Way contained within the solar circle. What is the result?
24. Briefly describe how, in principle, we can use the orbital characteristics of stars at many distances from the galactic center to determine the distribution of mass in the Milky Way. Why, in practice, do we determine the overall pattern of the galaxy's rotation primarily from clouds of atomic hydrogen gas?
25. What is a *rotation curve?* Describe the rotation curve of the Milky Way, and explain how it indicates that most of the Milky Way's mass lies outside the solar circle. What do we mean by *dark matter?*
26. How do we know that spiral arms are not simple "pinwheels" as they appear at first glance? What features do we see in images of spiral arms that suggest that they result from waves of star formation propagating through galaxies?
27. Describe a *spiral density wave.* What happens to gas clouds inside the crest of such a wave? How might such a wave account for the nature of spiral arms?
28. What is Sgr A*? What evidence suggests that it contains a massive black hole?

Discussion Questions

1. *Galactic Ecosystem.* This chapter has likened the star–gas–star cycle in our Milky Way to the ecosystem that sustains life on Earth. Here on our planet, water molecules cycle from the sea to the sky to the ground and back to the sea. Our bodies convert atmospheric oxygen molecules into carbon dioxide, and plants convert the carbon dioxide back into oxygen molecules. How are the cycles of matter on Earth similar to the cycles of matter in the galaxy? How do they differ? Do you think the term *ecosystem* is appropriate to discussions of the galaxy?
2. *Galaxy Stuff.* In the chapters on stars, we learned why we are "star stuff." Based on what you've learned in this chapter, explain why we are also "galaxy stuff." Does the fact that the entire galaxy was involved in bringing forth life on Earth change your perspective on the Earth or on life in any way? If so, how? If not, why not?

Problems

1. *Unenriched Stars.* Suppose you discovered a star made purely of hydrogen and helium. How old do you think it would be? Explain your reasoning.
2. *Enrichment of Star Clusters.* The gravitational pull of an isolated globular cluster is rather weak—a single supernova explosion can blow all the interstellar gas out of a globular cluster. How might this fact be related to observations indicating that stars ceased to form in globular clusters long ago? How might it be related to the fact that globular clusters are so deficient in elements heavier than hydrogen and helium? Summarize your answers in one or two paragraphs.
3. *Blue Nebulae.* Nebulae that scatter light are bluer than the stars that illuminate them. The Earth's sky is bluer than the Sun. Is this a coincidence? Explain why or why not.
4. *High-Velocity Star.* The average speeds of stars *relative* to the Sun in the solar neighborhood is about 20 km/s (i.e., the speeds at which we see stars moving toward or away from the Sun—*not* their orbital speeds around the galaxy). Suppose you discover a star in the solar neighborhood that is moving relative to the Sun at a much higher speed, say 200 km/s. What kind of orbit does this star probably have around the Milky Way? In what part of the galaxy does it spend most of its time? Explain.
5. *Mass from Rotation Curve.* Using velocities shown on the Milky Way's rotation curve in Figure 18.19c, along with the orbital velocity law (see Mathematical Insight 18.1), calculate the mass of the Milky Way

Galaxy within each of the following distances. Give your answers in both kilograms and solar masses.

a. 10,000 light-years.

b. 30,000 light-years.

c. 50,000 light-years.

6. *Dark Matter Beyond the Solar Circle.* In Mathematical Insight 18.1, we found that the mass of the Milky Way Galaxy within the solar circle is about 100 billion solar masses. At twice the Sun's distance from the galactic center, the orbital velocities of gas clouds still are about the same as the orbital velocity of the Sun (220 km/s). Use this fact to estimate the mass of the galaxy within a circle with twice the radius of the solar circle. Then compare the mass of the galaxy within the solar circle to the mass of the galaxy that lies *beyond* the solar circle (but within twice the radius of the solar circle). Given that most of the light in the Milky Way Galaxy comes from within the solar circle, explain how you can conclude that mass beyond the solar circle must be predominantly dark.

7. *Interstellar Pressures.* The ideal gas law from chemistry provides a simple formula relating the gas pressure P to density and temperature:

$$P = nkT \quad (k = 1.4 \times 10^{-23} \text{ joule/Kelvin})$$

where n is the number density of particles in atoms per cubic meter, T is the temperature in Kelvin, and k is a constant. Use the information in Table 18.1 to estimate the characteristic gas pressures in the various stages of the star–gas–star cycle. In what stage is the pressure significantly higher than in the others? Explain why the gas is more compressed in this stage. (*Hint:* Be sure to convert the densities in Table 18.1 from atoms per cubic centimeter to atoms per cubic meter before using the formula. The pressures should be in units of joules per cubic meter, which is equivalent to newtons per square meter.)

8. *Research: Discovering the Milky Way.* Humans have been looking at the Milky Way since long before recorded history, but only in the past century did we verify the true shape of the galaxy and our location within it. Learn more about how conceptions of the Milky Way developed through history. What names did different cultures give the band of light they saw? What stories did they tell about it? How have ideas about the galaxy changed in the past few centuries? Try to locate diagrams that illustrate these changes. Write a two- to three-page summary of your findings.

9. *Web Project: Images of the Star–Gas–Star Cycle.* Explore the web to find pictures of nebulae and other forms of interstellar gas in different stages of the star–gas–star cycle. Assemble the pictures into a sequence that tells the story of interstellar recycling, with a one-paragraph explanation accompanying each image.

10. *Web Project: The Galactic Center.* Search the web for recent images of the galactic center, along with information about whether the center hides a massive black hole. Present a two- to three-page report, with pictures, giving an update on current knowledge about the center of the Milky Way Galaxy.

19
Galaxies: from Here to the Horizon

FAR BEYOND THE MILKY WAY, WE SEE MANY OTHER GALAXIES—some similar to our own and some very different—scattered throughout space to the very limits of the observable universe. This fabulous sight inspires some fundamental questions: How far away are all these galaxies? How old is the universe? How big is it? Such questions might have seemed ridiculously speculative a century ago. Today, we believe we know the answers to all three questions with respectable accuracy.

Edwin Hubble, the man for whom the Hubble Space Telescope is named, provided the key discovery when he proved conclusively that galaxies exist beyond the Milky Way. The distances he measured revealed an astonishing fact: The more distant a galaxy is, the faster it moves away from us. Hubble's discovery dealt a mortal blow to the traditional belief in a static, eternal, and unchanging universe. The motions of the galaxies away from one another imply instead that the entire universe is expanding and that its age is finite. This chapter follows the steps leading to this profound revelation as we acquaint ourselves with the different types of galaxies in the universe and then learn how to measure their distances. From the distances to other galaxies, we can infer the age of the universe and determine how large the observable universe really is.

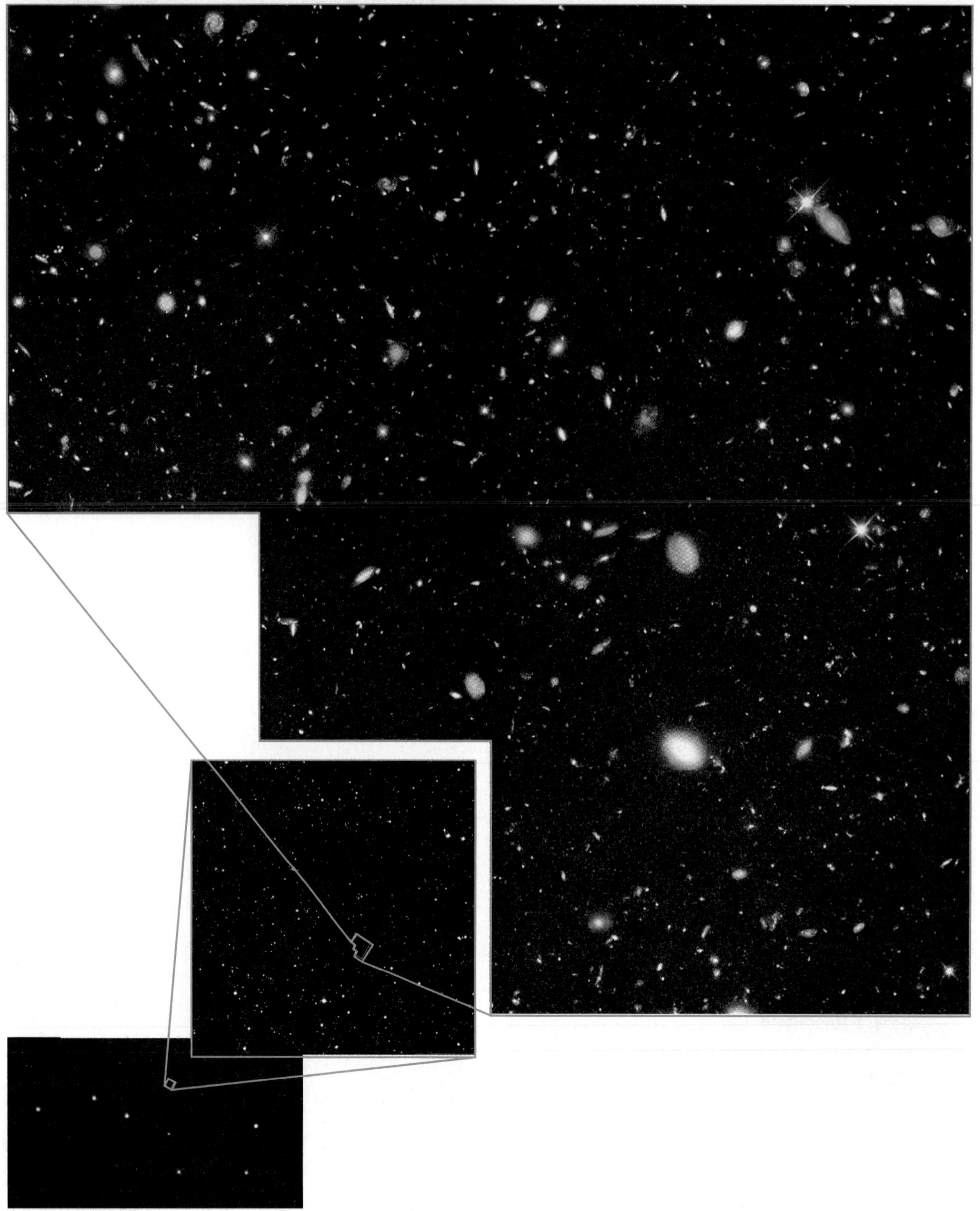

FIGURE 19.1 The Hubble Deep Field, an image composed of 10 days of exposures taken with the Hubble Space Telescope. Some of the galaxies pictured are three-quarters of the way across the observable universe. The field itself is located in the Big Dipper.

19.1 Islands of Stars

Figure 19.1 shows an amazing image of a tiny slice of the sky taken by the Hubble Space Telescope. The telescope pointed toward a single direction in the sky, collecting all the light it could for 10 days. If you held a grain of sand at arm's length, the angular size of the grain would match the angular size of everything in this picture. This slice of the sky is jam-packed with galaxies of many sizes, colors, and shapes. Some look large, some small. Some are reddish, some whitish. Some appear round, and some appear flat. Small galaxies greatly outnumber large ones, yet the large ones produce most of the light in the universe. Counting the galaxies in this slice of the sky and multiplying by the number of such slices it would take to make a montage of the entire sky, we find that the observable universe contains over 50 billion galaxies.

Astronomers classify galaxies into three major categories. **Spiral galaxies** look like flat white disks with yellowish bulges at their centers. The disks are filled with cool gas and dust, interspersed with hotter ionized gas as in the Milky Way, and usually display beautiful spiral arms. **Elliptical galaxies** are redder, more rounded, and often longer in one direction than in the other, like a football. Compared with spiral galaxies, elliptical galaxies contain very little cool gas and dust, though they often contain very hot, ionized gas. Galaxies that appear neither disklike nor rounded are classified as **irregular galaxies.** The sizes of all three types of galaxy span a wide range, from *dwarf galaxies* containing as few as 100 million (10^8) stars to *giant galaxies* with more than 1 trillion (10^{12}) stars. The differing colors of galaxies arise from the different kinds of stars that populate them, and we will see that these colors reflect their star-formation histories.

TIME OUT TO THINK *Take a moment and try to classify the larger galaxies in Figure 19.1. How many appear spiral? Elliptical? Irregular? Do the colors of galaxies seem related to their shapes?*

Spiral Galaxies

Like the Milky Way, other spiral galaxies also have a thin *disk* extending outward from a central *bulge* (Figure 19.2).[1] The bulge itself merges smoothly into a

[1] Objects carry "M" names (e.g., M 31) if they are among the 110 objects in the *Messier Catalog*, published by Charles Messier in 1781. "NGC" stands for the *New General Catalog*, a listing of more than 7,000 objects published in 1888.

FIGURE 19.2 Spiral galaxies: (**a**) M 83, a face-on spiral with a small bulge. (**b**) NGC 300, a face-on spiral with a larger bulge. (**c**) NGC 891, an edge-on spiral considered to be very similar to the Milky Way—note the dusty disk. (**d**) NGC 4594, also known as the Sombrero Galaxy, an edge-on spiral with a large bulge and a dusty disk.

halo that can extend to a radius of over 100,000 light-years (30 kpc). Together, the bulge and halo of a spiral galaxy make up its **spheroidal component,** so named because of its rounded shape. Although no clear boundary divides the pieces of the spheroidal component, astronomers usually consider stars within 10,000 light-years (3 kpc) of the center to be members of the bulge and those outside this radius to be members of the halo.

The **disk component** of a spiral galaxy slices directly through the halo and bulge. The disk of a large spiral galaxy like the Milky Way extends 50,000 light-years (15 kpc) or more from the center. The disks of all spiral galaxies contain an *interstellar medium* of gas and dust, but the amounts and proportions of the interstellar medium in molecular, atomic, and ionized forms differ from one spiral galaxy to the next. Spiral galaxies with large bulges generally have less interstellar gas and dust than those with small bulges.

Not all galaxies with disks are standard spiral galaxies. Some spiral galaxies appear to have a straight bar of stars cutting across the center, with spiral arms curling away from the ends of the bar. Such galaxies are known as *barred spiral galaxies* (Figure 19.3).

Other galaxies have disks but do not appear to have spiral arms (Figure 19.4). These are called *lenticular galaxies* because they look lens-shaped when seen edge-on (*lenticular* means "lens-shaped").

FIGURE 19.3 NGC 1365, a barred spiral galaxy.

FIGURE 19.4 NGC 5078, a lenticular galaxy. Most lenticular galaxies have less cool gas than this one, which displays a prominent black streak of interstellar dust.

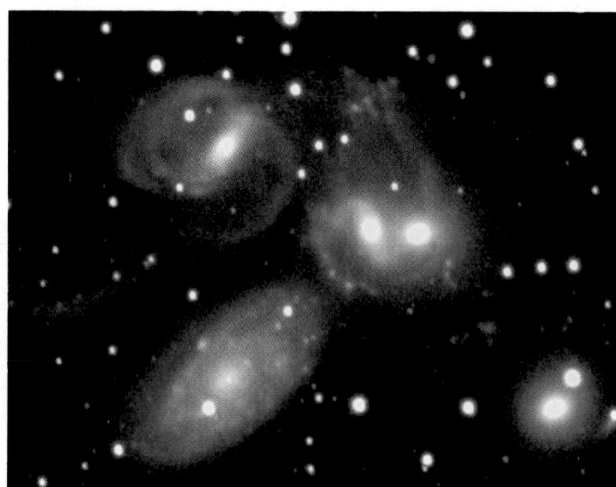

FIGURE 19.5 An unusually tightly packed group of galaxies called Stephan's Quintet. (The galaxy on the lower left is counted as part of the "quintet" but it is not a true member of the group. Instead it is a foreground galaxy that happens to lie along the line of sight.)

Because they have disks, lenticular galaxies are somewhat like spiral galaxies without arms. But they might be more appropriately considered an intermediate class between spirals and ellipticals—they tend to have less cool gas than normal spirals, but more than ellipticals.

About 75% to 85% of large galaxies in the universe are spiral or lenticular. Spiral galaxies are often found in loose collections of several galaxies, called **groups,** that extend over a few million light-years (Figure 19.5). Our Local Group is one example, with two large spirals: the Milky Way and the Great Galaxy in Andromeda (M 31) [Section 1.2]. Lenticular galaxies are particularly common in **clusters** of galaxies, which can contain hundreds and sometimes

THINKING ABOUT . . .

Hubble's Galaxy Classes

Edwin Hubble invented a system for classifying galaxies that remains widely used. It assigns the letter *E*, followed by a number, to each elliptical galaxy. The larger the number, the flatter the galaxy. An E0 galaxy is round, and an E7 galaxy is highly elongated. A spiral galaxy is assigned an uppercase S, or *SB* if it has a bar, and a lowercase *a*, *b*, or *c*. The lowercase letter indicates the size of the bulge and the dustiness of the disk. An Sa galaxy has a large bulge and a modest amount of dusty gas; an SBc galaxy has a bar, a small bulge, and lots of dusty gas. The lenticular galaxies are designated S0, signifying their intermediate spot between spirals and ellipticals. Irregular galaxies are designated Irr.

Astronomers had once hoped that the classification of galaxies might yield deep insights, just as the classification of stars did in the early twentieth century. The Hubble classification scheme itself was suspected for a time to be an evolutionary sequence in which galaxies flattened and spread out as they aged. Unfortunately for astronomers, galaxies turn out to be far more complex than stars, and classification schemes like this one have not led to easy answers about the nature of galaxies.

thousands of galaxies extending over more than 10 million light-years (Figure 19.6).

FIGURE 19.6 A portion of the Virgo cluster of galaxies, the nearest large galaxy cluster to the Milky Way.

Elliptical Galaxies

The major difference between elliptical and spiral galaxies is that ellipticals lack a significant disk component (Figure 19.7). Thus, an elliptical galaxy has only a spheroidal component and looks much like the bulge and halo of a spiral galaxy. (In fact, elliptical galaxies are sometimes called *spheroidal galaxies.*) Most of the interstellar medium in large elliptical galaxies consists of low-density, hot, X ray–emitting gas—rather like the gas in bubbles and superbubbles in the Milky Way [Section 18.2]. Elliptical galaxies usually contain very little dust or cool gas, although they are not completely devoid of either. Some have cold, relatively small gaseous disks rotating at their centers that might be the remnants of a collision with a spiral galaxy [Section 20.3].

Elliptical galaxies appear to be more social than spiral galaxies: They are much more common in large clusters of galaxies than outside clusters. Elliptical galaxies make up about half the large galaxies in the cores of clusters, while they represent only about 15% of the large galaxies found outside clusters. However, ellipticals are more common among small galaxies. Particularly small elliptical galaxies with less than a billion stars, called **dwarf elliptical galaxies,** are often found near larger spiral galaxies. At least 10 dwarf elliptical galaxies belong to the Local Group.

Irregular Galaxies

A small percentage of the large galaxies we see nearby fall into neither of the two major categories. This *irregular* class of galaxies is a miscellaneous class, encompassing small galaxies such as the Magellanic Clouds and "peculiar" galaxies that appear to be in disarray (Figure 19.8). These blobby star systems are usually white and dusty, like the disks of spirals. Telescopic observations probing deep into the universe show that distant galaxies are more likely to be irregular in shape than those nearby. Because the light of more distant galaxies was emitted longer ago in the past, these observations tell us that irregular galaxies were more common when the universe was younger.

a M 87, a giant elliptical galaxy in the Virgo Cluster.

b Leo I, a dwarf elliptical galaxy in the Local Group.

FIGURE 19.7 Elliptical galaxies.

a The Large Magellanic Cloud, a small companion to the Milky Way.

b The Small Magellanic Cloud, a smaller companion to the Milky Way.

FIGURE 19.8 Irregular galaxies.

Old Spheroids, Young Disks

The colors of galaxies provide a crucial clue to how they form and operate. Disks of spiral galaxies appear whitish with flecks of blue. This indicates that the stars in the disks of spiral galaxies, sometimes referred to as the *disk population,*[2] include hot, blue stars as well as cool, red ones. Because only short-lived, massive stars look blue, we conclude that star formation must be an ongoing process in galactic disks—just as it is in the disk of the Milky Way. In contrast, the *spheroidal population* of stars found in elliptical galaxies and in the bulges and halos of spiral galaxies looks reddish in color. The absence of blue stars among the spheroidal population indicates that most of the stars in these spheroids are old and that their star formation must have ceased long ago—just as is the case in the halo of the Milky Way [Section 18.3].

The characteristic motions of stars in disks and spheroids give another important clue to their nature (Figure 19.9). Disk stars generally orbit in orderly circles in the same direction, like the rotation of the Milky Way's disk [Section 18.4]. Stars of the spheroidal population in elliptical galaxies and in the bulges and halos of spiral galaxies tend to orbit in random directions. Astronomers reason that, because the motions of their stars are so different, disks and spheroids must have formed in very different ways.

According to the most basic model of galaxy formation, individual galaxies formed gradually from huge, collapsing **protogalactic clouds** of intergalactic gas (Figure 19.10). The stars of spheroidal populations formed first, while the cloud remained blobby in shape with little or no measurable rotation. The orbits of these stars around the center of the galaxy could have any orientation, accounting for the randomly oriented orbits of spheroidal-population stars that we see today. Later, at least in spiral galaxies, conservation of angular momentum caused the remaining gas to contract into a flattened, spinning disk. Stars that formed within this spinning gas disk were born on orbits moving at the same speed and in the same direction as their neighbors and thus represent the disk-population stars. In this basic model, elliptical galaxies are galaxies in which stars formed more quickly, using up most of the gas before it could settle into a rotating disk.

TIME OUT TO THINK *How does the disk of a spiral galaxy resemble a protoplanetary disk* [Section 8.3]*? How are the processes that form galactic disks and protoplanetary disks similar?*

We will see in Chapter 20 that this model is somewhat oversimplified, and parts of it may be just plain wrong. Collisions between neighboring galaxies

[2]Astronomers often refer to the younger, whitish stars in disks as *Population I* stars and to the older, reddish stars in spheroids as *Population II* stars. To keep these straight, you can use the mnemonic "*I* live in Population *I*," because the solar system sits in the Milky Way's disk.

d M 82, an irregular galaxy with lots of young stars.

c NGC 1313, an irregular galaxy with scattered patches of star formation.

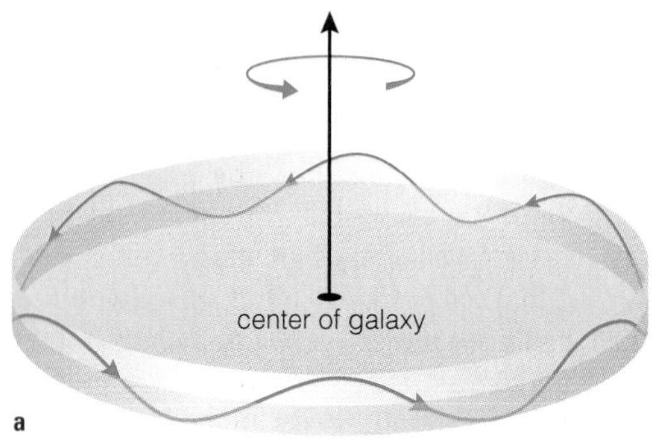

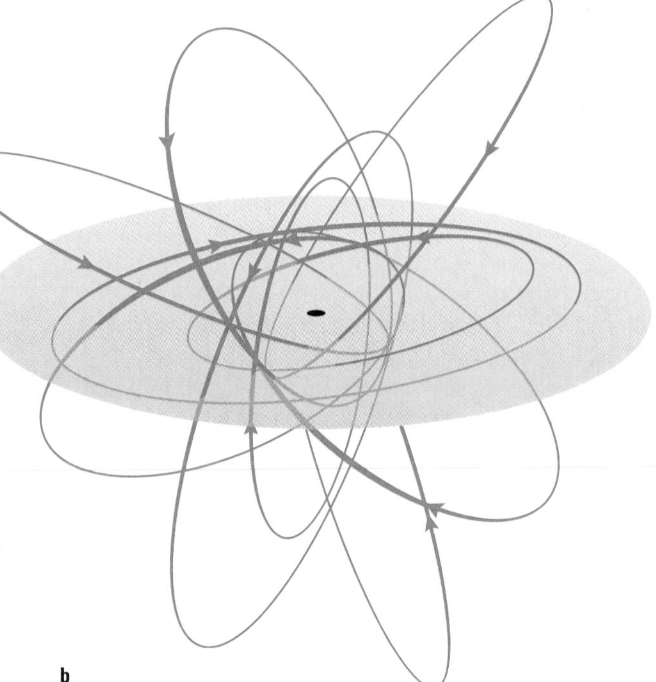

FIGURE 19.9 Characteristic stellar orbits in (**a**) disk component and (**b**) spheroidal component. Stars in the disk tend to orbit in the same direction around the center of the galaxy, with a bit of up and down motion. Stars in the bulge and halo also orbit the center of the galaxy, but the orientations of their orbits are more random.

and huge bursts of star formation can wreak havoc on the orderly progression from protogalactic cloud to blobby spheroid to spinning disk. Because of these interactions between galaxies and their neighbors, the evolution of a galaxy depends on the expansion and evolution of the universe as a whole. Our quest to understand the lives of galaxies thus depends on our learning how to measure the size, age, and expansion rate of the universe. We therefore devote the rest of this chapter to these topics, which will set the stage for us to study in more detail how galaxies evolve.

19.2 Measuring Cosmic Distances

Measuring cosmic distances is one of the most fundamental and challenging tasks we face when trying to understand galaxies and the universe as a whole. Our determinations of astronomical distances depend on a chain of methods that begins with knowing distances in our own solar system. At each link in this chain, we establish standards that help us forge the next link. Because each link depends on preceding links, any inaccuracies accumulate as we move up the chain. Nevertheless, we now can measure the

FIGURE 19.10 Schematic model of galaxy formation.

A protogalactic cloud contains only hydrogen and helium gas.

Halo stars begin to form as the protogalactic cloud collapses.

distances to the farthest galaxies to an accuracy of better than 30%. The distance chain begins with radar measurements of the solar system's size. The second link involves parallax measurements of the distances to nearby stars [Section 15.2]. We will now follow the rest of this chain, link by link, to the outermost reaches of the observable universe.

Standard Candles

Beyond the few hundred light-years within which we can measure stellar parallax, astronomical distance measurements become more difficult. To measure larger distances, we depend on a special set of standards. In our day-to-day lives, we use standards to judge distances without thinking much about it. Our standards are usually familiar things whose actual sizes we know. For example, you can probably guess the distance to a nearby house by its apparent size because you know the real size of a typical house.

In astronomy, the apparent sizes of things can easily deceive us. The actual sizes of galaxies differ, and the angular diameters of stars are generally too small to measure directly. Instead, we measure distances by comparing the brightness of distant objects. A simple formula, which we call the *luminosity–distance formula,* relates the apparent brightness, true luminosity, and distance of any object [Section 15.2]:

$$\text{apparent brightness} = \frac{\text{luminosity}}{4\pi \times (\text{distance})^2}$$

We can always measure an object's apparent brightness with a detector, so we can use this formula to calculate luminosity if we know the distance, or to calculate distance if we know the luminosity. For example, suppose we see a distant street lamp and know that all street lamps of its type put out 1,000 watts of light. Then, once we measure the apparent brightness of the street lamp (in units of watts/m^2), we can calculate its distance with the luminosity–distance formula.

An object such as a street lamp, for which we are likely to know the true luminosity, represents what astronomers call a **standard candle.** The term *standard candle* is meant to suggest a light source of a known, standard luminosity. Unlike light bulbs, astronomical objects do not come marked with wattage, so an astronomical object can serve as a standard candle only if we have some way of knowing its true luminosity without first measuring its apparent brightness and distance. Fortunately, many astronomical objects meet this requirement. For example, any star that is a twin of our Sun—that is, a main-sequence star with spectral type G2—should have about the same luminosity as the Sun. Thus, if we measure the apparent brightness of a Sun-like star, we can use the luminosity–distance formula to calculate its distance by assuming that it has the same luminosity we have measured for the Sun (3.8×10^{26} watts).

There is always some uncertainty in distances calculated by applying the luminosity–distance formula, because no astronomical object is a perfect standard candle. For example, H–R diagrams of star clusters show that, while Sun-like stars all have very similar luminosities, they do not all have *exactly* the

Conservation of angular momentum ensures that the remaining gas flattens into a spinning disk.

Billions of years later, the star-gas-star cycle supports ongoing star formation within the disk. The lack of gas in the halo precludes further star formation outside the disk.

same luminosity (Figure 19.11). Instead, individual Sun-like stars may be somewhat brighter or dimmer in luminosity than the Sun, so a distance that we calculate by assuming that a Sun-like star has precisely the Sun's luminosity may be somewhat different from its true distance.

In many ways, the challenge of measuring astronomical distances comes down to the challenge of finding the objects that make the best standard candles. The more confidently we know an object's true luminosity, the less uncertain the distance we calculate with the luminosity–distance formula.

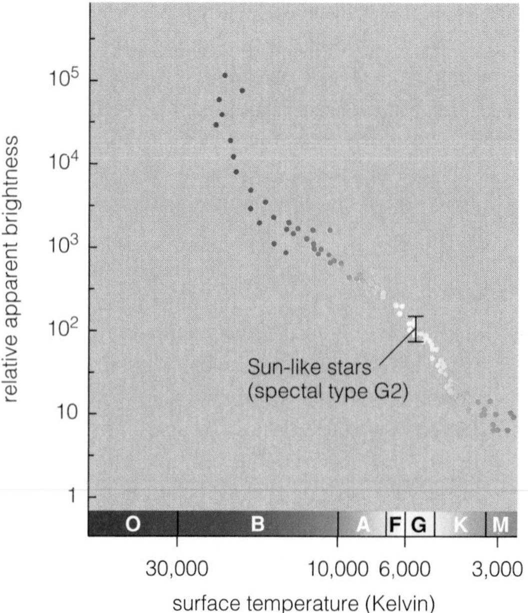

FIGURE 19.11 The H–R diagram for the Pleiades. Note that some Sun-like stars are slightly brighter or dimmer than others. The same is true for other main-sequence stars of a particular spectral type.

Main-Sequence Fitting

We can use Sun-like stars as standard candles because we know that they are similar to the Sun and because we can measure the Sun's luminosity quite easily. However, Sun-like stars are relatively dim, and we cannot detect them at great distances. To measure distances beyond 1,000 light-years or so, we need brighter standard candles. An obvious first choice is to use brighter main-sequence stars. However, before we can use any main-sequence star as a standard candle, we must first have some way of knowing its true luminosity. The key to understanding this process is to remember that we can use the luminosity–distance formula in two ways:

1. First, we identify a star cluster that is close enough for us to determine its distance by parallax and plot its H–R diagram. Because we know the distances to the cluster stars, we can use the luminosity–distance formula to establish their true luminosities from their apparent brightnesses.
2. Then we can look at stars in other clusters that are too far away for parallax measurements and measure their apparent brightnesses. If we assume that main-sequence stars in other clusters have the same true luminosities as their counterparts in the nearby cluster, we can calculate their distances with the luminosity–distance formula.

The nearest star cluster with a well-populated main sequence, the *Hyades Cluster* in the constella-

Mathematical Insight 19.1 Luminosity–Distance Relation Revisited

As we saw in Mathematical Insight 15.1, the luminosity–distance formula arises from the fact that apparent brightness goes down with the square of the distance to a light source. That is why we usually write it in the following form:

$$\text{apparent brightness} = \frac{\text{luminosity}}{4\pi \times (\text{distance})^2}$$

We can usually measure apparent brightness. Thus, we generally use this formula either to calculate luminosity when we know distance or to calculate distance when we know luminosity. We can solve the formula to find luminosity by multiplying both sides by $4\pi \times (\text{distance})^2$:

$$\text{luminosity} = 4\pi \times \text{apparent brightness} \times (\text{distance})^2$$

We can solve for the distance by dividing both sides of this version of the formula by ($4\pi \times$ apparent brightness) and then taking the square root of both sides:

$$\text{distance} = \sqrt{\frac{\text{luminosity}}{4\pi \times \text{apparent brightness}}}$$

tion Taurus, has long been crucial to this technique. We can measure the true distance of the Hyades Cluster through its parallax. Knowing this distance allows us to determine the true luminosities of all its stars with the luminosity–distance formula. We can find the distances to other star clusters by comparing the apparent brightnesses of their main-sequence stars with those in the Hyades Cluster and assuming that all main-sequence stars of the same color have the same luminosity (Figure 19.12). Recall that this technique of determining distances by comparing main sequences in different star clusters is called *main-sequence fitting* [Section 15.6].

The Hyades Cluster does not contain stars of every spectral type, so building a complete, standard H–R diagram required that astronomers use main-sequence fitting to find the distances to many nearby star clusters until every spectral type was represented. Over the past century, astronomers have compiled large catalogs listing the luminosity of every stellar type on the H–R diagram.

Note that, in principle, we need to measure the apparent brightness of only a single star in a cluster to determine the cluster's distance: Once we have determined the star's spectral type, we can look up its expected true luminosity in a catalog and apply the luminosity–distance formula to determine its distance.[3] This method is often used for measuring distances to isolated stars that are not members of clusters.

[3]This procedure is sometimes called *spectroscopic parallax,* a very confusing term because the method has nothing to do with true geometric parallax.

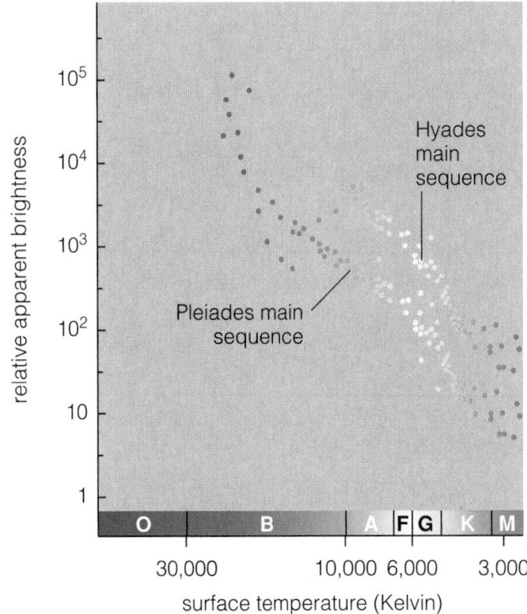

FIGURE 19.12 Comparisons of the apparent brightnesses of stars in the Hyades Cluster with the apparent brightness of those in the Pleiades Cluster show that the Pleiades are 2.7 times farther away because they are $2.7^2 = 7.3$ times dimmer.

The advantage of main-sequence fitting over using a single star in a cluster is that it reduces the uncertainty in the distance calculation. Remember that main-sequence stars of a particular spectral type may differ from one another somewhat in luminosity. However, the *average* luminosity of stars of a particular spectral type should hardly vary from one star cluster to another (assuming the clusters have similar ages and abundances of heavy elements). By

Example 1: You measure the apparent brightness of a particular star to be 2.5×10^{-10} watt/m^2. A parallax measurement shows the star's distance to be 42 light-years, or about 4×10^{17} meters. What is the luminosity of the star?

Solution: We simply substitute into the formula for the luminosity:

$$\text{luminosity} = 4\pi \times \left(2.5 \times 10^{-10} \frac{\text{watt}}{\text{m}^2}\right) \times (4 \times 10^{17}\ \text{m})^2 = 5 \times 10^{26}\ \text{watts}$$

The star's luminosity is 5×10^{26} watts, or about 25% greater than the Sun's luminosity of 3.8×10^{26} watts.

Example 2: Find the distance to a Sun-like star ($L = 3.8 \times 10^{26}$ watts) whose apparent brightness at Earth is 1.0×10^{-10} watt/m^2.

Solution: We substitute into the formula for the distance:

$$\text{distance} = \sqrt{\frac{3.8 \times 10^{26}\ \text{watts}}{4\pi \times \left(1.0 \times 10^{-10} \frac{\text{watt}}{\text{m}^2}\right)}} = 5.5 \times 10^{17}\ \text{m}$$

The distance to the star is 5.5×10^{17} meters, or about 59 light-years (18 parsecs).

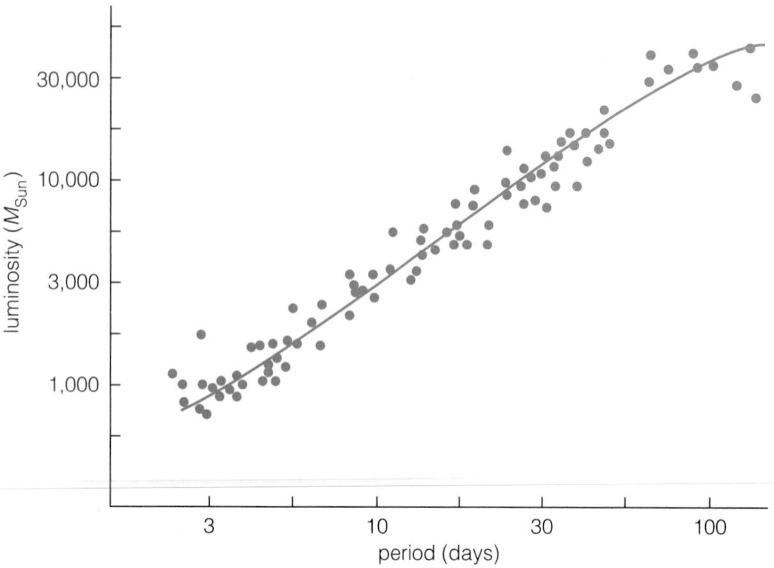

FIGURE 19.13 Cepheid period–luminosity relation. Note that all Cepheids of a particular period have very nearly the same luminosity.

comparing entire main sequences, we are essentially comparing average luminosities and thus achieve much more precise distance measurements.

TIME OUT TO THINK ***Suppose you measure the heights of students in many different astronomy classes and also calculate the average height of the students in each class. Should you expect to find greater variation in height among the students in a particular class or among the average heights that you compute for many different classes? Why? How does this example explain why main-sequence fitting gives better distance estimates than estimates based on individual main-sequence stars?***

Cepheid Variables

Main-sequence fitting works well for measuring distances to star clusters throughout the Milky Way, but not for measuring distances to other galaxies. Most main-sequence stars are too faint to be seen in other galaxies, even with our largest telescopes. Instead, we need very bright stars to serve as standard candles for distance measurements beyond the Milky Way, and the most useful bright stars are the *Cepheid variables* [Section 15.5]. Recall that Cepheids are *pulsating variable stars* that follow a simple *period–luminosity relation:* The longer the time period between peaks in brightness, the greater the luminosity of the Cepheid variable star (Figure 19.13). All Cepheids of a particular period have very nearly the same luminosity (within about 10%), so Cepheids effectively scream out their true luminosities when we measure their periods. "Hey, everybody, my luminosity is 10,000 solar luminosities!" is how we would translate the message of a Cepheid variable whose brightness peaks every 30 days. Thus, Cepheid variables are the primary standard candles used to determine distances to nearby galaxies.

Cepheids also played a key role in the discovery that the Milky Way is only one of billions of galaxies in the universe. The only galaxies visible to the naked eye are the Andromeda Galaxy and the Large and Small Magellanic Clouds (which appear as little more than fuzzy, glowing blobs). By the mid-1700s, telescopes revealed that the Andromeda Galaxy was really a much larger spiral-shaped structure and that many similar spiral-shaped structures littered the sky. Many astronomers, who thought these spirals were spinning gas clouds inside our own galaxy, called them "spiral nebulae." Around the same time, some creative thinkers guessed the truth about "spiral nebulae"—that they are galaxies like the Milky Way. Here is a 1755 quote from German philosopher Immanuel Kant, as translated by Edwin Hubble in his book *The Realm of the Nebulae:*

> *We see that scattered through space out to infinite distances, there exist similar systems of stars [i.e., galaxies], and that creation, in the whole extent of its infinite grandeur, is everywhere organized into systems whose members are in relation with one another. . . . A vast field lies open to discoveries and observation alone will give the key.*

Following Kant, some later writers referred to galaxies as "island universes," because they appeared to be vast, self-contained communities of stars, isolated from one another by chasms of empty space. However, these views could not be conclusively established with the evidence available at the time. As recently as the 1920s, many astronomers still held that the "spiral nebulae" were merely clouds of gas within the Milky Way—and therefore that the Milky Way represented the entire universe.

Edwin Hubble put this debate to rest in 1924 (Figure 19.14). Using the new, 100-inch telescope atop southern California's Mount Wilson (Figure 19.15)—the largest telescope in the world at the time—he discovered Cepheid variables in the Andromeda Galaxy by comparing photographs of the galaxy taken

a

b

FIGURE 19.14 (**a**) Edwin Hubble at the Mount Wilson Observatory. (**b**) The Hubble Space Telescope in orbit around Earth.

days apart. He used the period–luminosity relation to determine the luminosities of these stars and then used these luminosities in the luminosity–distance formula to compute their distances. His distance measurements proved that the Andromeda Galaxy sat far beyond the outer reaches of stars in the Milky Way, demonstrating that it is a separate galaxy.[4] This single stroke of scientific discovery dramatically changed our view of the universe. Rather than inhabiting a universe that ended with the Milky Way, we suddenly knew that we live in just one among billions of galaxies.

[4]At the time, Hubble was unaware that there actually are two different types of Cepheids, depending on their heavy-element content, with each type obeying a different period–luminosity relation. As a result, he underestimated the true distance of Andromeda by about half. But this underestimated distance was still great enough to prove that the Andromeda Galaxy is well outside the Milky Way.

Hubble's Law

Hubble's determination of the distance to the Andromeda Galaxy assured him a permanent place in the history of astronomy, but he didn't stop there. Hubble, an ambitious man who chose astronomy over a chance to box professionally,[5] proceeded to estimate the distances to many more galaxies. Within just a few years, Hubble made one of the most astonishing discoveries in the history of science: that the universe is expanding.

Astronomers had known since the 1910s that the spectra of most "spiral nebulae" tended to be *redshifted.* In other words, each of the emission and absorption lines in their spectra appeared at longer (redder) wavelengths than expected (Figure 19.16). Recall that redshifts occur when the object emitting the radiation is moving away from us [Section 7.5]. But, since Hubble had not yet proved that the "spiral

[5]Hubble declined the professional boxing offer in favor of studying law as a Rhodes Scholar, and then abandoned law in favor of astronomy after a few months as an attorney. He once fought an exhibition boxing match against the reigning world light-heavyweight champion.

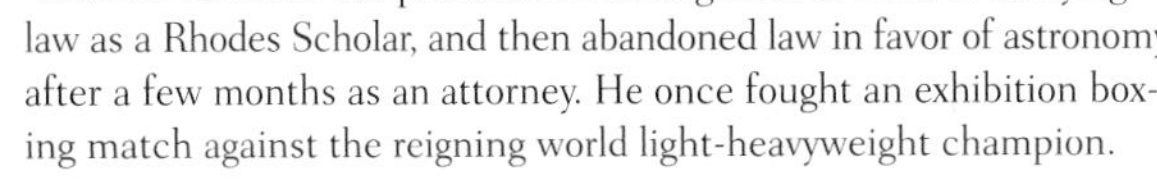

FIGURE 19.15 The 100-inch telescope on Mount Wilson, outside Los Angeles.

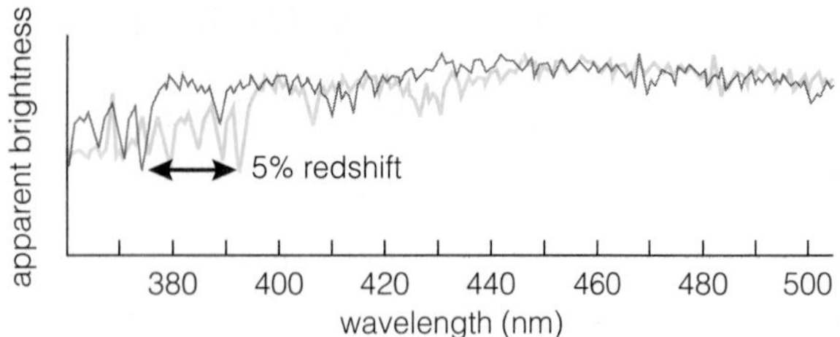

FIGURE 19.16 Redshifted galaxy spectrum. The grey line shows the spectrum of light originally emitted by the galaxy. The blue line shows the spectrum we observe, which is shifted by 5% to longer (redder) wavelengths, indicating that this galaxy is moving away from ours at 5% of the speed of light.

nebulae" were distant galaxies, no one understood the true significance of their motions.

Following his discovery of Cepheids in Andromeda in 1924, Hubble and his coworkers spent the next few years busily measuring the redshifts of galaxies and estimating their distances. Because even Cepheids were too dim to be seen in most of these galaxies, Hubble needed brighter standard candles for his distance estimates. One of his favorite techniques was to assume that the brightest object in each galaxy—which he assumed to be a bright star—had about the same luminosity, so he could use these objects as standard candles. In fact, these objects do not make very good standard candles, but Hubble still got the right general idea about the relationship between redshift and distance.

In 1929, Hubble announced his conclusion: The more distant a galaxy, the greater its redshift and hence the faster it moves away from us (Figure 19.17). Hubble's original assertion was based on an amazingly small sample of galaxies. Even more incredibly, he had grossly underestimated the luminosities of his standard candles. The "brightest stars" he had been using as standard candles were really entire *clusters* of bright stars. Fortunately, Hubble was both bold and lucky. Subsequent studies of much larger samples of galaxies showed that they are indeed receding from us, but they are even farther away than Hubble thought.

We can express the idea that more distant galaxies move away from us faster with a very simple formula, now known as **Hubble's law:**

$$v = H_0 \times d$$

where v stands for velocity (sometimes called a *recession velocity*), d stands for distance, and H_0 (pronounced "H-naught") is a number called **Hubble's constant.** Astronomers generally quote the value of Hubble's constant in strange-sounding units of *kilometers per second per megaparsec* (km/s/Mpc). To make sense of these units, suppose Hubble's constant had a value of $H_0 = 100$ km/s/Mpc. Then, according to Hubble's law, we would expect a galaxy located at distance $d = 10$ Mpc (32.5 million light-years) to be moving away from us at a speed of:

$$v = H_0 \times d = 100 \frac{\text{km/s}}{\cancel{\text{Mpc}}} \times 10 \cancel{\text{Mpc}} = 1{,}000 \text{ km/s}$$

Mathematical Insight 19.2 Redshift

The *redshift* of an object is the difference between the observed wavelength ($\lambda_{\text{observed}}$) of a line in the object's spectrum and the wavelength the line would have if the object were standing still (λ_{rest}). (The Greek letter λ is usually used to stand for wavelength.) Redshifts are best expressed as fractional differences:

$$\text{redshift} = z = \frac{\lambda_{\text{observed}} - \lambda_{\text{rest}}}{\lambda_{\text{rest}}}$$

We use the letter z to represent redshift. The value of z for every line in a moving object's spectrum is the same.

For a relatively nearby galaxy, with a redshift z that is much less than 1, we can find its velocity away from us with the following simple formula:

$$v = c \times z$$

where $c = 3.0 \times 10^5$ km/s is the speed of light. (A more complex formula can be used when the redshift is *not* much less than 1, but as we discuss later in this chapter, the idea of speed becomes hard to define for such distant galaxies.)

Example: Two emission lines from hydrogen in the visible spectrum have rest wavelengths of 656.3 nm and 486.1 nm. Suppose you see this pair of hydrogen

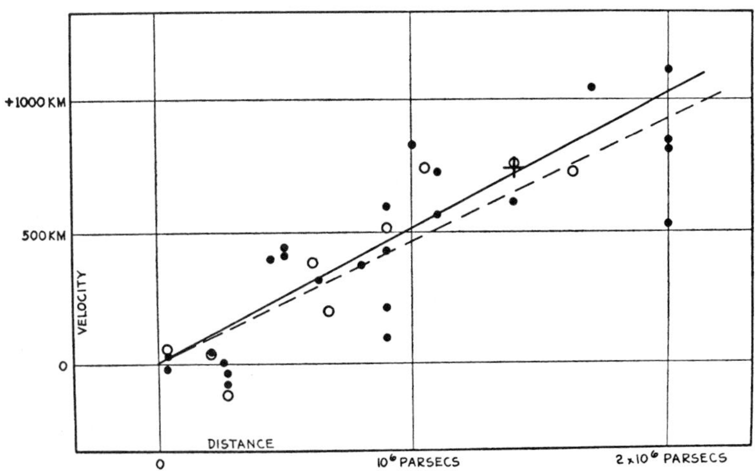

FIGURE 19.17 Hubble's original velocity–distance diagram. Although Hubble underestimated the true galactic distances, he still discovered the general trend (represented by the straight lines) in which more distant galaxies move away from us at higher speeds. He determined the galaxy speeds by measuring redshifts in the galaxy spectra.

Similarly, a galaxy located at a distance $d = 11$ Mpc (36 million light-years) would be moving away at a speed of:

$$v = H_0 \times d = 100 \frac{\text{km/s}}{\cancel{\text{Mpc}}} \times 11 \cancel{\text{Mpc}} = 1{,}100 \text{ km/s}$$

In other words, a value of $H_0 = 100$ km/s/Mpc would mean that, for each additional megaparsec of distance, a galaxy would be moving from us at a speed that is greater by 100 km/s.

TIME OUT TO THINK *Suppose H_0 is 50 km/s/Mpc (rather than 100 km/s/Mpc). In that case, what speed would Hubble's law predict for a galaxy located at a distance of 10 Mpc? Explain.*

Dividing both sides of Hubble's law by H_0 puts it into a form in which we can use a galaxy's velocity to determine its distance:

$$d = \frac{v}{H_0}$$

In principle, applying this law is one of the best ways to determine distances to galaxies. We find the velocity of a distant galaxy from its redshift; then we divide this velocity by H_0 to find its distance. However, we encounter two major practical difficulties in finding distances in this way:

1. Galaxies do not obey Hubble's law perfectly because they can have velocities relative to one another caused by gravitational tugs that Hubble's law does not take into account.
2. Even when galaxies do obey Hubble's law well, the distances we find with it are only as accurate as our best measurement of Hubble's constant.

The first problem makes Hubble's law difficult to use for nearby galaxies. Within the Local Group, for example, Hubble's law does not work at all: The galaxies in the Local Group are gravitationally bound together with the Milky Way and therefore are *not*

emission lines at 662.9 nm and 491.0 nm in the spectrum of a distant galaxy. What is the redshift of the galaxy? How fast is it moving away from us?

Solution: For the two different lines, we find a redshift or z of:

$$\text{656.3-nm line:} \quad z = \frac{662.9 \text{ nm} - 656.3 \text{ nm}}{656.3 \text{ nm}} = 0.010$$

$$\text{486.1-nm line:} \quad z = \frac{491.0 \text{ nm} - 486.1 \text{ nm}}{486.1 \text{ nm}} = 0.010$$

The redshift of the galaxy is $z = 0.010$. Note that, as we should expect, we find the same redshift from either line. (Astronomers generally try to measure the redshifts of at least two lines in any particular spectrum to make sure that they aren't mistaking a misidentified line for a shifted line.) Because this redshift is much less than 1, we can use the simple formula to find the galaxy's speed:

$$v = c \times z = (3 \times 10^5 \text{ km/s}) \times 0.01 = 3{,}000 \text{ km/s}$$

The galaxy is receding from us at 3,000 km/s, equivalent to 1% of the speed of light.

moving away from us in accord with Hubble's law. However, Hubble's law works fairly well for more distant galaxies: The recession speeds of galaxies at large distances are so great that any motions caused by the gravitational tugs of neighboring galaxies are tiny in comparison.

The second problem means that, even for distant galaxies, we can know only *relative* distances until we pin down the true value of H_0. For example, Hubble's law tells us that a galaxy moving away from us at 20,000 km/s is twice as far away as one moving at 10,000 km/s, but we can determine the actual distances of the two galaxies only if we know H_0. The quest to measure H_0 accurately is one of the main missions of the Hubble Space Telescope.

TIME OUT TO THINK *Suppose a galaxy is moving away from us at 10,000 km/s. Use Hubble's law in the form $d = v/H_0$ to calculate its distance (in Mpc) if $H_0 = 100$ km/s/Mpc. What is its distance if $H_0 = 50$ km/s/Mpc?*

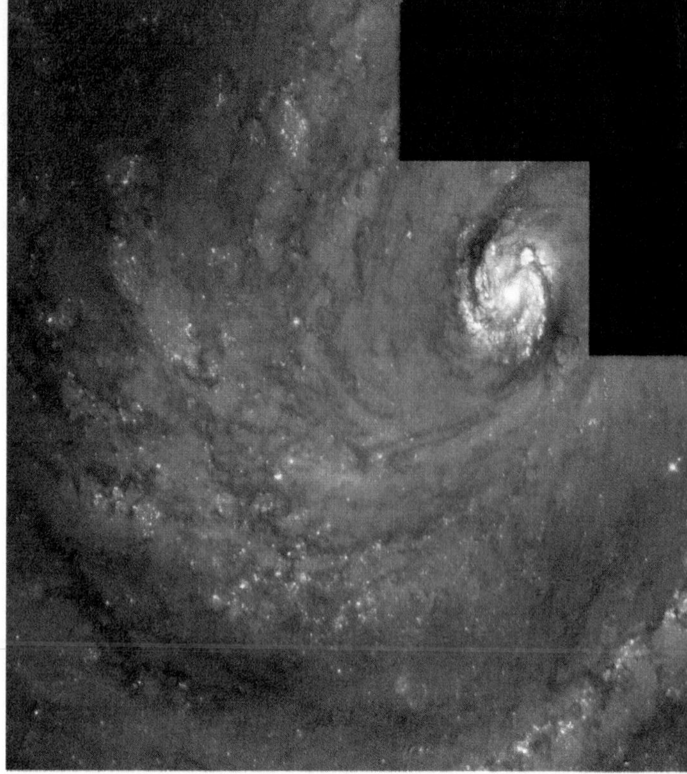

a

Seeking Distant Standards

The process of measuring H_0 is rather like the process of calibrating a scale. We can calibrate a scale by, for example, making sure that it reads 10 pounds when we place a 10-pound weight on it, 20 pounds when we place a 20-pound weight on it, and so on. Once it is properly calibrated, we can use it to weigh objects whose weights are not known in advance. In a similar way, astronomers calibrate Hubble's law by making sure that it gives correct distances for galaxies whose distances we already know from applying some other method. However, until recently, telescopic technology was insufficient for measuring distances accurately by other methods: We were simply unable to detect good standard candles, such as Cepheids, in galaxies much beyond the Local Group.

The superior resolving power of the Hubble Space Telescope now enables us to measure the periods and apparent brightnesses of Cepheid variable stars in galaxies as distant as the Virgo Cluster of galaxies (Figure 19.18), about 50 million light-years (15 Mpc) away. This distance may sound quite far, but it is still near enough that gravitational tugs cause noticeable deviations from Hubble's law. Nevertheless, these Cepheids can help us by enabling us to calibrate even brighter standard candles, such as supernovae.

Recall that white-dwarf supernovae are thought to represent exploding white-dwarf stars that have reached the 1.4-solar-mass limit and should all have nearly the same luminosity [Section 17.2]. Thus, white-dwarf supernovae should be good standard candles—once we know their true luminosities. Although only a few supernovae have been detected during the past century in galaxies within about 50 million light-years (15 Mpc) of the Milky Way, astronomers have kept careful records of these events. Today, we can look back at the light curves for these events to determine which of these past supernovae were white-dwarf supernovae (as opposed to massive star supernovae).

Once we have identified a white-dwarf supernova in the historical records, we can look for Cepheids in the same galaxy. We then use the Cepheids as standard candles to determine the distance to the galaxy in which the supernova occurred. Once we know the galaxy's distance, we can use the luminosity–distance formula to determine the supernova's true luminosity. This technique has allowed researchers using the Hubble Space Telescope to determine the true luminosities of several white-dwarf supernovae. As expected, they are all about the same, and we can now use them as reliable standard candles (Figure 19.19). Because white-dwarf supernovae are so bright—about 10 billion solar luminosities at their peak—we can use them to measure distances to galaxies billions of light-years away (Figure 19.20).

Although white-dwarf supernovae are excellent standard candles, most galaxies have not hosted supernovae during the time that humans have been watching them. Thus, it would be useful if we could use galaxies themselves as standard candles. Astronomers have discovered a close relationship among spiral galaxies between their total luminosities and the rotation speeds of their disks: The faster a spiral galaxy's rotation speed, the more luminous it is (Figure

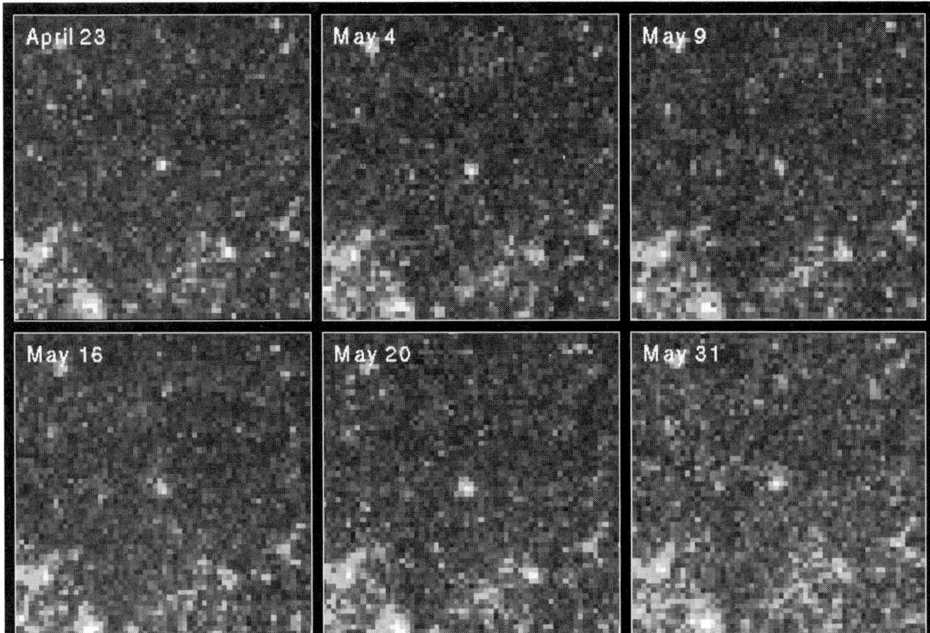

b

FIGURE 19.18 Hubble Space Telescope observations of (**a**) the center of galaxy M 100 in the Virgo Cluster and (**b**) one of its Cepheid variable stars varying in brightness over several weeks.

19.21). This relationship, called the **Tully–Fisher relation** (after its discoverers), holds because both luminosity and rotation speed depend on the galaxy's mass. A galaxy's luminosity depends on the number of stars it contains, which is related to the total amount of matter within it, and the total amount of matter determines a galaxy's rotation speed [Section 18.4].

We can measure the rotation speed of a spiral galaxy's disk by, for example, comparing the Doppler shifts of the portion of the disk rotating toward us and the portion rotating away from us. Once we have measured the rotation speed of a spiral galaxy, the Tully–Fisher relation tells us its true luminosity, which makes the galaxy itself a standard candle that we can use to determine its distance.

Despite the new calibrations of Hubble's law made possible by the Hubble Space Telescope, some uncertainties remain. These uncertainties should decrease as we measure the distances to more and more galaxies. As of 1998, the true value of H_0 appears to be somewhere between 50 and 80 km/s/Mpc.

Summary: The Distance Chain

Distance measurements are fundamental to our understanding of galaxies and the universe. Here we briefly summarize the chain of measurements that allows us to determine ever-greater distances (Figure 19.22). Note that, with each link in the distance chain, uncertainties become somewhat greater. Thus, although we know the Earth–Sun distance at the base of the chain extremely accurately, distances to

FIGURE 19.19 The relationship between the maximum brightness of a white-dwarf supernova and the galaxy in which it exploded is quite close, meaning that (1) white dwarf supernovae are good standard candles and (2) the galaxies in which they explode obey Hubble's law accurately. Thus, we can use white-dwarf supernovae to measure large distances in the universe.

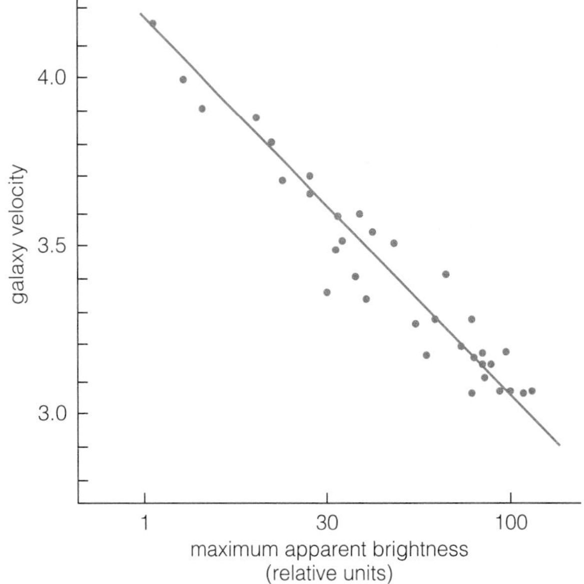

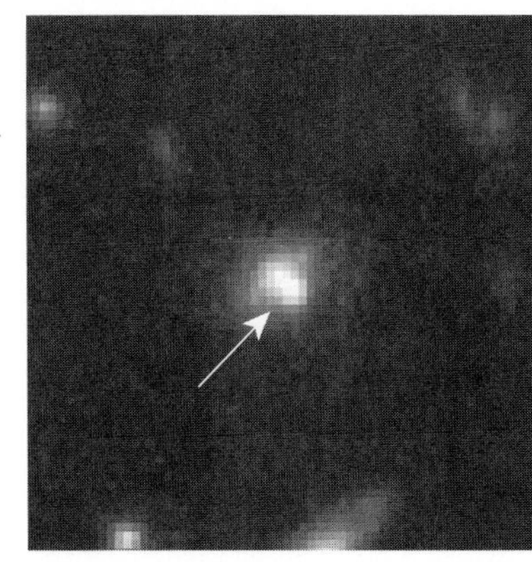

FIGURE 19.20 Photo of a white-dwarf supernova in a galaxy halfway across the observable universe, recorded in March 1996 by the Hubble Space Telescope.

the farthest reaches of the universe remain uncertain by about 30%.

- ***Radar ranging:*** We measure distances within the solar system by bouncing radio waves off planets. Then, with the aid of some geometry, we use these radar measurements to determine the Earth–Sun distance.
- ***Parallax:*** We measure the distances to nearby stars by observing how their positions change, relative to the background stars, as the Earth moves around the Sun. These distances thus rely on our knowledge of the Earth–Sun distance.
- ***Main-sequence fitting:*** We know the distance to the Hyades Cluster in our Milky Way Galaxy through parallax. Comparing the observed intensities of its main-sequence stars to the intensities of those in other clusters gives us the distances to these other star clusters in our galaxy.
- ***Cepheid variables:*** By studying Cepheids in star clusters with distances measured by main-sequence fitting, we learn the precise period–luminosity relation for Cepheids. When we find a Cepheid in a more distant star cluster or galaxy, we can determine its true luminosity by measuring the period between its peaks in brightness and then use this true luminosity to determine the distance.
- ***Distant standards:*** By measuring distances to relatively nearby galaxies with Cepheids, we learn the true luminosities of white-dwarf supernovae and the Tully–Fisher relation between the luminosities and rotation speeds of spiral galaxies. Thus, supernovae and entire galaxies become useful standard candles for measuring great distances in the universe.
- ***Hubble's law:*** Distances measured to galaxies with white-dwarf supernovae and the Tully–Fisher relation allow us to calibrate Hubble's law and measure Hubble's constant, H_0. Once we know H_0, we can determine a galaxy's distance directly from its redshift.

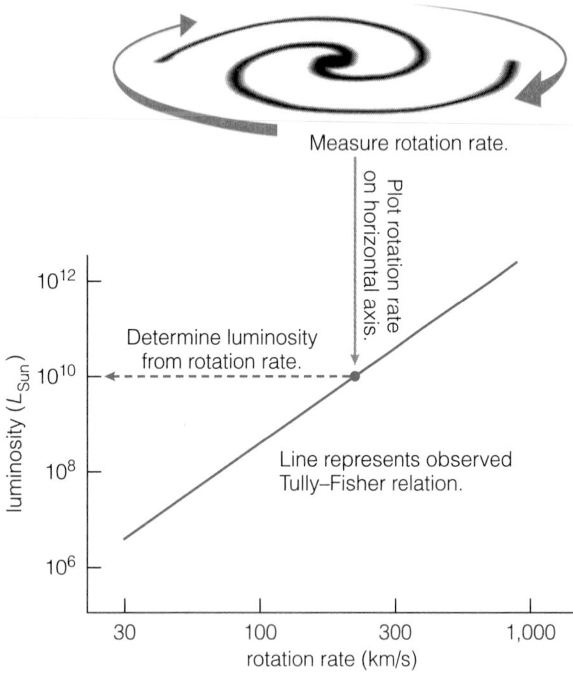

FIGURE 19.21 A schematic diagram of the Tully–Fisher relation. (The precise relation differs among different subclasses of spiral galaxies.)

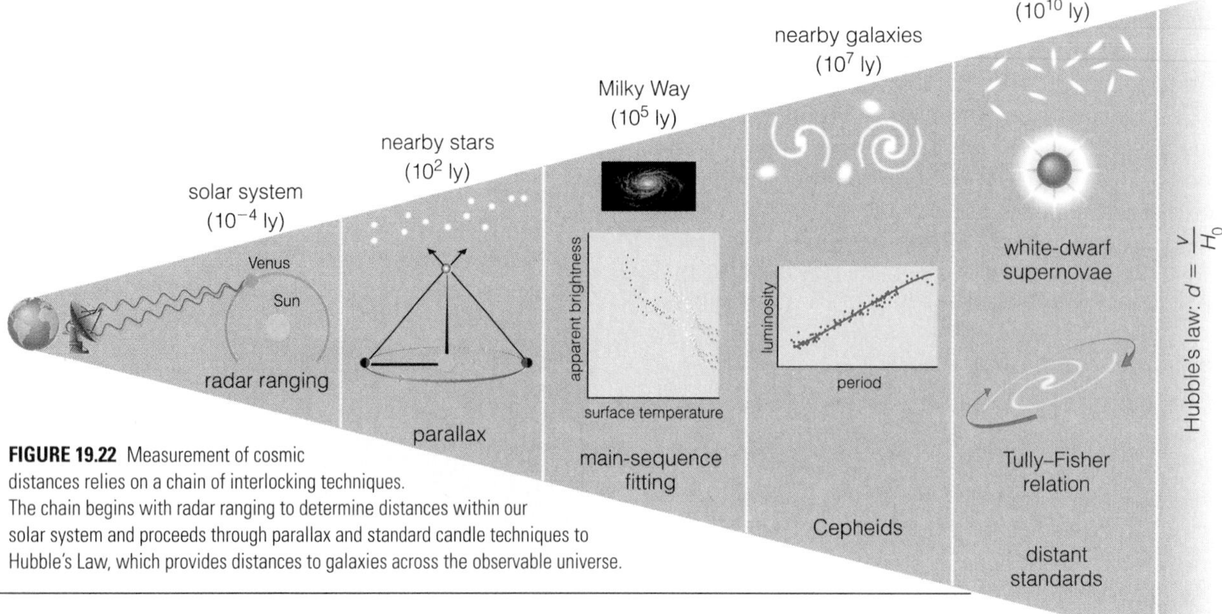

FIGURE 19.22 Measurement of cosmic distances relies on a chain of interlocking techniques. The chain begins with radar ranging to determine distances within our solar system and proceeds through parallax and standard candle techniques to Hubble's Law, which provides distances to galaxies across the observable universe.

19.3 The Age of the Universe

Galaxies all across the universe are moving away from one another in accordance with Hubble's law. This fact implies that galaxies must have been closer together in the past. Tracing this convergence back in time, we reason that all the matter in the observable universe started very close together and that the entire universe came into being at a single moment.

The universe has been expanding since its birth. All the while, the gravitational pull of each galaxy on every other galaxy has been working to slow the expansion rate. In the densest regions of the universe, where we find the most galaxies, gravity has locally won out over expansion. The Local Group no longer expands, nor do other clusters of galaxies. But the expansion continues on all larger scales.

Universal Expansion

When people first hear about the expansion of the universe, they are frequently tempted to think of the universe as a ball of galaxies expanding into a void. This impression is mistaken. The universe expands, but it is not expanding *into* anything. Each point moves away from every other point, and the universe looks about the same no matter where you stand.[6] Back in Chapter 3, we likened the expanding universe to a raisin cake rising as it baked [Section 3.5], but a cake has edges and a center. The universe has neither, so it is now time to deepen our metaphor.

A better analogy would be something that can expand but that has no center and no edges—like the surface of a balloon [Section S4.4]. Just as you'll never encounter an edge if you sail around the world, there is no edge to a balloon's surface. And, while a balloon has a center inside it, the center is not part of the *surface*. Thus, like the universe, the surface of a balloon has no center and no edges. The only significant problem with this analogy is that, because we want to represent the entire universe with the two-dimensional surface of a balloon, we must imagine flattening the universe into only two dimensions. We attach plastic polka dots to the surface of the balloon to represent flattened clusters of galaxies and blow into the balloon to represent the expansion of the universe (Figure 19.23).

[6]In this book, whenever we speak about the universe as a whole we are assuming that matter in the universe is evenly distributed on scales much larger than superclusters of galaxies. This assumption is known as the *Cosmological Principle.* It is impossible to prove that the Cosmological Principle holds outside the observable regions of the universe, but within the observable universe matter does appear to be distributed more and more evenly as we look at increasingly larger scales [Section 21.5].

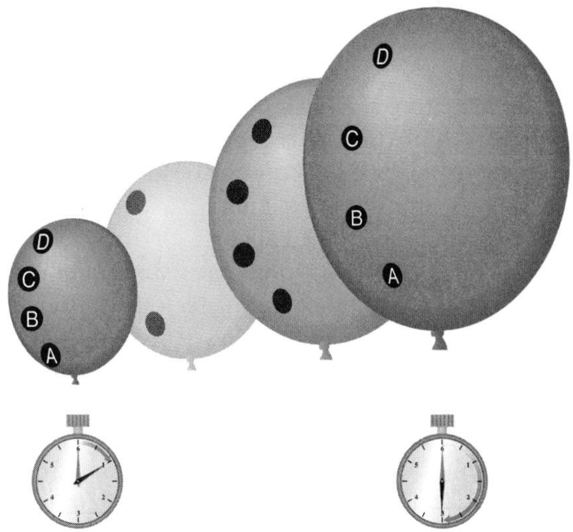

FIGURE 19.23 As the balloon expands, dots uniformly move apart.

As the real universe expands, the distances between clusters of galaxies grow. However, the clusters themselves do not grow because gravity binds each cluster's galaxies together. In a similar way, as the surface of the balloon expands, the polka dots move apart, but the dots themselves don't grow. Suppose that, 1 second after you begin blowing into the balloon, dots A, B, C, and D are spaced 1 cm apart along a line. After 2 seconds, the distances between neighboring dots grow to 2 cm, and after 3 seconds to 3 cm. If you recorded the distances of the other dots from dot B once each second, your observations would look like this:

Measuring Distances from Dot B

Time	Dot A Distance	Dot B Distance	Dot C Distance	Dot D Distance
1 s	1 cm	0 cm	1 cm	2 cm
2 s	2 cm	0 cm	2 cm	4 cm
3 s	3 cm	0 cm	3 cm	6 cm

Miniature scientists on Dot B might be tempted to think they are at the center of some explosion. After all, they see every other dot moving away from them, and the farthest dots move fastest. However, miniature scientists on Dot C would record a very similar set of observations:

Measuring Distances from Dot C

Time	Dot A Distance	Dot B Distance	Dot C Distance	Dot D Distance
1 s	2 cm	1 cm	0 cm	1 cm
2 s	4 cm	2 cm	0 cm	2 cm
3 s	6 cm	3 cm	0 cm	3 cm

The miniature scientists on Dot C would also be tempted to think they are at the center of some explosion. Eventually, the miniature scientists would realize that the balloon itself is expanding. Just as we cannot point to a single spot on a balloon's surface and say "the balloon is expanding from here," we cannot identify a single point in space from which the universe is expanding. The universe has no center. Likewise, neither the balloon nor the universe has any edge.

The balloon analogy, while helpful, has its limitations. For example, the balloon has a finite total surface area, but the universe might be infinite in extent. In addition, you might be tempted to say that the balloon is expanding away from a point *inside* the balloon or that the surface of the balloon has an inside edge and an outside edge. We notice these features of the balloon because the two-dimensional surface of the balloon exists within a three-dimensional space. But the third dimension that we perceive so easily would make no sense to a two-dimensional miniature scientist restricted to the surface of the balloon. Does our own universe have a center or an edge in some other dimension? We'll never know, so we might as well get used to the idea of a universe without a center or an edge.

Hubble Constant and Age

Let's return for a moment to the miniature scientists living on Dot B of the expanding balloon. If you look back at the table showing distances from Dot B, you'll see that the distance to Dot C grows by 1 cm every second. Thus, the scientists on Dot B will measure Dot C to be moving away from them at a speed of 1 cm/s. Similarly, they will see the distance to Dot D growing by 2 cm every second, so Dot D is moving away from them at a speed of 2 cm/s. If they made their measurements 3 seconds after the balloon began expanding, when Dot C was 3 cm away and Dot D was 6 cm away, they would conclude that the following formula relates the distances and velocities of other dots:

$$v = \left(\frac{1 \text{ cm/s}}{3 \text{ cm}}\right) \times d \quad \text{or} \quad v = \left(\frac{1}{3 \text{ s}}\right) \times d$$

where v and d are the velocity and distance of any dot, respectively.

TIME OUT TO THINK *Confirm that this formula gives the correct values for the speeds of dots C and D, as seen from Dot B, 3 seconds after the balloon began expanding. How fast would a dot located 9 cm from Dot B move, according to the scientists on dot B?*

If the miniature scientists think of their balloon as a bubble, they might call the number relating distance to velocity—the term $\frac{1}{3\text{s}}$ in the above formula—the "bubble constant." An especially insightful miniature scientist might flip over the "bubble constant" and find that it was exactly equal to the time since the balloon started expanding. That is, the "bubble constant" of $\frac{1}{3\text{s}}$ tells them that the balloon has been expanding for 3 seconds. Perhaps you see where we are heading.

Just as the inverse of the "bubble constant" tells the miniature scientists that their balloon has been expanding for 3 seconds, the inverse of the Hubble

Mathematical Insight 19.3 Age from Hubble's Constant

As discussed in the text, the reciprocal of Hubble's constant, or $1/H_0$, is directly related to the age of the universe. In particular, it tells us what the age of the universe is *if* the expansion rate has remained constant since the beginning. We can calculate this age from the measured value of H_0. The best current estimates put H_0 between 50 and 80 km/s/Mpc, so let's take the average and assume that $H_0 = 65$ km/s/Mpc.

Before we take the reciprocal, we will convert H_0 to more convenient units. Recall that a megaparsec is about 3.26 million light-years and that a light-year is about 9.5 trillion km. Putting these two facts together, we find that a megaparsec is equivalent to:

$$1 \text{ Mpc} \times \left(3.26 \times 10^6 \frac{\text{ly}}{\text{Mpc}}\right) \times \left(9.5 \times 10^{12} \frac{\text{km}}{\text{ly}}\right) = 3.1 \times 10^{19} \text{ km}$$

Substituting kilometers for megaparsecs in Hubble's constant, we find:

$$H_0 = 65 \text{ km/s/Mpc} = 65 \frac{\text{km}}{\text{s}} \times \frac{1}{1 \text{ Mpc}} = 65 \frac{\text{km}}{\text{s}} \times \frac{1}{3.1 \times 10^{19} \text{ km}} = \frac{65}{3.1 \times 10^{19} \text{ s}}$$

constant, or $1/H_0$, tells us something about how long our universe has been expanding.[7] For example, a bit of mathematical manipulation shows that a Hubble constant of $H_0 = 65$ km/s/Mpc is equivalent to $H_0 = 1/(15$ billion years). In this case, the inverse of the Hubble constant is 15 billion years. However, our universe would be 15 billion years old only if, as in our balloon analogy, the expansion rate has never changed.

The Elastic Universe

The expansion of the universe was faster in the past than it is now. Gravity slows the velocities at which galaxies are flying away from one another, just as it slows the velocity of a baseball thrown into the air. Will the expansion of the universe ever slow to a halt? Will gravity ever reverse the expansion and pull everything back together in a cataclysmic crunch? The ultimate fate of the universe comes down to a simple question: Does the universe have enough kinetic energy to escape its own gravitational pull?

Hubble's constant essentially tells us the kinetic energy of the universe, but we do not yet know the overall strength of the universe's gravitational pull. The strength of this pull depends on the *density* of matter in the universe. The greater the density, the greater the overall strength of gravity and the higher the likelihood that gravity will someday halt the expansion. Precise calculations show that gravity will win out over expansion if the current density of the universe exceeds a seemingly minuscule 10^{-29} gram per cubic centimeter—roughly equivalent to a few hydrogen atoms in a volume the size of your closet. But the universe might not be even this dense—we simply do not yet know [Section 21.6]. The precise density marking the dividing line between eternal expansion and eventual contraction is called the **critical density.** Figure 19.24 shows how the fate of the universe depends on density. Note that, in all cases, the average distance between galaxies has

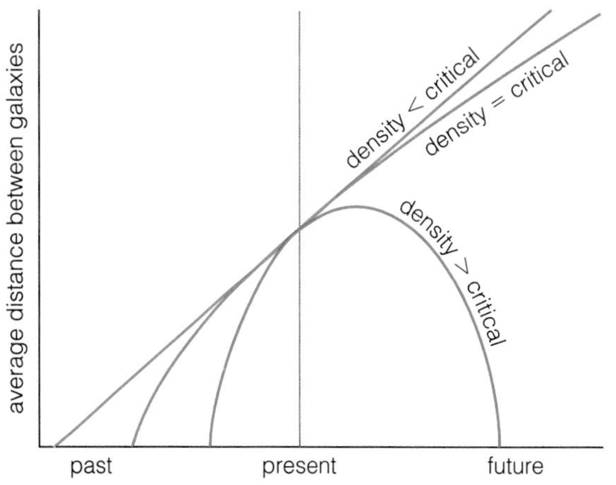

FIGURE 19.24 The expansion rate and fate of the universe depend on whether its density is greater than or less than the critical density. Note that the age of the universe we infer from its present expansion rate is smaller for higher densities than it is for lower densities.

[7]Note that the "bubble constant" for the balloon depends on when it is measured, but it is always equal to 1/(time since balloon started expanding). Similarly, the Hubble constant actually changes with time but stays roughly equal to 1/(age of the universe). We call it a constant because it is the same at all locations in the universe. Moreover, its value does not change noticeably on the time scale of human civilization.

If we divide both the top and the bottom of the fraction by 65, we get:

$$H_0 = \frac{65}{3.1 \times 10^{19}\ \text{s}} = \frac{\frac{65}{65}}{\frac{3.1 \times 10^{19}\ \text{s}}{65}} = \frac{1}{4.8 \times 10^{17}\ \text{s}}$$

Next, we convert the unit in the denominator from seconds to years:

$$H_0 = \frac{1}{4.8 \times 10^{17}\ \text{s} \times \frac{1\ \text{min}}{60\ \text{s}} \times \frac{1\ \text{hr}}{60\ \text{min}} \times \frac{1\ \text{day}}{24\ \text{hr}} \times \frac{1\ \text{yr}}{365\ \text{days}}} = \frac{1}{1.5 \times 10^{10}\ \text{yr}}$$

In this form, it is easy to see that the reciprocal of Hubble's constant is:

$$\frac{1}{H_0} = 1.5 \times 10^{10}\ \text{yr} = 15\ \text{billion years}$$

Thus, if the expansion rate has never changed, a value of $H_0 = 65$ km/s/Mpc implies that the age of the universe is 15 billion years. Since gravity has presumably slowed the expansion rate over time, the actual age of the universe is somewhat less than 15 billion years.

been increasing from the beginning of the universe to the present. If the actual density of the universe exceeds the critical density, then the universe will someday stop expanding and start contracting. If the actual density is equal to or below the critical density, then the universe will continue to expand forever.

Density and Age

We will return to questions concerning the fate of the universe in later chapters. Here our focus is on how the density of the universe affects our estimates of its age. If the actual density of the universe is much smaller than the critical density, then gravity is so weak that it can hardly slow the expansion at all. In that case, the age of the universe is simply the inverse of the Hubble constant, $1/H_0$—just as the "age" of the balloon was the inverse of the "bubble constant." Current estimates that place the value of Hubble's constant between 50 and 80 km/s/Mpc would put the age of the universe between about 12 and 20 billion years.[8]

In contrast, if the actual density exactly equals the critical density, then gravity should have substantially slowed the expansion of the universe since the beginning. In this case, the universe was expanding much faster in the past and hence reached its current size in a shorter amount of time than it would have if gravity were unimportant. Precise calculations show that the age of a universe with exactly the critical density would be $2/(3H_0)$, or somewhere between about 8 and 13 billion years. The age of the universe would be smaller still if the actual density exceeded the critical density.

Intriguingly, the oldest stars in globular clusters appear to be somewhere between 12 and 16 billion years old [Section 15.6]. These stars could not possibly have existed before the birth of the universe! Thus, if these stellar age estimates are correct, the universe

[8]This entire discussion of density, age, and expansion rate presumes that gravity is unopposed as it works to slow the expansion of the universe. However, as this book goes to press, some recent observations of white-dwarf supernovae in distant galaxies suggest that the expansion rate has stopped decreasing and is now increasing. This finding is very preliminary, but if it holds up under further scrutiny it would mean that the universe is actually older than the ages we predict by considering Hubble's constant and density alone. (This topic is discussed further in the Thinking About box in Section 22.3.)

Mathematical Insight 19.4 Lookback Time and Cosmological Redshift

Cosmological redshift is easy to measure (at least in principle): We observe spectral lines in a distant galaxy and compare their wavelengths to the rest wavelengths of these lines to determine the redshift (see Mathematical Insight 19.2). However, determining a lookback time from a redshift is not so easy because the lookback time depends on the age and expansion rate of the universe—values which we will not know precisely until we know precise values for Hubble's constant H_0 and the density of the universe.

Because cosmological redshift can be precisely measured and lookback time cannot, astronomers prefer to think about distant galaxies in terms of cosmological redshifts—at least when they talk among themselves. However, newspaper reporters usually want answers to questions about "how far away" or "how long in the past," so astronomers oblige by *estimating* lookback times.

The general formula for converting cosmological redshifts to lookback times is rather complicated, but below are two special versions of the formula that you can use to estimate lookback times yourself. The first version would be true if the mass density of the universe were much less than the critical density; astronomers refer to this case as a *low-density universe*. The second version would be true if the mass density of the universe turned out to be *equal* to the critical density; astronomers refer to this case as a *critical-density universe*.

$$\text{lookback time in a low-density universe} = \frac{z}{H_0(1+z)}$$

$$\text{lookback time in a critical-density universe} = \frac{2}{3H_0}\left[1 - \frac{1}{(1+z)^{3/2}}\right]$$

cannot be younger than 12 billion years old—which would rule out the possibility that the actual density of the universe greatly exceeds the critical density. Moreover, other evidence suggests that the actual density of the universe is also not much less than the critical density, which means its age is not much more than the 8- to 13-billion-year range found by assuming it has the critical density. All these facts together point to an age of the universe of around 13 billion years. However, it remains possible that we are overestimating the ages of the oldest stars. The best available evidence as of 1998 suggests that the universe is somewhere between 10 and 16 billion years old. If you are reading this book at a much later date, it is likely that the age of the universe is known within more tightly constrained limits.

19.4 The Horizon of the Universe

When we began our discussion of the expanding universe, we stressed that the universe does not have an edge. Yet the universe *does* have a *horizon,* a place beyond which we cannot see. This horizon is a boundary in time, not in space. It exists because we cannot see back to a time before the universe began.

Remember that we always see celestial objects as they were at some time in the past because light travels at a finite speed. If we see a supernova in a galaxy 400 million light-years away, we are seeing a supernova that occurred 400 million years ago. But what do we mean by the distance of 400 million light-years? Because the universe is expanding, the distance between Earth and the supernova is greater today than it was at the time of the supernova event. Are we talking about the distance from the supernova to Earth when the supernova exploded, or are we referring to the distance between the supernova and Earth when the light arrived here? Do we mean that photons traveled through 400 million light-years of space? When we start trying to measure large distances through the universe, our everyday notions of space-only or time-only become inadequate. The path light takes from the supernova to us is really a trajectory through *both* space *and* time, so we must think in terms of *spacetime.*

Figure 19.25 shows a *spacetime diagram* for the supernova and our observation of it [Section S4.3]. The horizontal axis shows distance (through space) from the Milky Way, and the vertical axis shows time. Photons of light from the supernova travel toward us at

---

If we further assume that H_0 = 65 km/s/Mpc = 1/(15 billion years), then these expressions become:

$$\text{lookback time in a low-density universe} = (15 \text{ billion years}) \times \frac{z}{(1+z)}$$

$$\text{lookback time in a critical-density universe} = (10 \text{ billion years}) \times \left[1 - \frac{1}{(1+z)^{3/2}}\right]$$

Example: Suppose you observe a galaxy whose spectral lines are redshifted by an amount z = 3. What is the lookback time to the galaxy if we live in a low-density universe? What is the lookback time if our universe has the critical density? Assume that H_0 = 65 km/s/Mpc.

Solution: In a low-density universe with H_0 = 65 km/s/Mpc, the lookback time to a galaxy with cosmological redshift z = 3 would be:

$$(15 \text{ billion years}) \times \frac{z}{(1+z)} = (15 \text{ billion years}) \times \frac{3}{(1+3)}$$
$$\approx 11 \text{ billion years}$$

In a critical-density universe, the lookback time to this galaxy would be:

$$(10 \text{ billion years}) \times \left[1 - \frac{1}{(1+z)^{3/2}}\right] = (10 \text{ billion years}) \times \left[1 - \frac{1}{(1+3)^{3/2}}\right]$$
$$\approx 9 \text{ billion years}$$

Current evidence suggests that we live in a universe that has a density of no more than the critical density. Thus, if H_0 = 65 km/s/Mpc, the lookback time to a galaxy with redshift z = 3 is somewhere between 9 and 11 billion years.

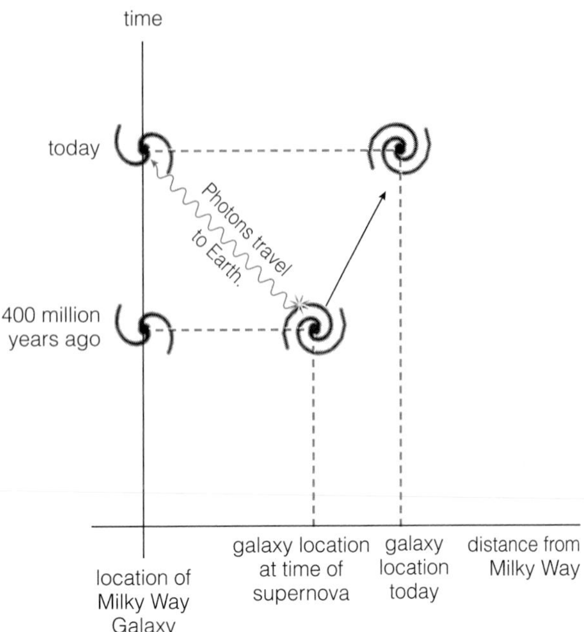

FIGURE 19.25 A spacetime diagram for a supernova in a distant galaxy. The lookback time to this supernova is 400 million years.

the speed of light, starting at the place and time of the supernova explosion. By the time these photons reach us, the galaxy in which the supernova occurred has moved farther away from us. By looking along the time axis, we see that it has taken 400 million years for the photons of light to reach us. We call this the **lookback time** to the supernova. In other words, a distant object's lookback time is the difference between the current age of the universe and the age of the universe when the light left the object. This quantity, which essentially combines both space and time, is much more meaningful than an actual distance when we discuss objects that are moving away from us with the expansion of the universe.

TIME OUT TO THINK *Explain why there is no question about meaning when we talk about distances to objects within the Local Group, but distances become difficult to define when we talk about objects much farther away.*

An object's lookback time is directly related to its redshift. We have seen that the redshifts of galaxies tell us how quickly they are moving away from us. In the context of an expanding universe, redshifts have an additional, more fundamental interpretation. Let's return one final time to the universe on the balloon. Suppose you drew wavy lines on the balloon's surface to represent light waves. As the balloon inflates, these wavy lines stretch out, and their wavelengths increase (Figure 19.26). This stretching closely resembles what happens to photons in an expanding universe. The expansion of the universe stretches out all the photons within it, shifting them to longer, redder wavelengths. We call this effect a **cosmological redshift.**

In a sense, we have a choice when we interpret the redshift of a distant galaxy: We can think of the redshift either as being caused by the Doppler effect as the galaxy moves away from us or as being caused

FIGURE 19.26 As the universe expands, photon wavelengths stretch like the wavy lines on this expanding balloon.

Common Misconceptions: Beyond the Horizon

Perhaps you're thinking there must be something beyond the cosmological horizon. After all, the horizon encircles more and more matter with each passing second. This new matter had to come from somewhere, didn't it? The problem with this reasoning is that the cosmological horizon, unlike a horizon on the Earth, is a boundary in *time,* not in space. At any one moment, in whatever direction we look, the cosmological horizon lies at the beginning of time and encompasses a certain volume of the universe. At the next moment, the cosmological horizon still lies at the beginning of time, but it encompasses a slightly larger volume. However, that does not mean that the matter at the outskirts of the observable universe used to lie "beyond the horizon." When we peer into the distant universe, we are looking back in both space and time. We cannot look "past the horizon" because we cannot look back to a time before the universe began.

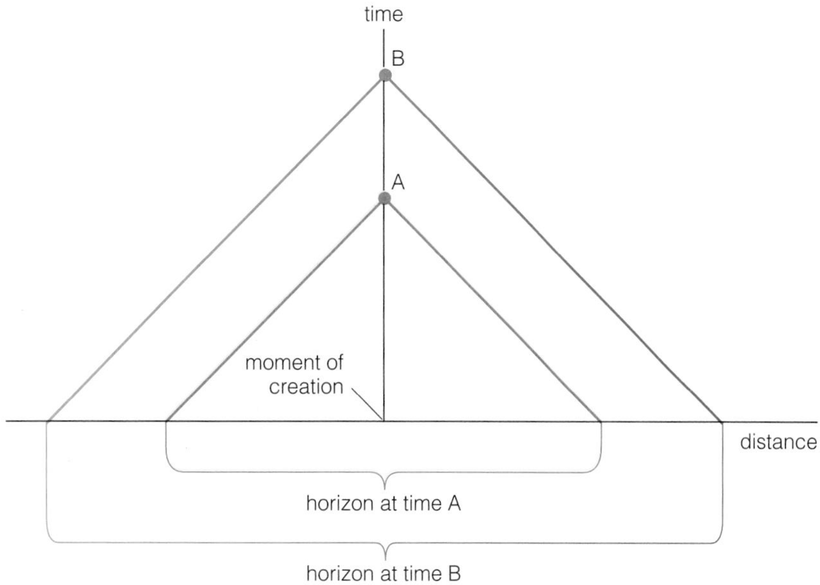

FIGURE 19.27 Spacetime diagram showing the cosmological horizon and how it grows. The horizon distance is greater at time B than at time A.

by a photon-stretching, cosmological redshift. However, as we look to very distant galaxies, the ambiguity in the meaning of *distance* also makes it difficult to specify precisely what we mean by a galaxy's *speed*. Thus, it becomes preferable to interpret the redshift as being due to photon stretching in an expanding universe. From this perspective, it is better to think of space itself as expanding, carrying the galaxies along for the ride, than to think of the galaxies as projectiles flying through a static universe. The cosmological redshift of a galaxy thus tells us how much space has expanded during the lookback time to that galaxy.

Looking back to galaxies with higher and higher redshifts, we see the universe as it appeared at earlier and earlier times. Ultimately, the age of the universe limits how far back we can look. At the **cosmological horizon,** the lookback time is equal to the age of the universe. Beyond this boundary in spacetime, we cannot see anything at all.

If we could see matter at the cosmological horizon, we would see it as it looked at the very beginning of time. One year from now, we would see this same matter as it looked 1 year after the beginning—which means the horizon will have moved somewhat farther from us. We would therefore be able to see some new matter next year that we cannot see today. In fact, our horizon is continually expanding outward at the speed of light. Our observable universe therefore grows larger in radius by 1 light-year with each passing year, continually encompassing more and more matter. Again, because we are dealing with both space and time, the growth of the horizon is best viewed on a spacetime diagram (Figure 19.27).

Our quest to study galaxies and measure cosmic distances has brought us to the very limits of the observable universe. The galaxies in Figure 19.1 at the beginning of this chapter extend from relatively nearby almost all the way to the horizon. Now it is time to examine what these galaxies have been doing for the past 10 billion or more years—the topic of our next chapter.

It is difficult beyond description to conceive that space can have no end; but it is more difficult to conceive an end. It is difficult beyond the power of man to conceive an eternal duration of what we call time; but it is more impossible to conceive a time when there shall be no time.

THOMAS PAINE, *The Age of Reason* (1796)

THE BIG PICTURE

The picture could hardly get any bigger than it has in this chapter. We have reached the limits of the observable universe in both space and time. If your head hasn't already exploded, try to make sure that these "big picture" ideas fit inside:

- The universe is filled with galaxies that come in a variety of shapes and sizes. The most fundamental distinctions in galaxies are between *disk components,* with stars of many ages and abundant gas for new star formation, and *spheroidal components,* which contain old stars and little gas. Both components are present in spiral galaxies, while elliptical galaxies generally lack disks.
- Our measurements of the distances to faraway galaxies rely on a chain of distance measurements that begins with the Earth–Sun distance measured with the aid of radar ranging and continues with parallax and various standard-candle techniques. The distance chain allowed Hubble to measure distances to galaxies and thus to discover Hubble's law and the expansion of the universe.
- When we say that the universe is expanding, we mean that *space itself* is expanding. The universe is not expanding "into" anything.
- As the universe expands, it carries the matter within it along for the ride. From the current expansion rate and from estimates of the overall strength of gravity in the universe and the ages of the oldest stars, we infer that the universe is somewhere between about 10 and 16 billion years old.
- When we look out into the universe, we are looking back in time. Just short of the cosmological horizon, we are seeing the universe shortly after the moment of creation. We can see no farther.

Review Questions

1. Distinguish between *spiral galaxies* and *elliptical galaxies* in terms of their shapes and colors. What are *irregular galaxies?*
2. Distinguish between the disk component and the spheroidal component of a spiral galaxy. Which component includes the galaxy's spiral arms? Which includes its bulge? Which includes its halo?
3. How does a *barred spiral galaxy* differ from a normal spiral galaxy? How does a *lenticular galaxy* differ from a normal spiral galaxy?
4. What is a *group* of galaxies? What is a *cluster* of galaxies?
5. What is the major difference between an elliptical galaxy and a spiral galaxy? How is an elliptical galaxy similar to the halo of the Milky Way?
6. Among large galaxies in the universe, about what proportion are spiral or lenticular? Elliptical? What do we mean by a *dwarf elliptical galaxy?*
7. Distinguish between disk-population stars and spheroidal-population stars, in terms of their ages, masses, colors, and orbits. Briefly describe how the gravitational contraction of a *protogalactic cloud* explains the different characteristics of these populations.
8. What do we mean by a *standard candle?* Briefly explain how, once we identify an object as a standard candle, we can use the luminosity–distance formula to find its distance.
9. Explain how, starting with measurements of stars in the Hyades Cluster, astronomers built up a catalog of the true luminosities of all types of main-sequence stars. Why is the Hyades Cluster so important to this process?
10. Describe the technique of main-sequence fitting. Why should we expect to get a better estimate of the distance to a star cluster through the technique of main-sequence fitting than by looking at only a single main-sequence star?
11. Why are Cepheid variables good standard candles? Briefly explain how Edwin Hubble used his discovery of a Cepheid in Andromeda to prove that the "spiral nebulae" were actually entire galaxies.
12. What is *Hubble's law?* How did Hubble discover it? Explain the meaning of Hubble's law when it is expressed mathematically as $v = H_0 \times d$. What are the units of *Hubble's constant,* H_0?
13. Describe how we can use Hubble's law to determine the distance to a distant galaxy. What practical difficulties limit the use of Hubble's law for measuring distances?
14. What makes *white-dwarf supernovae* good standard candles? Briefly describe how we have learned the true luminosities of white-dwarf supernovae.
15. What is the *Tully–Fisher relation?* How does it enable us to use entire spiral galaxies as standard candles?

16. What is our current best estimate of Hubble's constant? Summarize all the links in the distance chain that allow us to estimate distances to the farthest reaches of the universe.

17. In what ways is the surface of an expanding balloon a good analogy to the universe? In what ways is this analogy limited? Explain why a miniature scientist living in a polka dot on the balloon would observe all other dots to be moving away, with more distant dots moving away faster.

18. What do we mean by the *critical density* of the universe? Explain how the fate of the universe depends on whether the actual density is less than, equal to, or more than the critical density.

19. Based on current estimates of the value of Hubble's constant, how old is the universe if the actual density is much less than the critical density? How old is it if the density is equal to the critical density?

20. What is a *lookback time?* Why does it make more sense to talk about a distant object in terms of its lookback time than in terms of its distance?

21. What do we mean by a *cosmological redshift?* How does our interpretation of a distant galaxy's redshift differ if we think of it as a cosmological redshift rather than a Doppler shift?

22. What is the *cosmological horizon?* Why can't we see past it?

Discussion Questions

1. *Cosmology and Philosophy.* One hundred years ago, many scientists believed that the universe was infinite and eternal, with no beginning and no end. When Einstein first developed his general theory of relativity, he found it predicted that the universe should be either expanding or contracting. He believed so strongly in an eternal and unchanging universe that he modified the theory, a modification he would later call his "greatest blunder." Why do you think Einstein and others assumed that the universe had no beginning? Do you think that a universe with a definite beginning in time, some 15 billion or so years ago, has any important philosophical implications? Explain.

2. *Tired Light.* Hubble's law relates the redshifts of galaxies directly to their distances. The overwhelming majority of astronomers believe that these redshifts arise from the Doppler effect, an easily measurable and well-understood phenomenon that can be measured in the laboratory. However, a few astronomers argue that there might be some other explanation for galaxy redshifts based on physical effects that occur only on vast scales. For example, perhaps photons somehow get "tired" and lose energy as they cross immense intergalactic distances. Which explanation for redshifts do you prefer? Why is considering alternative explanations important in science?

Problems

1. *Classifying Galaxies.* Figure 19.1 shows a wide variety of galaxies. Devise your own scheme for classifying them by color, and describe it. Estimate the fraction of galaxies that falls into each color class. Devise and describe your own scheme for classifying galaxies by shape. Estimate the fraction that falls into each shape class. Describe any relationships you see between the shapes and colors of these galaxies.

2. *Counting Galaxies.* Estimate how many galaxies are pictured in Figure 19.1. Explain the method you used to arrive at this estimate. This picture shows about 1/30,000,000 of the sky, so multiply your estimate by 30,000,000 to obtain an estimate of how many galaxies like these fill the entire sky.

3. *Distance Measurements.* The techniques astronomers use to measure distances are not so different from the ones you use every day. Describe how parallax measurements are similar to your visual depth perception. Describe how standard-candle measurements are similar to the way you estimate the distance to an oncoming car at night.

4. *Standard Candles.* List at least three qualities that would tend to make a type of astronomical object useful as a standard candle. Explain how each quality is important if we hope to use objects of this type to measure distances in the universe.

5. *Cepheids as Standard Candles.* Suppose you are observing Cepheids in a nearby galaxy. You observe one Cepheid with a period of 8 days between peaks in brightness, and another with a period of 35 days. Estimate the luminosity of each star. Explain how you arrived at your estimate. (*Hint:* See Figure 19.13.)

6. *Cepheids in M 100.* Scientists using the Hubble Space Telescope have observed Cepheids in the galaxy M 100. Here are actual data for three Cepheids in M 100:
   - Cepheid 1: luminosity = 3.9×10^{30} watts, brightness = 9.3×10^{-19} watt/m^2.
   - Cepheid 2: luminosity = 1.2×10^{30} watts, brightness = 3.8×10^{-19} watt/m^2.
   - Cepheid 3: luminosity = 2.5×10^{30} watts, brightness = 8.7×10^{-19} watt/m^2.

Compute the distance to M 100 with data from each of the three Cepheids. Do all three distance computations agree? Based on your results, estimate the uncertainty in the distance you have found.

7. *Redshift and Hubble's Law.* Imagine that you have obtained spectra for several galaxies and have measured the observed wavelength of a hydrogen emission line that has a rest wavelength of 656.3 nm. Here are your results:
   - Galaxy 1: Observed wavelength of hydrogen line is 659.6 nm.
   - Galaxy 2: Observed wavelength of hydrogen line is 664.7 nm.
   - Galaxy 3: Observed wavelength of hydrogen line is 679.2 nm.

   a. Calculate the redshift, z, for each of the three galaxies.

   b. From their redshifts, calculate the speed at which each of the galaxies is moving away from us. Give your answers both in km/s and as a fraction of the speed of light.

   c. Estimate the distance to each galaxy from Hubble's law. Assume that $H_0 = 65$ km/s/Mpc.

8. *Age from Hubble's Constant.* In Mathematical Insight 19.3, we found that a value for Hubble's constant of $H_0 = 65$ km/s/Mpc implies an age for the universe of 15 billion years if the expansion rate has never changed, or 10 billion years if the actual density of the universe is equal to the critical density.

   a. Suppose $H_0 = 80$ km/s/Mpc. How old is the universe if the expansion rate has never changed? How old is it if the actual density is equal to the critical density?

   b. Repeat part (a) for the case where $H_0 = 50$ km/s/Mpc.

   c. Based on your answers to parts (a) and (b), what can you say about the range of possible ages for the universe? Explain.

9. *Lookback Time.* You have detected a supernova in a galaxy whose spectral lines are redshifted by an amount $z = 0.8$. Use the formulas given in Mathematical Insight 19.4 to answer the following questions.

   a. How long ago did the supernova explode, assuming we live in a low-density universe with $H_0 = 65$ km/s/Mpc?

   b. How long ago did the supernova explode if we live in a critical-density universe with $H_0 = 65$ km/s/Mpc?

   c. Repeat parts (a) and (b), but this time for a supernova in a galaxy with redshift $z = 2.5$.

   d. (*Challenge*) How do your answers to parts (a) through (c) change if $H_0 = 75$ km/s/Mpc, rather than 65 km/s/Mpc? Explain, without actually doing the calculations.

10. *Web Project: Galaxy Gallery.* Many fine images of galaxies are available on the web. Collect several images of each major type and build a galaxy gallery of your own. Supply a descriptive paragraph about each galaxy.

11. *Research Project: Hubble's Constant.* Measurements of Hubble's constant are rapidly growing more reliable. Find at least two measurements of H_0 determined in the past 2 years. Describe how these values were measured. What is the quoted uncertainty of each? By how much do they differ?

20

Galaxy Evolution

THE SPECTACLE OF GALAXIES STREWN LIKE BEAUTIFUL ISLANDS across the universe invites us to ponder their origins. Some galaxies have majestic spiral arms, some are elliptical, and some are irregular. What processes sculpted them into their present shapes?

If we look closely, we see even more spectacular sights. A few nearby galaxies are engaged in titanic collisions with other galaxies. Many of these colliding galaxies are full of star-forming clouds in which new stars are born and massive stars explode 100 times more frequently than in the Milky Way. Other galaxies appear to harbor enormous black holes surrounded by gigantic accretion disks that generate extraordinary luminosities. Narrow streams of matter jet from a few of these galaxies into intergalactic space at nearly the speed of light.

The origins of galaxies and the incredible phenomena they sometimes display puzzled astronomers for much of the twentieth century, but today we are assembling a rudimentary understanding of how galaxies evolve. In this chapter, we will sift through the fascinating clues that hint at how galaxies formed and developed, pausing now and again to admire the fantastic spectacles that this search for our origins has uncovered.

20.1 The Lives of Galaxies: An Incomplete Story

We would like to tell the story of a galaxy's life from beginning to end as completely as we told the life stories of stars (Chapter 16), but we cannot. Too many aspects of **galaxy evolution**—the formation and development of galaxies—remain mysterious. Galaxies are more challenging to study than stars both because they are far more complex and because most of them are so distant. Nevertheless, galaxy evolution is one of the most active research areas in astronomy, and each year brings new discoveries. Perhaps someday soon the story of galaxy evolution will be complete. For now we will have to discuss it in bits and pieces.

In broad terms, the story we seek should tell us how the hydrogen and helium gas in the early universe evolved into the galaxies we see today. As we'll discuss, even this general story still presents some puzzles, such as why some galaxies turned out to be spirals with gas-rich disks while others became gas-poor elliptical galaxies [Section 19.1]. But we also need to look beyond ordinary spiral and elliptical galaxies to assemble a complete story of galaxy evolution.

For example, observations show that some galaxies, called **starburst galaxies,** are currently forming stars at a tremendous rate. Such starbursts must be only a temporary phase in the lives of these galaxies, because otherwise they would have consumed all the galaxy's gas long ago. It is possible that many galaxies experience an episode of intense star formation at some point during their lives, and we will see that the effects of these starbursts on galaxy evolution can be profound.

A few galaxies show even more spectacular phenomena, such as extremely bright centers known as **active galactic nuclei** that sometimes outshine all the stars in the galaxy combined.[1] In some cases, these active galactic nuclei appear to be firing powerful jets of material millions of light-years into space. We now believe that the "engines" for active galactic nuclei are enormous black holes and that their tremendous power comes from gigantic accretion disks formed by gas spilling into the black holes.

Although we see active galactic nuclei in a few nearby galaxies, most are in much more distant galaxies. Because more distant galaxies are also those with greater lookback times [Section 19.4], we can conclude that active galactic nuclei were more common in the past than they are today. Indeed, the most luminous active galactic nuclei, which we call **quasars,** are generally also the most distant. This fact suggests that quasars represent an energetic, youthful phase in the lives of some galaxies and that nearby active galactic nuclei are rarer cases in which mature galactic centers show something akin to this youthful behavior.

Like starbursts, quasars and other active galactic nuclei appear to be only temporary phases in the lives of galaxies. If these phenomena really are powered by gigantic accretion disks around supermassive black holes, it may be that they die out once the black hole consumes most of the available gas near the galactic center. Quasars may be most common in young galaxies because the gas that feeds them runs out as the galaxies age. The black holes themselves do not go away, even though their accretion disks fade. It is quite likely that the centers of many, and perhaps most, large galaxies once shone brightly as quasars and then faded with time. Clearly, quasars and other active galactic nuclei are part of the puzzle of galaxy evolution, but their precise role in shaping the lives of galaxies remains mysterious.

In the rest of this chapter, we will discuss what we currently know about the lives of galaxies. We will begin by discussing the techniques astronomers use to study galaxy evolution and next examine the general question of why most galaxies end up either as spirals or as ellipticals. Then we will look in detail at the phenomena of starburst galaxies and active galactic nuclei and attempt to fit these dramatic stages into the overall puzzle of galaxy evolution. Finally, we will explore how very distant quasars are helping astronomers study galaxy evolution in a completely different way. The light from these distant quasars traverses billions of light-years on its way to Earth, and their spectra bear the signatures of many gaseous clouds that lie between the quasars and Earth. Some of these clouds probably represent galaxies in their most youthful state. Thus, the study of quasar spectra not only may teach us about some early phases in the evolution of galaxies, but also may soon help us understand the complete story of galaxy evolution.

20.2 Investigating Galaxy Evolution

We pieced together the life stories of stars by observing relatively nearby stars of many different ages, but we cannot do the same for galaxies. Unlike stars, galaxies do not continuously form and die. Most galaxies formed many billions of years ago, and those that we see nearby are now well into adulthood. Thus, we cannot simply take pictures of galaxies in our vicinity and assemble them into a birth-to-death sequence because these galaxies are all roughly the same age. Instead, we must try to weave several different lines of less direct evidence into a single coherent story

[1]Galaxies with active galactic nuclei are often called *active galaxies*.

of galaxy evolution. In this section, we investigate three important techniques that are useful for studying galaxy evolution: detailed studies of the Milky Way, observations of distant galaxies, and theoretical modeling of galaxy formation.

Clues from the Milky Way

A good starting point for studying how galaxies form and evolve is to search our own Milky Way Galaxy for clues to its past. Our study of the Milky Way in Chapter 18, which we extended to other galaxies in Chapter 19, suggested the following basic picture of galaxy evolution:

- A galaxy begins as a blobby *protogalactic cloud* of hydrogen and helium gas.
- Gravity causes the protogalactic cloud to collapse into a young galaxy. Meanwhile, smaller clouds of gas within the forming galaxy collapse to form stars.
- The *star–gas–star cycle* shapes the galaxy's interstellar medium, as gravity collects interstellar gas into new stars and other stars explode or drive winds that recycle matter back into interstellar space. Over time, this recycling gradually increases the galaxy's small fraction of heavy elements (elements besides hydrogen and helium).

This general sequence of events probably applies to all galaxies, but the details of each individual galaxy's life story certainly differ. Clearly, the histories of spiral and elliptical galaxies must differ substantially for these two types to look so different today. To understand more about how the differences between galaxies arose, we need to look far beyond the Milky Way.

A Family Album of Galaxies

Modern telescopes allow us to see galaxies at many different stages of development. Remember that the farther out we look into space the further back we look into the youth of the universe. When we look at a galaxy a billion light-years away, we are seeing it at a time when the universe was a billion years younger than it is today. When we look at a galaxy halfway to the cosmological horizon [Section 19.4], we see it as it was when the universe was half its present age. Even closer to the cosmological horizon, we see galaxies as they were within the first few billion years after the Big Bang.

Capitalizing on this ability to see back through time, astronomers are beginning to assemble "family albums" showing galaxies of many different ages (Figure 20.1). The photos in such an album are pictures of galaxies ranging in distance from a few million light-years to billions of light-years. Each picture shows one galaxy at a single stage in its life, so we can arrange these pictures to make a family album for galaxies of particular types. Pictures of the farthest galaxies show galaxies in their childhood, and pictures of the nearest show mature galaxies as they are today.

TIME OUT TO THINK *You may notice that we use the term "today" in a very broad sense. For example, if we look at a relatively nearby galaxy—one located, say, 20 million light-years away—we see it as it was 20 million years ago. In what sense is this "today"?* (Hint: *How does 20 million years compare to the age of the universe?)*

Our current instruments are not quite sensitive enough to reveal the whole history of galaxy evolution. The galaxies pictured in Figure 20.1 extend back to fairly young ages, but not all the way back to the infancy of galaxies. Because most galaxies formed long ago, we expect to find protogalactic clouds primarily at great distances from us, where we are seeing furthest into the past. The great distances of protogalactic clouds, along with the fact that they contain few if any stars, make them very difficult to detect. Young galaxies do have stars, but they still lie at extremely large distances from us, so they are also challenging to observe. The best images of young galaxies currently come from the Hubble Space Telescope, but even its superior resolving power and sensitivity cannot reveal all the details we would like to see.

Modeling Galaxy Birth

Because our telescopes cannot yet see back to the time when galaxies formed their first stars, we must use theoretical modeling to study the earliest stages in galaxy evolution. The most successful models for galaxy formation presume the following:

- Hydrogen and helium gas filled all of space more or less uniformly when the universe was very young—say, in the first million years after its birth.
- This uniformity was not quite perfect, and certain regions of the universe were ever so slightly denser than others.

Beginning from these assumptions, which are supported by a mounting body of evidence (discussed in Chapter 22), we can model galaxy formation using well-established laws of physics. The models show that the regions of enhanced density originally expanded along with the rest of the universe. However, the slightly greater pull of gravity in these regions

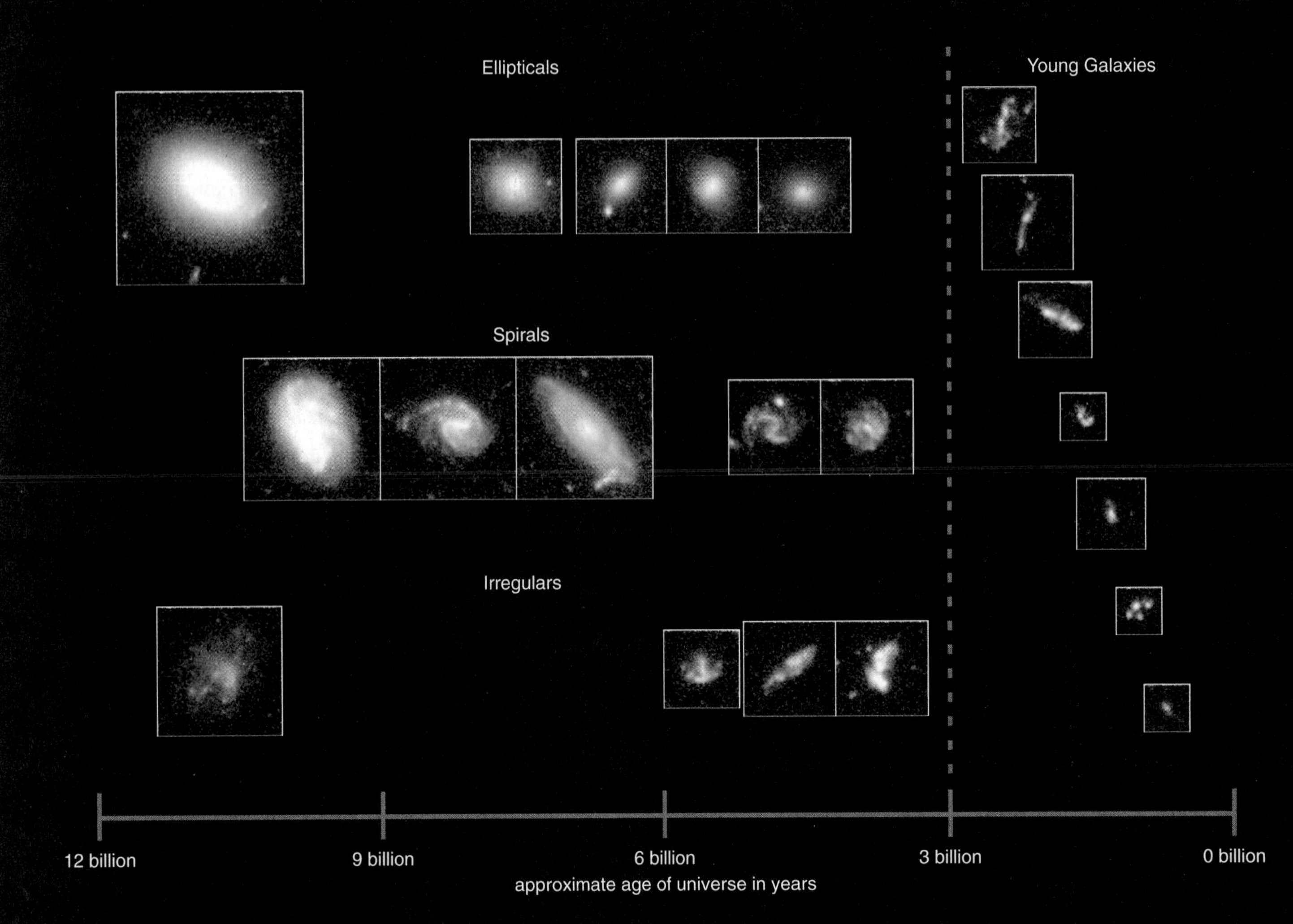

FIGURE 20.1 Family albums for spiral, elliptical, and irregular galaxies at different ages along with some very young, distant galaxies. (Younger galaxies appear smaller because they are more distant.)

gradually slowed their expansion. Within about a billion years, the expansion of these denser regions halted and reversed, and the material within them began to contract into protogalactic clouds.

Stars must have begun to form almost immediately, but current models suggest that explosions of the first stars slowed the rate of star formation shortly thereafter. Protogalactic clouds initially cooled as they contracted, radiating away their thermal energy, and the first generation of stars grew from the densest, coldest clumps of gas. The most massive of these stars lived and died within just a few million years, a short time compared to the time required for the collapse of a protogalactic cloud into a mature galaxy. The supernovae of these massive stars generated shock waves that heated the surrounding interstellar gas, slowing the collapse of young galaxies and the rate at which new stars formed within them.

On the whole, the idea that galaxies grow from contracting protogalactic clouds and form their stars over billions of years is consistent with what we know from studying the Milky Way and from our limited observations of very young galaxies. However, the lives of galaxies must become more complicated as they develop, for reasons we will now discuss.

20.3 Why Do Galaxies Differ?

Why do different people have different personalities? Some people argue that the primary cause is *nature*—that personality is ingrained in our genes. Others argue that the primary cause is *nurture*—that the experiences of our lives are more important. Astronomers have long pondered a similar "nature versus nurture" question about galaxies: Why do galaxies come in such a great variety of shapes and sizes? On the "nature" side, we might claim that a galaxy's current state depends only on the properties with which it was born. On the "nurture" side, we might claim that all galaxies are similar at birth but end up different because of environmental influences, such as collisions with other galaxies. In the billions of years since the initial collapse of protogalactic clouds, galaxies have had plenty of time to interact with their neighbors.

We now believe that both initial conditions and interactions with other galaxies affected the development of all types of galaxies. To help differentiate their roles, we consider a question about galaxy evolution that is still unanswered: Why do spiral galaxies have gas-rich disks while elliptical galaxies do not?

Initial Collapse

Two potential explanations for the differences between spirals and ellipticals fall into the category of "nature." That is, these explanations trace a galaxy's current appearance back to the properties of the protogalactic cloud from which it formed. The first explanation concerns angular momentum differences among protogalactic clouds (Figure 20.2). If a protogalactic cloud has enough angular momentum, it

FIGURE 20.2 One possible reason for the difference between spiral and elliptical galaxies may be their original angular momentum. The cloud on the left side of this artist's conception begins with little angular momentum and becomes an elliptical galaxy because it hardly rotates as it collapses. The cloud on the right begins with considerable angular momentum and becomes a spiral galaxy because it rotates faster and flattens as it collapses.

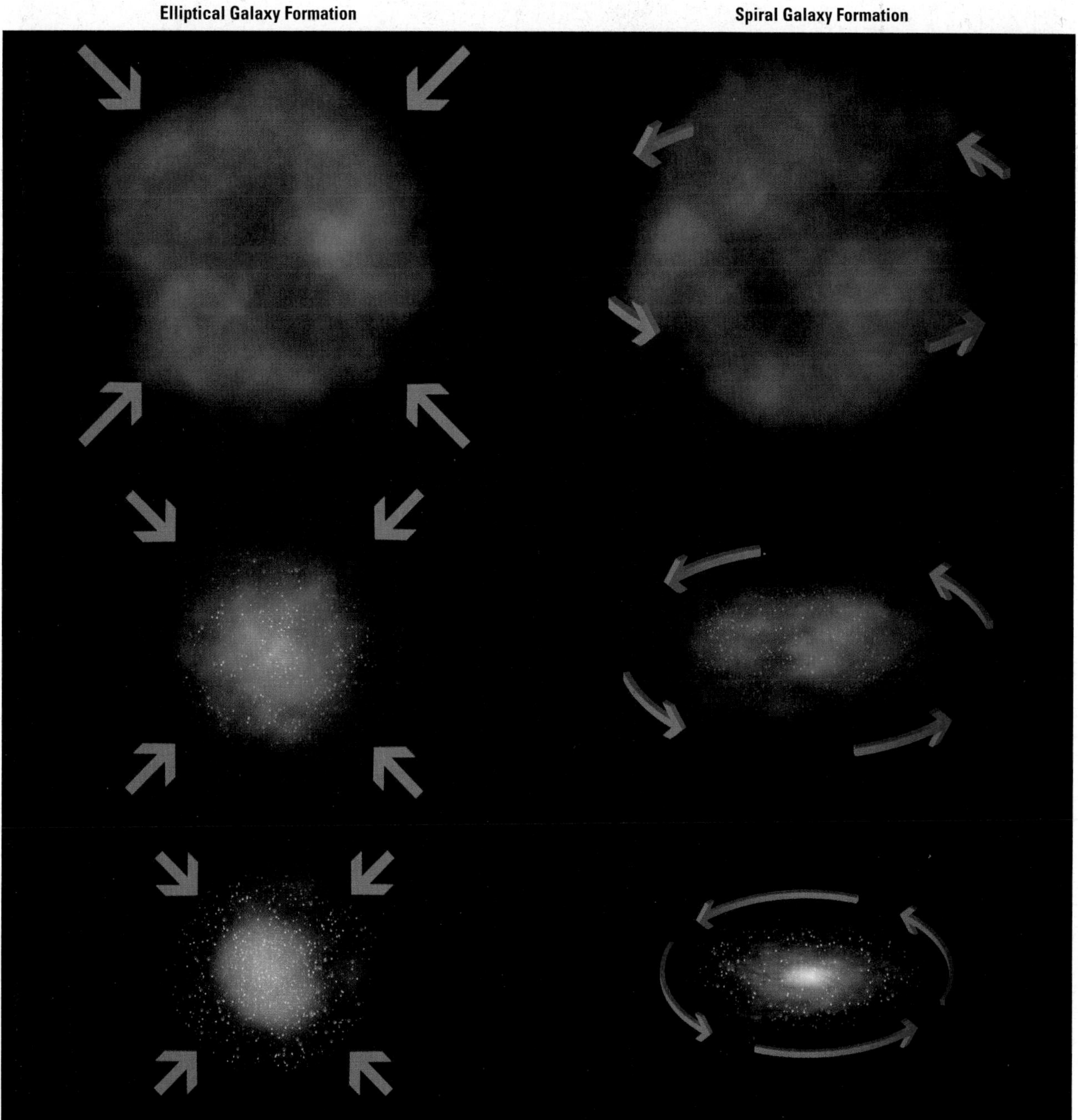

will rotate quickly as it collapses, and the galaxy it produces will tend to form a disk [Section 19.1]. The resulting galaxy would therefore be a spiral galaxy. If a protogalactic cloud has little or no net angular momentum, its gas might not form a disk at all, and the resulting galaxy would be elliptical.

The second "nature" scenario hinges on the ability of the protogalactic gas to cool and form stars (Figure 20.3). Remember that elliptical galaxies are most common in clusters of galaxies. The density of matter in these clusters was probably higher than elsewhere in the universe when galaxies first formed. Thus, the preponderance of elliptical galaxies in clusters suggests that elliptical galaxies formed from protogalactic clouds with higher densities than the clouds from which spirals formed. The higher gas

FIGURE 20.3 Another possible reason for the difference between spiral and elliptical galaxies may be their initial rate of star formation. The cloud on the left side of this artist's conception creates stars more rapidly than it collapses, becoming an elliptical galaxy because its gas is used up before it can flatten into a disk. The cloud on the right forms stars more slowly, becoming a spiral galaxy because it flattens before it forms most of its stars.

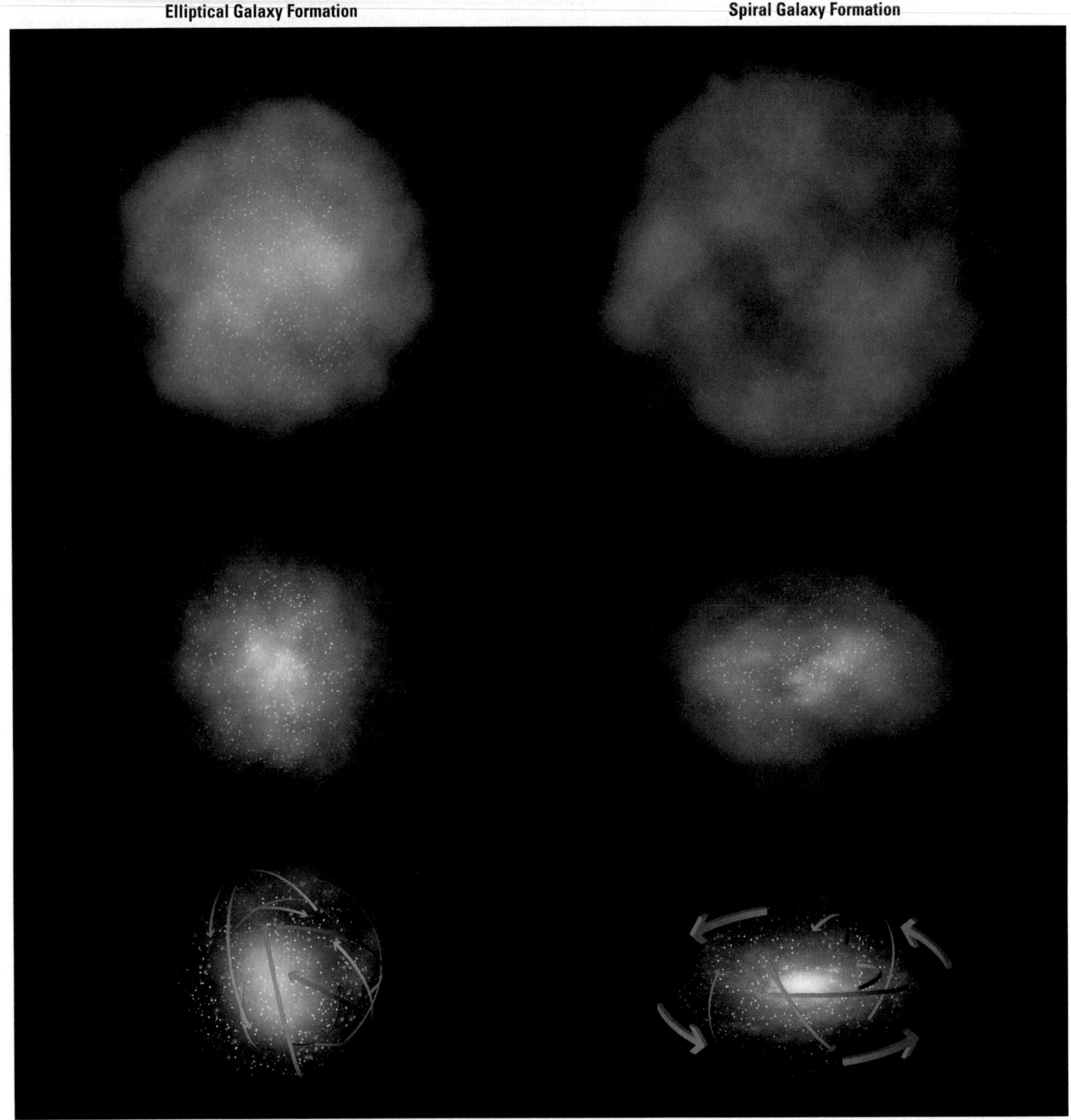

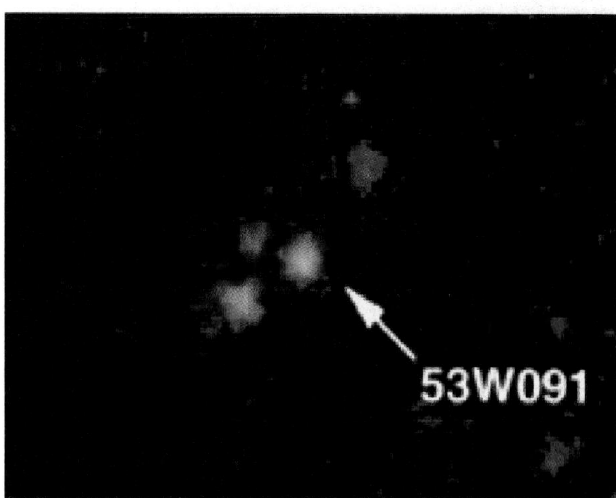

FIGURE 20.4 The redness of this distant elliptical galaxy, LBDS 53W091, indicates that its stars all formed during the first billion years or so after the birth of the universe. Its two red companions to the left are probably also elliptical galaxies with similar star-formation histories.

densities in these proto-elliptical clouds would have enabled them to radiate energy more efficiently and to cool more quickly. This cooling would have allowed gravity to collapse clumps of gas into stars more quickly, using up all the gas before the cloud had time to collapse into a disk. The resulting galaxy would look elliptical. In contrast, lower-density, proto-spiral clouds would have a slower rate of star formation, leaving plenty of gas to form a disk as the cloud collapsed. The resulting spiral galaxy would have a gas-filled disk in which the star–gas–star cycle could maintain ongoing star formation.

Some evidence for this latter scenario comes from a few giant elliptical galaxies at very large distances (Figure 20.4). These galaxies look very red even after we have accounted for their large redshifts. They apparently have no blue or white stars at all, indicating that new stars no longer form within these galaxies—even though we are seeing them as they were when the universe was only a few billion years old. This finding supports the idea that all the stars in these galaxies formed almost simultaneously, leaving no time for a disk to develop.

Subsequent Mergers

The two "nature" scenarios, in which the formation of a gas-rich disk depends on the angular momentum or density of the protogalactic cloud, probably describe important parts of the overall story. However, they ignore "nurture" effects: Galaxies rarely evolve in perfect isolation.

Recall that, in our scale-model solar system in Chapter 2, we made the Sun the size of a grapefruit. On this scale, the nearest star was like another grapefruit a few thousand kilometers away. Because the average distances between stars are so huge compared to the sizes of stars, direct star–star collisions are extremely rare. However, if we rescale the universe so that our entire *galaxy* is the size of a grapefruit, the Andromeda Galaxy is like another grapefruit only about a meter away, and a few smaller galaxies lie considerably closer. Thus, the average distances between galaxies are not tremendously larger than the sizes of galaxies, and collisions between galaxies are inevitable.

Such collisions are spectacular events, but they unfold over hundreds of millions of years. In our short lifetime, we can at best see an instantaneous snapshot of a collision in progress, as evidenced by the distorted shape of the colliding galaxies. Galactic collisions must have been even more frequent in the past, when the universe was smaller and galaxies were even closer together. Photographs of galaxies at a variety of distances confirm that distorted-looking galaxies—probably galaxy collisions in progress—were more common in the early universe than they are today (Figure 20.5).

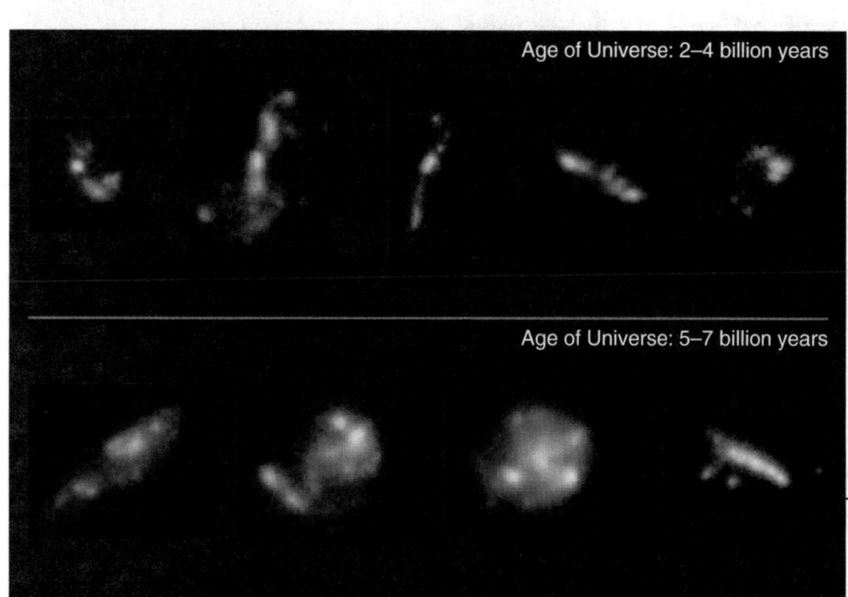

FIGURE 20.5 Hubble Space Telescope photos of distorted young galaxies.

We can learn much more about collisions between galaxies with the aid of computer models that simulate interactions between galaxies. Such models allow us to "watch" a collision that takes hundreds of millions of years in nature unfold in a few minutes on a computer screen.

These computer models show that a collision between two spiral galaxies can create an elliptical galaxy. (Look ahead to Figure 20.13.) Tremendous tidal forces between the colliding galaxies tear apart the two disks, randomizing the orbits of their stars. Meanwhile, a large fraction of their gas sinks to the center of the collision and rapidly forms new stars. Supernovae and stellar winds eventually blow away the rest of the gas. When the cataclysm finally settles down, the merger of the two spirals has produced a single elliptical galaxy; little gas is left for a disk, and the orbits of the stars have random orientations.

Observational evidence confirms the idea that at least some elliptical galaxies are the result of collisions and subsequent mergers. The fact that elliptical galaxies dominate the galaxy populations at the cores of dense clusters of galaxies, where collisions should be most frequent, may mean that some of the spirals once present became ellipticals through collisions. Stronger evidence comes from structural details of elliptical galaxies, which often attest to a violent past. Some elliptical galaxies have stars and gas clouds in their core that orbit differently from the other stars in the galaxy, suggesting that they are leftover pieces of galaxies that merged in a past collision. Other elliptical galaxies are surrounded by shells of stars that probably formed from stars stripped out of smaller galaxies that once strayed too close and were destroyed (Figure 20.6).

The most decisive evidence that collisions affect the evolution of some elliptical galaxies comes from observations of the **central dominant galaxies** found at the centers of many dense clusters. These giant elliptical galaxies apparently grew to huge sizes by consuming other galaxies through collisions. Central dominant galaxies frequently contain several tightly bound clumps of stars that probably once were the centers of individual galaxies later swallowed by the giant central galaxy (Figure 20.7). This process of *galactic cannibalism* can create galaxies over 10 times more massive than the Milky Way, making them the largest galaxies in the universe.

Clusters of galaxies might also turn a spiral galaxy into an elliptical galaxy simply by stripping the interstellar gas out of its disk. Hot plasma fills the centers of galaxy clusters [Section 21.3]. When a spiral galaxy cruises through the center of such a cluster, this plasma exerts drag forces that slow the galaxy's gas but not its stars. The galaxy's stars continue to move freely along their way, but the gas is left behind. If the disk of the spiral galaxy has not yet formed many stars, then the galaxy will evolve to look more like an elliptical galaxy as its massive stars die away. Its disk will fade, while its bulge and halo will remain prominent. In contrast, if the disk has already formed a large number of stars when its gas is stripped, the remaining galaxy is more likely to look *lenticular* in shape [Section 19.1].

More Clues from the Milky Way

The spiral galaxies that still exist must have avoided disruptive collisions that would have destroyed their

FIGURE 20.6 Shells of stars around the elliptical galaxy NGC 3923.

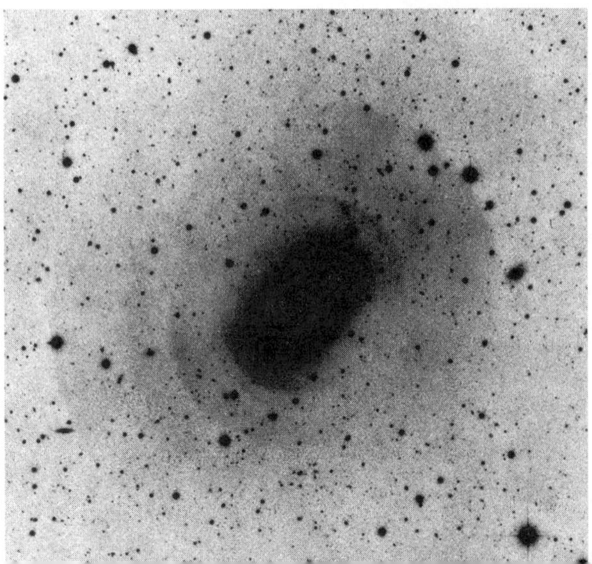

FIGURE 20.7 The central galaxy of the cluster Abell 3827, shown here in false color, has multiple clumps of stars that probably once were the centers of individual galaxies.

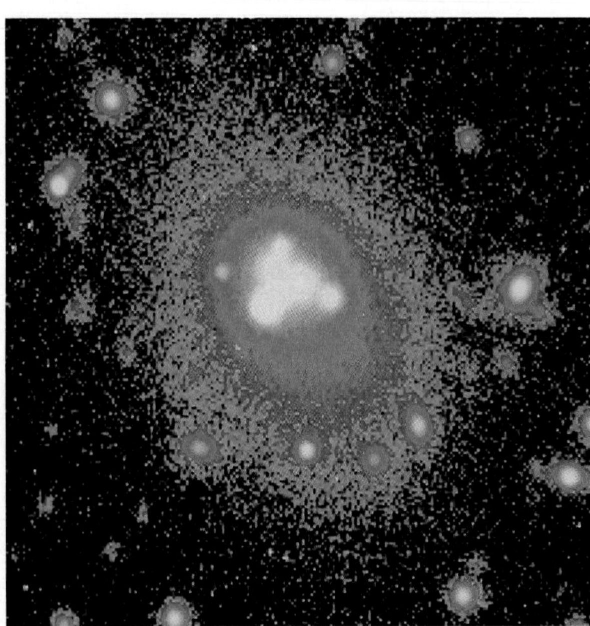

disks.[2] Otherwise, their disks would not be so thin and their rotation would not be so orderly. Yet the disorderly motions of the halo stars in spiral galaxies suggest that collisions may have disturbed the galaxies in the distant past, before their disks were fully formed. Because most spiral galaxies in today's universe look similar to our own Milky Way Galaxy, we can test our theories about the formation and evolution of spiral galaxies by studying the "fossil record" written within the stars of the Milky Way.

All available evidence confirms that the stars in the Milky Way's halo are indeed old. The main-sequence turnoff points in H–R diagrams of globular clusters show that their stars were born some 12–16 billion years ago [Section 15.6]. Individual halo stars (i.e., those not in globular clusters) and at least some of the bulge stars appear similarly old. Furthermore, the proportions of heavy elements in halo stars are much lower than those in the Sun, indicating that they formed before many generations of supernovae had a chance to enrich the Milky Way's interstellar medium [Section 18.2].

Variations in the heavy-element content of halo stars suggest that our galaxy formed from a few different gas clouds. If our Milky Way had formed from a single protogalactic cloud, it would have steadily accumulated heavy elements during its inward collapse as stars formed and exploded within it. In that case, the outermost stars in the halo would be the oldest and the most deficient in heavy elements, and the innermost stars would be the youngest and the richest in heavy elements. Halo stars in globular clusters do indeed differ in age and heavy-element content, but these variations do not seem to depend on the stars' distance from the galactic center. The easiest way to account for these variations is to suppose that the Milky Way's earliest stars formed in relatively small protogalactic clouds, each with a few globular clusters, and that these clouds later collided and combined to create the full protogalactic cloud that became the Milky Way (Figure 20.8).

[2]Collisions with small galaxies do not disrupt the disks of large spirals. On the far side of our galaxy, behind the galactic bulge, a dwarf galaxy in the constellation Sagittarius is currently passing directly through the Milky Way's disk without significantly disrupting it.

FIGURE 20.8 Schematic illustration of how the Milky Way's halo may have formed. The characteristics of stars in the Milky Way's halo suggest that several smaller gas clouds, already bearing some stars and globular clusters (top), may have merged to form the Milky Way's protogalactic cloud (bottom). These stars and star clusters then remained in the halo while the gas settled into the Milky Way's disk.

TIME OUT TO THINK *If the preceding scenario is true, it means that the Milky Way suffered several collisions early in its history but has not collided with a galaxy of comparable size since then. Explain why we should not be surprised that most collisions occurred in the distant past.* (Hint: *How did the average separations of galaxies in the past compare to their average separations today?*)

Once the full protogalactic cloud was in place, its collapse and heavy-element enrichment continued in a more orderly fashion. One result was a layer of stars intermediate between the disk and the halo. The heavy-element contents of stars and globular clusters in this intermediate layer *do* depend on their distances from the galactic center. These stars are roughly as old as halo stars, but they formed before the spinning protogalactic cloud finished flattening into a disk. Their proportions of heavy elements suggest that, at the time the disk formed, the Milky Way contained about 10% as much material in the form of heavy elements as it does today. Once the disk formed, generations of stars lived and died in the star–gas–star cycle that gradually increased the abundance of heavy elements. Thus, the ages of stars in the Milky Way's disk range from newly born to 10 billion or more years old. New stars will continue to be born as long as enough gas remains in the disk.

Summary: The Roles of Nature and Nurture

We have identified two potential "nature" scenarios that may determine whether a galaxy is born as an elliptical or a spiral galaxy: One scenario suggests that elliptical galaxies form when protogalactic clouds have low angular momentum, and the other suggests that elliptical galaxies form when high density leads to rapid star formation. But we have also found that "nurture" effects, such as collisions between galaxies, can influence the lives of galaxies and very likely cause some spiral galaxies to become elliptical. Each of these mechanisms probably plays a role in galaxy evolution, and we do not yet know whether any one is more important than the others. Irregular galaxies, which make up only a small percentage of large galaxies, probably arise from a variety of different and unusual circumstances, but they probably are affected by the same "nature" and "nurture" processes. For example, some galaxies may appear irregular because they are in the midst of a collision.

All in all, we seem to be well on our way to understanding how and why protogalactic clouds evolve into spiral and elliptical galaxies, and why we should expect a few galaxies to have irregular appearances. Nevertheless, a few puzzles remain. For example, while the oldest halo stars have very low fractions of heavy elements, these fractions are not zero. Where did the heavy elements in these stars come from? They probably came from an even earlier generation of stars, but where are the remnants of this first generation today? An even more fundamental question relates to the formation of the protogalactic clouds themselves. We said that these clouds formed in regions of slight density enhancement in the early universe, but what caused these density enhancements? This last question is one of the major puzzles in astronomy, and we'll revisit it in Chapters 21 and 22.

20.4 Starburst Galaxies

Most of the galaxies in the present-day universe appear comfortably settled, steadily forming new stars at modest rates. The Milky Way itself produces an average of about one new star per year. At this rate, the Milky Way won't exhaust the interstellar gas in its disk until long after the Sun has died. *Starburst galaxies* are not so thrifty. Some form new stars at rates exceeding 100 stars per year, more than 100 times the star-formation rate in the Milky Way. These galaxies cannot possibly sustain such a torrid pace of star formation for long: At their current rates of star formation, they would consume *all* their interstellar gas in just a few hundred million years—a relatively short time compared to the 10 billion or more years since galaxies first formed. Starburst galaxies are thus passing through an important but temporary phase in the evolution of at least some galaxies.

Starburst galaxies are relatively new subjects of study. These voracious consumers of interstellar gas look peculiar at visible wavelengths because they are filled with star-forming molecular clouds, but these clouds conceal much of the action. Dust grains in the molecular clouds absorb most of the visible and ultraviolet radiation streaming from the galaxy's many young stars. This radiation heats the dust grains to much higher temperatures than normal, and they ultimately reemit all the absorbed visible and ultraviolet energy as infrared light, with a peak wavelength of around 60,000 nanometers (60 micrometers). Because photons at these wavelengths cannot penetrate the Earth's atmosphere, the intensity of star-forming activity in starburst galaxies did not become fully apparent until the 1983 launching of the *Infrared Astronomy Satellite (IRAS)*. Figure 20.9 shows an especially luminous starburst galaxy, along with its spectrum from radio through visible wavelengths. Note that the visible output of this galaxy is about 10 billion solar luminosities ($10^{10}\ L_{\text{Sun}}$), not very different from the total luminosity of the Milky Way. However, its infrared output is a trillion times that of our Sun ($10^{12}\ L_{\text{Sun}}$), making it 100 times brighter in infrared light than in visible light.

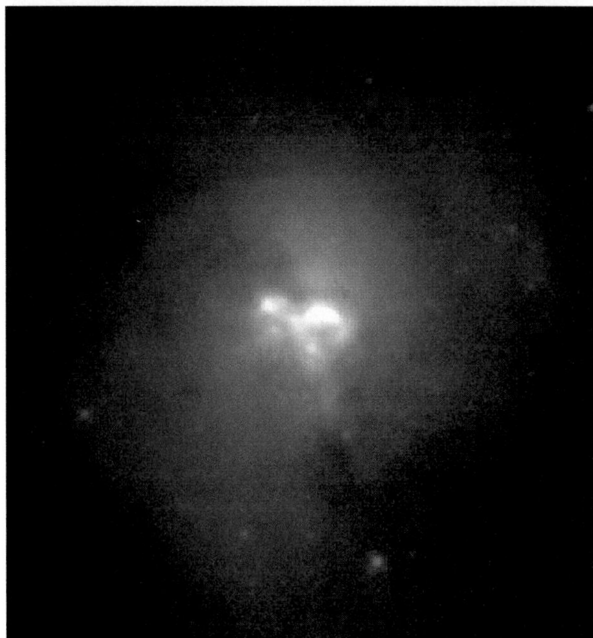

a

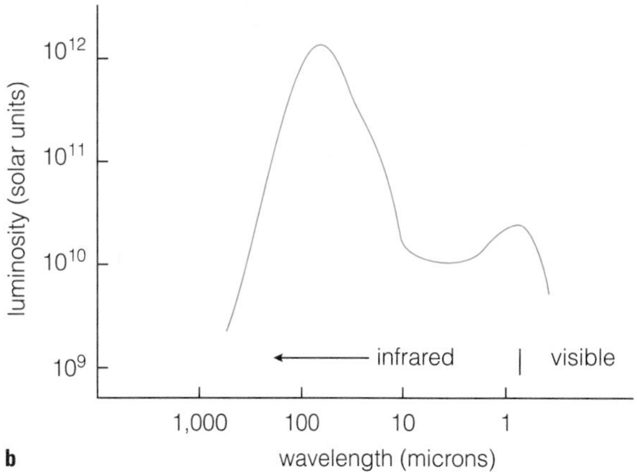

b

FIGURE 20.9 (**a**) An infrared photo of Arp 220, a large starburst galaxy. (**b**) The galaxy's spectrum shows that it emits most of its radiation as infrared light, rather than visible light.

Galactic Winds

A star-formation rate 100 times that of the Milky Way's also means an occurrence of supernovae at 100 times the Milky Way's rate. Just as in the Milky Way, each supernova in a starburst galaxy generates a shock wave that creates a *bubble* of hot gas [Section 18.2]. The shock waves from several nearby supernovae quickly overlap and blend into a much larger *superbubble*. The story usually ends here in the Milky Way, but in a starburst galaxy the drama is just beginning. Supernovae continue to explode inside the superbubble, adding to its thermal and kinetic energy. When the superbubble starts to break through the disrupted gaseous disk, it expands even faster. Hot gas erupts into intergalactic space, creating a **galactic wind.**

Galactic winds consist of low-density but extremely hot gas, typically with temperatures of 10–100 million Kelvin. They do not emit much visible light, but they do generate X rays. X-ray telescopes in orbit have detected pockets of X-ray emission surrounding the disks of some starburst galaxies, presumably coming from the outflowing galactic wind (Figure 20.10a). Sometimes we also see the glowing remnants of a punctured superbubble extending out into space (Figure 20.10b).

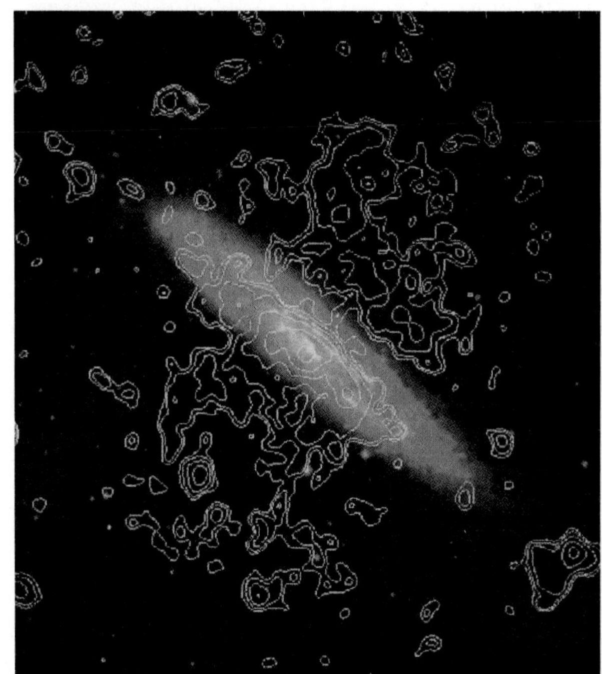

a

b

FIGURE 20.10 (**a**) This photographic image shows the visible light from the starburst galaxy NGC 253; the contour lines represent regions of strong X-ray emission. Note that the X rays appear on either side of the disk, suggesting that they come from hot gas blown out of the disk. (**b**) This photograph of the starburst galaxy M 82 shows violently disturbed gas (red) poking out from above and below the disk where hot gas is escaping.

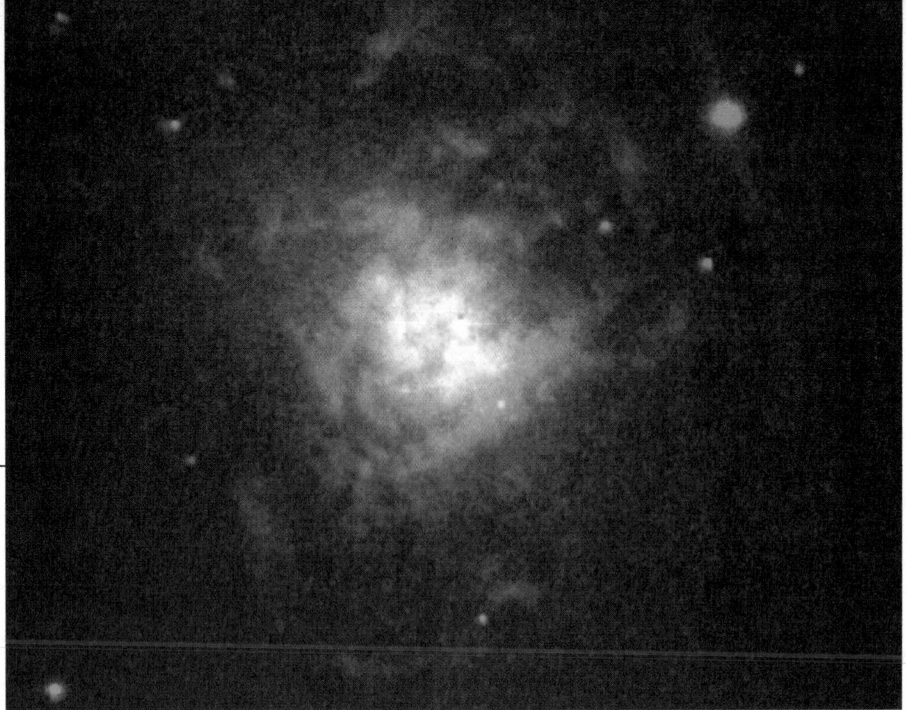

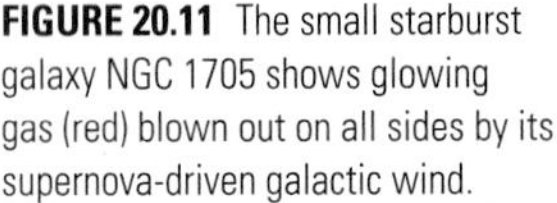

FIGURE 20.11 The small starburst galaxy NGC 1705 shows glowing gas (red) blown out on all sides by its supernova-driven galactic wind.

Supernova-driven galactic winds have an even more dramatic impact when they occur in small starburst galaxies (Figure 20.11). The winds can blow out of small galaxies on all sides, driving away virtually all their gas. As a result, star formation in these galaxies shuts down for billions of years. Some of the small elliptical galaxies in the Local Group apparently have burst like this at least twice during their lifetimes. In one example, about half the stars in the small galaxy were born 12–16 billion years ago, and the rest only 5–8 billion years ago. Presumably, each of these bursts of star formation ejected all of the galaxy's gas. Star formation was put on hold for several billion years until enough gas could reaccumulate within the galaxy for a new starburst to ignite.

TIME OUT TO THINK *Dwarf galaxies that have undergone bursts of star formation tend to have fewer heavy elements than large galaxies. Why do you think that is?* (Hint: *What happens to the heavy elements produced by the burst of star formation?*)

Collisions and Starbursts

Many of the most luminous starburst galaxies look violently disturbed (Figure 20.12). They are neither flat disks with symmetric spiral arms nor smoothly rounded balls of stars, like elliptical galaxies. Streamers of stars are strewn everywhere. Such galaxies are filled with dusty molecular clouds, and deep inside them we see two distinct clumps of stars that look like two different galactic centers. Their appearance suggests that such starbursts result from a collision between two gas-rich spiral galaxies. The collision apparently compresses the gas inside the colliding galaxies, leading to the burst of star formation.

Computer models of galaxy collisions convincingly reproduce the overall features of these starbursts (Figure 20.13, page 630). The simulations begin with two spiral galaxies on a collision course; the computer model then tracks the motions of stars and gas as the galaxies converge. As the galaxies approach each other, tidal forces [Section 6.5] start to tear off streamers of stars called *tidal tails*. Next, the gaseous disks begin to merge. The individual stars in the two disks don't interfere with one another (recall the huge distances between stars relative to their sizes), but the gas in the disks is greatly affected. The collision transfers angular momentum from the innermost gas clouds to the outer parts of the colliding galaxies. Because the inner clouds lose angular momentum, they drift closer to the center and merge. This central cloud grows so dense that the rate of star formation skyrockets, producing a tremendous starburst.

The computer models show that the starburst will die away over the course of a few hundred million years. The tidal tails slowly dissipate and fade, and the stars settle into the randomly oriented orbits characteristic of elliptical galaxies. What a sight it would be to watch colliding galaxies for another billion years or so, as they transform themselves into giant elliptical galaxies!

At least some large starbursts represent a short-lived phase in the life of galaxies that occurs when two spiral galaxies collide. The causes of smaller-scale starbursts are less clear: Some may result from

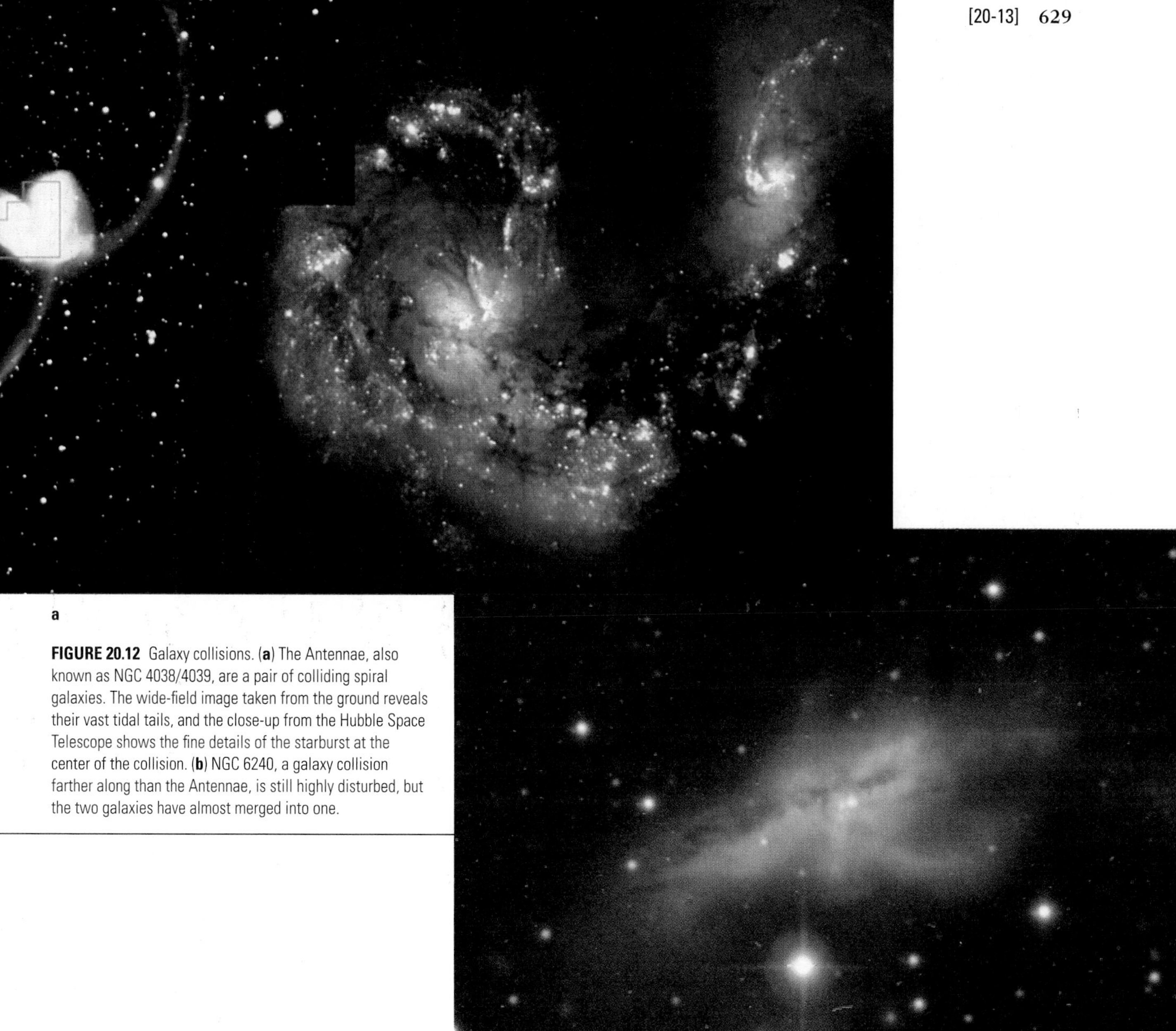

a

b

FIGURE 20.12 Galaxy collisions. (**a**) The Antennae, also known as NGC 4038/4039, are a pair of colliding spiral galaxies. The wide-field image taken from the ground reveals their vast tidal tails, and the close-up from the Hubble Space Telescope shows the fine details of the starburst at the center of the collision. (**b**) NGC 6240, a galaxy collision farther along than the Antennae, is still highly disturbed, but the two galaxies have almost merged into one.

direct collisions, while others may be caused by close encounters with other galaxies. The Large Magellanic Cloud is currently undergoing a period of rapid star formation, perhaps because of the tidal influence of the Milky Way. No matter what their exact causes turn out to be, starburst galaxies clearly represent an important piece in the overall puzzle of galaxy evolution.

20.5 Quasars and Other Active Galactic Nuclei

Starbursts may be spectacular, but some galaxies display even more incredible phenomena: extreme amounts of radiation, and sometimes powerful jets of material, emanating from deep in their centers. We generally refer to these unusually bright galactic centers as *active galactic nuclei* and reserve the term *quasar* for the very brightest of them. The brightest quasars shine more powerfully than 1,000 galaxies the size of the Milky Way.

The glory days of quasars are long past. The fact that we find quasars primarily at great distances tells us that these blazingly luminous objects arose during an early stage in the evolution of at least some galaxies. Because we find no nearby quasars and relatively few nearby galaxies with less luminous active galactic nuclei, we conclude that the objects that shine as quasars in young galaxies must become dormant as

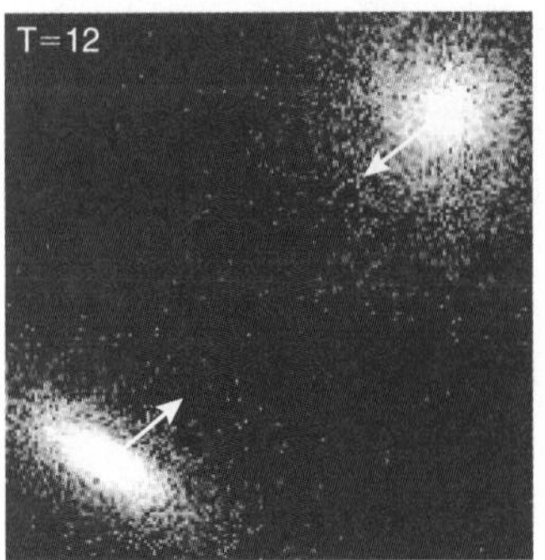

Two simulated spiral galaxies approach each other on a collision course.

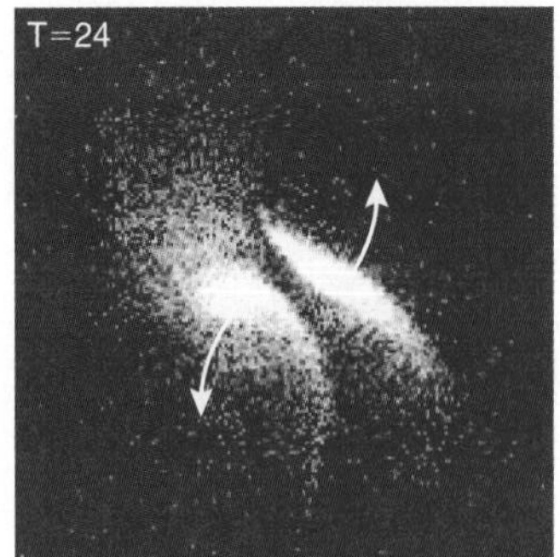

The first encounter disrupts the two galaxies and sends them into orbit around each other.

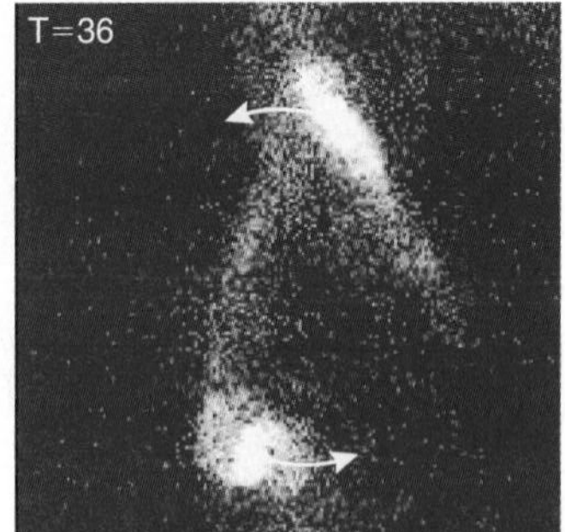

Gravitational forces between the galaxies tear out long streamers of stars.

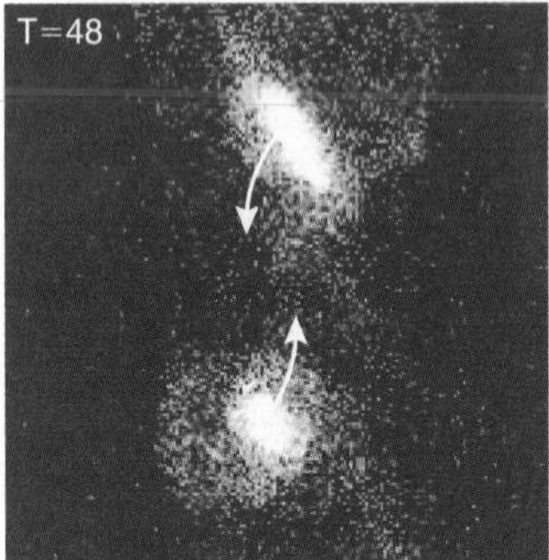

Because the first disruptive encounter saps kinetic energy from the galaxies, they cannot escape each other.

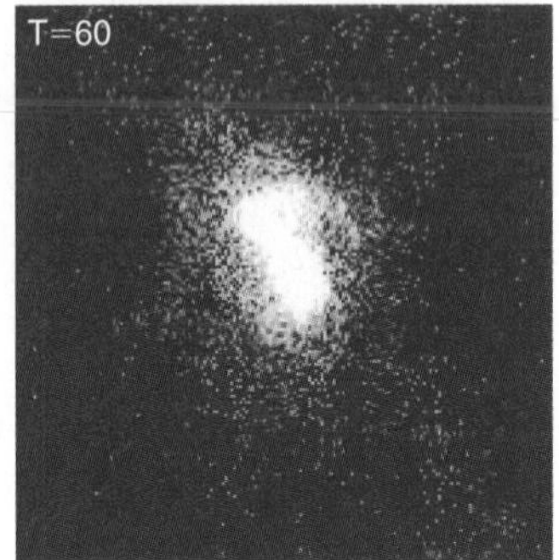

The second encounter is more direct than the first; the galaxies collide head-on and begin to merge.

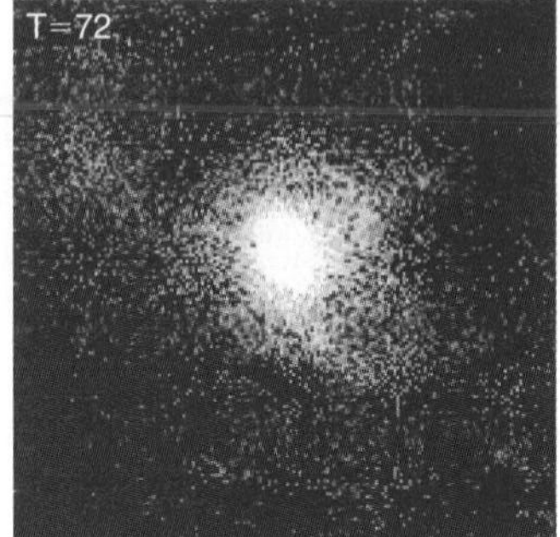

The single galaxy resulting from the collision and merger is an elliptical galaxy surrounded by debris.

FIGURE 20.13 Several stages in a supercomputer simulation of a collision between two spiral galaxies. The whole sequence spans about 1.5 billion years. (Each unit on the time counter "T=" corresponds to 25 million simulated years.)

the galaxies age. Thus, many nearby galaxies that now look quite normal—we don't yet know which ones—must have centers that once shone brilliantly as quasars.

What could possibly drive the incredible luminosities of quasars, and why did quasars fade away? A growing body of evidence points to a single answer: The energy output of quasars comes from gigantic accretion disks surrounding **supermassive black holes**—black holes with masses millions to billions of times that of our Sun. But before we study how these incredible powerhouses work, let's investigate the evidence that points to their existence.

The Discovery of Quasars

In the early 1960s, a young professor at the California Institute of Technology named Maarten Schmidt was busy identifying cosmic sources of radio-wave emission. Radio astronomers would tell him the coordinates of newly discovered radio sources, and he would try to match them with objects seen through visible-light telescopes. Usually the radio sources turned out to be normal-looking galaxies, but one day he discovered a major mystery. A radio source called 3C 273 looked like a blue star through a telescope but had strong emission lines at wavelengths that did not appear to correspond to any known chemical element.[3] After months of puzzlement, Schmidt suddenly realized that the emission lines were not coming from an unfamiliar element at all. Instead, they were emission lines of hydrogen hugely redshifted from their normal wavelengths (Figure 20.14). Schmidt calculated that the expansion of the universe was carrying 3C 273 away from us at 17% of the speed of light.

Schmidt computed the distance to 3C 273 using Hubble's law; then he plugged this distance into the

[3]The designation 3C 273 stands for 3rd Cambridge Radio Catalogue, object 273. A few other, similarly enigmatic radio sources were known at the time, but Schmidt's breakthrough came with 3C 273.

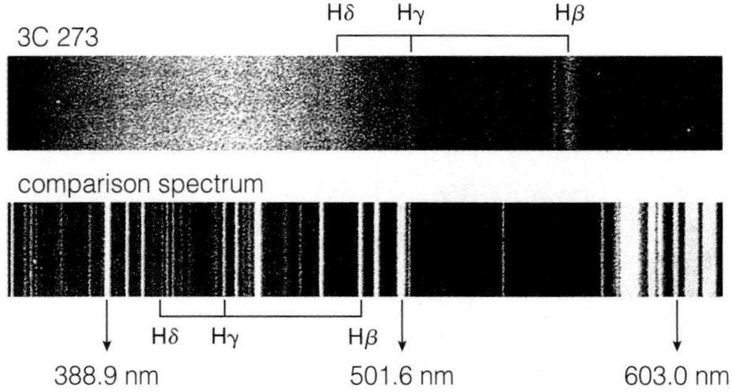

FIGURE 20.14 Spectrum of the quasar 3C 273. The lines labeled Hβ, Hγ, and Hδ are hydrogen emission lines. Note their significant redshift in the quasar spectrum relative to the "comparison spectrum" that shows them at their rest wavelengths.

luminosity–distance formula [Section 15.2]. What he found was astonishing: 3C 273 has a luminosity of about 10^{39} watts, or well over a trillion (10^{12}) times that of our Sun—making it hundreds of times more powerful than the entire Milky Way Galaxy. Discoveries of similar but even more distant objects soon followed. Because the first few of these objects were strong sources of radio emission that looked like stars through visible-light telescopes, they were named "quasi-stellar radio sources," or *quasars* for short. Later, astronomers learned that most quasars are not such powerful radio emitters, but the name has stuck.

For many years, a debate raged among astronomers over whether Hubble's law could really be used to determine quasar distances. Some argued that quasars might have high redshifts for other reasons and therefore might be much nearer to us than Hubble's law would suggest. The vast majority of astronomers now consider this debate settled. Improved images show that quasars are indeed the centers of extremely distant galaxies and often are members of very distant galaxy clusters. We will present more evidence for the large distances of quasars later in this chapter.

Most quasars lie more than halfway to the cosmological horizon. The lines in typical quasar spectra are shifted to more than three times their rest wavelengths, which tells us that the light from these quasars emerged when the universe was less than a third of its present age. The farthest known quasar as of 1998 has spectral lines shifted to 5.9 times their rest wavelengths[4] (Figure 20.15). The light we see from this distant quasar began its journey when the universe was only about a tenth of its present age.

The extraordinary energy output of quasars emerges across an unusually wide swath of the electromagnetic spectrum. Quasars radiate approximately equal amounts of power from infrared wavelengths all the way through to gamma rays (Figure 20.16); they also produce strong emission lines. By comparison, most stars and galaxies emit primarily visible light. The wide spread of photon energies coming

[4]In the Mathematical Insights, we talk about distant objects in terms of their redshifts, z. Lines in a redshifted spectrum have wavelengths $(1 + z)$ times their rest wavelengths. Thus, a quasar with lines at 5.9 times their rest wavelengths has a redshift $z = 4.9$.

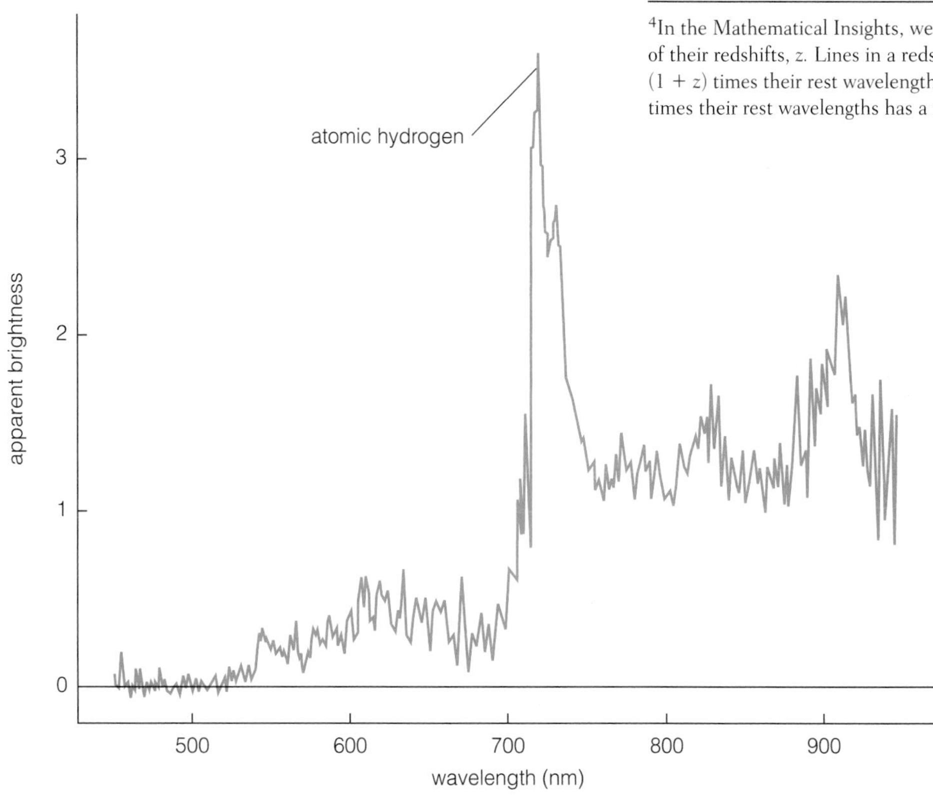

FIGURE 20.15 The tall peak at about 720 nm in this quasar spectrum is an emission line of atomic hydrogen that has a rest wavelength of 121.6 nm. Thus, the lines in this spectrum are redshifted to 720 ÷ 121.6 = 5.9 times their rest wavelengths.

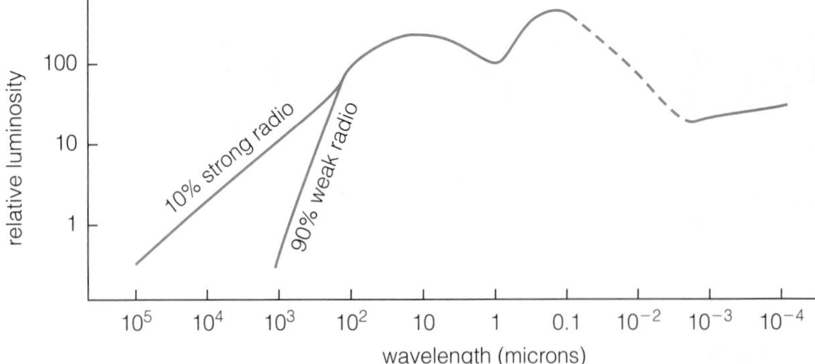

FIGURE 20.16 This schematic spectrum, representing the average of many quasar spectra, shows that all quasars emit strongly across most of the electromagnetic spectrum. However, only about 10% of quasars show strong radio emission. Quasar spectra also have strong emission lines, not shown in this diagram.

from quasars implies that they contain matter with a wide range of temperatures. How does this matter manage to radiate such a large luminosity?

Evidence from Nearby Active Galactic Nuclei

Quasars are difficult to study in detail because they are so far away. Luckily, some quasarlike objects are much closer to home. About 1% of present-day galaxies—that is, galaxies we see nearby—have less powerful active galactic nuclei that look very much like weak quasars.[5] The spectra of these active nuclei range from infrared to gamma rays, just like those of quasars. The only real distinction between nearby active galactic nuclei and quasars is their luminosity. Generally, the power outputs of quasars swamp those of the galaxies that contain them, making the surrounding galaxy hard to detect (Figure 20.17). That is why quasars look "quasi-stellar." The galaxies surrounding dimmer active galactic nuclei are much

[5]Because astronomer Carl Seyfert, in 1943, was the first to group galaxies with active galactic nuclei in a special class, nearby galaxies with active nuclei are often called *Seyfert galaxies.*

Mathematical Insight 20.1 Lookback Times to Quasars

In Mathematical Insight 19.4, we discussed how astronomers calculate the *lookback time* to a distant object from its redshift, z. We can use the same formulas to calculate lookback times to quasars. Recall that the lookback time we calculate depends on the values we assume for Hubble's constant and for the density of the universe. Thus, as the following examples show, we will need to know these values more precisely before we can fully interpret what distant quasars tell us about the early universe.

Example 1: The quasar whose spectrum is pictured in Figure 20.15 has a redshift $z = 4.9$. What is the lookback time to the quasar if we live in a low-density universe? How long after the Big Bang are we seeing this quasar? Assume that $H_0 = 65$ km/s/Mpc.

Solution: Mathematical Insight 19.4 gives the following formula for lookback time in a low-density universe with $H_0 = 65$ km/s/Mpc:

$$\text{lookback time} = (15 \text{ billion years}) \times \frac{z}{(1+z)}$$

Substituting $z = 4.9$, we find:

$$\text{lookback time} = (15 \text{ billion years}) \times \frac{4.9}{(1+4.9)} = 12.5 \text{ billion years}$$

If these assumptions about Hubble's constant and density are correct, the age of the universe is 15 billion years. Thus, we would be seeing this quasar as it was 12.5 billion years ago, or 2.5 billion years after the Big Bang. Note also

FIGURE 20.17 The bright quasar in the nucleus of the galaxy just left of center outshines the rest of the galaxy, which appears disturbed, as if it has recently undergone a collision.

easier to see because their centers are not so overwhelmingly bright.

The light-emitting regions of active galactic nuclei are so small that even the sharpest images do not resolve them. Our best visible-light images show only that active galactic nuclei must be smaller than 100 light-years (30 parsecs) across. Radio-wave images made with the aid of *interferometry* [Section S2.5] show that these nuclei are even smaller: less than 3 light-years (1 parsec) across. But rapid changes in the luminosities of some active galactic nuclei point to an even smaller size.

To understand how variations in luminosity give us clues about an object's size, imagine that you are a master of the universe and you want to signal one of your fellow masters a billion light-years away. An active galactic nucleus would make an excellent signal beacon, because it is so bright. However, suppose the smallest nucleus you can find is 1 light-year across. Each time you flash it on, the photons from the front end of the source reach your fellow master a full year before the photons from the back end. Thus, if you flash it on and off more than once a year, your signal will be smeared out. Similarly, if you find a source that is 1 light-day across, you can transmit signals that flash on and off no more than once a day. If you want to send signals just a few hours apart, you need a source no more than a few light-hours across.

Occasionally, the luminosity of an active galactic nucleus doubles in a matter of hours. The fact that we see a clear signal indicates that the source must be less than a few light-hours across. In other words,

that 2.5 billion years is one-sixth of 15 billion years, so we would be seeing this quasar at a time when the universe was only one-sixth its present age.

Example 2: Calculate the lookback time to the quasar in Example 1 if our universe instead has the critical density. Contrast the results with those of Example 1.

Solution: The formula from Mathematical Insight 19.4 for lookback time in a critical-density universe with H_0 = 65 km/s/Mpc is:

$$\text{lookback time} = (10 \text{ billion years}) \times \left[1 - \frac{1}{(1+z)^{3/2}}\right]$$

Substituting $z = 4.9$, we find:

$$\text{lookback time} = (10 \text{ billion years}) \times \left[1 - \frac{1}{(1+4.9)^{3/2}}\right] = 9.3 \text{ billion years}$$

In this case, our assumptions about density and Hubble's constant imply an age of 10 billion years for the universe. Thus, we would be seeing this quasar as it was 9.3 billion years ago, or 700 million years after the Big Bang. Note that 700 million years is only 7% of 10 billion years. Thus, the assumption of a critical density for the universe implies that the quasar existed when the universe was only 7% of its present age—relatively much earlier than the one-sixth ($\approx$ 17%) of present age implied by the assumption of a low-density universe in Example 1. You can see why measuring the density is so crucial to understanding the first few billion years of the universe.

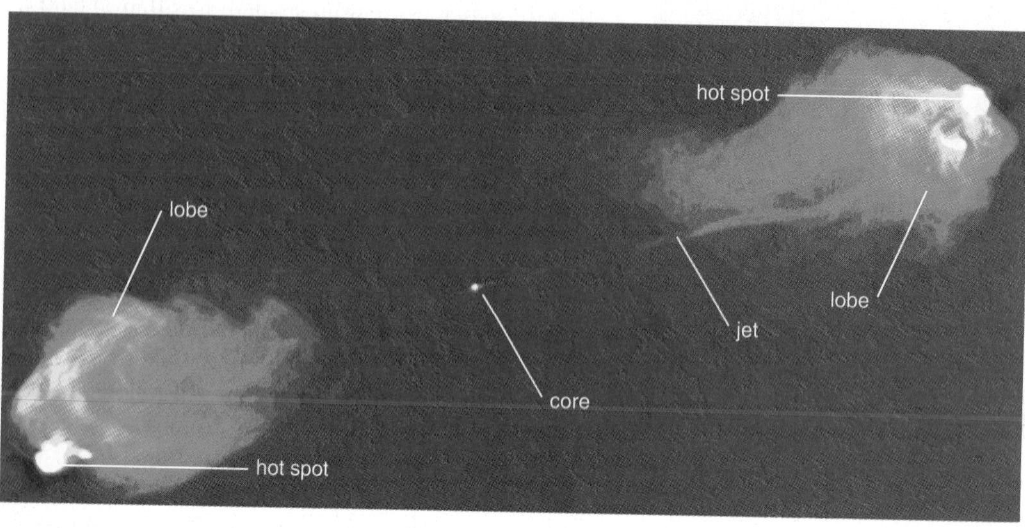

FIGURE 20.18 Radio image of the radio galaxy Cygnus A taken with the Very Large Array in New Mexico. The distance between the lobes is about 400,000 light years, several times larger than the extent of the galaxy in visible light.

the incredible luminosities of active galactic nuclei and quasars are apparently being generated in a volume of space not much bigger than our solar system.

Radio Galaxies and Jets

In the early 1950s, a decade before the discovery of quasars, radio astronomers noticed that certain galaxies emit unusually large quantities of radio waves. Today we believe that these **radio galaxies** are another class of celestial powerhouse that is closely related to quasars. Upon close inspection, we find that much of the radio emission comes not from the galaxies themselves, but rather from pairs of huge *radio lobes,* one on either side of the galaxy (Figure 20.18).

The radio waves from the lobes are produced by electrons and protons spiraling around magnetic field lines at nearly the speed of light. Astronomers of the 1950s guessed that this might be the source of the radio emission, but the implied amounts of energy in the lobes seemed implausibly large until the discovery of quasars. British theoretician Martin Rees led the way in showing that radio galaxies contain active galactic nuclei. He argued in 1971 that radio galaxies must have powerful **jets** spurting from their nuclei that transport energy to their lobes. That is, although the lobes lie far outside the visible galaxy, he asserted that the ultimate source of their energy would be found in the galaxy's tiny nucleus, which must therefore be as powerful as a quasar. A few years later, detailed radio images began to support his claims.

Today, radio telescopes resolve the structure of radio galaxies in vivid detail. At the center of a typical radio galaxy sits a tiny *radio core*—the active galactic nucleus of the radio galaxy—less than a few light-years across. Two jets of plasma shoot out of the core in opposite directions. Frequently, only the jet tilted in our direction is visible. Using time-lapse radio images taken several years apart, we can track the motions of various plasma blobs in the jets. Some of these blobs move at close to the speed of light. The lobes lie at the ends of the jets, sometimes as much as a million light-years from the core. The relative prominence of these three elements—cores, jets, and lobes—varies from radio galaxy to radio galaxy. Thus, a gallery of typical radio galaxies exhibits a wide variety of sizes and shapes (Figure 20.19).

Quasars and radio galaxies may be much more similar than they appear. Indeed, many quasars, including 3C 273, have core–jet–lobe radio structures reminiscent of radio galaxies. For example, Figure 20.20 shows a series of pictures of a jet in which a blob of plasma is moving outward from a quasar at close to the speed of light. Moreover, the active galactic nuclei of many radio galaxies seem to be concealed beneath donut-shaped rings of dark

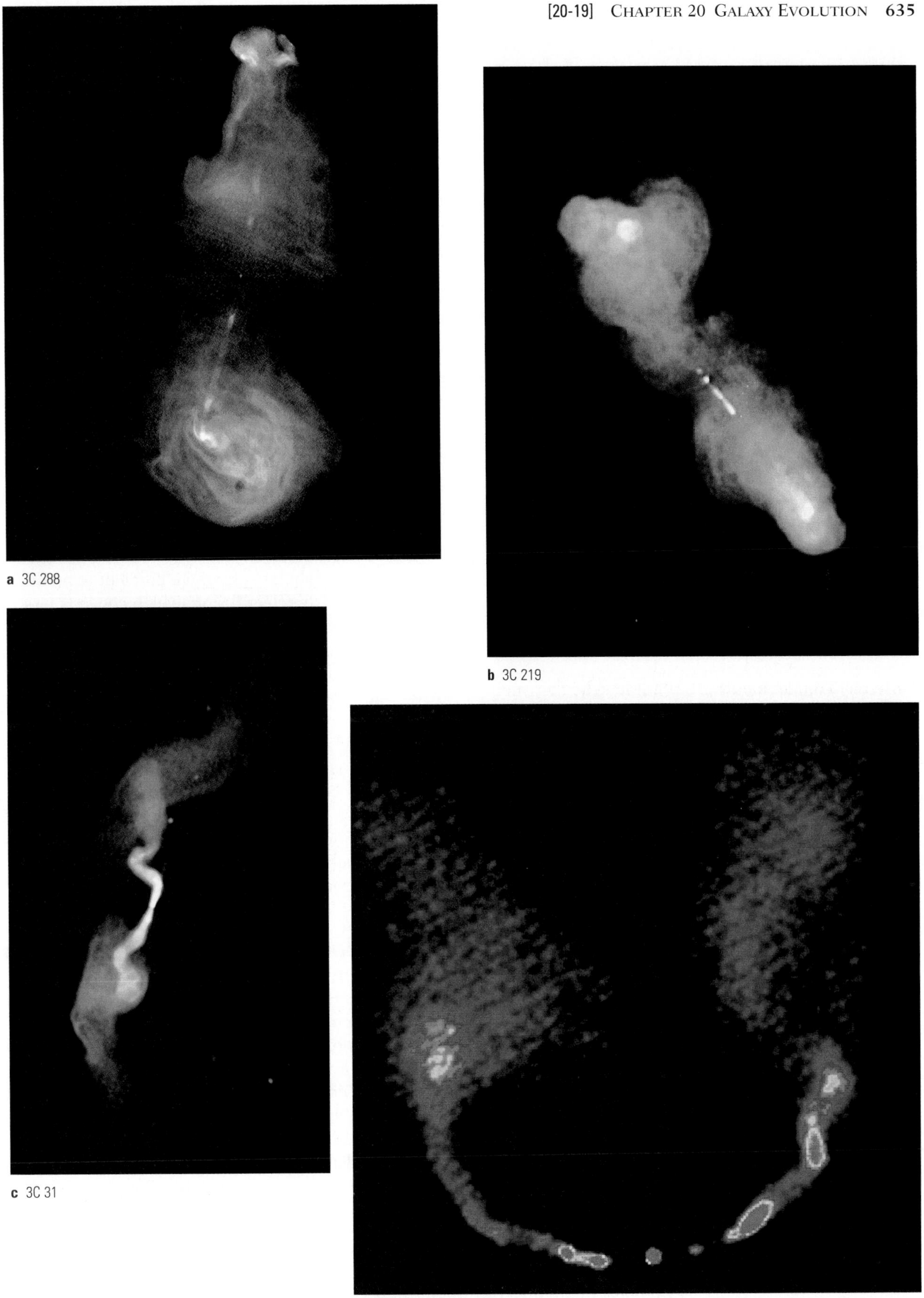

a 3C 288

b 3C 219

c 3C 31

d NGC 1265

FIGURE 20.19 Radio galaxy gallery. Note the double-lobe structure in all cases, but with very different shapes and sizes.

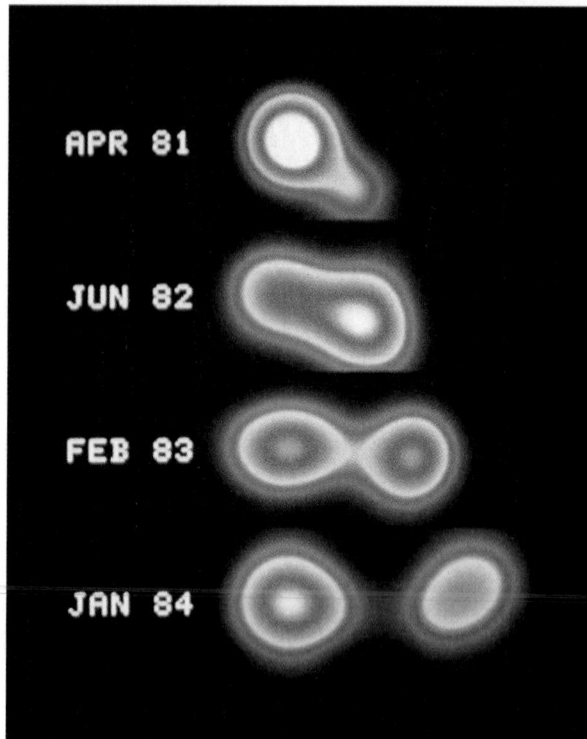

FIGURE 20.20 These images of the jet in the quasar 3C 345, taken over a period of several years, show a blob of plasma (right) moving away from the core (left) close to the speed of light.

molecular clouds (Figure 20.21). It may be that such structures look like quasars when they are oriented so that we can see the active galactic nucleus at the center, and like the cores of radio galaxies when the ring of dusty gas dims our view of the central object.[6]

[6]A subset of active galactic nuclei called *BL Lac objects* is likely to consist of the centers of radio galaxies whose jets point directly at us.

Supermassive Black Holes

Astronomers have worked hard to envision physical processes that might explain how radio galaxies, quasars, and other active galactic nuclei release so much energy within such small central volumes. Only one explanation seems to fit: The energy comes from matter falling into a supermassive black hole (Figure 20.22). Gravity converts the potential energy of the infalling matter into kinetic energy. Collisions between infalling particles convert the kinetic energy into thermal energy, and photons carry this thermal energy away. As in X-ray binaries, we expect that the infalling matter swirls through an accretion disk before it disappears beneath the event horizon of the black hole [Section 17.4].

This method of energy generation can be awesomely efficient. The gravitational potential energy lost by a chunk of matter falling into a black hole is equivalent to its mass-energy, $E = mc^2$. As much as 10–40% of this energy can emerge as radiation before the matter crosses the event horizon. Thus, accretion by black holes is far more efficient at producing light than is nuclear fusion, which converts less than 1% of mass-energy into photons. Note that, as with accretion into the black holes formed in supernovae, the light is coming not from the black hole itself but rather from the hot gas surrounding it.

A variety of mechanisms explain why quasars and other active galactic nuclei radiate energy across the electromagnetic spectrum. The hot gas in and above the accretion disk produces copious amounts of ultraviolet and X-ray photons. This radiation ionizes surrounding interstellar gas, energizing intense ionization nebulae that emit visible light. The emission lines produced by these nebulae are the same

Mathematical Insight 20.2 Feeding a Black Hole

As discussed in the text, 10–40% of the mass-energy of matter falling into a black hole can be radiated away as energy. (The precise value for a particular black hole depends on its rotation rate: Faster rotation allows more energy to be released.) Suppose 10% of the mass-energy is radiated away. Then the amount of energy radiated by mass m falling into a black hole is $E = \frac{1}{10}mc^2$. Equivalently, if we know the energy E radiated into space, then the amount of mass the black hole must accrete to radiate this energy is:

$$\text{accreted mass} = m = 10 \times \frac{E}{c^2}$$

Example: The most powerful quasars have luminosities of about 10^{40} watts. How much mass must the central black hole consume each second for its accretion disk to produce this luminosity? How many solar masses of material must it consume each year? Assume that 10% of the mass-energy of consumed mass is radiated away.

FIGURE 20.21 Artist's conception of the central region of a radio galaxy. The active galactic nucleus, obscured by a ring of dusty clouds, lies at the point from which the jets emerge. If viewed along a direction closer to the jet axis, the active nucleus would not be obscured and would look more like a quasar.

Solution: Recall that 1 watt = 1 joule/s, so the black hole must accrete enough mass to radiate 10^{40} joules each second. Thus, the amount of mass required to feed the black hole for 1 second is about:

$$m = 10 \times \frac{E}{c^2} = 10 \times \frac{10^{40}\ \frac{\text{kg} \times \text{m}^2}{\text{s}^2}}{\left(3 \times 10^8\ \frac{\text{m}}{\text{s}}\right)^2} = 1.1 \times 10^{24}\ \text{kg}$$

(Note that we replaced 1 joule with its equivalent units: 1 joule $= \frac{\text{kg} \times \text{m}^2}{\text{s}^2}$.)

The black hole must accrete 1.1×10^{24} kilograms of mass per second; we can convert this rate to kilograms per year as follows:

$$\frac{1.1 \times 10^{24}\ \text{kg}}{1\ \text{s}} \times \frac{1\ \text{solar mass}}{2.0 \times 10^{30}\ \text{kg}} \times \frac{3.1 \times 10^7\ \text{s}}{1\ \text{year}} = \frac{17\ \text{solar masses}}{1\ \text{year}}$$

Thus, the central black hole in a luminous quasar must consume about 17 times the mass of the Sun *every year!*

FIGURE 20.22 Artist's conception of an accretion disk surrounding a supermassive black hole. Note that this picture represents only the very center of an object like that shown in Figure 20.21.

ones Maarten Schmidt used to measure the first quasar redshifts. The infrared light may come from rings of molecular clouds that encircle the active galactic nucleus (see Figure 20.21). As in starburst galaxies, dust grains in these molecular clouds absorb the high-energy light and reemit it as infrared light. This ring of dense clouds could well be the reservoir for the accretion disk, supplying the matter that the giant black hole eventually consumes. Finally, the radio emission from active galactic nuclei comes from the fast-moving electrons that we sometimes see jetting from these nuclei at nearly the speed of light.

Although it's easy to understand why matter swirls inward through the accretion disk toward the black hole, explaining the powerful jets emerging from active galactic nuclei is more challenging. One plausible model for jet production relies on the twisted magnetic fields thought to accompany accretion disks. As an accretion disk spins, it pulls the magnetic field lines that thread it around in circles. Charged particles start to fly outward along the field lines like beads on a twirling string. The particles careening along the field lines accelerate to speeds near the speed of light, forming a jet that shoots out into space. Some jets blast all the way through the galaxy's interstellar medium and penetrate into the much less dense intergalactic gas. The *hot spots* at the ends of the lobes in radio galaxies (see Figure 20.18) are the places where the jets are currently ramming into the intergalactic gas. When particles traveling down a jet hit the hot spot, they are deflected into the surrounding radio lobe like water from a firehose hitting a wall. The particles then fill the lobe with energy, generating powerful radio emission.

The supermassive black hole theory explains many of the observed features of quasars and other active galactic nuclei, but it is incomplete. We do not yet know what would create such giant black holes, nor do we know why quasars eventually run out of gas and stop shining. The preponderance of quasars during the first several billion years of the universe suggests that the formation of supermassive black holes is somehow linked to galaxy formation, perhaps through collisions between galaxies. Some scientists have suggested that clusters of neutron stars resulting from extremely dense starbursts at the centers of galaxies might somehow coalesce to form an enormous black hole, but these speculations are still unverified. The origins of supermassive black holes remain mysterious.

Hunting for Monsters

Do supermassive black holes really drive the tremendous activity of radio galaxies, quasars, and other active galactic nuclei? The idea that such monsters even exist has been hard for some astronomers to swallow. Proving that we have found a black hole is tricky. Black holes themselves do not emit any light, so we need to infer their existence from the ways in which they alter their surroundings. In the vicinity of a black hole, matter should be orbiting at high speed around something invisible.

The case for supermassive black holes grew much stronger during the 1990s. The relatively nearby galaxy M 87 features a bright nucleus and a jet that emits both radio and visible light. Thus, it was already a prime black-hole suspect when astronomers pointed the Hubble Space Telescope at its core in 1994 (Figure 20.23). The spectra they gathered showed blueshifted emission lines on one side of the nucleus and redshifted emission lines on the other. This pattern of Doppler shifts is the characteristic signature of orbiting gas: On one side of the orbit the gas is coming toward us and hence is blueshifted, while on the other side it is moving away from us and is redshifted. The magnitude of these Doppler shifts shows that the gas, located up to 60 light-years (18 parsecs) from the center, is orbiting something invisible at a speed of hundreds of kilometers per second. This high-speed orbital motion indicates that the central object has a mass some 2–3 billion times that of our Sun. A supermassive black hole is the only thing we know of that could be so massive while remaining unseen.

Observations of NGC 4258, another galaxy with a visible jet, delivered even more persuasive evidence just 1 year later. A ring of molecular clouds orbits the nucleus of this galaxy in a circle less than 1 light-year in radius. We can pinpoint these clouds because they amplify the microwave emission lines of water molecules, generating beams of microwaves very similar to laser beams.[7] The Doppler shifts of these emission lines allow us to determine the orbits of the clouds very precisely. Their orbital motion tells us that the clouds are circling a single, invisible object—presumably a black hole—with a mass of 36 million solar masses.

We may never be 100% certain that we have discovered black holes in other galaxies. The best we can do is rule out all other possibilities. However, the hypothesis that gigantic black holes lie at the cores of quasars, nearby active galactic nuclei, and radio galaxies is withstanding the tests of time and thousands of observations.

20.6 Shedding Light on Protogalactic Clouds

The most mysterious part of galaxy evolution is the part we've never directly observed: the formation and early development of protogalactic clouds. But in recent years, the study of quasars has begun to shed light on even this very early stage of galaxy evolution. The most distant quasars inhabit the outskirts of the observable universe. Photons from some of these quasars began journeying to Earth when the universe was only 1 or 2 billion years old. Along the way, they have passed through *voids* that are virtually empty, as well as through numerous intergalactic hydrogen clouds. Some of these photons have even pierced

FIGURE 20.23 Doppler shifts of the emission lines from the gas disk in the elliptical galaxy M 87 indicate a 2–3-billion-solar-mass black hole.

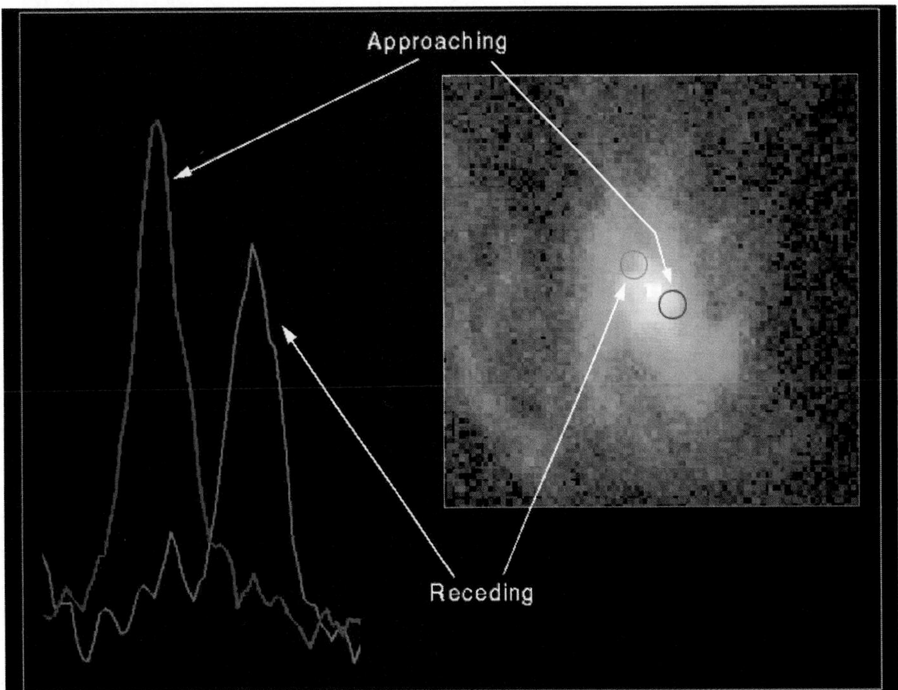

[7]The word *laser* stands for "light amplification by stimulated emission of radiation." These clouds contain *water masers,* with the word *maser* standing for "microwave amplification by stimulated emission of radiation."

the gaseous disks of spiral galaxies that are still in the process of forming.

The matter strewn across the paths of these photons leaves its mark in the spectra of quasars. Remember that atoms tend to absorb light at very specific wavelengths. Thus, every time a light beam from a quasar passes through an intergalactic or protogalactic cloud, some of the atoms in the cloud absorb photons from the beam, creating an absorption line. Studies of these absorption lines in quasar spectra can tell us what was happening in protogalactic clouds during the epoch of galaxy formation and provide unique clues about how galaxies evolve.

Distant Beacons

Our current interpretation of quasar redshifts—that they indicate vast distances—did not win general acceptance immediately. Recall that a small but vocal minority of astronomers argued for years that the huge luminosities of quasars were implausible. Instead of believing in supermassive black holes, these dissenters preferred to believe that quasar redshifts were misunderstood. The numerous hydrogen absorption lines scattered across the spectra of quasars are among the main reasons why the skeptics ultimately lost their case.

The most prominent emission line in the spectra of quasars is produced by hydrogen atoms in which the electron drops from energy level 2 to level 1 [Section 7.4]. (Astronomers refer to this line as *Lyman-alpha.*) The rest wavelength of this hydrogen line is 121.6 nm, putting it well into the ultraviolet portion of the spectrum. However, in the spectrum of a distant quasar, this line may be redshifted all the way into the visible band. Figure 20.24 shows the spectrum of a high-redshift quasar.

Note that the spectrum features hundreds of tiny absorption lines to the short-wavelength (bluer) side of the emission line from the quasar. Each of these absorption lines stems from the same transition that produces the quasar's emission line, but in reverse: The absorption line is produced when electrons in hydrogen atoms jump from energy level 1 to level 2. The fact that these lines are distinct tells us that each is produced in a distinct cloud of hydrogen gas; the fact that each has a different wavelength tells us that each cloud has a distinct redshift.

Hubble's law tells us that the quasar is more distant than any of the clouds because the quasar's emission line has the greatest redshift of any of these hydrogen lines. It also tells us that the clouds producing lines with smaller redshifts (shorter wavelengths) must lie closer to us, while the clouds producing lines with greater redshifts (longer wavelengths) must be more distant. Thus, we conclude that the many absorption lines represent signatures from hundreds of intergalactic clouds lying between us and the quasar.

Mathematical Insight 20.3 Weighing Supermassive Black Holes

We weigh supermassive black holes the same way we weigh almost everything else in the universe: by measuring the velocity v and orbital radius r of the matter circling the central black hole. Given these measurements, we can apply the orbital velocity law from Mathematical Insight 18.1 to find the mass M_r within a distance r of the galactic center:

$$M_r = \frac{r \times v^2}{G}$$

Example: Doppler shifts show that ionized gas in the nucleus of the active galaxy M 87 orbits at a speed of about 800 km/s at a radius of 18 parsecs = 5.6×10^{17} meters. Use these values to calculate the mass within 18 parsecs of the galactic center.

Solution: Substituting the given values into the orbital velocity law, we find:

$$M_r = \frac{(5.6 \times 10^{17} \text{ m}) \times \left(8.0 \times 10^5 \frac{\text{m}}{\text{s}}\right)^2}{6.67 \times 10^{-11} \frac{\text{m}^3}{\text{kg} \times \text{s}^2}} = 5.3 \times 10^{39} \text{ kg}$$

Converting kilograms to solar masses, we find:

$$M_r = (5.3 \times 10^{39} \text{ kg}) \times \frac{1 \text{ solar mass}}{2.0 \times 10^{30} \text{ kg}} = 2.7 \times 10^9 \text{ solar masses}$$

The ionized gas is orbiting around a mass of about 2.7 billion solar masses. Presumably, nearly all of this mass is in the central black hole.

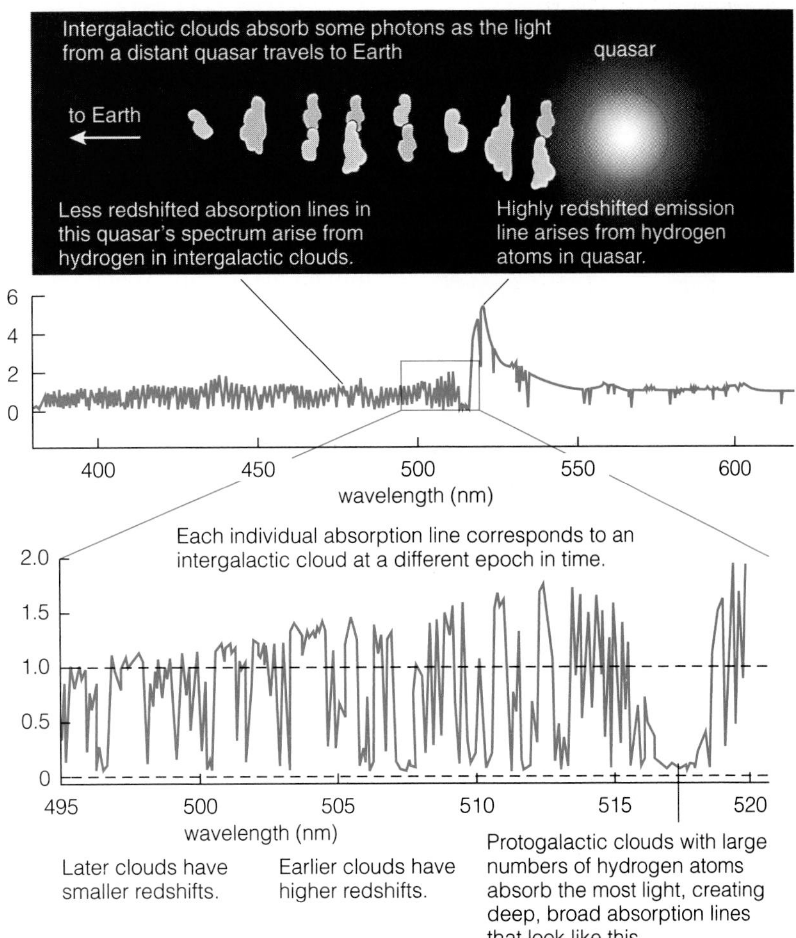

FIGURE 20.24 Intergalactic hydrogen clouds that would otherwise be invisible are revealed in the spectra of distant quasars. Top: Intergalactic clouds lying between Earth and a distant quasar absorb some of a quasar's light as it passes through. Middle: This spectrum of a quasar, taken at the Keck telescope, shows many different absorption lines. Bottom: A close-up of the spectrum reveals individual absorption lines, each corresponding to an individual intergalactic hydrogen cloud.

TIME OUT TO THINK *Why don't we see similar hydrogen absorption lines on the long-wavelength (redder) side of the emission line from the quasar?*

While almost all quasar spectra feature many hydrogen absorption lines from intergalactic clouds, a few also have highly redshifted absorption lines from elements such as carbon and magnesium. These elements are produced in stars, so the absorbing clouds must be near the galaxies that created their heavy elements. In some cases, we can now detect these galaxies directly even when they lie more than halfway across the universe (Figure 20.25). The quasars, which must lie even farther away, are really distant beacons from even earlier in time.

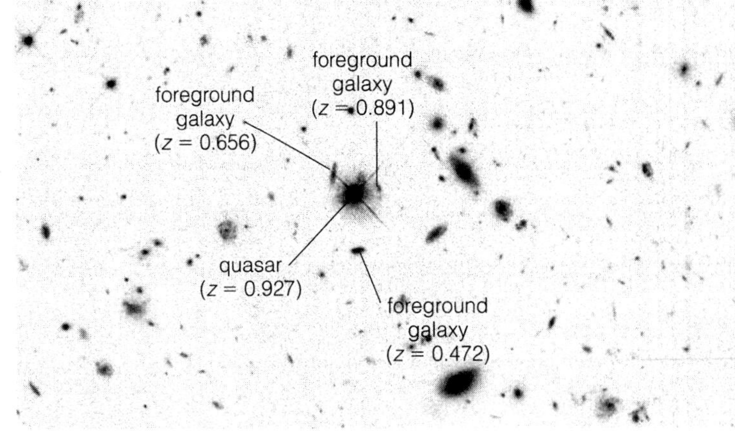

FIGURE 20.25 The galaxies marked by arrows have the same redshift as lines in the quasar spectrum. Thus, gas clouds in the outlying regions of these galaxies (too dim to be seen in this photograph) must lie directly in front of the quasar.

Layers in Space and Time

If you turn a quasar spectrum on its side, the absorption lines are reminiscent of the geological strata seen in the walls of the Grand Canyon (Figure 20.26).

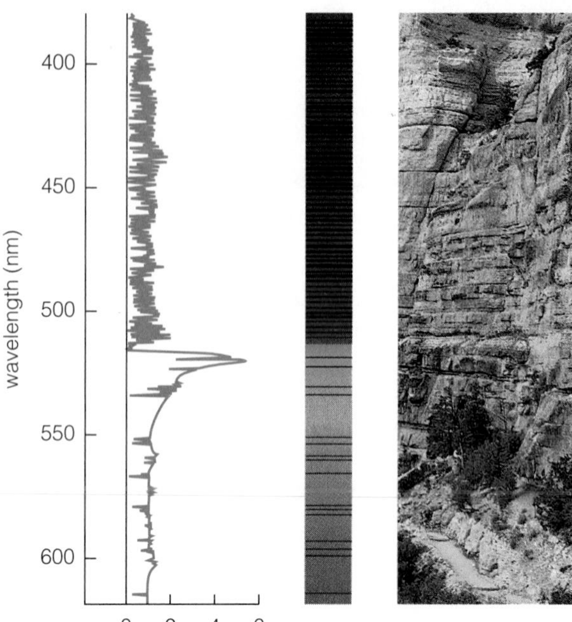

FIGURE 20.26 Interpreting the absorption lines in a quasar spectrum is similar to interpreting the geological layers laid down over the ages in a canyon wall. Each line corresponds to a different epoch in time.

Each stripe on the canyon wall corresponds to a different layer of sedimentation that has been deposited over the course of geological history. The topmost layers are the most recent; the deepest layers are the oldest. You can think of a quasar spectrum in a similar way. Each of the many hydrogen absorption lines in a quasar's spectrum has a different cosmological redshift that corresponds to a different lookback time [Section 19.4]. The higher the redshift, the further back in time (and the farther away in space) we are looking. Like layers of sediment, the multiple hydrogen absorption lines tell us about the state of the universe at different moments in its past.

Hydrogen absorption lines are most tightly bunched in the spectra of the highest-redshift quasars, indicating that many intergalactic clouds were present when the universe was young. Fewer absorption lines are present at shorter wavelengths and smaller redshifts, indicating that the intergalactic gas clouds have been vanishing as time progresses. Where all this gas is going is uncertain, but at least some of it forms stars and galaxies. Much of the rest is probably dispersing as the universe expands.

Clues to Galaxy Formation

The most prominent absorption lines in quasar spectra are of great interest to astronomers because they enable us to investigate what is happening in protogalactic clouds and young galaxies that have not yet formed many stars. When the light from a distant quasar passes through a large protogalactic cloud or the disk of a young galaxy, the resulting hydrogen absorption line is unusually wide and deep. The depths and widths of these lines tell us just how much atomic hydrogen gas the forming galaxies contain.

We are only just beginning to learn how to read the clues these huge absorption lines have etched into the spectra of quasars, but the evidence gathered so far supports our general picture of galaxy evolution. The deepest and widest lines typically lie at the highest redshifts, indicating that the youngest galaxies are made mostly of gas. In fact, they contain about the same amount of mass in the form of hydrogen gas as older galaxies contain in the form of stars. The widest lines at smaller redshifts, which arise in galaxies that are more mature, are not nearly as wide as those at higher redshifts, presumably because a greater fraction of the gas in these galaxies has already collected into stars.

Absorption lines from elements other than hydrogen corroborate this picture. The lines from these heavy elements are more prominent in mature galaxies than in the youngest galaxies, implying that the mature galaxies have experienced more supernovae, which have added heavy elements to their interstellar gas. This overall pattern of gradual heavy-element enrichment accompanied by the gradual diminishing of interstellar hydrogen agrees well with what we know about the Milky Way. Stars in the gaseous disks of spiral galaxies all across the universe appear to have been forming steadily for over 10 billion years.

With every new study of a quasar spectrum, we learn more about the galaxies and intergalactic clouds that have left their mark on the light from these distant beacons. Perhaps someday soon these observations will help us fit together all the pieces of the galaxy evolution puzzle, and we at last will understand the whole glorious history of galaxy evolution in our universe. Until that time, we will keep peering deep into space and back into time, searching for clues that will unlock the mysteries of cosmic evolution.

Reality provides us with facts so romantic that imagination itself could add nothing to them.

JULES VERNE

THE BIG PICTURE

We have not yet solved the whole puzzle of galaxy evolution, but in this chapter we have described some of its crucial pieces. As you look back, keep sight of these "big picture" ideas:

- Although we do not yet know the complete story of galaxy evolution, we are rapidly learning more. We know that galaxies grow from protogalactic clouds of gas, but collisions with neighboring galaxies have probably affected many galaxies.
- Galaxies in the present-day universe do not always evolve peacefully—many undergo temporary starbursts that can eject large fractions of their interstellar gas.
- The tremendous energy outputs of quasars and other active galactic nuclei, including those of radio galaxies, are probably powered by gas accreting onto supermassive black holes. The centers of many present-day galaxies must still contain the supermassive black holes that once enabled them to shine as quasars.
- Quasars are brilliant beacons whose light has crossed billions of light-years of space to reach us, passing through numerous intergalactic clouds and galaxies on the way. These clouds and galaxies leave their mark on quasar spectra, revealing how protogalactic clouds were behaving before they formed most of their stars.

Review Questions

1. What do we mean by *galaxy evolution?*
2. Briefly define *starburst galaxies, active galactic nuclei,* and *quasars.* Why do we think that all these phenomena represent temporary stages in galaxy evolution?
3. Describe three basic techniques we use to study galaxy evolution and what each technique enables us to learn.
4. What two assumptions underlie theoretical models of galaxy formation? According to these models, what processes eventually slow the collapse of protogalactic clouds?
5. Describe the two potential scenarios in which the properties of the original protogalactic cloud determine the difference between an elliptical galaxy and a spiral galaxy. What evidence supports these "nature" scenarios?
6. Briefly explain why we expect that collisions between galaxies should be relatively common, while collisions between stars are extremely rare. Why should galaxy collisions have been more common in the past than they are today?
7. Briefly describe how a collision between two spiral galaxies might lead to the creation of a single elliptical galaxy. What evidence supports this scenario for the formation of elliptical galaxies?
8. What is a *central dominant galaxy?* How do we think they formed?
9. Briefly describe how the hot gas in a galaxy cluster can help change a spiral galaxy into an elliptical or lenticular galaxy.
10. What evidence suggests that the protogalactic cloud that formed our own Milky Way resulted from several collisions among smaller clouds? How does this explain why the halo of our galaxy looks so much like an elliptical galaxy?
11. Briefly explain why starburst galaxies often appear ordinary when they are observed in visible light but extraordinary when they are observed in infrared light.
12. Briefly describe how a collision between two spiral galaxies could create a luminous starburst galaxy.
13. What is a *galactic wind?* What causes it? How is it similar to a superbubble in the Milky Way, and how is it different?
14. Why must starbursts cease after a relatively short time? What evidence suggests that small galaxies in our Local Group have undergone two or more starbursts in the past?
15. Briefly describe the discovery of quasars. What evidence convinced astronomers that the high redshifts of quasars really do imply great distances? Why can we learn more about quasars by studying nearby active galactic nuclei?
16. Briefly explain how we can use variations in luminosity to set limits on the size of an object's emitting region. For example, if an object doubles its luminosity in 1 hour, how big can it be?
17. What is a *radio galaxy?* Describe *jets* and *radio lobes.* Why do we think that their ultimate energy sources lie in quasarlike galactic nuclei?
18. Briefly explain the general picture of how *supermassive black holes* enable quasars to produce their huge luminosities. Summarize the evidence supporting the idea that supermassive black holes lie at the center of radio galaxies, quasars, and other active galactic nuclei.
19. Briefly explain how intergalactic clouds between a distant quasar and Earth leave distinctive marks in a quasar's spectrum.
20. Briefly explain how we can learn about the early history of galaxy evolution by studying quasar spectra.

Discussion Questions

1. *The Case for Supermassive Black Holes.* The evidence for supermassive black holes at the center of galaxies is strong. However, it is very difficult to prove absolutely that they exist because the black holes themselves emit no light. We can only infer their existence from their powerful gravitational influences on surrounding matter. How compelling do you find the evidence? Do you think astronomers have proved the case for black holes beyond a reasonable doubt? Defend your opinion.
2. *Life in Colliding Galaxies.* Suppose the Milky Way were currently undergoing a collision with another large spiral galaxy. Do you think this collision would affect life on Earth? Why or why not? How would the night sky look different if our galaxy was in the midst of such a collision?

Problems

1. *Life Story of a Spiral.* Imagine that you are a spiral galaxy. Describe your life history from birth to the present day. Your story should be detailed and scientifically consistent, but also creative. That is, it should be entertaining while at the same time incorporating current scientific ideas about the formation of spiral galaxies.
2. *Life Story of an Elliptical.* Imagine that you are an elliptical galaxy. Describe your life history from birth to the present. There are several possible scenarios for the formation of elliptical galaxies, so choose one and stick to it. Be creative while also incorporating scientific ideas that demonstrate your understanding.
3. *Jets.* Various kinds of astronomical objects produce jets. In two or three paragraphs, compare and contrast the jets produced by radio galaxies with those from protostars. How are they similar? How are they different?
4. *Distant Galaxies.* The most distant galaxies known have cosmological redshifts greater than $z = 4$. Suppose you observe a young galaxy with $z = 4$. In this problem, assume that $H_0 = 65$ km/s/Mpc.
   a. Calculate the lookback time to the galaxy under the assumption that we live in a low-density universe, in which case the current age of the universe is 15 billion years. How long after the Big Bang did this object form? What fraction of its current age was the universe when the light that we see left this object? (*Hint:* See Mathematical Insight 20.1.)
   b. Repeat part (a), but this time assume that we live in a critical-density universe, in which case the current age of the universe is 10 billion years.
   c. Briefly discuss how different assumptions about values of H_0 and density affect our interpretation of the lookback time to this galaxy.
5. *Your Last Hurrah.* Suppose you fell into an accretion disk that swept you into a supermassive black hole. Assume that, on your way down, the disk will radiate 10% of your mass energy, $E = mc^2$. (*Hint:* See Mathematical Insight 20.2.)
   a. What is your mass in kilograms? (*Hint:* Use the conversion 1 kg = 2.2 pounds.)
   b. Calculate how much radiative energy will be produced by the accretion disk as a result of your fall into the black hole.
   c. Calculate approximately how long a 100-watt light bulb would have to burn to radiate this same amount of energy.
6. *The Black Hole in NGC 4258.* The molecular clouds circling the center of the active galaxy NGC 4258 orbit at a speed of about 1,000 km/s, with an orbital radius of 0.15 parsec = 4.8×10^{15} meters. Use the orbital velocity law (see Mathematical Insight 20.3) to calculate the mass of the central black hole. Give your answer both in kilograms and in solar masses ($1M_{\text{Sun}} = 2.0 \times 10^{30}$ kg).
7. *Web Project: Future Missions.* The subject of galaxy evolution is a very active area of research. Using the web, look for information on current and future NASA missions involved in investigating galaxy evolution. How big are the planned telescopes? What wavelengths will they look at? When will they be launched? Assemble a journal from your web research, with pictures and a paragraph on each mission you find. If possible, make your journal a web page and include links to the relevant web sites.
8. *Research: Greatest Redshift.* As of early 1998, the quasar with the largest measured redshift is the quasar shown in Figure 20.15, with a redshift $z = 4.9$. Using the web or other resources, find the current record holder for the largest redshift. Write a one-page report describing the object and its discovery.
9. *Research: The Quasar Controversy.* For many years, some astronomers argued that quasars were not really as distant as Hubble's law implies. Research the history of the discovery of quasars and the debates that followed. What evidence led some astronomers to think that quasars might be nearer than Hubble's law suggested? Why did most astronomers eventually conclude that quasars really are far away? Write a one- to two-page report summarizing your findings.

21

Dark Matter

THE MAJORITY OF THE MATTER IN THE UNIVERSE, THE STUFF THAT binds galaxies together with the force of its own gravity, is too dark to see. We know that this matter exists because we can detect its gravitational influences on other things, but what is this so-called *dark matter*? Do armies of Jupiter-size bodies, too dim for us to see at a distance, populate the voids between the stars? Could black holes be responsible? Or is dark matter an entirely new form of matter, still undiscovered here on Earth? We don't yet know the answer. Incredibly, we still haven't identified the most common form of matter in the universe, making dark matter one of the greatest cosmic mysteries.

In this chapter, we will investigate why most astronomers believe that dark matter exists. We'll see how it affects structures the size of galaxies and larger. We'll investigate current hypotheses about the nature of dark matter. And we'll see how the fate of the universe rests on the question of just how much dark matter there is.

21.1 The Mystery of Dark Matter

We first encountered dark matter when we studied the Milky Way's rotation in Chapter 18. We saw that atomic hydrogen clouds lying farther from the galactic center than our Sun orbit the galaxy at unexpectedly high speeds. We concluded that much of our galaxy's mass must lie beyond the distance of the Sun's orbit around the galactic center, distributed throughout the galaxy's spherical halo [Section 18.4]. Yet most of the Milky Way's *light* comes from stars lying closer to the galaxy's center than does our Sun. Together, these facts imply that the halo contains large amounts of matter, but this matter emits so little light that we cannot see it. Thus, we call it *dark matter.*

Does dark matter really exist? To claim that the galaxy is filled with a form of matter that we cannot yet identify may seem strange, but only two possible explanations can account for the high orbital velocities of the atomic hydrogen clouds: Either their velocities are caused by the gravitational attraction of unseen matter, or we are doing something wrong when we apply the law of gravity to understand their orbits. The latter possibility implies that we do not understand how gravity operates on galaxy-size scales, but our theories of gravity successfully account for many other cosmic phenomena. Thus, the vast majority of astronomers believe that we correctly understand how gravity works and that dark matter must really exist.

The evidence that the Milky Way contains dark matter qualifies as an interesting surprise on its own, but it is also a glimpse into a more profound mystery. If our suspicions about dark matter are correct, then the luminous part of the Milky Way's disk must be rather like the tip of an iceberg, marking only the center of a much larger clump of mass (Figure 21.1). A more detailed analysis involving the motions of neighboring dwarf galaxies shows that the total mass of dark matter in the Milky Way might be 10 times greater than the mass of visible stars. Each galaxy in the universe is similar to our own in this respect, meaning that we probably cannot see most of the matter in the universe. This matter's darkness makes it extremely challenging to study, but its overwhelming dominance of the universe makes it extremely important. Dark matter, by virtue of its mighty gravitational pull, seems to be the glue that holds galaxies and clusters of galaxies together.

Dark matter has even more profound implications for the fate of the universe. Recall that the universe ultimately has two possible fates: It might continue to expand forever, or it might someday stop expanding and begin to collapse [Section 19.3]. We do not yet know which fate awaits our universe because the answer depends on the overall density of matter—and we cannot determine the overall density until we first determine how much dark matter is out there. Thus, the very fate of the universe hinges on the total amount of dark matter.

In this chapter, we will presume that we understand gravity correctly and consequently that dark matter pervades the universe. If our understanding of gravity is correct, then we already know quite a lot about dark matter, despite the fact that its composition remains a mystery. We do not know the total amount of dark matter in the universe, but we can measure the amounts in individual galaxies and clusters of galaxies from its gravitational effects. We can even make some reasonable guesses as to what it will turn out to be. We'll begin by discussing how we "discover" dark matter, and conclude by discussing what our current knowledge predicts about the ultimate fate of the universe.

FIGURE 21.1 Dark matter distribution relative to luminous matter in a spiral galaxy.

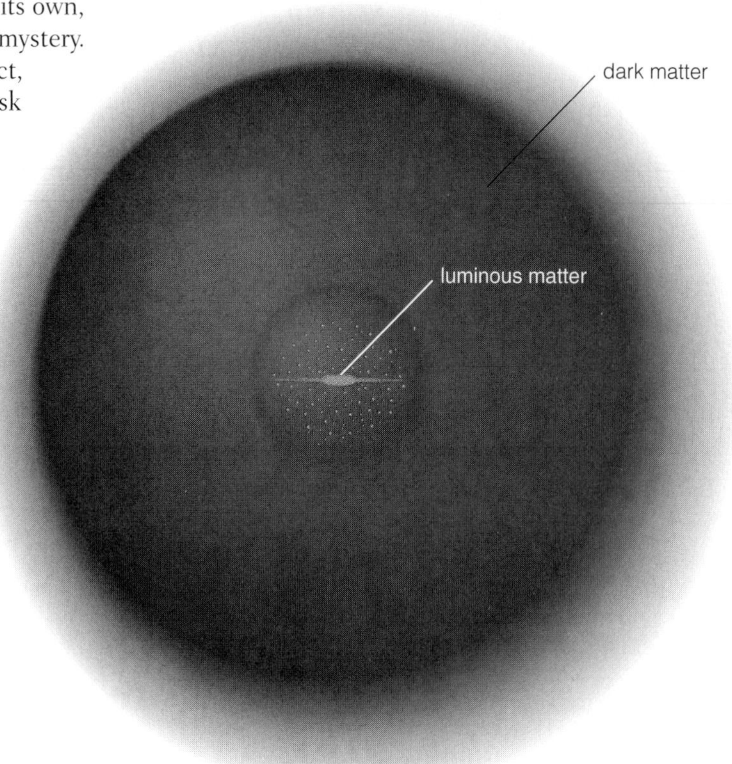

21.2 Dark Matter in Galaxies

The claim that dark matter far outweighs the visible matter in the universe might seem farfetched, but it rests on fundamental physical laws. Newton's laws of motion and gravity are among the most trustworthy tools in science. We have used them time and again to measure masses of celestial objects. We found the masses of the Earth and the Sun by applying Newton's version of Kepler's third law [Section 6.4]. We used this same law to calculate the masses of stars in binary star systems, revealing the general relationships between the masses of stars and their outward appearances. Newton's laws have also told us the masses of things we can't see directly, such as the masses of neutron stars in X-ray binaries and of black holes in active galactic nuclei. These laws have proved extremely reliable in many different applications. Thus, when Newton's laws tell us that dark matter exists, we are not inclined to ignore them.

We determine the amount of dark matter in a galaxy from the galaxy's mass and luminosity: We can infer the total mass in stars from the galaxy's luminosity, so whatever additional mass remains unaccounted for must be dark. Measuring a galaxy's luminosity is relatively easy. We simply point a telescope at the galaxy in question, measure its apparent brightness,[1] and calculate its luminosity from the luminosity–distance formula [Section 15.2]. Measuring the mass is more complicated. We would like to apply Newton's laws in the form of the orbital velocity law from Chapter 18 to matter orbiting as far from the galaxy's center as possible. However, the matter farthest from the center of a galaxy is extremely dim. For all we know, the matter at the edge of a galaxy might be completely dark.

Weighing Spiral Galaxies

We can weigh a spiral galaxy by measuring the gravitational effects of the galaxy's mass on the orbits of objects in its disk. Even beyond the point in the disk at which starlight fades into the blackness of intergalactic space, we can still see radio waves from atomic hydrogen gas. We can therefore use Doppler shifts of the 21-cm emission line of atomic hydrogen [Section 18.2] to determine how quickly this gas moves toward us or away from us (Figure 21.2). Galaxies beyond the Local Group have cosmological redshifts affecting all their spectral lines. However, on one side of a spiral galaxy the gas is rotating away from

[1]In practice, we must take pictures in a few different spectral bands because, for example, some galaxies emit much more red light than blue light.

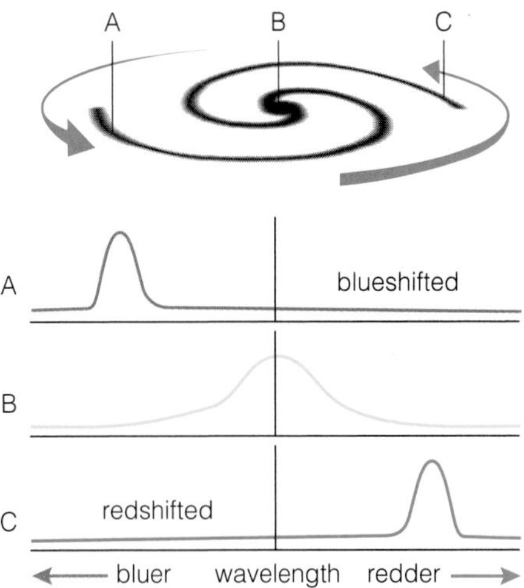

FIGURE 21.2 Measuring the rotation of a spiral galaxy with the 21-cm line of atomic hydrogen. Blueshifted lines on the left side of the disk show how fast that side is rotating toward us. Redshifted lines on the right side show how fast that side is rotating away from us.

us, so its 21-cm line is redshifted a little more than the redshift of the galaxy as a whole. On the other side, the 21-cm line is blueshifted relative to the redshift of the galaxy as a whole because the gas is rotating toward us. From the Doppler shifts of these clouds,[2] we can construct a *rotation curve*—a plot showing orbital velocities of gas clouds and stars—just as we did for the Milky Way (Figure 21.3) [Section 18.4]. A rotation curve contains all the information we need to measure the mass contained within the orbits of the outermost gas clouds.

TIME OUT TO THINK *As a brief review, draw a rotation curve for our solar system. How does the solar system rotation curve differ from the rotation curve of the Milky Way? What does that tell us about the distribution of matter in the Milky Way? Why?* (Hint: *See Chapter 18 for review.*)

The rotation curves of most spiral galaxies turn out to be remarkably flat as far out as we can see, which means that the orbital speeds of gas clouds remain roughly constant with increasing distance from the galactic center (Figure 21.4). We saw in Chapter 18 that our Milky Way's rotation curve remains more or less flat from well within the radius of the Sun's orbit (28,000 light-years, or 8.5 kpc) to beyond twice that radius. The rotation curves of some other spiral galaxies are flat to well beyond 160,000 light-years (50 kpc).

[2]Because the Doppler effect tells us only about the velocity of material directly toward or away from us, we must also take into account the tilt of the galaxy before we construct the rotation curve.

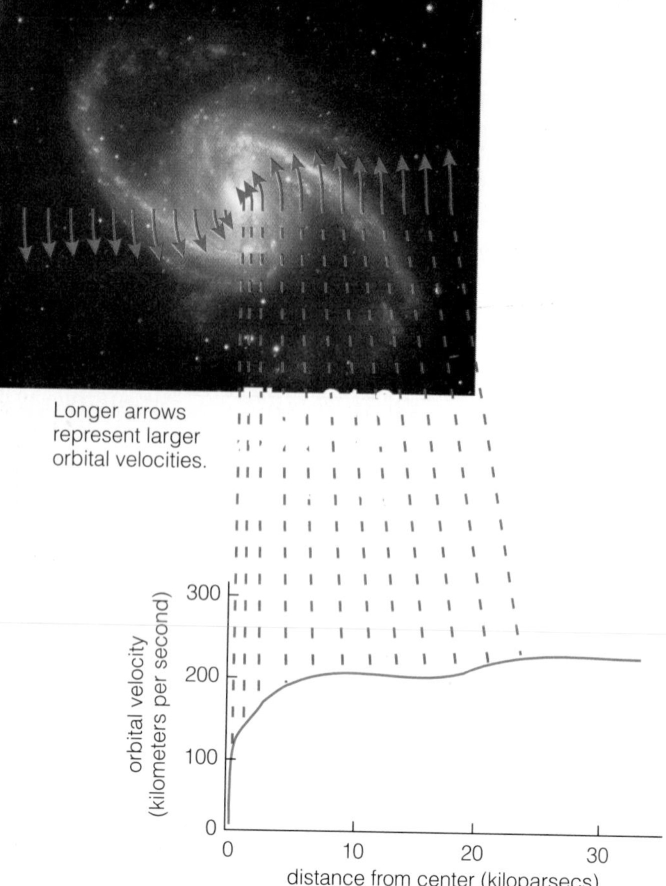

FIGURE 21.3 A rotation curve shows the orbital velocities of stars or gas clouds at different distances from a galaxy's center.

Because the orbital speeds of gas clouds tell us the amount of mass contained within their orbital paths, the flat rotation curves imply that a great deal of matter lies far from the galactic center [Section 18.4]. In particular, if we imagine drawing bigger and bigger circles around a galaxy, the flat rotation curves imply that we must keep encircling more and more matter. The mass of a spiral galaxy with gas clouds orbiting at 200 km/s at a distance of 160,000 light-years from its center must be at least 500 billion (5×10^{11}) solar masses.

Weighing Elliptical Galaxies

We must use a different technique to weigh elliptical galaxies, because most of them contain very little atomic hydrogen gas and hence do not produce detectable 21-cm radiation. Instead, we generally weigh the inner parts of elliptical galaxies by observing the motions of the stars themselves.

The motions of stars in an elliptical galaxy are disorganized, so we cannot assemble their velocities into a sensible rotation curve. Nevertheless, the velocity of each individual star still responds to the mass inside the star's orbit. At any particular distance from an elliptical galaxy's center, some stars are moving toward us and some away from us. Thus, the spectral lines from the galaxy as a whole tend to be smeared out: Instead of a nice sharp line at a particular wavelength, we see a *broadened* line spanning a range of wavelengths reflecting the various Doppler shifts of the individual stars (Figure 21.5). The greater the broadening of the spectral line, the faster the stars must be moving.

When we compare spectral lines from different regions of an elliptical galaxy, we find that the speeds of the stars remain fairly constant as we look farther from the galactic center. Thus, just as in spirals, most of the matter in elliptical galaxies must lie beyond the distance where the light trails off and hence must be dark matter. However, we cannot determine the *total* amount of dark matter in elliptical galaxies as well as we can in spirals, because we cannot measure their masses as far from their centers. In ellipticals we can study only the motions of stars, while in spirals we are able to detect the 21-cm radiation from gas clouds well beyond the radii at which we no longer see individual stars.

Mass-to-Light Ratio

We can determine how much dark matter a galaxy contains by comparing the galaxy's measured mass to its luminosity. For example, the Milky Way contains about 90 billion (9×10^{10}) solar masses of material within the Sun's orbit. However, the total luminosity of stars within this region is only about 15 billion (1.5×10^{10}) solar luminosities. Thus, on average, it takes 6 solar masses of matter to produce 1 solar luminosity of light in this region of the galaxy (15 billion $\times$ 6 = 90 billion). We therefore say that the Milky Way's **mass-to-light ratio** within the Sun's orbit is about 6 solar masses per solar luminosity. This fact tells us that most matter is dimmer than our Sun, which is not surprising since we know that most stars are smaller and dimmer than our Sun. We find similar mass-to-light ratios for the inner regions of most other spiral galaxies.

TIME OUT TO THINK *Note that, because the Sun has a mass of $1M_{Sun}$ and a luminosity of $1L_{Sun}$, its mass-to-light ratio is $1M_{Sun}/1L_{Sun}$ = 1 solar mass per solar luminosity. What is the mass-to-light ratio of a $1M_{Sun}$ red giant with a luminosity of $100L_{Sun}$? What is the mass-to-light ratio of a $1M_{Sun}$ white dwarf with a luminosity of $0.001L_{Sun}$?*

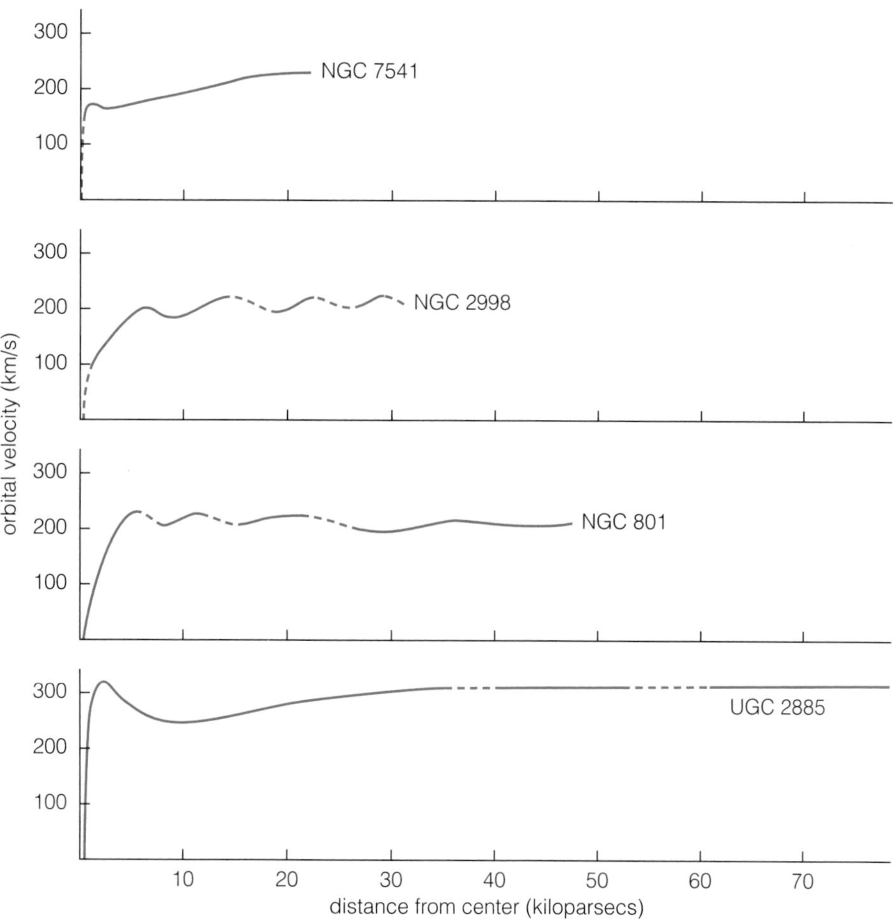

FIGURE 21.4 Actual rotation curves of four spiral galaxies.

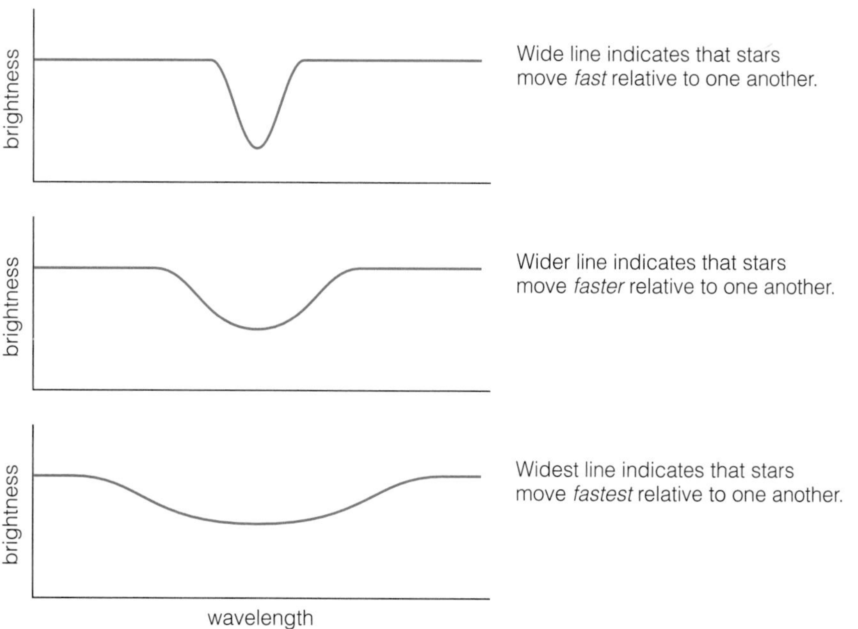

FIGURE 21.5 The broadening of absorption lines in an elliptical galaxy's spectrum tells us how fast its stars move relative to one another.

Elliptical galaxies contain virtually no luminous high-mass stars, so their stars are less bright on average than the stars in a spiral galaxy. In the inner regions of elliptical galaxies, the orbits of stars indicate a mass-to-light ratio of about 10 solar masses per solar luminosity—nearly double the mass-to-light ratio in the central region of the Milky Way. This is not surprising given the fact that elliptical galaxies have dimmer stars.

The surprise in mass-to-light ratios—and one of the key pieces of evidence for the existence of dark matter—comes when we look to the outer reaches of galaxies. As we look farther from a galaxy's center, we find a lot more mass but not much more light. The mass-to-light ratios for entire spiral galaxies can be as high as 50 solar masses per solar luminosity, and the overall ratios in some dwarf galaxies can be even higher. We are forced to conclude that stars alone cannot account for the amount of mass present in galaxies. For example, if a galaxy's overall mass-to-light ratio is 60 solar masses per solar luminosity and its stars account for only 6 solar masses per solar luminosity, then the remaining 90% of the galaxy's mass must be dark.

21.3 Dark Matter in Clusters

The problem of dark matter in astronomy is not particularly new. Back in the 1930s, astronomer Fritz Zwicky was already arguing that clusters of galaxies held enormous amounts of this mysterious stuff (Figure 21.6). Few of his colleagues paid attention.[3] Later observations supported Zwicky's claims. We can now weigh clusters of galaxies in three different ways: by measuring the speeds of galaxies orbiting the center of the cluster, by studying the X-ray emission from hot gas between the cluster galaxies, and

[3]At that time, Dr. Zwicky was also arguing that supernovae were exploding stars that left neutron stars behind, but his colleagues weren't buying that either.

Mathematical Insight 21.1 Mass-to-Light Ratio

Mathematically, an object's mass-to-light ratio is defined as its total mass in units of solar masses divided by its total luminosity in units of solar luminosities. Thus, by definition, the mass-to-light ratio of the Sun is:

$$\frac{1 \text{ solar mass}}{1 \text{ solar luminosity}} = 1 \text{ solar mass per solar luminosity}$$

A galaxy made entirely of stars just like the Sun would have a mass-to-light ratio identical to that of the Sun.

Example 1: What is the mass-to-light ratio of the matter inside the solar circle of the Milky Way?

Solution: To answer this question, we divide the 90 billion solar masses inside the solar circle by the 15 billion solar luminosities of the stars in the same region:

$$\frac{9 \times 10^{10} \text{ solar masses}}{1.5 \times 10^{10} \text{ solar luminosities}} = 6 \text{ solar masses per solar luminosity}$$

Because this mass-to-light ratio is larger than the Sun's ratio of 1, the average luminosity of objects in this region of the Milky Way must be less than that of the Sun. If not completely dark, these bodies must be dim.

Example 2: Suppose a galaxy contains 5×10^{11} solar masses within a radius of 150,000 light-years (50 kpc) of its center, but its total luminosity is only 1.5×10^{10} solar luminosities. What is its mass-to-light ratio? What does this imply?

Solution: Again we divide the total mass by the total luminosity. The mass-to-light ratio within 150,000 light-years of this galaxy's center is:

$$\frac{5 \times 10^{11} \text{ solar masses}}{1.5 \times 10^{10} \text{ solar luminosities}} = 33 \text{ solar masses per solar luminosity}$$

Taken as a whole, the mass in this galaxy is far darker, on average, than the Sun or than the objects in the inner region of the Milky Way.

THINKING ABOUT . . .

Pioneers of Science

Scientists always take a risk when they publish what they think are ground-breaking results. If their results turn out to be in error, their reputations may suffer. In the case of dark matter, the pioneers in its discovery risked their entire careers. A case in point is Fritz Zwicky, with his proclamations in the 1930s about dark matter in clusters of galaxies: Most of his colleagues considered him an eccentric who leapt to premature conclusions.

Another pioneer in the discovery of dark matter was Vera Rubin, an astronomer at the Carnegie Institute. Working in the 1960s, she became the first woman to observe under her own name at California's Palomar Observatory—then the largest telescope in the world. (Another woman, Margaret Burbidge, was permitted to observe at Palomar earlier but was required to apply for time under the name of her husband, also an astronomer.) Rubin first saw the gravitational signature of dark matter in spectra she recorded of stars in the Andromeda Galaxy. She noticed that stars in the outskirts of Andromeda moved at suprisingly high speeds, suggesting a stronger gravitational attraction than could be explained by the mass of the galaxy's stars alone. In other words, she found that the rotation curve for Andromeda is relatively flat to great distances from the center, just as we now know is also the case for the Milky Way. Working with a colleague, Kent Ford, she constructed rotation curves for the hydrogen gas in many other spiral galaxies (by studying Doppler shifts in the spectra of hydrogen gas) and discovered that flat rotation curves are common. Although Rubin and Ford did not immediately recognize the significance of the results, they were soon arguing that the universe must contain substantial quantities of dark matter.

For a while, other astronomers had trouble believing the results. Some astronomers suspected that the bright galaxies studied by Rubin and Ford were unusual for some reason. So Rubin and Ford went back to work, obtaining rotation curves for fainter galaxies. They found flat rotation curves—a signature of dark matter—even in these galaxies. By the 1980s, the evidence compiled by Rubin and Ford was so overwhelming that even their early critics came around. Either the theory of gravity was wrong, or they had discovered dark matter in spiral galaxies.

by observing how the clusters bend light as *gravitational lenses* [Section S4.5]. All three techniques indicate that clusters contain huge amounts of dark matter. Let's investigate each of these techniques more closely.

Orbiting Galaxies

Zwicky was one of the first astronomers to think of galaxy clusters as huge swarms of galaxies bound together by gravity. It seemed natural to him that galaxies clumped closely in space should all be orbiting one another, just like the stars in a star cluster. He therefore assumed that he could measure cluster masses by observing galaxy motions and applying the orbital velocity law.

Armed with a spectrograph, Zwicky measured the redshifts of the galaxies in a particular cluster and used these redshifts to calculate the speeds at which the individual galaxies are moving away from us. He

FIGURE 21.6 Fritz Zwicky.

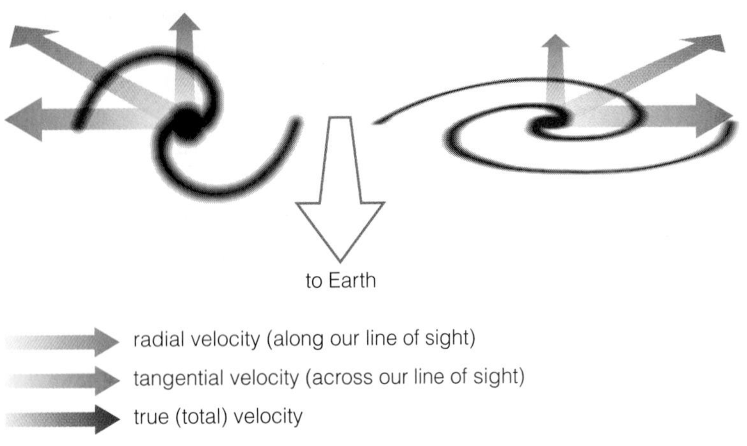

FIGURE 21.7 A galaxy's true velocity can be broken into a radial component along our line of sight and a tangential component across our line of sight. Doppler shifts tell us only the radial component; we cannot measure the tangential component for distant galaxies.

determined the velocity of the cluster as a whole by averaging the velocities of its individual galaxies. He could then estimate the speed of a galaxy around the cluster—that is, its orbital speed—by subtracting this average velocity from the individual galaxy's velocity. Finally, he plugged these orbital speeds into an appropriate form of the orbital velocity law [Section 18.4] to estimate the cluster's mass and compared this mass to the luminosity of the cluster. To his surprise, Zwicky found that clusters of galaxies have huge mass-to-light ratios: They hold hundreds of solar masses for each solar luminosity of radiation they emit. He concluded that most of the matter within these clusters must be almost entirely dark. Many astronomers disregarded Zwicky's result, believing that he must have done something wrong to arrive at such a strange answer.

Today, far more sophisticated measurements of galaxy orbits in clusters confirm Zwicky's original finding. We cannot measure the actual orbital velocity and position of a single galaxy. We can measure only its *radial velocity* along our line of sight (Figure 21.7) [Section 3.4]. However, if we measure the radial velocities of many galaxies in a cluster, we can estimate the average orbital velocity of all the cluster's galaxies. Cluster masses found in this way imply lots of dark matter. The visible portions of a cluster's galaxies represent less than 10% of the cluster's mass.

Intracluster Medium

A second method for weighing a cluster of galaxies relies on X-ray observations of the hot gas that fills

Mathematical Insight 21.2 Masses of Clusters

The first two of the three ways described in the text for measuring the masses of galaxy clusters rely on the orbital velocity law from Mathematical Insight 18.1 to find the mass M_r within a distance r of the galactic center:

$$M_r = \frac{r \times v^2}{G}$$

where v is the velocity of objects moving under the influence of gravity. This formula can also be applied to clusters of galaxies.

In the case where we measure the orbital speeds of galaxies directly, we simply use the average speed of the galaxies and the cluster radius in the orbital velocity law. In the case where we measure the temperature of hot, intracluster gas from its X-ray emission, we must first convert the gas temperature into an average speed for the gas particles. Although we will not present a derivation here, we can make this conversion with the following formula, which gives the approximate average speeds of the hydrogen nuclei in a gas of temperature T:

$$\text{average speed of hydrogen nuclei} \approx 100\,\frac{\text{m}}{\text{s}} \times \sqrt{T} \quad (T \text{ in Kelvin})$$

Once we find this speed, we can use it as v in the orbital velocity law.

Example: Suppose a cluster has a radius of 1 Megaparsec (3.1×10^{22} meters). Its galaxies orbit with an average speed of approximately 1,000 km/s (1×10^6 m/s), and its intracluster gas has a temperature of 9×10^7 K. Find the cluster's mass from both the galaxy speeds and the gas temperature. Do the results agree?

Solution: First, we find the mass from the galaxy speeds. We set $r = 3.1 \times 10^{22}$ m for the cluster radius, set $v \approx 1 \times 10^6$ m/s for the average speed of the galaxies,

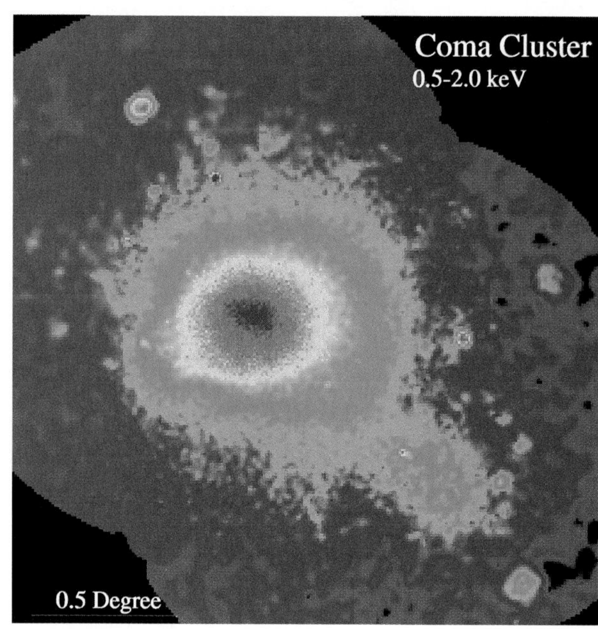

FIGURE 21.8 (**a**) The Coma Cluster of galaxies seen in visible light. Virtually every pictured object here is a galaxy. (**b**) A map of X-ray emission from the Coma Cluster. Hot, X-ray emitting gas bound to the cluster by gravity fills the spaces between the galaxies.

the space between the galaxies in the cluster (Figure 21.8). This gas, also known as the **intracluster medium** (*intra* means "within"), is so hot that it emits primarily X rays and therefore went undetected until the 1960s, when X-ray telescopes were finally launched above the Earth's atmosphere. The temperature of this gas is tens of millions of degrees in many clusters and can exceed 100 million degrees in the largest clusters. Even though the intracluster medium is invisible in optical light, its mass often exceeds the mass of all the visible stars in all the cluster's galaxies combined. The largest clusters of galaxies contain up to five times more matter in the form of hot gas than in the form of stars.

The temperature of the hot intracluster gas depends on the mass of the cluster itself, enabling us to use X-ray telescopes to measure the masses of galaxy clusters. The intracluster medium in most clusters is nearly in a state of *gravitational equilibrium*—that is, the outward gas pressure balances gravity's inward

---

and assign the gravitational constant G its usual value. Plugging these values into the orbital velocity law, we find:

$$\text{cluster mass} = M_r \approx \frac{(3.1 \times 10^{22}\ \text{m}) \times \left(1 \times 10^{6}\ \dfrac{\text{m}}{\text{s}}\right)^2}{6.67 \times 10^{-11}\ \dfrac{\text{m}^3}{\text{kg} \times \text{s}^2}} = 4.6 \times 10^{44}\ \text{kg}$$

To find the mass from the temperature of the X-ray emitting gas, we first use the given formula to find the average speeds of the hydrogen nuclei in the gas:

$$\text{average speed} \approx 100\ \frac{\text{m}}{\text{s}} \times \sqrt{T} = 100\ \frac{\text{m}}{\text{s}} \times \sqrt{9 \times 10^{7}} = 9.5 \times 10^{5}\ \frac{\text{m}}{\text{s}}$$

Plugging this velocity into the orbital velocity law, we find:

$$\text{cluster mass} = M_r \approx \frac{(3.1 \times 10^{22}\ \text{m}) \times \left(9.5 \times 10^{5}\ \dfrac{\text{m}}{\text{s}}\right)^2}{6.67 \times 10^{-11}\ \dfrac{\text{m}^3}{\text{kg} \times \text{s}^2}} = 4.2 \times 10^{44}\ \text{kg}$$

Note that the two methods have given two results that agree fairly well: 4.6×10^{44} kg and 4.2×10^{44} kg. Thus, we can be confident that we are in the correct range for the actual mass of the cluster. Taking 4.4×10^{44} kg as an intermediate value and recalling that the Sun's mass is 2.0×10^{30} kg, the cluster mass is about:

$$\text{cluster mass} \approx (4.4 \times 10^{44}\ \text{kg}) \times \frac{1\ \text{solar mass}}{2.0 \times 10^{30}\ \text{kg}} = 2.2 \times 10^{14}\ \text{solar masses}$$

pull [Section 14.1]. In this state of balance, the average kinetic energies of the gas particles are determined primarily by the strength of gravity and hence by the amount of mass within the cluster. Because the temperature of a gas reflects the average kinetic energies of its particles, the gas temperatures we measure with X-ray telescopes tell us the average speeds of the X ray–emitting particles. We can then use these speeds and the orbital velocity law to weigh the cluster.

The results obtained with this method agree well with the results found by studying the orbital motions of the cluster's galaxies. Again, we find that mass-to-light ratios in clusters of galaxies generally exceed 100 solar masses per solar luminosity. Even after accounting for the hot intracluster gas, it is clear that clusters must contain huge quantities of dark matter binding all the galaxies together.

FIGURE 21.9 Hubble Space Telescope picture of a galaxy cluster acting as a gravitational lens. The yellow elliptical galaxies are cluster members, and the odd, blue ovals are multiple images of a single galaxy that lies almost directly behind the cluster's center.

Gravitational Lensing

We have so far relied exclusively on methods derived from Newton's laws, such as the orbital velocity law, to measure galaxy and cluster masses. These laws keep telling us that the universe holds far more matter than we can see, a difficult result for some people to accept. Can we trust these laws to tell the truth? Orbital motions were our only tool for measuring masses until quite recently. Now we have another tool: *gravitational lensing*.

Gravitational lensing occurs because masses distort spacetime—the "fabric" of the universe [Section S4.3]. Massive objects can therefore act as **gravitational lenses** that bend light beams passing nearby. This prediction of Einstein's theory of general relativity was first verified in 1919 during an eclipse of the Sun [Section S4.5]. Because the light-bending angle of a gravitational lens depends on the mass of the object doing the bending, we can measure the masses of objects by observing how strongly they distort light paths.

Figure 21.9 shows a particularly striking example of how a cluster of galaxies can act as a gravitational lens. Many of the yellow elliptical galaxies concentrated toward the center of the picture belong to the cluster, but at least one of the galaxies pictured does not. At several positions on various sides of the central clump of yellow galaxies you will notice multiple images of the same blue galaxy. Each one of these images, whose sizes differ, looks like a distorted oval with an off-center smudge.

The blue galaxy seen in these multiple images lies almost directly behind the center of the cluster, at a much farther distance. Multiple images arise because photons traveling from this blue galaxy to Earth do not follow straight paths. Instead, the cluster's gravity bends their paths as they pass through the cluster, enabling light from this galaxy to arrive at Earth from a few slightly different directions. Each alternative path produces a separate, distorted image of the blue galaxy, making this one galaxy look like several galaxies (Figure 21.10).

Multiple images of a gravitationally lensed galaxy are rare, occurring only when a distant galaxy lies directly behind the lensing cluster, but single images of gravitationally lensed galaxies behind clusters are quite common. Figure 21.11 shows a more typical example. This picture shows numerous normal-looking galaxies and several arc-shaped galaxies. The oddly curved galaxies are not members of the cluster, nor are they really curved. They are normal galaxies lying far beyond the cluster whose images have been distorted by the cluster's gravity.

Careful analyses of the distorted images in pictures of clusters like these enable us to weigh the clusters without resorting to the orbital velocity law. Instead, Einstein's theory of general relativity tells us how massive these clusters must be to generate the observed distortions. It is reassuring that cluster masses derived in this way generally agree with those derived from galaxy velocities and X-ray temperatures. These completely different methods all indicate that clusters of galaxies hold very substantial amounts of dark matter.

TIME OUT TO THINK *Should the fact that we have three different ways of measuring cluster masses give us greater confidence that we really do understand gravity and that dark matter really does exist? Why or why not?*

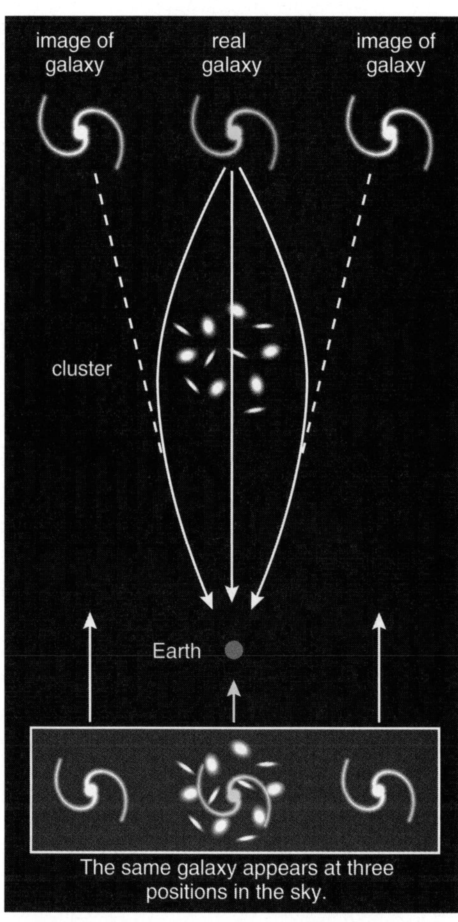

FIGURE 21.10 A cluster's powerful gravity bends light paths from background galaxies to Earth. If light can arrive from several different directions, we see multiple images of the same galaxy.

FIGURE 21.11 Hubble Space Telescope picture of the cluster Abell 2218. The thin, elongated galaxies are the images of background galaxies distorted by the cluster's gravity.

21.4 Dark Matter: Ordinary or Extraordinary?

We do not yet know what the dark matter in galaxies and clusters of galaxies is composed of, but our educated guesses fit into two basic categories. First, some or all of the dark matter could be *ordinary,* made of protons, neutrons, and electrons. In that case, the only unusual thing about dark matter is that it is dim; otherwise it is made of the same stuff as all the "bright matter" that we can see. The second possibility is that some or all of the dark matter is *extraordinary,* made of particles that we have yet to discover. A bit more terminology will be useful. Protons and neutrons belong to a category of particles called **baryons,** so ordinary matter is sometimes called **baryonic matter.**[4] By extension, extraordinary matter is called **nonbaryonic matter.**

[4]Technically, a baryon is a particle made from three quarks. (Recall that the fundamental particles are *quarks* and *leptons;* electrons and neutrinos are types of leptons [Section S5.2].)

Ordinary Dark Matter: MACHOs

Matter need not be extraordinary to be dark. In astronomy, "dark" merely means not as bright as a normal star and therefore not visible across vast distances of space. Your body is dark matter. Everything you own is dark matter. Earth and the planets are also dark matter. The process of star formation itself leaves behind dark matter in the form of *brown dwarfs,* those dim "failed stars" that are not quite massive enough to sustain nuclear fusion [Section 16.2]. Some scientists believe that trillions of faint red stars, brown dwarfs, and Jupiter-size objects left over from the Milky Way's formation still roam our galaxy's halo, providing much of its mass. They fancifully term these objects **MACHOs,** for *massive compact halo objects.*

Because MACHOs are too faint for us to see directly, astronomers who wish to detect them must resort to clever techniques. One innovative way to detect faint starlike objects in the Milky Way's halo takes advantage of gravitational lensing. If trillions of these MACHOs really exist, every once in a while a MACHO should drift across our line of sight to a more distant star. When a MACHO lies almost directly between us and the farther star, the MACHO's gravity will focus the star's light directly toward the Earth. The distant star thus will appear much brighter than usual for several days or weeks as the MACHO passes in front of it (Figure 21.12). We cannot see the MACHO itself, but the duration of the lensing event reveals its mass.

Gravitational lensing events such as these are rare; they happen to about one star in a million each

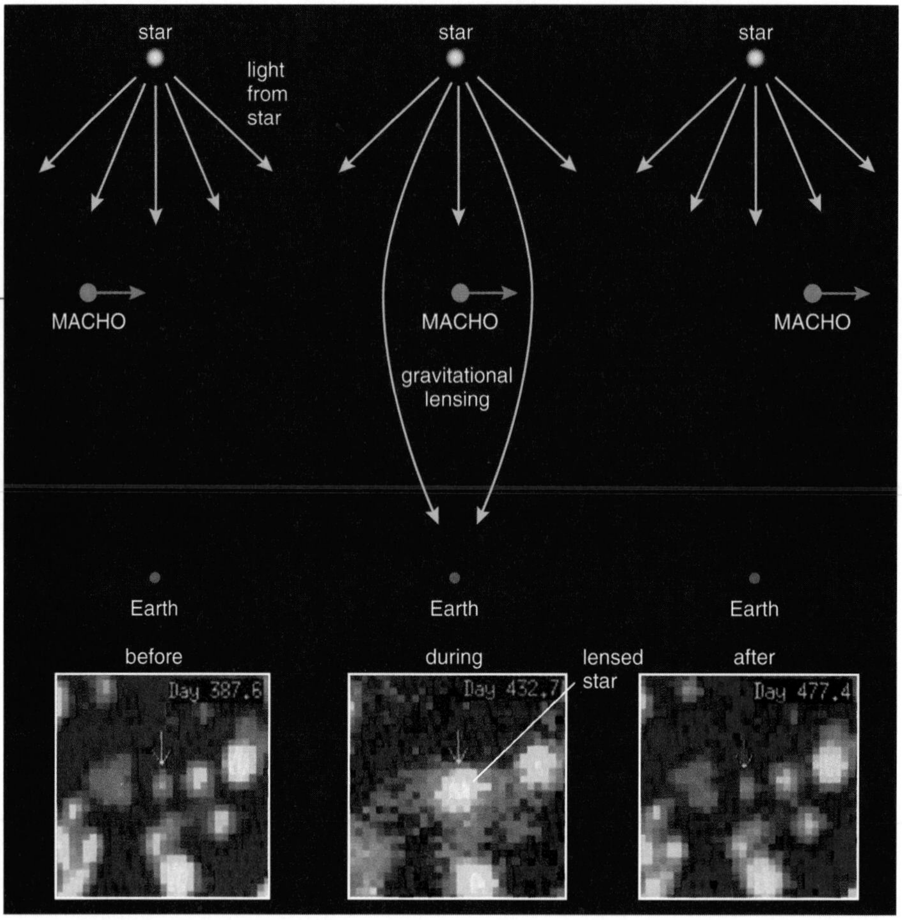

FIGURE 21.12 A lensing event in which a MACHO passing in front of a more distant star temporarily makes the star appear brighter, revealing the presence of the dark MACHO.

year. To detect MACHO lensing events, we therefore must monitor huge numbers of stars. Current large-scale monitoring projects now record numerous lensing events annually. These events demonstrate that MACHOs do indeed populate our galaxy's halo, but probably not enough of them to account for all the Milky Way's dark matter. Something else lurks unseen in the outer reaches of our galaxy. Maybe it is also normal matter made from baryons, but maybe not.

Extraordinary Dark Matter: WIMPs

A more exotic possibility is that most of the dark matter in galaxies and clusters of galaxies is not made of ordinary, baryonic matter at all. Let's begin to explore this possibility by taking another look at those nonbaryonic particles we discussed in Section 14.3: neutrinos. These unusual particles are dark by their very nature, because they have no electrical charge and hence cannot emit electromagnetic radiation of any kind. Moreover, they are never bound together with charged particles in the way that neutrons are bound in atomic nuclei, so their presence cannot be revealed by associated light-emitting particles. In fact, particles like neutrinos interact with other forms of matter through only two of the four forces: gravity and the *weak force* [Section S5.2]. For this reason, they are said to be *weakly interacting particles.* If you recall that trillions of neutrinos from the Sun are passing through your body at this very moment without doing any damage, you can see that the name *weakly interacting* fits well.

The dark matter in galaxies cannot be made of neutrinos, because these tiny particles travel through the universe at enormous speeds and can easily escape a galaxy's gravitational pull. (However, as we'll discuss shortly, neutrinos might contribute to the amount of dark matter outside galaxies.) But what if other weakly interacting particles exist that are similar to neutrinos but considerably heavier? They too would evade direct detection, but they would move more slowly and could collect into galaxies, adding mass without adding light. Such hypothetical particles are called **WIMPs,** for *weakly interacting massive particles.*[5] WIMPs could make up most of our

[5]WIMPs were the first form of hypothetical dark matter to be humorously named. The MACHOs were named later, to set them apart from the WIMPs and to add to the joke.

galaxy's mass, but they would be completely invisible in all wavelengths of light.[6]

It might surprise you that scientists would suspect the universe to be filled with particles they haven't yet discovered. However, WIMPs could also explain why dark matter doesn't behave like the visible matter in galaxies. During the early stages of the Milky Way's formation, the ordinary (baryonic) matter settled toward our galaxy's center and then flattened into a disk. Meanwhile, the dark matter stayed where it was, out in the galaxy's halo. This resistance of the dark matter to settling is exactly what we would expect from weakly interacting particles. Because WIMPs do not emit electromagnetic radiation, they cannot radiate away their energy. They are therefore stuck orbiting out at large distances and cannot collapse with the rest of the protogalactic cloud. By itself, the inability of dark matter to settle into the luminous regions of galaxies does not prove that dark matter is extraordinary and nonbaryonic. However, as we'll discuss in the next chapter, some other reasons lead us to believe that baryons represent only a minority of the universe's mass and hence that WIMPs really exist.

TIME OUT TO THINK *What do you think of the idea that much of the universe is made of as-yet-undiscovered particles? Can you think of other instances in the history of science in which the existence of something was predicted before it was discovered?*

21.5 Structure Formation

Dark matter remains enigmatic, but we are learning more about its role in the universe every year. Because galaxies and clusters of galaxies seem to contain much more dark matter than luminous matter, we believe that dark matter's gravitational pull must be the primary force holding these structures together. Thus, we strongly suspect that the gravitational attraction of dark matter is what pulled galaxies and clusters together in the first place.

Growth of Structure

Stars, galaxies, and clusters of galaxies are all **gravitationally bound systems,** meaning that their gravity is strong enough to hold them together. In most of the gravitationally bound systems we have discussed so far, gravity has completely overwhelmed the expansion of the universe. That is, while the universe as a whole is expanding, space is *not* expanding within our solar system, our galaxy, or our Local Group of galaxies.

Our best guess at how galaxies formed, briefly outlined in Section 20.2, envisions them as growing from slight density enhancements in the early universe. The expansion of space hindered galaxy formation for millions of years after the birth of the universe. No galaxies arose until gravity pulled the expanding hydrogen and helium gas that filled the universe into protogalactic clouds in which stars could begin to form. If dark matter is indeed the most common form of mass in galaxies, it must have provided most of this gravitational attraction. Regions with the highest densities of dark matter, and therefore the strongest gravity, would have gradually drawn in the surrounding gas. Some of this gas formed stars, but most of the galaxy remained dark. According to this model, the luminous matter in each galaxy must now be nestled inside the larger cocoon of dark matter that initiated the galaxy's growth.

The formation of a cluster of galaxies probably echoes the formation of a galaxy. Early on, all the galaxies that will eventually constitute the cluster are flying apart with the expansion of the universe. But the gravity of the dark matter associated with the cluster eventually reverses the trajectories of these galaxies. The galaxies ultimately fall back inward and start orbiting each other randomly, like the stars in the halo of our galaxy.

Some clusters of galaxies apparently have not yet finished forming: Their immense gravity continues to draw in new members. For example, the nearby Virgo Cluster of galaxies appears to be tugging on our own Local Group. Right now the Local Group is still moving away from the Virgo Cluster, but not as quickly as a simple application of Hubble's law would predict. Recall that Hubble's law lets us calculate the speed at which universal expansion carries us away from the Virgo Cluster [Section 19.3], but it does not account for any gravitational effects. The discrepancy between the predictions of Hubble's law and our actual velocity away from the Virgo Cluster is about 400 km/s. That is, we are moving away from the Virgo Cluster 400 km/s more slowly than we would be as a result of the expansion of the universe alone. Astronomers call this 400-km/s deviation from Hubble's law a **peculiar velocity,** but it's really not so peculiar: It is simply the effect of the gravitational attraction pulling us back toward Virgo against the flow of universal expansion. The overall effect is

[6]Weakly interacting particles that are slow-moving enough to collect into galaxies are sometimes called *cold dark matter* to set them apart from faster-moving *hot dark matter* particles such as neutrinos.

rather like that of swimming upstream against a strong current: You move "upstream" relative to other objects floating in the current, but you're still headed "downstream" relative to the shore because of the strong current. The speed of the current is like the speed of expansion of the universe, and your swimming speed "upstream" is your peculiar velocity.

Just as the Earth's gravity slows a rising baseball and eventually turns it around and pulls it back toward the ground, the gravitational tug of the Virgo Cluster may eventually turn the galaxies of our Local Group around and pull them into the cluster. Similar processes are taking place on the outskirts of other large clusters of galaxies, where we see many galaxies with large peculiar velocities pointing in the direction of the cluster. Their peculiar velocities indicate that the cluster's gravity is pulling on them. Eventually, some or maybe all of these galaxies will fall into the cluster. Thus, many clusters are still attracting galaxies, adding to the hundreds they already contain. On even larger scales, clusters themselves have surprisingly large peculiar velocities, hinting that they are parts of even bigger gravitationally bound systems, called **superclusters,** that are just beginning to form (Figure 21.13).

FIGURE 21.13 Peculiar velocities of galaxies flowing into superclusters. Each arrow shows the peculiar velocity of a galaxy inferred from a combination of observations and modeling. The Milky Way is at the center of the picture, and not all galaxies are shown. The area pictured is about 600 million light-years from side to side. Note how the galaxies tend to flow into regions where the density of galaxies is already high. These vast, high-density regions are probably superclusters in the process of formation.

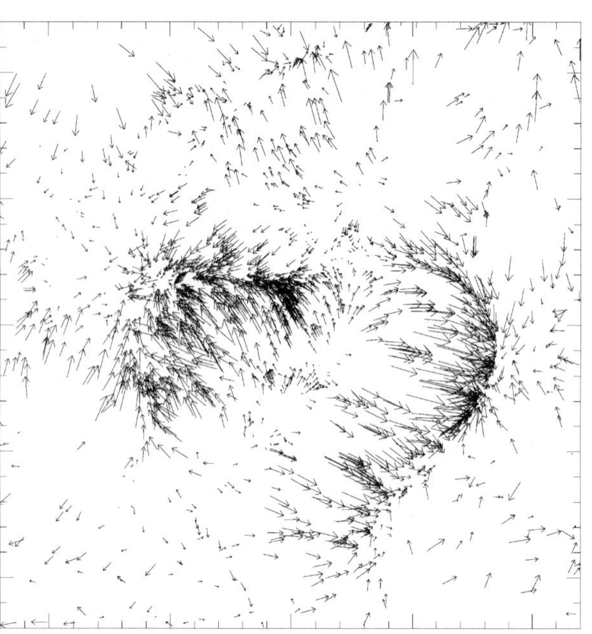

Large-Scale Structures

Beyond about 300 million light-years (100 Mpc), we can no longer measure distances accurately enough to determine peculiar velocities. We can estimate the distances to faraway galaxies only by measuring their redshifts and applying Hubble's law. Today we have redshift measurements for thousands upon thousands of such galaxies. When we convert the redshifts to distance estimates, we can create three-dimensional maps of the universe. Such maps have been made for only a few "slices" of the sky so far, but they are already revealing **large-scale structures** much vaster than clusters of galaxies.

One of the most famous depictions of large-scale structure is the "slice of the universe" pictured in Figure 21.14. Many astronomers collaborated to measure the redshifts of all the galaxies in this particular slice of the sky.[7] This project dramatically revealed the complex structure of our corner of the universe. Each dot in the picture represents an entire galaxy of stars. The arrangement of the dots reveals huge sheets of galaxies spanning many millions of light-years. Clusters of galaxies are located at the intersections of these sheets. Between the sheets of galaxies lie giant empty regions called **voids.**

Some of the structures we see in the universe are amazingly large. The so-called *Great Wall* of galaxies stretches across an expanse measuring some 180 million light-years (50 Mpc) side to side. Immense structures such as these apparently have not yet collapsed into randomly orbiting, gravitationally bound systems. They haven't had enough time. The universe may still be growing structures on ever larger scales, even though it is billions of years old (Figure 21.15).

Most astronomers believe that all these large-scale structures grew from slight density enhancements in the early universe, just as galaxies did. Galaxies, clusters, superclusters, and the Great Wall probably all started as mildly high-density regions of different sizes. The voids in the distribution of galaxies probably started as mildly low-density regions.

If this picture of structure formation is correct, then all the structures we see in today's universe mirror the original distribution of dark matter very early in time. Supercomputer models of structure formation in the universe can now simulate the growth of galaxies, clusters, and larger structures from very tiny density enhancements as the universe evolves. The results of these models look remarkably similar to the slice of the universe in Figure 21.14, bolstering our

[7]To keep the project manageable, the astronomers measured the redshifts of every galaxy they could find that was brighter than an agreed-upon brightness limit.

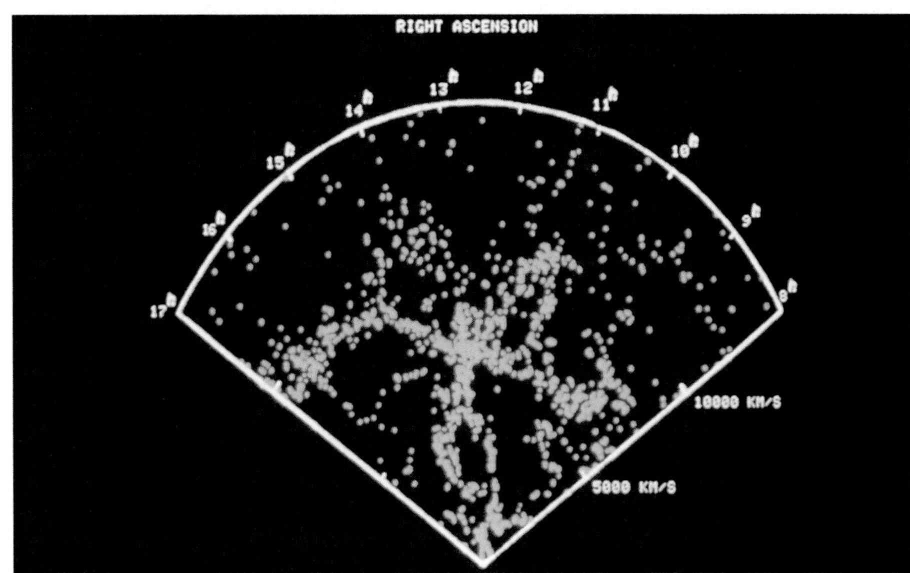

a

b

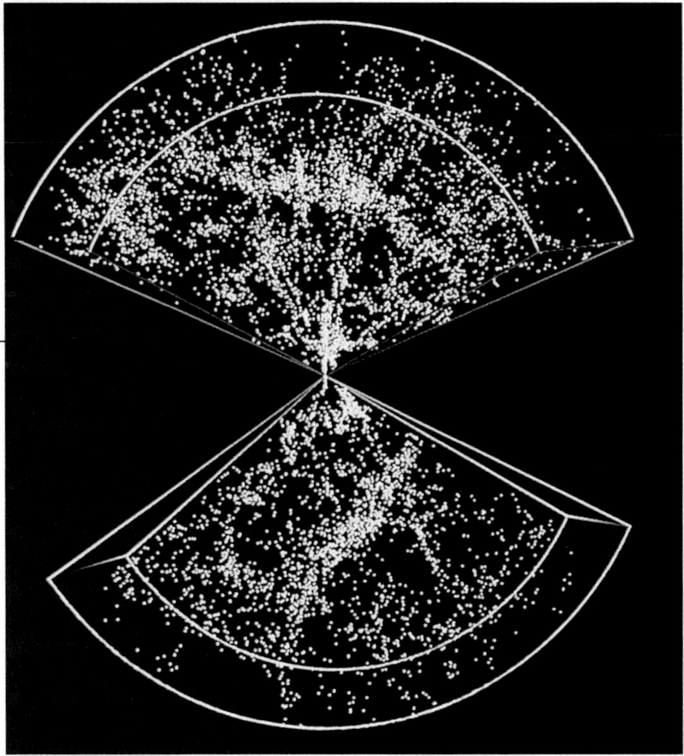

FIGURE 21.14 These "slices" show galaxies in a very thin, fanlike swath extending outward from the Earth and spanning 30° across our sky. Each dot represents a galaxy. Note that galaxies are not scattered randomly but instead seem to lie along sheets and strings interspersed with voids that contain very few galaxies. (**a**) The "Great Wall" is the sheet of galaxies that forms the arms of the stick figure at the center. (**b**) The upper swath is similar to (a) but thicker and thus contains more galaxies; the lower swath points in a different direction.

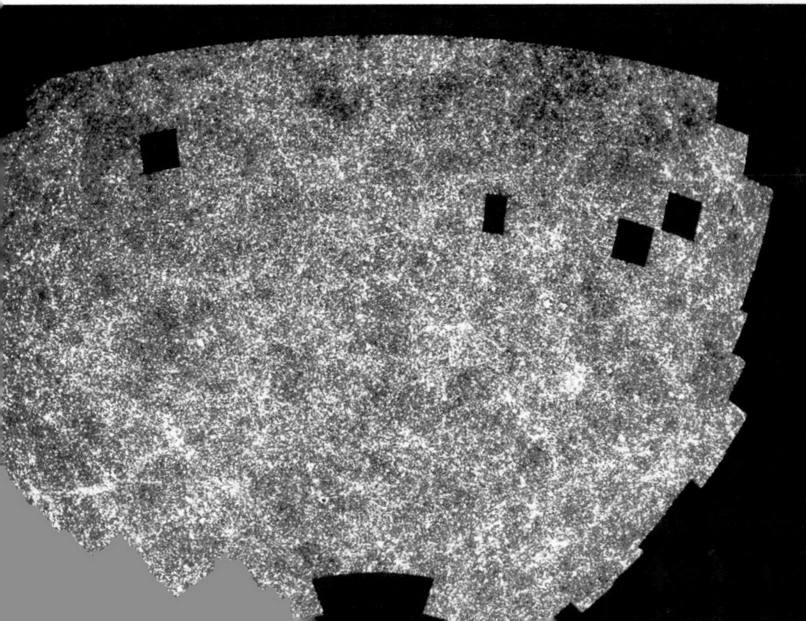

FIGURE 21.15 This picture shows the positions of over 3 million galaxies spread across 15% of the sky. Brighter areas contain more galaxies than darker areas, and the black rectangles are regions for which there is no data. Note that the universe is more uniform on large scales than on small scales.

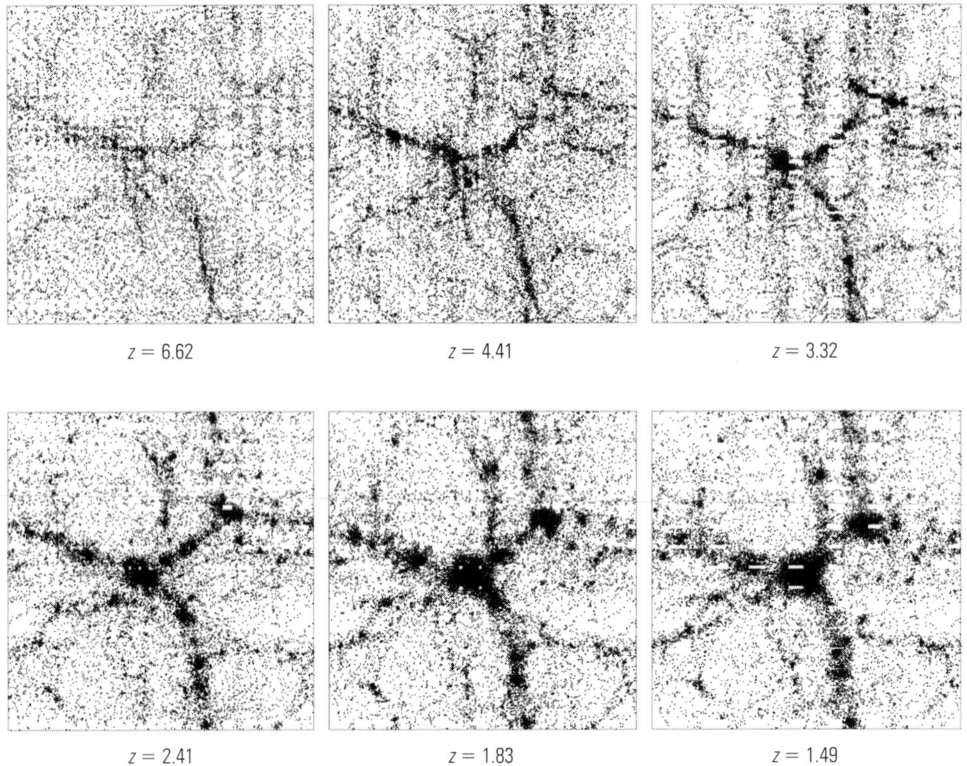

FIGURE 21.16 Frames from a supercomputer simulation of structure formation. As time progresses, the higher-density portions of the universe, shown in black, continue to attract more and more matter. Note how similar the filamentary structures in this simulation are to the slice of the universe in Figure 21.14.

confidence in this scenario (Figure 21.16). However, the models do not tell us *why* the universe started with these slight density enhancements—that is a topic for the next chapter. Nevertheless, it seems increasingly clear that these "lumps" in the early universe were the seeds of all the marvelous structures we see today.

21.6 Fire or Ice?

Some say the world will end in fire,
Some say in ice.
From what I've tasted of desire
I hold with those who favor fire.
But if it had to perish twice,
I think I know enough of hate
To say that for destruction ice
Is also great
And would suffice.

ROBERT FROST

We now confront one of the ultimate questions in astronomy: How will the universe end? This question basically comes down to asking whether or not the universe itself is a gravitationally bound system. Let's review the possibilities we discussed in Chapter 19:

- If the mass density of the universe is large enough, the collective gravity of all its matter will eventually halt and reverse the expansion. The galaxies will come crashing back together, and the entire universe will end in a fiery "big crunch." We call this type of universe a **closed universe.**
- If the mass density of the universe is too small, the collective gravity of all its matter cannot halt the expansion, and the universe will keep growing forever into the future. In this kind of universe, called an **open universe,** the galaxies will never come back together. The universe will cool into icy darkness.
- If the density of the universe equals the special value known as the *critical density* [Section 19.3], the collective gravity of all its matter is exactly the amount needed to balance the expansion. The universe will never collapse but will expand more and more slowly as time progresses. Such a precisely balanced universe is sometimes called a **flat universe,** but it is also open because its galaxies never come back together.

How the universe behaves in each of these scenarios is depicted in Figure 21.17.

TIME OUT TO THINK *Do you think that one of the potential fates of the universe is preferable to the others? If so, why? If not, why not?*

Counting all the luminous matter in the universe shows that the mass contained in stars falls far short of the critical density. The visible parts of galaxies contribute less than 1% of the mass density needed to halt the universe's expansion. The fate of the universe thus rests with the dark matter. Is there enough dark matter to halt the expansion of the universe?

Recall that we can estimate the amount of dark matter by looking at the mass-to-light ratio. It turns out that for dark matter to contribute enough mass for the universe to have the critical density, the average mass-to-light ratio must be approximately 1,000 solar masses per solar luminosity throughout the universe.[8] Clusters of galaxies have mass-to-light ratios of a few hundred solar masses per solar luminosity, still a few times less than the ratio needed for a closed universe. If the proportion of dark matter in the universe at large is similar to that in clusters, then the universe is open, with a density currently equal to about 25% of the critical density. For the universe to be closed, even more dark matter must lie beyond the boundaries of galaxy clusters.

Amazingly, tiny ghostlike neutrinos might supply enough mass to raise the universe's density to the critical density. The masses of individual neutrinos are still too small to measure, but what neutrinos lack in mass they might make up in sheer numbers—neutrinos are quite likely the most common particles in the universe other than photons. Because neutrinos zip through the universe so quickly, they do not necessarily collect into clusters of galaxies. However, they might gravitate into large-scale structures, where their influence would be measurable.

The peculiar velocities of galaxies allow us to probe the distribution of dark matter in large-scale structures, but studies of these departures from Hubble's law remain inconclusive. Some researchers have found that large-scale structures have even higher mass-to-light ratios than clusters of galaxies, indicating that the universe's mass density could be close to the critical value. As of 1998, however, most other studies hold the line near the 25% value we infer from clusters. The question of the universe's fate thus remains unanswered, with an open universe looking like the odds-on favorite.

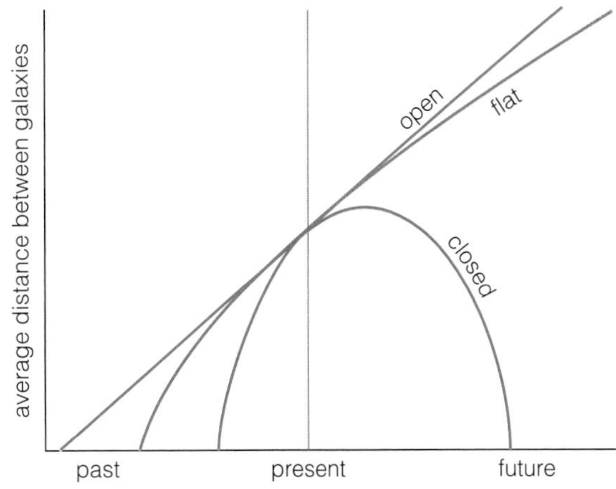

FIGURE 21.17 This graph shows how the average distance between galaxies changes with time for open, flat, and closed universes. A rising curve means the universe is expanding, and a falling curve means the universe is contracting.

This is the way the world ends
This is the way the world ends
This is the way the world ends
Not with a bang but a whimper.

FROM *The Hollow Men* BY T. S. ELIOT

[8]It's easy to see this fact: The "bright matter" in stars adds up to only about 1% of the critical density, so we need 100 times as much dark matter to reach the critical density. Because most stars are smaller and dimmer than the Sun, the average mass-to-light ratio of stars is on the order of 10 solar masses per solar luminosity. Multiplying this value by 100, we find that the mass-to-light ratio of the entire universe must be about 1,000 solar masses per solar luminosity if there is enough dark matter to reach the critical density.

THE BIG PICTURE

In this chapter, we have found that there may be much more to the universe than meets the eye. Dark matter too dim for us to see seems to far outweigh the stars. Here are some key "big picture" points to remember about this mysterious matter:

- Either dark matter exists, or we do not understand how gravity operates across galaxy-size distances. We have many reasons to have confidence in our understanding of gravity, so the majority of astronomers believe that dark matter is real.
- Measurements of the mass and luminosity of galaxies and galaxy clusters indicate that they contain far more mass in dark matter than in stars.
- Despite the fact that dark matter is by far the most abundant form of mass in the universe, we still have little idea what it is.
- Superclusters, walls, and voids much larger than clusters of galaxies extend many millions of light-years across the universe. Each of these structures probably began as a very slight enhancement in the density of dark matter early in time, and these enormous structures are still in the process of forming.
- Dark matter holds the key to the fate of the universe. If there is enough of it, its gravity will cause the expansion of the universe to halt and reverse someday. Current indications are that there is not enough dark matter to halt the expansion, in which case the universe will continue to expand forever.

Review Questions

1. In what sense is dark matter "dark"?
2. What evidence suggests that the Milky Way contains dark matter? What alternative explanation might be possible? Why do most astronomers prefer the idea of dark matter to the alternative explanation?
3. Briefly explain why it is more difficult to measure a galaxy's mass than to measure its luminosity.
4. Briefly describe how we construct rotation curves for spiral galaxies and how these curves lead us to conclude that spiral galaxies contain dark matter.
5. Why can't we apply the same techniques that we use for spirals to weigh elliptical galaxies? Briefly describe how we can infer the average orbital speeds of stars in an elliptical galaxy from the widths of its spectral lines.
6. What is a *mass-to-light ratio?* Explain why higher mass-to-light ratios imply more dark matter.
7. What are typical mass-to-light ratios for inner regions of spiral galaxies? For inner regions of elliptical galaxies? How do these mass-to-light ratios compare to the mass-to-light ratios we find when we look farther from a galaxy's center? What does that tell us about dark matter in galaxies?
8. Briefly describe the three different ways of measuring the mass of a cluster of galaxies. Do the results from the different methods agree? What do they tell us about dark matter in galaxy clusters?
9. What do we mean by the *intracluster medium* in galaxy clusters? Why wasn't it discovered prior to the 1960s? How does its temperature tell us about the mass of a cluster?
10. What is gravitational lensing? Why does it occur? How can we use it to estimate the masses of lensing objects?
11. Briefly explain what we mean by *baryonic* and *nonbaryonic matter.* Which type are you made of?
12. What do we mean by *MACHOs?* Describe several possibilities for the nature of MACHOs.
13. How can gravitational lensing events tell us about MACHOs? Based on current evidence, can MACHOs explain all the dark matter in the Milky Way? Why or why not?
14. Explain what we mean when we say that a neutrino is a *weakly interacting particle.* Why can't the dark matter in galaxies be made of neutrinos?
15. What do we mean by *WIMPs?* Why would we expect WIMPs to remain gravitationally bound to galaxies while neutrinos do not remain bound? Why would we expect WIMPs to be distributed throughout galactic halos, rather than settled into a disk?
16. What do we mean by a gravitationally bound system? Why isn't space expanding within systems such as our solar system or the Milky Way?
17. What are *peculiar velocities?* What causes them?

18. Briefly describe the appearance of the universe on the largest scales, including the distribution of galaxies in great sheets and the voids between the sheets. What is the *Great Wall* of galaxies?

19. Explain what we mean when we ask whether the universe is *closed, open,* or *flat.* How is the fate of the universe related to its density?

20. Explain why the total amount of dark matter in the universe holds the key to its final fate. According to current evidence about the amount of dark matter, what is the fate of the universe?

21. How might neutrinos affect the issue of the total amount of dark matter and the fate of the universe?

Discussion Questions

1. *Dark Matter or Revised Gravity.* One possible explanation for the large mass-to-light ratios we measure in galaxies and clusters is that we are currently using the wrong law of gravity to measure the masses of very large objects. If we really do misunderstand gravity, then many fundamental theories of physics, including Einstein's theory of general relativity, will need to be revised. Which explanation for these mass-to-light ratios do you find more appealing, dark matter or revised gravity? Explain why. Why do you suppose most astronomers find dark matter more appealing?

2. *Our Fate.* Scientists, philosophers, and poets alike have speculated on the fate of the universe. How would you prefer the universe as we know it to end, in a big crunch or through eternal expansion? Explain the reasons behind your choice.

Problems

1. *Mass-to-Light Ratio.* Suppose you discovered a galaxy with a mass-to-light ratio of 0.1 solar mass per solar luminosity. Would you be surprised? Explain why or why not. What would this measurement say about the nature of the stars in this galaxy?

2. *Weighing a Spiral Galaxy.* The top curve in Figure 21.4 shows the rotation curve for a spiral galaxy called NGC 7541.
   a. Use the orbital velocity law to determine the mass (in solar masses) of NGC 7541 enclosed within a radius of 10 kiloparsecs from its center. (*Hint:* 1 kpc = 3.1×10^{19} m.)
   b. Use the orbital velocity law to determine the mass of NGC 7541 enclosed within a radius of 20 kiloparsecs from its center.
   c. Based on your answers to parts (a) and (b), what can you conclude about the distribution of mass in this galaxy?

3. *More Spiral Galaxies.* Figure 21.4 shows the rotation curves for four spiral galaxies. Use the orbital velocity law and the given data to make the best possible estimate for the *total* mass of each galaxy. Explain your work.

4. *Weigh a Cluster of Galaxies.* Suppose a cluster of galaxies has a radius of about 2 Mpc = 6.2×10^{22} m and an intracluster medium with a temperature of 8×10^7 K. Estimate the mass of the cluster. Give your answer in both kilograms and solar masses.

5. *How Many MACHOs?* Suppose the rotation of a highly simplified galaxy whose stars are all identical to the Sun (1 solar mass per solar luminosity) shows that its overall mass-to-light ratio is equal to 30 solar masses per solar luminosity.
   a. What is the ratio of dark matter to luminous matter in this galaxy?
   b. Suppose all the dark matter consists of MACHOs similar to Jupiter, each with a mass of 0.001 solar mass. How many of these MACHOs must the galaxy contain for each ordinary star? Explain.

6. *Web Project: Gravitational Lenses.* Gravitational lensing occurs in numerous astronomical situations. Compile a catalog of examples from the Web. Try to find pictures of lensed stars, quasars, and galaxies. Supply a one-paragraph explanation of what is going on in each picture.

7. *Research Project: Fate of the Universe.* We are continually learning more about the amount of dark matter in the universe, and the question of whether the universe is open or closed is therefore constantly revisited. Search for the most recent information about dark matter and the fate of the universe, preferably by finding news articles on this topic published within the past year. Write a one- to three-page report on your findings. Be sure to state whether, according to the most recent evidence, the universe appears to be open or closed.

8. *Research Project: The Nature of Dark Matter.* Find and study recent reports on new ideas about the possible nature of dark matter. Write a one- to three-page report that summarizes the latest ideas about what dark matter is made of.

22
The Beginning of Time

THE UNIVERSE HAS BEEN EXPANDING FOR OVER 10 BILLION YEARS. During that time, matter collected into galaxies. Stars formed in those galaxies, producing heavy elements that were recycled into later generations of stars. One of these late-coming stars formed about 4.6 billion years ago, in a remote corner of a galaxy called the Milky Way. This star was born with a host of planets that formed in a flattened disk surrounding it. One of these planets soon became covered with life that gradually evolved into more and more complex forms. Today, the most advanced life-forms on this planet, human beings, can look back on this series of events and marvel at how the universe created conditions suitable for life.

To this point in the book, we have discussed how the matter in the early universe collected into galaxies, stars, planets, and ultimately people. We have seen that we are star stuff, and galaxy stuff as well. More to the point, we now recognize that we are the product of more than 10 billion years of cosmic evolution. We have one question left to address: Where did the *matter itself* come from?

To answer this ultimate question, we must go beyond the most distant galaxies and even beyond what we can see near the horizon of the universe. We must go back not only to the origins of matter and energy but to the beginning of time itself.

22.1 Running the Expansion Backward

Is it really possible to study the origin of the entire universe? Not long ago, questions about creation were considered unfit for scientific study. But that attitude began to change with Hubble's discovery that the universe is expanding. This discovery led to the insight that all things very likely sprang into being at a single moment in time, in an event that we have come to call the *Big Bang*. Today, powerful telescopes allow us to view how galaxies have changed over the last 10 billion years, and at great distances we see young galaxies still in the process of forming [Section 20.2]. These observations confirm that the universe is gradually aging, just as we should expect if the entire universe really was born some 10–16 billion years ago.

Unfortunately, we cannot see back to the very beginning of time. Light from the most distant galaxies shows us what the universe looked like when it was a few billion years old. Beyond these galaxies, we have not yet found any objects shining brightly enough for us to see them. But we face an even more fundamental problem. The universe is filled with a faint glow of radiation that appears to be the remnant heat of the Big Bang. This faint glow comes from a time when the universe was about 300,000 years old, and it prevents us from seeing directly to earlier times. Thus, just as we must rely on theoretical modeling to determine what the Earth or the Sun is like on the inside, we must use modeling to investigate what happened in the universe in its first 300,000 years.

Fortunately, calculating conditions in the early universe is straightforward if we consider some fundamental principles. We know that the expanding universe is cooling and becoming less dense as it grows. Therefore, the universe must have been hotter and denser in the past. Calculating exactly how hot and dense the universe must have been when it was more compressed is similar to calculating the temperatures and densities in a car engine as the pistons compress the fuel mix—except that the conditions become much more extreme. Figure 22.1 shows just how hot the universe was during its earliest moments, according to such calculations. To understand these early moments, we thus need to know how matter and energy behave under such extraordinarily hot conditions.

Advances made in modern physics during the latter half of the twentieth century enable us to predict the behavior of matter and energy under conditions far more extreme than in the centers of the hottest stars. When we apply our understanding of modern physics to the early universe, we find something so astonishing that it almost sounds preposterous: Today, we may understand the behavior of matter and energy well enough to describe what was happening in the universe just *one ten-billionth* (10^{-10}) of a second after the Big Bang.[1] And that's not all. While our understanding of physics is less certain under the more extreme conditions that prevailed even earlier, we have some ideas about what the universe was like when it was a mere 10^{-35} second old, and perhaps a glimmer of what it was like at the age of just 10^{-43} second. These tiny fractions of a second are so small that, for all practical purposes, we are studying the very moment of creation.

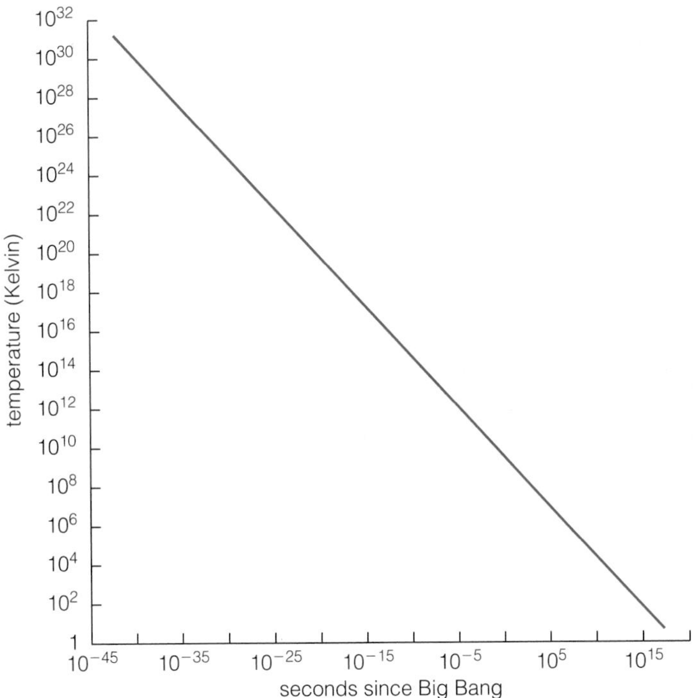

FIGURE 22.1 Temperature of the universe from the Big Bang to the present (10^{10} years $\approx 3 \times 10^{17}$ seconds).

22.2 A Scientific History of the Universe

The **Big Bang theory**—the scientific theory of the universe's earliest moments—presumes that all we see today, from Earth to the cosmic horizon, began as an incredibly tiny, hot, and dense collection of matter and radiation. It describes how expansion and cooling of this unimaginably intense mixture of particles and photons could have led to the present universe of stars and galaxies, and it explains several aspects

[1]Just as there is uncertainty in the exact age of the universe, there is also uncertainty in the exact time at which these conditions occurred. However, the times given here and in Figure 22.2 (e.g., 10^{-10} second) are probably accurate to the nearest power of 10.

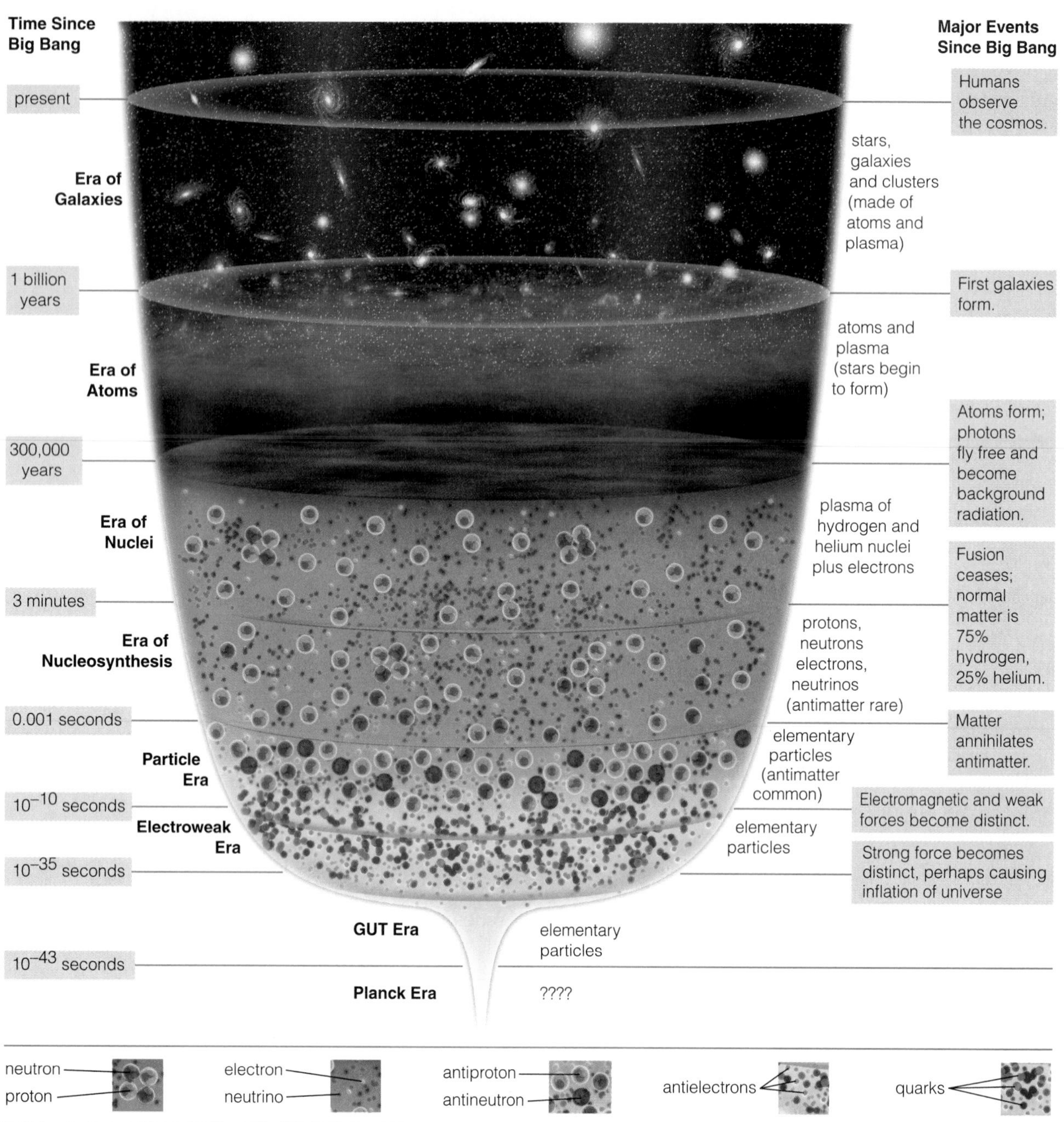

FIGURE 22.2 This diagram summarizes the eras of the universe. The names of the eras and their ending times are indicated on the left, and the state of matter during each era and the events marking the end of each era are indicated on the right.

of today's universe with impressive accuracy. Our ideas about the earliest moments of the universe are quite speculative because they depend on aspects of physics that we do not yet fully comprehend. However, the evidence supporting this theory grows stronger and stronger as we focus on progressively later moments in time. We will discuss the evidence supporting the Big Bang theory later in this chapter. First, in order to help you understand the significance of the evidence, let's examine the story of creation according to this theory.

Figure 22.2 summarizes the story by dividing the overall history of the universe into a series of *eras,* or time periods. Each era is distinguished from the next by some major change in the conditions of the universe. You'll find it easiest to keep track of the various eras if you refer back to this figure as we discuss each era in detail.[2]

Before we delve into the story of creation, we must discuss one important difference between the early universe and the present. The early universe was so hot that photons could transform themselves into matter, and vice versa, in accordance with Einstein's formula $E = mc^2$ [Section 5.2]. Physicists can produce reactions that create and destroy matter in their laboratories, but these reactions are now relatively rare in the universe at large.

One example of such a reaction is the creation or destruction of an *electron–antielectron pair* [Section S5.5]. When two photons collide with a total energy greater than twice the mass-energy of an electron (i.e., the electron's mass times c^2), they can create two brand-new particles: a negatively charged electron and its positively charged twin, the *antielectron,* also known as a *positron* (Figure 22.3). The electron is a particle of **matter,** and the antielectron is a particle of **antimatter.** The reaction that creates an electron–antielectron pair also runs in reverse: When an electron and an antielectron meet, they *annihilate* each other totally, transforming all their mass-energy back into photon energy.

Similar reactions can produce or destroy any particle–antiparticle pair, such as a proton and antiproton or neutron and antineutron. The early universe was therefore filled with an extremely dynamic blend of photons, matter, and antimatter, converting furiously back and forth. However, despite all these vigorous reactions, describing conditions in the early universe is straightforward because we can calculate the proportions of the various forms of radiation and matter from the universe's temperature and density at any time.

[2]The eras described in this book highlight some of the major transitions that took place in the early universe. In more specialized books on cosmology, these eras are subdivided into many more eras, providing additional details about the early universe not covered here.

FIGURE 22.3 Electron–antielectron creation and annihilation.

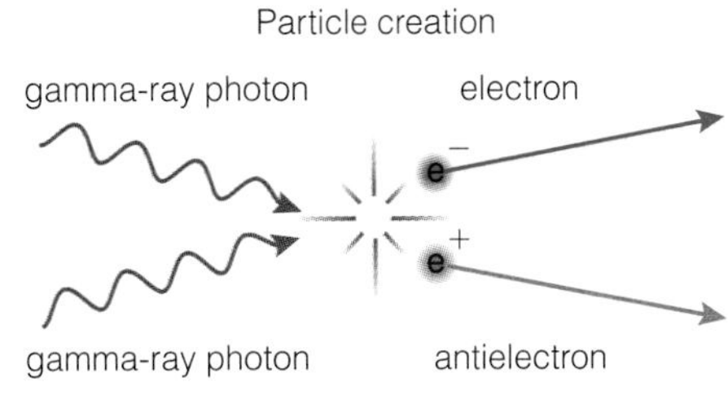

The First Instant

The scientific story of creation begins when the universe was an incomprehensibly tiny 10^{-43} second old. This instant in time is called the **Planck time** after physicist Max Planck, one of the founders of the science of quantum mechanics. According to the laws of quantum mechanics, there must have been substantial energy fluctuations from point to point in the very early universe. Because energy and mass are equivalent, Einstein's theory of general relativity tells us that these energy fluctuations must have generated a rapidly changing gravitational field that randomly warped space and time.

Prior to the Planck time, in the **Planck era,** these random energy fluctuations were so large that our current theories are powerless to describe what might have been happening. The problem is that we do not yet have a theory that links quantum mechanics and general relativity. In a sense, quantum mechanics is our successful theory of the very small (subatomic particles), and general relativity is our successful theory of the very big (the large-scale structure of spacetime). Perhaps someday we will be able to merge these theories of the very small and the very big into a single "theory of everything." Until that happens, science cannot describe the universe before the Planck time.

The GUT Era

Understanding the transition that marked the beginning of the next era requires thinking in terms of the *forces* that operate in the universe. Everything that happens in the universe today is governed by four distinct forces: *gravity, electromagnetism,* the *strong force,* and the *weak force* [Section S5.2]. Gravity dominates large-scale action, and electromagnetism dominates chemical and biological reactions. The strong and weak forces determine what happens within atomic nuclei: The strong force binds nuclei together, and the weak force mediates nuclear reactions such as fission or fusion.

These forces were not so distinct in the early universe. The four forces that act so differently today turn out to be separate aspects of a smaller number of more fundamental forces, probably only one or two (Figure 22.4). As an analogy, think about ice, liquid water, and water vapor. These three substances are quite different from one another in appearance and behavior, yet they are just different phases of the single substance H_2O. In a similar way, physicists have shown that the electromagnetic and weak forces lose their separate identities under conditions of very high temperature or energy and merge together into a single *electroweak force.* Physicists also believe that the electroweak force and the strong force ultimately lose their separate identities at even higher energies. The theories that predict this merger of the electroweak and strong forces are called **grand unified theories,** or **GUTs** for short; their merger is usually referred to as the *GUT force.* Many physicists believe that at even higher energies the GUT force and gravity merge into a single "super force" that governs the behavior of everything.[3]

Calculations from general relativity and quantum mechanics suggest that this unified "super force" may have reigned in the universe during the Planck era. If so, the Planck time (10^{-43} s) marks the instant when gravity became distinct from the other three forces, which were still merged as the GUT force. By analogy to ice crystals forming as a liquid cools, we say that gravity "froze out" at the Planck time. The universe subsequently entered the **GUT era,** when two forces operated in the universe: gravity and the GUT force.

The GUT era lasted but a tiny fraction of a second, coming to an end when the universe had cooled to 10^{27} K at an age of 10^{-35} second. Grand unified theories predict that the strong force froze out from the GUT force at this point, leaving three forces operating in the universe: gravity, the strong force, and the electroweak force. As we will discuss later in the chapter, it now seems possible that this freezing out of the strong force released an enormous amount of energy. This energy release would have caused the universe to undergo a sudden and dramatic expansion that we call **inflation.** In the space of a mere 10^{-33} second, pieces of the universe the size of an atomic

[3]The term *grand unified* describing theories that merge the strong and electroweak forces poses a bit of a crisis in terminology for theories that link all four forces together. If your theory is already grand, what should you call it when it becomes even better? How about "super"? Among the names you may hear for theories linking all four forces are *supersymmetry, superstrings,* and *supergravity.*

FIGURE 22.4 The four forces are distinct at low temperatures but merge at very high temperatures.

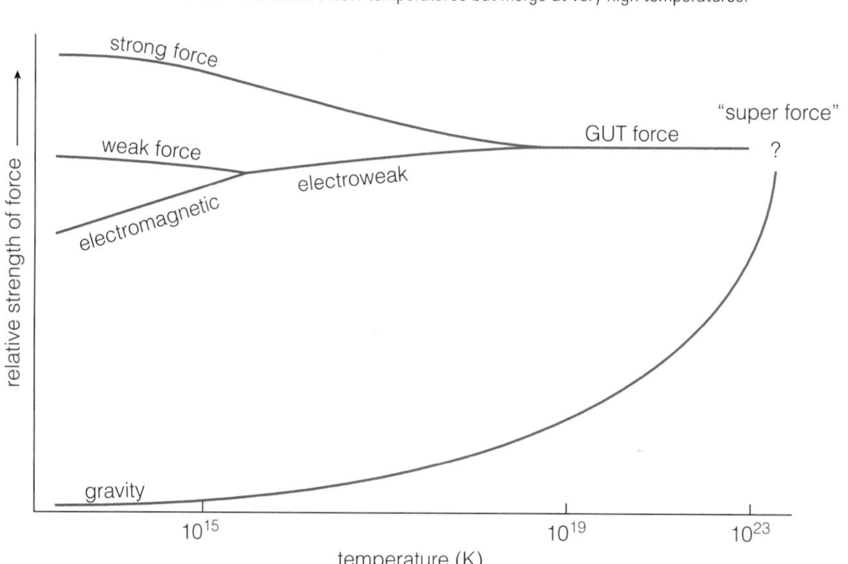

nucleus may have grown to the size of our solar system. Inflation sounds bizarre, but it might actually explain some puzzling features of today's universe, as we will see.

The Electroweak Era

The end of the GUT era marks the beginning of the **electroweak era,** so named because the electromagnetic and weak forces were still unified into the electroweak force. Intense radiation filled all of space, as it had since the Planck era, spontaneously producing matter and antimatter particles of many different sorts that immediately reverted back to energy. The universe continued to expand and cool throughout the electroweak era, dropping to a temperature of 10^{15} K when it reached an age of 10^{-10} second. This temperature is still 100 million times hotter than the temperature in the core of the Sun, but it was low enough for the electromagnetic and weak forces to freeze out from the electroweak force. After this instant (10^{-10} s), all four forces were forever distinct in the universe.

The end of the electroweak era marks an important transition not only in the universe, but also in human understanding of the universe. The theory that unified the weak and electromagnetic forces, developed in the 1970s, predicted the emergence of new types of particles (called the W and Z, or *weak bosons*) when temperatures rose above the 10^{15} K that pervaded the universe when it was 10^{-10} second old. In 1983, experiments performed in a huge particle accelerator near the French/Swiss border (CERN [Section S5.2]) reached energies equivalent to such high temperatures for the first time.[4] The new particles showed up just as predicted, produced from the extremely high energy in accord with $E = mc^2$. Thus, we have direct experimental evidence concerning the conditions in the universe at the end of the electroweak era. Prior to that time, we do not have any direct experimental evidence. Thus, our theories concerning the earlier parts of the electroweak era and the GUT era are much more speculative than our theories describing the universe from the end of the electroweak era to the present.

[4]In the words of Carlo Rubbia, who received the Nobel Prize for leading the 1983 experiments (along with Simon Van de Meer), the particle accelerators essentially produce "little bangs." That is, they produce in miniature the conditions that prevailed in the universe shortly after the Big Bang. The theory concerning the electroweak unification was proposed by Steven Weinberg, Sheldon Glashow, and Abdus Salam, who shared the Nobel Prize in 1979.

The Particle Era

Spontaneous creation and annihilation of particles continued throughout the next era, which we call the **particle era** because tiny particles of matter were as numerous as photons. These particles included electrons, neutrinos, and *quarks*—the building blocks of heavier particles such as protons, antiprotons, neutrons, and antineutrons [Section S5.2]. Near the end of the particle era, when the universe was about 0.0001 second old, the temperature grew too cool for quarks to exist on their own, and quarks combined in groups of three to form protons and neutrons. The particle era came to an end when the universe reached an age of 1 millisecond (0.001 s) and the temperature had fallen to 10^{12} K. At this point, it was no longer hot enough to spontaneously produce protons and antiprotons (or neutrons and antineutrons) from pure energy.

If the universe had contained equal numbers of protons and antiprotons (or neutrons and antineutrons) at the end of the particle era, all of the pairs would have annihilated each other, creating photons and leaving essentially no matter in the universe. In today's universe, photons outnumber protons by over a billion to one and appear to outnumber antiprotons by a much larger factor. Thus, we conclude that protons must have slightly outnumbered antiprotons at the end of the particle era: For every billion antiprotons, there were about a billion and one protons. Thus, for each 1 billion protons and antiprotons that annihilated each other, creating photons, a single proton was left over. This seemingly slight excess of ordinary matter over antimatter makes up all the matter in the present-day universe. That is, the protons (and neutrons) left over from when the universe was 0.001 second old are the very ones that make up our bodies.

The Era of Nucleosynthesis

So far, everything we have discussed occurred within the first 0.001 second of the universe's existence—a time span shorter than the time it takes you to blink an eye. At this point, the protons and neutrons left over after the annihilation of antimatter attempted to fuse into heavier nuclei. However, the heat of the universe remained so high that most nuclei broke apart as fast as they formed. This dance of fusion and break-up marked the **era of nucleosynthesis.**

The era of nucleosynthesis ended when the universe was about 3 minutes old. By this time, the density in the expanding universe had dropped so much that fusion no longer occurred, even though the temperature was still about a billion Kelvin (10^9 K)—

THINKING ABOUT . . .

Einstein's Greatest Blunder

Shortly after Einstein completed his general theory of relativity in 1915, he found that it predicted that the universe could not be standing still: The mutual gravitational attraction of all the matter would make the universe collapse. However, Einstein thought at the time that the universe should be eternal and static, so he decided to alter his equations. In essence, he inserted a "fudge factor," which he called the *cosmological constant,* that acted as a repulsive force to counteract the attractive force of gravity. Had he not been so convinced that the universe should be standing still, he might instead have come up with the correct explanation for why the universe is not collapsing: because it is still expanding from the event of its birth. After Hubble discovered that the universe is expanding, Einstein called his invention of the cosmological constant the greatest blunder of his career.

Despite Einstein's disavowal, variations both on the idea of an eternal universe and on the cosmological constant continued to crop up over the years. One of the cleverest ideas, developed in the late 1940s, was called the *steady state theory* of the universe. This theory accepted the fact that the universe is expanding but rejected the idea of a Big Bang and instead postulated that the universe is infinitely old. This idea may seem paradoxical at first: If the universe has been expanding forever, shouldn't every galaxy be infinitely far away from every other galaxy? The steady state theory suggested that new galaxies continually form in the gaps that open up as the universe expands, thereby keeping the average distance between galaxies the same at all times. In a sense, the steady state theory said that the creation of the universe is an ongoing and eternal process rather than having happened all at once with a Big Bang.

Two key discoveries ruled out the steady state theory as an alternative to the Big Bang. First, the 1965 discovery of the cosmic background radiation matched a prediction of the Big Bang theory but could not be adequately explained by the steady state theory. Second, the steady state theory predicts that the universe looks about the same at all times, which

much hotter than the temperature at the center of the Sun today. When fusion ceased, about 75% of the mass of the ordinary (baryonic) matter in the universe remained as individual protons, or hydrogen nuclei. The other 25% of this mass had fused into helium nuclei, with trace amounts of deuterium (hydrogen with a neutron) and lithium. Except for the small amount of matter that stars later forged into heavier elements, the chemical composition of the universe remains unchanged today.

The Era of Nuclei

At the end of the era of nucleosynthesis, the universe consisted of a very hot plasma of hydrogen nuclei, helium nuclei, and free electrons. This basic picture held for about the next 300,000 years as the universe continued to expand and cool. The fully ionized nuclei moved independently of electrons during this period (rather than being bound with electrons in neutral atoms), which we call the **era of nuclei.** Throughout this era, photons bounced rapidly from one electron to the next, just as they do deep inside the Sun today, never managing to travel far between collisions. Any time a nucleus managed to capture an electron to form a complete atom, one of the photons quickly ionized it.

The era of nuclei came to an end when the expanding universe was about 300,000 years old, at which point the temperature had fallen to about 3,000 K—roughly half the temperature of the Sun's surface today. Hydrogen and helium nuclei finally captured electrons for good, forming stable, neutral atoms for the first time. With electrons now bound into atoms, the universe suddenly became transparent. Photons, formerly trapped among the electrons, began to stream freely across the universe. We still see these photons today as the *cosmic background radiation,* which we will discuss shortly.

The Era of Atoms and the Era of Galaxies

We've already discussed the rest of the story in earlier chapters. The end of the era of nuclei marked the beginning of the **era of atoms,** when the universe consisted of a mixture of neutral atoms and plasma. Thanks to the slight density enhancements present in the universe at this time and the gravitational attraction of dark matter, the atoms and plasma slowly assembled into protogalactic clouds [Section 21.5]. Stars formed in these clouds, transforming is inconsistent with observations showing that galaxies at great distances look younger than nearby galaxies. As a result of these predictive failures, the steady state theory is no longer taken seriously by most astronomers.

More recently, astronomers have begun to take the idea of a cosmological constant more seriously. In the mid-1990s, a few observations suggested that the ages of the oldest stars are slightly older than the age of the universe derived from Hubble's constant [Section 19.3]. Clearly, stars cannot be older than the universe. If these observations were being interpreted correctly, the universe had to be older than Hubble's constant implies. Recall that we estimate the age of the universe by running the expansion backward and by assuming that the rate of expansion has either remained steady or slowed due to gravity. If instead the expansion rate has accelerated, so that the universe is expanding faster today than it was in the past, then the age of the universe would be greater than that ordinarily found from Hubble's constant. What could cause the expansion of the universe to accelerate over time? The repulsive force represented by a cosmological constant, of course.

Although further study of the troubling observations suggests that the stars probably are *not* older than the age of the universe derived from Hubble's constant, other observations also suggest a possible need for a cosmological constant. As of 1998, some preliminary measurements of distances to high-redshift galaxies (using white-dwarf supernovae as standard candles) suggest that the expansion *is* accelerating. A cosmological constant could account for this startling finding, but we'll need many more observations before we'll know whether it is correct. In either case, scientists are attempting to use variations of the cosmological constant to explain some of the remaining puzzles associated with the Big Bang theory, such as the precise nature of inflation. Einstein's greatest blunder, it seems, just won't go away.

the gas clouds into galaxies. The first full-fledged galaxies had formed by the time the universe was about 1 billion years old, beginning what we call the **era of galaxies.**

The era of galaxies continues to this day. Generation after generation of star formation in galaxies steadily builds elements heavier than helium and incorporates them into new star systems. Some of these star systems develop planets, and on at least one of these planets life burst into being a few billion years ago. And here we are, thinking about it all. Describing both the follies and the achievements of the human race, Carl Sagan once said, "These are the things that hydrogen atoms do—given 15 billion years of cosmic evolution."

22.3 Evidence for the Big Bang

Like any scientific theory, the Big Bang theory is a model designed to explain a set of facts. If it is close to the truth, it should be able to make predictions about the real universe that we can verify through observations or experiments. The Big Bang model has gained wide scientific acceptance for two key reasons:

- The Big Bang model predicts that the radiation that began to stream across the universe at the end of the era of nuclei should still be present today. Sure enough, we find that the universe is filled with what we call **cosmic background radiation.** Its characteristics precisely match what we expect according to the Big Bang model.
- The Big Bang model predicts that some of the original hydrogen in the universe should have fused into helium during the era of nucleosynthesis. Observations of the actual helium content of the universe closely match the amount of helium predicted by the Big Bang model. Fusion of hydrogen to helium in stars could have produced only about 10% of the observed helium.

The Cosmic Background Radiation

The first major piece of evidence supporting the Big Bang theory arrived in 1965. Arno Penzias and Robert Wilson, two physicists working at Bell Laboratories in New Jersey, were calibrating a sensitive microwave

FIGURE 22.5 Arno Penzias and Robert Wilson with the Bell Labs microwave antenna.

antenna designed for satellite communications (Figure 22.5). (*Microwaves* fall within the radio portion of the electromagnetic spectrum.) Much to their chagrin, they kept finding unexpected "noise" in every measurement they made with the antenna. Fearing that they were doing something wrong, they worked frantically to discover and eliminate all possible sources of background noise. They even climbed up on their antenna to scrape off pigeon droppings, on the off-chance that these were somehow causing the noise. No matter what they did, the microwave noise wouldn't go away. The noise was the same no matter where they pointed their antenna, indicating that it came from all directions in the sky and ruling out the possibility that it came from any particular astronomical object or from any place on the Earth. Embarrassed by their inability to explain the noise, Penzias and Wilson prepared to "bury" their discovery at the end of a long scientific paper about their antenna.

Meanwhile, physicists at nearby Princeton University were busy calculating the expected characteristics of the radiation left over from the heat of the Big Bang. They concluded that, if the Big Bang had really occurred, this radiation should be permeating the entire universe and should be detectable with a microwave antenna. On a fateful airline trip home from an astronomical meeting, Penzias sat next to an astronomer who told him of the Princeton calculations. The Princeton group soon met with Penzias and Wilson to compare notes. The "noise" in the Bell Labs antenna was not an embarrassment after all. Instead, it was the cosmic background radiation—and the first strong evidence that the Big Bang had really happened. Penzias and Wilson received the 1978 Nobel Prize in physics for their discovery of the cosmic background radiation.[5]

The cosmic background radiation consists of photons arriving at Earth directly from the end of the era of nuclei, when the universe was about 300,000 years old. Because neutral atoms could remain stable for the first time, they captured most of the electrons in the universe. With no more free electrons to block them, the photons from that epoch have flown unobstructed through the universe ever since (Figure 22.6). When we observe the cosmic background

[5]The dramatic story of the discovery of the cosmic background radiation is told in greater detail, along with much more scientific history, in Timothy Ferris, *The Red Limit* (New York: Quill, 1983).

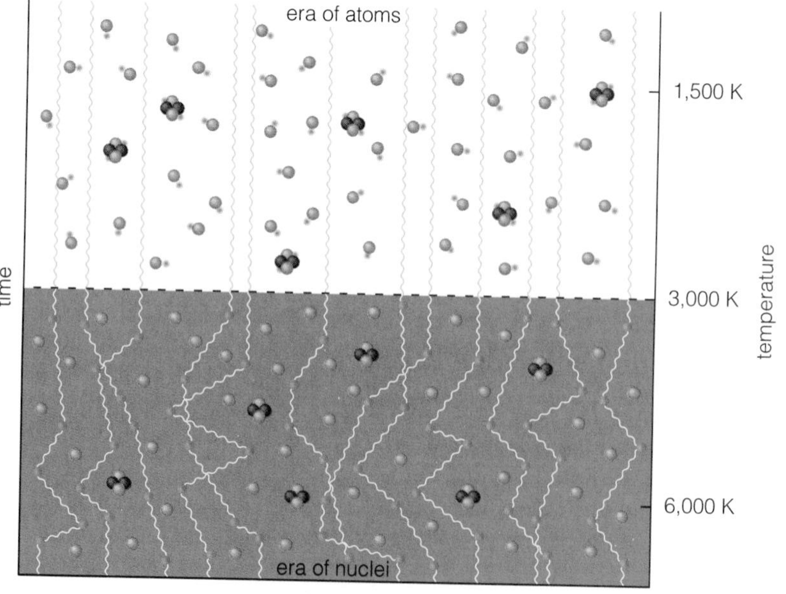

FIGURE 22.6 Photons frequently collided with free electrons during the era of nuclei and thus could travel freely only after electrons became bound into atoms. The photons released at the end of the era of nuclei make up the cosmic background radiation.

radiation, we essentially are seeing back to a time when the universe was only 300,000 years old. In that sense, we are seeing light from the most distant regions of the observable universe—only 300,000 light-years from our cosmological horizon [Section 19.4]. Surprisingly, it does not take a particularly powerful telescope to "see" this radiation. In fact, you can pick it up with an ordinary television antenna. If you set an antenna-fed television (i.e., *not* cable or satellite TV) to a channel for which there is no local station, you will see a screen full of static "snow." About 1% of this static is due to photons in the cosmic background radiation. Try it. If your friends ask why you are watching nothing, tell them that you are actually watching the most incredible sight ever seen on a television screen: the Big Bang, or at least as close to it as we'll ever get.

The cosmic background radiation came from the heat of the universe itself and therefore should have an essentially perfect thermal radiation spectrum [Section 7.4]. When this radiation first broke free 300,000 years after the Big Bang, the temperature of the universe was about 3,000 K, not too different from that of a red-giant star's surface. Thus, the spectrum of the cosmic background radiation originally peaked in visible light, just like the thermal radiation from a red star, with wavelengths of a few hundred nanometers. However, the universe has expanded by a factor of about 1,000 since that time, stretching the wavelengths of these photons by the same amount [Section 19.4]. Their wavelengths should therefore have shifted to about a millimeter, squarely in the microwave portion of the spectrum and corresponding to a temperature of a few degrees above absolute zero. In the early 1990s, a NASA satellite called the *Cosmic Background Explorer* (*COBE*) was launched to test these ideas about the cosmic background radiation. The results were a stunning success for the Big Bang theory. As shown in Figure 22.7, the cosmic background radiation does indeed have a perfect thermal radiation spectrum, with a peak corresponding to a temperature of 2.73 K. In a very real sense, the temperature of the night sky is a frigid 3 degrees above absolute zero.

TIME OUT TO THINK Suppose the cosmic background radiation did not really come from the heat of the universe itself but instead came from many individual stars and galaxies. Explain why, in that case, we would not expect it to have a perfect thermal radiation spectrum. How does the spectrum of the cosmic background radiation lend support to the Big Bang theory?

COBE achieved an even greater success mapping the temperature of the cosmic background radiation in all directions. It was already known that the cosmic background radiation is extraordinarily uniform throughout the universe. Conditions in the early universe must have been extremely uniform to produce such a smooth radiation field. For a time, this uniformity was considered a strike against the Big Bang theory because, as we discussed in Chapters 20 and 21, the universe must have contained some regions of enhanced density in order to explain the formation of galaxies. The COBE measurements restored confidence in the Big Bang theory because they showed that the cosmic background radiation is *not quite* perfectly uniform. Instead, its temperature varies very

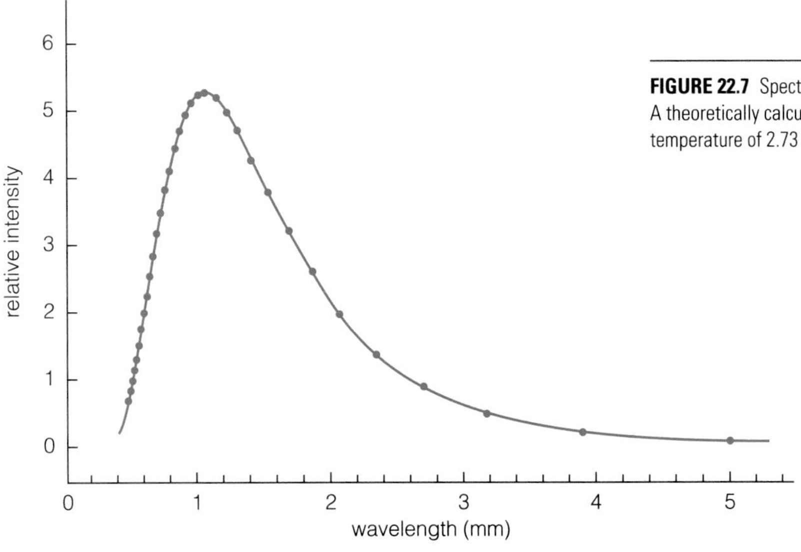

FIGURE 22.7 Spectrum of the cosmic background radiation from COBE. A theoretically calculated thermal radiation spectrum (smooth curve) for a temperature of 2.73 K perfectly fits the data (dots).

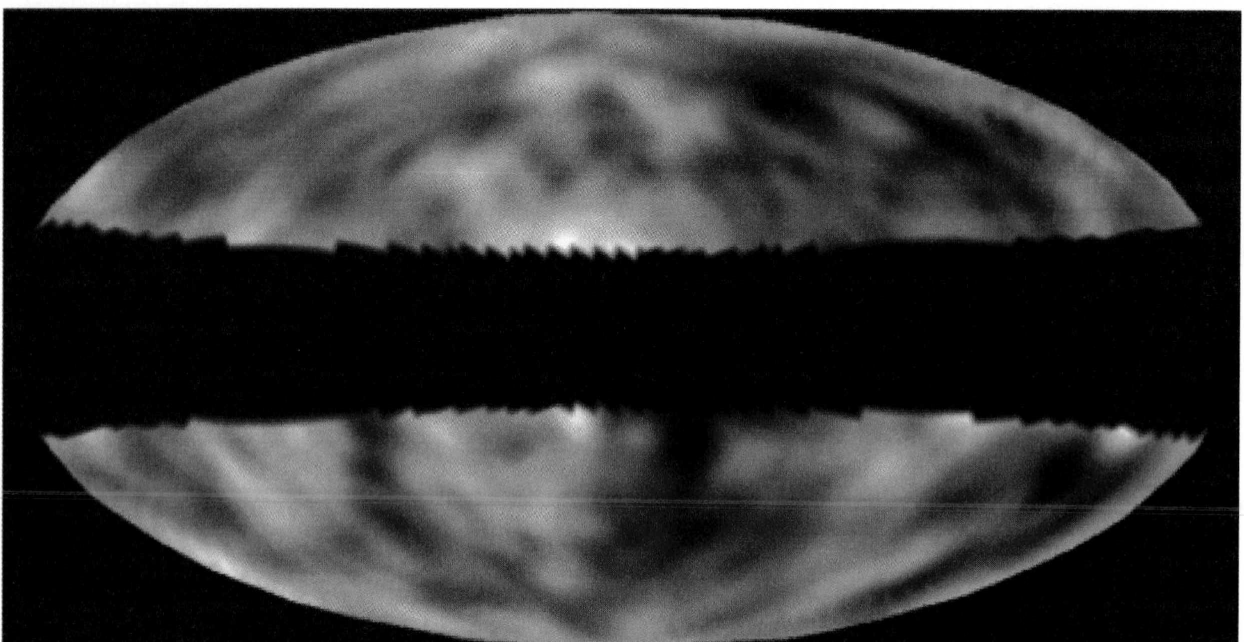

FIGURE 22.8 This all-sky map shows temperature differences in the cosmic background radiation measured by COBE. The background temperature is about 2.73 K everywhere, but the brighter regions of this picture are slightly less than 0.0001 K hotter than the darker regions—indicating that the early universe was very slightly lumpy. We are essentially seeing what the universe was like at the surface marked "300,000 years" in Figure 22.2. (The central strip of this map, which corresponds to the disk of the Milky Way, has been masked out because the brightness differences there stem primarily from radio noise in the Milky Way.

slightly from one place to another by a few parts in 100,000 (Figure 22.8).[6] These variations in temperature indicate that the density of the early universe really did differ slightly from place to place—the seeds of structure formation were indeed present during the era of nuclei.

COBE's discovery of density enhancements bolstered the idea that some of the dark matter consists of WIMPs (weakly interacting massive particles [Section 21.4]) that we have not yet identified and that the gravity of this dark matter drove the formation of structure in the universe. Regions of enhanced density can grow into galaxies because the extra gravity in these regions draws matter together, even while the rest of the universe expands. The greater the density enhancements, the faster matter should have collected into galaxies. Detailed calculations show that, to explain the fact that galaxies formed within a few billion years, the density enhancements at the end of the era of nuclei must have been significantly greater than the few parts in 100,000 suggested by the temperature variations in the cosmic background radiation. Because WIMPs are weakly interacting and do not interact with photons, we do not expect them to influence the temperature of the cosmic background radiation directly. However, the gravity of the WIMPs can collect ordinary baryonic matter into clumps that *do* interact with photons. Thus, the small density enhancements detected by COBE may actually echo the much larger density enhancements made up of WIMPs. Careful modeling of the COBE temperature variations shows that they are consistent with dark-matter density enhancements large enough to account for the structure we see in the universe today.

Synthesis of Helium

The discovery of the microwave background radiation in 1965 quickly solved another long-standing astronomical problem: the origin of cosmic helium. Everywhere in the universe, about one-quarter of the mass of ordinary matter (i.e., not dark matter) is helium. The Milky Way's helium fraction is about 28%, and no galaxy has a helium fraction lower than 25%. A small proportion of this helium comes from hydrogen fusion in stars, but most does not. The majority of the helium in the universe must already

[6]The Earth's motion (e.g., orbit of Sun, rotation of galaxy, etc.) means that we are moving relative to the cosmic background radiation. Thus, we see a slight blueshift (about 0.12%) in the direction we're moving, and a slight redshift in the opposite direction. We must first subtract these effects before analyzing the temperature of the background radiation.

have been present in the protogalactic clouds that preceded the formation of galaxies. In other words, the universe itself must once have been hot enough to fuse hydrogen into helium. The current microwave background temperature of 2.73 K tells us precisely how hot the universe was in the distant past and exactly how much helium it should have made. The result—25% helium—is another impressive success of the Big Bang theory.

A helium nucleus contains two protons and two neutrons, so we need to understand what protons and neutrons were doing during the era of nucleosynthesis in order to see why the primordial fraction of helium was 25%. Early in this era, when the universe's temperature was 10^{11} K, nuclear reactions could convert protons into neutrons, and vice versa. For example, a proton and an electron could combine to form a neutron and a neutrino. This reaction is one of several that change protons into neutrons. All such proton–neutron conversion reactions involve neutrinos or antineutrinos, which experience only the weak force. As long as the universe remained hotter than 10^{11} K, these reactions kept the numbers of protons and neutrons nearly equal.

As the universe cooled from 10^{11} K to 10^{10} K, neutron–proton conversion reactions began to favor protons. Neutrons are slightly heavier than protons, and therefore reactions that convert protons to neutrons require energy to proceed (in accordance with $E = mc^2$). Below 10^{11} K, the required energy for neutron production was no longer readily available, so the rate of these reactions slowed. In contrast, reactions that convert neutrons into protons release energy and thus are unhindered by cooler temperatures. At the time the temperature of the universe fell to 10^{10} K, protons began to outnumber neutrons because the conversion reactions ran only in one direction: Neutrons changed into protons, but the protons didn't change back.

At 10^{10} K, the universe was still hot and dense enough for nuclear fusion to operate. Protons and neutrons constantly combined to form *deuterium*—the rare form of hydrogen nuclei that contain a neutron in addition to a proton—and deuterium nuclei fused to form helium. However, during the early parts of the era of nucleosynthesis, the helium nuclei were almost immediately blasted apart by one of the many gamma rays that filled the universe.

Fusion finally created long-lasting helium nuclei when the universe was about 1 minute old, at which time the destructive gamma rays were gone (Figure 22.9). Calculations show that the proton-to-neutron ratio at this time should have been about 7 to 1. Moreover, virtually all the available neutrons should have become incorporated into nuclei of helium-4. Figure 22.10 shows that, based on the 7-to-1 ratio

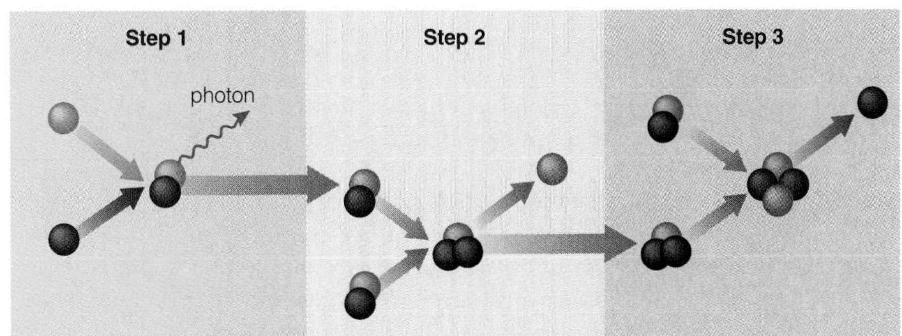

FIGURE 22.9 During the era of nucleosynthesis, virtually all the neutrons in the universe fused with protons to form helium-4. This figure illustrates one of several possible reaction pathways. In Step 1, a neutron and a proton fuse to form a deuterium nucleus, releasing a photon. In Step 2, two deuterium nuclei fuse to make hydrogen-3, releasing a proton. In Step 3, the hydrogen-3 nucleus fuses with deuterium to create helium-4, releasing one of the neutrons.

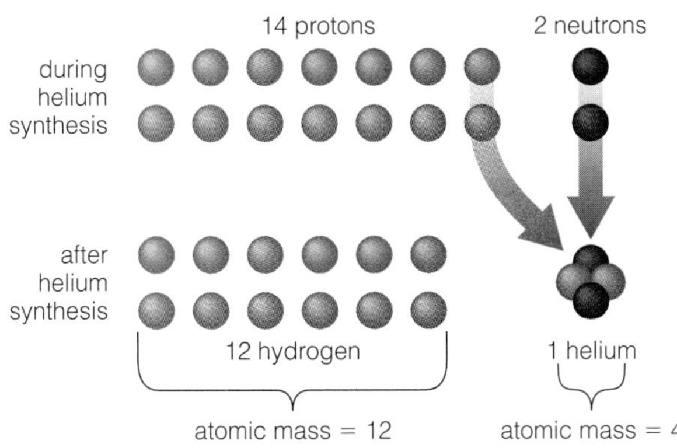

FIGURE 22.10 During helium synthesis, protons outnumbered neutrons 7 to 1, which is the same as 14 to 2. The result was 12 hydrogen nuclei (individual protons) for each helium nucleus. Thus, the hydrogen-to-helium mass ratio is 12 to 4, which is the same as 75% to 25%.

of protons to neutrons, the universe should have had a composition of 75% hydrogen and 25% helium by mass at the end of the era of nucleosynthesis.

Thus, the Big Bang theory makes a very concrete prediction about the composition of the universe: It should be 75% hydrogen and 25% helium by mass (at least for ordinary matter). The fact that observations confirm this predicted ratio of hydrogen to helium is another striking success of the Big Bang theory.

TIME OUT TO THINK *Briefly explain why it should not be surprising that some galaxies contain a little more than 25% helium, but it would be very surprising if some galaxies contained less.* (Hint: *Think about how the relative amounts of hydrogen and helium in the universe can be affected by fusion in stars.*)

Synthesis of Other Light Nuclei

Why didn't the Big Bang produce heavier elements? By the time stable helium nuclei formed, when the universe was about a minute old, the temperature and density of the rapidly expanding universe had already dropped too far for a process like carbon production (three helium nuclei fusing into carbon [Section 16.3]) to occur. Reactions between protons, deuterium nuclei, and helium were possible, but most of these led nowhere. In particular, fusing two helium-4 nuclei together results in a nucleus that is unstable and falls apart in a fraction of a second, as does fusing a proton to a helium-4 nucleus.

A few reactions involving hydrogen-3 (also known as *tritium*) or helium-3 *can* create long-lasting nuclei. For example, fusing helium-4 and hydrogen-3 produces lithium-7. However, the contributions of these reactions to the overall composition of the universe were minor because hydrogen-3 and helium-3 were so rare. Models of element production in the early universe show that, before the cooling of the universe shut off fusion entirely, such reactions generated only trace amounts of the next three lightest elements after helium: lithium, beryllium, and boron. Thus, aside from hydrogen, helium, and trace amounts of these other light nuclei, all other elements were forged much later in the nuclear furnaces of stars.

The Density of Ordinary Matter

Calculations made with the Big Bang model allow scientists to estimate the density of ordinary (baryonic) matter in the universe from the observed amount of deuterium in the universe today. The fact that some deuterium nuclei still exist in the universe indicates that the fusing of neutrons into helium nuclei in the early universe was incomplete. The amount of deuterium—roughly one hydrogen atom in 100,000 contains a deuterium nucleus—tells us about the density of protons and neutrons (baryons) during the era of nucleosynthesis. If this density had been higher in the early universe, protons and neutrons would have fused more efficiently into helium, and less deuterium would have been left over.

The results of such calculations show that the density of ordinary (baryonic) matter in the universe is somewhere between 1% and 10% of the critical density. (Recall that the critical density is the density required if the expansion of the universe is to stop and reverse someday [Section 19.3].) Similar calculations based on the observed abundance of lithium lead to the same conclusion, adding to our confidence in the Big Bang model and implying something very interesting about the fate of the universe: Unless extraordinary (nonbaryonic) matter outweighs ordinary matter by at least a factor of 10, the universe must be open. Moreover, if the universe's density is greater than 10% of the critical density—as appears to be the case [Section 21.6]—then extraordinary (nonbaryonic) matter may constitute the majority of the universe's mass.

One possible candidate for this extra mass might be the neutrinos left over after neutron–proton conversion reactions ceased. These ghostlike particles still swarm through the universe in enormous numbers: on average, about 100 per cubic centimeter throughout the universe. If these neutrinos weigh even as much as 10^{-34} kg, or one ten-thousandth of an electron's mass, then ghostlike neutrinos are the dominant form of matter in the universe and will determine the universe's ultimate fate.

TIME OUT TO THINK *The ideas discussed above point to a rather amazing fact: Although we do not yet know the mass of neutrinos and have yet to discover any WIMPs, we suspect that either neutrinos or WIMPs (or a combination of both) dominate the total mass of the universe. Briefly explain how this is possible, and comment on how confident we can be that weakly interacting particles make up the bulk of dark matter.*

22.4 Quandaries

The cosmic background radiation tells us what the universe was like when it was 300,000 years old, and the abundance of helium takes us all the way back to the era of nucleosynthesis, when the universe was just minutes old. The experiments that verified the unification of the weak and electromagnetic forces into the electroweak force provide at least some evidence that we understand conditions back to the end of the electroweak era (10^{-10} s). Can we find evidence to support the Big Bang model at even earlier times?

Prior to the early 1980s, scientists had identified four major aspects of our actual universe that were unexplained by the Big Bang model. This section briefly describes these four quandaries; the next section will describe a potential solution to all of them that may tell us what happened well before the electroweak era.

Why Matter and Not Antimatter?

Matter appears to be far more common than antimatter throughout the observable universe. In the solar system, where we can directly measure amounts of matter and antimatter, we find very little antimatter. We have good reason to believe that the same is true elsewhere in the universe. Whenever a particle of matter meets a corresponding particle of antimatter, the two particles annihilate each other, turning into pure photon energy. If a galaxy's interstellar medium held both matter and antimatter, the whole mixture would be extremely explosive. The entire galaxy would glow brightly as the matter and antimatter annihilated, quickly turning all the galaxy's gas into gamma-ray photons. Collisions between matter galaxies and antimatter galaxies would also shine brilliantly in gamma rays as the two galaxies annihilated each other. The fact that we never see entire galaxies being annihilated, either alone or in pairs, tells us that antimatter is relatively rare compared to matter.

The overwhelming prevalence of matter in our universe is somewhat puzzling in the context of the Big Bang theory. Particle accelerators on Earth can briefly re-create the highly energetic conditions characteristic of the universe at an age of 10^{-10} second, and these accelerators always produce equal amounts of matter and antimatter. We have never observed any reactions that produce even a tiny bit more matter than antimatter, yet somehow the early universe performed this trick. We will call this puzzle the *antimatter problem.*

Where Does Structure Come From?

According to the Big Bang theory, the process of galaxy formation required the density of the early universe to differ slightly from place to place. Otherwise, the pull of gravity would have been equally balanced everywhere, prohibiting the collapse of protogalactic clouds. The subtle temperature differences seen in the cosmic background radiation tell us that regions of enhanced density did indeed exist at the end of the era of nuclei, when the universe was 300,000 years old.

However, the standard Big Bang theory cannot explain where these density enhancements originally came from. Perhaps these slight lumps existed from the very beginning of time, as part of the initial blueprint for creation, but physicists would prefer an explanation that relies on a natural process to create such lumps. We will refer to the puzzle of where the density enhancements came from as the *structure problem.*

Why So Smooth?

Not only is the slight lumpiness of the early universe a problem, but its large-scale smoothness is also a problem. Observations of the cosmic background radiation show that, overall, the density of the universe at the end of the era of nuclei varied from place to place by no more than about 0.01%. The idea that different parts of the universe should be so similar shortly after the Big Bang may seem quite natural at first, but upon further inspection the smoothness of the universe becomes difficult to explain.

Imagine observing the cosmic background radiation in a certain part of the sky. You are seeing that region as it was only 300,000 years after it formed. The microwaves coming from it have taken some 10 billion years or more, almost the entire age of the universe, to travel all the way to Earth. Now imagine turning around and looking at the background radiation coming from the opposite direction. You are also seeing this region at an age of 300,000 years, and it looks virtually identical.

The two microwave-emitting regions are billions of light-years apart, but we are seeing them as they were when they were only 300,000 years old. They can't possibly have known about each other then. A signal traveling at light speed from one to the other would barely have started its journey (Figure 22.11). So how do they know to be at exactly the same temperature? This coincidence is as extraordinary as receiving the same letter, word for word, from two strangers living in two different countries who have never met each other and who grew up in completely different, isolated cultures.

A more physical example of such a coincidence would be to measure the temperatures of two completely different, totally isolated objects and find them to be precisely equal. If we measure the temperature of two bricks in the same room, for example, we might not be surprised to find their temperatures equal if both bricks are exposed to the same air

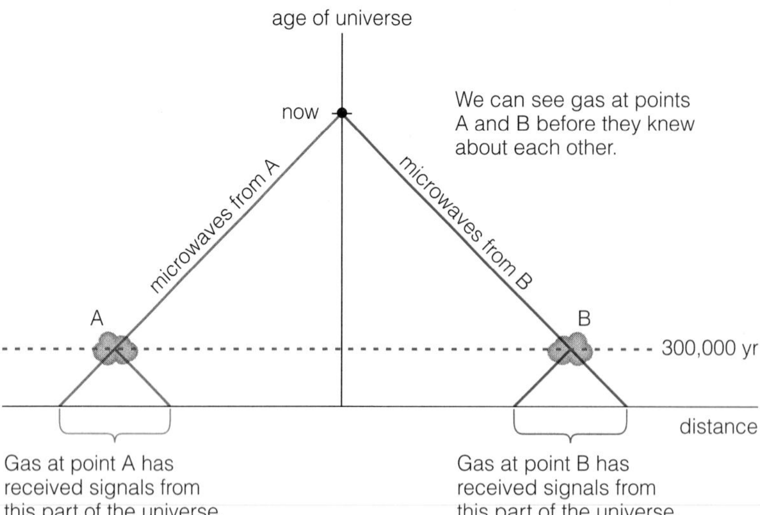

FIGURE 22.11 Light left the microwave-emitting regions we see on opposite sides of the universe long before they could have communicated with each other and equalized their temperatures, yet their temperatures are virtually identical.

temperature, the same lighting, and so forth. We would be even less surprised if the bricks had been touching each other for a long time, allowing heat to flow from the hotter one to the cooler one. But the two regions on opposite sides of the observable universe have not even had enough time for a single exchange of photons, let alone enough time to exchange heat energy. We will call this mystery the *smoothness problem*.

Why Almost Critical?

The last problem we will discuss verges on the philosophical, but it is worth mentioning. The density of matter in the universe is somewhere around 20–100% of the critical density [Section 21.6]—remarkably close to the critical density considering that the standard Big Bang theory doesn't say anything about what the density should be. Why isn't it 1,000 times the critical density, or 0.0000001 times the critical density?

Another way to state this problem is to say that the universe is remarkably flat. Recall that the density of a *flat universe* exactly equals the critical density. Its kinetic energy of expansion precisely balances its overall gravitational pull. Any deviation from this precise balance grows more severe as the universe evolves. For example, if the universe had been 10% denser at the end of the era of nuclei, it would have recollapsed long ago. On the other hand, if it had been 10% less dense at this time, galaxies would never have formed before expansion spread all the matter too thin. The universe had to start out remarkably balanced to be even remotely close to flat today. We will call this problem the *flatness problem*.

The reason the flatness problem verges on the philosophical is that our own universe is the only example we have, so we don't know whether its flatness is a coincidence or an absolute demand of nature's laws. Some creative thinkers have pointed out that our observation that the universe's density is close to critical might be biased because we happen to live in a universe that can bring forth life. A universe well above the critical density would not last long enough to produce the conditions for life, and a universe well below the critical density would never gather matter into galaxies, stars, planets, and humans. Therefore, no living beings could possibly observe a universe whose density differs radically from the critical density.[7]

Scientists generally pay little attention to such arguments, even though they are plausible, because they lead to intellectual dead ends. If they are true, then we needn't bother looking for scientific answers to puzzles like the flatness problem. And scientists prefer to have scientific answers for every observed aspect of the universe.

22.5 Inflation: A Solution to the Quandaries?

The Big Bang model relies heavily on our knowledge of particle physics. Current particle physics theories have been tested to temperatures of 10^{15} K, corre-

[7]This line of argument, pointing out why the physical conditions of the universe "coincidentally" favor the formation of life, is known as the *anthropic principle*.

sponding to the end of the electroweak era. Our knowledge of earlier times rests on weaker foundations, because we are less certain of the physical laws at work. In fact, the best laboratory for studying the laws of physics at such high temperatures is the Big Bang itself. Different guesses as to how matter might behave at such high energies predict different outcomes for the universe we see today. If a particular model predicts that our universe should look different from the way it really does, then that model must be wrong. On the other hand, if a new model of particle physics solves the four quandaries described above, then it may be on the right track.

Today, the grand unified theories that predict a sudden burst of expansion at the end of the GUT era seem to be on the right track. If these theories are correct, the universe may have expanded enormously in an episode of *inflation* that forever altered its history. Let's see how the GUTs and the associated idea of inflation address the four quandaries: the antimatter problem, the structure problem, the smoothness problem, and the flatness problem.

Broken Symmetry

Here is a one-sentence description of the universe's history: Everything started out highly symmetric and hot, and the universe grew less symmetric as it expanded and cooled. When we describe the early universe as *symmetric* in this context, we are saying that it looked virtually the same from any angle or vantage point. By this definition, a circle is more symmetric than a square. While a square looks the same each time you rotate it 90°, a circle looks the same no matter what rotation angle you choose. The early universe was highly symmetric because the soup of particles and energy that filled it was virtually the same everywhere. If the universe had remained this symmetric, it would be an extremely boring place today. The breaking of this symmetry as the various parts of the universe cooled and condensed is what made life possible—and interesting too.

According to the grand unified theories, one of the earliest episodes of symmetry breaking occurred when the strong force split from the GUT force at the end of the GUT era. This event was loosely similar to the freezing of water into ice. You can observe for yourself that water is more symmetric than ice: Water in a glass looks the same no matter how you rotate it, but ice cubes in a glass look different from different angles. Water is more symmetric than ice because its H_2O molecules can point in all different directions. In contrast, the H_2O molecules in ice crystals are organized in very particular ways. When water freezes into ice, substantial energy is released as the molecules settle into place.

TIME OUT TO THINK ***It takes heat to melt ice into water. Use this fact to explain why heat must be released when water freezes into ice.*** (Hint: *Remember the law of conservation of energy.*)

The grand unified theories predict that the "freezing out" of the strong force from the GUT force should have released enormous energy, causing the universe to expand dramatically. This dramatic expansion is what we call *inflation*. Amazingly, this expansion actually accelerated until the separation of the strong force was complete about 10^{-33} second later, at which point the period of inflation ended and the universe returned to a more gradual rate of expansion.

According to these same grand unified theories, particle reactions acting while the strong force was freezing out from the electroweak force could have created more matter than antimatter. The usual rules dictating that antimatter and matter must be created in equal measure were temporarily suspended. If these theories are correct, they naturally explain how matter particles came to outnumber antimatter particles by a relatively tiny amount at the dawn of the particle era.

This imbalance remained tiny until the temperature of the universe dropped to 10^{12} K at the end of the particle era, when particle reactions no longer produced proton–antiproton pairs. At this moment, roughly a billion and one protons were present for every billion antiprotons. The billion antiprotons quickly annihilated every proton they could, leaving a single proton behind. That explains why there are no antimatter galaxies. Only matter remained after this episode of annihilation, solving the antimatter problem. Without the original imbalance, there would not have been enough protons remaining to make stars, galaxies, or human beings.

Giant Quantum Fluctuations

To understand how the idea of inflation solves the structure problem, we need to recognize a special feature of energy fields. Laboratory-tested principles of quantum mechanics require that the energy fields at any point in space be always fluctuating (as a result of the *uncertainty principle* [Section S5.3]). Thus, the distribution of energy through space on very small scales is slightly irregular, even in a complete vacuum. These tiny quantum "ripples" can be characterized by a wavelength that corresponds roughly to their size and by an amplitude that corresponds to their strength.

In principle, quantum ripples in the very early universe could have been the seeds for density enhancements that later grew into galaxies. However, the wavelengths of the original ripples were far too small to explain density enhancements like those we see imprinted on the cosmic background radiation. Here again, inflation may provide the solution.

Inflation would have dramatically altered these quantum fluctuations. The fantastic growth of the universe during the period of inflation would have stretched tiny ripples in space to enormous wavelengths (Figure 22.12). Ultimately, inflation would have caused these quantum ripples to grow large enough to become the density enhancements that later formed large structures in the universe. Thus, inflation goes a long way toward solving the structure problem. However, it is important to note that the models invoking inflation do not entirely solve the structure problem. Although these models make reasonable predictions for the wavelengths of the quantum ripples, they do not predict their amplitudes. This problem points out the speculative nature of the idea of inflation. Many scientists are working hard to improve models of inflation, in hopes that they will soon be able to make more definitive predictions that can be tested observationally.

Equalizing Temperatures and Densities

Our description of the smoothness problem likened distant parts of the universe to vastly separated bricks with identical temperatures despite never having been in contact. The idea of inflation solves this problem because it says that the entire observable universe was less than 10^{-35} light-second across before the episode of enormous expansion. Radiation signals traveling at light speed would have had time to equalize all the temperatures and densities in this region. Inflation then pushed these equalized regions to much greater distances, far out of contact with one another (Figure 22.13). Temperature equalization followed by inflation thus rendered every part of the observable universe virtually identical, making the cosmic background radiation very smooth.

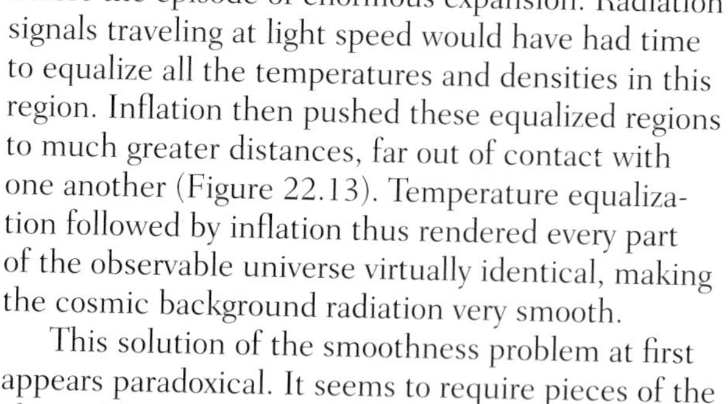

This solution of the smoothness problem at first appears paradoxical. It seems to require pieces of the observable universe to move faster than the speed of light during inflation. Yet we know that nothing can move faster than the speed of light [Section S3.2]. To understand why inflation does not violate this universal speed limit, remember that the expansion of the universe is really the expansion of *space itself*. Objects with large cosmological redshifts aren't moving at high speed so much as they are riding along with the expansion of the universe. The gap between two objects thus increases because the space between them is expanding. Photons can easily cross this gap before inflation, but during inflation the gaps expand much faster than photons can cross them, and communication between the two objects ceases. One object never detects the other moving faster than light speed. Communication between these objects finally resumes long after inflation, when photons have had time to cross the huge chasm in space that has opened.

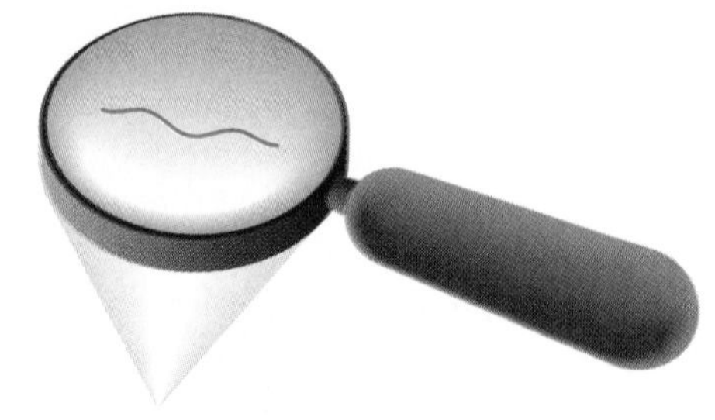

FIGURE 22.12 During inflation, ripples in spacetime would have stretched by a factor of perhaps 10^{30}.

From Curved to Flat

Inflation tends to flatten the curvature of space called for by Einstein's theory of general relativity, just as a balloon grows less tightly curved when you inflate it with your breath. Imagine a tiny insect on the balloon's surface. The insect might notice the curvature of the balloon when you begin to blow into it, but as the balloon expands, the surface flattens. The insect will hardly notice the curvature of the enlarged balloon (Figure 22.14). If the balloon were inflated to the size of the Earth, even *you* might think its surface was flat.

The flattening of space during the period of inflation would have been so enormous that it virtually would have eliminated any curvature the universe might have had previously. Inflation accomplishes this flattening effect by ensuring that the mass-energy providing the universe's gravity precisely balances its kinetic energy of expansion. Once this balance is set, it should never waver. The universe should remain finely balanced from the time of inflation to the present.

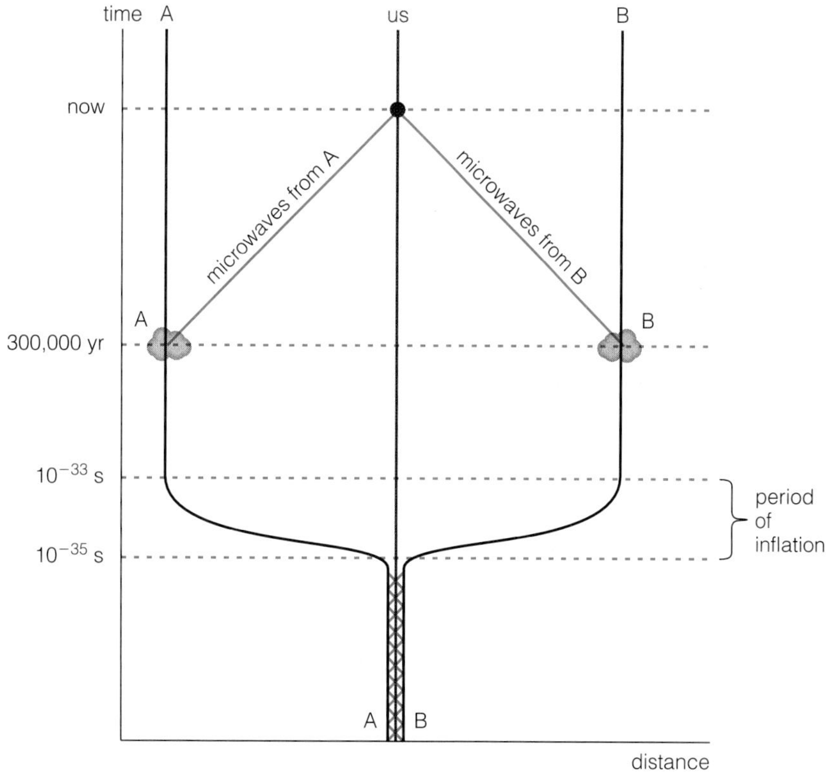

FIGURE 22.13 Before inflation, regions A and B were near enough to communicate and equalize their temperatures. Inflation pushed them far apart. Today, we can see both A and B, but they are too far apart to see each other.

This fine balance turns out to be both a success and a pitfall of the inflation theory. On one hand, it explains the flatness problem, enabling the universe to stay balanced long enough to bring forth galaxies. However, our studies of dark matter tend to show that the current universe is slightly out of balance: The best current estimates put the actual density of the universe at only about 25% of the critical density [Section 21.6], suggesting that the universe has about four times more kinetic energy than gravity can overcome. If these dark-matter measurements hold up, inflation theory will have a tough time explaining this slight imbalance.

The Bottom Line

All things considered, inflation does a remarkable job of addressing the problems inherent in the standard Big Bang theory. Many astronomers and physicists believe that some process akin to inflation did affect the early universe, but the details of the interaction between high-energy particle physics and the evolving universe remain unclear. If these details can be worked out successfully, we face an amazing prospect—a breakthrough in our understanding of the very smallest particles achieved by studying the universe on the largest observable scales.

FIGURE 22.14 As a balloon expands, its surface seems increasingly flat to an insect crawling along it.

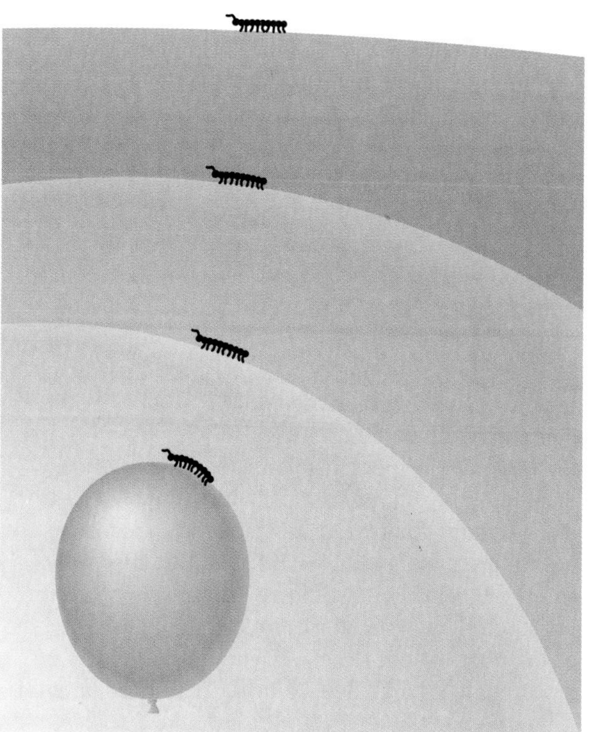

THINKING ABOUT . . .

The End of Time

According to the Big Bang theory, time and space have a beginning in the Big Bang. Do they also have an end? If the universe is closed, the answer seems to be a clear yes: Sometime in the distant future, the universal expansion will cease and reverse, and the universe will eventually come to an end in a fiery *Big Crunch*. But what if the universe continues to expand forever?

In an open universe, the end will come much more gradually. The star–gas–star cycle in galaxies cannot continue forever, because not all the material is recycled: With each generation of stars, more mass becomes locked up in planets, brown dwarfs, white dwarfs, neutron stars, and black holes. Eventually, about a trillion years from now, even the longest-lived stars will burn out, and the galaxies will fade into darkness. By this time, the expansion of the universe will have made the average distance between galaxy clusters about a hundred times greater than it is today.

At this point, the only new action in the universe will occur on the rare occasions when two objects—such as two brown dwarfs or two white dwarfs—collide within a galaxy. The vast distances separating star systems in galaxies make such collisions extremely rare. For example, the probability of our Sun (or the white dwarf that it will become) colliding with another star is so small that it would be expected to happen only once in a quadrillion (10^{15}) years. But forever is a long time, and even low-probability events will eventually happen many times. If a star system experiences a collision once in a quadrillion years, it will experience about 100 collisions in 100 quadrillion (10^{17}) years. By the time the universe reaches an age of 10^{20} years, star systems will have suffered an average of 100,000 collisions each, making a time-lapse history of any galaxy look like a cosmic game of pinball.

These multiple collisions will severely disrupt galaxies. As in any gravitational encounter, some objects lose energy and some gain energy. Objects that gain enough energy will be flung into intergalactic space, where they will drift ever-farther from all other objects with the expansion of the universe. Objects that lose energy will eventually fall to the galactic center, forming a gigantic black hole. The

22.6 Did the Big Bang Really Happen?

You might occasionally read an article in a newspaper or a magazine questioning whether the Big Bang really happened. We may never be able to prove with absolute certainty that the Big Bang theory is correct. However, no one has come up with any other model of the universe that so successfully explains so much of what we see. As we have discussed, the Big Bang model makes at least two specific predictions that we have observationally verified: the characteristics of the cosmic background radiation and the composition of the universe. It also explains quite naturally many other features of the universe. And, at least so far, we know of nothing that is absolutely inconsistent with the Big Bang model.

The Big Bang theory's very success has also made it a target for respected scientists, skeptical nonscientists, and crackpots alike. The nature of scientific work requires that we test established wisdom to make sure it is valid. A sound scientific disproof of the Big Bang theory would be a discovery of great importance. However, stories touted in the news media as disproofs of the Big Bang usually turn out to be disagreements over details rather than fundamental problems that threaten to bring down the whole edifice. Nevertheless, scientists must keep refining the theory and tracking down the disagreements, because every once in a while a small disagreement blossoms into a full-blown scientific revolution.

You don't need to believe all you have read without question. The next time you are musing on the universe's origins, try an experiment for yourself. Go outside on a clear night and look at the sky. Notice how dark it is, and then ask yourself *why* it is dark. If the universe were infinite, unchanging, and everywhere the same, then the entire night sky would blaze as brightly as the Sun. Johannes Kepler [Section 6.3] was perhaps the first person to realize that the night sky in such a universe should be bright, but this realization is now called **Olbers' paradox** after Heinrich Olbers, a German astronomer of the 1800s.

To understand how Olbers' paradox comes about, imagine that you are in a dense forest on a flat plain. If you look in any given direction, you'll likely see a tree. If the forest is small, you might be able to see through some gaps in the trees to the open plains, but larger forests have fewer gaps (Figure 22.15). An infinite forest would have no gaps at all—a tree trunk would block your view along any direction.

remains of the universe will consist of black holes with masses as great as a trillion solar masses widely separated from a few scattered planets, brown dwarfs, and stellar corpses. If the Earth somehow survives, it will be a frozen chunk of rock in the darkness of the expanding universe, billions of light-years from any other solid object.

If grand unified theories are correct, the Earth still cannot last forever. These theories predict that protons will eventually fall apart. The predicted lifetime of protons is extremely long: a half-life of at least 10^{32} years. However, if protons really do decay, then by the time the universe is 10^{40} years old the Earth and all other atomic matter will have disintegrated into radiation and subatomic particles, such as electrons and neutrinos.

The final phase in the future of an open universe is predicted to come with the disintegration of the giant, black-hole corpses of galaxies. Black holes are predicted to slowly evaporate through the process of *Hawking radiation*, finally disappearing in brilliant bursts of radiation [Section S5.5]. The largest black holes last longest, but even trillion-solar-mass black holes will evaporate sometime after the universe reaches an age of 10^{100} years. From then on, the universe will consist only of individual photons and subatomic particles, each separated by enormous distances from the others. Nothing new will ever happen, and no events will ever occur that would allow an omniscient observer to distinguish past from future. In a sense, the universe will finally have reached the end of time.

Lest any of this sound depressing, keep in mind that 10^{100} years is an indescribably long time. As an example, imagine that you wanted to write on a piece of paper a number that consisted of a 1 followed by 10^{100} zeros (i.e., the number $10^{10^{100}}$). It sounds easy, but a piece of paper large enough to fit all those zeros *would not fit in the observable universe* today. And, if that still does not alleviate your concerns, you may be glad to know that a few creative thinkers speculate about ways in which the universe might undergo rebirth, even after the end of time.

FIGURE 22.15 In a large forest (left) you'll see trees no matter where you look. In a small forest (below), you can see open spaces beyond the trees.

The universe is like a forest of stars in this respect. In an unchanging universe with an infinite number of stars, we would see a star in every direction, making every point in the sky as bright as the Sun's surface. Obscuring dust doesn't change this conclusion. The intense starlight would heat the dust over time until it too glowed like the Sun or evaporated away.

There are only two ways out of this dilemma: Either the universe has a finite number of stars, in which case we would not see a star in every direction, or it changes in time in some way that prevents us from seeing an infinite number of stars. For several centuries after Kepler first recognized the dilemma, astronomers leaned toward the first option. Kepler himself preferred to believe that the universe had a finite number of stars because it was finite in space, with some kind of dark wall surrounding everything. Astronomers in the early twentieth century preferred to believe that the universe was infinite in space, but that we lived inside a finite collection of stars; they thought of the Milky Way as an island floating in a vast black void. However, subsequent observations showed that galaxies fill all of space more or less uniformly. We are therefore left with the second option: The universe changes in time.

The Big Bang theory solves Olbers' paradox in a particularly simple way: It tells us that we can see only a finite number of stars because the universe began at a particular moment. While the universe may contain an infinite number of stars, we can see only those that lie within our cosmological horizon [Section 19.4]. There are other ways in which the universe could change in time and prevent us from seeing an infinite number of stars, so Olbers' paradox does not *prove* that the universe began with a Big Bang. But we must have some explanation for why the sky is dark at night, and no explanation besides the Big Bang also explains so many other observed properties of the universe so well.

> *A poet once said "The whole universe is in a glass of wine." . . . It is true that when we look in a glass of wine closely enough we see the entire universe. There are the things of physics: the twisting liquid which evaporates depending on the wind and weather, the reflections in the glass, and our imagination adds the atoms. The glass is a distillation of the earth's rocks, and in its composition we see the secrets of the universe's age, and the evolution of the stars. . . . If our small minds, for some convenience divide this glass of wine, this universe into parts—physics, biology, geology, astronomy, psychology, and so on—remember that nature does not know it! So let us put it all back together, not forgetting ultimately what it is for.*
>
> RICHARD FEYNMAN, (1918–1988), Physicist

THE BIG PICTURE

Our "big picture" is now about as complete as it gets. We've discussed the universe from the Earth outward, and from the beginning to the end. When you think back on this chapter, keep in mind the following ideas:

- Predicting conditions in the early universe is straightforward. The only real question is how matter and energy behave under such extreme conditions.
- Our current understanding of physics allows us to reconstruct the conditions that prevailed in the universe all the way back to the first 10^{-10} second. Our understanding is less certain back to 10^{-35} second. Beyond 10^{-43} second, we run up against the present limits of human knowledge.
- Although it may sound strange to talk about the universe during its first fraction of a second, our ideas about the Big Bang rest on a solid foundation of observational, experimental, and theoretical evidence. We cannot say with absolute certainty that the Big Bang really happened, but no other model ever proposed has so successfully explained how our universe came to be as it is.

Review Questions

1. Briefly explain why the early universe must have been much hotter and denser than the universe is today. How can we use this idea to determine the temperature and density conditions that prevailed at different times in the early universe?
2. What do we mean by the *Big Bang theory?*
3. What is *antimatter?* How were particle–antiparticle pairs created in the early universe? How were they destroyed?
4. Briefly summarize the characteristics marking each of the various eras shown in Figure 22.2.
5. What are the *Planck era* and the *Planck time?* Why can't our current theories describe the history of the universe during the Planck era?
6. What are *grand unified theories?* According to these theories, how many forces operated in the universe during the GUT era? Why?
7. What do we mean by *inflation?* Briefly describe why inflation might have occurred at the end of the GUT era.
8. What characterized the universe during the *electroweak era?* What evidence from particle accelerators supports the idea that the weak and electromagnetic forces were once unified as a single electroweak force?
9. What characterized the universe during the *particle era?* What happened to the *quarks* that existed freely in the universe early in the particle era?
10. How long did the *era of nucleosynthesis* last? Why was this era so important in determining the chemical composition of the universe forever after?
11. Briefly explain why radiation was trapped for 300,000 years during the *era of nuclei* and why the cosmic background radiation broke free at the end of this era.
12. What do we mean by the *era of atoms* and the *era of galaxies?* What era do we live in? Explain.
13. Briefly describe the two key pieces of evidence that support the Big Bang theory.
14. Briefly describe how the *cosmic background radiation* was discovered. How can you see the cosmic background radiation for yourself on your television set?
15. Why does the Big Bang theory predict that the cosmic background radiation should have a perfect thermal radiation spectrum? How did the *Cosmic Background Explorer* (COBE) support this prediction of the Big Bang theory? What is the temperature of the cosmic background radiation?
16. Briefly explain why we expect the cosmic background radiation to be almost, but not quite, the same in all directions. What did COBE find regarding the smoothness of the cosmic background radiation? Why does this finding suggest that some of the dark matter may consist of WIMPs?
17. Briefly describe how theoretical calculations allow us to predict the fraction of ordinary matter in the universe that should consist of helium as opposed to hydrogen. Do observations of the universe today agree with this theoretical prediction?
18. Why didn't the Big Bang produce elements heavier than helium, aside from trace amounts of lithium, beryllium, and boron?
19. How do measurements of the abundance of deuterium in today's universe allow us to estimate the density of ordinary matter during the era of nucleosynthesis? Explain why these measurements suggest that at least some of the dark matter in the universe must be extraordinary, consisting of either neutrinos that have mass or undiscovered WIMPs.
20. Briefly describe each of the four quandaries posed by the standard Big Bang theory: the *antimatter problem,* the *structure problem,* the *smoothness problem,* and the *flatness problem.*
21. What do we mean by *symmetry* in the context of the early universe? How might the idea of broken symmetry explain why the universe consists almost entirely of matter (as opposed to antimatter) today?
22. Why do we think that tiny quantum ripples should have been present in the early universe? How might inflation have caused these tiny ripples to become the seeds that later grew into galaxies?
23. How does inflation solve the smoothness problem? Briefly explain why, even though space grew by an enormous factor during inflation, this dramatic expansion does *not* mean that anything traveled faster than light and therefore does not violate the theory of relativity.
24. Briefly explain why the theory of inflation predicts that the universe ought to be *flat* and therefore have a density equal to the critical density. Do current measurements of the density of the universe support this prediction?
25. What is *Olbers' paradox?* Explain how this simple paradox provides at least some support for the idea that the universe began in a Big Bang.

Discussion Questions

1. *The Moment of Creation.* You've probably noticed that, in discussing the Big Bang theory, we never quite talk about the first instant. Even our most speculative theories at present take us back only to within 10^{-43} second of creation. Do you think it will *ever* be possible for science to consider the moment of creation itself? Will we ever be able to answer questions such as *why* the Big Bang happened? Defend your opinions.
2. *The Big Bang.* How convincing do you find the evidence for the Big Bang model of the universe's origin? What are the strengths of the theory? What does it fail to explain? Overall, do *you* think the Big Bang really happened? Defend your opinion.
3. *The End of Time.* According to our current understanding, the universe as we know it will eventually come to an end—either in a fiery Big Crunch if the universe is closed, or in a slow death as all the stars eventually die if the universe is open. However, some people speculate that the universe might then undergo some type of rebirth. In the case of a closed universe, the rebirth might consist of a new Big Bang; some people even speculate that the universe might repeatedly oscillate through cycles of Big Bangs and Big Crunches. Discuss some ideas about how the universe might undergo rebirth if it is open. (*Hint:* You may wish to read Isaac Asimov's short story *The Last Question,* published in 1956.)
4. *Forever.* If you could live forever, would you choose to do so? Suppose the universe will keep expanding forever, as the bulk of observations currently suggest. How would you satisfy your energy needs as time goes on? How would you grow food? What would you do with all that time?

Problems

1. *Life Story of a Proton.* Tell the life story of a proton from its formation shortly after the Big Bang to its presence in the nucleus of an oxygen atom you have just inhaled. Your story should be creative and imaginative, but it also should demonstrate your scientific understanding of as many stages in the proton's life as possible. You can draw on material from the entire book, and your story should be three to five pages long.
2. *Detecting Antimatter.* One way to detect the presence of antimatter at great distances is to look for photons created by matter–antimatter annihilation, such as by the annihilation of an electron and an antielectron. This process converts *all* the mass of the electron–antielectron pair to two photons of equal energy.
   a. Using the fact that the mass of an electron (and also of an antielectron) is 9.11×10^{-31} kg, calculate the energy, wavelength, and frequency of the photons created by electron–antielectron annihilation. In what portion of the electromagnetic spectrum do these photons lie?
   b. Suppose that, in the spectrum of an object suspected to harbor a massive black hole, we found a strong spectral line at the photon energy you calculated in part (a). Explain why this observation would mean that electron–antielectron pairs are being produced near the black hole. Does it seem reasonable that enough energy could be present in this vicinity to create such electron–antielectron pairs? Why or why not?
3. *The Cosmic Background Radiation.* The cosmic background radiation first began to stream freely across the universe at the end of the era of nuclei, when the universe was 300,000 years old and its temperature was about 3,000 K. Since that time, the universe has expanded by a factor of about 1,000.
   a. Using Wien's law (see Mathematical Insight 7.2), calculate the wavelength of maximum intensity for thermal radiation with a temperature of 3,000 K.
   b. The observed temperature of the cosmic background radiation today is about 2.73 K. Use Wien's law to calculate the wavelength of maximum intensity for the cosmic background radiation today.
   c. How much longer are the wavelengths of the cosmic background radiation today than they were at the time the radiation was first released? Is this change in wavelength consistent with the idea that the universe has expanded by a factor of about 1,000 since the end of the era of nuclei? Explain. (*Hint:* Look back at the discussion of cosmological redshifts in Chapter 19.)
4. *10^{100} Years.* In the box Thinking About . . . The End of Time, we found that the final stage in the history of an open universe will come about 10^{100} years from now. Such a large number is easy to write but difficult to understand. In this problem, we investigate some of the incredible properties of very large numbers.
   a. The current age of the universe is around 10^{10} years. How much longer is a trillion years than this current age? How much longer is 10^{15} years? 10^{20} years?
   b. Suppose protons decay with a half-life of 10^{32} years. When will the number of remaining protons be half the current amount? When will it be a quarter of its current amount? How many half-lives will have gone by when the universe reaches an age of 10^{34} years? What fraction of the original protons will remain at this time? Based on your answers,

is it reasonable to conclude that *all* protons in today's universe will be gone by the time the universe is 10^{40} years old? Explain.

c. Suppose you were trying to write 10^{100} zeros on a piece of paper and could write microscopically, so that each zero (including the thickness of the pencil mark) occupied a volume of 1 cubic micrometer—about the size of a bacterium. Could 10^{100} zeros of this size fit in the observable universe? Explain. (*Hints:* Calculate the volume of the observable universe in cubic micrometers by assuming it is a sphere with a radius of 15 billion light-years. The volume of a sphere is $4/3 \times \pi \times (\text{radius})^3$; 1 light-year $\approx 10^{15}$ meters; 1 cubic meter $= 10^{18}$ cubic micrometers.)

5. *Web Project: New Tests of the Big Bang Theory.* The Cosmic Background Explorer (COBE) satellite provided striking confirmation of several predictions of the Big Bang theory. But new satellites are already being designed that will test the Big Bang theory further, primarily by observing the subtle variations in the microwave background radiation with a much higher sensitivity than COBE. Use the web to gather pictures and information about the COBE mission and its planned successors. Write a one- to two-page report about the strength of the evidence compiled by COBE and how much more we might learn from the upcoming missions.

6. *Research Project: Decay of the Proton.* One of the most startling predictions of grand unified theories is that protons will eventually decay, albeit with a half-life of more than 10^{32} years. If this is true, it may be possible to observe an occasional proton decay, despite the extraordinarily long half-life. Several experiments to look for proton decay are under way or being planned. Find out about one or more of these experiments, and write a one- to two-page summary in which you describe the experiment(s), any results to date, and what these results (or potential results) mean to the grand unified theories.

7. *Research Project: New Ideas in Inflation.* The idea of inflation solves many of the puzzles associated with the standard Big Bang theory, but we are still a long way from finding strong evidence that inflation really occurred. Find recent news or magazine articles that discuss some of the latest ideas about inflation and how we might test these ideas. Write a two- to three-page summary of your findings.

S6 Interstellar Travel

ACCORDING TO SCIENCE FICTION, OUR DESCENDANTS WILL soon travel among the stars as routinely as we jet about the Earth in airplanes. They'll race around the galaxy in starships of all sizes and shapes, circumventing nature's prohibition on faster-than-light travel by entering hyperspace, wormholes, or warp drive. They'll witness incredible cosmic phenomena firsthand, including stars and planets in all stages of development, accretion disks around white dwarfs and neutron stars, and the distortion of spacetime near black holes. Along the way, they'll encounter numerous alien species, most of which will look a lot like us and share virtually the same level of technological advancement.

Real interstellar travel is likely to be limited by the speed of light. Moreover, given the billions of years over which the galaxy has existed, finding an alien species that shares our level of technological development would be an extraordinary coincidence; more likely, any other species we encounter will be either far ahead of us or far behind us.

Nevertheless, if our civilization survives long enough, it seems almost inevitable that our descendants will leave the confines of our solar system. In this final chapter, we will explore how that might happen. We will also consider whether other species might already have progressed beyond this point and, if so, how we might contact them.

S6.1 Starships: Distant Dream or Near-Reality?

Before our ancestors domesticated camels and horses, and before the invention of the wheel, the fastest way to travel was on foot. Traveling on foot, our ancestors could sustain speeds of no more than a few kilometers per hour. Today, our fastest interplanetary spacecraft travel through the solar system at speeds of a few *tens of thousands* of kilometers per hour—a factor of 10,000 faster than our ancestors traveled on foot. But we will need to go much faster to travel among the stars (at least if we are interested in making round-trip journeys within human lifetimes). At the speeds of our current interplanetary spacecraft, the journey to even the nearest stars (besides the Sun) would take more than 50,000 years.

The vast distances between the stars mean that practical starships would have to travel near the speed of light in order to complete a journey within a human lifetime. Our current spacecraft are nowhere near achieving such speeds. In fact, our seemingly fast interplanetary spacecraft travel less than 1/10,000 the speed of light—meaning that starships will need to go 10,000 times faster. From this standpoint, we are as far from interstellar flight as cavemen were from the space age.

From another standpoint, however, starflight might be just around the corner. Nearly all of the increase in speed we have achieved over our ancestors occurred within the past century or so, with the advent of flight and the dawn of the space age. If we can achieve a comparable speed advancement in the next century, our great-grandchildren may be building starships.

Regardless of how far in the future interstellar travel might be, we will have to overcome huge technological and social hurdles to achieve it. On the technological side, we will need entirely new types of engines to reach speeds close to the speed of light. We'll also need new types of shielding to protect crew members from instant death: When a starship travels through interstellar gas at near–light speed, ordinary atoms and ions will hit it like a flood of high-energy cosmic rays. And the construction of starships will require systems for engineering on an unprecedented scale: We will need enormous starships to hold the large crews and stores of food and supplies necessary for multi-year journeys across interstellar space.

The social hurdles may be even more difficult to overcome. Building starships will not be something we can do in a garage. It will almost certainly require construction facilities in space and probably cannot be achieved until we first learn to mine materials from the Moon or nearby asteroids. Thus, large-scale human colonization of space—including at least a few major space stations and major outposts on the Moon—is probably a prerequisite to any era of starship exploration. Given global political realities, with nations inherently suspicious of anything that might be construed as an attempt to gain military advantage, the colonization of space will probably not occur without far greater international trust and cooperation than we enjoy today.

Even if we overcome the technological and social hurdles, interstellar travel will pose further difficulties for starship crews. Einstein's theory of relativity offers high-speed travelers a "ticket to the stars": Time will run slow for the travelers, enabling them to make journeys across many light-years of space in relatively short amounts of time. For example, in a ship traveling at an average speed of 99.9% of the speed of light, the 50 light-year round-trip to Vega would take the travelers only about 2 years [Section S3.6]. But more than 50 years would pass on Earth while they were gone, which would make their lives very difficult when they returned. Family and friends would be older or gone, new technologies might have made their knowledge and skills obsolete, and it might take them many years just to understand the political and social changes that occurred in their absence.

Despite all these hurdles, the human race is progressing toward the achievement of interstellar travel. Technology continues to advance rapidly, and the potential for international cooperation is much greater today than it was just a couple of decades ago, when the United States and the Soviet Union seemed poised for World War III. Even the difficulties posed by long trips may not be insurmountable. Perhaps the trips will be one-way, with the travelers discovering and colonizing new planets. Or perhaps advances in medicine will extend human life spans to hundreds or even thousands of years, in which case being away from Earth for a few decades may be no more traumatic than going off to college is today.

All things considered, we seem bound to become interstellar travelers unless we *choose* otherwise. Such a choice is a very real possibility and may be either deliberate or a consequence of some catastrophe that halts our technological development. On the deliberate side, many people argue that money for space exploration would be better spent here on Earth, and others argue against space colonization on philosophical grounds. On the catastrophic side, we are fully capable of unleashing forces that could cause the collapse of our civilization, perhaps even our extinction. Despite the lessening of international tensions since the end of the Cold War, more than 10,000 nuclear weapons remain in the global arsenal, and many nations possess the technology to build chemical and biological weapons of mass destruction. Our civilization might also succumb to disasters brought on by human activity, such as overpopulation, epidemic disease, or global warming.

If an alien civilization were watching us, they might well conclude that we are poised on the brink of the most significant turning point in human history. If we destroy ourselves, all our achievements in science, art, and philosophy will be lost forever. But if we survive and choose to continue the exploration of space, we may be embarking on a path that will take us to the stars. Whether we like it or not, history has handed this decision to our generation first.

TIME OUT TO THINK ***While we could certainly destroy our civilization today, some people argue that a civilization with colonies spread among many different star systems would be essentially "extinction-proof." Because of the long travel times between star systems, no single event (such as war, disease, or environmental damage) could wipe out all the colonies. Do you agree that attaining large-scale interstellar travel would assure the long-term survival of the human species? Defend your opinion.***

S6.2 Starship Design

So far we have spoken of starships that could carry humans across vast interstellar distances. But if we expand the definition to include robotic spacecraft, then we have already launched our first starships. Four interplanetary probes—*Pioneers 10* and *11,* and *Voyagers 1* and 2—are traveling fast enough to escape from the solar system. It will take them more than 10,000 years to cover each light-year of distance, and their trajectories will not take them on close passes of any nearby stars. Nevertheless, because of the relatively low speeds at which they are traveling through the near-vacuum of interstellar space, these spacecraft should suffer little damage during their journeys and are likely to look almost as good as new for millions of years. Each therefore carries a greeting from Earth, in case someone comes across one of them someday (Figure S6.1).

Current spacecraft, including both robotic probes like Pioneer and Voyager and crew-carrying spacecraft such as the Space Shuttle, are launched by *chemical rockets.* As we discussed in Chapter 6, rockets work by taking advantage of Newton's third law of motion: Engines drive hot gas out the back of the rocket, which causes the rocket to accelerate forward. Chemical rockets are simply rockets in which the engine power comes from chemical reactions—which essentially means ordinary burning. For example, some chemical rockets burn kerosene, while others (including the Space Shuttle's main engine) tap the energy released when liquid hydrogen and liquid oxygen combine to make water.

Chemical rockets will never be practical for interstellar travel because a limiting process prevents them from reaching very high speeds: Achieving higher speeds requires more fuel, but the added weight of additional fuel makes it more difficult to increase the rocket's speed. Detailed calculations of this limiting process show that chemical rockets, even as propulsion systems for relatively small robotic spacecraft, cannot exceed speeds of about $0.001c$—in which case the journey to Alpha Centauri would take more than

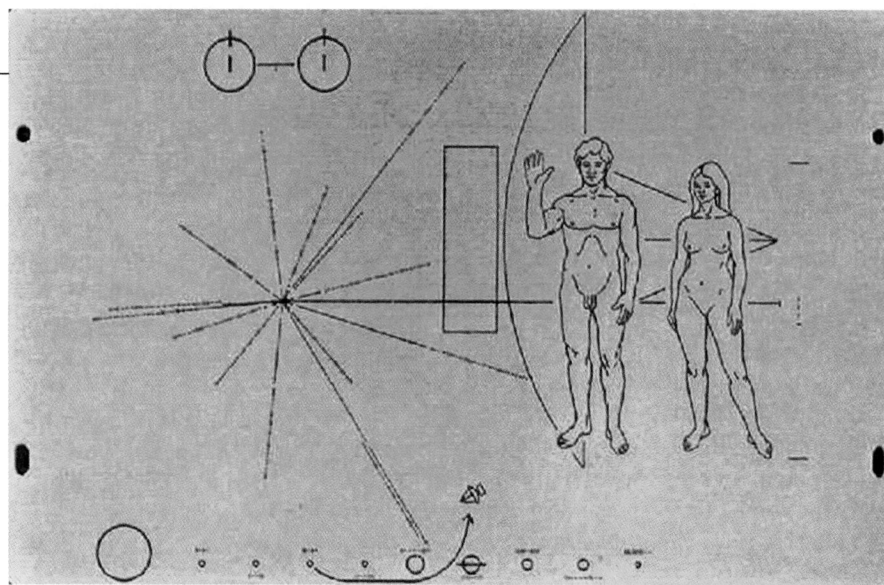

FIGURE S6.1 Messages aboard the Pioneer and Voyager spacecraft, which are bound for the stars.

a *Pioneers 10* and *11* carry a copy of this etching as a greeting to anyone who might find them.

4,000 years. For the much larger starships needed to support human crews, chemical rockets are completely out of the question. Fortunately, other technologies may allow for much faster travel.

Nuclear Rocket Propulsion

Rockets would be far more powerful if their engines used nuclear power rather than chemical power. The basic idea behind a nuclear-powered rocket is very similar to that of a chemical rocket, except that the particles expelled out the back are driven by nuclear energy instead of chemical energy. Nuclear reactions generate far more power than chemical reactions because they convert some of the mass in atomic nuclei into energy in accord with $E = mc^2$. (In contrast, chemical reactions generate power only by rearranging the energy levels of electrons in atoms or molecules.) Fusing a kilogram of hydrogen, for example, generates more than a million times more energy than chemical reactions in a kilogram of hydrogen and oxygen.

Fusion is somewhat more efficient at converting mass to energy than fission, but fission-powered rockets are probably closer to reality because we do not yet have the technology to build controlled fusion reactors. (In contrast, nuclear fission reactors are fairly commonplace and currently generate roughly 10% of the world's electricity.) In fact, during the 1960s, the United States government spent several billion dollars developing fission-powered rocket engines and successfully built and tested them on the ground. However, the nuclear rocket program was terminated in the early 1970s, before a nuclear rocket ever flew in space.

One way to realize the benefits of fusion power without the need for controlled fusion reactors would be to power starships with nuclear bombs. This idea, called *nuclear pulse propulsion,* involves generating energy with repeated detonations of relatively small H-bombs. One design, created under the name Project Orion,[1] envisions the explosions taking place a few tens of meters behind the space ship (Figure S6.2). The vaporized debris from each explosion would impact a "pusher plate" on the back of the spacecraft, propelling the ship forward. In principle, we could build an Orion spacecraft with existing technology, although it would be very expensive and would require an exception to the international treaty banning nuclear detonations in space.

Another design for a fusion-powered rocket, called Project Daedalus,[2] relies on generation of a continuous stream of energy from an on-board controlled nuclear fusion reactor (Figure S6.3). Daedalus is therefore beyond our current technological capabilities, but it might be achievable if and when we learn to control nuclear fusion.

Fusion-powered starships such as Orion or Daedalus could probably achieve speeds of about 10% of the speed of light, completing journeys to nearby stars in a few decades. Such speeds are certainly sufficient for sending colonists on one-way voyages, particularly if the crew can be put into some sort of hibernation that prevents them from aging during the trip.

---

[1]Project Orion lasted from 1958 to 1965 and involved about 40 engineers and scientists.

[2]Project Daedalus was designed in the 1970s by members of the British Interplanetary Society.

b *Voyagers 1* and *2* carry a phonograph record—a 12-inch gold-plated copper disk containing music, greetings, and images from Earth.

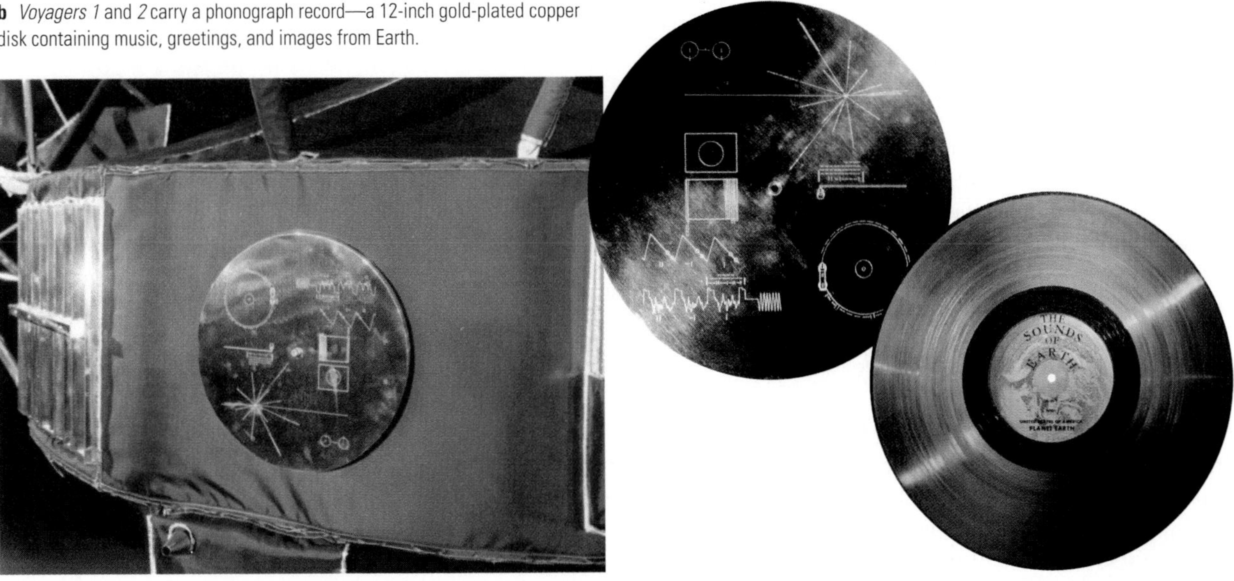

FIGURE S6.2 Artist's conception of the Project Orion starship, showing one of the small H-bomb detonations that propel it. Debris from the detonation impacts the flat disk, called the pusher plate, at the back of the spaceship. The central sections (enclosed in lattice) hold the bombs, and the crew lives in the front sections.

FIGURE S6.3 Artist's conception of a robotic Project Daedalus starship. The front section (right) holds the scientific instruments. The large spheres hold the hydrogen for the central fusion reactor.

TIME OUT TO THINK *From the sixteenth to the nineteenth century, many people left their homes in Europe on one-way journeys to the "new world" of America. In the future, similar one-way trips to colonies on Earth-like planets around other stars may be possible. If offered the opportunity, would you go? Do you know anyone who would? Why or why not?*

Matter–Antimatter Rocket Engines

In principle, there is an even more powerful energy source for rocket engines than fusion: matter–antimatter annihilation [Section S5.2]. Whereas fusion converts less than 1% of the mass of atomic nuclei into energy, matter–antimatter annihilation converts *all* the annihilated mass into energy. Starships with matter–antimatter engines could probably reach 90% or more of the speed of light. At these speeds, the slowing of time predicted by relativity becomes noticeable, putting many nearby stars within a few years' journey for the crew members.

However, while matter–antimatter engines may someday be the propulsion systems of choice for starships, this technology appears to be far in the future. The two major technological challenges to matter–antimatter engines are finding a way to produce sufficient quantities of antimatter to be used in the engines and finding a way to safely store the antimatter until it is needed.

FIGURE S6.4 Artist's conception of a spaceship propelled by a solar sail, shown as it approaches a forming planet in a young solar system. The sail is many kilometers across; the scientific payload is at the central meeting point of the four scaffold-like structures.

At present, we can produce antimatter only in huge particle accelerators that cost billions of dollars to build and require enormous amounts of energy to operate. With present technology, making just a single kilogram of antimatter would take more energy than the current annual energy use of all humanity. Moreover, even if we could produce the antimatter, we do not yet have the technology to store it safely. We could not put it in a normal container, because it would instantly annihilate with the matter in the container walls. We could potentially use a *magnetic container*—which uses magnetic fields to keep charged particles confined inside—but the container would have to be completely empty of ordinary matter in order to prevent annihilation.

Sails and Beamed Energy Propulsion

In principle, spacecraft with large, lightweight sails could be propelled by the solar wind and the radiation pressure of sunlight. Solar sailing may well prove to be a fairly inexpensive way of navigating within the solar system (Figure S6.4). Surprisingly, it may even be useful for interstellar travel. Energy flowing from the Sun would give an optimally designed sailing starship a small but continuous acceleration as it heads out of the solar system. By the time the ship reached the distance at which the solar wind could no longer provide acceleration, it might already be traveling at speeds of a few percent of the speed of light.

We might achieve even higher velocities by shining powerful lasers at the spacecraft sails. Most of the force accelerating solar sails actually comes from the momentum of photons of sunlight, rather than from particles in the solar wind. Thus, we could augment this force by building a powerful laser and directing its beam at the sail. Such *beamed energy propulsion* theoretically could accelerate ships to substantial fractions of the speed of light—but only if we could build enormous lasers. For example, accelerating a ship to half the speed of light within a few years would require a laser that uses 1,000 times more power than all current human power consumption. As a result, starships propelled by beamed energy probably remain far in the future. (Another challenge for beamed energy propulsion is figuring out how to slow down the ship when it reaches its destination.)

Interstellar Ramjets

All the propulsion systems we've discussed so far have limited capabilities: Rockets can carry only a limited amount of fuel no matter what their fuel source, and beamed energy becomes too weak to accelerate starships when they get very far from home. One way around these difficulties would be to build a starship that collects fuel as it goes. Such ships are known as *interstellar ramjets* (Figure S6.5).

An interstellar ramjet would essentially scoop up interstellar hydrogen and use it as fuel for a fusion engine. Because the density of interstellar space is so low, the giant scoops on the front of interstellar ramjets would have to be hundreds of kilometers across. Interstellar ramjets therefore seem at least as far in the future as any of the other designs we have discussed. Nevertheless, they have one major advantage over all the others: the ability to collect fuel continuously and therefore to accelerate continuously.

With a continuous acceleration of 1*g*, an interstellar ramjet could put virtually the entire universe within the range of its crew. Imagine that the ship accelerates at 1*g* for half the voyage to its destination and then turns around and decelerates at 1*g* until it arrives. The crew would find the trip quite comfortable, experiencing Earth-like gravity the entire way. During most of the journey, the ship would be traveling relative to Earth (and its destination) at a speed very close to the speed of light, so time on the ship would pass very slowly compared to time on Earth. Longer trips would mean reaching top speeds closer to the speed of light and therefore more extreme effects on time. For example, such a ship could make a trip to a star 500 light-years away in only about 12 years of ship time.[3] It could travel the 28,000 light-years to the center of the Milky Way Galaxy, where it could observe firsthand the mysterious galactic center [Section 18.5], in only about 21 years of ship time. It could make the trip to a star system in the Andromeda Galaxy in only about 29 years of ship time. Thus, the crew could travel to the Andromeda Galaxy, spend 2 years studying one of its star systems and taking pictures of the Milky Way Galaxy to bring home, and return in just 60 years of ship time. However, because the Andromeda Galaxy is about 2.5 million light-years away, the crew would find that about 5 million years had passed on Earth by the time they returned home.

Science Fiction

If you are a science fiction fan, this discussion of interstellar travel may be depressing. Interstellar tourism and commerce seem out of the question, even with interstellar ramjets that reach speeds very close to the speed of light. If we are ever to travel about the galaxy the way we now travel about the Earth, we will need spacecraft that can get us from here to there much faster than the speed of light.

The theory of relativity leaves little hope that we can ever find a way to travel *through* space faster than the speed of light, but science fiction writers have imagined all kinds of novel shortcuts that don't necessarily violate relativity or any other known laws of physics. We discussed some of these ideas (hyperspace, wormholes, and warp drive) in Chapter S4. The bottom line is that our present knowledge does not allow us to say one way or the other whether any of these technologies are possible.

S6.3 Radio Communication: Virtual Interstellar Travel

We cannot travel at the speed of light, but we can send information at the speed of light using radio waves, lasers, or other forms of light. Because radio waves penetrate interstellar gas and dust fairly easily,

FIGURE S6.5 Artist's conception of a spaceship powered by an interstellar ramjet. The giant scoop in the front (left) collects interstellar hydrogen for use as fusion fuel.

[3]You can calculate this and the other travel times for the crew using the formula given in problem 6 of Chapter S4.

they probably will prove to be the most useful form of interstellar communication. Although we usually think of radio communication as a way of sending and receiving fairly simple messages, there is really no limit to how much information we could transmit or receive. In principle, we could become *virtual* interstellar travelers. Imagine, for example, that we sent a robotic probe to Alpha Centauri at 10% of the speed of light. It would take 40 years to get there, but it could send data back to Earth via radio in just 4 years. Nearby stars, at least, could be explored with robotic spacecraft in the same way we now explore the planets of our own solar system.

An even better way to gather data about distant worlds would be to receive signals from civilizations in other star systems. If such civilizations exist, they may already be broadcasting messages to us. The *search for extraterrestrial intelligence,* or *SETI,* is a project in which scientists use large radio telescopes to scan the skies for signals from other civilizations. With enough data and the computer technology known as virtual reality, we could actually experience what it would be like to walk among the cities of an alien civilization.

The more immediate hope of SETI efforts is simply to learn whether anyone else is out there—and happens to be broadcasting radio messages. But how could we interpret these messages? After all, we could hardly expect alien civilizations to share any of our social customs, let alone to speak English. Fortunately, we know that all civilizations must share some things in common: the laws of mathematics and science that are the same everywhere. Imagine that we wanted to send a message to a distant civilization. As popularized in the movie *Contact,* we might begin by broadcasting radio pulses in the order of the prime numbers (1, 2, 3, 5, 7, 11, . . .)—clearly identifying the message as something sent by an intelligent species, since no known natural process can broadcast prime numbers. We could then represent the prime numbers with a binary code and use this same code to begin representing some of the laws of mathematics. Since alien civilizations would know these laws already, they could learn the intricacies of our code. Once they knew the full code, they could receive as much additional information about Earth and our civilization as we wished to send them.

Scientists working on SETI hope that someone out there is already broadcasting similar coded messages to Earth. Even if such messages are being continually broadcast to Earth, finding them may be rather like finding the proverbial needle in a haystack: There are billions of stars to search, and we don't know the radio frequencies at which other civilizations might choose to broadcast their messages. Nevertheless, we can make some guesses. For example, we might focus our attention on Sun-like stars and assume that the aliens would broadcast near some "natural" frequency, such as the frequency of 21-cm radio waves from hydrogen. In addition, new technologies allow us to search many frequencies simultaneously, increasing the odds of success. However, SETI efforts remain controversial because they require extensive time on powerful and expensive radio telescopes when we have no way of knowing whether anyone is even out there.

TIME OUT TO THINK ***SETI supporters argue that, despite the costs and the unknowns, the efforts are justified because contact with an extraterrestrial intelligence would be one of the most dramatic discoveries in human history. Do you agree? Defend your opinion.***

S6.4 Civilizations

Whether we consider the virtual travel of radio communication or actual ships bound for the stars, a recurrent theme of interstellar travel is contact with other civilizations. Thus, we are led to one of the most exciting questions in science: Is anyone else out there?

Unfortunately, we simply don't yet know. But we can speculate—and perhaps even learn something in the process. Let's try a systematic approach to guessing the number of alien civilizations that exist. We'll concentrate on the Milky Way Galaxy and then extend our results to the entire observable universe. For simplicity's sake, we'll also assume that alien life, like life on Earth, will need liquid water and therefore that it must evolve on a planet that lies within the *habital zone* around a star [Section 13.6].

How Many Civilizations Are Out There?

In principle, we could calculate the number of civilizations in the Milky Way Galaxy if we knew just a few basic facts. We would start with the number of planets that lie within habital zones, which we'll call N_P (for "number of planets"). Then we would determine the fraction of those planets that actually have life; let's use f_{life} to stand for this fraction. Multiplying N_P by f_{life} would tell us the number of life-bearing planets in the Milky Way. Next, we'd multiply this number by the fraction of life-bearing planets on which intelligent beings evolve and develop a civilization, which we'll call $f_{civilization}$. Finally, we'd multiply by the fraction of planets with civilizations that have those civilizations *now* (as opposed to the distant past or future), which we'll call f_{now}. Putting all these ideas together gives us the following simple equation for the number of civilizations now inhabiting the Milky Way Galaxy:[4]

$$\text{Number of civilizations} = N_P \times f_{life} \times f_{civilization} \times f_{now}$$

The only difficulty in using this equation is that we don't know the value of any of its terms! In fact, the only term for which we can even make a reasonable guess is the number of planets, N_P. Our theories of star formation tell us that planetary systems ought to be common, and recent discoveries of protoplanetary disks around nearby stars—and of a few full-fledged planets—support this idea [Sections 8.8 and 16.2]. Studies of our own solar system suggest that both Earth and Mars lie within the Sun's habitability zone [Section 13.6], and Venus is probably close to this zone. If other planetary systems are similar, we might expect the typical number of planets within the habitability zone to be between one and three. All in all, it seems likely that many, and perhaps even most, of the several hundred billion stars in the Milky Way have at least one planet within their habitability zone. In that case, we might reasonably guess that the number of potentially habitable planets in the Milky Way is in the neighborhood of 100 billion (10^{11}). The technology for detecting planets beyond our solar system is rapidly advancing, so we may be able to improve or confirm this estimate within the next one or two decades.

For the moment, we have no rational way to guess the fraction f_{life} of these 100 billion or so planets on which life actually arose. The fact that life arose quickly on Earth [Section 13.4] suggests that life develops fairly easily under the right conditions, but it is by no means conclusive. As far as we know today, f_{life} could be anywhere between 1 in 100 billion (it happened only on Earth) and 1 (it happened on all 100 billion planets suitable for life).

TIME OUT TO THINK *Suppose we discover life on Mars. Would this give us any greater insight into the value of f_{life}? Explain. Also explain what we would learn about f_{life} if we discovered that life had never existed on Mars.*

Similarly, we have little basis on which to guess the fraction $f_{civilization}$ of life-bearing planets that develop a civilization. However, the fact that life flourished on Earth for some 4 billion years before the rise of humans suggests that developing intelligent life is much more difficult than developing microbial life. The question is *how much* more difficult. Roughly half the stars in the Milky Way are older than our Sun, so if 4 billion years is a typical time for a civilization to develop then it should have happened already on about half the planets with life. On the other hand, if the Earth was "fast" and the typical time needed for intelligent life to develop is much longer, then most life-bearing planets may be covered with nothing more advanced than bacteria.

The huge uncertainties in f_{life} and $f_{civilization}$ make it impossible for us to determine the number of civilizations that have arisen in the Milky Way's history. If 1 in 1,000 suitable planets ends up with a civilization, then we are newcomers among a billion civilizations in the Milky Way. If 1 in 1 million develops a civilization, there would still have been some 100,000 civilizations out there in the Milky Way Galaxy alone. But if only 1 in 1 trillion planets ends up with a civilization, then we may be the first in the Milky Way—and the only civilization within millions of light-years.

TIME OUT TO THINK *Recall that the total number of stars in the observable universe is greater than the number of grains of sand on all the beaches on Earth* [Section 2.4]. *Do you think it's possible that we are actually alone in the* universe? *Defend your opinion.*

The final step in our calculation—determining the number of civilizations that exist *now*—requires knowing how long a civilization typically lasts in a planet's history. Here we must define what we mean by *civilization;* let's use a definition that presumes a civilization advanced enough to travel in space—and therefore advanced enough to be seeking life elsewhere. On the 4.6-billion-year-old planet Earth, the space age began only a little over 40 years ago. Thus, the fraction of Earth's history during which it has had a space-faring civilization is 40 in 4.6 billion, or about 1 in 100 million. If this fraction is typical, there would have to be some 100 million civilization-bearing planets in the Milky Way in order to have good odds of finding another civilization out there now. However, we'd expect this fraction to be typical only if we are on the brink of self-destruction, as the fraction will grow larger as long as our civilization continues to thrive. Thus, if civilizations are at all common, the key factor in whether any are out there now is their survivability. If most civilizations self-destruct shortly after achieving the technology for space travel, then we probably are alone in the galaxy at present. But if most survive and become interstellar travelers, the Milky Way may be brimming with civilizations—most far more advanced than ours.

[4] This equation is a slightly modified version of the *Drake equation,* originally proposed by astronomer Frank Drake.

TIME OUT TO THINK *What do you think is the likelihood that our civilization will survive to become interstellar travelers? Explain.*

A Paradox: Where Are the Aliens?

Imagine that we survive and become interstellar travelers and that we begin colonizing habitable planets around nearby stars. As the colonies grow at each new location, some of the people may decide to set out for other star systems. Even if our starships traveled at relatively low speeds—say, a few percent of the speed of light—we could have dozens of outposts around nearby stars within a few centuries. In 10,000 years, our descendants would be spread among stars within a few hundred light-years of Earth. In a few million years, we could have outposts scattered throughout the Milky Way and would likely have explored nearly every star system in the galaxy. We would have become a true galactic civilization.

Now, if we take the idea that *we* could develop a galactic civilization within a few million years and combine it with the reasonable (though unproved) idea that civilizations ought to be common, we are led to an astonishing conclusion: Someone else should already have created a galactic civilization. In fact, someone should have done it a long time ago. For argument's sake, let's suppose that civilizations arise around 1 in a million stars, in which case some 100,000 civilizations should have arisen in the Milky Way. Further, let's suppose that civilizations typically arise when their stars are 5 billion years old. Given that the galaxy is at least 10 billion years old, the first of these 100,000 civilizations would have arisen at least 5 billion years ago, and others would have arisen, on average, every 50,000 years. Under these assumptions, the youngest civilization besides ours would be some 50,000 years ahead of us technologically, and most would be millions or billions of years ahead of us.

Thus, we encounter a strange paradox:[5] Plausible arguments suggest that a galactic civilization should already exist, yet we have so far found no evidence of such a civilization.[6] There are many possible solutions to this paradox, but most fall into one of three basic categories:

1. There is no galactic civilization because civilizations are not common. Perhaps we are the first, or perhaps civilizations are so rare that many star systems, including ours, remain unexplored.
2. There is no galactic civilization because civilizations do not leave their home worlds—either because they are uninterested in interstellar travel or because they destroy themselves before achieving it.
3. There *is* a galactic civilization, but it has deliberately avoided revealing its existence to us.

We do not know which, if any, of these three explanations is the correct solution to the paradox, but each leads to intriguing possibilities. If the first is correct, then our civilization is an astonishing achievement—something that few if any other species have ever accomplished. From this point of view, humanity becomes all the more precious, and the collapse of our civilization would be all the more tragic. The second possible explanation is much less satisfying. Unless we "think differently" from other civilizations, it says that we will soon either lose our interest in interstellar travel or destroy ourselves—hardly a comforting thought. The third possibility is perhaps the most intriguing. It says that we are newcomers on the scene of a galactic civilization that has existed for millions or billions of years before us. Perhaps members of this civilization are deliberately leaving us alone for the time being and will invite us to join them when we prove ourselves worthy. If so, our entire species may be on the verge of beginning a journey every bit as incredible as that of a baby emerging from the womb and coming into the world.

---

[5] This paradox is often called *Fermi's paradox*, because during a conversation about extraterrestrial life in 1948 the scientist Enrico Fermi supposedly blurted out, "Where are they?"

[6] Note that we are neglecting reports of so-called UFOs. Although it is impossible to prove that reports of UFOs are *not* real, science demands hard evidence that can be seen and tested by anyone. So far, at least, UFO believers have not produced any such clear evidence.

We, this people, on a small and lonely planet
Travelling through casual space
Past aloof stars, across the way of indifferent suns
To a destination where all signs tell us
It is possible and imperative that we learn
A brave and startling truth.

MAYA ANGELOU, excerpted from *A Brave and Startling Truth*

THE BIG PICTURE

Throughout our study of astronomy, we have taken the "big picture" view of trying to understand how we fit into the universe. Here, at last, we have returned to Earth and examined the role of our own generation in the big picture of human history. Tens of thousands of human generations past have walked this Earth, but ours is the first generation with the technology to study the far reaches of our universe and to travel beyond the Earth. It is up to us to decide whether we will use this technology to advance our species or to destroy it.

Take a long view, a view down the centuries and millennia from this moment. Imagine our descendants living among the stars, having created or joined a great galactic civilization. They will have the privilege of experiencing ideas, worlds, and discoveries far beyond our wildest imagination. Perhaps, in their history lessons, they will learn of our generation—the generation that history placed at the turning point and that chose life over death and destruction.

Review Questions

1. How do speeds of current spacecraft compare to the speed of light?
2. Describe several hurdles, both technological and social, that we must overcome in order to achieve interstellar flight.
3. Why do we say that *Pioneers 10* and *11* and *Voyagers 1* and *2* are our first interstellar spacecraft? When will they be a light-year from Earth?
4. Why are chemical rockets inadequate for interstellar travel?
5. Briefly describe the prospects and limitations for each of the following alternative starship designs: nuclear propulsion, matter–antimatter engines, sails and beamed energy propulsion, and interstellar ramjets.
6. What is SETI? How do SETI scientists hope it will work?
7. Briefly describe and interpret the equation given in the text that allows us to estimate the number of civilizations in the Milky Way Galaxy. Why is the term f_{now} particularly important?
8. Describe several potential solutions to the paradox of "where are the aliens?"

Discussion Questions

1. *Distant Dream or Near-Reality?* Considering all the issues surrounding interstellar flight, when (if ever) do you think we are likely to begin traveling among the stars? Why?
2. *Where Are the Aliens?* Consider the paradox that arises from the fact that we do not yet have evidence of a galactic civilization. What do *you* think is the solution to this paradox? Why?
3. *Aliens in the Movies.* Choose a science fiction movie (or television show) that involves alien species, and discuss what aspects of it are realistic and what aspects are not. Pay particular attention to the depiction of alien species and encounters between different species. Do you think such encounters are portrayed realistically?

Problems

1. *Research: Pioneers of Flight.* Write a one- to two-page biography of one of the following pioneers of flight: Otto Lilienthal, the Wright brothers, Charles Lindbergh, Amelia Earhart.
2. *Research: Development of Rocketry.* Write a one- to two-page paper about the development of rocketry. Be sure to include the roles of Konstantin Tsiolkovsky and Robert Goddard.
3. *Research: Starship Design.* Find more details about one of the proposals for starship propulsion and design discussed in this chapter. How would the starship actually be built? What new technologies would be needed, and what existing technologies could be used? Summarize your findings in a one- to two-page research report.

APPENDIX A

A FEW MATHEMATICAL SKILLS

This appendix reviews the following mathematical skills: powers of 10, scientific notation, working with units, and the metric system. You should refer to this appendix as needed while studying the textbook, particularly if you are having difficulty with the Mathematical Insights.

A.1 Powers of 10

Powers of 10 simply indicate how many times to multiply 10 by itself. For example:

$$10^2 = 10 \times 10 = 100$$

$$10^6 = 10 \times 10 \times 10 \times 10 \times 10 \times 10 = 1{,}000{,}000$$

Negative powers are the reciprocals of the corresponding positive powers. For example:

$$10^{-2} = \frac{1}{10^2} = \frac{1}{100} = 0.01$$

$$10^{-6} = \frac{1}{10^6} = \frac{1}{1{,}000{,}000} = 0.000001$$

Table A.1 lists powers of 10 from 10^{-12} to 10^{12}. Note that powers of 10 follow two basic rules:

1. A positive exponent tells how many zeros follow the 1. For example, 10^0 is a 1 followed by no zeros, and 10^8 is a 1 followed by eight zeros.
2. A negative exponent tells how many places are to the right of the decimal point, including the 1. For example, $10^{-1} = 0.1$ has one place to the right of the decimal point; $10^{-6} = 0.000001$ has six places to the right of the decimal point.

Table A.1 Powers of 10

Zero and Positive Powers			Negative Powers		
Power	Value	Name	Power	Value	Name
10^0	1	One			
10^1	10	Ten	10^{-1}	0.1	Tenth
10^2	100	Hundred	10^{-2}	0.01	Hundredth
10^3	1,000	Thousand	10^{-3}	0.001	Thousandth
10^4	10,000	Ten thousand	10^{-4}	0.0001	Ten thousandth
10^5	100,000	Hundred thousand	10^{-5}	0.00001	Hundred thousandth
10^6	1,000,000	Million	10^{-6}	0.000001	Millionth
10^7	10,000,000	Ten million	10^{-7}	0.0000001	Ten millionth
10^8	100,000,000	Hundred million	10^{-8}	0.00000001	Hundred millionth
10^9	1,000,000,000	Billion	10^{-9}	0.000000001	Billionth
10^{10}	10,000,000,000	Ten billion	10^{-10}	0.0000000001	Ten billionth
10^{11}	100,000,000,000	Hundred billion	10^{-11}	0.00000000001	Hundred billionth
10^{12}	1,000,000,000,000	Trillion	10^{-12}	0.000000000001	Trillionth

Multiplying and Dividing Powers of 10

Multiplying powers of 10 simply requires adding exponents, as the following examples show:

$$10^4 \times 10^7 = \underbrace{10{,}000}_{10^4} \times \underbrace{10{,}000{,}000}_{10^7} = \underbrace{100{,}000{,}000{,}000}_{10^{4+7} = 10^{11}} = 10^{11}$$

$$10^5 \times 10^{-3} = \underbrace{100{,}000}_{10^5} \times \underbrace{0.001}_{10^{-3}} = \underbrace{100}_{10^{5+(-3)} = 10^2} = 10^2$$

$$10^{-8} \times 10^{-5} = \underbrace{0.00000001}_{10^{-8}} \times \underbrace{0.00001}_{10^{-5}} = \underbrace{0.0000000000001}_{10^{-8+(-5)} = 10^{-13}} = 10^{-13}$$

Dividing powers of 10 requires subtracting exponents, as in the following examples:

$$\frac{10^5}{10^3} = \underbrace{100{,}000}_{10^5} \div \underbrace{1{,}000}_{10^3} = \underbrace{100}_{10^{5-3} = 10^2} = 10^2$$

$$\frac{10^3}{10^7} = \underbrace{1{,}000}_{10^3} \div \underbrace{10{,}000{,}000}_{10^7} = \underbrace{0.0001}_{10^{3-7} = 10^{-4}} = 10^{-4}$$

$$\frac{10^{-4}}{10^{-6}} = \underbrace{0.0001}_{10^{-4}} \div \underbrace{0.000001}_{10^{-6}} = \underbrace{100}_{10^{-4-(-6)} = 10^2} = 10^2$$

Powers of Powers of 10

We can use the multiplication and division rules to raise powers of 10 to other powers or to take roots. For example:

$$(10^4)^3 = 10^4 \times 10^4 \times 10^4 = 10^{4+4+4} = 10^{12}$$

Note that we can get the same end result by simply multiplying the two powers:

$$(10^4)^3 = 10^{4\times 3} = 10^{12}$$

Because taking a root is the same as raising to a fractional power (e.g., the square root is the same as the 1/2 power, the cube root is the same as the 1/3 power, etc.), we can use the same procedure for roots, as in the following example:

$$\sqrt{10^4} = (10^4)^{1/2} = 10^{4\times(1/2)} = 10^2$$

Adding and Subtracting Powers of 10

Unlike with multiplication and division, there is no shortcut for adding or subtracting powers of 10. The values must be written in longhand notation. For example:

$$10^6 + 10^2 = 1{,}000{,}000 + 100 = 1{,}000{,}100$$

$$10^8 + 10^{-3} = 100{,}000{,}000 + 0.001 = 100{,}000{,}000.001$$

$$10^7 - 10^3 = 10{,}000{,}000 - 1{,}000 = 9{,}999{,}000$$

Summary

We can summarize our findings using n and m to represent any numbers:

- To *multiply* powers of 10, *add* exponents: $10^n \times 10^m = 10^{n+m}$
- To *divide* powers of 10, *subtract* exponents: $\dfrac{10^n}{10^m} = 10^{n-m}$
- To *raise* powers of 10 to other powers, multiply exponents: $(10^n)^m = 10^{n \times m}$

A.2 Scientific Notation

When we are dealing with large or small numbers, it's generally easier to write them with powers of 10. For example, it's much easier to write the number 6,000,000,000,000 as 6×10^{12}. This format, in which a number *between* 1 and 10 is multiplied by a power of 10, is called **scientific notation.**

Converting a Number to Scientific Notation

We can convert numbers written in ordinary notation to scientific notation with a simple two-step process:

1. Move the decimal point to come after the *first* nonzero digit.
2. The number of places the decimal point moves tells you the power of 10; the power is *positive* if the decimal point moves to the left and *negative* if it moves to the right.

Examples:

$$3{,}042 \xrightarrow[\text{3 places to left}]{\text{decimal needs to move}} 3.042 \times 10^3$$

$$0.00012 \xrightarrow[\text{4 places to right}]{\text{decimal needs to move}} 1.2 \times 10^{-4}$$

$$226 \times 10^2 \xrightarrow[\text{2 places to left}]{\text{decimal needs to move}} (2.26 \times 10^2) \times 10^2 = 2.26 \times 10^4$$

Converting a Number from Scientific Notation

We can convert numbers written in scientific notation to ordinary notation by the reverse process:

1. The power of 10 indicates how many places to move the decimal point; move it to the *right* if the power of 10 is positive and to the *left* if it is negative.
2. If moving the decimal point creates any open places, fill them with zeros.

Examples:

$$4.01 \times 10^2 \xrightarrow[\text{2 places to right}]{\text{move decimal}} 401$$

$$3.6 \times 10^6 \xrightarrow[\text{6 places to right}]{\text{move decimal}} 3{,}600{,}000$$

$$5.7 \times 10^{-3} \xrightarrow[\text{3 places to left}]{\text{move decimal}} 0.0057$$

Multiplying or Dividing Numbers in Scientific Notation

Multiplying or dividing numbers in scientific notation simply requires operating on the powers of 10 and the other parts of the number separately.

Examples:

$$(6 \times 10^2) \times (4 \times 10^5) = (6 \times 4) \times (10^2 \times 10^5) = 24 \times 10^7 = (2.4 \times 10^1) \times 10^7 = 2.4 \times 10^8$$

$$\frac{4.2 \times 10^{-2}}{8.4 \times 10^{-5}} = \frac{4.2}{8.4} \times \frac{10^{-2}}{10^{-5}} = 0.5 \times 10^{-2-(-5)} = 0.5 \times 10^3 = (5 \times 10^{-1}) \times 10^3 = 5 \times 10^2$$

Note that, in both these examples, we first found an answer in which the number multiplied by a power of 10 was *not* between 1 and 10. We therefore followed the procedure for converting the final answer to scientific notation.

Addition and Subtraction with Scientific Notation

In general, we must write numbers in ordinary notation before adding or subtracting.

Examples:

$$(3 \times 10^6) + (5 \times 10^2) = 3{,}000{,}000 + 500 = 3{,}000{,}500 = 3.0005 \times 10^6$$

$$(4.6 \times 10^9) - (5 \times 10^8) = 4{,}600{,}000{,}000 - 500{,}000{,}000 = 4{,}100{,}000{,}000 = 4.1 \times 10^9$$

When both numbers have the *same* power of 10, we can factor out the power of 10 first.

Examples:

$$(7 \times 10^{10}) + (4 \times 10^{10}) = (7 + 4) \times 10^{10} = 11 \times 10^{10} = 1.1 \times 10^{11}$$

$$(2.3 \times 10^{-22}) - (1.6 \times 10^{-22}) = (2.3 - 1.6) \times 10^{-22} = 0.7 \times 10^{-22} = 7.0 \times 10^{-23}$$

A.3 Working with Units

Showing the units of a problem as you solve it usually makes the work much easier and also provides a useful way of checking your work. If an answer does not come out with the units you expect, you probably did something wrong. In general, working with units is very similar to working with numbers, as the following guidelines and examples show.

Five Guidelines for Working with Units

Before you begin any problem, think ahead and identify the units you expect for the final answer. Then operate on the units along with the numbers as you solve the problem. The following five guidelines may be helpful when you are working with units:

1. Mathematically, it doesn't matter whether a unit is singular (e.g., meter) or plural (e.g., meters); we can use the same abbreviation (e.g., m) for both.
2. You cannot add or subtract numbers unless they have the *same* units. For example, 5 apples + 3 apples = 8 apples, but the expression 5 apples + 3 oranges cannot be simplified further.

3. You *can* multiply units, divide units, or raise units to powers. Look for key words that tell you what to do.
   - *Per* suggests division. For example, we write a speed of 100 kilometers per hour as

$$100 \frac{\text{km}}{\text{hr}} \quad \text{or} \quad \frac{100 \text{ km}}{1 \text{ hr}}$$

   - *Of* suggests multiplication. For example, if you launch a 50-kg space probe at a launch cost *of* \$10,000 per kilogram, the total cost is:

$$50 \cancel{\text{kg}} \times \frac{\$10{,}000}{\cancel{\text{kg}}} = \$500{,}000$$

   - *Square* suggests raising to the second power. For example, we write an area of 75 square meters as 75 m^2.
   - *Cube* suggests raising to the third power. For example, we write a volume of 12 cubic centimeters as 12 cm^3.

4. Often the number you are given is not in the units you wish to work with. For example, you may be given that the speed of light is 300,000 km/s but need it in units of m/s for a particular problem. To convert the units, simply multiply the given number by a *conversion factor*: a fraction in which the numerator (top of the fraction) and denominator (bottom of the fraction) are equal, so that the value of the fraction is 1; the number in the denominator must have the units that you wish to change. In the case of changing the speed of light from units of km/s to m/s, you need a conversion factor for kilometers to meters. Thus, the conversion factor is:

$$\frac{1{,}000 \text{ m}}{1 \text{ km}}$$

Note that this conversion factor is equal to 1, since 1,000 meters and 1 kilometer are equal, and that the units to be changed (km) appear in the denominator. We can now convert the speed of light from units of km/s to m/s simply by multiplying by this conversion factor:

$$\underbrace{300{,}000 \frac{\cancel{\text{km}}}{\text{s}}}_{\substack{\text{speed of light} \\ \text{in km/s}}} \times \underbrace{\frac{1{,}000 \text{ m}}{1 \cancel{\text{km}}}}_{\substack{\text{conversion from} \\ \text{km to m}}} = \underbrace{3 \times 10^8 \frac{\text{m}}{\text{s}}}_{\substack{\text{speed of light} \\ \text{in m/s}}}$$

Note that the units of km cancel, leaving the answer in units of m/s.

5. It's easier to work with units if you replace division with multiplication by the reciprocal. For example, suppose you want to know how many minutes are represented by 300 seconds. We can find the answer by dividing 300 seconds by 60 seconds per minute:

$$300 \text{ s} \div 60 \frac{\text{s}}{\text{min}}$$

However, it is easier to see the unit cancellations if we rewrite this expression by replacing the division with multiplication by the reciprocal (this process is easy to remember as "invert and multiply"):

$$300 \text{ s} \div 60 \frac{\text{s}}{\text{min}} = \underbrace{300 \cancel{\text{s}} \times \underbrace{\frac{1 \text{ min}}{60 \cancel{\text{s}}}}_{\text{invert}}}_{\text{and multiply}} = 5 \text{ min}$$

We now see that the units of seconds (s) cancel in the numerator of the first term and the denominator of the second term, leaving the answer in units of minutes.

More Examples of Working with Units

Example 1. How many seconds are there in 1 day?

Solution: We can answer the question by setting up a *chain* of unit conversions in which we start with 1 *day* and end up with *seconds*. We use the facts that there are 24 hours per day (24 hr/day), 60 minutes per hour (60 min/hr), and 60 seconds per minute (60 s/min):

$$\underbrace{1\ \cancel{\text{day}}}_{\substack{\text{starting}\\ \text{value}}} \times \underbrace{\frac{24\ \cancel{\text{hr}}}{\cancel{\text{day}}}}_{\substack{\text{conversion}\\ \text{from}\\ \text{day to hr}}} \times \underbrace{\frac{60\ \cancel{\text{min}}}{\cancel{\text{hr}}}}_{\substack{\text{conversion}\\ \text{from}\\ \text{hr to min}}} \times \underbrace{\frac{60\ \text{s}}{\cancel{\text{min}}}}_{\substack{\text{conversion}\\ \text{from}\\ \text{min to s}}} = 86{,}400\ \text{s}$$

Note that all the units cancel except *seconds,* which is what we want for the answer. There are 86,400 seconds in 1 day.

Example 2. Convert a distance of 10^8 cm to km.

Solution: The easiest way to make this conversion is in two steps, since we know that there are 100 centimeters per meter (100 cm/m) and 1,000 meters per kilometer (1,000 m/km):

$$\underbrace{10^8\ \text{cm}}_{\substack{\text{starting}\\ \text{value}}} \times \underbrace{\frac{1\ \text{m}}{100\ \text{cm}}}_{\substack{\text{conversion}\\ \text{from}\\ \text{cm to m}}} \times \underbrace{\frac{1\ \text{km}}{1{,}000\ \text{m}}}_{\substack{\text{conversion}\\ \text{from}\\ \text{m to km}}} = 10^8\ \cancel{\text{cm}} \times \frac{1\ \cancel{\text{m}}}{10^2\ \cancel{\text{cm}}} \times \frac{1\ \text{km}}{10^3\ \cancel{\text{m}}} = 10^3\ \text{km}$$

Alternatively, if we recognize that the number of kilometers should be smaller than the number of centimeters (because kilometers are larger), we might decide to do this conversion by dividing as follows:

$$10^8\ \text{cm} \div \frac{100\ \text{cm}}{\text{m}} \div \frac{1{,}000\ \text{m}}{\text{km}}$$

In this case, before carrying out the calculation, we replace each division with multiplication by the reciprocal:

$$\begin{aligned} 10^8\ \text{cm} \div \frac{100\ \text{cm}}{\text{m}} \div \frac{1{,}000\ \text{m}}{\text{km}} &= 10^8\text{cm} \times \frac{1\ \text{m}}{100\ \text{cm}} \times \frac{1\ \text{km}}{1{,}000\ \text{m}} \\ &= 10^8\ \cancel{\text{cm}} \times \frac{1\ \cancel{\text{m}}}{10^2\ \cancel{\text{cm}}} \times \frac{1\ \text{km}}{10^3\ \cancel{\text{m}}} \\ &= 10^3\ \text{km} \end{aligned}$$

Note that we again get the answer that 10^8 cm is the same as 10^3 km, or 1,000 km.

Example 3. Suppose you accelerate at 9.8 m/s^2 for 4 seconds, starting from rest. How fast will you be going?

Solution: The question asked "how fast?" so we expect to end up with a speed. Therefore, we multiply the acceleration by the amount of time you accelerated:

$$9.8\,\frac{\text{m}}{\text{s}^2} \times 4\ \text{s} = (9.8 \times 4)\,\frac{\text{m} \times \cancel{\text{s}}}{\text{s}^{\cancel{2}}} = 39.2\,\frac{\text{m}}{\text{s}}$$

Note that the units end up as a speed, showing that you will be traveling 39.2 m/s after 4 seconds of acceleration at 9.8 m/s^2.

Example 4. A reservoir is 2 km long and 3 km wide. Calculate its area, in both square kilometers and square meters.

Solution: We find its area by multiplying its length and width:

$$2\ \text{km} \times 3\ \text{km} = 6\ \text{km}^2$$

Next we need to convert this area of 6 km^2 to square meters, using the fact that there are 1,000 meters per kilometer (1,000 m/km). Note that we must square the term 1,000 m/km when converting from km^2 to m^2:

$$6\ \text{km}^2 \times \left(1{,}000\ \frac{\text{m}}{\text{km}}\right)^2 = 6\ \text{km}^2 \times 1{,}000^2\ \frac{\text{m}^2}{\text{km}^2} = 6\ \cancel{\text{km}^2} \times 1{,}000{,}000\ \frac{\text{m}^2}{\cancel{\text{km}^2}}$$

$$= 6{,}000{,}000\ \text{m}^2$$

The reservoir area is 6 km^2, which is the same as 6 million m^2.

A.4 The Metric System (SI)

The modern version of the metric system, known as *Système Internationale d'Unites* (French for "International System of Units") or **SI,** was formally established in 1960. Today, it is the primary measurement system in nearly every country in the world with the exception of the United States. Even in the United States, it is the system of choice for science and international commerce.

The basic units of length, mass, and time in the SI are:

- The **meter** for length, abbreviated m
- The **kilogram** for mass, abbreviated kg
- The **second** for time, abbreviated s

Multiples of metric units are formed by powers of 10, using a prefix to indicate the power. For example, *kilo* means 10^3 (1,000), so a kilometer is 1,000 meters; a microgram is 0.000001 gram, because *micro* means 10^{-6}, or one millionth. Some of the more common prefixes are listed in Table A.2.

Metric Conversions

Table A.3 lists conversions between metric units and units used commonly in the United States. Note that the conversions between kilograms and pounds are valid only on Earth, because they depend on the strength of gravity.

Example 1. International athletic competitions generally use metric distances. Compare the length of a 100-meter race to that of a 100-yard race.

Solution: Table A.3 shows that 1 m = 1.094 yd, so 100 m is 109.4 yd. Note that 100 meters is almost 110 yards; a good "rule of thumb" to remember is that distances in meters are about 10% longer than the corresponding number of yards.

Example 2. How many square kilometers are in 1 square mile?

Solution: We use the square of the miles-to-kilometers conversion factor:

$$(1\ \text{mi}^2) \times \left(\frac{1.6093\ \text{km}}{1\ \text{mi}}\right)^2 = (1\ \cancel{\text{mi}^2}) \times \left(1.6093^2\ \frac{\text{km}^2}{\cancel{\text{mi}^2}}\right) = 2.5898\ \text{km}^2$$

Therefore, 1 square mile is 2.5898 square kilometers.

Table A.2 SI (Metric) Prefixes

Small Values			Large Values		
Prefix	Abbreviation	Value	Prefix	Abbreviation	Value
Deci	d	10^{-1}	Deca	da	10^1
Centi	c	10^{-2}	Hecto	h	10^2
Milli	m	10^{-3}	Kilo	k	10^3
Micro	μ	10^{-6}	Mega	M	10^6
Nano	n	10^{-9}	Giga	G	10^9
Pico	p	10^{-12}	Tera	T	10^{12}

Table A.3 Metric Conversions

To Metric	From Metric
1 inch = 2.540 cm	1 cm = 0.3937 inch
1 foot = 0.3048 m	1 m = 3.28 feet
1 yard = 0.9144 m	1 m = 1.094 yards
1 mile = 1.6093 km	1 km = 0.6214 mile
1 pound = 0.4536 kg	1 kg = 2.205 pounds

APPENDIX B

THE PERIODIC TABLE OF THE ELEMENTS

12 — Atomic number
Mg — Element's symbol
Magnesium — Element's name
24.305 — Atomic mass*

*Atomic masses are fractions because they represent a weighted average of atomic masses of different isotopes—in proportion to the abundance of each isotope on Earth.

1 H Hydrogen 1.00794																	2 He Helium 4.003
3 Li Lithium 6.941	4 Be Beryllium 9.01218											5 B Boron 10.81	6 C Carbon 12.011	7 N Nitrogen 14.007	8 O Oxygen 15.999	9 F Fluorine 18.988	10 Ne Neon 20.179
11 Na Sodium 22.990	12 Mg Magnesium 24.305											13 Al Aluminum 26.98	14 Si Silicon 28.086	15 P Phosphorus 30.974	16 S Sulfur 32.06	17 Cl Chlorine 35.453	18 Ar Argon 39.948
19 K Potassium 39.098	20 Ca Calcium 40.08	21 Sc Scandium 44.956	22 Ti Titanium 47.88	23 V Vanadium 50.94	24 Cr Chromium 51.996	25 Mn Manganese 54.938	26 Fe Iron 55.847	27 Co Cobalt 58.9332	28 Ni Nickel 58.69	29 Cu Copper 63.546	30 Zn Zinc 65.39	31 Ga Gallium 69.72	32 Ge Germanium 72.59	33 As Arsenic 74.922	34 Se Selenium 78.96	35 Br Bromine 79.904	36 Fr Krypton 83.80
37 Rb Rubidium 85.468	38 Sr Strontium 87.62	39 Y Yttrium 88.9059	40 Zr Zirconium 91.224	41 Nb Niobium 92.91	42 Mo Molybdenum 95.94	43 Tc Technetium (98)	44 Ru Ruthenium 101.07	45 Rh Rhodium 102.906	46 Pd Palladium 106.42	47 Ag Silver 107.868	48 Cd Cadmium 112.41	49 In Indium 114.82	50 Sn Tin 118.71	51 Sb Antimony 121.75	52 Te Tellurium 127.60	53 I Iodine 126.905	54 Xe Xenon 131.29
55 Cs Cesium 132.91	56 Ba Barium 137.34		72 Hf Hafnium 178.49	73 Ta Tantalum 180.95	74 W Tungsten 183.85	75 Re Rhenium 186.207	76 Os Osmium 190.2	77 Ir Iridium 192.22	78 Pt Platinum 195.08	79 Au Gold 196.967	80 Hg Mercury 200.59	81 Ti Thallium 204.383	82 Pb Lead 207.2	83 Bi Bismuth 208.98	84 Po Polonium (209)	85 At Astatine (210)	86 Rn Radon (222)
87 Fr Francium (223)	88 Ra Radium 226.0254		104 Rf Rutherfordium (261)	105 Db Dubnium (262)	106 Sg Seaborgium (263)	107 Bh Bohrium (262)	108 Hs Hassium (265)	109 Mt Meitnerium (266)	110 Uun Ununnilium (269)	111 Uuu Unununium (272)	112 Uub Ununbium (277)						

Lanthanide Series

57 La Lanthanum 138.906	58 Ce Cerium 140.12	59 Pr Praseodymium 140.908	60 Nd Neodymium 144.24	61 Pm Promethium (145)	62 Sm Samarium 150.36	63 Eu Europium 151.96	64 Gd Gadolinium 157.25	65 Tb Terbium 158.925	66 Dy Dysprosium 162.50	67 Ho Holmium 164.93	68 Er Erbium 167.26	69 Tm Thulium 168.934	70 Yb Ytterbium 173.04	71 Lu Lutetium 174.967

Actinide Series

89 Ac Actinium 227.028	90 Th Thorium 232.038	91 Pa Protactinium 231.036	92 U Uranium 238.029	93 Np Neptunium 237.048	94 Pu Plutonium (244)	95 Am Americium (243)	96 Cm Curium (247)	97 Bk Berkelium (247)	98 Cf Californium (251)	99 Es Einsteinium (252)	100 Fm Fermium (257)	101 Md Mendelevium (258)	102 No Nobelium (259)	103 Lr Lawrencium (260)

APPENDIX C

PLANETARY DATA

Table C.1 Physical Properties of the Sun and Planets

Name	Radius (Eq) (km)	Radius (Eq) (Earth Units)	Mass (kg)	Mass (Earth Units)	Average Density (g/cm^3)	Surface Gravity (Earth = 1)
Sun	695,000	109	1.99×10^{30}	333,000	1.41	27.5
Mercury	2,440	0.382	3.30×10^{23}	0.055	5.43	0.38
Venus	6,051	0.949	4.87×10^{24}	0.815	5.25	0.91
Earth	6,378	1.00	5.97×10^{24}	1.00	5.52	1.00
Mars	3,397	0.533	6.42×10^{23}	0.107	3.93	0.38
Jupiter	71,492	11.19	1.90×10^{27}	317.9	1.33	2.53
Saturn	60,268	9.46	5.69×10^{26}	95.18	0.71	1.07
Uranus	25,559	3.98	8.66×10^{25}	14.54	1.24	0.91
Neptune	24,764	3.81	1.03×10^{26}	17.13	1.67	1.14
Pluto	1,160	0.181	1.31×10^{22}	0.0022	2.05	0.07

Table C.2 Orbital Properties of the Sun and Planets

Name	Distance from Sun* (AU)	Distance from Sun* (10^6 km)	Orbital Period (years)	Orbital Inclination† (degrees)	Orbital Eccentricity	Sidereal Rotation Period (Earth days)‡	Axis Tilt (degrees)
Sun	—	—	—	—	—	25.4	7.25
Mercury	0.387	57.9	0.2409	7.00	0.206	58.6	0.0
Venus	0.723	108.2	0.6152	3.39	0.007	−243.0	177.4
Earth	1.00	149.6	1.0	0.00	0.017	0.9973	23.45
Mars	1.524	227.9	1.881	1.85	0.093	1.026	23.98
Jupiter	5.203	778.3	11.86	1.31	0.048	0.41	3.08
Saturn	9.539	1,427	29.46	2.49	0.056	0.44	26.73
Uranus	19.19	2,870	84.01	0.77	0.046	−0.72	97.92
Neptune	30.06	4,497	164.8	1.77	0.010	0.67	29.6
Pluto	39.54	5,916	248.0	17.15	0.248	−6.39	118

*Semimajor axis of the orbit.

†With respect to the ecliptic.

‡A negative sign indicates retrograde rotation.

Table C.3 Satellites of the Solar System*

Planet Satellite	Radius or Dimensions† (km)	Distance from Planet (10^3 km)	Orbital Period (Earth days)	Mass** (kg)	Density** (g/cm^3)	Notes About the Satellites
Earth						**Earth**
Moon	1,738	384.4	27.322	7.349×10^{22}	3.34	*Moon* Probably formed in giant impact.
Mars						**Mars**
Phobos	$13 \times 11 \times 9$	9.38	0.319	1.3×10^{16}	2.2	*Phobos, Deimos* Probable captured asteroids.
Deimos	$8 \times 6 \times 5$	23.5	1.263	1.8×10^{15}	1.7	
Jupiter						**Jupiter**
Metis	20	127.96	0.295	9×10^{16}	—	*Metis, Adrastea, Amalthea, Thebe* Small moonlets within and near Jupiter's ring system.
Adrastea	$13 \times 10 \times 8$	128.98	0.298	1×10^{16}	—	
Amalthea	$135 \times 82 \times 75$	181.3	0.498	8×10^{18}	—	
Thebe	$55 \times ? \times 45$	221.9	0.6745	1.4×10^{17}	—	
Io	1,821	421.6	1.769	8.933×10^{22}	3.57	*Io* Most volcanically active object in the solar system.
Europa	1,565	670.9	3.551	4.797×10^{22}	2.97	*Europa* Possible oceans under icy crust.
Ganymede	2,634	1,070.0	7.155	1.482×10^{23}	1.94	*Ganymede* Largest satellite in solar system; unusual ice geology.
Callisto	2,403	1,883.0	16.689	1.076×10^{23}	1.86	*Callisto* Cratered iceball.
Leda	~8	11,094	238.7	4×10^{16}	—	*Leda, Himalia, Lysithea, Elara* Each probably a fragment of captured asteroid that broke apart.
Himalia	92.5	11,480	250.6	8×10^{18}	—	
Lysithea	~18	11,720	259.2	6×10^{16}	—	
Elara	~38	11,737	259.7	6×10^{17}	—	
Ananke	~15	20,200	−631	4×10^{16}	—	*Ananke, Carme, Pasiphae, Sinope* Each probably a fragment of captured asteroid that broke apart.
Carme	~20	22,600	−692	9×10^{16}	—	
Pasiphae	~25	23,500	−735	1.6×10^{17}	—	
Sinope	~18	23,700	−758	6×10^{16}	—	
Saturn						**Saturn**
Pan	10	133.570	0.574	4.2×10^{14}	—	*Pan, Atlas, Prometheus, Pandora, Epimetheus, Janus* Small moonlets within and near Saturn's ring system.
Atlas	$19 \times ? \times 14$	137.64	0.602	1.6×10^{16}	—	
Prometheus	$74 \times 50 \times 34$	139.35	0.613	5×10^{17}	—	
Pandora	$55 \times 44 \times 31$	141.7	0.629	3.4×10^{17}	—	
Epimetheus	$69 \times 55 \times 55$	151.42	0.694	5.6×10^{17}	—	
Janus	$97 \times 95 \times 77$	151.47	0.695	2.0×10^{18}	—	
Mimas	199	185.52	0.942	3.70×10^{19}	1.17	*Mimas, Enceladus, Tethys* Small and medium-sized iceballs, many with interesting geology.
Enceladus	249	238.02	1.370	1.2×10^{20}	1.24	
Tethys	530	294.66	1.888	6.17×10^{20}	1.26	
Calypso	$15 \times 8 \times 8$	294.66	1.888	4×10^{15}	—	*Calypso, Telesto* Small moonlets sharing Tethys's orbit.
Telesto	$15 \times 13 \times 8$	294.67	1.888	6×10^{15}	—	

Dione	559	377.4	2.737	1.08×10^{21}	1.44	*Dione* Medium-sized iceball, with interesting geology.
Helene	$18 \times ? \times 15$	377.4	2.737	1.6×10^{16}	—	*Helene* Small moonlet sharing Dione's orbit.
Rhea	764	527.04	4.518	2.31×10^{21}	1.33	*Rhea* Medium-sized iceball, with interesting geology.
Titan	2,575	1,221.85	15.945	1.3455×10^{23}	1.88	*Titan* Dense atmosphere shrouds surface; ongoing geological activity possible.
Hyperion	$180 \times 140 \times 112$	1,481.1	21.277	2.8×10^{19}	—	*Hyperion* Only satellite known not to rotate synchronously.
Iapetus	718	3,561.3	79.331	1.59×10^{21}	1.21	*Iapetus* Bright and dark hemispheres show greatest contrast in the solar system.
Phoebe	110	12,952	−550.4	1×10^{19}	—	*Phoebe* Very dark; material ejected from Phoebe may coat one side of Iapetus.
Uranus						**Uranus**
Cordelia	13	49.75	0.336	1.7×10^{16}	—	
Ophelia	16	53.77	0.377	2.6×10^{16}	—	
Bianca	22	59.16	0.435	7×10^{16}	—	
Cressida	33	61.77	0.465	2.6×10^{17}	—	
Desdemona	29	62.65	0.476	1.7×10^{17}	—	*Cordelia, Ophelia, Bianca, Cressida, Desdemona, Juliet, Portia, Rosalind, Belinda, Puck* Small moonlets within and near Uranus's ring system.
Juliet	42	64.63	0.494	4.3×10^{17}	—	
Portia	55	66.1	0.515	1×10^{18}	—	
Rosalind	29	69.93	0.560	1.5×10^{17}	—	
Belinda	33	75.25	0.624	2.5×10^{17}	—	
Puck	77	86.00	0.764	5×10^{17}	—	
Miranda	236	129.8	1.413	6.6×10^{19}	1.26	
Ariel	579	191.2	2.520	1.35×10^{21}	1.65	
Umbriel	584.7	266.0	4.144	1.17×10^{21}	1.44	*Miranda, Ariel, Umbriel, Titania, Oberon* Small and medium-sized iceballs, with some interesting geology.
Titania	788.9	435.8	8.706	3.52×10^{21}	1.59	
Oberon	761.4	582.6	13.463	3.01×10^{21}	1.50	
Neptune						**Neptune**
Naiad	29	48.2	0.296	1.4×10^{17}	—	
Thalassa	40	50.0	0.312	4×10^{17}	—	
Despina	74	52.5	0.333	2.1×10^{17}	—	*Naiad, Thalassa, Despina, Galatea, Larissa, Proteus* Small moonlets within and near Neptune's ring system.
Galatea	79	62.0	0.396	3.1×10^{18}	—	
Larissa	$104 \times ? \times 89$	73.6	0.554	6×10^{18}	—	
Proteus	$218 \times 208 \times 201$	117.6	1.121	6×10^{19}	—	
Triton	1,352.6	354.59	−5.875	2.14×10^{22}	2.0	*Triton* Probable captured Kuiper belt object—largest captured object in solar system.
Nereid	170	5,588.6	360.125	3.1×10^{19}	—	*Nereid* Small, icy moon, very little known.
Pluto						**Pluto**
Charon	635	19.6	6.38718	1.56×10^{21}	1.6	*Charon* Unusually large compared to its planet; may have formed in giant impact.

*Note: Authorities differ substantially on many of the values in this table.

†a × b × c values for the Dimensions are the approximate lengths of the axes for irregular moons.

‡Negative sign indicates retrograde rotation.

**Masses and densities are most accurate for those satellites visited by a spacecraft on a flyby.

APPENDIX D

STELLAR DATA

Table D.1 Stars Within 12 Light-Years

Star	Distance (ly)	Spectral Type		RA		Dec		Luminosity (L/L_{Sun})
				h	m	°	′	
Sun	0.000016	G2	V	—	—	—	—	1.0
Proxima Centauri	4.2	M5.5	V	14	30	−62	41	0.0006
α Centauri A	4.4	G2	V	14	40	−60	50	1.6
α Centauri B	4.4	K0	V	14	40	−60	50	0.53
Barnard's Star	6.0	M4	V	17	58	+04	42	0.005
Wolf 359	7.8	M6	V	10	56	+07	01	0.0008
Lalande 21185	8.3	M2	V	11	03	+35	58	0.03
Sirius A	8.6	A1	V	06	45	−16	42	26.0
Sirius B	8.6	DA2		06	45	−16	42	0.002
Luyten 726-8A	8.7	M5.5	V	01	39	−17	57	0.0009
Luyten 726-8B	8.7	M6	V	01	39	−17	57	0.0006
Ross 154	9.7	M3.5	V	18	50	−23	50	0.004
Ross 248	10.3	M5.5	V	23	42	+44	11	0.001
ε Eridani	10.5	K2	V	03	33	−09	28	0.37
Lacaille 9352	10.7	M1.5	V	23	06	−35	51	0.05
Ross 128	10.9	M4	V	11	48	+00	49	0.003
EZ Aquarii A	11.3	M5	V	22	39	−15	18	0.0006
EZ Aquarii B	11.3	M6	V	22	39	−15	18	0.0004
EZ Aquarii C	11.3	M6.5	V	22	39	−15	18	0.0003
61 Cygni A	11.4	K5	V	21	07	+38	42	0.15
61 Cygni B	11.4	K7	V	21	07	+38	42	0.09
Procyon A	11.4	F5	IV–V	07	39	+05	14	7.4
Procyon B	11.4	DA	—	07	39	+05	14	0.0005
Gliese 725 A	11.4	M3	V	18	43	+59	38	0.02
Gliese 725 B	11.4	M3.5	V	18	43	+59	38	0.01
Gliese 15 A	11.6	M1.5	V	00	18	+44	01	0.03
Gliese 15 B	11.6	M3.5	V	00	18	+44	01	0.003
DX Cancri	11.8	M6.5	V	08	30	+26	47	0.0003
ε Indi	11.8	K5	V	22	03	−56	45	0.26
τ Ceti	11.9	G8	V	01	44	−15	57	0.59
GJ 1061	11.9	M5.5	V	03	36	−44	31	0.0009

Note: These data were provided by the RECONS project, courtesy of Dr. Todd Henry. The luminosities are all total (bolometric) luminosities. The DA stellar types are white dwarfs. The coordinates are for the year 2000.

Table D.2 Twenty Brightest Stars

Star	Constellation	RA h	RA m	Dec °	Dec ′	Distance (ly)	Spectral Type		Apparent Magnitude	Luminosity (L/L_{Sun})
Sirius	Canis Major	6	45	−16	42	8.6	A1	V	−1.46	26
Canopus	Carina	6	24	−52	41	98	F0	Ib–II	−0.72	1,300
α Centauri	Centaurus	14	40	−60	50	4.4	G2 K0	V V	−0.01 1.3	1.6 0.53
Arcturus	Boötes	14	16	+19	11	36	K2	III	−0.06	170
Vega	Lyra	18	37	+38	47	26	A0	V	0.04	60
Capella	Auriga	5	17	+46	00	46	G8	III	0.05	200
Rigel	Orion	5	15	−08	12	820	B8	Ia	0.14	70,000
Procyon	Canis Minor	7	39	+05	14	11.4	F5	IV–V	0.37	7.4
Betelgeuse	Orion	5	55	+07	24	490	M2	Iab	0.41	52,000
Achernar	Eridanus	1	38	−57	15	65	B5	V	0.51	690
Hadar	Centaurus	14	04	−60	22	290	B1	III	0.63	30,000
Altair	Aquila	19	51	+08	52	17	A7	IV–V	0.77	10.5
Acrux	Crux	12	27	−63	06	390	B1 B3	IV V	1.39 1.9	33,000 11,000
Aldebaran	Taurus	4	36	+16	30	52	K5	III	0.86	220
Spica	Virgo	13	25	−11	09	260	B1	V	0.91	23,000
Antares	Scorpio	16	29	−26	26	390	M1	Ib	0.92	16,000
Pollux	Gemini	7	45	+28	01	39	K0	III	1.16	55
Fomalhaut	Piscis Austrinus	22	58	−29	37	23	A3	V	1.19	13
Deneb	Cygnus	20	41	+45	16	1,400	A2	Ia	1.26	52,000
β Crucis	Crux	12	48	−59	40	490	B0.5	IV	1.28	90,000

Note: Two of the stars on this list, α Centauri and Acrux, are binary systems with members of comparable brightness. They are counted as single stars because that is how they appear to the naked eye. All the luminosities given are total (bolometric) luminosities. The coordinates are for the year 2000.

GLOSSARY

21-cm line A spectral line from atomic hydrogen with wavelength 21 cm (in the radio portion of the spectrum).

absolute magnitude A measure of the luminosity of an object; defined to be the apparent magnitude of an object if it were located exactly 10 parsecs away.

absolute zero The coldest possible temperature, which is 0 K.

absorption (of light) The process by which matter absorbs radiative energy.

absorption-line spectrum A spectrum that contains absorption lines.

acceleration The rate at which an object's velocity changes. Its units are velocity divided by time; standard units are m/s^2.

acceleration of gravity The acceleration of a falling object. On Earth, the acceleration of gravity, designated by g, is $9.8\ m/s^2$.

accretion The process by which objects gather together to make larger objects.

accretion disk A rapidly rotating disk of material that orbits a starlike object as the material (e.g., white dwarf, neutron star, or black hole) gradually falls inward.

active galactic nuclei The unusually luminous centers of some galaxies, thought to be powered by accretion onto supermassive black holes. Some active galactic nuclei outshine all the stars in their galaxy combined, and some appear to be firing powerful jets of material millions of light-years into space. Quasars are the brightest type of active galactic nuclei; radio galaxies also contain active galactic nuclei.

adaptive optics A technique in which telescope mirrors flex rapidly to compensate for the bending of starlight caused by atmospheric turbulence.

albedo Describes the fraction of sunlight reflected by a surface; albedo = 0 means no reflection at all (a perfectly black surface); albedo = 1 means all light is reflected (a perfectly white surface).

altitude (above horizon) The angular distance between the horizon and an object in the sky.

amino acids The building blocks of proteins.

angular momentum Momentum attributable to rotation or revolution. The angular momentum of an object moving in a circle of radius r is the product $m \times v \times r$.

angular resolution (of a telescope) The smallest angular separation that two pointlike objects can have and still be seen as distinct points of light (rather than as a single point of light).

angular size (or **angular distance**) A measure of the angle formed by extending imaginary lines outward from our eyes to span an object (or between two objects).

Antarctic Circle The circle on the Earth with latitude 66.5°S.

antimatter Refers to any particle with the same mass as a particle of ordinary matter but whose other basic properties, such as electrical charge, are precisely opposite.

aphelion The point at which an object orbiting the Sun is farthest from the Sun.

apogee The point at which an object orbiting the Earth is farthest from the Earth.

apparent brightness The amount of light reaching us *per unit area* from a luminous object; often measured in units of $watts/m^2$.

apparent magnitude A measure of the apparent brightness of an object in the sky, based on the ancient system developed by Hipparchus.

apparent retrograde motion (of a planet) Refers to the period during which a planet appears to move westward relative to the stars as viewed from Earth.

apparent solar time Time measured by the actual position of the Sun in your local sky; defined so that noon is when the Sun is *on* the meridian.

arcminutes (or **minutes of arc**) One arcminute is 1/60 of 1°.

arcseconds (or **seconds of arc**) One arcsecond is 1/60 of an arcminute, or 1/3,600 of 1°.

Arctic Circle The circle on the Earth with latitude 66.5°N.

asteroid A relatively small and rocky object that orbits a star; asteroids are sometimes called *minor planets* because they are similar to planets but smaller.

asteroid belt The region of our solar system between the orbits of Mars and Jupiter in which asteroids are heavily concentrated.

astronomical unit (AU) The average (semimajor axis) distance of the Earth from the Sun, which is about 150 million km.

atmospheric pressure The surface pressure resulting from the overlying weight of an atmosphere.

atomic number The number of protons in an atom.

atomic weight (or **atomic mass**) The combined number of protons and neutrons in an atom.

atoms Consist of a nucleus made from protons and neutrons surrounded by a cloud of electrons.

aurora Dancing lights in the sky caused by charged particles entering our atmosphere; called the *aurora borealis* in the Northern Hemisphere and the *aurora australis* in the Southern Hemisphere.

azimuth (usually called *direction* in this book) Direction around the horizon from due north, measured clockwise in degrees. E.g., the azimuth of due north is 0°, due east is 90°, due south is 180°, and due west is 270°.

bar The standard unit of pressure, approximately equal to the Earth's atmospheric pressure at sea level.

baryonic matter Refers to ordinary matter made from atoms (because the nuclei of atoms contain protons and neutrons, which are both baryons).

baryons Particles, including protons and neutrons, that are made from three quarks.

basalt A type of volcanic rock that makes a low-viscosity lava when molten.

belts (on a jovian planet) Dark bands of sinking air that encircle a jovian planet at a particular set of latitudes.

Big Bang The event that gave birth to the universe.

binary star system A star system that contains two stars.

biosphere Refers to the "layer" of life on Earth.

black hole A bottomless pit in spacetime. Nothing can escape from within a black hole, and we can never again detect or observe an object that falls into a black hole.

black smokers Structures around seafloor volcanic vents that support a wide variety of life.

blueshift A Doppler shift in which spectral features are shifted to shorter wavelengths, caused when an object is coming toward the observer.

bosons Particles, such as photons, to which the exclusion principle does not apply.

bound orbits Orbits on which an object travels repeatedly around another object; bound orbits are elliptical in shape.

brown dwarf An object too small to become an ordinary star because electron degeneracy pressure halts its gravitational collapse before fusion becomes self-sustaining; brown dwarfs have mass less than $0.08M_{Sun}$.

bubble (interstellar) The surface of a bubble is an expanding shell of hot, ionized gas driven by stellar winds or supernovae; inside the bubble, the gas is very hot and has very low density.

Cambrian explosion The dramatic diversification of life on Earth that occurred between about 540 and 500 million years ago.

carbon stars Stars whose atmospheres are especially carbon-rich, thought to be near the ends of their lives; carbon stars are the primary sources of carbon in the universe.

carbonate rock A carbon-rich rock, such as limestone, that forms underwater from chemical reactions between sediments and carbon dioxide. On Earth, most of the outgassed carbon dioxide currently resides in carbonate rocks.

carbonate–silicate cycle The process that cycles carbon dioxide between the Earth's atmosphere and surface rocks.

CCD (charge coupled device) A type of electronic light detector that has largely replaced photographic film in astronomical research.

celestial coordinates The coordinates of right ascension and declination that fix an object's position on the celestial sphere.

celestial equator (CE) The extension of the Earth's equator onto the celestial sphere.

celestial navigation Navigation on the surface of the Earth accomplished by observations of the Sun and stars.

celestial sphere The imaginary sphere on which objects in the sky appear to reside when observed from Earth.

central dominant galaxy A giant elliptical galaxy found at the center of a dense cluster of galaxies, apparently formed by the merger of several individual galaxies.

Cepheid variables A particularly luminous type of pulsating variable star that follows a period–luminosity relation and hence is very useful for measuring cosmic distances.

charged particle belts Zones in which ions and electrons accumulate and encircle a planet.

chemical enrichment The process by which the abundance of heavy elements (heavier than helium) in the interstellar medium gradually increases over time as these elements are produced by stars and released into space.

chromosphere The layer of the Sun's atmosphere below the corona; most of the Sun's ultraviolet light is emitted from this region, in which the temperature is about 10,000 K.

circulation cells (also called Hadley cells) Large-scale cells (similar to convection cells) in a planet's atmosphere that transport heat between the equator and the poles.

circumpolar star A star that always remains above the horizon for a particular latitude.

climate Describes the long-term average of weather.

close binary A binary star system in which the two stars are very close together.

closed universe Refers to the case in which the density of the universe is greater than the critical density, so that the expansion will someday halt and the universe will begin to contract.

cluster of galaxies A collection of a few dozen or more galaxies bound together by gravity; smaller collections of galaxies are simply called *groups*.

CNO cycle The cycle of reactions by which intermediate- and high-mass stars fuse hydrogen into helium.

coma (of a comet) The dusty atmosphere of a comet created by sublimation of ices in the nucleus when the comet is near the Sun.

comet A relatively small, icy object that orbits a star.

comparative planetology The study of the solar system by examining and understanding the similarities and differences among worlds.

compound (chemical) A substance made from molecules consisting of two or more atoms with different atomic number.

condensates Solid or liquid particles that condense from a cloud of gas.

condensation The formation of solid or liquid particles from a cloud of gas.

conduction (of energy) The process by which thermal energy is transferred by direct contact from warm material to cooler material.

conservation of angular momentum (law of) The principle that, in the absence of net torque (twisting force), the total angular momentum of a system remains constant.

conservation of energy (law of) The principle that energy (including mass-energy) can neither be created nor destroyed, but can only change from one form to another.

conservation of momentum (law of) The principle that, in the absence of net force, the total momentum of a system remains constant.

constellation A region of the sky; 88 official constellations cover the celestial sphere.

convection The energy transport process in which warm material expands and rises, while cooler material contracts and falls.

convection cell An individual small region of convecting material.

convection zone (of a star) A region in which energy is transported outward by convection.

core (of a planet) The dense central region of a planet that has undergone differentiation.

core (of a star) The central region of a star, in which nuclear fusion can occur.

Coriolis effect Causes air or objects moving on a rotating planet to deviate from straight-line trajectories.

corona (solar) The tenuous uppermost layer of the Sun's atmosphere; most of the Sun's X rays are emitted from this region, in which the temperature is about 1 million K.

coronal holes Regions of the corona that barely show up in X-ray images because they are nearly devoid of hot coronal gas.

cosmic background radiation The remnant radiation from the Big Bang, consisting of photons that have traveled freely through the universe since the end of the era of nuclei.

cosmic rays Refers to electrons, protons, and atomic nuclei that zip through interstellar space at close to the speed of light.

cosmological horizon The boundary of our observable universe, which is where the lookback time is equal to the age of the universe. Beyond this boundary in spacetime, we cannot see anything at all.

cosmological redshift Refers to the redshifts we see from distant galaxies, caused by the fact that expansion of the universe stretches all the photons within it to longer, redder wavelengths.

critical density (of the universe) The precise average density for the entire universe that marks the dividing line between eternal expansion and eventual contraction; if the actual density is less than the critical density, the universe will continue to expand forever; if the actual density is greater than the critical density, the universe will someday stop expanding and begin to contract.

crust (of a planet) The low-density surface layer of a planet that has undergone differentiation.

dark matter Matter that we infer to exist from its gravitational effects but from which we have not detected any light; dark matter dominates the total mass of the universe.

daylight saving time Standard time plus 1 hour, so that the Sun appears on the meridian around 1 P.M. rather than around noon.

declination (dec) Analogous to latitude, but on the celestial sphere; it is the angular north-south distance between the celestial equator and a location on the celestial sphere.

degeneracy pressure A type of pressure unrelated to an object's temperature, which arises when electrons (electron degeneracy pressure) or neutrons (neutron degeneracy pressure) are packed so tightly that the exclusion and uncertainty principles come into play.

degenerate object An object in which degeneracy pressure is the primary pressure pushing back against gravity, such as a brown dwarf, white dwarf, or neutron star.

differential rotation Describes the rotation of an object in which the equator rotates at a different rate than the poles.

differentiation The process in which gravity separates materials according to density, with high-density materials sinking and low-density materials rising.

diffraction grating A finely etched surface that can split light into a spectrum.

diffraction limit The angular resolution that a telescope could achieve if it were limited only by the interference of light waves; it is smaller (i.e., better angular resolution) for larger telescopes.

dimension (mathematical) Describes the number of independent directions in which movement is possible; e.g., the surface of the Earth is two-dimensional because only two independent directions of motion are possible (north-south and east-west).

disk component (of a galaxy) The portion of a spiral galaxy that looks like a disk and contains an interstellar medium with cool gas and dust; stars of many ages are found in the disk component.

Doppler effect The effect that shifts the wavelengths of spectral features in objects that are moving toward or away from the observer.

dwarf elliptical galaxies Particularly small elliptical galaxies with less than a billion stars.

eccentricity A measure of how much an ellipse deviates from a perfect circle, which has zero eccentricity.

eclipse Occurs when one astronomical object casts a shadow on another or crosses our line of sight to the other object.

eclipse seasons Periods during which lunar and solar eclipses can occur because the nodes of the Moon's orbit are nearly aligned with the Earth and Sun.

eclipsing binary A binary star system in which the two stars happen to be orbiting in the plane of our line of sight, so that each star will periodically eclipse the other.

ecliptic The Sun's apparent annual path among the constellations.

ecliptic plane The plane of the Earth's orbit around the Sun.

ejecta (from an impact) Debris ejected by the blast of an impact.

electromagnetic field An abstract concept used to describe how a charged particle would affect other charged particles at a distance.

electromagnetic spectrum The complete spectrum of light, including radio waves, infrared, visible light, ultraviolet light, X rays, and gamma rays.

electromagnetic wave A synonym for light, which consists of waves of electric and magnetic fields.

electrons Fundamental particles with negative electric charge; the distribution of electrons in an atom gives the atom its size.

electron-volt (eV) A unit of energy equivalent to 1.60×10^{-19} joule.

electroweak era The era of the universe during which only three forces operated (gravity, strong force, and electroweak force), lasting from 10^{-35} second to 10^{-10} second after the Big Bang.

element (chemical) A substance made from individual atoms of a particular atomic number.

ellipse A type of oval that happens to be the shape of bound orbits. An ellipse can be drawn by moving a pencil along a string whose ends are tied to two tacks; the locations of the tacks are the foci (singular, focus) of the ellipse.

elliptical galaxies Galaxies that appear rounded in shape, often longer in one direction, like a football. They have no disks and contain very little cool gas and dust compared to spiral galaxies, though they often contain very hot, ionized gas.

emission (of light) The process by which matter emits energy in the form of light.

emission-line spectrum A spectrum that contains emission lines.

equivalence principle The fundamental starting point for general relativity, which states that the effects of gravity are exactly equivalent to the effects of acceleration.

era of atoms The era of the universe lasting from about 300,000 years to about 1 billion years after the Big Bang, during which it was cool enough for neutral atoms to form.

era of galaxies The present era of the universe, which began with the formation of galaxies when the universe was about 1 billion years old.

era of nuclei The era of the universe lasting from about 3 minutes to about 300,000 years after the Big Bang, during which matter in the universe was fully ionized and opaque to light. The cosmic background radiation was released at the end of this era.

era of nucleosynthesis The era of the universe lasting from about 0.001 second to about 3 minutes after the Big Bang, by the end of which virtually all of the neutrons and about one-seventh of the protons in the universe had fused into helium.

erosion The wearing down or building up of geological features by wind, water, ice, and other phenomena of planetary weather.

escape velocity The speed necessary for an object to completely escape the gravity of a large body such as a moon, planet, or star.

evaporation The process by which atoms or molecules escape into the gas phase from a liquid.

event Any particular point along a worldline represents a particular event; all observers will agree on the reality of an event but may disagree about its time and location.

event horizon The boundary that marks the "point of no return" between a black hole and the outside universe; events that occur within the event horizon can have no influence on our observable universe.

excited state (of an atom) Any arrangement of electrons in an atom that has more energy than the ground state.

exclusion principle The law of quantum mechanics that states that two fermions cannot occupy the same quantum state at the same time.

exosphere The hot, outer layer of an atmosphere, where the atmosphere "fades away" to space.

exposure time The amount of time for which light is collected to make a single image.

fall equinox (autumnal equinox) Refers both to the point in Virgo on the celestial sphere where the ecliptic crosses the celestial equator and to the moment in time when the Sun appears at that point each year (around September 21).

false-color image An image displayed in colors that are *not* the true, visible-light colors of an object.

fault (geological) A place where rocks slip sideways relative to one another.

fermions Particles, such as electrons, neutrons, and protons, that obey the exclusion principle.

filter (for light) A material that transmits only particular wavelengths of light.

fireball A particularly bright meteor.

flare stars Small, spectral type M stars that display particularly strong flares on their surfaces.

flat (or **Euclidean**) **geometry** Refers to any case in which the rules of geometry for a flat plane hold, such as that the shortest distance between two points is a straight line.

flat universe Refers to the case in which the density of the universe is equal to the critical density; as with an open universe, expansion will continue forever in this case.

focal plane The place where an image created by a lens or mirror is in focus.

focus (of a lens or mirror) The point at which rays of light that were initially parallel (such as light from a distant star) converge.

force Anything that can cause a change in momentum.

frame of reference (in relativity) Two (or more) objects share the same frame of reference if they are *not* moving relative to each other.

free-fall Refers to conditions in which an object is falling without resistance; objects are weightless when in free-fall.

free-float frame A frame of reference in which all objects are weightless and hence float freely.

frequency Describes the rate at which peaks of a wave pass by a point; measured in units of 1/s, often called *cycles per second* or *hertz*.

frost line The boundary in the solar nebula beyond which ices could condense; only metals and rocks could condense within the frost line.

fundamental forces The four fundamental forces in nature are gravity, the electromagnetic force, the strong force, and the weak force.

fundamental particles Subatomic particles that cannot be divided into anything smaller.

galactic fountain Refers to a model for the cycling of gas in the Milky Way Galaxy in which fountains of hot, ionized gas rise from the disk into the halo and then cool and form clouds as they sink back into the disk.

galactic wind A wind of low-density but extremely hot gas flowing out from a starburst galaxy, created by the combined energy of many supernovae.

galaxy A huge collection of anywhere from a few hundred million to more than a trillion stars, all bound together by gravity.

gamma rays Light with very short wavelengths (and hence high frequencies)—shorter than those of X rays.

gamma-ray burst A sudden burst of gamma rays from deep space; such bursts apparently come from distant galaxies, but their mechanism is unknown.

genetic code The "language" that living cells use to read the instructions chemically encoded in DNA.

geocentric universe (ancient belief in) The idea that the Earth is the center of the entire universe.

geology The study of surface features (on a moon, planet, or asteroid) and the processes that create them.

giants (luminosity class III) Stars that appear just below the supergiants on the H–R diagram because they are somewhat smaller in radius and lower in luminosity.

global positioning system (GPS) A system of navigation by satellites orbiting the Earth.

global wind patterns (or **global circulation**) Wind patterns that remain fixed on a global scale, determined by the combination of surface heating and the planet's rotation.

globular cluster A spherically shaped cluster of up to a million or more stars; globular clusters are found primarily in the halos of galaxies and contain only very old stars.

grand unified theories (GUTs) Theories that unify the strong, weak, and electromagnetic forces into a single force.

granulation (on the Sun) The bubbling pattern visible in the photosphere, produced by the underlying convection.

gravitational constant The experimentally measured constant G that appears in the law of universal gravitation;

$$G = 6.67 \times 10^{-11} \frac{\text{m}^3}{\text{kg} \times \text{s}^2}.$$

gravitational contraction The process in which gravity causes an object to contract, thereby converting gravitational potential energy into thermal energy.

gravitational encounters Occurs when two (or more) objects pass near enough so that each can feel the effects of the other's gravity and can therefore exchange energy.

gravitational equilibrium Describes a state of balance in which the force of gravity pulling inward is precisely counteracted by pressure pushing outward.

gravitational lensing The magnification or distortion (into arcs, rings, or multiple images) of an image caused by light bending through a gravitational field, as predicted by Einstein's general theory of relativity.

gravitational redshift A redshift caused by the fact that time runs slow in gravitational fields.

gravitational time dilation The slowing of time that occurs in a gravitational field, as predicted by Einstein's general theory of relativity.

gravitational waves Predicted by Einstein's general theory of relativity, these waves travel at the speed of light and transmit distortions of space through the universe. Although not yet observed directly, we have strong indirect evidence that they exist.

gravitationally bound system Any system of objects, such as a star system or a galaxy, that is held together by gravity.

great circle A circle on the surface of a sphere whose center is at the center of the sphere.

Great Red Spot A large, high-pressure storm on Jupiter.

greenhouse effect The process by which greenhouse gases in an atmosphere make a planet's surface temperature warmer than it would be in the absence of an atmosphere.

greenhouse gases Gases, such as carbon dioxide, methane, and water vapor, that are particularly good absorbers of infrared light but are transparent to visible light.

Gregorian calendar Our modern calendar, introduced by Pope Gregory in 1582.

ground state (of an atom) The lowest possible energy state of the electrons in an atom.

GUT era The era of the universe during which only two forces operated (gravity and the grand-unified-theory or GUT force), lasting from 10^{-43} second to 10^{-35} second after the Big Bang.

half-life The time it takes for half of the nuclei in a given quantity of a radioactive substance to decay.

halo (of a galaxy) The spherical region surrounding the disk of a spiral galaxy.

Hawking radiation Radiation predicted to arise from the evaporation of black holes.

heavy elements In astronomy, *heavy elements* generally refers to all elements *except* hydrogen and helium.

helium-capture reactions Fusion reactions that fuse a helium nucleus into some other nucleus; such reactions can fuse carbon into oxygen, oxygen into neon, neon into magnesium, and so on.

helium flash The event that marks the sudden onset of helium fusion in the previously inert helium core of a low-mass star.

helium fusion The fusion of three helium nuclei into one carbon nucleus.

hertz (Hz) The standard unit of frequency for light waves; equivalent to units of 1/s.

Hertzsprung–Russell (H–R) diagram A graph plotting individual stars as points, with stellar luminosity on the vertical axis and spectral type (or surface temperature) on the horizontal axis.

high-mass stars Stars born with masses above about $8M_{Sun}$; these stars will end their lives by exploding as supernovae.

horizon A boundary that divides what we can see from what we cannot see.

horizontal branch The horizontal line of stars that represents helium-burning stars on an H–R diagram for a cluster of stars.

hot spot (geological) A place within a plate of the lithosphere where a localized plume of hot mantle material rises.

Hubble's constant A number that expresses the current rate of expansion of the universe; designated H_0, it is usually stated in units of km/s/Mpc. The reciprocal of Hubble's constant is the age the universe would have *if* the expansion rate had never changed.

Hubble's law Mathematically expresses the idea that more distant galaxies move away from us faster; its formula is $v = H_0 \times d$, where v is a galaxy's speed away from us, d is its distance, and H_0 is Hubble's constant.

hydrogen compounds Compounds that contain hydrogen and were common in the solar nebula, such as water, ammonia, and methane.

hydrogen-shell burning Hydrogen fusion that occurs in a shell surrounding a stellar core.

hydrosphere Refers to the "layer" of water on the Earth consisting of oceans, lakes, rivers, ice caps, and other liquid water and ice.

hyperbola The precise mathematical shape of one type of unbound orbit (the other is a parabola) allowed under the force of gravity; at great distances from the attracting object, a hyperbolic path looks like a straight line.

hyperspace Any space with more than three dimensions.

ices (in solar system theory) Materials that are solid only at low temperatures, such as the hydrogen compounds water, ammonia, and methane.

imaging (in astronomical research) The process of obtaining pictures of astronomical objects.

impact basin A very large impact crater often filled by a lava flow.

impact cratering The excavation of bowl-shaped depressions (*impact craters*) by asteroids or comets striking a planet's surface.

inflation (of the universe) A sudden and dramatic expansion of the universe thought to have occurred at the end of the GUT era.

infrared light Light with wavelengths that fall in the portion of the electromagnetic spectrum between radio waves and visible light.

intensity (of light) A measure of the amount of energy coming from light of specific wavelength in the spectrum of an object.

interferometry A telescopic technique in which two or more telescopes are used in tandem to produce much better angular resolution than the telescopes could achieve individually.

intermediate-mass stars Stars born with masses between about $2–8M_{Sun}$; these stars end their lives by ejecting a planetary nebula and becoming a white dwarf.

interstellar cloud A cloud of gas and dust between the stars.

interstellar dust grains Tiny solid flecks of carbon and silicon minerals found in cool interstellar clouds; they resemble particles of smoke and form in the winds of red-giant stars.

interstellar medium Refers to gas and dust that fills the space between stars in a galaxy.

intracluster medium Hot, X ray–emitting gas found between the galaxies within a cluster of galaxies.

inverse square law Any quantity that decreases with the square of the distance between two objects is said to follow an inverse square law.

inversion (atmospheric) A local weather condition in which air is colder near the surface than higher up in the troposphere—the opposite of the usual condition, in which the troposphere is warmer at the bottom.

Io torus A donut-shaped charged-particle belt around Jupiter that approximately traces Io's orbit.

ionization The process of stripping an electron from an atom.

ionization nebula A colorful, wispy cloud of gas that glows because neighboring hot stars irradiate it with ultraviolet photons that can ionize hydrogen atoms.

ionosphere A portion of the thermosphere in which ions are particularly common (due to ionization by X rays from the Sun).

ions Atoms with a positive or negative electrical charge.

irregular galaxies Galaxies that look neither spiral nor elliptical.

isotopes Each different isotope of an element has the *same* number of protons but a *different* number of neutrons.

jets High-speed streams of gas ejected from an object into space.

joule The international unit of energy, equivalent to about 1/4,000 of a Calorie.

jovian nebulae The clouds of gas that swirled around the jovian planets, from which the moons formed.

jovian planets Giant gaseous planets similar in overall composition to Jupiter.

Julian calendar The calendar introduced in 46 B.C. by Julius Caesar and used until it was replaced by the Gregorian calendar.

Kelvin (temperature scale) The most commonly used temperature scale in science, defined such that absolute zero is 0 K and water freezes at 273.15 K.

Kepler's laws of planetary motion Three laws discovered by Kepler that describe the motion of the planets around the Sun.

kinetic energy Energy of motion, given by the formula $\frac{1}{2}mv^2$.

Kuiper belt The comet-rich region of our solar system that spans distances of about 30–100 AU from the Sun; Kuiper belt comets have orbits that lie fairly close to the plane of planetary orbits and travel around the Sun in the same direction as the planets.

latitude The angular north-south distance between the Earth's equator and a location on the Earth's surface.

length contraction Refers to the effect in which you observe lengths to be shortened in reference frames moving relative to you.

lenticular galaxies Galaxies that look lens-shaped when seen edge-on, resembling spiral galaxies without arms. They tend to have less cool gas than normal spiral galaxies but more gas than elliptical galaxies.

leptons Fermions *not* made from quarks, such as electrons and neutrinos.

life track A track drawn on an H–R diagram to represent the changes in a star's surface temperature and luminosity during its life; also called an *evolutionary track.*

light curve A graph of an object's intensity against time.

light gases (in solar system theory) Refers to hydrogen and helium, which never condense under solar nebula conditions.

light pollution Human-made light that hinders astronomical observations.

light-collecting area (of a telescope) The area of the primary mirror or lens that collects light in a telescope.

light-year The distance that light can travel in 1 year, which is 9.46 trillion km.

lithosphere The relatively rigid outer layer of a planet; generally encompasses the crust and the uppermost portion of the mantle.

Local Group The group of about 30 galaxies to which the Milky Way Galaxy belongs.

Local Supercluster The supercluster of galaxies to which the Local Group belongs.

longitude The angular east-west distance between the Prime Meridian (which passes through Greenwich) and a location on the Earth's surface.

lookback time Refers to the amount of time since the light we see from a distant object was emitted. I.e., if an object has a lookback time of 400 million years, we are seeing it as it looked 400 million years ago.

low-mass stars Stars born with masses less than about $2M_{Sun}$; these stars end their lives by ejecting a planetary nebula and becoming a white dwarf.

luminosity The total power output of an object, usually measured in watts or in units of solar luminosities ($L_{Sun} = 3.8 \times 10^{26}$ watts).

luminosity class Describes the region of the H–R diagram in which a star falls. Luminosity class I represents supergiants, III represents giants, and V represents main-sequence stars; luminosity classes II and IV are intermediate to the others.

luminosity–distance formula The formula that relates apparent brightness, luminosity, and distance:

$$\text{apparent brightness} = \frac{\text{luminosity}}{4\pi \times (\text{distance})^2}$$

lunar eclipse Occurs when the Moon passes through the Earth's shadow, which can occur only at full moon; may be total, partial, or penumbral.

lunar maria The regions of the Moon that look smooth from Earth and actually are impact basins.

lunar month The time required for a complete cycle of lunar phases, which is about $29\frac{1}{2}$ days (also called a *synodic month*).

lunar phase Describes the appearance of the Moon as seen from Earth.

MACHOs Stands for *massive compact halo objects* and represents one possible form of dark matter in which the dark objects are relatively large, like planets or brown dwarfs.

magma Underground molten rock.

magnetic braking The process by which a star's rotation slows as its magnetic field transfers its angular momentum to the surrounding nebula.

magnetic field Describes the region surrounding a magnet in which it can affect other magnets or charged particles in its vicinity.

magnetic-field lines Lines that represent how the needles on a series of compasses would point if they were laid out in a magnetic field.

magnetosphere The region surrounding a planet in which charged particles are trapped by the planet's magnetic field.

main sequence (luminosity class V) The prominent line of points running from the upper left to the lower right on an H–R diagram; main-sequence stars shine by fusing hydrogen in their cores.

main-sequence fitting A method for measuring the distance to a cluster of stars by comparing the apparent brightness of the cluster's main sequence with the standard main sequence.

main-sequence lifetime The length of time for which a star of a particular mass can shine by fusing hydrogen into helium in its core.

main-sequence turnoff A method for measuring the age of a cluster of stars from the point on its H–R diagram where its stars turn off from the main sequence; the age of the cluster is equal to the main-sequence lifetime of stars at the main-sequence turnoff point.

mantle (of a planet) The rocky layer that lies between a planet's core and crust.

Martian meteorites Meteorites found on the Earth's surface that apparently were chipped off the surface of Mars.

mass A measure of the amount of matter in an object.

mass extinction An event in which a large fraction of the living species on Earth go extinct in a very short period of time.

mass-energy The potential energy of mass, which has an amount $E = mc^2$.

massive-star supernova A supernova that occurs when a massive star dies, initiated by the catastrophic collapse of its iron core; often called a Type II supernova.

mass-to-light ratio The mass of an object divided by its luminosity, usually stated in units of solar masses per solar luminosity. Objects with high mass-to-light ratios must contain substantial quantities of dark matter.

matter–antimatter annihilation Occurs when a particle of matter and a particle of antimatter meet and convert all of their mass-energy to photons.

mean solar time Time measured by the average position of the Sun in your local sky over the course of the year.

meridian A half-circle extending from your horizon (altitude 0°) due south, through your zenith, to your horizon due north.

metals (in solar system theory) Elements, such as nickel, iron, and aluminum, that condense at fairly high temperatures.

meteor A flash of light caused when a particle from space burns up in our atmosphere.

meteor shower A period during which many more meteors than usual can be seen.

meteorite A rock from space that lands on Earth.

Metonic cycle The 19-year period, discovered by the Babylonian astronomer Meton, over which the lunar phases occur on the same dates.

Milky Way Used both as the name of our galaxy and to refer to the band of light we see in the sky when we look into the plane of the Milky Way Galaxy.

millisecond pulsars Pulsars with rotation periods of a few thousandths of a second.

molecular bands The tightly bunched lines in an object's spectrum that are produced by molecules.

molecular clouds Cool, dense interstellar clouds in which the low temperatures allow hydrogen atoms to pair up into hydrogen molecules (H_2).

molecular dissociation The process by which a molecule splits into its component atoms.

molecule Technically the smallest unit of a chemical element or compound; in this text, the term refers only to combinations of two or more atoms held together by chemical bonds.

momentum The product of an object's mass and velocity.

moon An object that orbits a planet.

mutations Errors in the copying process when a living cell replicates itself.

natural selection The process by which mutations that make an organism better able to survive get passed on to future generations.

nebula A cloud of gas in space, usually one that is glowing.

nebular capture The process by which icy planetesimals capture hydrogen and helium gas to form jovian planets.

nebular theory The detailed theory that describes how our solar system formed from a cloud of interstellar gas and dust.

net force The overall force to which an object responds; the net force is equal to the rate of change in the object's momentum, or equivalently to the object's mass × acceleration.

neutrino A type of fundamental particle that has extremely low mass and responds only to the weak force; neutrinos are leptons and come in three types—electron neutrinos, mu neutrinos, and tau neutrinos.

neutron star The compact corpse of a high-mass star left over after a supernova; typically contains a mass comparable to the mass of the Sun in a volume just a few kilometers in radius.

neutrons Particles with no electrical charge found in atomic nuclei, built from three quarks.

newton The standard unit of force in the metric system; 1 newton $= 1 \frac{\text{kg} \times \text{m}}{\text{s}^2}$.

Newton's laws of motion Three basic laws that describe how objects respond to forces.

nodes (of Moon's orbit) The two points in the Moon's orbit where it crosses the ecliptic plane.

nonbaryonic matter Refers to exotic matter that is not part of the normal composition of atoms, such as neutrinos or the hypothetical WIMPs.

north celestial pole (NCP) The point on the celestial sphere directly above the Earth's North Pole.

nova The dramatic brightening of a star that lasts for a few weeks and then subsides; occurs when a burst of hydrogen fusion ignites in a shell on the surface of an accreting white dwarf in a binary star system.

nuclear fission The process in which a larger nucleus splits into two (or more) smaller particles.

nuclear fusion The process in which two (or more) smaller nuclei slam together and make one larger nucleus.

nucleus (of an atom) The compact center of an atom made from protons and neutrons.

nucleus (of a comet) The solid portion of a comet, and the only portion that exists when the comet is far from the Sun.

observable universe The portion of the entire universe that, at least in principle, can be seen from Earth.

Olbers' paradox Asks the question of how the night sky can be dark if the universe is infinite and full of stars.

Oort cloud A huge, spherical region centered on the Sun, extending perhaps halfway to the nearest stars, in which trillions of comets orbit the Sun with random inclinations, orbital directions, and eccentricities.

opacity A measure of how much light a material absorbs compared to how much it transmits; materials with higher opacity absorb more light.

opaque (material) Describes a material that absorbs light.

open cluster A cluster of up to several thousand stars; open clusters are found only in the disks of galaxies and contain young stars.

open universe Refers to the case in which the density of the universe is less than the critical density, so that the expansion will continue forever.

optical quality Describes the ability of a lens, mirror, or telescope to obtain clear and properly focused images.

orbital resonance Describes any situation in which one object's orbital period is a simple ratio of another object's period, such as 1/2, 1/4, or 5/3. In such cases, the two objects periodically line up with each other, and the extra gravitational attractions at these times can affect the objects' orbits.

outgassing The process of releasing gases from a planetary interior, usually through volcanic eruptions.

oxidation Refers to chemical reactions, often with the surface of a planet, that remove oxygen from the atmosphere.

ozone The molecule O_3, which is a particularly good absorber of ultraviolet light.

ozone hole A place where the concentration of ozone in the stratosphere is dramatically lower than is the norm.

pair production The process in which a concentration of energy spontaneously turns into a particle and its antiparticle.

parabola The precise mathematical shape of a special type of unbound orbit allowed under the force of gravity; if an object in a parabolic orbit loses only a tiny amount of energy, it will become bound.

parallax angle Half of a star's annual back-and-forth shift due to stellar parallax; related to the star's distance according to the formula

$$\text{distance in parsecs} = \frac{1}{p}$$

where p is the parallax angle in arcseconds.

parsec (pc) Approximately equal to 3.26 light-years; it is the distance to an object with a parallax angle of 1 arcsecond.

particle era The era of the universe lasting from 10^{-10} second to 0.001 second after the Big Bang, during which subatomic particles were continually created and destroyed and ending when matter annihilated antimatter.

peculiar velocity (of a galaxy) The component of a galaxy's velocity relative to the Milky Way that deviates from the velocity expected by Hubble's law.

penumbra The lighter, outlying regions of a shadow.

perigee The point at which an object orbiting the Earth is nearest to the Earth.

perihelion The point at which an object orbiting the Sun is closest to the Sun.

period–luminosity relation The relation that describes how the luminosity of a Cepheid variable star is related to the period between peaks in its brightness; the longer the period, the more luminous the star.

photon An individual particle of light, characterized by a wavelength and a frequency.

photosphere The visible surface of the Sun, where the temperature averages just under 6,000 K.

pixel An individual "picture element" on a CCD.

Planck era The era of the universe prior to the Planck time.

Planck time The time when the universe was 10^{-43} second old, before which random energy fluctuations were so large that our current theories are powerless to describe what might have been happening.

planet An object that orbits a star and that, while much smaller than a star, is relatively large in size; there is no "official" minimum size for a planet, but the nine planets in our solar system all are at least 2,000 km in diameter.

planetary nebula The glowing cloud of gas ejected from a low-mass star at the end of its life.

planetesimals The building blocks of planets, formed by accretion in the solar nebula.

plasma A gas consisting of ions and electrons.

plate tectonics The geological process in which plates are moved around by stresses in a planet's mantle.

plates (on a planet) Pieces of a lithosphere that apparently float upon the denser mantle below.

potential energy Energy stored for later conversion into kinetic energy; includes gravitational potential energy, electrical potential energy, and chemical potential energy.

power The rate of energy usage, usually measured in watts (1 watt = 1 joule/s).

precession The gradual wobble of the axis of a rotating object around a vertical line.

primary mirror The large, light-collecting mirror of a reflecting telescope.

primitive meteorites Meteorites that formed at the same time as the solar system itself, about 4.6 billion years ago.

processed meteorites Meteorites that apparently once were part of a larger object that "processed" the original material of the solar nebula into another form.

protogalactic cloud A huge, collapsing cloud of intergalactic gas from which an individual galaxy formed.

proton–proton chain The chain of reactions by which low-mass stars (including the Sun) fuse hydrogen into helium.

protons Particles found in atomic nuclei with positive electrical charge, built from three quarks.

protoplanetary disk A disk of material surrounding a young star (or protostar) that may eventually form planets.

protostar A forming star that has not yet reached the point where sustained fusion can occur in its core.

protostellar disk A disk of material surrounding a protostar; essentially the same as a protoplanetary disk, but may not necessarily lead to planet formation.

protostellar wind The relatively strong wind from a protostar.

protosun The central object in the forming solar system that eventually became the Sun.

pulsar A neutron star from which we see rapid pulses of radiation as it rotates.

pulsating variable stars Stars that alternately grow brighter and dimmer as their outer layers expand and contract in size.

quantum mechanics The branch of physics that deals with the very small, including molecules, atoms, and fundamental particles.

quantum state Refers to the complete description of the state of a subatomic particle, including its location, momentum, orbital angular momentum, and spin to the extent allowed by the uncertainty principle.

quantum tunneling The process in which, thanks to the uncertainty principle, an electron or other subatomic particle appears on the other side of a barrier that it does not have the energy to overcome in . normal way.

quarks The building blocks of protons and neutrons, quarks are one of the two basic types of fermions (leptons are the other).

quasar The brightest type of active galactic nucleus.

radar ranging A method of measuring distances within the solar system by bouncing radio waves off planets.

radial motion The component of an object's motion directed toward or away from us.

radiation pressure Pressure exerted by photons of light.

radiation zone (of a star) A region of the interior in which energy is transported primarily by radiative diffusion.

radiative diffusion The process by which photons gradually migrate from a hot region (such as the solar core) to a cooler region (such as the solar surface).

radiative energy Energy carried by light; the energy of a photon is Planck's constant times its frequency, or $h \times f$.

radio galaxy A galaxy that emits unusually large quantities of radio waves; thought to contain an active galactic nucleus powered by a supermassive black hole.

radio waves Light with very long wavelengths (and hence low frequencies)—longer than those of infrared light.

radioactive dating The process of determining the age of a rock (i.e., the time since it solidified) by comparing the present amount of a radioactive substance to the amount of its decay product.

radioactive element (or **radioactive isotope**) A substance whose nucleus tends to fall apart spontaneously.

recession velocity (of a galaxy) The speed at which a distant galaxy is moving away from us due to the expansion of the universe.

red giant A giant star that is red in color.

red-giant winds The relatively dense but slow winds from red-giant stars.

redshift (Doppler) A Doppler shift in which spectral features are shifted to longer wavelengths, caused when an object is moving away from the observer.

reflecting telescope A telescope that uses mirrors to focus light.

reflection (of light) The process by which matter changes the direction of light.

refracting telescope A telescope that uses lenses to focus light.

rest wavelength The wavelength of a spectral feature in the absence of any Doppler shift or gravitational redshift.

retrograde motion Motion that is backward compared to the norm; e.g., Venus is said to rotate retrograde because it rotates in the opposite direction of most other planets.

revolution The orbital motion of one object around another.

right ascension (RA) Analogous to longitude, but on the celestial sphere; it is the angular east-west distance between the vernal equinox and a location on the celestial sphere.

rings (planetary) Consist of numerous small particles orbiting a planet within its Roche zone.

Roche zone The region within two to three planetary radii (of any planet) in which the tidal forces tugging an object apart become comparable to the gravitational forces holding it together; planetary rings are always found within the Roche zone.

rocks (in solar system theory) Material common on the surface of the Earth, such as silicon-based minerals, that are solid at temperatures and pressures found on Earth but typically melt or vaporize at temperatures of 500–1,300K.

rotation The spinning of an object around its axis.

rotation curve A graph that plots rotational (or orbital) velocity against distance from the center for any object or set of objects.

runaway greenhouse effect A positive feedback cycle in which heating caused by the greenhouse effect causes more greenhouse gases to enter the atmosphere, which further enhances the greenhouse effect.

saddle-shaped (or **hyperbolic**) **geometry** Refers to any case in which the rules of geometry for a saddle-shaped surface hold, such as two lines that begin parallel eventually diverge.

saros cycle The period over which the basic pattern of eclipses repeats, which is about 18 years $11\frac{1}{3}$ days.

satellite Any object orbiting another object.

scattered light Light that is reflected into random directions.

Schwarzschild radius A measure of the size of the event horizon of a black hole.

secondary mirror A small mirror in a reflecting telescope, used to reflect light gathered by the primary mirror toward an eyepiece or instrument.

sedimentary rock A rock that formed from sediments created and deposited by erosional processes.

seismic waves Earthquake-induced vibrations that propagate through a planet.

semimajor axis Half the distance across the long axis of an ellipse; in this text, it is usually referred to as the *average* distance of an orbiting object, abbreviated a in the formula for Kepler's third law.

shield volcano A shallow-sloped volcano made from the flow of low-viscosity basaltic lava.

shock wave A wave of pressure generated by gas moving faster than the speed of sound.

sidereal day The time of 23 hours 56 minutes 4.09 seconds between successive appearances of any particular star at its highest point on the meridian; essentially the true rotation period of the Earth.

sidereal month About $27\frac{1}{4}$ days, the time required for the Moon to orbit the Earth once (as measured against the stars).

sidereal year The time required for the Earth to complete exactly one orbit as measured against the stars; about 20 minutes longer than the tropical year on which our calendar is based.

silicate rock A silicon-rich rock.

singularity The place at the center of a black hole where, in principle, gravity crushes all matter to an infinitely tiny and dense point.

solar activity Refers to short-lived phenomena on the Sun, including the emergence and disappearance of individual sunspots, prominences, and flares; sometimes called *solar weather.*

solar circle The Sun's orbital path around the galaxy, which has a radius of about 28,000 light-years.

solar day Twenty-four hours, which is the average time between appearances of the Sun on the meridian.

solar eclipse Occurs when the Moon's shadow falls on the Earth, which can occur only at new moon; may be total, partial, or annular.

solar flares Huge and sudden releases of energy on the solar surface, probably caused when energy stored in magnetic fields is suddenly released.

solar maximum The time during each sunspot cycle at which the number of sunspots is the greatest.

solar minimum The time during each sunspot cycle at which the number of sunspots is the smallest.

solar nebula The piece of interstellar cloud from which our own solar system formed.

solar neutrino problem Refers to the disagreement between the predicted and observed number of neutrinos coming from the Sun.

solar prominences Vaulted loops of hot gas that rise above the Sun's surface and follow magnetic-field lines.

solar system (or **star system**) Consists of a star (sometimes more than one star) and all the objects that orbit it.

solar wind A stream of charged particles ejected from the Sun.

sound wave A wave of alternately rising and falling pressure.

south celestial pole (SCP) The point on the celestial sphere directly above the Earth's South Pole.

spacetime The inseparable, four-dimensional combination of space and time.

spacetime diagram A graph that plots a spatial dimension on one axis and time on another axis.

spectral resolution Describes the degree of detail that can be seen in a spectrum; the higher the spectral resolution, the more detail we can see.

spectral type A way of classifying a star by the lines that appear in its spectrum; it is related to surface temperature. The basic spectral types are designated by a letter (OBAFGKM, with O for the hottest stars and M for the coolest) and are subdivided with numbers from 0 through 9.

spectroscopic binary A binary star system whose binary nature is revealed because we detect the spectral lines of one or both stars alternately becoming blueshifted and redshifted as the stars orbit each other.

spectroscopy (in astronomical research) The process of obtaining spectra from astronomical objects

speed The rate at which an object moves. Its units are distance divided by time, such as m/s or km/hr.

speed of light The speed at which light travels, which is about 300,000 km/s.

spherical geometry Refers to any case in which the rules of geometry for the surface of a sphere hold, such as that lines that begin parallel eventually meet.

spheroidal component (of a galaxy) The portion of any galaxy that is spherical (or football-like) in shape and contains very little cool gas; generally contains only very old stars. Elliptical galaxies have only a spheroidal component, while spiral galaxies also have a disk component.

spin angular momentum Often simply called *spin,* it refers to the inherent angular momentum of a fundamental particle.

spiral density waves Gravitationally driven waves of enhanced density that move through a spiral galaxy and are responsible for maintaining its spiral arms.

spiral galaxies Galaxies that look like flat, white disks with yellowish bulges at their centers. The disks are filled with cool gas and dust, interspersed with hotter ionized gas, and usually display beautiful spiral arms.

spreading centers (geological) Places where hot mantle material rises upward between plates and then spreads sideways.

spring equinox (vernal equinox) Refers both to the point in Pisces on the celestial sphere where the ecliptic crosses the celestial equator and to the moment in time when the Sun appears at that point each year (around March 21).

standard candle An object for which we have some means of knowing its true luminosity, so that we can use its apparent brightness to determine its distance with the luminosity–distance formula.

standard time Time measured according to the internationally recognized time zones.

star A large, glowing ball of gas that generates energy through nuclear fusion in its core. The term *star* is sometimes applied to objects that are in the process of becoming true stars (e.g., protostars) and to the remains of stars that have died (e.g., neutron stars).

starburst galaxy A galaxy in which stars are forming at an unusually high rate.

stellar parallax The apparent shift in the position of a nearby star against distant objects that occurs as we view the star from different positions in the Earth's orbit of the Sun each year.

stellar wind A stream of charged particles ejected from the surface of a star.

stratosphere An intermediate-altitude layer of the atmosphere which is warmed by the absorption of ultraviolet light from the Sun.

stratovolcano A steep-sided volcano made from viscous lavas that can't flow very far before solidifying.

stromatolites Large bacterial "colonies."

subduction zones Places where one plate slides under another.

subgiant A star that is between being a main-sequence star and a giant; subgiants have inert helium cores and hydrogen-burning shells.

sublimation The process by which atoms or molecules escape into the gas phase from a solid.

summer solstice Refers both to the point on the celestial sphere where the ecliptic is farthest north of the celestial equator and to the moment in time when the Sun appears at that point each year (around June 21).

sunspot cycle The period of about 11 years over which the number of sunspots on the Sun rises and falls.

sunspots Blotches on the surface of the Sun that appear darker than surrounding regions.

superbubble Essentially a giant interstellar bubble, formed when the shock waves of many individual bubbles merge to form a single, giant shock wave.

supercluster Superclusters consist of many clusters of galaxies, groups of galaxies, and individual galaxies and are the largest known structures in the universe.

supergiants (luminosity class I) The very large and very bright stars that appear at the top of an H–R diagram.

supermassive black hole Giant black hole, with a mass millions to billions of times that of our Sun, thought to reside in the centers of many galaxies and to power active galactic nuclei.

supernova The explosion of a star.

Supernova 1987A A supernova witnessed on Earth in 1987; it was the nearest supernova seen in nearly 400 years and helped astronomers refine theories of supernovae.

supernova remnant A glowing, expanding cloud of debris from a supernova explosion.

synchronous rotation Describes the rotation of an object that always shows the same face to an object that it is orbiting because its rotation period and orbital period are equal.

tangential motion The component of an object's motion directed across our line of sight.

tectonics The disruption of a planet's surface by internal stresses.

temperature A measure of the average kinetic energy of particles in a substance.

terrestrial planets Rocky planets similar in overall composition to Earth.

theories of relativity (*special* and *general*) Einstein's theories that describe the nature of space, time, and gravity.

thermal emitter An object that produces a thermal radiation spectrum; sometimes called a "blackbody."

thermal energy Represents the collective kinetic energy, as measured by temperature, of the many individual particles moving within a substance.

thermal escape The process in which atoms or molecules in a planet's exosphere move fast enough to escape into space.

thermal pressure The ordinary pressure in a gas arising from motions of particles that can be attributed to the object's temperature.

thermal pulses The predicted upward spikes in the rate of helium fusion, occurring every few thousand years, that occur near the end of a low-mass star's life.

thermal radiation The spectrum of radiation produced by an opaque object that depends only on the object's temperature; sometimes called "blackbody radiation."

thermosphere A high, hot X ray–absorbing layer of an atmosphere, just below the exosphere.

tidal force A force that is caused when the gravity pulling on one side of an object is larger than that on the other side, causing the object to stretch.

tidal friction Friction within an object that is caused by a tidal force.

tidal heating A source of internal heating created by tidal friction. It is particularly important for satellites with eccentric orbits such as Io and Europa.

time dilation Refers to the effect in which you observe time running slower in reference frames moving relative to you.

timing (in astronomical research) The process of tracking how the light intensity from an astronomical object varies with time.

torque A twisting force that can cause a change in an object's angular momentum.

transmission (of light) The process in which light passes through matter without being absorbed.

transparent (material) Describes a material that transmits light.

Trojan asteroids Asteroids found within two stable zones that share Jupiter's orbit but lie 60° ahead of and behind Jupiter.

tropic of Cancer The circle on the Earth with latitude 23.5°N.

tropic of Capricorn The circle on the Earth with latitude 23.5°S.

tropical year The time from one spring equinox to the next, on which our calendar is based.

troposphere The lowest atmospheric layer, in which weather occurs.

Tully–Fisher relation A relationship among spiral galaxies showing that the faster a spiral galaxy's rotation speed, the more luminous it is; it is important because it allows us to determine the distance to a spiral galaxy once we measure its rotation rate and apply the luminosity–distance formula.

turbulence Rapid and random motion.

ultraviolet light Light with wavelengths that fall in the portion of the electromagnetic spectrum between visible light and X rays.

umbra The dark central region of a shadow.

unbound orbits Orbits on which an object comes in toward a large body only once, never to return; unbound orbits may be parabolic or hyperbolic in shape.

uncertainty principle The law of quantum mechanics that states that we can never know both a particle's position and its momentum, or both its energy and the time it has the energy, with absolute precision.

universal law of gravitation The law expressing the force of gravity (F_g) between two objects, given by the formula

$$F_g = G\frac{M_1 M_2}{d^2}$$

$$\left(G = 6.67 \times 10^{-11} \frac{\text{m}^3}{\text{kg} \times \text{s}^2}\right).$$

universal time (UT) Standard time in Greenwich (or anywhere on the Prime Meridian).

universe The sum total of all matter and energy.

velocity The combination of speed and direction of motion; it can be stated as a speed in a particular direction, such as 100 km/hr due north.

virtual particles Particles that "pop" in and out of existence so rapidly that, according to the uncertainty principle, they cannot be directly detected.

viscosity Describes the "thickness" of a liquid in terms of how rapidly it flows; low-viscosity liquids flow quickly (e.g., water), while high-viscosity liquids flow slowly (e.g., molasses).

visible light The light our eyes can see, ranging in wavelength from about 400 to 700 nm.

visual binary A binary star system in which we can resolve both stars through a telescope.

voids Huge volumes of space between superclusters that appear to contain very little matter.

volatiles Refers to substances, such as water, carbon dioxide, and methane, that are usually found as gases, liquids, or surface ices on the terrestrial worlds.

volcanism The eruption of molten rock, or lava, from a planet's interior onto its surface.

wavelength The distance between adjacent peaks (or troughs) of a wave.

weather Describes the ever-varying combination of winds, clouds, temperature, and pressure in a planet's troposphere.

weight The net force that an object applies to its surroundings; in the case of a stationary body on the surface of the Earth, weight = mass × acceleration of gravity.

weightless A weight of zero, as occurs during free-fall.

white-dwarf limit (also called the *Chandrasekhar limit*) The maximum possible mass for a white dwarf, which is about $1.4M_{\text{Sun}}$.

white-dwarf supernova A supernova that occurs when an accreting white dwarf reaches the white-dwarf limit, ignites runaway carbon fusion, and explodes like a bomb; often called a Type Ia supernova.

white dwarfs The hot, compact corpses of low-mass stars, typically with a mass similar to the Sun compressed to a volume the size of the Earth.

WIMPs Stands for *weakly interacting massive particles* and represents a possible form of dark matter consisting of subatomic particles that are dark because they do not respond to the electromagnetic force.

winter solstice Refers both to the point on the celestial sphere where the ecliptic is farthest south of the celestial equator and to the moment in time when the Sun appears at that point each year (around December 21).

worldline A line that represents an object on a spacetime diagram.

wormholes The name given to hypothetical tunnels through hyperspace that might connect two distant places in our universe.

X rays Light with wavelengths that fall in the portion of the electromagnetic spectrum between ultraviolet light and gamma rays

X-ray binary A binary star system that emits substantial amounts of X rays, thought to be from an accretion disk around a neutron star or black hole.

X-ray burster An object that emits a burst of X rays every few hours to every few days; each burst lasts a few seconds and is thought to be caused by helium fusion on the surface of an accreting neutron star in a binary system.

zenith The point directly overhead, which has an altitude of 90°.

zodiac The constellations on the celestial sphere through which the ecliptic passes.

zones (on a jovian planet) Bright bands of rising air that encircle a jovian planet at a particular set of latitudes.

Acknowledgments

Illustrations by Joe Bergeron
Figures 1.1, 1.3, 1.11, 3.14, 3.20, 3.22, 3.23, 3.24, 8.6, 8.7, 8.12, 8.13, 8.14, 11.19c, 12.18g, 16.3, 16.10b, 16.23, 16.24, 17.3, 17.9, 17.15b, 18.1, 18.18, 19.10, 20.2, 20.3, 20.8, 20.21, 20.22, 22.2, S2.11. ©Joe Bergeron S6.2, S6.3, S6.4, S6.5.

Part Opener and Chapter Opener Photos—***All part opener observatory photographs by Roger Ressmeyer/ ©Corbis; Part I / Chapter opener 1–3, S1:*** Jerry Lodriguss; ***Part II / Chapters 4–7, S2:*** ©Anglo-Australian Observatory, photography by D. Malin; ***Part III / Chapters 8–13***: ©Michael Horn 1998; ***Part IV / Chapters S3–S5*** HST/MIRLIN/Lindsay King (Univ. of Manchester); ***Part V / Chapters 14–17:*** ©Anglo-Australian Observatory/Royal Observatory Edinburgh, photography by D. Malin; ***Part VI / Chapters 18–22, S6:*** Robert Williams and the Hubble Deep Field Team (STScI) and NASA.

Chapter 1 **1.2** NASA **1.4** Jerry Lodriguss **1.8** Andrea Dupree (Harvard-Smithsonian CFA), Ronald Gilliland (STScI), ESA, and NASA **1.10** Gordon Garradd **1.13** Frank Zullo **1.15** David Nunuk **1.20** Akira Fujii **1.22** Jan Curtis (Geophysical Institute, Fairbanks, AK) **1.23** Niescja Turner and Carter Emmart.

Chapter 2 **2.1** Stan Maddock **2.3** National Optical Astronomy Observatories/NSO, Sacramento Peak **2.4** NASA/USGS **2.5** NASA, courtesy NSSDC **2.6** (**a**) NASA; (**b**) ©1994 Hansen Planetarium Publications, Salt Lake City, UT, and W. T. Sullivan, III, Univ. of Washington-Seattle, reproduced with permission **2.7** (**a**) NASA/USGS, courtesy NSSDC; (**b**) NASA **2.8** (**a**) NASA/JPL/USGS and Cornell Univ., courtesy NSSDC; (**b**) Brad Snowder **2.9** (**a**) NASA/JPL/USGS; (**b**) NASA/ JPL/DLR **2.10 (a,b), 2.11** (**a**), **2.12** NASA/JPL **2.13** (**a**) Dr. R. Albrecht (ESA/ESO) and NASA **2.14** (**a**) Hal Weaver, P. D. Feldman (Johns Hopkins Univ.), and NASA; (**c**) NASA **2.15** Megan Donahue (STScI).

Chapter 3 **3.11**(**moons**) Akira Fujii; (**earth**) NASA **3.16** (*top, bottom*) Akira Fujii; (*center*) Dennis diCicco **3.17, 3.18** Akira Fujii **3.19** Art by Stan Maddock. Adapted from map by Fred Espenak (NASA/GSFC).

Chapter S1 **S1.17** ©Husmo-foto **S1.18** ©Bernd Wittich/Visuals Unlimited **S1.19** Dennis diCicco **S1.20** Stan Maddock **S1.23** (**a**-*left*) Corbis-Bettmann; (**a**-*right*) The Bettmann Archive; (**b**) Corbis-Bettmann; (**c**) ©Science VU/Visuals Unlimited.

Chapter 4 **4.1** Adapted from *Ancient Astronomers,* Anthony Aveni, page 96 **4.2** ©Erich Lessing/Art Resources, NY **4.3** (**a**) ©N. Pecnik/ Visuals Unlimited (**b**) Stan Maddock **4.4** ©Kenneth Garrett **4.5** (**a**) ©Wm. E. Woolam/Southwest Parks; (**b**) Richard A. Cooke, III/Tony Stone Images, Inc. **4.6** ©1987 by Margaret R. Curtis **4.7** Richard A. Cooke, III **4.8** ©Loren McIntyre/Woodfin Camp & Assoc. **4.9** (*left*) John A. Eddy/Visuals Unlimited; (*right*) ©Jeff Henry/Peter Arnold, Inc. **4.10** ©Oliver Strewe (Wave Productions Pty, Ltd.) **4.11** Werner Forman Archive/Art Resources, NY **4.12** Stan Maddock **4.13** Courtesy of Carl Sagan Productions, Inc., from *Cosmos* (Random House) © 1980 Carl Sagan **4.14** Corbis-Bettman.

Chapter 5 **5.1** Dimitri Lundt/Tony Stone Images **5.5** ©Science VU/Visuals Unlimited.

Chapter 6 **Page 142** Giraudon/Art Resources, NY **Page 143** (*left*) The Granger Collection, NY; **143** (*right*) Archive Photos **Page 144** ©Erich Lessing/Art Resources, NY **Page 147** (*left*) Corbis-Bettmann **6.12** Jerry Lodriguss **Page 149** Corbis-Bettmann.

Chapter 7 **7.1** Runk/Schoenberger from Grant Heilman.

Chapter S2 **S2.6** Dennis diCicco **S2.7** (**b**) Yerkes Observatory **S2.8** (**b**) Palomar Observatory/Caltech **S2.10** (**a**) ©Richard Wainscoat; (**b**) Russ Underwood (W. M. Keck Observatory) **S2.12** Mark Voit (STScI) **S2.13** ©Anglo-Australian Observatory, photograph by D. Malin **S2.14** N. Levenson et al. 1997, Ap J.484,304 **S2.18** Jodi Schoemer **S2.19** ©Richard Wainscoat **S2.20** NASA/Ames Research Center **S2.21** Canada-France-Hawaii Telescope Corporation, Hawaii **S2.22** (**a**) Stan Maddock; (**b**) NASA; (**c**) Don Foley/National Geographic Image Collection **S2.23** (**c**) Eastman-Kodak **S2.24** NASA **S2.25** David Parker, 1997/Science Library. *The Arecibo Observatory is part of the National Astronomy and Ionosphere Center, which is operated by Cornell Univ. under a cooperative agreement with the National Science Foundation* **S2.27** ©Joel Gordon Photography **S2.30** NASA/JPL; painting by Ken Hodges **S2.31** (**a**) NASA/JPL **S2.31**(**b**) and **S2.32** "Sojourner™, Mars Rover™, and spacecraft design and images copyright

(©) 1996–97, California Institute of Technology. All rights reserved. Further reproduction prohibited." **S2.33** (**b**) NASA/JPL **S2.35** NASA/JPL/Caltech.

CHAPTER 8 **8.2** (*left*) Johns Hopkins Univ./APL/NASA **8.3** (*left*) ©1997 H. Mikuz and B. Kambic (Crni Vrh Observatory, Slovenia) **8.4** C. R. O'Dell (Rice Univ.) and NASA **8.8** (**a-***left*) NASA/STScI, courtesy of Alfred Schultz and Helen Hart; (**a-***right*) JPL/Caltech, Franklin & Marshall College, and NASA at W. M. Keck Observatory; (**b-***left*) M. J. McCaughrean (MPIA), C. R. O'Dell (Rice Univ.), and NASA; (**b-***right*) J.Bally, D. Devine, and R. Sutherland **8.11** Meteorite specimen courtesy of Robert Haag Meteorites **8.15** NASA/JPL, courtesy NSSDC **8.16** ©William K. Hartmann **8.18** (**a**) S. Terebey (Extrasolar Research Corp.) and NASA; (**b**) T. Nakajima and S. Kulkarni (Caltech), S. Durrance, D. Golimski (JHU), and NASA.

CHAPTER 9 **9.1**(**moon**) Akira Fujii; (**others**) NASA **9.2** (**a**) NASA/USGS; (**b**) NASA/JPL **9.3** (**a**) ©Don Davis; (**b**) NASA, courtesy NSSDC; (**c**) ©Barrie Rokeach; (**d**) Artis Planetarium/The Netherlands; (**e**) NASA/JPL **9.4** Roger Ressmeyer/©Corbis **9.9** (**a**) ©Jules Bucher/Photo Researchers, Inc. **9.11** Don Gault, Peter Schultz, and NASA Ames Research Center **9.12** (**a**) NASA/USGS; (**b**) NASA, courtesy of LPL **9.13** (**a**) NASA/JPL (Viking Orbiter); (**b**) NASA/USGS; (**c**) NASA/ JPL (Viking Orbiter) **9.14** NASA **9.16** (**b**) ©Paul Chesley/Tony Stone Images **9.17** (*top-right*) NASA, courtesy of the Astronomy Society of the Pacific; (*center-right*) NASA/JPL; (*bottom-right*) ©Forest Buchanan/Visuals Unlimited **9.19** NASA/USGS **9.22** NASA **9.23** (**a,c**) NASA/JPL; (**b**) NASA, courtesy of Mark Robinson **9.24** (**a**) NASA, courtesy of Mark Robinson **9.25** Lowell Observatory Photographs **9.26** NASA/USGS **9.27** NASA/JPL **9.28** NASA/JPL and Malin Space Science Systems **9.29** NASA/JPL.

CHAPTER 10 **10.1**(**Moon**) *globe,* Akira Fujii, *surface,* Artis Planetarium, The Netherlands; (**Mercury**) *globe,* NASA, courtesy NSSDC, *surface,* ©Don Davis; (**Venus**) *globe,* NASA/JPL, *surface* ©Don Davis; (**Earth**) *globe,* NASA, *surface* ©Barrie Rokeach; (**Mars**) *globe,* NASA, courtesy of Calvin Hamilton, *surface,* NASA/JPL/Caltech **10.2** NASA **10.10** (**b**) Dr. L. A. Frank, The Univ. of Iowa; (**c**) NASA **10.12** and **10.14** (**a**) ©Tom Van Scant/The Geosphere Project, The Stock Market **10.14** (**b**) GOES-10 satellite image. Colorization by ARC Science Simulations. Copyright © 1998 **10.15** (**a, b**) © Tom Van Scant/The Geosphere Project, The Stock Market **10.16** (**a,c**) NASA/JPL; (**b**) J. Bell III (Cornell Univ.), T. Clancy (STScI), P. James, M. Wolff (Univ. of Toledo), S. Lee (Univ. of Colorado), L. Martin (Lowell Observatory), and NASA **10.18** NASA, courtesy NSSDC **10.19** NASA/JPL and Malin Space Science Systems **10.21** ©D. Cavagnaro/Visuals Unlimited **10.24** (**a**) D. Potter, T. Morgan, and R. Killen; (**b**) M. Mendillo, J. Baumgartner, and J.Wilson of Boston Univ.

CHAPTER 11 **11.1** NASA **11.3** (**b**) NASA/JPL **11.6** (**a**) ©Michael Carroll **11.7** (**b**) STScI, R. Beebe and A. Simon (NMSU); (**c**) NASA Infrared Telescope Facility and Glenn Orton **11.8** (**a**) NASA/JPL **11.11** (**Uranus**) Dr. Heidi Hammel (MIT) and NASA; (**others**) NASA/JPL **11.12** (**b**) J. Clarke and G. Ballester (Univ. of Michigan), J. Trauger, R. Evans (JPL), and NASA **11.13** Nick Schneider (CU/LASP) and John Trauger (JPL) **11.14** Data provided by Fran Bagenal **11.15** Courtesy Tim Parker/JPL **11.16** NASA/JPL **11.18** (**a,d,e**) NASA/USGS; (**c**) ©David Matherly/Visuals Unlimited; (**f**) NASA/JPL/Ames Research Center; (**g**) NASA/JPL/Univ. of Arizona **11.20** (**a**) DLR/NASA/JPL; (**b,c**) NASA/JPL/Arizona State Univ. **11.21** (**a**) NASA/JPL, courtesy of Calvin Hamilton; (**b**) NASA/JPL and Brown Univ.; (**c**) NASA/JPL **11.22** (**a**) NASA/JPL; (**b**) NASA/JPL and Arizona State Univ. **11.23** (**a,inset**) NASA/ JPL/Caltech; (**b**) ©Don Davis **11.24, 11.25** NASA/JPL **11.26** (**a**) NASA/ USGS; (**b**) NASA/JPL **11.27** (**a**) S. Larson (Univ. of Arizona/LPL); (**b**) NASA/JPL; (**c**) ©William K. Hartmann **11.28** NASA/JPL/Caltech **11.29** NASA/JPL, courtesy of M. Showalter (Plantary Data Systems Ring Node) **11.30** (**a,b**) NASA/JPL **11.31** (**b**) NASA/JPL, courtesy of M. Showalter (Plantary Data Systems Ring Node) **11.32** (**Uranus**) Erich Karkoschka (Univ. of Arizona/LPL) and NASA; (**others**) NASA/JPL.

CHAPTER 12 **12.1, 12.2** Data provided by Dave Tholen (Univ. of Hawaii) **12.3** Discovery photograph made January 7, 1976 by Eleanor F. Helin/JPL (Helin then associated with Caltech/Palomar Observatory) **12.4** (**a**) NASA/USGS; (**b**) NASA/JPL; (**c**) Johns Hopkins Univ./APL/NASA **12.7** JPL (DIAL) and NASA **12.8** Courtesy of Calvin J. Hamilton **12.9** (**a**) Walt Radomski/nyrockman.com; (**b**) PEANUTS reprinted by permission of United Feature Syndicate, Inc. **12.10** Meteorite specimens courtesy of Robert Haag Meteorites **12.11** (**a**) ©Peter Ceravolo; (**b**) Tony and Daphne Hallas **12.12** (*left*) Astuo Kuboniwa; March 9, 1997; 19:25:00–19:45:out; BISTAR Astronomical Observatory, Japan; telescope—d=125mm, f=500mm refractor; film—Ektracrome E100S; (*right*) Halley Multicolour Camera Team, Giotto, ESA, ©MPAE **12.15** (**a**) U.S. Naval Observatory; (**b**) Dr. R. Albrecht, ESA/ESO, and NASA **12.16** A. Stern (SRI), M. Buie (Lowell Observatory), ESA, and NASA **12.17** (**a**) Hal Weaver, T. E. Smith (STScI), and NASA; (**b**) Courtesy of Paul Schenk (Lunar and Planetary Institute) **12.18** (**a**) NASA/JPL/Caltech; (**b**) HST Jupiter Imaging Science Team; (**c**) MSSO, ANU/Science Library/Photo Researchers, Inc.; (**d**) Courtesy of Richard Wainscoat et al. (Univ. of Hawaii); (**e**) H. Hammel (MIT) and NASA; (**f**) HST Comet Team and NASA **12.19** (**c**) G. Emerson (E. E. Barnard Observatory) **12.20** Kirk Johnson (Denver Museum of Natural History) **12.21** Image courtesy of Dr. Virgil Sharpton (Lunar and Planetary Institute) **12.22** ©William K. Hartmann **12.23** TASS/Sovfoto.

CHAPTER 13 **13.1** NASA **13.2** (**Earth**) NOAA, (**Venus**) NASA/ Magellan data courtesy Peter Ford (MIT), (**Mars**) NASA/Viking data courtesy Mike Mellon (Univ. of Colorado) **13.5** Geological Survey of Canada **13.7** (**a**) ©Gene Ahrens/Bruce Coleman, Inc.; (**b**) J. Messerschmidt/Bruce Coleman, Inc.; (**c**) Biological Photo Service; (**d**) ©Martin G. Miller/Visuals Unlimited **13.8** Stan Maddock **13.9, 13.10** Digital image by Dr. Peter W. Sloss (NOAA/NESDIS/NGDCO) **13.12** Stan Maddock **13.13** Stan Maddock. Adapted from maps in *Modern Physical Geology,* Thompson & Turk, Saunders **13.14** NASA **13.15** Earth Satellite Corp./Science Photo Library/Photo Researchers, Inc. **13.16** (**a**) Lloyd Cluff/Corbis; (**b**) Mike Yamashita/ Woodfin Camp & Assoc. **13.17** (*clockwise from top*) ©Philip Rosenberg; ©Philip Rosenberg; Univ. of Hawaii; R. Shallenberger (Midway Atoll National Wildlife Refuge) Art by Stan Maddock **13.19** ©James L. Amos/Photo Researchers, Inc. **13.20** ©Jeff Greenberg/Visuals Unlimited **13.21** (**a, c**) Biological Photo Services; (**b**) ©Kevin Collins/Visuals Unlimited **13.22** ©Nih. R. Feldman/Visuals Unlimited **13.24** (**a**) Woods Hole Oceanographic Institute; (**b**) ©Barrie Rokeach **13.25** ©Ken Lucas/Visuals Unlimited **13.28** NASA/GSFC **13.29** NASA.

CHAPTER S4 **S4.23(a)** ESA/NASA; (**b**) HST/MIRLIN/Lindsay King (Univ. of Manchester) **S4.26** Taken from *Black Holes and Time Warps* by Kip Thorne.

CHAPTER S5 **S5.1** Fermilab Visual Media Service **S5.5** ©Harold & Esther Edgerton Foundation, 1998, courtesy of Palm Press, Inc.

CHAPTER 14 **14.1** Photo by Corel **14.3** National Optical Astronomy Observatories/NSO, Sacramento Peak **14.8** National Optical Astronomy Observatories **14.9** Courtesy of Dr. Kenneth Lande (Univ. of Pennsylvania) **14.10** ICRR (Institute for Cosmic Ray Research), The Univ. of Tokyo **14.13** W. Livingston (National Solar Observatory) **14.14** NOAO/National Solar Observatory. **14.17, 14.18** (**a**) National Optical Astronomy Observatories; (**b**) Courtesy of the SOHO/SOI-MDI project of Stanford Univ. *Soho is a project of international cooperation between the European Space Agency and NASA* **14.19** Courtesy of B. Haisch and G, Slater (Lockheed Palo Alto Research Laboratory).

CHAPTER 15 **15.3** Constellation Orion, defocused star trails, copyright D. Malin **15.4, 15.5** Harvard College Observatory **15.6** Lowell Observatory Photograph **15.16** ©Anglo-Australian Observatory/Royal Observatory Edinburgh, photography by D. Malin **15.17** ©Anglo-Australian Observatory, photograph by D. Malin.

CHAPTER 16 **16.1** ©Anglo-Australian Observatory, photograph by D. Malin **16.2** IPAC (Infrared Processing and Analysis Center) and Caltech/JPL **16.4** (**b,c**) C. Burrows and J. Morse (STScI), J. Hester (Arizona State Univ.), and NASA **16.13** (**a**) Nordic Optical Telescope, La Palma; (**b**) ©Anglo-Australian Observatory, photograph by D. Malin; (**c**) J. P. Harrington, K. J. Borkowski (Univ. of Maryland) and NASA; (**d**) B. Balick (Univ. of Washington), V. Icke (Leiden Univ.), G. Mellema (Stockholm Univ.), and NASA; (**e**) R. Sahai, J. Trauger (JPL), the WFPC2 Science Team, and NASA **16.21** From plates taken in 1956 with the Hale telescope, ©Malin/Pasachoff/Caltech **16.22** ©Anglo-Australian Observatory, photograph by D. Malin.

CHAPTER 17 **17.4** (**b**) M. Shara, B. Williams, and D. Zurek (STScI); R. Gilmozzi (ESO); D. Prialnik (Tel Aviv Univ.); and NASA **17.8** N. A. Sharp with Kitt Peak 4-meter telescope (NOAO) **17.10** Naval Research Laboratory **17.16, 17.17** NASA **Pages 556-557** (*Bell*) Copyright is held by CWP/Regents of the Univ.of California (http://www.physics.ucla.edu/~cwp/Phase2/Burnell,_Jocelyn_Bell@841234567.html); (*others*) Courtesy of Astronomical Institute of Bonn Univ.

CHAPTER 18 **18.2** (**clockwise from top**) NASA's NSSDC; S. L. Snowden; D. Malin; D. Malin; J. Hester and P. Scowen; J. Bally **18.3** H. Bond (STScI) and NASA **18.4** ©Anglo-Australian Observatory, photography by D. Malin **18.5** S. L. Snowden (NASA/GSFC) **18.6** (**a**) Tony and Daphne Hallas; (**b**) Jeff Hester (Arizona State Univ.) and NASA **18.7** (**a**) J. Keuhane, B. Koralesky, M. Anderson, L. Rudnick (Univ. of Minnesota), and R. Perley (NRAO); (**b**) J. W. Keohane, E. V. Gotthelf, R. Petre (NASA/GSFC) **18.8** Simulation by D. Strickland and I. Stevens **18.10, 18.11** John Bally (Univ. of Colorado) **18.12** Jeff Hester and Paul Scowen (Arizona State Univ.) and NASA **18.13** Courtesy of NASA's National Space Science Data Center at the Goddard Space Flight Center **18.14** ©Anglo-Australian Observatory/Royal Observatory Edinburgh, photography by D. Malin **18.16, 18.17, 18.20** ©Anglo-Australian Observatory, photography by D. Malin **18.23** (**a**) David Talent at the Cerro Tololo Inter-American Observatory, Chile; (**b**) AFRL/BMDO, courtesy of Dr. Mehradad Moshir; (**c**) NRAO/AUI, courtesy of Cornelia Lang (NRAO/UCLA) and Mark Morris (UCLA).

CHAPTER 19 **19.1** ©Sky Publishing Corporation, reproduced with permission **19.2** (**a,b,d**) ©Anglo-Australian Observatory, photography by D. Malin; (**c**) ©IAC photo from plates taken with the Isaac Newton telescope, photo by D. Malin **19.3, 19.4** ©Anglo-Australian Observatory, photography by D. Malin **19.5** W. C. Keel and R. E. White, III **19.6** ©Anglo-Australian Observatory/Royal Observatory Edinburgh, photography by D. Malin **19.7** ©Anglo-Australian Observatory, photography by D. Malin **19.8** (**a,b**) ©Anglo-Australian Observatory/Royal Observatory Edinburgh, photography by D. Malin; (**c**) ©Anglo-Australian Observatory, photography by D. Malin; (**d**) U.S. Naval Observatory **19.14** (**a**) The Observatories of the Carnegie Institution of Washington; (**b**) NASA/STScI **19.15, 19.17** The Observatories of the Carnegie Institution of Washington **19.18** (**a**) J.Trauger, JPL, and NASA; (**b**) Dr. W. L. Freedman (The Observatories of the Carnegie Institution of Washington), and NASA **19.20** Megan Donahue (STScI).

CHAPTER 20 **20.1** Robert Williams and the HDF Team (STScI) and NASA **20.4** Hyron Spinard, Univ. of California at Berkely et al. **20.5** Robert Williams and the HDF Team (STScI) and NASA **20.6** ©Anglo-Australian Observatory, photography by D. Malin **20.7** Dr. Michael J. West (Saint Mary's Univ.) **20.9** (**a**) R. Thompson, M. Rieke, G. Schneider (Univ. of Arizona); N. Scoville (Caltech); and NASA **20.10** (**a**) MPE (Max Plank Institute); (**b**) W. C. Keel (Univ. of Alabama) **20.11** Data obtained on the 3.9m Anglo-Australian Telescope by G. R. Meurer (JHU) **20.12** (**a**) Brad Whitmore (STScI) and NASA; (**b**) W. C. Keel (Univ. of Alabama) **20.13** Chris Mikos (Case Western Reserve Univ.) **20.14** Maarten Schmidt (Palomar Observatory/Caltech) **20.17** John Bahcall (Institute for Advanced Study) and NASA **20.18** Courtesy of C. Carilli and R. Perley (NRAO) **20.19** (**a-c**) Alan Bridle (NRAO/AUI); (**d**) NRAO/AUI **20.20** John Biretta (STScI) **20.23** H. Ford (STScI/Johns Hopkins Univ.); L. Dressel, R. Harms, A. Kochhar (Applied Research Corp.); Z. Tsvetanov, A. Davidsen, G. Kriss (Johns Hopkins Univ.); R. Bohlin, G. Hartig (STScI); B. Margon (Univ. of Washington-Seattle); and NASA **20.25** Mark Dickinson (STScI) **20.26** (*right*) ©Frank T. Awbrey/Visuals Unlimited.

CHAPTER 21 **21.4** Redrawn from Vera C. Rubin, *Scientific American,* June 1983, vol. 248:6, page 96. **21.6** Courtesy of Caltech **21.8** (**a**) Omar Lopez-Cruz and Iam Shelton; (**b**) Steve Snowden, NASA/GSFC/USRA **21.9** W. N. Colley, E. Turner (Princeton Univ.), J. A. Tyson (Bell Labs, Lucent Technologies), and NASA **21.11** W. Couch (Univ. of New South Wales), R. Ellis (Cambridge Univ.), and NASA **21.12** (*inset photos*) Charles Alcock/Lawrence Livermore National Laboratory **21.13** Michael Strauss, Princeton Univ. **21.14** Harvard-Smithsonian Center for Astrophysics **21.15** S. Maddox, G. Efstathiou, and W. Sutherland, MNRAS, 242, 43P **21.16** Frank Summers (American Museum of Natural History).

CHAPTER 22 **2.5** Roger Ressmeyer/©Corbis **22.8** E. Bunn (Bates College). **22.15** (**a**) Joel Gordon; (**b**) John Kieffer/Peter Arnold, Inc.

CHAPTER S6 **S6.1** NASA/JPL.

INDEX

Italic typeface refers to illustrations.
The letter t preceding page numbers refers to tables.
Where several page references are given, **bold** type is used to denote a main discussion.